Lecture Notes in Electrical Engineering

Volume 574

Series Editors

The book series *Lecture Notes in Electrical Engineering* (LNEE) publishes the latest developments in Electrical Engineering - quickly, informally and in high quality. While original research reported in proceedings and monographs has traditionally formed the core of LNEE, we also encourage authors to submit books devoted to supporting student education and professional training in the various fields and applications areas of electrical engineering. The series cover classical and emerging topics concerning:

- Communication Engineering, Information Theory and Networks
- Electronics Engineering and Microelectronics
- Signal, Image and Speech Processing
- Wireless and Mobile Communication
- Circuits and Systems
- Energy Systems, Power Electronics and Electrical Machines
- Electro-optical Engineering
- Instrumentation Engineering
- Avionics Engineering
- Control Systems
- Internet-of-Things and Cybersecurity
- Biomedical Devices, MEMS and NEMS

For general information about this book series, comments or suggestions, please contact leontina.dicecco@springer.com.

To submit a proposal or request further information, please contact the Publishing Editor in your country:

China

Jasmine Dou, Associate Editor (jasmine.dou@springer.com)

India

Aninda Bose, Senior Editor (aninda.bose@springer.com)

Japan

Takeyuki Yonezawa, Editorial Director (takeyuki.yonezawa@springer.com)

South Korea

Smith (Ahram) Chae, Editor (smith.chae@springer.com)

Southeast Asia

Ramesh Nath Premnath, Editor (ramesh.premnath@springer.com)

USA, Canada:

Michael Luby, Senior Editor (michael.luby@springer.com)

All other Countries:

Leontina Di Cecco, Senior Editor (leontina.dicecco@springer.com)
Christoph Baumann, Executive Editor (christoph.baumann@springer.com)

**** Indexing: The books of this series are submitted to ISI Proceedings, EI-Compendex, SCOPUS, MetaPress, Web of Science and Springerlink ****

More information about this series at http://www.springer.com/series/7818

China Society of Automotive Engineers
(China SAE)
Editor

Proceedings of China SAE Congress 2018: Selected Papers

Springer

Editor
China Society of Automotive Engineers
(China SAE)
Beijing, China

ISSN 1876-1100 ISSN 1876-1119 (electronic)
Lecture Notes in Electrical Engineering
ISBN 978-981-13-9720-2 ISBN 978-981-13-9718-9 (eBook)
https://doi.org/10.1007/978-981-13-9718-9

This Springer imprint is published by the registered company Springer Nature Singapore Pte Ltd.
The registered company address is: 152 Beach Road, #21-01/04 Gateway East, Singapore 189721, Singapore

Contents

Performance Simulation Study of Vehicle Engine Cooling System

Xin Li, Liang Zhu, Yongcheng Zhu, Zhixin Zeng and Jujiang Liu

Abstract The cooling system is an important subsystem of the engine, and its performance directly plays a key role in engine performance. In this paper, the vehicle cooling system model is built up with AMESim, which taken the CFD simulation results of the wind speed field and temperature field as input, to simulate the cooling system under the limit steady-state conditions such as low-speed and high-load, high-speed climbing, and NEDC transient conditions. We can calculate the water temperature rise curve of the transient condition, and balanced water temperature of steady-state conditions, and then compared with the test. The result verifies the correctness of the cooling system model. Simulation and test were carried out on a car with I-type radiator and U-flow radiator respectively. The final result shows that the I-type radiator is equipped with better cooling performance.

Keywords Cooling system · Heat transfer coefficient · Steady-state condition · NEDC condition · Optimization design

List of Symbols in the Text

Latin Alphabet

A	Heat exchange area
C	Mass heat flow
c	Specific heat capacity
C_p	Constant pressure specific heat capacity
C_r	Mass heat flow ratio
d_h	Hydraulic diameter
$\mathrm{d}pA$	Pressure component
E	Test heat transfer efficiency

X. Li (✉) · L. Zhu · Y. Zhu · Z. Zeng · J. Liu
Guangzhou Automobile Group Co., Ltd. Automotive Engineering Institute, Guangzhou, China
e-mail: lixin@gaei.cn

China SAE (ed.), *Proceedings of China SAE Congress 2018: Selected Papers*,
Lecture Notes in Electrical Engineering 574,
https://doi.org/10.1007/978-981-13-9718-9_1

error	Error
f	Coefficient of friction
g	Gravity acceleration
H	Head
m	Quality
NTU	Number of heat transfer units
Nu	Nusselt number
Pr	Prandt number
P	Heat exchange
P	Pressure
Q	Flow
Q	Heat
Re	Reynolds number
S	Circulation area
T	Temperature
U	Heat transfer coefficient
V	Flow rate
z	Height above the datum

Greek Alphabet

γ	Severe
Δ	Change
ε	Heat exchange efficiency
η	Pump efficiency
λ	Thermal conductivity
μ	Dynamic viscosity
ρ	Density
τ	Shear stress

Subscript

1	Water-side parameter
2	Wind-side parameter
in	Fluid inlet
out	Fluid outlet
simu	Simulation value
exp	Test value
laminar	Laminar flow
turbulent	Turbulent flow
min	Minimum
max	Maximum

1 Introduction

The cooling effect of the engine has a great impact on the fuel economy and emission performance of the whole vehicle. With higher and higher requirements of national emission regulations and fuel consumption regulations for engine performance, some new technologies such as small displacement and high-power turbocharged direct injection are widely used, which led to the heat load of the cooling system also getting higher and higher, so that the cold start warming process is paid more attention. However, the research and development of the vehicle cooling system are facing a large technical challenge. Especially, the precise matching of the engine cooling system is an important technical problem that can be solved urgently [1]. The traditional logarithmic mean temperature difference method and the efficiency heat transfer unit number method are cumbersome to calculate [2], and it is also necessary to find the heat exchanger performance chart. So these methods are less direct, less accurate, and less efficient. In the literature [3], the one-dimensional simulation model of the cooling system has calculated the engine performance under the steady point operating conditions such as low-speed and high-load conditions and studied the influence of various main components' parameters on the engine cooling performance, which provides a basis for the design or selection of the cooling system. This paper establishes a mathematical model of the radiator based on the convection heat transfer theory of ε-NTU, the heat transfer coefficient of the radiator is fitted through the single performance data, the quasi-three-dimensional model of the cooling module is established by the AMESim software. At the same time, the vehicle dynamic model and engine heat capacity model are established. The temperature rise curve of dynamic condition and engine balanced temperature under the steady-state conditions can be obtained through changing the wind speed calculation mode and pressure calculation mode of the cooling module. This method can significantly improve the design efficiency of cooling system and obtain more accurate performance results of steady or dynamic cooling system.

2 Create Models and Set Parameters

2.1 Fluid and Pump Fan Modeling

The kinetic energy (flow) of the actual fluid under the steady-state uncompressible condition is mainly affected by the pressure component, the gravity component and the fluid friction (viscous) component, as shown in Eq. (1). Simultaneously, the continuity equations and energy loss equations, as shown in Eqs. (2–3), the flow and pressure distribution of the corresponding component, can be calculated [4]. In the engine water cooling system model, the parameters of radiator fan, grille, heater and engine water jacket are inputted according to the flow pressure loss data obtained from the single-product test. The pressure loss parameter of the cooling pipeline is input according to the pipeline resistance coefficient. The coolant is defined as an equal volume mixture of ethylene glycol and water. The oil is defined as 10W-40. The gas is defined as dry air.

$$-\mathrm{d}pA - \rho g A\mathrm{d}z - \tau P\mathrm{d}s = (\rho \mathrm{d}s A)V\frac{\mathrm{d}V}{\mathrm{d}s} \tag{1}$$

$$A_1V_1 = A_2V_2 = Q \tag{2}$$

$$h_f = f\frac{L}{D}\frac{V^2}{2g} \tag{3}$$

Above the equations, $\mathrm{d}pA$ is the pressure component. $\rho g A\mathrm{d}z$ is the gravity component. $\tau P\mathrm{d}s$ is the viscosity component. $(\rho \mathrm{d}s A)V\frac{\mathrm{d}V}{\mathrm{d}s}$ is the increase of kinetic energy.

By measuring the pressure of the inlet and outlet pipes, combining with the flow rate of the pump, thus the curve of flow capacity head of the pump is obtained, as shown in Eq. (4). The efficiency of the pump is shown in Eq. (5).

$$h = H_{\mathrm{in}} - H_{\mathrm{out}} = \left\{\frac{P_{\mathrm{in}}}{\gamma} + \frac{V_{\mathrm{in}}^2}{2g} + Z_{\mathrm{in}}\right\} - \left\{\frac{P_{\mathrm{out}}}{\gamma} + \frac{V_{\mathrm{out}}^2}{2g} + Z_{\mathrm{out}}\right\} \tag{4}$$

$$\eta = \frac{P_{\mathrm{in}}}{P_{\mathrm{out}}} \tag{5}$$

In Eq. (4–5), h is the water head. H_{in} is the inlet pressure. H_{out} is the outlet pressure. Q is the flow rate. η is the pump efficiency.

The fan model only works in pressure mode, and its input is a static pressure curve characteristic of the air volume at different speeds, as shown in Fig. 1. The fan rotation area is equivalent to the air volume pressure field. The wind speed (vehicle speed) is converted into the inlet pressure before the cooling module, as shown in Eqs. (6–7), wherein the pressure outlet after the cooling module is assumed to be one atmosphere.

$$\Delta P = \frac{1}{2} \cdot K \cdot \rho \cdot v^2 \tag{6}$$

In Eq. (6), K is the drag coefficient. ρ is the air density. v is the wind speed.

$$P_{\mathrm{out}} = P_{\mathrm{in}} - \mathrm{dP} \tag{7}$$

In Eq. (7), P_{in} is the inlet pressure of the cooling module. dP is the pressure drop of the cooling module. P_{out} is the outlet pressure of the cooling module.

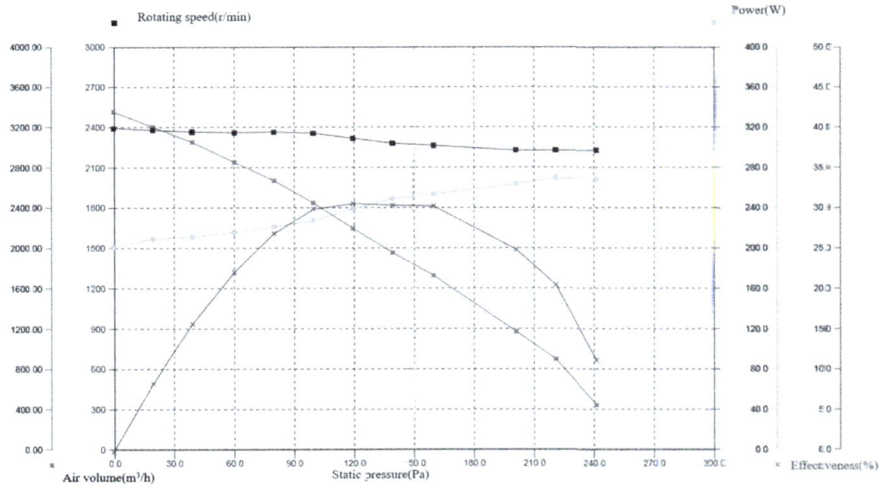

Fig. 1 Performance curve of the fan

2.2 *Modeling of Radiator and Fitting of Heat Transfer Coefficient*

It is difficult to discretize the physical model of the radiator sink into a mathematical model. The mathematical model of radiator is established based on the ε-NTU method [5], and it is necessary to determine the one smaller heat capacity of two fluids mass flow, as shown in Eq. (8). Theoretically, while assuming the heat transfer efficiency is 100%, only the smaller fluid of mass flow heat capacity can reach the maximum temperature difference. Actually, as shown in Eq. 9, the key calculation amount of heat exchange is turn to obtain heat transfer efficiency ε, which can be defined as a function of NTU and C_r, as shown in Eq. (10), wherein C_r is the ratio of the heat capacity of two fluid mass flows, as in Eq. (11). The NTU is the number of heat transfer units, as shown in Eq. (12), and it is related to heat transfer coefficient, heat transfer area *A,* and heat capacity of mass flow of a smaller fluid, which reflects three directions of heat transfer optimization. When the radiator is shaped, *A* is calculated from the geometrical dimensions, and *U* is a function of Nusselt number *Nu*, thermal conductivity λ and equivalent diameter *d* are shown in Eq. (13). When the radiator material and dimensions are determined, the key to the solving of the heat transfer efficiency is transformed into the fitting of Nusselt number *Nu*. According to the test data of the radiator bench, this paper fit the *Nu* with Reynolds number *Re* and Prandtl number *Pr*, as shown in Eqs. (14–16). The simulated heat transfer efficiency ε_{simu} and heat transfer amount P_{simu} are calculated by the fitting heat transfer coefficient and then compared analyzes with the heat transfer efficiency ε_{exp} and heat transfer amount P_{exp} to obtain the error, as shown in Eq. (18).

$$C_{\min} = \min\left(\left|m_1 \cdot C_{p1}\right|, \left|m_2 \cdot C_{p2}\right|\right) \tag{8}$$

$$P = C_{\min} \cdot |T_{\text{in1}} - T_{\text{in2}}| \cdot \varepsilon = UA\Delta T_m \tag{9}$$

$$\varepsilon = f(\text{NTU}, C_r) = 1 - \mathrm{e}^{\frac{\mathrm{e}-C_r*\text{NTU}-1}{C_r}} \tag{10}$$

$$C_r = \frac{C_{\min}}{C_{\max}} \tag{11}$$

$$\text{NTU} = \frac{UA}{C_{\min}} \tag{12}$$

$$UA = \frac{1}{\frac{d_1}{Nu_1\lambda_1} + \frac{1}{Gwall} + \frac{d_2}{Nu_2\lambda_2}} * \frac{1}{\frac{1}{A}} = \frac{1}{\frac{d_1}{Nu_1\lambda_1 A_1} + \frac{1}{Gwall} + \frac{d_2}{Nu_2\lambda_2 A_2}} \tag{13}$$

$$Re = \frac{m \cdot d_h}{\mu S} \tag{14}$$

$$Pr = \frac{\mu \cdot c_p}{\lambda} \tag{15}$$

$$Nu = \alpha \cdot Re^{\beta} \cdot Pr^{\gamma} \tag{16}$$

$$P_{\text{simu}} = C_{\min} \cdot |T_{\text{in1}} - T_{\text{in2}}| \cdot \varepsilon_{\text{simu}} \tag{17}$$

$$\text{error} = 100 \cdot \left|\frac{P_{\text{simu}} - P_{\exp}}{P_{\exp}}\right| \tag{18}$$

Above equations, m_1 is the mass flow of air. C_{p1} is the heat capacity at constant pressure of air. m_2 is the mass flow of the coolant. c_{p2} is the specific heat capacity of the coolant. T_{in1} is the inlet temperature of the air. T_{in2} is the inlet temperature of the coolant. ΔT_m is the average temperature difference between air and coolant. d_{h} is the hydraulic diameter. μ is the dynamic viscosity. S is the flow area. λ is the thermal conductivity. Nu_1 is the water-side Nusselt number. A_1 is the water-side heat exchange area. Nu_2 is the Nusselt number of the wind side. A_2 the heat exchange area of the wind side. P_{simu} is the heat transfer amount calculated by the simulation. $P_{\exp}$ is the heat transfer amount of the radiator bench test. The error is the error of simulation and test.

The geometrical dimensions of the radiator are shown in Table 1. The heat transfer area, flow area, and equivalent diameter of the water side and the wind side of the radiator can be calculated from the geometrical dimensions.

From the data of single character energy from radiator stand test, as shown in the column (1–4) of Table 2, v_1 is the wind speed, q_2 is the coolant flow, $P_{\exp}$ is the test heat transfer amount, $\varepsilon_{\exp}$ is the test heat transfer efficiency, $T_{\text{in1}} = 20$ °C is the test wind temperature, $T_{\text{in2}} = 80$ °C is the test inlet water temperature, and

Table 1 Radiator geometry

Core characteristic	Size (mm)
Flow type	I-type
Length * width * thick	673 * 411.3 * 16
Flat pipe height	1.5
Flat tube material thickness	0.22
Fin wave height	6.5
Fin distance	2.4
Fin material thickness	0.07
Fin efficiency	0.85

the maximum inlet temperature difference of the standard liquid gas is 60 °C. The Reynolds number, the Prandtl number, and the Nusselt number can be calculated in column 5–10 of Table 2; according to these data, the equation of the Nusselt number can be fitted, as shown in Eqs. (19–22).

Then, the heat transfer efficiency ε of the radiator is calculated to simulate heat dissipation. The error can be analyzed by comparing between the test data and the simulated data of heat transfer, as shown in column (11–13) of Table 2. The Nusselt number equation is obtained by fitting the test data of the radiator single item, which can be used for calculation of heat exchange amount of other working conditions.

$$Nu_{1_{\text{laminar}}} = 0.0126 \cdot Re^{1.3313} \cdot Pr^{0.3333} \tag{19}$$

$$Nu_{1_{\text{turbulent}}} = 1.1205 \cdot Re^{0.9102} \cdot Pr^{0.3333} \tag{20}$$

$$Nu_{2_{\text{laminar}}} = 0.1679 \cdot Re^{0.3683} \cdot Pr^{0.3333} \tag{21}$$

$$Nu_{2_{\text{turbulent}}} = 2.6116 \cdot Re^{8.9329} \cdot Pr^{0.3333} \tag{22}$$

Above equations, $Nu_{1_{\text{laminar}}}$ is the wind-side laminar Nusselt number. $Nu_{1_{\text{turbulent}}}$ is the wind-side turbulent Nusselt number. $Nu_{2_{\text{laminar}}}$ is the water-side laminar Nusselt number. $Nu_{2_{\text{turbulent}}}$ is water-side turbulence Nusselt number.

From Table 2, the error of each bench test condition under standard temperature difference is within 1.1%, and the results indicate that the Nusselt number fitting formula has high precision, conforms to the design requirements, which can be used for checking various limit working conditions.

Table 2 Results of heat transfer coefficient fitting of radiator

v_1/m/s	q_2/L/min	P_{exp}/kW	ε_{exp}/%	Nu_1	Nu_2	Re_1	Re_2	Pr_1	Pr_2	ε_{simu}/%	P_{simu}/kW	Error/%
2	40	31.4	73.00	12.33	41.28	355.91	278.68	0.71	35.37	73.16	31.47	0.22
4	40	47.3	53.35	19.67	39.31	718.65	246.95	0.72	39.17	53.92	47.81	1.07
6	40	56.2	52.58	22.72	38.14	1083.81	229.14	0.72	41.74	52.43	56.05	0.27
2	80	34.5	77.70	12.31	69.70	355.10	626.74	0.71	32.07	78.35	34.79	0.83
4	80	57.2	64.47	19.61	67.26	714.77	574.31	0.71	34.49	64.86	57.55	0.61
6	80	70.5	53.01	22.66	65.83	1078.15	544.86	0.72	36.05	53.08	70.59	0.13
2	120	35.7	80.39	12.30	93.60	354.63	978.57	0.71	31.02	80.99	35.97	0.75
4	120	61.6	69.41	19.59	91.14	713.06	916.94	0.71	32.74	69.25	61.46	0.23
6	120	77.3	58.11	22.63	89.65	1075.48	880.72	0.72	33.86	57.88	77.00	0.38
2	150	36.2	81.51	12.29	105.42	354.44	1243.73	0.71	30.59	82.00	36.42	0.60
4	150	62.7	70.64	19.58	103.48	712.63	1179.84	0.71	31.96	71.00	63.02	0.50
6	150	79.9	60.05	22.62	102.23	1074.46	1139.42	0.71	32.90	59.89	79.68	0.28

2.3 Modeling of Vehicle Cooling System

The vehicle dynamic model is build up according to the methods in vehicle dynamics and engine principles [6]. The running resistance models are calculated as Eqs. (23–27). The road conditions are divided into steady-state conditions and transient conditions. The steady-state conditions are a stable vehicle speed condition with a certain slope and an ambient temperature 40 °C. The transient condition is NEDC with an ambient temperature 25 °C.

$$P_{\text{roll}} = m \cdot g \cdot xkr \cdot v \tag{23}$$

$$xkr = K_1 + \frac{K_2}{P_{\text{tire}}} + \frac{K_3 \cdot v^2}{P_{\text{tire}}} \tag{24}$$

$$P_{\text{incl}} = m \cdot g \cdot \sin(\alpha) \cdot v \cdot k \tag{25}$$

$$P_{\text{aero}} = 0.5 \cdot \rho_{\text{air}} \cdot scx \cdot v^3 \tag{26}$$

$$P_{\text{accel}} = m \cdot \text{accel} \cdot v \tag{27}$$

Above equations, P_{roll} is the rolling resistance. P_{incl} is the slope resistance. P_{aero} is the air resistance. P_{accel} is the acceleration resistance.

An engine heat capacity model divides into two parts with the heat capacity of the cylinder and the cylinder head and also defines the heat exchange area, so that it can be used to calculate the temperature rise curve of the transient condition. The engine's universal characteristics and heat balance test data are inputted into the model for calculation. The heat release of the engine is equal to the heat dissipation of the radiator plus the heat capacity of the engine and the heat absorption of the coolant, as shown in Eq. (28). The transient cooling system model is established as shown in Fig. 2, and it contains the vehicle cooling system model, the radiator fan model, the dynamic model, and the operating conditions and physical parameters.

Meanwhile, the wind-side 3D model of grille, radiator, fan, and other components is established according to the geometric dimensions and layout data, as shown in Fig. 3.

$$Q_{\text{out}} = Q_{\text{in}} = \text{cm}\Delta T_{\text{coolant}} \tag{28}$$

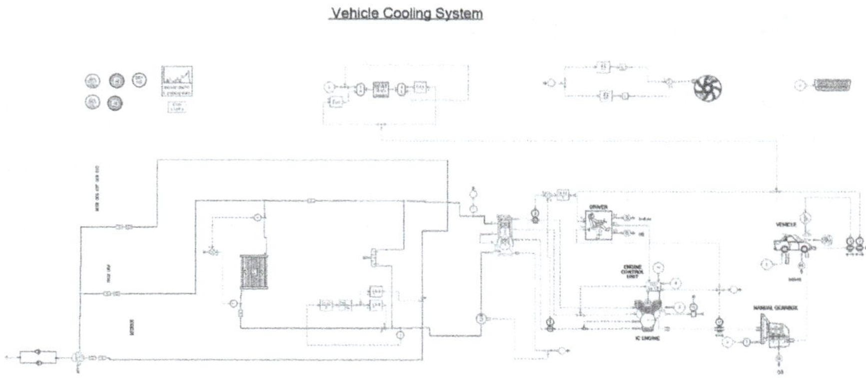

Fig. 2 Vehicle transient cooling system model

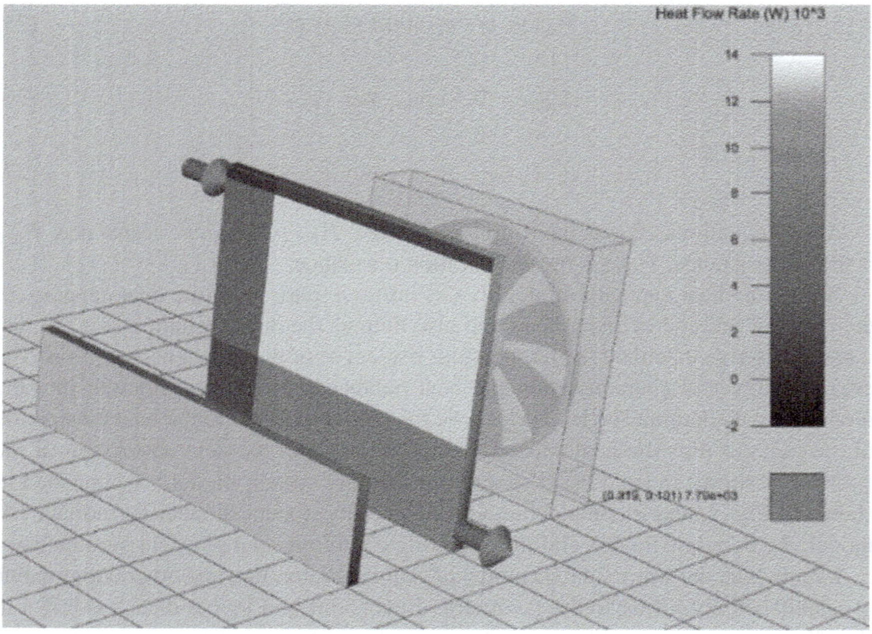

Fig. 3 Wind-side quasi-three-dimensional model

Table 3 Limit steady-state conditions' test conditions

Working condition name	Low-speed and high-load	High-speed climbing
Ambient temperature (°C)	40	45
Speed (km/h)	40	120
Road gradient (%)	10%	3%
Engine speed (r/min)	2048	3730
Transmission gear	3	5
Engine heat release (kW)	31	36

3 Calculation and Result Analysis of Limit Working Condition

The assessing is quite severe for vehicle cooling system under the low-speed and high-load conditions or the high-speed climbing condition, as shown in Table 3. These two operating conditions are steady-state conditions that means the vehicle speed and engine speed are all stable, so the flow rate of the radiator side is calculated by speed mode. The input values of wind speed and wind temperature in front of the radiator are obtained by the three-dimensional CFD simulation method [7], as shown in Fig. 4.

The simulation results of the engine outlet water temperature are shown in Table 4. Compared with the vehicle test results, the low-speed climbing water temperature error is 0.4%, the high-speed climbing condition error is 0.8%, and it indicates that the simulation results are relatively accurate.

4 NEDC Working Condition Calculation and Result Analysis

NEDC working condition is a transient condition, where the speeds are constantly changing. Therefore, the wind-side flow of the radiator is simulated in the pressure mode. The pressure source and fan of the cooling module are defined as a differential pressure flow element, and at the same time, the quasi-three-dimensional model of cooling module and the heat capacity of engine model are simulated calculation according to the vehicle cooling system model established above.

The simulation of temperature rise curve is shown in Fig. 5. The engine outlet water temperature rise curves of test and simulations are good consistency. When the cycle is carried out to 342 s, the warming up is completed, and when reaches 80.16 °C, the engine balanced water temperature is 88 °C. The water temperature difference between the test and the simulation is within 2 °C.

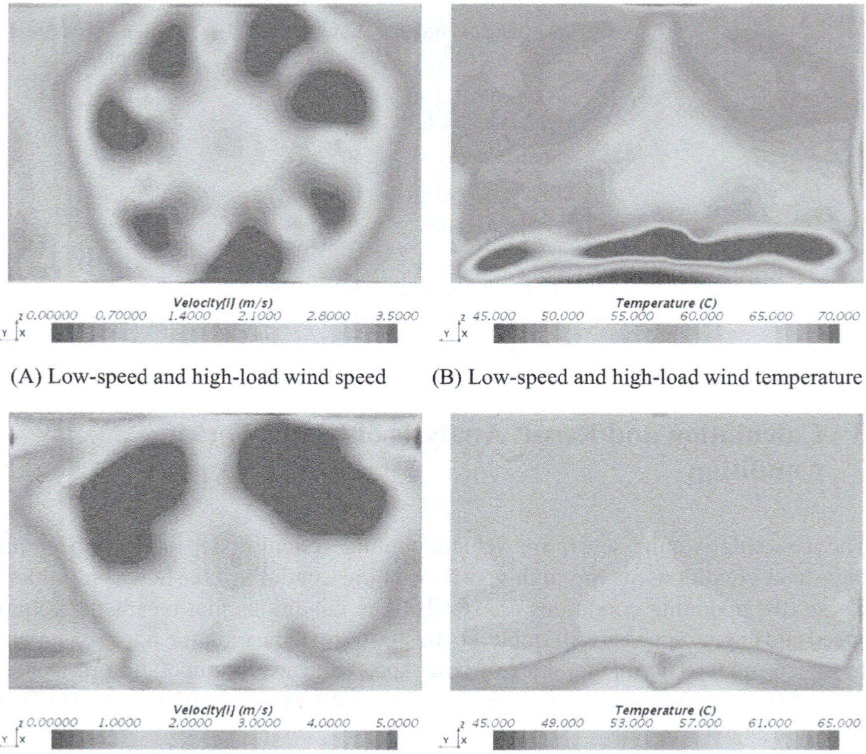

(A) Low-speed and high-load wind speed (B) Low-speed and high-load wind temperature

(C) High speed climbing wind speed (D) High-speed climbing wind temperature

Fig. 4 Wind speed field and wind temperature field in front of the radiator

Table 4 Comparison between simulation and experiment of engine outlet balanced water temperature

Working condition	Simulation value (°C)	Experimental value (°C)	Error (%)
Low-speed climbing	114.1	114.5	0.4
High-speed climbing	110.2	109.3	0.8

5 Optimization Design of U-Type and I-Type Radiator

The original scheme of a certain car radiator is a U-type radiator. The improvement is I-type radiator. So the balanced water temperature of the engine is decreased in limit working condition and achieved a better cooling effect. The dimensions of U-type and I-type radiators of the same size are shown in Table 5. Heat transfer performance bench test is shown in Fig. 6. The obtained heat dissipation, wind resistance, and water resistance are shown in Figs. 7 and 8. It can be known from the results of the

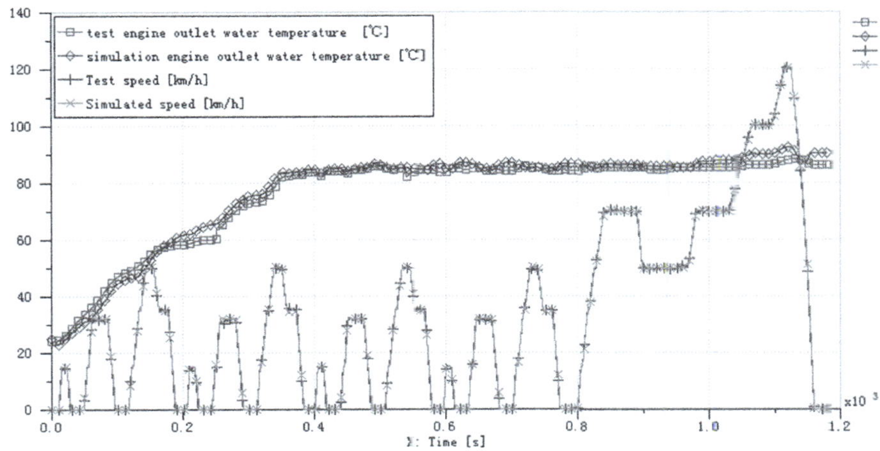

Fig. 5 Comparison between simulation and experiment of engine temperature rise curve

Table 5 Radiator geometry

Size	I-type (mm)	U-type (mm)
Length * width * thick	703 * 438.25 * 26	703 * 438.25 * 26
Flat pipe height	1.75	1.75
Flat tube material thickness	0.22	0.22
Fin wave height	6.25	6.25
Fin distance	2.9	2.9
Fin material thickness	0.07	0.07
Fin efficiency	0.85	0.85

A. U-type B. I-type

Fig. 6 U-type and I-type single-product performance test

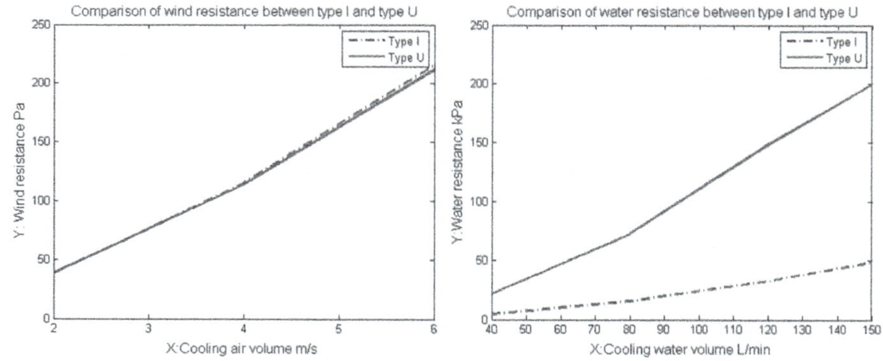

(A) Wind resistance (B) Water resistance

Fig. 7 Comparison of wind resistance and water resistance of two radiators

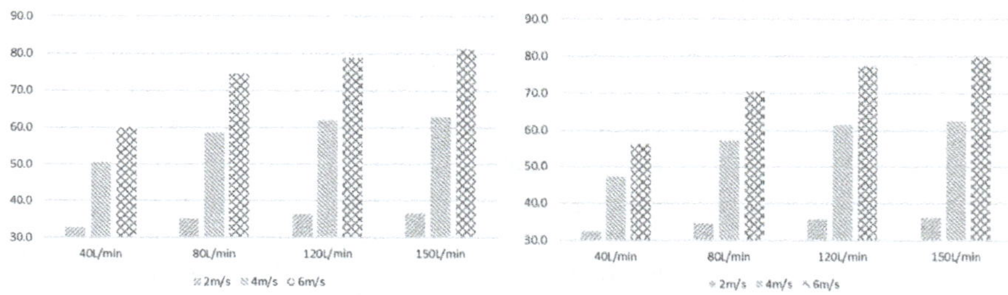

(A) U-type dispersion thermal performance (B) I-typedispersion thermal performance

Fig. 8 Heat dissipation performance comparison of two types of radiators

single product that the heat dissipation performance of U-type is slightly better than that of I-type at the same water flow and air volume, but the water resistance is greater than that of I-type, and the wind resistances are almost the same. The heat transfer is simulated calculation in low-speed climbing and high-speed climbing conditions for the cooling system model and radiator model of the vehicle model, as shown in Figs. 9 and 10.

From Tables 6 and 7, comparing with the U-type radiator, the balanced water temperature of the I-type radiator in the low-speed climbing condition reduces 2.64 °C, and the water flow increases 20.93 L/min. But in the high-speed climbing condition, the balanced water temperature decreases 2.5 °C, and the radiator water flow increases 31.12 L/min. So concluded that the I-type can reduce the balanced temperature of the engine under limit working conditions, which is conducive to the engine combustion and the fuel consumption reducing. At the same time, the structure of the I-type chamber is simple, which reduces the risk of internal leakage of the radiator during long-term use.

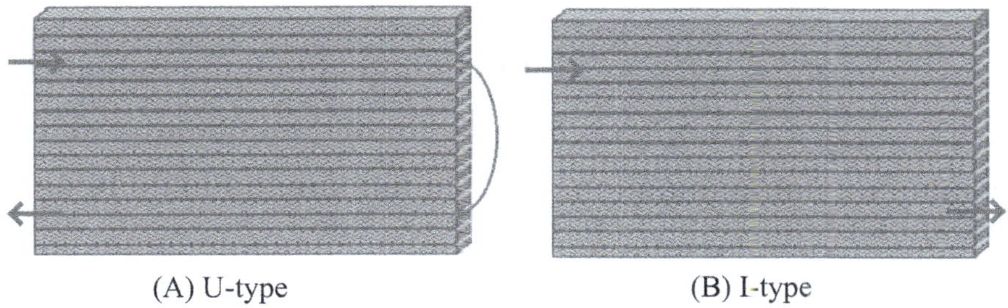

Fig. 9 Modeling of U-type and I-type radiators

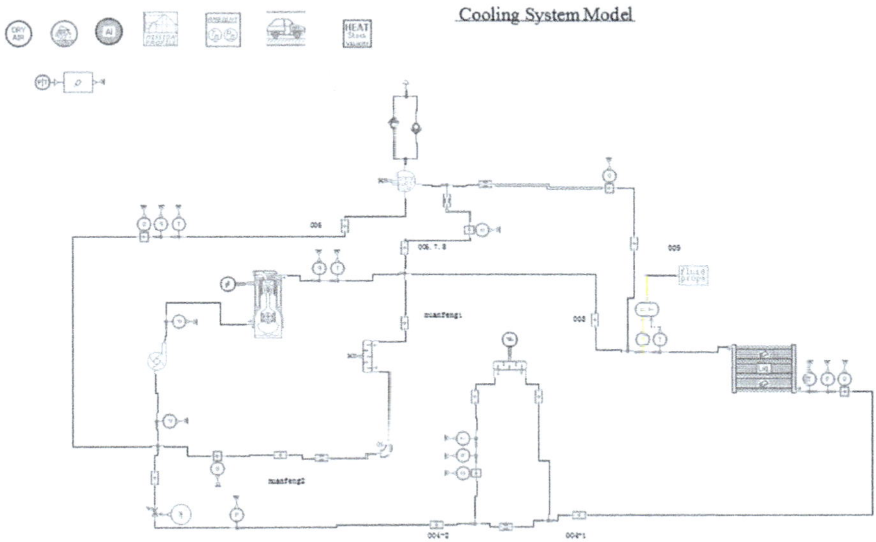

Fig. 10 Vehicle steady-state cooling system model

Table 6 Simulation results of low-speed climbing conditions

	I-type	U-type	Difference
Engine outlet temperature (°C)	114.36	117	2.64
Radiator flow (L/min)	54.2	33.27	20.93
Radiator average wind speed (m/s)	2.79	2.80	0.01
Heat exchange (kW)	35.61	35.52	0.09

Table 7 Simulation results of high-speed climbing conditions

	I-type	U-type	Difference
Engine outlet temperature (°C)	109.5	112.0	2.5
Radiator flow (L/min)	95.3	64.18	31.12
Radiator average wind speed (m/s)	4.65	4.65	0
Heat exchange (kW)	48.12	48.11	0.01

6 Conclusion

Firstly, the heat transfer coefficient of the radiator was fitted by the data of radiator bench test data. The vehicle cooling system model established in this paper is able to accurately evaluate the cooling system performance. It is conducive to matching precisely of the cooling system, optimizing the development cycle and saving the test cost, and also playing a positive role in the fuel consumption of the whole vehicle.

Secondly, through the establishment of a vehicle model and radiator simulation model, the I-type radiator was selected for development which is advantageous to the engine combustion and reducing fuel consumption.

Thirdly, the cooling system model established in this paper can be easy to couple with the models of vehicle thermal management or air conditioning, which has a strong expansibility.

References

1. Zhu L, Li X et al (2017) The study of thermal management and control strategy based on the transient system model. In: 2017 China automotive engineering society proceedings. China Machine Press, Beijing, pp 1467–1478
2. Yao Z, Wang X (2001) Vehicle cooling heat transfer. Beijing Institute of Technology Press, Beijing, pp 179–193
3. Zhang B, Zhao J et al (2016) Parameter matching simulation analysis of engine cooling system based on AMESim. Mach Des Manuf 12:190–193
4. John Finnemore E, Franzini JB (2015) Fluid mechanics and its engineering applications. China Machine Press, Beijing, pp 101–102
5. Yang W, Zhou J et al (2002) Three variables characterize the heat transfer characteristics of heat exchangers. CIESC J 53(4):436–439
6. Mitschke M, Wallentowiz H (2009) Vehicle dynamics. Tsinghua University Press, Beijing, pp 74–85
7. Liang X, Yuan X (2010) Analysis of vehicle thermal management by one-dimensional/three-dimensional combined simulation. Automot Eng 32(9):793–798

Study on Thermal Simulation Technology of Liquid-Cooled Power Battery Pack Based on Coupling of Equivalent Circuit and Heat Flow

Weimin Wang, Yisu Ye, Jinpeng Yu, Kun Zhou, Yang Xia, Hongtao Li and Renhe Xu

Abstract In the process of electric vehicle operation, the heat generated by the battery charge and discharge will cause the temperature of the battery to rise, and the temperature will affect the performance, service life and even combustion explosion of the battery. In the development of power battery pack, it is necessary to simulate and analyze the temperature field in the battery pack and combine the test of the battery bed to ensure that the thermal performance meets the requirements in the operation process. In this paper, the thermal simulation analysis model of the battery pack based on the coupling of equivalent circuit and the heat flow was established, and its simulation results were an agreement with test results. It is shown that the simulation method is correct, which can provide advices and suggestions for the design and development of EV power battery pack.

Keywords CFD · Power battery pack · Thermal simulation

1 Introduction

As the source of power, the battery pack is one of the most important components in electric vehicles. The temperature requirement of the battery is very strict, and higher or lower temperature will have a great negative impact on the battery performance, such as affecting the charging time, accelerating the life decay and even thermal runaway happening. During the development of the battery pack, thermal management measures must be taken into account to control the battery working temperature within a suitable range. Simulation method can be used to predict the thermal performance in the early stage of development, evaluating whether the battery pack design meets the requirements in various operating conditions, and then test work can be reduced to some extent. In terms of battery heat generating power, it is affected by many factors including charge and discharge rate, temperature, SOC, etc. In fact, it is difficult to obtain accurate heating power in the actual operating process. This

W. Wang (✉) · Y. Ye · J. Yu · K. Zhou · Y. Xia · H. Li · R. Xu
Dongfeng Motor Corporation Technical Center, Wuhan 430058, Hubei, China
e-mail: 1206833496@qq.com

China SAE (ed.), *Proceedings of China SAE Congress 2018: Selected Papers*, Lecture Notes in Electrical Engineering 574,
https://doi.org/10.1007/978-981-13-9718-9_2

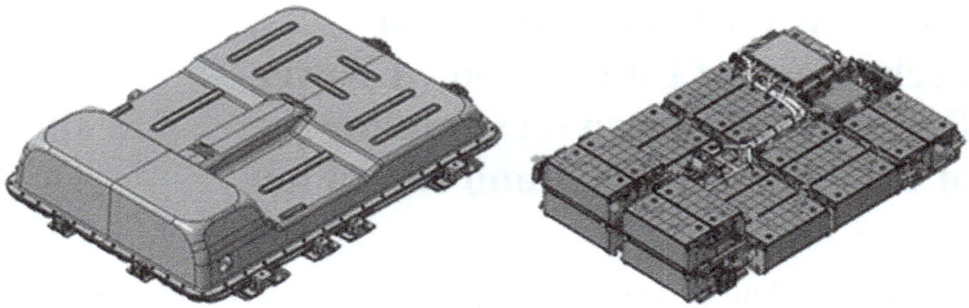

Fig. 1 Battery pack structure

paper described the coupling method of equivalent circuit and heat flow, which can obtain accurate heating power in real time. Compared with the traditional method, precision has been improved.

2 Brief Description of Pack Structure

A battery pack is composed of several modules, and the module is composed of several battery cells. The battery pack also includes key components such as BMS, explosion-proof valves, wiring harnesses, brackets and thermal management system, such as water-cooled panels.

2.1 Pack

The power source device of an electric vehicle is called a battery pack. Figure 1 shows a battery pack structure diagram of an electric vehicle. The battery pack includes battery module, control system and other components. As an important part of the water-cooled battery pack, the water-cooling device is used to dissipate heat and warm the battery pack when the battery pack temperature is too high or too low, ensuring the temperature inside the battery pack is within the normal operating temperature range.

2.2 Battery Module

The battery module is an important part of the pack. The bracket of the module is a box made of an insulating and heat-conductive material, in which battery cells are arranged by means of series and parallel connection. These battery cells are separated by heat-conducting plates. Figure 2 shows the module diagram of a battery

Fig. 2 Module assembly diagram

Fig. 3 Exploded perspective view of a battery module

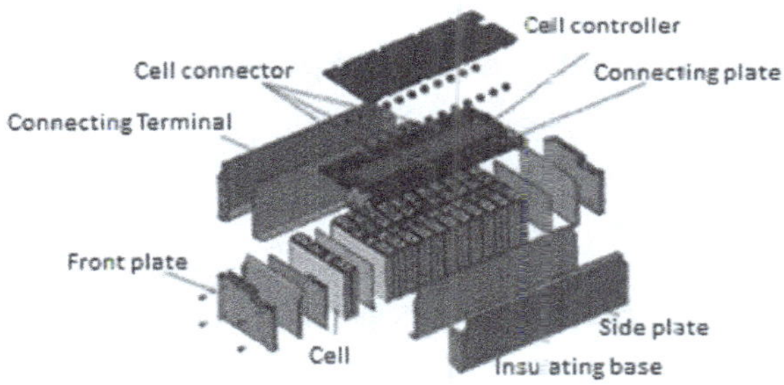

pack. There are insulating plates and damping plates around the box. Figure 3 is an exploded perspective view of a battery module.

2.3 Battery Cell

In terms of Li-ion battery cell, there are two types including cylindrical and square structure. The battery cell described in this paper is square type, as shown in Fig. 4. The shell is made of aluminum, and the core is composed of positive electrode material, a separator, a negative electrode material, an electrolyte, which is the main part of the heat generation of the battery. Then, the battery core material is wound and stacked into a square battery according to the chemical reaction mechanism [1].

Fig. 4 Battery cell

3 Thermal Characteristic Analysis of Power Battery

3.1 Heat Generation Mechanism of the Battery Cell

It is difficult to obtain the expression of the heat generation rate q of the battery cell in the field of engineering accurately. At present, engineers obtain q mainly by means of theoretical calculation and experimental measurements. Direct measurement is very difficult due to the high requirements of the test condition. So when estimating the heat generation rate of the battery, the traditional method is described in the following Formula (1), which was proposed by Dr. Bernardi from the University of California, Berkeley [2].

$$q = \frac{I}{V_b}\left[(V_0 - V_1) - T\frac{\mathrm{d}V_0}{\mathrm{d}T}\right] \tag{1}$$

In Formula 1, V_b is volume of battery cell, I is current when charge or discharge, V_0 is open-circuit voltage, V_1 is terminal voltage, T is temperature, $I * (V_0 - V_1)$ represents Joule heat, $I * T\ (\mathrm{d}V_0)/\mathrm{d}T$ is reversible reaction heat, which represent correlation quantity of electrochemical reaction. For a specific battery, $I * T\ (\mathrm{d}V_0)/\mathrm{d}T$ can be assumed constant, and it is weaker compared with Joule heat.

The RCR equivalent circuit model is used to calculate the heating power of the cell in this paper, as shown in Fig. 5. In this figure, V_0 is open-circuit voltage, R_0 is ohmic internal resistance, R_p and capacitor C constitute the polarization internal resistance of the battery. Battery heat generation model will be created by investigating these four parameters (V_0, R_0, R_p, τ) within RCR circuit at different temperature and different (SOC) state of charge.

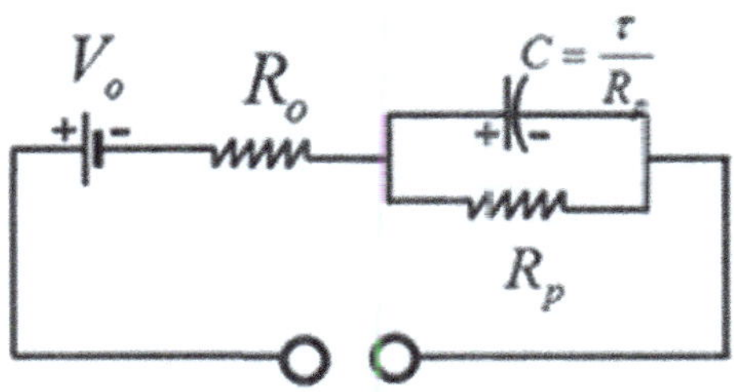

Fig. 5 Equivalent circuit diagram

Considering specific operating conditions on the vehicle, the heat generation rate and temperature of battery cells are changed in real time. With the traditional method, heat generation rate was assumed to be distributed uniformly in space, so the influence of the temperature change on the heating generation rate cannot be reflected in time. Based on the coupling of equivalent circuit and thermal flow CFD simulation, the uneven temperature distribution can be obtained due to the difference of the core internal structure, and then heat generation rate updated subsequently. Such a real-time coupling method can be used to predict temperature distribution and SOC accurately under various operating conditions.

3.2 Heat Transfer Path Inside Pack

The heat transfer between the cells and the solid components in the battery module is conducted by means of heat conduction and the heat transfer between the module and inside air by means of natural convection. When the battery temperature reached the limit, the water-cooling system was activated. The extra heat will be taken away by coolant so as to lower the battery temperature, ensuring battery cell working within a suitable temperature range.

4 Thermal CFD Simulation

Following simplification was considered: the battery core is divided into two parts, a shell and a core, and the core is the only heat source. Simulation analysis flow is shown in Fig. 6. Firstly, the battery pack (except the core) mesh model was created in STAR-CCM+ software; meanwhile, battery heating generation model and circuit file were built in BDS software using HPPC test data and battery operating conditions. And then, the BSM module was enabled in STAR-CCM+ to assemble the core and pack meshing model, calculating the temperature field and SOC within the real-time coupling. Finally, we can evaluate whether cooling system design reaches the requirements based on detailed temperature analysis by post-processing in STAR-CCM+.

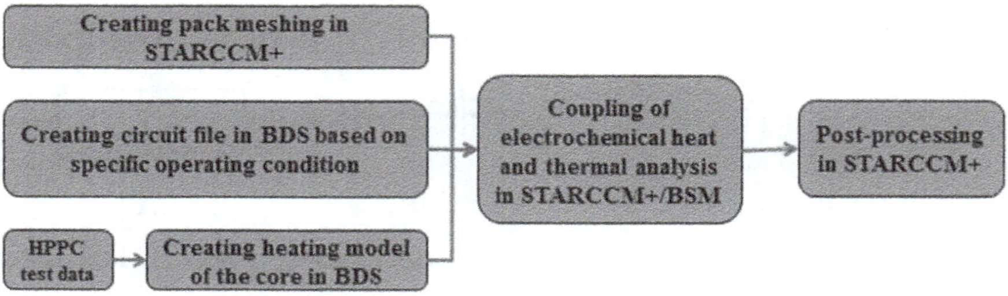

Fig. 6 Pack thermal simulation process

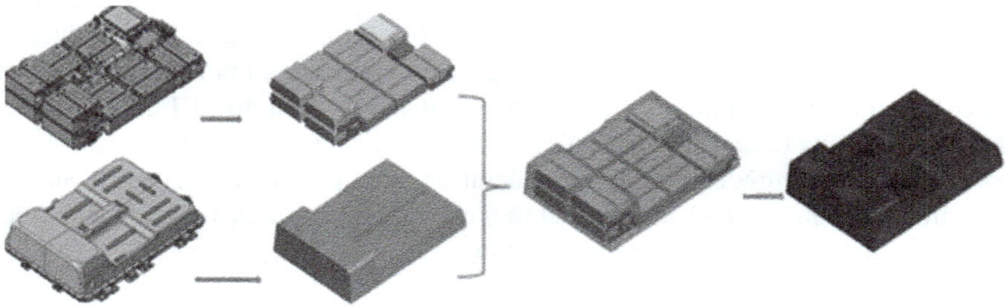

Fig. 7 Meshing generation process

4.1 Creating Simulation Model

4.1.1 Meshing

The CAD model of the battery package was simplified appropriately, and then imported into the simulation software STAR-CCM+, as shown in Fig. 7. The wire harness, bolts, brackets and other small parts are removed, and the small gap between the parts is simplified, and the chamfer of the parts is removed. The heat transfer between the components is connected through the interface. Regarding the mesh, the battery cell, the end plate and the fluid space adopt a polyhedral mesh, and the thin-walled parts such as the heat-conductive plate and the insulating plate adopt a hexahedral mesh, and the total number of meshes is 24 million.

4.1.2 Physical Model Setup

Heat transfer between the fluid and solid is mainly convective, and the k-epsilon turbulence model is selected according to the engineering experience [3]. The fluid–solid coupling heat transfer surface is set for the contact surface between the fluid and the solid, and the wall surface is set to the non-slip temperature wall boundary

Table 1 Properties of solid materials

Part name	Density (kg m^{-3})	Specific heat capacity (J kg^{-1} K^{-1})	Thermal conductivity (W m^{-1} K^{-1})
Battery cell	2980	1000	Anisotropy
Insulating plate	1300	1600	0.2
End plate	2710	714	3.917
Damping plate	240	1000	0.04

condition. Considering the high temperature of the air close to the module, select the ideal gas as the air, taking into account the buoyancy of the air [4].

4.1.3 Properties of the Solid Materials

Material properties of important several solid components are shown in Table 1.

4.1.4 Boundary and Initial Conditions

During the experiment, the battery pack was placed in an incubator. The heat transfer between the battery pack and the outside can be classified as convective. In the simulation, the outer surface of the battery pack case was set to the temperature and natural convection boundary, as shown in Fig. 8. Before the experiment, the battery pack should be left in the incubator for a period of time, so that the initial temperature of the battery pack is the same as the preset experimental initial temperature. During the experiment, the inlet flow and water temperature of the water-cooling system were controlled by the water-cooling unit and kept at a constant state. The inlet and outlet boundary of the water-cooled plate were shown in Fig. 9. With respect to the initial condition, the temperature of the battery pack was set to 45 °C, which was consistent with test data.

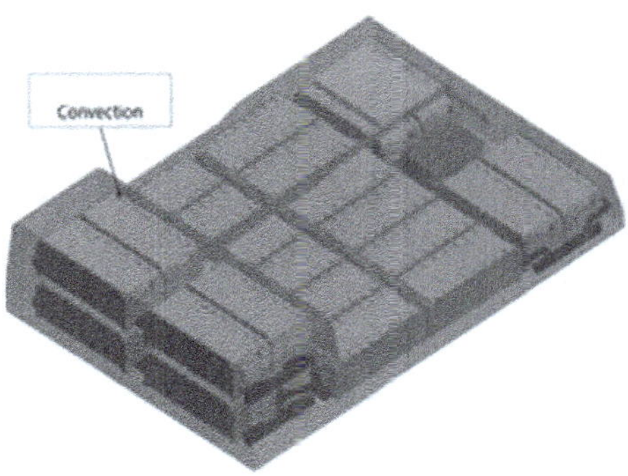

Fig. 8 Shell boundary condition of the pack

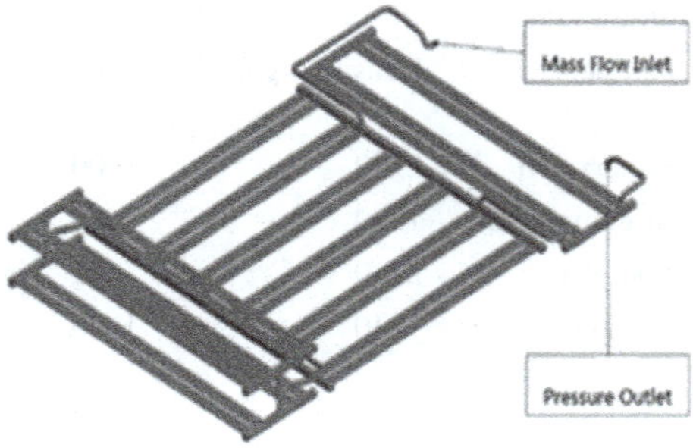

Fig. 9 Boundary condition of the water-cooling plate

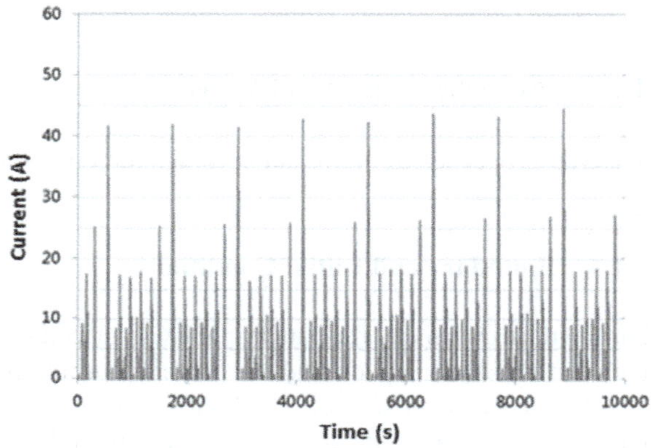

Fig. 10 Current change under NEDC condition

The NEDC condition (before 10,000 s) was simulated according to the test setup. Each NEDC operating condition was 1180 s, and the initial SOC value was set to 1. The current was shown in Fig. 10. For cooling system conditions, inlet flow was 10L/min, inlet temperature was 20 °C, coolant is a mixture of ethylene glycol and water, and the volume fraction each accounted for 50%.

4.1.5 Solver Setup

In this paper, the SIMPLE algorithm was used to solve the coupling field. The heat transfer process in different regions was combined into a whole heat transfer process [5]. The same numerical discrete rule was used for the iterative calculation of the fluid domain and the solid domain.

4.2 Simulation Results

At the end of the NEDC condition, the temperature distribution was shown in Fig. 11. It can be seen from the figure that the battery temperature increases along the direction of the water flow, and the battery temperature is lowest at inlet area of the liquid cooling plate, highest near outlet area. This is due to inlet temperature is the lowest, and coolant in the water-cooled plate is heated by the battery pack from upstream to downstream, so the cooling effect decreased. Since the coolant is disposed at the bottom of the cell, the temperature of these areas was lower than upside.

Due to the temperature difference between the battery shell and the internal components of the battery pack, the natural convection phenomenon occurs inside the battery pack. The velocity vector diagram at the cross section was shown in Fig. 12. It can be seen that the velocity of the air in the simulation results is greater than 0, indicating that natural convection has occurred.

Fig. 11 Pack temperature distribution at the end of operating condition

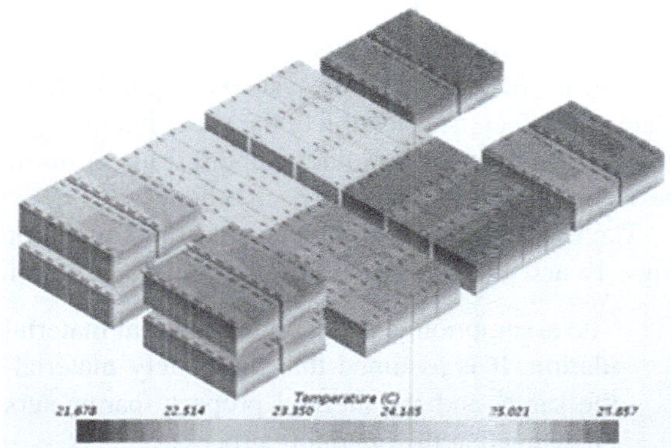

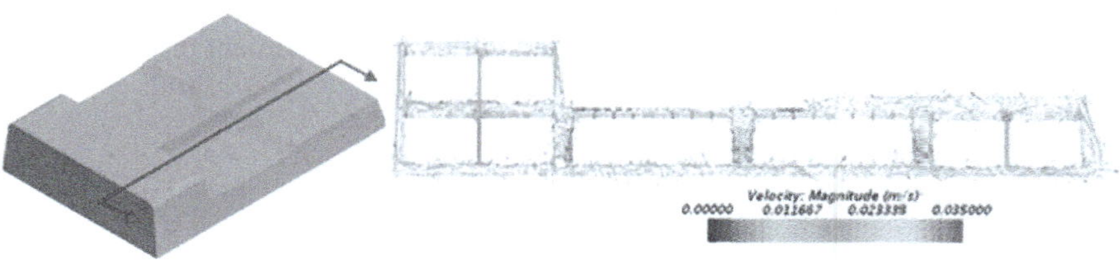

Fig. 12 Velocity vectors in key areas inside the pack

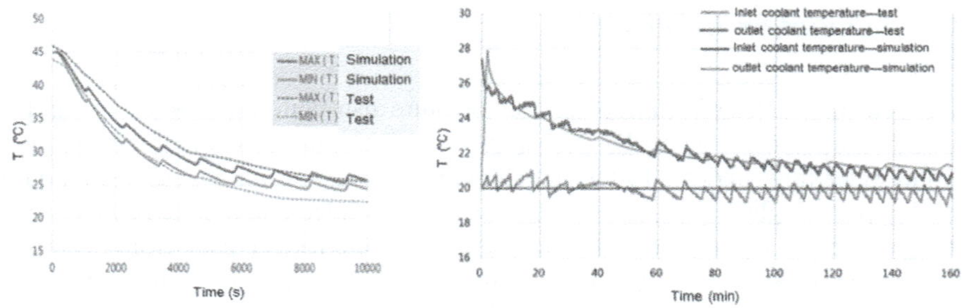

Fig. 13 Comparison between simulation and test results about temperatures

4.3 Comparison Between Simulation and Test Results

Regarding the maximum and minimum temperature inside the battery pack, inlet and outlet coolant temperature, the comparison between simulation and test results was shown in Fig. 13. It can be seen that the simulation results are basically consistent with the experimental results.

In terms of water-cooling system design, the flow resistance of the water-cooled plate will affect the selection of the water pump. Figure 14 shows the flow resistance comparison between simulation and test under normal ambient temperature. It can be seen that they are close to each other.

There are still deviations between the simulation and test results as shown in Figs. 13 and 14. The main reasons could be summarized as follows:

(1) The assumption is made on the internal material of the battery during the simulation. It is assumed that the battery material is uniform and the density is the same, and the thermal property parameters of the Li-ion battery remain unchanged at the operating process.
(2) The initial value of the battery pack temperature in the test deviates from the simulation set value.
(3) The inlet coolant temperature fluctuated during the experiment process.

Fig. 14 Comparison between simulation and test results about flow resistance

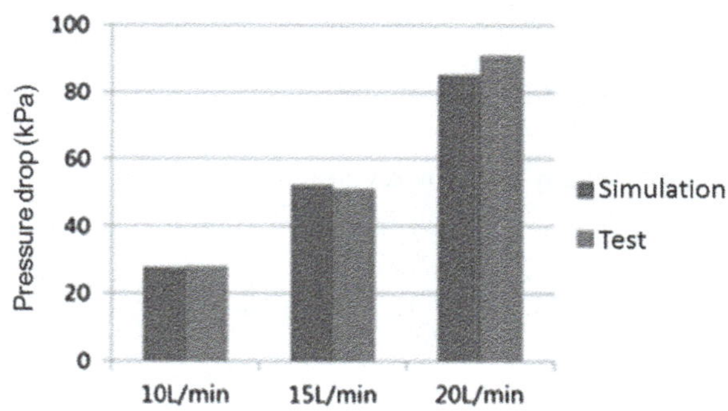

By improving the input difference between the above simulation and test, it can be expected that the simulation accuracy can be further improved.

5 Summary and Conclusions

The necessity of thermal analysis for power battery pack in electric vehicles is introduced at first in this paper, and the traditional battery heat generation rate model and its limitations are explained. And then a new battery heating model by using coupling of equivalent circuit and thermal flow is introduced. The physical properties and boundary conditions of the heat transfer model were determined based on the specific conditions of the bench experiment.

The simulation software was used to calculate the thermal characteristic of the battery pack. By comparing some simulation and test results, simulation accuracy can be guaranteed to a certain degree, indicating that the thermal simulation method can predict the temperature change of the battery pack in the actual working process and provide theoretical support for the thermal management performance development of power battery pack of the electric vehicle.

References

1. Xing N (2012) Thermal characteristic analysis and simulation study of lithium-ion power pack in EV. Jilin University, Changchun
2. Luo Y (2012) Thermal dynamic simulation of power lithium-ion battery pack based on driving conditions (English). J Automot Saf Energy Saving 3(1):51–57
3. Wang F (2004) Computational fluid dynamics analysis-CFD software principles and applications. Tsinghua University Press, Beijing
4. Yang J, Zhang Z, Jiu X, Tan J (2010) Engineering fluid mechanics. Peking University Press, Beijing
5. Lin G (2011) Research on the temperature field of lithium battery pack and heat dissipation structure of EV. Chongqing University

Crack Analysis and Structure Optimization of Torsion Beam Mounting Point

Yan Li, Junfei Yan, Ming Gao, Lizhong Mao and Hongsheng Tian

Abstract The torsion beam mounting point, one of the important connection points between the car body and the chassis, is very important for the normal running and safety of passenger cars. The rear torsion beam structure is a very mainstream chassis-rear suspension system. The reliability of torsion beam mounting point is related to passenger's safety, which should not be ignored. So, the study of the factors for the strength and durability of the torsion beam installation point, ensuring the performance meeting the durability and strength demand is very important. In this paper, by analyzing the stress and strain of the structure and using structure comparison and topology optimization method, we summarized the affecting factors and the optimization ways of the durability of the torsion beam mounting point, which provided guidance for the torsion beam mounting point design of further vehicles.

Keywords Torsion beam · Durability · Strength · Structure optimization

1 Introduction

Torsion beam suspension system, invented in the 1970s, is still a mainstream in the rear suspension system and widely used on the rear suspension system of the small and medium family cars. The suspension system is the collective name of all the connections between the car body and wheels or axle, which conveys the force and torque between the car body and wheels, and weakens the impact force and vibration caused by road irregular excitation to ensure the stability, ride comfort and safety of the vehicles. With the rise of the importance of the vehicle for daily life, vehicle has become an important part of modern life, and its safety is more and more significant to the customers. As one of the most important mounting points between the chassis and car body, the strength of the torsion beam mounting point is vital for safety and comfort. So, the study of the crack of the torsion beam and the structure optimization to solve the problem are especially important.

Y. Li (✉) · J. Yan · M. Gao · L. Mao · H. Tian
BYD Auto Industry Company Limited, Shenzhen 518118, China
e-mail: Liyanhero@163.com

China SAE (ed.), *Proceedings of China SAE Congress 2018: Selected Papers*, Lecture Notes in Electrical Engineering 574,
https://doi.org/10.1007/978-981-13-9718-9_3

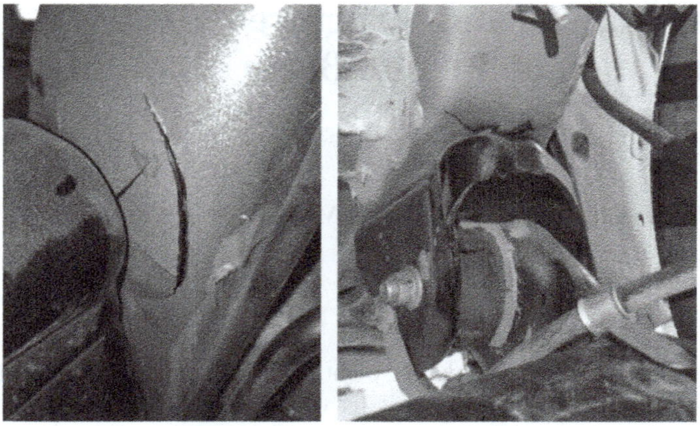

Fig. 1 Cracking of torsion beam mounting point

2 Problem Description

The rear suspension structure of a vehicle is a torsion beam structure, which cracked severely during the road test besides the torsion beam mounting point. To solve the problem, we simulated the force status of the torsion beam mounting point by dynamics analysis, established a static analysis model according to the real condition, and used topology optimization method to optimize the structure of the torsion beam mounting point. Limited by the tooling cost, time, the number of the mounting points and mounting space, we adopted the high-cost bolting scheme to meet the durability requirement of 300 thousand kilometers in the end. The road tests proved that the scheme could meet the durability requirement. Afterward, due to chassis adjustment, the torsion beam mounting point moved 35 mm in the z direction of the body coordinate system, which increased the force on the torsion beam mounting point. After moving, the torsion beam mounting point could not meet the durability requirement of 300 thousand kilometers. Therefore, the study and topology optimization of the torsion beam mounting points continued. After several rounds of simulation and optimization, the scheme which met the durability requirement of 300 km was formed (Fig. 1).

3 Analysis and Structure Optimization

3.1 Structure of the Torsion Beam Suspension System

In general, the chassis torsion beam suspension system is composed of beam, vertical arm, spring, shock absorber, hubs and mounting bracket, which is shown in Fig. 2.

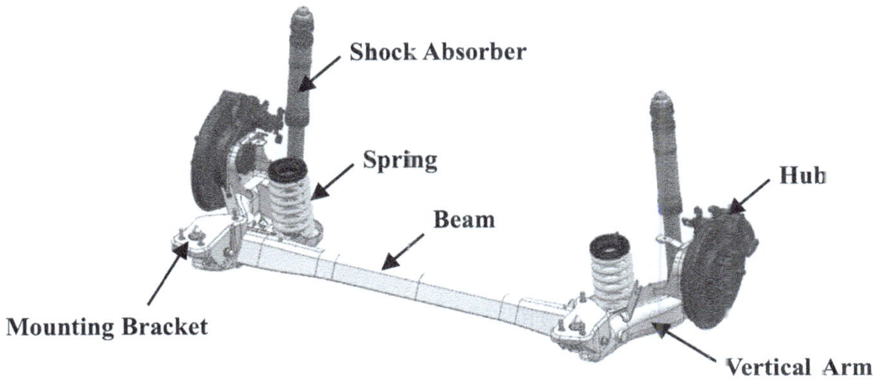

Fig. 2 Torsion beam

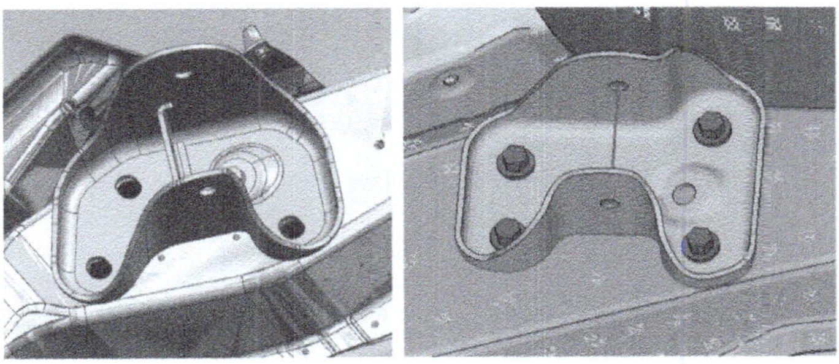

Fig. 3 Torsion beam mounting bracket

3.2 General Types of the Torsion Beam Mounting Point

There are generally two mounting forms that connect the bracket of the torsion beam and the carling, which, respectively, has three and four mounting points, as shown in Fig. 3.

3.3 Analysis and Simulation Verification

3.3.1 Failure Analysis of Original Structure Cracking

The first road test model adopted the form of mounting bracket with three mounting points, as shown in Fig. 4.

The nuts at the mounting points of this car were directly welded to the reinforced plate attached to carling inside the carling. In this form, only the carling and reinforced

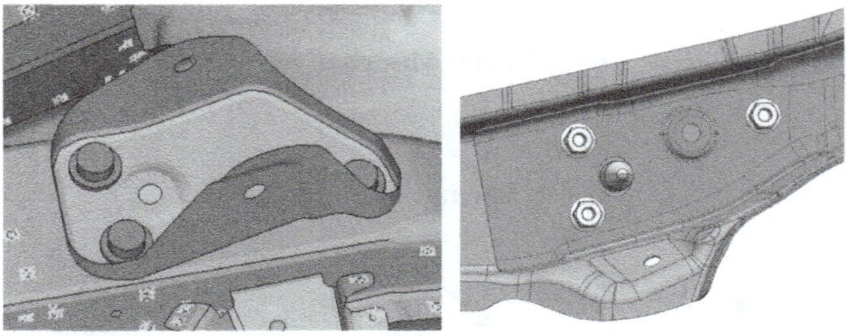

Fig. 4 Torsion beam mounting point of the road test model

Fig. 5 Cross section drawn of the torsion beam mounting point

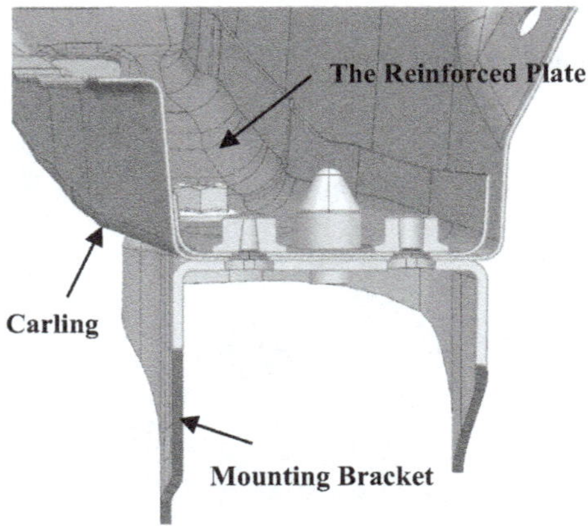

plate directly born the forces and torques of the torsion beam, thus the mounting points had weak stiffness and poor bearing capacity and caused stress concentration near the mounting point, which caused poor durability and led to the crack of the carling. The cross-sectional drawn of the structure is illustrated in Fig. 5.

In the simulation model, we chose the forces and torques when vehicles passing speed bumpers by one side, which is the worst operating condition during the road test, and forced on the plastic strain near the crack of carling. The plastic strain cloud of the carling was shown as Fig. 6. The plastic strain which is 1.7% near the mounting holes is largest, whose position consists of the crack position in the road test. According to static strength evaluation criteria requires that the largest strain is less than 0.2%. Thus, the carling would certainly fail and crack due to not meeting durability requirement, which is consistent with the fact.

3.3.2 Analysis of the Bolt Sleeve Structure

Limited by the retooling cost, process cycle, the number of the mounting points, position and mounting space, only increasing the thickness or adding some stiffen-

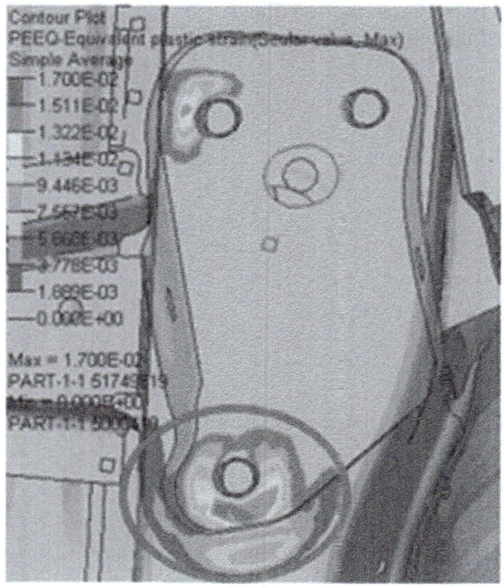

Fig. 6 Plastic strain cloud of the carling in original structure of a cracked car

ing plates cannot make the largest strain less than 0.2% by topology optimization methods. This means it cannot meet the durability requirement. After overall consideration, the original nuts were replaced with the bolt sleeves. The bolt sleeves are connected to the carling reinforcing plate and the bolt sleeve fixed plate, which adds the bolt sleeve fixed plate to bear the forces and torques compared to the original structure that has only the carling and stiffening plate to bear the loads. The new structure is shown in Figs. 7 and 8. The strain of the new structure is shown in Fig. 9,

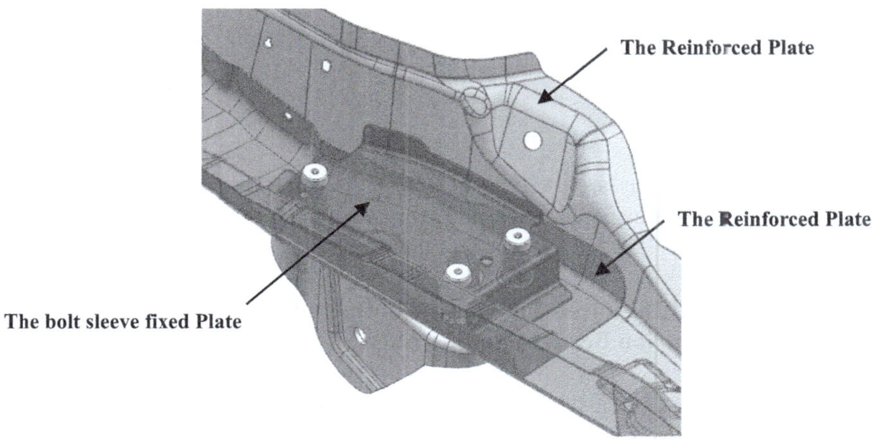

Fig. 7 Bolt sleeve structure

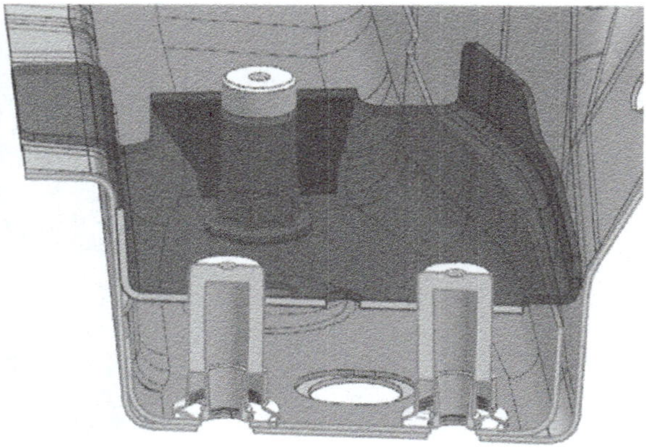

Fig. 8 Cross section drawn of the bolt sleeve structure

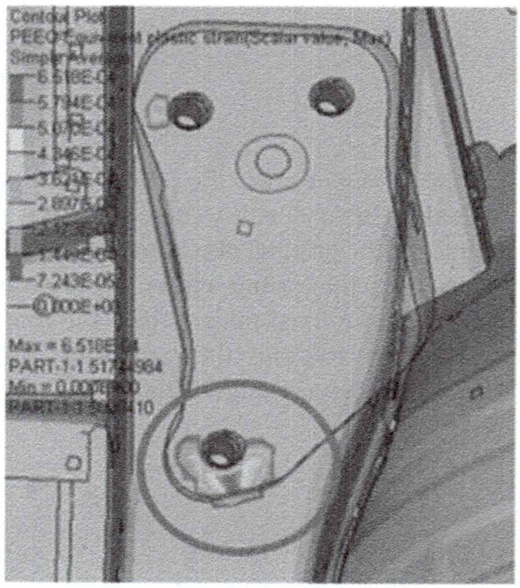

Fig. 9 Plastic strain cloud of the carling in bolt sleeve structure

which indicates that the largest strain of 0.065% is much less than 0.2%. After the new road test, there was no cracking problem at the torsion beam mounting point.

3.3.3 Downward Displacement Analysis of the Bolt Sleeve Structure and the Torsion Beam Mounting Axis

Due to the chassis tuning, the torsion beam mounting axis has a displacement of 35 mm in the z-axis, which changes the distance between the torsion beam mounting axis and the carling from 59 to 94 mm, as shown in Fig. 10, i.e., lengthen the arm of force by 60%.

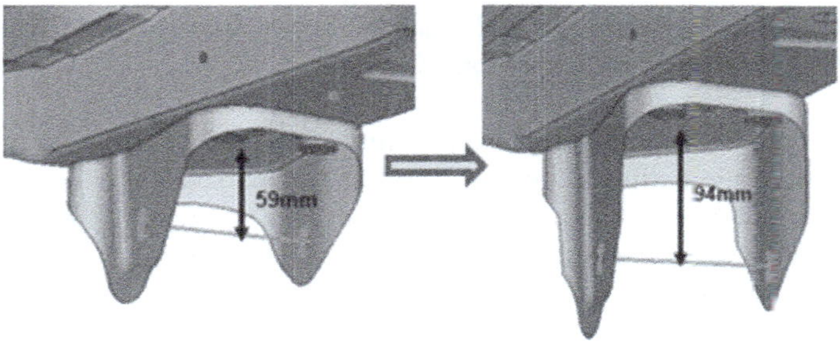

Fig. 10 Downward displacement of the torsion beam mounting axis

Longer arm of force results in larger forces and torques. The simulation result of strain of the carling is shown in Fig. 11, which illustrates that the largest strain of 0.17% is less than 0.2%. So, the structure was safe according to the simulation analysis. However, the structure still cracked after the road test of more than 200 thousand km. The reason should be the simulation error. After all, 0.17% is close to the limit of 0.2%. So, the structure still needs to be optimized to lower the strain and solve the crack problem.

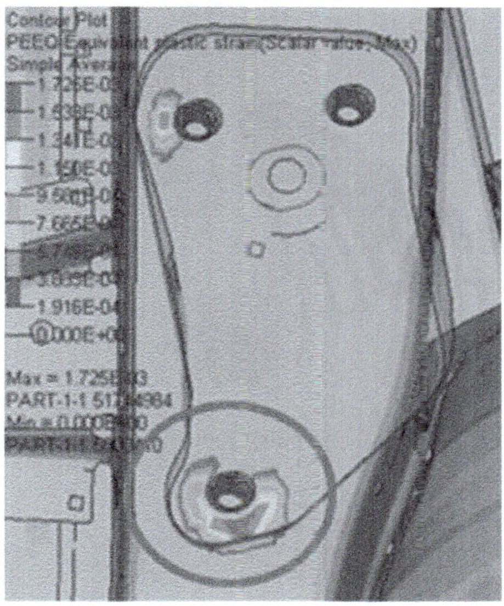

Fig. 11 Plastic strain cloud of the carling after the torsion beam mounting axis down

3.3.4 Structure Optimization and Analysis Verification After Moving Down the Torsion Beam Mounting Axis

To further decrease the strain, topology optimization method was used to optimize the structure of the carling cavity on the basis of the scheme using bolt sleeve. The optimization result is shown in Fig. 12, and the red regions are the effective distribution of materials in the carling cavity. The scheme in consideration of the optimization result is shown in Fig. 13.

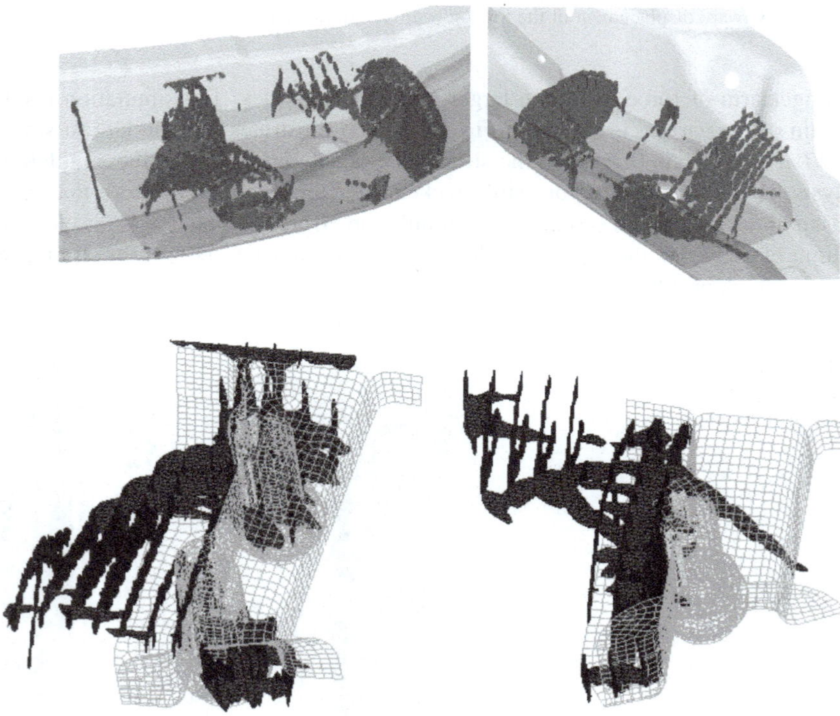

Fig. 12 Topology optimization result of the carling cavity

Fig. 13 Optimized result

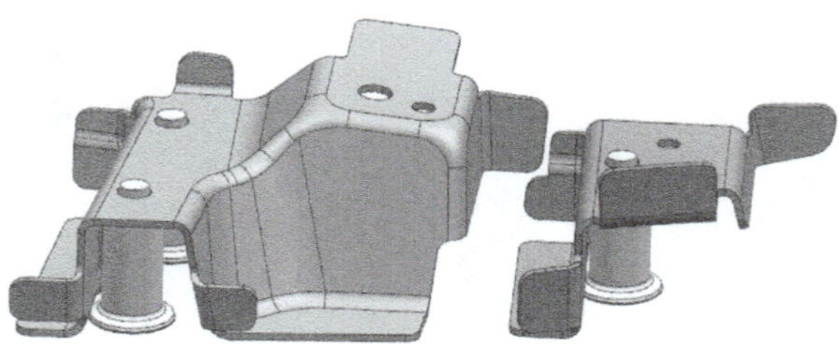

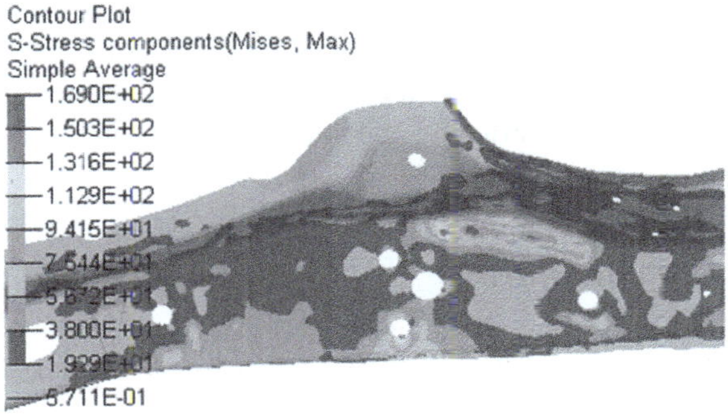

Fig. 14 Stress cloud of the carling after optimizing structure

Put the optimization scheme to the mechanical model. The simulation result showed that there was no plastic strain in the carling, and the maximum stress was 169 MPa, as shown in Fig. 14. So, the new structure can fully meet the durability requirement of the car body.

4 Scheme Verification

The scheme after topology optimization is realized by welding the sheet metal by hand to the car. The torsion beam mounting point had no crack problem after both test bench of 24 channels road simulator and road test of 300 thousand kilometers. The test results provided that the scheme meets the demand for strength and durability of car body. The crack problem was completely solved.

5 Summary of the Experience of the Torsion Beam Mounting Point Design

After the study of the crack problem of the torsion beam mounting point and the real vehicle test, design experience of the structure of the torsion beam mounting point was gained. As a matter of fact, there are two ways to improve durability. First, improve the strength grade of materials and increase the thickness of components. Second, optimize the car body structure. Improving the strength grade of materials would result in higher costs and a higher difficulty of stamping. Increasing thickness would result in higher costs and weight. The optimization of structure generally takes a longer time, but it can ensure less cost and weight with a better performance. So during the design of the structure of the torsion beam mounting point, unless there are limits of the mounting or layout space, directly improving the strength grade and increasing the thickness are not recommended. The current criteria are that the plastic strain should be lower than 0.2%. In consideration of error, the maximum stress of the components should be as low as possible to the yield strength of the material. Otherwise, the vehicle verification is required to confirm whether the car body meets the durability requirement. The following is design points from the two aspects of the structure and material specifications.

5.1 Structure

(In all of the following examples, the material type of the carling was 340/590DP and the thickness was 1.5 mm. The material type of the mounting nut plates was SAPH440 and the thickness is 2.5 mm.)

(1) If there are three mounting points for the mounting of torsion beam, the scheme of bolt sleeves must be used to ensure that the loads can be bore by the bolt sleeves and the bolt sleeve fixed plate. The strength grade of the material cannot be lower than SAPH440, and the thickness cannot be less than 2.5 mm to decrease the stress and strain of the carling and avoid cracking due to poor durability.

(2) If the space allows, mounting type of four torsion beam mounting points is recommended, and the layout of the mounting point should be similar to a rectangle, as shown in Fig. 15. It is recommended that the distance between point 1 and point 4, and the distance between point 2 and point 3 are both above 60 mm. And under this circumstance, nuts can be welded directly on the stiffness plate, and the bolt sleeve fixed plate and the bolt sleeves can be removed from the structure, which not only reduced the body weight by 2 kg but also reduced the cost of per car by about 20 Yuan. This kind of mounting structure is recommended for further designed vehicles.

    Another vehicle is in the design stage, whose mounting brackets between the bracket of the torsion beam and carling has four bolts, as shown in Fig. 16.

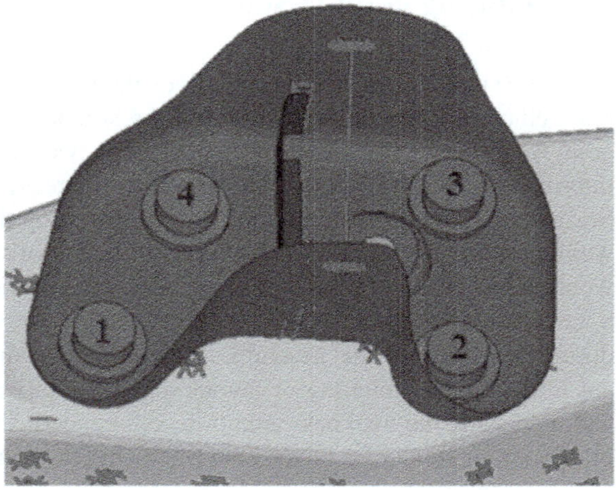

Fig. 15 Mounting point location

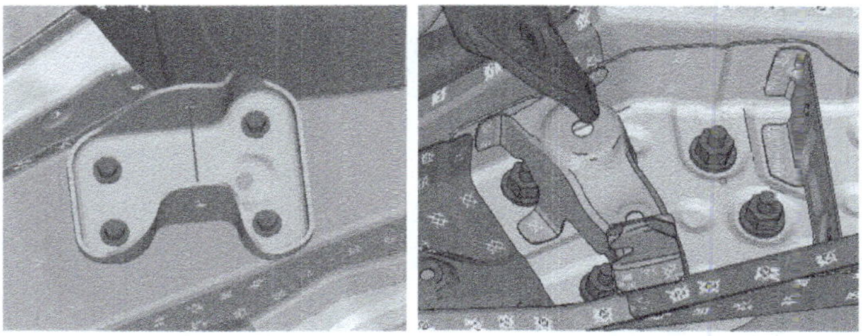

Fig. 16 Torsion beam mounting point

The simulation result of the carling is shown in Fig. 17, which indicates that there is no plastic strain, and the maximum stress is only 107.8 MPa, which proves that the type of four mounting points has excellent durability and strength.

(3) More stiffeners or bowl structure that has flanging on its whole border should be used on the mounting nut plate of the torsion beam.

a. Verification of stiffeners effect on the mounting nut plate of the torsion beam.

As shown in Fig. 18, on the basis of the structure that has four mounting points, two mounting nut plate structures were designed. The structure on the left had stiffeners and bosses, while the structure on the right had no features. Put the two structures into simulation analysis, and the results showed that both structures had no plastic strain. The stress distribution of the carling is shown in Fig. 19, and the stress distribution of the mounting nut plate is shown in Fig. 20. More specific results are listed in Table 1. The results demonstrate that add stiffeners on the mounting nut plate can lower the stress near the torsion beam mounting point and improve the durability and strength.

b. Verification of stiffeners effect on the mounting nut plate of bowl structure that has flanging on its whole border

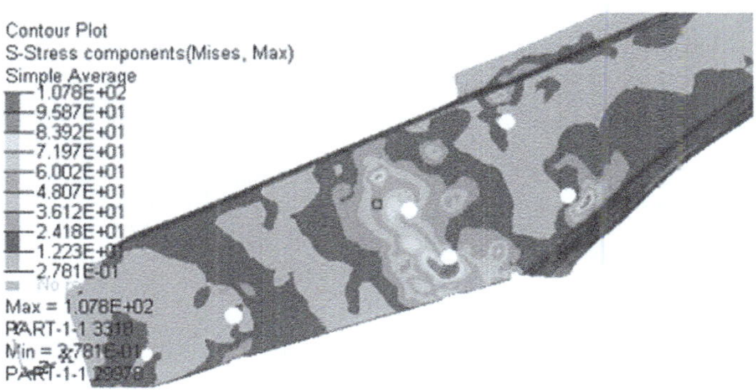

Fig. 17 Stress cloud of the carling

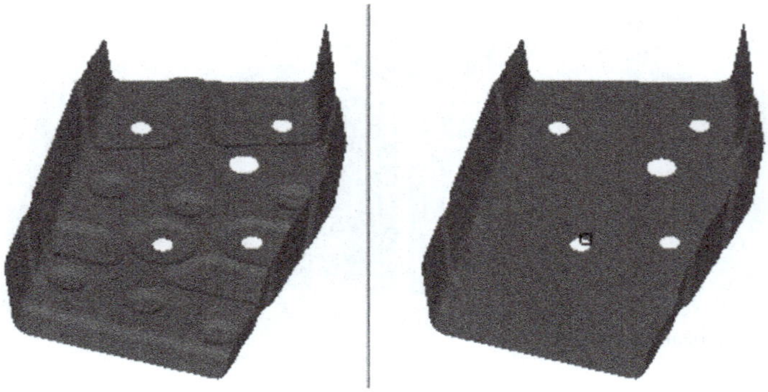

Fig. 18 Different structures of the mounting nut plate

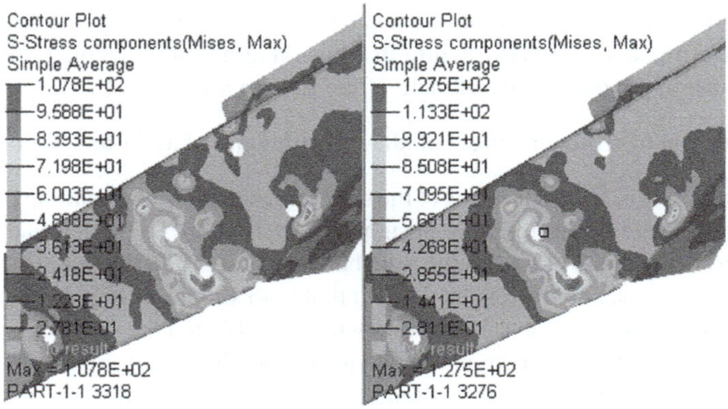

Fig. 19 Stress cloud of the carling of different structures

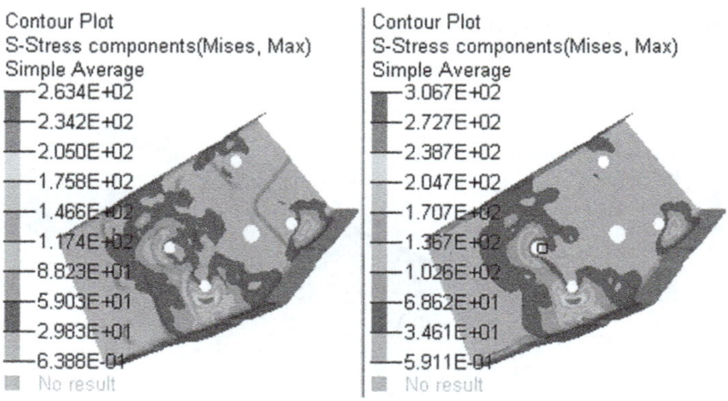

Fig. 20 Stress cloud of the fixed plate of different structures

Table 1 Comparison table of different structural stress values

Type	Maximum stress of the carling (Mpa)	Maximum stress of nuts (Mpa)
Nut plate with stiffeners	107.8	263.4
Nut plate without stiffeners	127.5	305.7
Difference	−19.7	−43.3

As shown in Fig. 21, on the basis of the structure that has four mounting points, two mounting nut plate structures were designed. The structure on the left is bowl structure that has flanging on its whole border, and the structure on the right is structure that has flanging only on its front and rear border. Put the two structures into simulation analysis, and the results show that both structures have no plastic strain, and the stress distributions of the carling and the mounting nut plate are, respectively, shown in Figs. 22 and 23. More specific results are listed in Table 2. The results demonstrate that bowl structure that has flanging on its whole border plays a great role in lowering the stress near the torsion beam mounting point.

(4) Decrease the distance between the mounting axis of the torsion beam and the carling

While the structure has four mounting points, a sink feature could be put at the bottom of the torsion beam mounting bracket and sink into the carling hole, which can decrease the distance between the mounting axis of the torsion beam and the carling,

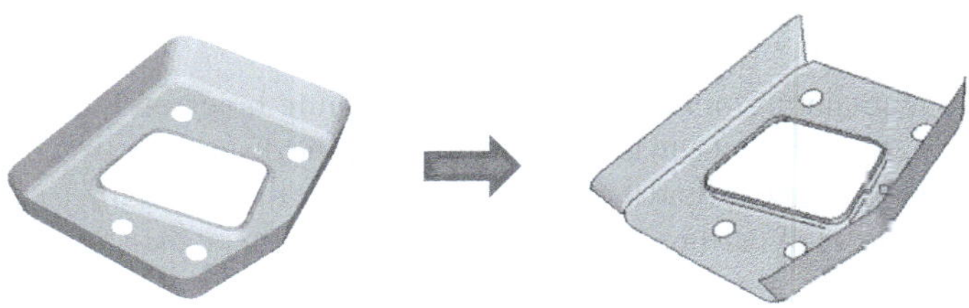

Fig. 21 Different structures of the mounting nut plate

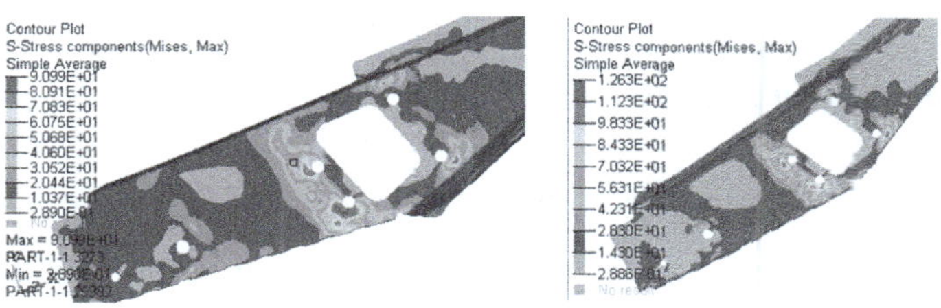

Fig. 22 Stress cloud of the carling of different structures of the mounting nut plate

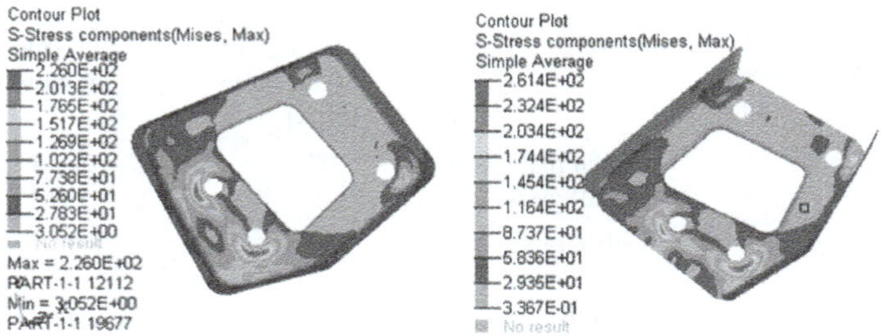

Fig. 23 Stress cloud of the mounting nut plate of different structures

Table 2 Comparison table of different structural stress values

Type	Maximum stress of the carling (Mpa)	Maximum stress of nuts (Mpa)
Fully enclosed bowl structure	90.99	226
Front and rear openings' structure	126.3	261.4
Difference	−35.31	−35.4

as shown in Fig. 24. According to a reference car, the distance can be decreased by 18 mm, as shown in Fig. 25.

The results of the simulation analysis show that there is no plastic strain on both the carling and the mounting nut plate. The stress is listed in Table 3, and the stress distribution of the carling and the mounting nut plate are, respectively, shown in Figs. 26 and 27.

Using four mounting points' structure and opening holes on the mounting nut plates can not only lessen the mass but also decrease the distance between the mounting axis of the torsion beam and the carling. This reduce the arm of force and decrease the strain and stress of carling and mounting nut plate, make the strength and durability of the torsion beam mounting point as good as possible under the limit of cost and mass, totally avoid crack problem caused by poor durability, and greatly improve the efficiency of car body development and reliability of car body.

5.2 *Material and Thickness*

(1) Improve the strength grade of the material

It is the poor strength and durability of the structure that cause the crack of the carling, so the most direct way to solve the problem is to improve the strength grade of the material. Taking one car as an example, improve its strength grade of carling from

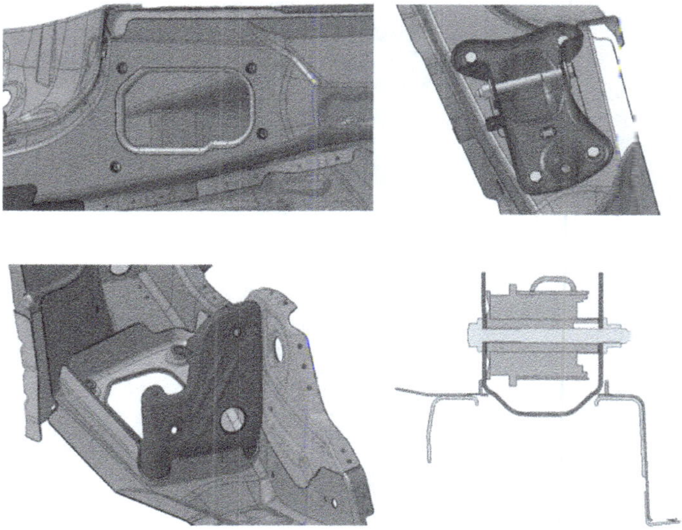

Fig. 24 Opening structure of the torsion beam mounting point

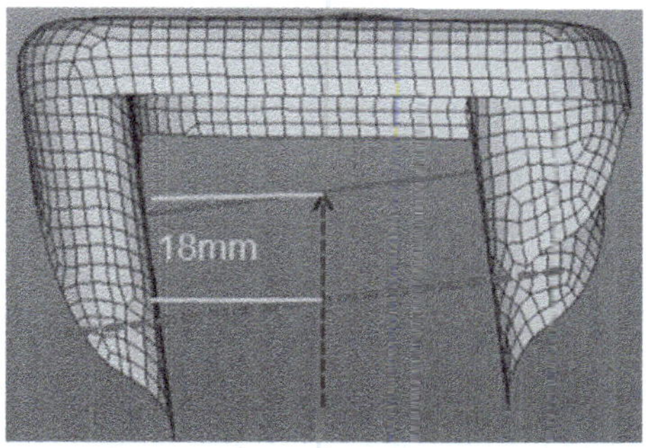

Fig. 25 Opening structure of the torsion beam mounting point (axis up)

Table 3 Comparison table of stress values at different distances between the mounting axis and the carling

Type	Maximum stress of the carling (Mpa)	Maximum stress of the mounting nut plate (Mpa)
Original distance	90.99	226
Distance reduced by 18 mm	68.56	169.7
Difference	22.43	56.3

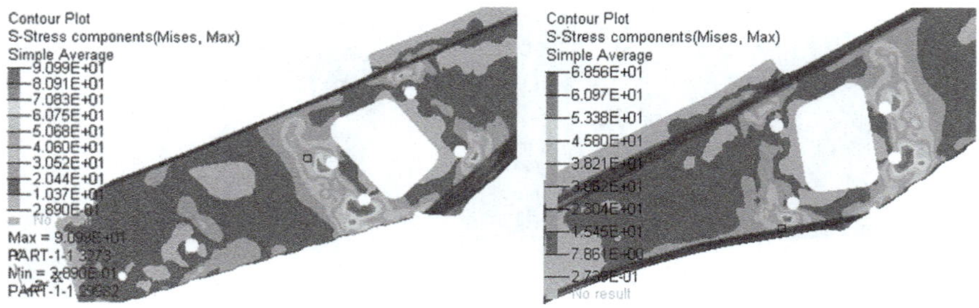

Fig. 26 Stress cloud of the carling

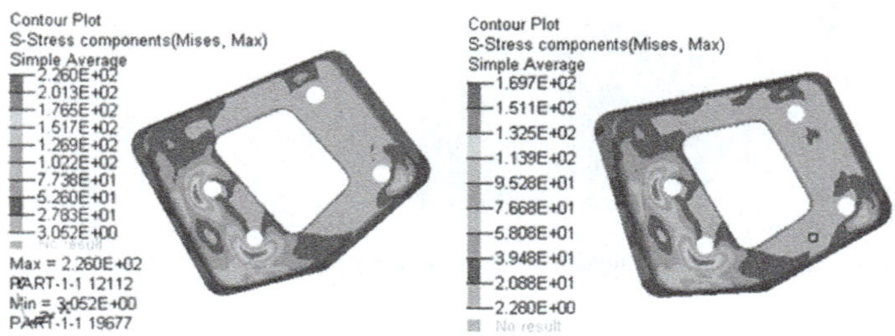

Fig. 27 Stress cloud of the mounting nut plate

280VK to 340/590DP while keeping the thickness the same. As Fig. 28 shows, the plastic strain of the carling decreases from 0.186 to 0.102%, which greatly decreases the risk of the crack of the carling.

(2) Increase the thickness of the mounting nut plate

Mounting nut plates not only provide nuts for mounting the torsion beam but also strengthen the carling nearby, and disperse the forces and torques. So, to lower

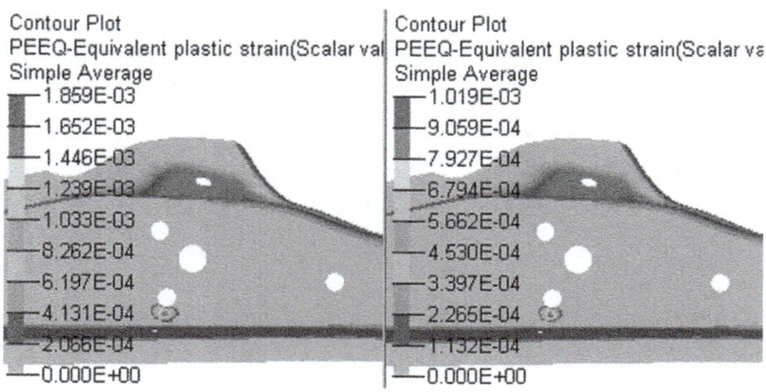

Fig. 28 Clouds of stress and plastic strain of the carling

the cost and reduce the weight, thickness of small mounting nut plates should be increased to decrease the strain and stress of the carling. One car used the 2-mm-thick mounting plates. To prove the influence of thickness for the strain and stress of the carling, the thickness was changed to 2.0 mm. According to the simulation analysis of two conditions, the stress of the carling and the mounting nut plate are, respectively, shown in Figs. 29 and 30, and more specific data of the stress is listed in Table 4.

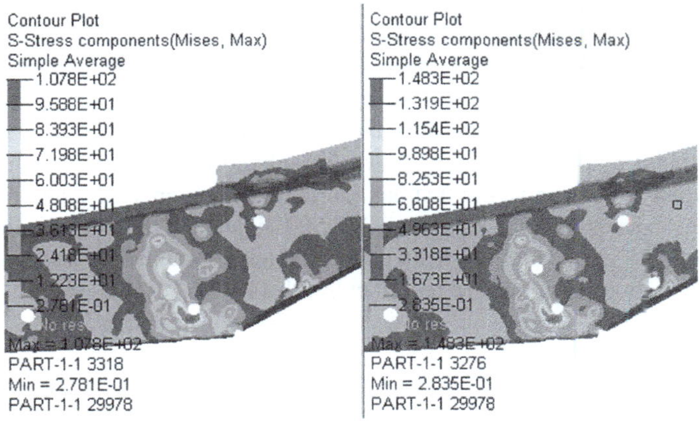

Fig. 29 Clouds of stress of the carling

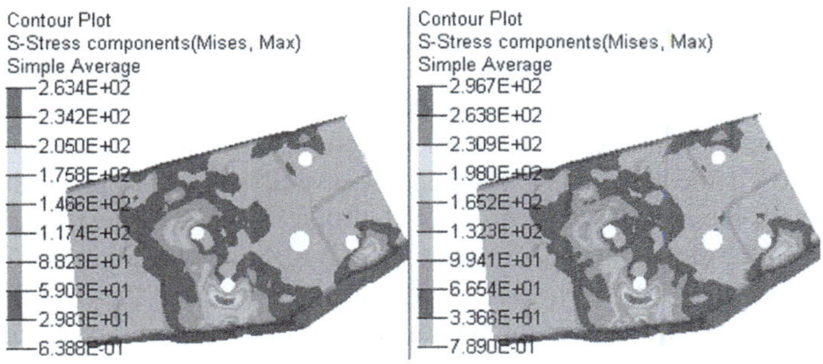

Fig. 30 Stress cloud of the mounting nut plate

Table 4 Comparison table of stress values at different thicknesses of the mounting nut plate

Type	Maximum stress of the carling (Mpa)	Maximum stress of the mounting nut plate (Mpa)
2.5-mm thick	107.8	263.4
2.0-mm thick	148.3	296.7
Difference	−40.5	−33.3

According to the simulation analysis, after decreasing the thickness from 2.5 to 2.0 mm, the maximum stress of the carling increased by 37.5%, and the maximum stress of the mounting nut plate increased by 12.64%. So, it is obvious that the thickness of the mounting nut plate has a great effect on the stress near the torsion beam mounting point. The thicker the mounting nut plate is, the smaller the maximum stress of the carling is. As a matter of fact, by our verification, taking one car whose torsion beam mounting point cracks in the road test as an example, when the thickness of its mounting nut plate was changed to 5 mm and the other components kept the original thickness, the maximum stress of the carling would become so small that the crack problem could be completely solved. However, the mounting nut plates were too thick to be welded following the normal welding process. So, the thickness of the mounting nut plates is limited to less than 3.0 mm.

By analysis and verification, no matter how the structure of the torsion beam mounting point is optimized, it should be considered first that the structure should be the rational mounting layout that has four mounting points. Second, the strength grade of the carling cannot be less than 340/590DP and the thickness cannot be less than 1.5 mm. As for the mounting nut plate, its thickness cannot be less than 2.5 mm and the strength grade cannot be less than SAPH440.

6 Conclusion

In this paper, CAE mechanical analysis model was established for solving the crack problem that occurs to the torsion beam mounting point of a car during the durability road test, and the reason of the crack problem was analyzed. After many times of comparison, the effective simulation model and the judging criteria are finally found. Then, on the basis of the results of the mechanical analysis, by using the topology optimization method and the structure of the reference car for comparison, the final optimization scheme was finally formed after several rounds of optimization. And, it was through the double verification of the road test and the test bench of road simulator. Furthermore, the experience of the optimization was summarized, and the contribution values of different factors are given. It indicated the direction for the future design of the structure of the torsion beam mounting point. Although bearing large forces and torques, the torsion beam mounting points are prone to failure and cracking due to poor strength and durability, the feasible optimization schemes that are economical and meet the demand for strength and durability can be made by using the means of topology optimization and simulation analysis, analyzing and sorting the effects of the different factors, and considering the lightweight design, cost, feasibility of stamping and welding. The main conclusions are summarized as below.

(1) The main reason that the torsion beam mounting point cracks is that the forces and torques between the car body and the chassis are too severe, and the dura-

bility and strength of the torsion beam mounting point are too poor to bear the loads repeatedly.

(2) If there are three mounting points on the torsion beam mounting point, the bolt sleeve structure must be used, and the strength grade of the bolt sleeve fixed plate cannot be less than SAPH440, and its thickness cannot be less than 2.5 mm.
(3) The mounting layout that has four mounting points should be considered first.
(4) The mounting axis of the torsion beam should be as close as possible to the carling to shorten the arm of force.
(5) The mounting nut plate of the torsion beam should include some stiffeners or be bowl structure that has flanging on its whole border, to decrease the strain and stress near the torsion beam mounting point and greatly decrease the risk of crack. The thickness of the mounting nut plate should be no less than 2.5 mm, and its strength grade should not less than SAPH440.
(6) The strength grade of the carling cannot be less than 340/590DP and its thickness cannot be less than 1.5 mm.
(7) Aiming at solving the conflict between the problem that the torsion beam mounting point cracks during the road test and the fact that the number of the mounting points of the torsion beam mounting point cannot be added to 4, a bracket similar to 'Ω' is designed by topology optimization, which weld the base, the top and the facade of the bolt sleeve on the bracket by two shielded welding to change the stress of the bracket from shearing stress to both shearing stress and the tension stress, and spread the stress of different components. Putting the optimized scheme to the road test and the test bench road simulator for verification, the results showed that the car body met the durability and strength requirements.

Measurement and Modeling of Hysteresis in Pneumatic Actuator Under Different Loading Rate

Yunbin Gong and Henry Guo

Abstract A hysteresis model under different loading rate is proposed based on measurement results with simple polynomial method. This model will be used for better control and compensation of pneumatic actuator in actual application. The hysteresis model is achieved using simple polynomial method based on measurement data. Six groups of testing data are collected under different loading rate from quasi-static (2 kPa/s) to 25 kPa/s at growth rate of 5 kPa/s. History independence research is also conducted in order to understand more detailed hysteresis feature of pneumatic actuator. Based on measurement results, it is concluded that the hysteresis model is effective to describe the effect of stroke and loading rate on the hysteresis. The history independence of hysteresis study validates the model is robust enough

Keywords Turbocharger · Pneumatic actuator · Hysteresis · Measurement · Modeling · Loading rate

1 Introduction

Turbocharging technology is important for automotive spark-ignition engines to support downsize concept in order to reduce fuel consumption and exhaust emissions. Free-floating turbocharger boost is controlled by turbine size, but it is at the expense of lower speed boost. A smart wastegate turbocharger controls boost by bypassing turbine flow, so it allows smaller turbine to increase low speed boost. Wastegate allows optimizing both low and high speed engine performance [1]. It is widely used in passenger vehicle due to downsizing design. Pneumatic actuator is a key component for wastegate turbocharger since it controls the position of bypassing valve and controls the boost as required.

Hysteresis is the time-based dependence of a system's output on present and past inputs. The dependence arises because the history affects the value of an internal state [2]. Pneumatic actuator hysteresis often leads to problems in the wastegate

Y. Gong · H. Guo (✉)
Honeywell Integrated Technology (China) Co., Ltd, Shanghai, China
e-mail: guohezhong2000@sina.com

China SAE (ed.), *Proceedings of China SAE Congress 2018: Selected Papers*,
Lecture Notes in Electrical Engineering 574,
https://doi.org/10.1007/978-981-13-9718-9_4

turbocharger with internal friction, pressure loading rate, material types, actuator size because it creates tracking errors, limit cycles and undesired slip motions. The demand for the measurement and modeling of actuator hysteresis is rising. There is a need for hysteresis model: a simple and physically meaningful model that is suitable for control purposes. Lack of this knowledge will leave a difficulty for the control system afterward.

Currently, there are several hysteresis models, simple polynomial model, Preisach model, Lapshin model, Bouc–Wen model, etc.

Preisach model shows hysteresis is nonlinear. In this model, hysteresis is as a linear superposition of square loops called non-ideal relays. Many other complex hysteresis models were arising from the simple parallel connection, or superposition, of elementary carriers of hysteresis termed hysterons [3]. In the Lapshin model of hysteresis, a simple parametric description of various hysteretic loops may be found [4]. Along with the classical loop, substitution of rectangle, triangle or trapezoidal pulses instead of the harmonic functions also allows piecewise-linear hysteresis loops frequently used in discrete automatics to be built in the model. The Bouc–Wen model of hysteresis is used to describe nonlinear hysteretic systems very often. This model was introduced by Bouc [5, 6] and extended by Wen [7], who demonstrated its versatility by producing a variety of hysteretic patterns. These hysteresis models have been widely used in engineering applications [8–12].

In this paper, the simply polynomial model will be used to describe the actuator hysteresis based on the testing data. Six groups of testing data are collected under different loading rate from quasi-static (2 kPa/s) to 25 kPa/s at growth rate of 5 kPa/s. History independence research is also conducted in order to understand more detailed hysteresis feature of pneumatic actuator.

2 Hysteresis Characteristics Study for Pneumatic Actuator

2.1 Pneumatic Actuator Geometry Introduction

The geometric model of the actuator is shown in Figs. 1 and 2 shows the scheme of the actuator. There is an elastic diaphragm. Generally, the diaphragm is made of rubber (a) which is fixed between two disks (b) in the closed case (c) and it is mounted on a steel core (d). This core is supported by the steel spring (e) on the one side. Spring is heat set after winding for stability and it has special coatings in order to have cracking and corrosion resistance. This assembly connects a control rod. The rubber diaphragm is reinforced by the textile on one side (f). A pressure difference on the one and the other side of the diaphragm causes movement of the rod in the axial direction of the actuator. Final position of the control rod depends on the mechanical properties of the diaphragm, on the spring stiffness, and on the pressure difference inside the actuator. This type of actuator is derived by pneumatic which is so-called pneumatic actuator. The actuator is subjected many types of tests: salt spray test,

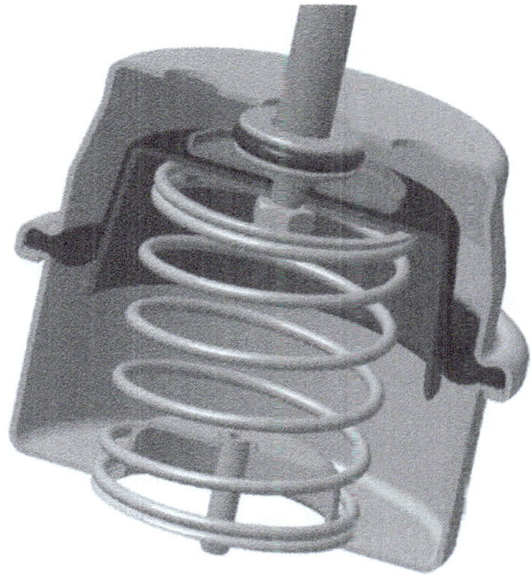

Fig. 1 Partially cut model of the actuator

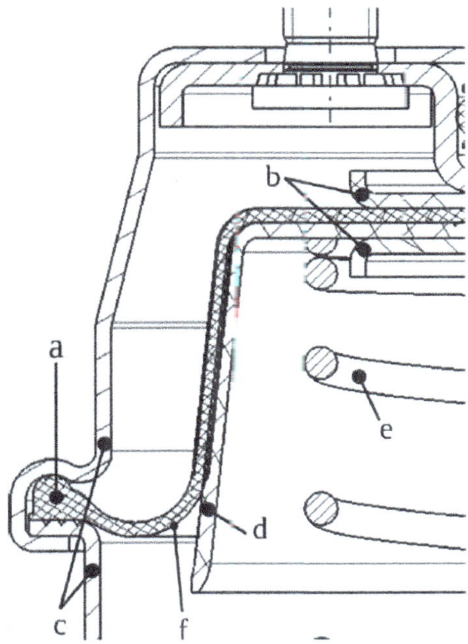

Fig. 2 Scheme of the actuator structure

dust test, chemical test, cyclic test of diaphragms and spring and hysteresis test. In this paper, hysteresis test and study are mainly discussed.

2.2 Experiments and Setup

Since there is no analytical analysis and solution to describe pneumatic actuator pressure-stroke relationship, a variety of testing configurations has been chosen to illustrate part of this relationship. The hysteresis characteristics for the pneumatic actuator used in this study were studied and measured under different loading rate. A layout of experiment system is shown in Fig. 3. The actuator (h) is fixed in the testing bench by adaptor (i). The pressure source (j) applies pressure loading to the actuator and the displacement sensor (g) detects the displacement change. The pressure inside the actuator increases, and as a result, the rod extends. Computer (k) can set the pressure increasing and decreasing rate and record pressure-displacement curves.

The stiffness of spring in actuator is 5.43 N/mm and the free length is 52.58 mm. The spring is compressed to 28.21 mm after installing in the actuator. The effective diameter for pressure loading is 57.5 mm which is depended on the actuator size. The thickness of diaphragm is 0.5 mm and the material is fluorosilicone elastomer (FVMQ).

Three actuators are prepared and each actuator is tested two times under same condition. The pressure rate is set as 2, 5, 10, 15, 20 and 25 kPa/s to study the effect of loading rate on hysteresis. Once the setup is finished and motion starts, computer

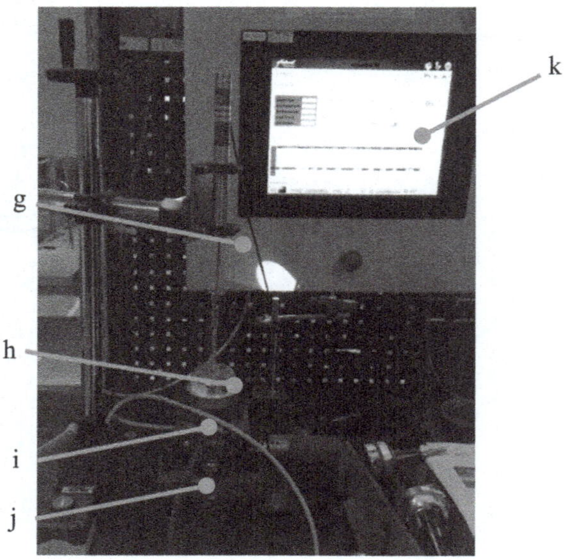

Fig. 3 Picture of pneumatic actuator setup

can record the forward and backward pressure-stroke curve automatically. The gap of two curves is considered as hysteresis value at specified stroke.

2.3 *Experiments Results Under Different Loading Rate*

The average data is taken for each pressure rate. Figures 4, 5, 6, 7, 8 and 9 show the relationship between internal pressure and stroke under different pressure rate.

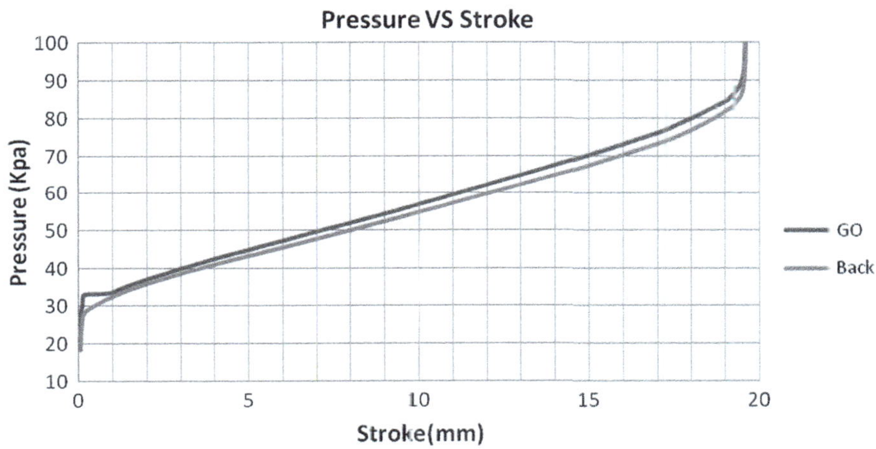

Fig. 4 Internal pressure and stroke curve (2 kPa/s)

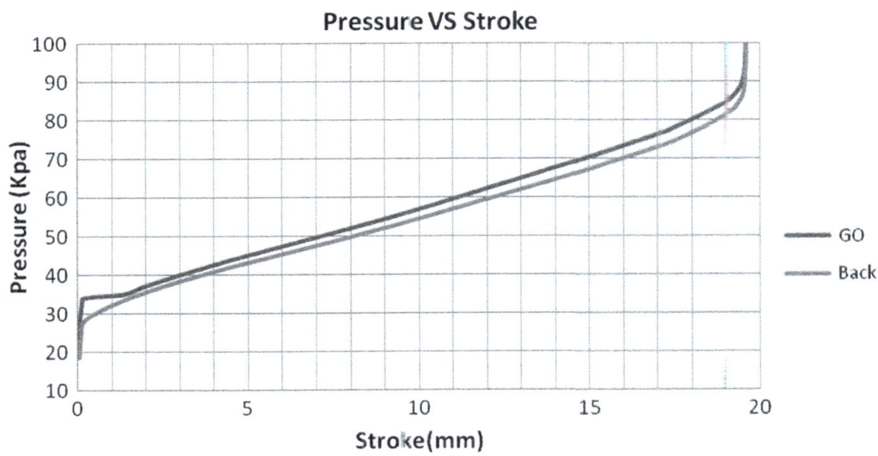

Fig. 5 Internal pressure and stroke curve (5 kPa/s)

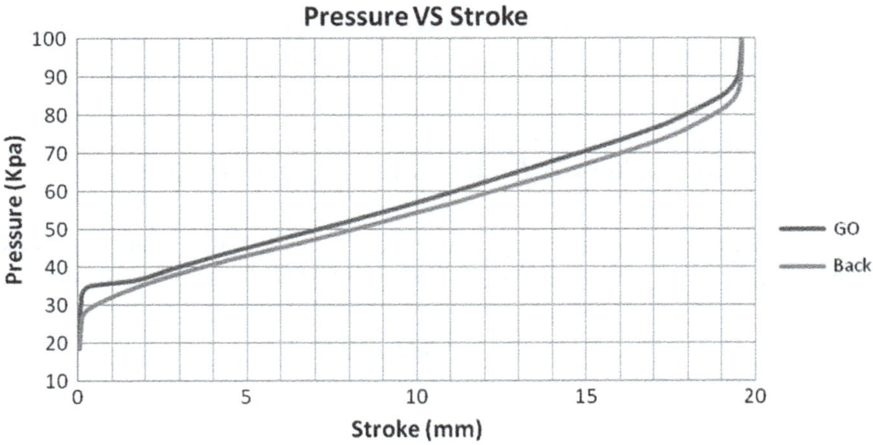

Fig. 6 Internal pressure and stroke curve (10 kPa/s)

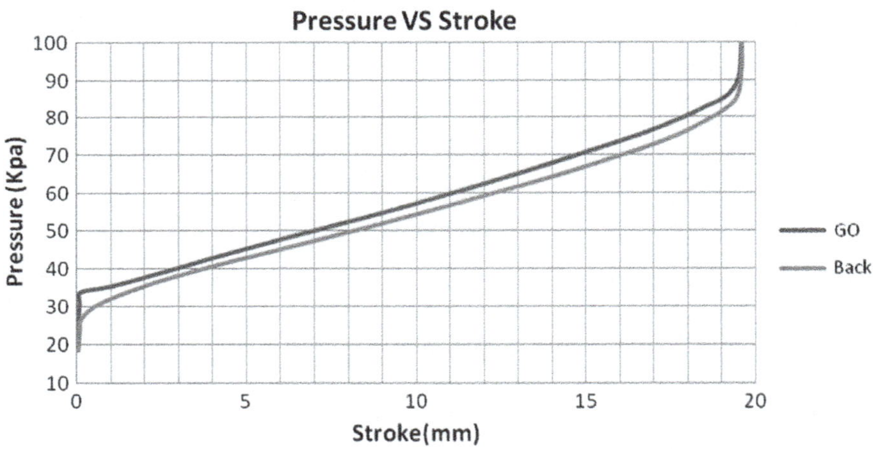

Fig. 7 Internal pressure and stroke curve (15 kPa/s)

2.4 Hysteresis Modeling

By fitting these lines with a linear equation, the slope and the intercepts are obtained as functions of the rod stroke. These lines look very straight. This brings us to formulate the hysteresis model as a linear function of the rod stroke. It can have:

$$Y_1 = k_1 * X + b_1 (\text{forward}), \quad Y_2 = k_2 * X + b_2 \text{ (backward)}$$

As described before, the hysteresis is the gap of both lines. It can be expressed as

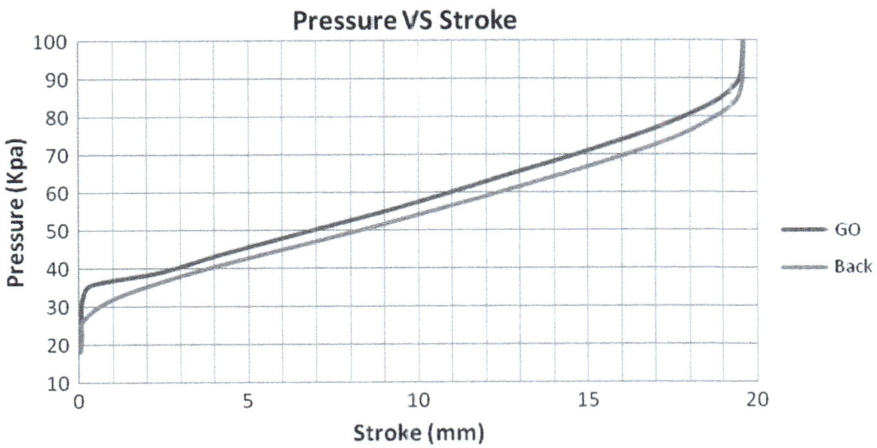

Fig. 8 Internal pressure and stroke curve (20 kPa/s)

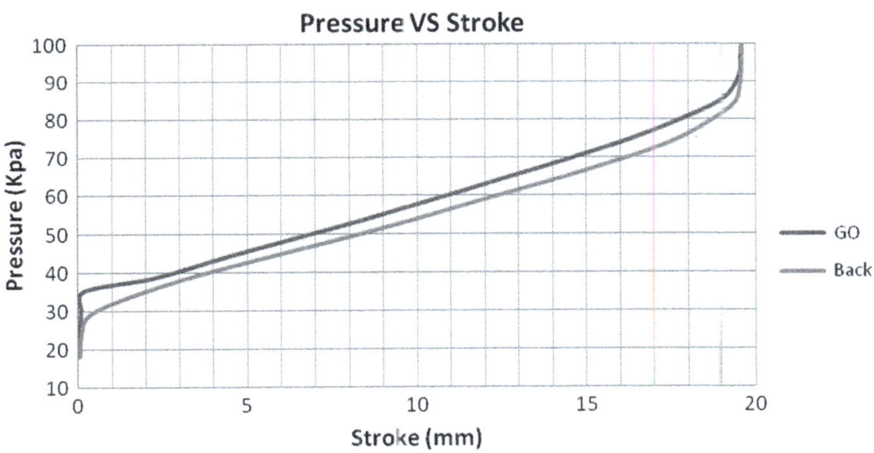

Fig. 9 Internal pressure and stroke curve (25 kPa/s)

$$H = Y_1 - Y_2 = k * X + b,$$

where $k = k_1 - \mathrm{k}_2$, $b = b_1 - b_2$.

All tested data are regressed as below, see Table 1.

Regression analysis is done to describe k and b with the function of loading rate v.

$$k = m_1 * v^3 + n_1 * v^2 + p_1 * v - t_1 \quad b = m_2 * v^3 + n_2 * v^2 + p_2 * v + t_2$$

where:

Table 1 Parameters of hysteresis function under different loading rate

Loading rate	2 kPa/s		5 kPa/s		10 kPa/s		15 kPa/s		20 kPa/s		25 kPa/s	
$k1, k2$	2.549	2.437	2.554	2.432	2.558	2.430	2.559	2.428	2.563	2.416	2.574	2.412
$b1, b2$	31.875	30.793	31.966	30.776	32.043	30.589	32.208	30.443	32.334	30.316	32.417	30.236
R^2	0.999	0.999	0.999	0.999	0.999	0.999	0.999	0.999	0.999	0.999	0.999	0.999
Hysteresis: k	0.112		0.122		0.128		0.132		0.147		0.162	
Hysteresis: b	1.082		1.190		1.454		1.765		2.018		2.181	

$$m_1 = 7.11\text{e}{-06} \quad n_1 = -2.60\text{e}{-4} \quad p_1 = 3.28\text{e}{-3} \quad t_1 = 2.61 \quad R^2 = 0.996$$

$$m_2 = -6.03\text{e}{-05} \quad n_2 = 2.34\text{e}{-3} \quad p_2 = 1.23\text{e}{-3} \quad t_2 = 31.34 \quad R^2 = 0.999$$

Totally it is expressed:

$$H = \left(m_1 * v^3 + n_1 * v^2 + p_1 * v + t_1\right) * x + \left(m_2 * v^3 + n_2 * v^2 + p_2 * v + t_2\right)$$

$$\text{or } H = (m_1 * x + m_2) * v^3 + (n_1 * x + n_2) * v^2 + (p_1 * x + p_2) * v + (t_1 * x + t_2)$$

where the unit for loading rate v is ka/s and for stoke x is mm.

R^2 is very high that shows the data fitness very good. It is easy to see from measured pressure-stroke curves and the hysteresis model that hysteresis is related to rod stroke and loading rate. And it suggests the hysteresis is much obvious with longer stroke and higher loading rate.

2.5 Experiments Results Under Different Operating Stroke

In order to understand the hysteresis of actuator is history dependent or independent, one actuator is performed testing under different operating stroke. The same loading rate is set as 10 kPa/s and three different operating ranges: 20–100 kPa, 20–80 kPa and 20–50 kPa. Two groups of testing data are plotted including forward and backward paths, see Figs. 10 and 11.

It can be easily seen the forward and backward paths remain the same. This indicates the hysteresis is history independent.

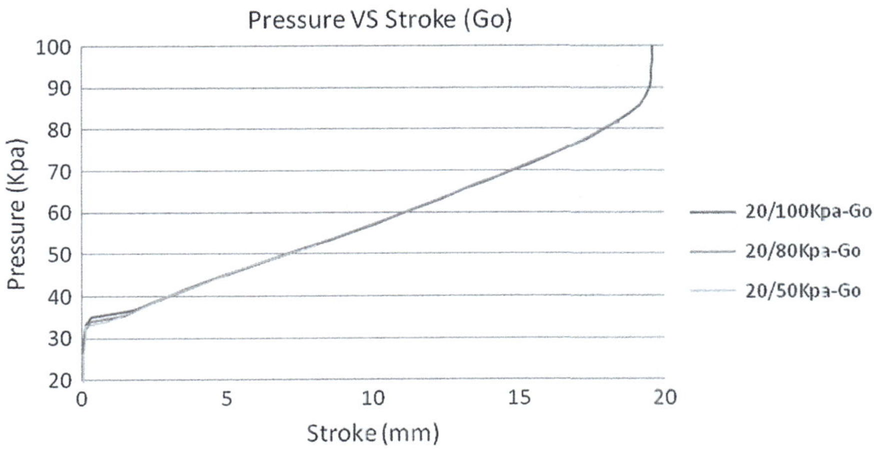

Fig. 10 Forward paths under 3 pressure ranges

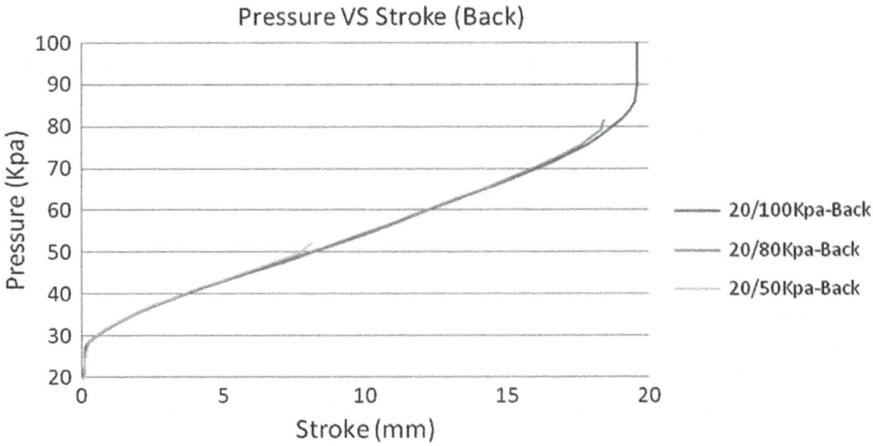

Fig. 11 Backward paths under 3 pressure ranges

3 Conclusion and Further Work

The simply polynomial hysteresis model gave us simple yet accurate description of the actuator and pneumatic behaviors. This model can be used in simulation of actuator system and for better control and compensation of pneumatic actuator in actual application.

Based on measurement results, it is concluded that the hysteresis model is effective to describe the effect of stroke and loading rate on the hysteresis. The history independence of hysteresis study validates the model is robust enough.

However, the hysteresis model in pneumatic model and system, which effects control, compensation and power to the actuator, still needs a lot of study and further improvement. For example, it needs to find the relationship between hysteresis and spring stiffness, actuator size and diaphragm types. The further research in this direction is highly recommended.

References

1. Honeywell Turbo Technologies internal document
2. https://en.wikipedia.org/wiki/Hysteresis
3. Mayergoyz ID (2003) Mathematical models of hysteresis and their applications, 2nd edn. Academic Press. ISBN 978-0-12-480873-7 (Electromagnetism)
4. Lapshin RV (1995) Analytical model for the approximation of hysteresis loop and its application to the scanning tunneling microscope. Rev Sci Instrum 66(9):4718–4730. Bibcode:1995RScI...66.4718L (USA: AIP)
5. Bouc R (1967) Forced vibration of mechanical systems with hysteresis. In: Proceedings of the fourth conference on nonlinear oscillation. Prague, Czechoslovakia, p 315

6. Bouc R (1971) Modèle mathématique d'hystérésis: application aux systèmes à un degré de liberté. Acustica 24:16–25 (in French)
7. Wen YK (1976) Method for random vibration of hysteretic systems. J Eng Mech 102(2):249–263
8. Gupta R, Reitsma SY (1987) Measurement and analysis techniques for hysteresis of the dairy cows' teat. Can Agric Eng 29(1):51–57
9. Wojewoda J et al (2008) Hysteretic effects of dry friction: modelling and experimental studies. Philos Trans R Soc Lond A Math Phys Eng Sci 366(1866):747–765
10. Tan X, Baras JS (2004) Modeling and control of hysteresis in magnetostrictive actuators. Automatica 40(9):1469–1480
11. Vo-Minh T et al (2011) A new approach to modeling hysteresis in a pneumatic artificial muscle using the Maxwell-slip model. IEEE/ASME Trans Mechatron 16(1):177–186
12. Chou C-P, Hannaford B (1996) Measurement and modeling of McKibben pneumatic artificial muscles. IEEE Trans Robot Autom 12(1):90–102

Airflow–Resistance Analysis of Cooling Module Based on CFD Method

Yao Zhang, Bin Liu, Bo Liu, Hongtao You and Yan Zhao

Abstract To explore the relationship between the supply air rate and the resistance of the cooling module itself, this paper provides a simulation method for cooling front module and single cooler supply air rate and wind resistance. Using the matching method, the monomer with the same characteristic structure is simplified to porous medium, and its characteristic parameters are obtained, the CFD is used to simulate the problem, and the accuracy is verified by the test; and the wind speed–resistance characteristics are simulated by the superposition of multiple monomers, and the results are verified. This method can accurately predict the wind resistance of the cooling system during the design stage and provide a basis for the selection of the scheme.

Keywords Cooling system · Radiator · CFD simulation analysis · Wind resistance curve

Foreword

With the rapid growth of automobile technology, new energy, hybrid energy, and other new power cars have emerged to the market which also got great support from national policies. Therefore, high-power-density engine, super-power density generators, and wheel motors which require intensive radiating devices have arisen, and in the meantime, to achieve lightweight and miniaturization, it is a great challenge to the design of the cooling system. Therefore, it is particularly important to optimize the design of the cooling system and excavate the cooling system from all aspects. The cooling capacity of the whole vehicle is inevitably related to its air intake, so how to optimize the allocation of cooling components to improve its air intake coefficient is an effective way to optimize the cooling system. Many domestic scholars and experts have done a lot of research; for example, Xianjun Hou and his colleague have using GT-SUITE, STAR-CMM+ to build a vehicle cooling matching analysis model [1], Feng Lv has been studying on commercial vehicle cooling matching method [2], and

Y. Zhang (✉) · B. Liu · B. Liu · H. You · Y. Zhao
Dongfeng Motor Corporation Technical Center, Wuhan, Hubei 430058, China
e-mail: braveheartzy@126.com

China SAE (ed.), *Proceedings of China SAE Congress 2018: Selected Papers*,
Lecture Notes in Electrical Engineering 574,
https://doi.org/10.1007/978-981-13-9718-9_5

Zhengming Tong and his colleague using wind tunnel test and simulation analysis to analyzing how radiating tunnel numbers would effect radiating performance [4]. This article mainly using wind tunnel test and CFD simulation method to study on wind resistance of cooling module.

1 Single Module Analysis

1.1 Calculation of Flow Characteristic Parameters of Radiator

Because the radiator core contains a large number of turbulent belt structure, complete modeling calculation ability is limited, so it is simplified as a porous medium to solve. In star-ccm+, the porous medium is described with two parameters of inertia resistance coefficient and viscous resistance coefficient [5], so we determine the characteristic coefficient of the radiator according to the experimental data and determine its characteristic coefficient in star according to Formula [1] in star:

$$\frac{\Delta p}{L} = -(p_{\mathrm{i}}|v| + p_{\mathrm{v}})v \tag{1}$$

where

- Δp Import and export pressure drop;
- L Porous media region length;
- v Inlet liquid velocity;
- p_{i} Inertia resistance coefficient;
- p_{v} Viscous resistance coefficient.

According to the above method, the flow characteristic parameters of radiator A can be solved, and the exterior profile characteristic size of radiator A is as shown in Table 1.

In the wind tunnel to complete the radiator monomer wind tunnel test, obtain the experimental data as shown in Table 2.

The fitting curve of velocity and unit length pressure drop in the table conforms to quadratic curve:

$$y = 123.88x^2 + 430.36x;$$

Table 1 Radiator A outline size

Structure parameter of core body (mm)	Width	High	Thick
Radiator core dimensions (mm)	834	730	50

Table 2 Radiator A wind speed–wind resistance test results

Va (m/s)	4	6	8	10
ΔPa (Pa)	178	359	569	833
Pressure drop per unit length (Pa/m)	3560	7180	11,380	16,660

The formula can be obtained from analogy with (1): $p_{\mathrm{i}} = 123.88\ \mathrm{kg/m^3}$; $p_{\mathrm{v}} = 430.36\ \mathrm{kg/(m^3\ s)}$.

1.2 Simulating Calculation of CFD

The radiator B with the same internal characteristics of type A radiator is selected to simulate and analyze the B model to obtain its airflow–resistance curve. To establish the B-type radiator flow area geometric parameters:

The 3D model of the radiator flow area is established according to the ratio of 1:1, and the cutting body grid is divided into star in Fig. 1.

CFD Solver settings. Since the flow is a low-speed flow, select the separation flow model and the specific boundary type set as in Table 3 [5].

According to the above setup, the radiator monomer is solved, and according to the characteristic parameters derived from Model A in step 1.1 as the simulation parameters of the porous media (radiator) in the star-ccm+, the wind resistance of the radiator under different wind speed is solved; the results are as shown in Fig. 2.

1.3 Model B Test Verification and Comparison of Results

The Model B is put into the wind tunnel to complete the radiator monomer wind tunnel test and obtain the test airflow–resistance curve [3] (Fig. 3).

The following experimental data are obtained, and the simulation results are compared with the experimental results as shown in Table 4.

In order to verify the reliability of the method, a set of verification data is proposed, then a set of simulation and experiment are done according to the above method, the simulation test process is no longer repeat, and the experimental results are as shown in Table 5.

Through the comparison between the above two sets of test results and simulation results, the radiator with the same internal structure (heat dissipation pipe, dispersed tropical characteristic size) which is fitted to the parametric porous media simulation result monomer error rate is less than 2.5%, within acceptable range; therefore, it is considered that the method modeling simulation results are reliable.

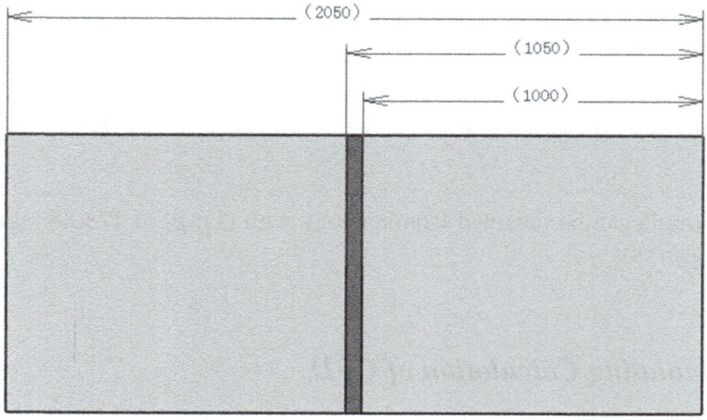

(a) geometrical model

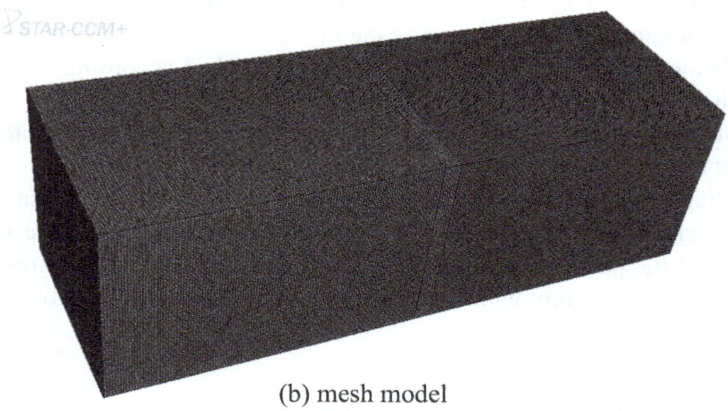

(b) mesh model

Fig. 1 Region model of heat sink solution

Table 3 Solver boundary settings

Inlet	Velocity inlet
Outlet	Pressure outlet
Wall	Treatment of two-layer boundary layer *y*+ wall surface
Solving model	K-Epsilon
Radiator area	Porous media
Mesh type	HEXA
Mesh number	1,687,449

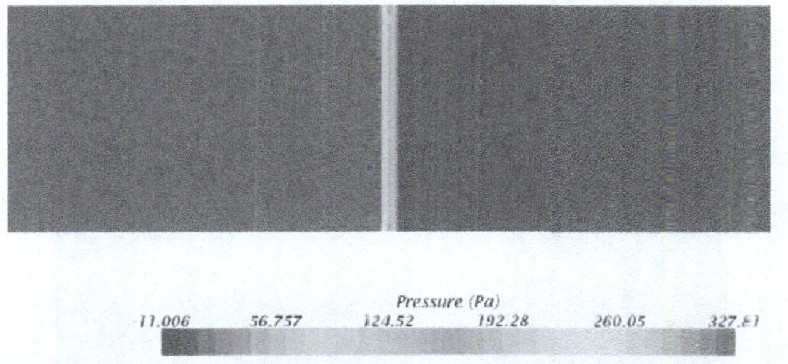

(a) Section Y=0 pressure distribution

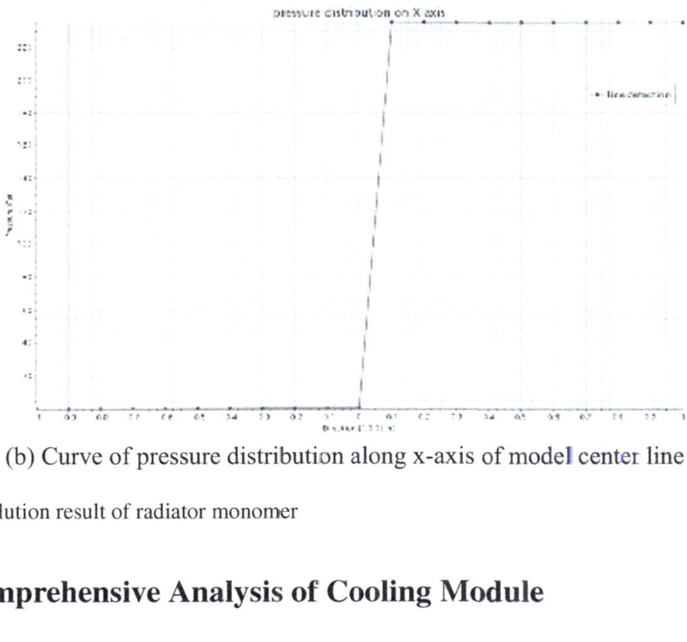

(b) Curve of pressure distribution along x-axis of model center line

Fig. 2 Solution result of radiator monomer

2 Comprehensive Analysis of Cooling Module

2.1 *Solution of Flow Characteristic Parameters of the Remaining Single Modules*

According to the conclusion of the first part, the model with the same external characteristics can be simulated and analyzed by using porous media model, and the results are reliable. The relevant experimental data of the remaining monomers (intercooler, ATF oil cooler, condenser) are obtained by means of experiments. According to Eq. (1), the porous medium characteristic parameters of intercooler, ATF oil cooler, and condenser with the same internal characteristic size are fitted and solved. The results are as follows:

Fig. 3 Wind tunnel test of radiator

Table 4 Comparison of results of radiator test and simulation

Va (m/s)	4	5	6	7	8
The simulation value (Pa)	177.0	263.1	358.9	460.3	570.4
Experimental value (Pa)	178	260.2	350.5	460	571.4
Rate of deviation (%)	0.56	−1.10	−2.34	−0.07	0.18

Table 5 Comparison of results of radiator test and simulation (Group II)

Va (m/s)	4	5	6	7	8
The simulation value (Pa)	163.2	231	314.5	427.3	557.6
Experimental value (Pa)	161.2	231.4	320.2	425.6	562.3
Rate of deviation (%)	−1.23	0.17	1.81	−0.40	0.84

The experimental data of three monomer modules were fitted as shown in Tables 6 and 7.

Table 6 Pressure drop test value of unit length of intercooler, ATF cooler, condenser monomer

Va (m/s)	4	5	6	7	8
Intercooler ΔPa (Pa/m)	1397.36	2142.10	2992.10	3897.3	4992.10
ATF cooler ΔPa (Pa/m)	3031.25	4375	5968.75	7843.75	9937.5
Condenser ΔPa (Pa/m)	5937.5	8250	10,937.5	14,187.5	17,375

2.2 Overall Simulation of Cooling Module

The porous media simulation parameters of each module are solved above; next, we will fit to solve the pressure drop of cooling package superimposed by multiple modules.

As the cooling package is composed of cooling modules arranged in series, we only know the inlet air velocity, the pressure drop characteristic of each module, but the pressure drop of the cooling package is not a simple superposition of the corresponding wind speed pressure drop of each monomer. Because the structure of each cell is not the same, and the back pressure of each module is different, this leads to the inevitable error if the simple pressure drop is superimposed. Theoretically, the flow process conforms to the continuity equation and the momentum conservation equation. Since the process can be simplified to a constant density fluid, the equation can be simplified to the following form:

$$\frac{\partial u}{\partial x} + \frac{\partial v}{\partial x} + \frac{\partial w}{\partial x} = 0 \tag{2}$$

$$\delta_F = \delta_m \cdot \frac{\mathrm{d}v}{\mathrm{d}t} \tag{3}$$

where u, v, and w represent the velocity components along the x-, y-, and z-directions, respectively.

Set up the complete cooling module of each monomer core, according to the actual relative position and distance for assembly, and import star-ccm+ to divide the grid. The porous characteristic parameters of the above components are taken as the modules of each monomer, and the integrated simulation of the cooling module is carried out.

The geometrical model is established as shown in Fig. 4.

Import the model into star-ccm+, divide hexahedral grid, and increase local (cooling module location) grid density, and a total of 2,667,427 meshes are generated. The boundary conditions of the solver are set in accordance with the radiator monomer. The wind resistance under the general test wind speed (3, 4, 5, 6, 7, 8) is simulated and calculated. The results are as shown in Fig. 5.

Table 7 Flow characteristic parameters of intercooler, cooler, and condenser

Component	Inertia resistance coefficient (kg/m^3)	Viscous resistance coefficient (kg/(m^3 s))
Intercooler	109.54	342.41
ATF cooler	122.44	262.6
Condenser	174.23	786.52

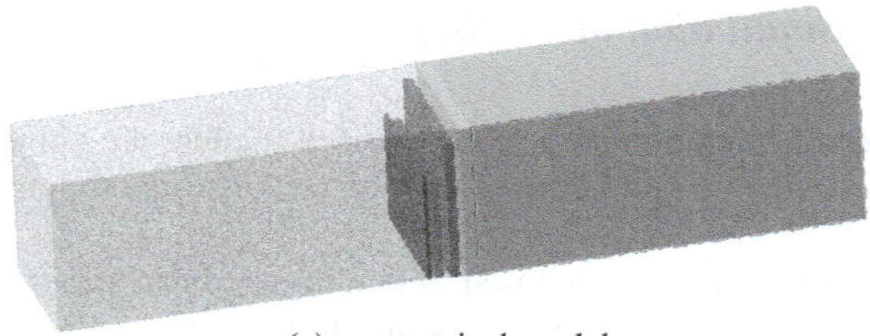

(a) geometrical model

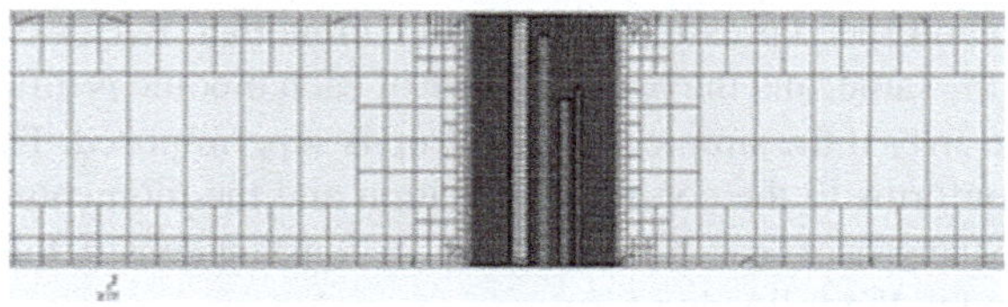

(b) mesh model Y=0 (Solving regional Local encryption)

Fig. 4 Region model for cooling module solving

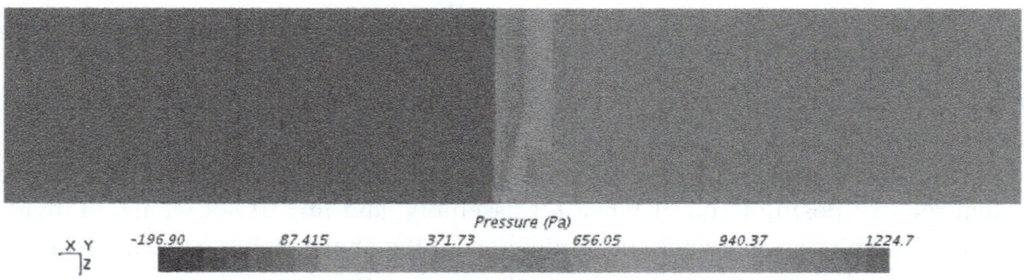

Fig. 5 $y = 0$ pressure distribution cloud chart of cooling module

2.3 Overall Wind Tunnel Test of Cooling Module and Comparison of Results

The physical cooling module which is identical to the data in the simulation is tested in the wind tunnel, the pressure drop at different wind speed is measured, and the wind speed–resistance curve is obtained [3] (Fig. 6).

The following experimental data are obtained, and the simulation results are compared with the experimental results as shown in Table 8.

In order to show the resistance characteristic of the cooling module more intuitively, the resistance curve is compared according to the experimental results and simulation results, as shown in Fig. 7. According to the comparison results, the maximum error of the simulation results is 2.86%, the simulation results are in good agreement with the experimental values, and the results are reliable.

Fig. 6 Wind tunnel test of cooling module

Table 8 Simulation and test results of cooling module comparison

Va (m/s)	4	5	6	7	8
The simulation value (Pa)	338	488	670	869	1103
Experimental value (Pa)	335.7	482.7	654.7	850	1071.5
Rate of deviation (%)	0.68	1.09	2.28	2.19	2.86

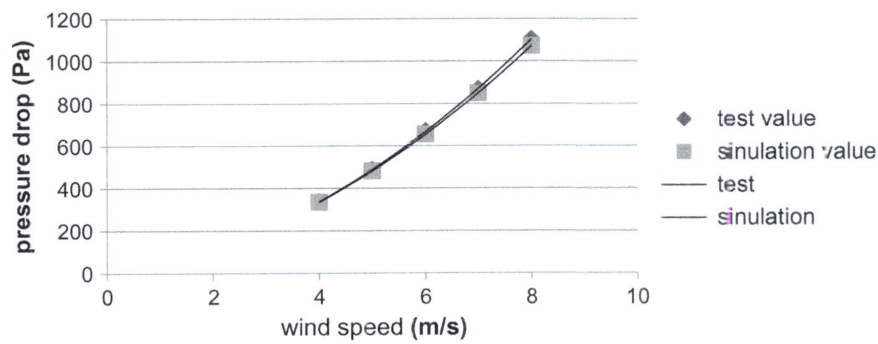

Fig. 7 Wind resistance–pressure drop curve of cooling module

3 Ending

a. Building CFD multi-hole radiating medium model, simulating radiating device with same structure feature but different outer size and conducting the test, error rate is less than 3%, is a rather accurate.
b. Building cooling module model, using the combination of multiple composite, each composite using CFD simulating multi-hole medium, solving pressure loss

feature in cooling module, and conducting the test, rate of deviation is less than 3%, and the model is rather accurate.

c. Based on this model, it has guiding significance for the design of the cooling system. (1) Guide the cooling module unit selection on the basis of having a certain cooling module monomer data accumulation. (2) Cooling module unit combination configuration scheme optimization. The different unit configuration positions in the design stage can accurately calculate the wind resistance, which provides a design basis for the scheme determination.
d. The resistance of cooling module is not simple adding up different device, and using traditional experience algorithm is not accurate enough (cooling power is weak or redundant) not consistent with modern design methodology; using the method mentioned in this article, we will no longer need to use experiment to iterate which will save a lot of human resource and time and have higher accuracy.

References

1. 候献君, 马将森, 等 (2015) 基于 CFD 方法的整车冷却系统匹配分析.汽车技术 (11):11–14
2. 吕峰 (2011) 商用车冷却模块匹配设计方法研究. 浙江大学
3. JB/T2293-1978.《汽车、拖拉机散热器风筒试验方法》
4. 童正明,陈丹,等 (2013) 管带式汽车散热器试验及数值模拟研究.汽车技术 (10):51–54
5. 《STAR-CCM+ Usr Guide》CD-adapco 2011

Numerical Simulation and Experimental Study of a Vehicle Exhaust Heating Heat Exchanger

Yan Zhao, Wenjie Hu, Jianjun Meng, Hongtao You and Yao Zhang

Abstract Shell-and-tube heat exchanger is a product with simple structure, mature technology, and convenient use. The heat exchanger mentioned in this study is used in a vehicle exhaust heating system, which uses the exhaust heat of the engine to heat the air inside the vehicle, improve the internal heating effect of the vehicle in the cold season, realize waste turning into treasure, save energy, and reduce emissions. In this study, CATIA three-dimensional software is used to carry out detailed modeling of heat exchangers. The flow path which is extracted from ANSYS software and simplified is divided into mesh. The appropriate boundary conditions are used to calculate the CFD three-dimensional numerical simulation, calculating the velocity field, temperature field, and pressure field inside the fluid heat exchanger at different temperature inlets. The test platform is used to test the temperature field of the heat exchanger. The simulation results are in good agreement with the experimental test data, which provides a good research method for heat exchanger design and performance prediction.

Keywords Exhaust heating · Heat exchanger · CFD · Fluent

Foreword

The rapid development of modern industry has made people's demand for energy grow. Due to the limited storage of non-renewable energy such as oil, natural gas, and coal, countries are also striving to save energy and reduce emissions while trying to develop new energy. Therefore, energy conservation and emission reduction are also the long-term development theme of the automotive industry in the future. In this study, the shell-and-tube heat exchanger is applied to the automobile exhaust system, and the waste heat in the exhaust gas is recycled to solve the problems caused by the low temperature of the engine cooling fluid or the poor effect of the warm air in the car due to the large volume and insufficient energy of the car, so as to achieve a good supplementary heating effect.

Y. Zhao (✉) · W. Hu · J. Meng · H. You · Y. Zhang
Technical Center of Dongfeng Motor Group Co., Ltd., Wuhan
430058, Hubei, China
e-mail: zhaoyan@dfmc.com.cn

China SAE (ed.), *Proceedings of China SAE Congress 2018: Selected Papers*,
Lecture Notes in Electrical Engineering 574,
https://doi.org/10.1007/978-981-13-9718-9_6

Since the 1960s, the American Heat Transfer Research Inc. (HTRI) and the British Heat Transfer and Fluid Flow Service (HTFS) had achieved a lot of research results in heat exchangers. The heat calculation micro-computing software, network optimization software, and process design software developed by them had high calculation accuracy, which not only saved a lot of manpower and material resources, improved work efficiency, but also brought high economic benefits. In China, the traditional methods of early research on heat exchangers are mainly based on theoretical calculations and experimental methods. Due to the complex internal structure and flow heat transfer of the heat exchanger, it is difficult to satisfy the research on flow and heat transfer performance through simple theoretical calculation. The experimental research relies on the physical and experimental equipment and cannot meet the needs of large-scale, high-speed, fast-paced development of heat exchangers, and the singularity of test results can not provide improved direction for the design and development and internal optimization of high-efficiency heat exchangers.

With the continuous advancement of computer hardware technology and the rapid development of computational fluid dynamics (CFD) and numerical heat transfer (NHT), numerical simulation of flow and heat transfer has become an effective means of heat exchanger research. Fluent is an international leader in commercial CFD software, which can be used in industries related to fluid, heat transfer, and chemical reaction [1].

In this study, Fluent is used as the solution tool, and the flow field, temperature field, and pressure field inside the heat exchanger are studied by numerical simulation method, and the flow and heat transfer inside the heat exchanger were demonstrated in detail. It provides the theoretical basis for the optimization design of the internal structure of the heat exchanger. Compared with the traditional test method, the numerical simulation does not depend on the test model and the test site, which saves a lot of manpower, material, and test costs, and greatly shortens the development cycle.

1 Numerical Simulation

1.1 System Principle

The shell-and-tube heat exchanger belongs to a type of partition wall heat exchanger, the fluid passage in the tube is called a tube process, and the fluid passage outside the tube is called a shell side. When the tube and the shell flow two different temperatures of fluid, respectively, the higher temperature fluid transfers heat to the lower temperature fluid through the heat exchange tube wall, the higher temperature fluid is cooled, and the lower temperature fluid is heated. Thus, the heat transfer between the two fluids is achieved.

In this study, the cold air inside the vehicle enters the rear part of the heat exchanger through the air inlet duct under the action of the fan. The high-temperature exhaust gas from the engine enters the front part of the heat exchanger, and multiple sets of

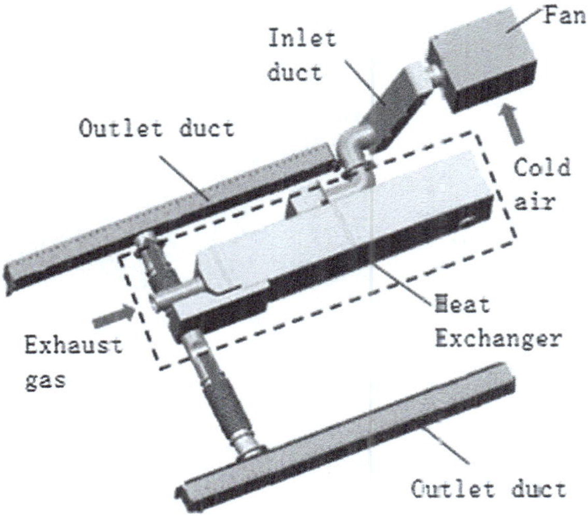

Fig. 1 Schematic diagram of the exhaust heating system

heat exchange tubes are arranged in a rectangular shape inside the heat exchanger. The heat is exchanged, and the finally heated hot air flows back into the vehicle through the left and right air outlet duct to complete the air circulation circuit. The system schematic is shown in Fig. 1. The main part of this study is the heat exchanger (inside the dotted line), and the rest is outside the scope of this study.

1.2 Problem Simplification and Assumptions

In order to eliminate the interference of secondary factors, reduce the complexity of the mesh, and save computing time, this study makes the following simplifications and assumptions:

(1) This paper studies the flow and heat transfer characteristics of heat exchanger with stable flow, using Fluent steady-state solution.
(2) The ideal gas is incompressible and is solved by pressure-based solver (pressure based).
(3) The effect of gravity on the fluid is not considered.
(4) The wall surface of the casing is a non-slip and thermal insulation wall, and it does not exchange heat with the outside [1].
(5) The wall of the heat exchange tube is a constant temperature wall, and the temperature value depends on the temperature of the exhaust gas in the tube.

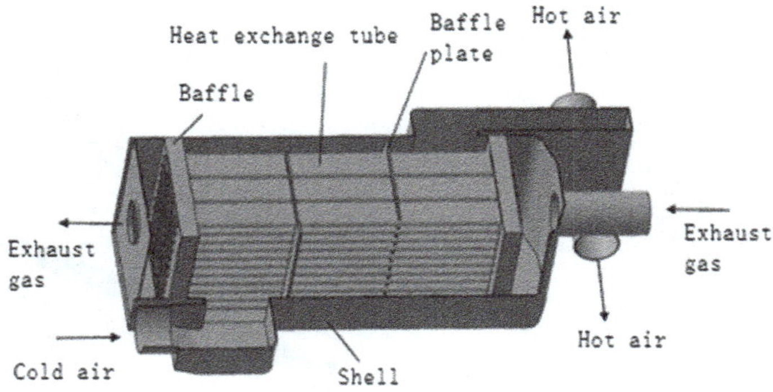

Fig. 2 Schematic diagram of heat exchanger structure

1.3 Calculation Model

The heat exchanger in this study (shown in Fig. 2) mainly consists of shell, baffle, baffle plate, end cap, etc. The heat exchange tube bundle is arranged in a rectangular shape in parallel, and the wall surface of the tube bundle is the heat exchange surface. The baffle plate not only prevents fluid short circuit, increases fluid velocity, but also forces the fluid to flow through the heat exchange tube bundle multiple times according to a prescribed path, so that the turbulence intensity is greatly increased and the heat exchange efficiency is improved.

1.4 Mesh Model

Extract the fluid domain, specifying the inlet, outlet (outlet1, outlet2), gwall, and qwall boundaries. Divide with a tetrahedral mesh. The total mesh element in the entire mesh model (shown in Fig. 3) is 7.40 million, the node is 1.36 million, and the average mesh quality is 0.83, which satisfies the Fluent computing requirements.

1.5 Solving Settings

(1) Mesh import: Import the mesh model into Fluent, set unit, check the mesh quality, and smooth the mesh[2].
(2) Solving model: Select based on pressure implicit solver, steady flow, standard k-ε turbulence model, energy equation, fluid parameter setting, SIMPLE algorithm, sub-relaxation factor and convergence residual which are all default. The result convergence is judged by the quality flow difference between import and export (i.e., quality conservation) monitoring [2].

Fig. 3 Grid model

(3) Boundary condition setting [3]:

   (a) The entrance boundary condition (inlet) selects the mass flow inlet, the inlet flow rate is 0.078 kg/s, the direction is perpendicular to the boundary, and the turbulence mode selects the turbulence intensity and the hydraulic diameter. After calculation, in this example, the turbulence intensity = 4%, the hydraulic diameter = 76.6 mm, and the inlet temperature range is 0–15 °C (the value is repeated every 1 °C interval).
   (b) The gwall is set as a constant temperature wall, and the wall material is defined as stainless steel, regardless of the wall thickness, and the specific value is measured according to the exhaust temperature.
   (c) The shell wall (qwall) is set to adiabatic and non-slip wall.
   (d) The exits (outlet1, outlet2) are set to outflow, and the ratios are each 0.5.

(4) Setting up the monitor and calculations: Repeat the calculation by taking different inlet temperatures, in turn, and analyze the convergence curve.

1.6 Results and Analysis

The inlet air temperature is taken as 0 °C for analysis, and the plane ($z = -56$) at the center of the two outlets of the heat exchanger is taken as a special plane.

(1) Pressure field

The fluid creates a large pressure drop at the inlet and outlet, while the pressure drop across each baffle plate is relatively small. Especially in the case where the inlet airflow is bent, due to the limitation of the structure, the flow direction changes abruptly and the pressure distribution is uneven. It can also be seen from Fig. 4 that

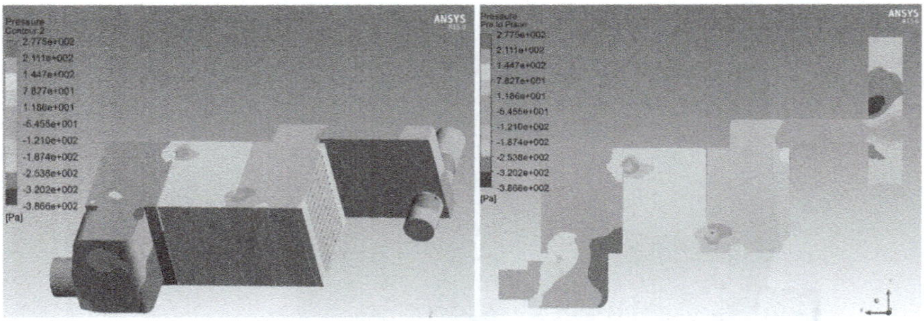

Fig. 4 Pressure field

the pressure in the whole fluid region is decreasing along the flow direction of the fluid.

(2) Velocity field

The eccentrically arranged structure of the inlet generates a strong vortex region at the inlet after fluid entry, which increases the turbulence intensity and also has the adverse effect of a large pressure drop. Due to the existence of the baffle plates, the fluid is more in contact with the heat exchange wall surface, and the velocity periodically changes direction, resulting in a "Z"-type reentry flow. There is a region with a lower flow rate near the bottom of the leeward side of each baffle plate, called the "speed dead zone"; the fluid flowing through the opening of the baffle plate creates a region with a higher flow rate. Figure 5 (left) velocity streamline diagram clearly depicts the flow path of the fluid throughout the heat exchanger.

(3) Temperature field

As shown in Fig. 6 (middle), along the flow direction of the fluid, as the contact time of the fluid with the heat exchange wall surface increases, the amount of heat transferred also increases, and the temperature of the fluid continuously increases. However, there is a "heat transfer dead zone" near the "speed dead zone" position

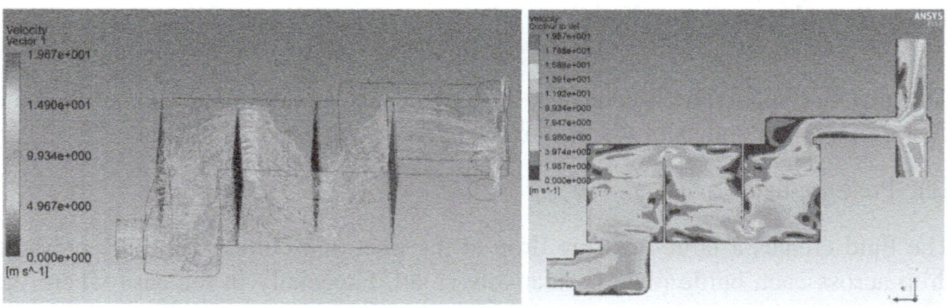

Fig. 5 Velocity field

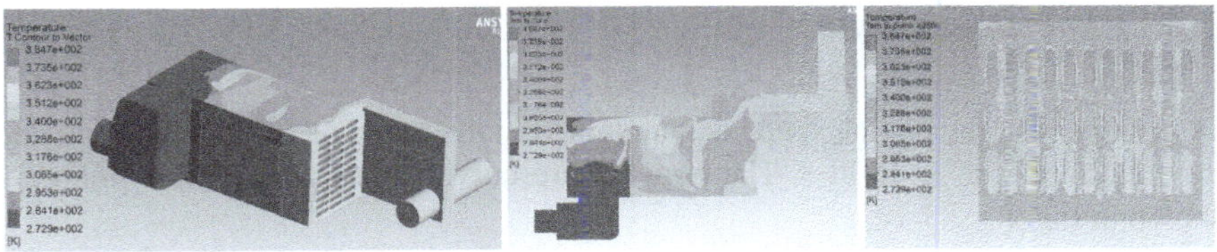

(Temperature field) (z= −56 Temperature field) (x=2800 Temperature field)

Fig. 6 Temperature field

on the leeward side of the baffle plate. The temperature in this zone is significantly higher than the ambient temperature. This is because the fluidity of this zone is poor or no flow, and it is not very good to flush the heat exchange tubes. Cross-sectional temperature cloud diagram in Fig. 6 (right) shows that the temperature gradient is mainly concentrated in the thermal boundary layer; that is, the temperature rise of the fluid near the heat exchange wall is the most rapid and obvious. The turbulent motion of the fluid allows the heat to diffuse sufficiently, and the overall temperature of the fluid in the shell rises rapidly and remains stable.

2 Experimental Research

2.1 Experiment Method

The heat exchanger under study is equipped with a vehicle exhaust system and placed in a low-temperature test chamber for testing. The exhaust system pipes and heat exchanger are wrapped with heat insulation cotton to reduce heat dissipation. The heat exchanger air inlet mass flow rate is controlled by the fan during the test. The temperature sensors record the exhaust temperature, the air inlet, and the left duct and right duct in real time. The test system is shown in Fig. 7.

Fig. 7 Test system diagram

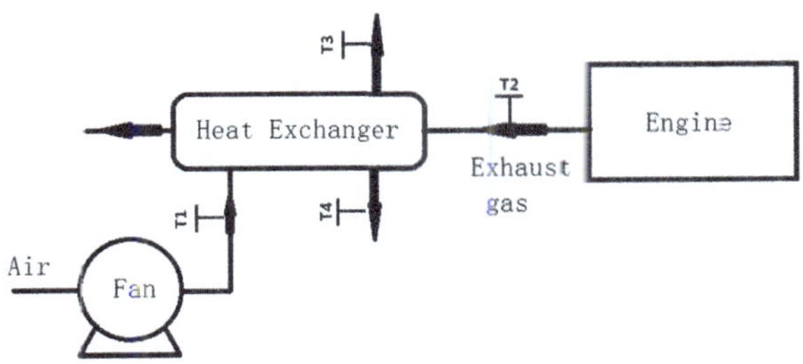

2.2 *Experiment Data*

During the test, the engine speed is maintained in the range of 1200–1400 rpm, and the total test time is 1800 s. The temperature curves of each measured point are shown in Fig. 8.

2.3 *Data Analysis*

It can be seen from Fig. 8 (air inlet temperature T1) that as the test continues, the cold air in the vehicle enters the heat exchanger under the action of the fan, and the heated air gradually flows back into the vehicle after the heat exchange. The temperature inside the car gradually rises, and the temperature at the air inlet of the heat exchanger also tends to rise. In the first few tens of seconds of the test, the hot air has not yet completely filled the interior of the vehicle, and the fan draws continuous cold air, causing a brief drop in temperature at the air inlet point of the heat exchanger.

It can be seen from Fig. 8 (engine exhaust temperature T2) that the temperature of the engine exhaust gas entering the heat exchanger during the test has a small range fluctuation and is basically maintained at 100–110 °C.

From Fig. 8 (heat exchanger left outlet temperature T3, heat exchanger right outlet temperature T4), it can be seen that the temperature at the left and right outlets of the heat exchanger rise rapidly and reach a steady state, which is basically maintained between 60 and 72 °C.

3 Comparative Analysis

The test conditions of the heat exchanger in this study are that the air inlet mass flow rate is constant and the temperature is constantly rising. Considering that the temperature inside the vehicle is not uniform at the beginning of the test, the whole system has not yet reached a steady state. Therefore, the air inlet temperature is only numerically simulated at a range of 0–15 °C (1 °C per interval) and compared with the test data in the same range of air inlet temperature (as shown in Table 1). The left-side outlet simulation value T3 of the heat exchanger deviates from the test value by 0.07 to −2.54%, and the right outlet simulation value T4 deviates from the test value by 0.02 to −2.54%. Figures 9 and 10 show the comparison between the simulation results and the test data.

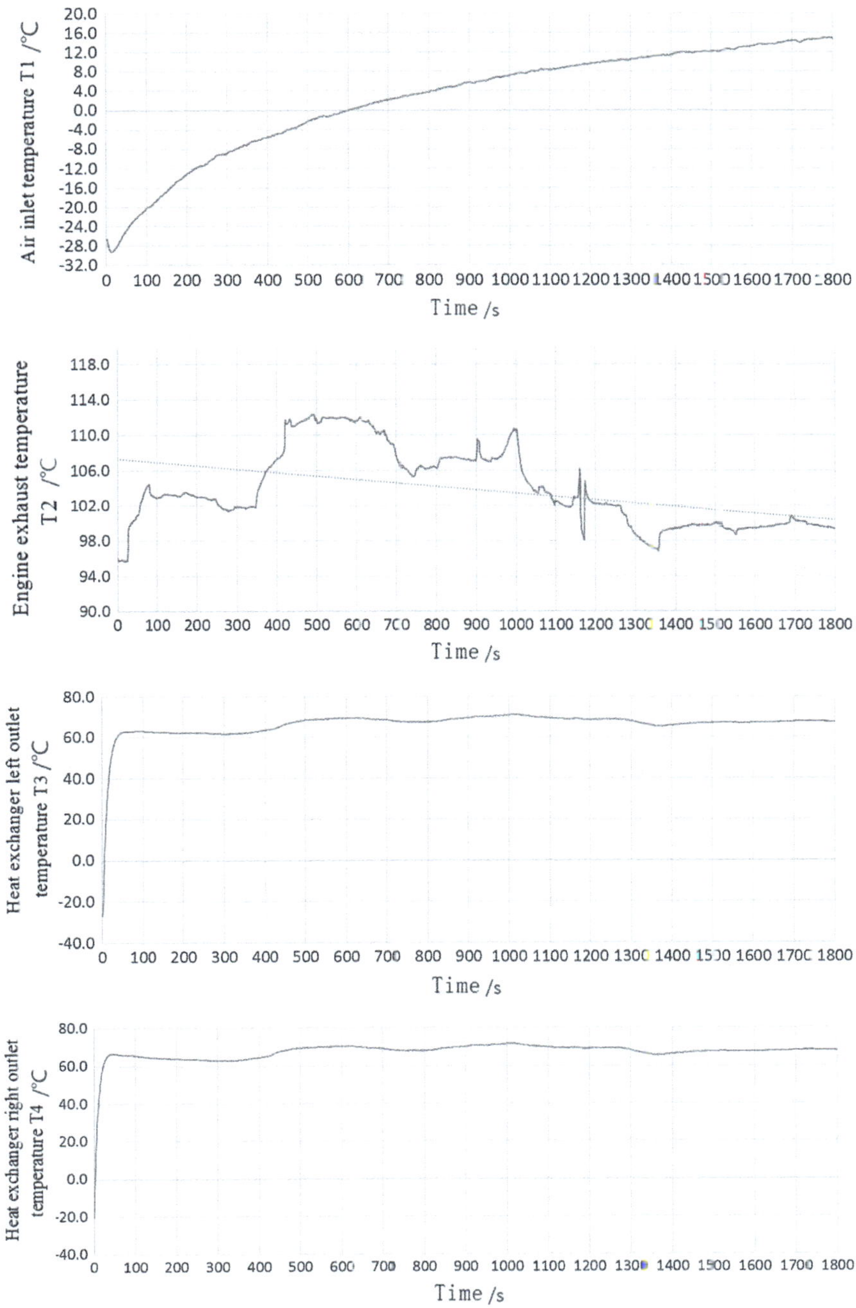

Fig. 8 Temperature curves

Table 1 Comparison table of heat exchanger simulation results and test data

Time/s	T1/°C	T3 test/°C	T3 simulation/°C	Error (%)	T4 test/°C	T4 simulation/°C	Error (%)
600	0	69.4	70.11	1.02	70.3	70.43	0.18
644	1	69.2	70.02	1.18	69.9	70.32	0.60
688	2	68.5	69.3	1.17	69.2	69.61	0.59
737	3	67.5	67.55	0.07	68.1	67.83	−0.40
805	4	67.3	68.7	2.08	67.8	68.97	1.73
870	5	68.9	69.53	0.91	69.5	69.81	0.45
931	6	70.1	69.73	−0.53	70.7	70.05	−0.92
986	7	70.5	71.96	2.07	71	72.25	1.76
1051	8	70.2	68.42	−2.54	70.5	68.71	−2.54
1150	9	68.8	68.4	−0.58	69.1	68.67	−0.62
1230	10	68.6	68.31	−0.42	69	68.59	−0.59
1349	11	65.6	65.49	−0.17	65.7	65.71	0.02
1428	12	66.2	67.51	1.98	66.6	67.77	1.76
1574	13	67	67.79	1.18	67.3	68.09	1.17
1675	14	67.4	68.54	1.69	67.8	68.82	1.50
1788	15	67.6	68.62	1.51	67.9	68.89	1.46

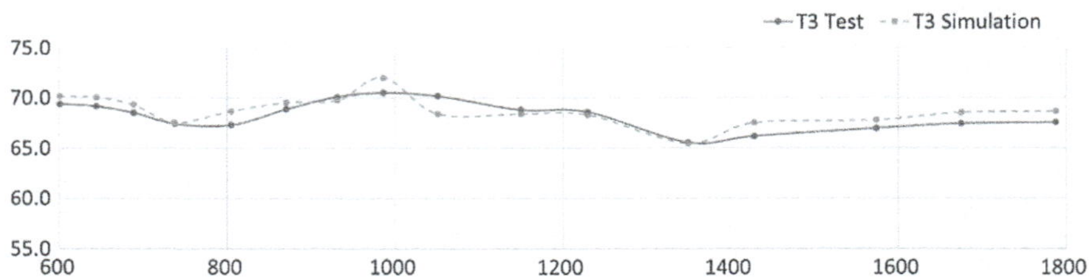

Fig. 9 T3 test compared with T3 simulation

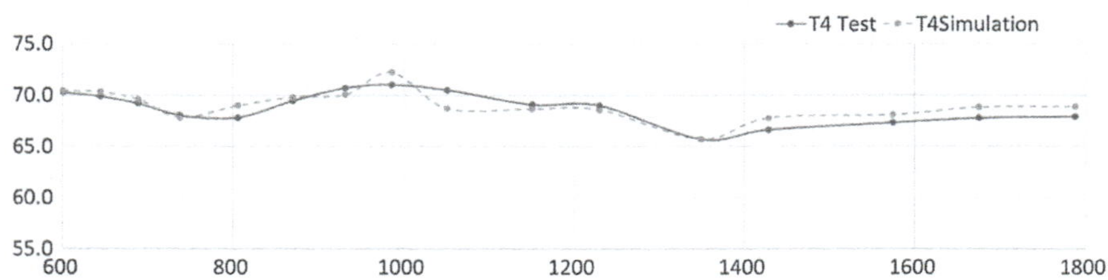

Fig. 10 T4 test compared with T4 simulation

4 Conclusions and Gains

In this paper, the shell-side flow and heat transfer performance of the shell-and-tube heat exchanger for vehicles are studied by means of numerical simulation and experimental research. The main conclusions and gains are as follows:

(1) This paper uses Fluent software to simulate the shell-side fluid of the heat exchanger and uses reasonable problem simplification, hypothesis, and solution settings to simulate the internal pressure field, velocity field, and temperature field of the heat exchanger under different temperature inlet conditions. The simulation results are consistent with actual flow and heat transfer.

(2) The temperature data of heat exchanger outlet are recorded by a suitable test system, and the simulation results are in good agreement with the test data, which proves the reliability of the numerical simulation method in this study. It provides a good engineering research method for the performance prediction of heat exchanger in the early stage of design.

(3) The numerical simulation method can obtain detailed data such as pressure field, velocity field, and temperature field inside the heat exchanger, which cannot be obtained by the test method. In addition, the influence of the baffle plate on the flow and heat transfer in the shell side of the heat exchanger can be reflected. It provides the guidance and reference for the arrangement of internal heat transfer tube bundle, the optimum design of baffle plate, the improvement of shell shape, and the further improvement of heat transfer efficiency.

References

1. Liu L, Huang W (2006) FLUENT simulates the three-dimensional flow field of shell-and-tube heat exchanger. Chem Equip Technol 27(3):54–57
2. Han Z, Wang J, Lan X (2004) FLUENT fluid engineering simulation calculation examples and applications. Beijing Institute of Technology Press, Beijing
3. Yu Y (2008) FLUENT introduction and advanced tutorials. Beijing Institute of Technology Press, Beijing

Research on Road Simulation Accelerated Durability Test Method for Front Swingarm

Zhenliang Yang, Songbo Zhang, Jianke Jia, Junjun Fu and Jiangtao Zhai

Abstract Ability to reproduce the fault of proving ground and user use is poor, which uses the traditional durability test method of the front swingarm bench. However, the road simulation test method is able to accurately reproduce the real road load spectrum of the swingarm, which has good validation results. The article focuses on accelerated editing the target response signal of swingarm, a new method for compressing and reconstructing the time-domain load spectrum, is put forward based on the intensified extrapolation and superposition of the load rainflow matrix according to the experimental conditions so as to further shorten the bench test period. The mean cycles of the load rainflow matrix were corrected, according to the Goodman method. And the S-N curve of the swingarm is calculated according to the nominal stress method. The damage degree of the measuring point load of the swingarm is compared before and after accelerated editing. The equivalence relation between the reconstructed spectrum of rainflow matrix and the original load spectrum was verified. Finally, the results showed that the speed-up ratio of the simulation iterative target response signal of the swingarm reaches 78.8 times compared with the road test at the proving ground, which is 2.4 times compared with the traditional bench test standard. A bench test was performed to verify that this accelerated editing method is effective.

Keywords Road simulation · Rainflow reconstruction · Nominal stress method · Accelerated editing · Equivalent relation

1 Introduction

As the key component of the whole vehicle, the front swingarm, is subjected to the vibration of the road roughness feedback and the dynamic load of the vehicle acceleration, braking, steering, roll, and other driving conditions, the working environment is extremely bad. The swingarm often has failures such as cracking of the bushing and

Z. Yang (✉) · S. Zhang · J. Jia · J. Fu · J. Zhai
OuShang Automotive Division, Chongqing Changan Automobile Co., Ltd., Chongqing, China
e-mail: 282570329@QQ.com

China SAE (ed.), *Proceedings of China SAE Congress 2018: Selected Papers*, Lecture Notes in Electrical Engineering 574,
https://doi.org/10.1007/978-981-13-9718-9_7

cracking of the solder joints. At present, the reliability verification methods for components such as swingarm mainly include road durability test on proving ground and indoor bench durability test. In the current fierce market competition environment, the advantages of bench test in terms of time and cost are very obvious compared with road test. It has become the trend of future automobile component reliability verification that the bench test gradually replaces the road test. The ultimate goal of the swingarm bench test is equivalent to the failure of user usage. Therefore, it is especially important to use the actual road spectrum to load the bench test.

The test results of the current swingarm bench mainly using sine wave of constant amplitude for low-frequency loading often deviate greatly from the results of actual road test, since the influence of load phase and loading sequence was not considered. Therefore, it has become the main research and application direction of the current component bench test using the simulation of the real road load spectrum to obtain the driving spectrum of the test bench. The exploration of accelerated test methods based on simulation iterative tests has become a development trend to adapt to market demand in the future.

2 Swingarm Load Spectrum Acquisition and Calibration

In the swingarm ball pin close to the ball pin seat, the bending half bridge was formed using the resistance strain gauge in the *X*-direction and the *Y*-direction of the real vehicle. The real-time strain load received by the swingarm ball pin fixed point feedback was measured in both directions. In the western Chongqing Automobile Proving ground, the actual load spectrum of the swingarm was collected for all test conditions according to the Changan Company Road Durability Test Specification and the loading requirements.

The swingarm bench test uses the ball pin as the loading end and fixes the other two bushings to simulate the *X*-direction and *Y*-direction loading of the real vehicle. The positive direction of the force must conform to the right-hand rule. The bench of the swingarm durability test (Fig. 1a) was made to calibrate the swingarm after the proving ground test, which must be pre-tensioned before calibration. The corresponding strain value was recorded when the discrete load was used for loading the bench. Afterward, the quantitative relationship between the load and the strain was fitted. It is necessary to keep the coupling error within 10% between the *X*-direction and the *Y*-direction calibration. It was seen from Fig. 1b that the calibration curve is very linear and satisfies subsequent analysis requirements. Compared with other vehicle data, it is found that the overall rigidity of the swingarm material is small from the calibration coefficient.

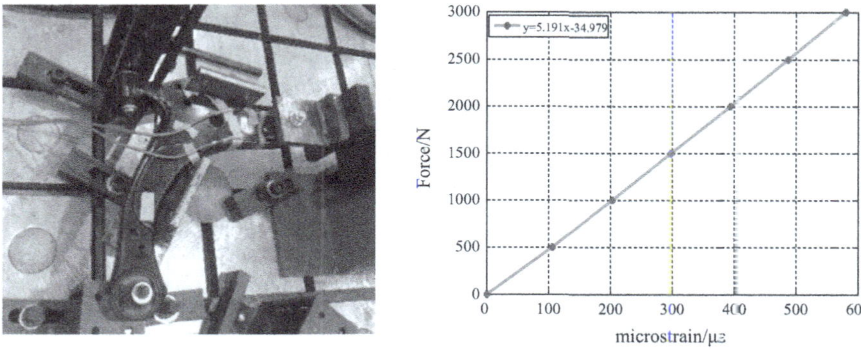

(a) Swing arm durability test bench (b) Swing arm calibration curve

Fig. 1 Swingarm calibration test

3 Load Spectrum Compression Based on Rain Flow Matrix

3.1 Rainflow Count of Measured Load Spectrum

Due to its good mechanical properties, the rainflow counting method perfectly matches the stress–strain hysteresis loop of the elastoplastic zone of the material, but it is also applicable to the parts with elastic range. The two-parameter rainflow counting method is widely used, in which the amplitude represents the load dynamic strength and the mean value represents the static strength. The two-parameter statistical method perfectly reproduces the random dynamic load characteristics of the components during the actual running of the vehicle.

Using the improved "three-point" [1] rainflow counting method, the two-dimensional mean-amplitude frequency matrix of the sample load of the swingarm in bad road condition was obtained (Fig. 2). It is found that the large load area of the sample load is relatively discrete due to the short time history of the sample load. Most of the circulation is in the middle and low-amplitude range from the rainflow diagram, indicating that a small amount of discrete large load is the cause of the damage of the swingarm measure point. The load rainflow statistics process of other working conditions is similar and will not be described again.

3.2 Time-Domain Reconstruction Based on Rainflow Matrix

A series of rainflow counting cycles of the desired target response signal of swingarm were counted by the rainflow counting method. The time-domain sequence of the swingarm was reconstructed by setting higher cycle frequency for these load cycles. Comparing the original load spectrum with the rainflow matrix reconstruction spec-

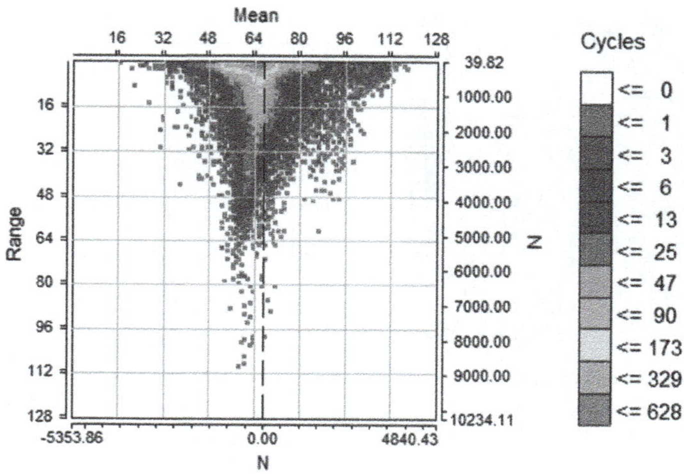

Fig. 2 Swingarm load rainflow matrix diagram

trum (Fig. 3), it can be found that the original spectrum was reduced from 1857 to 144.9 s, which was accelerated by 12.8 times. This reconstructed time-domain spectrum only removes the rainflow cycles below 200N, which will be demonstrated later in the ineffective damage interval, so that nearly 100% of the damage intensity is retained and almost no energy is lost.

3.3 Load Spectrum Compression Based on Rainflow Matrix

The time-domain load from the rainflow reconstruction of the sample load spectrum of swingarm was edited again. The small amplitude load cycles in the rainflow matrix were appropriately deleted, which can significantly reduce the load frequency and accelerate the test process. Engineering believes that 10% of the maximum amplitude of the rainflow cycle [2] is basically equivalent to the symmetrical fatigue limit of the material. The comparison between the amplitude–frequency matrix after deleting 10% of maximum amplitude and the original matrix is shown in Fig. 4. The curve 1 indicates the total frequency before the deletion is 31,397 times, and the curve 2 indicates the total frequency after the deletion is 11,269 times, and the acceleration ratio is 2.8 times.

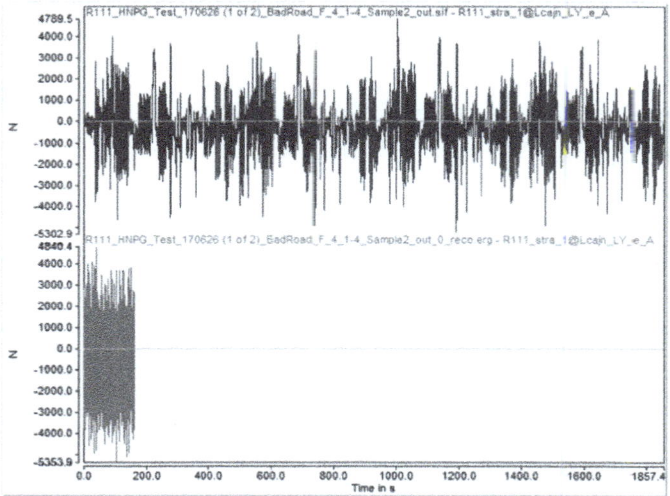

(a) Time-domain comparison before and after load reconstruction

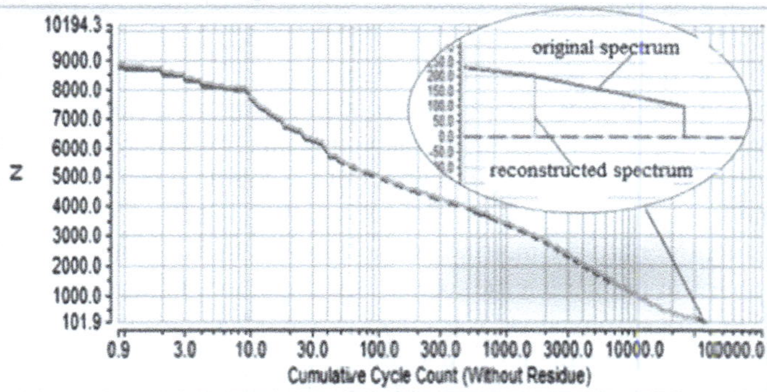

(b) Comparison of rainflow matrix before and after reconstruction

Fig. 3 Comparison of the original spectrum and reconstruction spectrum of the bad road

4 Enhanced Extrapolation and Superposition of Load Spectrum

4.1 Enhanced Extrapolation of Swingarm Load

The load amplitude distribution in the whole life region of the road test specification cannot be fully represented by the sample load spectrum of swingarm. The occurrence of some extreme load is related to the probability, which is difficult to reproduce in the short test time of the sample load. The overall acceleration editing idea is briefly described for the sample load analysis, but the input condition for the bench test must

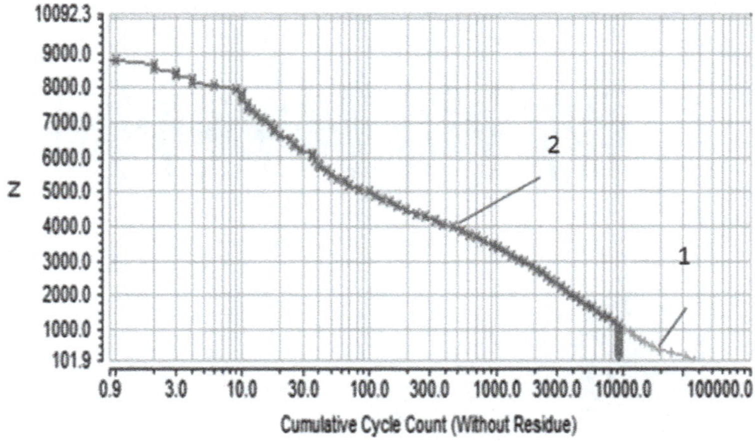

Fig. 4 Rainflow matrix compression process

be the acceleration target response spectrum which was obtained by re-editing after extrapolating all sample load cycles to the million level.

At present, there are four extrapolation methods: parameter extrapolation, rainflow matrix extrapolation, time-domain extrapolation, and mile–percentile extrapolation [3]. Considering the good prediction effect of the rainflow matrix extrapolation on the extreme load, the extended rainflow frequency matrix of swingarm load was obtained using the rainflow matrix extrapolation method. Rainflow matrix extrapolation is a typical nonparametric extrapolation method, first proposed by Dessler. The distribution of extreme load was effectively predicted using an appropriate kernel probability density estimation function. The discrete probability density of any point in the rainflow matrix is analyzed, and then the discrete probability density is transformed into a continuous probability density function according to Eqs. 1 and 2. Due to the densely frequency distribution of small load areas, the variability of which is small, and the large load areas is just opposite. It is necessary to use different bandwidths for different areas to predict the probability density. According to the Weibull fitting result, the Epanechnikov kernel function with adaptive variable bandwidth is used to estimate the probability density [4], which has higher prediction accuracy and more engineering applications.

$$\hat{f}(x, y) = \frac{1}{n}\sum_{i=1}^{n}\frac{1}{(h\lambda_i)^2}K\left[\left(\frac{x - X_i}{h\lambda_i}, \frac{y - Y_i}{h\lambda_i}\right)\right] \tag{1}$$

$$K\left(\frac{x - X_i}{h\lambda_i}, \frac{y - Y_i}{h\lambda_i}\right) = \begin{cases} \frac{2}{\pi}\left(1 - \frac{\left(\left\|\sqrt{(X-X_i)^2+(Y-Y_i)^2}\right\|\right)^2}{(h\lambda_i)^2}\right), & \\ 0 & \end{cases}$$

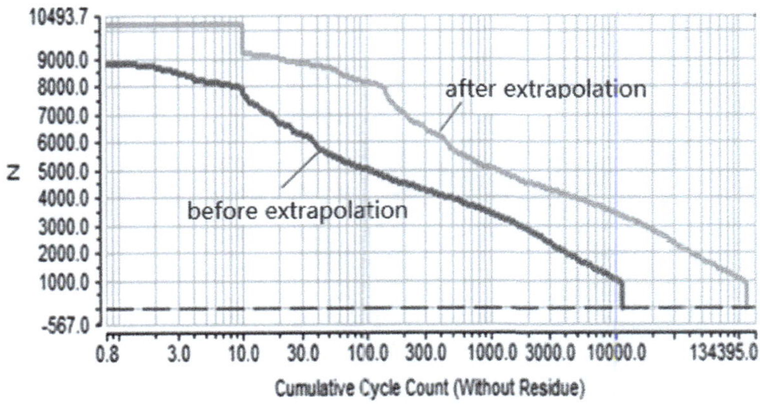

Fig. 5 Comparison of 10 times rainflow extrapolation results

$$\begin{cases} \dfrac{\left(\left\|\sqrt{(X-X_i)^2+(Y-Y_i)^2}\right\|\right)^2}{(h\lambda_i)^2} < 1 \\ \text{other} \end{cases}$$

$\|.\|$---Euclidean distance; h---Optimal fixed bandwidth;

λ_i---Adaptive coefficient; n---Total cycles (2)

Due to the abundant typical road surface of the proving ground, the swingarm has experienced excitation of various road surface irregularities in a short time, which result in the amplitude distribution of response load relatively comprehensive. Therefore, taking a conservative 10 times rainflow matrix extrapolation, it is found that the extreme amplitude of the sample load in bad road is increased from 4428N to 5097N, and the cumulative cycles are increased from 11,269 to 112,820 (Fig. 5). The amplitude and cycles of the extreme load are significantly increased, which indicate that the amplitude distribution of extreme load is effectively predicted.

4.2 *Full-Life Condition Superposition of Swingarm Load*

Based on the requirements of the road test specification, the load rainflow matrix under the total mileage range of the whole working condition was established (Fig. 6) by means of superimposing extended rainflow matrix of each test condition, according to the statistical cycle information of conditions (Table 1). It can be seen that the amplitude distribution of the full-life rainflow matrix of the swing arm load is more uniform from the extreme load, the median load to the small amplitude load, eliminating the dispersion phenomenon of load points of the sample load spectrum. And the overall load cycles have reached million level.

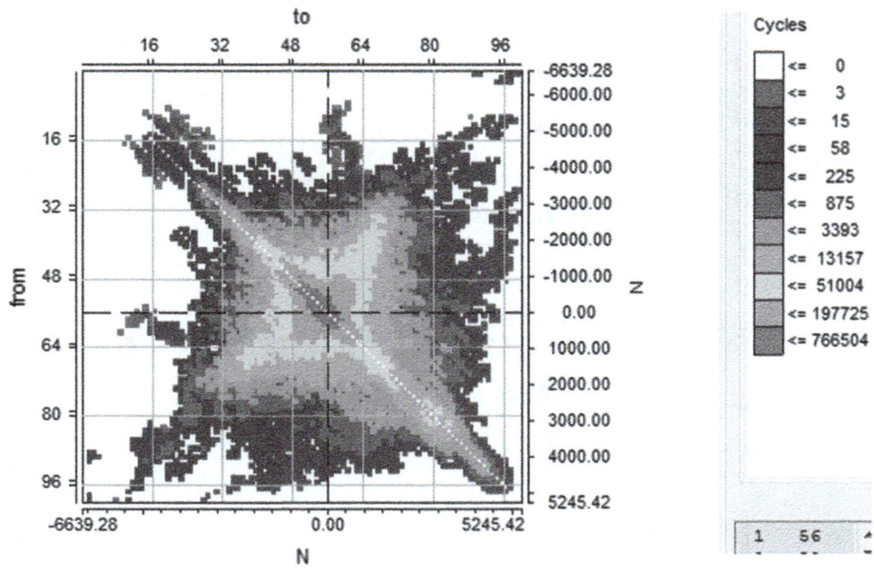

Fig. 6 Full-life cycle rainflow matrix diagram

Table 1 Total cycles count of specification conditions

Condition	Stowage	Total cycles
Bad road	Full load	900
City	Full load	134
ASD	Full load	15
Shift gear	Full load	120
High speed	Full load	90
Snake	Full load	80
Bad road	Half load	300
City	Half load	150
ASD	Half load	57
Shift gear	Half load	280
High speed	Half load	210
Snake	Half load	187

According to the principle of rainflow matrix compression described above, the interval cycles equivalent to 10% of maximum amplitude or less were deleted to obtain the final extended rainflow matrix (Fig. 7).

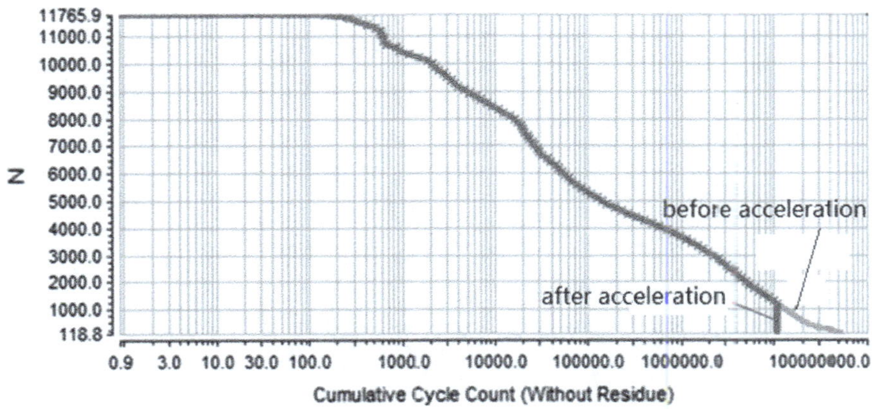

Fig. 7 Accelerated full-life rainflow matrix

4.3 Swingarm Acceleration Target Response Spectrum Rainflow Reconstruction

The desired target response signal required for the final bench test was reconstructed in the time domain, after concentrating the frequency of the original rainflow cycle in the higher-frequency band. The time-domain target response signal containing all the test conditions was finally obtained according to the rainflow reconstruction method in the time domain (Fig. 8). The total length of the target response signal obtained by superposing all conditions after multiplying the original test data omitted the transitional road surface of each working condition by the corresponding number of cycles is 930 h. However, the overall target response signal obtained by compressing and reconstructing the rainflow matrix is only 11.8 h, and the acceleration ratio is reached 78.8 times. In addition, it is assumed that the conventional bench test of the swingarm is loaded with a total of 300,000 times for 27.8 h in this direction, even if it is loaded with a frequency of 3 Hz. So, the acceleration ratio of the rainflow reconstruction spectrum is 2.4 times compared to the conventional bench test specification. The time-domain history created by rainflow reconstruction preserves the true amplitude, frequency, and phase information of the load, and the verification accuracy is better than the traditional steady-state loading method.

Comparing the power spectral density curves of the reconstructed spectrum with the original load spectrum (Fig. 9), it is found that the power spectral density is more gradual in the 0–50 Hz band after reconstruction, which is beneficial to the smooth loading of the bench during the road simulation test. The violent jitter is improved in the low- and medium-frequency bands when the bench of swingarm is loading.

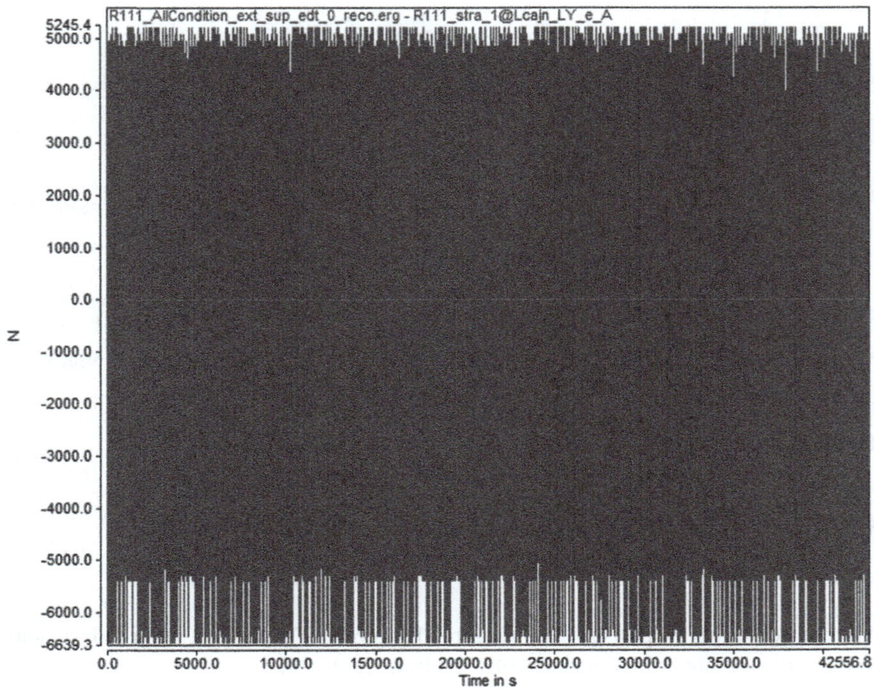

Fig. 8 Full-life interval expectancy target response signal

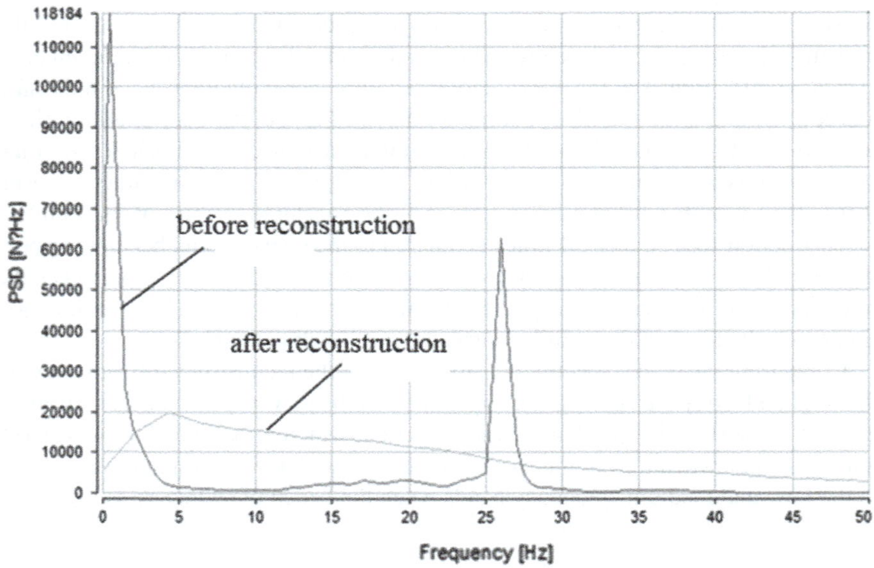

Fig. 9 PSD comparison before and after time-domain reconstruction

5 Acceleration Target Spectrum Damage Equivalent

5.1 Average Stress Correction

Based on the rainflow cycle distribution, the equivalent acceleration spectrum and the original load spectrum are analyzed from the perspective of the damage occurrence principle. In order to obtain more intuitive comparison results, the damage results are further compared and analyzed. The mean stress of the previous superimposed extended rainflow matrix is corrected based on the Goodman model. Since the measured load spectrum is a kind of random load, the load cycle distribution is irregular and the mean stress contains positive and negative values. The cyclic stress ratio *R* (Fig. 10) which is greater than 1 or less than −1 is the case where the mean stress is less than 0. The cyclic stress ratio *R* which is greater than or equal to −1 and less than or equal to 0 as well as greater than 0 and less than 1 is the case where the mean stress is greater than 0. The mean stress cycle of the measured load spectrum is distributed in four intervals, two transitions limitR and limitR2 are set in the interval where the cyclic stress ratio *R* is greater than 0 and less than 1. The purpose is to improve the theoretical damage of load cycles with both large mean and small-amplitude characteristics, because the real damage caused by the small-amplitude cycle (below material fatigue limit) of the actual load is very weak, but the calculated damage corrected by this slope M_1 will be too large. Therefore, the transition value is set in the interval where the cyclic stress ratio *R* is greater than 0 and less than 1 to maintain a more realistic damage process. The one-dimensional amplitude–frequency matrix corrected by the extended rainflow matrix according to Formula (4) is shown in Fig. 11. It can be seen that the mean cycle in the rainflow matrix has been equivalently superimposed to the corresponding amplitude cycle, and the effective damage of the mean stress cycle is better considered

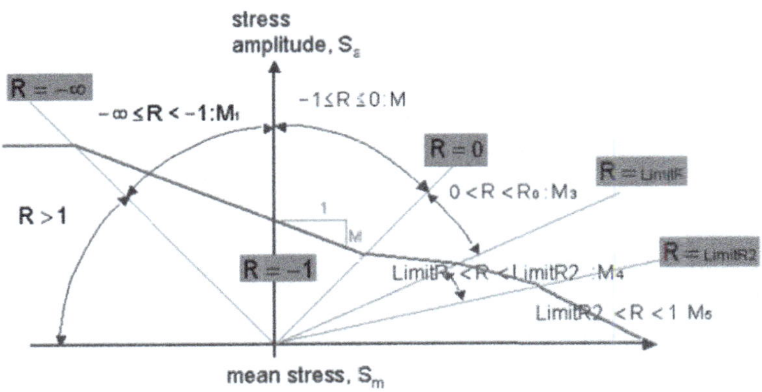

Fig. 10 Improved Goodman mean stress correction model

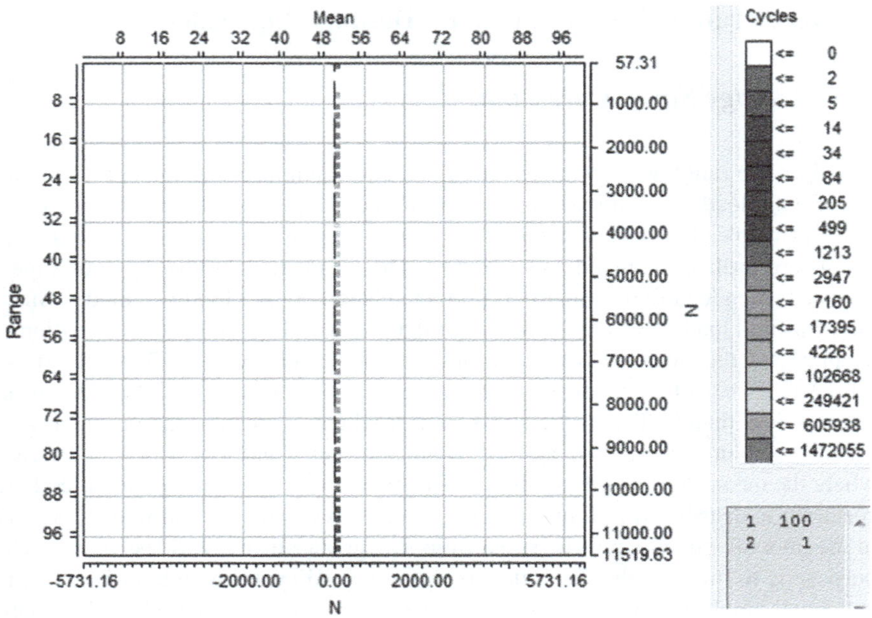

Fig. 11 Swingarm load one-dimensional amplitude–frequency matrix

$$M = \frac{S_{R=-1}}{S_{R=-0}} - 1 \tag{3}$$

In the formula, M is the slope of the corrected curve, $S_{R=-1}$ is the symmetrical cyclic fatigue limit of the material, and $S_{R=-0}$ is the material fatigue limit under the tensile mean stress.

$$\frac{S_a}{S_{a,\text{eq}}} + \frac{S_m}{S_{\max}} = 1 \tag{4}$$

In the formula, S_a is the amplitude of the load, $S_{a,\text{eq}}$ is the equivalent amplitude of the load, S_m is the mean value of the load, and $S_{\max}$ is the material tensile strength.

5.2 *Pseudo-Damage Analysis*

The pseudo-damage of the swingarm was analyzed using the S-N curve with both 50% survival and $R = -1$ characteristics in this direction. The material of the swing ball pin seat is usually 45# steel with quenched and tempered treatment. The following are basic material characteristic parameters: tensile strength (σ_b) is 735 MPa, elastic modulus (E) is 200,000 MPa, and material symmetrical bending fatigue limit (σ_{-1}) is 340 MPa. According to the anti-fatigue design manual [5], the load factor (C_l), surface quality coefficient (C_s), size factor (C_d), reliability level coefficient (C_r), and fatigue notch coefficient (k_f) which will influence the fatigue strength were estimated, and the bending fatigue limit of the material was corrected according to Formula 5, which is 255 MPa. The slope and intercept of the S-N curve are calculated according to the corrected material parameters by Eq. 6, the slope of −0.0657 in double logarithmic coordinates with an intercept of 735 were obtained, respectively. The S-N curve of the swingarm pin shown in Fig. 12 was drawn by MATLAB.

$$\sigma_{-1d} = \sigma_{-1} c_l c_d c_s c_r / k_f \tag{5}$$

In the formula, σ_{-1d} is corrected symmetrical bending fatigue limit of material, and other parameters have been mentioned above.

$$\sigma^m N = C \rightarrow m \lg \sigma + \lg N = \lg C \tag{6}$$

In the formula, m is the constant, C is the constant, N is the cycle life, and σ is the equivalent stress amplitude.

It can be seen from the damage distribution of the one-dimensional load at the ball pin of Fig. 13 that the magnitude of the damage between the rainflow reconstruction load and the original load is basically the same, and the overall distribution trend is basically coincident. The total damage results in Table 2 can prove that the damage of the rainflow compression–reconstruction spectrum retains 93% of the original load, and the damage consistency effect is very good, which is in accordance with the equivalent relation. It proves that the method of using the rainflow matrix compression and reconstruction to obtain the acceleration target response spectrum for the bench test is theoretically effective.

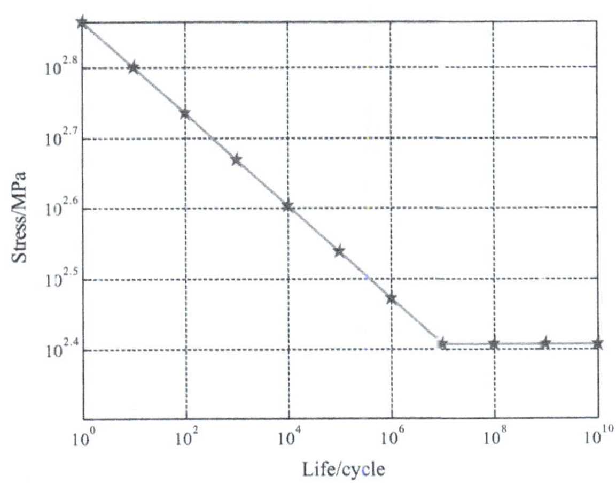

Fig. 12 S-N curve of the swingarm ball pin

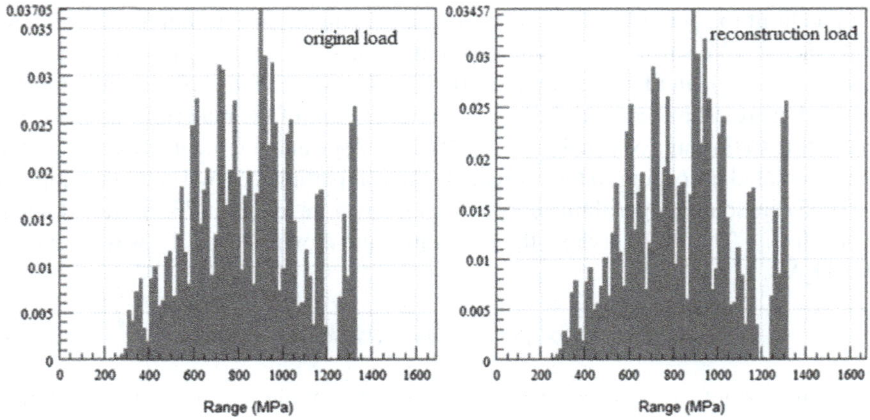

Fig. 13 Swingarm load pseudo-damage distribution matrix

Table 2 Comparison of specification pseudo-damage

Target response signal	Pseudo-damage
Original load	0.87
Rainflow reconstruction load	0.81

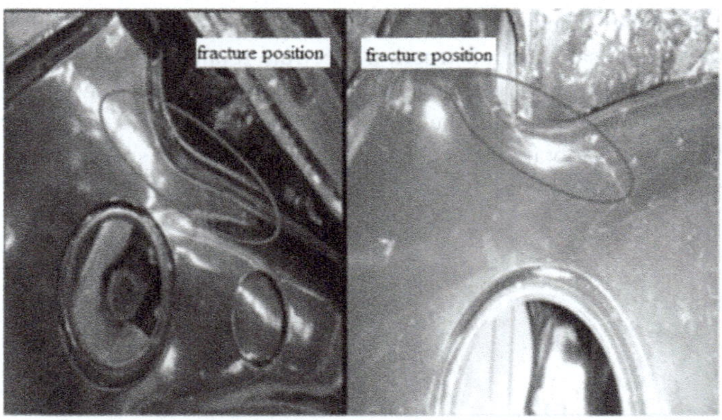

(a) Road simulation accelerated test (b) Traditional durability test

Fig. 14 Comparison of two types of swingarm durability bench test

6 Bench Test Verification

Set up the durability test bench for the swingarm (Fig. 1a), and the durability test was carried out using the accelerated edited target spectrum of swingarm according to the road simulation test method. And the specification full-life loading was completed.

The final test result is shown in Fig. 14a. Compared with the test results (Fig. 14b) after the full load of the traditional swingarm durability test, it was found that the fracture position of the swingarm was basically the same after the two tests, which proved that the acceleration method basically keeps the failure mode of the swingarm consistent and effective.

7 Conclusion

(1) The small cyclic load of the original load rainflow matrix of all working conditions for the swingarm is reduced based on the equivalent damage principle, and the time-domain load history is reconstructed. The duration is only 11.8 h. On the one hand, the verification acceleration ratio is 78.8 times, compared with the 930 h of the road test specification. On the other hand, the acceleration ratio is 2.4 times, compared with the loading time of the swingarm bench using traditional specification.

(2) According to the comparison of the pseudo-damage between the reconstructed load and the original load of the swingarm ball pin, it is found that the reconstructed load retains 93% of the damage strength of the original load. Further, based on consistency with the comparison results of the two bench tests, it is proved that this reconstruction acceleration method is effective.

References

1. Chen L, Liu X (1990) Application of improved rain-flow method to estimation of tractor components fatigue life. J Jiangsu Inst Technol 2:18–24
2. Ping A, Wang D, Xu H (1993) Study on determining criterion for omission of small loads. J Agric Mach 3:64–69
3. Liu Y, Zhang X, Wang Z (2011) Contrast of extrapolations in compiling load pectrum. Mod Manuf Eng 11:8–11
4. Adrian G (2009) Kernel bandwidth estimation for nonparametric modeling. IEEE Trans Syst 39(6):1543–1555
5. Zhao S, Wang Z (2015) Kangpilao sheji shouce. China Machine Press, Beijing

Optimization and Investigation on SNR of Engine Mass Airflow Sensor

Zhaojing Wang, Xiaomao Zhang, Ming Chen and Zheng Xu

Abstract In this paper, aiming at the problem of the signal-to-noise ratio exceeding the target of the mass airflow sensor in an engine air intake system test, it is analyzed to the overall flow state of the air intake system via the CFD analysis software STAR CCM+. Based on the $K-\varepsilon$ turbulence model, the reason for the signal-to-noise ratio exceeding the target is studied through numerical analysis. The rationality and feasibility of various optimization proposals are explored. Finally, the optimization proposal of the rectifier grid is adopted. The optimized results are tested and verified that signal-to-noise ratio of the optimized intake system meets the design requirements.

Keywords CFD · MAF of the air intake system · Single-to-noise ratio · Optimization

Foreword

At present, automobile emission regulations are becoming more and more stringent, engine technology tending to be electronically controlled, and the electronic control system controls the injection quantity and the control of the EGR rate under different conditions. It is necessary to accurately measure the amount of fresh air sucked into the cylinder through the intake system. The mass airflow sensor (MAF) is used to measure the air intake quantity of the engine. If the airflow distribution in front of the airflow sensor is not uniform, even turbulence, the airflow signal measured by the airflow sensor will be inaccurate, and the engine electronic control unit will not get the correct intake air signal, so it cannot accurately control the fuel injection amount and EGR rate. When the engine electronic control unit receives a continuously unstable intake air amount signal, or the unstable fuel injection amount or EGR rate, which may cause the engine to wobble, idle speed, and problems such as high fuel consumption and poor exhaust emissions, and therefore, the airflow at the installation of the airflow

Z. Wang (✉) · X. Zhang · M. Chen · Z. Xu
SAIC Motor Technical Center, Shanghai, China
e-mail: wangzhaojing@saicmotor.com

China SAE (ed.), *Proceedings of China SAE Congress 2018: Selected Papers*, Lecture Notes in Electrical Engineering 574,
https://doi.org/10.1007/978-981-13-9718-9_8

sensor must be sufficiently uniform to minimize the turbulence intensity and ensure the accuracy of the airflow sensor measurement of the intake system.
In this paper, the excessive signal-to-noise ratio (SNR) of mass airflow sensor in a certain engine intake system test is carried out. The CFD analysis software STAR CCM+ performs numerical simulation analysis on the overall flow of the intake system without MAF. Based on the turbulence model and the standard wall function method, the reasons for the excessive signal-to-noise ratio are studied through numerical analysis. The rationality and feasibility of various optimization schemes are explored. Finally, the optimization scheme of the rectifier grid is determined. Based on the optimization scheme, this paper analyzes the flow field of the MAF intake system before and after optimization and solves the problem that the MAF signal-to-noise ratio of the intake system exceeds the standard.

1 SNR Indicator of Mass Airflow Sensor Exceeding the Standard

The mass airflow sensor signal-to-noise ratio indicator is expressed by statistical concept. The sensor signal is sampled at a specific flow rate. 100 points of data are collected for the flow test, and the confidence is 99%. The data that the percentage of 2.581σ in the average value of the sample data is used to estimate the signal-to-noise ratio at a given flow rate, where σ is the standard deviation of the sample, the smaller the percentage value, the more accurate the sensor signal. The intake system requires a measurement flow range of 5–500 kg/h, and the signal-to-noise ratio indicator is controlled below the target line. Figure 1 is a schematic diagram of a test equipment for measuring the accuracy of the intake system MAF. Figure 2 shows the results of the MAF signal-to-noise ratio (confidence level is 99%) of the intake system and the target line. The different color curves are different MAFs for the same set of intake systems' measurement results. It can be seen from the results that under the flow rate of 100 kg/h, most of the signal-to-noise ratio results are beyond the target line, especially when the flow rate is 10 kg/h, the signal-to-noise ratio indicator exceeds the target line by about 7.5%. Therefore, the intake system structure or the MAF installation position needs to be optimized.

2 Establishment of Mathematical Model

The existing CFD analysis calculations follow the three laws of mass conservation, energy conservation, and momentum conservation [1, 2]. The calculation of the air filter is a single-phase flow calculation; the calculation conditions are carried out at normal temperature; and the temperature change is small, so the thermodynamic model of the gas is considered. Generally speaking, the airflow movement inside the

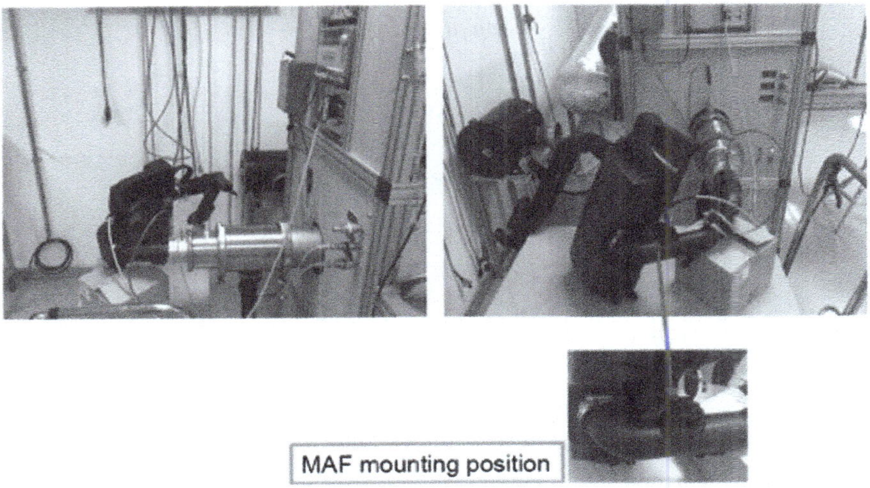

Fig. 1 Mass airflow sensor SNR measurement equipment

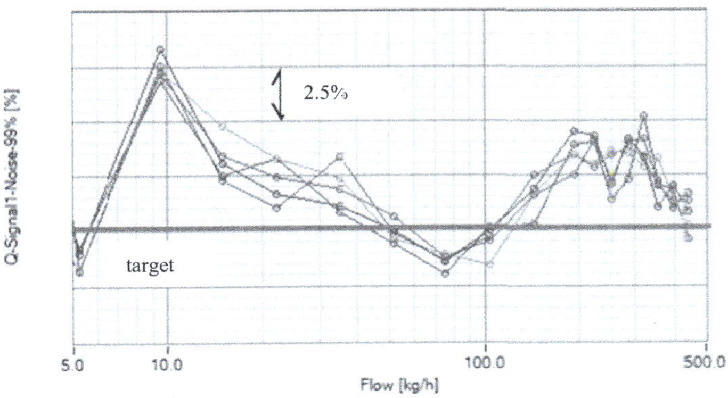

Fig. 2 MAF signal-to-noise ratio of the intake system at a 99% confidence

air filter belongs to the turbulent flow, so the turbulence model needs to be set. This paper adopts the more mature $k-\varepsilon$ double equation model. The relevant mathematical equations are as follows:

Mass conservation equation

$$\frac{\partial \rho}{\partial t} + \frac{\partial}{\partial x_j}(\rho u_j) = s_m \tag{1}$$

Momentum conservation equation

$$\frac{\partial \rho u_i}{\partial t} + \frac{\partial}{\partial x_j}(\rho u_j u_i - \tau_{ij}) = -\frac{\partial p}{\partial x_i} + s_i \tag{2}$$

Energy conservation equation

$$\frac{\partial \rho h}{\partial t} + \frac{\partial}{\partial x_j}(\rho h u_j + F_{h,j}) = \frac{\partial p}{\partial t} + u_j \frac{\partial p}{\partial x_j} + \tau_{i,j} \frac{\partial u_i}{\partial x_j} + s_h \tag{3}$$

Turbulent k equation

$$\begin{aligned} &\frac{\partial}{\partial t}(\rho k) + \frac{\partial}{\partial x_j}\left[\rho u_j k - \left(\mu + \frac{\mu_t}{\sigma_\varepsilon}\right)\frac{\partial k}{\partial x_j}\right] \\ &= \mu_t(P + P_B) - \rho\varepsilon - \frac{2}{3}\left(\mu_t \frac{\partial u_i}{\partial x_i} + \rho k\right)\frac{\partial u_i}{\partial x_i} + \mu_t P_{\mathrm{NL}} \end{aligned} \tag{4}$$

Turbulent ε equation

$$\begin{aligned} &\frac{\partial}{\partial t}(\rho\varepsilon) + \frac{\partial}{\partial x_j}\left[\rho u_j \varepsilon - \left(\mu + \frac{\mu_t}{\sigma_\varepsilon}\right)\frac{\partial \varepsilon}{\partial x_j}\right] = C_{\varepsilon 1}\frac{\varepsilon}{k}\left[\mu_t P - \frac{2}{3}\left(\mu_t \frac{\partial u_i}{\partial x_i} + \rho k\right)\frac{\partial u_i}{\partial x_i}\right] \\ &+ C_{\varepsilon 3}\frac{\varepsilon}{k}\mu_t P_B - C_{\varepsilon 2}\rho\frac{\varepsilon^2}{k} + C_{\varepsilon 4}\rho\varepsilon\frac{\partial u_i}{\partial x_i} + C_{\varepsilon 1}\frac{\varepsilon}{k}\mu_t P_{NL} \end{aligned} \tag{5}$$

In Eq. (1), ρ is the fluid density, t is the time, x_i is the Cartesian coordinate system ($i = 1, 2, 3$), u_j is the absolute velocity component of the fluid in the x_i direction, and s_{m} is the mass source term; In (2), τ_{ij} is the shear stress, p is the surface pressure, and s_{i} is the momentum source term; in Formula (3), h is the static enthalpy, which is the sum of the enthalpy chemical reaction enthalpy; in Formula (4), the first term on the right is the turbulent term caused by tangential, normal stress and buoyancy, the second term represents the viscous dissipation term, and the third term represents the enhancement or attenuation term of the fluid due to compression, in Eq. (5), σ_ε, $C_{\varepsilon 1}$, $C_{\varepsilon 2}$, $C_{\varepsilon 3}$, and $C_{\varepsilon 4}$ are empirical constants.

The calculation is solved by the simple method, and the speed and pressure separation solver is used. Since the grid is large and scattered, the second-order upwind-style discrete processing is adopted. It is assumed that the gas temperature during the intake process is constant, normal temperature, ignoring the influence of temperature change on the air density, so the gas adopts a constant density; since the k–ε turbulence model is only applicable to turbulent regions that are a certain distance away from the wall, the standard wall function method is used for the region near the wall.

2.1 Geometric Models and Mesh Models

The geometric model of the air filter is built by CATIA software. In order to thoroughly understand the flow characteristics inside the air filter and ensure the accuracy of the calculation, the constructed air filter geometry completely retains the local details of the actual components without any simplification. The model includes the dirty pipe of the intake system, the upper and lower casings of the air filter, the clean pipe, and the mass airflow sensor installed on the clean pipe. Firstly, the CATIA model is imported into the preprocessing software ANSA for geometric cleaning to obtain the surface mesh of the fluid domain, and then the surface mesh of the air filter is imported into the STAR CCM+ software for mesh division, which is mainly divided by a polyhedral mesh. In order to capture detailed geometric features and consider the calculation time, the total number of volume meshes is finally controlled at around 1.5 million. It is worth that in order to calculate the pressure loss of the inlet more accurately, a large enough hemisphere is added at the inlet of the dirty pipe. Figure 3 shows the geometric model and mesh model of the initial intake system.

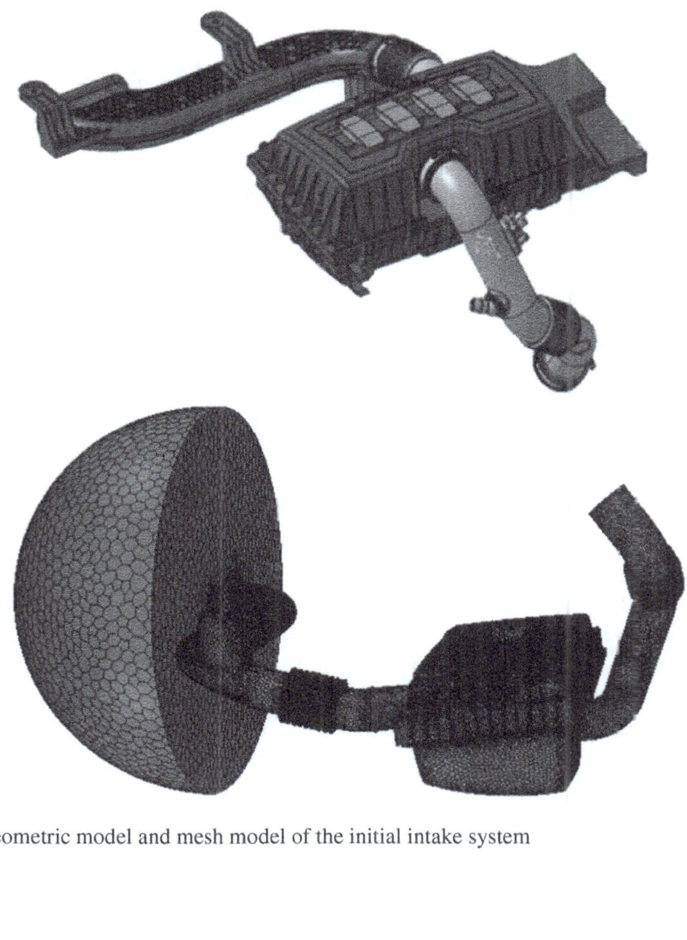

Fig. 3 Geometric model and mesh model of the initial intake system

2.2 Boundary Conditions and Model Parameter Settings

The model entrance adopts the stagnation inlet, which is generally the standard atmospheric pressure, and the outlet adopts the mass airflow inlet. The research flow rate is 10 kg/h, and the wall surface is set to the nonslip boundary. The calculated medium is air. Ambient atmospheric pressure is 101,325 Pa, and temperature is normal temperature 300 K; air density is 1.18415 kg/m^3, and air viscosity is 1.85508×10^{-5} Pa s.

In model settings, the filter element is assumed to be a porous medium model, and its inertial resistance coefficient and viscous resistance coefficient are expressed by Eq. (6) [3]:

$$\frac{\Delta P}{L} = -(P_{\mathrm{i}}|v| + P_{\mathrm{V}})v \tag{6}$$

where P_{i} is the inertial resistance coefficient, kg/mA; P_{v} is the viscous resistance coefficient, kg/m^{3}s.

3 Application of CFD in Optimizing the MAF SNR of Intake System

3.1 Flow Field Analysis of Intake System Without MAF

According to the requirements of the uniformity of the intake airflow at the front end of the airflow sensor, it is necessary to know the flow in the intake system in detail, calculate the intake system without MAF, and understand the flow and key section of the intake system, and velocity distribution, etc. The velocity uniformity index γ of a section defined in the software is as shown in Formula (7) [4], where V_{mean} is the average velocity of the section, V_{i} is the velocity of the unit surface, and A_{i} is the area of the unit surface.

$$\gamma = 1 - \frac{\sum_{\mathrm{i}} |V_{\mathrm{i}} - V_{\mathrm{mean}}| A_{\mathrm{i}}}{2|V_{\mathrm{mean}}| \sum_{i} A_{\mathrm{i}}} \tag{7}$$

The schematic diagram of the clean pipe structure of the intake system and the research section is shown in Fig. 4. The red arrow indicates the flow direction in the clean pipe of the intake system. Three research sections S1, S2, and S3 are selected to show the airflow uniformity state after the pipe bend. S1 is the cross section just after the end of the elbow of the clean pipe, S2 is the cross section of the farthest point where the MAF can be installed, and S3 is the cross section of the end of the straight section after the elbow. For simpler description, the side close to the core of

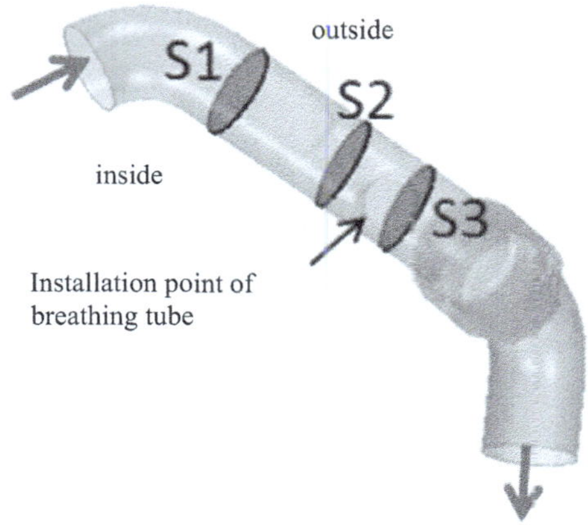

Fig. 4 Structure and cross section of the clean pipe of the intake system

the curved pipe is defined as inside, and as the outside on the side away from the core of bend pipe.

Table 1 shows the velocity uniformity index of different sections after clean pipe at the flow rate of 10 kg/h. The velocity uniformity indexes of sections S1, S2, and S3 after bending are 84.0, 87.6, and 91.6%, respectively. With the effect of straight pipe, the uniformity index of the section along the flow direction gradually increased, and the maximum increases to 91.6%. However, since the MAF must be installed upstream of the breathing tube, the MAF cannot be installed at the S3.

Figure 5 shows the velocity distribution (Fig. 5a) and the turbulence intensity distribution (Fig. 5b) at a flow rate of 10 kg/h. As shown in the figure, the inner and outer sides of the pipe have obvious velocity gradients on the three sections by the influence of the straight pipe section structure, wherein the S1 section is the closest to the elbow exit, the velocity gradient is the largest. The length of the straight pipe before the S3 section is the longest, and the velocity gradient is significantly reduced. The distribution of turbulence intensity has similar characteristics. A high turbulence intensity region appears on the inside of the elbow on the S1 section. This appearance of this area is the flow passage on the inner side of the elbow suddenly expands, and the fluid generates a reverse pressure gradient, forming a coiling of the main flow and a vorticity. Under the rectification of the straight tube, the swirling effect of the roll vortex gradually decreases, the S3 cross section has almost completely disappeared, and the turbulence intensity tends to be uniform in all directions. The flow distribution on the three sections has a distinct velocity gradient between the inner and outer sides, and the flow state is asymmetrical under the symmetry of

Table 1 Velocity uniformity index of different sections after clean pipe bending (flow rate: 10 kg/h)

Section	Mass airflow (kg/h)	UI_V (γ)
S1	10	84.0%
S2	10	87.6%
S3	10	91.6%

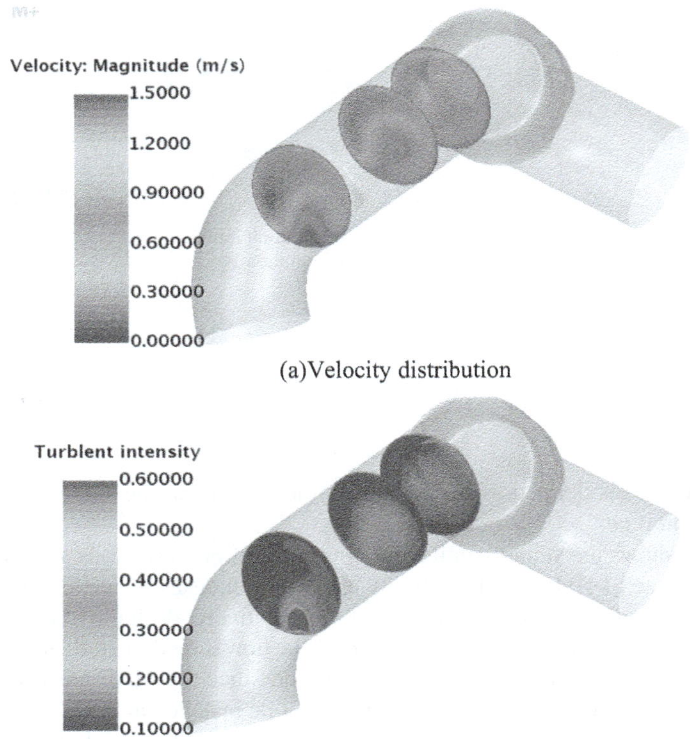

Fig. 5 Schematic diagram of velocity distribution and turbulence intensity distribution

the left- and right-side structures. When the speed of the clean pipe inlet is evenly distributed, such a phenomenon does not occur. Therefore, in this paper, the inlet flow of the clean pipe is also non-uniform in the entire intake system due to the upstream link of the upper casing chamber. The non-uniformity causes a left–right asymmetry of the flow and also certainly influences the uniformity of the flow.

Figure 6 shows a vector diagram of the gas flow velocity after a bend pipe in a clean pipe. A similar conclusion can be drawn from this figure. In the straight pipe after bending, the outer velocity is large while the inner velocity is small, but a high-velocity region appears at the inner side of the pipe, about at the position of 1 times pipe diameter downstream and that breaks the gradient distribution. Combined with the turbulence intensity, the phenomenon is caused by the eddy blending in the direction of secondary flow, which is enhanced by the unevenness of the intake air, and is unfavorable for flow uniformity.

According to the flow analysis and flow uniformity index of the S1, S2, and S3 sections, the main schemes can be proposed from the following three optimization directions.

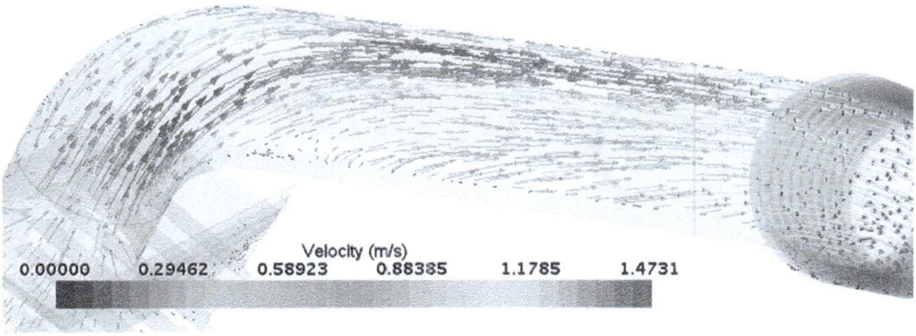

Fig. 6 Schematic diagram of gas velocity vector of axial section of clean pipe elbow

1. Scheme 1 is to change the arrangement of the upstream of the clean pipe. By adjusting the length of the straight pipe section after the air filter outlet, the non-uniformity flow of the clean pipe inlet can be changed, thereby achieving the purpose of improving the flow uniformity of the MAF measurement position. However, the implementation of this program is very difficult for the lack of actual engine compartment and vehicle space.
2. Scheme 2 is adjusting the measurement position of the MAF, but it is limited by the installation position of the breathing tube (Fig. 4). The MAF front flow uniformity index can only reach 87.6% and still cannot meet the relevant requirements.
3. Scheme 3 is adding a rectifying grid. The effect of this solution to improve flow uniformity is more pronounced, but it will increase the system resistance. According to the comprehensive analysis, there are problems that cannot be arranged in both scheme 1 and scheme 2. Therefore, scheme 3 is adopted, a rectifying grid is added in the clean pipe after the elbow, and the flow state is adjusted to improve the flow uniformity and reduce the signal-to-noise ratio.

3.2 *Flow Field Analysis of Intake System Based on Optimized Scheme*

In this section, case 1 refers to the initial intake system structure, and case 2 refers to the optimized intake system structure. The MAF structure was considered in the calculation, and the influence of the added rectifying grid on the flow uniformity and turbulence intensity was studied.

As shown in Fig. 7, the added rectifying structure is a honeycomb hexagonal grid structure disposed at a position close to the elbow before the MAF behind the elbow. Table 2 gives a comparison of the velocity uniformity index of case 1 and case 2. In case 1, the HFM front-end section velocity uniformity index is 83.8%, and the case 2 velocity uniformity index rises to 89.2%, an increase of 5.4%.

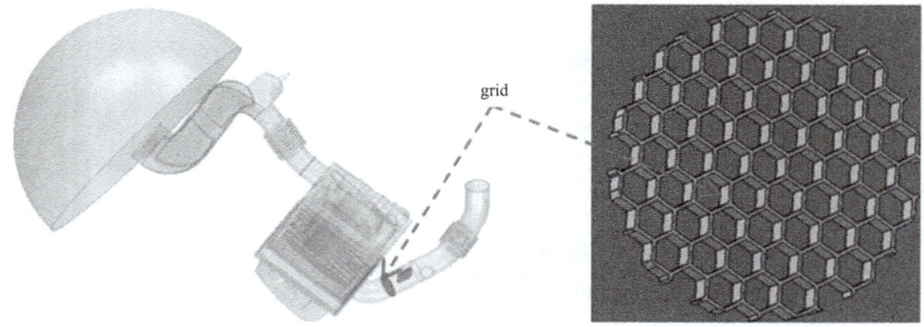

Fig. 7 Intake system with rectifying grid and rectifier grid

Table 2 Comparison of velocity uniformity index between initial scheme (case 1) and optimized scheme (case 2)

Cases	Mass airflow (kg/h)	UI_V (γ) (%)
Case 1	10	83.8
Case 2	10	89.2

Figure 8 shows the velocity distribution and turbulence intensity comparison of the front axial section of MAF between case 1 and case 2. Compared with case 1, case 2 has significantly improved flow uniformity due to the rectification of the grid structure, and the section velocity gradient is significantly reduced, eliminating vortices due to uneven intake air. It can be seen from the turbulence intensity diagram of case 2 that the influence of the bend pipe structure upstream of the rectifying grid is significantly weakened. Figure 9 shows the gas flow velocity vector of the case 2 clean pipe. It can be seen from the figure that the rectification effect of the rectification

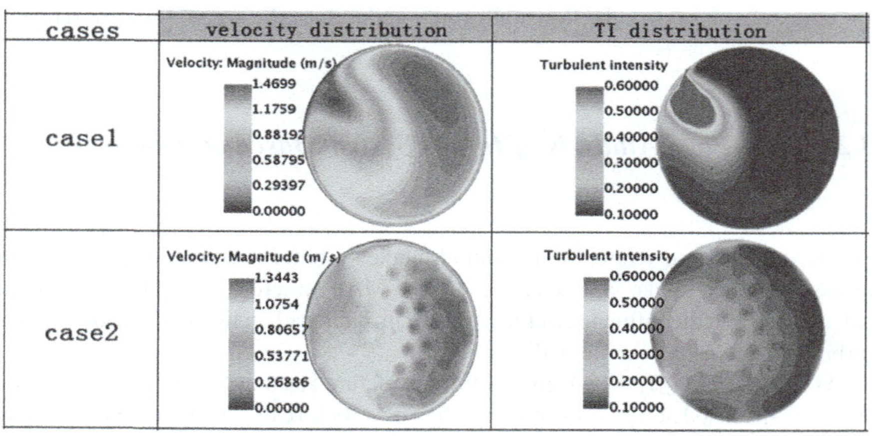

Fig. 8 Comparison of velocity distribution and turbulence intensity distribution between initial scheme (case 1) and optimized scheme (case 2)

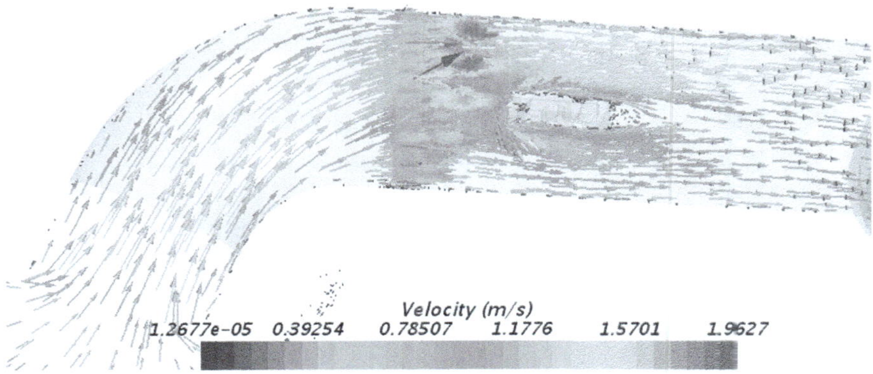

Fig. 9 Internal velocity vector in case 2 clean pipe

grid is obvious. The velocity vector direction is rectified from the outer side of the elbow to the axial direction of the straight pipe. However, since the size of the grid is the same, the overall velocity distribution law is still that the high outside of the elbow and the low inner.

4 Verification of SNR Test of Optimized Intake System

Figure 10 is a diagram of the signal-to-noise ratio test equipment for the optimized intake system. Figure 11 shows the signal-to-noise ratio of the optimized intake

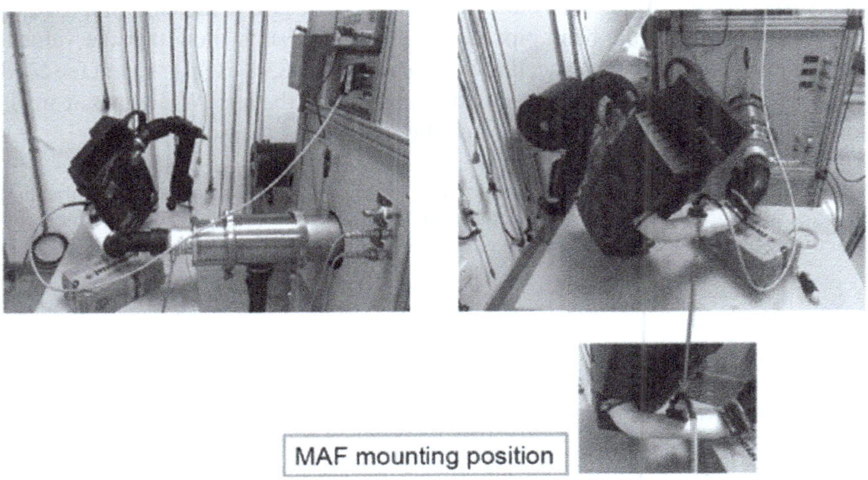

Fig. 10 Mass airflow sensor measurement equipment diagram

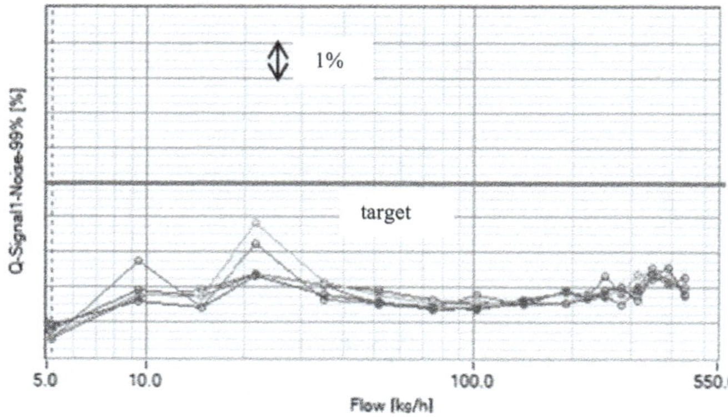

Fig. 11 SNR of MAF of the optimized intake system at a 99% confidence

system MAF. The signal-to-noise ratio is below the target line in the flow range of 5–500 kg/h. Especially, under the condition of the air intake rate of 10 kg/h, the signal-to-noise ratio decreased by 9.5%.

5 Conclusion

This paper is aimed at the problem that the MAF SNR test index exceeds the standard in a certain intake system. Based on Star CCM+ software, the flow field of the intake system without MAF is analyzed, and various schemes for improving the flow uniformity of the front end of the MAF of the intake system are proposed. The uniformity scheme is determined that the rectifier grid is added to improve the flow uniformity after the bend pipe. The simulation results show that the velocity uniformity index of the target section is increased from 83.8 to 89.2%, and the MAF signal-to-noise ratio index is also greatly improved in the test. While the air intake flow rate is 10 kg/h, the signal-to-noise ratio indicator decreased by 9.5%.

Acknowledgements This project is supported by the Science and Technology Commission of Shanghai Municipality and a research grant from the Shanghai Key Laboratory of Automobile Powertrain (contract no. 14DZ2260600).

References

1. Li J, Hao P, Zhang X et al (2002) Non-uniformity of flow in curved pipe and its rectification. J Mech Eng 38(12):146–148
2. Li J, Liu Z, Liu Z et al (2012) Simulation and experimental analysis of air filter flow process. J Zhejiang Univ (Eng Sci) 46(2):327–332
3. Bell J (1983) Porous media fluid dynamics. China Building Industry Press, Beijing
4. CD-Adapco, STAR-CCM+ Documentation Version 10.06:7736–7737

Luggage Retention Optimization Design for Rear Seat of Vehicle

Junhui Liu, Jianjun Huang, Miao He, Zhaoliang Gou, Chi Dai and Zhou Zhang

Abstract This paper focuses on a vehicle with rear seat adjustable backrest, analysis of validation phrase rear luggage retention test failure, to find the weak position of the rear seat frame structure. Optimize the backrest frame from two aspects, collide energy transfer direction and frame structure strength, and release reasonable structure improvement proposal from the perspective of cost reduction and lightweight, combining simulation analysis and physical verification. Optimized structure of the rear seat frame can meet the design specification, saving project cost, shorten the development cycle, reduce development risk.

Keywords Rear seat backrest adjustable · Luggage retention · CAE · Seat frame optimization

1 Introduction

Seats, as the major components of the vehicle, mainly provide support, occupant position determination, driving comfort experience, and protect occupants. In emergency conditions, together with the restraint system to protect the occupants reduce damage. The seat design is based on the premise of safety, then improves the comfort and functional design. The rear seat, as one of the main safety components, and its skeleton strength plays a vital role in the safety of the whole vehicle [1].

In order to obtain better comfort, the rear backrest angle adjustable function has been applied in some vehicle models. The backrest structure is fixed on the floor by the recliners. Compared with the traditional rear seat, the upper end of the backrest is not fixed by locks, which requires higher structure strength. The luggage impact test is the most stringent test for the strength of the rear seat frame, when the middle seat belt retractor is deployed on the roof.

J. Liu (✉) · J. Huang · M. He · Z. Gou · C. Dai
Dongfeng Motor Corporation Technical Center, Wuhan 430058, China
e-mail: liujunh@dfmc.com.cn

Z. Zhang
Dongfeng Lear Technical Center, Wuhan 430058, China

China SAE (ed.), *Proceedings of China SAE Congress 2018: Selected Papers*, Lecture Notes in Electrical Engineering 574,
https://doi.org/10.1007/978-981-13-9718-9_9

According to GB15083-2006, two luggage blocks of 300 mm × 300 mm × 300 mm mass 18 kg are placed on the floor of the luggage compartment of the car body according to Fig. 1. The body is firmly fixed on the sled, and the car body is decelerated, the deceleration pulse is shown in Fig. 2. Before the deceleration, the original speed of the car body should be 50_{0}^{+2} km/h. The seat needs to meet, during the test process, the front profile of the backrest and/or headrest [shore (A) hardness above 50] cannot be move beyond the specified scope [2].

This paper mainly introduces the improvement of a rear seat with recliner to achieve backrest angle adjustment, failed to meet the luggage retention test, with the aid of finite element analysis, find out the control points of the rear seat backrest frame, through structure optimization design to get low cost and light weight proposals.

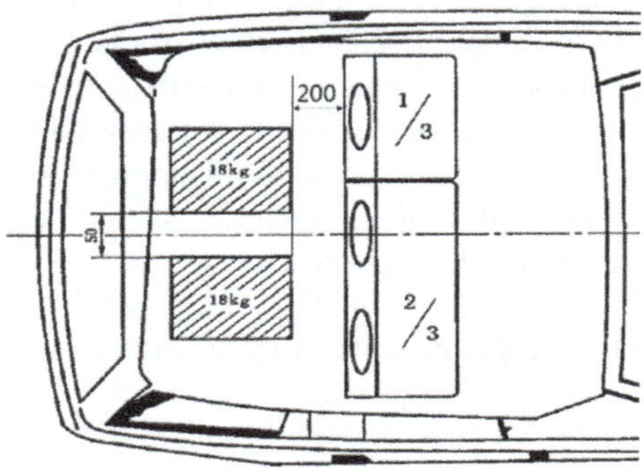

Fig. 1 Luggage block layout

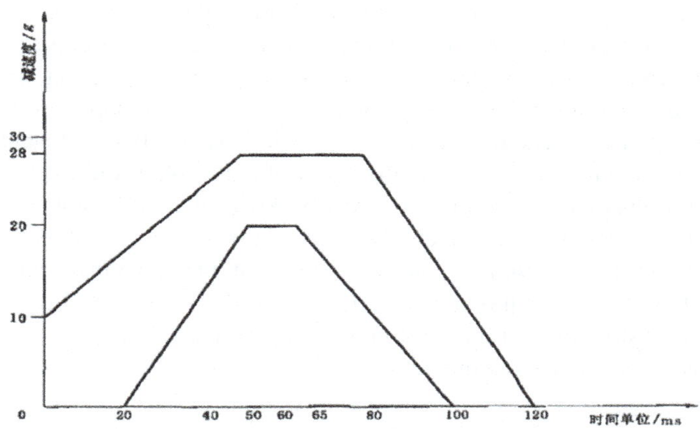

Fig. 2 Collide pulse region

2 Failure Model Description

The rear seat backrest of the vehicle is divided into 40 and 60% sides, and the backrest angle adjustment function is achieved through the angle recliner, as shown in Fig. 3. The backrest skeleton is connected to the lower bracket through the angle recliners and fixed to body floor by eight bolts, as shown in Fig. 4. The impact energy of the luggage block received by the backrest is passed to the angle recliner through the backrest skeleton and then transmitted to the brackets of the angle recliners, and finally to the floor of the vehicle body. The parts related to test strength include backrest tube, upper bracket of the recliner, angle recliner, lower bracket of the recliner, mounting bracket, and vehicle floor. The simulation analysis model is shown in Fig. 5.

In the data design stage, the FEA is carried out with luggage blocks deployed according to the requirements of the GB 15083, and also the test pulse, which is shown in Fig. 6.

The reception criteria of the rear seat luggage retention test are as follows: the front profile of the backrest and/or the headrest [shore (A) hardness above 50] part cannot move forward a lateral vertical plane, and this plane passes through:

(a) a point which is 150 mm front of the seat *R* point (for seat headrest part);
(b) a point which is 100 mm front of the seat *R* point (for seat backrest part).

Then, it is judged that the luggage retention strength of the rear seats meets the requirements of GB15083.

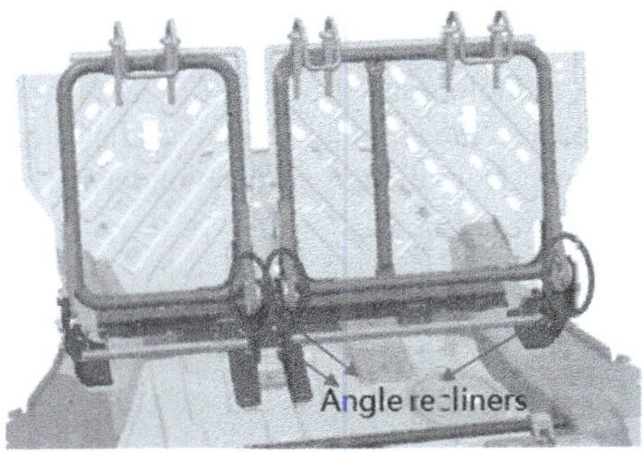

Fig. 3 Seat structure diagram

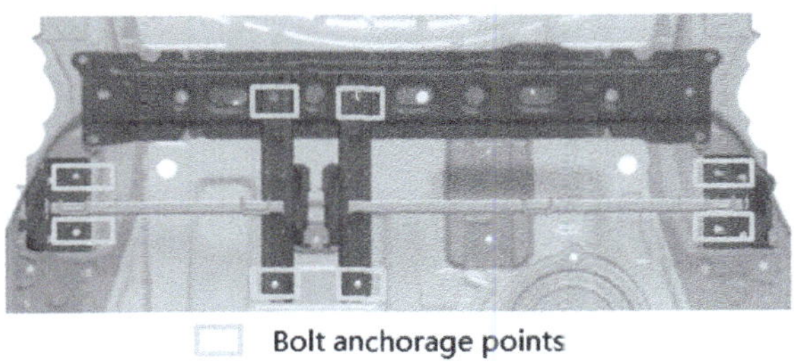

Fig. 4 Seat anchorage diagram

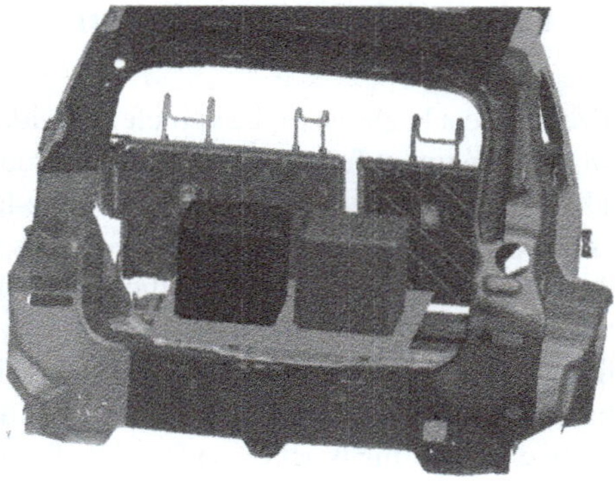

Fig. 5 Simulation analysis module

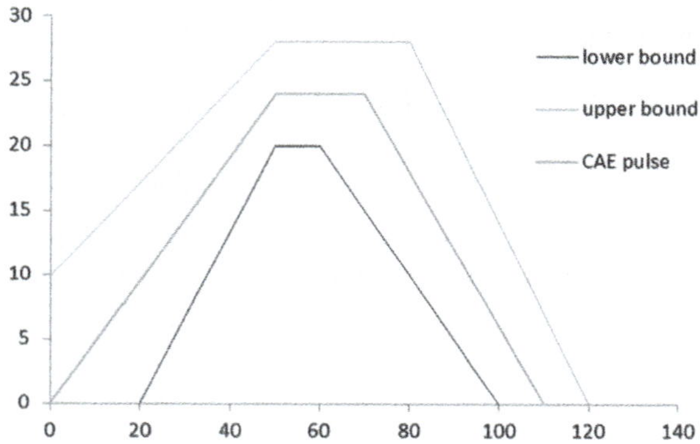

Fig. 6 Simulation load pulse (图片更新)

A vehicle seat (R point: $X = 2100$ mm) is shown in Fig. 7. With several optimizations, the simulation analysis has a certain safety margin on the backrest or the headrest from the target plane under the specified loading. The CAE analysis result is shown in Fig. 8.

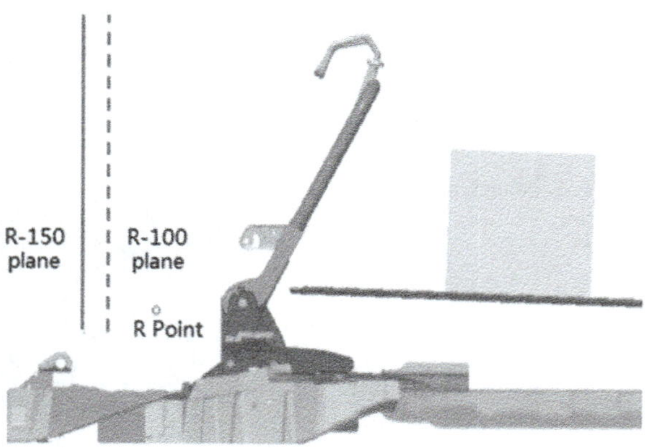

Fig. 7 Acceptance criteria of GB

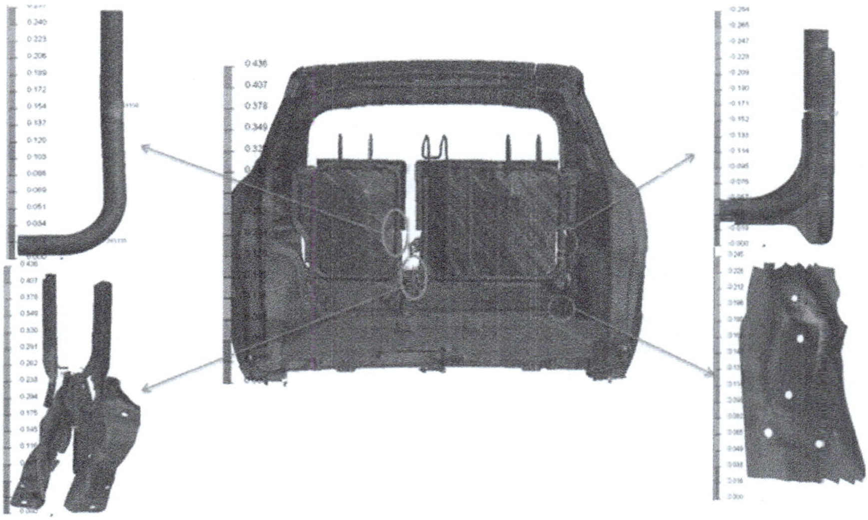

Fig. 8 Seat luggage retention analysis

Produce seats according to the data and carry out the seat sled test [3]. During the test, the dynamic displacement of 60% side seat headrest exceeds the plane of R-150 mm plane, and the test failed.

The results of CAE analysis and physical test are shown in Table 1

Table 1 CAE and physical test results

	Sled CAE analysis	Vehicle CAE analysis	Sled test
Diagram		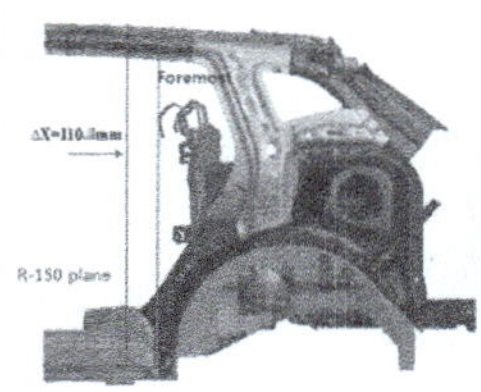	
Test pulse	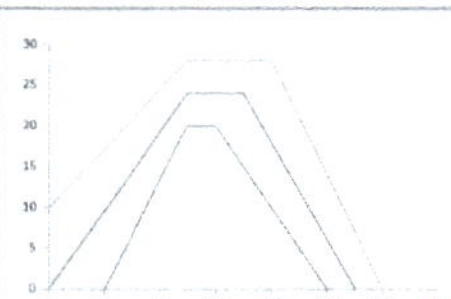	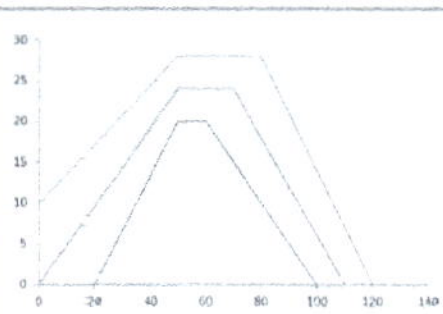	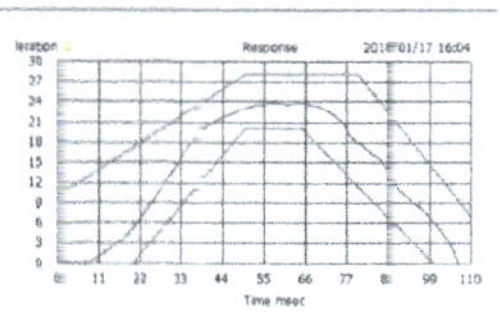
Result	OK (safety margin 124 mm)	OK (safety margin 111 mm)	**Fail** (exceed about 25 mm)

3 Analysis

In the sled test, the deformation of the seat beyond the target plane mainly comes from the 60% side backrest skeleton, for the parts in the energy transmission path, compared the CAE analysis and the physical test results (Fig. 9), to find the cause of the failure [4].

3.1 Inner Upper Connecting Plate and Tube

The tube is severely bent in FEA, but is not bending in a physical test, which indicates that the energy of the backrest skeleton is not effectively absorbed, but transfers down to the recliner.

3.2 Inside Recliner

In the CAE analysis, the recliner is set as a rigid part and records the torque, and the default setting is not fail; but for physical test, after disassembling the recliner, the lock plate is severely sheared, leading to the lean of the backrest, which is the main reason of the failure.

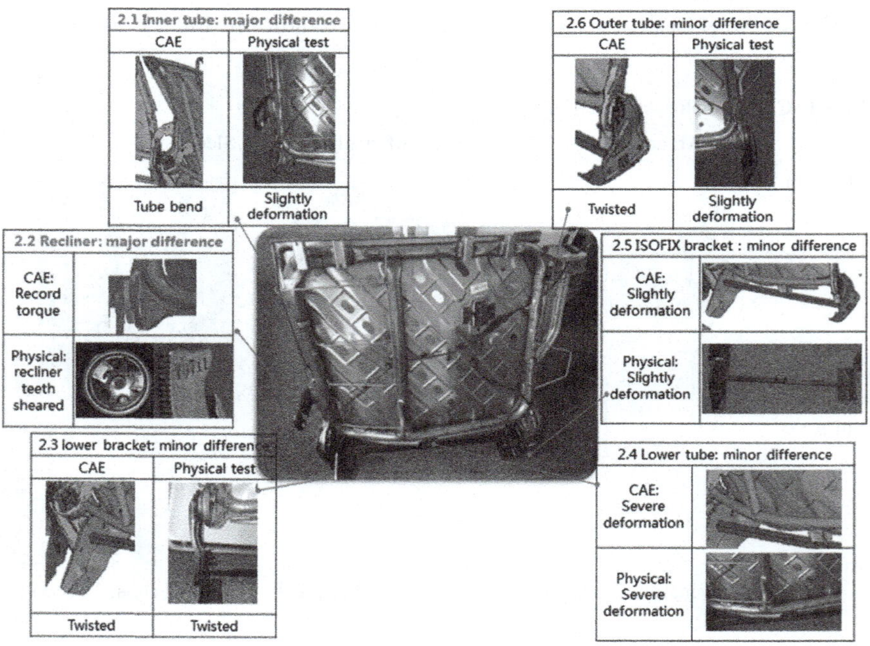

Fig. 9 Seat luggage retention CAE and physical comparison

3.3 Inside Recliner Lower Connecting Plate

The CAE analysis shows that the lower connecting plate is twisted outwards, with little difference from the physical test.

3.4 Backrest Lower Transverse Tube

The forward deformation of the lower transverse tube in the CAE analysis is consistent with the physical test condition.

3.5 ISOFIX Mounting Bracket

The CAE analysis shows slight deformation and is consistent with the physical test condition.

3.6 Outer Bracket and Tube

The CAE analysis is distorted outwards, and the physical test is slightly distorted, with little difference.

Because the grid precision of the vehicle collision cannot reflect the deformation of the internal tooth of the recliner, it is generally set as a rigid part in the CAE analysis, and the torque of the 60% side backrest inner recliner is recorded at 2075 Nm less than the rated torque 2100 Nm of the recliner, but the safety margin is small.

In this case, due to the torque of the angle recliner cannot be measured in the physical test, it cannot be compared to the FEA result. The failure of the recliner leads to the deviation between the result of FEA and the result of the physical test. The optimization proposals need to monitor the torque of the recliner and also set the safety margin suggestion of the recliner torque in FEA.

Through the above analysis, the direct reason for the failure of the physical test is that the load of the 60% side backrest inside recliner exceeds the rated torque 2100 Nm, resulting recliner teeth slippage.

4 Proposal Optimization

According to the comparison between CAE results and physical tests, the angle recliner cannot hold the load is the main reason of the failure.

Option 1, by changing an inside recliner with a higher rated torque, or

Option 2, by adjusting the load distribution of the 60% side backrest skeleton, to reduce the load of the inside recliner.

Because of recliner change would influence the parts matching interface and cost increase, the analysis direction of this case is option 2: to reduce the load of the inside angle recliner.

According to the load transfer path, optimization measures can be carried out in the following two directions:

(1) Transfer path optimization: transferred the load of the inside recliner to the outside recliner. With topological analysis, a variety of schemes is analysed, and the transverse steel tube is added by consideration of parts modification and technological feasibility.
(2) Structural strength optimization: reduces the load transferred to the inside recliner during the transfer process, such as increasing the energy absorption of the backrest tube and the connecting plate [5, 6].

According to the above analysis, carry out the improvement first on the seat system level, then on the vehicle level.

4.1 First Round Optimization Design

Optimize the load transfer path: According to the topological analysis method, considering a variety of change proposals, combined with manufacturing process and change impact, select the proposal: based on the 60% side backrest skeleton, increase the lateral tube, transfer the collision energy to the outside recliner ③, reduce the load of the inside recliner ②. The proposal descriptions are shown in Table 2.

According to the above optimization proposal, CAE analysis is carried out, and the result is shown in Table 3.

The proposal A meets the luggage retention design target, and the safety margin increases by 23 mm; meanwhile, the load of the 60% side backrest inside recliner decreases by 86 Nm.

By comparing proposal A with the original one, it is shown that adding the lateral tube can transfer the load of the inside recliner to the outside one, effectively dispersing the load. After adding the tube, the safety margin is further increased, and the weight and cost also increased.

4.2 Second Round Optimization Design

Optimize the strength of the skeleton structure: reduce the strength of the backrest tube and the connection plate of the recliner, increase the energy absorption of the backrest skeleton, reduce the load transferred to the recliner, meanwhile, reduce the

Table 2 First round proposal optimize

	Original failed proposal	Proposal A
Description	60% side backrest skeleton structure: a rectangular shape with a vertical reinforcement tube Tube specification: QSTE500, Φ30 mm, T1.5 mm	60% side backrest skeleton structure: material specification unchanged, add horizontal tube
Diagram	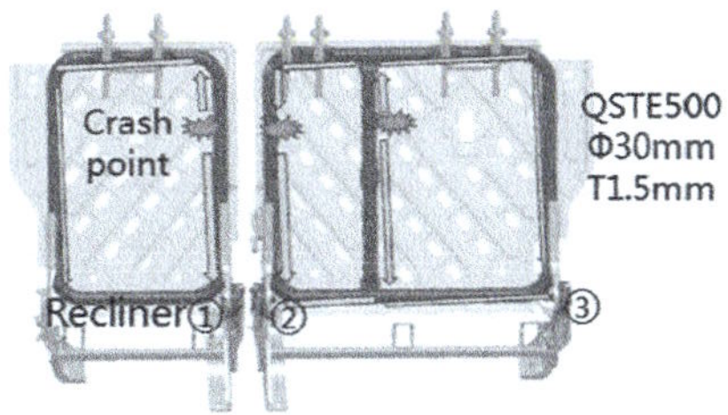	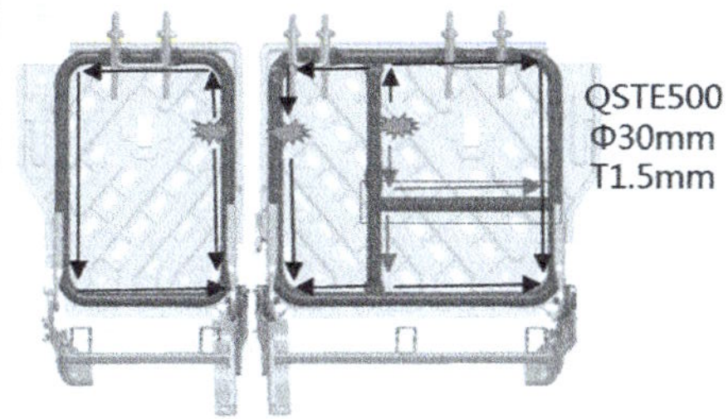

Table 3 First round proposal optimize result

		Original failed proposal	Proposal A
CAE analysis		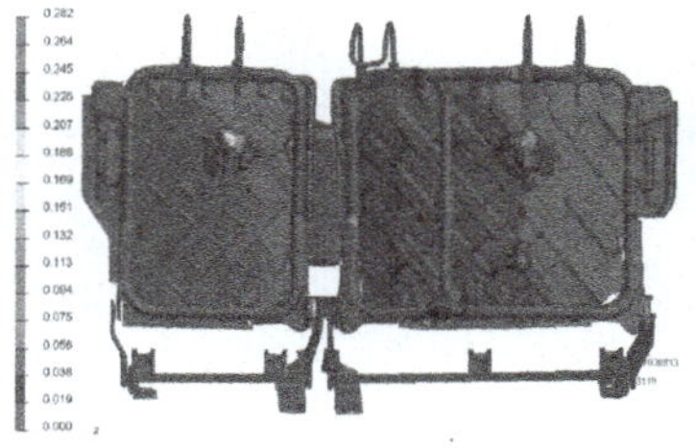	
CAE result		OK (safety margin: 145 mm)	OK (safety margin: 159 mm)
Recliner torque Nm	①	1853	1849
	②	2075	1989
	③	1358	1497

weight and optimize the cost, to find the best solution. The proposal descriptions are shown in Table 4.

According to the two sets of proposals above, the CAE analysis is carried out, and the results are shown in Table 5.

The proposal B only reduces the material grades of the backrest tube. The simulation test shows that the skeleton performance is lower than the original one, but still with a 127 mm safety margin. Through the FEA analysis result, the deformation of the tube at the 60% side backrest is larger than that of the original, the energy absorption effect has increased, and the torque of the recliner is reduced by 134 Nm.

Proposal C, based on proposal B, reduces the strength of the recliner connection plate and the reinforce plate. The simulation results show that the skeleton performance is slightly reduced, but there is still a 115 mm safety margin. The load on the 60% side back angle recliners has further reduced, and the result meets the design target.

4.3 *Vehicle Level CAE Analysis*

According to the results of the stimulation analysis, the three sets of optimization proposals are compared and summarized as shown in Table 6.

Through the simulation analysis of the luggage retention test, it is found that the main factors affecting the strength of the backrest skeleton are the balance distribution of the skeleton structure strength and the recliners' load, a reasonable framework design can effectively disperse the load, to achieve the design target of safety, low cost and lightweight. The optimization proposal C can meet the CAE simulation target, and the weight is reduced by 0.4 kg and the cost also reduced.

Table 4 Second round proposal optimize

	Original failed proposal	Proposal B	Proposal C
Description	Tube specification:QSTE500, Φ30 mm, T1.5 mm	Tube specification: material change from QSTE500 to Q345	1. Tube material change to Q345 2. Recliner upper bracket (green): Height reduced from 208 mm to 158 mm 3. Upper reinforce bracket (blue): thickness reduced from 2.5 to 2.0 mm
Diagram	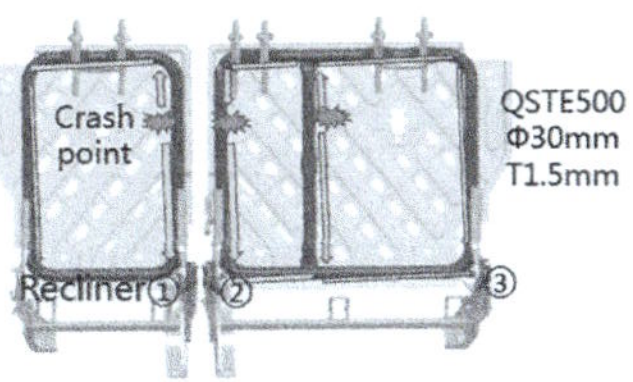	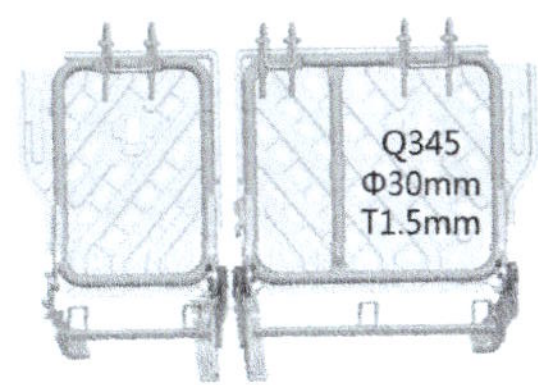	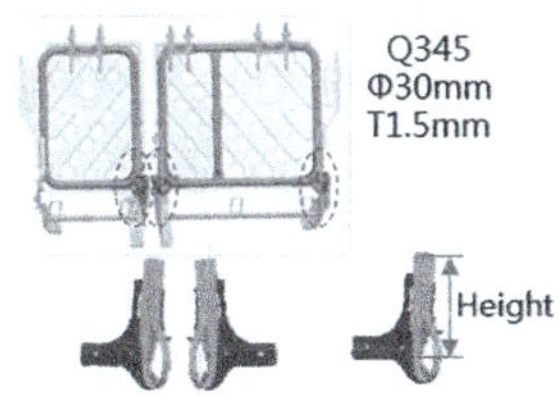

Table 5 Second round proposal optimize result

		Original failed proposal	Proposal B	Proposal C
CAE analysis		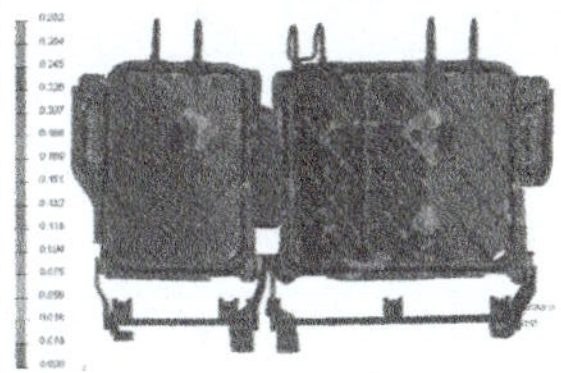		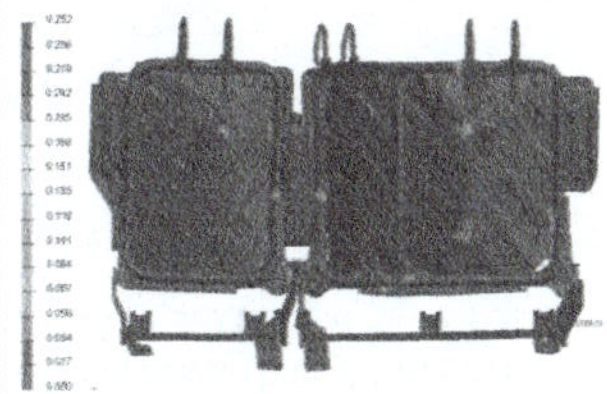
CAE result		OK (safety margin: 145 mm)	OK (safety margin: 127 mm)	OK (safety margin: 115 mm)
Recliner torque Nm	①	1853	1706	1667
	②	2075	1941	1873
	③	1358	1244	1048

Table 6 Optimize proposal comparison

	Original failed proposal	Proposal A	Proposal B	Proposal C
Description	Base	Add tube	Tube specification: material change from QSTE500 to Q345	1. Tube material: Q345 2. Recliner upper bracket: Height reduced 50 mm 3. Upper reinforce bracket thickness reduced 0.5 mm
		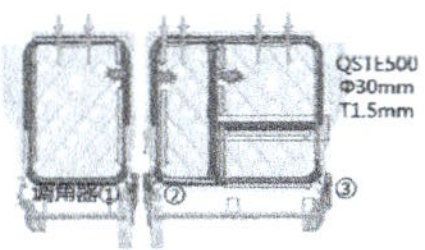	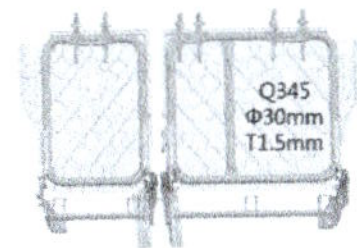	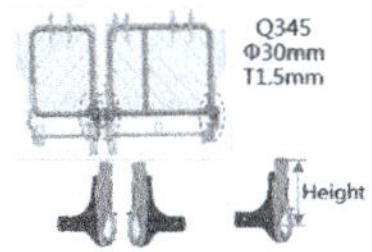
CAE safety margin	145 mm	159 mm	127 mm	115 mm
Recliner ② torque	2075 Nm	1989 Nm	1941 Nm	1873 Nm
Mass	Base	+0.3 kg	0	−0.4 kg
Cost	Base	Increased	Decreased	Decreased

Table 7 Vehicle test improve result

		Original failed proposal	Proposal C
CAE analysis		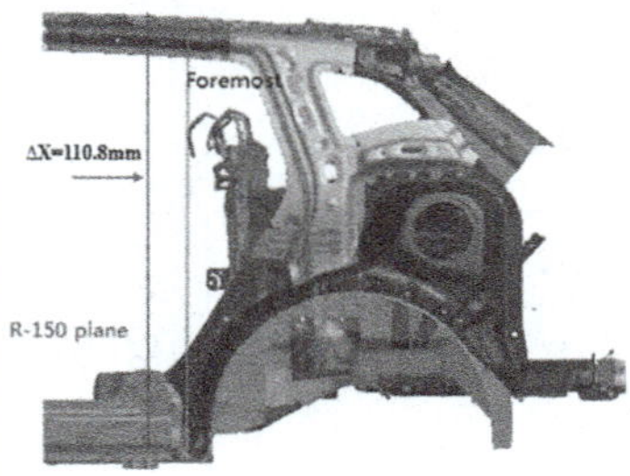	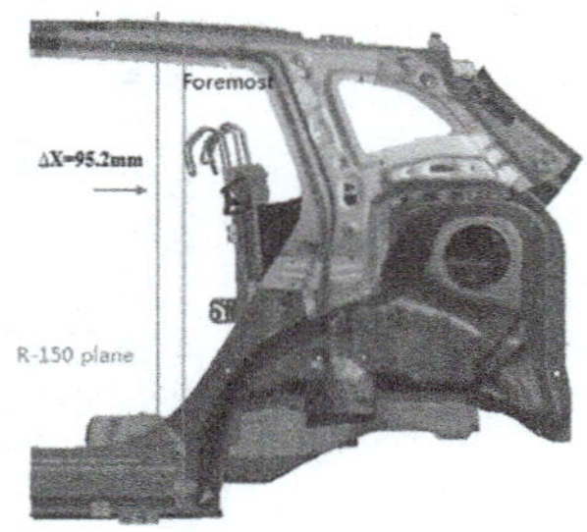
CAE result		OK (safety margin: 111 mm)	OK (safety margin: 95 mm)
Recliner torque Nm	①	1769	1617
	②	2051	1850
	③	1196	1016

Proposal C is selected to carry out vehicle CAE analysis, and the result is shown in Table 7.

In the vehicle level, the performance of the proposal C meets the design target and has a 95 mm safety margin; at the same time, the load of the 60% side backrest inside recliner is sharply reduced to 1853 Nm, with a 11.8% safety margin from the rated torque 2100 Nm.

In summary, the parts are produced according to proposal C.

5 Test Verification

According to the requirements of the luggage retention test in GB 15083, carry out the sled test of proposal C. The recliner does not fail, and the displacement of the backrest and headrest skeleton meets the design requirements.

Then, carry out the physical test of the vehicle level, the recliner has no failure, and the displacement of the backrest and headrest skeleton meets the requirements of the regulations.

The results of the tests are shown in Table 8.

Table 8 Physical test result

	Original failed proposal—sled test	Proposal C—sled test	Proposal C—vehicle test
Test picture			
Test pulse	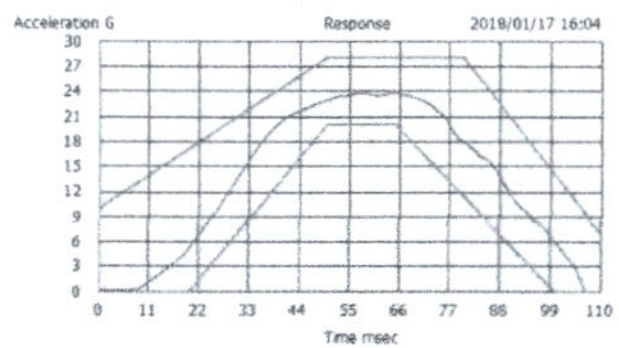	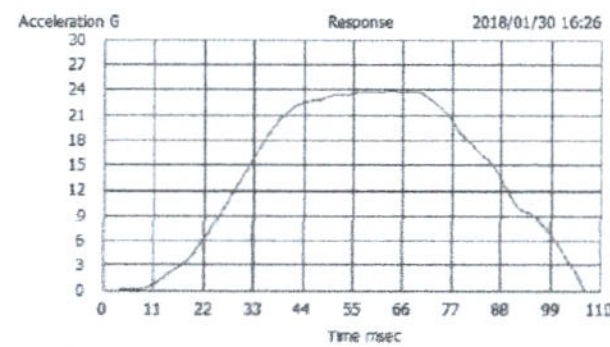	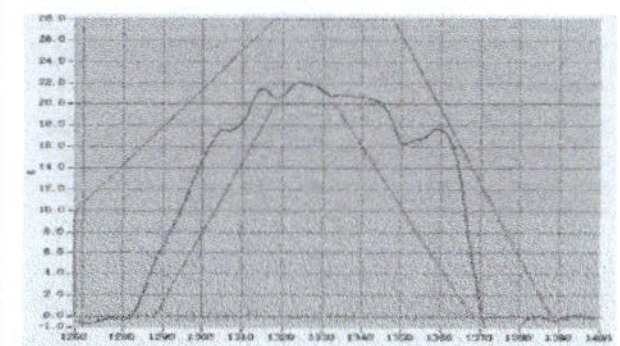
Test result	Fail	OK (safety margin 105 mm)	OK (safety margin 78 mm)
Recliners	Slippage	OK	OK

6 Conclusion

In this paper, an analytical method is provided to the design of the rear seat structure strength with a seat backrest angle adjustable, the strength of the angle recliner and the strength of the backrest skeleton structure should be considered comprehensively.

In this case, by analysing and comparing a variety of optimization proposals, the conclusions are as follows:

(1) The seat structure performance of the rear adjustable backrest is both determined by the strength of the angle recliner and the strength of the backrest skeleton structure.
(2) Because the vehicle collision simulation precision cannot analyse the internal deformation of the recliner, the load of the recliner recorded from the simulation analysis is suggested with a safety margin of 12% (compared to the rated torque) be reserved.
(3) Increasing the energy absorption of the backrest skeleton structure can effectively reduce the load on the angle recliner.

References

1. Yao W, Sun D (2002) Summarizes for the safety of automobile seat. Automobile Technol 8:5–8
2. GB 15083-2006 (2007) Strength requirement and test of automobile seats, their anchorages and any bead restraints. Standards Press of China
3. Wang H, Zhang D (2004) Modelling and simulating with finite element method in vehicle seats. J Tongji Univ 32(7):947–951
4. Xu Z, Hao W, Zhang Z, Jin J (2009) Analyses on stress tensor and crash worthiness simulation of an automobile seat. J Chongqing Univ 32(5):512–515
5. Vangipuram R, Long L, Truong H (2009) Parameter design based FEA correlation studies on automotive seat structures. SAE paper vol 1, issue 1, pp 175–183
6. Park SD, Pyun JK, Choi BY (2010) Seat common frame design optimization. SAE paper 2010-01-0390

Analyzing and Optimization the Resistance Welding Spots of a Trunk Lid Hinge Reinforcement Plate

Zhiqiang Huang, Li Han, Yujie Tian, Kai Kang and Junping Qiao

Abstract Based on the finite element method (FEM), this research focused on the trunk lid hinge reinforcement plate welding spot crack problem. The number of the resistance spot welding (structure adhesives) and the structure of the hinge reinforcement plate are analyzed, and we found the method to solve the trunk lid hinge reinforcement plate welding spot crack problem. The stress of trunk lid inner panel was reduced from 128 to 80 Mpa, the fatigue life was increased to more than 10,000,000 from 248,000 by CAE simulations, and the CAE simulations have been effectively verified with the prototype test.

Keywords Trunk lid hinge reinforcement plate · FEM · Welding spot cracking

1 Introduction

Resistance spot welding (RSW) is a time-saving and very effective way to join two or more metal workpieces together, and it is widely used in automotive industry since thousands of spot welds were applied in the auto body, so the number and layout of the RSW have an important influence on the performance of the body in white (BIW). The unreasonable RSW layout or structure will increase the number of RSW, which provides adverse impact on the body structure stiffness improvement and will reduce the jobs per hour (JPH). In this study, an optimization scheme is proposed for the RSW fatigue cracking of a trunk lid hinge reinforcement plate in the normal temperature durability test. And the finite element method (FEM) is used to evaluate the RSW fatigue cracking. Finally, the design was verified in the high- and low-temperature durability test.

Z. Huang (✉) · L. Han · Y. Tian · K. Kang · J. Qiao
Beijing Automotive Technology Center, BAIC Motor Co., Ltd, Beijing 101300, China
e-mail: huangzq09@163.com

China SAE (ed.), *Proceedings of China SAE Congress 2018: Selected Papers*, Lecture Notes in Electrical Engineering 574,
https://doi.org/10.1007/978-981-13-9718-9_10

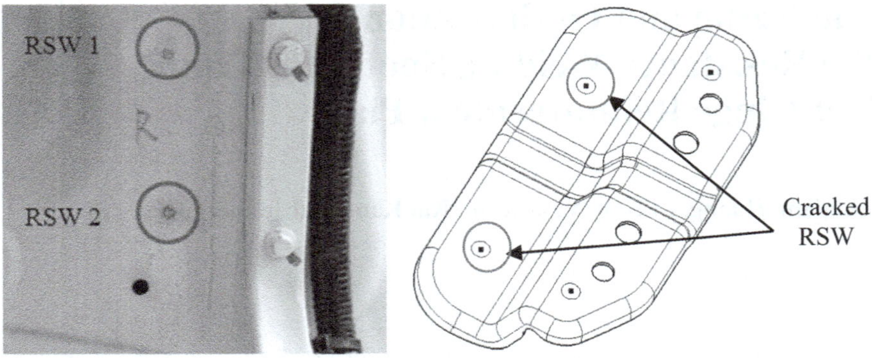

Fig. 1 RSW fatigue crack of trunk lid hinge reinforcement plate

2 Description and Analysis of the Issue

The target of a trunk lid opening and closing durability test, which is implemented at the normal temperature, has no RSW damage in 30,000 cycles and no RSW cranking in 25,000 cycles. But when we finished the 30,000 cycle opening and closing durability test, the RSW of the trunk lid hinge reinforcement plate right was cracked, as shown in Fig. 1. And the two welding spots are named RSW 1 and RSW 2, respectively.

According to the design experience, the main reasons for the RSW fatigue cracking of BIW are as follows: (1) improper materials of the welding sheet metals. When the yield strength gap between the welding sheet metals is large [1] or the thickness ratio between the welding sheet metals is less than 1/3, the RSW crack will arise in the sheet metal with lower strength or thinner thickness. (2) The imbalance stress of the local structure is another reason for the RSW cracking [1]. When the local stress exceeds the limit yield strength of the material, the alternating load causes the RSW cracking.

3 FEM of the Trunk Lid Hinge Reinforcement Plate RSW Cracking

3.1 Trunk Lid FEM Model

The trunk lid finite element method model is established, as shown in Fig. 2. The joints between the trunk lid and BIW are bolts, and the constraint of the joints in the BIW is set with full constraint model. And constrain the Z direction displacement of the Trunk lid in the center of the Trunk lid bumper. Loading 415 N (Simulate the load of the latch strike) to the Trunk lid latch along the trunk lid close direction.

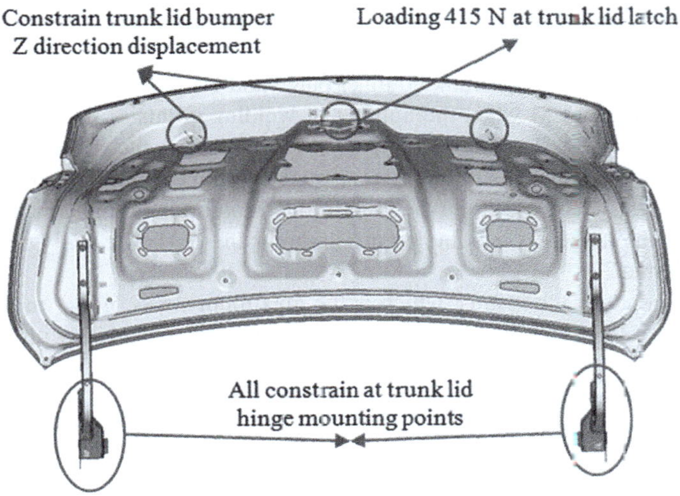

Fig. 2 FEM model and constraint diagram of the trunk lid

Analyzing the stress of the trunk lid inner panel with FEM, and simulating the RSW fatigue life [3] with the FEMFAT [2].

3.2 *Trunk Lid Hinge Reinforcement Plate RSW FEM Scheme*

To solve the problem of the RSW fatigue cracking of a trunk lid hinge reinforcement plate, this paper analyzes the main factors of the welding spots cracking by CAE and includes the number of the hinge reinforcement plate welding spots (structural adhesive) and the structure of the hinge reinforcement, as shown in Table 1.

Table 1 CAE validation protocol of the trunk lid hinge reinforcement plate RSW cracking

Hinge reinforcement plate	RSW or structural adhesive		
	4 RSW	6 RSW	Remarks
Original design	Original proposal	Proposal one	
Hinge reinforcement plate 1	–	Proposal two	
Hinge reinforcement plate 2	–	Proposal three	

3.2.1 The Number of the Hinge Reinforcement Plate RSW

The maximum stress of the trunk lid inner panel and the fatigue life of the RSW are analyzed with the FEM when the number of welding points of the trunk lid hinge reinforcement plate is 4 and 6, respectively. Figure 3 shows the stress cloud chart of the trunk lid inner panel with the original trunk lid hinge reinforcement plate, four welding points. The stress cloud diagram of the trunk lid inner panel with the hinge reinforcement plate 1, six welding points, is shown in Fig. 4. The analysis results are shown in Table 2.

Table 2 contains the results, which is the maximum stress of the trunk lid inner panel and the RSW fatigue life of the hinge reinforcement plate with a different number of welding points. According to the CAE analysis results in Table 2, under the condition that we only changed the number of hinge reinforcement plate welding

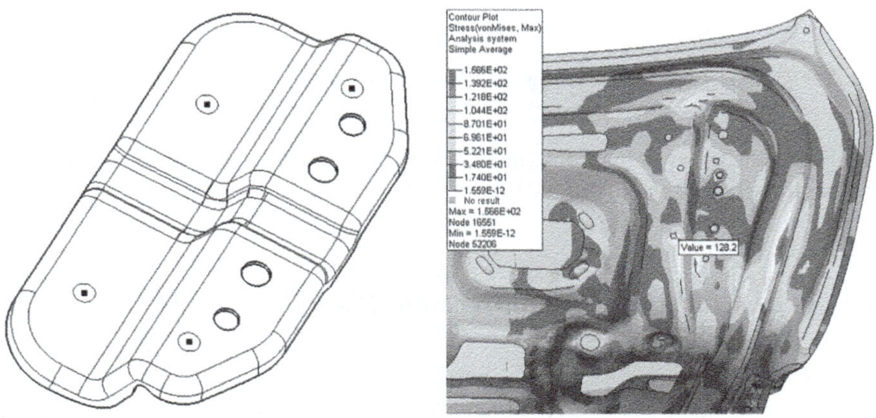

Fig. 3 Stress cloud chart of the trunk lid inner panel with the original proposal

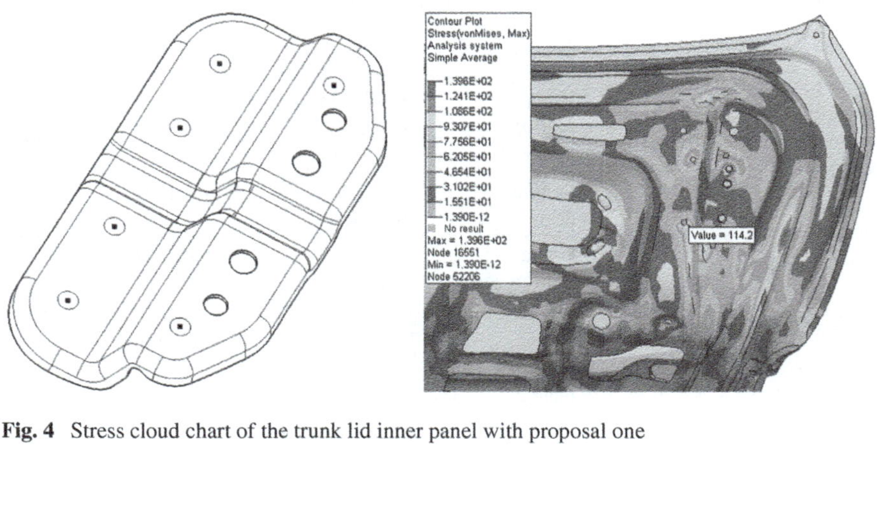

Fig. 4 Stress cloud chart of the trunk lid inner panel with proposal one

Table 2 CAE analysis results of different hinge reinforcement plate welding point numbers

Proposal	Index			
	Maximum stress (Mpa)	Life of RSW (cycles)		Remarks
		RSW 1	RSW 2	
Original proposal	128	248,000	514,000	
Proposal one	114	294,000	618,000	

points from 4 to 6, the maximum stress of the trunk lid inner panel was reduced by 10.9%. And the life cycles of the RSW 1 and RSW 2 increase by 18.4 and 20.2%, respectively. Therefore, increasing the number of the trunk lid hinge reinforcement plate welding points will be helpful to reduce the maximum stress of the trunk lid inner panel and improve the fatigue life of the RSW.

3.2.2 The Structure of the Hinge Reinforcement Plate

The structure of the trunk lid hinge reinforcement plate is a key factor of the hinge reinforcement plate RSW fatigue life. And we analyzed the proposal of the hinge reinforcement plate with a different structure and same number of the welding points. Figures 5, 6, and 7, respectively, describe the stress of the trunk lid inner panel with different hinge reinforcement plates.

Table 3 contains the maximum stress of the trunk lid inner panel and the RSW fatigue life of the hinge reinforcement plate of the proposal 1, proposal 2, and proposal 3. According to the results of the CAE analysis in Table 3, (1) with the same number of welding points, the maximum stress of the trunk lid inner panel was reduced and the RSW fatigue life was increased with the increase of the welding contact area. (2) Compared with proposal one, the maximum stress of the trunk lid inner

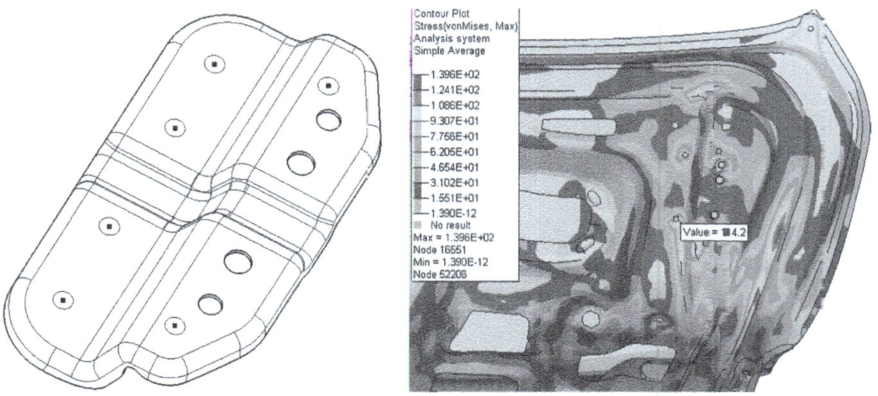

Fig. 5 Stress cloud chart of the trunk lid inner panel with the hinge reinforcement plate 1

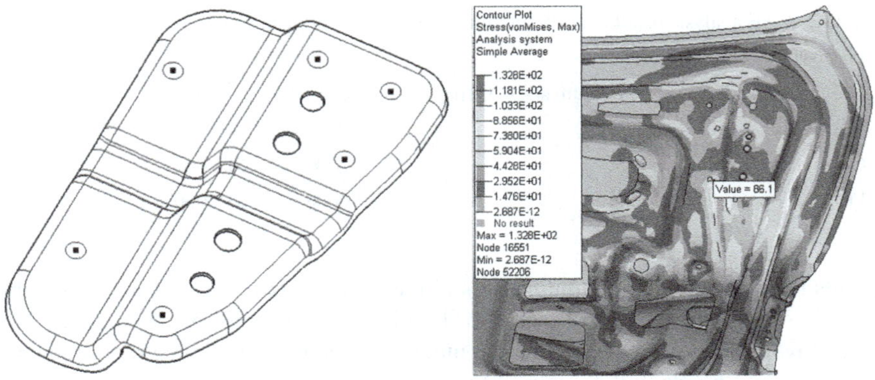

Fig. 6 Stress cloud chart of the trunk lid inner panel with proposal two

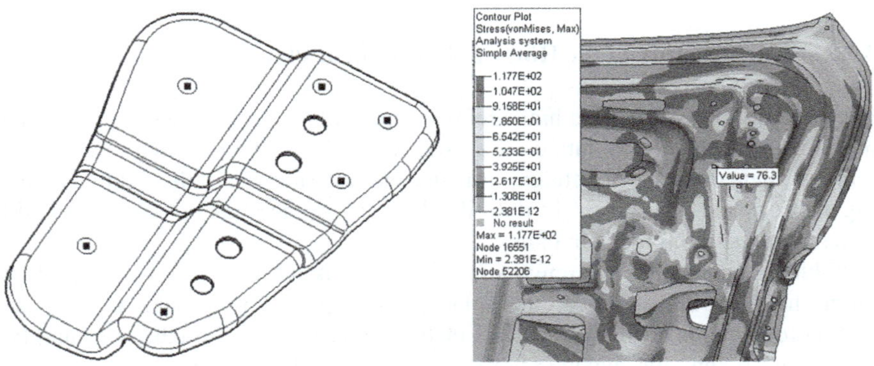

Fig. 7 Stress cloud chart of the trunk lid inner panel with proposal three

Table 3 CAE analysis results of different hinge reinforcement plates

Proposal	Index			
	Maximum stress (Mpa)	Life of RSW (cycles)		Remarks
		RSW 1	RSW 2	
Proposal one	114	294,000	618,000	
Proposal two	86	388,000	646,000	
Proposal three	76	415,000	857,000	

panel from proposal two is reduced by 24.6%, and the fatigue life of the RSW 1 and RSW 2 increased by 32% and 4.5%, respectively. (3) In the same hinge plate welding point layout proposal, the maximum stress from the trunk lid inner panel of the proposal three was reduced by 24.6%, and the RSW 1 and RSW 2's fatigue life of the proposal three increased by 7.0 and 32.7%, respectively, compared with the proposal two. Therefore, increasing welding contact area will be helpful to reduce the maximum stress of the trunk lid inner panel and improve the fatigue life of the RSW.

4 Solutions and Verification

4.1 Solutions

The preliminary analysis in Sect. 3 shows that increasing the welding contact area and the welding point number of the hinge reinforcement plate is helpful to reduce the maximum stress of the trunk lid inner panel and the RSW fatigue life. In order to solve the problem of RSW cracking, proposal four and proposal five, which are shown in Table 4, were presented according to the FEM analysis results of Sect. 3. Figures 8 and 9 are results, which are the stress cloud chart of the trunk lid inner panel, of the proposal four and proposal five.

Table 5 contains the maximum stress of the trunk lid inner panel and the RSW fatigue life of the hinge reinforcement plate, of the original proposal, proposal four, proposal five, and the benchmarking (BM). According to the CAE analysis results in Table 5, compared with the original proposal, the maximum stress of the trunk lid inner panel from proposal four and proposal five was reduced by 40.5 and 37 5%, respectively, and was similar with that of the benchmarking. The RSW fatigue life from proposal four and proposal five was obviously increased and was consistent with that of the benchmarking.

The cost and design cycles of the solutions are shown in Table 6. Considering the cost, cycles, and factory, proposal five is chosen.

Table 4 CAE validation protocol of the trunk lid hinge reinforcement plate RSW cracking

Hinge reinforcement plate	RSW or structural adhesive		
	9 RSW	Structural adhesive	Remarks
Original design	–	Proposal five	
Hinge reinforcement plate 3	Proposal four	–	

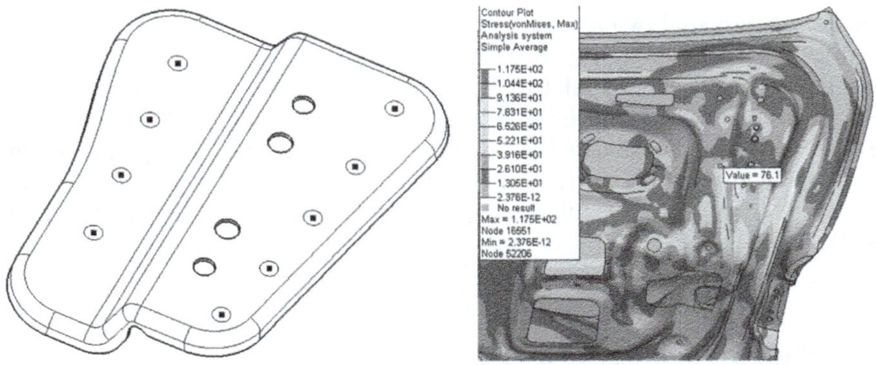

Fig. 8 Stress cloud chart of the trunk lid inner panel with proposal four

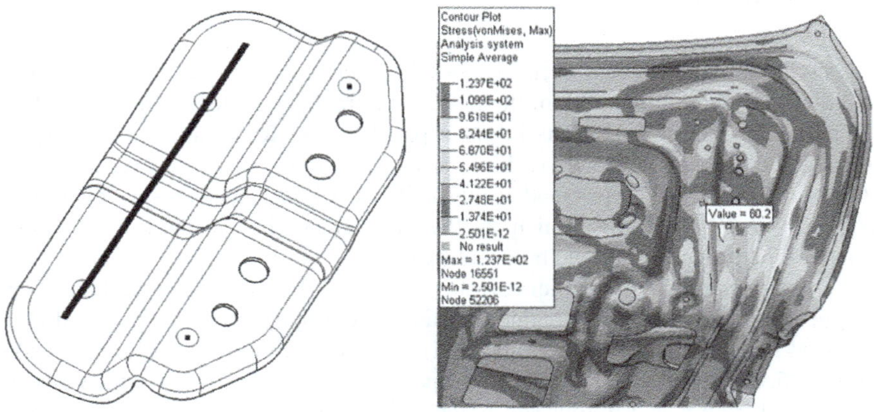

Fig. 9 Stress cloud chart of the trunk lid inner panel with proposal five

Table 5 CAE analysis results of different hinge reinforcement plates

Proposal	Index			
	Maximum stress (Mpa)	Life of RSW (cycles)		Remarks
		RSW 1	RSW 2	
Original proposal	128	248,000	514,000	
Proposal four	76	2,025,000	3,332,000	
Proposal five	80	>10,000,000	>10,000,000	
The benchmarking	81	>10,000,000		

Table 6 Cost and design cycles of the solutions

	Design proposal	Cost (¥)	Cycles (days)	Remarks
Proposal four	1. Redesign the trunk lid hinge reinforcement plate 2. Increase the number of the welding points from 4 to 9	Part cost: +0.3 Retooling fee: 50,000	60	1. Extra robots are needed to meet the JPH requirement, but no space was available for the extra robot
Proposal five	1. Structural adhesive is needed between the trunk lid hinge reinforcement plate and the trunk lid inner panel only, which is 120 mm in length	Part cost: +1	90	

4.2 *Test Verification*

4.2.1 Prototype Test

Theoretical calculation cannot fully verify the improvement effect, and it still needs to be verified by prototype test finally. And a prototype was made as proposal five, as shown in Fig. 10. Thirty thousand cycle high and low-temperature opening and closing durability test is carried out by the third test organization (SGS). The RSW cracking problem did not arise until the end of the test. Figure 11 shows the durability test results of the proposal five.

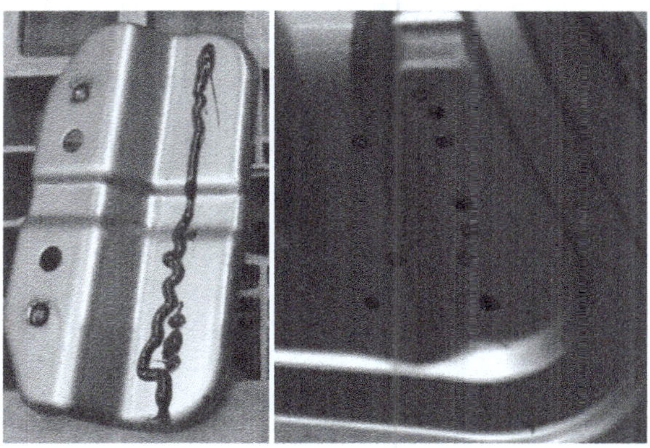

Fig. 10 Prototype

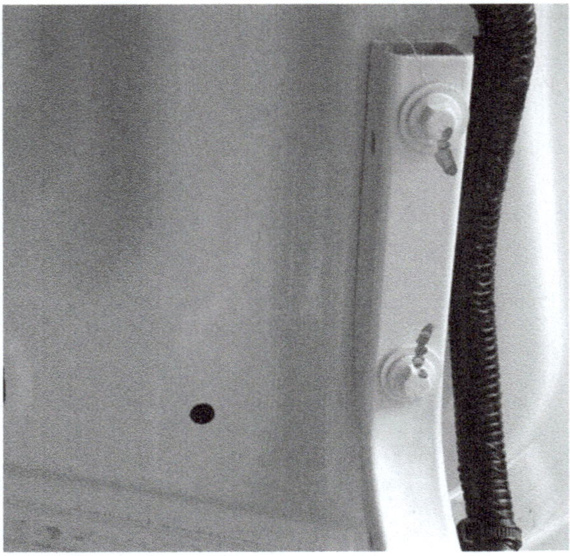

Fig. 11 Durability test results

5 Conclusion

Researching on the trunk lid hinge reinforcement plate welding spot tear problem, the number of the hinge reinforcement plate welding spots (structural adhesive) and the structure of the hinge reinforcement are analyzed, and it can be concluded:

(1) Optimizing the number of the hinge reinforcement plate welding spots (structural adhesive) and the structure of the hinge reinforcement plate can effectively reduce the maximum stress of the trunk lid inner panel and improve the fatigue life of the hinge reinforcement plate welding points.
(2) The structural adhesive's rational use can effectively reduce the maximum stress of the trunk lid inner panel and improve the fatigue life of the hinge reinforcement plate welding points.

From the above analysis, the RSW 1 and RSW 2 fatigue lives were increased from 248 thousand cycles to more than 10 million cycles by CAE simulations. And the prototype test has verified the effectiveness of proposal five.

References

1. Yang J, Zhang D, Duan Z (2016) Stiffness analysis and optimization of welding spots of a vehicle backdoor lock ring reinforcement plate. In: SAECCE, 2016
2. Zhao S (2014) The fatigue life analysis of BIW based on FEMFAT. Automobile Appl Technol 4:84–86
3. Zhu T, Gao F, Liu G (2006) Pre-estimation of fatigue life of welding spots of BIWs. Automobile Technol 2:37–39

Cross-Sectional Optimal Design of 6082 Aluminium Alloy Anti-collision Beam Based on the Crash Simulation Analysis Using LS-DYNA

Zhimin Liu, Jingchen Huo, Chunxiang Liu, Bensheng Xiong, Bo Gao, Junping Qiao and Zhuqing Cheng

Abstract Four cross sections of 6082 aluminium alloy anti-collision beam were designed for one car. Based on the crash simulation analysis using LS-DYNA, the deformation and energy absorption parameters of 6082 aluminium alloy beam were obtained.The results of crash simulation for a rear low-speed impact indicated that the invasion of the collider was minimum and the effect of energy absorption was optimum, when the tangent angle of cross section between arc rib and horizontal plane of the anti-collision beam was twenty degree.Finally, the pendulum test revealed that the performance of this 6082 aluminium alloy beam satisfied with performance requirements, when the tangent angle of arc rib cross section of the anti-collision beam was designed to the twenty degree.

Keywords Aluminium alloy · Anti-collision beam · The crash simulation · Lightweight · Cross section

1 Introduction

Automotive lightweight technology is an effective means to solve the problem of automobile fuel consumption and emission. The realization of automotive lightweight is mainly from three aspects such as material, structure and processing of technique. And the lightweight effect of using lightweight materials instead of steel is significant among them. As the light structural material, aluminium alloys have high potential application in automobile body manufacture for which can satisfy with the safety performance requirements and realize lightweight effectively. With the improvement of safety regulations, anti-collision beam is playing a more and more important role as a key part of safety in vehicle collision. A number of studies have shown that the aluminium alloy anti-collision beam can not only reduce the weight by 40–70% compared with the steel anti-collision beam, but also greatly improve the collision absorption energy [1, 2]. As the price of aluminium alloy is

Z. Liu (✉) · J. Huo · C. Liu · B. Xiong · B. Gao · J. Qiao · Z. Cheng
BAIC Motor Co. Ltd. R&D Center, Beijing, China
e-mail: 007liuzhimin@163.com

China SAE (ed.), *Proceedings of China SAE Congress 2018: Selected Papers*, Lecture Notes in Electrical Engineering 574,
https://doi.org/10.1007/978-981-13-9718-9_11

decreasing gradually, aluminium alloy anti-collision beam will be more and more widely used in car body in the future.

Generally, the anti-collision beam assembly part is made up of the anti-collision beam and the welding parts of the left and right energy-absorbing boxes. The energy absorption of anti-collision beam assembly part for steel material mainly depends on the left and right energy absorption boxes, but for the aluminium alloy anti-collision beam assembly part, the aluminium alloy anti-collision beam body has also the better energy absorption effect [3, 4]. Therefore, for the design of aluminium alloy anti-collision beam, not only the strength of the anti-collision beam body should be considered, but also the energy absorption effect can be improved by optimizing the cross section of the anti-collision beam body reasonably, so as to improve the crashworthiness of the anti-collision beam and reduce the weight of the part.

In this paper, an aluminium alloy anti-collision beam assembly part is developed for one certain car, and four kinds of aluminium alloy anti-collision beams with four sections are designed. According to GB 17354, LS-DYNA software was used to simulate the protection for the rear end of the car body in order to get the optimal cross section. Finally, the optimized rear anti-collision beam assembly part of aluminium alloy was tested by pendulum test. The testing results revealed that the optimized aluminium alloy anti-collision beam assembly part was consistent with the requirements of the collision performance.

2 Cross-Sectional Design of Aluminium Alloy Anti-collision Beam

In this paper, 6082 aluminium alloy was adopted on the body of anti-collision beam. The thickness of front and rear plates, upper and lower plates and middle rib plate of the anti-collision beam body is 2 mm. Four cross sections of the beam are shown in Fig. 1, there is no rib plate in the first cross section, the middle rib in the second cross

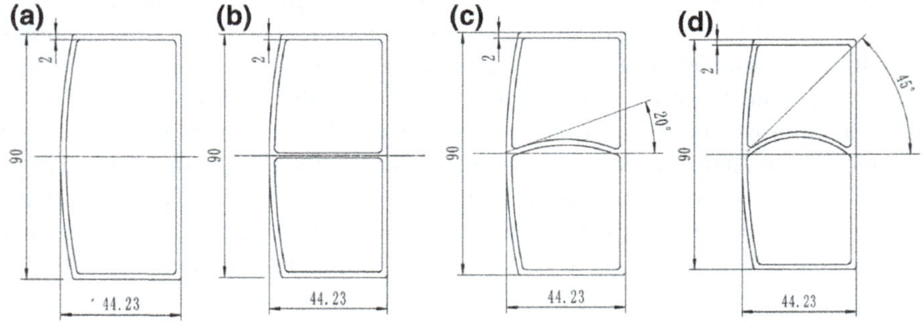

Fig. 1 Four cross sections of 6082 aluminium alloy anti-collision beam: **a** the cross section with mouth shape; **b** the cross section with plane rib; **c** the cross section with 20-degree arc rib; **d** the cross section with 45-degree arc rib

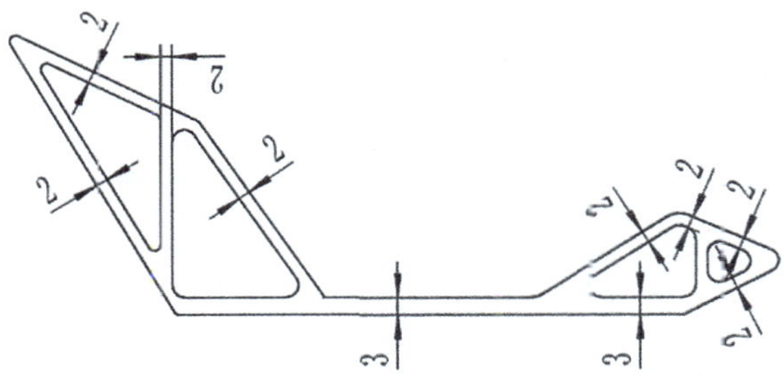

Fig. 2 Cross section of the installation bracket of 6061-T6 aluminium alloy

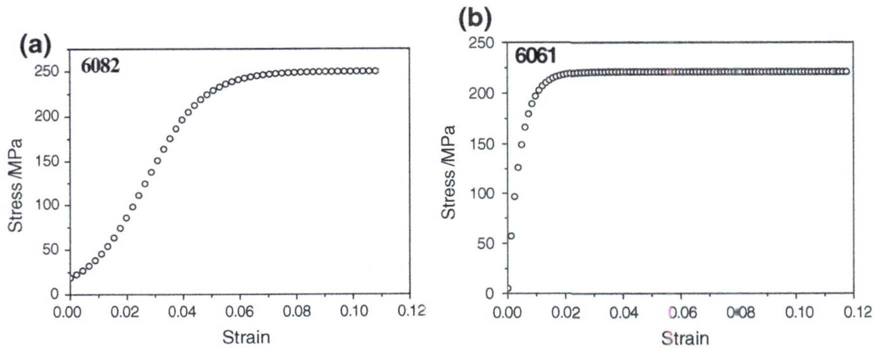

Fig. 3 Stress-strain curves of aluminium alloy: **a** 6082; **b** 6061-T6

section is a plane, the middle rib in the third cross section is a arc surface, and the tangent angle between the arc surface and the horizontal plane is 20°, and the tangent angle of arc rib in the fourth cross section is 45 degrees. As the rear collision is a low-speed collision, the requirements for collision resistance are not high, and the aluminium alloy anti-collision beam has one good energy absorption effect, so the left and right energy absorption box and the installation blank are integrated into an integral installation bracket. The cross sections of the designed installation bracket are shown in Fig. 2, and the material of 6061-T6 adopted and the thickness is 2 mm. Figure 3 shows the stress-strain curves of aluminium alloy 6061-T6 and 6082.

3 Establishment of Finite Element Model

Firstly, the geometric model of the vehicle had been simplified. For the rear longitudinal beam, assembly part under the rear floor is made of high strength steel with high yield strength, and the energy-absorbing box of the rear anti-collision beam is in alignment with the rear longitudinal beam, so the force flow generated by the collider during the collision of the rear collision beam can be directly passed along the rear longitudinal beam to the body frame. Since the frame of body in white (BIW)

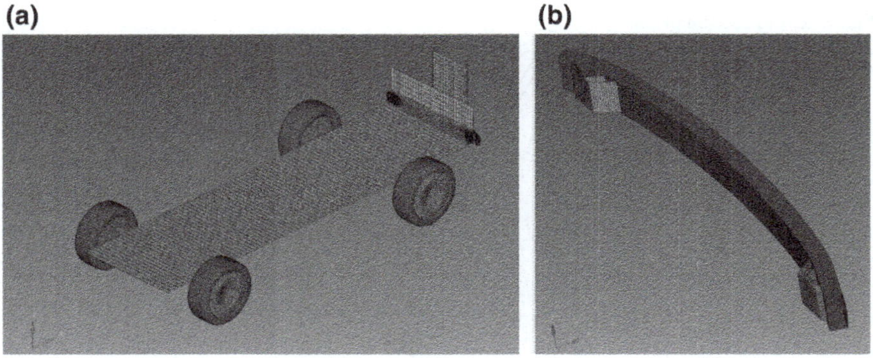

Fig. 4 Finite element model of the whole vehicle: **a** simplified BIW and chassis; **b** aluminium alloy anti-collision beam

is made of high strength steel, and the cross-sectional area is large enough, it can be considered that the BIW hardly deformed during the collision. Therefore, the BIW can be further simplified as a rigid body, and the software of hyper-mesh is used for finite element modelling. As shown in Fig. 4, in the finite element model of the whole vehicle, the BIW is set as a rigid body, the aluminium alloy anti-collision beam assembly part is a plastic deformation body, the tire is elastic body, and the ground is rigid wall. The friction coefficient between the internal parts of the vehicle system is 0.15, and the friction coefficient between the tire and the ground is 0.015. In the finite element model, the grid is divided by BT shell element. The grids of BIW and chassis are 30 mm, and the grid of the aluminium alloy anti-collision beam is 5 mm. Collision simulation is conducted with the regulation of GB 17354.

4 Results of Simulation

4.1 Safe Distance of the Collision

Figure 5 illustrates that the displacement cloud diagram of aluminium alloy anti-collision beam after frontal collision with no-load. As seen in Fig. 5, the anti-collision beams of the four cross sections have undergone different degrees of deformation and bending. The safety distance between the anti-collision beam and the BIW is the vertical distance which is between the maximum bending point of the anti-collision beam and the BIW. Among those four sections, the safety distance of anti-collision beam with 20-degree arc rib section was the largest, and the anti-collision beam with mouth shape section had collided with the BIW. In addition, during the initial period of collision, that is, at 0.0025 s, the anti-collision beam with plane rib section was cut and torn along the rib plate, as shown in Fig. 6. Therefore, it can be considered that the collision resistance with the 20-degree arc rib section is the best.

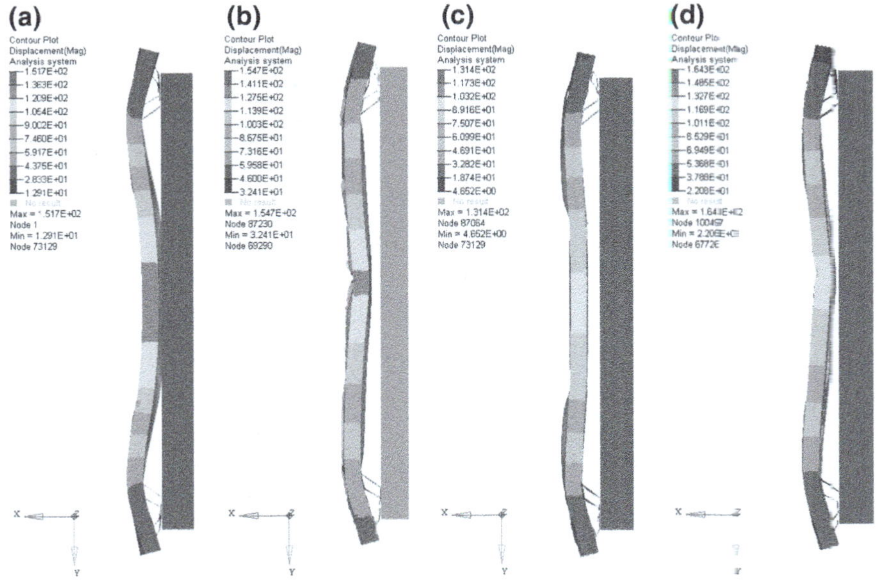

Fig. 5 Displacement cloud diagrams of aluminium alloy anti-collision beam after front collision with no-load: **a** the cross section with mouth shape; **b** the cross section with plane rib shape, **c** the cross section with 20-degree arc rib; **d** the cross section with 45-degree arc rib

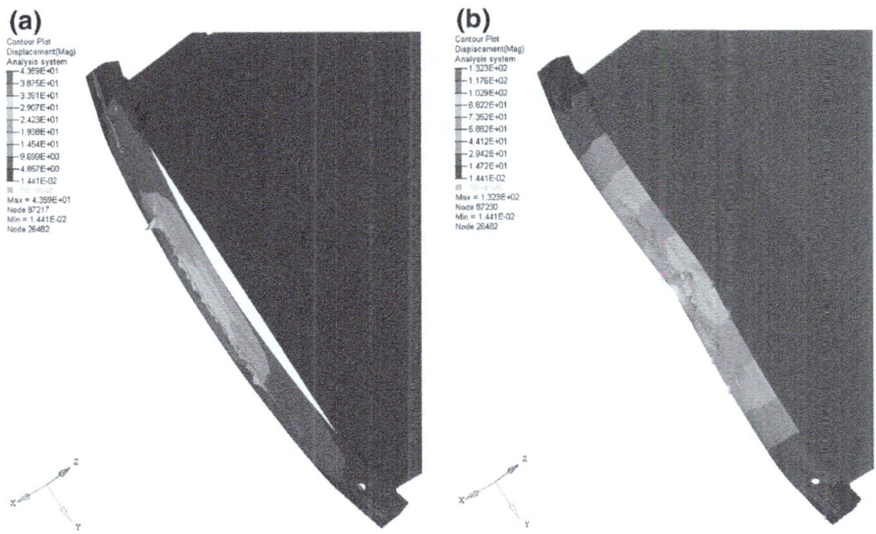

Fig. 6 The anti-collision beam with plane rib section was cut and torn during the initial period of collision: **a** $t = 0.0025$ s; **b** $t = 0.1$ s

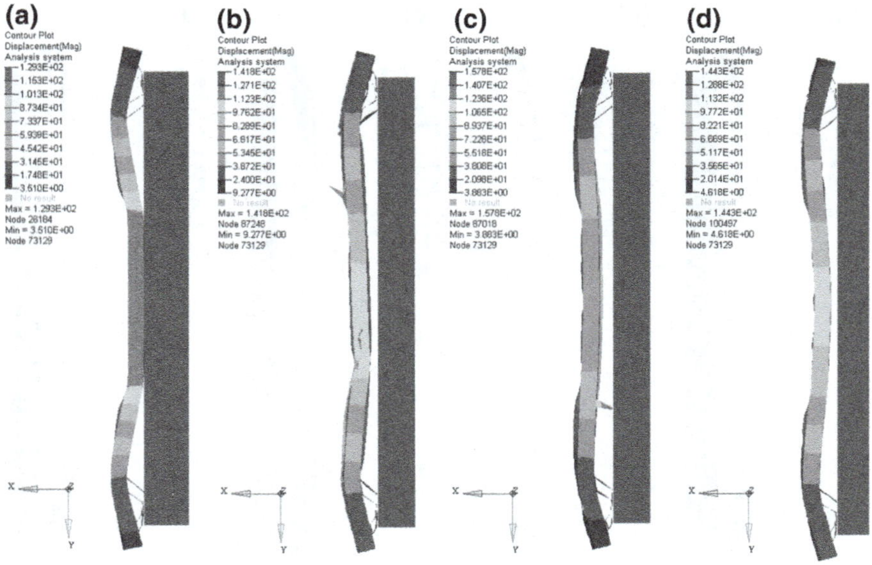

Fig. 7 Displacement cloud diagram of aluminium alloy anti-collision beam after front collision with half-load: **a** the cross section with mouth shape; **b** the cross section with plane rib shape; **c** the cross section with 20-degree arc rib; **d** the cross section with 45-degree arc rib

Figure 7 illustrated the displacement cloud diagram of aluminium alloy anti-collision beam after front collision with half-load. It can be clearly seen that the collision resistance with 20-degree arc rib section was the best, and the anti-collision beam with mouth shape section had collided with the BIW.

4.2 Curves of Maximum Invasion and Energy Absorption

The maximum invasion curve of anti-collision beams after the front collision was shown in Fig. 8a. In the case of no-load, the maximum invasion of anti-collision beam with mouth shape is 103 mm, and then the maximum invasion of anti-collision beams with the sections of plane rib shape and 45-degree arc rib is 74, 75 mm, respectively, and which do not fit with the performance requirements. The minimum invasion is the anti-collision beam with the cross section of 20-degree arc rib, which is 61 mm, and the result fits with the performance requirement. Figure 8b indicated that in the case of half-load, the invasion of the anti-collision beam with the sections of mouth shape, the 45-degree arc rib and the plane rib are respectively 104, 80 and 71 mm, and which do not satisfied with the performance requirements of 70 mm. The minimum invasion curve is the anti-collision beam with the cross section of 20-degree arc rib, which is 68 mm, and the result meets the performance requirement.

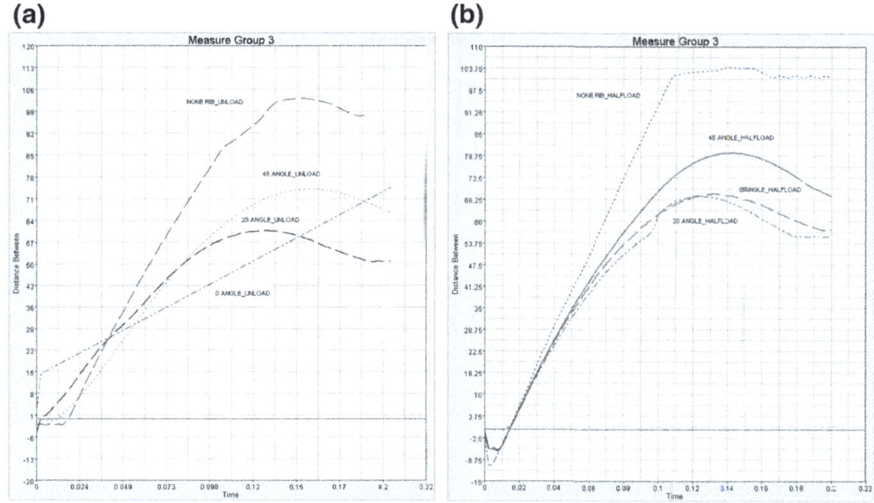

Fig. 8 Maximum invasion curves after the front collision of anti-collision beams: **a** the front collision with no-load; **b** the front collision with half-load

Figure 9 illustrated the curves of energy absorption in the case of un-load and half-load; it is evident that the peak value of energy absorption with plane rib shape section is the highest, while the peak value of anti-collision beam with 20-degree arc rib, 45-degree arc rib and mouth shape sections decreases successively. This phenomenon revealed that plastic deformation of the anti-collision beam with the planar rib section is the largest, and the energy absorption is the most. In addition, the anti-collision beam with the 20-degree arc rib section has good effect on energy absorption, while the anti-collision beam with mouth shape section has the worst effect due to the lack of rib structure.

To sum up, in the case of front collision, the anti-collision beam with the 20-degree arc rib section has the largest safety distance from the car body and the least invasion. Although the anti-collision beam with plane rib section has the best effect of absorbing energy during the collision. In the case of no-load, due to the low coincidence degree between the collider and the anti-collision beam, serious cut-off and tear occurred at the initial stage of the collision and then quickly broke off. Therefore, the 6082 aluminium alloy anti-collision beam had been manufactured in the cross section with 20-degree arc rib. Figures 10 and 11 show the curve of displacement cloud diagram and maximum invasion of the whole vehicle, when the collision is under no-load of angular collision and half-load of angular collision. As shown in Fig. 10, in the no-load condition, it is obvious that the beam is easy to be torn along the rib plane because of a low degree of overlap between the collider and the anti-collision. Figure 11 indicated that the maximum invasion under no-load of angular collision and half-load of angular collision is 31 and 29 mm, respectively, and the maximum invasion fit with the requirements of 40 mm under angular collision.

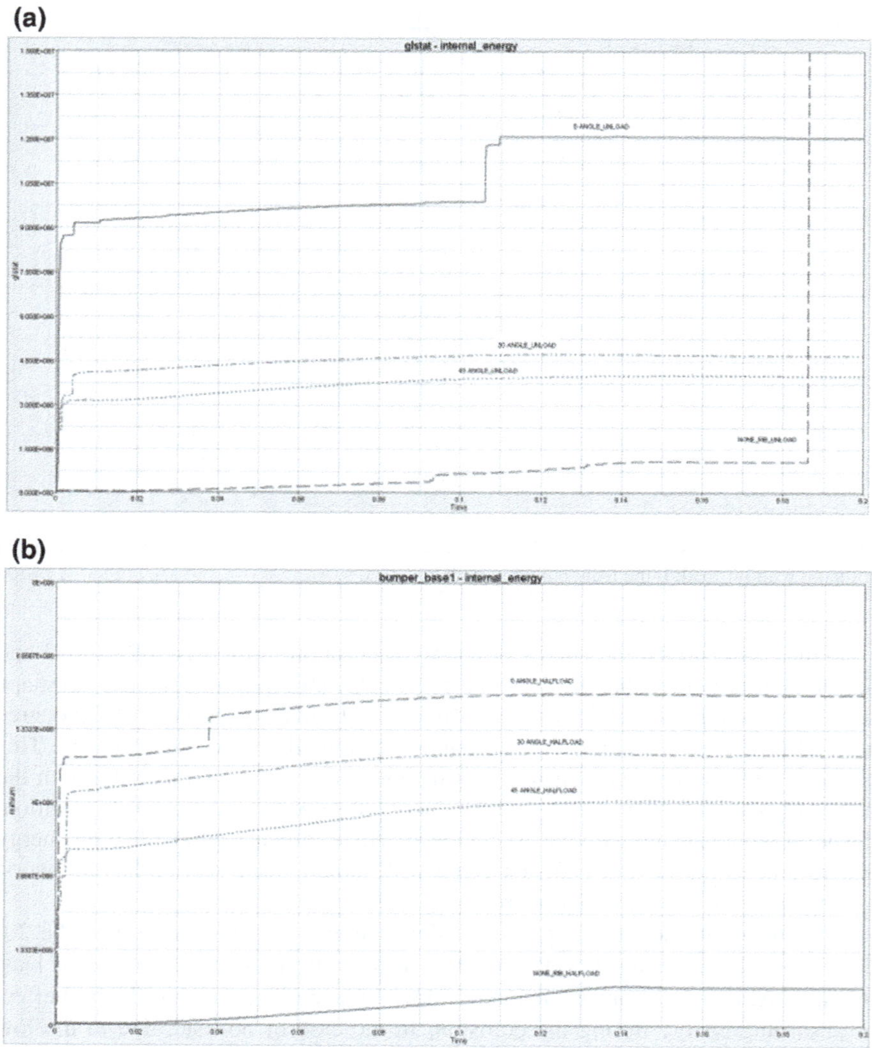

Fig. 9 Energy absorption curves after the front collision of anti-collision beams: **a** the front collision with no-load; **b** the front collision with half-load

5 Pendulum Test

According to the optimized section, the aluminium alloy anti-collision beam was manufactured; the extrusion die was developed, and the aluminium alloy anti-collision beam assembly part was prepared through process of extrusion forming, artificial ageing, bending forming and welding. The pendulum test was conducted with GB 17354. Figure 12 illustrated the comparison in deformation behaviours of

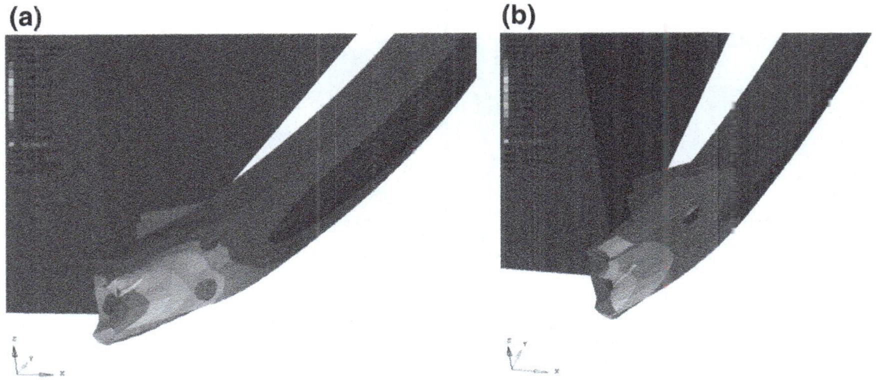

Fig. 10 Curve of displacement cloud diagram: **a** the angular collision under no-load; **b** the angular collision under half-load

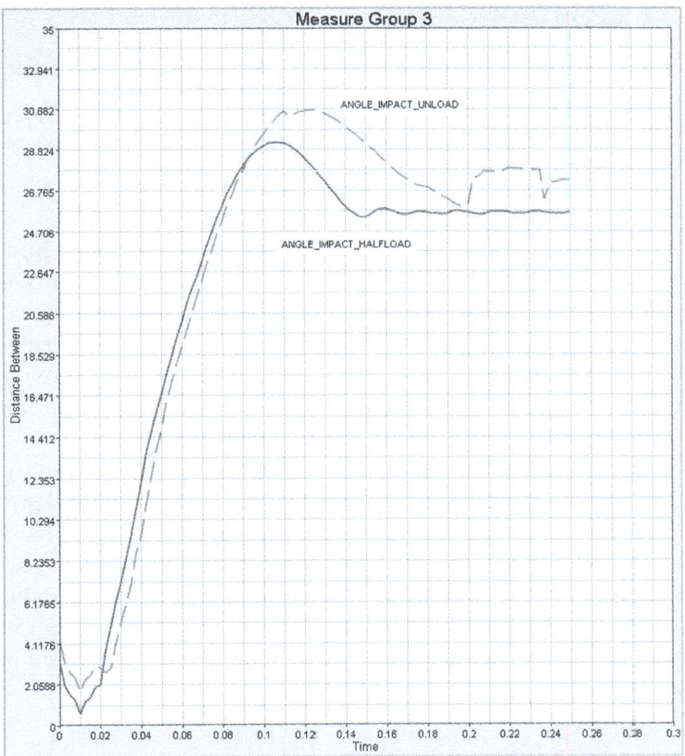

Fig. 11 Maximum invasion curves of angular collision

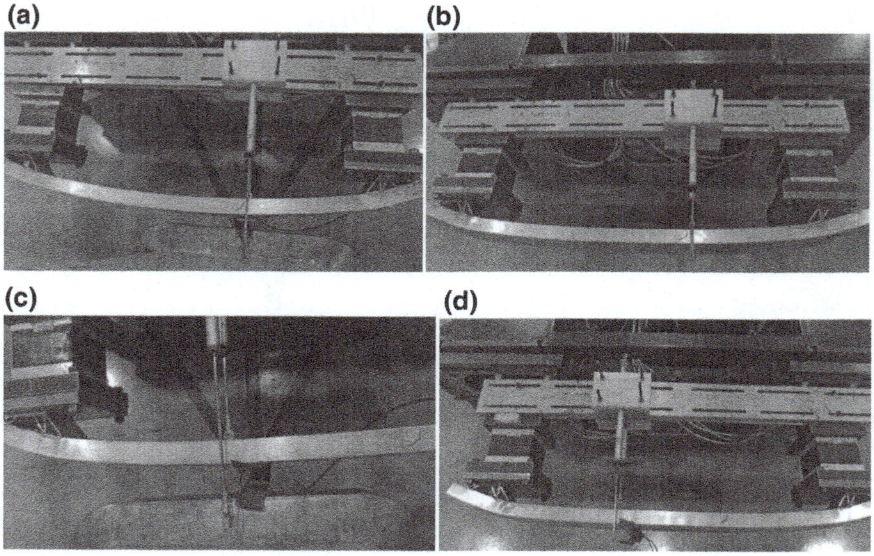

Fig. 12 Comparison in deformation of the anti-collision beam before and after the pendulum test: **a** before pendulum test under no-load; **b** after pendulum test under no-load; **c** before pendulum test under half-load; **d** after pendulum test under half-load

the anti-collision beam before and after the pendulum test under no-load and half-load. It can be seen that the deformation of the anti-collision beam is small after the collision, and which there is no obvious damage phenomenon. Figure 13 plotted the maximum invasion curves of the anti-collision beam after the pendulum test, among which the maximum invasion peak value is 35.8 mm under no-load, and the maximum half-load invasion peak value is 48.3 mm under half-load. It was found that the results satisfied with the invasion requirement of 70 mm.

Figure 14 revealed the deformation of the anti-collision beam assembly part before and after the angular impact of the pendulum. It can be seen that the deformation of the aluminium alloy anti-collision beam and energy-absorbing box is small after the pendulum test, and the anti-collision beam there is no obvious damage. Figure 15 illustrated the curve of the maximum invasion measured in the pendulum angular test. The results indicated that the maximum invasion under no-load is 15.8 mm and the maximum invasion under half-load is 18.3 mm, and the maximum invasion fit with the performance requirement of 40 mm.

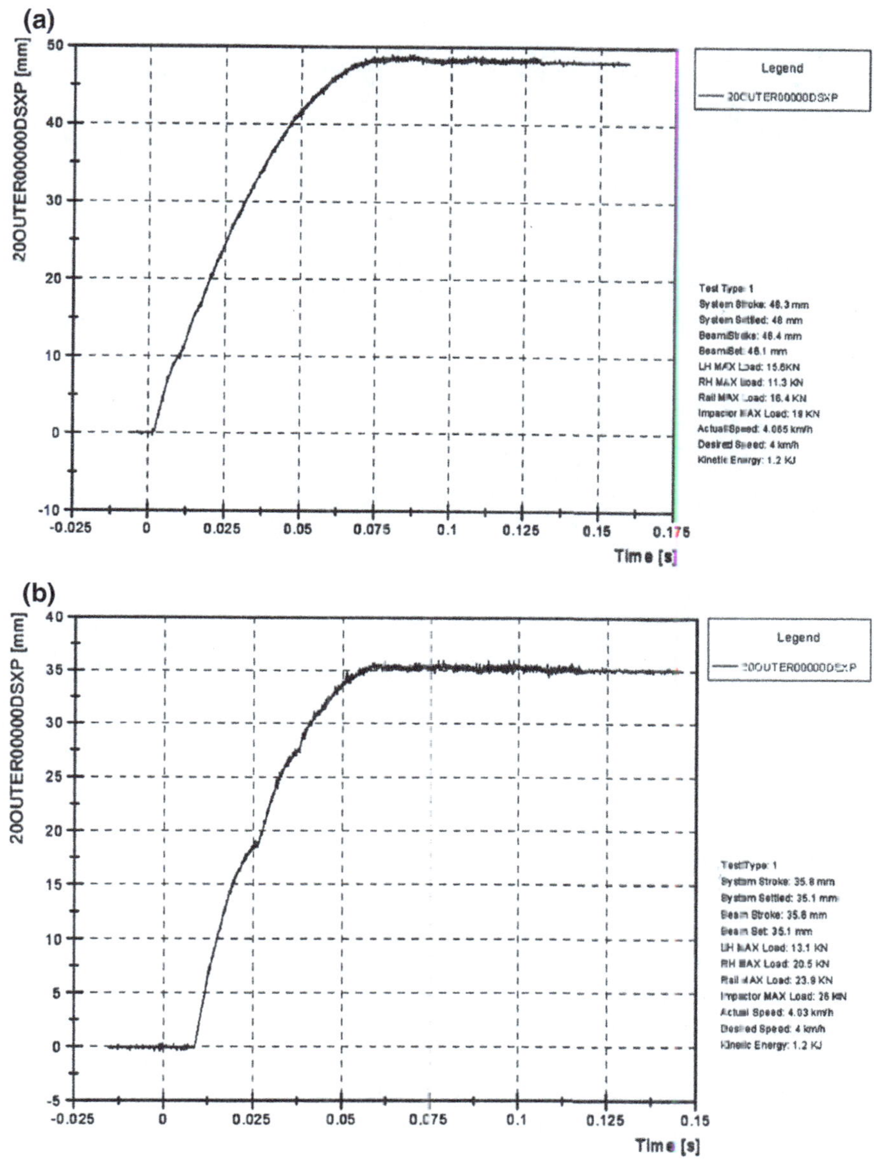

Fig. 13 Maximum invasion curves of the anti-collision beam after the pendulum test: **a** pendulum test under no-load; **b** pendulum test under half-load

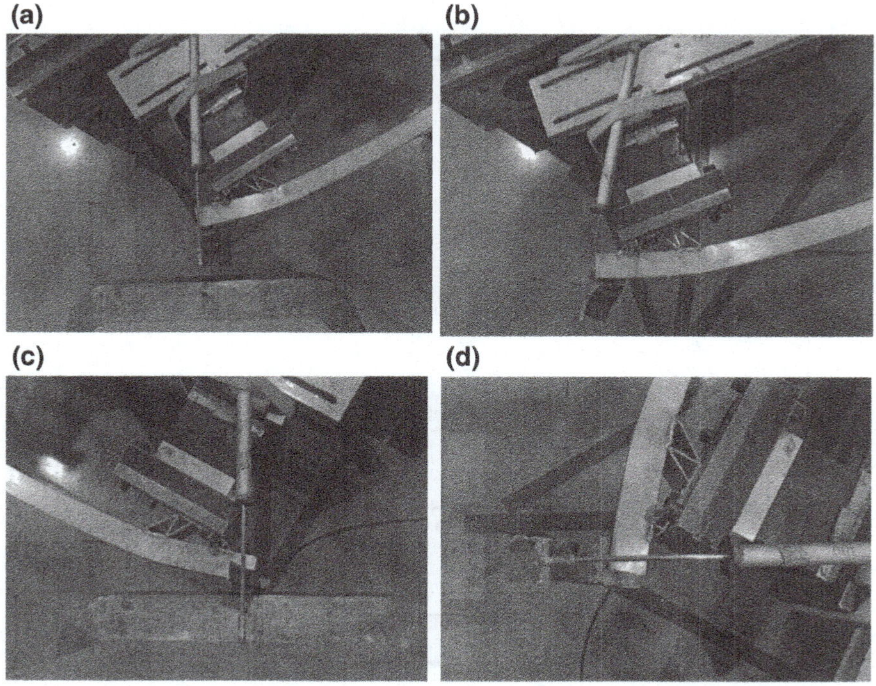

Fig. 14 Deformation of the anti-collision beam assembly parts before and after the angular impact of the pendulum: **a** before pendulum test under no-load; **b** after pendulum test under no-load; **c** before pendulum test under half-load; **d** after pendulum test under half-load

6 Conclusions

Four cross-sectional structures were designed for 6082 aluminium alloy anti-collision beam. According to GB 17354, the collision simulation was carried out, and the strength, invasion and energy-absorbing performance were further studied, and key evaluation parameters were obtained.

The simulation results indicated that different cross-sectional parameters have different effects on improving the collision performance of parts, among which the anti-collision beam with 20-degree arc rib section has the best collision resistance.

It was found that the safety of the anti-collision beam after the optimization of the section parameters there is better performance in pendulum test. This optimization method of cross section has some referential significance for the structural design of the aluminium alloy anti-collision beam in the future.

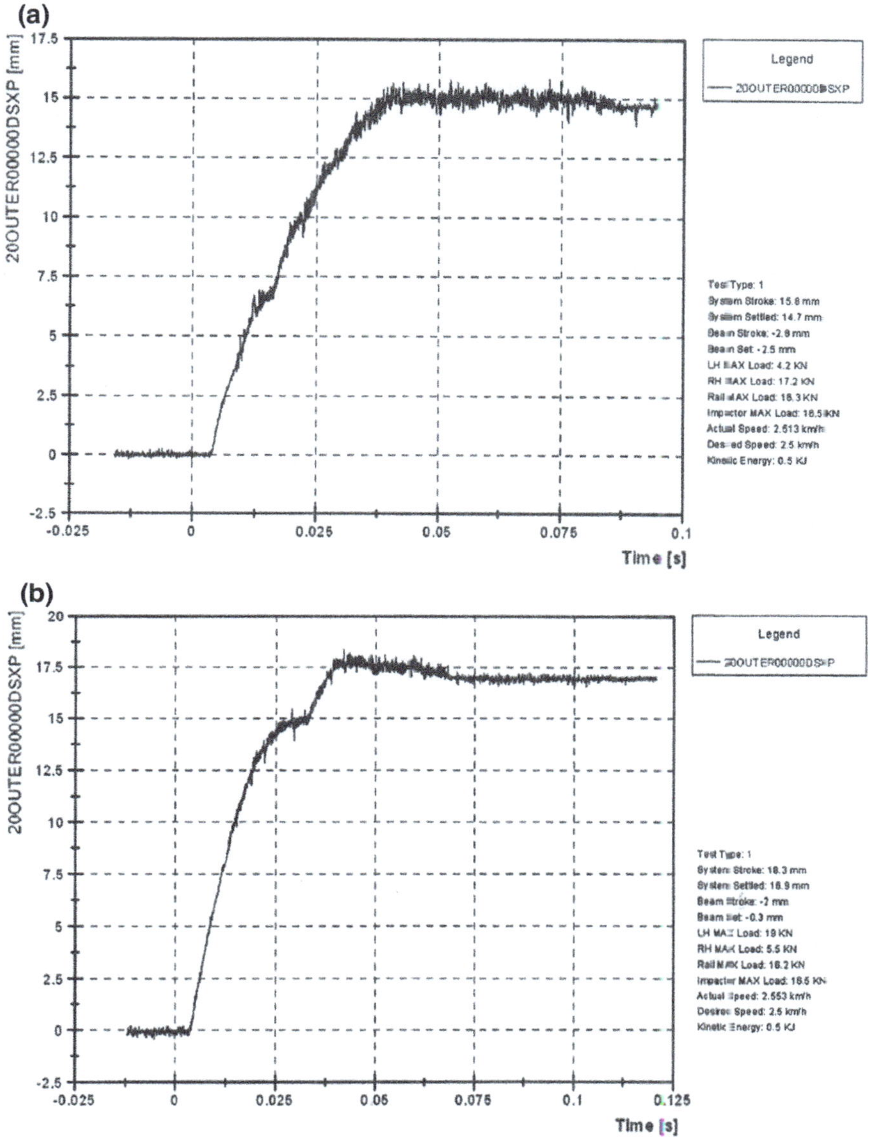

Fig. 15 Curves of the maximum invasion measured in angular impact of the pendulum: **a** pendulum test under no-load; **b** pendulum test under half-load

References

1. Long Q, Zhou C, Xiang X et al (2018) Development and application of anti-collision beam aluminum alloy profiles for high performance anti-impact vehicle. Alum Fabr 1(240):20–24
2. Liu Z, Wu X, Zhang D et al (2014) Improvement of formability of Y-shaped tubular part of 6016 aluminium alloy by pulsating hydroforming. In: Proceedings of SAE-China Congress 2014: selected papers, pp 221–225
3. Xu Z, Xu X, Wan X et al (2013) Structure optimal design of aluminum alloy bumper anti collision beam. J Mech Eng 49(8):136–142
4. An Z, Zha Y (2016) The structural strength design of the light weight of the car crash beam of anti-collision beam. Comput Simul 7(33):206–284

Management and Innovation of Robot in Automobile Production Line

Feng Chang, Maowei Jie and Wei Liu

Abstract Automobile automatic production cannot be separated from a large number of industrial robots, but robots will have a variety of problems in practical applications. Aiming at this problem, this paper makes a breakthrough from innovation of robot management, innovation of maintenance technology and skilled talent training, lists the innovative methods used in the robot management, shares the technological innovation and inventions in the maintenance work, and shows several items for the innovation and reuse of the scrapped robot. In the face of mass use of industrial robots, the equipment management methods and special talents training ways of automobile manufacturing companies are discussed in all directions.

Keywords Industrial robot · Innovation · Management · Talent training

Foreword

With the rapid development of the China's automobile manufacturing industry, the automation rate of automobile manufacturing is becoming higher and higher in recent years. It has gradually become the leader of the manufacturing industry. When we talk about the automation of automobile manufacturing, an important equipment robot is always mentioned. With the increasing requirements for handling, welding, spraying, and testing of automotive parts, the robot is demanded to deal with this kind of high-volume, high-strength, reproducible, and high-precision tasks. With the popularity of the automotive industry, a new direction, which is called the industrial robot repair and management, has been extended from the robot industry. It is new in China and is also lack of relevant experience exchanges and discussions. Based on more than ten years' experiences on the industrial robot maintenance and management of the authors, the experiences of robot equipment management and related personnel training will be analyzed, summarized, and finally improved in this article.

Robot is a device that works automatically. It can not only accept the manipulation of human, but also run under the guidance of pre-defined programs. Industrial robots are originated and developed from the automotive industry. After long-term accu-

F. Chang (✉) · M. Jie · W. Liu
Beijing Hyundai Motor Company, Beijing, China
e-mail: changf@bhmc.com.cn

China SAE (ed.), *Proceedings of China SAE Congress 2018: Selected Papers*,
Lecture Notes in Electrical Engineering 574,
https://doi.org/10.1007/978-981-13-9718-9_12

mulation of technology and experience, the industrial robots have been integrated into the automotive assembly line well.

1 Application of Industrial Robots in Automobile Production

1.1 Development Status of China's Industrial Robot

In 2014, the installed capacity of industrial robots in China reached 57,000 units. As a result, China has become the world's largest market for industrial robots. With the increase in labor costs, the demand for robots in domestic enterprises is increasing year by year. However, the annual sales of self-owned brand robots are less than 20,000 units, and they are mainly low-end and medium-end robots. At present, the four major brands of industrial robots ABB, KUKA, FANUC, and Yaskawa, which are all foreign brands, occupy about 70% of the market share of Chinese industrial robots. The dilemmas faced by domestic robots include the lack of robot core technology and high-tech talents and is consistent with the difficulties of domestic cars in previous years. It is clear that the development of the domestic robot industry is still very difficult.

In June 2014, General Secretary Xi Jinping mentioned in the report of the Academician of the two academies that "robot is the jewel on the top of the manufacturing industry, and its development, manufacturing and application are an important sign for measuring the level of technological innovation and manufacturing in a country." At the national level, we also began to notice the importance of industrial robots for Industry 4.0 and Smart Manufacturing 2025. Therefore, with the investment in science and technology and the cultivation of talents, as the country with the world's largest market, the future of domestic robots is both brilliant and challenging [1].

1.2 The Applications of Industrial Robots in Beijing Hyundai

As early as 2002, Beijing Hyundai built a factory, which pays special attention on the automation of automobile production. Due to the use of high-volume industrial robots, Beijing Hyundai has become the first auto factory in China with 100% welding automation rate. Nowadays the scale of five factories in three places in Beijing Hyundai has been formed. Corresponding with the extremely high automation rate, a total of 2320 industrial robots were installed in those five factories. These robots replacing heavy labor were used in various functions such as welding, handling, coating, and inspection in automobile production (Fig. 1).

Fig. 1 Robot on the car production line

1.3 Problems and Countermeasures in the Application of Industrial Robots

In Beijing Hyundai, due to the early use of industrial robots, the first batch of robots has been worked for nearly 15 years. More than 5000 days and nights work, many robots are dealing with the problem of component aging. Since automobile production is a pipeline operation, once any robot equipment failure occurs, the entire workshop will be discontinued. The discontinuation of the workshop will result in a large number of direct and indirect economic losses. If the production line is shut down for one minute, it will result in economic loss of more than 700 yuan. Therefore, the stability of robot operation has gradually become an important aspect of vehicle cost control in automobile production.

Another important issue for robotics application is the lack of sophisticated personnel for robot repair. Due to the technical monopoly, robot manufacturers can take advantages of the technology to provide maintenance services to customers, which saves a large amount of maintenance costs. Therefore, learning robot maintenance technology is facing the challenges of no data and no experience to learn. Most of the maintenance workers in the production line can only simply debug the robot. Once the robot has complicated electrical and mechanical faults, it is often helpless. In this case, the only solution is inviting the expert of equipment manufacturer for maintenance, and it will result in not only long line-stop time, but also high maintenance costs.

According to the experience accumulated in the management and maintenance of robots for more than ten years, the problems existing in the application of robots in Beijing Hyundai are summarized and analyzed and summarized one by one. The main breakthroughs in three aspects are the management innovation of robots, the innovation of maintenance technology, and the cultivation of highly sophisticated talents. Through five years of difficult research, Beijing Hyundai has realized the complete autonomous management, repair, and maintenance of the robot and got rid of the shackles of robot manufacturers completely by making the engineers master of principles of the robot maintenance. Beijing Hyundai has also built a robot talent training process. Many robot improvement cases are used as references by robot manufacturers. Those strategies not only cultivate a large number of talents for the company, but also reduce greatly the maintenance cost of each workshop robot.

2 Innovation of Robotic Equipment Management

2.1 Establishment of "ID Card"

Due to the factory's high automation rate, the application of a large number of robots must be managed uniformly. Because there is no relevant experience of the management of a large number of robots in China, innovation of equipment management method is necessary to manage every robot.

According to the specific situation of the production line robot, we set up a management account for each robot, which means an individual ID card for each robot. The identity card contains details of the robot's production line, installation date, functional use, component replacement records, regular maintenance records, and its fault resume. Once the robot has broken down, it can quickly check whether the other robots in the factory have the similar faults and find out the problems that once occurred in the fault handling by the management ledger with the advantages of speeding up the fault diagnosis processing.

The innovation and improvement started in 2011. After years of experiences accumulation, according to the management ledger, this system can calculate the service life of each part of the robot, handle the common faults, and accumulate experience of the vulnerable parts and so on, so that it can carry out the preventive maintenance pertinently. By the overall production of the ledger, all robots are brought into the management state, some experience in maintenance can be summed up in a quantitative way. This management method of ID card is a management method that is initially explored in robot management and it will be further promoted and optimized (Fig. 2).

2.2 Self-maintenance and Self-manufacturing Management of Robot Components

Because of the mass use of robots, it is impossible to avoid the aging and damage to the products. The purchase of the robot's parts is also a big expense in the use of robots every year. According to the calculation of the maintenance data of robots for many years, more than 4 million yuan will be spent on the maintenance of a factory with 400 robots, which means that every robot needs 10 thousands yuan for its maintenance (Table 1).

In order to reduce the purchase cost of the robots, we are gradually promoting self-production and self-maintenance of the robots' parts. At present, the main advance

NO.	LINE		STATION	USE'S	ROBOT MODEL		PRODUCT		ENC. CABLE		SPOT NUMBER			BJ CABLE		BALANCE		MOTO & REDUCER						7 AXIS	
	LINE	NUM			TYPE	NO.	ROBOT	CONTROL	FIRST	2th	RC	XD	RC5	BJ1	BJ2	H	V	S	H	V	R2	B	R1	CABLE	M
1			110-1	H/CA	HS220	-02	11/06	11/06			17	18	17												
2			111-1	H	HX165	-02	03/08	03/08	12'01		50	34	50	14'02	14'02							13'02 R			
3			111-3	H/Vision	HX200L	-00	05/01	05/01	12'01		17	14	17	13'10	13'10			13'08 M	13'02 M						14'09
4			111-4	H/Vision	HX200L	-00	05/01	05/01	12'01		17	14	17	13'10	13'10			13'08 M	13'07 M						
5			112A-1	S(S)	HX165	-02	03/06	03/07	15'05		11	6	11	15'05	15'05									15'05	15'05
6			112A-2	S(S)	HX165	-02	03/06	03/07	15'05		12	6	12	15'05	15'05									15'05	15'05
7			112A-3	S(S)	HX165	-02	03/06	03/07	15'05		11	7	11	15'05	15'05							13'02 R		15'05	15'05
8	FRT FLR (15)	15	112A-4	SPOT	HX165	-02	03/06	05/01	13'04			6													
9			112A-5	SPOT	HX165	-02	03/07	03/07	13'04			6													
10			112B-1	S(S)	HX165	-02	04/05	04/05	15'05		11	7	11	15'05	15'05									15'05	15'05
11			112B-2	S(S)	HX165	-02	04/05	04/05	15'05		12	6	12	15'05	15'05									15'05	15'05
12			112B-3	S(S)	HX165	-02	04/05	04/05	15'05		11	6	11	15'05	15'05									15'05	15'05
13			112B-4	SPOT	HX165	-02	04/05	04/05	13'04			6													
14			112B-5	SPOT	HX165	-02	04/05	04/05	13'04			6													
15			113-1	H/CA	HS200S	-10	11/06	11/06	15'03		30	31	30												
16			159-1	H/CA	HX200	-00	05/01	05/01	15'05		26	24	26	13'10	13'10										14'05
17			160-1	H/CA	HX200	-00	05/01	05/01	15'05		28	23	28	13'10	13'10										
18			161-1	H/Vision	HX200	-00	05/01	05/01	15'05		18	17	18	13'10	13'10										
19			161-2	H/Vision	HX200	-00	05/01	05/01	12'01		18	16	18												
20			161-3	H/Vision	HX165	-04	04/11	04/11	12'01		23	19	33	13'10	13'10										
21			162A-1	S(S)	HX165	-04	04/12	04/12	12'01		11	14	13												
22			162A-2	S(S)	HX165	-04	04/12	04/12	12'01		11	14	11											14'04	
23			162A-3	S(S)	HX165	-04	05/01	05/01	12'01		18	16	18												
24	F/C (18)	18	162A-4	S(S)	HX165	-04	04/12	04/12	12'01		18	16	18												
25			162A-5	S(S)	HX165	-04	04/12	04/12	12'01		23	24	23												
26			162A-6	S(S)	HX165	-04	05/01	05/01	12'01		26	24	26												
27			162B-1	S(S)	HX165	-04	04/12	04/12	12'01		11	14	13												
28			162B-2	S(S)	HX165	-04	04/12	04/12	12'01		11	14	11											16'08	
29			162B-3	S(S)	HX165	-04	05/01	05/01	12'01		18	16	18												
30			162B-4	S(S)	HX165	-04	05/01	05/01	12'01		18	16	18												15'02
31			162B-5	S(S)	HX165	-04	04/12	04/12	12'01		23	24	23												
32			162B-6	S(S)	HX165	-04	05/01	05/01	12'01		26	24	26												
33			163-1	H/CA	HX400S	-05	11/06	11/06	12'01		29	30	29												

Fig. 2 Robot management ledger (partial)

Table 1 Comparison of robot components purchase and independent maintenance costs

No.	Parts	Average purchase price (¥)	Annual usage	Maintenance cost (¥)
1	Cable	20,000	30	1000
2	Robot Board	16,000	40	300
3	Reduce	40,000	5	–
4	Servomotor	30,000	20	500
5	Servo drive unit	50,000	5	1500
6	Fan	300	50	–

is in the following aspects. First, by measuring the cable sequence of the original robot, the cable can be self-made by requisitions of cable plugs and special welding tools. The service life of the self-made cable is comparative with the life of original cable. The second point is the independent maintenance of the circuit board. By employing electronic maintenance experts from the factory, the intimate knowledge of the construction and principle of robot circuit board can be acquired. The fault maintenance account of the circuit board can thus be set up with the capacities of rapid identification of wear electronic components for various circuit boards. It may make an abandoned circuit board repaired by replacing a small electronic component sometimes. The third point is that by the analysis of damaged and disassembled servomotor in the factory, the root cause of the damage can be found. Dealing with the main reason of the memory failure of the encoder, we made a special maintenance tool for the servomotor. The fourth point is that, we try to figure out the corresponding control unit of each axis of six-axis motor, which is integrated in one control module, so an error can be repaired by replacing an control unit (Fig. 3).

In order to facilitate the maintenance and production of robot parts better, Beijing Hyundai has set up an IT workstation belongs to the robot studio, which main jobs are the self-maintenance and the production of robots. The IT workstations can save more than 300 yuan for the company every year.

2.3 *Robot Rapid Repair Method*

Because of the large number of the robots, the preventive maintenance and regular maintenance operation cannot make sure that the faults are eliminated, when the robot's failure cannot be avoided, how to deal with the failures quickly and resume the operation of production lines in time has become an important research direction. After summarizing the failure of the robots, it is found that the robot fault occurs mainly on four components, namely cable, motor, reducer, and drive unit. According to the four most vulnerable parts, a quick solution was set up and makes standardized training methods for each maintenance worker (Table 2).

Fig. 3 Robot self-maintenance and cable self-made

Table 2 Robot rapid repair method

No.	Fault location	Rapid repair method	Remarks
1	Internal cable	Install external junction box, Temporary repair	Tool invention
2	Servomotor	Use the repair box to clear the data, Set new data	Tool invention
3	Reducer	Reduce the speed of the robot (100% → 70%)	Change parameters
4	Servo drive unit	Use compressed air to cool down	Temporary plan

Robot troubleshooting: first inspection and maintenance, easy first and then difficult, temporary countermeasures first and then final improvement

3 Technological Innovation of Robot Maintenance

3.1 The Invention of Servomotor Maintenance Box

With the analysis of the failure, it is found that, in the bodywork factory with the largest number, 25.1% of the shutdown failures are caused by robots and related equipment. Fault analysis of the robot found that the servomotor fault accounted for 38.5% of the total time of the robot equipment failure (Fig. 4).

When there is a fault of the servomotor, the traditional maintenance method is to disassemble the bad motor and replace a new one. Replacing servomotors is a complex operation, requiring multiple maintenance personnel to work together. The weight of servomotor is more than 20 kg, and there are some hidden dangers in its replacement. After the replacement of the servomotor, it is necessary to adjust the

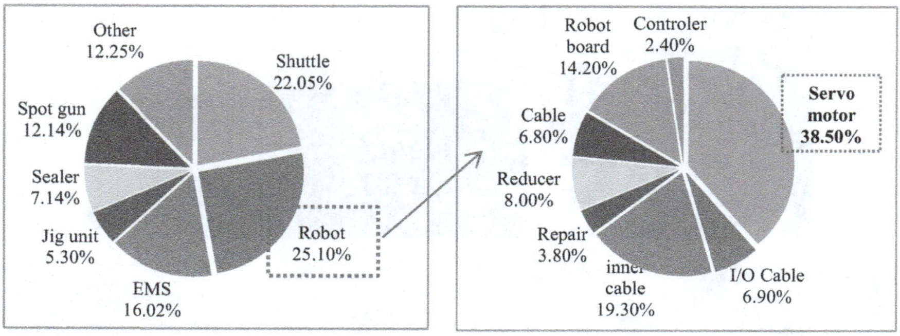

Fig. 4 Production line equipment failure analysis table

working trajectory of all the robot models, and the whole maintenance process will take more than 2 h.

In order to find out the cause of the failure and formulate countermeasures, the servomotor is disassembled and studied. By analysis, it is found that the stator and rotor parts of the servomotor are not worn, and the fault is caused by the encoder storage data overflow. The data overflow makes the encoder crash, so the robot cannot start work. The environment in the production workshop does not meet the requirements for on-site replacement of the encoder, so the original scheme for replacing the encoder was negated. By chance, an idea come out with some electronic devices such as routers and MP4, they all have a RESET button. By pressing the RESET button, the device can clear all the external settings data and restore the factory settings.

The servomotor encoder is analyzed and tested repeatedly in the experimental field. Finally, a set of external RESET circuits is also made to the servomotor of the robot. When the servomotor of the robot causes the fault due to the overflow of encoder data, just insert the plug of the repair box into the encoder of the fault motor, start the power switch on the repair box and all the data of the fault motor will be emptied after 5 s. The robot's fault can be repaired quickly by setting up the required data through the pedagogue. It took 120 min to replace the motor operation in the past, but now, the time reduces to 20 min (Fig. 5).

Servomotor repair box was put into use in October 2014, reduced the maintenance costs, and has been promoted and applied in more than 20 Hyundai factories worldwide. The project won the first prize of Beijing modern excellent proposal. Bronze Award of Beijing scientific and technological innovation the second award of the Beijing staff innovation award and obtained the national patent.

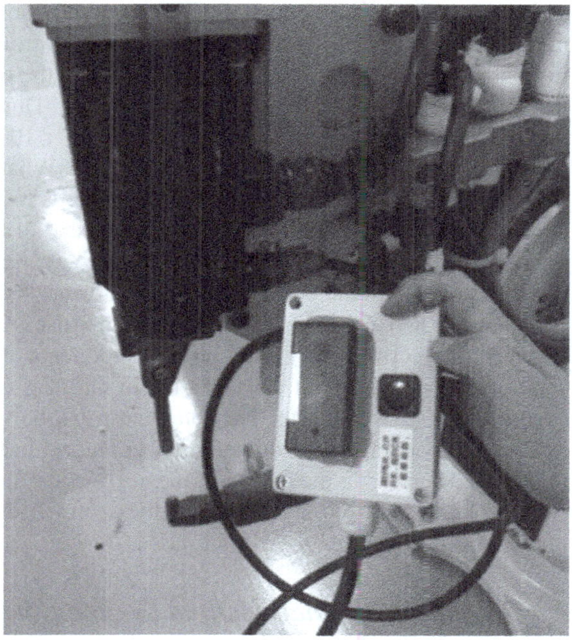

Fig. 5 Use of the repair box

3.2 Circuit Modification of the Brake of the Robot Control Cabinet

When the robot in the factory is running, burn out the 20 V output terminals of transformer often burn out. The works cannot understand why it happened. The concrete condition is shown in Fig. 6.

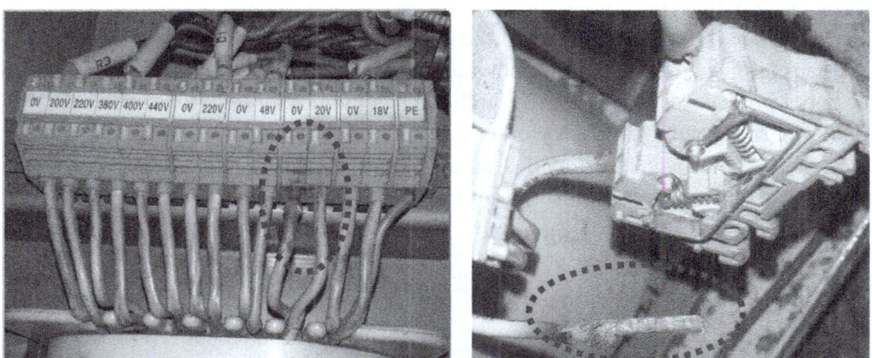

Fig. 6 Burning diagram of transformer terminal in control cabinet

In light of the condition that the 20 V terminal transformer burn down frequently, measure the circuit of 20 V terminal of robot control cabinet is measured, the corresponding control circuit diagram is made, and the operation of robot 20 V circuit is simulated with components in the laboratory. During the trial run, I found that the circuit had an instantaneous high current over 100 A passing through the transformer terminal at the moment of power failure. What causes such a large current is the capacitance in circuit design. By analysis, it is found that the robot designer has applied the filtering function of capacitor instead of the capacitor's energy storage function. When the system is powered off, the capacitor will instantly discharge, which will result in the burnout of the transformer terminals or circuit boards.

After repeated experiments, it was decided to use DC power supply module instead of capacitor, rectifier bridge and other components of the original circuit. After the experiment, it was rebuilt and installed in the factory. No similar failure occurred during the three years after the completion of the transformation. The improved project and content were adopted after being fed back to the robot manufacturing factory and applied to the design of the next-generation robot. By improvement, not only huge cost has been saved, but also the confidence in robot maintenance has been improved.

3.3 Angle Difference Method to Calculate the Original Position of the Robot

The origin reference data losses occur when the robot is running, which will cause the robot trajectory changing in the working procedures and welding grinding procedures of all the models stored. Once the situation happens, each procedure trajectory of the robot needs re-teaching, which will result in a long-term shutdown failure.

In response to this phenomenon, the control theory of robot position is analyzed. A fast method for data calculation has eventually been designed, named "Axis Difference Method for Calculating Axis Data." The specific method is to set a reference point. Once data error or trajectory offset happens in the robot, a difference is calculated between the new position and the original position, and then the difference is compensating calculated in the robot data. By this way, the trajectory problem of all the robot procedures will repair through the teaching of a point.

In order to ensure that each worker can operate the method, a standardized operation method has been developed. And uniform training has been carried out for the production line workers to ensure that each repairer masters the rapid maintenance skill.

4 Transformation and Reuse Innovation of Retirement Robot

Robot is a kind of production equipment, which composed of complex mechanical and electrical control components, and the robot has its own service life The accuracy of robot affects the final quality of coating, sound insulation, welding quality, assembly precision, and so on. The accuracy of a newly installed robot is below 0.3 mm, but with the constant employing, the accuracy will decline. When the accuracy is more than 2 mm, robots are also not suitable for the automated vehicle production. When robots meet the requirements of scrapping, scrapping directly will waste resources, but continuing to use cannot conducive to ensuring the quality of the whole vehicle (Fig. 7).

4.1 Difficulties of Scrapped Robots' Reutilization

There are four difficulties in the reutilization of scrapped robots:

① Lack of standards, lack of standardized methods for processing flow, and accuracy requirements for different working conditions.

② Specialization of technology, industrial robots have three layers of data systems: the use of data layer, maintenance data layer, and design data layer. If the old robot needs to be rebuilt, you must master the skill of how to use the underlying design data layer, well this layer of use involves industry secrets and lack of training materials and related information, so the explore independently is necessary.

③ Accuracy of technology, it is necessary to overhaul its electrical and mechanical components and adapt to the new environment.

④ Lack of experience, before the project begins, it has not carried out the design contents of robot application and function independently.

Fig. 7 Scrapped industrial robot

4.2 Innovative Reuse of Scrapped Robots

By analyzing, four aspects for reutilization of the scrapped robot are determined, they are construction of professional training center; replace manual labor with the less precise work; training of mechanical component disassembly; and display of robot-related innovations showcase.

In order to regulate the recycling of the scrapped robots and improve efficiency, a standardized process is developed. With the promotion of the project, 16 scrapped robots were transformed and reused within a half year. These scrapped robots are reapplied to the automatic cleaning of stamping die, training center construction, robot making planed noodles, robot calligraphy display, robot lion dance performance, robot playing table tennis, and so on. By the exploration of robots, we mastered the use of the underlying data of the robot, which break the technical constraints of equipment manufacturers. The project has improved the understanding of robots for all employees, broadened their horizons, narrowed the distance between ordinary employees and intelligent manufacturing, and improved the design skills and innovative thinking of robots.

The picture below shows six cases for scrapped robots in practical innovation applications (Fig. 8).

Reduced purchase of three robots after robotic automatic cleaning of the mold, and save is 1.05 million Yuan. The maintenance cost of the robot has also been reduced from 400,000 per year to 700,000 per year per 100 units. The project not only makes the team improves their skills, but also explore and master various skills about robots. The project has accumulated experience in robot design. The scrap robot

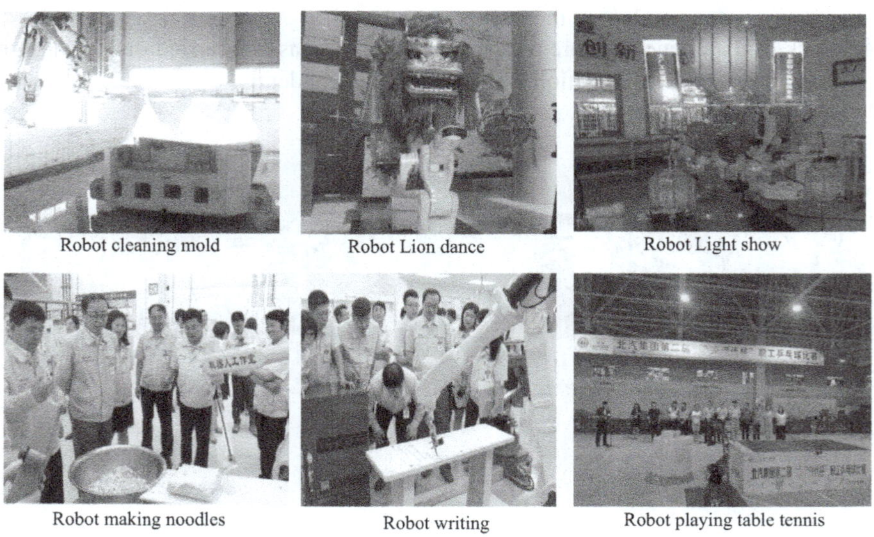

Fig. 8 Reconstruction of scrapped robots

is also planned for the company's auto parts life test and quality-related repeatability experiments, etc.

5 Innovative Talent Training Mode

5.1 Innovative Studio Mode

In order to promote the cultivation of robot professionals and innovative talent training mode, Beijing Hyundai established the first robot innovation studio in the automotive industry. The studio has seven training robots: HHI robot, KUKA robot, YASKAWA robot, and so on. The studio has built a platform for the robot research, repair and maintenance personnel of Beijing Hyundai since it was found.

The studio's operation mainly depends on two aspects: One is the promotion of the key projects, and the other is skills competition of employee. These two main aspects are the keys to promote the operation of the studio. The studio regularly takes three months to complete the off-line training for robot technicians of various factories and improve the worker's skill level. One of the trained talents won the championship in the national robot skill contest, which improved the popularity of the enterprises, and affirmed the training mode.

5.2 Application of WeChat Information Platform

In order to improve the overall quality of robot-related talents, an elite group of maintenance talents was set up. Robot maintenance workers from five factories will join the group to give full play to the role of WeChat. If robot equipment fails in any factory, the fault phenomena and alarm contents will be immediately sent to the group. In this way, all personnel can timely link their work experience to provide their similar troubleshooting experience and the skills and experience of all maintenance personnel can be improved. All staffs can display their knowledge, exchange business, publish some of the latest robot-related news in the world on WeChat to improve the interests on robot of team members pursuing better invest of robot research.

6 Conclusion

Robot is one of the main equipment of automobile production enterprise exploration of autonomous management, autonomous maintenance robots has begun in various enterprises. Mastering the related technology of robot will greatly reduce the cost of robot using and also make the enterprise get a breakthrough in training automation talents (Fig. 9).

With the innovation of robot management and technology, robots will be more standardized and systematized in maintenance. The aging robot is closely monitored.

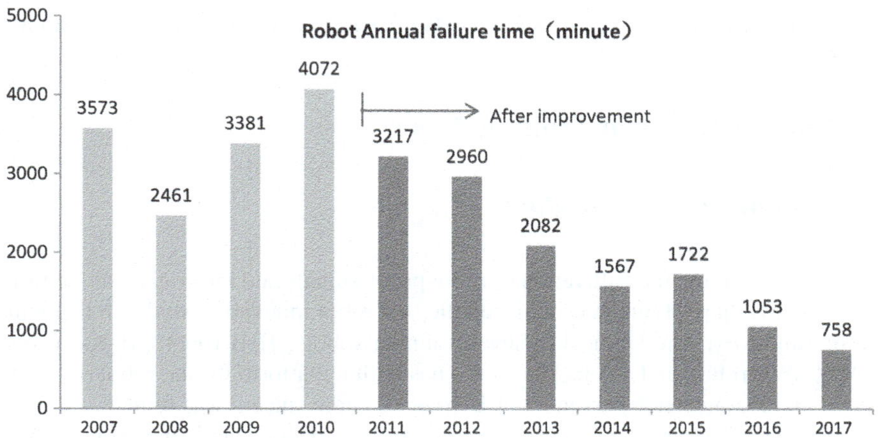

Fig. 9 Time comparison of robot fault

Once there is something different, the preventive maintenance work will be targeted. With the development of robot skill talents and innovative maintenance methods, the frequency and failure time of robot decreases year by year.

The effectiveness of robot management:

① A professional robot studio has been set up to improve skills, cultivate talents, and promote innovation.
② Write two robot troubleshooting manuals to improve skills.
③ Develop 25 standardization processes of robot operation and maintenance.
④ Develop a professional robot management ledger to make it systematic and targeted.
⑤ The application of computerized PDA inspection mode reduces the incidence of poor spot inspection.
⑥ Develop an APP of troubleshooting to assist robot troubleshooting.
⑦ The failure time and the maintenance costs significantly reduced.

The robots that have reached its service life have been managed and maintained reasonably and continue to operate steadily. In a word, equipment aging is not terrible, the terrible is no management, once there is no scientific management, New equipment will be damaged quickly. As a new device, the robot is becoming popular. Its management and maintenance are still at the initial stage of exploration. Continuous summarization and analysis will boost the development of the national industrial robot industry.

Reference

1. Xi Jinping's speech at the academician conference (2014)

Design of Automatic Feeding System for White Body Welding Production Line

Lin Yang, Xiaobo Wang and Xuefeng Zhu

Abstract In order to solve the problem of automatic feeding of multi-model and heterogeneous body sheet metal parts, this paper designs a system for automatic assembly of top cover sheet metal parts. The system adopts the key technologies such as reliable picking, precise feeding and automatic replacement of material rack. Moreover, the special system is incorporated into the automated welding line. Finally, a comprehensive comparison is made between the technological process before and after the improvement. From the efficiency, economy, reliability and other comparison results, it can be seen that the new technology has a significant advantage, to enhance the white body welding production capacity.

Keywords BIW · Welding · Production line · Automatic feeding system

1 Introduction

At present, body engineering is the most active and rapidly developing field in the world automotive industry, in which BIW (body-in-white) welding production line is regarded as one of the four fundamental processes in automobile body manufacturing [1]. In order to ensure the production capacity and market competitiveness of the products, advanced automobile manufacturers have widely adopted robot-centered high-efficiency, reliable and flexible BIW welding production line [2], which has greatly improved the body welding process rhythm control and quality stability. However, for the welding production line with multi-type and heterogeneous body sheet metal parts, man-machine cooperation mode is often adopted in feeding operation, which causes a great degree of waiting waste or even leading to work stagnation.

L. Yang (✉) · X. Wang
Dongfeng-Nissan Dalian Branch Company, Ruigang Road 8, Double D District, Dalian City 116600, Liaoning Province, China
e-mail: yang-lin@dfl.com.cn; gordonyl@outlook.com

X. Zhu
Dalian University of Technology, Linggong Road 2, Dalian City, Liaoning Province, China

China SAE (ed.), *Proceedings of China SAE Congress 2018: Selected Papers*, Lecture Notes in Electrical Engineering 574,
https://doi.org/10.1007/978-981-13-9718-9_13

For example, the IE (Industry Engineering) method was used to analyze the man-machine operation of a vehicle roof feeding workstation. It was found that the equipment utilization rate was 65.74% in a production cycle time, while the operator's work plumpness was only 40.74%. In order to meet the demand of uninterrupted supply of materials, some automakers add the frequency of storage rack replacement, such as "one-feeding with two-supply storage rack" mode. The practical production shows that the production cycle time is occupied by a considerable number of low value-added operations or even no value operations. Some other automakers or researchers have made improvements in the scheduling of feeding operations to balance the operation time of processes and workers' operations. However, it is always difficult to be integrated when the feeding work stations are isolated with each other. Therefore, it is urgent to break through the feeding automation of body sheet metal parts in order to realize the operation integration and efficiency improvement of welding production line.

Automakers and experts in the field of automobile manufacturing all over the world have done a lot of meaningful research work to improve the level of automation and intelligence of BIW welding production line. The BIW welding production line abroad has been developing with high automation and high flexibility. Open Robogate assembly system, a patent of COMAU in Italy, combines its own flexible concept with high-precision BIW assembly, which can adapt to mixed-flow production of multi-products and also can raise efficiency to 60 JPH. Aiming at the problem of precision control in BIW process, S. M. Wu Research Center of Michigan University and Lin's team of Shanghai Jiao Tong University put forward the "2 mm project" of automobile body manufacturing and established the clamping and positioning technology system of automatic welding production line for body sheet metal parts [3]. Jianqiang Wang from HeFei University of Technology has made research on multi-PLC coupling and multi-level PROFIBUS bus control technique for B class vehicles BIW mixed-flow production line in China, which improved the standardization and modularization of the whole line control system [4]. Mingyuan Tao from Huazhong University of Science and Technology, made his study based on the characteristics of body sheet metal parts and its requirement of clamping and positioning, analyzed characteristics of the pneumatic system and its composition of BIW production line [5]. These research results provide theoretical and technical support for our research on automatic feeding system for automatic feeding research of body sheet metal parts.

Although much research has been made on automatic feeding system, there is still a big gap on the automation of automobile manufacturing between China and abroad. This paper, taking the welding process of SUV roof parts in Dongfeng-Nissan Dalian branch company as a typical research object, shows our research on automatic feeding system for BIW production line, which can greatly improve efficiency of the welding production line and reduce the labor intensity of manipulators.

2 System Design

According to the overall process planning of body manufacturing, BIW welding is the next step after sheet metal stamping, which is the process of splicing and assembling the stamped sheet metal parts into the whole body. Feeding operation is to pick up the stamped sheet metal parts from the storage rack and place them in the specified welding station, which is an important auxiliary process for the automatic welding production line. To this end, the actual geometric characteristics and material properties of the sheet metal parts of the manufactured models must be taken into consideration for automatic loading planning and system design.

Taking the roofs of Dalian Nissan SUV model as an example. According to the sunroof types, this model mainly includes three roof types: fully metal roof, standard roof with sunroof and roof with a panoramic sunroof, as shown in Fig. 1. Meanwhile, in order to ensure the structural strength and rigidity, a certain number of reinforcing ribs are also designed under the roof. The basic parameters of the roof are: the maximum width is 1200 mm; the maximum length is 2000 mm; the thickness is about 1 mm; and alloy steel material. Most sheet metal parts of the body roof are flattened and streamline curvature with relatively heavyweight and high surface quality requirements.

In order to realize the automatic loading of various types of roofs at one loading station, the constraints to be considered in the design of the feeding system mainly include quick recognition of different roof models, automatic picking and moving and automatic loading continuity of roof sheet metal parts. This paper firstly designs a process flow that can achieve the automatic loading of passenger car roofs of different sizes and models. As shown in Fig. 2, the basic process mainly includes: the production line system communicates through Ethernet, sending the loading command package including the car model, roof model and other information to the loading subsystem according to the production cycle; coding and correlating racks

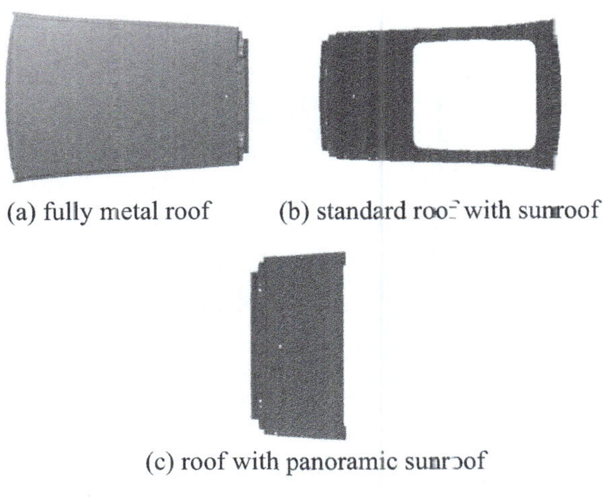

Fig. 1 Roof schematic diagram of Dalian Nissan SUV model

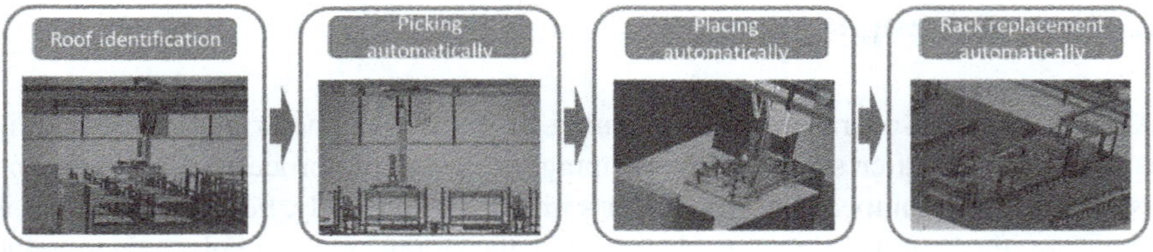

Fig. 2 Process flow of automatic roof feeding

and body roofs to ensure that the loading subsystem rapidly identifies the body roof required by the command; the special manipulator picks up and loads the body roof; the automatic swapping of the standby rack and the empty rack is completed to ensure the loading continuity; the multi-sensor online integration is used for information feedback and process monitoring of the automatic loading.

According to the loading process above, the automatic roof loading system is designed accordingly (as shown in Fig. 3). The main functions and features are as follows:

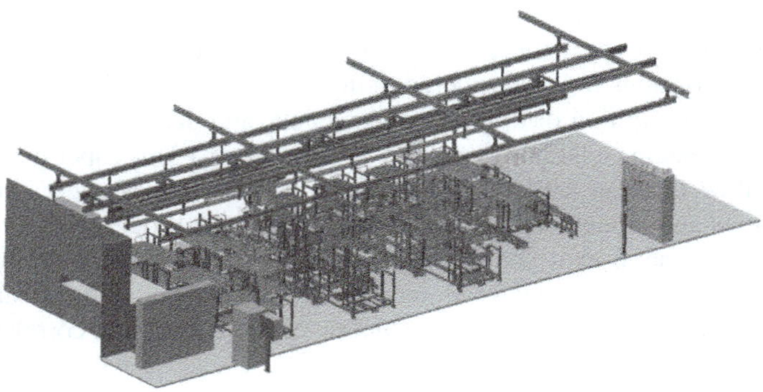

(a) Master plan of the system

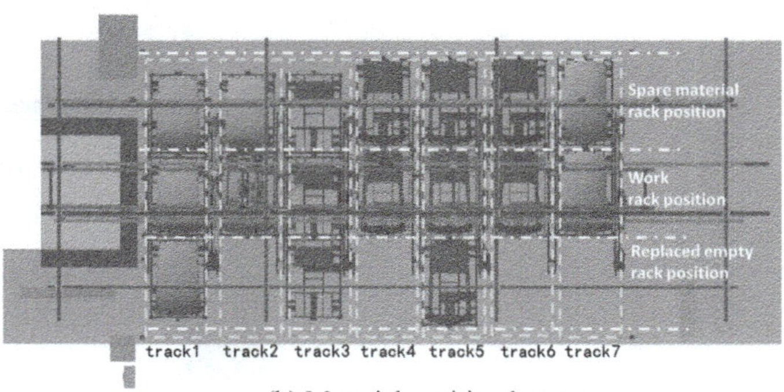

(b) Material position layout

Fig. 3 Automatic loading system

(1) The system is generally composed of three rows of material positions: a standby material position, a working material position and a changeover material position. To this end, it is necessary to ensure that the standby material position always remains ready with materials and the changeover material position is filled with an empty rack;
(2) Each vertical row of the standby material position, working material position and changeover material position constitutes a material position swap group. The system is designed with X material position swap groups and with expandability;
(3) The special manipulator installed at the end of the drag system can realize the reciprocating motion, the rotational motion, the up and down motion, etc. along the track above the working material position, to complete part identification, pickup and loading;
(4) Laser, visual and other sensors are used for material position detection, material cart status monitoring, roof position and posture detection, etc.

3 Key Technologies

3.1 Reliable Picking Based on Vacuum Adsorption

The vacuum adsorption clamp has the technical advantages of large controllable suction, soft contact and no pollution and also has a low price and good applicability. Therefore, it is often used in automobile body production lines.

3.1.1 Design of Vacuum Adsorption Clamp

According to the surface quality, material, shape and model of the roof cover, a special suction fixture with independent and controllable multi-channel sucker is designed (as shown in Fig. 4). The vacuum sucker is straight, and each sucker is controlled by a separate air path to ensure the adsorption reliability. The eight suckers are arranged in two rows as a rectangular pattern. In order to prevent the roof from falling off due to the unstable grasping, single-hook and double-hook safety hooks are added to the fixture to assist the sucker to carry out the roof picking.

Among all the external plates of the roofs, the difference between the high-configured and the middle- or low-configured plates is large. For the external plate of the panoramic roof, the absorbable area is about one-third of the middle- or low-external plate, and the left four suckers have no absorbable area on the roof, making the grasping difficult. The sucker is matched with the pneumatic hook for grasping, and the safety hook must be turned to the bottom of the roof under the action of cylinder. When arranging the suction point of the sucker, it is necessary to ensure

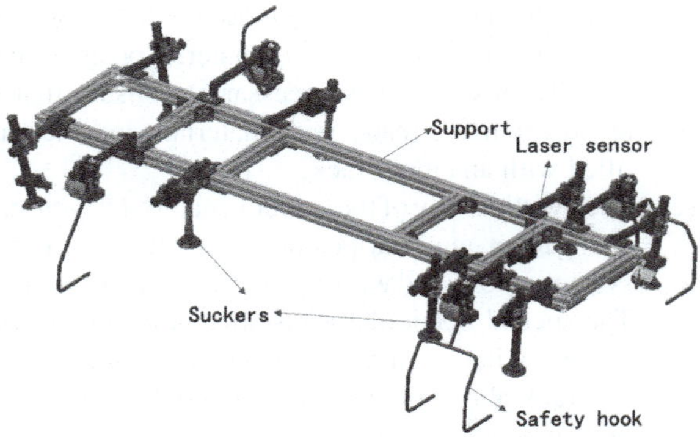

Fig. 4 Special vacuum adsorption clamp

that the resultant suction force is close to the center of gravity of the component. When the gripping member is a middle- or low-configured roof, all the eight suckers act on the plate, and the action point of the resultant suction force should be close to the center of gravity of the plate.

The suction force of the vacuum sucker is calculated according to the configuration of the clamp, the number of suckers and the gravity of the part.

$$\begin{cases} F_i = P \times A \\ F_S = \sum_{i=1}^{n} F_i = \text{mg} \end{cases} \tag{1}$$

where F_i is the adsorption force of a single sucker, with the unit of N, and all the sucker adsorption force are consistent; i is the sucker number, $1,\ldots,n$; P is the vacuum (relative pressure) inside the sucker, with the unit Pa; A is the effective adsorption area, with the unit of m^2; F_s is total adsorption force, with the unit of N; m is the mass of the part; g is gravitational acceleration, with the unit of g/s^2.

The adsorption safety factor f should be considered when calculating the sucker diameter. During the picking and transportation of the roof, the parts are basically horizontally adsorbed, and f is taken as 4. The vacuum P in the sucker should be selected in the range of 63–95% of the maximum vacuum degree pV of the vacuum generator, to improve the vacuum adsorption capacity without causing excessive adsorption response.

$$D = \sqrt{\frac{4}{\pi} \times \frac{F_s \times f}{P \times n} \times 1000} \tag{2}$$

For example, a certain model of roof weighs 120 N, the number of suckers is 8, the sucker diameter D is 30 mm, and the vacuum generator has a maximum vacuum of −88 kPa. The German SCHMALZ sucker is used, the gas supply pipe diameter should not be smaller than 1/2 in., the switch valve should be provided at the interface, the gas supply pressure should be stable and not less than 5 bar, and the gas is dry and clean compressed air.

3.1.2 Design of Picking Motion Mechanism

Install the special fixture designed in Sect. 3.1.1 to the end of the lifting mechanism to construct the roof picking device (as shown in Fig. 5). In order to achieve the picking action, the picking device needs to carry out query motion (*Y*-direction) of the material level and the position seeking motion (*Z*-direction), and the double-direction motion is designed to adopt the SEW servo control system and the EEPOS aluminum alloy track. The *Z*-direction movement of the lifting mechanism is driven by a servo belt. The *Y*-direction motion track is equipped with a special servo reducer and a driving mechanism of rack and pinion for precise positioning and adjustable speed. In order to shorten the interval time, all the materials to be picked up are arranged in the longitudinal direction coincident with the middle vertical line in the feeding port turntable, and the material level is sorted according to the expected production plan (the current production ratio of each model), that is, no *X*-direction movement is required during the picking process (as shown in Fig. 6). With the plates in the material rack are taken out one by one, the mechanical arm senses the position of the plate in the lower layer through its own set of laser range finder (DT35-B15551), getting the order of downward *Z*-direction motion and the height value of plate need to lift, and the installation position of the detecting device is shown in Fig. 4.

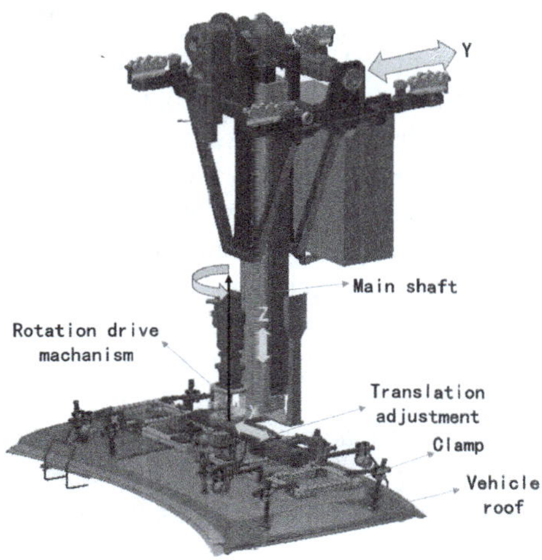

Fig. 5 Picking device

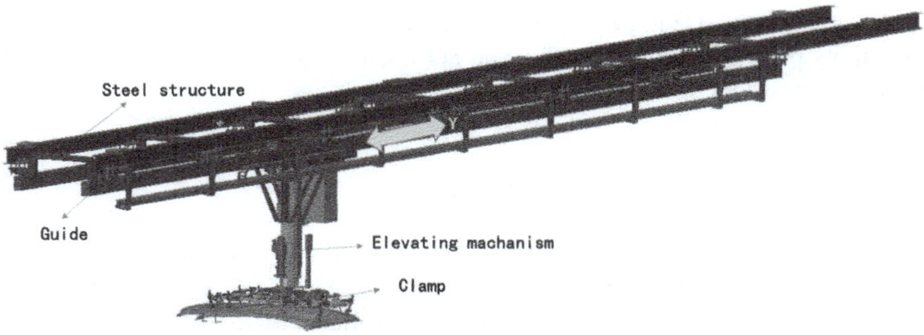

Fig. 6 Y-axis moving of picking device

3.2 Accurate Placing Based on Vision System

The picking device needs to place the roof sheet metal parts to the specified welding preparation station by matching the pinholes of roof with the guide pins. While, the non-uniformity and irregularity of the placement of sheet metal parts in the material rack would lead to misplacement of the parts between layers and layers. Also, there is position deviation between the floor and the material rack.

The vision system is used to identify the shape and location of roof sheet metal parts and the position of the guide pins, which can assist assemble action complete accurately. 2D vision system can be used to complete the contour figure identification in the horizontal plane because of the large area and the curvature of the sheet metal parts, so we can obtain the model of the roof as well as the vehicle type and roof height. 3D coordinate can be calculated out by checking assembly information. According to the camera installation height and the sample of FANUC intelligent vision system which is called iRVision, 25 mm focal length is adopted, and the resolution of camera is 130 million (1280 $\times$ 1024) based on the assembly accuracy ($\pm$1 mm).

Under the condition that position accuracy of the rack is reliable, the primary function of vision system is placing accuracy, and if the camera is installed on the picking device and moves with it, the scope of camera cannot cover the whole parts because of the height limit and also occupies the cycle time for waiting non-vibration state of picking device. Thus, the assembly position detection scheme is adopted by installing the camera above the assembly fixture instead of installing it on the picking device.

To guarantee the accuracy of loading attitude, there are translation and rotation adjustment drive on the picking device (as shown in Fig. 5). Distance (*X*-direction) and angle trimming (*Z*-direction) are adjusted by transmission of ball screw and turbine worm via position detection of vision system. The adjustable range of *X*-direction is $\pm$100 mm, and the adjustment accuracy of *Z*-direction rotation is $\pm$1 mm; thus, it can place the roof on the assembly fixture successfully and accurately.

3.3 Automatic Replacement of Material Storage Rack

As for automatic equipment by erecting guideways in the air, the protective enclosure and grating sensors are necessary for safety reasons. These protective devices can prevent personnel from entering the area when the equipment is working. While, the logistics staff have to enter the operation area and shut down the equipment to carry out material rack replacement, which will affect production cycle time. To minimize the influence time of reloading, the automatic rack replacement project has been proposed in this paper. The logistics staff only need to put the spare rack into the entrance of standby position track, the spare rack can replace the working rack in time accurately through precise positioning of guide and transmission mechanism when the working rack is empty. The process can complete automatically without any machine halt.

Each workpiece on site corresponds to a unique rack, and the relative position of the same workpiece on the rack needs to be consistent. The position deviation of the workpiece relative to the rack is ±3 mm. The rack is arranged strictly according to Fig. 7, and the ground limit device is set up.

In order to obtain the spare state information of the material rack effectively and ensure that the rack is not obstructed, the Omron E3Z-D82 photoelectric switch is embedded into the ground plane, and the existence information of the plate is obtained through the hollow position of the bottom layer of the material rack. With the gradual consumption of the same type of plate in the material rack, the automatic replacement program of the rack can be triggered immediately when the plate cannot be detected at this position. The automatic switching mechanism of the ground material rack is equipped with a special reducer. When the equipment obtains the replacement signal, the motor drives the chain, uses the cylinder to locate and limit the position and drives the rack to move to replace the rack whose material has been exhausted. At this time, the empty material rack is pushed to the conversion position, waiting for the logistics staff to pull.

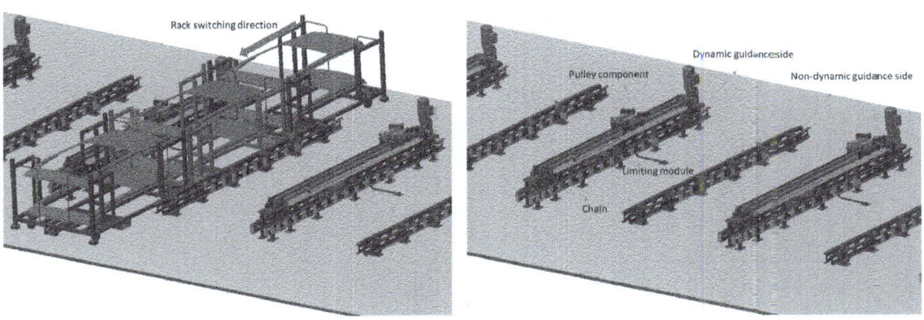

Fig. 7 Rack ground location mechanism

Table 1 Comprehensive evaluation table

Type	Item	Calculation method	Proportion			Evaluation standard					
			Strategy	Quality	Environment	5	4	3	2	1	0
C	Investment recovery period	Calculation based on the income scale	85%	65%	65%	<2			2–2.5		>2.5
Q	DPU decline	(Current value-target-value)/target value		30%		>50%	50–30%	30–20%	20–10%	10–5%	<5%
T	Mode of production	Production mode after improving	5%			LOT → JIT & IS	Batch → JIT & IIS	LOT → Batch	Batch (unchanged)	LOT (unchanged)	Deterioration
	LT	(Current value- target-value)/target value	5%			>60%	60–50%	50–40%	40–25%	25–10%	<10%
E	Man-machine engineering	Tough job improvement	5%	5%	5%	A → 0	A → C	B → 0	A → B; B → C	C → 0	Unchanged
M	Carbon decline	(Current value-target-value)/target value			30%	>50%	50–30%	30–20%	20–10%	10–5%	<5%
Total			100%	100%	100%	–					

(continued)

Table 1 (continued)

Value	Scores			
	Strategy	Quality	Environment	Total
2.42	1.70	1.30	1.30	2
15.0%	0.00	0.60	0.00	2
Batch (unchanged)	0.10	0.00		2
100%	0.25	0.00		5
A → 0	0.25	0.25	0.25	5
	0.00%		0.00	0
	2.3	2.15	1.55	

4 Comprehensive Evaluation

The solution became one of the first one to realize the unmanned operation with utilizing multi-sensor communication technology, combining model recognition, automatic feeding and intelligent replacement of storage. This solution makes an achievement of reduction of two labors per day in the planning of 30JPH production capacity. The expected investment recovery period is 2.4 years. The decline of operation time such as loading, processing, unloading, checking, storing, transportation and assembly is totally 20 h/day with effects of 100%. Meanwhile, in the aspect of man-machine engineering, it also makes progress in eliminating the tough job. Table 1 shows that this program, which is one of the key innovation subjects in iFA (integrated Factory Automatic) improvement activities in Dongfeng-Nissan Dalian branch company, gets 6 points of the comprehensive evaluation.

5 Conclusion

This paper presents an approach of automatic feeding for BIW sheet metal parts, in which multi-sensors and manipulator are integrated online. Taking welding process of SUV roof in Dongfeng-Nissan Dalian branch company as the typical object, we designed automatic feeding system dedicated for three typical types of roofs, which makes breakthrough in key technology such as reliable pickup, accurate feeding, automatic replacement of rack comprehensive evaluation shows that the approach can improve the productivity and reduce production costs remarkably.

Acknowledgements Foundation project: This work was funded by the Natural Science Foundation of China (No.11302041). These supports are gratefully acknowledged.

References

1. Chen J, Luo H, Zhang W (1997) Automobile body manufacturing technology. Chongqing University Press, Chongqing
2. Yang J (2008) Application of robot in automobile body manufacture. Aeronaut Manuf Technol 3:36–37
3. Wang J, Wang C, Sun C, Jianlin W (2008) Application of multi PLC and multilevel field bus in robot welding line. J Shanghai Jiaotong Univ 42:20–24
4. Zhang Y, Chen S, Chen G, Lai X (2002) Research on pattern recognition method of BIW assembly size deviation. Mech Sci Technol 21(5):689–691
5. Tao M, Cao B, Wu C (2002) Application of pneumatic technology in automobile body welding production line. Chin Hydraul & Pneum Chin Hydrau Pneum 12:21–22

Lin Yang female, born in 1987, graduated from Dalian University of Technology, and was awarded Master of Engineering in Mechanical Manufacturing and Automation, IE Engineer of Dongfeng-Nissan Dalian branch company, professional title, the Research Direction is production efficiency improvement, body welding process rhythm control and quality stability, efficient human-machine operation.

ODISSEE Test of ABS Anti-lock Braking System

Ruicheng Zhang

Abstract The hardware composition and software testing principle of ODISSEE communication equipment for ABS anti-lock braking system of car are studied. According to the function test requirement of ABS, ABS initialization test and ABS start test, ABS static test, ABS dynamic test, ABS stop test, ABS write process, ABS result test, WSS wheel speed sensor test, etc are carried out in turn. According to the value of ABS dynamic test, the braking force of the four wheels is calculated to meet the requirements of the regulations; the operator interface during the testing process shows the test steps, the operation of the test requirements and the test results in time, which is convenient for the operator to operate the test. This paper can be used as reference for equipment maintenance technicians and automobile testing technicians.

Keyword ODISSEEN ABS

1 ODISSEE Testing and Communication (1) Page 61

The dialogue tool series vehicle electronic system, using portable computer and automobile diagnostic dialogue series console (PC group), is a car diagnostic dialogue tool developed by France PSA Automobile Company and French company ACTIA. As shown in Fig. 1 is the ODISSEE test device. ODISSEE test testing automotive electronics such as air conditioning and ABS systems.

The software of hub test bed adopts ODISSEE, which is a kind of equipment for communicating with automobile electronics.

The hardware of the system is as follows: motherboard (CPU, RAM, ROM, flash disk), I/O interface module (vehicle communication interface, USB input interface, serial interface, super input, the ODBII vehicle communication input interface, the keyboard interface and the screen I/O interface), wireless Bluetooth interface

R. Zhang (✉)
Dongfeng Peugeot Citroen Automobile Co., Ltd. Wuhan Factory, Wuhan
430056, Hubei, China
e-mail: 888zrc@sina.com

China SAE (ed.), *Proceedings of China SAE Congress 2018: Selected Papers*,
Lecture Notes in Electrical Engineering 574,
https://doi.org/10.1007/978-981-13-9718-9_14

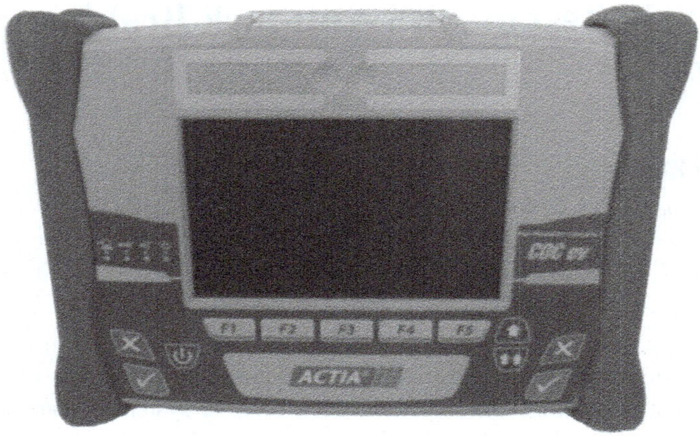

Fig. 1 ODISSEE equipment

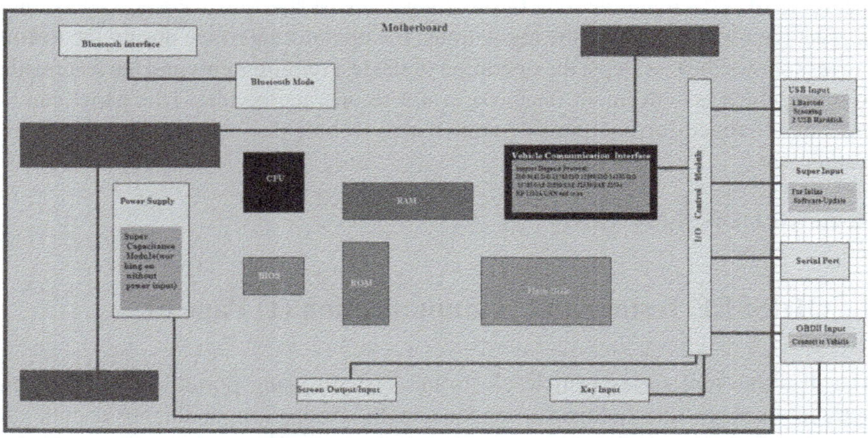

Fig. 2 Structure of ODISSEE system

(wireless Bluetooth module), wireless communication processing module (antenna), power supply module, etc. The structure of the ODISSEE system is illustrated in Fig. 2.

1.1 Synchros

PSA developed ODISSEE test program and established synchronizer based on the same mechanism. These synchros are 2-byte words in a common memory. The bench PC can launch a routine by putting the synchro value of the corresponding routine on

0 × FF00. The ODISSEE executes the routine and writes the state of the routine in the same synchro. For instance, 0 × 0002 means that the routine is being executed.

1.2 Swap Area

When parameters have to be transmitted to an ODISSEE routine, they are first written to a swap area (in French: zone d'échange'). This area is located in the common memory of the bench PC and the ODISSEE. When the parameters are written to the area, the routine is launched by means of a synchro.

When a routine makes results available, these results will be in the swap area after the routine is finished. PC reads results from swap area.

1.3 Jbus Communication

Reading and writing in the common memory are done via a serial link. The Jbus protocol is used for this. The PC of the roll test machine is the master. The common memory is physically located in the ODISSEE PC.

1.4 ODISSEE Communication Problem

Problem: Serial communication between PC and ODISSEE fails, Fig. 3.

If a communication error occurs between the PC and the ODISSEE, all further tests will not be executed. Thus, the cycle has come to an end.

Of all the tests that have been performed before the communication error occurred, the results will appear on the print report. Of the tests that were not performed, there will be no results nor mentions.

Furthermore, there will be a message on the print report to inform that there was an ODISSEE communication error:

******* Cycle stopped *******
******* ODISSEE COMMUNICATION FAILED *******

Due to this communication error, the global result at the top of the print report will be 'CAR NOK.'

Fig. 3 Serial communication between PC and ODISSEE

2 ABS Initialization and Start Test (2) Page 62, 64, 65, 67

2.1 ABS Initialization Test

In ABS initialization test, bar code is transmitted to ODISSEE by scanning gun on hub test bench. Accordingly, the ODISSEE knows what type of car this barcode corresponds to.

The bench will do a software reset of the ODISSEE, write the bar code to the swap area of the ODISSEE and inform the ODISSEE about this by means of a synchro.

This sequence will be repeated until the ODISSEE has correctly processed the bar code or until the operator has stopped the cycle.

2.2 ABS Initialization Test PC Cycle

During the PC cycle phase, the PC will perform the following actions:

(1) The PC will do a software reset of the ODISSEE.
(2) The PC writes the bar code in the swap area of the ODISSEE.
(3) The PC informs the ODISSEE by means of SynchroNumber that the bar code is ready.
(4) The PC checks if the ODISSEE has processed the code. If not, the PC will start all over again from stage 1.

2.3 ABS Start Test

This test starts the communication with the ABS system. The ABS system is put in test mode (the ABS error indicator light on the dashboard goes on).

If the car has not come to a standstill, place a message on the screen in which the operator is asked to brake to a standstill.

The PC waits until the car has come to a standstill and then starts the ODISSEE initialization routine.

2.4 ABS Start Test PC Cycle

During the PC cycle phase, the PC will perform the following actions:

(1) If the car has not come to a standstill, a message on the operator screen will be shown, in which the operator is asked to brake to a standstill.

(2) Wait until the car completely stands still.
(3) Execute the initialization routine (SynchroNumber) of the ODISSEE Every PollingInterval msec, a result is asked from the ODISSEE.

3 ABS Static and Dynamic Testing (3) Page 68, 71–74, 76, 79

3.1 ABS Static Test

In this test, several checks are done in a static way, such as:

(1) Check the ECU identification number of the ABS/ESP system
(2) ABS/ESP system detection brake switch of the brake pedal.

3.2 ABS Static Test Description

(1) The PC asks for the ECU identification number via the ODISSEE. After reading the number, it is compared with a preset list.
(2) To test the brake pedal, the operator is first invited to push the brake pedal. The ODISSEE will then check if the ABS/ESP block detects that the brake pedal is pushed. Then the same procedure is performed for release of the brake pedal. This is done by asking the operator the question and having the ODISSEE perform the check. According to this, brake pedal brake switch detection is realized.

3.3 ABS Static Test PC Cycle

During the PC cycle phase, the PC will perform the following actions:

3.3.1 Reading the ECU Identification Number Stage

(1) The ODISSEE is asked to transmit the identification number of the ABS/ESP system (Synchro ECU ID). Every PollingInterval msec, a result is asked from the ODISSEE.

(2) Check if the number is on the list ECU Group. The following is mentioned in the print report: 'CONFORMITE OK/NOK,' depending on the presence of the number in the list. The read identification number will also be printed.

3.3.2 Stage to Check the Working of the Brake Pedal

(1) Place a message on the screen in which the operator is asked to brake.
(2) Launch the check routine (SynchroBrakePress) of the ODISSEE. Every Polling-Interval msec, a result is asked from the ODISSEE.
(3) If this routine is finished, place the message on the screen, in which the operator is asked to release the brake.
(4) Launch the check routine (SynchroBrakeRelease) of the ODISSEE. Every PollingInterval msec, a result is asked from the ODISSEE.
(5) The ODISSEE reckons with the fact that the operator needs some time to operate the brake pedal. The operator has not necessarily pushed or released the brake pedal at the moment that the check routines are launched.

3.4 ABS Dynamic Test

This test checks the proper functioning of the valves of the ABS/ESP system.

This test comes after a brake test. The operator is still asked to brake with a certain force. The operator can see the brake force on the screen.

During braking, the brake forces applied on the wheels are continuously measured.

One by one, the valves of the ABS/ESP system are commanded:

(1) Activation of the valve → releasing the brake force
(2) 'Hold' of the valve → brake force is maintained
(3) Deactivation of the valve → increasing the brake force again

After performing the test sequence, the 'reduction' and the 'recovery' of the corresponding wheel and the 'stability' of the other wheels are calculated.

Reduction is a measure for the quantity of brake force that remains on a wheel when commanding the ABS valve (remove brake force).

Recovery is a measure for the quantity of brake force that is active again after stopping the command of the valve.

Stability is a measure for the quantity of the brake force that remains on a wheel when commanding the ABS valve of another wheel.

The test will first check whether each valve is connected to its corresponding wheel: 'valve cross check.' Secondly, the test checks if reduction and recovery are within the preset limits.

3.5 ABS Dynamic Test Calculation

The brake force is calculated as described in the brake test. For each valve activation, the following graphic is obtained, as shown in Fig. 4.

For each valve, after the activation, three values are obtained:

(1) StartValue: brake force at StartIndex
(2) MinimumValue: minimal brake force between Start and EndIndex
(3) EndValue: brake force at EndIndex

In which: StartIndex = start of the activation; EndIndex = end of the activation
Note that (EndIndex)Valve(N) = (StartIndex)Valve(N + 1)
Calculation of reduction/recovery/stability for every brake:
For every X = FL; FR; RL; RR and for every commanded valve:

$$\begin{cases} (\text{Reduction})X = \frac{100\times(\text{MinimumValue})X}{(\text{StartValue})X} \\ (\text{Recovery})X = \frac{100\times(\text{EndValue})X}{(\text{StartValue})X} \end{cases}$$

Test algorithm:

(1) Checks when the FrontLeft valve is activated:

For the commanded wheel:
ReductionFL $\leqq$ MaximumFrontReduction
RecoveryFL $\geqq$ MininumFrontRecovery

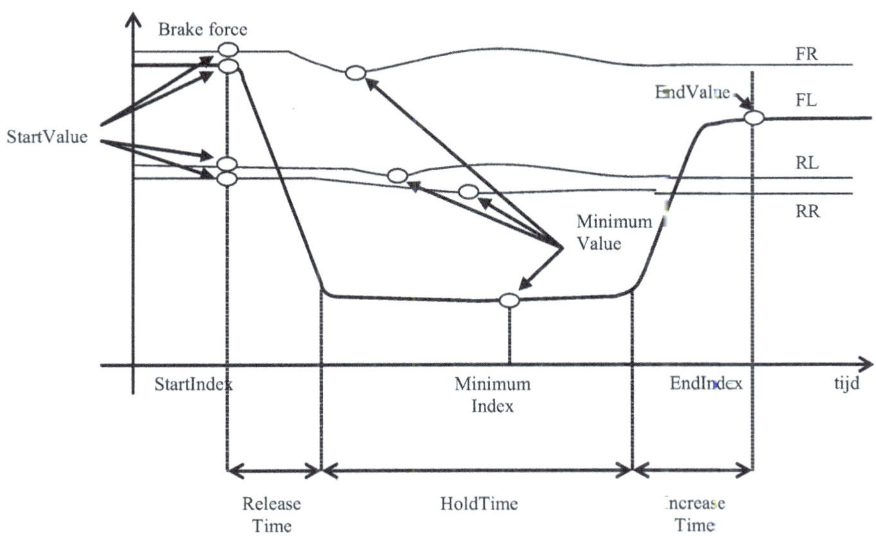

Fig. 4 ABS dynamic test braking force curve

For the non-commanded wheels:
StabilityFL-FR ≧ StabilityEdgeFront
StabilityFL-RL ≧ StabilityEdgeRear
StabilityFL-RR ≧ StabilityEdgeRear

(2) Checks when the FrontRight valve is activated:

For the commanded wheel:
ReductionFR ≦ MaximumFrontReduction
RecoveryFR ≦ MininumFrontRecovery

For the non-commanded wheels:
StabilityFR-FL ≧ StabilityEdgeFront
StabilityFR-RL ≧ StabilityEdgeRear
StabilityFR-RR ≧ StabilityEdgeRear

(3) Checks when the RearLeft valve is activated:

For the commanded wheel:
ReductionRL ≦ MaximumRearReduction
RecoveryRL ≧ MininumRearRecovery

For the non-commanded wheels:
StabilityRL-FL ≧ StabilityEdgeFront
StabilityRL-FR ≧ StabilityEdgeFront
StabilityRL-RR ≧ StabilityEdgeRear

(4) Checks when the RearRight valve is activated:

For the commanded wheel:
ReductionRR ≦ MaximumRearReduction
RecoveryRR ≧ MininumRearRecovery

For the non-commanded wheels:
StabilityRR-FL ≧ StabilityEdgeFront
StabilityRR-FR ≧ StabilityEdgeFront
StabilityRR-RL ≧ StabilityEdgeRear

3.6 Actia ABS Dynamic Test Parameters

ABS dynamic test parameters are shown in Table 1.

Table 1 ABS dynamic test parameters

Description	Type	Limits/Choice	Units
ID	Text		
Timeout	Integer		sec
Retries	Integer	≥0	sec
Blocking	Integer	≥0	
Dump Front	Integer	[0, 100]	%
Dump rear	Integer	[0, 100]	%
Recovery front	Integer	[0, 100]	%
Recovery rear	Integer	[0, 100]	%
Stability front	Integer	[0, 100]	%
Stability rear	Integer	[0, 100]	%
MinSpeed start	Float	[0, 200]	Km/h
MinSpeed end	Float	[0, 200]	Km/h
Enable valve actuations	Integer	≥0	
Actuation FL	Integer	≥0	
Actuation FR	Integer	≥0	
Actuation RL	Integer	≥0	
Actuation RR	Integer	≥0	
PollingInterval	Integer	≥0	Msec

3.7 ABS Dynamic Test PC Cycle

During the PC cycle phase, The PC will perform the following actions:

(1) Display the speeds of the four rollers on the screen during the whole test.
(2) Check the current speed: Because this always comes after the brake test, the current speed still is sufficient enough to perform this 'dynamic test.'
(3) If the current speed is lower than MinimumSpeedBegin, then the test is aborted.
(4) The ODISSEE routine is launched (SynchroNb) to check the valves. Every PollingInterval msec, the system checks if the ODISSEE is ready to perform the valve check.
(5) The operator was already braking for the brake test. He continues braking according to the brake bar.
(6) Start registering the speeds and brake forces every 20 ms.
(7) The valves are activated one by one. The sequence to activate the valves is determined by FL, FR, RL and RR. The activation of the valves is done by means of the ODISSEE routines (SynchroFL, SynchroFR, SynchroRL and SynchroRR). Every PollingIntervalActivations msec, the system checks if activation is finished.
(8) If—after the activations of the valves—the actual speed is lower than MinimumSpeedEnd, the test fails and is aborted.

(9) Analysis of the measured data: see calculations.
(10) Perform the following calculations per wheel, from the registered data (see part calculations):

Activation FrontLeft:
ReductionFL
RecoveryFL
ReductionFR
ReductionRL
ReductionRR

Activation FrontRight:
ReductionFR
RecoveryFR
ReductionFL
ReductionRL
ReductionRR

Activation RearLeft:
ReductionRL
RecoveryRL
ReductionFL
ReductionFR
Reduction-RR

Activation RearRight:
ReductionRR
RecoveryRR
ReductionFL
ReductionFR
ReductionRL

(11) Evaluate the obtained results (see test algorithms):

Check if two or more valves have been swapped. Check if the measured values are within the preset limits.

4 ABS Stop Test and Write Process Byte (4) Page 80, 82, 83, 85

4.1 ABS Stop Test

This test stops the communication with the ABS/ESP system, so that the ABS/ESP goes in normal operation (ABS/ESP indicator light goes out). If the car has not come to a standstill, the operator is asked to brake to a standstill.

Then the communication with the ABS/ESP system is stopped.
The error indicator light of the ABS/ESP system in the car must go out.

4.2 ABS Stop Test PC Cycle

During the PC cycle phase, the PC will perform the following actions:

(1) If the car has not come to a standstill, a message is displayed on the screen in which the operator is asked to brake to a standstill.
(2) Wait until the car completely stands still.
(3) Perform the 'Fin de dialogue' routine (SynchroNumber) of the ODISSEE. Every PollingInterval msec, a result is asked from the ODISSEE.

4.3 ABS Write Process Byte

During this test, the global result of the preprogrammed ABS/ESP tests is written to the memory of the ABS/ESP block.

When all the ABS/ESP tests preceding this test in the gamme were OK, the process byte OK is written. If one of the tests was NOK, the process byte NOK will be written.

4.4 ABS Write Process Byte PC Cycle

During the PC cycle phase, the PC will perform the following actions:

When all the preceding ABS/ESP tests were positive, the routine SynchroTestOK will be performed, otherwise it will be the routine SynchroNbTestNOK. Every PollingInterval msec, a result is asked from the ODISSEE.

5 ABS Results Test and WSS Test (5) Page 86, 88, 92, 94

5.1 ABS Results Test

This test asks for the ODISSEE results.

On request, the ODISSEE puts the results of its tests available as an ASCII string. OK is 'AA,' NOK is 'EE.'

5.2 ABS Results Test PC Cycle

During the PC cycle phase, the PC will perform the following actions:

(1) The routine 'Send results' (SynchroNumber) is performed. Every PollingInterval milliseconds, the system checks if the routine is finished.
(2) When the routine was performed correctly, the results are read from the swap area of the ODISSEE.

5.3 WSS Test

WSS test checks the proper functioning of the wheel speed sensors.

Initially, motor spins the rolls up to 2 km/h speed with a force of 200 N. After attaining the initial speed (2 km/h) motor spins the roll up to target speed with the target force and once the target speed is attained, the motor is stopped and the rolls are set to free running mode. Execute the initialization routine (SynchroNumber) of the ODISSEE. Every PollingInterval msec, a result is asked from the ODISSEE.

5.4 WSS Test PC Cycle

During the PC cycle phase, the PC will perform the following actions:

(1) Initially, motor spins the rolls up to 2 km/h speed with a force of 200 N.
(2) After attaining the initial speed (2 km/h) motor spins the roll up to target speed with the target force and once the target speed is attained, the motor is stopped and the rolls are set to free running mode.
(3) Perform the 'Fin de dialogue' routine (SynchroNumber) of the ODISSEE. Every PollingInterval msec, a result is asked from the ODISSEE.

6 Ending Language

The ODISSEE test of ABS anti-lock braking system includes ABS initialization test, ABS start test, ABS static test, ABS dynamic test, ABS stop test, ABS write process byte test, ABS result test, WSS wheel speed sensor test, etc. It covers all the testing contents of ABS system. Meet the test requirements of ABS anti-lock braking system. Only the vehicle with qualified test results can get off the line, and the unqualified vehicle cannot get off the line, which can guarantee the safety and quality of the car.

Reference

1. BEP EUROPE. [520844_R&B_tests_v1.c_en] Test manual 2015. No published information: (1) page 61, (2) page 62, 64, 65, 67, (3) page 68, 71–74, 76, 79, (4) page 80, 82, 83, 85, (5) page 86, 88, 92, 9

Design, Development, and Application of Industrial Engineering in New Factory Assembly Workshop

Yangwen Zhu

Abstract Established on October 18, 2002, Beijing Hyundai Motor Co., Ltd. It is the first Sino-foreign joint venture project in the field of automobile production approved after China's accession to the WTO. Beijing Hyundai Motor Co. Ltd needs to have a production capacity of 1.8 million units by 2020 to ensure its market share of Beijing Hyundai Motor Co., Ltd as passenger vehicles in China have been growing at a growth rate of more than 7% in recent years. Therefore, on the basis of the original three factories, two complete vehicle manufacturing plants with an average annual capacity of 300,000 capacity will be added. As the department with the most complex production process and the largest number of operators, the final assembly workshop, its scheme in the design stage of the new factory will directly affect the operating efficiency of the lean production of the whole factory. Therefore, this thesis to Beijing modern new factory in assembly workshop in the industrial engineering design improvement project of systematic application as the research object, using all kinds of basic design method of industrial engineering discipline and philosophy, emphatically analyzes the old assembly workshop design, internal logistics operation man–machine design defects existing in the operation, at the same time, combining with the plant design in front of the stage as a whole, put forward the improvement in the material system, reduce the man–machine operation fatigue strength, improve production efficiency, and so on comprehensive application of basic industrial engineering work method research strategy, action analysis and economic improve study method; based on the application of industrial engineering expertise, the optimized system design scheme is obtained, which provides systematic basis for the planning decision of final assembly workshop material design.

Keywords Industrial engineering · Efficiency · Optimization · Logistics

---

Y. Zhu (✉)
Beijing Hyundai Motor Co. Ltd, Beijing, China
e-mail: zhuyw@bhmc.com.cn

China SAE (ed.), *Proceedings of China SAE Congress 2018: Selected Papers*,
Lecture Notes in Electrical Engineering 574,
https://doi.org/10.1007/978-981-13-9718-9_15

1 Analysis of Automobile Logistics and Assembly Line of Old Factory

1.1 Analysis of Side Logistics in Assembly Workshop

Automobile logistics refers to the integrated organization and planning activities of automobile parts storage, movement, transportation, and distribution in production. The logistics management of the final assembly workshop of Beijing modern old factory mainly includes the receiving, storage and supply of parts, and at the same time, the rational and optimized supply route of parts is formulated, as shown in Fig. 1. In the whole vehicle assembly manufacturing process, whether the parts are put online in time and whether the coordination between operators and material transportation will affect the speed and quality of the whole vehicle assembly in the production line. Due to the poor supply of materials, parts are not delivered to the production line in time, which will cause the vehicle assembly stop line, or slow down the speed of group assembly.

In order to ensure the smooth flow of parts, the old factory assembly workshop adopted centralized component unit management measures and set up logistics containers and key parts sequence supply mode at the production line. Wire-edge logistics refers to specially made pallet loading devices, or the use of standard plastic boxes, boxes, and storage of a large number of parts, as shown in Fig. 2.

1.2 Serial Feeding Mode

Sequence feeding is the supply of the necessary parts in accordance with the production sequence of the assembly line, as shown in Fig. 3. The production manager strictly classifies the parts according to the input instruction information system of the production line. Through this system, the logistics distribution part can get real-

Fig. 1 Logistics transportation

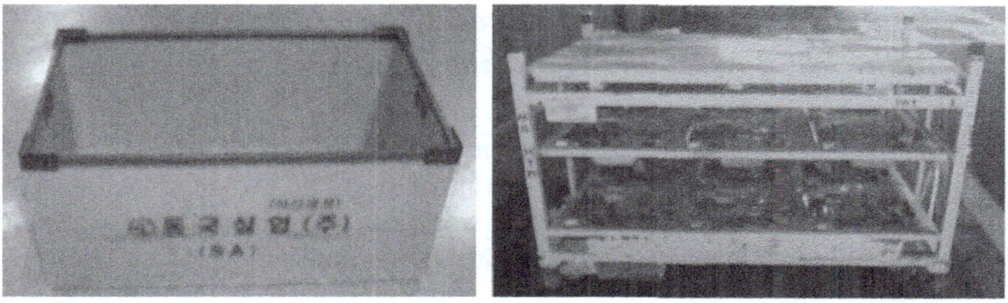

Fig. 2 Material box

Fig. 3 Ordinary material appliances

time information about the input configuration, quantity, and model of the vehicle in the final assembly workshop. Within the specified time, the parts are placed in the supply container in order and the operators only need to take the materials in order from the number to the number, and the assembled vehicles correspond exactly with the sequence number, as shown in Fig. 4.

Fig. 4 Ergonomic Material Frame

1.3 Analysis on the Status of Material Sorting and Feeding in the Final Assembly Workshop

Because Beijing Hyundai model's material components shape differences, so in addition to the modular components (instrument desk, ECU, before and after the suspension, etc.), the old factory assembly workshop manual sorting classification, sweep material subordinate to warehouse after system code, ordinary material stored in pallet or container, and sequence of the material before the outbound, sweep pattern in sequence order by specialized personnel to mobile rack. Materials in the form of AGV trolley need logistics personnel to move the materials to the conveyor trolley and push the trolley to the outlet manually, as shown in Fig. 5.

From the point of view of automobile assembly workers in the final assembly workshop, the way of sequence loading is more intuitive and convenient than the way of linear material picking. Parts are arranged according to the order of 1-n, which is completely corresponding to the current vehicle model that is actually assembled, saving a large number of parts and part's number check, and the time expenditure confirmed. It can be said that under the limited space of the old factory, it is more reasonable to adopt the integrated approach of linear logistics and warehousing, one part of direct supply of parts, and the other part of AGV transplanting materials of the production line. However, from the perspective of man–machine engineering, the labor intensity in the process of moving parts from the shelf or material table is still very high, and the operator needs to constantly adjust the posture to grab parts, which causes the waste of time and labor consumption in this process, as shown in Fig. 6.

Fig. 5 Sequence equipment on trolley

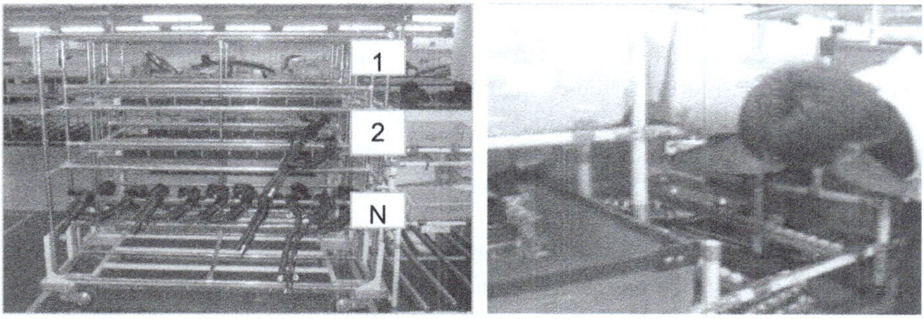

Fig. 6 Improve the contrast between before and after

2 Analysis of Man–Machine Engineering Relationship in Assembly and Sorting Process

2.1 Analysis of Production Line Assembly Process

Automobile assembly process is the final link of all automobile manufacturing process. It is a combination of modules and complete vehicles with tens of thousands of various parts that have been tested and qualified according to the specified precision standards and technical requirements. Final assembly is the last process of automobile production. The quality of assembly is directly related to the quality of the whole vehicle. Assembly operators adopt mobile assembly activities to carry out assembly work. The station is set on both sides of the conveyor line, sharing space system with a material rack, part box, and tool placement box, as shown in Fig. 7. Operators move in different positions of vehicles according to assembly sequence and different process contents. Due to the low degree of synchronization and less mutual restriction of human assembly, a certain degree of freedom can be considered in the programming process.

Fig. 7 On-site operation status

According to different types of vehicles, various types of internal, external, and electric parts assembled in the general assembly workshop are different in appearance and size. These parts are distributed in different forms in the material storage positions on both sides of the production line. As shown in Fig. 8, the first type of material boxed supply mode is online storage and supply, the second type of supply line is in line with the sub-type (SUV, Sedan), and the third type is AGV full-model supply mode.

Taking the assembly process of refueling caliber as an example, this paper compares three kinds of operations and material supply methods in the old factory and obtains the analysis conclusion of action time evaluation and labor intensity expenditure through the process and analysis technology of industrial engineering. After the unified calculation of the number of moving steps of personnel, the horizontal comparison was conducted according to the moving consumption time, as shown in Fig. 9.

It can be seen from the time and distance ratio that the transportation state of AGV trolley is the least step for production line operators. No matter what the model type is, operators only need to take materials from the rack of one AGV, but the

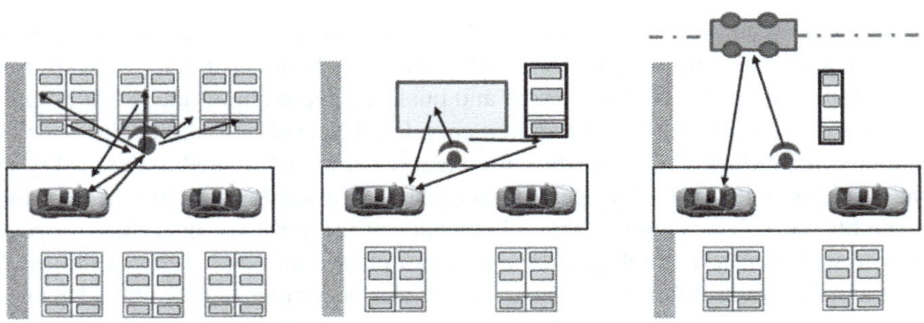

Fig. 8 On-site operation status

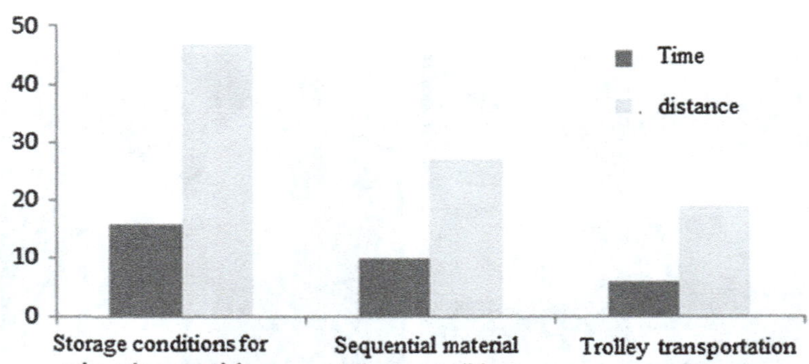

Fig. 9 Comparison of working hours and walking distance

corresponding trolley cannot be close to the production line, resulting in the longest walking distance for personnel. Through the method of taking the materials by the linear edge, it can be seen that for different models, people need to grasp and move the materials from different positions, resulting in increased walking distance and time consumption of the head line. Under the condition of serial supply, operators can take materials according to the sequence of vehicles, but still need to adopt the moving mode of crossing between different material racks between SUV and Sedan.

2.2 Analysis of Personnel Operation in the Delivery Yard

There are five employees in the material pickup yard of the final assembly workshop. They take the materials to the AGV loading rack according to the electronic vehicle number by grabbing the materials on one side. As shown in Fig. 10, the advantage is that the maintenance and manufacturing cost are extremely low, and it can be stopped at any time when the human is manually propelled. As the quantity of materials gradually increases, the weight of the trolley becomes heavier and the fatigue of the operator increases, as shown in Fig. 10.

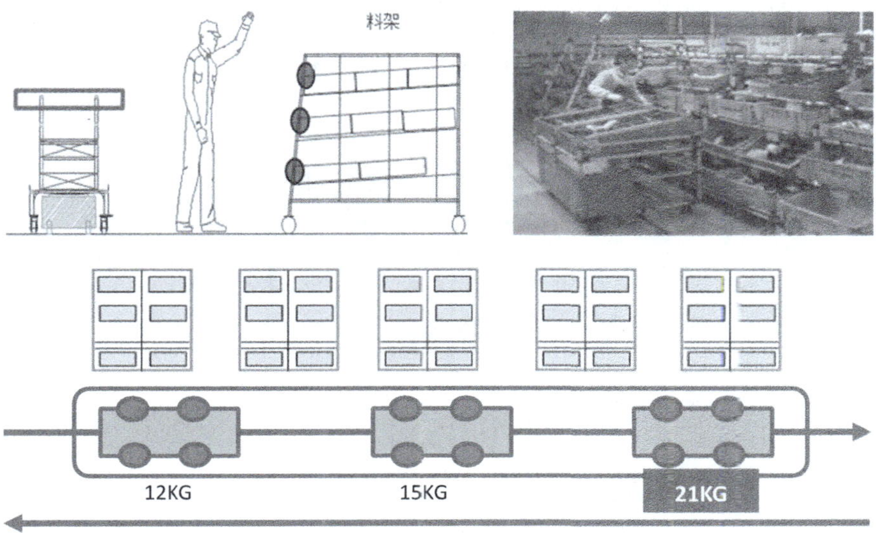

Fig. 10 Old Sorting Way and Load

3 Optimization Design of Production Logistics and Operation Process in New Plant Assembly Shop

3.1 Optimization of Logistics Transportation Scheme

Integrating the advantages and disadvantages of three existing logistics and material picking methods, the optimization plan of the new factory assembly workshop in production logistics and operation process focuses on the principle of "reducing the strength of human action." AGV transportation to the edge of the production line is the solution with the minimum labor consumption of personnel for comprehensive operation, but it is far away from the production line. Therefore, it is necessary to first pull the logistics trolley to the edge of the assembly line to meet the requirements of material collection design for operators. The final transmission position of the feed rack is confirmed through the simulation position, as shown in Fig. 11.

The AGV can safely drive to set goals based on a given starting point and end point without human intervention to complete a variety of given tasks, but it is costly to buy. Materials in assembly workshop line edge device design process, the car through the linear rail series, as shown in Fig. 12, through the chain and crawler, unified power, AGV car only used in material sorting factory and transport links, finally reduced the operation of the whole AGV car weight and quantity, save the overall operating costs, for the following carrier rack space module development design work to create positive conditions.

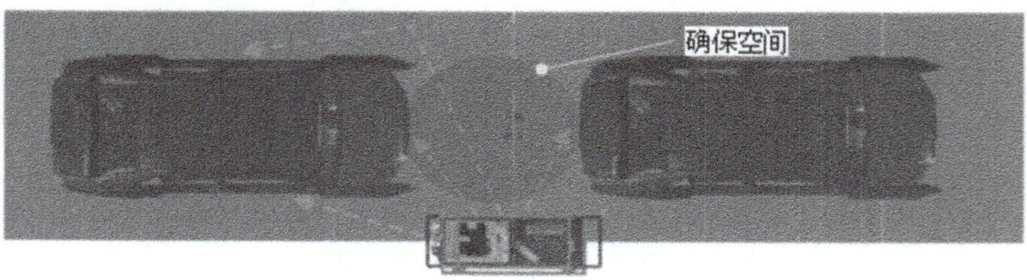

Fig. 11 Unilateral feeding area

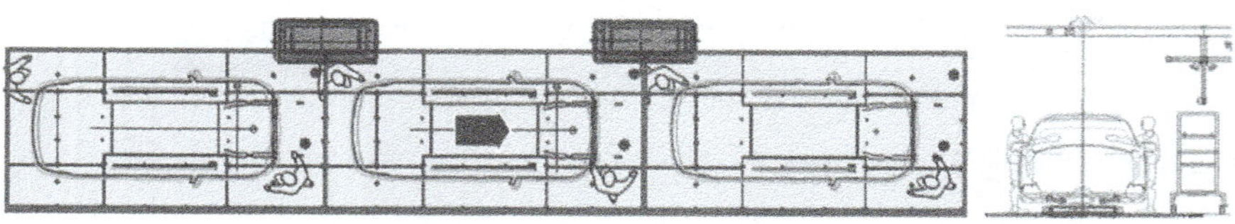

Fig. 12 Personnel Operating Location Design Drawings

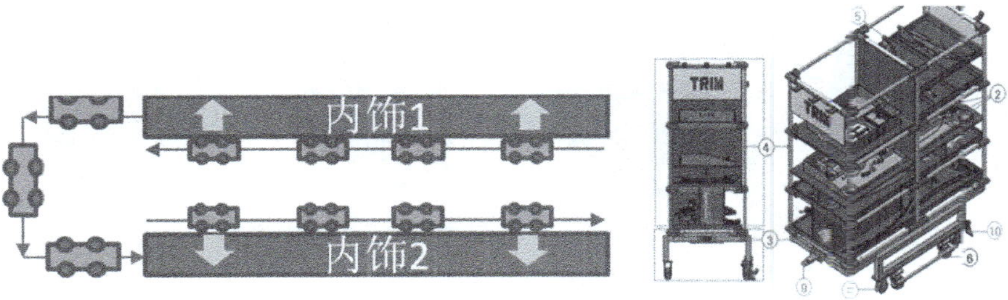

Fig. 13 Design of interior material vehicle

3.2 *Common Platform Development Between Adjacent Production Lines*

The AGV trolley of the old factory directly undertakes the conveying work of the material rack and parts. The materials are placed on the material rack horizontally according to the requirements of the current cycle style. After the operators grab the parts and install them on the body, the material rack goes forward gradually in an empty state.

But the horizontal placement of materials is beneficial with small volume and difficult to transform the old factory. In the early stage of the process design of the new factory, the space of the trolley can be utilized to the maximum, so as to accommodate all the operating stations of the final assembly line for the purpose of material collection at hand. Therefore, a new design is made for the rack, which is symmetrical on the left and right, and a similar material storage device is added on the three-dimensional area so that the bicycle will automatically turn the side of the complete rack to the workers after the rotation of two production lines so that the trolley can reach the state of sharing of two production lines, as shown in Fig. 13.

3.3 *Differential Design of Sequential Trolley*

The process settings of the final assembly workshop are divided according to the door line D, the interior line T, and the final line F. For example, the interior 1–2 lines are mainly for the installation of small parts according to the design requirements. They are divided into indoor/front/rear parts according to the working position. At this time, the material rack is separated into multiple areas for the installation of bicycle parts. As for the final 1–2 lines, according to the design requirements, they are mostly heavy parts with large volume and heavyweight. Therefore, multiple spaces need to be merged in the material rack and table of the final line movement. At the same time, protective grille should be added to eliminate the possibility that the weight will fall during the moving process, as shown in Fig. 14.

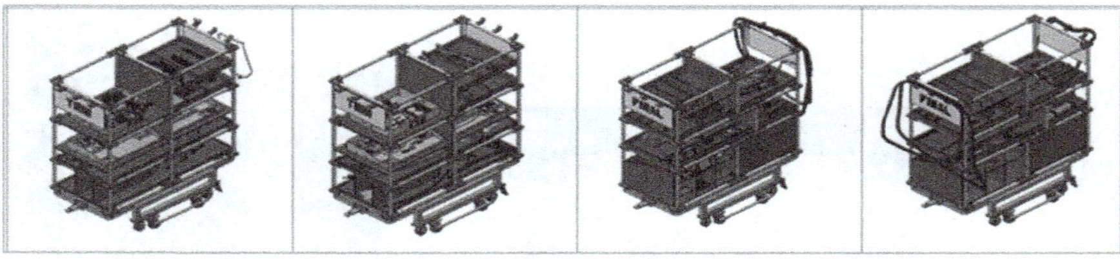

Fig. 14 Design of final line material trolley

3.4 Human Factor Engineering Simulation for Production Line Operators

In the process of operation, operators will produce the deterioration of working skills, which is accompanied by subjective symptoms such as fatigue. Through the testing of human engineering simulation software, the three-dimensional human body model is applied to simulate the operation process and attitude inspection of human movement. As shown in Fig. 15, operators in the different parts in the process of taking and out its location and vision have different influence results, such as the high take material fatigue of operators, the greater the thus design layout for the center with the operators of product material is placed, the weight of different parts according to the average best grasping position arrangement, reduce the boredom of the operators, so as to improve the operators work efficiency. Smaller volume and weight of component parts in the process of the operators to grasp will be more difficult, so small parts are to take out the height, in the chest of a operators to take place, and in order to improve the visibility, the height of the final parts' design should be in the position of the operators knee, designed to tilt angle at the same time, can make the parts more easily into assembly. Door decorative strips and door wire harness, and other strap items can be hung on both sides of the material rack with hooks, which can effectively maintain its original image and prevent the deformation and crimp failure of repeated winding.

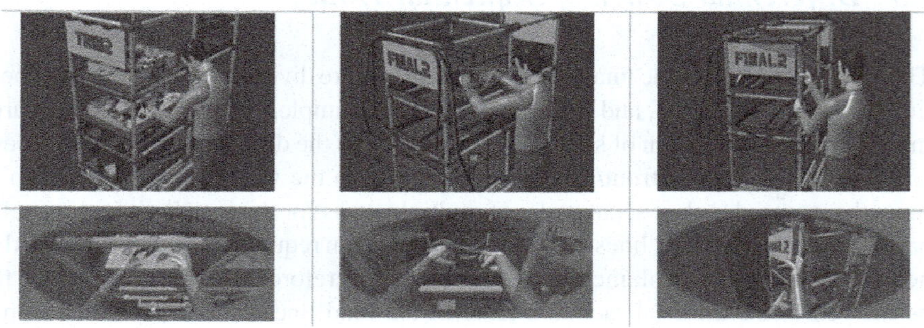

Fig. 15 Optimum fetching posture

3.5 *Material Sorting Plant Sequence Area Design*

(1) **Design of human–machine system sequence fault-prevention device**

In the basic industrial engineering technology application concept, a man–machine system is a system that has different functions and can interact with each other. At the same time, it compensates the deficiency of both systems and forms a reliable relationship in the use process. In the monitoring process of material handling by mistake, information acceptance is realized by automatic monitoring means, and part number identification is able to judge and intervene the wrong action of personnel at any time. Such systems have low workload requirements for personnel, which can reduce their work stress. In the process of extracting the material sequence of various parts of the new factory, logistics personnel take the material sequence of the internal product boxes corresponding to the parts benchmark. In the process of taking and placing, in order to ensure that the pattern is completely corresponding, the sequence bin light prompt and the desk lamp synchronous indicator are adopted. The process of operation is shown in Fig. 16. The material of the same color lamp can be moved, and only two lights can be turned on in a single grasp moving step. In case of a wrong placement, the automatic high-speed camera device will be detected, and the alarm device will immediately carry out voice control prompt operation.

(2) **Optimized design of working space of sequence shelves**

In general, the job space contains two main spatial scopes. The working range of the body movement when the human body is operating in the specified position is known as the working contact space, most of the work requiring production assembly and movement falls into this range. And in the assembly work or other bend, turn the movement, the human body's free movement of the standard scope, known as the activity space. The degree of freedom of operators in these two kinds of contact and activity space is the priority factor of this series shelf design. Referring to the design experience of the mobile trolley on the production line, the classification of small parts and heavy objects is standardized for multi-layer placement, and the tilting angle is gradually increased from the second floor. After the empty boxes are used,

Fig. 16 Sequence sorting yard

they are put back from the upper tilting position to facilitate the operation of the material recovery personnel, as shown in figure.

To sum up, the material and production line optimization design of the new factory's final assembly workshop is mainly composed of two modules, one is the online material transfer process, and the other is the material sorting process. In the transportation process, the biggest innovation feature of the factory optimization program is to shorten the process of AGV from fully conveying materials to work station, and only let it bear the power carrier of the goods shelf from the material pickup yard to the entrance of the production line, which greatly reduces the total amount of use.

By using the basic knowledge of industrial engineering to optimize the design of materials and production system in the new plant, the labor intensity of grasping parts is reduced, and the satisfaction of operators is improved. This design experience will be used more widely as new plant construction continues to increase in the future, further reducing the overall manufacturing and operating costs of the plant.

References

1. Zhang X (July, 2007) Introduction to industrial engineering, 3rd edn. Tsinghua university press
2. Han G (2009) Production logistics system can meet the demand of flexible manufacturing, p 12
3. Yishuping GF (2013) Basic industrial engineering, p 8
4. Zhou D (2011) Due to engineering, p 8

Design of Rear Floor of Automobile Flexible Automatic Welding Production Line

Zhiyang Wan

Abstract With the increasing production capacity of DongFeng Honda second factory, the cost of labor, workers labor intensity, quality, and process stability faces enormous challenges. DongFeng Honda no. 2 welding department promotes to sustain economic restructuring automation project step by step to achieve goals to reduce manual labor costs and improve the quality of body. The no. 2 welding department promotes to carry out the manual line renovation project of rear frame and rear panel to flexible automatic production line. Through the optimization of design by installing 10 welding robots, 3 handle robots, 1 gluing robot, and 1 stud welding equipment to realize 6 models of vehicle production in the same flexible line. Through the automation renovation project, we can realize cutting 22 workers and welding points automation rate increase 4.25%,and it cut down labor cost 2.48424 million RMB per year. It is not only achieved to reduce manual labor cost and improve the quality of the body, and the purpose of ensuring process stability, but also the automation can provide strong quality guarantee and cost reduction for the follow-up new model introduction.

Keywords Automobile · Welding · Flexible automatic production line

1 Preface

1.1 Necessity of Develop Flexible Production Line

With the development of science and technology, the cycle of product renewal is getting shorter and shorter, and the traditional mass production is being challenged because flexibility and productivity contradict each other. So in order to improve both manufacturing industry's flexibility and productivity, shorten the production cycle

Z. Wan (✉)
DongFeng Honda Automobile Co., LTD, Wuhan, Hubei Province, China
e-mail: wanzhiyang@wdhac.com.cn

China SAE (ed.), *Proceedings of China SAE Congress 2018: Selected Papers*, Lecture Notes in Electrical Engineering 574,
https://doi.org/10.1007/978-981-13-9718-9_16

and reduce the cost of products on the premise of ensuring product quality, flexible automation system came into being [1, 2].

Welding production is one of the most important four processes in automobile production; it guarantees the whole body strength and vehicle's safety performance. For welding process design, the chiefly goal is to guarantee the whole body's quality, and meanwhile to meet the production capacity. Welding production is a complex and professional manufacturing process, facing numerous fixture, welding, and carrier equipment, and at the same time, it must meet changed market demand under the premise of constant total production, which can adjust the production proportion of each car model at will, so implement flexible production process is the key to ensure the flexibility of welding production and stable performance. Flexible production line is broadly defined as permanent production line, when importing a new model, we just need to make appropriate adjustment or transformation of work station and equipment, and then we can manufacture new car model with smaller investment in the shortest period and minimize the investment of equipment.

1.2 Advantages of Flexible Production Line

Flexible production line is a technical complexity and highly automatic system; it combines technology such as microelectronics, computers, system engineering, and so on, which ideally and satisfactorily solved the contradiction between high automation and flexibility in mechanical manufacturing.

Characteristics as follows:

First, high equipment utilization rate. The equipment in the flexible production line has the ability to produce all kinds of car models, its flexibility is higher than single product line, in a sense, which made the equipment universal and greatly improved the utilization rate of equipment.

Second, WIP is reduced. Because of production line's flexibility, each product is produced on a general production line so that the stock number is drastically reduced.

Third, production capacity is relatively stable.

Fourth, high-quality product. In the parts processing, handing complete at once, it had high process accuracy and stable processing.

Fifth, the product has high strain capacity and adjustable production equipment, and reasonable system layout, which is convenient to increase or decrease equipment and meet market needs [3].

2 Project Brief

Flexible design of automatic production line in welding field has always been an important concept in production process planning of DongFeng Honda automobile Co., LTD, and from the first factory's expanding construction to the second factory's planning construction, this idea has been well practiced. Since the second factory begins production, the capacity keeps climbing quickly from 400 cars per day to 530, and then to 800, and until 2015 reached to 1060 cars per day. Each sub-assembly flexible automatic welding production line is the important focus of DongFeng Honda automobile Co., LTD. With the increasing labor cost, production capacity, labor intensity of workers, and quality and process stability faced enormous challenge, so welding department 2 carried out automatic transformation on the manual production line of the rear floor frame/panel sub-assembly of six car models, and gradually achieve the goal of reducing labor cost and improving the body's quality.

3 Design of Flexible Automatic Welding Line

3.1 Layout Design of Rear Floor Automatic Line

The rear floor of the body is mainly composed of the rear frame and the rear panel. The precision and welding strength of the floor directly affect the assembly, safety, stability, and other important functions of the product. This design combined the two parts together and realized a comprehensive and integrated flexible automatic production line through the methods of layout optimization, sharing of electrical system and display of the main console with the same screen. The automatic line can produce at the same time, or the other sideline can be excluded, and the single-line production can be realized, which provide conditions for the standby production to the mass production. The automatic line consists of preinstalled station, welding station, gluing station, stud welding station, and outputting station, etc., which comprises 14 robots, including 10 welding robots, 3 handling robots, and 1 gluing robot. Figure 1 is the layout of CAD design. The welding robot completes spot welding to ensure the body strength and assembly precision. The handling robot is mainly responsible for repair welding and finishing the transfer of parts and the installation of the rear floor assembly station. The gluing robot is mainly responsible for the spot welding structure gluing of the rear frame coordinate position.

3.2 Simulation Design of Rear Floor Frame/Panel Automatic Line

According to the shape and solder spot distribution of six vehicle models, combined with on-site confirmation of production site, the CAD drawing should reflect the overall layout of line body, the number and general position of robots, and the distance of transmission track, etc. Then, 3D simulation is carried out to confirm whether the layout of robot and device is reasonable and relevant dimension data are available. The static layout is simulated by ROB CAD simulation software, as shown in Fig. 2,

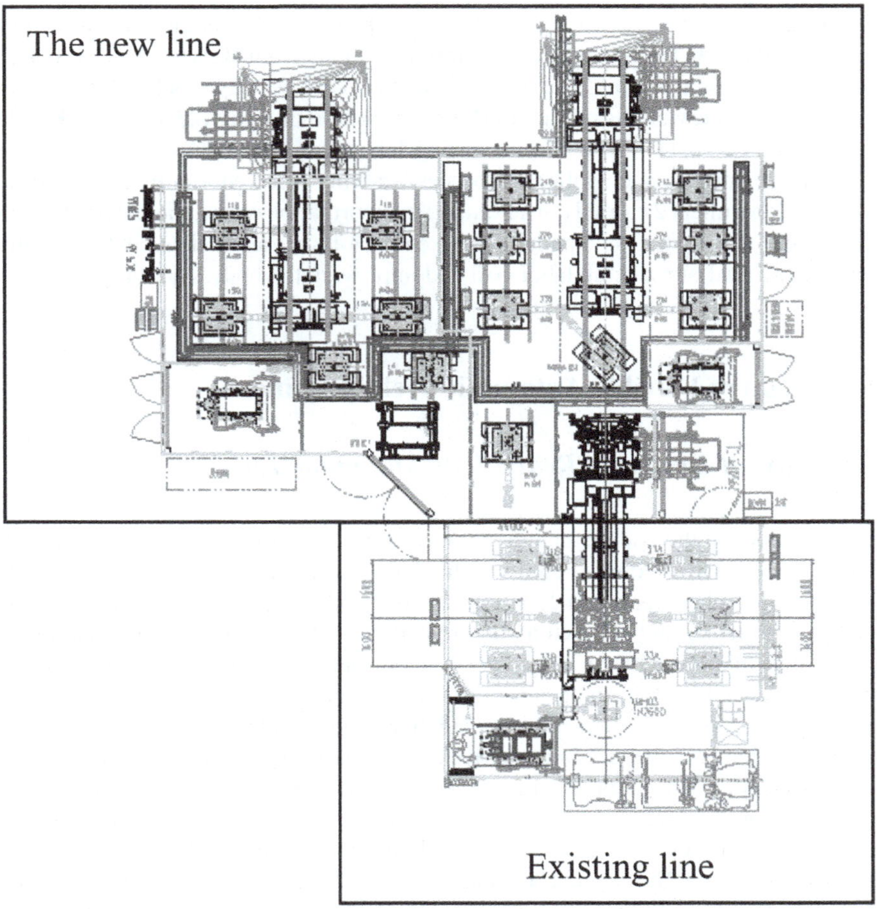

Fig. 1 CAD design drawing of rear floor line layout (The new line)

after off-line simulation and teaching, interlock and interference are set up to allow the robot to move in accordance with the welding and gluing trajectories, and by simulating the same trajectory as the actual production, the rationality of the layout and the selection of robot equipment is finally determined. Meanwhile, the cycle time is simulated, which need to be within the setting range. After this step, the elaboration design stage can be entered.

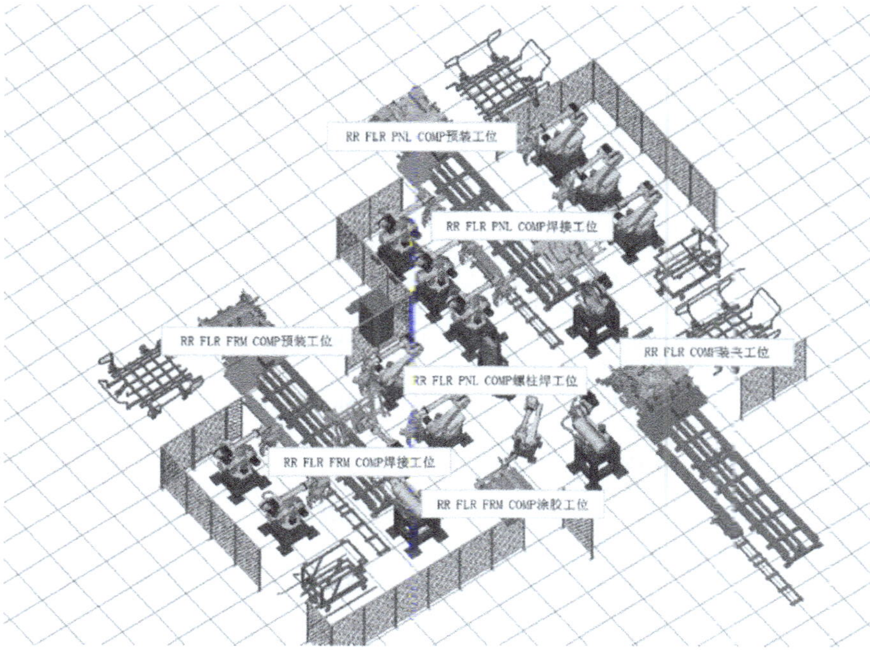

Fig. 2 Simulation design drawing of rear floor line

3.3 *Robot Simulation Selection*

The automatic line of the rear floor frame/panel is composed of preinstalled station, welding station, gluing station, stud welding station, outputting station, etc. Through simulation, 14 FANUC robots shall be set up for the line, including 10 welding robots, 3 handling robots, and 1 gluing robot. The model selection of robot is mainly based on the accessibility and the load of robot.

After the simulation review, the robot selection table is shown in Table 1.

The welding robot's configuration diagram selection is shown in Fig. 3, the handling robot is shown in Fig. 4, and the gluing robot is shown in Fig. 5.

The FANUC LVC function is specially configured on the selected robot, which is used to optimize the robot's motion action, allow the robot to learn the optimized action by itself, and carry out shock reduction and adjustment, reduce redundant trajectory, and improve the comprehensive cycle of the automatic line. The comparison before and after optimization is shown in Fig. 6.

Table 1 Robot selection list

No.	Robot No.	Robot model	ATC	Welding gun	Type	Load (kg)
1	21A	FANUC R-2000iC/210F	No	SRTX series	WELD	210
2	21B	FANUC R-2000iC/210F	No	SRTX series	WELD	210
3	22A	FANUC R-2000iC/210F	No	SRTX series	WELD	210
4	22B	FANUC R-2000iC/210F	No	SRTX series	WELD	210
5	23A	FANUC R-2000iC/210F	No	SRTX series	WELD	210
6	23B	FANUC R-2000iC/210F	No	SRTX series	WELD	210
7	MH01A	FANUC R-2000iC/210R	Yes	–	HAND	210
8	11A	FANUC R-2000iC/210F	No	SRTX series	WELD	210
9	11B	FANUC R-2000iC/210F	No	SRTX series	WELD	210
10	12A	FANUC R-2000iC/210F	No	SRTX series	WELD	210
11	12B	FANUC R-2000iC/210F	No	SRTX series	WELD	210
12	MH01B	FANUC R-2000iC/210R	Yes	–	HAND	210
13	MH02	FANUC R-2000iC/210R	Yes	–	HAND	210
14	13C	FANUC M-710iC/50	No	–	GLUE	50

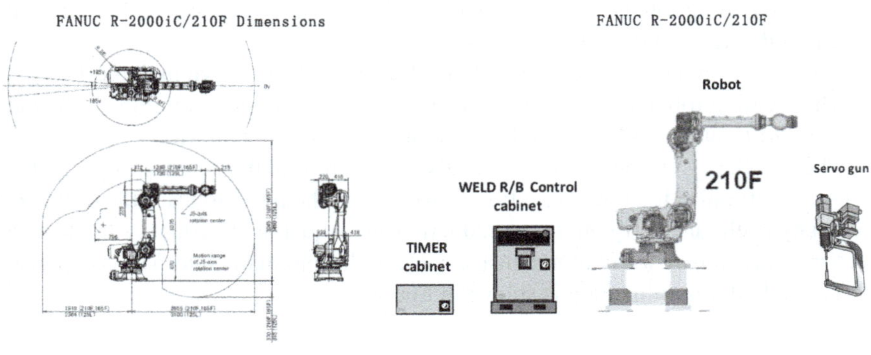

Fig. 3 Selection of the welding robot

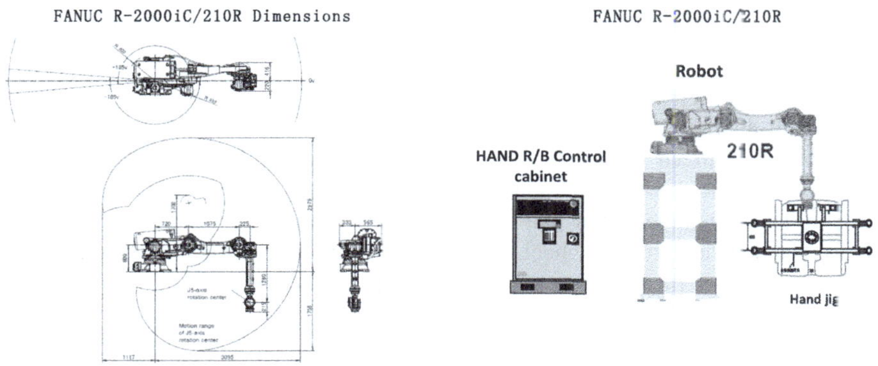

Fig. 4 Selection of the handing robot

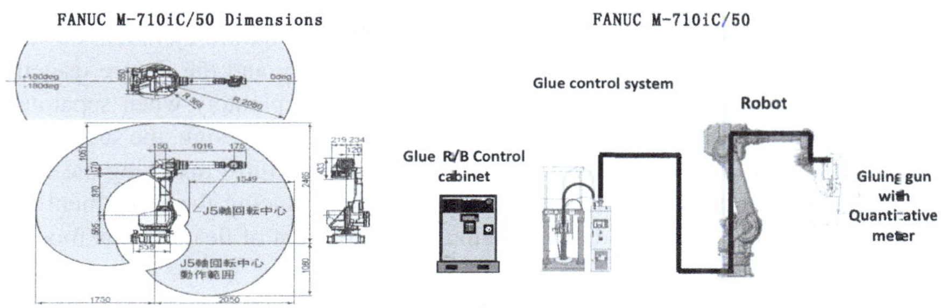

Fig. 5 Selection of the gluing robot

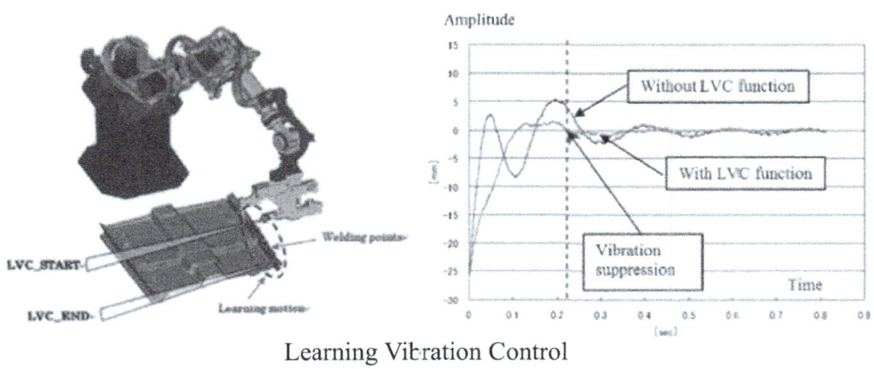

Fig. 6 FANUC LVC functional effect comparison diagram

3.4 Flexible Design of Jig Positioning and MSR Conveying

The jig base design includes two parts: clamping switch and sliding platform car.

The jig changing part is special jig for different models, and flexible integral switching is realized when vehicle model is switched. Each jig includes a special transfer trolley for switch and transfer jig. The special trolley can be connected with the simple guide of the sliding platform trolley and the jig can slide smoothly between them. The dedicated trolley wheel adopts the 150-mm rubber wheel. The two wheels at the push end of the trolley are in a universal form and a brake lock is set which is shown in Fig. 7.

MSR conveyor is a general part which 6 models jig set up on it. MSR conveyor configured with servo motor, reciprocating between the pre-installation station and the welding station, and the jig and the sliding trolley are connected in the form of aviation joints. The actuators of the cylinders on the fixture are all placed in the sliding carriage, and all the air paths are connected to the fixture cylinders through the aviation joints. The combination of the sliding trolley and the fixture uses the PASCAL mechanism, and the base can lift the clamp separately when separated. Because both the MSR and the conveying track are universal design, the subsequent introduction of new models only needs to make special fixtures matching the MSR base, instead of revamping the equipment itself, thus reducing the investment and cycle of new model introduction and realizing the addition of flexible new models. See Fig. 8.

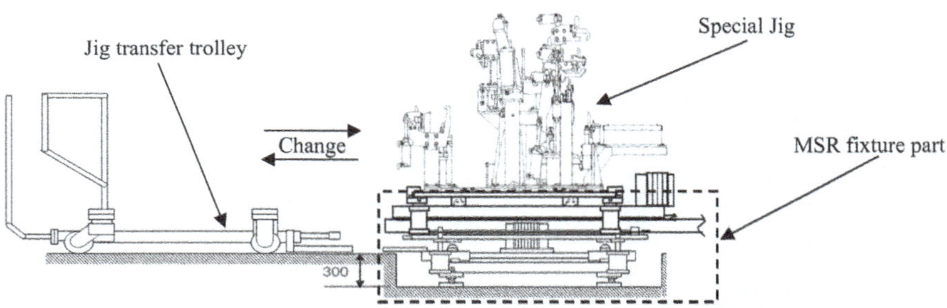

Fig. 7 Design of jig and MSR change

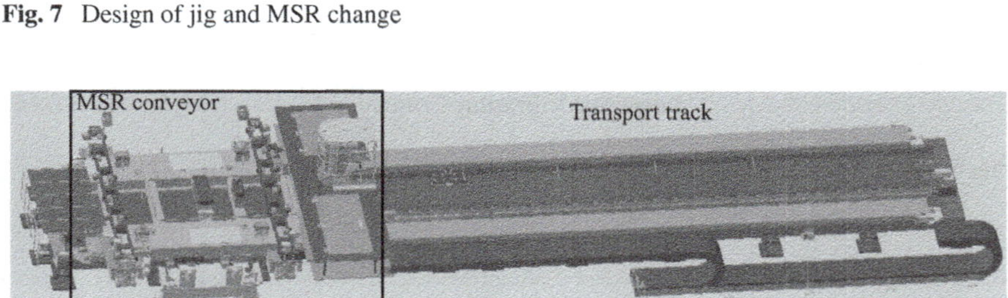

Fig. 8 Rear floor transport track and MSR conveyor design drawing

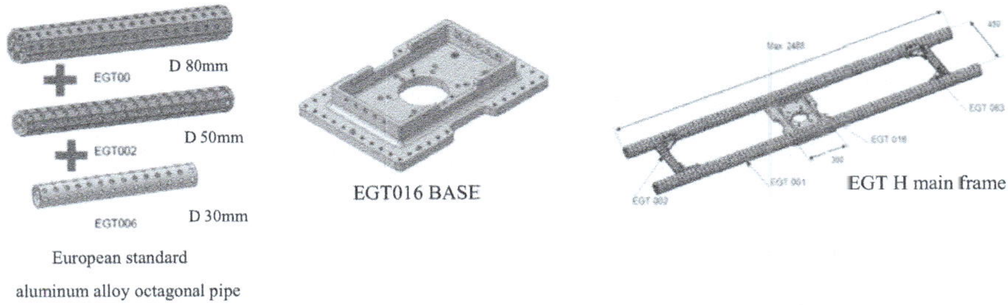

Fig. 9 Lightweight aluminum alloy octagonal pipe material

3.5 *Flexible Design of Gripper*

This design considered special grippers. Each model is equipped with a handing gripper for the handing robot. The purpose of handing gripper is to move the workpiece from the welding station to the gluing station or to the subsequent welding station. Considering that it is a fixed stud welding equipment, the rear panel grippers also play the role of grasping the workpiece for stud welding when targeting the three models of XR-V, GREIZ, and GIENIA. In the design, the technology needs to be taken into full account to meet the positioning function of grasping part welding, the non-deformation of part, the guarantee of welding strength and quality, and the non-falling position and non-deviation of part when welding abnormality occurs in the process of welding. It is the first time to introduce aluminum alloy octagonal pipe structure in the company (as shown in Fig. 9), which is used to make the body frame of the gripper. Compared with the original square steel pipe frame structure, the same kind of gripper can reduce weight by about 25% on average, which effectively reduced the load of the robot and increased the durability of the robot. In addition, the aluminum alloy octagonal pipe structure is composed of different octagonal pipe and some round pipe profiles with the European standard, which is convenient for equipment engineer to prepare standard spare parts. All the structural connections are connected by the socket nut without welding. When the octagonal gripper collide and deform, the equipment engineer only needs to replace the original damaged octagonal pipe and replace it with the new standard octagonal pipe, then quick recovery is complete, which provide a strong guarantee for efficient production. And, octagonal pipe itself has better strength and deflection.

3.6 *General Design of Welding Gun*

Selection of robot welding gun: the main part of the welding gun arm is made of lightweight and high-strength alloy, and the hollow-type lightweight servo gun is adopted, which is matched with FANUC standard servo motors to realize the application of lightweight welding gun. The form and size of the welding gun repeat-

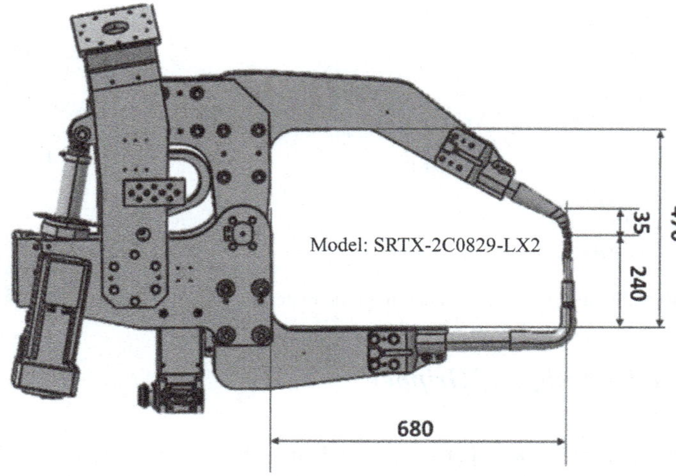

Fig. 10 Design drawing of welding gun simulation model selection

edly consider the clamping fixture positioning requirements and gun gesture. After repeated simulation, the technological accessibility of each welding spot of six models has been verified in detail.

During the process of review and design of servo welding gun, taking into account that the structural differences between rear frame and rear panel of all Honda models are small, the rear frame line and rear panel line are configured four and six general X-type servo welding gun which can completely meet the demand of all models, and fully realize the general production of existing models and future models, as shown in Fig. 10. The grinding device matching with the welding gun is an all-in-one machine that can automatically replace the electrodes and the caps, which can automatically replace the upper and lower electrode and grinding, reduce the time for the operator to change the electrodes and improve the utilization rate of the equipment. The all-in-one machine is equipped with high carbon steel blade and tool holder, which can be used for dressing and grinding various materials of electrodes, such as chromium zirconium copper and alumina copper. Figure 11 is the illustration of the automatic cap changer all-in-one machine.

3.7 Type Selection Design of Stud Welding Equipment

This project is the first time for our company to introduce the automation of stud welding into the automatic line, mainly for the three models of XR-V, GREIZ, and GIENIA. First of all, there are two forms of single-side stud welding and double-side stud welding, each of which has advantages. According to the comparative analysis of workpiece, site and quality stability, etc., as shown in Table 2, the "single-side

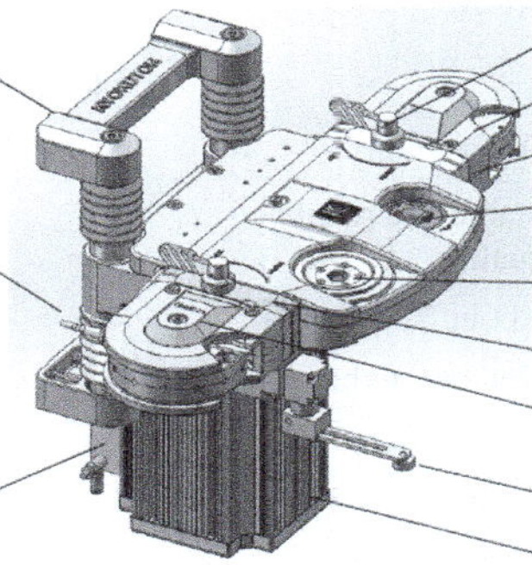

Fig. 11 Description of automatic cap changing and electrode cap grinding machine

Table 2 Comparison between "single-side stud welding" and "double-side stud welding"

Comparison item	Score			
	Single-side stud welding		Double-side stud welding	
	Score	Description	Score	Description
Quality reliability	5	Stable and reliable quality	5	Stable and reliable quality
Time	4	About 3 s/spot	2	About 4.5 s/spot
Failure rate	3	Not very easy to happen	4	Not very easy to happen
Area covered	4	1.5–2 m^2	2	3–4 m^2
Investment	4	High cost performance	2	Low cost performance
Spare parts cost	4	More reasonable	2	high cost
Workability of the workpiece	5	Good	2	Bad
After-sale service	4	Fast response	1	Basically no service
Total	33		20	

welding" mode had been chosen. Secondly, there are two welding methods in the construction of stud-welding equipment. One is the robot with stud-welding gun, and the workpiece is fixed on the jig, which is welded by the robot with gun, that is "movable welding gun" type. The other way is that the stud welding gun is fixed in a certain position, and the robot grabs the work piece for welding, that is "fixed welding gun" type. After the analysis of the workpiece and equipment, as shown in Table 3, we decided to adopt the "fixed welding gun" type for the design. Figure 12 shows the configuration of the "fixed welding gun" stud-welding machine. The welding time of each screw column takes 3.2 s, the failure rate of the equipment is less than 6 parts per million, and the equipment covers only 1.2 m^2. Compared with manual stud welding, automatic stud welding is more efficient and stable.

Table 3 Comparison between "fixed gun" and "movable gun"

Comparison item	Score			
	Fixed gun		Movable gun	
	Score	Description	Score	Description
Quality reliability	5	Stable and reliable quality	5	Stable and reliable quality
Time	4	About 3.5 s/spot	4	About 4 s/spot
Failure rate	4	Not easy to happen	4	Not easy to happen
Area covered	4	About 1.5 m^2	2	About 2.7 m^2
Investment	4	High cost performance	1	More expensive
Spare parts cost	4	More reasonable	3	More spare parts
Workability of the workpiece	5	Good	5	Good
After-sale service	4	Fast response	4	Fast response
Total	34		28	

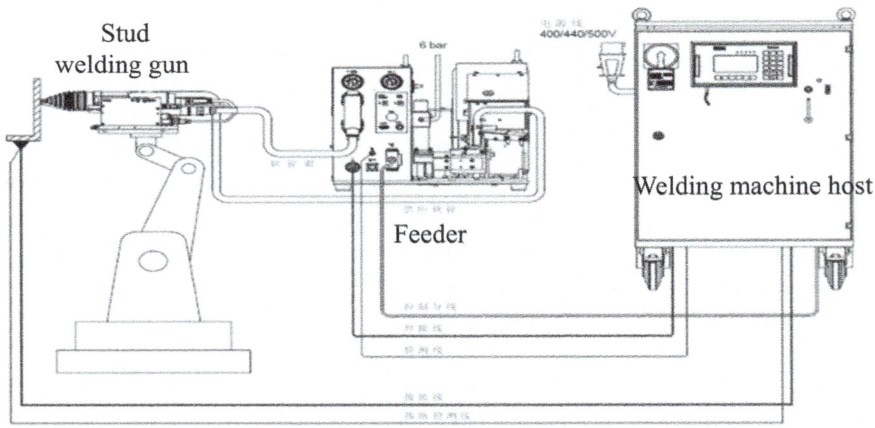

Fig. 12 Configuration of the "fixed welding gun" stud-welding machine (TUCKER DCE1800)

3.8 Selection of Gluing Equipment and Design of Gluing Platform

The gluing equipment adopted robot with glue gun, SCA precision quantitative system and GRACO C68 200L large glue pump. Through the PID temperature controller, the whole process of heat preservation and heating ensured the smoothness of glue output in low temperature environment. See Fig. 13.

When manual line is used, the rear frame is manually glued. Due to the existence of artificial deviation factor in manual gluing, sometimes glue breaking and glue leakage

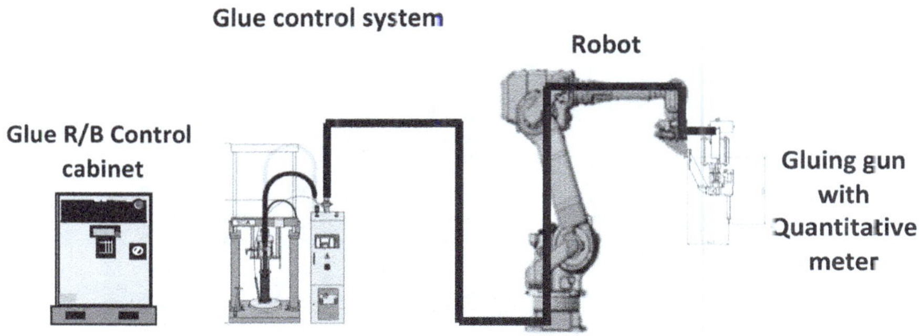

Fig. 13 Gluing system configuration diagram

will lead to unqualified gluing, which will lead to subsequent leakage, abnormal sound, and other quality problems. This project is designed for the automatic gluing system of the all kinds of vehicle rear frame, whose repetitive positioning accuracy reached 0.8 mm, and equipped with glue—coated vision system. At the same time of robot glue coating, real-time monitoring and judgment will be carried out. If there were quality problems such as glue leakage, glue breaking, glue diameter deviation, alarm would appear, then the operator will perform manual processing as shown in Fig. 14.

Since all models use the same solder glue of S412-H, only one glue pump device is required to solve the problem, but the difference is only to apply the glue at different positions of each type of vehicle frame. This time we considered to equip glue-coating robot with glue gun, the work piece is placed on the jig of the universal gluing platform, different models with different supports and clamps, the supports and clamps all adopted flexible plug- and pull form. When it is necessary to process

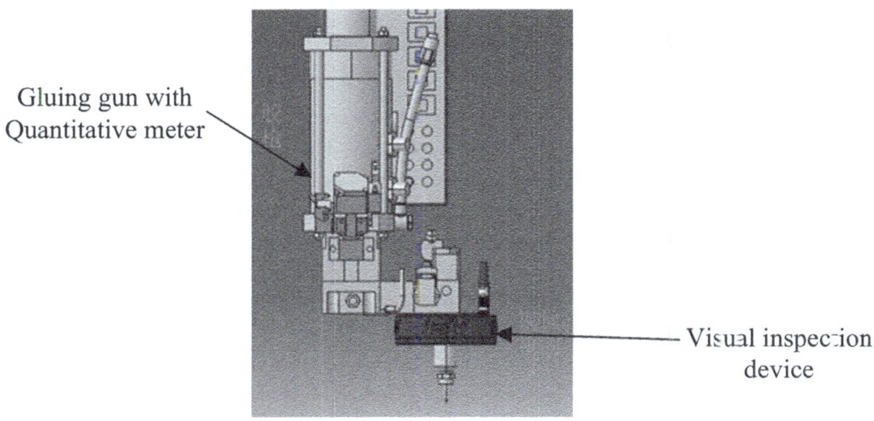

Fig. 14 Glue visual inspection device

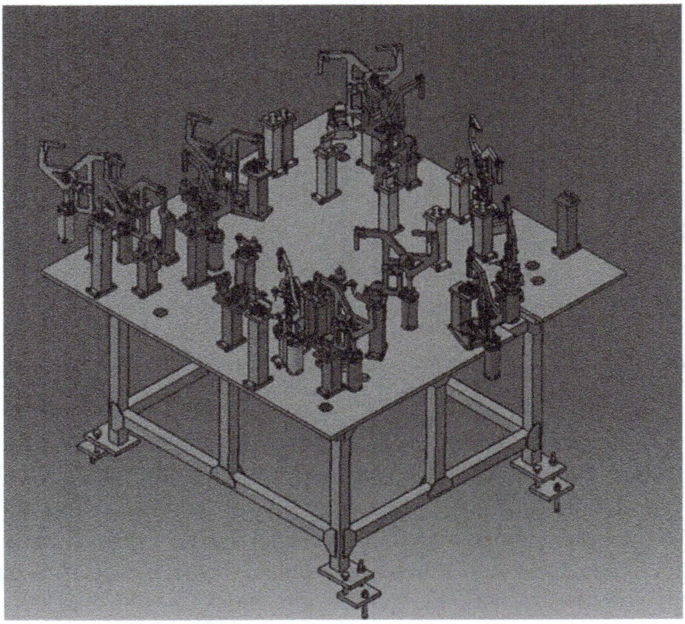

Fig. 15 Design drawing of gluing platform

one car model, its plug and pull support and chuck are installed on the base, and the rest of the car's support and chuck are placed on the adjacent hook. The universal gluing platform and special insert and pull pieces can meet the gluing production of six models as shown in Fig. 15.

3.9 Design of Electrical System

In this project, the original two manual wire bodies were designed intensively, and the two wire bodies will be merged into one wire body. Through the sharing of electrical system, MASTER cabinet/SERVO cabinet, master control cabinet and bridge frame, the purpose of small floor place, brief electrical equipment, simple operation, and compact layout will be realized. According to the use function and operation steps, and referring to the electrical control system diagram of past achievements, we drew the control system diagram of Fig. 16, the network system diagram (Fig. 17), and the servo control schematic diagram (Fig. 18).

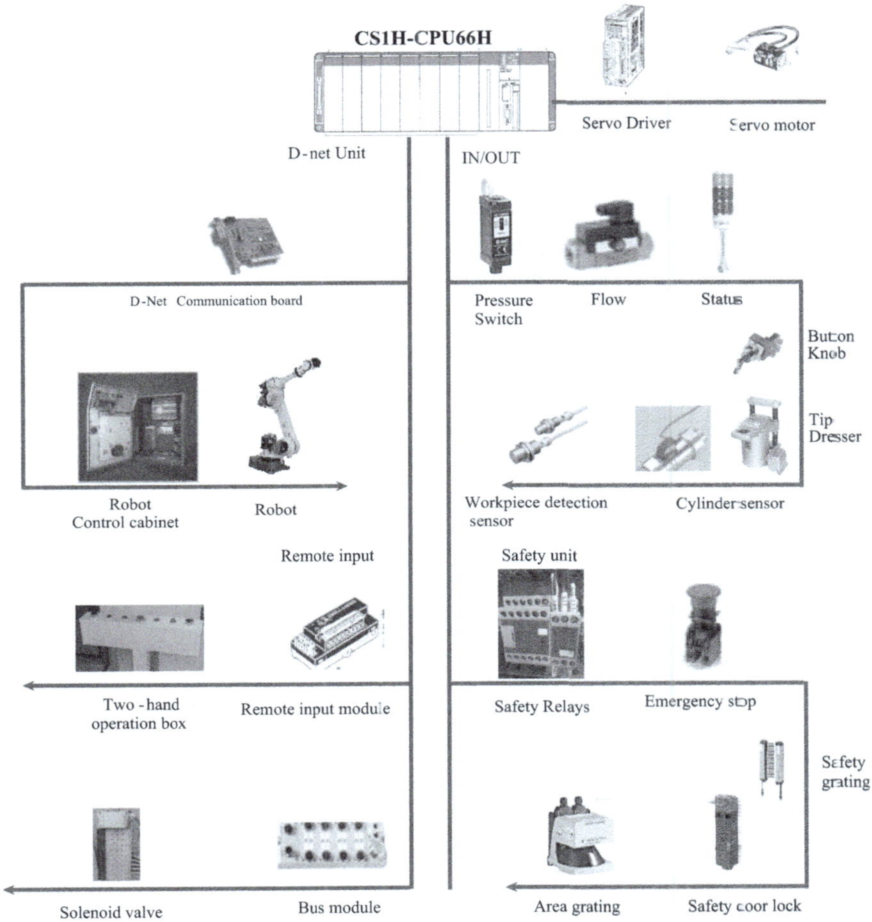

Fig. 16 Control system configuration diagram

4 Conclusion and Application of the Actual Production

The project entered into mass production in April 2017. Since mass production, the comprehensive operation rate of equipment had reached to 87.6%, and the comprehensive quality achievement rate is 98.2%. We achieved all the design and application objectives of this project and following effects: (1) Cutting down 22 operators, which saved 2484.24 thousand RMB per year for the company. (2) After automation, the labor intensity of operator is greatly reduced, and the comprehensive labor intensity is reduced by about 35%. (3) This automatic line not only had high utilization rate of equipment but also saved cost for company, and the comprehensive utilization rate of equipment was 87.49%. (4) The welding automation rate is improved. Welding

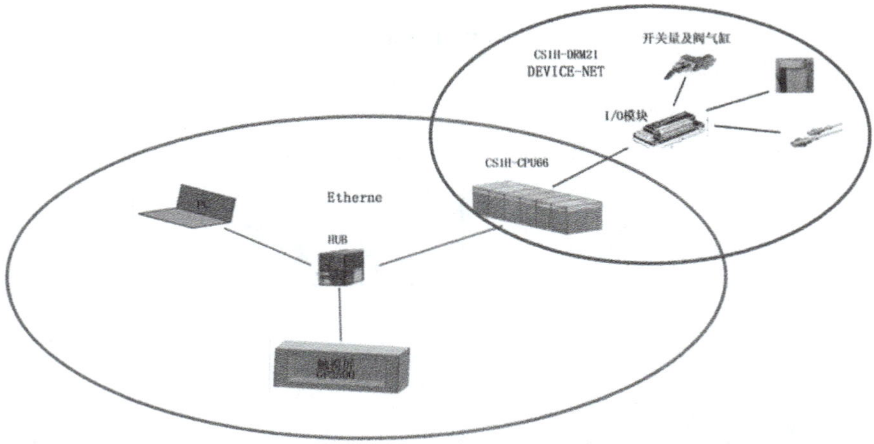

Fig. 17 Network system composition diagram

Fig. 18 Servo control schematic diagram

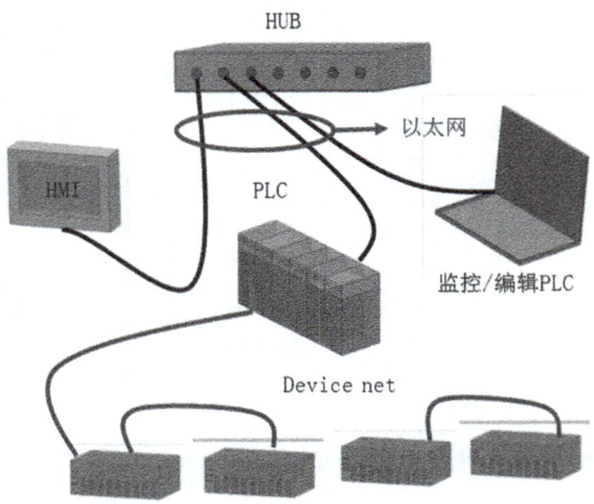

spot automation rate of A-zone increased by 11.96%, since which A-zone implement 100% welding spot automation. The automation rate of welding spots was increased 4.25% to 81.32%, and the comprehensive labor efficiency was further improved. (5) The flexible production of 6 + N (N is the follow-up new model) models is realized, which is convenient for the subsequent introduction of new models.

5 Epilogue and Future Prospect

China's car market is becoming increasingly competitive, about how to grasp the market dynamics for automobile enterprises, get a head start on the market and adjust production to fit market, the flexibility of production is particularly important. Automobile production is a complex and continuous system process, and flexible production mode emphasizes the flexibility and integrity of production system. So when importing a device, the current model and subsequent model should be fully considered, and even the corresponding method when the process of vehicle model is updated. The design of the automatic flexible welding production line should reflect the concept of one-time planning, step-by-step implementation, and rapid development in accordance with the market In each specific application of production flexibility, the rational design planning should be developed according to the process characteristics of each enterprise and the prospect of the industry, so as to give full play to the role of production flexibility.

References

1. Wei J (2009) Flexible manufacturing system (Chap. 10). Chemical Industry Press, Beijing
2. Wang X, Zhang P (2008) Efficient automatic processing and manufacturing system (Chap. 7). National Defense Industry Press, Beijing
3. Ma J (2012) Application of flexible production line in automobile manufacturing. East China Sci Technol 4:25–26

White Body Flexible Welding Line Development and Application

YongSheng Fu and Xiang Li

Abstract Dongfeng Renault's multi-type welding line project in the Chinese city of wuhan will be summarized in this article. And we will focus on the new process scheme, process layout and its program features applied in the Dongfeng Renault project and the realization of multi-variety model mixed-flow production. A variety of solutions was introduced, and the features of each line in the project were explained. It will provide a high reference value for the development of welding lines for multi-variety models of similar welding equipment projects in the future.

Keywords Overall scheme · Vehicle switching · Mixed-flow production

1 Introduction

Foreword

With the global economic and technological integration process, it has become a trend to strengthen capital and capacity cooperation among key auto companies in the global automotive industry and improve resource utilization efficiency. The establishment of Dongfeng Renault Automotive Co., Ltd. (DRAC) has enabled Renault to enter the world's largest automotive market through the localization of joint ventures. DRAC plan's production capacity is based on the annual production capacity of 300,000 complete vehicles. The first phase will achieve an annual production capacity of 150,000 vehicles.

Dongfeng Machine Plant Co., Ltd. has strong R & D and design strength, strong processing and manufacturing capabilities, superb installation and commissioning level, won the trust of the group and customers and assumed the heavy responsibility of the overall planning and construction of DRAC Automotive welding production line.

Y. Fu (✉) · X. Li
Dongfeng Machine Tool Plant Co., Ltd., Shiyan, China
e-mail: fuyongsheng@dfl.com.cn

China SAE (ed.), *Proceedings of China SAE Congress 2018: Selected Papers*, Lecture Notes in Electrical Engineering 574,
https://doi.org/10.1007/978-981-13-9718-9_17

2 Overview

2.1 Introduction

Dongfeng Renault welding workshop adopted a flexible welding production line adapted to the production of multiple models. The whole welding line can produce four types of models, including 133 robots. The production line is divided into engine compartment line, floor main line, body main line, body side line, left and right front door line, left and right back door line, engine cover/back door line and sub-assembly line of each area. In the early stage, HZH and HZJ models should be put into production.

2.2 Technical Requirements

2.2.1 Production Beat

Double shift production

Meet HZH, HZJ and future production of two models, a total of four models of mixed-flow production.

Meet the 30JPH production capacity, the production cycle is 107 s/vehicle.

2.2.2. It can accurately position the vehicle to meet the process requirements, accuracy requirements and beat requirements.
2.2.3. Reserve the vehicle switching space to meet the equipment addition of the subsequent models.
2.2.4. Adopting the form of roller bed conveying, the conveying speed is high and reliable.
2.2.5. The control system meets the requirements of the Renault standard and is safe, stable and efficient.

3 Overall Plan

3.1 Overall Program Development

3.1.1 Layout and Process

Firstly, according to the components, the whole welding line is divided into engine compartment line, floor main line, body main line, body side line, front floor line, rear floor line, rolling hem line and corresponding sub-assembly fixture.

Secondly, according to the process requirements of automation and parts, and the welding difficulty, the engine compartment line, floor main line, body main line, body side line and rolling hem line are designed as automatic lines. The front floor line, the real floor line, the opening fixture, sub-installed jig are artificial lines.

Finally, according to the plant area of the workshop, adjust the layout of the line and arrange each assembly and sub-assembly line reasonably. The floor line and the body line adopt a circular layout. In the form of trolley transport, a belt transport trolley is used on the longitudinal side, and a rotating roller bed transport trolley is used on the other side, and a body is arranged between the two rotating roller beds. Cover the wire and the welding station to make reasonable use of the space.

The EC assembly is transported to the floor line through the storage. The front floor and rear floor assembly are transported to the floor line by the transfer trolley and manually placed on the floor line station by the gripper. The body side is manually transported by the transfer trolley to the side of the body main line and pre-assembled by the gripper. The top cover beam is pre-assembled after being manually glued The roof is grabbed by the robot and placed on the body main line, and the robot performs welding. After the body assembly is welded, the body assembly is moved from the body main line to the AGV trolley of the adjustment line by the transfer machine.

In general, the welding shop carries out the overall process layout according to the principle of top-down, parts and sub-assembly gradual upgrading. The parts are transferred between the production lines by elevators, and the assembly is transferred by AGV transmission mode. People are automatically transported as shown in Fig. 1.

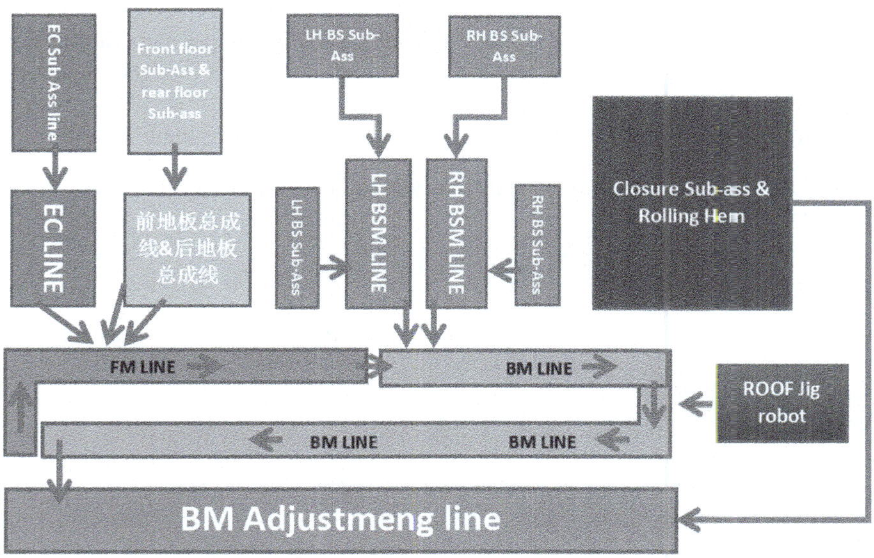

Fig. 1 Overall process plan layout

3.1.2 Features of the Program

3.1.2.1. According to the investment plan and meeting the production tact requirements, the combination of an artificial line and automation line is adopted. The EC sub-assembly, the front and rear floor sub-total and assembly lines, the body side sub-assembly line, the opening part sub-assembly jig, the roof assembly jig, etc., are in the form of manual welding, the EC assembly line and the BSM assembly line. The floor main line and the body main line use an automated line for robot welding.

3.1.2.2. In the overall layout, centering on the floor main line and the body main line, each sub-assembly line is on one side of the main line, and the adjustment line is on the other side of the main line. After each sub-assembly is welded, it is transported to the main line through various forms of logistics. With this layout, the logistics distance of each sub-assembly to the main line is short, the transit time is short, and fewer transfer devices are required.

3.1.2.3. On the basis of the original endless conveying roller bed, innovation was carried out, and the longitudinal conveying on the line body was carried out at one end. The conventional longitudinal conveying was mostly belt-type conveying, which satisfied the storage needs of the trolley. At the other end, a rotating roller bed is used, and between the two rotating roller beds, a top cover welding station is added, which effectively utilizes the line space.

3.1.2.4. In the body main forming welding station, the front station pre-assembled body side, the top cover beam and the rear wall plate are used, and the welding process is performed at the forming station. The body forming station is operated in the form of a robot changeable OG, and four OGs are shared to position the left and right front, rear and upper parts of the body. Moreover, OG can be replaced according to different models to meet the production of multiple models.

3.1.2.5. Comparison between Renault welding shop and other manufacturers.

The Renault welding shop is compared with a factory welding shop as shown in Table 1.

Table 1 Comparison table

	Tact time (JPH)	Occupied area (m^2)	Internal welding points	POE	Automation rate (%)
Phase I project of other factory	30	25,200	3088	2142	40
Renault Wuhan Plant Phase I Project	30	23,250	3520	1830	42

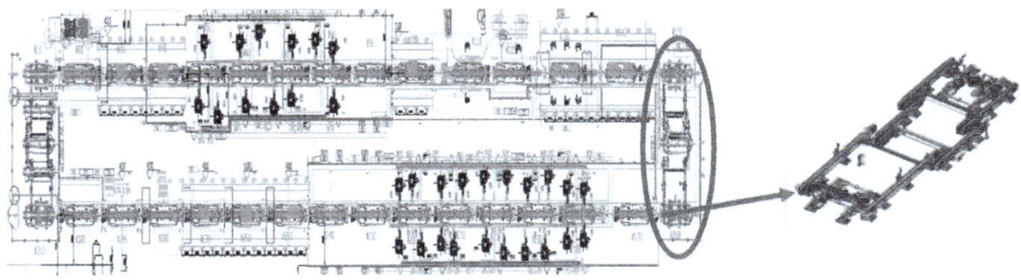

Fig. 2 Traditional roller bed ring line

4 Main Technical Content

4.1 Innovation of the Overall Layout of Each Welding Area

4.1.1 Innovation in the Layout of the Body Mainline

In the arrangement of the ring line of the usual trolley roll, at both ends of the circulation line, the mechanism for longitudinally transporting the trolley is to use a belt or a chain for longitudinal conveyance. Such a circular line, a longitudinally transported station, cannot be arranged for welding. The on-line station of the station or workpiece reduces the effective use of the ring line and also limits the layout of the assembly fixture as shown in Fig. 2.

In the main line of Dongfeng Renault, the main line layout was innovated. Instead of using the belt to transport vertically, a rotating roller bed was developed to increase the overhead line and welding station in the longitudinal direction as shown in Fig. 3.

By adopting an innovative main line layout, the space for longitudinal handling is rationally utilized, the welding station is increased, and the robot island of the roof assembly can be more easily arranged on the side of the body main line, shortening the roof assembly to the body main line.

4.1.2 Innovation of Line Layout for Various Types of EC Lines

Dongfeng Renault's EC line, although also an automated line, has innovated in the form of welding fixtures and panel logistics in the workstation. It adopts the form of annular layout, four-position rotary table and multi-vehicle common fixture combination. Robots are arranged outside the rotary table, responsible for welding and handling. With this layout, there are only two porting robots, and the rest are welding robots, reducing the number of porting robots and grippers as shown in Fig. 4.

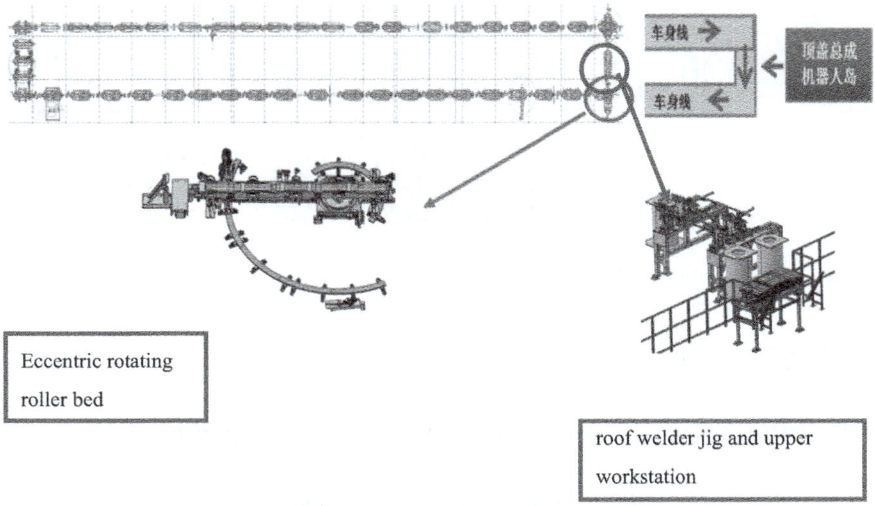

Fig. 3 Renault's roller ring line

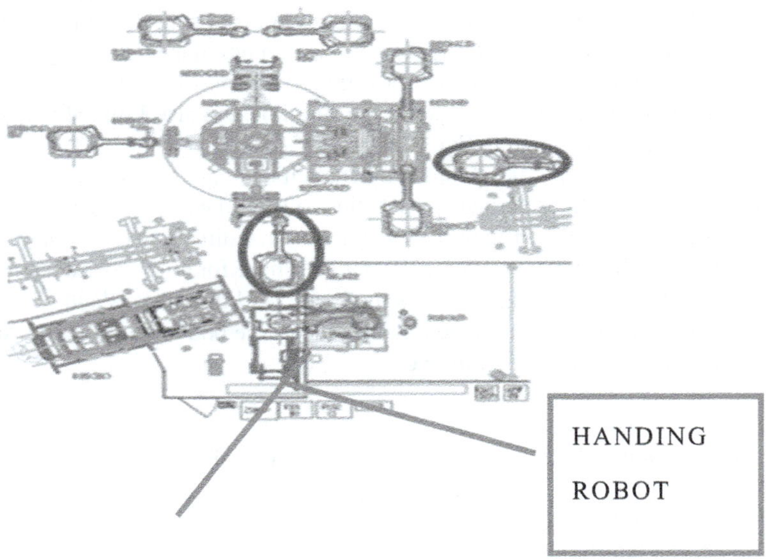

Fig. 4 Renault's EC line

Table 2 Robot usage comparison

Factory	Occupied area (m^2)	Tact time (JPH)	Number of robots	Welding robot	Handling robot	Number of welding points
Renault EC line	205	30	8	6	2	91
Other EC line	244	30	8	4	4	75

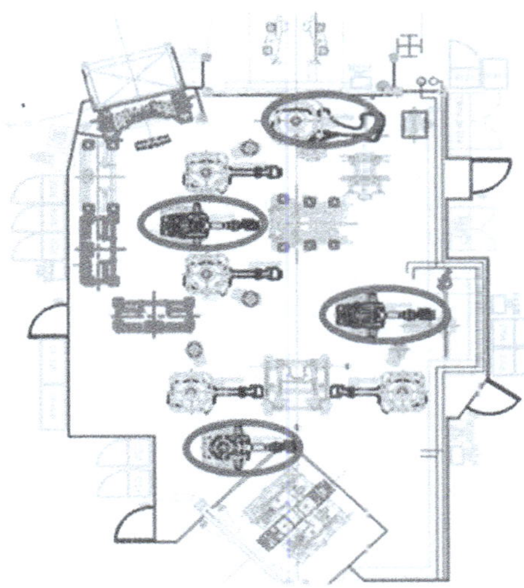

Fig. 5 EC line of other factory layout

The layout of Dongfeng Renault's cabin line has the following advantages over that of other automobile factories:

1. less floor space.
2. the robot has a high utilization rate and few auxiliary robots as shown in Table 2, as shown in Fig. 5.

4.1.3 Innovation of the Layout of the Closure Rolling Hem Line

The closure is divided into three rolling hem islands, which are left/right front door rolling islands, left/right rear door rolling islands, hood and back door rolling islands. In a rolling hem island, there are two rotary tables. Each revolving table can place four models of rolling tire film, the center of each revolving table is arranged with one robot, and between the two revolving tables, two robots are arranged. At work, three edge-rolling robots work at the same time. After the right edge-rolling is completed, two robots will turn to the left and roll along with the robot arranged on the left as shown in Fig. 6.

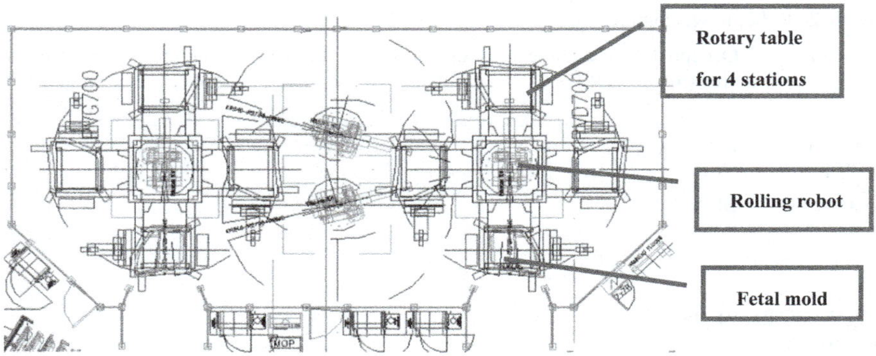

Fig. 6 Layout of Renault rolling island

Comparing with other rolling islands, the robot has the following characteristics as shown in Table 3.

By comparison, Renault's rolling island robot uses less amount, has fewer auxiliary facilities, does not need the tire film storage rack, the grip storage rack and the robot walking shaft, occupies less area, the equipment investment is relatively small. The layout is shown in Fig. 7.

Table 3 Rolling hem comparison

Renault's rolling island	Explain	Other factory rolling islands	Explain
There are two four-position turntables and four models can be placed on each rotating table	No need to replace the die	Two two-position A-type rotary tables with two tires of one type on each turntable can hold the front and rear doors of one type of car. The mold needs to be switched	It is necessary to switch the rolling edge mold, a special membrane switching robot and moving device, and the mold storage rack and the field are also needed
It can be done manually, and each tire mold has a part pressing device	There is a special positioning clamping device on the tire die	The robot and the gripper, as well as the moving device, are needed to place the gripper on the tire mold and clamp the parts to meet the requirements of the upper part	Special positioning gripper is needed, special hand hold storage rack and robot mobile device are needed
A total of four robots	All rolling robots	A total of six robots, four roller robots, two handling robots	The handling robot is a large robot, which has high cost and large investment

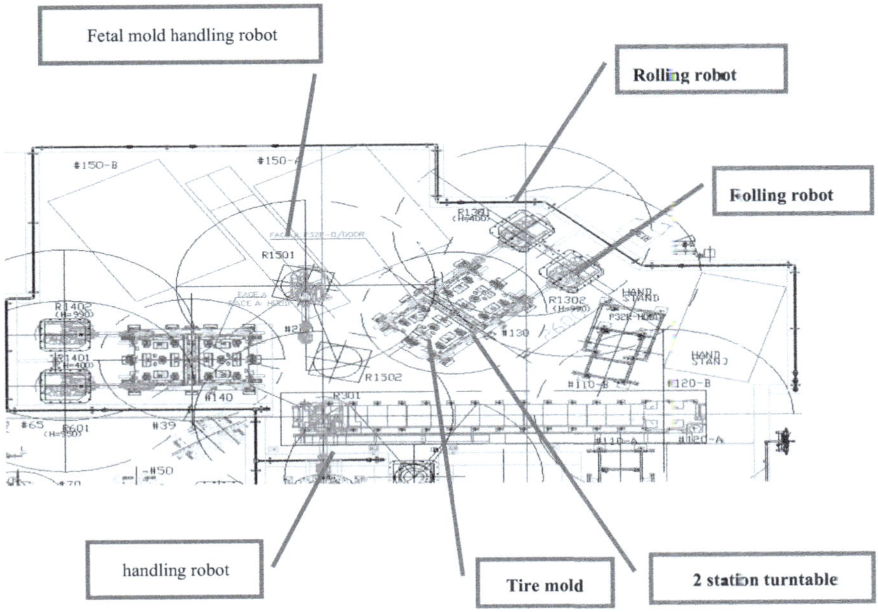

Fig. 7 Other factory rolling hem layout

4.2 *Development of Body Side Line*

4.2.1 Overall Layout of Welding Position for Body Side Line Forming

The body side line forming station should be able to realize the mixed-flow production of four models, and all of them are robot welding, which is fully automatic control. The body side panels of different models in the A-pillar to C-pillar of the product, the shape change is not very large, the difference between the different models in the rear wheel cover and the taillight seat part of the side wall is very large. According to the characteristics of the product, the design of the fixture is adopted, and the front part of the body side is in the form of pneumatic positioning of the vehicle type. At the rear of the BSM complex, a single overall OG positioning mechanism is employed. The positioning OG of the rear part of each side of the vehicle adopts a switching device that combines the translational transmission and the rotation to switch the vehicle to meet the production requirements of different models as shown in Fig. 8.

By adopting the body side front vehicle switch, the rear different vehicle model OG rotation switch, the side assembly fixture structure of five robots welding in the two sides of the fixture and in the air, it provides a new way to meet the multi-vehicle model. High-speed beat solution for body side molding stations.

Fig. 8 Layout of BSM line

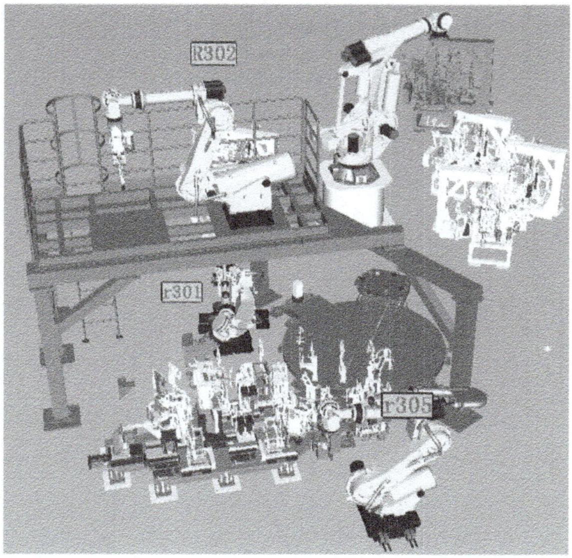

4.2.2 Switching Device Development

In order to use multi-vehicle model positioning, OG positioning mechanism is adopted in the rear of body side circumference. However, for the switching mode of OG, a rotating table switching device is designed, which combines translation transportation and rotation. The device comprises a OG working fixed position, a rotating platform and a translation mechanism as shown in Fig. 9.

At the same time, in the fixed working position and the storage fixed position, the rear side of the two models is placed separately. When the switch is needed, the translation mechanism of the turntable moves the two OGs from the storage table to the middle of the turntable at the same time, and then the turntable rotates. After the

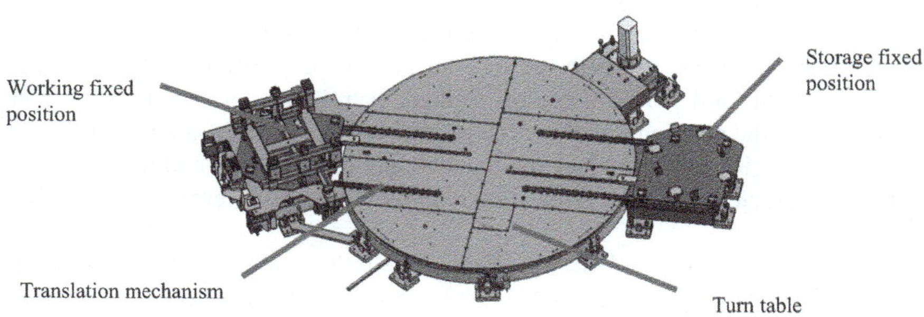

Fig. 9 Overall switching device for OG at the rear of the BSM

turntable rotates 180 degrees, the translation mechanism moves two OGs from the turntable to the working position and the storage position, respectively.

4.2.3 Development of the Connection Structure of the Gas Path and Control Signal of the Switching Device

In the rear part of the side wall, the overall switching positioning OG is adopted. Electrical, pneumatic components and control signals of the detection switch on the OG, we are required when working with the electrical fixture coupling control, and to ensure the operation of the control OG. The air control connection adopts the form of a telescopic pair interface, and the form of the wireless signal connection is adopted in the electronically controlled connection form. The connection and communication of control signals are realized by the wireless communication module of the OG and the fixture fixed table.

Compared with the pin-type docking method, it has the following advantages (as shown in Fig. 10):

1. Easy to install and low installation requirements.
2. The failure rate is low. Through the wireless connection, the pin that is not prone to the pin type is not in place, the contact is poor, and the signal transmission is stable.

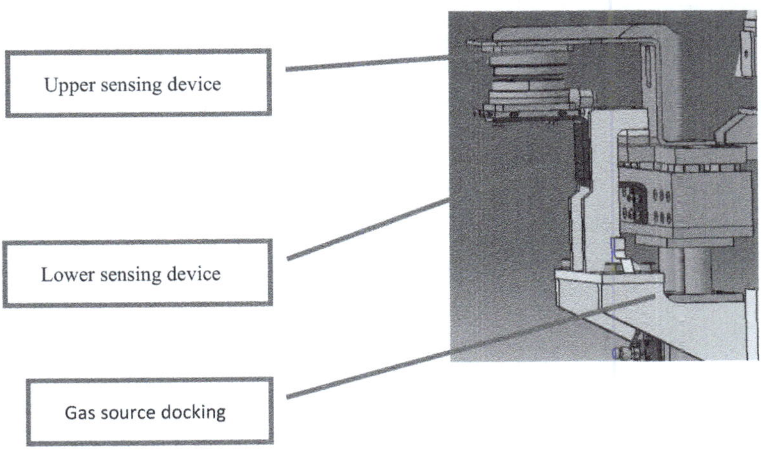

Fig. 10 Connection structure of gas path and control signal

4.3 Development of the Closure Rolling Hem and Fixture

4.3.1 Development of the Edge Cutting Mechanism

Due to the product structure, the flange angle of the workpiece is relatively large, greater than 110°, and it is impossible to achieve the edging by the robot piping. Therefore, according to the demand of the product, a set of piping edge mechanism has been developed. After the hood is positioned, the edge of the rim is extended, and the burr is bent to less than 100° by squeezing and then retracted after completion. Then, the robot performs the piping to complete the hood of the hood as shown in Fig. 11.

Through the development of the edge cutting mechanism, the problem that the robot cannot roll when the flange is at a large angle is solved, and the application range of the robot piping is expanded.

A comparison of the use of a flanged edge with a non-rolled edge is shown in Table 4.

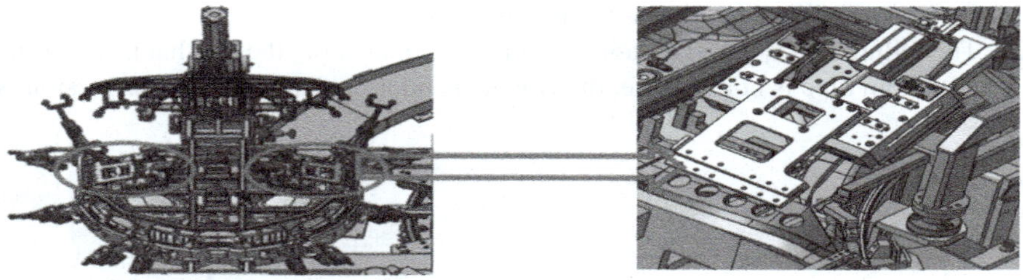

Fig. 11 Application of the edge cutting mechanism on the hood rolling die

Table 4 Comparison

Content	Explain	Feature
Hobbing edge	Can complete all hemming work with robots	The robotic hemming process, which can adapt to the large angle of the part flanging, reduces the investment and does not require additional equipment
No edge cutting	It is necessary to add a special pre-edge jig to bend the flange to less than 100°, and then place the workpiece on the piping film for robot rolling, which increases the station and increases the transfer logistics between the stations	Pre-wrapped fixtures have been added to meet robot piping requirements. Increased inter-station transfer Increased equipment and investment, increased site area

4.3.2 Development of Flexible Switching Device for Fixture Models

The sub-assembly fixtures for the closure are all fixtures for a single model. In order to reduce the investment, a solution for sharing the welding tongs for each type of vehicle fixture is adopted. However, instead of using conventional clamps with wheels, when using any model, the fixture is pushed to the welding station. Instead, a set of a rotating platform fixed to the floor and capable of rotating 270 degrees is used. The vehicle fixture is placed by the forklift. On the rotating platform, the fixture can be controlled by the docking device, the clamp and the control air circuit and the circuit are connected. For four models, only one set of electrical control cabinets and button control system are required. The push-type fixtures are avoided, each with its own control and operating system, saving equipment investment. And the vehicle-type fixture adopts shelf-type storage, which is stored by the forklift and takes up a small area of the factory as shown in Figs. 12 and 13.

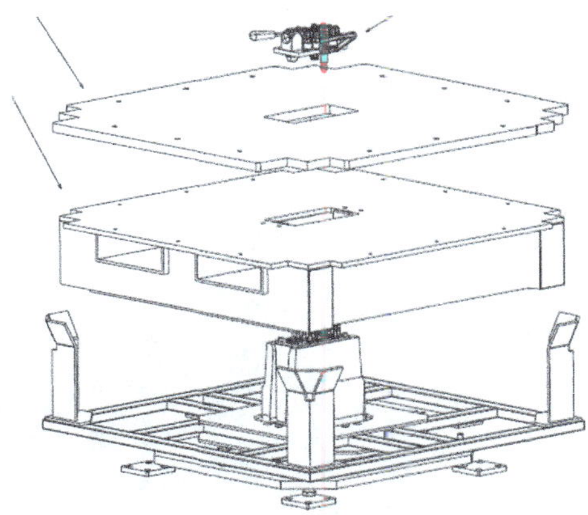

Fig. 12 Ground fixed turntable

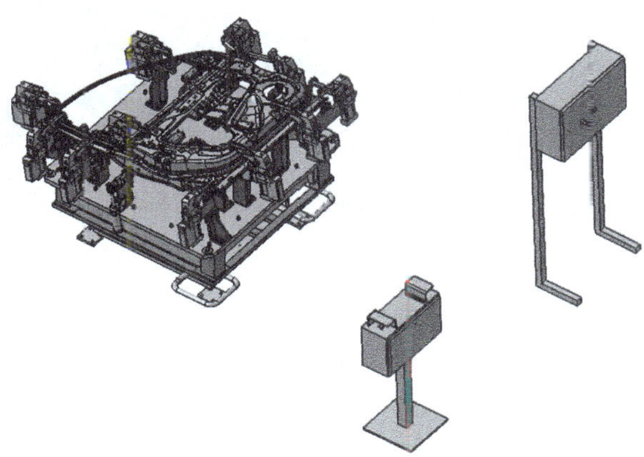

Fig. 13 Complete set of vehicle fixtures

4.4 Research and Development of Body Main Lines

4.4.1 Design of the Body Forming Station

Before the body main assembly is transported to the body forming welding station, the side wall assembly, the roof beam and the rear wall are pre-assembled. After being transported to the body assembly station, the side wall assembly, the roof beam and the rear panel are not grounded, but are positioned by the OG of the robot. The OG is divided into a left/right body side OG, a top cover beam OG and a rear OG. The left/right body side OG mainly locates the left/right front part. The top cover beam OG mainly locates the front cover front beam, strengthens the beam and the side cover upper part and ensures the opening size of the front wind window. The rear OG mainly locates the rear cover cross member, the rear coaming plate and the rear side of the side wall to ensure the opening size of the back door opening as shown in Fig. 14.

By adopting the form of partition block OG positioning, the design of the body assembly fixture is innovated, and the structure of the body assembly fixture is simplified, and the purpose of vehicle switching production is quickly realized.

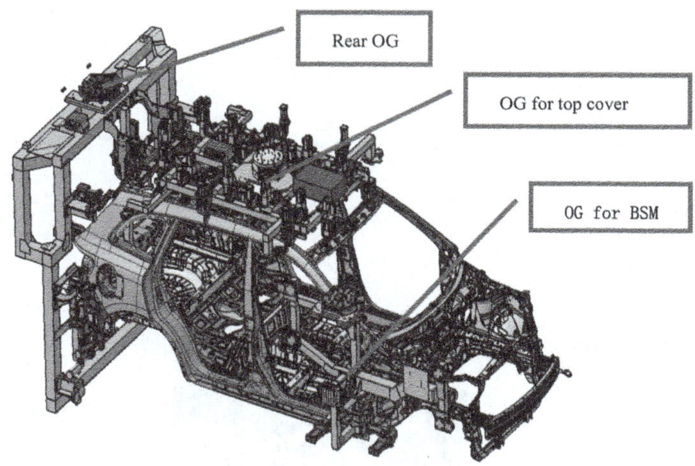

Fig. 14 OG fixture is positioned in the body assembly

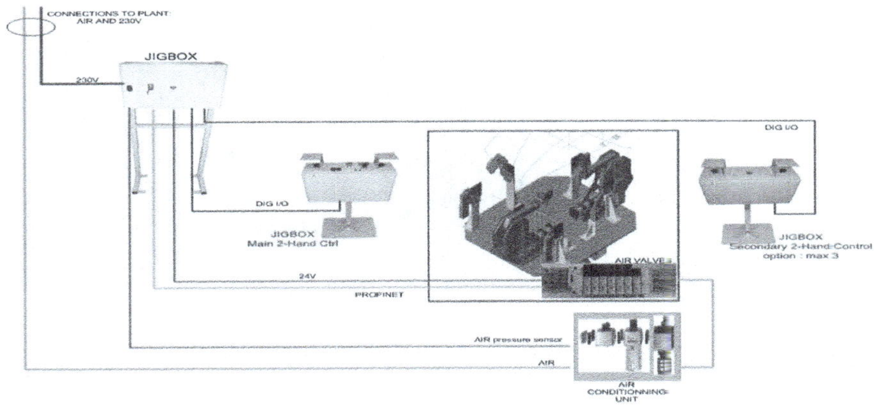

Fig. 15 JIGBOX control system

4.5 *Design of the Control System*

4.5.1 Design of JIGBOX Control System for Sub-assembly Jigs

Renault's sub-assembly A/J fixture, also known as JIGBOX, is an electronically controlled pneumatic fixture system controlled by Siemens PLC. It is manufactured according to Renault standard regardless of network layering or hardware configuration. The main control unit is Siemens ET200S PLC and FESTO valve island, which realizes the solution of electric control gas and gas feedback into electricity. In addition, due to the expandability of the valve island, the valve can also be expanded into a single Siemens ET200S control cabinet. Control multiple sets of simple fixtures individually. When programing, according to the functional requirements of the fixture, only input conditions are required, and it is filled in EXCL, and the control program can be automatically generated. The programing is simple, the work efficiency is improved, and human programing errors do not occur. JIGBOX has the advantages of high safety, easy operation, reliable anti-dead treatment and simple maintenance as shown in Fig. 15.

4.5.2 Development of Renault JIGBOX Dedicated Signal Controller

Renault's electrical requirements require the proximity switch, cylinder-in-position switch and other signals to be connected to the input module of the valve terminal, but the European company's dedicated signal controller is purchased and only one of them provides high procurement costs. In order to reduce the cost, the JIGBOX fixture dedicated signal controller was independently developed The function of four signal controllers of foreign products is realized on one controller, which is equivalent

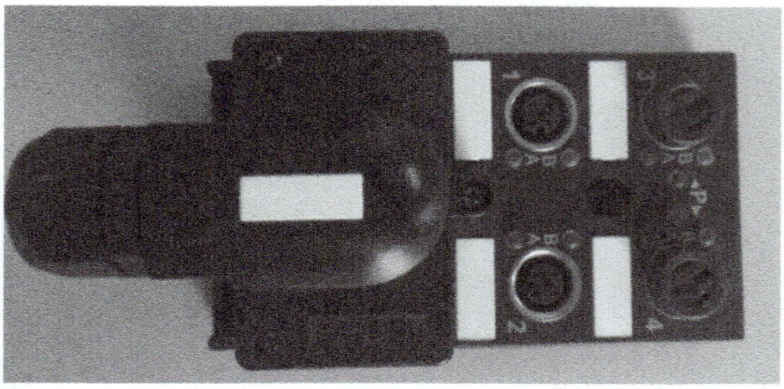

Fig. 16 Self-developed JIGBOX fixture dedicated signal controller

Table 5 Comparison of independent development and foreign manufacturers' products

Supplier	Performance
Self-development	Four signal access functions, adapt to a variety of situations and have strong scalability
Foreign manufacturers	Only one signal access function, single function, no scalability

to focusing the functions of four foreign models on one product, powerful, scalable and inexpensive as shown in Fig. 16 and as shown in Table 5.

4.5.3 SCUBE System for Main Line Control

Dongfeng Renault welding main line is based on the French Renault SCUBE standard welding line. The standard successfully automates, standardizes and modularizes hardware and software design. It adopts powerful Siemens PLC and is compatible with mainstream robots such as KUKA, ABB and FANUC. Solved the problem of design, production and maintenance requirements of the automotive manufacturer's welding shop.

4.5.3.1. Network Shelf

Based on the excellent flexibility and openness of PROFINET, users can arbitrarily design the system architecture of the machine or factory according to actual needs. It includes NETWORK N0—PROFINETIO RT–field device layer, NETWORK N1—industrial Ethernet (HMI man-machine interface, LED car-type display and backup layer), NETWORK N2—connected to the shop floor server as shown in Fig. 17.

4.5.3.2. SCUBE Standard Hardware Design Features—Modular Design SCUBE

Prior to modular standardization, the equipment in the workshop was completely determined by the supplier. Different suppliers may have different standards. The layout of the control unit is also diverse. This will increase the difficulty of learning for maintenance personnel and operators in the workshop management. The advantages of modular design are standardization of design, standardization of wiring,

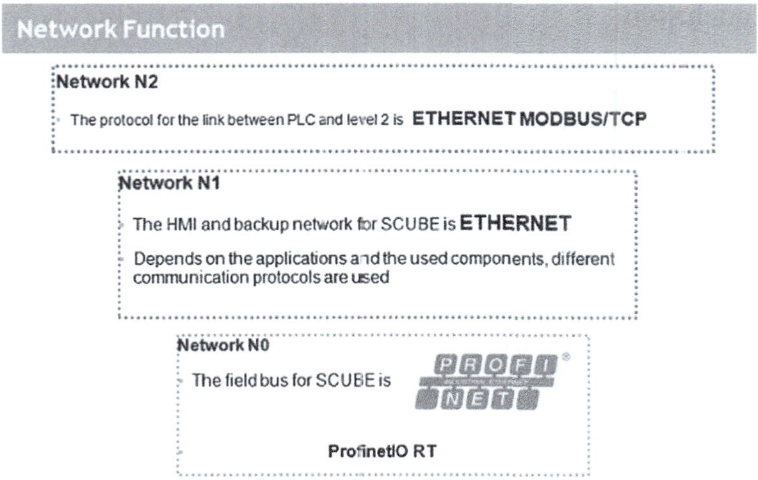

Fig. 17 Network shelf

standardization of maintenance of standardized equipment capacity and modular design to standardize the design and use of welding equipment throughout the shop as shown in Fig. 18.

4.5.3.3. SCUBE man–machine interface and programing features—The human—machine interface uses ODIL software, and ODIL is a powerful software. Realize the design operation of the screen, to manual operation and alarm processing. With special handling, the STEP7 program framework can also be generated, including the HW configuration. ODIL automates design, including interface design automation and automatic generation of STEP7 programs and hardware configurations.

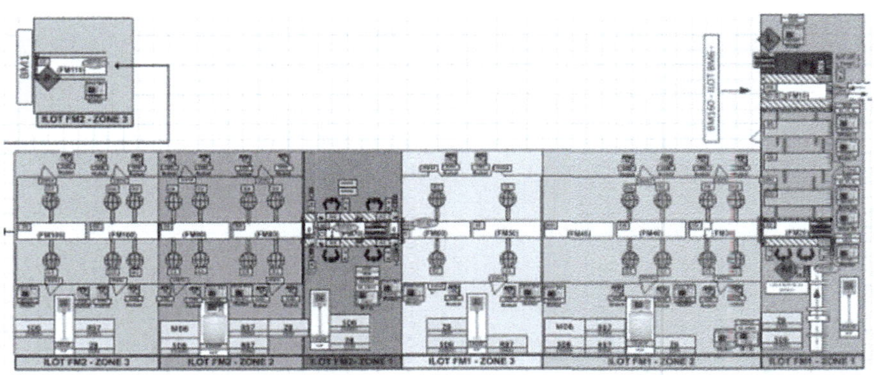

Fig. 18 Modular design schematic

5 Conclusion

This project has developed and integrated a large number of new technologies, such as the body main line, the innovation of the overall layout of the engine compartment line and the rolling hem, the four-station turn table of the engine compartment line and the multi-model OG positioning technology of the body main work station. The rotating platform and jig switching technology and edge cutting edge technology of the body side line jig storage further improve the ability of integrating advanced welding production technology and improve the technical level of welding equipment in an all-round way.

After two years of development , the overall layout of innovation and the new technology equipment has been put into use, which meets the needs of multi-variety mixed-flow production of the white body in Dongfeng Renault welding workshop and meeting the user's requirements for use. In accordance with the relevant standards of Renault Company, it provides a technical accumulation for the development and application of similar wire bodies in the future and lays a solid foundation for the application of intelligent welding workshop.

Probability Density Evolution Method for Vehicle Dynamics Analysis Under Uncertainty

Xiaokai Chen, Hao Lei, Cheng Zhang and Jiahui Liu

Abstract Because of the uncertainties and highly nonlinear characteristics of intelligent vehicles, how to control the motion of intelligent vehicles more accurately and effectively has become the key to the development of intelligent vehicles. In this paper, based on the nonlinearity and parametric uncertainty of the suspension system, the probability density evolution method is used to analyze the variation of the probability density of the suspension dynamic response over time, in order to facilitate more effective motion control. Combining the dynamic random state equation of the suspension and the probability conservation principle in the conservative stochastic system, the generalized probability density evolution equation for vehicle random vibration is established. Considering the nonlinear stiffness of the hydro-pneumatic spring in the suspension system and the uncertainties of the parameters, such as the suspension and the tire, the stochastic vibration response analysis was performed with the sprung-mass acceleration response as an example. The random probability space is discretely selected by the number-theoretic method. Finite-difference method such as Lax-Wendroff scheme and TVD (Total Variation Diminishing) scheme is used to numerically solve the generalized probability density evolution equation, and the variation of the probability density of the response is obtained. The probability density evolution method has good calculation accuracy and can give specific probability distribution information, which lays a foundation for subsequent motion control.

Keywords Suspension nonlinearity · Uncertainty · Probability density evolution method

---

X. Chen (✉) · H. Lei · C. Zhang · J. Liu
Beijing Institute of Technology, Beijing 100081, China
e-mail: chenxiaokai@263.net

China SAE (ed.), *Proceedings of China SAE Congress 2018: Selected Papers*,
Lecture Notes in Electrical Engineering 574,
https://doi.org/10.1007/978-981-13-9718-9_18

1 Introduction

The intelligentization of vehicles is a new direction in the field of vehicle engineering and the development of the future automotive industry. Its motion control is a key technical issue for intelligent vehicles to achieve autonomous driving. Intelligent vehicles are typical multi-input–multi-output complex systems with parameters uncertainty, time lag and high nonlinearity [1]. Therefore, the key to intelligent vehicles' motion control is how to deal with the effects of its inherent random uncertainty and highly nonlinearity on the vehicle motion control system. The random uncertainty of the parameters will inevitably lead to the uncertainty of the system response. The nonlinearity effect of the system can make the response of the Gaussian-excited system significantly deviate from the Gaussian distribution, and this deviation often has a significant impact on precise control of the system [2]. Under the coupling of random uncertainty and nonlinearity, the motion control of intelligent vehicles is difficult to calculate and control precisely.

Stochastic dynamics began in 1905 with the study of Brownian motion by Einstein et al., and the study of nonlinear stochastic dynamics has been developed for more than 50 years [3]. At present, the research on stochastic systems has formed various analysis methods such as Monte Carlo simulation method, equivalent linearization method, polynomial chaotic expansion method, random perturbation method and probability density evolution method. The Monte Carlo simulation (MCS) method is almost regarded as an omnipotent method for dealing with random problems. The simulation of the random field is the core technology of the MCS. The biggest shortcoming of the MCS method is that the amount of calculation is huge, and the calculation results are randomly converged, which makes it difficult to be applied to practical engineering problems. The equivalent linearization method approximates the nonlinear problem to a linear problem [4], but since the response of the nonlinear system is often a non-stationary process, the computational accuracy of the equivalent linearization method tends to be poor. Polynomial Chaos Expansion (PCE) approximates the input and output of an uncertain system into a series of expanded forms of standard random variables [5, 6], but PCE has insufficient accuracy for solving strong nonlinear problems. In the field of structural stochastic response analysis, the stochastic perturbation method uses the Taylor formula to expand the random function to achieve random analysis [7, 8]. At present, the random perturbation method has been used for random static problems with small parameter variation and is not suitable for dynamic problems. The probability density evolution method (PDEM) is derived from the Fokker-Planck-Kolmogorov (FPK) equation in statistical mechanics. There are many numerical solutions to the FPK equation. Zhang [9] uses the finite-difference method and the multigrid method to solve the high-dimensional FPK equation. Huang et al. [10] used the path integration method to solve the FPK equation and applied it to the ship field. However, the FPK equation only considers the evolution law of the probability that a deterministic system responds under random excitation and does not consider the system with parameter uncertainty [11]. The generalized density evolution equation established by the principle of conservation of probability is decoupled, which is proposed by Li and Chen [11–13]. Therefore, the probability density function (PDF) of random response of structures and its evolution law can be directly obtained. The PDEM has good solution efficiency and calculation accuracy and has been widely used in stochastic dynamics of structures.

This paper applies the PDEM to the random vibration problem of vehicles and lays a foundation for subsequent motion control. Taking the suspension system of a car as an example, this paper studies the dynamic response of the stochastic systems with uncertainties and nonlinearities and uses the PDEM to obtain the PDF and its evolution law of the suspension response.

2 The Probability Density Evolution Method

2.1 The Principle of Conservation of Probability

The principle of conservation of probability is a general law of conservative stochastic systems, similar to the principle of conservation of energy. Whether it is the probability density analysis of random functions in probability theory or the derivation of FPK equations, it implies the basic idea of the principle of conservation of probability [2]. Referring to the description of the conservation principle in physics, the principle of conservation of probability can be expressed as: If the random factors involved in a stochastic system are retained, the probability will be preserved in the evolution process of the system [11]. The conservative stochastic system means that during the evolution of the stochastic system, neither the existing random factors disappeared nor new random factors were added.

Consider a general dynamical system with the state equation

$$\dot{\boldsymbol{Y}} = \boldsymbol{A}(\boldsymbol{Y}, t) \quad \boldsymbol{Y}(t_0) = \boldsymbol{Y}_0 \tag{1}$$

where $\boldsymbol{Y} = (Y_1, Y_2, \ldots, Y_n)^{\mathrm{T}}$ is the n-dimension state vector; $\boldsymbol{Y}_0 = (Y_{0,1}, Y_{0,2}, \ldots, Y_{0,n})^{\mathrm{T}}$ is the initial value vector; $\boldsymbol{A}(\cdot)$ is the deterministic operator vector. The state Eq. (1) essentially establishes a mapping from $\boldsymbol{Y}_0$ to $\boldsymbol{Y}(t)$. Since $\boldsymbol{Y}_0$ is a random vector, $\{\boldsymbol{Y}_0 \in \Omega_0\}$ is a random event. Here, Ω_0 is any arbitrary domain in the distribution range of $\boldsymbol{Y}_0$. According to the stochastic state Eq. (1), $\boldsymbol{Y}_0$ will be changed to $\boldsymbol{Y}(t)$ at time t. The domain Ω_0 to which $\boldsymbol{Y}_0$ belongs at time t_0 is accordingly changed to Ω_t to which $\boldsymbol{Y}(t)$ belongs at time t.

Because the probability is preserved in the mapping of any arbitrary element event, we have

$$\int_{\Omega_{t_0}} p_{\boldsymbol{Y}_0}(\boldsymbol{y}, t_0)\mathrm{d}\boldsymbol{y} = \int_{\Omega_t} p_{\boldsymbol{Y}}(\boldsymbol{y}, t)\mathrm{d}\boldsymbol{y} \tag{2}$$

where $p_Y(y, t)$ is the joint probability density function of $Y(t)$. Evidently, the above equation also holds at $t + \Delta t$, which will then result in

$$\frac{D}{Dt}\int_{\Omega_t} p_Y(\boldsymbol{y}, t)\mathrm{d}\boldsymbol{y} = 0 \tag{3}$$

where $D(\cdot)/Dt$ denotes the total derivative. Equation (3) is the embodiment of the principle of conservation of probability in a stochastic dynamical system.

2.2 The Generalized Density Evolution Equation (GDEE)

Consider the general stochastic dynamical system

$$\dot{\boldsymbol{Y}} = \boldsymbol{G}(\boldsymbol{\Theta}, \boldsymbol{Y}, t) \quad \boldsymbol{Y}(t_0) = \boldsymbol{Y}_0 \tag{4}$$

where $\boldsymbol{Y} = (Y_1, Y_2, \ldots, Y_m)^{\mathrm{T}}$ is the state vector; m is the dimension of the system; $\boldsymbol{Y}_0$ is the initial value vector; and $\boldsymbol{\Theta} = (\Theta_1, \Theta_2, \ldots, \Theta_s)^{\mathrm{T}}$ is an s-dimensional random vector characterizing the randomness involved with known joint PDF $p_{\boldsymbol{\Theta}}(\theta)$.

In general, most dynamic problems in engineering are well-posed problems. The solution of Eq. (4) exists and is unique, and the solution's behavior changes continuously with the random parameter vector $\boldsymbol{\Theta}$. Let $\mathbf{Z} = (Z_1, Z_2, \ldots, Z_m)^{\mathrm{T}}$ be the solutions of Eq. (4), then $\mathbf{Z}$ can be expressed as:

$$\boldsymbol{Z}(t) = \boldsymbol{H}(\boldsymbol{\Theta}, t) \quad \dot{\boldsymbol{Z}}(t) = \boldsymbol{h}(\boldsymbol{\Theta}, t) \tag{5}$$

Then the augmented system ($\mathbf{Z}(t)$; $\boldsymbol{\Theta}$) is probability preserved because all the random factors are involved, thus according to Eq. (5) we have

$$\frac{D}{Dt}\int_{\Omega_t \times \Omega_\theta} p_{z\Theta}(z, \boldsymbol{\theta}, t)\mathrm{d}z\mathrm{d}\boldsymbol{\theta} = 0 \tag{6}$$

where $p_{z_\Theta}(z, \theta, t)$ is the joint probability density function (PDF) of ($\mathbf{Z}(t)$; $\boldsymbol{\Theta}$). $\Omega_t \times \Omega_\theta$ is the domain in the augmented state space $\Omega_{\boldsymbol{Z}} \times \Omega_{\boldsymbol{\Theta}}$ corresponding to the $\Omega_0 \times \Omega_\theta$ at time t, where $\Omega_0 \times \Omega_\theta$ is any arbitrary domain in the initial space $\Omega_{z_0} \times \Omega_{\boldsymbol{\Theta}}$. After a series of mathematical manipulations from Eq. (6), we have

$$\int_{\Omega_t \times \Omega_\theta} \left(\frac{\partial p_{z\Theta}(z, \boldsymbol{\theta}, t)}{\partial t} + \sum_{j=1}^{m} \dot{z}_j(\boldsymbol{\theta}, t) \frac{\partial p_{z\Theta}(z, \boldsymbol{\theta}, t)}{\partial z_j} \right) \mathrm{d}z\mathrm{d}\boldsymbol{\theta} = 0 \tag{7}$$

Because of the arbitrariness of $\Omega_t \times \Omega_\theta$, we have

$$\frac{\partial p_{Z\Theta}(z, \boldsymbol{\theta}, t)}{\partial t} + \sum_{j=1}^{m} \dot{Z}_j(\boldsymbol{\theta}, t)\frac{\partial p_{Z\Theta}(z, \boldsymbol{\theta}, t)}{\partial z_j} = 0 \tag{8}$$

Equation (8) is referred as the generalized density evolution equation (GDEE). Specifically, as $m = 1$, i.e. only one physical quantity is considered, the GDEE becomes

$$\frac{\partial p_{Z\Theta}(z, \boldsymbol{\theta}, t)}{\partial t} + \dot{Z}(\boldsymbol{\theta}, t)\frac{\partial p_{Z\Theta}(z, \boldsymbol{\theta}, t)}{\partial z} = 0 \tag{9}$$

Equation (9) is a one-dimensional partial differential equation, and the boundary condition and the initial condition can be written as

$$p_{Z\Theta}(z, \boldsymbol{\theta}, t)\big|_{z \to \pm\infty} = 0 \tag{10}$$

$$p_{Z\Theta}(z, \boldsymbol{\theta}, t)|_{t=t_0} = \delta(z - z_0)p_{\Theta}(\boldsymbol{\theta}) \tag{11}$$

where z_0 is the deterministic initial value; $\delta(\cdot)$ is Dirac-delta function. Solving the GDEE (8), the instantaneous PDF of $\boldsymbol{Z}(t)$ can be obtained by

$$p_Z(z, t) = \int_{\Omega_\Theta} p_{Z\Theta}(z, \boldsymbol{\theta}, t)\mathrm{d}\boldsymbol{\theta} \tag{12}$$

3 Computational Algorithm for the PDEM

In the probability density evolution method, solving the stochastic dynamic problem needs to incorporate physical equations and the GDEE. Equation (8) is the GDEE with initial and boundary conditions specified by Eqs. (10) and (11). Such a computational algorithm can be carried out with the following steps:

Step (1): *Select representative points in the probability space and determine the assigned probability.*

Select a set of representative points in the distribution domain Ω_θ. Denote them by $\boldsymbol{\theta}_q = (\theta_{q,1}, \theta_{q,2}, \ldots, \theta_{q,s})^{\mathrm{T}}$, $q = 1, 2, \ldots, n_{\mathrm{sel}}$, where n_{sel} is the number of the selected points. Simultaneously, determine the assigned probability of each point according to the following equation

$$p_q = \int_{V_q} p_{\boldsymbol{\Theta}}(\boldsymbol{\theta})d\boldsymbol{\theta} \quad q = 1, 2, \ldots, n_{\text{sel}} \tag{13}$$

where V_q denotes a representative volume (domain) in the random parameter space.

Step (2): *Solve deterministic dynamical systems.*

For the specified $\boldsymbol{\Theta} = \boldsymbol{\theta}_q, q = 1, 2, \ldots, n_{\text{sel}}$, solve the deterministic dynamics equation to obtain time rate (velocity) of the physical quantities $\mathbf{Z} \cdot {}_j\ (\boldsymbol{\theta}_q, t), j = 1, 2, \ldots, m$.

Step (3): *Solve the GDEE with the finite-difference method.*

After *Step* 1, the GDEE becomes

$$\frac{\partial p_{z\boldsymbol{\Theta}}(z, \boldsymbol{\theta}_q, t)}{\partial t} + \sum_{j=1}^{m} \dot{z}_j(\boldsymbol{\theta}_q, t)\frac{\partial p_{z\boldsymbol{\Theta}}(z, \boldsymbol{\theta}_q, t)}{\partial z_j} = 0 \tag{14}$$

Correspondingly, the initial condition becomes

$$p_{z\boldsymbol{\Theta}}(z, \boldsymbol{\theta}, t)|_{t=t_0} = \delta(z - z_0)p_q \tag{15}$$

Take $\mathbf{Z}_j\ (\boldsymbol{\theta}_q, t), j = 1, 2, \ldots, m$, which we have obtained in *Step* 2, into the Eq. (14), and then solve the partial differential equation with the finite-difference method. With different difference schemes, the stability and convergence conditions of the solution are also different. Taking $m = 1$ (i.e. only one physical quantity is considered) as an example, the expressions of two difference schemes are given. For convenience, the GDEE is abbreviated as follows

$$\frac{\partial p(z, t)}{\partial t} + a(t)\frac{\partial p(z, t)}{\partial t} = 0 \tag{16}$$

where

$$p(z, t) = p_{\mathbf{Z}\boldsymbol{\Theta}}(z, \theta_q, t); \quad a(t) = \dot{Z}_j(\theta_q, t).$$

Take the z-axis grid as Δz and the t-axis grid as Δt, then the grid ratio is $r = \frac{\Delta t}{\Delta z}$. The Lax-Wendroff difference scheme with second-order accuracy constructed using Taylor series expansion is as follows

$$p_{j,k+1} = \frac{1}{2}(r^2a_k^2 - ra_k)p_{j+1,k} + (1 - r^2a_k^2)p_{j,k} + \frac{1}{2}(r^2a_k^2 + ra_k)p_{j-1,k} \tag{17}$$

Add a flux limiter based on the Law-Wendroff scheme, then we have the TVD (Total Variation Diminishing) scheme as follows

$$
\begin{aligned}
p_{j,k+1} = p_{j,k} - r\left[\frac{1}{2}(a_k + |a_k|)(p_{j,k} - p_{j-1,k}) + \frac{1}{2}(a_k + |a_k|)(p_{j+1,k} - p_{j,k})\right] \\
- \frac{1}{2}(1 - |ra_k|)|ra_k| \left[\varphi\left(r^+_{j+1/2}, r^-_{j-1/2}\right)\left(p_{j+1,k} - p_{j,k}\right)\right. \\
\left. - \varphi\left(r^+_{j+1/2}, r^-_{j-1/2}\right)\left(p_{j,k} - p_{j-1,k}\right)\right]
\end{aligned}
\tag{18}
$$

where $\varphi\left(r^+, r^-\right)$ is Roe-Sweby flux limiter with relatively small dissipation.

$$
\varphi\left(r^+, r^-\right) = \begin{cases} \max\left[0, \min(2r^+, 1), \min\left(r^+, 2\right)\right] & a_k < 0 \\ \max\left[0, \min(2r^-, 1), \min\left(r^-, 2\right)\right] & a_k \geq 0 \end{cases} \tag{19}
$$

where

$$
r^+_{j+1/2} = \frac{p_{j+2,k} - p_{j+1,k}}{p_{j+1,k} - p_{j,k}}, \quad r^-_{j+1/2} = \frac{p_{j,k} - p_{j-1,k}}{p_{j+1,k} - p_{j,k}}, \quad r^+_{j-1/2} = \frac{p_{j-1,k} - p_{j,k}}{p_{j,k} - p_{j-1,k}},
$$

$$
r^-_{j-1/2} = \frac{p_{j-1,k} - p_{j-2,k}}{p_{j,k} - p_{j-1,k}}
$$

The convergence conditions of the above two difference schemes read

$$
|ra_k| \leq 1 \tag{20}
$$

Step (4): *Synthesize the results.*

Sum up the joint probability density function $p_{\mathbf{Z\Theta}}(z, \theta_q, t)$, $q = 1, 2, \ldots, n_{sel}$, which we have got in *Sept* 3, to obtain the numerical solution of the PDF of $\mathbf{Z}(t)$.

$$
p_{\mathbf{Z}}(z, t) = \sum_{q=1}^{n_{\text{sel}}} p_{\mathbf{Z\Theta}}\left(z, \boldsymbol{\theta}_q, t\right) \tag{21}
$$

4 Case Study

4.1 Nonlinear Stiffness Characteristics of Hydro-pneumatic Spring

The hydro-pneumatic spring used in a suspension is a single chamber type, and its schematic diagram is shown in Fig. 1.

Assuming that the cylinder is fixed, the pressure of the gas in the static equilibrium position accumulator is p_0, and the volume is V_0. The cross-sectional area of the main

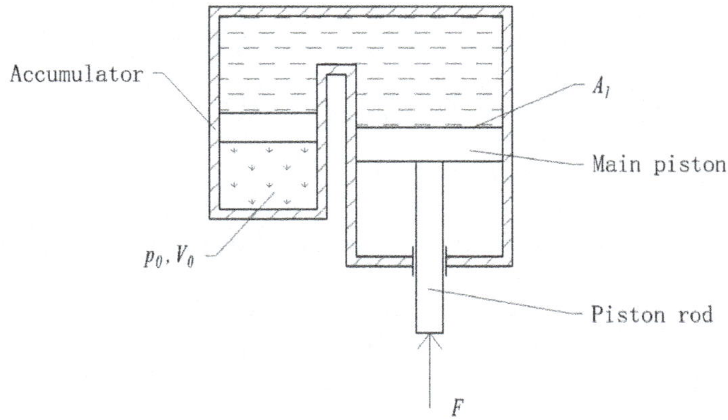

Fig. 1 Schematic diagram of the single chamber hydro-pneumatic spring

piston is A_1. The converted height of the gas at the static equilibrium position is $H_0 = V_0/A_1$. The external load acting on the main piston rod is F. Taking the static equilibrium position as the origin, the displacement of the main piston is x, and the symbol of x is positive during the compression stroke. The nonlinear stiffness characteristics of the hydro-pneumatic spring can be derived from the thermodynamic equation of the gas [14].

$$F = \frac{p_0 A_1}{(1 - x/H_0)^r} \tag{22}$$

Let the initial load $F_0 = p_0 A_1$, and the dynamic load acting on the main piston rod can be expressed as

$$\Delta F = \frac{p_0 A_1}{(1 - x/H_0)^r} - p_0 A_1 \tag{23}$$

where r is polytropic index. If the piston moves slowly, the gas will be in isothermal process, and $r = 1$. If the piston moves quickly, the thermodynamic process will become an adiabatic process, and $r = 1.4$. In general, take $r = 1.3$.

4.2 The 2DOF Nonlinear Model of Suspension

This paper simplifies the suspension into a 2 degree-of-freedom dual-mass vibration model, as shown in Fig. 2.

According to Newton's second law of motion, taking the static equilibrium position as the initial state, the dynamic equations are as follows:

Fig. 2 Schematic diagram of 2DOF vibration model

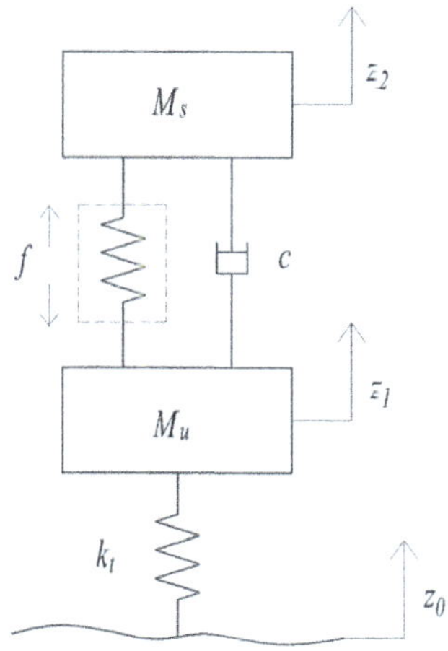

$$\begin{cases} M_s\ddot{z}_2 = c(\dot{z}_1 - \dot{z}_2) + f \\ M_u\ddot{z}_1 = k_t(z_0 - z_1) - c(\dot{z}_1 - \dot{z}_2) - f \end{cases} \tag{24}$$

where f denotes the force of the suspension spring; M_s is the sprung mass; M_u is the unsprung mass; k_t is the vertical stiffness of the tire; c is the damping coefficient of the damper. If Eq. (23) is used instead of f in Eq. (24), then:

$$\begin{cases} M_s\ddot{z}_2 = c(\dot{z}_1 - \dot{z}_2) + \frac{p_0A_1}{[1-(z_1-z_2)/H_0]^r} - p_0A_1 \\ M_u\ddot{z}_1 = k_t(z_0 - z_1) - c(\dot{z}_1 - \dot{z}_2) - \frac{p_0A_1}{[1-(z_1-z_2)/H_0]^r} + p_0A_1 \end{cases} \tag{25}$$

In this paper, the integral white noise method is used to simulate the random excitation from the road, and the road excitation is

$$\dot{z}_0(t) = -2\pi f_0 z_0(t) + 2\pi\sqrt{G_0 u}w(t) \tag{26}$$

where f_0 is the lower cut-off frequency; G_0 is the road roughness coefficient; u denotes the velocity of the vehicle; z_0 is the vertical displacement of road input; w is unit Gaussian white noise signal.

4.3 Numeral Simulation

In the numerical simulation of this paper, the uncertainties of three variables are considered, which are the vertical stiffness of the tire, the damping coefficient of the damper and the initial converted height of the gas. It is assumed that the three random variables are independent of each other and obey the normal distribution with

Table 1 Suspension parameter

Parameter (unit)	Symbol	Value	COV
Road roughness coefficient (m^3)	G_0	1×10^{-7}	–
Lower cut-off frequency (Hz)	f_0	0.01	–
Velocity of vehicle (km h^{-1})	u	70	–
Sprung mass (kg)	M_s	317.5	–
Unsprung mass (kg)	M_u	45.4	–
Gas pressure in equilibrium position (MPa)	p_0	1.5	–
Cross-sectional area of the main piston (m^2)	A_1	2×10^{-3}	–
Converted height of the gas (m)	H_0	0.2	0.08
Damping coefficient of the damper (N s m^{-1})	c	1500	0.08
Vertical stiffness of tire (N mm^{-1})	k_t	192	0.08

a coefficient of variation (COV) of 0.08. The remaining variables are deterministic variables. The specific parameters are shown in Table 1.

The response selected in this paper is the vertical acceleration of the sprung mass, and the simulation time is 3 s. The 135 representative points were selected by the number theory selection method [15]. The deterministic analysis was used to obtain the time derivative information of the vehicle body acceleration and then substituted into the GDEE. The two differential schemes, the Lax-Wendroff scheme (hereinafter referred to as LW scheme) and TVD scheme, were used to solve the GDEE, and then the evolution of PDF over time was obtained. The calculated results were compared with the Monte Carlo method (10^4 times).

The probability distribution of the response at any time can be obtained by PDEM. Figure 3 shows the PDF of the vertical acceleration of the sprung mass at $t = 1.0$ s, $t = 1.5$ s, $t = 2.0$ s and $t = 3.0$ s. The red dashed line with "+" is obtained by Monte Carlo method, and the black and blue lines are both obtained by PDEM, where the black is for LW scheme and the blue is for TVD scheme.

It can be seen that at the same time, the two schemes have the same trend as the PDF obtained by the Monte Carlo method, and the LW scheme has better precision than the TVD scheme. However, the LW scheme has oscillations on both sides of the probability density peak, and the probability density may be negative. This is caused by the characteristics of the LW difference scheme itself. The LW scheme cannot guarantee the non-negativity of the solution. The TVD scheme adds a flux limiter based on the LW, which increases the dissipation effect of the algorithm and ensures the non-negativity of the solution, but the calculation accuracy is lower than that of the LW format.

The mean and standard deviation (STD) of the response in time domain can be obtained from the PDF. The comparison of the results, which are obtained by Monte Carlo method and PDEM in LW and TVD scheme, is shown in Figs. 4 and 5. It can be seen from Fig. 4 that the mean response curves of the two difference schemes are both well matched with the mean curve of the Monte Carlo method and have

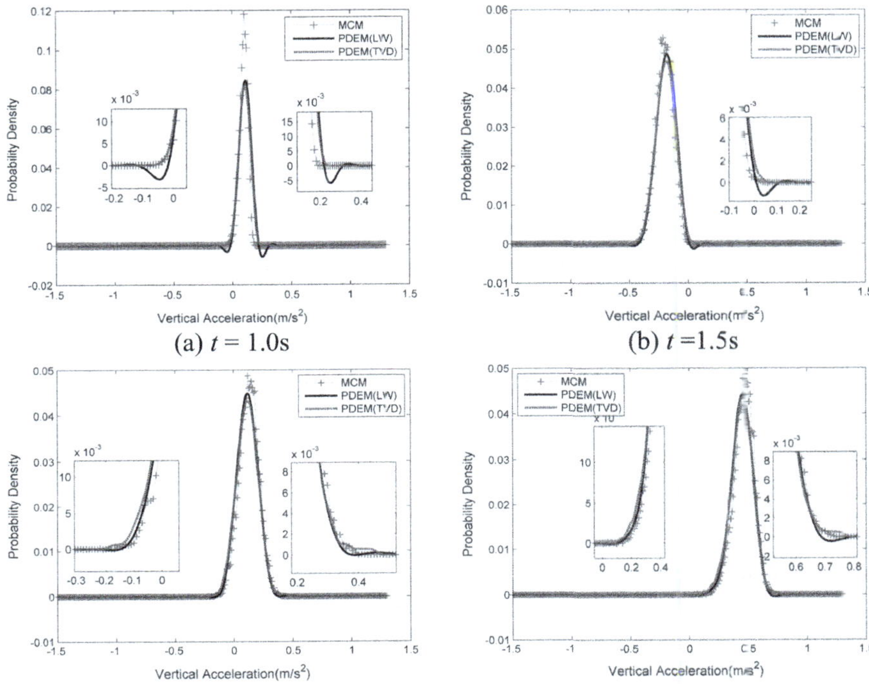

(a) $t = 1.0$s (b) $t = 1.5$s

(a) $t = 2.0$s (b) $t = 3.0$s

Fig. 3 PDF at some time instants

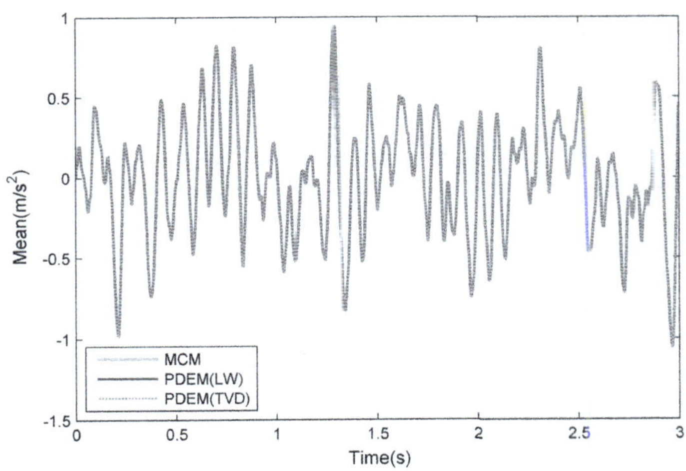

Fig. 4 Mean comparison of PDEM and Monte Carlo

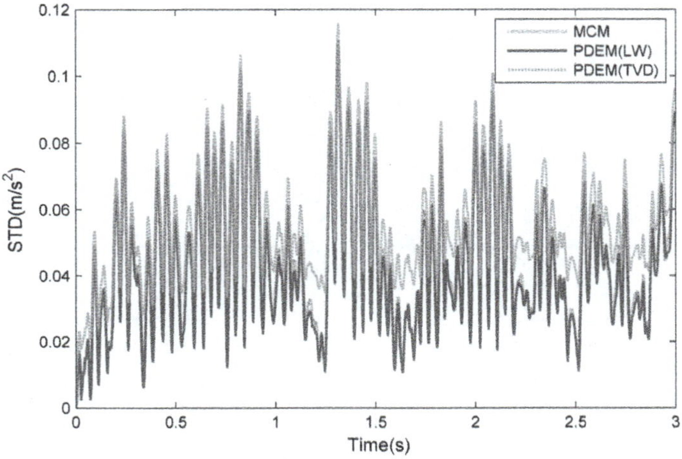

Fig. 5 STD comparison of PDEM and Monte Carlo

good calculation accuracy. Figure 5 shows the STD response curve obtained by LW scheme is also almost consistent with the Monte Carlo method and has good accuracy, but the curve obtained by TVD scheme has an error with the Monte Carlo method. The average absolute error is 0.011 m/s^2, and the maximum error is 0.026 m/s^2. The STD of the TVD scheme is generally larger, because the TVD scheme increases the dissipation effect, making the probability peak slightly wider and the standard deviation larger.

The PDEM can also visually demonstrate the time-varying process of its probability distribution. Figure 6 shows the variation of the PDF over time in the PDEM of TVD scheme. Figure 7 shows the contour of the PDF, which reflects the evolution of the probability distribution of the body acceleration response over time from 1.5 to 2.0 s.

As shown in Fig. 6, the PDF of the response changes with time just like the mountain stretching to the distance. For a nonlinear system, the PDF of the response does not obey the normal distribution, and some complex systems may even show an irregular or multi-peak PDF. As for these systems, if we only study the mean and standard deviation of the response, we may not be able to describe its true probability distribution. However, the PDEM can get more probability distribution information, which is the premise of realizing more precise motion control of the system.

The contour of the PDF surface in Fig. 7 seems like the water flowing in a river with a finite boundary, and the GDEE (Eq. 9) can also be analogized to the continuity equation in a hydrodynamic problem. This analogy can be understood as: The evolution of probability density reflects the flow of probability in the response space. For this non-stationary stochastic process, the PDEM can visually represent the change process of the response and can give more specific probability density information, which lays a foundation for the motion control of the system.

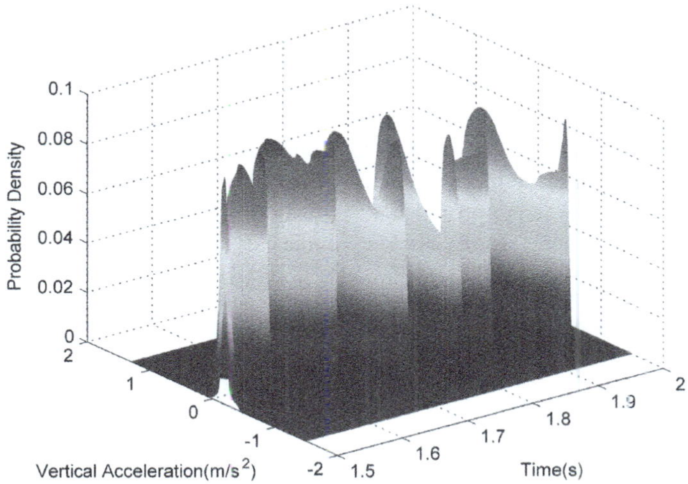

Fig. 6 PDF surface varying with time

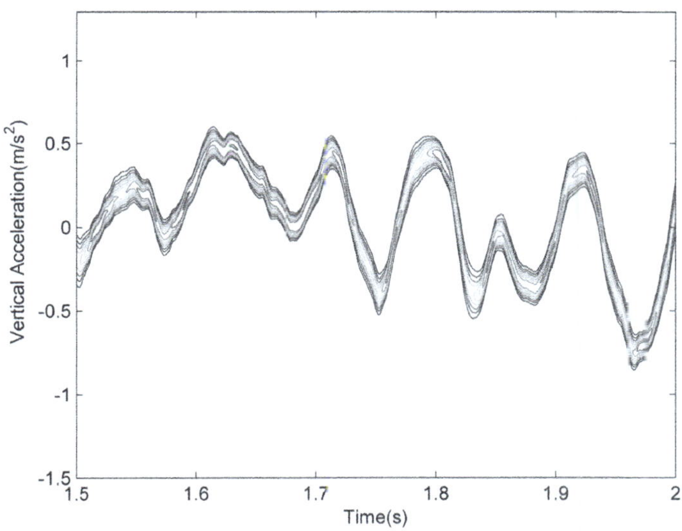

Fig. 7 Contour of the PDF surface

5 Conclusion

Based on the probability density evolution method, the nonlinear 2-degree-of-freedom vibration model is established, and the random vibration analysis of the suspension is carried out. Based on the state equation of stochastic system, the generalized density evolution equation of vehicle suspension vibration is derived from the principle of conservation of probability. The probability space is divided by the number theory method, and the GDEE is numerically solved by 2 finite-difference schemes (LW and TVD scheme). The results are compared with the traditional Monte Carlo method. The results show that the PDEM has good calculation accuracy, and it can intuitively reflect the time-varying process of the probability density of the response, so more probability distribution information of the response can be obtained. This paper provides a new way of thinking and a new method for stochastic response analysis of vehicle dynamics and nonlinear dynamic systems.

Acknowledgements This work was supported by the National Key R&D Program of China (No. 2017YFB0103704) and National Natural Science Foundation of China (Grant No. 51675044).

References

1. Guo JH, Luo YG, Li KQ (2015) Collaborative design of a motion control system for intelligent vehicles. J Tsinghua Univ (Sci Technol) 55(7):761–768
2. Zhu WQ, Cai GQ (2017) Introduction to stochastic dynamics. Science Press, Beijing
3. Zhu WQ (2005) Recent developments in nonlinear stochastic dynamics and control. World Sci- Tech R&D 27(1):1–4
4. Fang JJ, Fang T (1991) A weighted equivalent linearization technique for random vibration analysis. Chin J Appl Mech 3:114–120
5. Kewlani G, Crawford J, Iagnemma K (2012) A polynomial chaos approach to the analysis of vehicle dynamics under uncertainty. Veh Syst Dyn 50(5):749–774
6. Sandu C, Sandu A, Ahmadian M (2006) Modeling multibody dynamic systems with uncertainties. Part II: numerical applications. Multibody Syst Dyn 15(3):241–262
7. Zhang YM, Chen XY, Zhou ZP, Liu TQ (1995) Generalized probabilistic perturbation method for static analysis. Appl Math Mech 16(8):709–714
8. Zhang YM, Liu QZ, Wen BB (2003) Probability perturbation finite element method for response analysis of multi-degree-of-freedom nonlinear vibration systems with random parameters. Chin J Comput Mech 20(1):8–11
9. Zhang LQ (2006) Numerical solution of high-dimensional FPK equation. Zhejiang University, Hangzhou
10. Huang XL, Zhu XX (2001) Calculation of ship capsize probability by using path integration method. J Ship Mech 5(4):7–16
11. Li J, Chen JB (2010) Advances in the research on probability density evolution equations of stochastic dynamical systems. Adv Mech 40(02):170–188
12. Li J, Chen JB, Sun WL, Peng YB (2012) Advances of the probability density evolution method for nonlinear stochastic systems. Probab Eng Mech 28:132–142
13. Chen JB, Li J (2005) Dynamic response and reliability analysis of non-linear stochastic structures. Probab Eng Mech 20:33–44
14. Zou Y, Yu F, Sun T (2004) Simulation of nonlinear hydro-pneumatic suspension for comfort. Comput Simul 21(10):157–269
15. Li J, Chen JB (2009) Stochastic dynamics of structures. Wiley, Singapore

A Study on the Influence of Mass Change on Vibration Characteristic of ISD Suspension

Xiaofeng Yang, Hongchang Li, Yanling Liu, Yi Yang and Wentao Zhao

Abstract In order to study the influences of mass variation on vibration response characteristics of ISD suspension, the ISD suspension composed of four elements is established. Body acceleration, suspension travel and dynamic tire load as the assessment criteria are studied in the time and frequency-domain simulation with sprung and unsprung mass increased and decreased by 50%, respectively. The results show that the root mean square (RMS) of body acceleration reduces and the RMS of suspension deflection increases with the increase of the sprung mass. The RMS of dynamic load increases with the increase of unsprung mass. In the low frequency (0–5 Hz), the peak value of low-frequency resonance of suspension dynamic travel and dynamic tire load increases significantly. In the high frequency (5–15 Hz), the resonant frequency decreases and amplitude increases with the increase of the unsprung mass. Compared with the traditional passive suspension, when the sprung mass is increased by 50%, the RMS of body acceleration, suspension dynamic travel and dynamic tire load are decreased by 6.92, 4.48 and 4.02%, respectively. When the unsprung mass is increased by 50%, three factors are decreased by 6.51, 20.32 and 5.96%, respectively. It indicates that ISD suspension possesses better ride comfort and handling stability when the mass varies due to the coupled vibration effect of inerter.

Keywords ISD suspension · Mass change · Vibration response · Coupled vibration

1 Introduction

Vehicle suspension is the general term of the power transfer device between the wheel and the axle. The performance of the suspension has a great influence on the ride comfort, handle and stability of the vehicle [1, 2]. The traditional suspension cannot alleviate the contradiction between the three performance indicators, so that the vibration isolation performance is limited [3–5]. Active suspension [6] and semi-

X. Yang (✉) · H. Li · Y. Liu · Y. Yang · W. Zhao
School of Automotive and Traffic Engineering, Jiangsu University, Zhenjiang, China
e-mail: 1057898390@qq.com

China SAE (ed.), *Proceedings of China SAE Congress 2018: Selected Papers*,
Lecture Notes in Electrical Engineering 574,
https://doi.org/10.1007/978-981-13-9718-9_19

active suspension [7] system has complex structure, and it has shortcomings in its effectiveness, real-time, reliability, energy consumption and so on. It is difficult to be popularized in a large scale. Therefore, the new vibration isolation theory is applied to fully exploit the potential of the traditional suspension and to improve the performance of the suspension.

Inerter is a two-terminal mechanical device proposed by Smith in 2002 [5, 8], which has the property that the applied force at the terminals is proportional to the relative acceleration across the terminals. The two terminals should be moved independently, and the equation for the ideal inerter is given by

$$F = b(\dot{v}_2 - \dot{v}_1) \tag{1}$$

In Eq. (1), F is the force of the two terminal of the inerter. b is the inertance of the inerter, and the unit is kg. v_1, v_2 are the velocities of the two terminals of the inerter.

In recent years, performance advantages in many mechanical systems employing inerter have been researched and investigated [9], such as vehicle, train, motorcycle and building suspensions. It is shown that, by incorporating the inerter element, the new "Inerter-Spring-Damper" (ISD) network can effectively improve the suspension performance [10, 11].

Nowadays, new energy vehicles and vehicle lightweight are two major directions of vehicle research. According to the existing technology, the quality of the new energy vehicle is higher than that of the traditional gasoline engine, and it will have a certain influence on the performance of the suspension. Meanwhile, the optimization of the quality for the body and wheels by the lightweight of the automobile is bound to put forward higher requirements for the vehicle suspension system [12]. At the same time, for passenger cars or light goods vehicles, the number of passengers and cargo loading are different, and the quality parameters of suspension have also changed. At present, the influence of quality parameter change on the performance of ISD suspension has not been widely concerned.

Therefore, based on the "Inerter-Spring-Damper" suspension structure system, the 1/4 model of vehicle is established, and the influence of the change of mass parameters on the vibration response characteristics of ISD suspension and traditional passive suspension is analyzed and compared, which lays a theoretical foundation for the follow-up study of controllable ISD suspension.

2 ISD Suspension Structure

Inerter as a new type of two endpoint mechanical components is replaced by the "single end point" quality component, and the suspension structure based on the classical vibration isolation theory of "Mass-Spring-Damper" is changed. The inerter, spring and damper can be freely combined. The vehicle ISD suspension structure, such as Fig. 1, is studied in this paper according to the document [13].

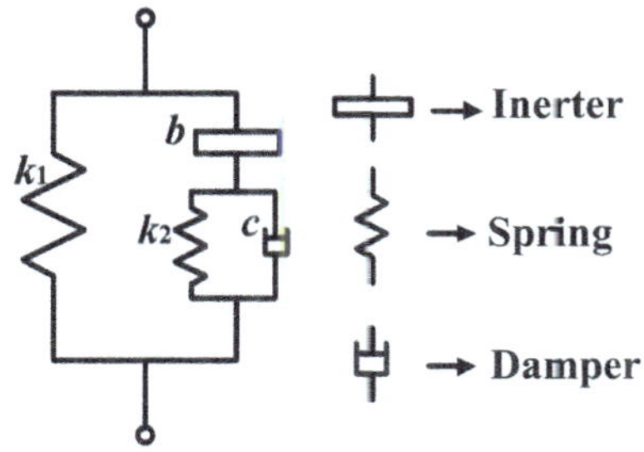

Fig. 1 ISD suspension structure

As shown in Fig. 1, the ISD suspension structure consists of two spring elements (main and auxiliary springs), one damping element and one inerter.

3 Dynamic Modeling of 1/4 ISD Suspension

A vehicle 1/4 ISD suspension model is established, as shown in Fig. 2.

According to Newton's second law, the sprung mass and unsprung mass of ISD suspension are analyzed, and the differential equation of motion is obtained.

$$\begin{cases} m_2\ddot{z}_2 + k_1(z_2 - z_1) + F = 0 \\ m_1\ddot{z}_1 + k_t(z_1 - z_0) - k_1(z_2 - z_1) - F = 0 \\ F = b_2(\ddot{z}_2 - \ddot{z}_b) = k_2(z_b - z_1) + c_2(\dot{z}_b - \dot{z}_1) \end{cases} \tag{2}$$

In Eq. (2), m_1 is the unsprung mass and m_2 is sprung mass, while z_1, z_2, z_b and z_0 are the vertical displacements of the tire, the vehicle body, the inerter and the road surface, respectively. Furthermore, k_1 is the main spring stiffness of the suspension, k_2 is the secondary spring stiffness of the suspension and kt is the tire stiffness, respectively. b_2 is inertance, c_2 is the damping coefficient of the damper and F is the force of the inerter or damper, respectively.

The random road input model is given by Eq. (3).

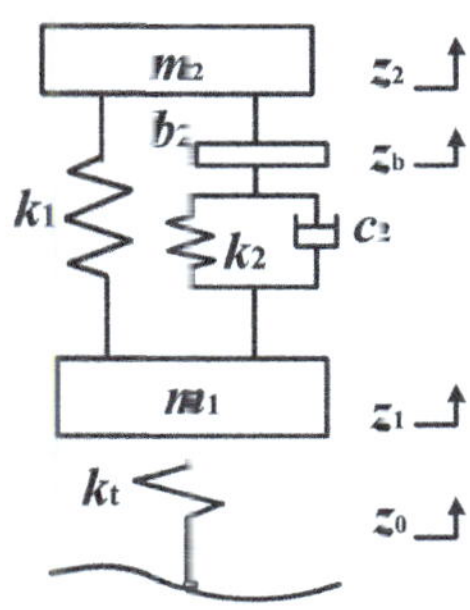

Fig. 2 1/4 ISD suspension model

$$\dot{z}_0(t) = -2\pi f_0 z_0(t) + 2\pi\sqrt{G_0 v}w(t) \quad (3)$$

In Eq. (3), G_0 is the road roughness coefficient, f_0 is the cut-off frequency, $w(t)$ is Gauss white noise with a mean of 0 and v is the speed.

Thus, the state-space equation is established as Eq. (4).

$$\begin{cases} \dot{X} = AX + Bu \\ Y = CX + Du \end{cases} \quad (4)$$

The state variable is

$$X = \begin{bmatrix} \dot{z}_2 \ \dot{z}_1 \ \dot{z}_b \ z_2 \ z_1 \ z_b \end{bmatrix}^{\mathrm{T}}$$

The input variable is

$$u = z_0$$

The output variables are the body acceleration, suspension deflection and dynamic tire load, respectively.

$$Y = \begin{bmatrix} \ddot{z}_2 \ z_2 - z_1 \ k_t(z_1 - z_0) \end{bmatrix}^{\mathrm{T}}$$

In Eq. (4)

$$A = \begin{bmatrix} 0 & \frac{c_2}{m_2} & -\frac{c_2}{m_2} & -\frac{k_1}{m_2} & \frac{k_1+k_2}{m_2} & -\frac{k_2}{m_2} \\ 0 & -\frac{c_2}{m_1} & \frac{c_2}{m_1} & \frac{k_1}{m_1} & -\frac{k_1+k_2+k_3}{m_1} & \frac{k_2}{m_1} \\ 0 & \left(\frac{c_2}{m_2} + \frac{c_2}{b_2}\right) & -\left(\frac{c_2}{m_2} + \frac{c_2}{b_2}\right) & -\frac{k_1}{m_2} & \left(\frac{k_1+k_2}{m_2} + \frac{k_2}{b_2}\right) & -\left(\frac{k_2}{m_2} + \frac{k_2}{b_2}\right) \\ 1 & 0 & 0 & 0 & 0 & 0 \\ 0 & 1 & 0 & 0 & 0 & 0 \\ 0 & 0 & 1 & 0 & 0 & 0 \end{bmatrix}$$

$$B = \begin{bmatrix} 0 \ \frac{k_t}{m_1} \ 0 \ 0 \ 0 \ 0 \end{bmatrix}^{\mathrm{T}}$$

$$C = \begin{bmatrix} 0 & \frac{c_2}{m_2} & -\frac{c_2}{m_2} & -\frac{k_1}{m_2} & \frac{k_1+k_2}{m_2} & -\frac{k_2}{m_2} \\ 0 & 0 & 0 & 1 & -1 & 0 \\ 0 & 0 & 0 & 0 & k_t & 0 \end{bmatrix}$$

$$D = \begin{bmatrix} 0 \ 0 \ -k_t \end{bmatrix}^{\mathrm{T}}$$

4 Simulation Analysis

In order to study the influence of the sprung mass change and the unsprung mass change on the vibration response characteristics of the traditional passive suspension and ISD suspension, the reference values of the sprung mass m_2 and the unsprung mass m_1 are increased by 50% and reduced by 50%, respectively, and the other suspension parameters remain unchanged, as shown in Table 1.

4.1 *Time-Domain Response Simulation*

Matlab/Simulink is used to simulate in the time-domain, and the specific parameters are shown in Table 2.

In this paper, the traditional passive suspension and ISD suspension are simulated and analyzed with the sprung mass m_2 and the unsprung mass m_1. The three performance indexes of the body acceleration mean square root value (BA), the suspension deflection mean square root value (SWS) and the dynamic tire load mean square root value (DTL) are used as evaluation criteria, and the vibration response characteristics of the suspension are analyzed in time-domain.

Table 1 Quality parameters

Quality parameters	Sprung mass m_2 (kg)	Unsprung mass m_1 (kg)
Reference values	320	45
Reference values +50%	480	67.5
Reference values −50%	160	22.5

Table 2 Model parameters

Parameters	Value
Deputy spring stiff k_2 (KN m^{-1})	10
Main spring stiff k_1 (KN m^{-1})	22
Inertance b_2 (kg)	217
Damper coefficient c_2 (N s m^{-1})	1067
Tire stiff k_t (kN m^{-1})	190
The road roughness coefficient G_0 (m^3 cycle^{-1})	5×10^{-6}
Speed v (km h^{-1})	20

4.1.1 Influence of the Change of Sprung Mass on the RMS of Suspension

When the sprung mass m_2 changes, the change trends of BA, SWS and DTL of traditional passive suspension and ISD suspension are shown in Figs. 3, 4 and 5.

As shown in Figs. 3, 4 and 5, when the reference value of the sprung mass m_2 increases by 50% and reduced by 50%, the RMS of the three performance indicators of the traditional passive suspension and the ISD suspension are shown in Table 3.

It can be seen from Figs. 3, 4, 5 and Table 3, the RMS of body acceleration is inversely proportional to the sprung mass, and the RMS of suspension deflection and dynamic tire load are directly proportional to the sprung mass. The RMS of body acceleration and the suspension deflection of the ISD suspension are more sensitive to the change of the sprung mass. By comparing with the traditional passive suspension, the RMS of the body acceleration for the ISD suspension always keeps the same as that of the traditional passive suspension. The RMS of the dynamic load and the dynamic load of the suspension are always lower than that of the traditional passive suspension. When the sprung mass is 480 kg, the RMS of the body acceleration is reduced by 6.92%, the RMS of the suspension deflection is reduced by 4.48%, and

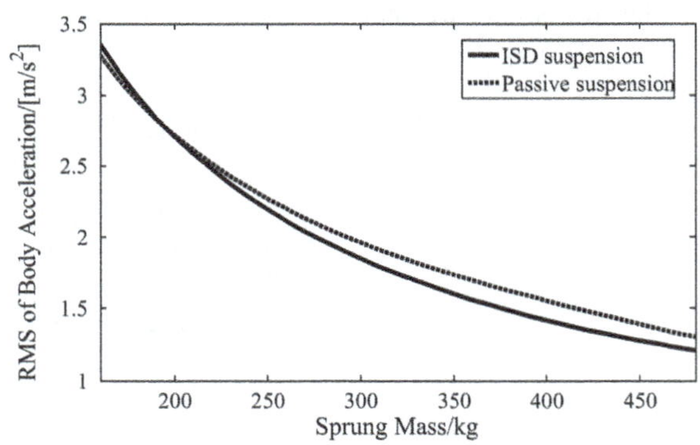

Fig. 3 RMS of body acceleration

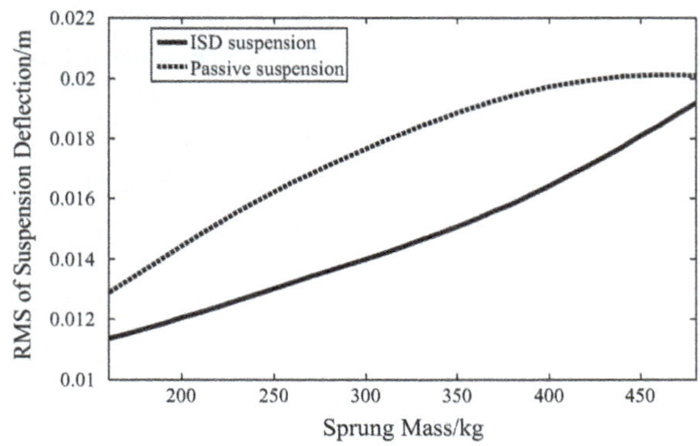

Fig. 4 RMS of suspension deflection

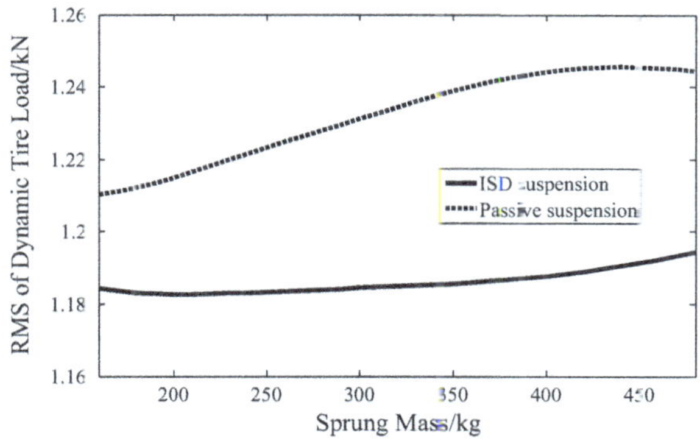

Fig. 5 RMS of dynamic tire load

Table 3 Suspension performance index when the sprung mass is changed

Suspension structure	m_2 (kg)	BA (m/s^2)	SWS (m)	DTL (kN)
Passive suspension	160	3.2720	0.0129	1.2104
	320	1.8657	0.0182	1.2344
	480	1.3026	0.0201	1.2444
ISD suspension	160	3.3536	0.0114	1.1844
	320	1.7418	0.0144	1.1849
	480	1.2125	0.0192	1.1944

the RMS of the dynamic tire load is reduced by 4.02% than that of the traditional suspension. It is shown that the change of the sprung mass has great influence on the RMS of the body acceleration and the suspension deflection. In addition, the coupling effect of the ISD suspension with the inerter can effectively improve the impact on vehicle ride comfort and handling stability caused by the change of the sprung mass.

4.1.2 Influence of the Change of Unsprung Mass on the RMS of Suspension

When the unsprung mass m_1 changes, the change trends of BA, SWS and DTL of traditional passive suspension and ISD suspension are shown in Figs. 5, 7 and 8.

As shown in Figs. 6, 7 and 8, when the reference value of the unsprung mass m_1 increases by 50% and reduced by 50%, the RMS of the three performance indicators of the traditional passive suspension and the ISD suspension are shown in Table 4.

It can be seen from Figs. 6, 7, 8 and Table 4, the RMS of the three performance indexes are directly proportional to the unsprung mass. The dynamic tire load of ISD suspension is more sensitive to the variation of unsprung mass. Compared with passive suspension, under the condition of different unsprung mass, the three per-

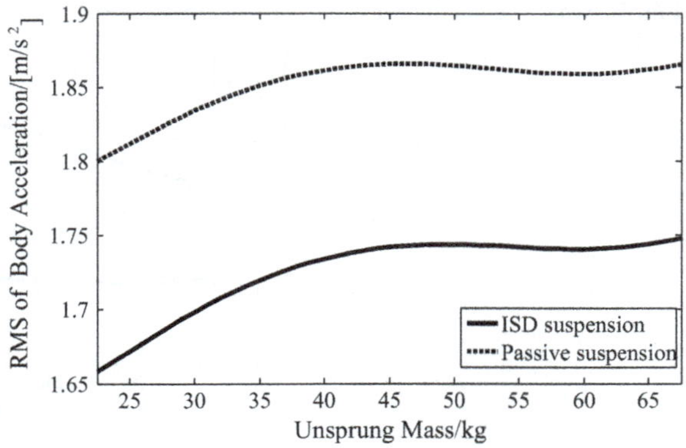

Fig. 6 RMS of body acceleration

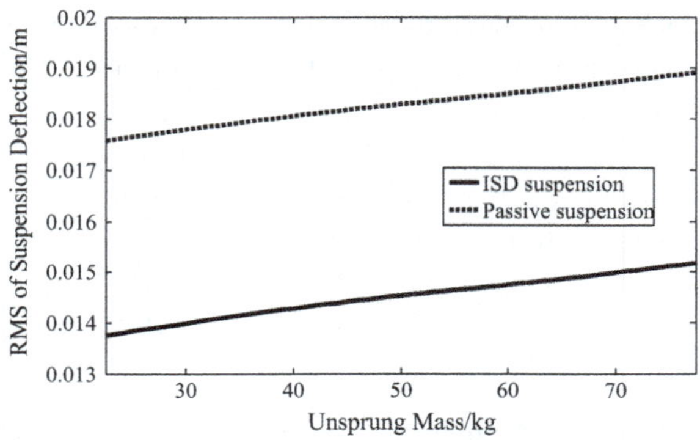

Fig. 7 RMS of suspension deflection

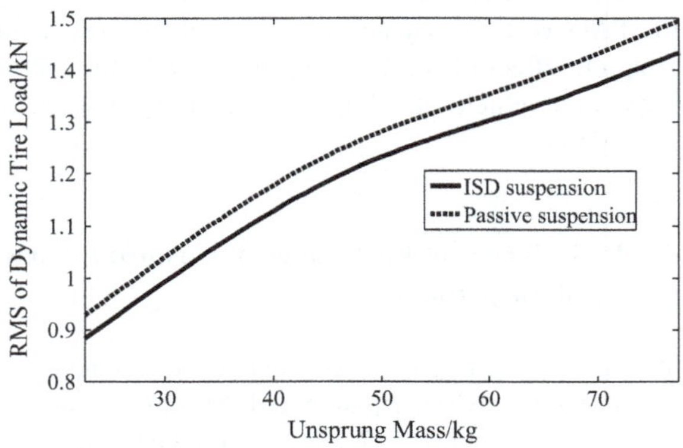

Fig. 8 RMS of dynamic tire load

Table 4 Suspension performance index when the unsprung mass is changed

Suspension structure	m_1 (kg)	BA (m/s^2)	SWS (m)	DTL (kN)
Passive suspension	22.5	1.8001	0.0176	0.9296
	45	1.8657	0.0182	1.2344
	67.5	1.8658	0.0187	1.4394
ISD suspension	22.5	1.6585	0.0138	0.8993
	45	1.7418	0.0144	1.1849
	67.5	1.7480	0.0149	1.3536

formance indexes of ISD suspension are obviously lower. When the unsprung mass is 67.5 kg, the RMS of the body acceleration is reduced by 6.31%, the RMS of the suspension deflection is reduced by 20.32% and the RMS of the dynamic tire load is reduced by 5.96% than that of the traditional suspension. It is shown that with the increase of unsprung mass, the effect of vibration energy on the body and tire is enhanced, the comfort of the vehicle becomes worse, so that the ride comfort and the handling stability are reduced. In addition, ISD suspension can effectively improve vehicle performance deterioration caused by unsprung mass change and improve handling stability and ride comfort.

4.2 *Frequency-Domain Response Simulation*

Based on the dynamic vibration absorption principle [14–17], the vibration transfer characteristics of the 1/4 suspension model in the frequency domain are analyzed by transfer function method. In order to study the vibration transmission characteristics of the ISD suspension, the isolator part of the vehicle 1/4 suspension model shown in Fig. 2 is considered as a mechanical impedance transfer function, as shown in Fig. 9.

Thus, the dynamic equation is established as Eq. (5)

$$\begin{cases} m_2\ddot{z}_2 + (\dot{z}_2 - \dot{z}_1)T(s) = 0 \\ m_1\ddot{z}_1 + k_t(z_1 - z_0) - (\dot{z}_2 - \dot{z}_1)T(s) = 0 \end{cases} \tag{5}$$

Fig. 9 1/4 suspension model

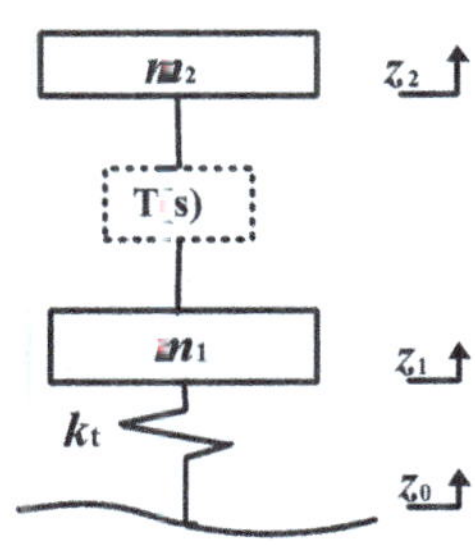

After Laplace transformation, as shown in Eq. (6)

$$\begin{cases} m_2 s^2 Z_2 + sT(s)(Z_2 - Z_1) = 0 \\ m_1 s^2 Z_1 + k_t(Z_1 - Z_0) - sT(s)(Z_2 - Z_1) = 0 \end{cases} \tag{6}$$

In Eq. (6), Z_0, Z_1 and Z_2 are Laplace transform of the road surface input, the unsprung mass and the sprung mass displacement, respectively. $T(s)$ is the velocity type impedance of the ISD suspension structure.

In Eq. (6)

$$\begin{aligned} T(s) &= \frac{k_1}{s} + \frac{1}{\frac{1}{bs} + \frac{1}{\frac{k_2}{s} + c}} \\ &= \frac{bcs^3 + b(k_1 + k_2)s^2 + ck_1 s + k_1 k_2}{bs^3 + cs^2 + k_2 s} \end{aligned}$$

Based on Table 2, the frequency-domain model of ISD suspension is established. The PSD of body acceleration, suspension deflection and dynamic tire load compared with passive suspension when the sprung mass or unsprung mass changed.

4.2.1 Effect of Sprung Mass on Suspension Performance

When the sprung mass m_2 changes, the changing trend for the PSD of body acceleration, suspension deflection and dynamic tire load of the passive suspension and the ISD suspension is shown in Figs. 10 and 11.

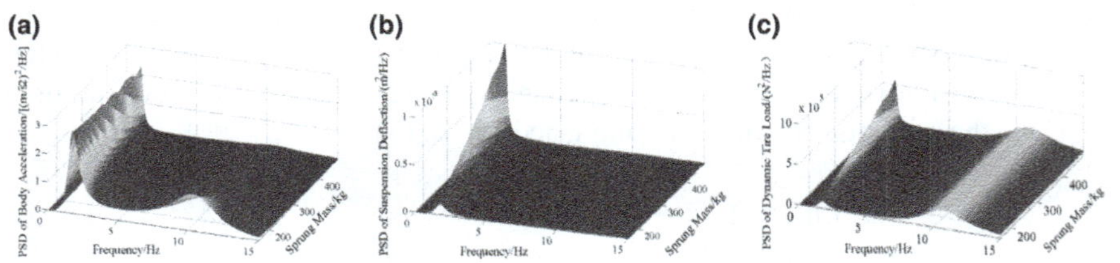

Fig. 10 Trend of three performance indexes of passive suspension with sprung mass

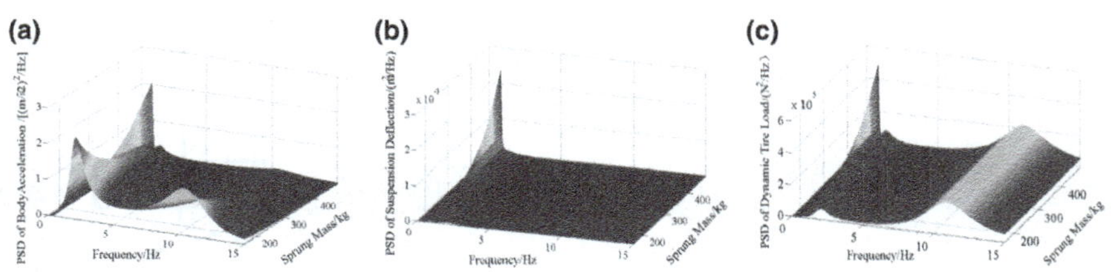

Fig. 11 Trend of three performance indexes of ISD suspension with sprung mass

The comparison between Figs. 10a and 11a is known, and the PSD of body acceleration's peak of the ISD suspension is lower than that of the passive suspension under the same sprung mass in low frequency. It shows that the ISD suspension can reduce the peak of the body acceleration in the low frequency through the coupling of the inerter and the sprung mass and can effectively improve the ride comfort of the vehicle in the low frequency.

The comparison between Figs. 10b and 11b is known, the PSD of suspension deflection does not change significantly at high frequency and the peak value at low-frequency resonance increases with the sprung mass. Meanwhile, the peak value of low-frequency resonance of ISD suspension is lower than that of passive suspension. It is shown that the coupling vibration between the inerter and the sprung mass can help to suppress the resonance of the suspension deflection at low frequencies and improve the handling stability of the vehicle.

The comparison between Figs. 10c and 11c is known, at high frequency, the resonance of the dynamic tire load' PSD is not affected by the change of the sprung mass, and the peak value at the low-frequency resonance peak increases with the sprung mass and the resonance frequency decreases. ISD suspension can reduce the resonance of the dynamic tire load in the low frequency through the coupling vibration of the inerter and the sprung mass. It can effectively improve the ride comfort of the vehicle and improve the safety.

4.2.2 Effect of Unsprung Mass on Suspension Performance

When the unsprung mass m_1 changes, the changing trend for the PSD of body acceleration, suspension deflection and dynamic tire load of the passive suspension and the ISD suspension is shown in Figs. 12 and 13.

The comparison between Figs. 12a and 13a is known, the formant peak value of the high frequency increases with the unsprung mass and the resonance frequency moves forward gradually. The formant peak value of ISD suspension in low frequency and high frequency is obviously lower than that of passive suspension. It is shown that the resonance performance of the body acceleration PSD at low frequency is not affected by the change of unsprung mass. With the increase of unsprung mass, the high-frequency resonance is aggravated, and the resonance frequency decreases, so

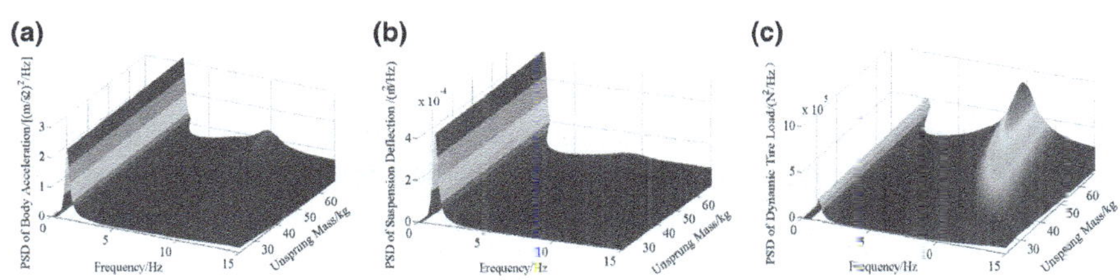

Fig. 12 Trend of three performance indexes of passive suspension with unsprung mass

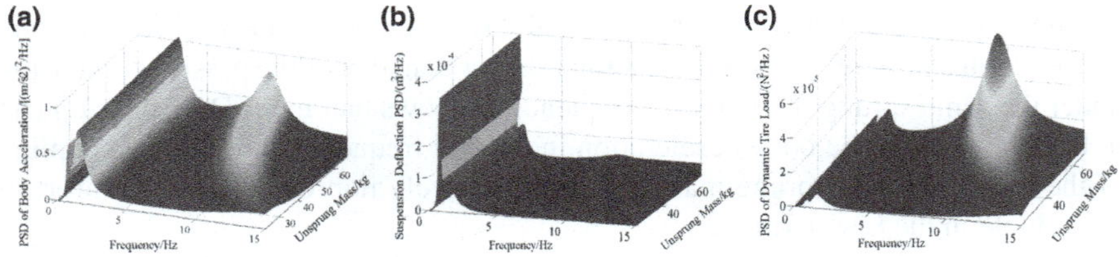

Fig. 13 Trend of three performance indexes of ISD suspension with unsprung mass

that the ride comfort is reduced. But the ISD suspension has a slight improvement on the vibration characteristics at high-frequency resonance.

The comparison between Figs. 12b and 13b is known, and the peak value of the ISD suspension at low frequencies is lower than that of the passive suspension. It shows that the performance of the low-frequency resonance is not affected by the change of the unsprung mass. With the increase of the unsprung mass, the high-frequency resonance peak is slightly increased and the resonance frequency decreases, which reduces the ride comfort of the vehicle. The performance of ISD suspension is similar to that of passive suspension at high frequency.

The comparison between Figs. 12c and 13c is known, and the change of the dynamic tire load PSD is similar to that of the body acceleration PSD, so that the ride comfort of the vehicle is reduced, the ground performance of the tire is deteriorated and the safety is reduced. Due to the addition of mass impedance, the ISD suspension has slowed down the increase of the formant peak value and improved the deterioration of the vehicle performance by the coupling of the inerter and the unsprung mass.

5 Conclusion

The main conclusions that may be derived from this research are following:

(1) In time domain, for the ISD suspension, with the increase of the sprung mass, the RMS of body acceleration decreases and the RMS of the suspension deflection increases, with the increase of the unsprung mass, the RMS of dynamic tire load increases significantly. As a result, in order to make the vehicle have good ride comfort and stability, the sprung mass should be increased and the unsprung mass should be reduced within the allowable range.

(2) In frequency domain, for the ISD suspension, with the increase of the sprung mass, the peak value of the resonance peak at low frequency of the suspension deflection and the dynamic tire load increases, the resonance frequency moves forward, the ride comfort decreases, with the unsprung mass increases, the high resonance peak of the body acceleration and the dynamic tire load increases, the resonance frequency decreases, the ride comfort of the vehicle is reduced,

and the tire grounding performance is bad at the same time so that reduce the security.

(3) Compared with passive suspension, when the sprung mass increased by 50%, the body acceleration, suspension deflection and dynamic tire load of ISD suspension decreased by 6.92, 4.48 and 4.02%, respectively. When the unsprung mass increased by 50%, and the three indexes decreased by 6.31, 20.32 and 5.96%, respectively. The results show that the ISD suspension with mass impedance can effectively improve the suspension performance when quality parameters change, taking into account the ride comfort and handling stability of the vehicle better and laying a theoretical foundation for the study of the follow-up controllable ISD suspension.

References

1. Wang R, Chen L, Zhang X et al (2012) Design and test of semi-active air suspension system of vehicle. Trans Chin Soc Agric Mach 43(4):6–9
2. Sun S, Zhan C, Chen K et al (2014) Air suspension based on hybrid fuzzy-suboptimal control strategy. J Chongqing Univ Technol 28(7):21–25
3. Chen L, Yang X, Wang R et al (2013) Design and performance study of vehicle passive suspension based on two-element inerter-spring-damper structure vibration isolation mechanism 32(6):90–95
4. Sun X, Chen L, Wang S et al (2013) Analysis of vibration isolation performance for nolinear inerter-spring-damper suspension. Trans Chin Soc Agric Eng 29(23):38–45
5. Smith MC, Wang F (2004) Performance benefits in passive vehicle suspensions employing inerters. Veh Syst Dyn 42(4):235–257
6. Zhang C, Zhang Z, Zhao H (2015) Analysis of dynamic characteristics and ride performance of automobile active suspension. J Chin Agric Mech 36(2):176–179
7. Huang C, Chen L, Yuan C et al (2014) Hybrid fuzzy control of semi-active suspension system. Automot Eng 36(8):999–1018
8. Smith MC (2002) Synthesis of Networks: the inerter. IEEE Trans Autom Control 47(10):1648–1662
9. Hu YL, Chen MZQ, Shu Z (2014) Passive vehicle suspensions employing inerters with multiple performance requirements. J Sound Vib 333(8):2212–2225
10. Sun X, Chen L, Wang S, Zhang X, Yang X (2016) Performance investigation of vehicle suspension system with nonlinear ball-screw inerter. Int J Autom Technol 17(3):399–408
11. Brzeski P, Perlikowski P (2016) Effects of play and inerter nonlinearities on the performance of tuned mass damper. Nonlinear Dyn 88(2):1–15
12. Li Y, Li Y, Lou M et al (2012) Lightweighting of car body and its challenge to joining technologies 48(18):44–54
13. Long C, Yujie S, Xiaofeng Y et al (2014) Design and experiment of vehicle suspension based on inerter-spring structure. J Vib Shock 33(22):83–87
14. Li L, Ding J, Tian Y (2016) Frequency-widening control design and optimization with Simulink simulation for MRE vibration absorbers. J Vib Shock 35(17):171–176
15. Zheng G, Weirui W (2017) Vehicle active ISD-DVA suspension system. J Vib Shock 36(1):167–174
16. Yang X, Shen Y, Chen L et al (2014) Design and performance analysis of vehicle ISD suspension based on dynamic vibration theory. Automot Eng 36(10):1262–1277
17. Chen L, Liu C et al (2016) Network synthesis and parameter optimization for vehicle suspension with inerter. Adv Mech Eng 8(12):1–7

A Study on the Mechanical Properties of Tire in the High-Speed Frontal Impact Condition

Fangxia Yang, Taisong Cui, Bo Dian and Bo Wei

Abstract The simulation accuracy of the tire model has an important effect on ensuring the simulation accuracy of the vehicle model in vehicle frontal crash simulation. A tire sub-model is set up based on the data of a Changan vehicle tire. and dynamic impact analysis for the tire is performed with a certain velocity. Tires' acceleration, speed, and deformation parameters of mechanical properties are acquired and compared with the tire impact test data. Then, the tire finite element model is optimized through comparing the tire performance parameters with the test results. After that, it is applied to the simulation model of high-speed frontal crash condition to analyze the effect of tire simulation method on the consequences. Comparative analysis is performed between simulation and test The results show that the vehicle acceleration and tire deformation in the vehicle crash model are more consistent with the test after optimization of the tire model, thus improved the simulation precision of tire model in simulation analysis.

Keywords Tire model · Mechanical properties · High-speed frontal impact condition

1 Introduction

With the development of vehicle technology, the people have the greater depth of vehicle safety knowledge, vehicle safety performance is been paid more and more attention, and crash safety standards more stringent. In 2012 July, the C-NCAP increased the 40% offset impact test velocity from 56 to 64 km/h; In addition, Insurance Institute for Highway Safety (I I H S) issued the test method of small overlapping frontal impact in 2012, that is, car at a speed of 64 km/h crash rigid barrier of 25% overlaps [1]. From January 1, 2018, 25% overlap test was implemented in the China Insurance Automobile Safety Index (C-IASI). All of these have greatly improved

F. Yang (✉) · T. Cui · B. Dian · B. Wei
State Key Laboratory of Vehicle NVH and Safety Technology, Chongqing 401120, China
e-mail: yfx3224@163.com

Changan Auto R&D Center, Changan Automobile Co., Ltd., Chongqing 401120, China

China SAE (ed.), *Proceedings of China SAE Congress 2018: Selected Papers*, Lecture Notes in Electrical Engineering 574,
https://doi.org/10.1007/978-981-13-9718-9_20

the requirements of vehicle safety performance, and it shows that the future vehicle crash safety will develop in a more severe direction.

In the vehicle development process, crash safety performance is an indispensable evaluation criteria of vehicle performance. But due to the high test cost, test will be performed at a later stage in the development process, and is not conducive to the early data optimization etc., at the early stage of the vehicle development, most of the simulation analysis method to simulation optimization vehicle safety performance. Therefore, the simulation accuracy is a crucial impact on the development of the whole car.

In the frontal offset crash condition, one side of the car body is forced, the front end of the car body deformed, tire squeezed contact with A-pillar, and the crash force through A-pillar transmitted to the threshold beam. So, the tire as an important force transmission path is in frontal offset crash condition. The simulation accuracy of the tire model has much effect on the 40 and 25% offset crash condition of 64 km/h. But the accuracy of the current simulation tire model is not high, tire parameters cannot be real reaction, and a direct impact on vehicle simulation accuracy. Therefore, in order to improve the accuracy of the current tire model, we should collect the dynamic test data of tire by the sled test and compare the tire mechanical properties data (such as acceleration, speed, and deformation) with simulation data. Then, the modified tire model is applied to the whole vehicle model and compared with the test data.

2 The Establishment of Tire Dynamic Impact Test Method and Simulation Model

2.1 The Establishment of Tire Dynamic Impact Test Method

Tire is fixed on the rigid barrier with fixture, using a rail car with a certain speed impacted the tire (see Fig. 1). Then tires' acceleration, speed, and deformation param-

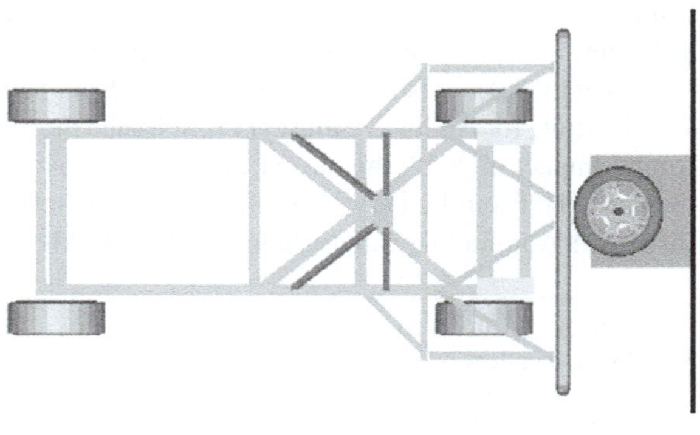

Fig. 1 Tire impact method

eters of mechanical properties are acquired. To analyze the dynamic mechanical characteristics of tire in the process of crash.

2.2 The Establishment of Tire Simulation Model

The tire sub-model in this paper is set up based on the data of a Changan vehicle A tire. Based on the anisotropic characteristics of the tire material [2], tire was simplified and divided into the tread rubber, sidewall rubber, airbag and rim by the Software Hyper Mesh (see Fig. 2).

We use shell element for tire model, common node connection between tread rubber and sidewall, and the material as *MAT_001 for simulation Airbag was composed of the inner surface of the tread, sidewall and rim which formed a closed cavity. Tire initial pressure is 210 kPa, the airbag parameter simulation using *AIRBAG_SIMPLE_PRESSURE_VOLUME [3]. Tire pressure is affected by the coefficient factor beta, CN, and the relative volume. In this analysis model, CN = 1.0 e−4, BEAT = 2.1, the specific pressure formula as shown in (type 1):

$$\text{Pressure} = \beta \frac{\text{CN}}{\text{Relative Volume}} \quad \text{(type 1)}$$

$$\text{Relative Volume} = \frac{\text{Current Volume}}{\text{Initial Volume}} \quad \text{(type 2)}$$

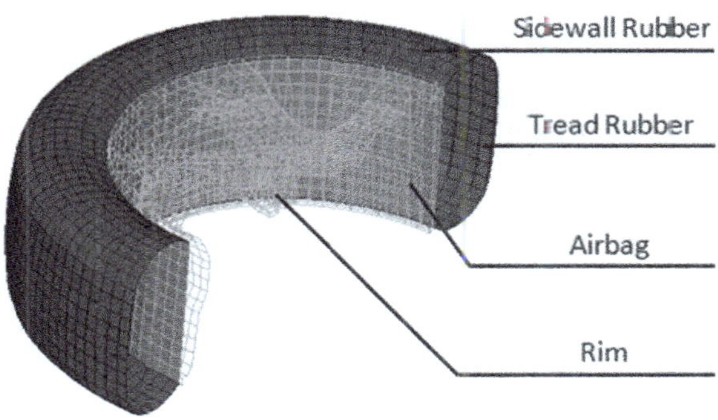

Fig. 2 Tire model

2.3 *Tire Dynamic Impact Analysis Results Compared with the Test Data*

According to the simulation analysis, the acceleration, velocity, displacement, and deformation of tire were acquired and compared with the tire impact test results, as shown in Figs. 3, 4, 5, and 6.

By comparison with the test and simulation, the tire deformation, acceleration curve and speed recovery time in analysis are different from the test data, indicating that the current tire simulation method is not accurate, cannot simulate the real tire reaction.

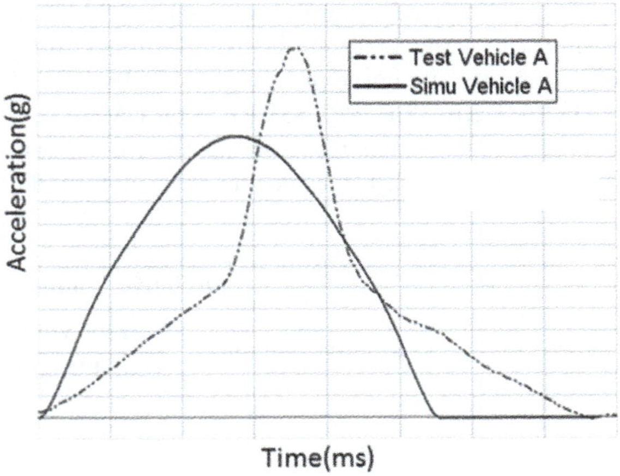

Fig. 3 Tire test and simulation acceleration curve of vehicle A

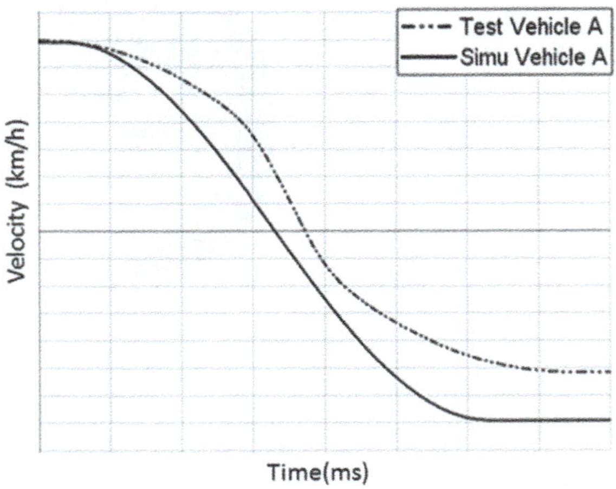

Fig. 4 Tire test and simulation velocity curve of vehicle A

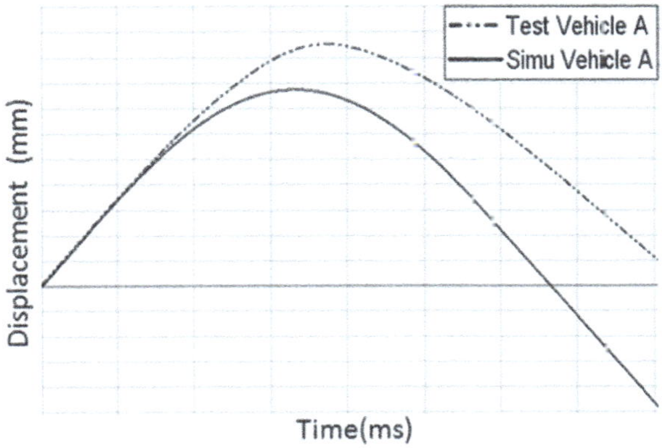

Fig. 5 Tire test and simulation displacement curve of vehicle A

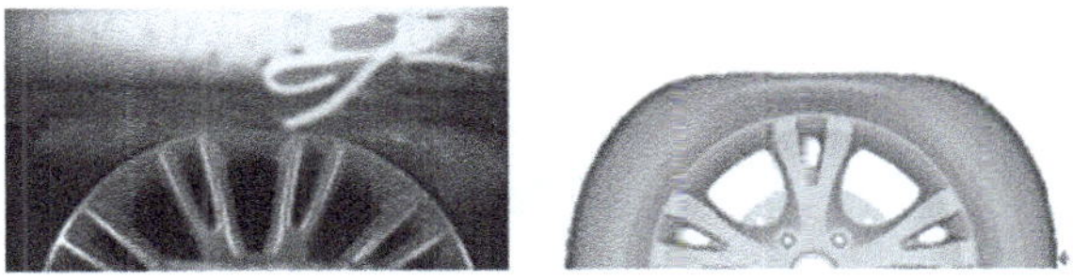

Fig. 6 Tire test and simulation deformation of vehicle A

3 Tire Simulation Model Optimization and the Analysis Results

3.1 *The Influence Factors of Tire Mechanical Properties*

In practice, tire rubber materials and tire pressure were the major influence factors of tire mechanical properties. The tire is composed of fiber, wire, cord, and rubber and has obvious nonlinear characteristic [4]. Effect under force, the stress–strain curve is a nonlinear function.

Tire stiffness is mainly decided by the internal gas pressure of airbag, usually use * AIRBAG_SIMPLE_PRESSURE_VOLUME simulation. This method put the over-all stiffness (tire stiffness and internal pressure) equivalent to the internal pressure, but in the actual use, process need to correct tire pressure and has some limitations.

3.2 *Tire Simulation Model Optimization*

With the analysis of the tire mechanics performance and influence factors, tread rubber and sidewall rubber use the set of parameters in *MAT_ORTHOTROPIC_ELASTIC for simulation, such as young's modulus a, b, c, Poisson's ratio and shear modulus, and so on. Tire pressure use the airbag simulation,

Fig. 7 Airbag inflated curve

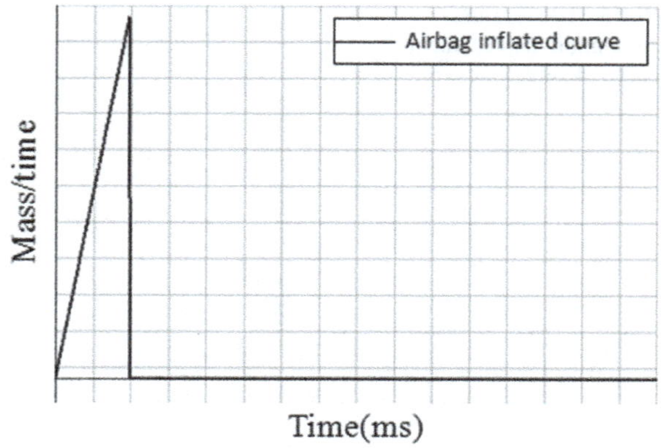

its keyword is: *AIRBAG_SIMPLE_AIRBAG_MODEL, and set the airbag inflated curve, shown in Fig. 7.

3.3 *Optimized Results Compared with the Test*

We optimized the tire material parameters and airbag simulation method to improve the tire stiffness performance, then compared with the test results. From Figs. 8, 9, and 10, it shows that the rail car simulation analysis results are basically in accord with the test data, but the acceleration, velocity, and displacement curve has some differences, the main reasons are as following:

Tire with geometric structure complexity and the diversity of materials, and the simulation model of tire is by simplifying, too idealistic.

Fig. 8 Test and simulation acceleration curve of vehicle A after optimization

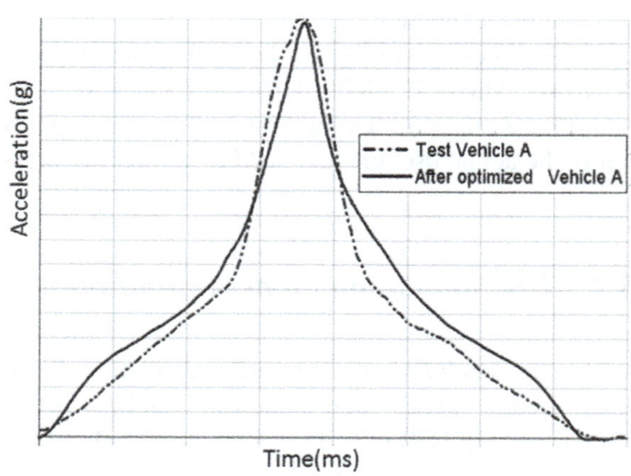

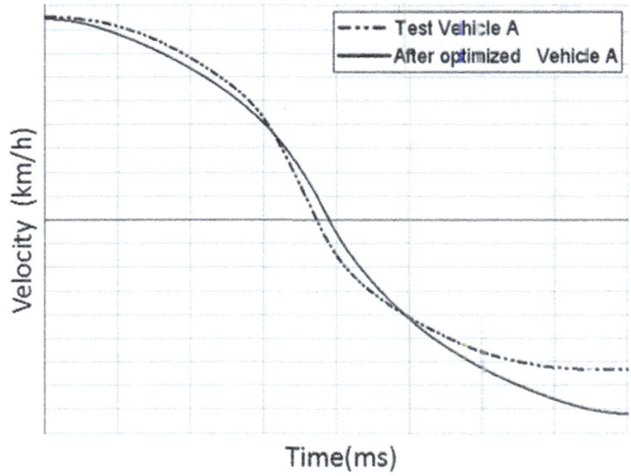

Fig. 9 Test and simulation velocity curve of vehicle A after optimization

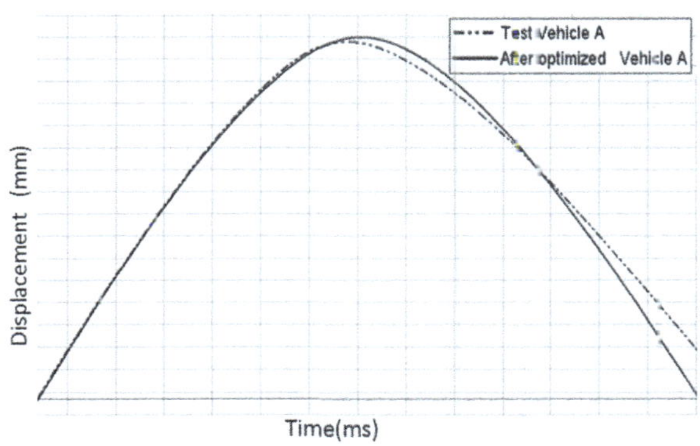

Fig. 10 Test and simulation displacement curve of vehicle A after optimization

In the process of tire impact test, rail car is affected by some situation, such as the friction coefficient, deflection of car and tire, and so on. This all also are the reason for the difference.

To solve the above problems, we will further study the simulation method of tire model in the subsequent R&D process to further improve the simulation accuracy.

4 Tire Simulation Method Validation

4.1 Tire Simulation Analysis and Test Contrast of Vehicle B

Established the vehicle B tire model used the optimized simulation method, and compared the analysis results with test, the parameters such as rail cars, acceleration, velocity, and displacement are more consistent with the test data, but the problems

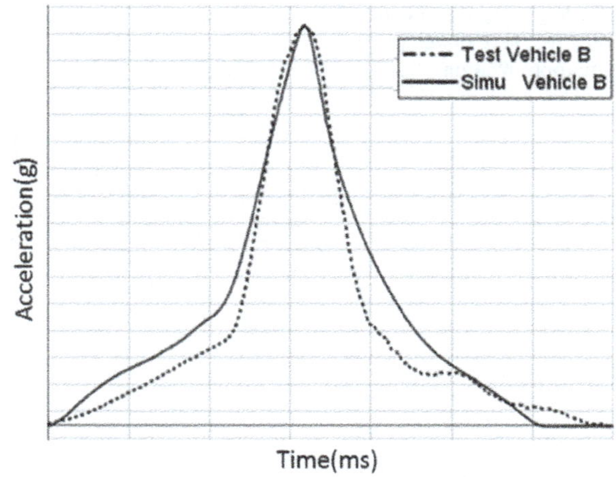

Fig. 11 Tire test and simulation acceleration curve of vehicle B

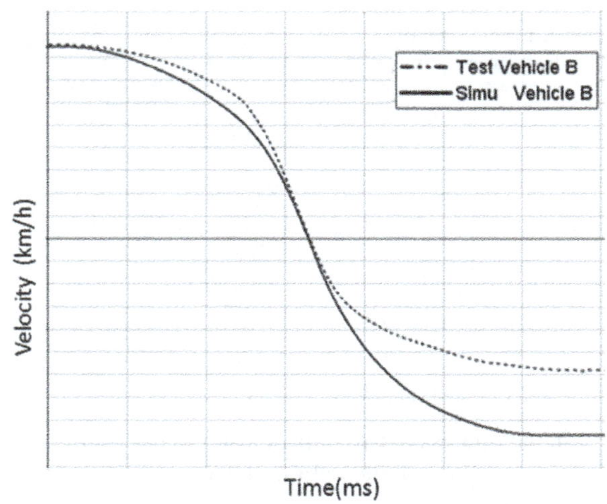

Fig. 12 Tire test and simulation velocity curve of vehicle B

as well as arising from the vehicle A, however, still appear some differences, verified the effectiveness and generality of the tire simulation method (Figs. 11, 12 and 13).

4.2 The Vehicle Simulation Analysis Results Compared with Test Data

In the high-speed frontal impact condition, we used the tire simulation model to analyze the 40% offset impact condition with the speed of 64 km/h, then the vehicle acceleration, firewall deformation, tire deformation and the threshold beam section

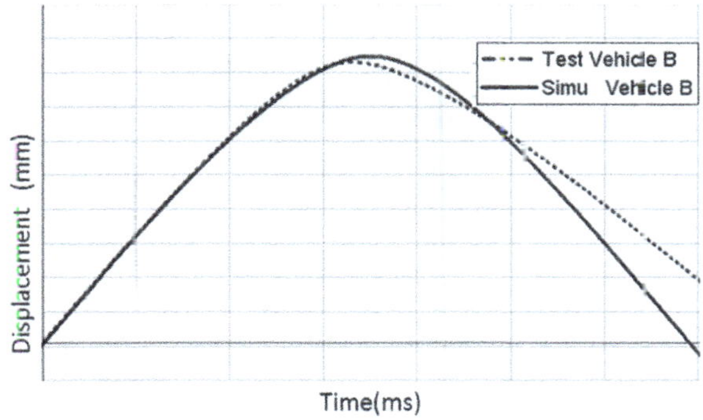

Fig. 13 Tire test and simulation displacement curve of vehicle B

force were used to verify the tire influence on the vehicle crash performance [5] (Figs. 14, 15, 16, 17 and 18).

These results indicated that after optimizing the tire simulation method, in the vehicle collision, the pulse width of vehicle acceleration curve in the simulation

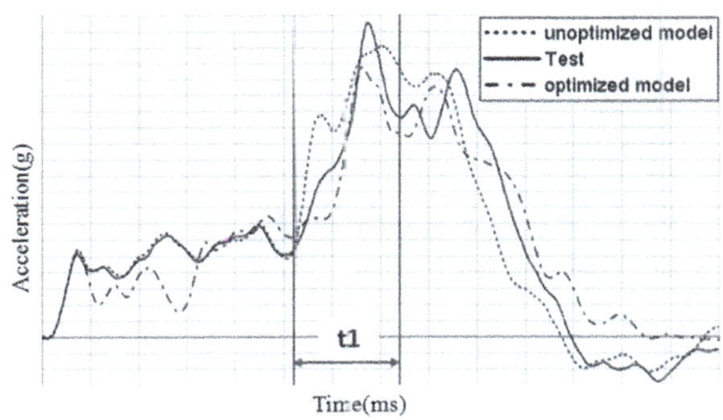

Fig. 14 Vehicle acceleration curve comparison

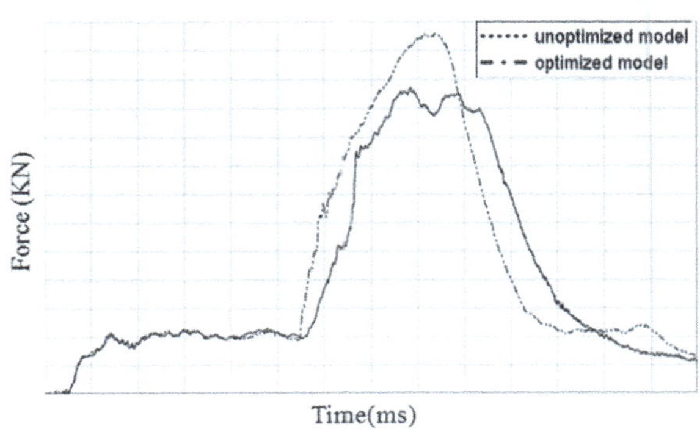

Fig. 15 Comparison before and after optimization of threshold section force

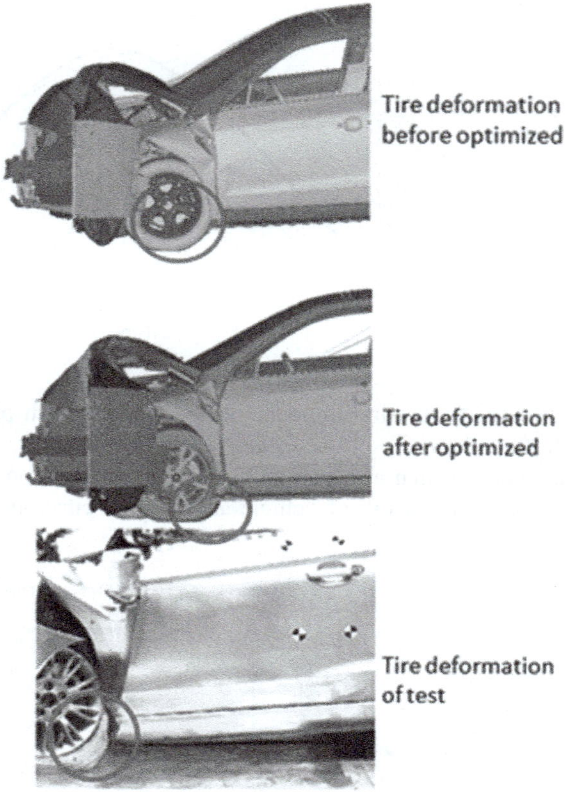

Fig. 16 Tire deformation comparison of vehicle simulation and test

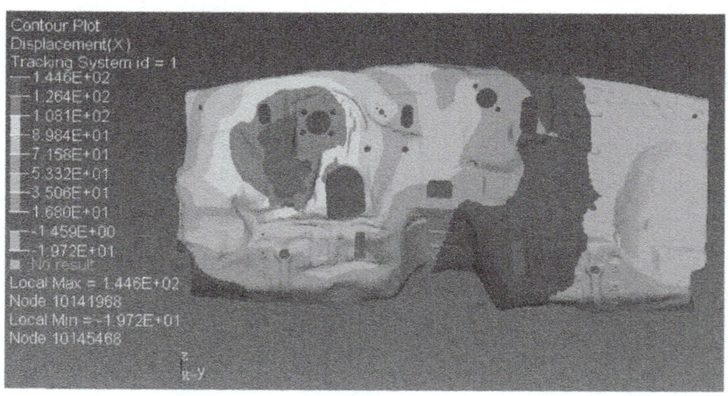

Fig. 17 Firewall deformation before optimization

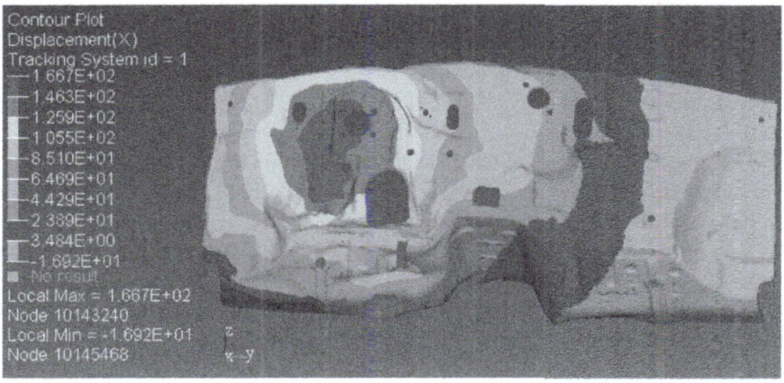

Fig. 18 Firewall deformation after optimization

model increased and its trend becomes closer to the test results. Especially in the t1 time period, reduced the acc curve differences between simulation and tests. Simultaneously, comparing the tire deformation before and after optimization, we could see the tire deformation is much more synchronized to the test results. It also shows that the tire stiffness is larger in the base model. Meanwhile, the threshold beam section force and the firewall deformation also show that, threshold beam section force and deformation of car body is bigger, tire deformation smaller; But when the tire stiffness is low, tire deformation become larger, threshold beam section force and firewall deformation are reduced.

In the process of high-speed offset impact condition, the tire in most cases by extrusion deformation, car body contact with the rim and its stiffness have great influence on the whole crash results. But not considering the influence of the rim in this paper, and will further study in the follow-up research.

5 Conclusion

1. Tire model simulation accuracy is affected by the factors of rubber materials and tire pressure. To improve the simulation accuracy, in this paper, we established the tire dynamic impact test method, and by means of the combination of simulation and test, to optimize the simulation model and then applied to the vehicle model to validate it.
2. The simulation analysis results before optimization cannot simulate the real tire situation. After the tire optimization, the tire model, identical degree of vehicle acceleration, tire deformation in the simulation model agree with the test results more higher, more responsive to the actual situation of tires and effectively improved simulation accuracy.
3. This paper has not considered the tire rotation, rim stiffness, and other factors, and the next step will be to study this.

References

1. Insurance Institute for Highway Safety (IIHS) (May 2014) Small Overlap Crash Test Protocol (Version III), IIHS secretary
2. Si-ba-gen H, Ling Z, Qin S, Lei Z (2015) Optimization of finite element modeling of tire and simulation of its stiffness characteristics. J Hefei Univ Technol (Natural Science)
3. Dyna LS, Keyword User's Manual Version 971
4. Zichen X, Jianyun H, Shitao D, Hua Y, Zhiwei J (2016) Structure design and mechanical properties of new radial-bias tire. Tire Industry
5. Dong C, Shugang X, Xiaoxiao Y, Yanjun H (2013) The differences between IIHS 40% overlap crash and small overlap crash in the requirement of body structure. China Automotive engineers Society

A Method for Estimating the Vertical Load of Tires

Kaifeng Zheng, Zhaoyang Mi, Chao Wang, Yuanxiang Liu and Xiaoce Sun

Abstract The vertical load of tire affects the lateral force and longitudinal force of the tire, thus affecting the longitudinal dynamic characteristics and lateral dynamic characteristics of the vehicle. In vehicle dynamics control, real-time and accurate estimation of wheel vertical load is the basis for calculating other important vehicle motion state parameters. However, there are many shortcomings in the formula of estimating the vertical load of tires at present. One is strong pertinence and poor universality; two is too much simplification and low estimation precision; three is different expression form, poor readability, and poor normalization. Aiming at these problems, the formula for calculating the vertical load of tire is given in detail and verified by CarSim. The simulation results confirm the advantages of the formula.

Keywords Vertical load of the wheel · Lateral dynamics · Roll motion · Real-time estimation

Foreword

In many vehicle active control systems, real-time and accurate estimation of the vehicle's motion state becomes a key and difficult point in designing the controller. However, estimates of the state of these vehicles are based on the known vertical load of the tire. For example, in the automotive electronic stability control system [1], in order to identify whether the vehicle is in a normal steering state or in a limit side slip instability state, it is necessary to estimate the vehicle sideslip angle in real time. However, the estimate of vehicle sideslip angle is calculated on the premise of knowing the vertical load of the tire. In the identification of road friction coefficient and control technology of ABS [2, 3], by estimating the road peak friction coefficient, ABS system can achieve better braking effect under different road conditions. The controller is required to estimate the ground-braking force and the ground normal load in real time. In the evaluation and design of the roll-warning control system [4, 5], it is necessary to know the lateral transfer rate of the automobile tire force,

K. Zheng (✉) · Z. Mi · C. Wang · Y. Liu · X. Sun
China North Vehicle Research Institute, Beijing, China
e-mail: zhengkaifeng1111@163.com

China SAE (ed.), *Proceedings of China SAE Congress 2018: Selected Papers*,
Lecture Notes in Electrical Engineering 574,
https://doi.org/10.1007/978-981-13-9718-9_21

and it is obvious that the vertical load of the tire is required to be estimated in real time. In addition, in electronic differential control [6], four-wheel steering control [7], active suspension control [8], ramp assist control [9], etc., it is necessary to estimate the vertical load of the tire. So, it is very important to estimate the vertical load of the tire in real time and accurately. However, as far as the current literature is concerned, there are many shortcomings in the formula for estimating the vertical load of the wheel. Firstly, it is highly targeted and has poor universality. The tire load calculation formula given in paper [10] is only applicable to lateral dynamics control, not applicable to vehicle acceleration and braking conditions. The formula given in paper [11] is only applicable to longitudinal dynamics control. It is not suitable for the vehicle to turn. Obviously, when developing the chassis integrated control system, the two calculation methods are necessarily combined into one. Secondly, the conditions are to simplify too much [12, 13], leading to the estimation accuracy is low. In order to calculate the effectiveness, the influence of airlift on the vertical load of wheel is usually ignored, which leads to low calculation accuracy. Most of the calculation formulas ignore the forward rake angle of the vehicle body in static balance state. Some vehicles have a certain forward rake angle when the vehicle is stationary, and the forward rake angle is not reflected in the vertical load of the wheel. In addition, most of the formulas ignore the lateral and longitudinal load transfer caused by roll stiffness and pitch stiffness, which will also reduce the accuracy of estimation. Thirdly, the expressions are different. Readability and standardization are poor. It can be seen from the paper [7, 14] that the vertical load calculation formulas of the four wheels seem similar, but differ in the symbols, so it is easy to have many program parameters, poor standardization, and make it easy to make mistakes during the editing process. Based on the above three problems, this paper proposes a universal, high-precision tire vertical load estimation formula, and the formula is expressed in a unified form, which is easy to understand and program.

1 Wheel Vertical Load Calculation

The vertical load of the wheel mainly consists of the following aspects. One is the static load. The other is the airlift caused by the speed of the vehicle relative to the speed of the wind. The third is the lateral load transfer caused by the lateral force. The fourth is the longitudinal load transfer of the wheel caused by the longitudinal force. Figure 1 is a vehicle dynamics model from which the vertical load calculation formula for each wheel can be analyzed.

Static load: when the vehicle is stationary, the vertical load of each wheel includes the corresponding unsprung mass (hereinafter referred to as the wheel mass) and the sprung mass (hereinafter referred to as the body mass). According to the torque balance equation, the static vertical load of the wheel is:

$$F_{\mathrm{zi}}^{A} = \frac{m_{\mathrm{s}} g (L - |L_{\mathrm{xi}}|)}{2L} + m_{\mathrm{wi}} g \tag{1}$$

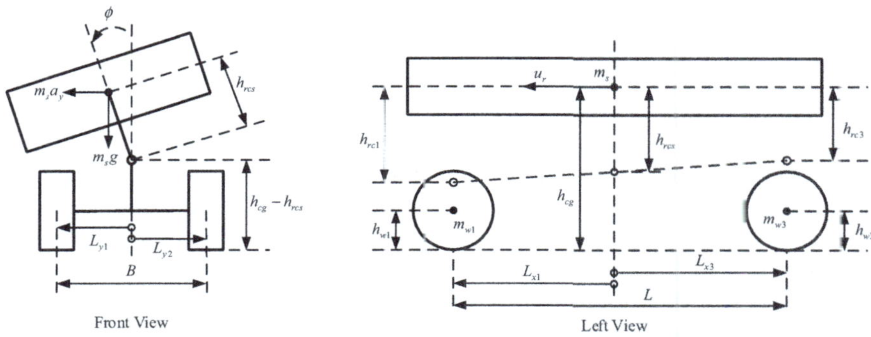

Fig. 1 Vehicle dynamics model

Here, m_s is the body mass. g is gravity acceleration, which is a constant. L is the wheel-track. m_{wi} is the wheel mass. What needs to be highlighted is the subscript of all the parameters and variables in this paper, $i = [1\ 2\ 3\ 4]$ representing the left front wheel, the right-front wheel, the left-rear wheel and the right-rear wheel, respectively. L_{xi} is the distance from the center of each wheel to the center of mass of the car (this value is positive or negative), and $L = L_{x1} - L_{x3} = L_{x2} - L_{x4}$.

Airlift: the lift generated by the flow of air through the top and bottom of the car when the vehicle is moving. Decompose the airlift into the airlift acting on the grounding point of each wheel. The formula is:

$$F_{zi}^{B} = -\frac{C_{ai} A \rho u_r^2}{4} \tag{2}$$

Here, C_{ai} is the equivalent airlift coefficient at each wheel. A is the frontal area. ρ is the air density. u_r is the relative speed, which is the speed of the car when there is no wind.

Wheel load transfer under lateral force: when the car turns, the external wheel load increases due to the lateral moment, and the inner wheel load decreases. This is called load transfer. The torque caused by the load transfer around the longitudinal axis of the vehicle is equal to the torque caused by the lateral force, but the transfer of the front and rear axles is closely related to the structure of the front and rear suspensions. The load transfer of the inner and outer wheels generated by the inertial force of the vehicle is caused by the couple of the lateral inertial force and the ground reaction force not at the same height. The roll torque generated by the lateral inertial force consists of two parts.

Part of the torque generated from the center of mass of the vehicle to the center of the vehicle's roll (rotation around the center of the roll) is transmitted to the ground through the spring, resulting in load transfer. Assume that the equivalent roll stiffness of the front and rear axles is $K_{\phi f}$ and $K_{\phi r}$, converting it to the equivalent stiffness at each wheel, which is $K_{\phi i} = \left[K_{\phi f}\ K_{\phi f}\ K_{\phi r}\ K_{\phi r}\right]/2$, then $C_{\phi i} = \left[C_{\phi f}\ C_{\phi f}\ C_{\phi r}\ C_{\phi r}\right]/2$. Therefore, the equivalent vertical stiffness of the suspension and the equivalent vertical damping of the wheel are:

$$F_{\text{zi}}^{C1} = -\frac{2(K_{\phi i}\phi + C_{\phi i}\dot{\phi})}{B}\text{sign}(L_{\text{yi}}) \tag{3}$$

Here, ϕ is the body roll angle. $\dot{\phi}$ is the body roll angle speed. $K_{\phi i}$ and $C_{\phi i}$ are the equivalent roll stiffness and damping at each wheel. L_{yi} is the distance to the longitudinal center plane of each wheel (this value is positive or negative). B is the wheel-track, $B = L_{y1} - L_{y2} = L_{y3} - L_{y4}$.

The other part is the torque generated by the lateral inertial force ($m_s a_y$) moving from the roll center to the ground, which is transmitted directly from the guidance system to the ground, resulting in load transfer, and its distribution at the four wheels is

$$F_{\text{zi}}^{C2} = -a_y\left[\frac{m_s(L - |L_{\text{xi}}|)(h_{\text{cg}} - h_{\text{rci}})}{\text{LB}} + \frac{2m_{\text{wi}}h_{\text{wi}}}{B}\right]\text{sign}(L_{\text{yi}}) \tag{4}$$

Here, h_{cg} is the distance from the center of mass of the car to the ground. h_{rci} is the distance between the center of the roll at each wheel from the center of mass of the car. h_{wi} is the distance between the wheel centroid from the ground.

Wheel load transfer under longitudinal force: when the car accelerates or brakes, the front and rear axle load transfer will occur. Similarly, the front and rear load transfer will also be two parts, one part is the torque generated after moving from the center of mass of the vehicle to the center of the vehicle's pitch. This part of the torque is transmitted to the ground through the spring, resulting in load transfer.

$$F_{\text{zi}}^{D1} = \frac{(K_{\theta i}\theta + C_{\theta i}\dot{\theta})}{L}\text{sign}(L_{\text{xi}}) \tag{5}$$

Here, θ is the body pitch angle. $\dot{\theta}$ is the body pitch rate. $K_{\theta i}$ is the equivalent pitch stiffness at each wheel. $C_{\theta i}$ is the equivalent pitch damping at each wheel.

The other part is the moment generated by the longitudinal inertial force $m_s a_x$ moving from the pitch center to the ground, which is transmitted directly from the guiding system to the ground, resulting in load transfer, and its distribution at each wheel is

$$F_{\text{zi}}^{D2} = -a_x\left[\frac{m_s h_{\text{cg}} + \sum(m_{\text{wi}}h_{\text{wi}})}{2L}\right]\text{sign}(L_{\text{xi}}) \tag{6}$$

Combining the above aspects, the formula for calculating the wheel load is

$$\begin{aligned} F_{\text{zi}} &= F_{\text{zi}}^{A} + F_{\text{zi}}^{B} + F_{\text{zi}}^{C1} + F_{\text{zi}}^{C2} + F_{\text{zi}}^{D1} + F_{\text{zi}}^{D2} \\ &= \frac{m_s g(L - |L_{\text{xi}}|)}{2L} + m_{\text{wi}}g - \frac{C_{\text{ai}}A\rho u_r^2}{4} \end{aligned}$$

$$- a_y\left[\frac{m_s(L-|L_{xi}|)(h_{cg}-h_{rci})}{LB}+\frac{2m_{wi}h_{wi}}{B}\right]\text{sign}(L_{yi})$$
$$-\frac{2(K_{\phi i}\phi+C_{\phi i}\dot{\phi})}{B}\text{sign}(L_{yi})-a_x\left[\frac{m_sh_{cg}+\sum(m_{wi}h_{wi})}{2L}\right]\text{sign}(L_{xi})$$
$$+\frac{(K_{\theta i}\theta+C_{\theta i}\dot{\theta})}{L}\text{sign}(L_{xi}) \quad (7)$$

2 Wheel Vertical Load Simulation

To verify the correctness of Eq. (7), compare it to the output of commercial vehicle dynamics software CarSim, as shown in Fig. 2. The inputs quantity are the speed of the car-u_r, vehicle longitudinal acceleration-a_x, vehicle lateral acceleration-a_y, body roll angle-ϕ, roll rate-$\dot{\phi}$, body pitch angle-θ, and pitch angle rate-$\dot{\theta}$. The outputs are the vertical load of four wheels-F_{zi}. The parameters used in the simulation are shown in Table 1. In real vehicle applications, these inputs in Fig. 2 are directly available through navigation and sensors.

Simulation case 1. The vehicle travels at a constant speed and the speed is 100 km/h. The simulation results are shown in Fig. 3. From the simulation results, Formula (7) can accurately estimate the change of wheel load. In contrast, taking the left front wheel as an example, considering the influence of airlift, the vertical load of the wheel is different from the reference value by 10 N. When the influence of airlift is neglected, the difference between the value and the reference value is 100 N. Compared with the static value of 1.3%, it can be seen that the airlift has a certain influence on the vertical load of the wheel, but the influence is small. When the vehicle speed is low, the effect of airlift is completely ignored.

Simulation case 2. First, accelerate the vehicle to 90 km/h, then brake at 0.8 MPa pressure until the end of the simulation. The simulation results are shown in Fig. 4.

Taking the left front wheel as an example, it can be seen from Fig. 4c that the vertical load of the wheel calculated by Eq. (7) differs from the reference value by a maximum of 100 N, which is relative to 1.3% of the static value. It can be seen that Formula (7) can accurately estimate the change of wheel load in the accelerated

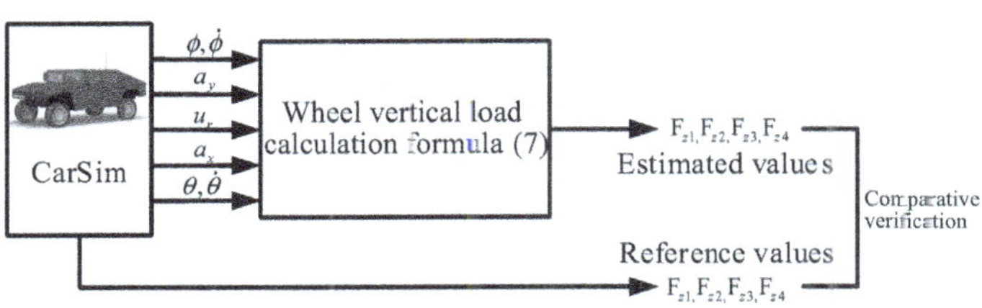

Fig. 2 Simulation comparison verification

Table 1 Vehicle simulation parameters

Parameters	Values	Meanings
m_s	2210 kg	Sprung mass
L_{xi}	[1.07 1.07 −2.232 −2.232] m	The distance from the spring loaded center of mass
L_{yi}	[0.95 −0.95 0.95 −0.95] m	The distance between the wheel and the longitudinal plane of the vehicle
$K_{\phi i}$	[50 50 50 50] kN/rad	Equivalent rollover stiffness at each wheel
$C_{\phi i}$	[5 5 5 5] kN/rad/s	Equivalent roll damping at each wheel
$K_{\theta i}$	[70 70 80 80] kN/rad	Equivalent pitching stiffness at each wheel
$C_{\theta i}$	[10 10 10 10] kN/rad/s	Equivalent pitching damping at each wheel
k_ϕ	180 kN/rad	Total equivalent tilting stiffness
c_ϕ	16 kN/rad/s	Total equivalent roll damping
k_θ	400 kN/rad	Total equivalent pitching stiffness
c_θ	40 kN/rad/s	Total equivalent pitching damping
I_{xxb}	1243	Body roll moment of inertia
I_{yyb}	4332	Body pitch moment of inertia
m_{wi}	[62.5 62.5 62.5 62.5] kg	Unsprung mass at each wheel
h_{wi}	[0.5 0.5 0.5 0.5] m	The height of the center of the unsprung mass at each wheel
h_{rci}	[0.52 0.52 0.54 0.54] m	Distance between the centerline of the roll at each wheel from the centroid of the sprung
h_{rcs}	0.53 m	Distance of the corrugated center of mass from the centerline of the roll
h_{cg}	0.66 m	Reed centroid height
θ_0	0.23°	Initial pitch angle of body
C_{ai}	[0.2 0.2 −0.05 −0.05]	Equivalent airlift coefficient at each wheel
A	2.4 m^2	Frontal area
ρ	1.2 kg/m^3	Air density

braking condition. In contrast, when ignoring the influence of the equivalent stiffness and damping of the suspension, i.e., the influence of Eq. (5) is neglected, the vertical load of the wheel is different from the reference value by 300 N, which is 3.8% relative to the static value. Equation (5) has little effect on the wheel load and can be ignored in the case where the vertical load accuracy of the wheel is not high.

Simulation case 3: double-lane change condition, the speed is 100 km/h. The simulation results are shown in Fig. 5. Taking the left front wheel as an example, it can be seen from Fig. 5c that the maximum difference between the vertical load of the

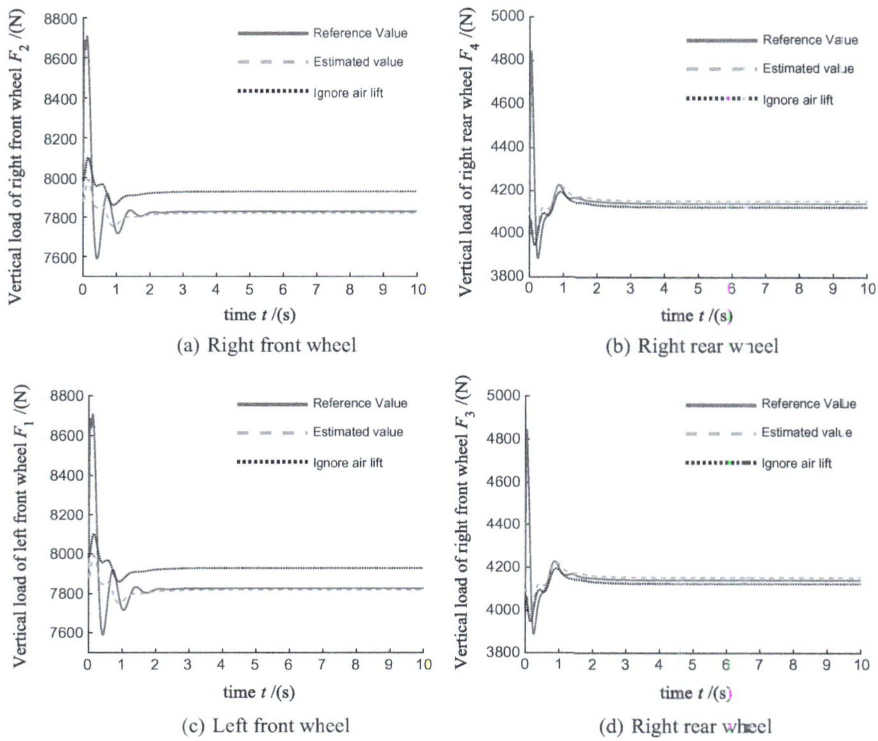

Fig. 3 Simulation comparison of uniform liner driving

wheel calculated by Eq. (7) and the reference value is 150 N, which is 1.9% relative to the static value. It is confirmed that the Formula (7) can well estimate the change of the wheel load under the double-lane change condition (steering condition). In contrast, when only the influence of the equivalent damping of the suspension is ignored, that is, when the damping in Eq. (3) is ignored, it has little effect on the result. While ignoring the equivalent stiffness and damping of the suspension, i.e., when Eq. (3) is ignored, the result varies greatly. It can be seen that the stiffness term in Eq. (3) has a great influence on the vertical load of the wheel and cannot be ignored. The damping term has little effect on the vertical load of the wheel and can be ignored.

3 Vehicle Roll and Pitch Motion State Estimation

From the above analysis, it can be concluded that the vertical load of the wheel can be estimated by using Eq. (7), whether in a uniform driving condition, an accelerated braking condition, or a steering condition. And, the estimation accuracy is very high.

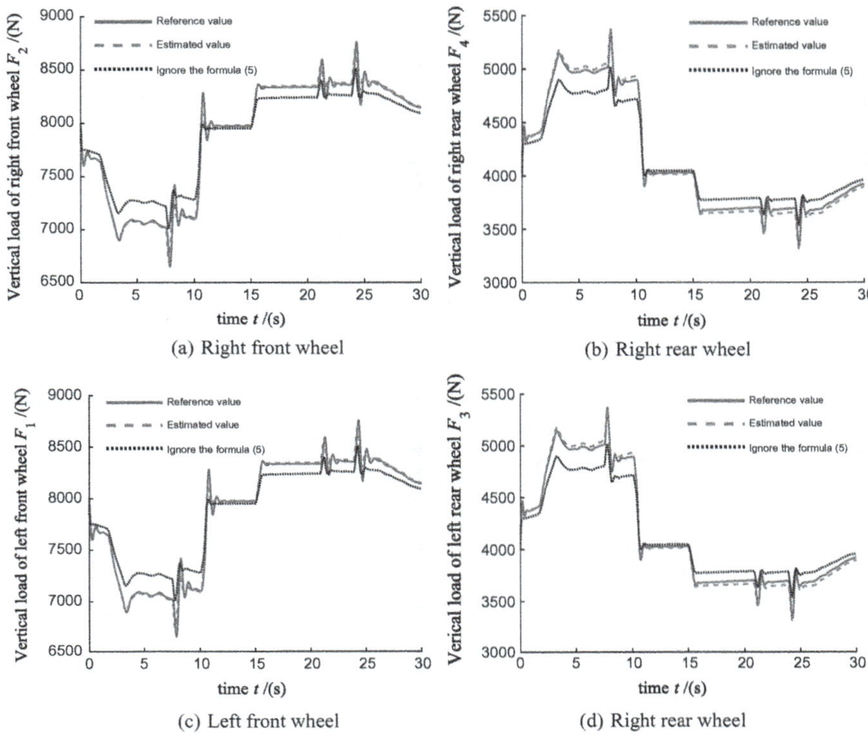

Fig. 4 Simulation comparison of braking conditions after acceleration

However, we can also see from Fig. 2, the calculation of the vertical load of the tire using Eq. (7) requires real-time acquisition of the roll and pitch motion states of the vehicle body by the sensor, which undoubtedly increases the complexity and cost of the system. Therefore, it is very important to estimate the roll and pitch motion state of the vehicle body by other known quantities.

According to the analysis of Fig. 1, the roll moment of the body includes these aspects: (1) the body mass biasing torque generated by the roll; (2) the resistance torque of the equivalent roll stiffness of the suspension, including the moment of the suspension spring and the stabilizer bar; (3) the resistance torque generated by the equivalent angular damping of the suspension; (4) the torque produced by the active suspension control force. According to the moment balance, the differential equation of the body roll motion can be obtained as

$$\left(I_{xxb} + m_s h_{\mathrm{rcs}}^2\right)\ddot{\phi} + c_\phi \dot{\phi} + \left(k_\phi - m_s g h_{\mathrm{rcs}}\right)\phi = m_s h_{\mathrm{rcs}} a_y \tag{8}$$

When ignoring the influence of roll acceleration and speed, the algebraic expression of the body roll angle can be directly obtained according to the above formula [5]

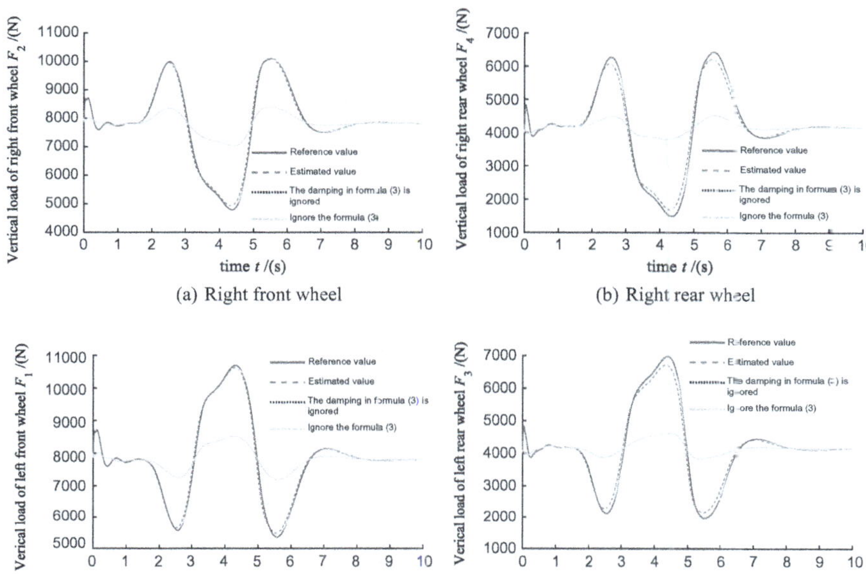

(a) Right front wheel (b) Right rear wheel

(c) Left front wheel (d) Right rear wheel

Fig. 5 Comparison of double-lane change simulation conditions

$$\phi = \frac{m_s h_{rcs} a_y}{k_\phi - m_s g h_{rcs}} \tag{9}$$

Similarly, the pitching moment of the body includes these aspects: (1) the body mass offset torque generated by the pitch; (2) the resistance torque of the equivalent pitch stiffness of the suspension; (3) the resistance torque generated by the equivalent angular damping of the suspension; (4) the pitch torque produced by the active suspension. According to the moment balance, the differential equation of the vehicle pitch motion is

$$\left(I_{yyb} + m_s h_{rcs}^2\right)\ddot{\theta} + c_\theta \dot{\theta} + (k_\theta - m_s g h_{rcs})\theta = -m_s h_{rcs} a_x \tag{10}$$

With the longitudinal acceleration and the lateral acceleration as inputs, the body roll and pitch motion states can be estimated by Eqs. (8) and (10), and the vertical load of the wheel is calculated by Eq. (7). The principle is shown in Fig. 6. The simulation analysis results are shown in Figs. 7, 8 and 9. It can be seen from Fig. 7 that the body roll angle and the roll angular velocity calculated by the Formula (8) are basically consistent with the reference value, and the maximum difference between the roll angle and the reference value is only 0.2°. In addition, the body roll angle calculated by the differential Eq. (8) and the algebraic Eq. (9) is basically equal, and the error is small, that is, when the vehicle body transient change is not violent, the Formula (9) can be substituted for the Formula (8) to estimate the body roll angle. It

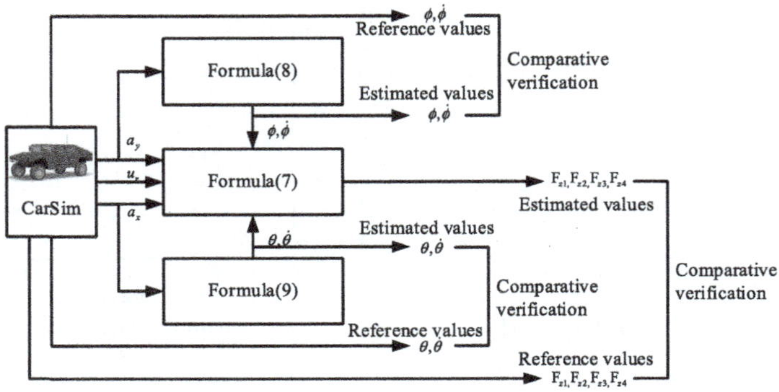

Fig. 6 Calculate the vertical load of the wheel from the estimated body motion state

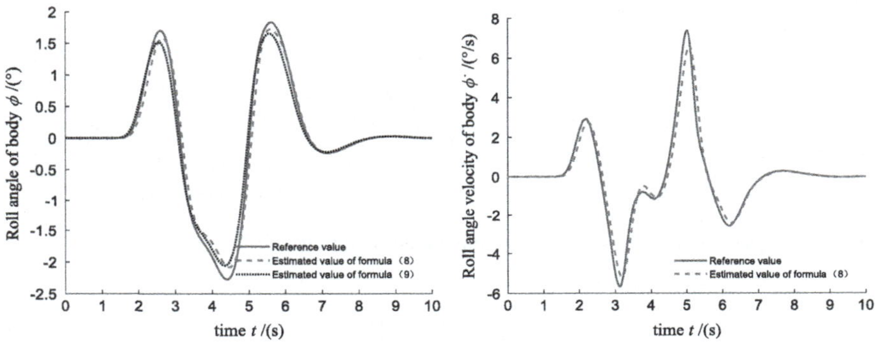

(a) body roll angle (b) body roll angle velocity

Fig. 7 Estimation of the roll motion state of the body under double-lane change condition

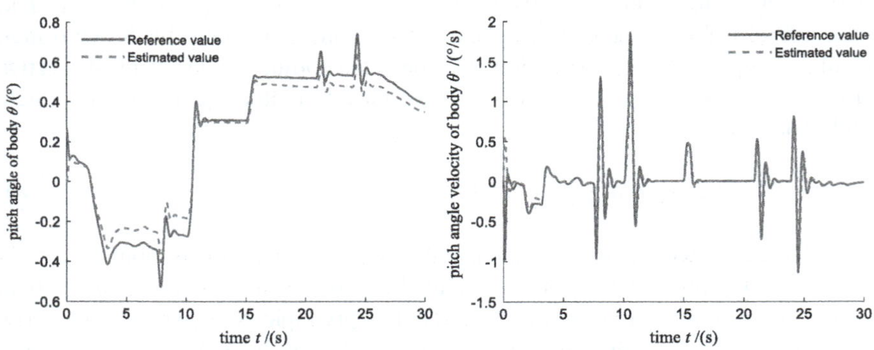

(a) body pitch angle (b) body pitch angle velocity

Fig. 8 Estimation of pitch motion state of accelerated braking conditions

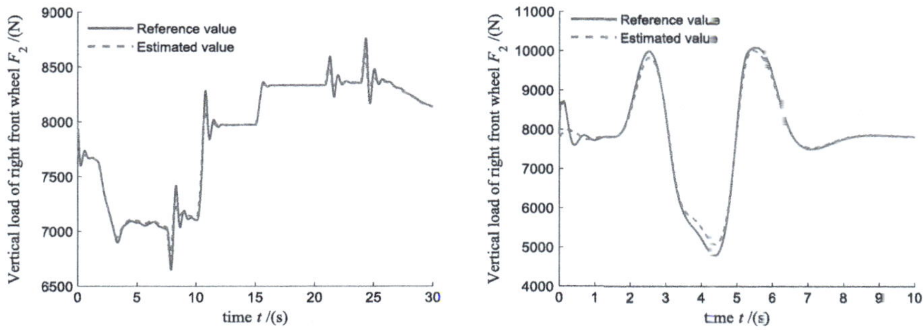

(a) The vertical load of the left front wheel under the accelerated braking condition (b) The vertical load of the left front wheel under double lane change

Fig. 9 Calculate the vertical load of the wheel based on the estimated body motion state

can be seen from Fig. 8 that the body pitch angle and the pitch angle speed calculated by Eq. (10) are basically consistent with the reference value, and the maximum pitch angle difference is only 0.25°. Figure 9 shows the vertical load of the vehicle using the estimated body motion state under double-lane change and accelerated braking conditions. It can be seen from Fig. 9 that the estimated value and the reference value are substantially coincident, and the estimation accuracy is high thereby confirming that the body roll and pitch motion states can be estimated by using Eqs. (8) and (10). And, calculating the vertical load of the wheel according to this formula is feasible and accurate.

4 Simplification of the Formula for Calculation of the Vertical Load of the Wheels

It can be seen from the above analysis that the damping terms in the airlift Eqs (2), (5), and (3) have little effect on the vertical load of the wheel. In addition, in the case where the transient change of the vehicle body is not violent, the body roll angle can be estimated by the Formula (9), so the Formula (7) can be simplified as

$$\begin{aligned} F_{\text{zi}} = {} & \frac{m_{\text{s}} g(L - |L_{\text{xi}}|)}{2L} + m_{\text{wi}} g \\ & - a_y \left[\frac{m_{\text{s}}(L - |L_{\text{xi}}|)(h_{\text{cg}} - h_{\text{rci}})}{\text{LB}} + \frac{2m_{\text{wi}} h_{\text{wi}}}{B} \right] \text{sign}(L_{\text{yi}}) \\ & - \frac{2m_{\text{s}} h_{\text{rcs}} a_y K_{\phi i}}{B(K_\phi - m_{\text{s}} g h_{\text{rcs}})} \text{sign}(L_{\text{yi}}) \\ & - a_x \left[\frac{m_{\text{s}} h_{\text{cg}} + \sum (m_{\text{wi}} h_{\text{wi}})}{2L} \right] \text{sign}(L_{\text{xi}}) \end{aligned} \tag{11}$$

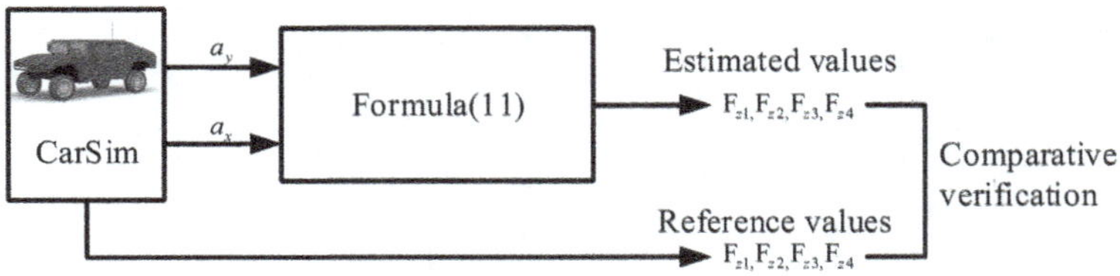

Fig. 10 Simplified estimation of wheel vertical load

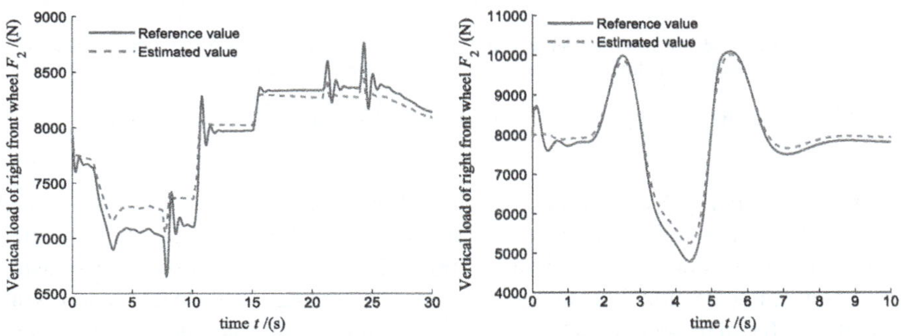

(a) The vertical load of the left front wheel under the accelerated braking condition

(b) The vertical load of the left front wheel under double lane change

Fig. 11 Simulation results of the simplified formula of the wheel vertical load

The correctness of the verification Formula (11) is verified by simulation analysis. The simulation principle is shown in Fig. 10. The simulation conditions are the same as above, and the simulation results are shown in Fig. 11. It can be seen from Fig. 11 that the method can correctly estimate the change of the wheel load regardless of the acceleration braking condition or the double shifting line condition. Although the estimated accuracy is not as high as that of Eq. (7), this method reduces the sensor configuration and reduces the computational complexity of the single-chip microcomputer. It has good practicability in vehicle control that does not require high vertical load calculation.

5 Conclusion

In this paper, a formula for estimating the vertical load of a tire with universality and high precision is proposed. The formula is unified and easy to understand and program. The validity of the formula is verified by simulation analysis. On this basis, by introducing the estimation of the roll and pitch motion state of the vehicle body and then estimating the vertical load of the tire, the number and cost of the sensors used in the system can be reduced. The simulation analysis verifies the correctness and accuracy of the method. Finally, on the basis of the above analysis, a simplified form of the tire load calculation formula is given by ignoring some items with smaller influence factors. Therefore, a compromise scheme is given in terms of calculation accuracy, real-time performance, system cost, and reliability.

References

1. Guo J (2011) Research on control algorithm and performance evaluation. Jilin University, Changchun
2. Zhang B (2008) Research on control technology and road identification of ABS. Beijing Institute of Technology, Beijing
3. Qi Z (2004) Research on control technology of antilock braking system in integrated ABS/ASR/ACC system for motor vehicles. Beijing Institute of Technology, Beijing
4. Zhou X (2013) Research on vehicle stability modeling and control strategy based on yaw and roll. Chongqing University of Technology, Chongqing
5. Wang Z, Liao M et al (2013) The study of the way to roll control. M&E Eng Technol 42(7):117–120
6. Zhao Y, Zhang J (2008) Study on electronic differential control system of independent in-wheel motor drive electric vehicle. J Syst Simul 20(18):4767–4775
7. Du F (2009) Simulation research on control strategies for active 4WS vehicle based on the steer-by-wire technology. Chang'an University, Xi'an
8. Lu F (2014) Study on vehicle vibration state observation algorithm based on the nonlinearity of suspension. Beijing Institute of Technology, Beijing
9. Li Y (2012) Research on the identification of mass for heavy vehicle and the estimation of road gradient. Jilin University, Changchun
10. Jin Z (2008) Rollover stability and anti-roll control for sport utility vehicles. Nanjing University of Aeronautics and Astronautics, Nanjing
11. Villagra J, D'Andrea-Novel B, Fliess M et al (2011) A diagnosis-based approach for tire-road fores and maximum friction estimation. Control Eng Pract 19:174–174
12. Zhang Y (2010) Research on state parameters estimation and control algorithm of electronic stability control system for passenger car. Jilin University, Changchun
13. Li G (2013) Study on stability and energy saving control for X-by-wire electric vehicle with four-wheel independent drive in-wheel motors. Jilin University, Changchun
14. Shim T, Ghike C (2007) Understanding the limitations of different vehicle models for roll dynamics studies. Veh Syst Dyn 45(3):191–216

The Characteristic of Transient HC Emissions During Cold Start on A Port-Fuel-Injection Gasoline Engine

Huan Chen, Chuanhui Cheng, Honglin Xu, Tao Wu, Zheng Xu and Yunchao Wang

Abstract The percentage of HC emission before catalyst light-off is beyond 90% in WLTC driving cycle. It is a big challenge to reduce the HC emission during cold start phase due to poor mixture preparation from port injection spark ignition engines. This study focuses on cold start period to investigate HC emissions based on cycle-by-cycle method, using fast FID HC analyzer (HFR500) and cylinder pressure indicating system. The results have been compared with two different types of fuel injectors installed on a 1.5-L port-fuel-injection gasoline engine. The transient HC emission of first 10 cycles has been analyzed. The test results also show that HC emissions of the type B injector (8-hole) are 11% lower than that of the type A injector (6-hole) under same fuel injection strategy.

Keywords Port fuel injection · Gasoline engine · Cold start · Transient HC emissions · Injector

1 Introduction

More stringent China 6 emission regulation for light-duty vehicles will be implemented based on new test cycle—WLTC (world light test cycle) in the year of 2020, which brings new challenges for the vehicle emissions control of the gasoline engines. HC emissions during cold start and subsequent engine warm-up contribute more than 90% of the WLTC test, before three-way catalytic converters reaching "light-off" temperature [1–3]. Optimizing engine control strategies and combustion system, to reduce HC emissions during this portion of the test is necessary to meet the more stringent China 6 emissions regulations.

It is known that high cold start emissions come from the unfavorable mixture preparation process. During this process, a large portion of excess injection fuel does not vaporize and enters the cylinder as liquid because intake port, valves, and chamber

H. Chen (✉) · C. Cheng · H. Xu · T. Wu · Z. Xu · Y. Wang
Shanghai Key Laboratory of Automobile Power-Train, SAIC Motor Technical Center, Shanghai 201804, China
e-mail: chenhuan01@saicmotor.com

China SAE (ed.), *Proceedings of China SAE Congress 2018: Selected Papers*, Lecture Notes in Electrical Engineering 574,
https://doi.org/10.1007/978-981-13-9718-9_22

wall are in low temperature. This fuel existing in combustion chamber survives as unburned during the subsequent combustion cycles and comes out into the exhaust gas, which contributes significantly to HC emissions. Furthermore, the aggressive transient behavior of engine speed and throttle during cold start also causes air and fuel mixture insufficient homogeneous, which leads to poor combustion quality and increases further level of HC emissions. Vehicle manufacturers and researchers focus on the investigations of reducing gasoline engine cold start emissions for decades. Horie et al. [4] in Honda technical center used fast response HC detector to study the first 30 cycles of cold start process and optimized injection strategies to get better air/fuel ratio (AFR), which reduced 14% HC emissions at Federal Test Procedure (FTP) 75 cycle. In 2006, Lee et al. [5] researched the effects of various intake charge motion control valves on mixture preparation, combustion, and HC emissions during the first 20 s of cold start process. The results indicate that the moderate swirl and tumble intensities will increase combustion stability and burning rate and reduce HC emissions. Li et al. [6] investigated the characteristic of mixture formation and transient HC emissions of the first firing cycle on a liquefied petroleum gas (LPG) motorcycle engine, which showed that the spark timing and AFR greatly affected the HC emissions of cold start process. In 2009, Yu et al. [7] researched transient HC emissions on a 1.5-L PFI hybrid electrical vehicle engine and implemented cycle-by-cycle control strategy for emission reduction at both cold and hot start condition. In 2016, Rodriguez et al. [8] studied the characteristic of HC and PM/PN emissions of the first 3 cycles on a 2.0-L GDI engine through the cycle-by-cycle analysis, and the HC emission of each cycle shows varied sensibility to AFR, injection, and spark timing.

More and more cycle-by-cycle combustion analysis research indicates that cycle-by-cycle control strategy is a promising method to optimize fuel injection and reduce the cold start emissions. At the initial phase of engine development project, cycle-by-cycle combustion analysis with the fast response emissions analyzer can effectively estimate the cold start emission level, which is useful tool/technology for hardware development of combustion system. In this study, the characteristic of HC emissions of the first 10 firing cycles has been investigated with two different kinds of fuel injectors based on cycle-by-cycle combustion analysis with fast response HC analyzer.

2 Experimental Setup

The experiment was carried out on a 1.5-L, 4-cylinder, PFI gasoline engine. The engine specifications are given in Table 1. The engine has a masking at intake valve seat, which can improve mixture preparation due to increased tumble intensities, big intake cam duration to realize Atkinson cycle reducing fuel consumption, and an intelligent thermal management split cooling system to accelerate the engine warming process. The engine was started up in no-load condition by disconnecting

Table 1 Engine specification

Item	Engine parameters
Engine model	PFI, 4-cylinder, 4-stroke, DOHC
Displacement (L)	1.498
Bore/stroke (mm)	75/84.8
CR	11.5
Injector pressure (bar)	3.8
Fuel	RON 92#

to dynamometer, then adjusting the accessory system load to simulate the operation of engine fast idle running in vehicle.

To keep the engine run under same conditions during the tests, the air/fuel/coolant has been well controlled. Figure 1 shows the schematic diagram of the test bench. The engine was instrumented to measure the transient process, including the transient HC and combustion analyzers. All experiment data have been taken from cylinder 1#. The cylinder pressure has been collected with a Kistler 6125c quartz pressure transducer mounted on the cylinder head. A Cambustion fast response HC analyzer (Cambustion HFR500, $t_{10\text{–}90\%}$ ~ 1 ms) was used to measure the transient HC emissions from the exhaust port of cylinder 1#, as shown in Fig. 2. A hot film air mass flowmeter

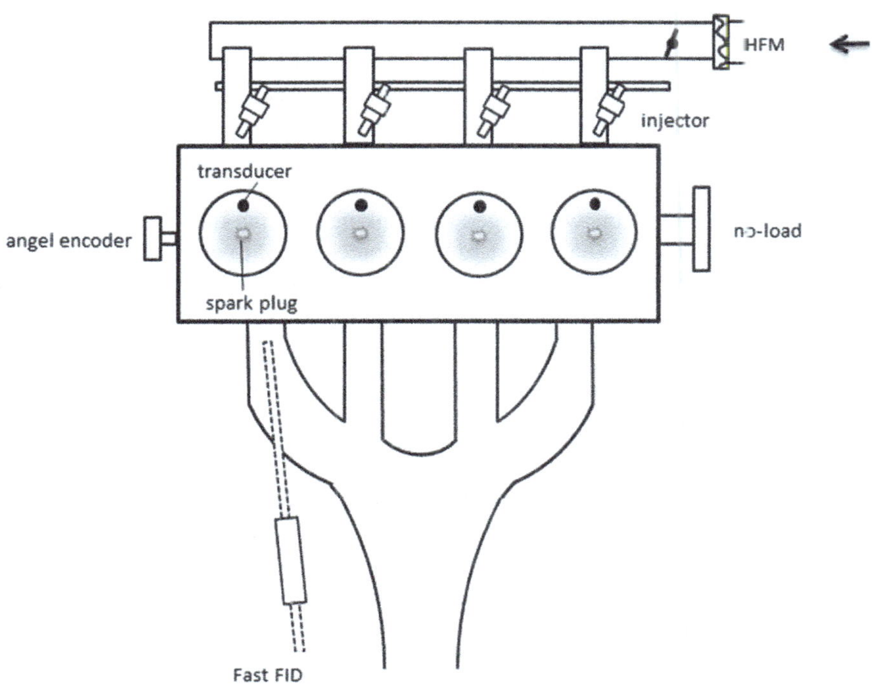

Fig. 1 Schematic diagram of the test bench

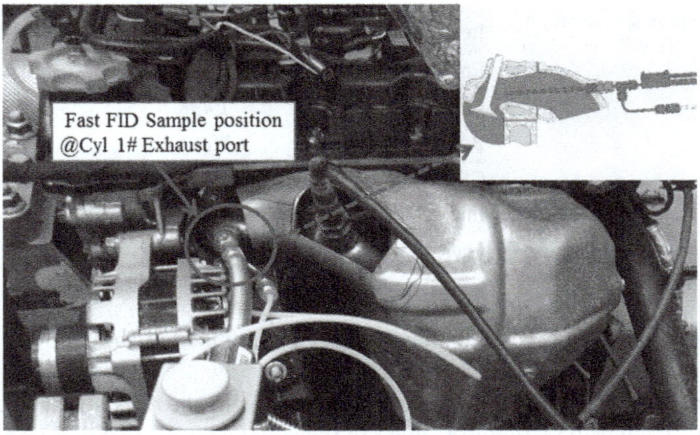

Fig. 2 Fast FID HC sampling position

was used to measure engine intake air mass flow. Cylinder pressure and transient engine out HC have been simultaneously recorded by AVL IndiModul 622 system, and the data acquisition was triggered by the optical encoder at every 1 °CA. Coolant temperature and pressure, fuel temperature and pressure, etc., have been controlled and recorded with Horiba STARS bench operating system. All start-up tests were conducted at approximately 25 °C environmental temperature. The coolant temperature was initially controlled at 25 °C by the external heat exchanger. Moreover, before each test, the engine was cranked manually to ensure the piston of cylinder 1# at 90 °CA ATDC of the expansion stroke to keep consistency of experiment condition. After each test, the engine was warmed up to 90 °C coolant temperature and run at steady idle for 30 min to bleed residual fuel building-up in the intake port. After shutdown, the engine was purged by cranking with the fuel cut off for 2 min, and then the engine and exhaust system were force-cooled to ambient temperature. Afterward, the engine was restarted for another test and data recording was triggered at the same time.

3 Research Method

In this work, the accurate bench simulation of vehicle test over cold start process was investigated, Figs. 3 and 4 show the bench simulation results compared with actual fast idle speed of engine onto the vehicle. Upon the first cylinder firing, the engine experiences a speed flare, resulting in a rapid increase in intake air mass flow rate. The engine then decelerates to the normal idle speed; therefore, intake air mass flow rate also transits to the idle demanding state. The results indicate that the engine speed and intake air mass flow in actual vehicle test can be precisely simulated by the engine bench test, especially for the first 2 s.

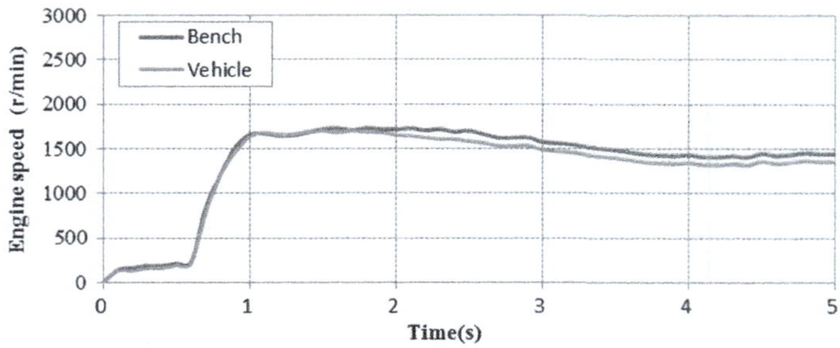

Fig. 3 Engine speeds of test bench and vehicle at cold start

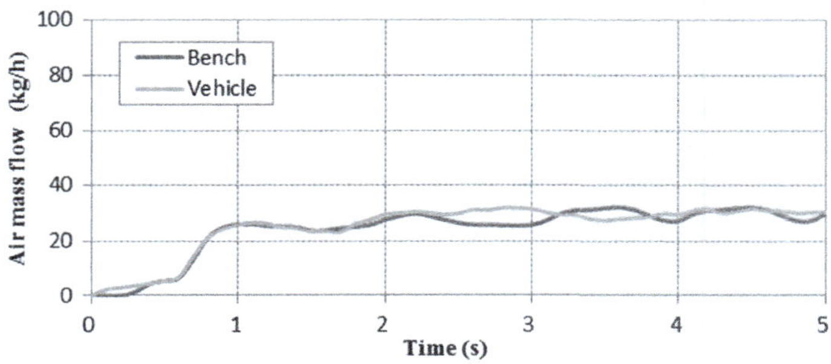

Fig. 4 Intake air mass flow rate of test bench and vehicle at cold start

Figure 5 reveals the engine out HC emissions (measured by Horiba MEXA-7100DEGR) of both bench test result and actual vehicle, and they are almost the same during first 2 s cold start process. Thus, the bench test can be used to study the first 10 combustion cycles (about 1.5 s) during engine cold start. This study focuses on the characteristics of transient HC emissions during first 10 firing cycles at cold start process and the HC emissions have been compared with two kinds of fuel injector.

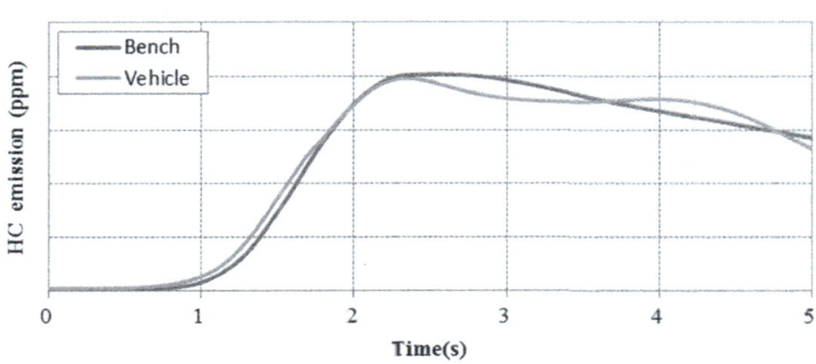

Fig. 5 Cold start HC emissions of test bench and vehicle at cold start

4 Results and Discussion

The two kinds of injector have been studied on test bench. Table 2 gives some features of fuel injector A and B. The parameters of fuel spray pattern, static flow, and setup position are the same. The fuel spray SMD of injector B is 2.6 μm less than injector A, due to its two more injection holes. The fuel injection mass of first 10 cycles during cold start is the same for injector A and B, as shown in Fig. 6.

The behavior of combustion in the first several cycles is very important for cold start period, and it is widely known that the first several cycles have great relationship with each other and effect on the mixture formation, cylinder, and port wall temperature as well as emission characteristics due to the complex transient process and fuel film evaporation characteristics in intake port wall. Since only a small portion of injected fuel vaporizes and contributes to a combustible mixture in a cold engine, the first combustion cycle generally involves significant overfueling. Figure 6 indicates that the fuel injection amount of first cycle is 4 times that of normal combustion cycle. This kind of fuel enrichment provides a sufficient amount of vaporized fuel so that an acceptable degree of drivability can be assured. The fuel injection amount decreases rapidly from the second cycle, while the residual fuel coming from the first cycle injection was still slowly evaporating and gradually entering succedent several combustion cycles, which affects the proportion of fuel and gas mixing in the latter several combustion cycles.

Table 2 Injector parameters

Injection pressure 3.8 bar	Injector A	Injector B
Hole numbers	6 holes	8 holes
Photograph	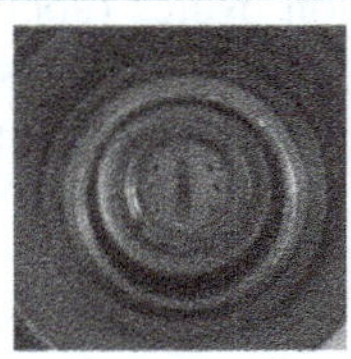	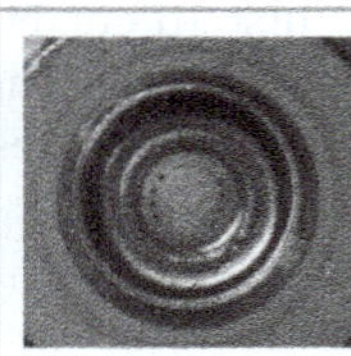

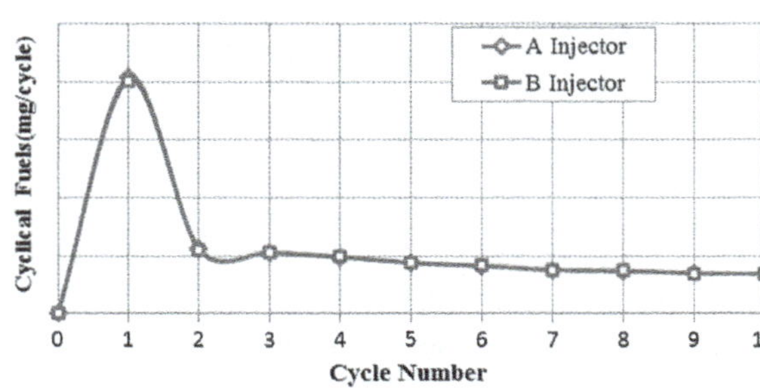

Fig. 6 Cyclic fuel mass of the two injectors during first 10 cycles

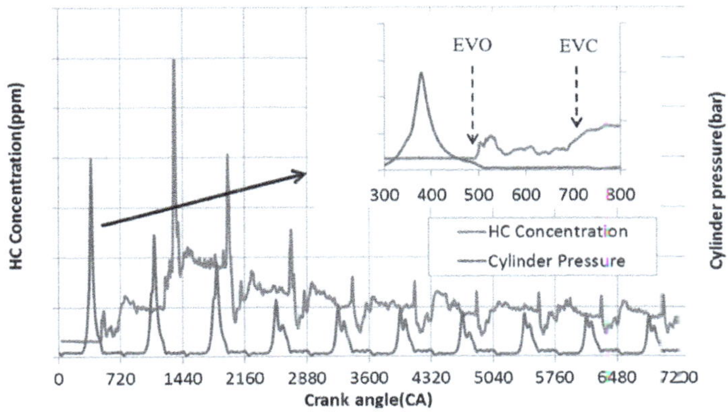

Fig. 7 Transient characteristics of the injector A during the first 10 combustion cycles

Figure 7 demonstrates the first 10 combustion cycles and HC emission of injector A during cold start period. The first cycle has good combustion and low HC emission, because the calibration setting was already optimized for injector A and the fuel and gas mixture formation in-cylinder is nearly stoichiometric mixing. When coming to the second combustion cycle, due to rising intake port wall temperature, the residual fuel of the first cycle injection is accelerated to evaporate and remained into the second combustion cycle,which formed a rich mixture and contributed to a high HC concentration after exhaust valve open. The fuel and air mixture of the latter third and fourth combustion cycles is similar to that of second combustion cycle; i.e., the AFR inside cylinder varies from rich to stoichiometric as wall fuel film evaporation gradually decreased. The HC emission keeps a lower level after the fifth combustion cycle due to the wall fuel film is nearly vaporized totally.

Figure 8 gives the first 10 combustion cycle and HC emission of injector B during cold start period. The fuel evaporation rate is faster than injector A since injector

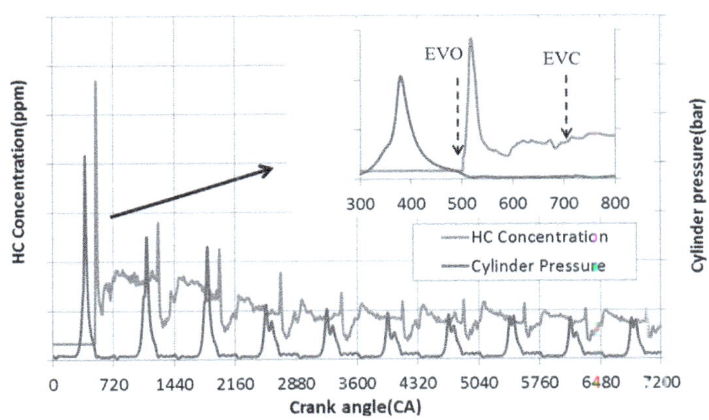

Fig. 8 Transient characteristics of the injector B during the first 10 combustion cycles

B has a small fuel spray SMD. Therefore, a large amount of fuel vapor enters the cylinder in the first cycle, which results in rich mixture and a large portion of fuel not burning completely; then, a large amount of HC emissions come out after the exhaust valve opened. From the second to fourth cycles, the unvaporized fuel of the first cycle keeps vaporizing, so rich air–fuel mixture is formed leading to relatively high HC emissions. After the fifth cycle, the fuel film formed by the first combustion cycle is evaporated completely; therefore, the air–fuel mixture in-cylinder is stabilized near the stoichiometric state, and the combustion is stable and produces low HC emission.

The HC emission mass out of each combustion cycle can be calculated through the parameters of the intake air mass flow rate, the in-cylinder pressure during the exhaust process, and the transient HC concentration at the exhaust port. Cyclical HC emissions comparison during the first 10 cycles between the two injectors is given in Fig. 9. The cyclical HC emission of injector A achieved the maximum value in the third cycle and then decreased slowly, while the cyclical HC emission of injector B reached the highest point at the first cycle and then declined rapidly. Compared to injector A, the HC emissions of injector B are 180% higher in the first cycle of cold start, but HC emissions are significantly lower than injector A during the following cycles, especially decreased 53% in the third cycle.

The cumulative HC emissions of the two injectors during first 10 combustion cycles are shown in Fig. 10. Cumulative HC emissions of the injector B are 11% lower than that of the injector A based on the same fuel injection strategy. The fuel injector B has a smaller SMD, but still caused a high HC emission in the first cycle, mainly because the original calibration data used in the study was optimized on the injector A. In addition, the HC emissions of the injector B may be further improved by calibration optimization such as reducing the amount of fuel injection in the first cycle.

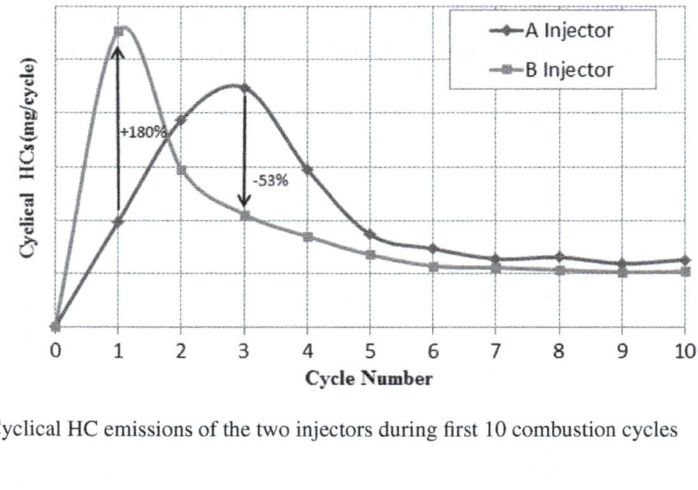

Fig. 9 Cyclical HC emissions of the two injectors during first 10 combustion cycles

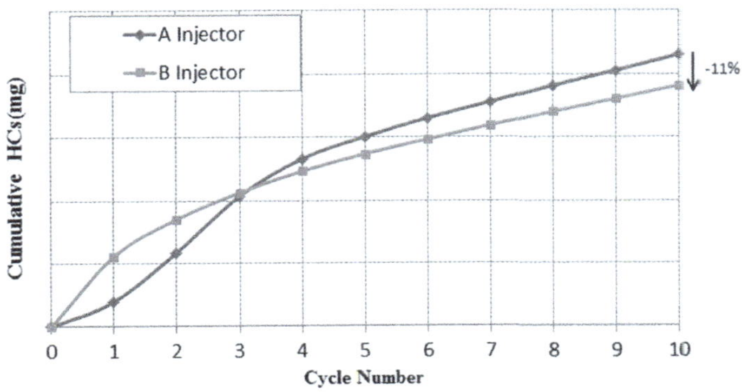

Fig. 10 Cumulative HC emissions of the two injectors during first 10 combustion cycles

5 Conclusions

In this work, the accurate simulation of vehicle test over the cold start process was performed on a no-load engine test bench. This is a low-cost and effective test method to investigate cold start emissions and can be used on hardware development of combustion system and the subsequent vehicle emission calibration work. The first 10 cycles combustion and emission characteristic based on cycle-by-cycle combustion analysis has been investigated with two different types of injectors using fast FID HC analyzer, and the test results reveal that HC emissions of 8-hole fuel injector are 11% lower than those of 6-hole fuel injector under the same fuel injection strategy.

References

1. Takeda K, Yaegashi T, Sekiguchi K, Saito K (1995) Mixture preparation and HC emissions of a 4-valve engine during cold starting and warm-up. SAE Paper 950074
2. Santoso H, Cheng WK (2002) Mixture preparation and hydrocarbon emissions behaviors in the first cycle of SI engine cranking. SAE Paper 2002-01-2805
3. Zheng C, Liu X, Mu J, Li F (2016) Emission correlation test of WLTC and NEDC cycles. China Measurement & Test, 2016-12-0022
4. Horie K, Takahasi H, Akazaki S (1995) Emissions reduction during warm-up period by incorporating a wall-wetting fuel model on the fuel injection strategy during engine starting. SAE Paper 952478
5. Lee D, Heywood JB (2006) Effects of charge motion control during cold start of SI engines. SAE Paper 2006-01-3399
6. Li L, Qiu D, Liu Z (2006) The characteristic of trasient HC emissions of the first firing cycle during cold start on an LPG SI engine. SAE Paper 2006-01-3403
7. Yu S, Zhang Y, Dong G, Li L (2009) Fuel injection optimization during engine quick start by means of cycle-by-cycle control strategy for HEV application. SAE Paper, No. 2009-01-2718
8. Rodriguez J, Cheng W (2016) Cycle-by-cycle analysis of cold crank-start in a GDI engine. SAE Paper 2016-01-0824

Research on Hybrid Electric Vehicle Engine Heating-Up Process and Fuel Consumption by Applying Auxiliary Heaters

Ping Sun, Chuanzhao Yao, Wei Dong, Shaozhen Liu, Ling He, Xiumin Yu and Huichao Zhao

Abstract Based on an engine of hybrid electric vehicle, the effect of auxiliary heating system on heating-up time and fuel consumption during the heating process of engine cooling system is investigated by the combination of experiment and simulation. The experimental results show that the auxiliary heating can obviously shorten the heating-up time and reduce the fuel consumption of system, regardless of synchronous heating or preheating. The fuel heater has a shorter heating-up time, while the PTC heater is more economical and green. The simulation results show that the body temperature has a great influence on the heating-up time of the cooling system, and the heating-up time of the coolant can be shortened obviously by using the auxiliary heater. In this research, the experiment and simulation are complementary and related to each other. The influence of auxiliary heating on the heating-up time and the fuel consumption are quantified. The purpose of this research is to find a way to shorten the heating-up time and improve fuel economy of hybrid power system.

Keywords Auxiliary heating · Engine coolant · Heating-up time · Fuel consumption

1 Introduction

The essential role of the vehicle's cooling system is to establish a thermal balance between coolant and engine body so that the engine can run well [1]. When the body is overheated, coolant takes away the excess heat, conversely, when the body is cold, coolant transfers heat to the body, so that the body can reach the appropriate

P. Sun · C. Yao (✉) · W. Dong · S. Liu · L. He · X. Yu
State Key Laboratory of Automotive Simulation and Control, Jilin University, Changchun 130022, China
e-mail: 2458671847@qq.com

W. Dong
e-mail: dwei@jlu.edu.cn

H. Zhao
Institute of New Energy Development, FAW, Changchun 130011, China

China SAE (ed.), *Proceedings of China SAE Congress 2018: Selected Papers*, Lecture Notes in Electrical Engineering 574,
https://doi.org/10.1007/978-981-13-9718-9_23

working temperature quickly. In northern China, the ambient temperature is very low in winter. In the cold start stage, low temperature causes bad fuel atomization. In order to ensure the engine can start smoothly, it usually needs to supply excessive fuel, but this will cause high fuel consumption and emission at cold start stage [2–5]. Experiments by Gordon W. Taylor et al. show that the fuel consumption of cold start process accounts for 25% of the whole test cycle [6]. Meanwhile, the incomplete combustion of fuel causes a large amount of pollutants, such as HC, and the lower the ambient temperature is, the more the pollutants exhaust [7, 8]. Low temperature causes high viscosity of lubricating oil, and it is bad for the life of the engine [9–11]. Some scholars have improved the initial combustion in cylinder by means of intake heating, fuel heating, and blending new fuels [12–14]. Some teams use active HC trap, electric heating catalytic converter, quick ignition three effect catalytic converter and post-combustion heating catalytic conversion system to reduce the emission of cold start process, and the results are satisfactory [15–17].

The idea of this study is to use the auxiliary heating method to heat the coolant rapidly before or during the warm-up process, so as to reduce the pollutant discharge during the cold start and warm-up process. Meanwhile, it's also propitious to reduce the wear of the engine, improve economy, and make the cockpit heat up quickly for driving comfort. In this study, the effect of auxiliary heating on the temperature rise and fuel consumption of the engine under different conditions is studied with the means of experiment and simulation, so as to provide some reference for the development and design of the engine cooling system for hybrid electric vehicle.

2 Test Equipment Setup and Scheme

2.1 Test Equipment Setup

The main technical parameters of the engine selected in this study are shown in Table 1.

The dynamometer is a CW160 eddy current dynamometer produced by Luoyang Kai Mechatronics Co., Ltd., and the Lambda Meter LA4 produced by ETAS is used for Lambda measuring. The fuel flowmeter is the DF-2420 volume fuel flowmeter

Table 1 Main technical parameters of the engine

Parameters	Value
Displacement (L)	1.798
Cylinder number	4
Compression ratio	9.6
Rated power (kW)	118
Maximum torque (N m)	250

Table 2 Main experimental equipments

Item	Test instruments	Type	Resolution
1	Dynamometer	CW160	±0.1 r/min
2	Lambda measuring instrument	Lambda Meter LA4	0.01
3	Temperature sensor	PT100	±0.1 °C
4	Turbine flowmeter	DN15	0.5 level
5	Fuel flowmeter	Onosokki DF2420	±0.5%
6	DC power	**	±0.5%
7	Fuel heater	Thermo Top Evo	**
8	PTC	**	Maximum power 5 kw

The asterisk indicates UNKNOWN. Some instruments are customized, so the type or resolution can't be provided.

produced by Onokazu Company of Japan. Customized machine is used to supply stable DC, and the fuel heater is a Thermo Top Evo fuel heater manufactured by Webasto, Germany. The PTC heater is produced by Jiangsu Chaoli Electric Manufacture Co., Ltd, and the thermometer is PT100. The liquid turbine flowmeter is used to measure coolant flow. The specific parameters are shown in Table 2.

The arrangement of the test cooling system is shown in Fig. 1.

2.2 *Test Scheme*

This study mainly explores the effects of fuel heater and PTC heater on the temperature rise time and fuel consumption of the engine. During the test, the engine's initial body temperature was always 12 °C, and intake temperature was −7 °C which was introduced from the outside by an aluminum foil drawing tube. The temperature rise time and fuel consumption of the engine were recorded when temperature rose from 12 to 80 °C. In the experiment, the coolant was heated by means of synchronous heating and preheating using fuel heater and PTC heater. The engine speed was controlled at 1000 rpm during the test. The excess air ratio was maintained at 1, and the throttle opening was 9%.

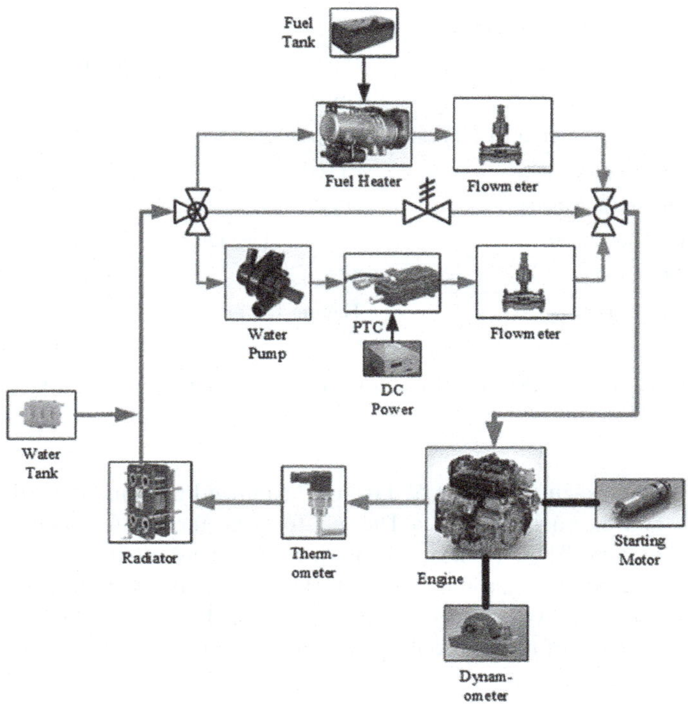

Fig. 1 Engine cooling system

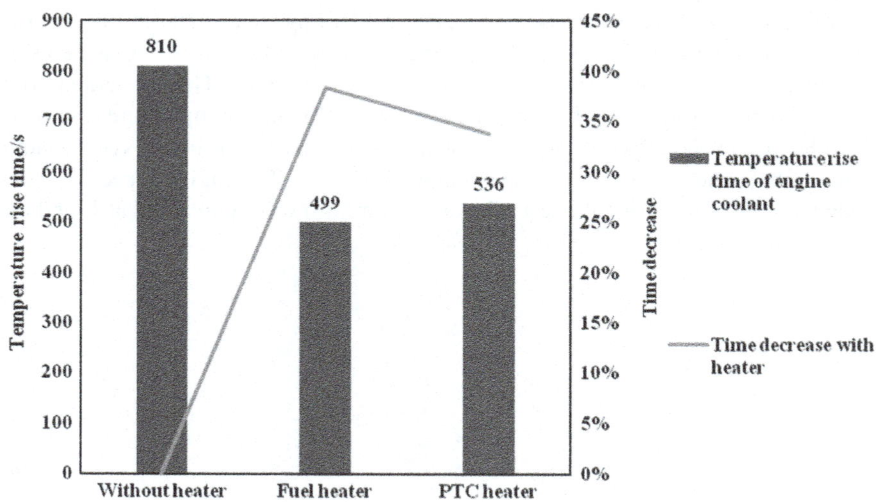

Fig. 2 Temperature rise time of coolant at synchronous heating

3 Results and Discussions

3.1 Synchronous Heating

3.1.1 Temperature Rise Time of Coolant

As shown in Fig. 2, the heating power of both fuel heater and PTC heater is 4.2 kW during synchronous heating, and the temperature rise time decreases due to the additional heat sources. The temperature rise time decreases by 38.4% when using fuel heater and it is 33.8% when PTC heater is used. Therefore, the synchronous heating method can obviously reduce the engine temperature rise time, and the reduction is more than 30%. Moreover, fuel heater occupies less time at synchronize heating.

3.1.2 Fuel Consumption

While reducing the temperature rise time, the fuel consumption of the engine also decreases at synchronous heating. As shown in Fig. 3, the fuel consumption reduction at synchronous heating is 36.9% when using fuel heater, and it is 34.26% when PTC heater is used. Therefore, the synchronous heating method cannot only reduce the temperature rise time, but also reduce the fuel consumption of the engine. Meanwhile, fuel heater is stronger than PTC heater to reduce the fuel consumption of the engine.

Although the heaters will reduce the temperature rise time and engine fuel consumption, they will also consume additional energy. Therefore, the costs of fuel and electricity consumed by engine, fuel heater, and PTC heater are converted into total costs, as shown in Table 3.

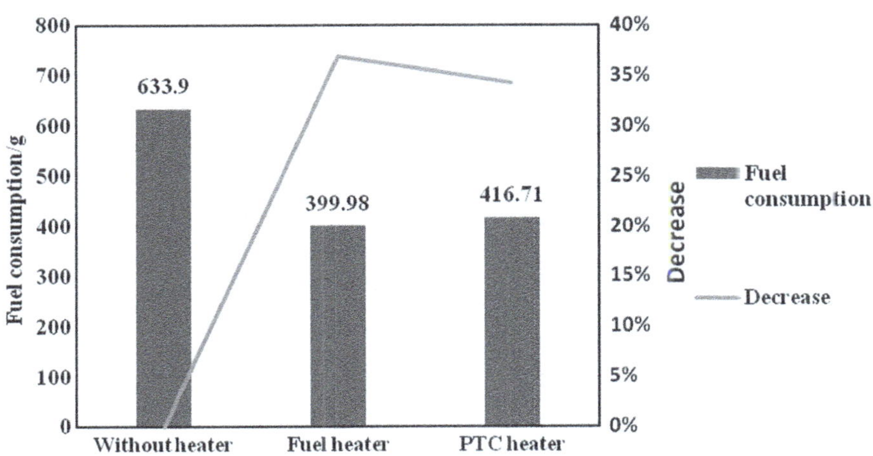

Fig. 3 Fuel consumption of the engine at synchronous heating

Table 3 Costs of fuel heater and PTC heater

Heating mode	95#Gasoline (¥/L)	Electricity (¥/kW h)	Fuel consumption (g)	Electricity consumption (kW h)	Energy consumption (kJ)	Total costs after compromise (¥)	Costs reduction (%)
Fuel heater	7.24	–	474.13	–	20,421	4.64	25.16
PTC heater	7.24	0.525	416.71	0.621	20,183	4.4	29.03
Engine only	7.24	–	633.9	–	27,302	6.2	0

In conclusion, considering the heating-up time, the temperature rise time of fuel heater is shorter, which can reduce the heating-up time and abrasion of the engine. However, from an economic point of view, the PTC heater is more economical and green.

3.2 *Preheating*

The preheating in this study refers to starting the heaters before the engine, and when the coolant rises to a certain temperature, stop the heaters and start the engine. Until coolant temperature rises to 80 °C, stop the engine. Both heaters heat coolant to 36, 42, 48, 54, and 60 °C with a power of 4.2 kW.

3.2.1 Preheating with Fuel Heater

As shown in Fig. 4, when preheating with fuel heater, the total temperature rise time will be longer than engine only, and the higher the preheating temperature is, the longer temperature rise time become. The main reason is that the heating power of the fuel heater is lower than that of engine. Therefore, the higher preheating temperature is, the longer time the fuel heater works, and the longer total temperature rise time of

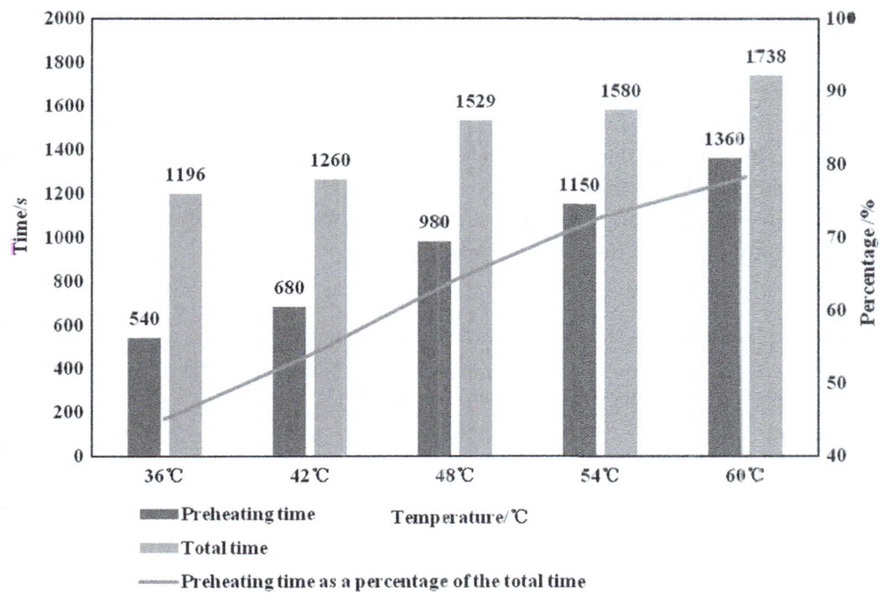

Fig. 4 Preheating time and total time at different preheating temperatures

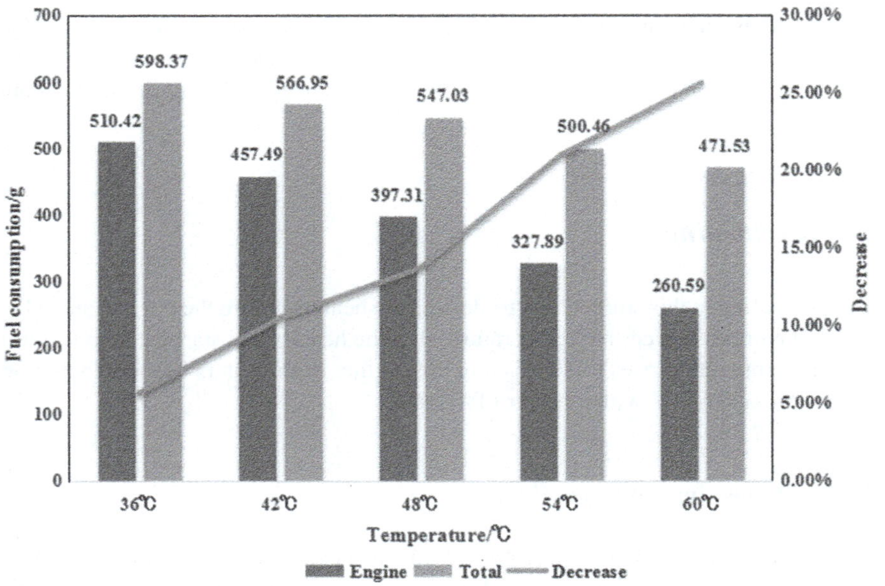

Fig. 5 Fuel consumption and its reduction of fuel heater preheated to different temperatures

coolant becomes. However, compared with engine only, the preheating can obviously shorten the running time of the engine. And the longer preheating time is good for shortening the engine running time and starting condition.

As can be seen from Fig. 5, the higher preheating temperature of the fuel heater leads to more fuel consumption of the heater and less fuel consumption of the engine. Moreover, the higher the preheating temperature is, the less the total fuel consumption becomes. The reason is that most of the energy produced by the heater is transferred to the coolant, and the energy produced by engine is also transferred to the engine body besides the coolant. And exhaust also takes away most of the energy; so it is more economical to use fuel-assisted heating.

Table 4 Replacement costs of fuel heater preheated to different temperatures

Preheating temperature (°C)	Fuel consumption (g)	Energy consumption (kJ)	Total costs (¥)	Costs reduction (%)
Engine only	633.9	27302	6.2	0
36	598.37	25,772	5.85	5.65
42	566.95	24,419	5.55	10.48
48	547.03	23,561	5.35	13.71
54	500.46	21,555	4.90	20.97
60	471.53	20,309	4.61	25.65

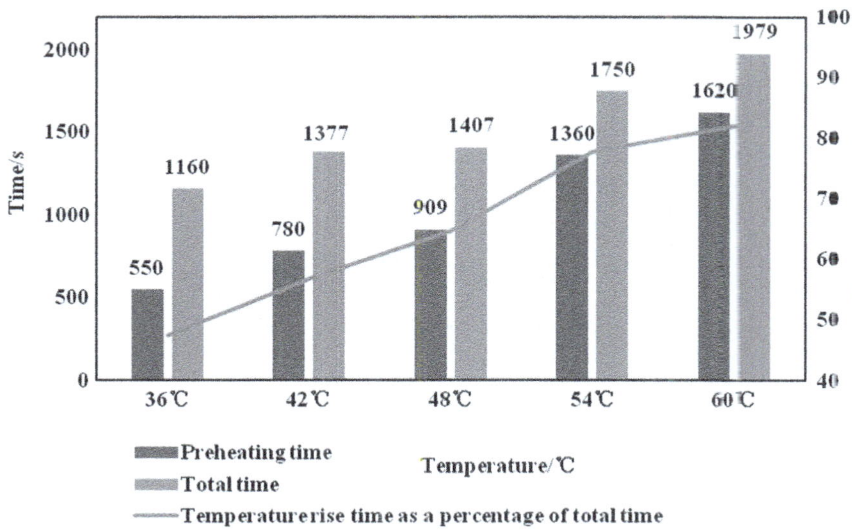

Fig. 6 Preheating time and total time of PTC heater preheated to different temperatures

Table 4 is the total cost of using fuel heater preheated to different temperatures. Higher preheating temperature leads to less total cost, and the maximum can reach 25.65%, which indicates that using fuel heater is more economical.

3.2.2 PTC Heater Preheating

From Fig. 6, it is known that with the increase of preheating temperature, the preheating time and the total temperature rise time are gradually lengthened, and the percentage of preheating time is higher. However, the running time of the engine decreases gradually, which helps to reduce the engine's wear and improve its life.

As shown in Fig. 7, the higher the preheating temperature of PTC heater is, the less fuel consumption and warm-up time of the engine become, which is beneficial to reduce the wear of the engine.

From Table 5, it is known that higher preheating temperature of PTC heater leads to more electricity consumed by PTC heater and less fuel consumed by engine, but the total cost is gradually reduced and the maximum drop can reach 40.43%. Compared with fuel heater, the PTC heater is more efficient, economical, and green.

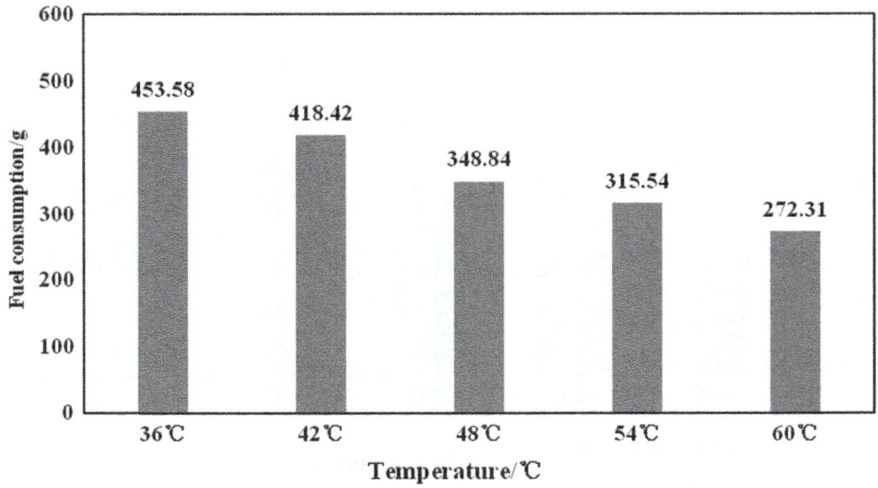

Fig. 7 Fuel consumption of PTC heater preheated to different temperatures

Table 5 Conversion cost of PTC heater preheated to different temperatures

Preheating temperature (°C)	Electricity consumption (kW h)	Fuel consumption (g)	Energy consumption (kJ)	Total costs (¥)	Costs reduction (%)
Engine only	–	633.9	27,302	6.2	0
36	0.642	453.58	21,847	4.77	23.06
42	0.910	418.42	21,297	4.57	26.29
48	1.062	348.84	18,848	3.97	35.97
54	1.528	315.54	19,091	3.89	37.26
60	1.948	272.31	18,741	3.69	40.48

4 Simulation Calculation and Analysis of Cooling System

4.1 Parameters and Model Building

The temperature of the engine has a great influence on temperature rise time and fuel consumption; however, the temperature of the engine body cannot be decreased to the ambient temperature during the actual test in this study. Therefore, the simulation method is used to realize the situations those cannot be realized in the experiment, which can greatly reduce the research time and cost. In this study, one dimensional simulation software FloMASTER is used to explore the temperature rise time of the engine coolant with or without auxiliary heaters.

The engine module is the most important module in the system. The heat transfer from the fuel to the coolant is Q_W, which is very difficult to calculate precisely due to some complex factors. Therefore, the empirical formula is usually used to estimate the energy of the engine. The formula is as follows:

$$Q_W = \frac{A g_e N_e h_n}{3600} \text{ (kJ/s)} \tag{1}$$

A is the percentage of the heat transferred to the cooling system The heat of general gasoline engine is 0.23–0.3. In this study, when A is 0.28, the temperature rise curve of the engine coolant is closer to the test value, so A is 0.28. g_e is the fuel consumption rate of ICE (kg/kW h), N_e is the power of ICE (kW), and h_n is the low calorific value of the fuel (kJ/kg).

The FloMASTER software is used to build the transient simulation model of the engine cooling system, as shown in Fig. 8, based on the circulating coolant route and actual operating parameters of the engine. The radiator is not involved in heat transfer in this model, which is equivalent to the short-circuit condition.

Table 6 shows the parameters of the engine block.

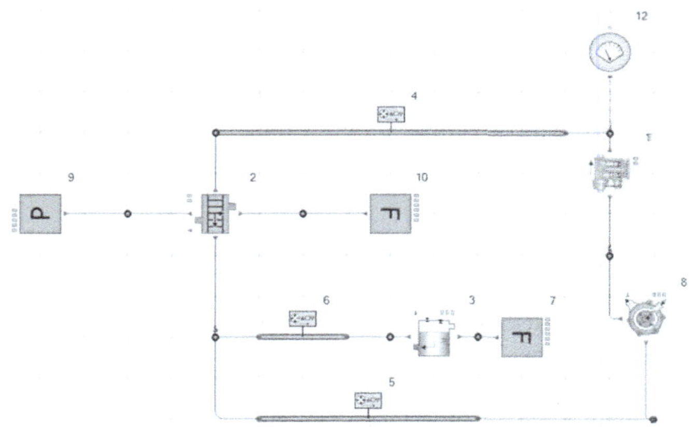

1 Engine 2 Radiator 3 Expansion Tank 4 Waterway 5 Waterway 6 Waterway 7 Flow Source 8 Water Pump 9 Pressure Source 10 Flow Source 11 Water Temperature Monitoring Meter

Fig. 8 Simulation model of the cooling system

Table 6 Parameter setting of the engine block

Material	Aluminum
Mass (kg)	90
Body temperature (°C)	12
Resistance coefficient	15
Density (kg/m^3)	2698
Heat conduction coefficient (kW/m K)	0.239
Specific heat capacity (kJ/kg K)	0.917

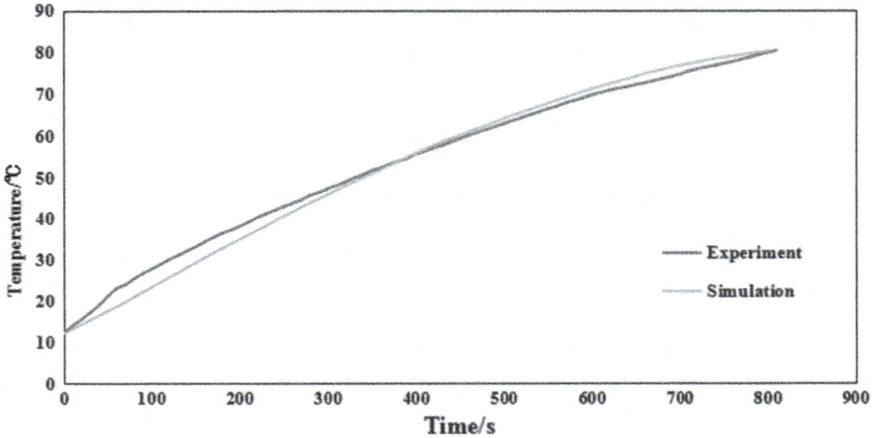

Fig. 9 Comparison between simulation and experiment of temperature rise process

4.2 *Analysis of Simulation Results*

In order to verify the correctness of the simulation model, the simulation and the test results are compared. As shown in Fig. 9, the test and the simulation curve are basically the same, and the correctness of the simulation model can be considered.

4.2.1 Temperature Rise Process of Different Body Temperature

In the simulation, the external temperature is set at 0, −5, −10, −15, −20, −25 °C, respectively, and the engine body temperature is consistent with the ambient temperature. The temperature rise time of engine coolant is shown in Table 7. It can be seen that the lower ambient temperature causes the longer temperature rise time of engine coolant, which will have a greater influence on the engine cold start and warm-up process. Therefore, it is necessary to adopt the auxiliary heating method to shorten the temperature rise time of engine coolant.

Table 7 Warm-up process under different ambient temperatures

Ambient temperature (°C)	Temperature rise time (s)	Time increase (%)
0	920	0
−5	968	5.22
−10	1020	10.87
−15	1070	16.30
−20	1122	21.96
−25	1181	28.37

4.2.2 Effect of Auxiliary Heating on Temperature Rise Time of Engine Coolant

In order to investigate the effect of auxiliary heating on temperature rise time of the coolant, the auxiliary heater module is added. As shown in Fig. 10, the label 7 in the figure is the auxiliary heater model. The others are the same as in Fig. 8.

In order to verify the correctness of the simulation model, the simulation results are compared with the experiments. As shown in Fig. 11, the experimental results are close to the simulations, which verify the correctness of the simulation model.

In order to investigate the influence of auxiliary heating on the temperature rise time of coolant, the results with or without auxiliary heater were compared under different ambient temperatures. As shown in Fig. 12, it can be seen that when the ambient temperature is 0 °C, the temperature rise time decreases by 27.72% with auxiliary heater, and when the ambient temperature is −25 °C, the reduction is 23.29%. It can be seen that the effect of the auxiliary heating is remarkable no matter how low the ambient temperature is. The decrease of the temperature rise time reduces as the ambient temperature decreases. The main reason is that the effect of low temperature on the temperature rise time is greater than that of auxiliary heater as the lower temperature.

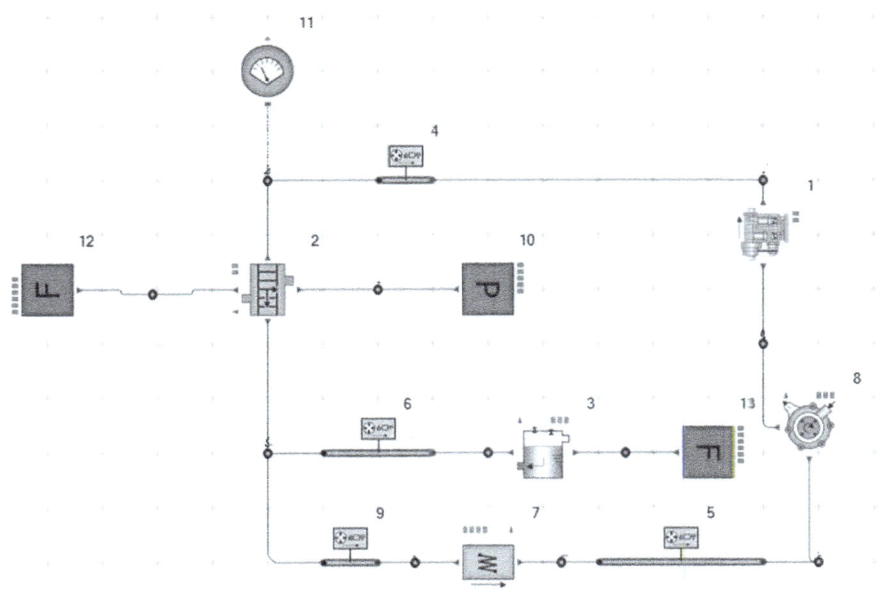

Fig. 10 Engine cooling system with auxiliary heater

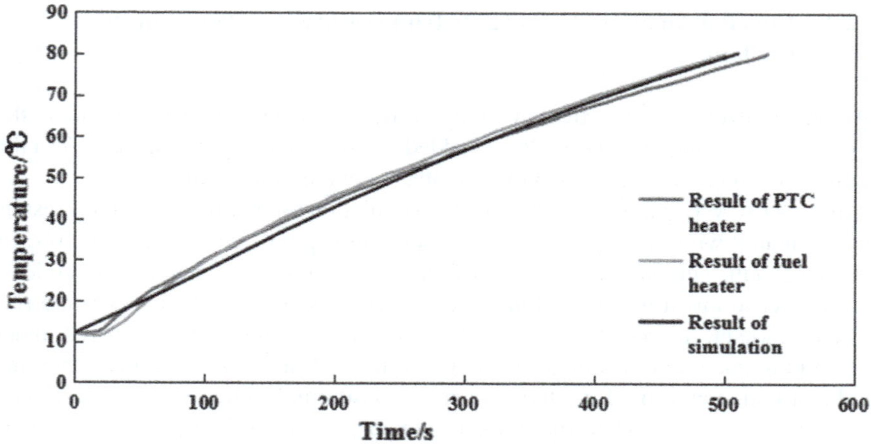

Fig. 11 Temperature rise time of simulation and experiment

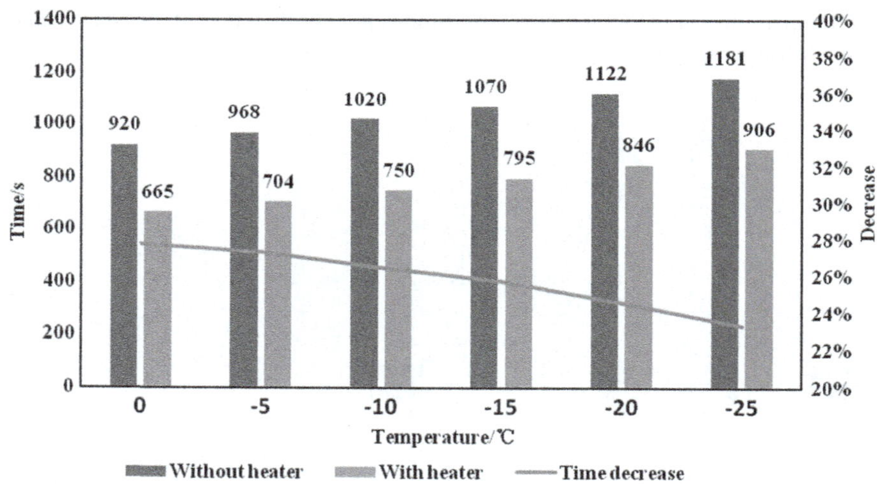

Fig. 12 Heating time at different ambient temperatures with or without auxiliary heater

5 Conclusion

The purpose of this study is to investigate the effects of auxiliary heating on the temperature rise time and fuel consumption of the engines at low temperature. The effects of auxiliary heaters on temperature rise time and fuel consumption of the engine cooling system are investigated by means of experiment and simulation. The following conclusions are drawn:

1. When the synchronous heating method is adopted, temperature rise time and fuel consumption of the engine can be obviously reduced by auxiliary heating. The decrease of temperature rise time with fuel heater is 38.4%, which is higher than that with PTC heater, 33.8%. Also, the fuel consumption of the engine can be reduced by 36.9% with fuel heater, which is higher slightly than that with PTC heater, 34.26%. However, fuel heater and PTC heater need to consume extra fuel and electricity, and the total cost is reduced by 25.16 and 29.03%, respectively. Although the temperature rise time of fuel heater is shorter than that of PTC heater, PTC heater is more economical and green.
2. The use of preheating will lead to an increase in the total temperature rise time, but the engine running time is significantly reduced, which can reduce the engine's wear. The preheating of the fuel heater can reduce the total cost by 25.65%, less than 40.48% of the PTC heater. Similar to synchronous heating, the temperature rise time of fuel heater is shorter, and the PTC heater is more economical and green.
3. According to the simulation results, the lower the temperature of the body is the longer the temperature rise time becomes. With auxiliary heating, the temperature rise time is obviously shortened, with a drop of more than 20%.

In summary, auxiliary heating can obviously reduce the temperature rise time and fuel consumption of the engine, reduce the wear of the engine, and improve the fuel economy of the vehicle, which has the value of popularization and utilization.

Acknowledgements This research was funded by the National Natural Science Foundation of China (51276079) and the Jilin Provincial Science and Technology Department (20150101036JC).

References

1. Chen J (2005) Automobile structure. Machinery Industry Press, Beijing
2. Junjun H, Huang Z, Zhou L et al (2003) An analysis on combustion characteristics of gasoline direct injection engine. Automot Engine 25(6):550–552
3. Zhang D, Li S, Li K et al (2010) Influence of coolant temperature on diesel engine emissions. Trans CSICE 28(6):510–513
4. Whelan I, Smith W, Timoney D et al (2012) The effect of engine operating conditions on engine-out particulate matter from a gasoline direct-injection engine during cold-start. SAE Technical Paper
5. Rodriguez JF, Cheng WK (2015) Effect of operation strategy on first cycle CO, HC, and PM/PN emissions in a GDI engine. SAE Int J Engine 8(2015-01-0887):1098–1106
6. Taylor GWR, Stewart S (2001) Cold start impact on vehicle energy use. SAE Technical Paper
7. Su Y, Liu Z, Han Y, Wang Y, Du B (2007) Effect of intake air temperature on combustion and emissions of direct injection diesel engine during preliminary start phase. Chin Int Combust Eng Eng 06:28–32
8. Zhou S, Xiao Y, Zhu Y (2010) Emission and pollution control of internal combustion engines. Beijing University of Aeronautics and Astronautics Press, Beijing
9. Miao L (2015) Study on thermal management system of distributed hybrid electric special vehicle. Beijing Institute of Technology

10. Schmidt S, Rieger P, Trattner A et al (2011) A methodical approach for thermal management simulation of hybrid powersport applications. SAE Technical Paper 32-0586
11. Liu Z, Wang J (2001) Automobile engine fundamentals. Tsinghua University Press, Beijing
12. Ni P (2007) Reduction of cold start emissions from gasoline engine by preheating intake air and related technologies. Tianjin University
13. She J, Zhang Y, Chen Z, Wang L (2012) Improvement of performance and emissions of DI diesel engine using preheated diesel fuel. J Huazhong Univ Sci Tech (Natural Science Edition) 40(01):113–118
14. Niu R, Zhang Y, Yu X, Du Y, Wu H, Song B (2017) Effect of hydrogen injection timing on combustion and emission characteristics in a hydrogen-blended gasoline engine. Trans CSICE 35(01):32–37
15. Li L, Shuai S, Xiao J, Ma W, Wang J (2005) Experimental study of zeolite effect on trap of hydrocarbon from gasoline engine during cold-start. Veh Engine 04:38–41
16. Qian R (2012) Electric heating catalytic converter for extended-range hybrid vehicle engines. Automob Parts 33:24–27
17. Wang J, Xiao J, Li J, Zhu Y (2000) A research on light-off performance and test method of automotive catalytic converter. Automot Eng 01:25–28+32

Adaptive Steering Stability Control for A Four In-Wheel-Motor Independent-Drive Electric Vehicle

Rufei Hou, Li Zhai and Tianmin Sun

Abstract This paper proposes an adaptive steering stability control strategy for a four in-wheel-motor independent-drive electric vehicle (4MIDEV) to improve its stability on the road with different adhesion coefficients, such as on the joint road. The proposed strategy is designed as a hierarchical structure whose upper control level takes the road adhesion coefficient into account in the yaw moment control and realizes the integrated control of the yaw rate and sideslip angle. The lower control level adopts a weight-based optimal allocation algorithm to achieve different weight control of each motor torque according to road adhesion coefficient. The proposed stability control strategy was validated in a co-simulation of the Matlab/Simulink and Carsim, the results of which indicate that the proposed strategy can effectively adapt to different adhesion coefficients and achieve good steering stability control effect.

Keywords Electric vehicles · In-wheel motor · Steering stability · Adhesion coefficient

1 Introduction

As the electric vehicle and in-wheel motor technology develops rapidly, a four in-wheel-motor independent-drive electric vehicle (4MIDEV) has attracted lots of attention [1, 2]. Compared with the traditional vehicle, the 4MIDEV is directly driven by four motors integrated into wheel hubs without the need for additional mechanical transmission components, such as drive shafts. Since the torque of each drive wheel can be independently controlled, this distributed drive form brings more possibilities for achieving better stability and maneuverability of the vehicle [3]. In recent years,

R. Hou (✉) · L. Zhai
Beijing Institute of Technology, Beijing, China
e-mail: hrfbit@163.com

T. Sun
Beijing Electric Vehicle CO. LTD, Beijing, China

China SAE (ed.), *Proceedings of China SAE Congress 2018: Selected Papers*,
Lecture Notes in Electrical Engineering 574,
https://doi.org/10.1007/978-981-13-9718-9_24

a great deal of valuable work has been carried out on the steering stability control of the 4MIDEV.

The stability control strategy of the 4MIDEV generally adopts a hierarchical structure, including an upper control level to compute the virtual force commands and a lower control level to realize the control of each wheel torque [4]. However, the previously proposed control strategy rarely considers the effects of changes in the adhesion coefficient, which may introduce additional control errors. As the road conditions become more complex, there is a need for a new stability control strategy adapted to road adhesion, so as to ensure good steering stability control effect. Thus, the requirements for the upper control level and the lower control level are much higher.

The direct yaw moment control (DYC) has been widely used in the upper control level and proved to be very effective [4–6]. Most of the DYCs proposed so far only control the yaw rate without the consideration of the sideslip angle [7] or take the sideslip angle as an intermediate variable in the control of the yaw rate [8, 9]. Although accurate control of the yaw rate can be achieved, these methods have poor consideration of the sideslip angle or require a high accuracy for estimating the sideslip angle [2]. Furthermore, the impact of adhesion coefficient is seldom taken into account in the DYC, which will greatly affect the final stability control results when the adhesion coefficient changes. In the lower control level, the optimal torque allocation algorithm is preferred over other algorithms such as average allocation and dynamic-load allocation [10, 11], whose optimal objective usually selects the tire workload usages to reflect steering stability [1, 12]. However, the different adhesion conditions at each wheel are always neglected in the torque allocation. The tire workload usages are generally considered to be equally important for stability, i.e., their weight coefficients were equal or independent of the adhesion coefficient, which may result in extreme loads on the wheels at low adhesion while other wheels still have a relatively high stability margin.

To solve the above problems, this paper proposes an adaptive stability control strategy to accommodate different adhesion coefficients. The proposed strategy adopts a hierarchical structure which takes the road adhesion coefficient into account in both the upper control level and lower control level. The adaptive control of the yaw rate and the sideslip angle is realized in the upper control level to ensure a better DYC control result. A weight-based optimal torque allocation algorithm is also developed in the lower control level to achieve different weight control of each motor torque according to road adhesion coefficient. The adaptive upper control level and the lower control level can effectively improve the steering stability of the 4MIDEV under different road adhesion coefficients.

2 Design of Adaptive Steering Stability Control Strategy for the 4MIDEV

The proposed adaptive stability control strategy, as shown in Fig. 1, consists of a reference output level with reference dynamic model, an upper control level based on the yaw moment control and speed control, and a lower control level with the weight-based optimal allocation algorithm.

2.1 Reference Output Level

The reference output level adopts a reference model to compute the desired dynamics parameters, such as the sideslip angle and the yaw rate, for the further control of the upper control level. To investigate the planar motion of the vehicle, a three-degree-of-freedom (3-DOF) model is constructed by ignoring the roll and pitch motions, as shown in Fig. 2, which can be described as:

$$\begin{bmatrix} \dot{V}_x \\ \dot{V}_y \\ \dot{\gamma} \end{bmatrix} = \begin{bmatrix} V_y\gamma - \frac{\rho C_d}{2m} A V_x^2 \\ -V_x\gamma \\ 0 \end{bmatrix} + \mathbf{B}_x \begin{bmatrix} F_{xfl} \\ F_{xfr} \\ F_{xrl} \\ F_{xrr} \end{bmatrix} + \mathbf{B}_y \begin{bmatrix} F_{yfl} \\ F_{yfr} \\ F_{yrl} \\ F_{yrr} \end{bmatrix} \tag{1}$$

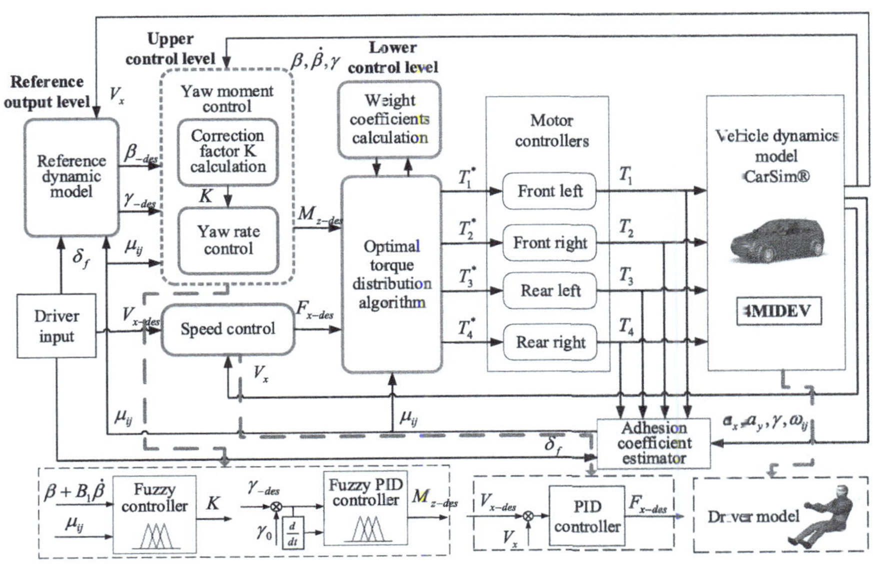

Fig. 1 Configuration of adaptive stability control strategy

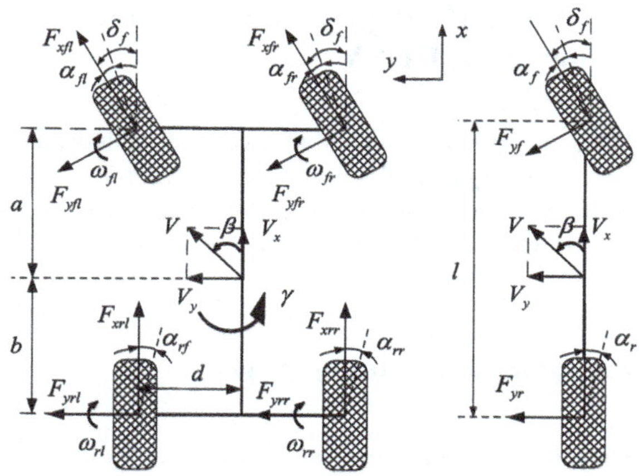

Fig. 2 Vehicle dynamics model

(a) 3-DOF model (b) 2-DOF model

where V_x and V_y denote the longitudinal velocity and the lateral velocity of vehicle, respectively, and $\dot{V}_x$ and $\dot{V}_y$ are their derivatives. γ and $\dot{\gamma}$ denote the yaw rate and its derivative, respectively. ρ denotes the air density, C_d is the air resistance coefficient, and A represents the maximum cross-sectional area. m stands for the mass of vehicle. F_{xij} and F_{yij} are the longitudinal and lateral forces of the wheels, respectively. $i \in \{f, r\}$ stands for the front or the rear and $j \in \{l, r\}$ stands for the left or the right.

The coefficient matrices $\mathbf{B}_x$ and $\mathbf{B}_y$ can be formulated as:

$$\mathbf{B}_x = \begin{bmatrix} \frac{\cos\delta_f}{m} & \frac{\cos\delta_f}{m} & \frac{1}{m} & \frac{1}{m} \\ \frac{\sin\delta_f}{m} & \frac{\sin\delta_f}{m} & 0 & 0 \\ \frac{a\sin\delta_f - d\cos\delta_f}{I_z} & \frac{a\sin\delta_f + d\cos\delta_f}{I_z} & -\frac{d}{I_z} & \frac{d}{I_z} \end{bmatrix} \tag{2}$$

$$\mathbf{B}_y = \begin{bmatrix} -\frac{\sin\delta_f}{m} & -\frac{\sin\delta_f}{m} & 0 & 0 \\ \frac{\cos\delta_f}{m} & \frac{\cos\delta_f}{m} & \frac{1}{m} & \frac{1}{m} \\ \frac{a\cos\delta_f + d\sin\delta_f}{I_z} & \frac{a\cos\delta_f - d\sin\delta_f}{I_z} & -\frac{b}{I_z} & \frac{b}{I_z} \end{bmatrix} \tag{3}$$

where I_z denotes the yaw inertia of the vehicle, and δ_f is the front-wheel steering angle. d is half of the tread. a is the distance between the center of gravity and the front axle, and b is the distance to the rear axle.

To more clearly and directly reflect the driver's steering intention, the above vehicle dynamics model is simplified to two degrees of freedom, i.e., the later motion and yaw motion, as shown in Fig. 2. The target values of sideslip angle and yaw rate can be obtained from the 2-DOF model:

$$\begin{cases} \gamma' = \dfrac{1}{1 - AV_x^2} \cdot \dfrac{V_x}{l} \cdot \delta_f \\ \beta' = \dfrac{1 - \left(\frac{m}{L}\right)\left(\frac{a}{bK_r}\right)V_x^2}{1 + AV_x^2} \cdot \dfrac{b}{l} \cdot \delta_f \end{cases} \tag{4}$$

where $A = \frac{m}{l^2} \cdot \frac{aK_f - bK_r}{K_f K_r}$, K_f and K_r stand for the sum of the tire cornering stiffnesses of the front axle and the rear axle, respectively.

Considering the different road adhesion conditions, it is necessary to limit the above target values [2]:

$$\begin{cases} \gamma_{\text{limit}} = \frac{0.85}{V_x} \mu g \\ \beta_{\text{limit}} = \tan^{-1}(0.02\ \mu g) \end{cases} \tag{5}$$

In summary, the desired yaw rate $\gamma_{-\text{des}}$ and sideslip angle $\beta_{-\text{des}}$ can be expressed as:

$$\begin{cases} \gamma_{-\text{des}} = \min\{ |\gamma'|, \gamma_{\text{lim}}\} \cdot \text{sgn}(\delta_f) \\ \beta_{-\text{des}} = \min\{ |\beta'|, \beta_{\text{lim}}\} \cdot \text{sgn}(\delta_f) \end{cases} \tag{6}$$

2.2 Upper Control Level

The upper control level consists of speed control and yaw moment control, as shown in Fig. 1, which computes the virtual commands of traction force and yaw moment to control the longitudinal and lateral motions of the vehicle based on the desired dynamics parameters form the reference output level.

2.2.1 Speed Control

In order to reduce the calculation cost and ensure the real-time performance, the speed control, as shown in Fig. 1, adopts the PID method to control the longitudinal traction force $F_{x-\text{des}}$ of the vehicle according to the error between the actual speed V_x and the desired speed $V_{x-\text{des}}$ from the "preview-follow" driver model [13], so as to meet the driving demand for speed. The speed V_x can be estimated by the driver's input to the acceleration/brake pedal [14].

2.2.2 Yaw Moment Control

The yaw moment control selects both the yaw rate and the sideslip angle corresponding to maneuverability and stability as control variables, as shown in Fig. 1, which control the lateral motion of vehicle through a yaw moment. Unlike the yaw rate, the sideslip angle and the yaw moment are related by the nonlinear tire forces, and the estimation accuracy of the sideslip angle is limited. Thus, the influence of the sideslip angle is considered in the control of the yaw rate by introducing a correction coefficient K. Then, the nominal yaw rate γ_0 can be obtained:

$$\gamma_0 = K\gamma \tag{7}$$

The correction coefficient K is studied by a fuzzy controller, as shown in Fig. 1. Considering the influence of sideslip angle rate $\dot{\beta}$, the integrated control quantity $\beta + B_1\dot{\beta}$ is designed as the input of the fuzzy controller based on the β – Method theory [15] and the phase plane method [16], where B_1 is a coefficient related to the adhesion coefficient. Furthermore, since the impact of the sideslip angle is different for the stability under different adhesion coefficients, the adhesion coefficient μ is also introduced as the other input of the fuzzy controller. The fuzzy rules for the correction coefficient K are shown in Table 1, where seven fuzzy subsets are designed to define the values of both inputs and outputs, such as Zero (Z), Positive-Small (PS), Positive-Medium (PM), Positive-Big (PB), Negative-Small (NS), Negative-Medium (NM), and Negative-Big (NB).

In order to adapt to the nonlinear system and minimize the calculation cost, the fuzzy PID method is adopted in the yaw moment control to make the nominal yaw rate γ_0 follow the desired value $\gamma_{-\text{des}}$, as shown in Fig. 1. The inputs of the fuzzy PID controller are the error between the γ_0 and $\gamma_{-\text{des}}$ and its rate. The output is the desired yaw moment $M_{z-\text{des}}$. The fuzzy rules of the fuzzy PID controller are shown in Table 2, where Δk_p, Δk_i and Δk_d are the correcting values of the proportional, integral, and differential coefficients, respectively.

Table 1 Fuzzy rules for coefficient K

K		$\beta + B_1\dot{\beta}$						
		NB	NM	NS	Z	PS	PM	PB
μ	NB	PB	PB	PB	PM	PB	PB	PB
	NM	PB	PB	PM	PM	PM	PB	PB
	NS	PM	PM	PS	PS	PS	PM	PM
	Z	NM	NS	NS	Z	PS	PS	PM
	PS	NM	NM	NM	NS	NS	NS	NM
	PM	NB	NB	NM	NM	NM	NB	NB
	PB	NB	NB	NB	NM	NB	NB	NB

Table 2 Fuzzy rules for yam moment control

Δk_p		$e(s)$						
		NB	NM	NS	Z	PS	PM	PB
$\Delta e(s)$	NB	PB	PB	PM	PS	PS	NS	NS
	NM	PB	PB	PM	PS	PS	NS	NS
	NS	PB	PB	PM	PS	NS	NM	NM
	Z	PB	PB	PM	Z	NS	NM	NB
	PS	PM	PM	PS	NS	NM	NB	NB
	PM	PS	PS	NS	NM	NM	NB	NB
	PB	PS	PS	NS	NM	NM	NB	NB
Δk_i		$e(s)$						
		NB	NM	NS	Z	PS	PM	PB
$\Delta e(s)$	NB	NB	NB	NM	NM	NS	Z	Z
	NM	NB	NM	NM	NS	NS	Z	Z
	NS	NM	NM	NS	NS	Z	PS	PS
	Z	NM	NS	NS	Z	PS	PS	PM
	PS	NS	NS	Z	PS	PS	PM	PM
	PM	Z	Z	PS	PM	PM	PB	PB
	PB	NB	NB	NM	NM	NM	Z	Z
Δk_d		$e(s)$						
		NB	NM	NS	Z	PS	PM	PB
$\Delta e(s)$	NB	PS	PS	Z	Z	Z	PS	PB
	NM	NB	NM	NM	NS	PM	PB	PM
	NS	NB	NB	NM	NS	PS	PS	PM
	Z	NS	NS	NS	NS	Z	PS	PB
	PS	NB	NM	NM	NS	PS	PB	PB
	PM	NB	NB	NM	NS	PM	PB	PB
	PB	PS	PS	Z	Z	Z	PS	PS

2.3 Lower Control Level

The lower control level converts the desired yaw moment and traction force from the upper control level to the driving/brake torque required for in-wheel motors based on the torque allocation algorithm. The relationship between the lower level and upper level can be summarized as:

$$\mathbf{v} = \mathbf{B}\mathbf{u} \tag{8}$$

where $\mathbf{u} = [T_{fl}\ T_{fr}\ T_{rl}\ T_{rr}]^{\mathrm{T}}$, $\mathbf{v} = [F_{x-\mathrm{des}}\ M_{z-\mathrm{des}}]^{\mathrm{T}}$,

$$\mathbf{B} = \frac{1}{R}\begin{bmatrix} \cos\delta_f & \cos\delta_f & 1 & 1 \\ -d\cos\delta_f + a\sin\delta_f & d\cos\delta_f + a\sin\delta_f & -d & d \end{bmatrix}.$$

For torque allocation, the friction ellipse constraint needs to be fully considered to ensure the feasibility of control:

$$F_{xij}^2 + F_{yij}^2 \le \left(\mu_{ij} F_{zij}\right)^2 \tag{9}$$

where μ_{ij} denote the road adhesion coefficient at each wheel, and F_{zij} stands for the vertical force. The lateral force F_{yij} can be obtained from the simplified tire model proposed in [2]:

$$F_{yij} = -K_{\alpha ij}\sqrt{1 - \left(\frac{F_{xij}}{\mu_{ij} F_{zij}}\right)^2}\frac{\mu_{ij}}{k}\tan^{-1}\left(\frac{K_{\alpha ij}\pi}{2\mu_{ij} F_{zij}}\alpha_{ij}\right) \tag{10}$$

where α_{ij} and $K_{\alpha ij}$ are the slip angle and lateral stiffness of each tire, respectively. Since there is a certain delay in the change of vertical load, the lateral tire force in the next sampling period can be estimated approximately according to the vertical load in the current sampling period and the longitudinal tire force in the next sampling period. The adhesion coefficient μ_{ij} can be estimated in real time by the second-order nonlinear state observer proposed in [17]. The longitudinal tire force F_{xij} is related to the torque T_{ij} of the in-wheel motor:

$$\dot{\omega}_{ij} = \frac{1}{J_c}\left(T_{ij} - F_{xij}R - M_{fij}\right) \tag{11}$$

where M_{fij}, $\dot{\omega}_{ij}$, R and J_c are the rolling resistance moment, angular acceleration, rolling radius, and moment of inertia of the wheel, respectively.

The lower control level develops a weight-based optimal allocation algorithm, which selects the tire workload usages as the main optimization objective to reflect the stability margin. As this objective goes smaller, the stability margin of vehicle will be increased [1, 2]. The tire workload usage Ω_{ij} can be described as

$$\Omega_{ij} = \frac{F_{xij}^2}{\mu_{ij}^2 F_{zij}^2} \tag{12}$$

In previous studies, the tire workload usage of each wheel was often simply added as an overall optimization goal, regardless of the differences in adhesion condition at each wheel. However, this may result in some low-attachment wheels being unstable, while the torque of other wheels cannot be utilized to the maximum. Thus, the weight factor ψ_{ij} for Ω_{ij} is investigated to suit different adhesion coefficients as follows:

$$\psi_{ij} = \frac{\Omega_{ij}}{\sum \Omega_{ij}} \tag{13}$$

In conclusion, the proposed torque allocation algorithm can be formulated as:

$$\min J_1 = \|\boldsymbol{\Gamma}\,\mathbf{u}\|_2 \tag{14}$$

where $\mathbf{u} = [T_{fl}\ T_{fr}\ T_{rl}\ T_{rr}]^{\mathrm{T}}$, $\boldsymbol{\Gamma} = \mathrm{diag}\left(\frac{\psi_{ij}}{\mu_{ij} F_{zij} R}\right)$, R denotes the rolling radius. The constraints of optimal allocation can be described as:

$$\begin{cases} F_{xij}^2 + F_{yij}^2 \le \left(\mu_{ij} F_{zij}\right)^2 \\ \mathbf{B}\mathbf{u} = \mathbf{v} \\ \mathbf{u}_{\lim -} \le \mathbf{u} \le \mathbf{u}_{\lim +} \end{cases} \tag{15}$$

where $\mathbf{u}_{\lim -}$ and $\mathbf{u}_{\lim +}$ are the lower limit and upper limit of $\mathbf{u}$, respectively. Considering that the constraint strength of the equation is too strong to solve, it can be converted into a penalty term and introduced into (11), then:

$$\min J_2 = \|\boldsymbol{\Gamma}\,\mathbf{u}\|_2 + \xi\,\|\mathbf{W}_v(\mathbf{B}\mathbf{u} - \mathbf{v})\|_2 \tag{16}$$

where ξ and $\mathbf{W}_v$ denote the allocation weight coefficient and matrix, respectively. In particular, the $\mathbf{W}_v$ reflects the control priority of the yaw moment and traction force. For example, the control priority of the yaw moment is much higher than that of the traction force when the adhesion coefficient is low, so as to ensure good lateral stability.

The (16) can be reformulated as a general form:

$$J_2 = \arg \min_{\underline{\mathbf{u}} \le \mathbf{u} \le \bar{\mathbf{u}}} \left\| \begin{pmatrix} \xi^{\frac{1}{2}} \mathbf{W}_v \mathbf{B} \\ \boldsymbol{\Gamma}_1 \end{pmatrix} \mathbf{u} - \begin{pmatrix} \xi^{\frac{1}{2}} \mathbf{W}_v \mathbf{v} \\ 0 \end{pmatrix} \right\|_2^2 = \|\mathbf{A}\mathbf{u} - \mathbf{C}\|_2^2 \tag{17}$$

The proposed optimization problem can be solved by the active set method [2].

3 Simulation Results

The adaptive steering stability control strategy proposed for the 4MIDEV was validated in the co-simulation with a strategy model in the Matlab/Simulink and a vehicle model in the Carsim, where the double-lane-change (DLC) maneuver was respectively carried out on the joint road and the μ-split road to further verify the adaptability of the strategy to the adhesion coefficient. In addition, the other two strategies, referred to as "ordinary control" and "speed control," were compared with the proposed control strategy, referred to as "adaptive control." The "ordinary control" proposed in [1] also developed a hierarchical control structure, but did not consider the influence of the adhesion coefficient. The "speed control" includes only

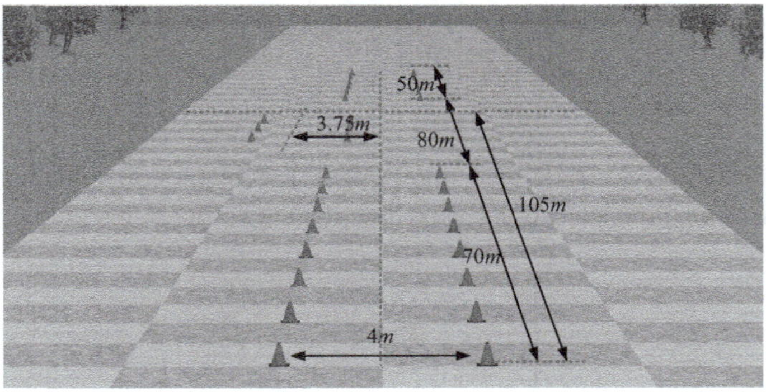

Fig. 3 Route of the DLC maneuver

Table 3 The parameters of the vehicle and in-wheel motors

Name	Symbol	Value
Vehicle mass	m	1411 kg
Moment of inertia about the yaw axis	I_z	2031.4 kgm^3
Height of the center of mass	hg	0.54 m
Length from the center of gravity (CG) to the front wheel axis	a	1.04 m
Length from the CG to the rear wheel axis	b	1.56 m
Tread width	d	1.48 m
Tire radius	R	0.3 m
Rated power	P_e	14 kW
Maximum power	P_m	28 kW
Rated speed	n_e	800 rpm
Maximum speed	n_m	1200 rpm
Rated torque	T_e	170 Nm
Maximum torque	T_m	340 Nm

the PID speed controller and a driver model to control lateral stability. The route of the DLC maneuver is shown in Fig. 3. The main parameters of the vehicle and hub motors for simulation are shown in Table 3.

3.1 μ-Split Road

In the simulation test for the μ-split road, the DLC maneuver was conducted at a constant speed of 60 km/h, where the adhesion coefficient was initially set to 0.75 and then changed to 0.75 on the right side and 0.2 on the left side at a distance of 105 m from the start, as shown in Fig. 4. Figure 5a–c show the track, yaw rate, and

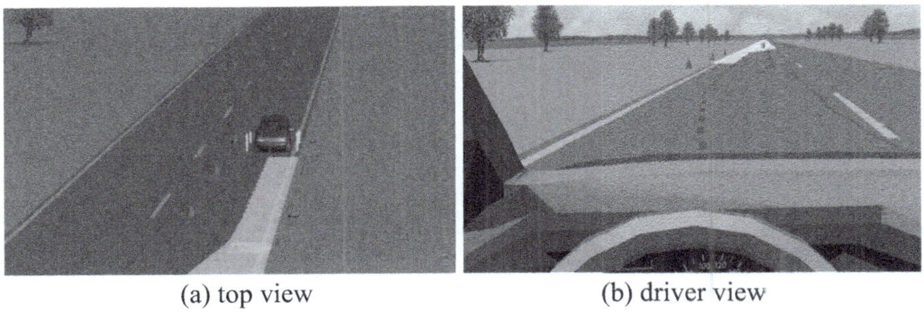

Fig. 4 μ-Split road

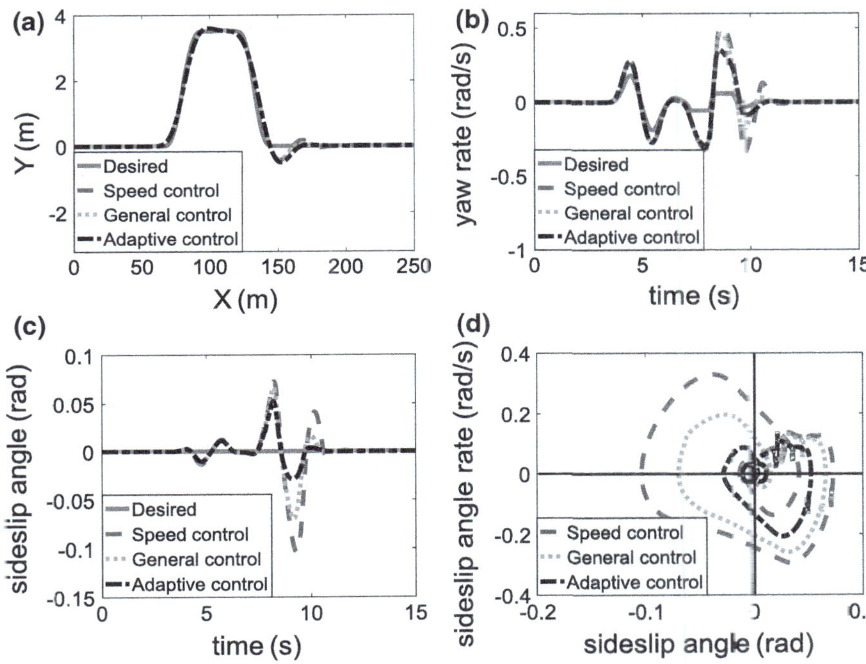

Fig. 5 Simulation results of the DLC maneuver under the μ-split road

sideslip angle responses of the vehicle under aforementioned strategies, respectively. Although the adhesion coefficient changed, the adaptive control could better follow the desired trajectory, with the minimum error and the fastest stable speed. In addition, a good control effect of the yaw rate and sideslip angle was also guaranteed by the adaptive control. Compared with the general control, the error of these two quantities could be reduced by up to 55 and 58.8% respectively under the adaptive control. Figure 6f shows the phase plane of the sideslip angle and the sideslip angle rate. The curves of the speed control, general control, and adaptive control were gradually

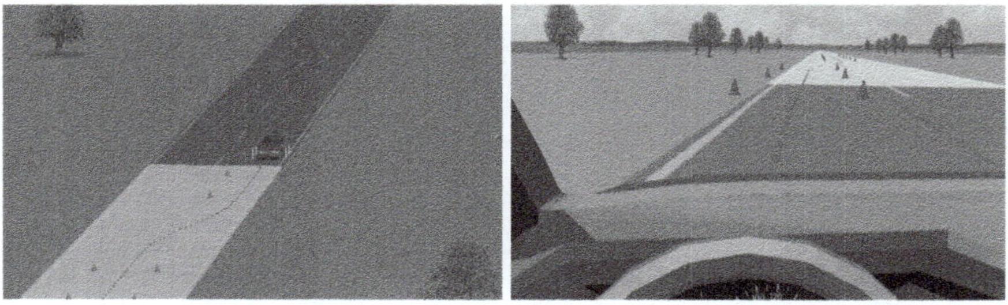

(a) top view (b) driver view

Fig. 6 Joint road

more concentrated on the origin, which means that the stability control effects of the above strategies were getting better [1].

3.2 Joint Road

The DLC maneuver was then conducted on the joint road for further verification, where the vehicle kept a constant speed of 60 km/h. The setting of joint-road surface is shown in Fig. 6, the adhesion coefficient of which was initially set to 0.75 and then changed to 0.2 at the middle of the maneuver. Figure 7a–c show the track, yaw rate, and sideslip angle responses of the vehicle under above three strategies, respectively. The adaptive control showed the best control effect in the following of the target trajectory and could effectively reduce the control error of the yaw rate and sideslip angle after the change of adhesion coefficient. Figure 7f shows the phase plane of the sideslip angle and the sideslip angle rate. The curve of the adaptive control is most concentrated on the origin compared to those of the other two strategies, indicating that its stability control effect is the best.

4 Conclusion

In this paper, an adaptive stability control strategy is proposed for the 4MIDEV to accommodate different adhesion coefficients. A new yaw moment control adapted to the adhesion coefficient is investigated in the upper control level to realize the comprehensive control of yaw rate and sideslip angle. The lower control level develops the optimization algorithm based on different weight coefficients, so as to realize different control of each motor torque according to the adhesion conditions at each wheel. The final simulation results show that the proposed control strategy can effectively improve the steering stability of the 4MDIEV on the road with different adhesion

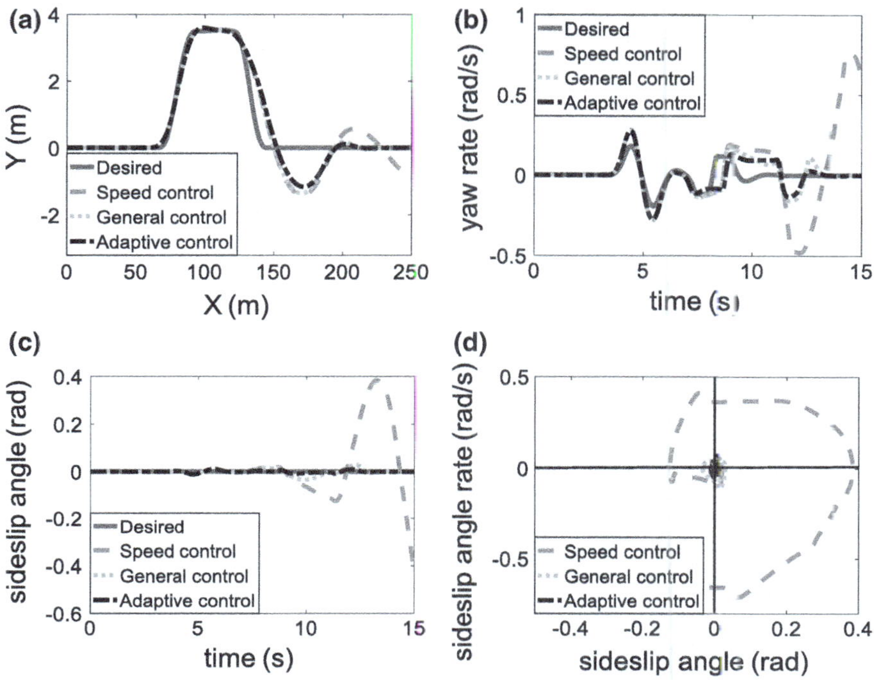

Fig. 7 Simulation results of the DLC maneuver under the joint road

coefficients, such as on the joint road. In the future work, this strategy will verified by the hardware-in-the-loop simulation, and finally the real-car experiment will be also carried out.

Acknowledgments This work was supported by the National Natural Science Foundation of China under Grant 51475045.

References

1. Zhai L, Sun T, Wang J (2016) Electronic stability control based on motor driving and braking torque distribution for a four in-wheel motor drive electric vehicle. IEEE Trans Veh Technol 65(6):4726–4739
2. Zhai L, Hou R, Sun T et al (2018) Continuous steering stability control based on an energy-saving torque distribution algorithm for a four in-wheel-motor independent-drive electric vehicle. Energies 11(2):350
3. Fallah S, Khajepour A, Fidan B et al (2013) Vehicle optimal torque vectoring using state-derivative feedback and linear matrix inequality. IEEE Trans Veh Technol 62(4):1540–1552
4. Shuai Z, Zhang H, Wang J et al (2014) Combined AFS and DYC control of four-wheel-independent-drive electric vehicles over CAN network with time-varying delays. IEEE Trans Veh Technol 63(2):591–602

5. Chen Y, Wang J (2012) Design and evaluation on electric differentials for overactuated electric ground vehicles with four independent in-wheel motors. IEEE Trans Veh Technol 61(4):1534–1542
6. Kim DH, Kim CJ, Kim SH et al (2011) Development of adaptive direct yaw-moment control method for electric vehicle based on identification of yaw-rate model. In: Intelligent vehicles symposium. IEEE, New York, pp 1098–1103
7. Tahami F, Kazemi R, Farhanghi S (2003) A novel driver assist stability system for all-wheel-drive electric vehicles. IEEE Trans Veh Technol 52(3):683–692
8. Zou G, Luo Y, Lian X et al (2008) A research of DYC for independent 4WD EV based on control target dynamic regulated. In: IEEE international conference on vehicular electronics and safety. IEEE, New York, pp 1–7
9. Xiong L, Yu Z (2009) Control allocation of vehicle dynamics control for a 4 in-wheel-motored EV. In: International conference on power electronics and intelligent transportation system. IEEE, New York, pp 307–311
10. Yamakawa J, Kojima A, Watanabe K (2007) A method of torque control for independent wheel drive vehicles on rough terrain. J Terramech 44(5):371–381
11. Kang J, Yoo J, Yi K (2011) Driving control algorithm for maneuverability, lateral stability, and rollover prevention of 4WD electric vehicles with independently driven front and rear wheels. IEEE Trans Veh Technol 60(7):2987–3001
12. Li B, Goodarzi A, Khajepour A et al (2015) An optimal torque distribution control strategy for four-independent wheel drive electric vehicles. Veh Syst Dyn 53(8):1172–1189
13. Guo K, Ma F (2003) Preview optimized artificial neural network driver model. Chin J Mech Eng 39(1):26–28
14. Ma Y, Guo H, Wang F et al (2016) An modular sideslip angle and road grade estimation scheme for four-wheel drive vehicles. In: Chinese control conference, pp 8962–8967
15. Shimada K, Shibahata Y (1994) Comparison of three active chassis control methods for stabilizing yaw moments. In: International Congress & exposition
16. Inagaki S (1994) Analysis on vehicle stability in critical cornering using phase-plane method. Avec 16(2):216–216(1)
17. Zhao L, Deng N, Ge Z et al (2014) Real-time road condition estimation for four-wheel-drive vehicle. J Harbin Inst Technol 46(11):42–46

Comparative Study on the Performance Prediction of Fuel Cell Using Support Vector Machine with Different Kernel Functions

Hanqi Ye, Xiaojun Ma, Tianxin Yang and Yongping Hou

Abstract The performance attenuation characteristics of fuel cell stack show a strong nonlinearity, but there is still no good model to predict its nonlinear attenuation characteristics. Support vector regression (SVR) is used as linear regression algorithm and realizes nonlinear regression function by introducing kernel function. Based on the test data of 4 kW fuel cell stack running 600 h on the bench, the results of using the SVR model with Gaussian radial basis kernel function (G-RBF), sigmoid kernel function, polynomial kernel function, and mixed kernel function to predict fuel cell performance are compared. The results show that the SVR model with polynomial kernel function has higher prediction precision.

Keywords Fuel cell · Performance prediction · Support vector regression machine · Kernel function

1 Introduction

Fuel cell vehicle, as one of the new energy vehicles, characterizes zero emission, high efficiency, and long range. The short lifetime of fuel cell stack which is the core component of fuel cell vehicle has limited the development of fuel cell vehicle. At present, the research on the life of fuel cell stack is based on the road test or durability bench test, which takes long time, high manpower cost, and capital cost. A reliable performance prediction method however can reduce the test time and cost of durability test by accurately predicting results consistent with conventional durability test data based on relatively few durability test data. At the same time, the prediction method can optimize the strategy of operating fuel cell stack and improve the life span.

H. Ye (✉) · Y. Hou
Tongji University, Shanghai, China
e-mail: hqye@tongji.edu.cn

X. Ma · T. Yang
Shanghai Greenbus Power System Corporation Ltd., Shanghai, China

China SAE (ed.), *Proceedings of China SAE Congress 2018: Selected Papers*,
Lecture Notes in Electrical Engineering 574,
https://doi.org/10.1007/978-981-13-9718-9_25

In the early study of performance prediction, Pei et al. [1] presented a lifetime prediction equation of fuel cell. He claimed that the lifetime had linear relationship with load changing cycle times, start-stop cycle times, idle times, and high load power times. With the further research on this problem, that the performance attenuation of fuel cell stack is nonlinear is widely accepted by many scholars. In recent years, Wu et al. [2–4] established the performance prediction model of proton exchange membrane fuel cell (PEMFC) based on the improved relevance vector machine (RVM), which predicted the voltage attenuation trend of the fuel cell, and verified the model by using several sets of experimental data, and compared it with the classical support vector machine (SVM) prediction model. It is found that the improved RVM model is more effective in the prediction of fuel cell performance attenuation. Kheirandish et al. [5] proposed the performance prediction method of PEMFC system based on SVM and compared it with the multilayer perception artificial neural network model. They concluded that the error of SVM was smaller. Han et al. [6] established the SVM model combining empirical equation to predict the polarization curve of fuel cell under different operating conditions. Silva et al. [7] established the predictive model of fuel cell voltage based on adaptive neuro-fuzzy system, and the model was validated by two sets of 1000 h experimental data. Javed et al. [8] established the PEMFC prediction model based on data-driven method, and the experimental data of two fuel cell stacks was used for model verification. Bae et al. [9] carried out the accelerated attenuation experiment of fuel cell and used a none-parameter smoothing regression algorithm to predict lifetime under non-accelerated conditions.

However, the current research is only to predict the voltage under single current density point, and to validate the prediction model by durability test data, which cannot reflect the comprehensive performance change of fuel cell stack. In order to make up for this shortcoming, this paper builds a model based on SVR, predicts the polarization curve of fuel cell stack in the future, and compares the prediction performance of the SVR with four different kernel functions.

The performance attenuation characteristic of fuel cell is strongly nonlinear. Only the nonlinear prediction model can be established to make accurate prediction of the polarization curve. The nonlinear characteristic of SVR is determined by kernel function, so this paper compares the prediction performance of the SVR with different kernel functions, including G-RBF kernel function, sigmoid kernel function, polynomial kernel function and a mixed kernel function to find which kernel function can be most suitable for fuel cell performance prediction. The hyperparameters in SVR are optimized by genetic algorithm (GA).

2 Introduction to Durability Test of Fuel Cell Stack

2.1 Test Bench and Duty Cycle

The test data of this paper is obtained from a 4 kW fuel cell stack durability bench test. The test bench is the FCATSTMG500 from Canadian Greenlight Innovation Company. It can complete the performance characterization test of 2–30 kW fuel cell stack. It mainly comprises the gas supply system, cooling system, load system, control system, data sampling system, security system and other subsystems and can accurately control the gas flow, gas temperature, gas humidity, gas pressure, etc. The test bench is shown in Fig. 1.

The fuel cell stack is operated on the test bench according to the durability duty cycle of Tongji University for the vehicular fuel cell stack, and the duty cycle is shown in Fig. 2. The operating conditions are based on the characteristics of the New European Driving Cycle (NEDC) and the actual vehicle driving conditions. It is adapted to the durability test condition of vehicular fuel cell stack, including idle, acceleration, deceleration, constant speed and other common conditions, as well as the rated working conditions [10]. The durability bench test lasts 600 h. At the beginning of the durability bench test, the polarization curve was tested per 90 cycles. As the running time was increased, the attenuation of the fuel cell stack was accelerated. Then the polarization curve was measured per 60 cycles. In conclusion, 28 polarization curves were measured, which reflected the performance change of the fuel cell stack during the 600 h durability test.

Fig. 1 Fuel cell stack test bench

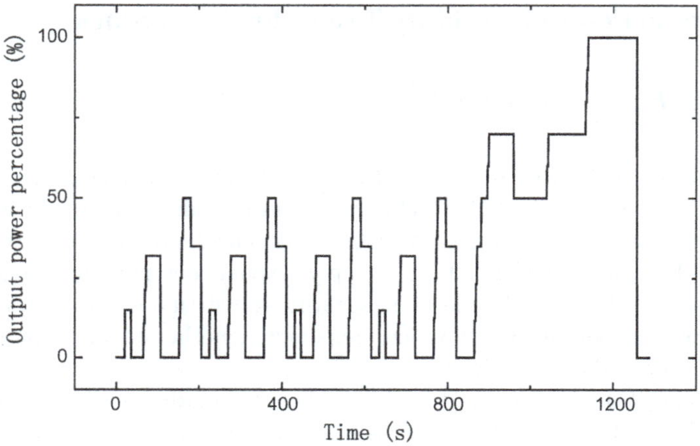

Fig. 2 Durability duty cycle of Tongji University for the vehicular fuel cell stack

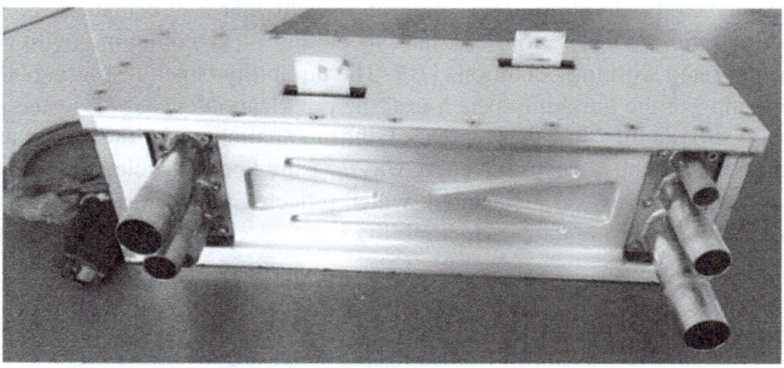

Fig. 3 Fuel cell stack for durability cycle test

2.2 *Test Object*

The test object is a 4 kW fuel cell stack, which is 488 mm length, 155 mm width, 145 mm height, composed of 25 pieces of single cell, 312 cm^2 of active area, uses deionized water as cooling medium, composite material as bipolar plate, and fuel cell stack as shown in Fig. 3.

3 The Principle of SVR Model Based on GA Optimization

SVM is a learning algorithm based on statistical learning theory, proposed by Vapnik et al. in 1995, which can overcome the traditional difficulties such as "dimensionality disaster" and "over-fitting" in some extent, and has been successfully applied in the fields of classification and prediction. SVR is essentially linear two classification algorithm. By introducing the kernel function, the low dimensional input data is mapped to the high dimensional space. The classification hyperplane is solved in the high dimensional space, and the classification hyperplane in high dimensional space shows non-linear in the low dimensional input space, thus the non-linear classification is realized. The regression function is realized by the introduction of ε insensitive loss function.

In the SVR model, the training dataset is given as $\boldsymbol{T} = \{(\boldsymbol{x}_1, y_1), (\boldsymbol{x}_2, y_2), \ldots, (\boldsymbol{x}_n, \boldsymbol{y}_n)\}$, $\boldsymbol{x}_i \in R^n$, $y_i \in R$, n as the quantity of training group samples. By nonlinear mapping function $\boldsymbol{\Phi}(\boldsymbol{x})$, the data in the input space is mapped to the high dimensional feature space, and the linear regression is carried out in the high dimensional feature space. The optimal linear regression function is:

$$f(\boldsymbol{x}) = \boldsymbol{\omega} \cdot \Phi(\boldsymbol{x}) + b \tag{1}$$

In Eq. (1): $\boldsymbol{\omega}$ is the weight, $\boldsymbol{\Phi}(\boldsymbol{x})$ is non-linear mapping function, and b is bias term.

According to SVR's structural risk minimization principle, weight $\boldsymbol{\omega}$ and bias term b can be obtained by solving the following objective functions:

$$\begin{aligned} &\min\left\{\frac{1}{2}\|\boldsymbol{\omega}\|^2 + C\sum_{i=1}^{n}\left(\xi_i + \xi_i^*\right)\right\} \\ &\text{s.t.}\begin{cases} y_i - \boldsymbol{\omega}\cdot\Phi(\boldsymbol{x}) - b \le \varepsilon + \xi_i \\ \boldsymbol{\omega}\cdot\Phi(\boldsymbol{x}) + b - y_i \le \varepsilon + \xi_i^* \\ \xi_i, \xi_i^* \ge 0, \quad i = 1, 2, \ldots, n \end{cases} \end{aligned} \tag{2}$$

In Eq. (2): ξ_i and ξ_i^* are the introduced nonnegative relaxation variables, which are used to allow regression error, and C is the penalty parameter, which is used to punish the regression error.

If the penalty parameter C is large, the allowable regression error is small, which may lead to the over-fitting phenomenon, which leads to the decline of prediction ability. Oppositely, if the penalty parameter is small, the allowable regression error is larger, which may cause the under-fitting phenomenon.

In order to establish the dual problem of the original problem, nonnegative Lagrange multiplier α_i and α_i^* are introduced to construct the Lagrange equation of the original problem. The partial derivative to ω, b, ξ_i, ξ_i^* of Lagrange equation is solved and are equal to zero. The dual principle is used to transform the original problem to the dual problem:

$$\min\left\{\frac{1}{2}\sum_{i,j=1}^{n}\left(\alpha_i - \alpha_i^*\right)\left(\alpha_j - \alpha_j^*\right)\Phi(\boldsymbol{x}_i)\cdot\Phi(\boldsymbol{x}_j) + \sum_{i=1}^{n}\alpha_i(\varepsilon - y_i) + \sum_{i=1}^{n}\alpha_i^*(\varepsilon + y_i)\right\}$$

$$\text{s.t.}\begin{cases}\sum_{i=1}^{n}(\alpha_i-\alpha_i^*)=0\\ \alpha_i,\alpha_i^*\in[0,C],\quad i=1,2,\ldots,n\end{cases}\tag{3}$$

As can be seen from Eq. (3), the input data in the high dimensional feature space exists in the form of the inner product $\Phi(\boldsymbol{x}_i)\cdot\Phi(\boldsymbol{x}_j)$. If Eq. (3) is directly solved in the situation of that the specific equation, parameter and dimension of nonlinear mapping function $\Phi(\boldsymbol{x})$ are given, the "dimensionality catastrophe" may occur when the dimension is very high. To deal with the "dimensionality catastrophe", SVR introduced kernel function $\kappa(\boldsymbol{x}_i,\boldsymbol{x}_j)$ to transform the calculation of inner product in high dimensional feature space into calculation of kernel function in low dimensional input space. Finally, the original problem becomes the solution to the following extremum problem:

$$\min\left\{\frac{1}{2}\sum_{i,j=1}^{n}(\alpha_i-\alpha_i^*)\left(\alpha_j-\alpha_j^*\right)\kappa(\boldsymbol{x}_i,\boldsymbol{x}_j)+\sum_{i=1}^{n}\alpha_i(\varepsilon-y_i)+\sum_{i=1}^{n}\alpha_i^*(\varepsilon+y_i)\right\}$$

$$\text{s.t.}\begin{cases}\sum_{i=1}^{n}(\alpha_i-\alpha_i^*)=0\\ \alpha_i,\alpha_i^*\in[0,C],\quad i=1,2,\ldots,n\end{cases}\tag{4}$$

In Eq. (4), the penalty parameter C and the insensitive loss parameter ε are the SVR parameters, which need to be given previously. The GA is used to determine the optimal penalty parameter C and insensitive loss parameter ε.

Finally, the optimal regression function is:

$$f(\boldsymbol{x})=\sum_{i=1}^{n}(\alpha_i-\alpha_i^*)\kappa(\boldsymbol{x}_i,\boldsymbol{x}_j)+b\tag{5}$$

In Eq. (5):

$$b=y_j-\sum_{i=1}^{n}(\alpha_i^*-\alpha_i)\kappa(\boldsymbol{x}_i,\boldsymbol{x}_j)+\varepsilon,\quad j\in\{m|0<\alpha_m<C\}\tag{6}$$

In Eq. (6): The training set points corresponding to $\alpha_i^*\neq\alpha_i$ are support vectors.

In this paper, the prediction performance of SVR with different kernel functions is compared to find the kernel function which is mostly suitable for the prediction of fuel cell performance. The common single kernel function mainly has the following three kinds:

(1) Polynomial kernel function:

$$\kappa(\boldsymbol{u},\boldsymbol{v})=(g\times\boldsymbol{u}\cdot\boldsymbol{v}+r)^d\tag{7}$$

In Eq. (7), g, r and d are the kernel function parameters. When $d=1$, it is the linear kernel function; when d is very large, high dimensional feature space dimension is also very high, which can increase the complexity of the model and the computation volume surges. To consider the nonlinear attenuation characteristic of fuel cell and reduce calculation, this paper chooses $d = 3$.

(2) G-RBF kernel function:

$$\kappa(\boldsymbol{u}, \boldsymbol{v}) = \exp\left(-\sigma \times \|\boldsymbol{u} - \boldsymbol{v}\|^2\right) \tag{8}$$

In Eq. (8): σ is the kernel function parameter.

(3) Hyperbolic tangent sigmoid kernel function:

$$\kappa(\boldsymbol{u}, \boldsymbol{v}) = \tanh(g \times \boldsymbol{u} \cdot \boldsymbol{v} + r) \tag{9}$$

In Eq. (9), g and r are the kernel function parameters.

Depending on the nature of the kernel function, a mixed kernel function can also be constructed from single kernel functions. If $\kappa_1(\boldsymbol{u}, \boldsymbol{v})$ and $\kappa_2(\boldsymbol{u}, \boldsymbol{v})$ are kernel functions, the following functions are also kernel functions, named as mixed kernel functions:

$$\kappa(\boldsymbol{u}, \boldsymbol{v}) = a \times \kappa_1(\boldsymbol{u}, \boldsymbol{v}) + b \times \kappa_2(\boldsymbol{u}, \boldsymbol{v}) \tag{10}$$

$$\kappa(\boldsymbol{u}, \boldsymbol{v}) = a \times \kappa_1(\boldsymbol{u}, \boldsymbol{v}) \cdot \kappa_2(\boldsymbol{u}, \boldsymbol{v}) \tag{11}$$

$$\kappa(\boldsymbol{u}, \boldsymbol{v}) = \exp(\kappa_1(\boldsymbol{u}, \boldsymbol{v})) \tag{12}$$

In Eqs. (10 and 11), a and b are constant.

In this paper, one kind of mixed kernel function is established by combining the polynomial kernel function with the weighted summation of G-RBF kernel function, and the equation is as followed:

$$\kappa(\boldsymbol{u}, \boldsymbol{v}) = \lambda \times (g \times \boldsymbol{u} \cdot \boldsymbol{v} + r)^d + (1 - \lambda) \times \exp\left(-\sigma \times \|\boldsymbol{u} - \boldsymbol{v}\|^2\right) \tag{13}$$

In Eq. (13): λ is the weighted coefficient, $\lambda \in [0, 1]$. According to repeated tests, this paper selects $\lambda = 0.95$. g, r and σ are the kernel function parameter.

The parameters of each kernel function need to be given previously and are determined by GA in this paper.

3.1 The Principle of GA

GA, a computational model that simulates the phenomenon/mechanism of 'survival of the fittest' in the natural world, makes the alternative solutions generate better offspring after selection, crossover and mutation, thereby obtaining the optimization solution by cyclical iteration.

In general, the solution to a problem is a number or set of numbers. Firstly, the solution is encoded into chromosomes. The common method is to convert the solution into binary encoding. If the solution is a set of numbers, each number can be converted to binary encoding and then spliced into a column encoding. GA produces the initial population, which is randomly generated consisting of several alternative solutions. Each solution is called an individual and a number of individuals form a population. Combining the sample data with the fitness function, we can calculate the fitness of each individual, and the fitness reflects the quality of each individual. The selection is to select some individuals with large fitness in the population. Then the crossover is to select the partial binary number of two individuals to exchange. Finally, the mutation operation refers to a random change in the binary encoding, for example, 0–1 or 1–0. After the original population has been selected, crossovered and mutated, a new population is produced. The new population is generated by cyclical iteration until the optimal target is achieved.

4 Prediction Model Establishment

4.1 SVR Model Based on GA Optimization

In this paper, the input data of SVR model is time t and current density i, the output data is the voltage V corresponding to time t and current density i. The dataset is $\boldsymbol{T} = \{([t_1, i_1], V_1), ([t_2, i_2], V_2), \ldots, ([t_n, i_n], V_n)\}$. The kernel functions are selected as the polynomial kernel function, the G-RBF kernel function, the sigmoid kernel function, and the mixed kernel function composed of the polynomial kernel function and the G-RBF kernel function. These four kernel functions are then compared and analyzed. The SVR model with four kernel functions requires given hyperparameters as shown in Table 1.

The hyperparameters in Table 1 are put into the GA to find the optimal solutions. After repeated tests, the initial population quantity of GA is 20, the genetic algebra is 50, the crossover probability is 0.8, and the mutation probability is 0.01. The fitness function is defined as the reciprocal of k-folder cross-validation error of the training data, and the k-fold cross-validation process is as follows:

Table 1 The hyperparameters of four kernel functions

Type	Hyperparameters				
Polynomial	C	ε	g	r	
G-RBF	C	ε	σ		
Sigmoid	C	ε	g	r	
Mixed	C	ε	g	r	σ

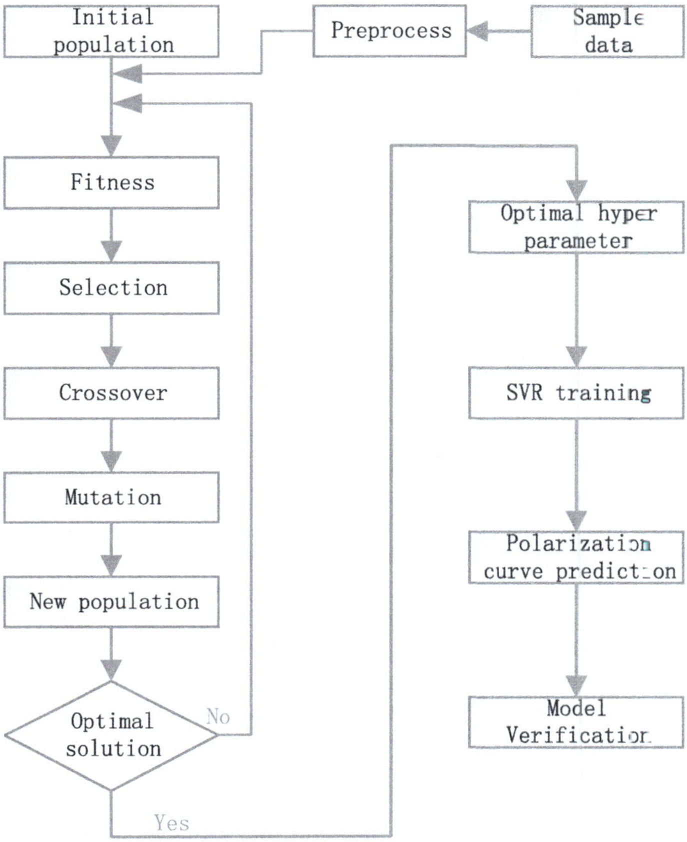

Fig. 4 The process of SVR optimized by GA

Step 1: Divide the training data randomly into k groups;
Step 2: Remove one group of data from k groups, and use the remaining k-1 group data training model;
Step 3: Regress the results of the removed group in Step 2;
Step 4: Calculate the square root of regression error of the results in step 3;
Step 5: Step 2–4 are repeated k times, and the mean error of k square root of regression error is obtained.

According to the characteristic of sample data, k is selected as 5, that is, 5-folder cross-validation method is used.

The process of SVR optimized by GA is shown in Fig. 4.

Table 2 Training and prediction set

Training set (h)	Prediction set (h)
180, 210, 230, 250, 270, 290, 310, 330, 350, 370, 390	410, 430, 450, 470, 490, 510, 530, 550, 570, 590, 600

Table 3 Random group of 5-folder cross-validation method

Group	1st	2nd	3rd	4th	5th
Time (h)	330, 290	310, 210	350, 250	390, 270	370, 230, 180

4.2 Data Preprocess and Grouping

Data preprocess refers to the normalization of data. Because of the different dimension and magnitude of time, current density and voltage in the sample data, we need to normalize the sample data before input the SVR model.

From the 28 polarization curve data measured by the fuel cell bench cycle test, the former 150 h is the activation period of the fuel cell stack. In this stage, the performance of the fuel cell is increased. In order to reduce the influence of the activation period on the prediction model, this paper removes the former 150 h data and select the polarization curve data of 150–600 h.

The former 50% of the remaining data were used as SVR training dataset, and the performance of the latter 50% was predicted, as shown in Table 2.

The training set consists of 11 polarization curves and Sect. 3.1 uses 5-folder cross-validation method to calculate the regression error of the training set. To divide the 11 polarization curve data into five groups randomly, it is impossible to achieve strict average distribution, so four groups contain two polarization curves data, one group contain three polarization curves data, and the random group is shown in Table 3.

5 Prediction Results and Comparative Analysis

5.1 Comparison of Prediction Results of SVR with Different Kernel Functions

The optimal parameters and optimum fitness of the SVR model with four kernel functions are shown in Table 4.

The regression results of the polarization curve of 230 and 330 h in the training set, and the prediction results of the polarization curves of 450, 530 and 600 h in the prediction set were shown in Figs. 5, 6, 7, 8 and 9.

Table 4 Optimal hyperparameter and optimum fitness

Type	C	ε	g	r	σ	Best fitness
Polynomial	142.0137	0	5.6051	0.8773		52.1942
G-RBF	199.8358	0.0151			2.3605	111.9737
Sigmoid	183.9078	0.2882	0.0883	0.0003		31.3617
Mixed	71.0899	0.0060	4.9851	0.3397	929.0487	70.4177

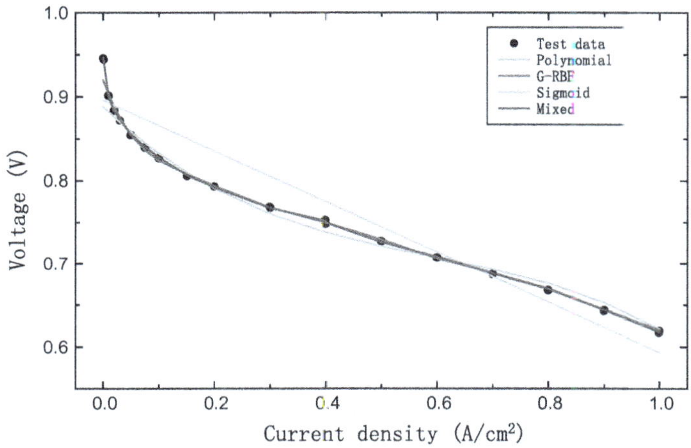

Fig. 5 230 h regression results

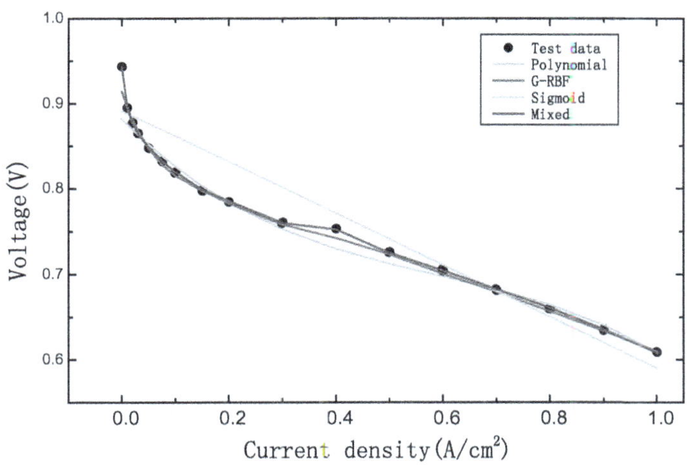

Fig. 6 330 h regression results

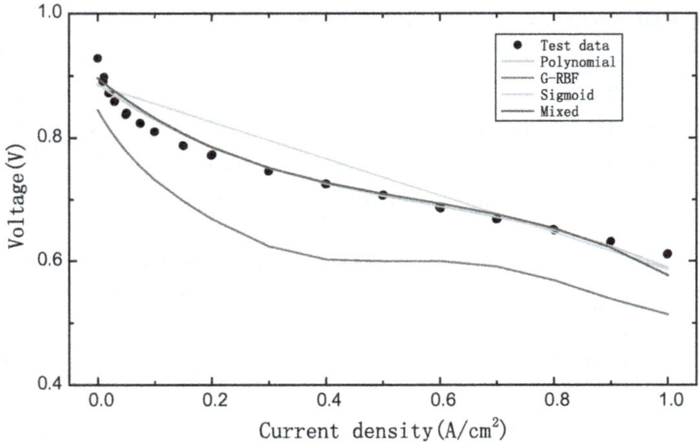

Fig. 7 450 h prediction results

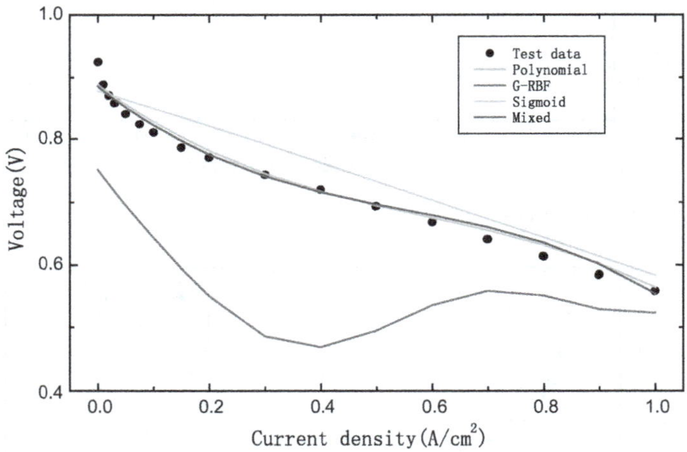

Fig. 8 530 h prediction results

5.2 Comparative Analysis of Prediction Results

It can be concluded from the Figs. 5 and 6 that regression precision of the training set from four kinds of SVR with different kernel functions is different. The SVR with sigmoid kernel function has the worst regression precision, the regression results of the other three kernel functions are almost same, among which the SVR with mixed kernel function is the most accurate.

It can be seen from the Figs. 7, 8 and 9 that the prediction results of SVR with polynomial kernel function are very close to that of SVR with mixed kernel function, and the prediction results are also better, comparing with that of two other kernel

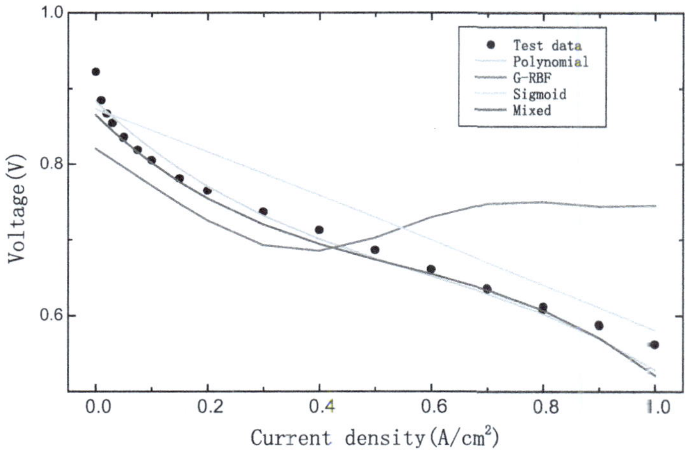

Fig. 9 600 h prediction results

functions. The prediction precision of the SVR with G-RBF kernel function is the lowest, and the shape of the polarization curve cannot accord with normal shape of polarization curve. It indicates that the G-RBF kernel function characterizes strong regression ability and poor generalization ability. SVR with sigmoid kernel function predicts that the polarization curve is a straight line and does not meet the requirement of prediction precision. Therefore, we exclude G-RBF and sigmoid kernel functions in the following error analysis, and only analyze the regression and prediction errors of SVR with polynomial kernel function and mixed kernel functions.

The error analysis mainly uses the maximum error, the mean absolute error (MAE), the root mean square error (RMSE) to compare and analyze, shown as Eqs. (14), (15) and (16).

$$\text{MaxError} = \max_{i=1,2,\ldots,n} \left| v_{p,i} - v_{r,i} \right| \tag{14}$$

$$\text{MAE} = \frac{1}{n} \sum_{i=1}^{n} \left| v_{p,i} - v_{r,i} \right| \tag{15}$$

$$\text{RMSE} = \sqrt{\frac{1}{n} \sum_{i=1}^{n} \left(v_{p,i} - v_{r,i} \right)^2} \tag{16}$$

In Eqs. (14–16): $v_{p,i}$ is predictive or regressive voltage; $v_{r,i}$ is the test voltage; n is the quantity of dataset.

The error results are shown in Table 5.

Firstly, it can be concluded from the regression error that the SVR with mixed kernel function is superior to that of the polynomial kernel function, whether in the aspect of the max error, MAE or RMSE. From the prediction error, it can be concluded

Table 5 Comparation of regression and prediction error of SVR with polynomial kernel function and mixed kernel function

Time (h)	Max error		MAE		RMSE	
	Polynomial	Mixed	Polynomial	Mixed	Polynomial	Mixed
230	0.0575	0.0031	0.0099	0.0010	0.0163	0.0011
330	0.0619	0.0037	0.0099	0.0010	0.0166	0.0012
Mean regression error	0.0597	0.0034	0.0099	0.0010	0.0164	0.0012
450	0.0413	0.0350	0.0092	0.0119	0.0128	0.0151
530	0.0373	0.0401	0.0120	0.0113	0.0146	0.0145
600	0.0395	0.0575	0.0125	0.0136	0.0151	0.0186
Mean prediction error	0.0394	0.0442	0.0112	0.0123	0.0141	0.0161

that the max error of the SVR with polynomial kernel function for 450 h is larger, the max error of that for 530 and 600 h is smaller, and the MSE of that for 530 h is greater, and the MSE of that of 450 and 600 h is smaller. The RMSE of that for 530 h is larger, while RMSE of that for 450 and 600 h is smaller. In a comprehensive perspective, the regression ability of the SVR with mixed kernel function is better, and the prediction ability of the SVR with polynomial kernel function is better.

6 Conclusions

1. The prediction ability of SVR with polynomial kernel function is the best, of which the average maximum prediction error is 0.0394, the average prediction error of MAE is 0.0112, and the average prediction error of RMSE is 0.0141, among polynomial kernel function, G-RBF kernel function, sigmoid kernel function, and mixed kernel function.
2. The prediction performance of the SVR with mixed kernel function is slightly poor, but it can still satisfy the prediction precision requirement. The average prediction error of max error, MAE, and RMSE is 0.0442, 0.0123, and 0.0161, and the regression ability of it is very strong.
3. The regression error of the SVR with G-RBF kernel function is small, but the prediction error is very large. If regression is carried out using the SVR with G-RBF kernel function, the accuracy requirement can still be satisfied; if the prediction is carried out, it is not suitable for the fuel cell stack. The regression precision and prediction precision of SVR with sigmoid kernel function are very poor, which indicates that SVR with sigmoid kernel function is not suitable for fuel cell performance prediction.

4. To sum up, the SVR with polynomial kernel function is recommended in the fuel cell performance prediction.

Acknowledgements The work was financially supported under the Science and Technology Commission of Shanghai Municipality, Project No. 16DZ1204202.

References

1. Pei P, Chang Q, Tang T (2008) A quick evaluating method for automotive fuel cell lifetime. Int J Hydrogen Energy 33(14):3829–3836
2. Wu Y, Breaz E, Gao F et al (2016) A modified relevance vector machine for PEM fuel-cell stack aging prediction. IEEE Trans Ind Appl 52(3):2573–2581
3. Wu Y, Breaz E, Gao F et al (2016) Nonlinear performance degradation prediction of proton exchange membrane fuel cells using relevance vector machine. IEEE Trans Energy Convers 31(4):1570–1582
4. Wu Y, Breaz E, Gao F et al (2015) Prediction of PEMFC stack aging based on relevance vector machine. In: Transportation electrification conference and expo. IEEE, New York, pp 1–5
5. Kheirandish A, Shafiabady N, Dahari M et al (2016) Modeling of commercial proton exchange membrane fuel cell using support vector machine. Int J Hydrogen Energy 41(25):11351–11358
6. Han IS, Chung CB (2017) A hybrid model combining a support vector machine with an empirical equation for predicting polarization curves of PEM fuel cells. Int J Hydrogen Energy
7. Silva RE, Gouriveau R, Jemeï S et al (2014) Proton exchange membrane fuel cell degradation prediction based on adaptive neuro-fuzzy inference systems. Int J Hydrogen Energy 39(21):11128–11144
8. Javed K, Gouriveau R, Zerhouni N et al (2016) Prognostics of proton exchange membrane fuel cells stack using an ensemble of constraints based connectionist networks. J Power Sources 324:745–757
9. Bae SJ, Kim SJ, Park JI et al (2012) Lifetime prediction of a polymer electrolyte membrane fuel cell via an accelerated startup–shutdown cycle test. Int J Hydrogen Energy 37(12):9775–9781
10. Yanwei H, Yongping H, Jianwen Z et al (2017) Comparison of duty cycles for fuel cell durability bench tests. Hunan Province: Battery Bimonthly, 6

Magnetic Field Distribution of Resonance Coupling Coils of 7.7 kW Wireless Charging System

Guangyuan Zhong, Li Zhai, Guixing Hu, Yu Cao and Liwen Lin

Abstract This paper designs a 7.7 kW power resonance wireless charging system for electric vehicles. The resonant frequency of this system is 85 kHz, and the coupling coil adopts LCC compensation circuit. A wireless charging system circuit model was built and simulated by MATLAB/Simulink to obtain the currents on the primary and secondary coils. The 3D FEA software ANSYS Maxwell was used to model circular coils with ferrites to predict the magnetic field distribution in the vehicle environment. According to the SAE J2954, four points on the body and the surrounding area and four points on the driver's seat were selected as markers, respectively. Studying three working conditions of the alignment and offset of the coupling coil, the magnetic field distribution of the entire coupled coil and each marker are observed. A pair of aluminum shielding shells was designed for the magnetic field distribution of the coupling coils. The magnetic field strength of the coils with or without shielded aluminum shells was simulated and compared. Then, the inhibitory effect of the shielding shell on the leakage magnetic field of the resonance coupling coils was verified. Safety of magnetic field distribution of a wireless charging system with a shielded aluminum shell was explained.

Keywords Electric vehicle · Wireless charging · Resonance coupling coils · Magnetic field

---

G. Zhong (✉) · L. Zhai · G. Hu · Y. Cao
Beijing Institute of Technology, 100081 Beijing, China
e-mail: z_g_yuan@163.com

L. Zhai
e-mail: ZhaiLi26@126.com

Collaborative Innovation Center of Electric Vehicles in Beijing, 100081 Beijing, China

L. Lin
State Grid Corporation of China, 350000 Fujian, China

China SAE (ed.), *Proceedings of China SAE Congress 2018: Selected Papers*,
Lecture Notes in Electrical Engineering 574,
https://doi.org/10.1007/978-981-13-9718-9_26

1 Introduction

In recent years, with the promotion of new energy industry in China, the sales and retention of electric vehicles (EVs) have continued to increase. At the same time, higher requirements have been placed on the convenience, safety, and environmental protection of electric vehicle charging. Compared with conductive charging, the main advantage of wireless charging is that the system can work without exposed conductors and interlocks. Thereby, wireless charging can effectively reduce the risk of electric accident to the system. It can be used in wet and rainy environments and just needs a small footprint. In addition, it has less wearing and aging problems. What is more, wireless charging can be highly integrated with smart electric vehicles to achieve unattended charging and bring a better user experience [1]. Therefore, it is necessary to develop the wireless charging industry.

With the deeply development of the system, the power transferred by the wireless system is increasing, and the power transmitted through the resonance coupling of the magnetic field has reached a kilowatt level. The probability that people and animals will be exposed to a dangerous radiation frequency (RF) electromagnetic field is also greatly increased. At the same time, leakage of high-frequency magnetic fields from resonance coupling coils may cause electromagnetic interference (EMI) on some low-voltage sensitive components [such as tire pressure sensors, battery management systems (BMS)] and on-vehicle radio equipment installed on or under the vehicle chassis. It can also affect the normal function of electric vehicles [2]. Therefore, it is necessary to analyze and estimate the mutual magnetic field and the leakage magnetic field distribution of the coupling coil of the wireless charging system.

Many researches have proposed the design method of the coupling coil and the core structure and analyzed the magnetic field distribution characteristics of the coupling coil [3]. The purpose is to improve the transmission efficiency of the wireless charging system under different charging levels. However, electromagnetic field (EMF) noise and EMI effects due to magnetic resonance coupling have not been described in detail. Many studies have analyzed and simulated the magnetic field distribution of coupling coils, but did not consider the influence of vehicle shell on the leakage magnetic field [4].

At present, there are generally three shielding methods that can reduce the EMF noise and EMI noise caused by resonance coupling: shielding shell, passive shield using conductive or ferromagnetic material, and coil design and resonant reactive shield [5]. The latter two types of shield structures are too complex and difficult to apply. Therefore, aluminum shell shields with simple structures and effective effects are usually selected to attenuate leakage magnetic fields and suppress EMI noise.

2 Design and Analysis of the Wireless Charging System

2.1 Design of Wireless Charging System

According to the SAE J2954 regulation of wireless charging system, a WPT2 level 7.7 kW wireless charging system is selected here. The design indicators are shown in Table 1. The input parameters of the system are 220 V, 50 Hz alternating current of the public power grid, the output is direct current 400 V, and the output voltage ripple is less than 1%, the current ripple is less than 3%. The maximum output power of the system is 7.7 kW [6].

The system includes primary rectifier bridges, power factor correction (PFC), inverter, coupling coils, and secondary compensation network structures, secondary rectifier bridges, high-frequency filter circuit and battery systems (Fig. 1 and Table 2).

The single-phase interleaved PFC that can increase the power factor of the system, the full-bridge inverter with higher power, and the LCC compensation structure with better resonance characteristics are selected [7].

Table 1 Design indicators of Wireless charging system

Parameter	Symbol	Value
Power	P	7.7 kW
Resonant frequency	f	85 kHz
Gap between coils	h	150 mm
Grid voltage	U_{ac}	220 V
Grid frequency	f_1	50 Hz
Charging voltage	U_c	300–400 V
Charging current	I_c	<20 A

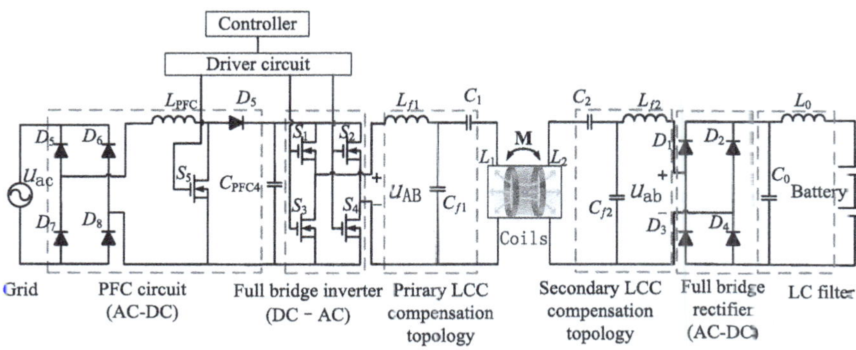

Fig. 1 Circuit structure of wireless charging system

Table 2 Parameter of wireless charging system

Parameter	Symbol	Value
Primary coil	L_1	226.44 μH
Secondary coil	L_2	226.44 μH
Mutual inductance of coupling coils	M	79.82 μH
Rectifier inductance	L_0	0.1319 mH
Rectifier capacitor	C_0	34.3 μF
PFC inductance	L_{PFC}	274.6 μH
PFC capacitor	C_{PFC}	1300 μF
Primary compensated inductance	L	42.4 μH
Secondary compensated inductance	L_{f2}	42.4 μH
Primary compensated capacitor 1	C_{f1}	82.7 nF
Secondary compensated capacitor 1	C_{f2}	82.7 nF
Primary compensated capacitor 2	C_1	19.1 nF
Secondary compensated capacitor 2	C_2	19.1 nF

2.2 Analysis of Wireless Charging System

The system adopts LCC compensation topology network because the characteristics of the LCC circuit cannot be affected by the coil coupling factor and secondary side load conditions.

The equivalent circuit of the LCC compensation topology network is shown in Fig. 2, where U_{AB} and U_{ab} are the input and output voltages of the compensation topology circuit, respectively. L_1 and C_1 are equivalent to inductance L_{e1}; L_2 and C_2 are equivalent to inductance L_{e2}. Based on the above equivalent circuit, the primary coil input current $\dot{I}_1$ can be expressed as

$$\dot{I}_1 = \frac{U_{ab}}{2\pi f L_{f1e}}\dot{\delta} \tag{1}$$

where $\dot{\delta}$ is the unit vector. Secondary coil input current $\dot{I}_2$ can be expressed as

Fig. 2 Equivalent circuit of LCC compensation topology network

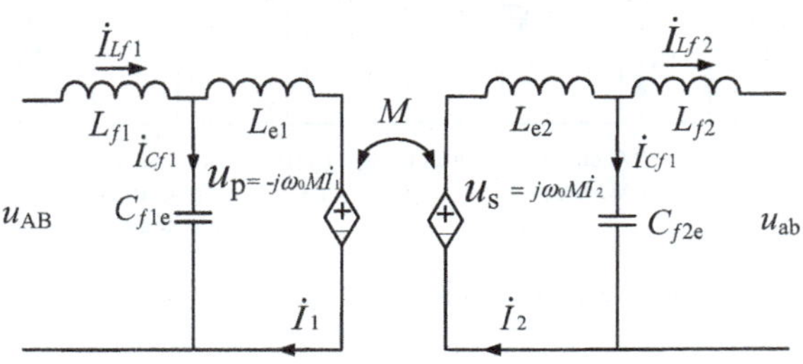

$$\dot{I}_2 = \frac{U_{ab}}{2\pi f L_{f2e}} \dot{\delta} \tag{2}$$

The power delivered by the system can be expressed as follows:

$$P = \dot{U}_{AB} \cdot \dot{I}_{Lf1} = \frac{U_{AB} U_{ab} M}{2\pi f L_{f1} L_{f2}}. \tag{3}$$

The LCC circuit on both sides will have a magnetic resonance at a fixed frequency. The distribution of the magnetic field is closely related to the current in the circular coil. The flux of mutual inductance can be expressed as follows

$$\varphi_{21} = \frac{2u_0 \pi a_1^2 a_2^2 N_1 I_1}{(a_1^2 + d^2)^{\frac{3}{2}}} \tag{4}$$

where a_1 is the primary coil radius, a_2 is the secondary coil radius, N_1 is the primary coil turns, and d is the length of the magnetic path. From formula (4), it can be seen that the magnetic field is mainly determined by the current in the coil [8].

2.3 Simulation of Wireless Charging System

The main circuit model of the wireless charging system is constructed by the control software MATLAB/Simulink, as shown in Fig. 2. The PFC control circuit model is used to control the power switch to achieve the desired DC voltage. The inverter control circuit model is used to control the inverter to generate the required U_{AB} and a constant sinusoidal current in the primary coil as excitation current. The simulation results of the high-voltage battery output charging voltage and charging current are 400 V and 19.25 A, respectively, as shown in Fig. 3, and the output power reaches the design value of 7.7 kW (Fig. 4).

The simulation results of the primary coil current and secondary coil current are obtained through the circuit model in MATLAB/Simulink, as shown in Fig. 3.

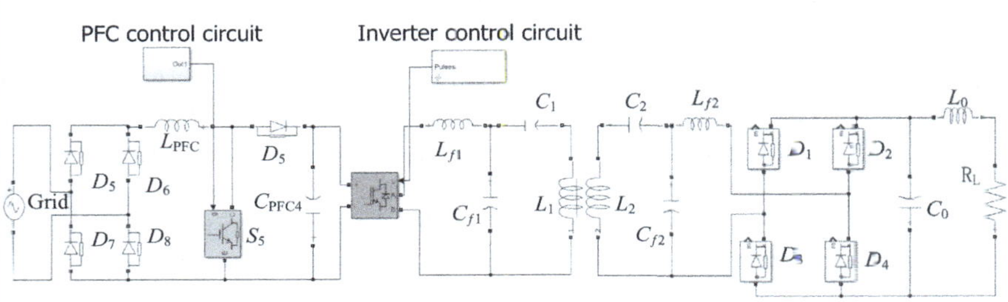

Fig. 3 Wireless charging system circuit model

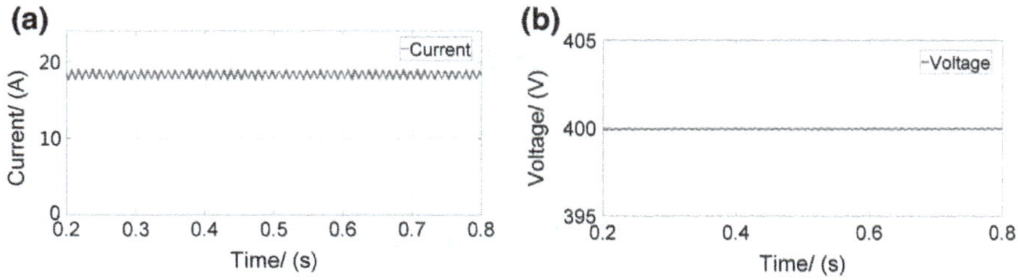

Fig. 4 Simulation results of the main circuit: **a** Output charging current. **b** Output charging voltage

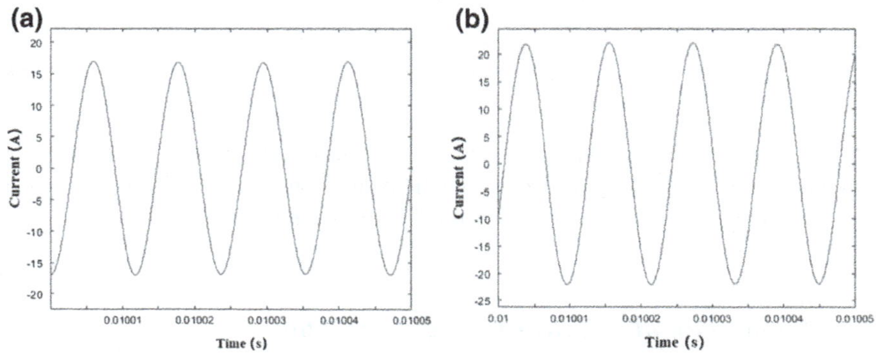

Fig. 5 Simulation results: **a** Current of primary coil. **b** Current of secondary coil

As can be seen from Fig. 5, the currents of the primary coil and secondary coil are 17.0 and 22.5 A which are consistent with the calculation. Two signals for two analog current sources are extracted for field distribution simulation.

3 Modeling and Simulation Analysis

3.1 Modeling for Resonance Coupling Coils

In this paper, primary coils and secondary coils of the same size and shape are used. The shape of the coil is selected here to have a circular arrangement with good anti-offset characteristics, and the ferrite is chosen to have a long and short staggered spoke arrangement. This arrangement can guarantee a large extent magnetic field distribution and save the amount of ferrite. The three-dimensional model of the coupling coil and ferrite is shown in Fig. 6, and its size is suitable for general passenger cars.

The 3D model of the coupling coil is established in ANSYS Maxwell. The electric parameters of the whole coupling coil model can be extracted conveniently and effec-

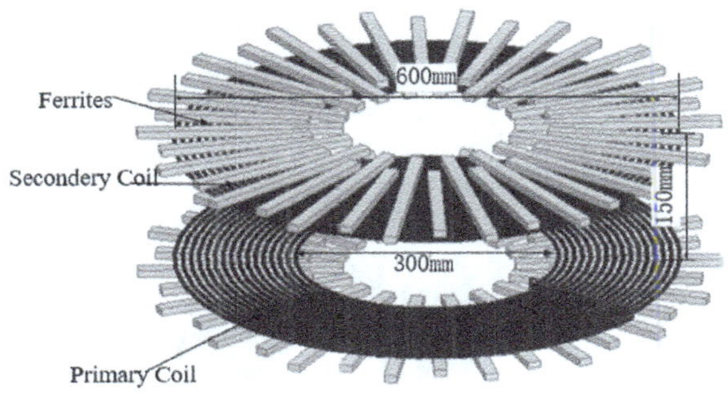

Fig. 6 3D model of the coupling coil

Table 3 Parameters of coils and ferrite

Circle coil attribute name	Value/Type	Ferrites attribute name	Value/Type
Coil geometry	Planar spiral	Ferrite material	Ferroxcube 3C94
Outer diameter	600 mm	Long ferrite size	240 mm × 15 mm × 9 mm
Inner diameter	300 mm	Short ferrite size	180 mm × 15 mm × 9 mm
Number of turns	16	Long ferrite quantity	18
Conductor diameter	3.9 mm	Short ferrite quantity	18
Litz conductor	II(16AWG4*5X24/36)	Adjacent ferrite angle	10°

tively after the coil and ferrite materials are correctly valued. The specific parameters of the coupling coil and ferrite are shown in Table 3.

3.2 *Modeling for Vehicle*

The general size SUV model is selected as the vehicle model. The model is built according to the actual vehicle size in the 3D modeling software CATIA and then also imported into Maxwell. Finally, the coupling coil and the vehicle model are arranged according to the standard as shown in Fig. 7. Three-dimensional size of the model are 3.8 m × 1.7 m × 1.6 m. The body is made of iron, and the chassis is a fully enclosed in iron plate [9]. The coil is located at the right of the driver's seat and close to the front axle of the vehicle (Fig. 8).

Fig. 7 3D vehicle model

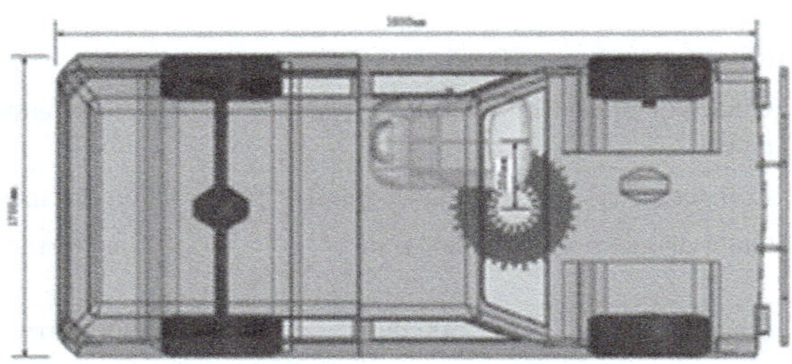

Fig. 8 Coupling coil and driver seat relative position

3.3 Simulation Analysis of Magnetic Field Strength

3.3.1 Marker Selection

In order to clarify the distribution of the magnetic field around the vehicle equipped with coupling coils, the definition of SAE J2945 is used to divide the vehicle into four physical areas 1, 2a, 2b, and 3. As shown in Fig. 9, area 1 is the entire area under the vehicle, including the periphery of the wireless power unit; area 2 is around the vehicle, where area 2a is 700 mm below the ground, and area 2b is 700 mm above the ground; 3 is the interior of the vehicle.

The four points m1, m2, m3, and m4 in different regions are selected as our markers, which can effectively reflect the magnetic field relationships in different regions. The four points are located in the center section of the coupling device, and the selection of the four points strictly follows the guidelines of SAE J2954. The relative positions are as shown in Figs. 9 and 10.

Since the driver stays in area 3 for a long time while driving, it is necessary to protect the driver's magnetic field in area 3. The SAE J2954 also gives makers for the field distribution of the driver's seat in area 3. It is necessary to record the position of the head (m5), chest (m6), seat cushion (m7), and foot (m8) in each seat position. In addition, the space peaks and their positions should be recorded to determine the electric and magnetic fields at each seat location. Measurements should be repeated

Fig. 9 Area division

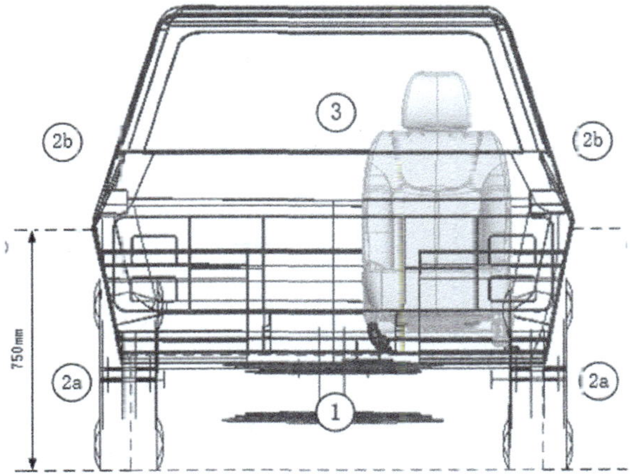

Fig. 10 Regional markers distribution

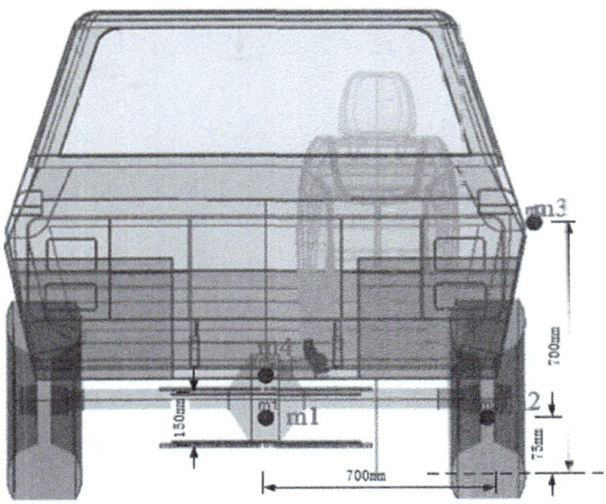

for each combination of alignment and gap conditions. The location of the measuring points is shown in Fig. 11. Each measurement point is 10 mm from the seat surface (Fig. 12).

3.3.2 Magnetic Field Limit Selection

The limits for magnetic field exposure in public places are given in the SAE J2954 and meet the requirements of ICNIRP 2010 (the International Commission on Non-Ionizing Radiation Protection). The limits are shown in Table 4.

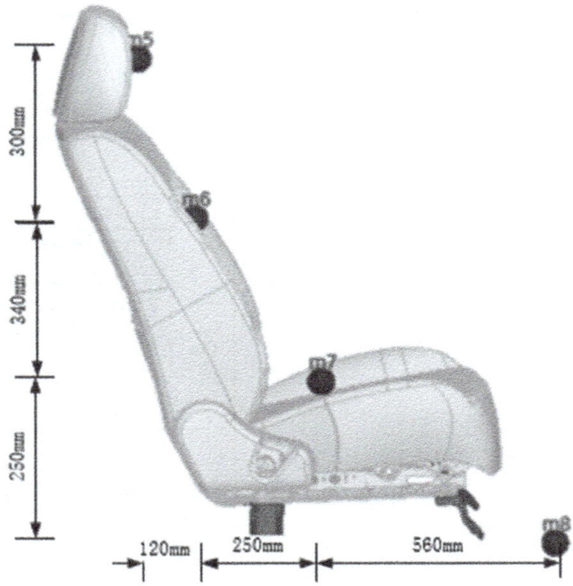

Fig. 11 Markers distribution of driver seat

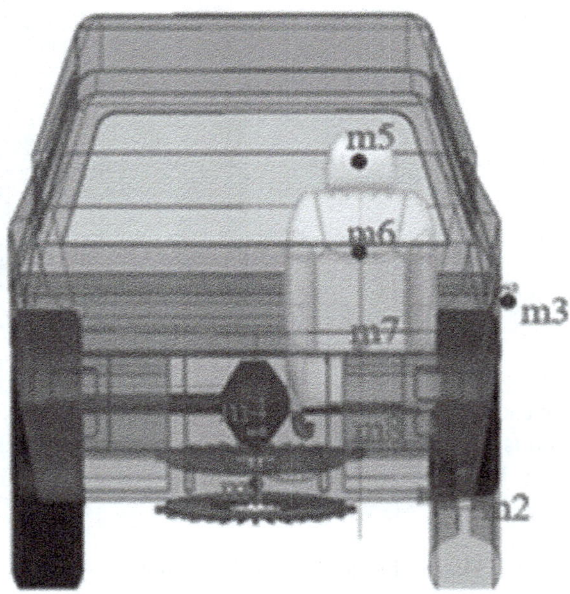

Fig. 12 Markers distribution

Table 4 Magnetic field limit reference

Reference	Region	Magnetic field mean limit (RMS)	Magnetic field peak limit (PEAK)
EMF exposure standards	2a, 2b, 3	27 uT or 21.4 A/m	38.2 uT or 30.4 A/m
Pacemaker/IMD limit	3, 2b	15 uT or 11.9 A/m	21.2 uT or 16.9 A/m
Pacemaker/IMD limit	2a	29.4 uT or 23.4 A/m	41.6 uT or 33.1 A/m

Considering the low-frequency modulation of the field or the possibility of contact currents (e.g., due to incomplete mains supply filtering) and compliance with the peak exposure limits can also ensure that RMS exposure limits are met, peak detection is evaluated.

The table can be seen in the maximum magnetic field and electric field operating conditions, and the peak value of the exposed magnetic field should be less than 30.4 A/m, but for those with pacemakers, long exposure is higher than the limits specified in area 3 and area 2. At this time, possible pacemaker effects are expected to limit the temporary return during the exposure (asynchronous pacing). To ensure the safety of special populations equipped with sensitive devices such as pacemakers and implantable nerve stimulators, the EMF exposure criteria and pacemaker/IMD limits are used here as a reference.

3.3.3 Magnetic Field Limit Selection

Selecting eight markers specified in SAE J2954, the current excitation when the power of the wireless charging system reaches 7.7 kW is loaded into the primary coil and the secondary coil. The magnetic field is simulated to obtain the magnetic field strength of each marker. The magnetic field is shown in Fig. 13.

Considering the limits in SAE J2954 for all normal operating conditions of the wireless charging system, including the coupling coil offset, the field exposure limits should be met in area 2a, 2b, and 3. Coupling coil deflection conditions are also

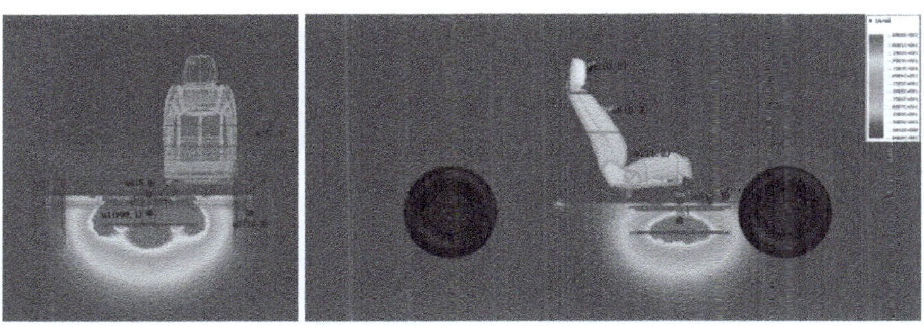

Fig. 13 Magnetic field distribution when coupling coils are aligned

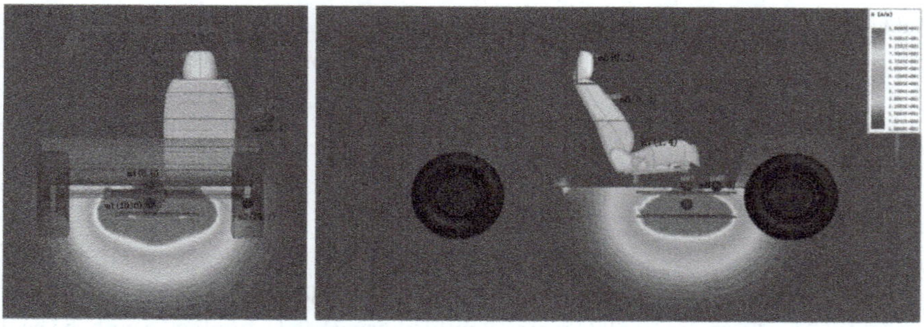

Fig. 14 Magnetic field distribution of 50 mm offset

considered here. The wireless coupling coil allows a certain amount of offset while guaranteeing certain working efficiency. This value must meet the manufacturer's deviation from the vehicle's implementation. Only the 100 mm offset specified in SAE J2954 is considered here. The 50 mm tolerance is used to divide the offset conditions into 50 and 100 mm offset. Simulation is performed according to the above-mentioned non-offset conditions to obtain the following simulation results (Figs. 14, 15 and Table 5).

It can be seen from the simulation results that the magnetic field is mainly distributed in the chassis area of the vehicle and gradually diverges toward the surroundings. The magnetic field in the vehicle is at a low level due to the shielding effect of the metal body. From the magnetic field distribution of each marker, it can be seen that m1 located at the center of the coupling coil in area 1 transmits energy between the coils, and its magnetic field can be as high as 1000 A/m. The m2 marker located in area 2a has a magnetic field of several tens of amperes per meter due to its proximity to the working area of the chassis. In particular, when the coupling coil is offset by 100 mm, the magnetic field intensity at m2 is further increased due to the distance. Increased to 37.2 A/m, exceeding the standard limit of 30.4 A/m for exposed areas in public areas, will cause some harm to organisms. The magnetic field at m3 located in the vicinity of the door 2b is in the range of 0.9–2.4 A/m, which is much smaller

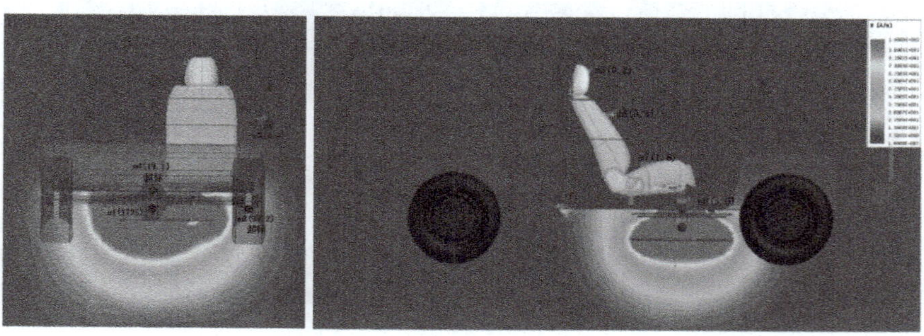

Fig. 15 Magnetic field distribution of 100 mm offset

Table 5 Magnetic field strength at markers under three conditions

Marker number	Alignment (A/m)	50 mm offset (A/m)	100 mm offset (A/m)	Limit value (A/m)
m1	998.1	1030	1725	None
m2	16.8	20.7	37.2 (Excessive)	30.4
m3	2.3	2.4	0.9	16.9
m4	8.9	8.6	19.1 (Excessive)	16.9
m5	0.2	0.2	0.2	16.9
m6	0.3	0.3	0.3	16.9
m7	1.2	1.4	1.6	16.9
m8	3.3	3.6	5.9	16.9

than the pacemaker/IMD limit. The point m4 located in the vehicle interior area 3 is directly above the coupling coil and is close to the working area. Due to the shielding effect of the vehicle body, the magnetic field value is not large but there is still a risk, especially when the offset reaches 100 mm. Values up to 19.1 A/m may adversely affect special people equipped with sensitive equipment and electrical equipment at the chassis. The four makers around the seat, due to the shielding effect of the vehicle body, whose magnetic field strength is within the normal range, and the distance from the marker and the coupling coil gradually decreases.

4 Wireless Charging System Magnetic Field Suppression

4.1 Design of Shielding Device

According to the simulation results above, it can be seen that the markers m2 and m4 exceed the specified magnetic limit when the coupling coil is offset by 100 mm. This may cause some damage to the human body exposed to the magnetic field. In addition, the magnetic field at the lower side of the vehicle chassis and doors is also at a high level, so it is necessary to perform magnetic field suppression on the coupling device. A shielded aluminum housing with a simple structure is used as a shielding device to facilitate the protection and installation of the coupling coil on the vehicle chassis. The thickness of entire housing is 5 mm, and the coil surrounds the coupling coil. The three-dimensional simulation model of the vehicle's magnetic field after installing a shielded shell is shown in Fig. 16.

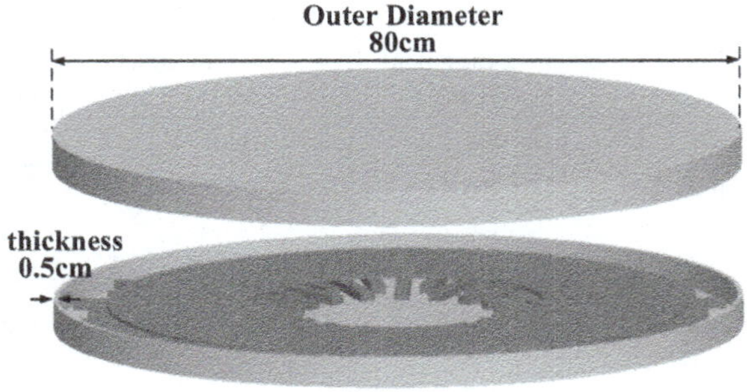

Fig. 16 Coupling coil with shielding aluminum plate

4.2 *Simulation Contrast*

In the same way, simulation verification was carried out for the three conditions of fully aligned, offset 50 and 100 mm coupling devices (Fig. 17 and Table 6).

The magnetic field distribution can be obtained from the simulation. After the shield shell is installed, the distribution of the magnetic field generated by the reso-

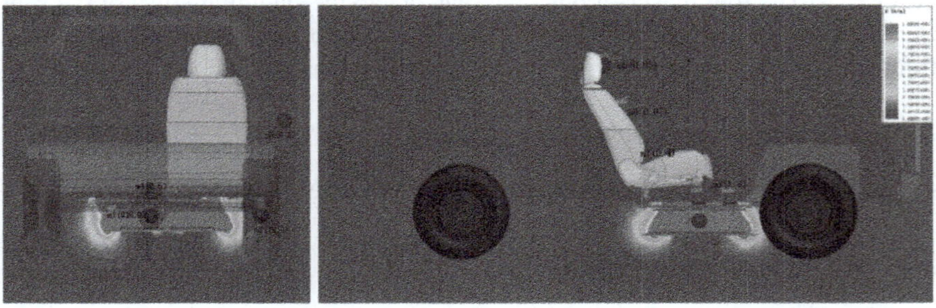

Fig. 17 Magnetic field distribution of 100 mm offset (with shielding aluminum shell)

Table 6 Magnetic field strength at markers under three conditions (with shielding aluminum shell)

Marker number	Alignment (A/m)	50 mm offset (A/m)	100 mm offset (A/m)	Limit value (A/m)
m1	672.0	721.0	836.9	None
m2	4.9	6.9	10.3	30.4
m3	0.5	0.7	0.8	16.9
m4	0.3	0.4	0.5	16.9
m5	0.02	0.04	0.05	16.9
m6	0.05	0.06	0.07	16.9
m7	0.2	0.3	0.4	16.9
m8	0.8	1.1	1.8	16.9

nance coupling coil is better improved. The magnetic field is concentrated between the two shielded aluminum shells, and the leakage magnetic field is significantly reduced. From the obtained numerical analysis, the m2 located in area 2a decreased from the maximum magnetic field of 16.2–37.2 A/m to 4.9–10.3 A/m. The amplitude of the drop was significant, which played a very good shielding effect, especially when the coil was at 100 mm offset. Under the moving conditions, compared with 37.2 A/m before shielding, the drop was 26.9 A/m, which effectively solved the problem of excessive magnetic field before shielding. The m3 located in area 2b also has a significant drop. The intensity of the magnetic field at the measurement point inside the vehicle and near the seat is further reduced, and it is always in a relatively safe range. It shows that the wireless charging system has less interference with the magnetic field inside the vehicle.

5 Conclusion

A 7.7 kW resonance wireless charging system is designed and built for electric vehicles. Circular coils with ferrite and the environment of the vehicle are modeled based on ANSYS Maxwell. The magnetic field distribution of the vehicle region under different working conditions was studied. Eight markers according to the SAE J2954 standard can be obtained. The magnetic field strengths of the markers m2 and m4 are 37.2 and 19.1 A/m, respectively, when the coupling coil is deflected by 100 mm, which exceeds the EMF exposure criteria and the pacemaker/IMD limit. By designing a pair of aluminum shield shells for the magnetic field distribution of the coupled coils, the magnetic field of the shielded aluminum shells are compared by simulation, and the magnetic field of the markers m2 and m4 are effectively suppressed, which are 10.3 and 0.5 A/m. And the magnetic field of other markers has been further suppressed. The shielding effect of the shield shell on the leakage magnetic field of the resonant coupling coil is verified that the electronic equipment and the body can be effectively prevented from being exposed to the EMF. Furthermore, the magnetic field distribution safety of a wireless charging system with a shielded aluminum shell was explained.

References

1. Lee W, Hong YK, Park JH, Lee J, Baek IS et al (2017) A simple wireless power charging antenna system: evaluation of ferrite sheet. IEEE Trans Magn 53(7):1–5
2. Ho SL, Wang J, Fu WN, Sun M (2011) A comparative study between novel witricity and traditional inductive magnetic coupling in wireless charging. IEEE Trans Magn 47(5):1522–1525
3. Kim S, Park HH, Kim J, Kim J, Ahn S (2014) Design and analysis of a resonant reactive shield for a wireless power electric vehicle. IEEE Trans Microw Theor Tech 62(4):1057–1066
4. Zeng H, Liu Z, Hou Y, Tong H, Zhou B (2017) Optimization of magnetic core structure for wireless charging coupler. IEEE Trans Mag 53(6):1–4

5. Ahmad A, Alam MS, Chabaan R, A comprehensive review of wireless charging technologies for electric Vehicles. IEEE Trans Transp Electrification 4(1):38–63
6. Cai C, Wang J, Fang Z, Zhang P, Hu M, Zhang J, Li L (2018) Design and optimization of load-independent magnetic resonant wireless charging system for electric vehicles. IEEE Access (6):17264–17274
7. Hwansoo M, Sungkyu K, Hyun HP, Seungyoung A, Design of a resonant reactive shield with double coils and a phase shifter for wireless charging of electric vehicles. IEEE Trans Magn 51(3):1–4
8. Jerdvisanop C, Kanako W, Takuji A, Soichi W, Toru U, Exposure evaluation of an actual wireless power transfer system for an electric vehicle With near-field measurement. IEEE Trans Microw Theory Tech 66(3):1543–1552
9. Guidelines for limiting exposure to time-varying electric and magnetic fields (1 Hz to 100 KHz): erratum. Health Phys 100(1):112

Fatigue Design of Rubber Mount for Automobile Powertrain

Xiaoqiang Deng, Haoju Hu, Junjie Qiu and Shubin Ye

Abstract In this paper, taking the vehicle powertrain anti-torsion mount as the research object, the stress and strain curves of its rubber material were first tested, and then the Ogden hyperelastic constitutive model was used to simulate its mechanical performance. Based on the accurate simulation of its stiffness characteristics, the finite element model in Abaqus was calibrated. Based on the theory of strain fatigue of rubber material, the ε-N fatigue curve of its rubber material was determined by the experiments, and its rig fatigue life was calculated and optimized. The rubber material fatigue curve has been corrected according to the durability test results of the rubber structure, and then its fatigue life on the vehicle driving endurance test is forecasted eventually.

Keywords Powertrain anti-torsion mount · Rubber material ε-N fatigue curve · The finite element · Fatigue calculation

1 Introduction

In a vehicle design, the reasonable design of powertrain mount system can reduce the vibration of the powertrain to the frame and body, improve ride comfort and reduce vibration noise [1]. The powertrain mount is the key link between the engine and automobile body, and it consists of a rubber mount and bracket fittings; the rubber parts used for vibration isolation should not only provide good vibration isolation performance, but also meet the durability and safety requirements of the system. How to design the rubber mount parts that meet the fatigue durability target has become an urgent problem for every manufacturer. The rubber components research mainly depends on test, if the fatigue characteristics of rubber can be effectively controlled at the initial stage of structural design, the development level and quality of rubber parts will be greatly improved.

X. Deng (✉) · H. Hu · J. Qiu · S. Ye
Guangzhou Automobile Group Co., Ltd, Automotive Engineering Institute, 511434 Guangzhou, China
e-mail: dengxiaoqiang@gaei.cn

China SAE (ed.), *Proceedings of China SAE Congress 2018: Selected Papers*, Lecture Notes in Electrical Engineering 574,
https://doi.org/10.1007/978-981-13-9718-9_27

Rubber, as a kind of high molecular engineering material, is different from elasto-plastic mechanical properties of metal materials. It has hyperelastic, viscoelastic, amplitude correlation, temperature correlation and Mullin characteristics [2]. The fatigue performance of rubber vibration isolator is mainly determined by many factors, such as material, working environment and structural design. This paper discusses the fatigue design of rubber structure which does not involve the rubber material formula and vulcanization technology and other material fields.

The general design process of a vehicle's rubber mount structure is shown as follows: (1) performance target input; (2) three-dimensional structure design of rubber; (3) stiffness checked by CAE; (4) fatigue endurance checked by CAE; (5) physical sample preparation; (6) rig test verification; (7) carrying out the vehicle verification. According to the general process of rubber structure design, the static tensile test of rubber material is carried out, and then the Ogden constitutive model is adopted to simulate the mechanical properties of rubber structure, then the stiffness characteristics are calculated. After achieving the ε-N fatigue curve via the rubber material fatigue test, its fatigue life is predicted, calculated and optimized, and then the final design of rubber mount structure could satisfy the requirement of durability will be achieved. The experience of rubber structure design can provide a good reference for other rubber structure design.

2 Rubber Material Features

2.1 Static Mechanical Properties and Constitutive Models of Rubber Materials

Different from the linear elasticity of metal materials, the static performance of rubber materials is characterized by the nonlinearity of stress and strain, and it is commonly called hyperelasticity. The mechanical properties of rubber materials are complex, and the hyperelastic model is usually used to describe them. At present, the constitutive models representing hyperelasticity mainly include Mooney Rivlin model, van der Waals model, Ogden model, Yeoh model, etc. [3, 4]. The constitutive model can be fitted with experimental data of rubber material static foundation test including uniaxial tensile, biaxial tensile, plane tensile and volume compression test, etc. [5, 6].

In this paper, the vehicle powertrain anti-torsion mount is made of HA55-degree natural rubber. The standard test rubber samples are prepared according to GB/T 528-528, and then the stress and strain curves are obtained via uniaxial tension, biaxial stretching and plane tensile test, as shown in Fig. 1.

Abaqus software provides the abundant material constitutive model and strong nonlinear analysis capability. Therefore, Abaqus is selected as a rubber stiffness and strain calculation tool in this paper [7]. The strain range involved in the analysis of rubber in this paper is large, and the calculation accuracy can be obtained by using

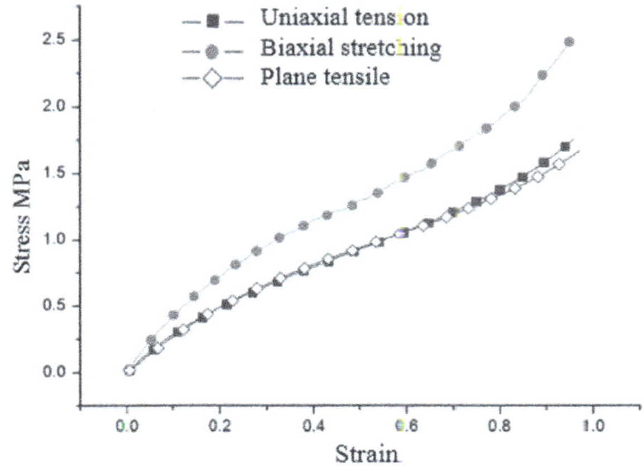

Fig. 1 Tensile curve of rubber material

the third-order Ogden model [5]. The form of the third-order Ogden strain energy potential is as follows:

$$U = \sum_{i=1}^{3} \frac{2\mu_i}{a_i^2}\left(\overline{\lambda}_1^{ai} + \overline{\lambda}_2^{ai} + \overline{\lambda}_3^{ai} - 3\right) + \sum_{i=1}^{3} \frac{1}{D}(J_{el} - 1)^{2i} \qquad (1)$$

where μ_i, a_i, D_i are the material parameter; λ_i is prime elongation; J_{el} is the elastic Jacobian. The third-order Ogden constitutive model parameters of its rubber materials were obtained as follows: $\mu_1 = 0.8126$, $\mu_2 = 0.0045$, $\mu_3 = 0.0027$, $\alpha_1 = 0.7833$, $\alpha_2 = 8.0915$, $\alpha_3 = -4.9914$, $D_i = 0$.

2.2 *The Durability of Rubber Material and the Determination of Its ε-N Fatigue Curve*

The fatigue fracture process of rubber material can be divided into two stages: the first stage, from no crack to the formation stage of small crack; the second stage, the crack extends from the small crack to the fracture stage. Studies on fatigue life are mainly carried out from these two aspects, and the two methods of life analysis and prediction are called crack initiation and crack propagation [8]. There are no obvious boundaries between crack initiation stage and crack growth stage. Generally, for the structure with an obvious gap or crack, the calculation method of fatigue life is based on the crack extension method, and for those with no obvious gap or crack, the calculation method of life is mostly based on the crack initiation method. The fatigue crack initiation method assumes that the fatigue life of rubber material has a one-to-one correspondence with some mechanical parameters of the material. This kind of mechanical parameter is often called fatigue prediction factor or fatigue damage parameter. The reasonable selection of fatigue damage parameters is the key point for predicting the fatigue life of rubber materials by crack initiation method. The early rubber fatigue life prediction theory aimed at establishing an empirical relationship between the fatigue crack initiation life and fatigue damage parameters, such as the commonly selected power law relationship [8]:

$$P = KN_f^b \tag{2}$$

where P is the fatigue damage parameter; N_f is crack initiation life; K and b are material constants.

In order to obtain the relationship between strain and life of the rubber material, the tensile test with seven groups of different tensile ratios (l/l_0) was prepared as shown in Table 1, each of which has 24 dumb-bell-type samples. According to GB/T 531.1-2008, uniaxial tensile fatigue test was carried out by the frequency of 1.7 Hz with the equipment shown in Fig. 2. Tensile fatigue tests were carried out with seven groups of rubber specimens, respectively, and the value of the test life of each group of rubber specimens was recorded in the test. Table 2 shows the fatigue test results of the 24 rubber specimens in group 3, and the fatigue failure of the samples in this group is shown in Fig. 3.

By analyzing the rubber fatigue life data of different elongation-ratio (strain level) specimens, and calculating each rubber specimens under 50% survival rate test lifetimes, the relational expression between the maximum principal strain ε and the sample life N_f is obtained by fitting as shown in Eq. 3, and the curve is shown in Fig. 4.

Table 1 Fatigue test samples of rubber materials

The sample group	1	2	3	4	5	6	7
Tensile ratios (l/l_0)	1.6	1.8	2.0	2.2	2.4	2.6	2.8
Logarithmic strain	0.470	0.588	0.693	0.788	0.875	0.956	1.030
Number of samples	24	24	24	24	24	24	24

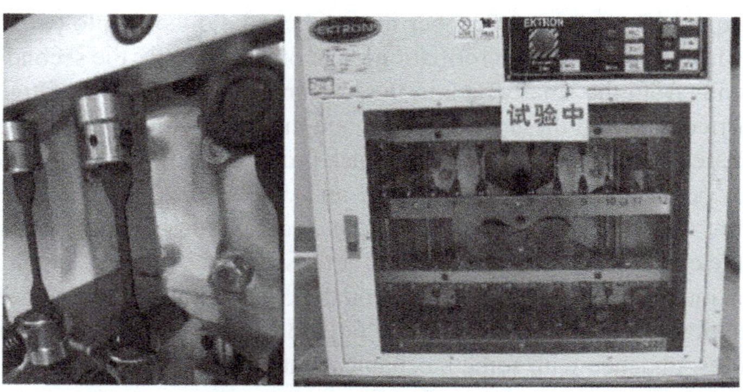

Fig. 2 Fatigue test diagram of rubber material

Table 2 Fatigue test life of group 3 rubber specimens

The specimens	Fatigue life	The specimens	Fatigue life
1#	91,885	13#	68,114
2#	108,854	14#	84,252
3#	57,796	15#	91,435
4#	89,767	16#	82,806
5#	77,700	17#	62,266
6#	80,286	18#	72,480
7#	26,717	19#	86,894
8#	58,015	20#	66,494
9#	56,704	21#	63,093
10#	–	22#	56,948
11#	50,201	23#	55,942
12#	30,985	24#	55,688

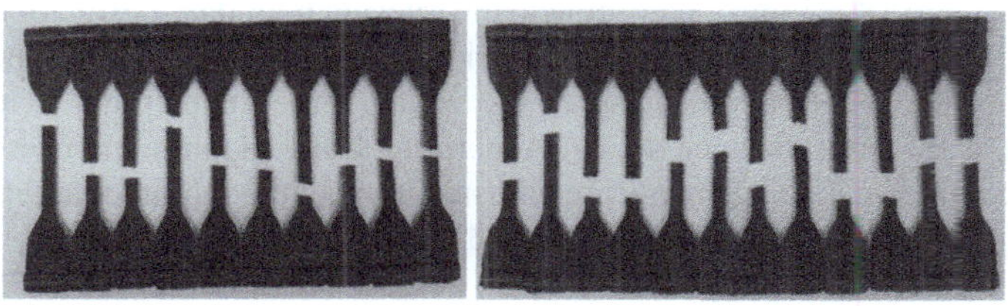

Fig. 3 Fatigue failure of 24 samples in group 3

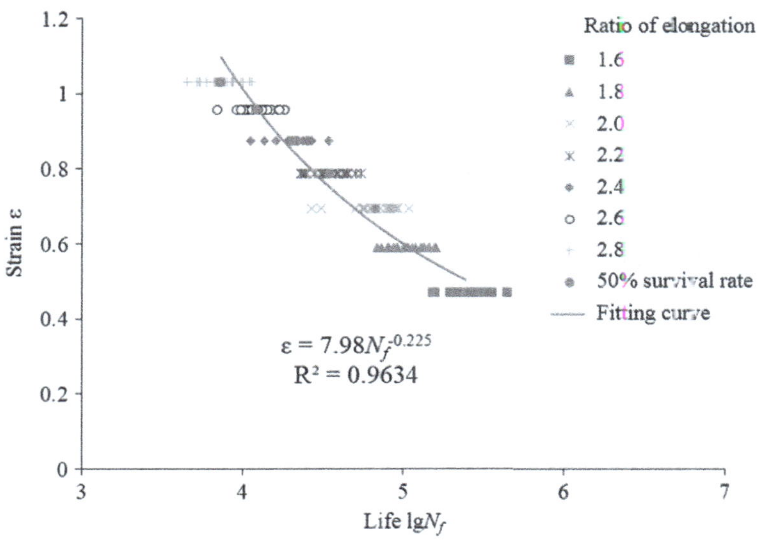

Fig. 4 Strain-life curve of rubber material

$$\varepsilon = A(N_f)^{-B} \tag{3}$$

where ε denotes the maximum principal strain, which is the fatigue evaluation parameter. N_f is fatigue life. A and B are fatigue parameters for fatigue life prediction. The fatigue parameters $A = 7.98$ and $B = 0.225$ of the rubber specimens are studied in this paper, and the fitting correlation coefficient $R^2 = 0.96$.

3 Rubber Mount Structure Design

3.1 Rubber Mount Stiffness Design

After the related performance objective was decomposed, the main vertical loading direction (X) linear stiffness of the powertrain anti-torsion mount was obtained and its value is set to 180 N/mm. According to the relevant boundary design of the rubber structure shown in Fig. 5, the major design element of the rubber mount includes the angle of the rubber main spring, θ, the width of the main spring, d, the height h, as well as the limit distance d_1, d_2. According to the requirement of its stiffness target, the related design elements are adjusted to meet the required target, and then the rubber preliminary structure is proposed to be determined.

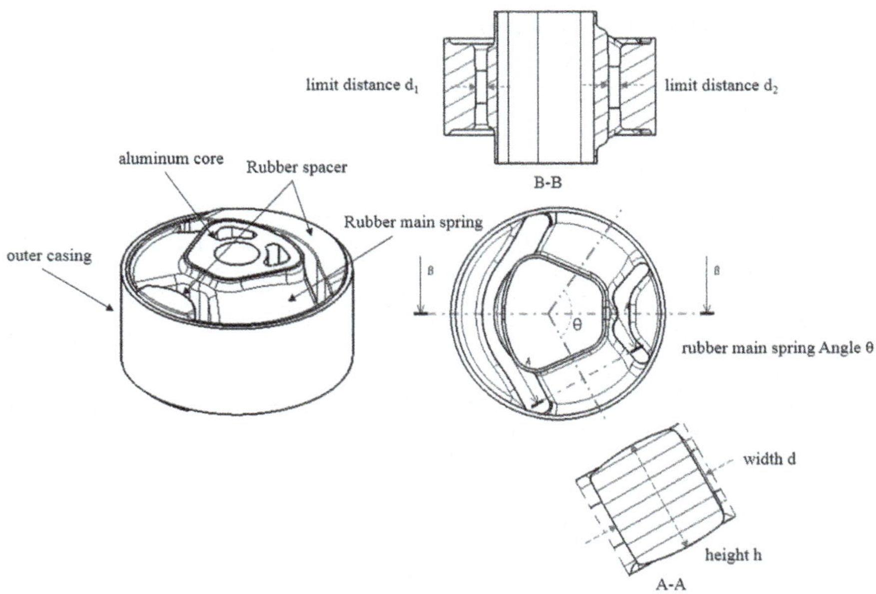

Fig. 5 3D structure of rubber mount

After completing the 3D structure data, the stiffness of the rubber structure needs to be checked via CAE method. Firstly, the finite element model was set up as shown in Fig. 6. Hexahedron element model is adopted, and its type is C3D8H. As the rigidity of the outer casing and the aluminum core is much higher than that of rubber, the rigid body simulation is adopted for the inner and outer casing. The third-order Ogden model was used to simulate rubber materials.

After several structure optimizations, the mainspring rubber angle, width and height had been adjusted, and the simulation result of x-direction stiffness is 192 N/mm while the design target is 180 N/mm. The error is +6.7% which is considered to meet the design target. Finally, this structure was selected to make samples, and the measured stiffness curve is shown in Fig. 7, and the linear stiffness is 175 N/mm. The error between the measured value and the calculated value is 9.7%, and the error between the measured value and the design target is 2.8%. Moreover, the CAE calculation curve of the nonlinear section is in good agreement with the test

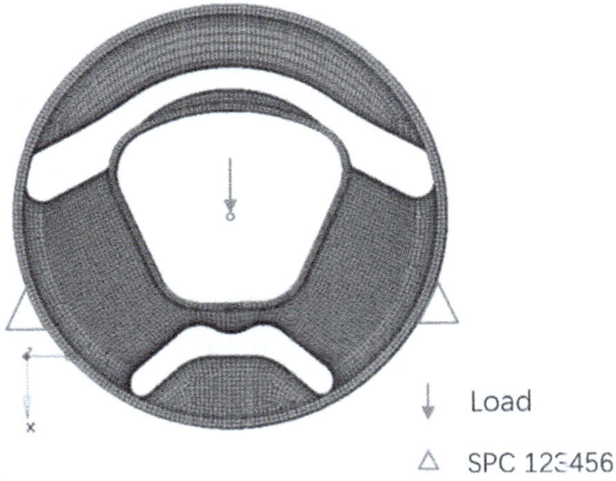

Fig. 6 Finite element model of rubber mount

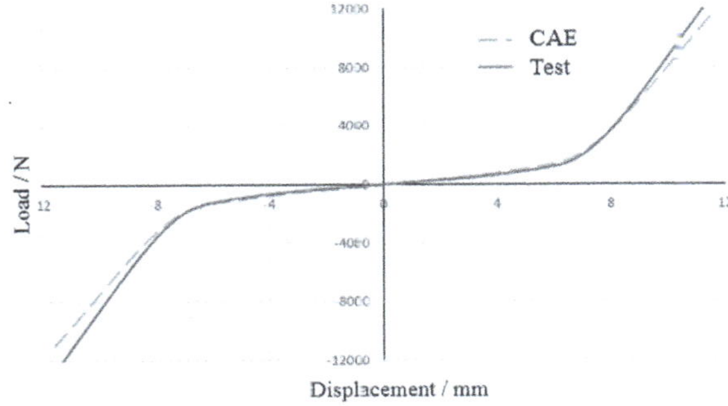

Fig. 7 CAE calculation and test stiffness curve

curve. The results show that the CAE method can predict the stiffness characteristics of rubber mount well.

3.2 *Rubber Mount Fatigue Life*

3.2.1 Rig Fatigue Life Optimization

In addition to meeting the stiffness design requirements, the rubber structure of rubber mount also needs to meet the corresponding fatigue durability. According to the relevant design standards, the durability target of this rubber mount is 380,000 times, and its fatigue endurance condition is shown in Table 3.

Fatigue test was carried out for the rubber mount in accordance with the fatigue condition in Table 3; during the test, the 1/10 of Level 1 cycle (34,000 times) was first loaded, and then the 1/10 of Level 2 cycle (4000 times) was loaded, and the total test lifetimes were recorded for ten times. The fatigue life of the rubber structure was finally measured as 338,000 times when the root of the rubber mainspring was damaged and cracked, as shown in Fig. 8. According to the evaluation standard of its durability target, 380,000 times, the fatigue durability design of this rubber structure does not meet the design requirement and needs structure optimization.

Table 3 Fatigue design objectives of the rubber mount

Load case	+*X*	−*X*	Cycles
Level 1	+3940	0	340,000
Level 2	+6400	−5800	40,000

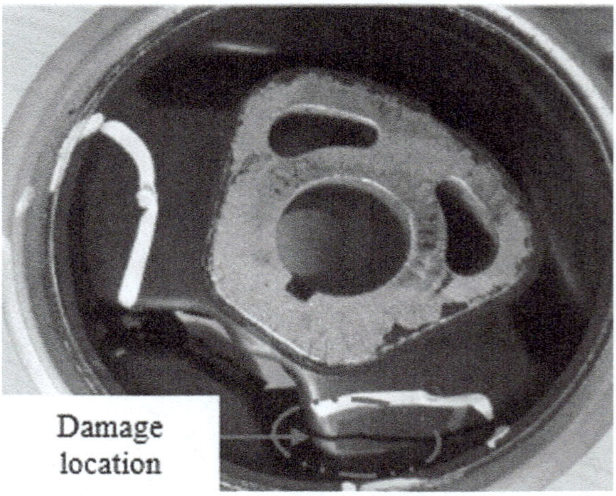

Fig. 8 Fatigue test failure results of the rubber mount

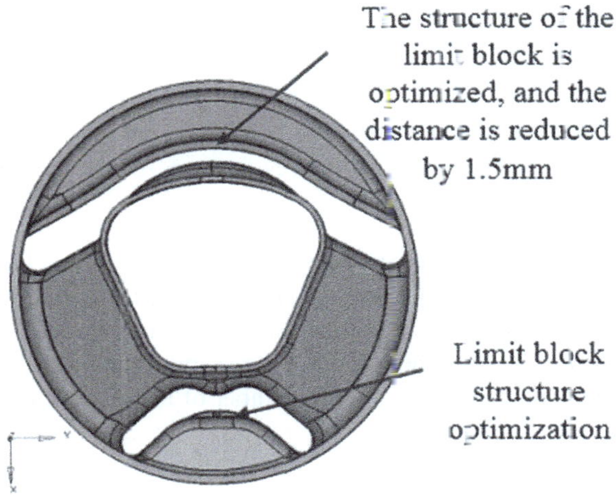

Fig. 9 Structure comparison before and after durability optimization

Since the fatigue life of this rubber structure has been close to the design target and durable optimization design must ensure the original stiffness, in the optimization of structural durability, the structure of rubber mainspring should remain unchanged as much as possible and changes should be made in other positions to achieve the requirement. The structural optimization is shown in Fig. 9, including changing the structure of the limit block in the *x*-direction and adjusting the limit distance slightly (with less impact on the stiffness of the nonlinear segment).

In order to investigate whether the optimized structure durability meets the design requirements, the finite element method is used to calculate and predicts its durability. The finite element model of the structure before and after optimization is established according to the method in Sect. 3.1, as shown in Fig. 10. According to the fatigue linear accumulation Miner theory, the load conditions in Table 3 were divided into three symmetric loads for strain calculation simulation, respectively, as shown in Table 4. During strain calculation simulation, rubber materials were simulated using a third-order Ogden constitutive model, whose parameters were obtained by fitting in Sect. 2.1. The maximum strain contours of each hot spot area of rubber structure

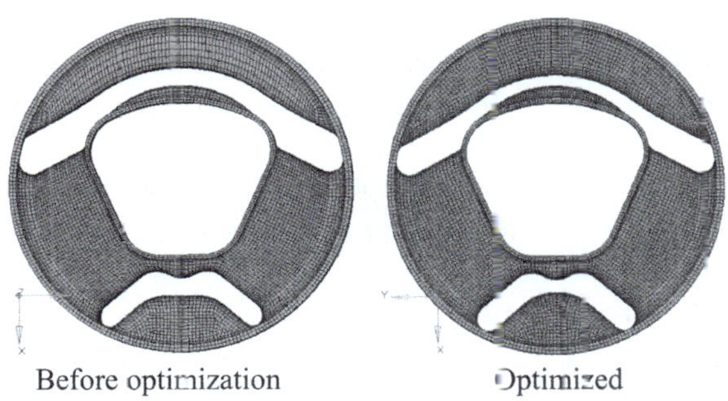

Fig. 10 Finite element model before and after rubber structure optimization

Table 4 Strain calculation values of each loading condition before and after optimization

Load	Hot point 1 (Maximum strain)			Hot point 2 (Maximum strain)		
	3940 N	6400 N	−5800 N	3940 N	6400 N	−5800 N
Before optimization	0.410	0.441	0.743	0.406	0.540	0.407
Optimized	0.405	0.436	0.680	0.401	0.538	0.384

under various load conditions are shown in Fig. 11. The comparison of the calculated strain values of the original and optimized structures is shown in Table 4.

Based on the ε-N fatigue curve of the rubber material, the damage $1/N_{f1}$, $1/N_{f2}$, $1/N_{f3}$ are obtained under every symmetric load. Then, the total damage D was calculated according to the number of loading cycles N_1, N_2 and N_3.

$$D = \frac{1}{N_{f1}} \times N_1 + \frac{1}{N_{f2}} \times N_2 + \frac{1}{N_{f3}} \times N_3 \tag{4}$$

And the total life

$$N_f = \frac{1}{D} \tag{5}$$

The comparison between the calculated fatigue life of the rubber structure before and after optimization is shown in Table 5. After optimization, the calculated life

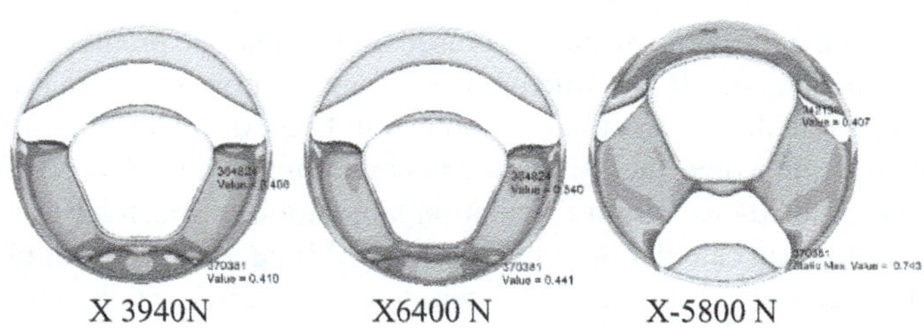

Fig. 11 Cloud diagram of strain calculation in each test load condition

Table 5 Comparison between the calculated life and the test

Category	Life of hot point 1 (location of test failure)			Life of hot point 2		
	Test	Calculation 1	Calculation 2	Test	Calculation 1	Calculation 2
Before optimization	0.89	0.587	0.92	–	1.09	1.71
Optimized	1.14	0.721	1.13	–	1.15	1.82

Note Life calculation 1 is obtained from the calculation of the material ε-N curve, while life calculation 2 is obtained from the modified e-n calculation

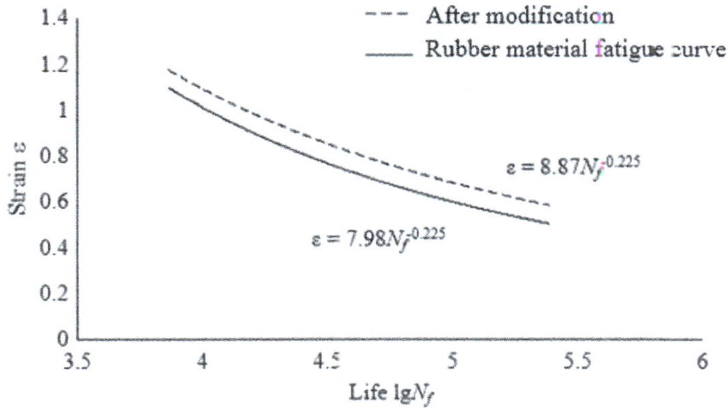

Fig. 12 Rubber material fatigue curve before and after modification

has increased from 0.587 to 0.721 (life 1.0 represents 380,000 times), and the life has been increased by 22.8%. The optimized rubber mount structure was used to make samples, and the fatigue test was carried out. The test life had increased from 338,000 times to 433,000 times which was increased by 28.1% to meet the fatigue durability design requirement.

From the comparison of fatigue calculation and test results, it is found that there is a certain error between calculation and test, the reason of which may be that the ε-N curve is a pure unidirectional tensile fatigue curve, and it is different from the actual rubber structure. So, the calculation is relatively conservative [9].

Therefore, the ε-N curve needs to be corrected based on the rubber fatigue test results, and the revised curve is shown in Fig. 12. During the correction, the value of A was changed and the value of B remained the same. If the value of A changed from 7.98 to 8.87, the ε-N fatigue curve will be translated overall to the right to make the revised life calculation be consistent with the experimental results as shown in Table 5, which indicates that the revised ε-N curve can be used for life prediction.

3.2.2 Vehicle Drive Durability Prediction

The rubber mount needs to meet the requirements of the driving durability test. The total cycle target of driving durability test is 400 cycles. In order to accurately predict its drive durability, the road load spectrum of the powertrain anti-torsion mount under a single cycle of drive durability was collected. The test was shown in Fig. 13.

The road load spectrum collected in the test field is a kind of random load spectrum, and the main factors causing fatigue damage of parts are load peak-valley value, mean value and load cycle times. Therefore, in the fatigue analysis of rubber parts, it is necessary to use the counting method to make statistics and analysis for random loads.

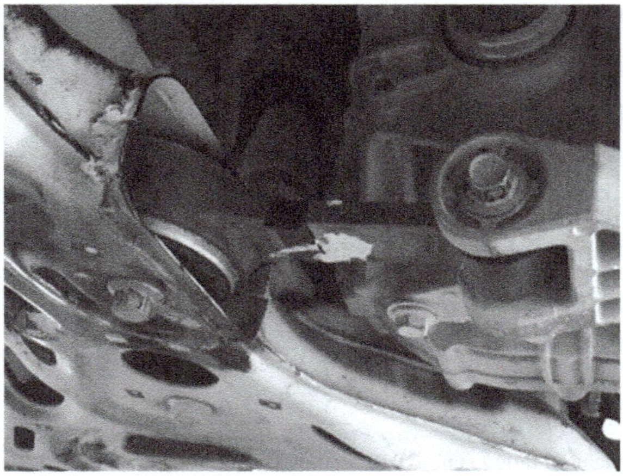

Fig. 13 Drive durability road load spectrum collected and strain gauge layout location

The main x-direction load of the anti-torsion mount is shown in Fig. 14. The force data is divided into eight stages of loading as shown in Table 6. Firstly, calculate the optimized rubber mount structure under various loads and the strain values are

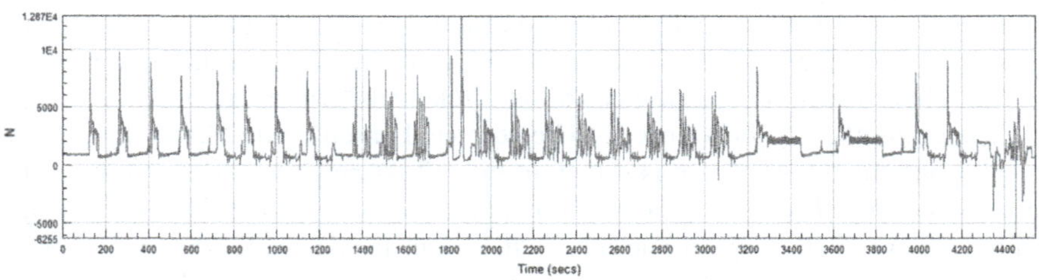

Fig. 14 The x-direction load of the anti-torsion mount under a single cycle of drive durability

Table 6 Calculation results of strain and life under driving endurance conditions

Drive durability (400 cycles)			Hot point 1		Hot point 1	
Load level		Cycles	Strain value	Life	Strain value	Life
Load1	12870N	400	0.951	1.18	0.981	1.31
Load2	9658N	2800	0.805		0.85	
Load3	7511N	6000	0.511		0.512	
Load4	5365N	10,800	0.43		0.429	
Load5	3219N	33,600	0.39		0.389	
Load6	1073N	1.24E6	0.207		0.200	
Load7	−3890N	800	0.41		0.221	
Load8	−6250N	400	0.78		0.395	

shown in Table 6. The modified rubber materials ε-N fatigue curve and load cycles at various levels were used to forecast the driving endurance life. The total life is shown in Table 6. Its minimum life is 472 cycles, and it is greater than 400, the target value, which meets the requirements of driving durability target. Finally, the vehicle drive durability test with the new rubber mount structure was carried out, it passed the test of vehicle driving durability, and there had been no rubber cracking problems, which indicate that the life prediction results of the rubber structure were reliable.

4 Conclusion

According to the static tensile test data of rubber material, the third-order Ogden constitutive model parameters were obtained by fitting, which was used to calculate and predict the stiffness characteristics of the rubber mount structure. The test results were in good agreement with the analytical results.

By combining the fatigue test of rubber material sample and finite element analysis for rubber mount, the fatigue life prediction model of rubber mount structure with high accuracy was obtained.

According to the rubber material ε-N fatigue curve, the fatigue durability of the rubber mount structure can be well predicted, which can save time and improve efficiency.

References

1. Min Q, Yi L, Tieli M et al (2001) A study on dynamic characteristics of hydraulic mount system. Automot Eng 33(6):381–384
2. Ye B, Duan X, Huang X et al (2012) Structure fatigue design on rubber mount of automobile power train. Comput Aided Eng 21(5):61–64
3. Li X, Li X et al (2005) Hyperelastic constitutive model of rubber material. China Elastomerics 15(1):50–58
4. Li X, Guizhong W et al (1999) Some forms of rubber strain energy function in finite element analysis. China Rubber Ind 46(12):707–711
5. Chen Z, Shi W, Wang Q et al (2010) Finite element analysis of engine rubber mount based on material test. Automot Tech 7:28–31
6. Deng X, Deng X, Qiu J et al (2016) Research on stiffness features analysis method of automotive suspension rubber bushing. Bus Tech Res 38(3):1–3
7. Zuang Z, Zhang F, Yin S et al (2005) Nonlinear finite element analysis and example of Abaqus. Beijing Science Press, Beijing
8. Wang H, Wei Y (2015) Study on rubber fatigue. Tire Ind 35(10):576–585
9. Fang M, Tan J, Guang X et al (2013) Random fatigue analysis of vehicle rubber bushing. Automot Eng 35(10):949–954

Study on the Green Wave Optimal Speed Guidance System Based on V2X

Bo Yang, Ying Zhang, Zhijun Cai, S. Ted Huang and Qigao Feng

Abstract In order to improve the city's traffic efficiency, this paper presents a green wave optimal speed guidance system based on V2X (Vehicle-to-Everything) and the method to calculate the optimal velocity. All vehicles of the system should be equipped with V2X-Boxes which can broadcast the ID, time, location and speed of the vehicle while receiving the messages from other V2X-Boxes and the traffic light status from roadside unit (RSU) in real time. The V2X-Box calculates the optimal speed by acceleration guidance method and deceleration guidance method, which not only increases the number of vehicles passing through the green light, but also provides green wave optimal speed guidance for vehicles entering the arterial road from the bypass. This system guarantees all road-running vehicles through traffic lights without stopping.

Keywords Green wave · Optimal speed · Guidance · V2X-Box · Arterial road

1 Introduction

With the development of automobile industry, more and more vehicles have appeared in cities. However, the speed of the vehicles in the center of metropolises is generally slow while the traffic delays are increasing rapidly. Prolonged traffic jams not only affect the drivers' travel efficiency, but also increase the pollution to the environment. Therefore, how to improve the efficiency of urban traffic has attracted significant attention of traffic participants.

B. Yang (✉) · Y. Zhang · Z. Cai · S. T. Huang · Q. Feng
GAC Engineering, Guangzhou Automobile Group Co. Ltd., Guangzhou, China
e-mail: yangbo_bit@126.com

China SAE (ed.), *Proceedings of China SAE Congress 2018: Selected Papers*,
Lecture Notes in Electrical Engineering 574,
https://doi.org/10.1007/978-981-13-9718-9_28

2 Traditional Green Wave Speed Guidance Method

Nowadays, green wave speed guidance method is widely used in arterial road control. The driver who drives at the guidance speed can pass through the intersections in arterial road without stopping because the traffic light would always be green when the driver reaches the intersection. Therefore, the method could improve traffic efficiency and reduce traffic congestion [1, 2].

Figure 1 shows the traditional green wave speed guidance method, in which the green light duration is T_1 and the red light duration is T_2 for each traffic light cycle on the arterial road while the distance between intersection I0 and intersection I2 is L. If the driver drives at the speed of $V_{\alpha 1}$ and arrives at I0 at the time of T_3, he can pass through intersection I1 without stopping. Similarly, the driver can pass through both intersection I1 and intersection I2 without stopping with the velocity of $V_{\alpha 2}$ [3–5].

$$V_{\alpha 1} = \tan \alpha_1 \tag{1}$$

$$V_{\alpha 2} = \tan \alpha_2 \tag{2}$$

However, the traditional green wave speed guidance method still has the following three problems. First of all, with the increasing number of intersections, the bandwidth of green wave will become smaller and smaller, which means fewer and fewer vehicles will be allowed to pass without stopping. In addition, the green wave cannot

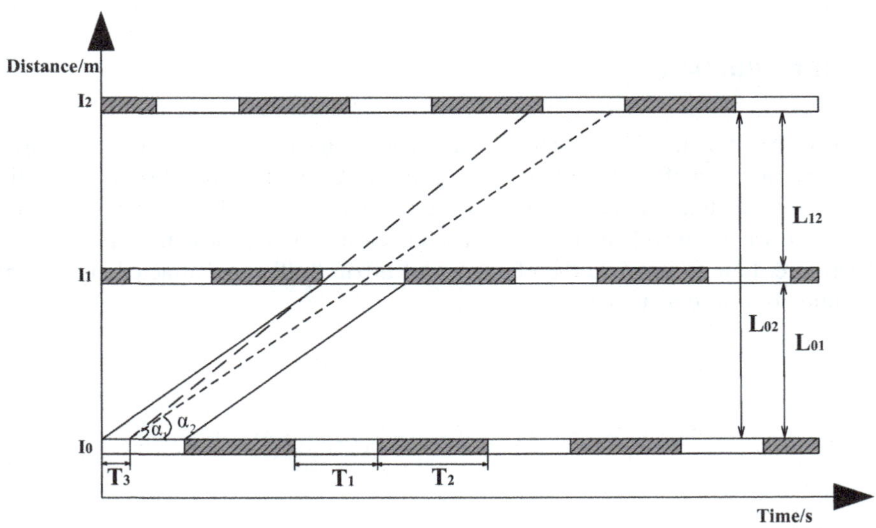

Fig. 1 Traditional green wave speed guidance method

always maintain full-load traffic which will waste the traffic efficiency. Last but not least, vehicles arriving at the arterial road from the bypass will affect the vehicles on the arterial road [6, 7].

3 Green Wave System Based on V2X

As shown in Fig. 2, the traffic lights of each intersection on the arterial road alternate in changing states with constant periods. All the states at intersection I1 are included in Table 1.

All the vehicles of the system have to be equipped with V2X-Box which can broadcast the ID, time, location and speed of the vehicle and receive the messages from other V2X-Boxes and the traffic light status from roadside unit (RSU) in real

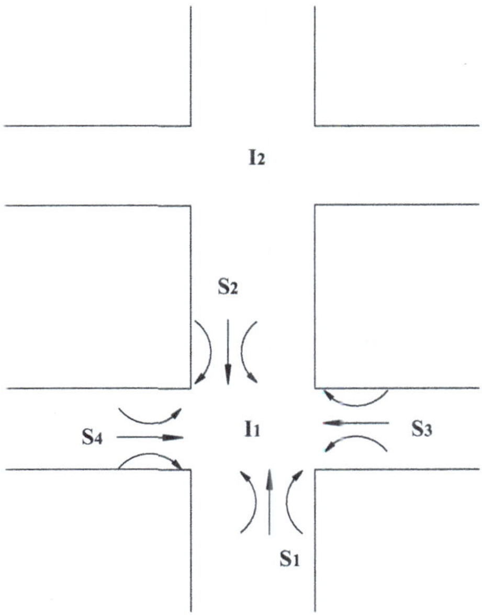

Fig. 2 Traffic light states of the system based on V2X

Table 1 Traffic light states

State	Upward direction of I1	Downward direction of I1	Left direction of I1	Right direction of I1
S1	Green	Red	Red	Red
S2	Red	Green	Red	Red
S3	Red	Red	Green	Red
S4	Red	Red	Red	Green

time. The traffic lights of the system are connected with RSU to broadcast the light status. V2X-Box will filter out the nearest front vehicle on the same lane and receive the messages from it to calculate the optimal speed and send it to the instrument cluster module through CAN bus for display.

4 Speed Guidance Method

Taking the arterial road upward direction as an example, all vehicles enter the arterial road from intersection I0 and some of them will exit at intersection I1. Some other vehicles will enter the arterial road from the left and the right bypass at intersection I1 when the traffic light of arterial road upward direction is red [1, 4, 5].

4.1 Acceleration Guidance

Figure 3 illustrates the acceleration guidance method. Define L_{12} as the distance between intersection I1 and I2, T_1 as the duration of the green light in a cycle, T_2 as the duration of the red light in a cycle, T as the phase difference between the traffic light of I1 and the traffic light of I2 and B as the safety distance between the nearest two vehicles. Moreover, t_1 is the difference between the time when the first vehicle through I1 without leaving the arterial road reaches I1 and the time when the

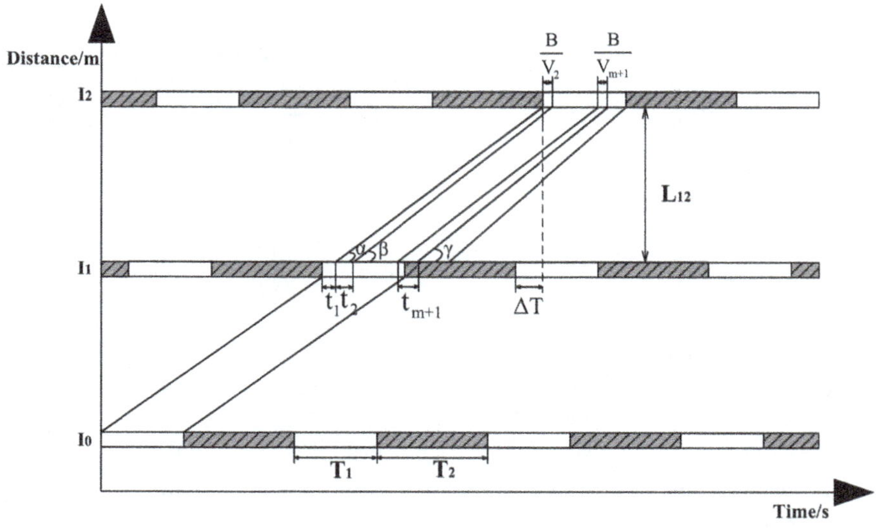

Fig. 3 Acceleration guidance method

traffic light turns from red to green of the current cycle. Similarly, t_2 is the difference between the time when the second vehicle through I1 without leaving the arterial road on the same lane reaches I1 and the time when the first vehicle through I1 without leaving the arterial road reaches I1. Likewise, t_{m+1} is the difference between the arrival times at I1 of the first vehicle entering the same lane of arterial road from the bypass and the last vehicle through I1 without leaving the arterial road of this cycle. t_n is the difference between the arrival times at I1 of the last vehicle and the last but one vehicle entering the arterial road from the bypass of this cycle.

Some vehicles accelerate to a new speed after passing through intersection I1. We can calculate the velocity of the first vehicle by Eq. (3).

$$V_1 = \tan\alpha = \frac{L_{12}}{T_1 + T_2 + \Delta T - t_1} \tag{3}$$

The velocity of the second vehicle can be determined using Eq. (5).

$$V_2 = \tan\beta = \frac{L_{12}}{\frac{L_{12}}{V_1} + \frac{B}{V_2} - t_2} \tag{4}$$

Therefore, we can get the speed of the second vehicle as:

$$V_2 = \frac{L_{12} - B}{\frac{L_{12}}{V_1} - t_2} \tag{5}$$

The speed of the last vehicle can be given by Eq. (6)

$$V_m = \frac{L_{12} - B}{\frac{L_{12}}{V_{m-1}} - t_m} \tag{6}$$

The speed of the first vehicle entering the same lane of arterial road from the bypass can be determined by Eq. (8).

$$V_{m+1} = \tan\gamma = \frac{L_{12}}{\frac{L_{12}}{V_m} + \frac{B}{V_{m+1}} - t_{m+1}} \tag{7}$$

$$V_{m+1} = \frac{L_{12} - B}{\frac{L_{12}}{V_m} - t_{m+1}} \tag{8}$$

Likewise, the velocity of the last vehicle entering the arterial road from the bypass of this cycle can be calculated as:

$$V_n = \frac{L_{12} - B}{\frac{L_{12}}{V_{n-1}} - t_n} \tag{9}$$

In which,

$$T_1 \geq \frac{B}{V_2} + \cdots + \frac{B}{V_m} + \cdots + \frac{B}{V_n} \tag{10}$$

4.2 Deceleration Guidance

As illustrated in Fig. 4, the remaining vehicles entering the arterial road from the bypass have to slow down and pass through intersection I2 until the next green light. Define t_1' as the difference between the time when the first vehicle decelerating into the arterial road from the bypass reaches intersection I1 and the time when the traffic light turns from green to red of this cycle. In the same way, t_2' is the difference between the time when the second vehicle decelerating into the arterial road from the bypass reaches intersection I1 and the time when the first vehicle decelerating into the arterial road from the bypass reaches intersection I1. Likewise, t_{m+1}' is the difference between the time when the first vehicle through I1 without leaving the arterial road of the next cycle reaches I1 and the time when the last vehicle decelerating into the arterial road from the bypass reaches I1. Similarly, t_n' is the difference between the arrival times at I1 of the last vehicle and the last but one vehicle through I1 without leaving the arterial road of the next cycle.

Employing the same method of accelerating guidance, we can calculate the velocity of the first vehicle decelerating into the arterial road from the bypass by Eq. (11).

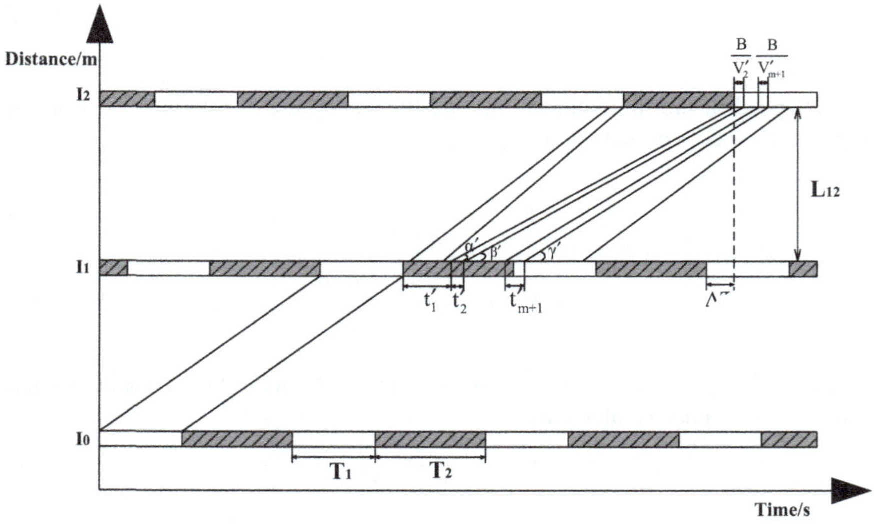

Fig. 4 Deceleration guidance method

$$V_1' = \tan\alpha' = \frac{L_{12}}{T_1 + 2T_2 + \Delta T - t_1'} \tag{11}$$

Similarly, the speed of the second vehicle decelerating into the arterial road from the bypass can be determined by Eqs. (12) and (13).

$$V_2' = \tan\beta' = \frac{L_{12}}{\frac{L}{V_1'} + \frac{B}{V_2'} - t_2'} \tag{12}$$

$$V_2' = \frac{L_{12} - B}{\frac{L_{12}}{V_1'} - t_2'} \tag{13}$$

Likewise, the speed of the last vehicle decelerating into the arterial road from the bypass can be given by Eq. (14).

$$V_m' = \frac{L_{12} - B}{\frac{L_{12}}{V_{m-1}'} - t_m'} \tag{14}$$

The velocity of the first vehicle passing through I1 without leaving the arterial road of the next cycle can be calculated as:

$$V_{m+1}' = \tan\gamma' = \frac{L_{12}}{\frac{L_{12}}{V_m'} + \frac{B}{V_{m+1}'} - t_{m+1}'} \tag{15}$$

$$V_{m+1}' = \frac{L_{12} - B}{\frac{L_{12}}{V_m'} - t_{m+1}'} \tag{16}$$

In the same way, the speed of the last vehicle passing through I1 without leaving the arterial road of the next cycle can be determined as:

$$V_n' = \frac{L_{12} - B}{\frac{L_{12}}{V_{n-1}'} - t_n'} \tag{17}$$

In which,

$$T_1 \geq \frac{B}{V_2'} + \cdots + \frac{B}{V_m'} + \cdots + \frac{B}{V_n'} \tag{18}$$

5 Test Verification

Construct a virtual road according to the road model described in Sect. 2, in which, the distance L_{12} between I1 and I2 is 1500 m. Additionally, the duration of red light is 60 s while the duration of green light is 30 s. The phase difference between the traffic light of I1 and the traffic light of I2 is 10 s while the safety distance B is 20 m. Thus, the green wave speed between I0 and I1 is 15 m/s, and there are 21 vehicles passing through I0 on the same lane during one cycle. Assuming five of them drive off the arterial road at intersection I1 while i vehicles enter the arterial road during the red light period, and the time interval between each vehicle is two seconds.

If $i \le 5$, the average travel time of traditional green wave speed guidance system can be calculated by Eq. (19).

$$t = \frac{16 \times 100 + 160 + \left(160 - \frac{16}{15}\right) \times (i - 1)}{16 + i} \tag{19}$$

The average travel time of green wave optimal speed guidance system based on V2X can be determined by Eq. (20).

$$t' = \frac{\frac{1500}{V_1} + \frac{1500}{V_2} + \cdots + \frac{1500}{V_{16+i}}}{16 + i} \tag{20}$$

If $i > 5$, the average travel time of traditional green wave speed guidance system can be computed as:

$$t = \frac{16 \times 100 + 160 + (160 - \frac{16}{15}) \times (i - 1)}{16 + i} \tag{21}$$

Similarly, we can compute the average travel time of green wave optimal speed guidance system based on V2X by Eq. (22).

$$t' = \frac{\frac{1500}{V_1} + \frac{1500}{V_2} + \cdots + \frac{1500}{V_{21}} + \frac{1500}{V_1'} + \cdots \frac{1500}{V_{i-5}'}}{16 + i} \tag{22}$$

From Table 2, we can see that when the number of vehicles entering the arterial road during the red light period is equal to the number of vehicles driving off the arterial road during the green light, the system based on V2X has the shortest average time and the highest improved traffic efficiency.

Table 2 Test result

Amount of the vehicles entering arterial road from bypass	Average time of the traditional system (s)	Average time of the system based on V2X (s)	Saved time (s)	Improved efficiency (%)
One vehicle	103.53	86.37	17.16	16.57
Two vehicles	106.61	84.78	21.83	20.48
Three vehicles	109.36	83.98	25.38	23.20
Four vehicles	111.84	83.17	28.67	25.63
Five vehicles	114.08	82.34	31.74	27.82
Six vehicles	116.12	89.16	26.96	23.22
Ten vehicles	122.71	94.87	27.84	22.69
Fifteen vehicles	128.55	106.83	21.72	15.90

6 Conclusion

This paper analyzes the deficiency of the traditional green wave speed guidance method, introducing a green wave optimal speed guidance system based on V2X and the method for calculating guidance speed. The system can effectively take advantage of the remaining light space to ensure the vehicles from bypass get smooth access to the arterial road, which will reduce the number of vehicle stops and improve the traffic efficiency.

References

1. Abu-Lebdeh G (2002) Integrated adaptive-signal dynamic-speed control of signalized arteries. J Transp Eng 128(5):447–451
2. Torok J, Kertesz J (1996) Green wave model of two-dimensional traffic: transitions in the flow properties and in the geometry of the traffic jam. Stat Mech Appl 231(4):515–533
3. Gartner NH, Stamatiadis C (2004) Progression optimization featuring arterial and route-based priority signal networks. Intell Transp Syst 8(2):77–84
4. Nagatani T (2007) Vehicular traffic through a sequence of green-wave lights. Stat Mech Appl 380(2):503–511
5. Corman F, D'Ariano A, Pacciarelli D, Pranzo M (2009) Evaluation of green wave policy in real-time railway traffic management. Transp Res Part C Emerg Technol 17(5):607–616
6. Berezhnoy A, Grakovsky A, Nesterov A (2007) The "green wave" mode production on the two-lane highways during the construction works time period. Transport 22(4):263–268
7. Wong CK, Wong SC (2005) Optimization methods for off-line traffic signal settings: recent advances and prospective future research. J Transp Syst Eng Inf Technol 5(2):36–55

Automatic Driving Joint Simulation Technology and Platform Design

Ji Song, Rongjun Huang, Wei Zhang and Qinghe Liu

Abstract In order to improve the production development of automatic driving technology ability, this paper puts forward a hybrid platform to establish virtual joint simulation technology. This platform includes sensor module, vehicle dynamics module, scenario module and control algorithm module in order to solve the automatic questions including roads of all weather conditions, sensor-type selection design, automatic control algorithm, driving performance testing and other difficulties which are complex to validate. Through the virtual joint simulation platform, autonomous driving performance can be improved in the simulation, and it is a reasonable method to realize automatic mass production.

Keywords Automatic mass production · SIL · Joint simulation platform · HD maps · Vehicle model

1 Introduction

Since autonomous driving technology becomes more and more popular recently, researchers focus on proposing in-depth research applicable to a certain technology in autonomous driving on the theoretical level [1]. However, this research method does not adopt a design and verification method with good vehicle interaction in China [2, 3]. In traditional road tests, although the real vehicle verification is true and effective, there are also many limited test conditions such as the testing difficulty, low efficiency and high cost [4]. Therefore, an effective method is needed to design

J. Song (✉) · Q. Liu
Harbin Institute of Technology, Weihai, Shandong, China
e-mail: 14694934@qq.com

R. Huang
BAIC MB-Tech Development Center Co., Ltd, Beijing, China

W. Zhang
China Automotive (Beijing) Intelligent Network Union Automotive Research Institute Co., Ltd, Beijing, China

China SAE (ed.), *Proceedings of China SAE Congress 2018: Selected Papers*, Lecture Notes in Electrical Engineering 574,
https://doi.org/10.1007/978-981-13-9718-9_29

and verify the research related to autonomous driving by integrating various modules to build a comprehensive simulation platform.

In China, some Internet companies have launched a simulator based on simulation in the loop (SIL). In terms of the dynamic characteristics, the complex vehicle model is represented by the dynamic model with less freedom, which is difficult to simulate the dynamic characteristics of the entire vehicle, especially the accuracy of suspension and tire is relatively low. In the scene simulation part, it is difficult to carry out the effective simulation, and it is hard to carry out a real simulation on complex road conditions or real road conditions.

The automatic driving virtual joint simulation platform proposed in this paper can complete the vehicle model with more than 25-(DOFs) degrees of freedom, importing real high-precision maps in time, building complex working conditions and completing the design and test of each module.

2 System Overview

The virtual joint simulation platform of autonomous driving proposed in this paper is built for the design of automatic driving mass production, which can build a foundation for automatic driving design and test. The simulation platform is mainly divided into four modules: sensor module, vehicle dynamics module, simulation scene module and control algorithm module. This simulation technology is able to combine the big data importing high-precision maps in batches and build test conditions close to real roads. The vehicle dynamics module completes the simulation construction and feedback of vehicle dynamics by combining real driving habits and driving comfortable requirements. The dynamic characteristics of the vehicle are integrated through integrated sensor simulation and control algorithm. Finally, the automatic driving performance design and software loop test are completed under the feedback of the external operating environment. The overall framework of the simulation platform is shown in Fig. 1.

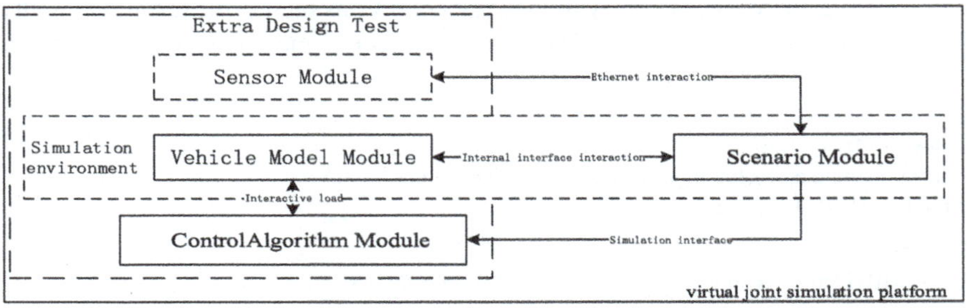

Fig. 1 Overall framework of the platform

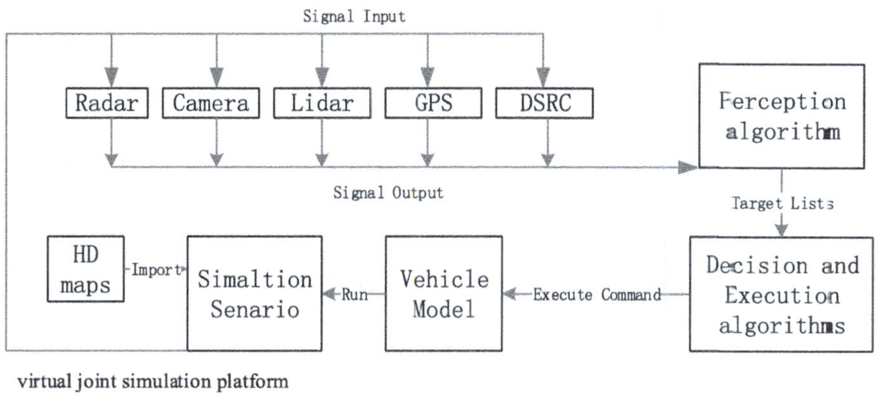

Fig. 2 Functional schematic of sensor module design

The platform takes the simulation scenario module as the main body, docking with the vehicle dynamics parameters by calling back the library parameters inside the vehicle module, and the sensor senses parameters through Ethernet connection. The control algorithm module interacts with the simulation scenario module through the CPU shared memory.

The functional principle diagram of the virtual simulation platform is shown in Fig. 2. Firstly, the scenario module is completed through the import of cloud-related traffic data. Secondly, the sensing module is parameterized by the type, quantity and location of the on-board sensor, and the sensor group model is built which acquires getting the dynamic driving environment signals from the simulation scenario. These signals are collected and timely identified and processed through the sensing algorithm. After that, the control algorithm module which can be loaded from other programs is carried out in the input algorithm to control the vehicle. The algorithm instruction is completed as a whole loop in the simulation scenario via vehicle dynamics module.

3 Platform Module Design

Modular design and testing is the core of this technology and the only way to realize SIL. The details are following.

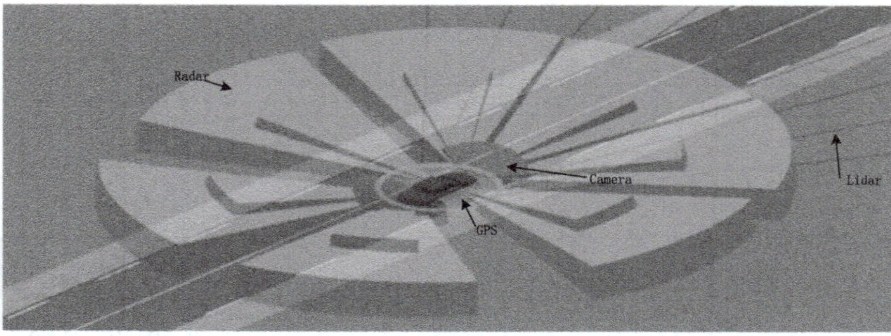

Fig. 3 Multi-sensor model building

3.1 Sensor Module

Sensor sensing module can simulate millimeter wave radar, camera, laser radar and other sensors. This platform provides two design methods of sensor sensing modules: Advanced sensor and physical sensor.

Advanced sensor simulation technology is a partial simulation of the perception sensor process, but it ignores the perception algorithm, so as to directly read all kinds of information from the environment. The information data without error and delay is obtained. The signal can be used perfectly as a control algorithm module combination, and it can also be completed without the delay testing of control algorithm accuracy. The simulation technology can ignore the sensor's perception accuracy and focus on the optimization and debugging of the control execution algorithm.

Physical sensor simulation technology simulates the real process of the signal source and signal acquisition, such as radar signal of millimeter wave radar, cameras, point cloud data, the GPS signals, etc. The processing of signal source information by sensing algorithm, and surrounding environment related information data, for the control algorithm to make decisions. The difference between the advanced sensor and physical sensor is that the physical sensor can simulate the signal source better. The more reflection and error signals in the simulation are produced in a physical way; the more realistic simulation process and data will be obtained.

Through the sensor simulation technology, the sensor module of this platform is parameterized. As shown in Fig. 3, the platform completed the construction of 7 mm wave radar, 2 single-eye cameras, 1 GPS and 1 16-line laser radar sensor model.

3.2 Vehicle Model Module

In order to reflect the characteristics of the vehicle better, the platform establishes vehicle dynamics models with more than 25-DOFs, including the simulation of spring, suspension, engine, chassis, tire, etc. Even though the automatic vehicle is mainly focused on the vehicle's deceleration and steering characteristic, the vehicle model is established in lower DOFs will lead to a high differences of results in simulation contrasting with the real vehicle, so the control algorithm which is based on this model lets the passengers feel discomfort, and driving way also can appear deviation from the normal driver. Furthermore, for the study of one DOF such as the acceleration and deceleration characteristics of vehicles, the characteristic curve should be extracted from the complete vehicle model to let the result better and more realistic. Otherwise, the lower DOFs will also affect accuracy.

This platform completes the appearance of the vehicle through 3D-Max. Figure 4 shows the external characteristic curve of the dynamic model with 28-DOFs. Figure 5 shows a real vehicle appearance.

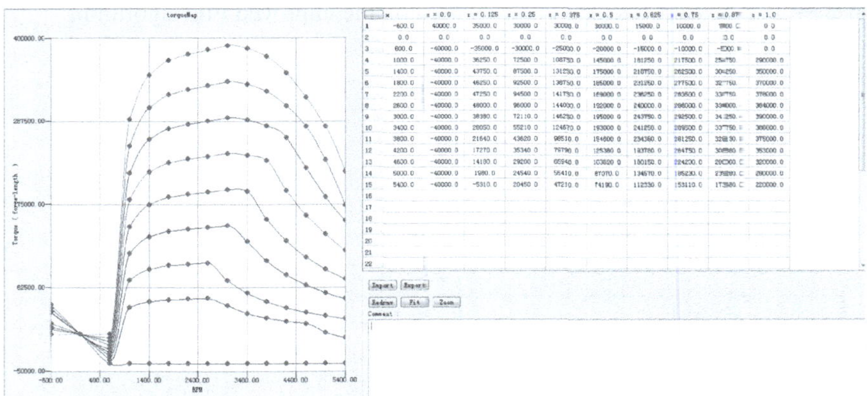

Fig. 4 Dynamic external characteristic curve

Fig. 5 Appearance of a certain type of vehicle

3.3 Control Algorithm Module

The control algorithm module based on sensor module to control the vehicle is the core technology in automatic driving. The API is used to overlap the simulation scenario module, including Python, Simulink, C++, etc. By reading the pre-loaded information of the vehicle appearance, maps information, vehicle dynamics model and signals output from the sensor module, it can be tested for all kinds of situations perfectly. Through the integrated platform output, the control algorithm is evaluated by the control effect of various working conditions. Figure 6 shows the flow chart of the platform overlapping other control algorithms.

After completing the joint of the control algorithm and the system, relevant design tests can be. Figure 7 shows the path planning under the high-order polynomial based on C++ and the path tracking control based on the improved PID algorithm.

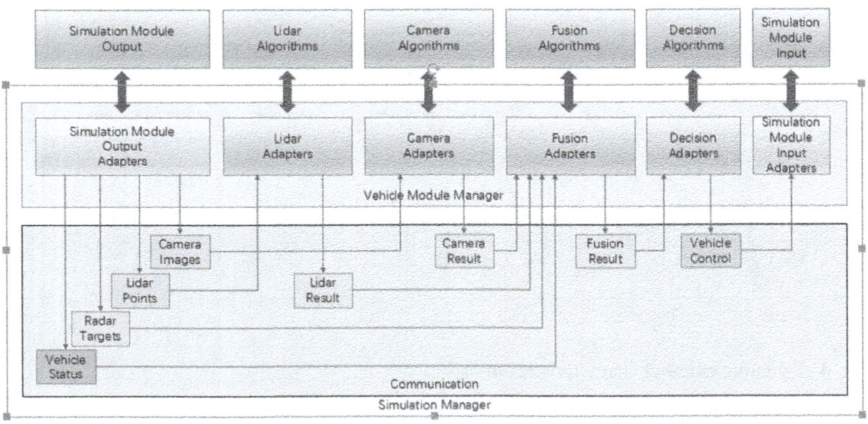

Fig. 6 Lapping control algorithm for simulation platform

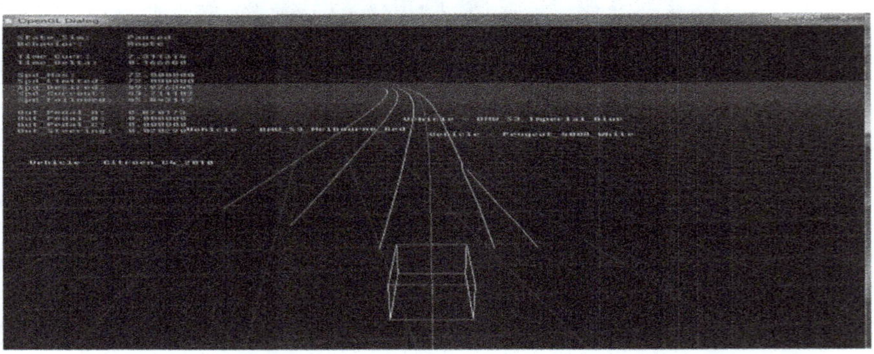

Fig. 7 OPENGL for simulation platform routing

3.4 Scenario Module

The simulation scenario module provides serial assembly and signal input for all other modules, which is the core module of the automatic driving simulation platform. Its functions contain the introduction of high-precision maps, the introduction and correction of map elements within buildings, trees, street lamps, etc., the setting of traffic logic including traffic lights, signal plates, etc., and the editing of scene elements including vehicles, pedestrians, automatic vehicles, etc.

The joint simulation platform proposed in this paper can be imported in high-precision maps, but in the process of importing, a lot of map problems are generated. High-precision map information includes the coordinates of the roads, the height, the intersection, the road signs, the traffic lights, etc. Due to the fuzziness of the OpenDrive rules, it leads to high-precision map appeared in the process of importing lack of roads, such problems as lack of logic, the definition of the sidewalk. The platform modifies the simulation scene for OpenDrive adaptation, to solve the above problems. In addition, the longitude and latitude information adapted by GPS needs to be converted into planar three-dimensional information. This platform completes the equiangular tangent cylindrical projection of the earth through the Mercator coordinate projection (UTM coordinate system), which can complete the transformation of longitude and latitude in Chinese road into three-dimensional coordinates better.

Scenario information can be modeled in various ways, including automatic import of the OSM model, automatic generation of cloud data and construction of third-party map background, so as to improve the accuracy of scenario simulation. Figure 8 shows the information elements of a simulated scenario of the platform, and Fig. 9 shows the schematic diagram of its programmable mode.

In the simulation scenario module, the simulation of the working condition is another important part. The substantial content and core point of the automatic simulation platform lie in the difficulty in completing and verifying such problems as multi-operating condition changing scene design, multi-control strategy comparative study and self-driving assembly test by means of software in the loop before realizing

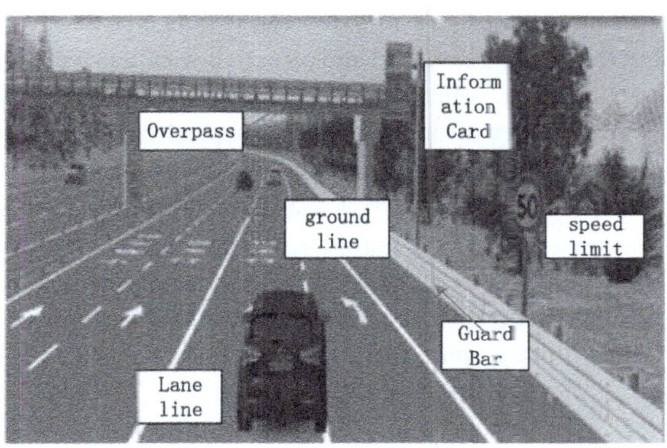

Fig. 8 Simulation platform schematic scene elements

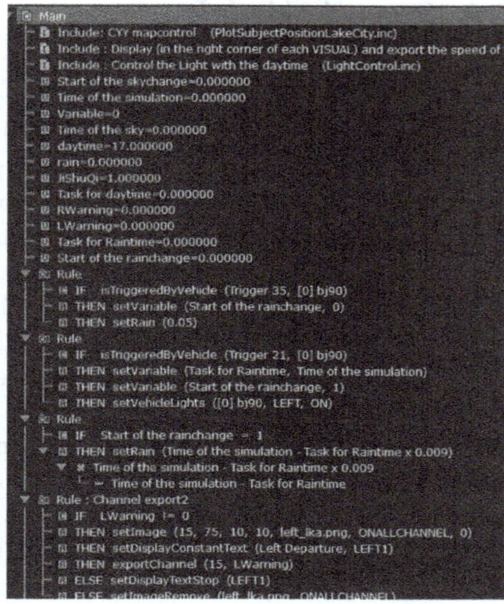

Fig. 9 Programmatic condition schematic

the self-driving mass production design. Working condition simulation will directly affect the simulation effect.

This platform has designed three working conditions to complete the simulation of real road conditions: ordinary working conditions, dangerous working conditions and accident working conditions. Most of the simulation scenarios can be restored in ordinary working conditions, including road condition simulation of common road sections such as city, expressway and parking lot. Dangerous working conditions extract possible dangerous working conditions in real traffic flow and virtual road test simulation and simulate the strain and adjustment made by self-driving vehicles when facing danger. In the case of an accident, the safety is verified through the reconstruction of accident data from the (CIDAS) China In-Depth Accident Study.

Safety is the core issue in autonomous driving. How to verify the feasibility of autonomous driving has become a hot topic in the field of autonomous driving. The vast majority of research and test is concentrated in L3 and below. In the face of L4, but there was no unified and effective way, it is well-known that most of the automatic transmission failure and accident are in extreme conditions. However, the extreme conditions are difficult to control or predict. Faced with the test of L3 level and below, the platform proposes to complete the conventional test based on ordinary working conditions. In the face of the test at level L3.5 or L4 of autonomous driving, the platform proposes to complete the special test in the complex scenes generated based on the fusion of real traffic data in hazardous and accident conditions. Therefore, the development and test of the automatic driving control algorithm and decision-making algorithm are completed.

4 Testing and Results

This platform tested the visual recognition in the sensor algorithm, and the simulation condition was that the Hough transform was used to detect the lane line under the condition of the urban expressway. In the Hough transform, the discretization line of polar parameters is shown in Eq. 1:

$$x\cos(\theta) + y\sin(\theta) = R \tag{1}$$

In Eq. 1, x is the independent variable of the locus, y is the dependent variable of the locus, R is the vertical distance of the straight line, and is the angle between R and the X-axis.

In the Hough transform, the circular equation with parameters is shown in Eq. 2:

$$\begin{cases} x = x_0 + R\cos(\theta) \\ y = y_0 + R\sin(\theta) \end{cases} \tag{2}$$

In Eq. 2, x_0, y_0 are the x and y coordinates of the center pixel points, respectively. R is the circular radius, θ is the rotatable angle.

Simulation lane line detection in sensor module completed by Hough transform is shown in Fig. 10.

The virtual simulation platform system has carried out the functional test of the general operating mode of autopilot and the safety verification test of the accident operating mode, and both have achieved good simulation test results.

Functional test results of ordinary working conditions: The vehicle dynamics module adopts a 28-DOF vehicle model, and the sensor sensing module is modeled as a physical sensor, which is tested in the urban road simulation scene. The simulation effect of the artificial potential field method for the low-speed ACC control strategy is shown in Figs. 11 and 12. LKA, AEB and other functional designs have also completed the simulation test.

Fig. 10 Highway lane line detection

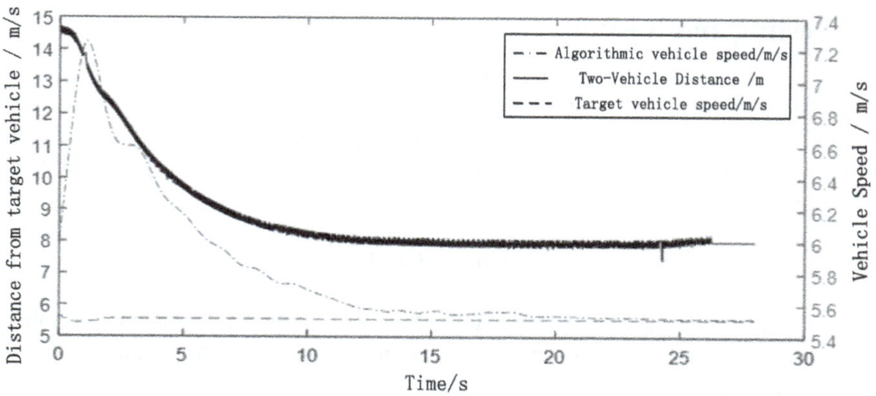

Fig. 11 Slow speed Acc tracking rendering

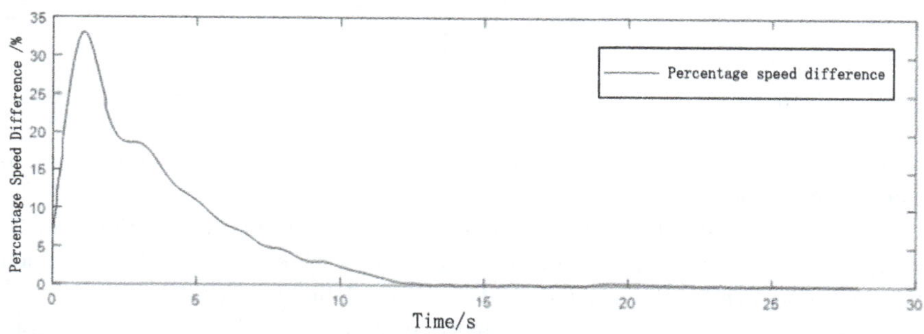

Fig. 12 Low speed Acc speed tracking differences

Safety verification result of accident condition: The system conducted reduction and safety test for a CIDAS bicycle accident. In this CIDAS traffic accident, the comparison of the minimum safe distance after applying the self-driving technology is shown in Fig. 13. Therefore, it can be concluded that in the real accident condition, the control algorithm can complete the safety task well and avoid the accident.

5 Conclusion and Prospect

In this paper, the virtual joint simulation platform for autonomous driving is set up, which can replace the vehicle to complete the automatic driving test with large tasks and high complexity. The test results are convincing. Further development is the following:

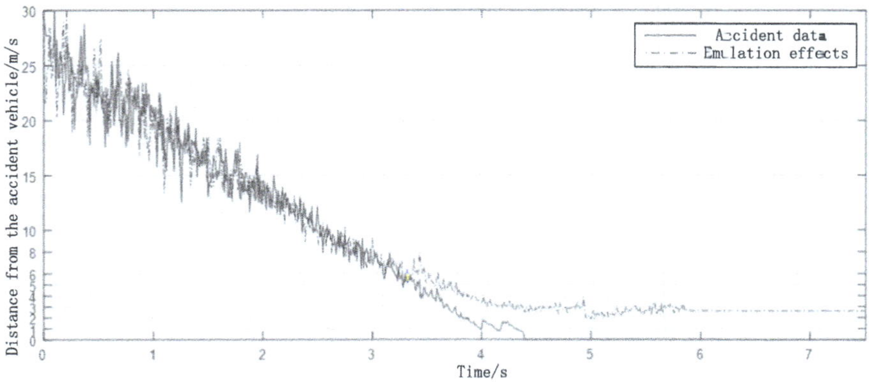

Fig. 13 CIDAS bike accident simulation

1. Be able to complete relevant design and test of current automatic driving technology. It can complete point test including signal, vehicle and control algorithm, upper test based on the combination of software, software and hardware and the advanced test of real vehicle simulation in various working conditions.
2. The simulation scenario database takes the risk conditions into account, and tests such as danger condition and accident condition are completed for CIDAS accident data, which can better verify the safety of self-driving vehicles under different conditions.
3. The simulation platform based on the dynamics of high DOFs vehicles can better provide security for the road on the autonomous vehicle and provide more test conditions and times.

In the future, this simulation platform can import real traffic flow in the simulation scenario and complete the survival of the test vehicle in the real traffic flow In terms of vehicle dynamics, electric and hybrid models will be added to complete the development and testing of autonomous driving technology for more types of cars. In the aspect of simulation scenario modeling, aiming at many problems of high-precision map import, the protocol docking between high-precision map and simulation system will be improved, so as to improve the automatic modeling ability of simulation scenario.

More importantly, through the combination of this virtual simulation platform system and real road driving, it will continuously explore and study the safe driving rules and requirements of autonomous driving, so as to promote the implementation of self-driving technology and mass production as soon as possible.

Acknowledgments Our deepest gratitude goes to the anonymous reviewers for their careful work and thoughtful suggestions that have helped improve this paper substantially.

References

1. Liu J (2017) Research on automatic driving system for unmanned vehicles based on human simulated intelligent control and fractional PID control. Int J Comput Eng 4
2. Xiu C, Chen H (2013) A study of local planning of unmanned vehicle based on improved artificial potential field method. Automot Eng 35(9):808–811
3. Khoshnejad M, Demirli K (2005) Autonomous parallel parking of a car-like mobile robot by a neuro-fuzzy behavior-based controller. Nafips 2005 Meeting of the North American. IEEE, pp 2876–2891
4. Zhang W (2014) Research on unmanned ground vehicle autonomous driving system based on human-like intelligent control. University of science and technology of China, AnHui Province

Research on the Decision Control Methods and Ride Comfort of Autonomous Vehicle

Pingfan Jin, Guilong Li, Caijing Xiu, Guoqi Zhong and Jishun Guo

Abstract In this paper, the comfort of automatic driving vehicle is improved by optimizing the decision control algorithm, and the comfort of the vehicle in different acceleration conditions is analyzed by experiment and simulation. According to the standard of ISO2613, ISO15622 and passenger somatosensory response, the riding sense grade of automatic driving vehicle is divided, and the rationality of ride sense classification is verified by real car test.

Keywords Autonomous · Ride · Decision and planning · Vehicle control · Assessment method

1 Introduction

Vehicle vibration caused by external traffic environment, vehicle running state, and vehicle component movement will directly affect the ride comfort of autonomous vehicle. When evaluating the ride comfort of automobile in dynamic environment, it is necessary to obtain the response characteristics of ride comfort to lateral and longitudinal accelerations, especially the response characteristics of vibration to different acceleration levels. The acceleration and lateral acceleration of vehicles are closely related to the decision-making and control methods of the automatic driving system and are also affected by the execution subsystem. However, the existing research has not yet seen the real-time riding comfort based on the characteristics of the automatic driving system in the dynamic riding environment and is used to optimize the decision-making, planning, and control algorithms. For this reason, based on the analysis of the interaction between passengers, vehicles and automatic driving system and the classification standard of ride comfort, an automatic driving vehicle ride model is established based on ISO2613 and ISO15622. A simulation model is constructed by using IPG CarMaker and MATLAB/Simulink simulation

P. Jin (✉) · G. Li · C. Xiu · G. Zhong · J. Guo
Guangzhou Automobile Group Motor Co., Ltd, 668 East of Jinshan Road,
Panyu District, Guangzhou, People's Republic of China
e-mail: kimpingfan@163.com

China SAE (ed.), *Proceedings of China SAE Congress 2018: Selected Papers*,
Lecture Notes in Electrical Engineering 574,
https://doi.org/10.1007/978-981-13-9718-9_30

tools. Real-time calculation and fast prediction of vibration response characteristics under different acceleration environments can be realized, and then, the magnitude and variation rate of longitudinal acceleration output can be optimized by feedback control to improve ride comfort. After the simulation, a real vehicle experiment is carried out to verify the accuracy of the theory and to correct it.

2 Technology Research Status

2.1 The Relationship Between Ride Sense and Vibration

2.1.1 Research Progress of Ride Comfort

Riding comfort is usually defined as the feeling state of passengers in vehicle dynamic environment. A variety of factors affect this state, including subjective sensory evaluation (such as expectations, personal sensitivity to the vibration environment) and objective sensory evaluation (such as vibration, temperature, noise, ventilation, space, changes in ear pressure, visual field, seat structure). There are several evaluation methods of traditional vehicle comfort in the world, but there is no uniform evaluation standard for the comfort of an automobile. At present, the mainstream evaluation methods are as follows: ISO263—Evaluation Indicators of Human Vibration Exposure, European Interim Standard ENV12299 and Japanese Standard JISE4023. In this paper, the riding comfort grade of self-driving vehicles will be classified according to the most applicable ISO263 standard (Table 1).

2.1.2 Adaptive Cruise Control Strategy

Vehicle adaptive cruise control system (ACC) controls the vehicle, cruising at a fixed speed when there is no front vehicle. When the target vehicle appears in the millimeter

Table 1 Standard statistics for vehicle comfort evaluation

Standard number	Name
ISO 2361—1 Original edition	Evaluation of human exposure to whole-body vibration—the first part: general discussion
ISO 2361—1 Revised edition	Evaluation of human exposure to whole-body vibration—the first part: general discussion
ISO 2361—4	Assessment of human exposure to systemic vibration—Part 4: Assessment of the effects of vehicle passengers, crew vibration and rotational vibration
ENV12299 (1999)	Comfort for passengers: Measurement and evaluation
JISE4023 (1990)	Vibration characteristics of vehicles—measurement method

Magnitude of overall vibration total value	Discomfort response
Less than 0.315 m/s^2	Not uncomfortable
0.315 m/s^2 to 0.63 m/s^2	A little uncomfortable
0.5 m/s^2 to 1.0 m/s^2	Fairly uncomfortable
0.8 m/s^2 to 1.6 m/s^2	Uncomfortable
1.25 m/s^2 to 2.5 m/s^2	Very uncomfortable
Greater than 2.0 m/s^2	Extremely uncomfortable

Fig. 1 ISO2613-1's criterion of ride acceleration for a given vibration acceleration

wave radar detection range, the output torque and braking power of the vehicle are calculated and adjusted to make the driving system and braking system coordinate operation. Following control: ACC detects the vehicle and the front obstacles in real time. The distance and relative speed of the object follow the front car to accelerate and decelerate automatically. If ACC decelerates the vehicle following the vehicle to a stop state, ACC will continue to control the vehicle after a short stop. If the vehicle remains in a stop state for a long time, the driver can step on the accelerator pedal or operate the ACC switch to make the vehicle rerun. After a certain period of time, the vehicle will automatically stop. The whole can be divided into: constant speed cruise: When there is no front car, the speed of the vehicle will be adjusted to reach the set speed of the driver. However, the existing ACC system has no intelligent control technology which can adjust vehicle acceleration according to the external driving environment and driver response. It has problems with the matching of automatic driving vehicles. Therefore, this paper analyzes the interaction between passengers, vehicles and automatic driving system, and the ride comfort classification. On the basis of the standard, the acceleration control strategy suitable for automatic driving vehicles is formulated, which enables the vehicle to intelligently adjust acceleration and deceleration in a reasonable acceleration range (Fig. 1).

3 Experimental Background

3.1 The Environment of Experiment and Automatic Driving Vehicle System

3.1.1 Simulation Environment

See Fig. 2.

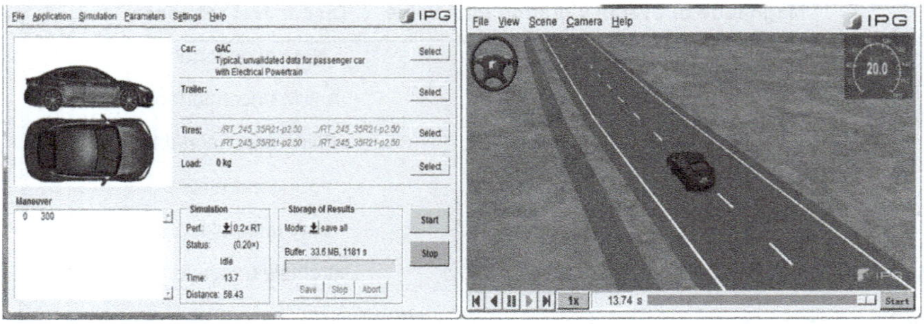

Fig. 2 Automatic driving simulation environment based on CarMaker and Simulink development

3.1.2 Automatic Driving Test Vehicle and System

The GAC Intelligent Driving Vehicle is modified based on the GA6 vehicle platform of Guangzhou Automobile Chuanqi. The appearance and additional external sensors are shown below. Vehicle steering (including steering lights), braking, and EPB can be controlled automatically (Fig. 3).

Compared with the traditional vehicle, the VCU also communicates with the upper intelligent terminal SCU to control the vehicle in the automatic mode. VCU sends a signal of gear position, mode, current torque, and EPB status to SCU, receives a request of control mode, torque, and braking from SCU, and controls vehicle mode and torque output according to current vehicle status. In manual mode, the driver can control the normal running of the vehicle by accelerator pedal, brake pedal and steering wheel, just like driving an ordinary automatic vehicle; in automatic control mode and remote control mode, VCU controls the output of torque and energy recovery of the vehicle according to the target torque and brake request of SCU. The vehicle is controlled by active braking through the iBoost system. IBS (NBS) directly detects braking pressure and controls the whole vehicle braking. In automatic mode, VCU sends expected braking pressure to IBS by receiving the signal of SCU about braking pressure, combining with internal logic calculation and fault state of the whole vehicle.

4 Experimental Process

4.1 The Experiment Simulation Modeling System

4.1.1 Behavior Decision System

Behavior decision-making module is an enabling subsystem, which can only run after receiving the information of perception, navigation, and CPT. Otherwise, even in

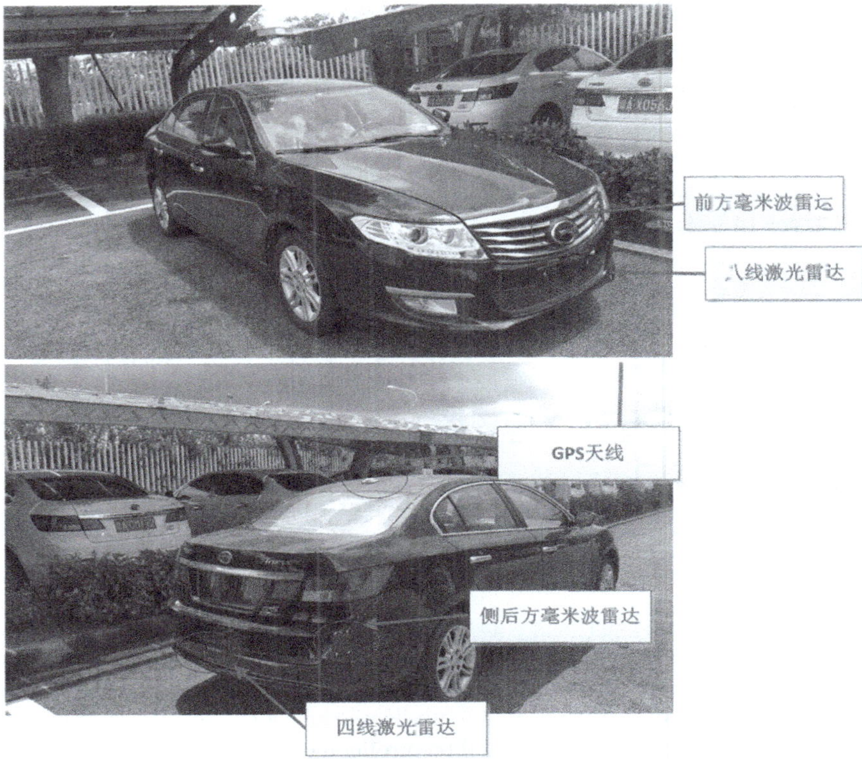

Fig. 3 Vehicle appearance and external sensor location

automatic driving mode, the vehicle cannot start, thus avoiding danger. The behavior decision system is composed of several modules, such as RoadEnvironment, PassMissionPoint, Lane, ApproIntersection, and Intersection. Each module implements certain functions. The behavior decision model is shown in the following Fig. 4.

In the decision system and information of perception system, navigator and CPT are received by UDP interruption. When the decision-making model judges the input data of the driver's control, the control strategy will be updated, and the deceleration of the vehicle will be judged according to the relative position of the vehicle and the obstacle to brake comfortably, normally or urgently. Acceleration will be judged according to the target speed in motion mode, general mode, and comfortable mode into three states to adopt different acceleration for acceleration.

4.1.2 Motion Control System

The motion control system is decomposed into the lateral control system and the longitudinal control system to realize the given trajectory following and unmanned

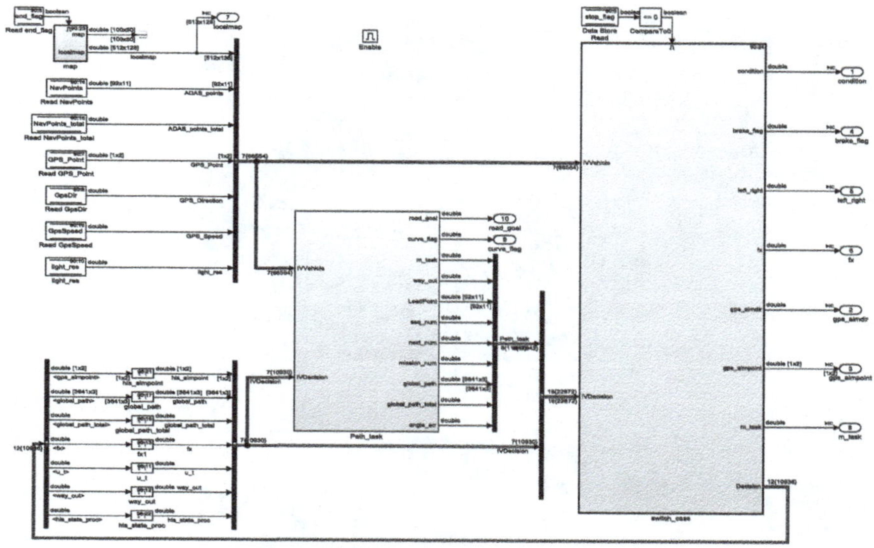

Fig. 4 GAC intelligent driving vehicle behavior decision model structure

longitudinal and lateral motion control. Lateral motion control system is divided into nearest point search module, preview distance calculation module, preview point search module, and steering angle PID controller module; longitudinal motion control system is divided into driving torque PID controller module and braking driving state conversion module (Fig. 5).

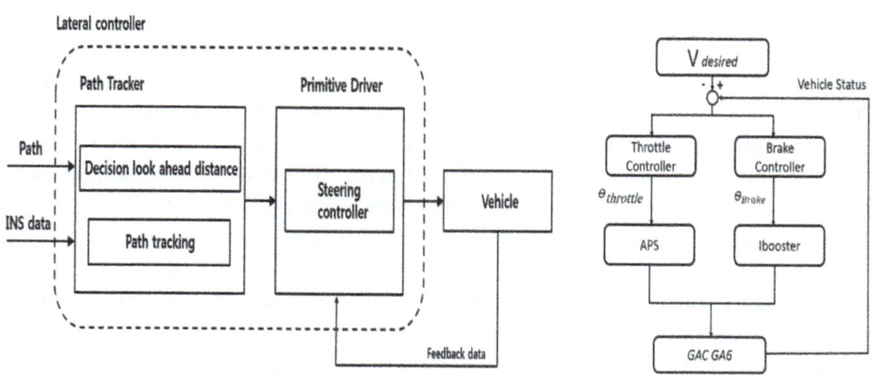

Fig. 5 Structure of horizontal and vertical control model for intelligent driving vehicle of GAC

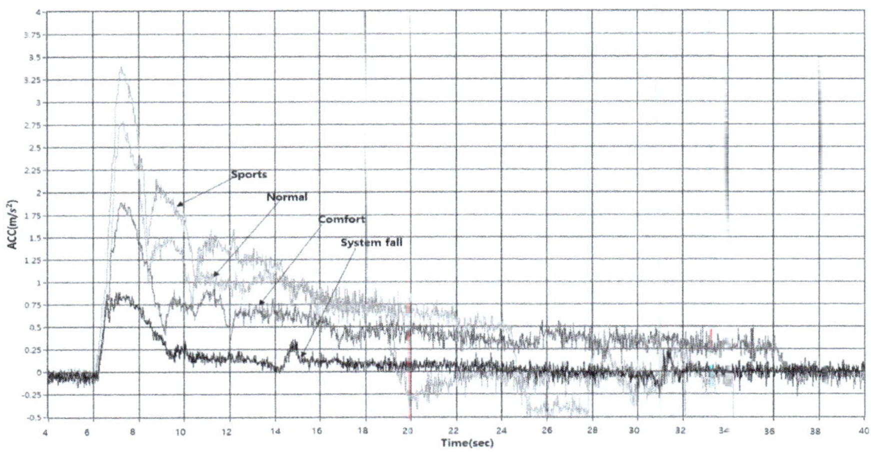

Fig. 6 Comparison of vehicle acceleration in different modes

4.2 Real Vehicle Experiment Results and Analysis

4.2.1 Accelerated Experiment of Ride Comfort and Ride Comfort of Vehicle

- The initial vehicle speed is 0 km/H
- The target speed is 100 km/H (Fig. 6).

Experimental results:

The definition of comfort satisfying ISO 15622 and acceleration, a less than 2.0 m/s;

The definition of comfort satisfying ISO 15622 and acceleration, a less than 2.0 m/s;

The definition of ISO 2613-1, a little uncomfortable (Fig. 7; Table 2).

4.2.2 Deceleration Experiment of Ride Comfort and Ride Comfort of Vehicle

- The initial vehicle speed is 80 km/H;
- Target speed: 75, 70, 60, 40, 20 km/H (Fig. 8).

Experimental result:

It satisfies the definition of ISO 15622, deceleration $a \leq 3.5$ m/s (average more than 2 s) (Fig. 9).

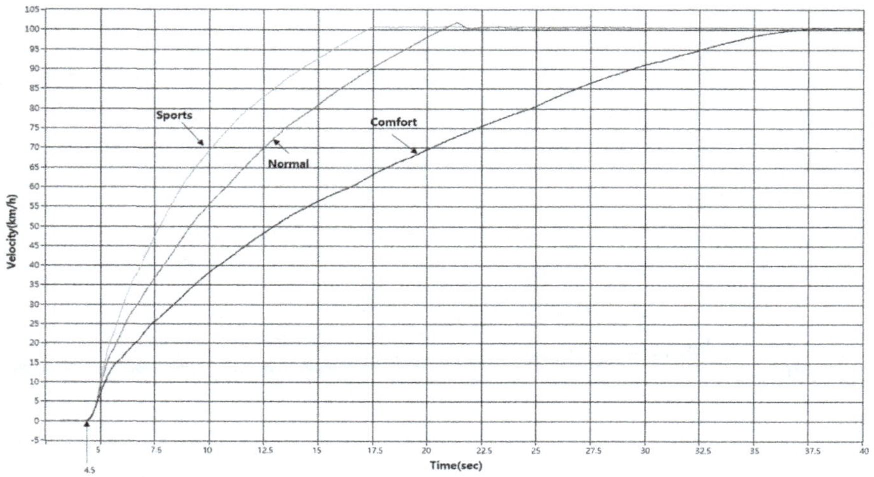

Fig. 7 Comparison of vehicle speed in different modes (0–100 km/H)

Table 2 Performance evaluation of vehicle acceleration results

Performance index	Result		
	Comfort mode	Normal mode	Sports mode
Overresponse (%)	0.9	2	1.3
Time consuming (s)	33.5	17.5	12.5

5 Conclusion

Based on ISO2613 and ISO15622, the ride model of automatic driving vehicle is established. The system simulation model is constructed by using IPG CarMaker and MATLAB/Simulink. The model realizes real-time calculation and fast prediction of vibration response characteristics under different acceleration environments and then improves the ride comfort of the passengers in the automotive vehicle by optimizing the decision-making control method of feedback control. After the simulation, a real vehicle experiment is carried out to verify the accuracy of the theory and to correct it. At the same time, the analysis results provide a basis for further objective experiments and simulation analysis.

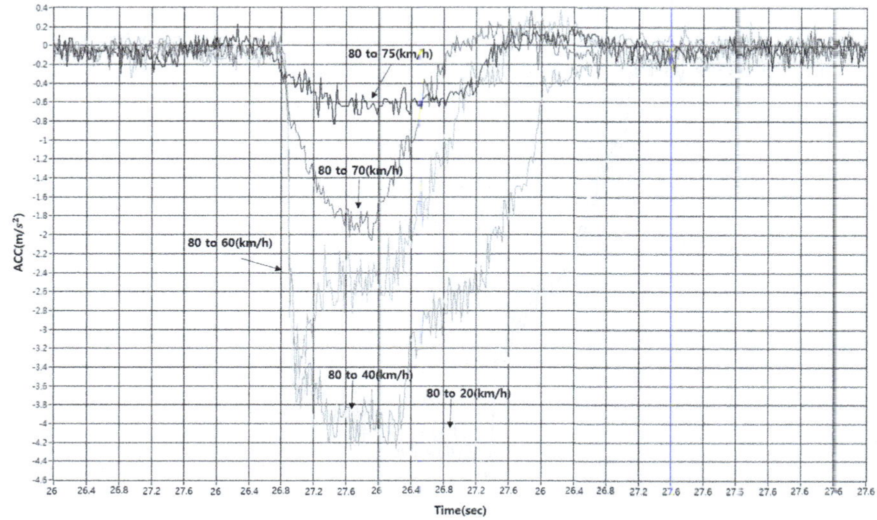

Fig. 8 Comparison of vehicle deceleration results under different modes

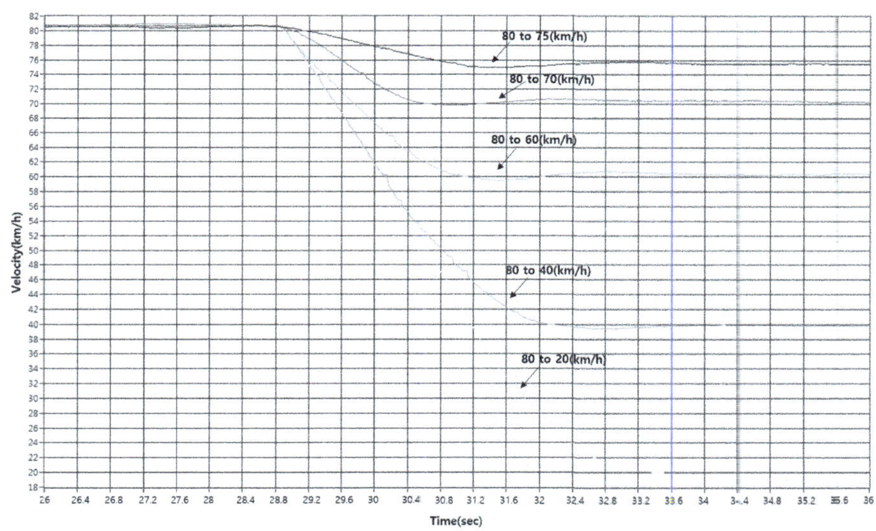

Fig. 9 Comparison of vehicle deceleration results at different target speeds

A Research on the Adaptive Cruise Controller for Electric Bus

Xiaoliang Li, Xudong Zhang, Yuan Zou, Tao Zhang and Shouyang Wei

Abstract The intelligent trend of vehicles became irresistible, and the application of adaptive cruise control systems to different vehicles, especially to electric vehicles, became a trend. The application of adaptive cruise control systems played a significant role in environmental protection, road-traffic efficiency and safety. In this paper, we proposed a hierarchical controller for the adaptive cruise system of the electric bus. Based on millimeter wave radar, target tracking and recognition algorithms for straight and curved conditions had been performed. In the upper controller, a proportional differential control algorithm had been adopted based on the vehicle spacing with variable time-headway, and its stability had been proved. In the lower level controller, the lookup table for acceleration and braking had been designed based on test data and a switching rule for acceleration and braking had been formulated. The results of offline simulation and hardware-in-loop (HIL) simulation showed that the adaptive cruise controller designed with mature control strategy can achieve faster and more accurate control response, real-time performance and was suitable for application on real vehicles.

Keywords Adaptive cruise control · Electric bus · Object identification · Hardware-in-loop

1 Introduction

With the advancement of technology and changes in people's needs, vehicles are more than just transporting people from place A to B. Electrification, intelligence, interconnectedness, and sharing have become the development direction of vehicles and have attracted extensive attention from vehicle manufacturers, Internet companies, and research institutions. As one of the typical applications in driving assistance, the

X. Li · X. Zhang · Y. Zou (✉) · T. Zhang · S. Wei
School of Mechanical Engineering, Beijing Institute of Technology, Beijing, China
e-mail: zouyuanbit@vip.163.com

X. Li
e-mail: xiao_liang_li@126.com

China SAE (ed.), *Proceedings of China SAE Congress 2018: Selected Papers*, Lecture Notes in Electrical Engineering 574,
https://doi.org/10.1007/978-981-13-9718-9_31

adaptive cruise control (ACC) system can make the vehicle accelerate or decelerate automatically according to the real-time traffic environment so that the vehicle can maintain a certain safe distance to the preceding one. It can reduce driving burden [1], so more and more ACC system has been equipped in various vehicles, and it has become a standard configuration. At present, the researches on ACC mainly focus on passenger vehicles. The researches on ACC for commercial vehicles, especially for electric buses, are rather rare. At the same time, "Safety specifications for commercial bus" (JT/T 1094-2016) stipulates that commercial bus which is longer than nine meters must install forward collision warning (FCW) system from April 2018. Killing two birds with one stone, we can add FCW function based on the hardware of ACC system. The application of ACC system on electric bus is of great significance for reducing the burden on drivers of commercial bus, ensuring the safety of passengers' lives, and improving the efficiency of traffic.

The current research on ACC system can be divided into four parts: environment perception, drivers' characteristics, vehicle dynamics model, and control algorithm. At present, research institutions, experts, and scholars have carried out relevant research on the above four parts and have gained valuable results. In terms of environmental perception: literature [2] used radar to detect and track target vehicle ahead and used Kalman filtering to test its effectiveness; literature [3] combined radar and camera to identify the front target so that the type and motion state of target can be detected well. With regard to drivers' characteristics: literature [4] considered the variable time-headway based on the host vehicle speed; literature [5] proposed a variable time-headway based on the relative vehicle speed; literature [6] proposed a nonlinear expected time-headway. With respect to vehicle dynamics model construction: literature [7] combined the theoretical analysis model with the vehicle experimental data and established a longitudinal dynamic model for simulating the complex driving conditions of the vehicle in the forward collision avoidance system. Literature [8] established inverse vehicle model to construct the ACC control system; the whole system consists of drive/brake mode switching, desired torque calculation, and engine inverse model brake inverse model. About control algorithm: literature [9] combined PI control and feedforward to design the lower layer controller, which made the system response faster, but its robustness was poor. Considering various constraints such as minimum error, driver perception, and fuel consumption, literature [10] used the quadratic programming method to obtain the optimal control law based on the model predictive control (MPC) theory and established the multi-objective ACC system. The current research mainly focuses on the comparative study of different algorithms. There are few studies on the whole process of ACC controller development used in real vehicles, especially in electric buses. In this paper, the research object is ACC system used in electric buses. The object identification algorithm of millimeter wave radar is designed for straight and curve road, and its effectiveness and reliability have been tested through experiments on actual vehicle. We analyzed control strategies and established offline controller simulation model with Simulink and TruckSim, hardware-in-loop simulation model with dSPACE and RT-ECU (the real-time controller). The results of simulation indicate that the algorithm we proposed can achieve fast and accurate response, and its real-time performance and stability can be guaranteed. We accomplished the design of ACC controller in addition to the vehicle test and verified its rationality.

2 Target Recognition Tracking Based on Millimeter Wave Radar

Radar is one of the most important sensors in ACC system, whose accuracy and stability of detection determine the performance of the whole system. At present, the type of radars used in vehicles includes ultrasonic radar, laser radar, and millimeter wave radar. Compared with others, stable performance, long detection distance, small weather impact, mature technology, and high-cost performance are the advantages of millimeter wave radar. Considering this, Delphi's electronically scanning radar (ESR) millimeter wave radar is our good choice. The radar uses frequency-modulated continuous wave, and its detection process is shown in Fig. 1.

2.1 Main Sensor-Millimeter Wave Radar

The frequency-modulated continuous wave of radar is generated by the oscillator and the wave is transmitted through the antenna. When the wave encounters an obstacle, a part of wave is reflected back. Through calculating the time difference between the transmitted wave and the echo wave, the distance between the obstacle and radar can be obtained. Due to relative motion, the Doppler effect for radar wave is obvious, and the relative velocity can be measured. There are so many multiple antennas in millimeter wave radar; generally that we can measure the azimuth of the target using the phase difference between two echo signals received by different antennas.

The frequency of millimeter wave is 76–77 GHz. The medium distance measurement mode of radar can detect obstacle within the azimuth range of ±45° and maximum distance range of 60 m. The remote measurement mode can detect azimuth range of ±10° and maximum distance range of 100 m. Specific radar detection performance in the two modes is shown in Table 1.

It is shown that the two modes work simultaneously to detect targets can not only ensure the accuracy of the long-distance measurement but also provide a larger

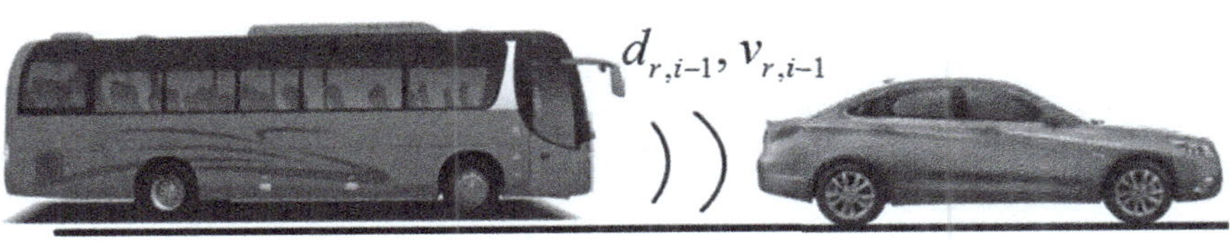

Fig. 1 Radar detection schematic

Table 1 Radar detection performance in two modes

Category	Remote mode	Medium distance mode
Refresh rate (Hz)	20	20
Relative speed detection range (m/s)	−100 to +25	−100 to +25
Distance detection accuracy (m)	±0.5	±0.25
Relative speed detection accuracy (m/s)	±0.12	±0.12
Azimuth detection range	±0.5°	±1°

range and more types. The target update frequency of ESR radar is 20 Hz, and the information of two working modes for 64 targets are integrated and output in each detection cycle. The relative distance, relative velocity, and azimuth information of the detected target are output through CAN bus.

2.2 *Target Tracking and Recognition*

A. Tracking algorithm in straight road condition

Firstly, we should transform the information of the radar output from rectangular coordinate system into polar coordinate. Secondly, we can filter out the redundant targets by using the effective detection area so that the computational burden can be reduced and computational speed can be improved. According to the radar performance standard 3-m-wide lane, the effective detection area is set as a rectangular area in which the host vehicle forward distance is 70 m and the left and right side widths are both 4.5 m (Fig. 2).

Fig. 2 Effective detection area schematic

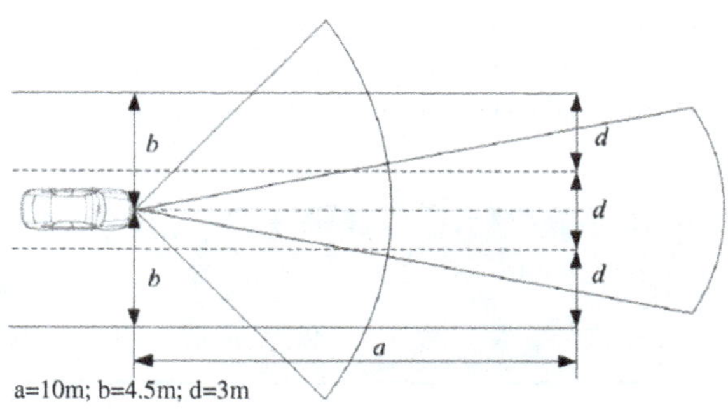

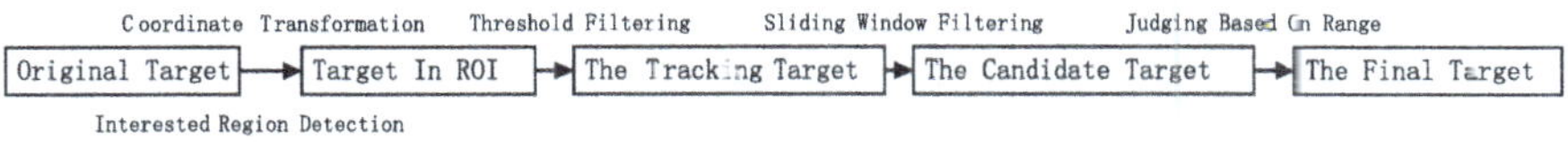

Fig. 3 Tracking and filtering process schematic

Because the stationary target in the side lane is invalid target that does not affect the driving safety, it needs to be filtered out. Since ACC system needs to do forward collision prevention warning, the target which lateral distance falls into the range [−1.5, 1.5] m needs to be reserved.

What's more, in order to make ACC system stable, it is still necessary to track and filter the remaining targets to determine the following target. The tracking and filtering process is shown in Fig. 3.

The tracking and filtering of targets are based on historical consistency. Since the vehicle is an inertial system and the sampling time of ACC system is very short, the motion state of vehicles will not be abrupt, so the next motion state can be inferred by the current state. The comparison of the measured motion state at the next moment and the inferred motion state can determine whether the target is the same or the data is distorted. The measured value of the next cycle of the target is compared with the predicted value as shown in formula (1):

$$\begin{bmatrix} d'_{x(n+1)} \\ d'_{y(n+1)} \\ v'_{x(n+1)} \\ v'_{y(n+1)} \end{bmatrix} - \begin{bmatrix} d_{x(n+1)} \\ d_{y(n+1)} \\ v_{x(n+1)} \\ v_{y(n+1)} \end{bmatrix} \leq \begin{bmatrix} K_{xd} \\ K_{yd} \\ K_{xv} \\ K_{yv} \end{bmatrix} \tag{1}$$

In the formula: the data with the superscript is the predicted value of the data for the motion state at the next moment; K is the threshold value, its subscript represents the corresponding variable of motion state. If the formula (1) is satisfied, the corresponding target information existing in the cache will be updated with the latest target information, and the tracking flag is set to 1 at the same time, otherwise is set to 0. If it is impossible to find the target that meets formula (1) in the existing target, this target will be stored in the cache as a new target. By this method, the existing target information can be updated in time, and the new one cannot be missed.

Finally, the final candidate target is selected from the cached targets according to sliding window detection as shown in Fig. 4.

Where S is set as 10 sampling periods. If in the last three consecutive sampling periods, and at least 7 in all 10 sampling periods, the flag is 1. The goal can be seen as a candidate. Therefore, even if the millimeter wave radar accidentally loses the target due to road undulation and body vibration, the target tracking can still perform effectively and continuously. The jump of the target motion information and data distortion can be avoided.

ACC system will filter two targets in the side lanes and one target in the host lane closest to the host vehicle from all candidates for control.

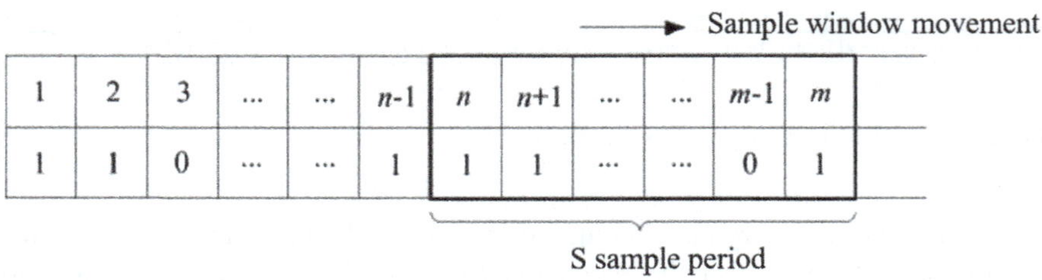

Fig. 4 Sliding window detection schematic

B. Tracking algorithm in curved road condition

The curved road conditions are more complicated than the straight road. For example, there may be other front vehicles in the adjacent lane as shown in Fig. 5. At this time, if the tracking process still bases on the algorithm used in original straight road, it is easy to determine the side lane target erroneously as the target vehicle for tracking.

Therefore, it is necessary to design the target tracking algorithm separately. According to international standards, the radius of the curve for ACC system is generally greater than 500 m, which is much larger than the longitudinal detection distance of the ESR radar. Therefore, it can be considered that when the host vehicle and the target vehicle are in the same corner, the curvature of the roads is equal. Therefore, it can be judged the two vehicles are in the same curve by calculating whether the front vehicle is in a position that the host vehicle can reach with its current road curvature.

When the centroid side deflection angle is ignored and the steering wheel angle is small, the original steering radius of the host vehicle can be written as:

$$R_0 = \frac{L_b}{\delta} \tag{2}$$

Fig. 5 Interference caused by front vehicle in adjacent lane schematic

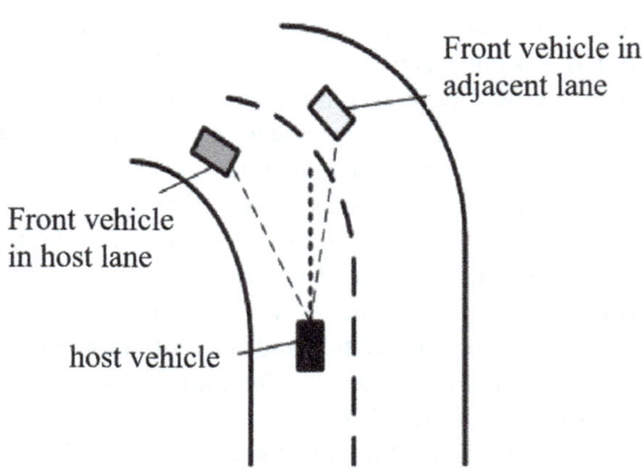

In the equation: R_0 is the original steering radius of the host vehicle, m; L_b is the wheelbase of the host vehicle, m; δ is the rotation angle of the front wheel, rad.

From the steady-state response of the vehicle at the step rotation angle of front-wheel input [11]:

$$\frac{\omega_r}{\delta} = \frac{v_h/L_b}{1 + K v_h^2} \tag{3}$$

In the equation: ω_r is the yaw rate of the host vehicle, rad/s; v_h is the host vehicle speed, m/s; K is the vehicle stability factor, s^2/m^2; the rest of the variables have the same meaning as before.

It can be derived from Eq. (3) that the relationship between the real steering radius R and the original steering radius R_0 of the host vehicle at a certain vehicle speed is as follows:

$$R = \frac{v_h}{\omega_r} = \left(1 + K v_h^2\right) R_0 \tag{4}$$

In the formula: the meaning of each parameter is the same as before. The host vehicle speed v_h is transmitted to the controller through CAN bus. The front-wheel rotation angle δ of the host vehicle can be calculated by collecting the rotation angle of steering wheel, and then we can get the road radius R through the formula (2)–(4).

As shown in Fig. 6, the host vehicle runs on the curve C. A front vehicle is in the point D. The central angle of the road can be expressed as follows:

$$\phi_d = \frac{1}{2} \arctan \frac{L + \rho \cos\theta}{R - \rho \sin\theta} \tag{5}$$

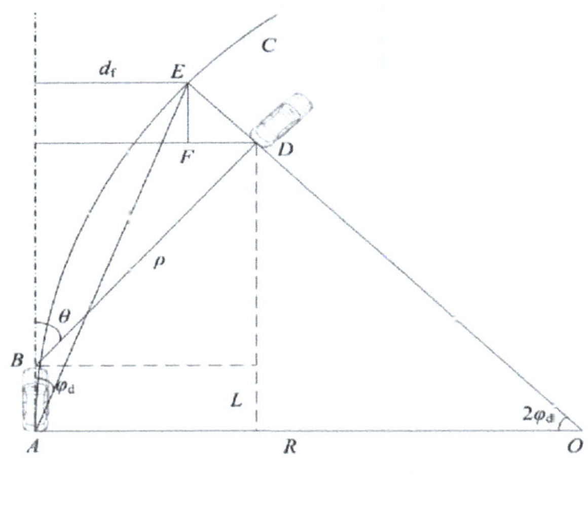

Fig. 6 Calculation in curved road schematic

In the formula: φ_d is a half of the central angle, rad; L is the longitudinal distance from the rear axle of the vehicle to the radar, m; R is the radius of the curved road, m; the remaining variables are the same as before.

The front vehicle is located in the adjacent lane, and the point in host lane corresponding to the same central angle is E, and the lateral distance d_f from the host vehicle can be calculated by formula (6):

$$d_f = 2R\sin^2\phi_d \tag{6}$$

At this time, the following relationship can be used to determine whether the target vehicle and the host vehicle are traveling in the same lane.

$$\left|\rho\sin\theta - d_f\right| < d_{f\,\mathrm{lim}} \tag{7}$$

In the formula: $d_{f\mathrm{lim}}$ is a threshold, which takes 0.5 m to prevent false due to sensor measurement errors and vehicle jitter.

2.3 Experiment and Verification in Real Vehicle

The results of experiment and verification in straight road conditions are shown in Figs. 7 and 8.

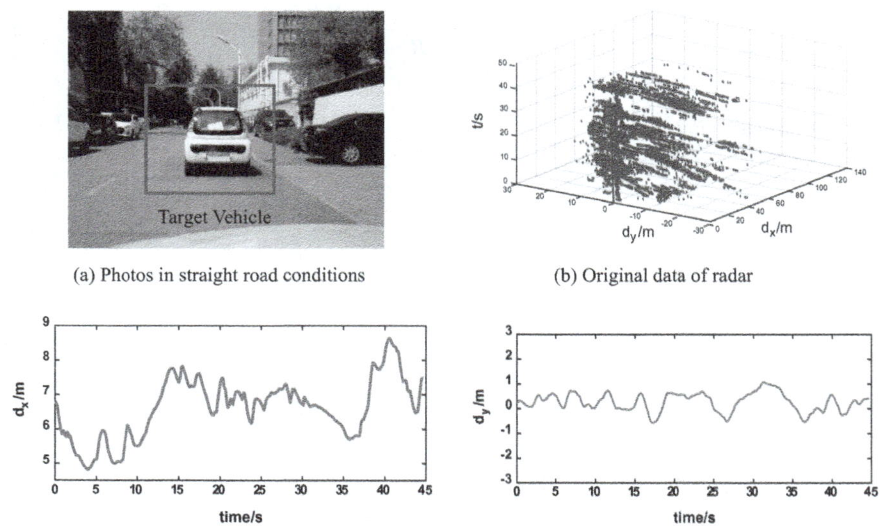

(a) Photos in straight road conditions

(b) Original data of radar

(c) The longitudinal distance of target

(d) The lateral distance of target

Fig. 7 Experiment and verification in straight road conditions

(a) Photos in curved road conditions

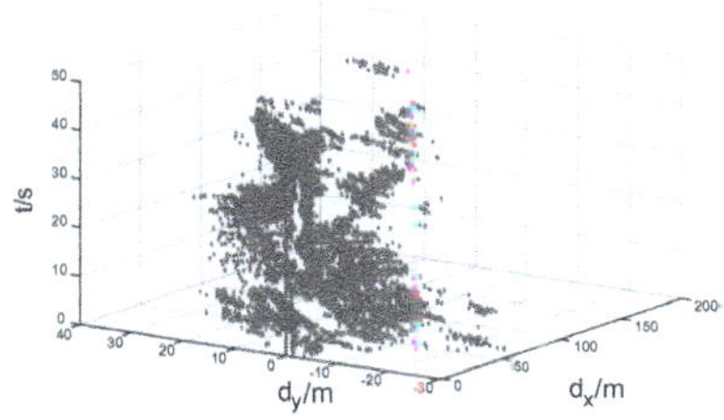

(b) Original data of radar

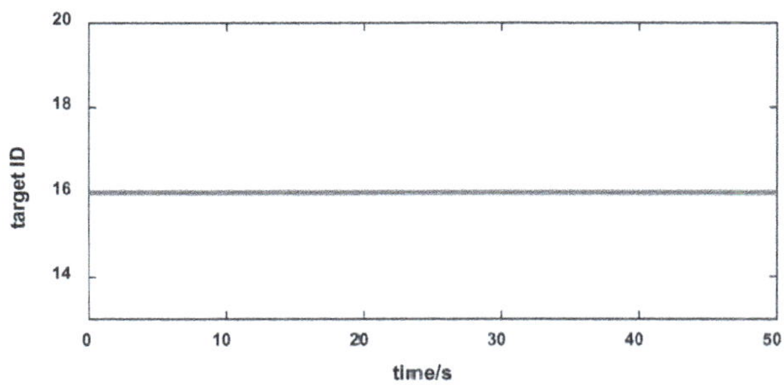

(c) The target ID

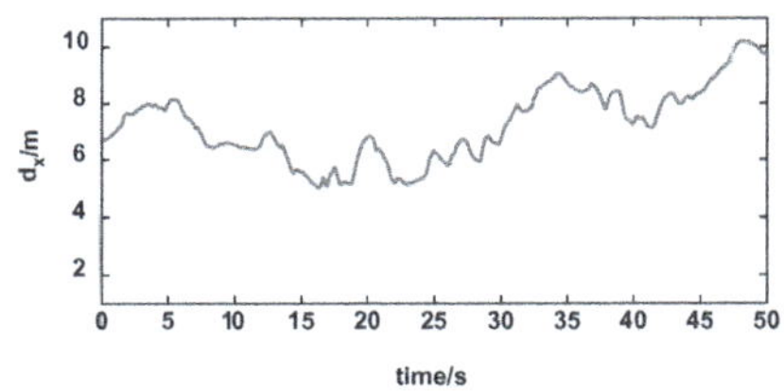

(c) The longitudinal distance of target

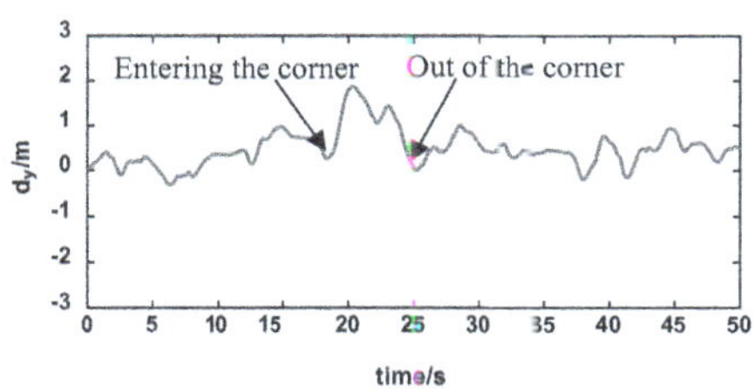

(d) The lateral distance of target

Fig. 8 Experiment and verification in curved road conditions

Through the experiment and verification in real vehicle, the proposed target recognition algorithm is able to meet the requirements. It can also track the vehicle target well in the cornering curved road conditions without losing, which can meet the actual needs.

3 Adaptive Cruise Control System Design

The ACC system adopts a hierarchical structure. The upper controller is a solver for desired acceleration. According to the host vehicle speed, the distance between the host vehicle and the target vehicle, what's more, the relative speed, we can calculate the desired acceleration of the host vehicle. The lower controller is used

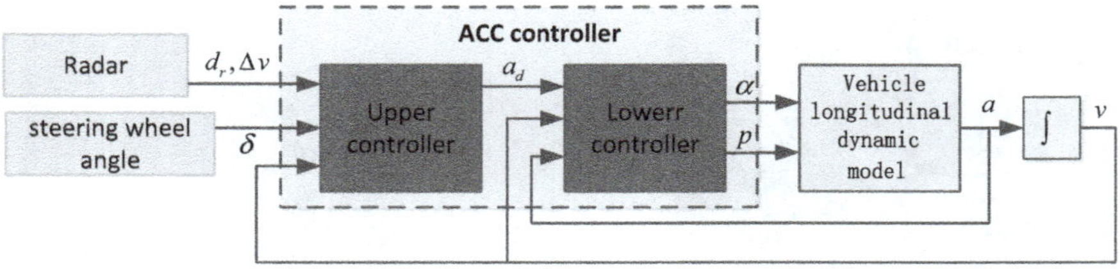

Fig. 9 Architecture of ACC system. a_{des}: Desired acceleration, α: Acceleration signal, p: Brake signal, a_i: Host vehicle acceleration, v_i: Host vehicle speed, δ: Steering wheel angle

to control actuator, which is responsible for calculating the acceleration and braking commands so that the actual acceleration accurately follows the desired acceleration. The structure of the control system is shown in Fig. 9.

3.1 Upper Controller Design

There are three distance control strategies—fixed distance, fixed time-headway, and variable time-headway. Considering the practicality and comfort under complicated conditions, the variable time-headway strategy is adopted in this paper. The desired distance between the host vehicle and the target vehicle is:

$$d_{\mathrm{d}} = v_{\mathrm{h}} h + d_0 \tag{8}$$

$$h = h_0 - c\Delta v \tag{9}$$

In the formula: d_{d} is the desired distance, m; v_{h} is the host vehicle speed, m/s; h is the time-headway, s; d_0 is the safety distance, m; h_0 is the time-headway we set, s; c is a constant greater than zero; Δv is the relative speed, m/s. According to driver characteristics, h_0 is set to 1.8 s, and d_0 is set to 3 m. At the same time, this paper sets the following saturation function:

$$h = \begin{cases} h_{\max} & h_0 - c\Delta v > h_{\max} \\ h_0 - c\Delta v & \text{others} \\ h_{\min} & h_0 - c\Delta v < h_{\min} \end{cases} \tag{10}$$

The vehicle following error is:

$$\begin{cases} d_{\text{error}} = d_r - d_d \\ \Delta v = v_p - v_h \end{cases} \tag{11}$$

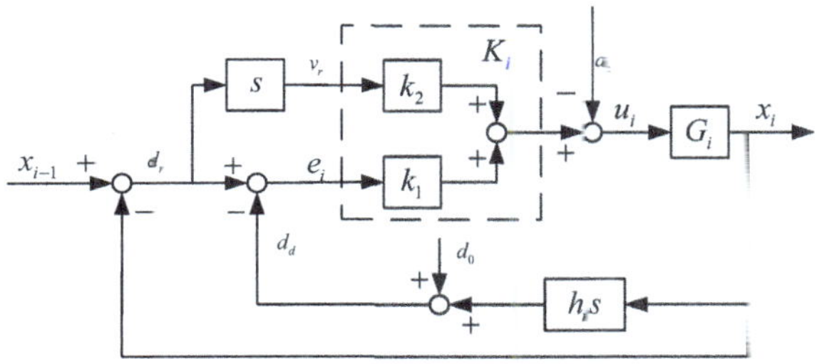

Fig. 10 Block diagram of the control algorithm in upper layer

where: d_{error} for the distance error, v_p for the target vehicle speed; d_r for the actual distance.

We used PID control to minimize the vehicle following error, but the existence of the integral is easy to make the control system unstable, thus PI control is finally selected. According to PID control algorithm, the vehicle's desired acceleration can be calculated as follows:

$$u = k_1 \Delta v + k_2 d_{\text{error}} - a_z \tag{12}$$

u is the desired acceleration, k_1 and k_2 are the parameters of the PI controller; a_z is the compensation amount, which is provided by the vehicle acceleration adaptive adjustment mechanism, which is used for the change of the main vehicle parameters and the external environment. The deviation is the correction.

In the real control process, the input of the upper ACC controller is d_{error} and Δv. The controller outputs the desired acceleration, which is converted into the final accelerator opening and brake pressure control in the lower layer control. The block diagram of the control algorithm in upper layer is shown in Fig. 10.

3.2 *Stability of Distance Error Proof*

For the distance control of the adaptive cruise control system, if the host vehicle running stably at a constant vehicle speed, the distance control strategy must ensure that the distance error is converged.

The requirement of stability can be expressed as:

$$\Delta v + k_d d_{\text{error}} = 0 \tag{13}$$

We can get from formulas (8) and (11):

$$d_{\text{error}} = d_r - v_h h - d_0 \tag{14}$$

Differentiate the formula (8) to get:

$$\dot{d}_{\text{error}} = \Delta v - v_h \dot{h} - \dot{v}_h h \tag{15}$$

$$\dot{h} = \begin{cases} -c\Delta\dot{v} & h_{\min} < h < h_{\max} \\ 0 & \text{others} \end{cases} \tag{16}$$

Construct the following function:

$$a_t = \begin{cases} 1 & h_{\min} < h < h_{\max} \\ 0 & \text{others} \end{cases} \tag{17}$$

Then, Eq. (16) can be expressed as:

$$\dot{h} = -a_t c \Delta\dot{v} \tag{18}$$

Differentiate the formula (13) to get:

$$\Delta\dot{v} = -k_d \dot{d}_{\text{error}} \tag{19}$$

Substituting the above formulas (18) and (19) into the formula (15), when the host vehicle acceleration is 0, the following formula is established:

$$\dot{d}_{\text{error}} = -\frac{k_d}{1 + (h + a_t c v_h) k_d} d_{\text{error}} \tag{20}$$

In the formula, $h > 0$, $a_t > 0$, $c > 0$, $v_h > 0$, $k_d > 0$. Therefore, the distance error converges to zero, the variable time-headway strategy adopted in this paper satisfies stability requirement.

3.3 Lower Controller Design

Due to the characteristics of electric buses, the lower controller can control the torque directly by controlling the motor. So that the actual acceleration of the host vehicle can accurately follow the change of the desired acceleration. The structure of lower controller is shown in Fig. 11.

According to the vehicle dynamics equation, the desired torque of motor when the vehicle is running on a windless straight road is as follows:

$$T_d = \frac{r}{i_0}\left(mgf + \frac{C_D A}{21.15} v_i^2 + \delta a_{\text{des}}\right) \tag{21}$$

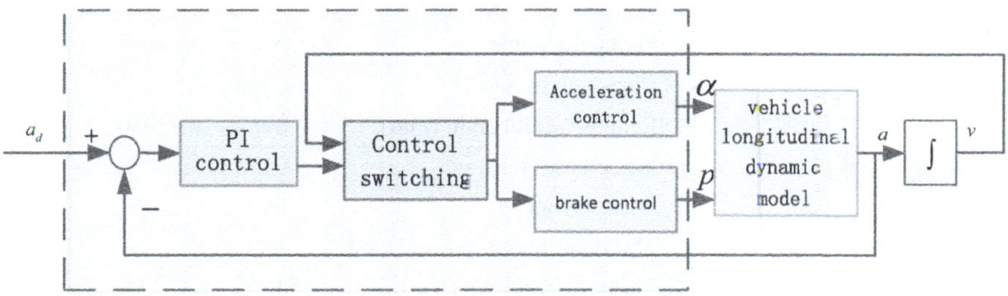

Fig. 11 Architecture of actuator controller

where: r for the wheel radius, i_0 for the transmission ratio, m for the curb weight, f for the coefficient of rolling resistance, C_D for the drag coefficient, A for the windward area, δ for the rotational mass conversion coefficient.

The relationship between motor torque and speed under different acceleration levels is shown in Fig. 12a. The acceleration command is obtained by looking up the curve. In this paper, the vehicle dynamics modeling software TruckSim is used to simulate the braking characteristics. The brake control inquiry table can be constructed by measuring the braking deceleration generated by different brake pressures at different speeds. The basic parameters of the electric bus model we established in TruckSim are shown in Table 2.

We convert the brake pressure into a percentage signal and measure the corresponding deceleration at different speeds in the range of 0–80 km/h. The result is made into the lookup table as shown in Fig. 12b. We can look up the table according to the speed and the desired acceleration to obtain the desired brake pressure.

It is necessary that the designed controller can avoid frequent switching between acceleration and braking operations. Referring to the method in literature [1], the strategy is established to avoid frequent switching between acceleration and braking.

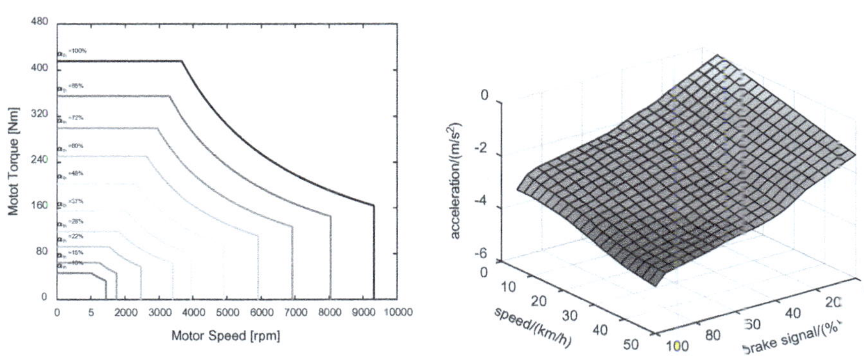

(a) Acceleration-torque-speed relationship (b) Acceleration-brake-speed relationship

Fig. 12 Lookup table about acceleration and brake control

Table 2 Basic parameters of vehicle model

Parameter name	Numerical value
Curb weight (kg)	6360
Length/width/height (mm)	8982/1794/2950
Wheelbase (mm)	6150
Drive layout	RR
Maximum motor power (kW)	120
Motor peak torque (Nm)	480
Highest speed (km/h)	90
Type of transmission	Single gear transmission
Wheel radius (mm)	520

Table 3 Switching strategy

Switching boundary	Control strategy
$a_{\text{des}} > -0.025v + 0.288$	Accelerate
$-0.037v - 0.225 \le a_{\text{des}} \le -0.025v + 0.288$	No control
$a_{\text{des}} < -0.037v - 0.225$	Brake

It can be fitted to a linear expression as follows:

$$a_s = -0.028v + 0.164 \tag{22}$$

In the formula:a_s is the reference acceleration. The switching boundary is set above and below the reference acceleration as shown in Table 3.

In addition, the desired acceleration is saturated as formula (23):

$$a_{\text{des},i} = \begin{cases} 1.2 & a_{\text{des},i} \ge 1.2 \\ a_{\text{des},i} & -4 < a_{\text{des},i} < 1.2\ (\text{m/s}^2) \\ -4 & a_{\text{des},i} \le -4 \end{cases} \tag{23}$$

4 Offline Simulation and Hardware-in-Loop Simulation

4.1 Offline Simulation

TruckSim is often used to simulate complex nonlinear vehicle dynamics models. Co-simulation with MATLAB/Simulink and TruckSim not only makes establishing vehicle model simple but also makes designing controller easy.

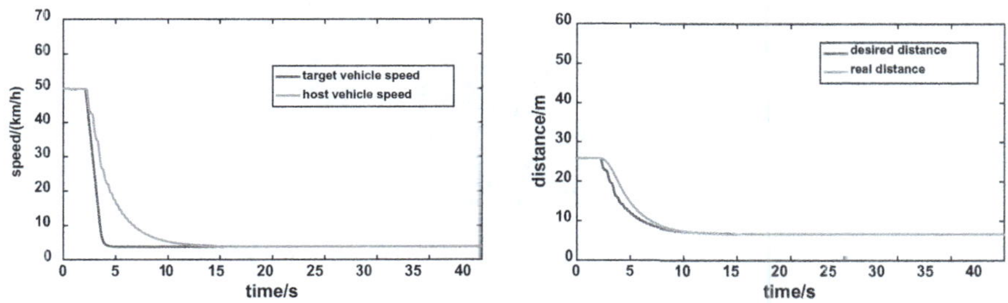

(a) $h = 1\text{s}$, emergency braking conditions

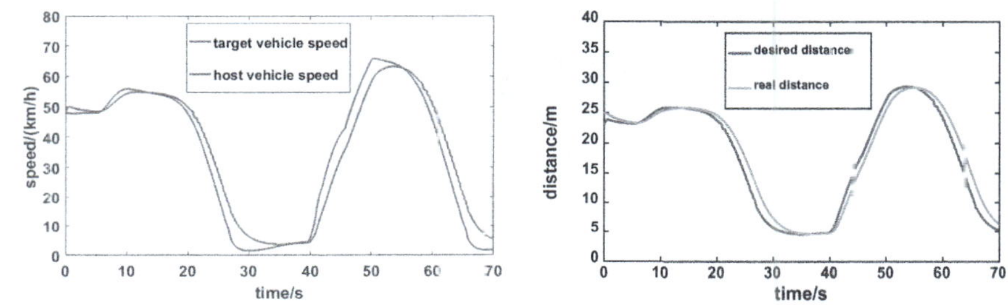

(b) $h = 1\text{s}$, target vehicle speed is shifting

Fig. 13 Result of offline simulation

Then, through the offline simulation, the effect of control algorithm is tested under typical conditions. Because the dynamics of acceleration and brake are different, different parameters are selected for controller.

By analyzing the simulation results of the emergency braking condition shown in Fig. 13a and the condition that target vehicle speed is shifting shown in Fig. 13b, the controller can control the host vehicle to keep up with the target vehicle and make the distance within safe range quickly and stably.

4.2 Hardware-in-Loop Simulation

After completing the offline simulation, the HIL simulation platform shown in Fig. 14 is built.

This paper chooses the self-developed real-time electronic control system RT-ECU with multiple data interfaces as the core controller. The dSPACE AutoBox and the ControlDesk are also used in the simulation. The 7-DOF vehicle dynamics model is imported as the controlled object in dSPACE, the control parameters are set in the ControlDesk, and the communication is through CAN bus.

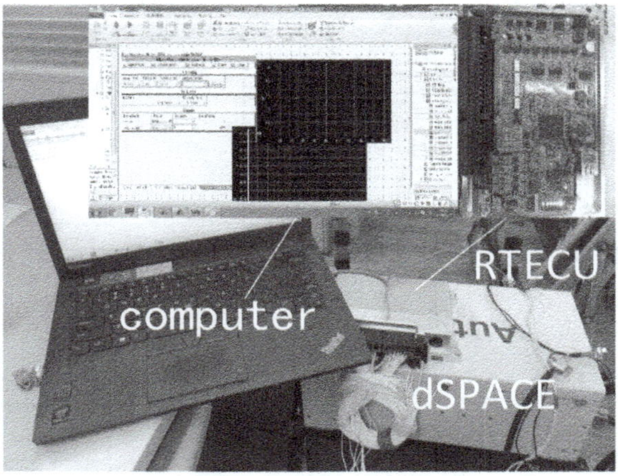

Fig. 14 Platform of hardware-in-loop simulation

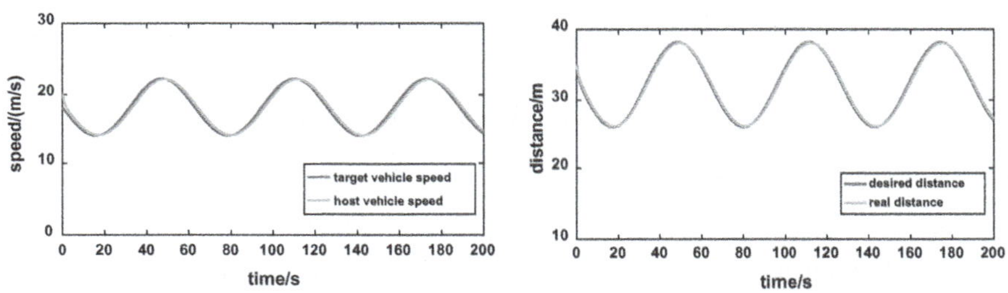

Fig. 15 $h = 1.75$ s, the result of hardware-in-loop simulation

By analyzing the results of the hardware-in-loop simulation shown in Fig. 15, the controller can also control the host vehicle to follow the target vehicle at a higher speed quickly and stably.

4.3 Real-Vehicle Experiment

In addition, we also carried out an experiment in real vehicle. The results are shown in Fig. 16, the effectiveness of the proposed control strategy had been verified.

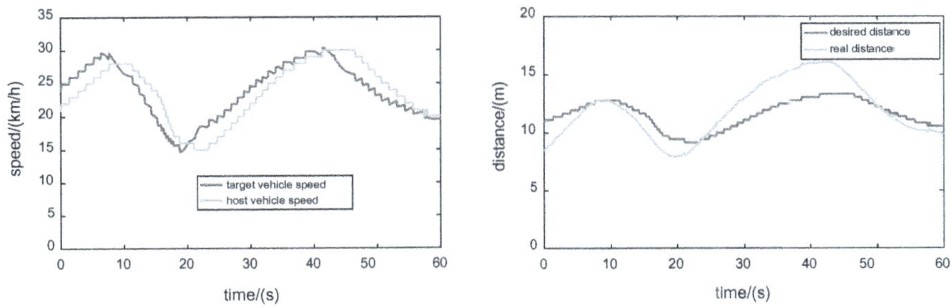

Fig. 16 Result of real-vehicle experiment

5 Conclusion

This paper carries out the researches about ACC system based on the electric bus. The results are listed as follows:

1. Target recognition and tracking algorithm based on millimeter wave radar for straight and curved conditions are designed and verified by real-vehicle experiments.
2. The upper controller which uses PI control algorithm based on variable time-headway and the lower layer actuator controller are designed.
3. Through the offline simulation, hardware-in-loop simulation and real-vehicle experiment, the results show that the control strategy can achieve faster and more accurate control response. It is suitable to use in real vehicles.
4. With the development of the vehicle networking, collaborative adaptive cruise will become the mainstream. In the future, the vehicle networking module can be added to the electric bus to research collaborative adaptive cruise.

References

1. Pei X, Liu Z, Ma G, Qi Z et al (2013) An adaptive cruise control system based on throttle/brakes combined control. Automot Eng 35(4):375–380
2. Liu Z, Wang J, Li K (2008) Robust vehicular radar target determination. J Tsinghua Univ (Science and Technology) 48(5):875–878
3. Shimomura N, Fujimoto K, Oki T et al (2002) An algorithm for distinguishing the types of objects on the road using laser radar and vision. IEEE Trans Intell Transp Syst 3(3):189–195
4. Broqua F (1991) Cooperative driving: basic concepts and a first assessment of "Intelligent Cruise Control" strategies. In: Drive conference advanced telematics in road transport
5. Yanakiev D, Kanellakopoulos I (1995) Variable time headway for string stability of automated heavy-duty vehicles. In: Proceedings of the 1995 IEEE conference on decision and control. IEEE Xplore, vol 4, pp 4077–4081
6. Yanakiev D, Kanellakopoulos I (1998) Nonlinear spacing policies for automated heavy-duty vehicles. IEEE Trans Veh Technol 47(4):1365–1377

7. Hou D, Gao F, Li K et al (2004) Vehicle longitudinal dynamic model for vehicle collision avoidance systems. J Tsinghua Univ (Science and Technology) 44(2):258–261
8. Li P, Wei M, Hou X (2012) Modeling and co-simulation of adaptive cruise control system. Automot Eng 34(7):622–626
9. Yi K, Hong J, Kwon YD (2001) A vehicle control algorithm for stop-and-go cruise control. In: Proceedings of the intelligent transportation systems. IEEE Xplore, pp 478–482
10. Li S, Li K, Rajamani R et al (2011) Model predictive multi-objective vehicular adaptive cruise control. IEEE Trans Control Syst Technol 19(3):556–566
11. Yu Z (2009) Vehicle theory. China Machine Press, Beijing

Combustion System Design and Development on 1.4L Miller Gasoline Engine

Zhanfeng Wang, Yaodong Liu, Haie Chen, Pinghui Huang, Yunfeng Han and Ziming Liu

Abstract In order to meet fuel consumption regulation China Phase 4 and emission regulation China 6, China FAW (First Automobile Work) developed a 1.4L GDI miller engine. The cam profile and turbocharger were optimized and designed by one-dimensional thermodynamic analysis. A delicate design of high compression ratio, spray target, and intake port was realized by Computational Fluid Dynamics (CFD) analysis of in-cylinder flow, atomization and mixing. The combustion system was finally settled based on optical and thermodynamic engine with the primary target: The specific power is 79 kW/L, the specific torque is 168 N m/L, the Brake-specific fuel consumption (BSFC) is 225 g/kW h and the maximum thermal efficiency is 37.5%.

Keywords Gasoline · Miller cycle · Fuel consumption · Emission · Combustion · Performance development

1 Introduction

With the requirement of energy saving and emission reduction, the regulations of fuel consumption and emission are becoming stricter. Therefore, OEMs launch the new generation gasoline engines one after another [1–5], which has the features of high compression ratio (>11), Miller/Atkinson cycle, CVVL/two-step VVL, 25/35 MPa spray system and optimized intake port and combustion system. In order to meet the fourth phase fuel consumption regulation and national six emission regulation, FAW developed a 1.4L GDI miller engine with 35 Mpa spray system and electronic control valve type turbocharger in consideration of cost reduction. Based on optical and thermodynamic engine with the primary target: The specific power is 79 kW/L, the specific torque is 168 N m/L, the Brake-specific fuel consumption (BSFC) is 225 g/kW h and the maximum thermal efficiency is 37.5%.

Z. Wang (✉) · Y. Liu · H. Chen · P. Huang · Y. Han · Z. Liu
China FAW Corporation Limited General R&D Institute, No. 8899 Dongfeng Street, Changchun, Ji Lin, China
e-mail: wangzhanfeng@faw.com.cn

China SAE (ed.), *Proceedings of China SAE Congress 2018: Selected Papers*, Lecture Notes in Electrical Engineering 574,
https://doi.org/10.1007/978-981-13-9718-9_32

Table 1 Main parameters and performance targets

Main parameters	
Bore (mm)	76.5
Stroke (mm)	75.6
Displacement (L)	1.39
Cylinder number	4
Firing order	1–3–4–2
Compression ratio	11.5
Max. cylinder pressure (MPa)	12
Fuel	RON 92
Power/Speed (kW/r/min)	110/4500–5500
Max. torque/speed (N m/r/min)	235/1600–4350
Specific power (kW/l)	79
Specific torque (N m/l)	168
Min. BSFC (g/kW h)	225
Emission level	China 6a/b

2 Main Parameters and Performance Target of 1.4L Miller Gasoline Engine

This engine will be matched on an A class SUV with curb weight 1475 kg. The main parameters and performance targets are shown in Table 1.

3 Combustion System Design

3.1 Combustion Chamber and Port

Compared with traditional Otto cycle, intake valve of Miller cycle closes earlier, which causes that the real compression ratio is smaller than the expansion ratio. Therefore, the temperature and pressure at TDC are significantly reduced, which is shown in Fig. 1 and so as knock tendency at high load conditions. At the same time, the combustion efficiency is improved and fuel consumption is reduced.

Miller cycle realizes the possibility of improving compression ratio in gasoline engine, which is one of the useful ways to improve the thermoefficiency. However, high compression ratio causes a tall stage on the surface of the piston top, which has a disadvantage on in-cylinder flow and spray design. Therefore, in order to maintain a high compression ratio and a low stage on the piston top, volume of the chamber should be kept smaller. The research shows that, smaller angle between in- and ex-valve and lower chamber height which can reduce the volume, as shown in Fig. 2.

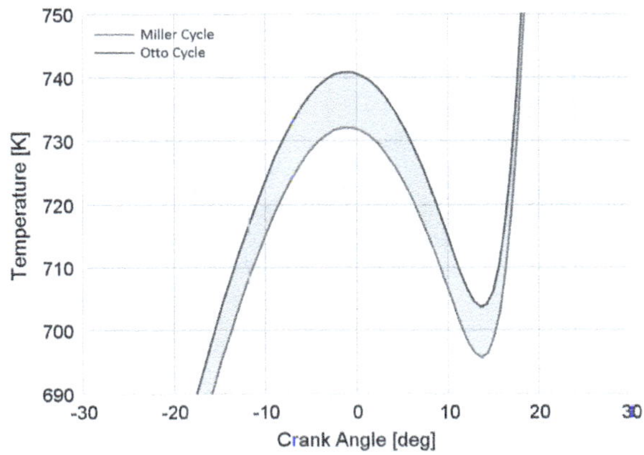

Fig. 1 Comparison of cylinder temperature between Otto and Miller cycle

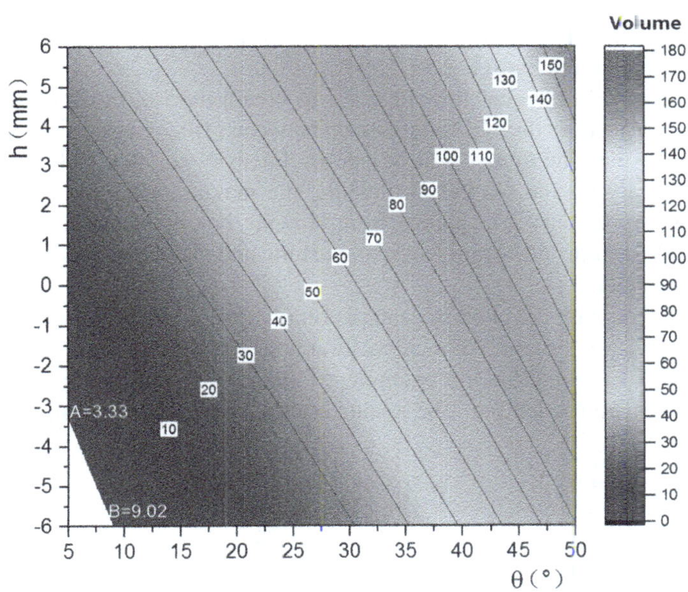

Fig. 2 Relationship among chamber volume, valve angle and chamber height

The design of intake port is very important in Miller cycle engine. Due to the reduced wrap angle and lift of intake valve, the flow strength and Turbulence Kinetic Energy (TKE) at TDC are lower than traditional Otto cycle. In order to compensate this disadvantage, straight intake port and 180° masking structure are designed, as shown in Fig. 3.

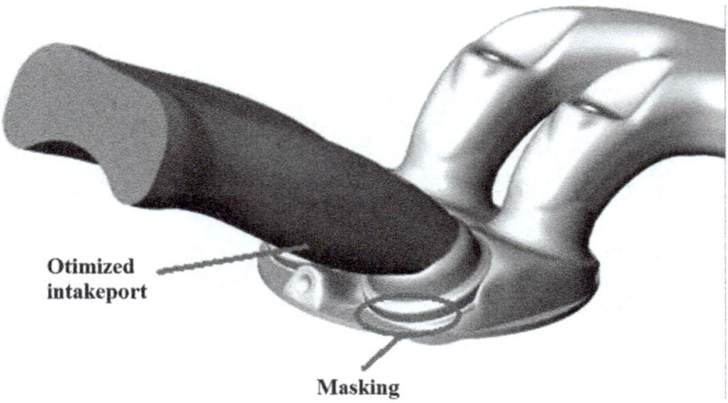

Fig. 3 Design of intake port and masking

Subsequently, the experimental results of different valve lifts are obtained by blowing box of intake port, as shown in Fig. 4. The average flow coefficient is 0.278 and the average tumble ratio is 2.71.

The design of in-cylinder flow target is to obtain the level of traditional valve profile. Figure 5 shows the CFD comparison between Miller engine and Otto engine. It can be seen that, the tumble and turbulence intensity are all equal to even better than the primary engine. At the meanwhile, the maximum tumble ratio is nearly 3.0, which is more than 2.71 gained from blowing box of intake port. The results show that, the design and optimization of masking and piston are very suitable to the combustion system.

In connection with the chamber and piston matching, two schemes are designed for the combustion system. CFD calculation and analysis are then followed. Take 3000 r/min-12 bar condition, for example, evaluation on ω tumble, ω swirl, tumble ratio and TKE are carried out. The results are shown in Fig. 6. It can be seen that,

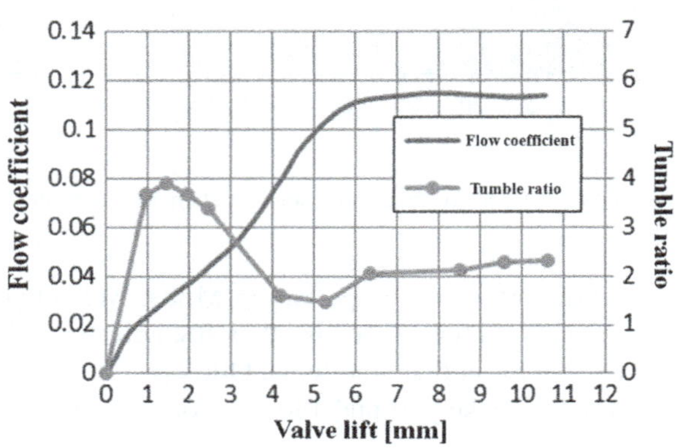

Fig. 4 Result of blowing box of intake port

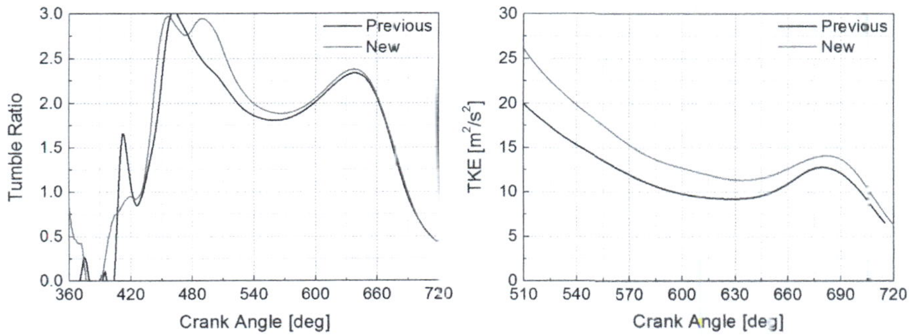

Fig. 5 Comparison of tumble ratio and TKE between previous and new system

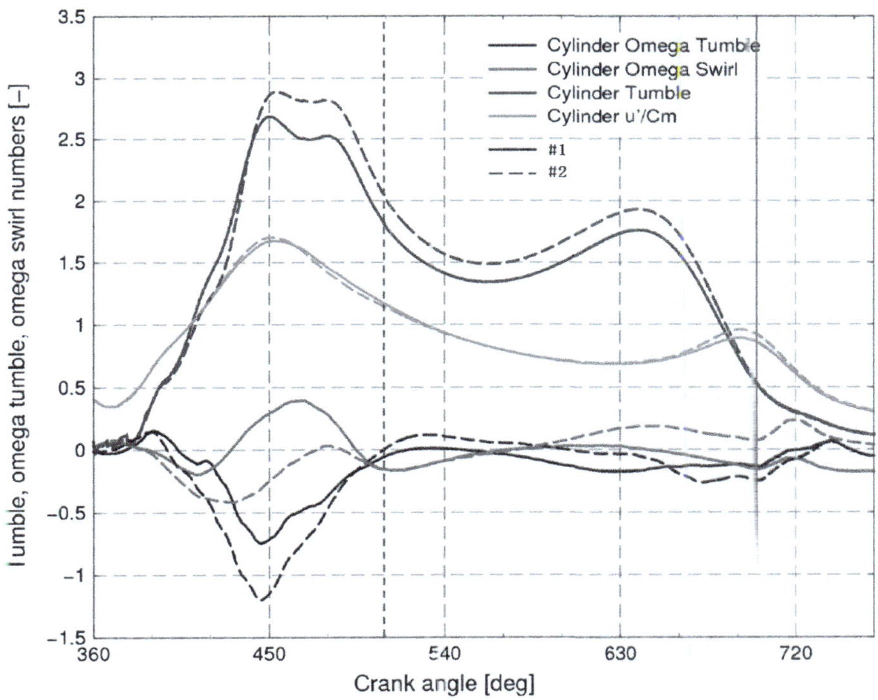

Fig. 6 Comparison between two schemes in CFD results

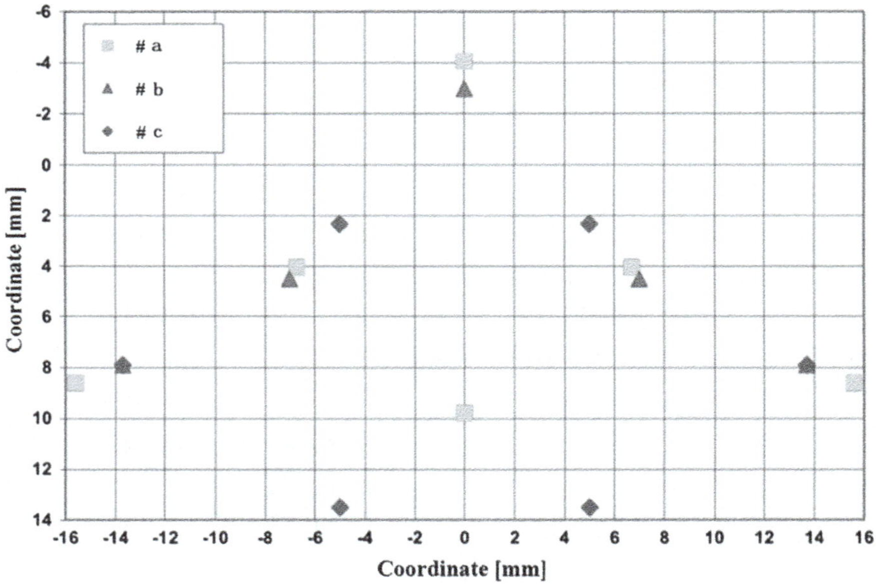

Fig. 7 Schemes of spray target

tumble ratio and TKE of #2 are all better than #1. At 700 °CA, ω tumble and ω swirl are all under the threshold value of 0.5, which means #2 is better than #1. Due to limited space, evaluation of other conditions is omitted. The #2 is chosen as the final scheme for the combustion system.

3.2 Spray Target

Due to the combustion system and existing database, three schemes of spray target on 35 MPa injector system are designed. The details are shown in Fig. 7. Also, CFD calculation and analysis are then followed, as shown in Fig. 8. It can be seen that, for the condition of 2000 r/min-2 bar, the amount of wetting is #b < #c < #a, for the condition of 1500 r/min-WOT (Whole Open Throat), the amount of wetting is #b < #a < #c, which means #b is the best. However, the final choice will be in the optical engine test.

3.3 Valve Profile

Compared with traditional Otto cycle, Miller cycle can reduce pumping loss at part load conditions. As shown below in Fig. 9, pumping loss is negative work at the

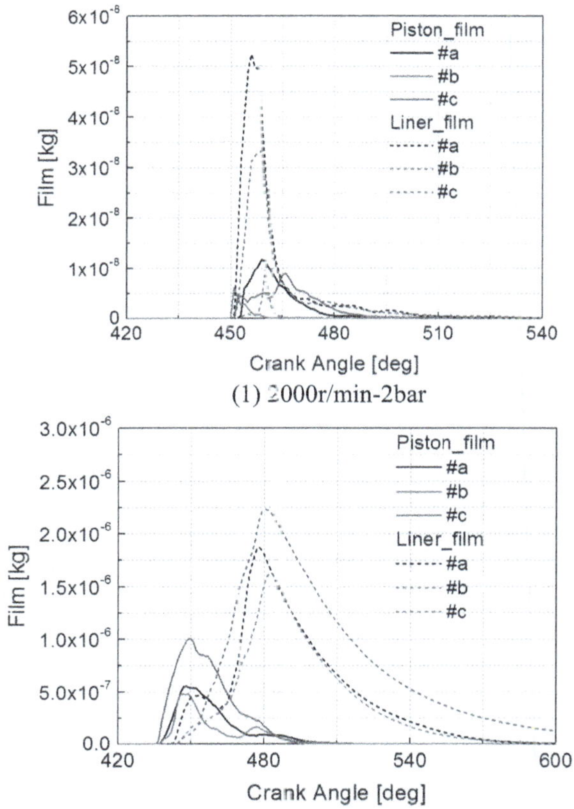

(1) 2000r/min-2bar

(2) 1500r/min-WOT

Fig. 8 Comparison among three schemes of spray target

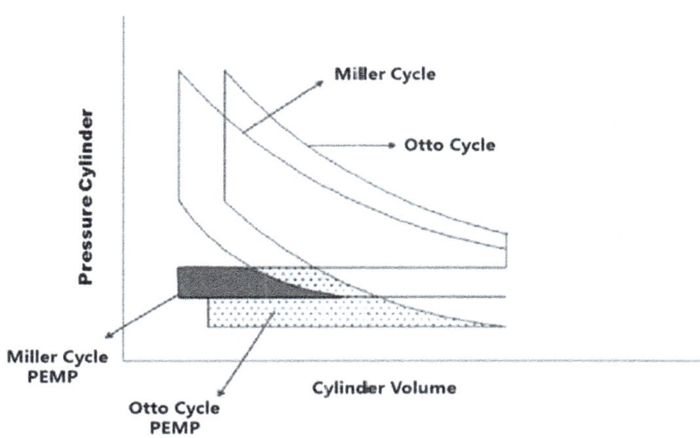

Fig. 9 Comparison of pumping losses between Miller cycle and Otto cycle

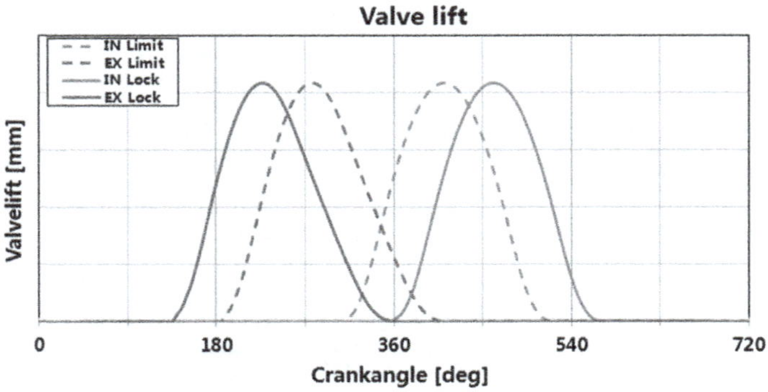

Fig. 10 Profile of in- and ex-valve

lower left corner. The dotted area is less than the blue area, namely pumping loss of Miller cycle is less than Otto cycle.

The temperature can be decreased significantly at the end of compression by using smaller intake valve duration. At the meanwhile, pumping loss at the part load conditions can also be reduced. Miller engine needs larger boost pressure in order to maintain the amount of in-cylinder air. Many works have been done to balance the demand of boost pressure, intake valve duration and intake valve lift with 1D thermodynamic simulation. Finally, the intake valve duration is confirmed of 170 °CA. The detailed profile curve is shown in Fig. 10. Based on the optimization design of the valve profile, the Miller cycle gasoline engine with high mechanical compression ratio can achieve almost the same temperature with traditional Otto cycle gasoline engine. At the same time, the Miller engine can achieve performance target.

4 Combustion System Development

4.1 Development in Optical Engine

The experimental tests are carried out on the above three schemes of spray target. The main evaluation value is spray penetration, amount of spray wetting, and soot emission, etc. Take 2000 r/min-0.2 Mpa and 1500 r/min for example to give the results.

Figure 11 shows the optical results of spray at 2000 r/min-0.2 MPa condition. It can be seen that, the penetration of #a and #c is too large to impinge the cylinder wall and piston, respectively, which is identical with the results of CFD (Fig. 8).

Figures 12 and 13 show the optical test results of single injection and triple injection, respectively, at 1500 r/min-WOT condition. It can be seen that, under the both injection strategies, the soot emission of #a and #c is larger than that of #b, which is

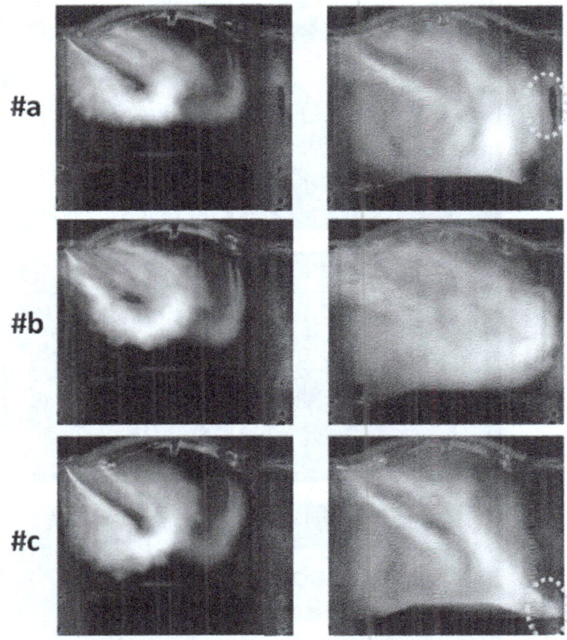

Fig. 11 Comparison of optical results of spray at 2000 r/min-2 bar condition

also identical with the results of CFD. Moreover, the soot emission of triple injection is better than that of single injection. According to the results of CFD and optical engine test, #b is chosen as the spray target scheme.

4.2 Development of Full Load Performance

Optimization of intake and exhaust Variable Valve Timing (VVT) is one of the most effective methods to get more power. For Miller cycle engine, it can achieve higher power with larger VVT phasing and higher boost pressure.

Figure 14 is the comparison of torque with different VVT phasing at 4500 r/min-WOT condition. It can be seen that, higher torque can be achieved in a large area when intake VVT is above 30 °CA.

Figure 15 is the comparison of combustion center with different VVT at 4500 r/min-WOT condition. It can be seen that, temperature at TDC decreases promisingly when advancing intake valve opening phasing, which has a suppressive effect on knocking. Therefore, the combustion center can be further advanced.

Figure 16 is the comparison of boost pressure with different VVT at 4500 r/min-WOT condition. It can be seen that, the degree of Miller cycle is greater with the advance of intake valve opening, thus, the fresh air will be more difficult to be

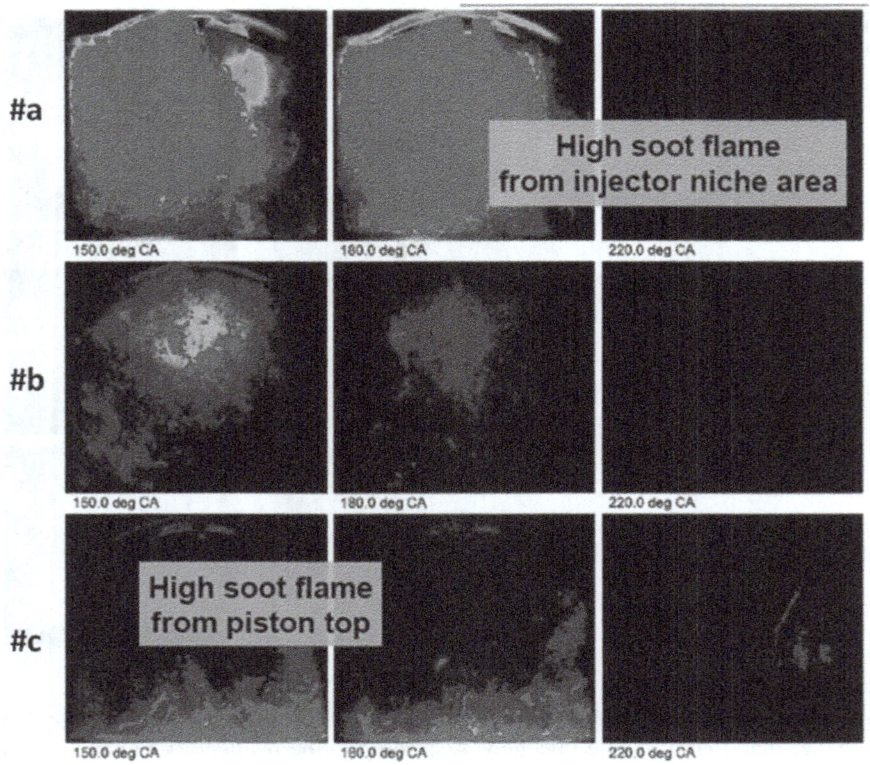

Fig. 12 Comparison of optical results of emission at 1500 r/min-WOT condition, single injection

introduced into the cylinder. Miller cycle engine requires greater boost pressure in order to reach the same torque in Otto cycle engine.

Turbocharger could not provide required boost pressure and enough air in time when engine is changing the operating condition, because it has a lag. The larger the lag is, the worse the acceleration capability appears. Lagging phenomenon is more remarkable because of the small inertia and wide speed range of gasoline engine.

Figure 17 is the comparison of transient capability with different VVT. It can be seen that, with the advance of intake valve opening, the greater the Miller cycle is, the more difficult the fresh air is introduced into cylinder, the higher the boost pressure is required, the worse the transient performance will be.

Figure 18 shows the transient response results of turbocharger. It can be seen that, the transient response of this Miller cycle gasoline engine is even better than traditional Otto cycle gasoline engine due to the advanced control strategy.

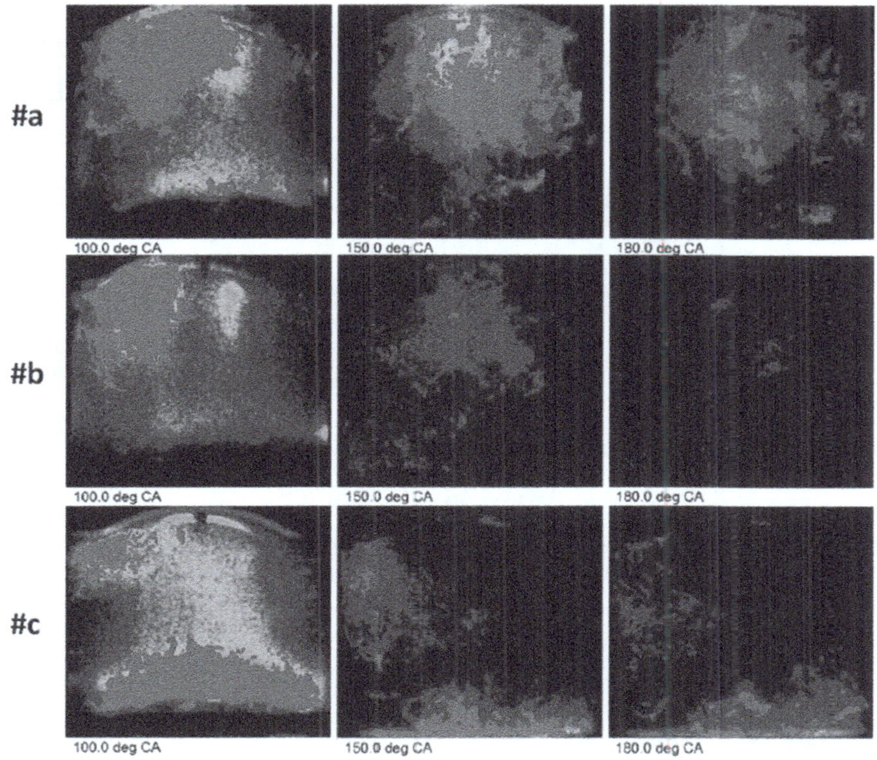

Fig. 13 Comparison of optical results of emission at 1500 r/min-WOT condition, triple injection

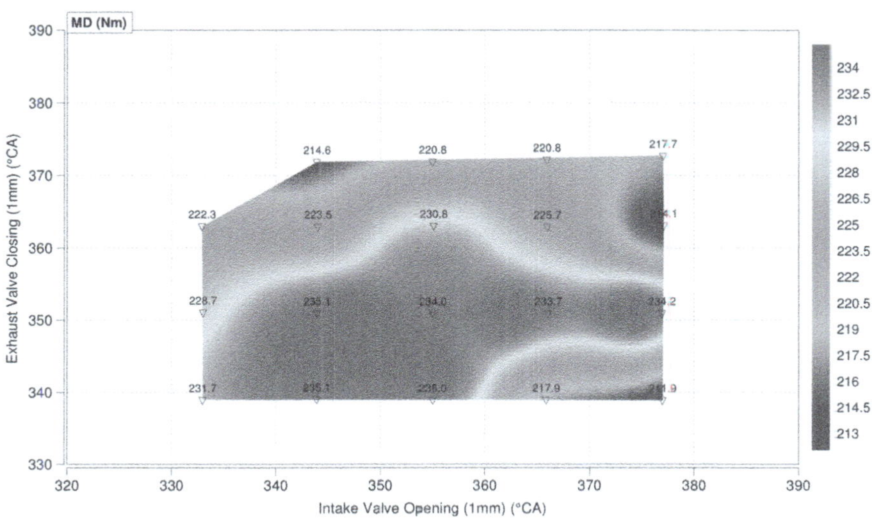

Fig. 14 Comparison of torque with different VVT

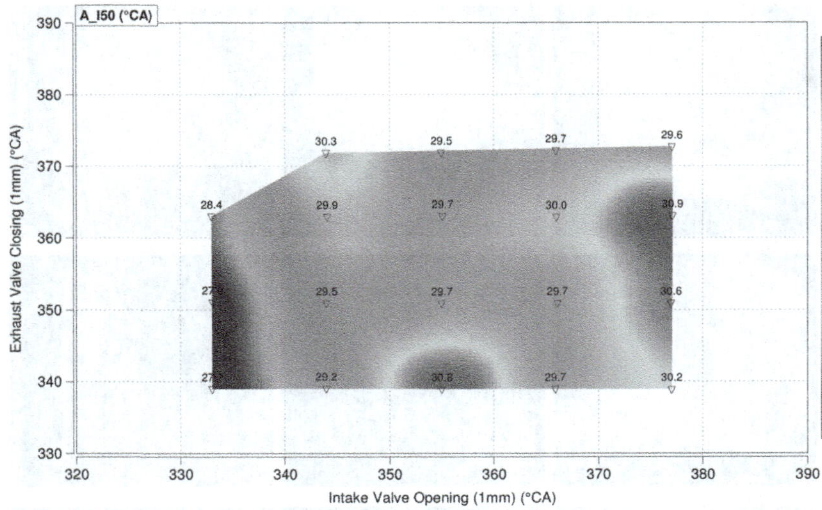

Fig. 15 Comparison of combustion center with different VVT

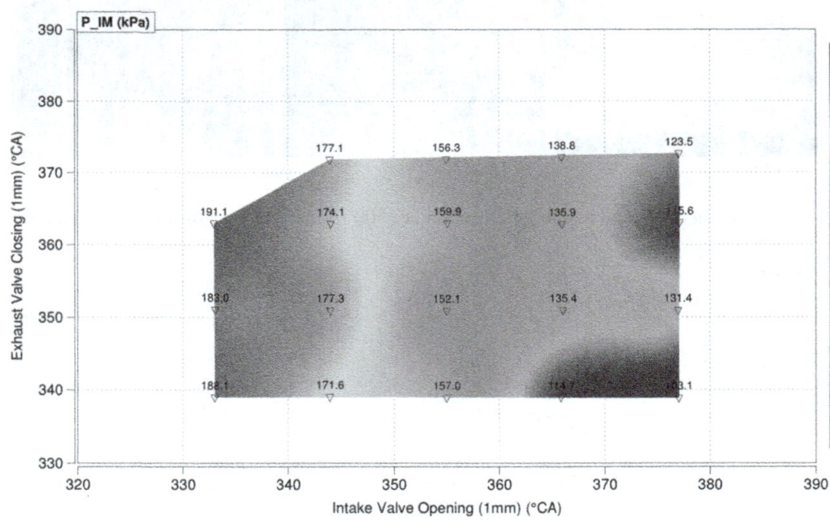

Fig. 16 Comparison of boost pressure with different VVT

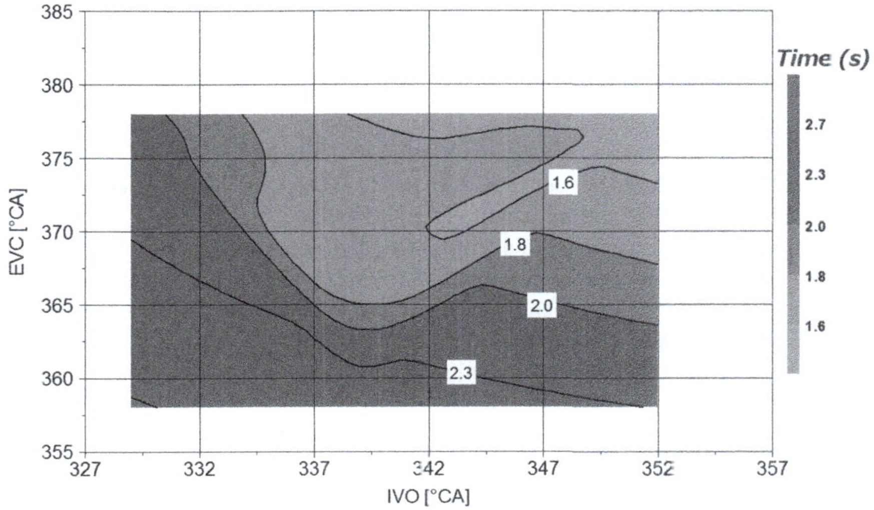

Fig. 17 Comparison of transient capability with different VVT

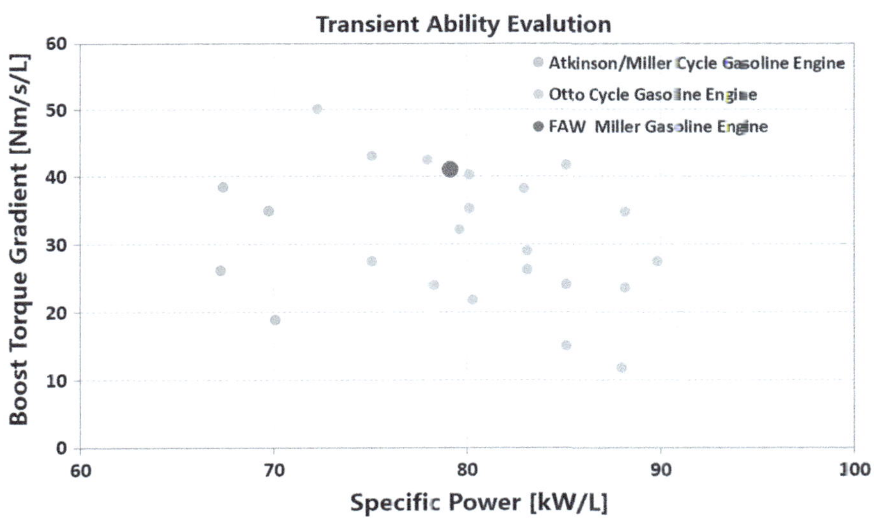

Fig. 18 Transient capability evaluation of Miller gasoline engine

4.3 Development of Emission Performance

The particle number limit is more stringent in China 6a/6b emission regulation. In order to realize lower particulate number emission, 35 MPa injector system is adopted. Figure 19 shows the comparison of particulate number emission with different fuel pressure. It can be seen that, the particle number of 35 MPa injector system can be reduced by 70% compared with that of 25 MPa.

Optimization test is carried out at cold start condition during performance development by adjusting fuel pressure, injection timing, injection ratio, lambda, etc. on the target of (IMEPC) Coefficient Of Variation (COV) of Indicated Mean Effective Pressure (IMEP), HydroCarbon (HC) emission and exhaust Particle Number (PN). Figure 20 is effect of different first injection ratio under the same third injection

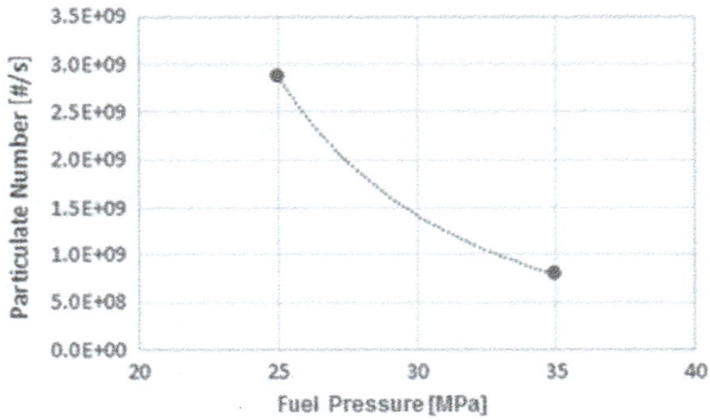

Fig. 19 Particle number comparison with different fuel pressure

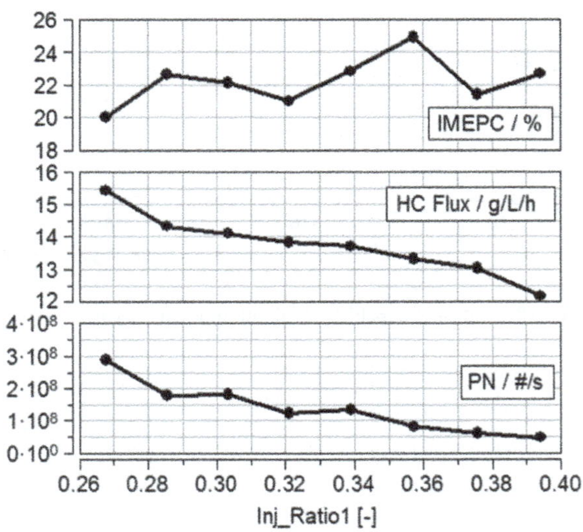

Fig. 20 Performance effect with different first injection ratio

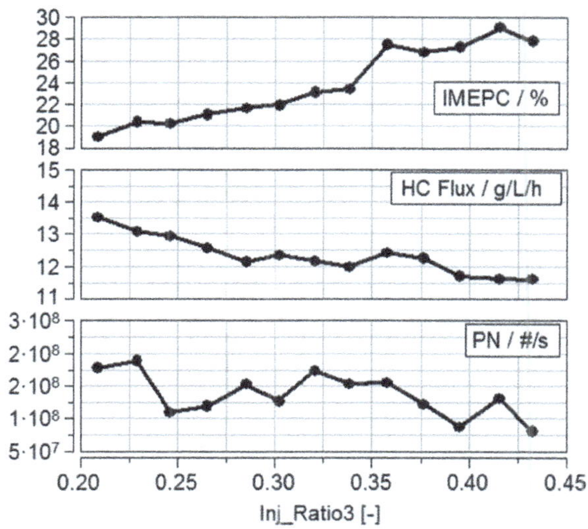

Fig. 21 Performance effect with different third injection ratio

ratio. It can be seen that, particle number and HC emission are decreased with first injection ratio increased.

Figure 21 is performance effect with different third injection ratio. It can be seen that, with the third injection ratio increased, HC emission is decreased and combustion stability is worse.

4.4 Development in Pre-ignition

Pre-ignition should be eliminated in gasoline engine of high power density. Optimization on VVT is first carried out at the condition of 1500 r/min-WOT which is followed by the optimization of spray timing and ratio. Table 2 shows the test

Table 2 Test results of pre-ignition

Times of pre-ignition (1/h)				
Optimization target	Cyl-1	Cyl-2	Cyl-3	Cyl-4
SOI [290,240,125], Ratio [6:3:1]	1.3	0	1.7	0.3
SOI [290,240,90], Ratio [6:3:1]	0	0	6	4
SOI [290,240,115], Ratio [4.5:4.5:1]	2.7	1.3	2.7	0
SOI [290,240,115], Ratio [2:1:1]	1	1	1	1
SOI [290,240,90], Ratio [2:1:1]	0.7	0.3	1	1.3
SOI [300,260,90], Ratio [1:1:1]	0	0	0	0

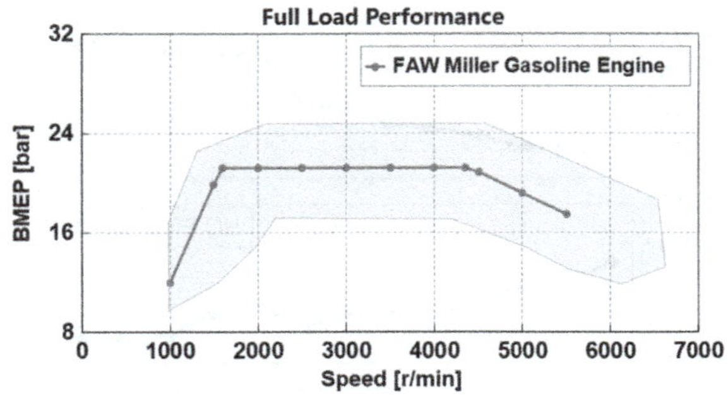

Fig. 22 Full load curve

results of pre-ignition. It can be seen that, when the spray ratio is the traditional one of 6:3:1 and 2:1:1, varying the spray timing could not reduce the frequency of pre-ignition substantially. Along the thought of 6:3:1, keeping 1/10 of the third injection and changing the spray ratio into 4.5:4.5:1 can neither gain the satisfactory result. According to the analysis of spray ratio, it can be seen that, from 6:3:1 and 4.5:4.5:1 to 2:1:1, increasing the percentage of the third injection can reduce the frequency of pre-ignition a little. Therefore, the spray ratio of 1:1:1 is chosen for the test and it turns out that the phenomenon of pre-ignition disappears. The results tell us that, it should reduce or balance the quantity of the first and second injection to avoid the cylinder wetting and the quantity of the third injection is not a key factor if it is in an appropriate SOI window.

4.5 Final Results of Combustion Development

Figure 22 shows the results of performance development and Fig. 21 shows the results of fuel consumption development. The specific power of this FAW Miller gasoline engine can reach 79 kW/L, and the specific toque can reach 168 N m/L. The minimum BSFC can achieve 225 g/kW h, and the maximum thermal efficiency can achieve 37.5% (Fig. 23).

4.6 Comparison with Traditional Otto Cycle Engine

Figure 24 shows the comparison of combustion between Miller cycle and Otto cycle. It can be seen that, ignition timing and combustion center of Miller cycle are more delayed than that of Otto cycle. Due to the delayed ignition timing, ignition delay of Miller cycle is longer, but burn duration is equivalent to Otto cycle.

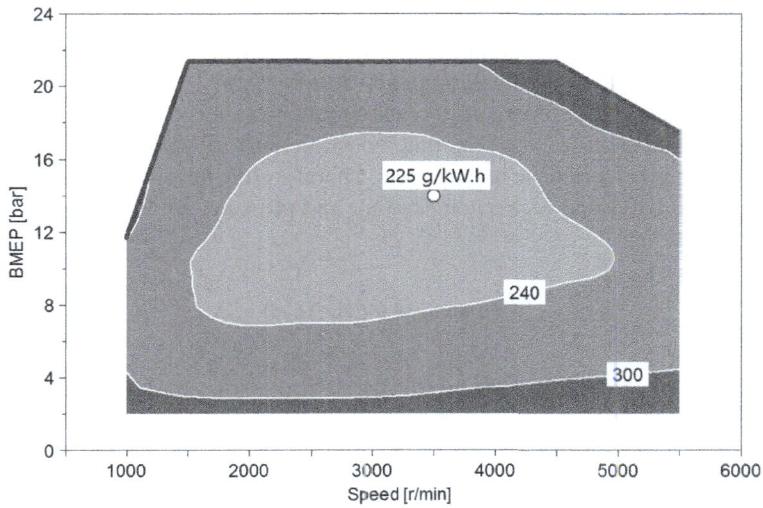

Fig. 23 BSFC map

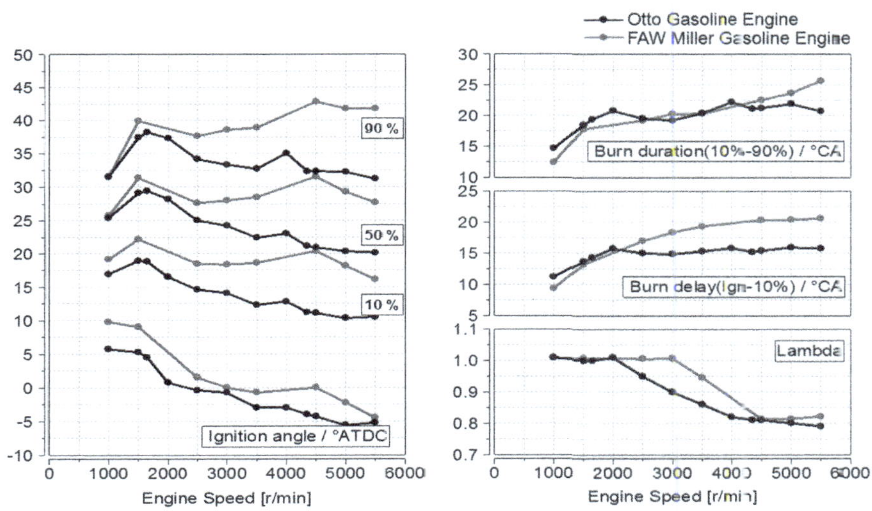

Fig. 24 Comparison of combustion between Miller cycle and Otto cycle

Figure 25 shows the comparison of pumping losses between Miller cycle and Otto cycle. It can be seen that, the pumping loss at WOT of Miller cycle is larger than that of Otto cycle because of the smaller intake valve duration.

Figure 26 shows the comparison of boost temperature and pressure between Miller cycle and Otto cycle. It can be seen that, boost pressure of Miller cycle is higher to achieve the same performance in Otto cycle. Therefore, the inlet pipe system of Miller cycle needs to withstand higher temperature and pressure, so does the capacity of the intercooler.

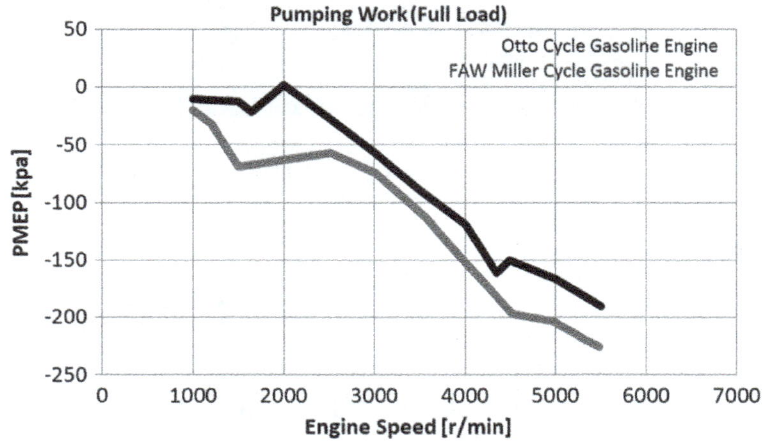

Fig. 25 Comparison of pumping losses between Miller cycle and Otto cycle

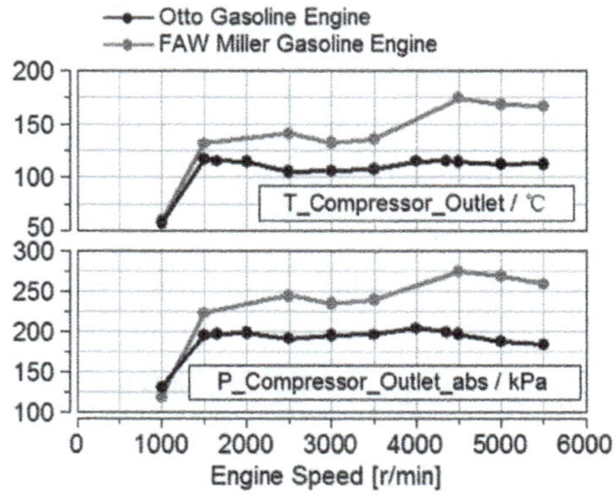

Fig. 26 Comparison of boost temperature and pressure between Miller cycle and Otto cycle

5 Conclusion

This paper discussed the design and development of FAW's 1.4L GDI Miller engine. The intake valve lift of 170 °CA well balances the need of inlet air at WOT and pumping loss reduction at part load. The optimization of intake port, masking and piston design realizes the need of in-cylinder flow and TKE. The power target is achieved just by choosing a normal turbocharger. The results of development are as follows: The specific power is 79 kW/L, the specific torque is 168 N m/L, BSFC is 225 g/kW h and the maximum thermal efficiency is 37.5%.

References

1. Wurms R, Budack R, Grigo M et al (2015) The new Audi 2.0 l engine with innovative rightsizing—a further milestone in the TFSI technology. In: 36th International wiener motor symposium
2. Eichler F, Demmelbauer-Ebner W, Theobald J et al (2016) The new EA211 TSI® evo from Volkswagen. In: 37th international wiener motor symposium
3. Wada Y, Nakano K, Mochizuki K et al (2016) Development of a new 1.5L I4 turbocharged gasoline direct injection engine. SAE 2016-01-1020
4. Murase E, Shimizu R (2016) Innovative gasoline combustion concepts for toyota new global architecture. In: 25th aachen colloquium automobile and engine technology
5. Björnsson H, Adkin P, Johannesson T et al (2016) The new volvo drive-E 3-cylinder engines. In: 25th Aachen colloquium automobile and engine technology

Diesel Engine Cylinder Deactivation Experimental Research on Asymmetric Exhaust System

Peng Zhou, Jianrui Zhang, Zhengxing Zhang, Baoyu Hao and Yun Lu

Abstract The exhaust temperature could be increased to heat DOC and promote DPF regeneration by part of cylinder deactivation in low load condition. In the article, with asymmetric exhaust system and EGR valve, cylinder deactivation can be achieved. The advantages and disadvantages of diesel engine cylinder deactivation are analyzed by experimental and modeling methods. Besides the keys factors are given. For test engine, cylinder deactivation can raise the exhaust temperature by 160 °C. Totally, the fuel consumption drops by 3%. In theoretical extreme situation, the airflow ratio of half deactivated cylinder and normal condition is located from 1.29 to 1.34, which is influenced by booster efficiency in greatest degree. The fuel consumption and NO_x emission increase in some conditions for lower air-fuel ratio and not optimized injection parameters, and further calibration and booster efficiency promotion are helpful.

Keywords Diesel engine cylinder deactivation · Asymmetric exhaust system · Exhaust temperature increase · Key factors

1 Introduction

Engine cylinder deactivation originates from variable displacement, namely the effective displacement changes in engine operation to match different conditions [1]. For one thing, pump loss and friction loss decrease with cylinder deactivation, for another less total superficial area lower heat transfer loss in combustion. The high load of single cylinder improves combustion efficiency. Most HC and CO can be converted in DOC for low mass flow and high temperature of exhaust gas [2, 3].

Common methods that deactivate diesel engine cylinder include: fuel cutoff, valve stop, inner circle forming [4] by leading exhaust gas into cylinder or by asymmetric exhaust structure. Fuel cutoff can be easiest implemented by electronic control, but limited by exhaust temperature government, air-fuel ratio, EGR ratio and fuel

P. Zhou (✉) · J. Zhang · Z. Zhang · B. Hao · Y. Lu
Engine Department, FAW Group Co. Jiefang, 130011 Changchun, China
e-mail: zhoupeng@rdc.faw.com.cn

China SAE (ed.), *Proceedings of China SAE Congress 2018: Selected Papers*,
Lecture Notes in Electrical Engineering 574,
https://doi.org/10.1007/978-981-13-9718-9_34

economy. By valve stop method, the valves of deactivated cylinders close and fuel cutoff. In recent years, many OEMs and consulting firms released related valve control system [5–9]. The merits include fuel-efficient, less throttle loss, high exhaust temperature in cylinder, but structure is complex. The method leading exhaust gas into cylinder can keep temperature in cylinder that is benefit for engine lifespan. The method of inner circle formed by asymmetric exhaust structure combines EGR valve and flap that controls asymmetric degree. It can connect the deactivated cylinder to intake manifold via external EGR. For one thing, asymmetric design prompts all exhaust gas from active cylinder flow into turbo to enhance energy utilization. The compressed fresh charge in deactivated cylinder flows into intake manifold which brings about no fresh air waste, less pump loss and easy implementation.

Some limit factors should be taken into account in cylinder deactivation condition. First is NVH limit that results from low fire frequency and stronger torque vibration of crankshaft [10]. Then highest torque decreases to a great extent for displacement reduction by half; besides turbo efficiency is affected for low exhaust flow mass, which reduces the maximum fuel injection quantity. In the article, asymmetric exhaust structure accompanied by EGR valve is applied to implement cylinder deactivation. Experiment and modeling method is taken to analyze the merits and drawbacks of diesel engine cylinder deactivation. In addition, key factors and improvement methods are given.

2 Cylinder Deactivation Experiment on Asymmetric Exhaust Structure

2.1 Experiment Object

The research object is an 11 L diesel engine with asymmetric exhaust structure, six cylinders (Table 1; Fig. 1).

Table 1 Engine parameters

Fuel feeding system form	High-pressure common rail
Boost form	Waste gate
Maximum net torque	2100 Nm
Injection pressure	200 MPa
EGR form	High-pressure EGR

Fig. 1 Test diesel engine of symmetric exhaust

2.2 *Experimental Devices*

Experimental devices include FEV eddy current dynamometer, FEV fuel consumption meter, Horiba emission analyzer, AVL combustion analyzer, LMS NVH testing equipment, temperature pressure sensor, airflow meter, LAMBDA sensor and CANAPE and so on.

3 Cylinder Deactivation and Experimental Principle

Asymmetric exhaust structure impels exhaust gas in 4-5-6 cylinder to enter the turbo, exhaust gas in 1-2-3 cylinder to enter the turbo and enter the exhaust gas recirculation system, respectively. A custom-designed throttle in 1-2-3 cylinder exhaust manifold accompanied by EGR valve impels whole charge in 1-2-3 cylinder to enter intake manifold via EGR, meanwhile fuel cutoff in 1-2-3 cylinder. In accordance with smoke limit and NVH limit, the operation range of available 3 cylinder deactivation mode is circled. The intake mass flow, fuel consumption and emissions of cylinder deactivation mode and normal mode are recorded at the same speed and torque. Three direction vibration sensors are attached and rotate speed signal from crankshaft speed sensor is divided to measuring instrument to get speed fluctuation.

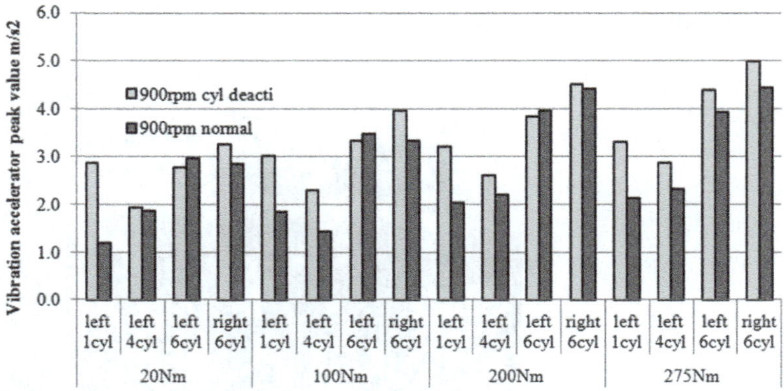

Fig. 2 Vibration comparison

4 Experimental Result and Analysis

4.1 Cylinder Deactivation Operation Range

NVH is monitored to find out proper operation range and protect engine bench. The vibration acceleration amplitude is shown in Fig. 2. The vibration increases as the load increases. When the speed is less than 900 rpm, vibration of cylinder deactivation in the small load is multiplied by several times compared to the normal working, especially at the side of pulley, and altogether cylinder deactivation will result in an increase in vibration.

As shown in Fig. 3, the operation range of cylinder deactivation and normal mode is circled on basis of NVH and smoke limit monitoring. In cylinder deactivation mode, the maximum torque that the engine can reach is quite limited for little intake quantity and low air-fuel ratio. It is hard to control in low speed for vibration, besides the test bench generates resonance at 1100 rpm and thus few tests are carried out.

4.2 Influence of Cylinder Deactivation on Intake Quantity

As the 1-3 cylinder fresh air goes back to the intake manifold through EGR, the fresh gas is only burned in the 4-6 cylinder and then through the turbine, and the same amount of fuel produces the similar exhaust energy. When the cylinder is deactivated, the intake quantity is less than that of the normal mode. Under the same load, the intake manifold pressure is slightly larger, the exhaust back pressure is lower, and the exhaust temperature of the cylinder is higher than that of the normal mode when the cylinder is deactivated. Some conclusions can be drawn from the following conclusions concerning the specific relationship of intake quantity between cylinder

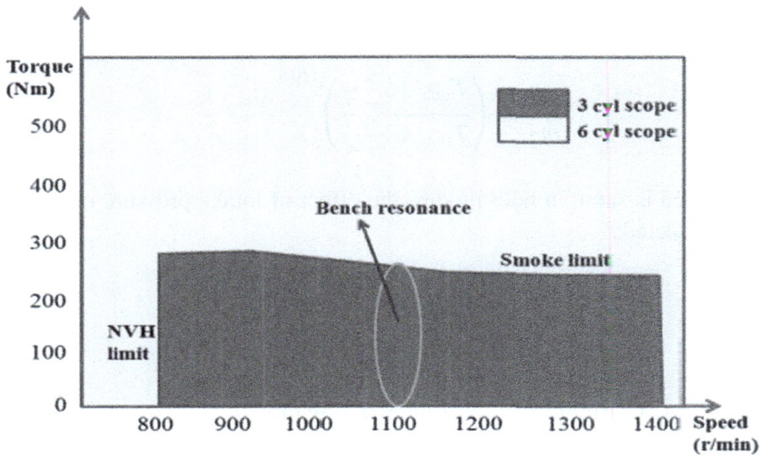

Fig. 3 Operation range

deactivation and normal mode. The shaft work of the compressor is derived from the turbine, which is partly converted into the enthalpy of the air intake. Assuming the same shaft work of compressor, the relationship among the pressure downstream compressor, compressor power and efficiency can be obtained according to formula 1 [11].

$$p_{\text{im}} = \left(\frac{T_{\text{amb}} + \Delta T\eta}{T_{\text{amb}}}\right)^{1.4/0.4} p_{\text{amb}} = \left(\frac{T_{\text{amb}} + \frac{W}{c_p \dot{m}}\eta}{T_{\text{amb}}}\right)^{1.4/0.4} p_{\text{amb}} \tag{1}$$

where p_{im} is intake manifold pressure, T_{amb} is ambience temperature, p_{amb} is ambience pressure, and W and η are compressor energy and efficiency, respectively. The mass flow is given according to mean value model.

$$\dot{m}(t) = \rho_m(t)\tilde{\lambda}_l(p_m, \omega_e)\frac{V_d}{N}\frac{\omega_e(t)}{2\pi} = \frac{p_m}{RT_m}(t)\tilde{\lambda}_l(p_m, \omega_e)\frac{V_d}{N}\frac{\omega_e(t)}{2\pi} \tag{2}$$

where λ is intake efficiency, ρ is density, V_d is displacement, and ω is engine speed. Intake quantity is the function of intake manifold pressure, temperature and engine speed. The displacement of cylinder deactivation is half of normal mode. When the engine speed is same:

$$\frac{\dot{m}_1}{\dot{m}_2} = \frac{\left(\frac{T_{\text{amb}} + \frac{W_1}{c_p \dot{m}_1}\eta_1}{T_{\text{amb}}}\right)^{1.4/0.4} p_{\text{amb}}}{\left(\frac{T_{\text{amb}} + \frac{W_2}{c_p \dot{m}_2}\eta_2}{T_{\text{amb}}}\right)^{1.4/0.4} p_{\text{amb}}} \frac{\frac{W_2}{c_p \dot{m}_2} + T_{\text{amb}}}{\frac{W_1}{c_p \dot{m}_1} + T_{\text{amb}}} \frac{\dot{\lambda}_{l1}}{\dot{\lambda}_{l2}} \cdot \frac{1}{2} \tag{3}$$

When compressor efficiency of both modes is equal to 1:

$$\frac{\dot{m}_1}{\dot{m}_2} = \left(\frac{T_{\text{amb}} + \frac{W_1}{c_p \dot{m}_1}}{T_{\text{amb}} + \frac{W_2}{c_p \dot{m}_2}} \right)^{1/0.4} \frac{\dot{\lambda}_{l1}}{\dot{\lambda}_{l2}} \cdot \frac{1}{2} \tag{4}$$

When speed is same in both modes, the effect of intake pressure on charge efficiency is neglected:

$$\frac{\dot{m}_1}{\dot{m}_2} = \frac{1}{2} \left(\frac{T_{\text{amb}} + \frac{W_1}{c_p \dot{m}_1}}{T_{\text{amb}} + \frac{W_2}{c_p \dot{m}_2}} \right)^{1/0.4} \tag{5}$$

When $W_1 = W_2$, suppose k_1, k_2 is temperature ratio of compressor in both modes, respectively:

$$\begin{aligned} &\frac{W}{c_p \dot{m}_1} = k_1 T_{\text{amb}}, \frac{W}{c_p \dot{m}_2} = k_2 T_{\text{amb}} \\ &\Rightarrow \frac{k_1}{k_2} = \frac{m_2}{m_1} \Rightarrow \frac{k_2}{k_1} = \frac{1}{2} \left(\frac{k_1 + 1}{k_2 + 1} \right)^{1/0.4} \end{aligned} \tag{6}$$

The flow mass ratio is equal to the reciprocal of the compressor outlet temperature ratio. The outlet temperature is related to the ratio of the power to flow of the turbocharger, which is based on the engine system and the turbocharger application. When both the effective power of turbocharger and air charging efficiency are same, 2 times displacement and both the turbocharger efficiency are 1, the flow ratio will be between 1.29 and 1.34 for the ratio upstream and downstream compressor is not more than 2 in general. In fact, even if the exhaust energy is approximately the same under the same operating conditions, the efficiency of the turbine and compressor is different, in that the exhaust flow and the exhaust temperature is different. Besides, lower intake mass results in lower exhaust pressure in deactivation mode; thus, turbine can extract less and the formula of flow ratio approximates: where the efficiency denotes a synthetical one.

$$\frac{k_1}{k_2} = \frac{m_2}{m_1} \Rightarrow \frac{k_2}{k_1} = \frac{1}{2} \left(\frac{\eta_{c1} k_1 + 1}{\eta_{c2} k_2 + 1} \right)^{1.4/0.4} \left(\frac{k_2 + 1}{k_1 + 1} \right) \tag{7}$$

The intake flow and temperature upstream and downstream compressor are measured through experiments. The calculation results of the intake airflow ratio according to temperature ratio upstream and downstream the compressor and turbine efficiency are compared with the experimental results as shown in Table 2. The intake quantity ratio of normal mode to cylinder deactivation has the most direct relationship with the efficiency of the turbocharger. By improving the efficiency of the turbocharger, the air-fuel ratio in the cylinder deactivation mode can be greatly increased

Table 2 Intake flow comparison

Speed/rpm-torque/Nm	Experiment	Calculation
900-100	1.58	1.44
900-200	1.544	1.436
1200-100	1.58	1.62
1200-200	1.585	1.62

Table 3 Experimental mean intake air comparison

Speed/rpm	6 cyl/kg/h	3 cyl/kg/h	Relative decrease (%)
900	266.545	168.33	36.8
1000	293.15	199.375	32
1200	367.75	253. 92	31
1300	405.94	284.315	30
1400	437.4	318.59	27.1
Average	354.157	244.906	31.4

and operation range is extended. In the same operating condition, the average intake quantity in the normal and deactivation model is compared in Table 3.

4.3 *The Effect of Cylinder Deactivation on Turbocharger*

High-performance turbocharger can bring more air intake, increase air-fuel ratio, thermal efficiency, fuel saving and emission improvement. Turbocharger can work in efficient range. In the thesis, the effect of exhaust temperature and flow rate on turbocharger is analyzed in the case of similar exhaust energy.

The temperature, pressure upstream and downstream turbine (the bypass valve unopened) and exhaust flow are measured; meanwhile, the one of the compressors and intake flow are measured. The turbine and compressor efficiency are calculated as shown in Fig. 4. According to the turbine characteristics, the flow range under the fixed turbine speed is narrow to keep high efficiency. When flow decreases or increases in the fixed turbo speed, the efficiency will reduce rapidly in theory.

In the case of the same exhaust energy, the increase of exhaust temperature and decrease of exhaust flow will result in lower turbocharger speed compared to the normal mode. The decrease of the turbocharger speed will result in a significant reduction in the efficiency. In the same way, the compressor does not work in the high-efficiency range, so the compressor power in deactivation mode is lower than that of normal mode as Fig. 5. A reasonable supercharger applied can effectively improve the efficiency when cylinders are deactivated and high-performance supercharger can extend the working condition range. In fact, the cylinder deactivation condition

Fig. 4 Turbine efficiency MAP

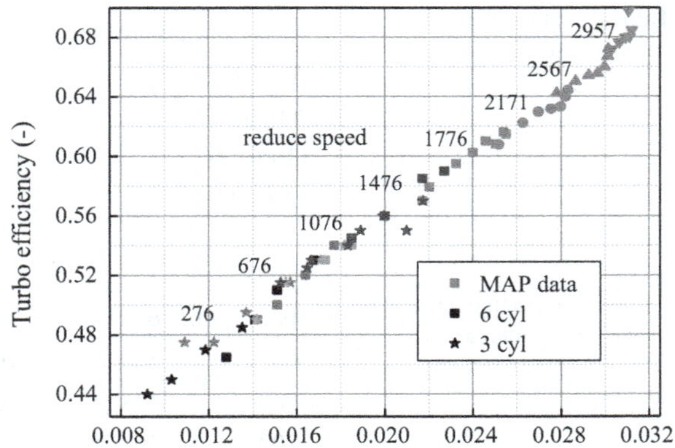

Fig. 5 Compressor efficiency comparison

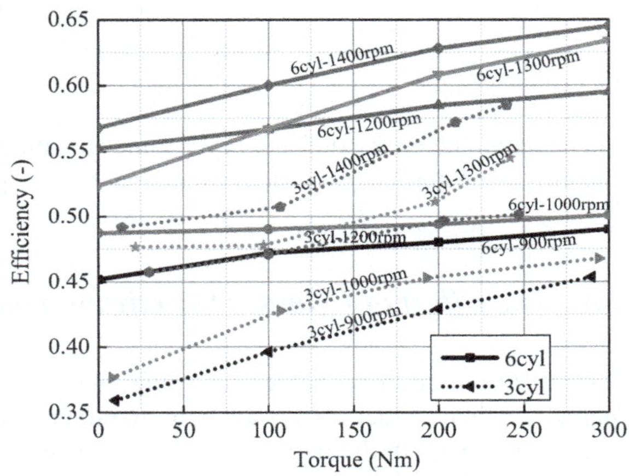

is rarely used if working condition range is limited. When smoke limit locates in air-fuel ratio 1.5, the engine maximum torque with different booster efficiency is shown in Fig. 6.

4.4 *Fuel Consumption Comparison*

As shown in Fig. 7, when the torque output is 0, the fuel consumption of normal and deactivation are quite similar. With the load increase, the fuel consumption in deactivation mode will be gradually higher than normal mode, which is mainly due to the air-fuel ratio decrease.

Empirically, the best combustion center is about 8 degrees after the top dead. Faster burning brings about better constant volume, so the combustion center and duration are mainly investigated. Combustion analyzer shows that the combustion

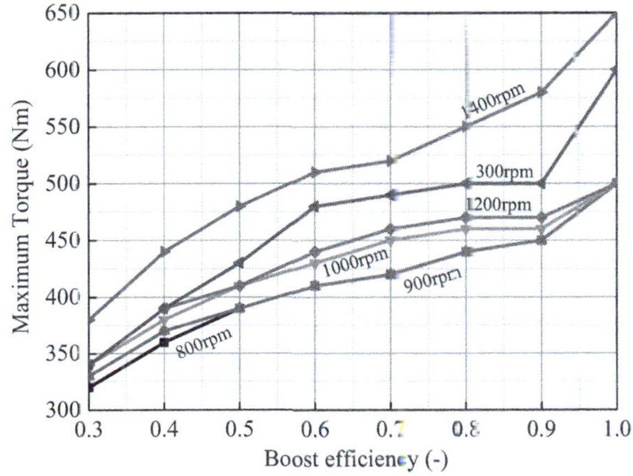

Fig. 6 Turbocharger efficiency and maximum torque

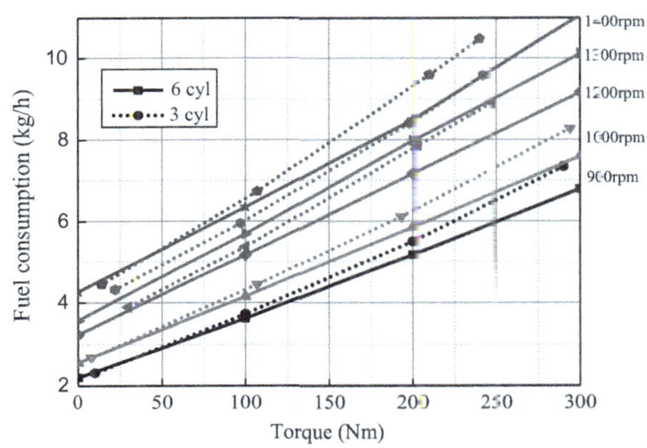

Fig. 7 Fuel consumption comparison

center (average AI50 = 3) near top dead brings more heat transfer loss in normal mode, but the lower air-fuel ratio in cylinder deactivation mode produces long combustion duration, as a result of no optimization to injection parameters (mainly the rail pressure). Basically, the long combustion duration in cylinder deactivation causes additional 5% fuel consumption compared to normal mode. The fuel consumption can be reduced by increasing the intake quantity and adjusting the injection parameters.

4.5 *Energy Distribution*

More fuel injection rate of single cylinder reduces the air-fuel ratio (Table 4) and brings a significant increase of exhaust temperature (Fig. 8) in cylinder deactivation

Table 4 Air fuel ratio comparison

Torque/Nm	AFR cyl Deacti	AFR normal
0	3.562	6.148
100	2.655	4.235
200	1.926	3.188
300	1.632	2.62
Average	2.443	4.047

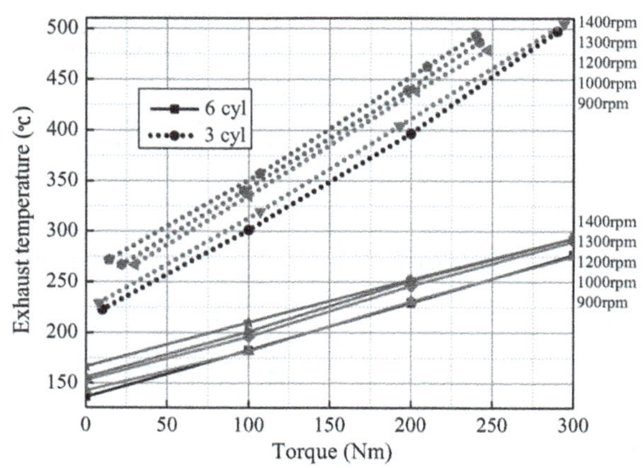

Fig. 8 Exhaust temperature comparison

mode. With load from small to large, the exhaust temperature increases from 80 to 263 °C due to the decrease of the air-fuel ratio. The 1-3 cylinders have low temperature of 70–80 °C due to the absence of the combustion process.

The fuel combustion chemical energy is converted into three parts: indicated work, heat transfer and exhaust energy. Chemical energy is given by product of fuel quality and calorific value of the diesel; the indicated work is calculated by the combustion analyzer; exhaust energy = exhaust flow * specific heat capacity * (the exhaust temperature − intake temperature); pressure ratio heat capacity is the function of the component and the temperature. As only half of cylinders fire, the heat transfer area is half when half of cylinders deactivation, but air-fuel ratio is lower, so the combustion temperature is higher. The result of the comparison of the exhaust energy and heat transfer loss between normal and deactivation mode is as shown in Figs. 9 and 10 respectively.

The exhaust energy in cylinder deactivation mode is higher, but the efficiency of the turbocharger is lower. Obviously, this test turbine is not fit for cylinder deactivation condition. Reasonable turbocharger for cylinder deactivation contributes to better use of exhaust energy. The heat transfer in normal mode is significantly beyond that in deactivation mode as engine speed increases, as a result that heat transfer area is prominent and effect of combustion is weak increasingly. Convective heat transfer of cylinder deactivation is gradually lower compared to normal mode. The saving heat transfer can be converted to exhaust energy or power output.

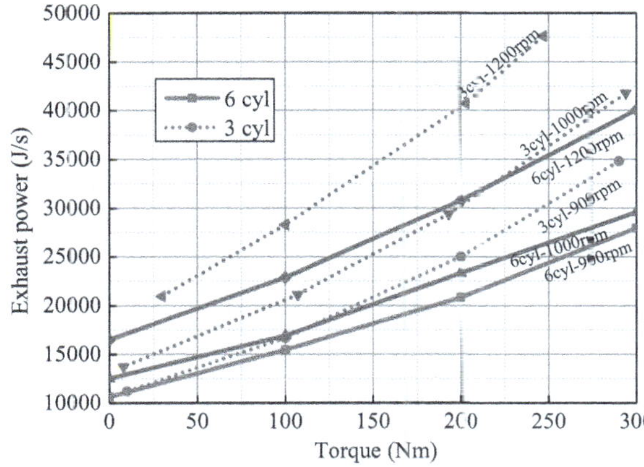

Fig. 9 Exhaust energy comparison

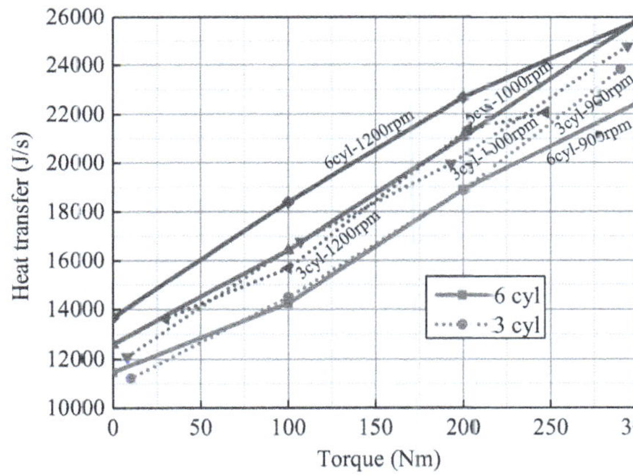

Fig. 10 Cylinder heat transfer comparison

4.6 Effect of Cylinder Deactivation on After Treatment

Injection quantity per cylinder increases, and rich mixture produces higher combustion temperature, which increases NO_x emission significantly, and CO remains roughly the same. The NO_x emission can be improved by optimizing the fuel injection parameters and selecting reasonable operation condition. In addition, more efficient turbine can bring more air intake and expand the operation condition. The problem of row emission increase can be tackled by high-efficiency SCR. The increased exhaust temperature in deactivation mode can be applied in DOC heating, DPF regeneration and SCR heating. The additional exhaust energy is converted to fuel quantity, and the proportion of the fuel consumed is calculated to reach the same exhaust temperature under DPF regeneration condition. The total fuel saving is shown as Table 5 (Fig. 11).

Table 5 Fuel saving effect in generation condition

Speed-torque	Fuel addition (%)	Exhaust energy (%)	Total fuel saving (%)	Speed-torque	Exhaust energy (%)	Total fuel saving (%)	Fuel addition (%)
900-0	−1.1	0.426	1.53	1000-200	6.8	9.22	2.43
900-100	2.7	2.65	−0.05	1000-300	0.9	4.67	3.77
900-200	6.38	6.3	−0.07	1200-0	0.9	4.67	3.77
900-300	7.87	9.27	1.4	1200-100	4.08	8.19	4.11
1000-0	−0.27	2.31	2.59	1200-200	7.2	10.78	3.58
1000-100	3.6	6.68	3.09	1200-300	11.2	13.6	2.4

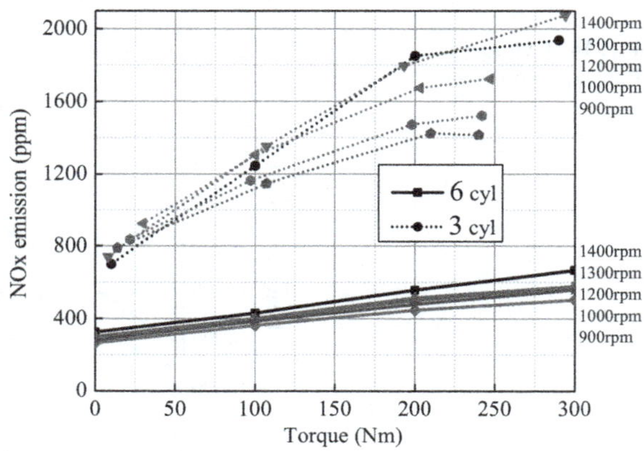

Fig. 11 NO_x emission comparison

As a result, even if the fuel injection parameters are not further optimized, fuel saving is quite obvious compared to normal mode. The high nitrogen oxide emission leads stress to the SCR. A balance between the fuel and urea consumption will be found out to get lowest cost.

5 Conclusions

The merits and drawbacks of cylinder deactivation via asymmetrical exhaust structure are studied, and the key factors that affect the application are found out and conclude:

(1) The exhaust temperature can be improved by 160 °C via cylinder deactivation for after treatment heating, and the fuel consumption reduces by about 3% compared with normal mode in DPF generation or SCR heating.

(2) When the speed is less than 900 rpm, vibration of cylinder deactivation in the small load is multiplied by several times compared to the normal working, especially at the side of pulley.
(3) Theoretically, intake charge ratio between normal and deactivated cylinder mode is 1.29–1.34 and the intake quantity ratio of normal mode to cylinder deactivation has relationship with the efficiency of the turbocharger, exhaust pressure and temperature. By improving the efficiency of the turbocharger, the air-fuel ratio in the cylinder deactivation mode can be greatly increased and the working condition range is extended.
(4) Rich mixture produces higher combustion temperature, so the emission values increase significantly. More efficient turbine can bring more fresh air and increase the air-fuel ratio to reduce NO_x emission.

References

1. Albertson WC, Bolander T, Chen JS et al (2004) Displacement on demand for improved fuel economy without compromising performance in GM S high value engine. In: 2004 Global powertrain conference
2. Pillai S, LoRusso J, Van Benschoten M (2015) Analytical and experimental evaluation of cylinder deactivation on a diesel engine. SAE Paper 2015-01-2809
3. Bharath AN, Yang Y, Reitz RD, Rutland C (2015) Comparison of variable valve actuation, cylinder deactivation and injection strategies for low-load RCCI operation of a light duty engine. SAE 2015-01-0843
4. Douglas KJ, Milovanovic N, Turner JWG, Blundell D (2005) Fuel economy improvement using combined CAI and cylinder deactivation(CDA)—An initial study. SAE 2005-01-0110
5. Kitabatake R, Minato A, Inukai N, Shimazaki N (2011) Simultaneous improvement of fuel consumption and exhaust emissions on a multi-cylinder camless engine. SAE Paper 2011-01-0937
6. Flierl R, Lauer F, Breuer M, Kolbenschmidt Pierburg AG, Hannibal W (2010) Cylinder deactivation with mechanically fully variable valve train. SAE Paper SAE 2010-01-0160
7. Rebbert M, Kreusen G, Lauer S (2008) A new cylinder deactivation by FEV and Mahle. SAE 2008-01-1354
8. Fujiwara M, Kumagai K, Segawa M, Sato R, Tamura Y (2008) Development of a 6-cylinder gasoline engine with new variable cylinder management technology. SAE 2008-01-0610
9. Flierl R, Gollasch D, Knecht A (2006) Improvements to a four cylinder gasoline engine through the fully variable valve lift and timing system UniValve. SAE 2006-01-0223
10. Leone TG, Pozar M (2001) Fuel economy benefit of cylinder deactivation sensitivity to vehicle application and operating constraints. SAE Paper 2001-01-3591
11. Guzzella L, Onder CH (2009) Introduction to modeling and control of internal combustion engine systems, 2nd edn. Springer, Berlin

Combustion and Emission Characteristics of Soybean Oil Biodiesel (SOB) and Gasoline Blends in a Partially Premixed Compression Ignition Engine

Qian Li, Jing Lang and Xianghong Nie

Abstract In the previous studies, partially premixed compression ignition (PPCI) combustion can obtain good combustion controllability, performance and emissions in diesel engines with gasoline. This article focuses on achieving soybean oil biodiesel (SOB)/gasoline PPCI combustion in a single cylinder diesel engine through moderate and low operating loads. Combustion and emission characteristics of SOB and gasoline blends in a PPCI engine have been studied. The results show that low temperature ignition property of gasoline can be improved by doping SOB. The best SOB addition is 30% in volume fraction which has the highest indicated thermal efficiency: 52%, and better emission characteristic compares to pure gasoline. Further experiments on PPCI combustion of 30% SOB/gasoline blend with different fuel injection timings and exhaust gas recirculation (EGR) ratios have been carried out to find out impacts of fuel injection timing and EGR on PPCI combustion. The results show that both the fuel injection timing and EGR can affect performance and emissions of the PPCI engine. For the consideration of engine emissions and efficiency, 12 °CA BTDC fuel injection timing is the best choice. 26% hot EGR (EGR without cooling) and 40% Quasi-EGR (CO_2 as an EGR simulation gas) are the best for low load (0.158 MPa BMEP) and moderate load (0.395 MPa BMEP), separately. When EGR ratio increases to 50%, both hot EGR and Quasi-EGR could reduce NO_x emission more than 70% while the fuel consumption increases slightly.

Keywords Partially premixed compression ignition (PPCI) · Soybean oil biodiesel (SOB) · Injection timing · Exhaust gas recirculation (EGR) · Combustion and emission

---

Q. Li (✉) · X. Nie
Powertrain R&D Institute, Changan Automobile Global R&D Center, Chongqing Changan Automobile Co. Ltd, 401133 Chongqing, China
e-mail: liqian6@changan.com.cn

Q. Li · J. Lang
School of Energy and Power Engineering, Huazhong University of Science and Technology, 430074 Wuhan, China

China SAE (ed.), *Proceedings of China SAE Congress 2018: Selected Papers*, Lecture Notes in Electrical Engineering 574,
https://doi.org/10.1007/978-981-13-9718-9_35

1 Introduction

Along with intensifying of urban environment pollution, the expansion of the energy consumption pressure and increasingly stringent CO_2 emissions, the development of efficient and low-emission engines becomes the urgent matter. As we know, formation processes of NO_x versus soot, HC and CO versus CO_2 are "trade-off" in conventional internal combustion engines. The introduction of homogenous charge compression ignition (HCCI) seemed to be the solution to reduce both NO_x and soot emissions and obtain high efficiency at the same time. However, HCCI also exists narrow operation range problem [1], and another kind of low temperature combustion partially premixed compression ignition (PPCI) can solve the defects of HCCI because of its more controllable ignition [2].

2 Partially Premixed Combustion

PPC can be seen as a combination of the classical diesel combustion and HCCI. In the PPC process, fuel is injected during the compression stroke, and then the fuel is mixed with air to some extent before combustion starts. The fuel should be injected before combustion starts while the fuel injection should not be too early to generate violent combustion once reactions start [3]. Through PPC, we can avoid the formation of the soot produced by over rich mixture in local area inside cylinder and drop the combustion temperature through the control of EGR, while lower combustion temperature and oxygen concentration is good for NO_x emission (Fig. 1).

Kalghatgi studied gasoline PPC concept both in heavy and in light duty engines [5, 6]. Both soot and NO_x emissions reduced while achieving 46% indicated efficiency. Fuel design is also under research combined with new combustion mode of engines [7–9].

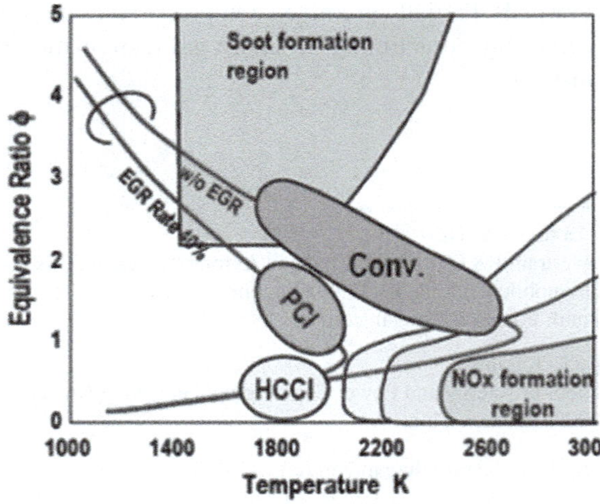

Fig. 1 NO_x-soot map with the locations of conventional Diesel, PPC (PCI), and HCCI combustion [4]

Table 1 FAME composition in SOB

Fame	Molecular geometry	Mass fraction (%)
Linoleic acid methyl ester ($C_{19}H_{34}O_2$)		49.65
Methyl oleate ($C_{19}H_{36}O_2$)		33.93
Methyl palmitate ($C_{17}H_{34}O_2$)		11.60
Methyl stearate ($C_{19}H_{38}O_2$)		4.82

This paper focuses on achieving gasoline PPCI combustion in a single cylinder diesel engine under moderate and low operating loads. Effects of soybean oil biodiesel (SOB), fuel injection timing, and exhaust gas recirculation (EGR) ratio on PPCI combustion were studied (Table 1).

3 Materials and Methods

3.1 SOB Preparation

Biodiesel used in this paper is produced from esterification of soybean oil with methanol. A proper amount of catalyst sodium hydroxide will speed up the reaction. The reaction is shown below:

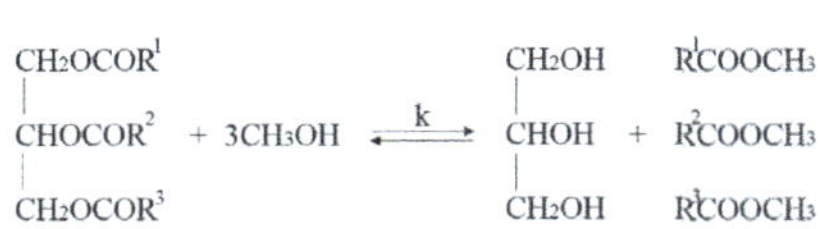

Sheng et al. [10] found that the optimal molar ratio of methanol to oil was 6:1, and 1% mass (oil basis) fraction catalyst amount was effective for esterification reaction. After ten minutes reaction under 60 °C, the biodiesel yield reached 88%. In consideration of excessive catalyst addition could result in a larger degree of emulsion production [11], the amount of catalyst in this paper is fixed at 0.5%. And one hour reaction time under 60 °C reaction temperature and 1000 r/min stirring speed can ensure a complete reaction. Generally, biodiesel production is composed of three aspects: esterification reaction; separation and cleaning; and drying.

Table 2 Main properties of gasoline and SOB

	Gasoline	SOB
Molecular	C_5–C_{12}	FAME
Octane number	93	–
Cetane number	13–17	62
Auto-ignition temperature (°C)	420	–
Boiling point (°C)	25–210	182–338
Reid vapor pressure (kPa@37.8 °C)	55–103	–
Latent heat of evaporation (kJ kg^{-1})	297	–
Low heating value (MJ kg^{-1})	43	38
Dynamic viscosity [CP (°C)]	0.34 (20 °C)	3.5 (37.8 °C)
Density (g mL^{-1})	0.730–0.780	0.842

Finally, the component testing of SOB was carried out through gas chromatography–mass spectrometer (GC-MS) machine Agilent 7890A. The main composition of SOB was fatty acid methyl esters (FAME): linoleic acid methyl ester, methyl oleate, methyl palmitate, and methyl stearate. And SOB's high cetane number has been proved in the primary testing: At least 5% volume fraction SOB is needed to realize compression ignition of SOB and gasoline blends in a PPCI engine without intake heating (Table 2).

3.2 *Experimental Setup and Procedure*

A water-cooling natural inspiration four-stroke single cylinder diesel engine is modified and used in this research. Basic engine properties can be seen in Table 3. Engine speed measuring and engine load loading are realized by Nanfeng CW25-2400/10000 electric eddy current dynamometer equipped with CMU3 dynamometer measuring and control device. The cylinder pressure data record system consists of KISTER 6125B cylinder pressure sensor, 5011B charge-amplifier, magnetoelectric tachometric transducer, ADlink PCI-9112 multi-function data acquisition card, and a self-developed cylinder pressure data acquisition program. Gas exhausts are measured by HORIBA MEXA-584L portable emission analyzer and soot emission is measured by FBY-200 smokemeter. The fuel consumption is measured by FCM-D automatic measuring instrument for fuel consumption and engine speed. The test bed is equipped with an intake heating system and EGR pipelines.

Table 3 Engine properties

Displacement	1.592 L
Stroke	120 mm
Bore	130 mm
Compression ratio	17:1
Connecting rod length	185 mm
Fuel supply advance angle	20 °CA
Injection pressure	18 MPa
Injection timing	−12 °CA
Number of valves	2
IVO	BTDC 12 °CA
IVC	ABDC 38 °CA
EVO	BBDC 55 °CA
EVC	ABDC 12 °CA

4 Results and Discussion

Experiments are conducted with SOB and gasoline blends, and the volume fractions of SOB are 10, 20, 30, 40, 50, 80, and 100%, 10% SOB blend is marked as SOB10, and so on (SOB0: pure gasoline). The results are compared with the baseline of pure gasoline. The optimum SOB blending fraction was selected as the fuel in the following fuel injection timing and EGR researches. Five kinds of injection timing: 4, 6, 8, 10, 14 are compared with the initial injection timing 12°CA BTDC (before top dead center). Two kinds of exhaust gas recirculation (EGR) are analyzed separately in two different loads. 13, 26, 38, 40, 45% EGR without cooling are compared with 0% EGR under 0.158 MPa BMEP; at 0.395 MPa BMEP load, CO_2 was added into the intake air to simulate the EGR effect; and it was referred to as Quasi-EGR. 26, 36, 40, 48, 51% EGR are compared with 0% EGR in this moderate load (Table 4).

Table 4 Engine operation condition

Fuel	Biodiesel/93# gasoline
Engine speed (rpm)	1400
BMEP (MPa)	0.079–0.631
Intake temperature (°C)	35 (Pure Gasoline:50)
Intake pressure (MPa)	0.1
Injection pressure (MPa)	17
Injection timing (°CA BTDC)	4–19
Volume ratio of biodiesel (%)	0–100
Hot EGR ratio at 0.158 MPa BMEP (%)	0–45 (65 ± 5 intake temperature)
Quasi-EGR ratio at 0.395 MPa BMEP (%)	0–51

4.1 Effects of SOB/Gasoline Blends on the PPCI Combustion Mode

SOB addition to gasoline increases the cetane number of the blend fuels, thus allowing the PPCI engine operating steadily at low load. Although 5% SOB addition could make the blends ignition under ambient intake temperature, but emissions of the engine were extremely worse in this condition. So SOB addition is above 10% through the whole research.

As shown in Fig. 2, peak cylinder pressure of combustion increased with SOB addition increasing, and the phase of peak cylinder pressure of combustion moved close to the top-dead-center (TDC), while there existed a little difference that the peak pressure reduced a bit at higher loads when the fraction of SOB exceeded 40%. The reason is that concentrated heat release tended to decrease when the percentage of gasoline in fuel blends reduced, because heat release of gasoline is violent and rapid when combustion starts. The more SOB addition, the higher percentage of diffusion combustion, so the peak pressure decreased. Heat release occurred earlier with the increase of SOB addition. And heat release of pure SOB (SOB100) was close to TDC. The maximum cylinder temperature point got close to TDC gradually with the increase of SOB addition. When the engine operated at low load, the higher the percentage of SOB, the higher the cylinder temperature. However, the cylinder temperature was close to each other except for pure gasoline (SOB0) under moderate and high loads. The reason is that combustion was more complete and closer to TDC with higher SOB addition fuels under low load. Relatively, the differences of complete combustibility among different fuels became smaller under higher operation loads.

Figure 3 gives curves of crank angle of 10% of total heat release (CA10: ignition timing), crank angle of 50% of total heat release (CA50: combustion phase center), crank angle of 10–90% of total heat release (combustion duration), and the maximum pressure rise rate varies with the change of loads.

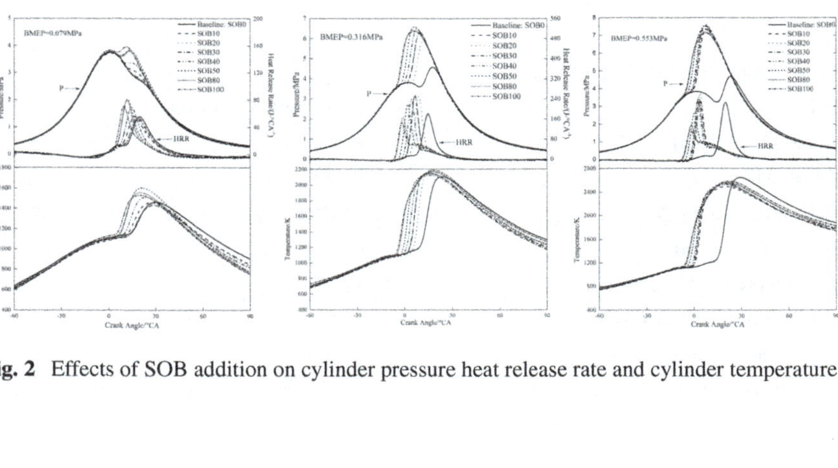

Fig. 2 Effects of SOB addition on cylinder pressure heat release rate and cylinder temperature

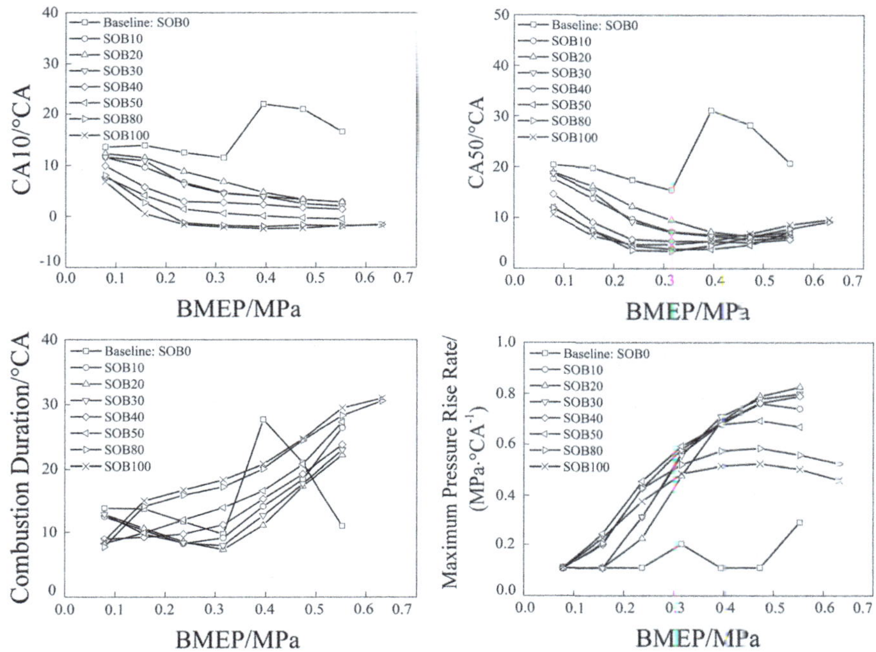

Fig. 3 Effects of SOB addition on combustion parameters

As a whole, ignition point and combustion phase of SOB/gasoline blends got close to TDC, gasoline ignition and combustion properties at low load with ambient temperature intake gas were improved by SOB, but it was easy to generate knocking with load increasing, and had bad impacts on high load condition expanding which can be easy to see from the maximum pressure rise rate figure.

Figure 4 gives curves of emissions of different SOB/gasoline blends under different loads. HC (hydrocarbon) and CO emissions reduced with the increase of SOB addition. This is because that SOB has improved ignition property of gasoline at low intake temperature, combustion phase center was advanced, the maximum combustion temperature increased, and the combustion efficiency of fuel has been enhanced. Besides, the ignition delay period reduced, fuel distribution around cylinder wall became less, the generation of HC from 'cylinder effect' and 'gap effect' decreased. Oxidation rate of intermediates CO of hydrocarbon fuels were speeded up with SOB addition. Also, longer residence time of CO provided sufficient oxidation time for CO, so the emission of CO reduced. However, CO emission increased at high loads because of the huge cycle fuel injection quantity which led to more combustion intermediates.

In general, NO_x emission increased with the increase of SOB addition. The reason is that with increasing of SOB proportion, combustion phase center was advanced, and the ignition delay period became short which led to concentrated heat release and

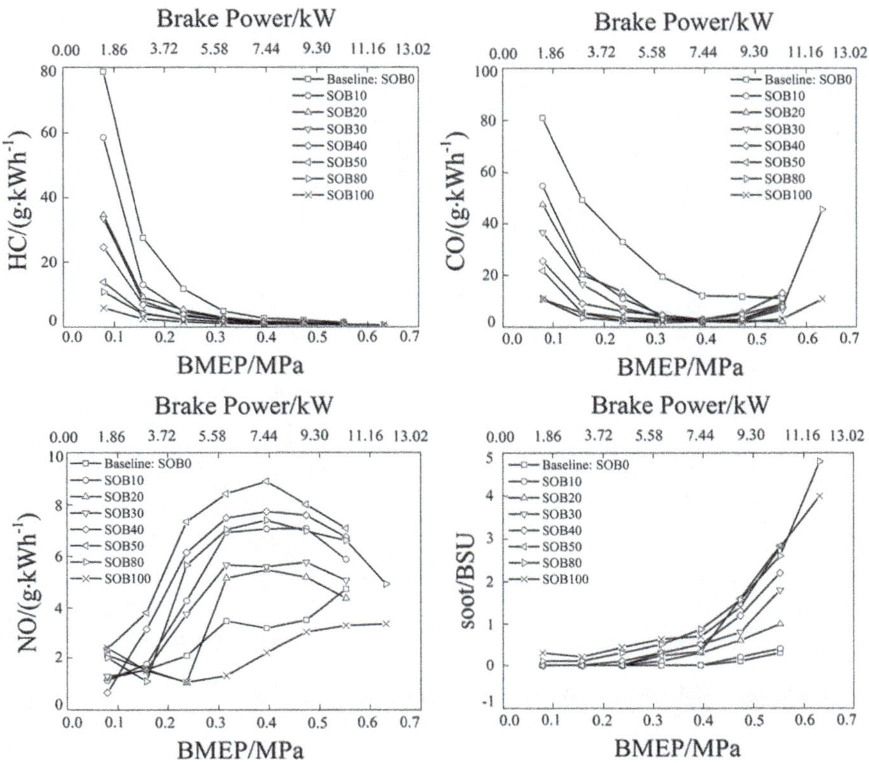

Fig. 4 Effects of SOB addition on emissions

high cylinder temperature. These promoted the generation of NO_X. With the increase of load, the amount of fuel circulation injection increased, formation of mixture in cylinder increased, burning rate became fast, heat releasing is concentrated, isopycnal of combustion was high in this case, and the rise of combustion temperature was faster, all those led to NO_X increased sharply, also knock was easy to occur under high loads. When the engine operated up to higher load, the increase of SOB fraction led to drop of NO_X. This is because the higher the addition of SOB, the lower the percentage of premixed combustion, so concentrated heat release weakened, and local high-temperature area diminished.

Soot emission increased with increasing of SOB addition. The higher the SOB addition, the higher the percentage of diffusion combustion. As we know, the formation of soot is in connection with local low oxygen concentration and high temperature. It was easy to form over rich mixture in local areas under high proportion SOB condition. Compared with diesel, SOB is a kind of alternative fuels with oxygen content which has significant function in decreasing soot.

Figure 5 gives comparison curves of fuel consumption and thermal efficiency of different SOB/gasoline blends. Due to pure gasoline's long ignition delay period,

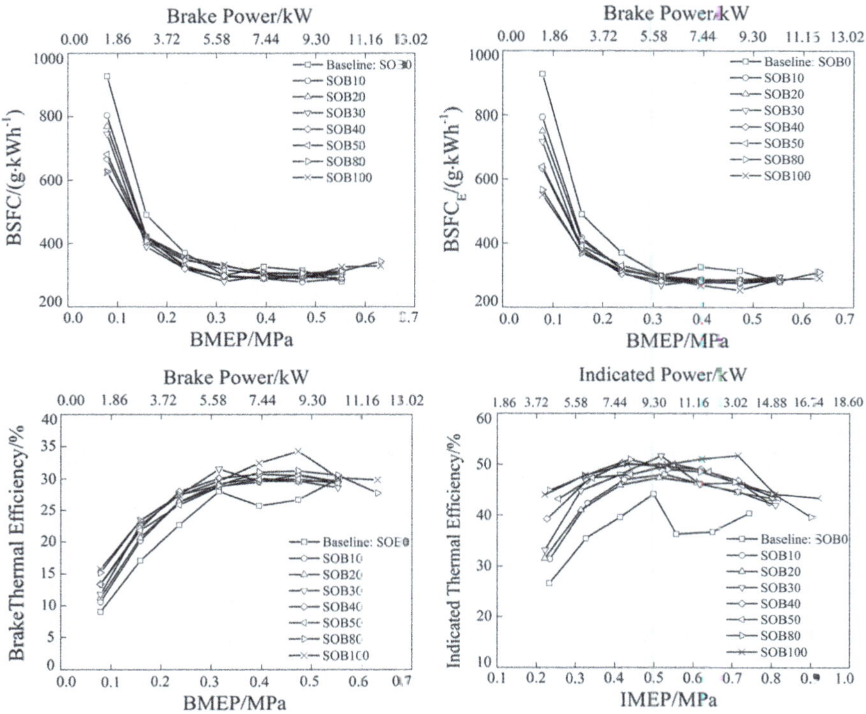

Fig. 5 Effects of SOB addition on fuel economy characteristics

gasoline without SOB addition had lower brake thermal efficiency and higher BSFC (brake specific fuel consumption) under low and moderate loads. BSFC under low and moderate loads could be decreased with advancing of combustion starting point and combustion phase through doping biodiesel into gasoline. When the engine operated up to high loads, the differences of BSFC among different SOB/gasoline blends became small. As the low heating value of SOB is lower than gasoline, the equivalent fuel consumption will be more accurate. Equivalent brake specific fuel consumption ($BSFC_E$) can be calculated by the following formula:

$$BSFC_E = (BSFC \times H_L)/H_{LG} \quad (1)$$

$$H_L = [(100 - B) \times \rho_G \times H_{LG} + B \times \rho_B \times H_{LB}]/[(100 - B) \times \rho_G + B \times \rho_B] \quad (2)$$

$BSFC_E$	Equivalent Brake Specific Fuel Consumption, g/kWh
BSFC	Brake Specific Fuel Consumption, g/kWh
H_L	Low heating value of mixed fuel; MJ/kg
H_{LG}	Low heating value of gasoline; MJ/kg

H_{LB}	Low heating value of biodiesel; MJ/kg
B	Volume fraction of biodiesel; %
ρ_G	Density of gasoline; kg/m^3
ρ_B	Density of biodiesel; kg/m^3.

$BSFC_E$ decreased with the increase of SOB addition under the whole operation range. SOB is a kind of clean renewable bioenergy and regarded as an excellent substitute for diesel fuel. From this thesis's fuel economic characteristic, we can see its great potential in fuel market. From the indicated thermal efficiency figure, we can see clearly that biodiesel was an effective way to improve thermal efficiency of PPCI engine with high compression ratio. The maximum indicated thermal efficiency was 52% of pure SOB and SOB30 PPCI combustion.

For the consideration of engine emissions and efficiency, SOB30 may be the best choice: In the load of BMEP 0.316 MPa, HC emission is 2.2 g/kWh, CO emission is 3.6 g/kWh, NO_x emission is 5.6 g/kWh, soot emission is 0.3BSU (Bosch Smoke Unit), equivalent brake specific fuel consumption is 268 g/kWh because of 31.2% brake thermal efficiency (52% indicated thermal efficiency). Compared to pure gasoline PPCI, HC emission reduced 56%, CO emission reduced 82%, NO_x emission increased 60%, soot emission increased 0.3BSU (still very small), equivalent brake-specific fuel consumption reduced 11%. In order to obtain better performance and emission characteristics of the PPCI engine, further researches of fuel injection timing and EGR on PPCI combustion mode were carried out based on SOB30 fuel.

4.2 Effects of Fuel Injection Timing on the PPCI Combustion Mode

Figure 6 gives curves of emissions of SOB30 PPCI combustion mode vary with loads under different fuel injection timings. From the figure, we can see that brake-specific emission of unburned HC increased with delaying of fuel injection timing. The reason is that, with delaying of fuel injection timing, start timing of combustion and combustion phase delayed, the maximum combustion temperature in cylinder dropped, and ignition delay period was extended. Atomization of gasoline and SOB were better, proportion of premixed mixture in cylinder increased, and ignition temperature was low, which led to the increase of incomplete combustion fuels. In addition, temperature near the engine cylinder wall was low, which made nearby combustion mixture quenched, also more fuel spread to cylinder wall because of longer ignition delay period which would result in incomplete combustion. Also brake-specific emission of CO increased with delaying of fuel injection timing. The reason, on the one hand, is decreasing of the maximum combustion temperature in cylinder and, on the other hand, is shorter resistance time of CO in cylinder caused by ignition delay.

Unlike HC and CO emissions, brake-specific emission of NO_x decreased with delaying of fuel injection timing. This is because the formation ratio of NO_x reduced with the decrease of the maximum combustion temperature in cylinder.

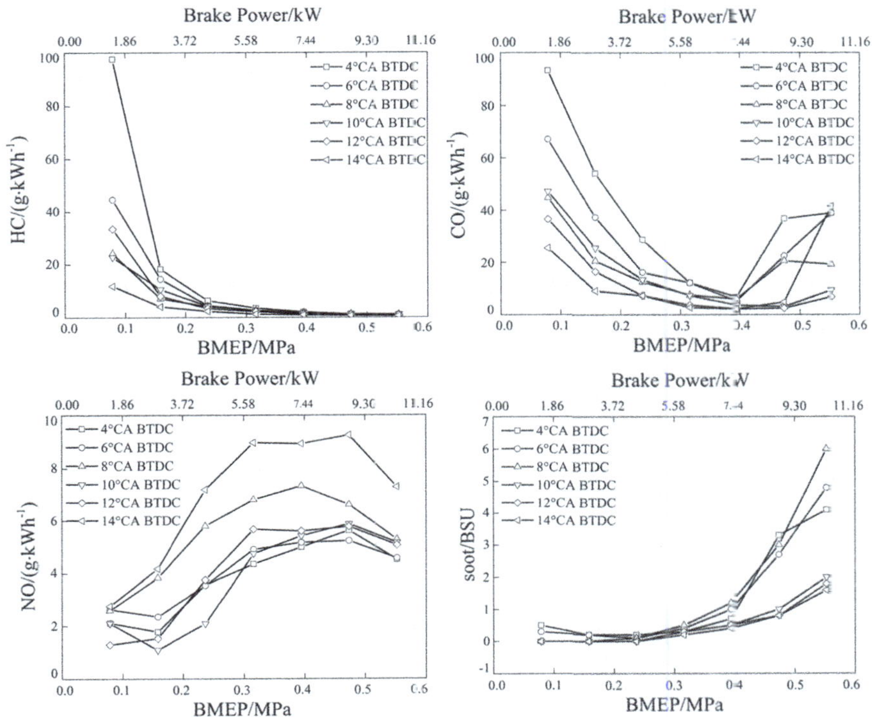

Fig. 6 Effects of fuel injection timing on emissions

NO_X increased with the increase of loads with the same fuel injection timing under low and moderate loads. And there was a drop at high loads. This is because with load increasing, the amount of fuel circulation injection increased, which led to rise of combustion temperature in cylinder. NO_X was also in an upward trend; however, oxygen-enriched area decreased with huge cycle fuel injection quantity at high load which led to a little drop of NO_X emission.

As shown in Fig. 6, soot emission remained low under low and moderate loads, and there was a dramatic increase when the engine operated up to high loads. Under low loads, atomization of gasoline and biodiesel in cylinder was better, proportion of premixed mixture in cylinder was high, and percentage of diffusion combustion in the whole combustion process was low; in addition, SOB improved oxygen content of modified gasoline, which had significant effects on decreasing soot emission. When the engine operated up to high loads, cylinder temperature was high, ignition delay period was short, anoxic area increased relatively, so soot emission increased.

Figure 7 gives the curves of fuel economic parameters of SOB30 fuel vary with loads under different fuel injection timings. Obviously, delaying of fuel injection timing would cause increasing of BSFC and $BSFC_E$ under low and moderate loads. It is mainly because start timing of combustion and combustion phase center was far-

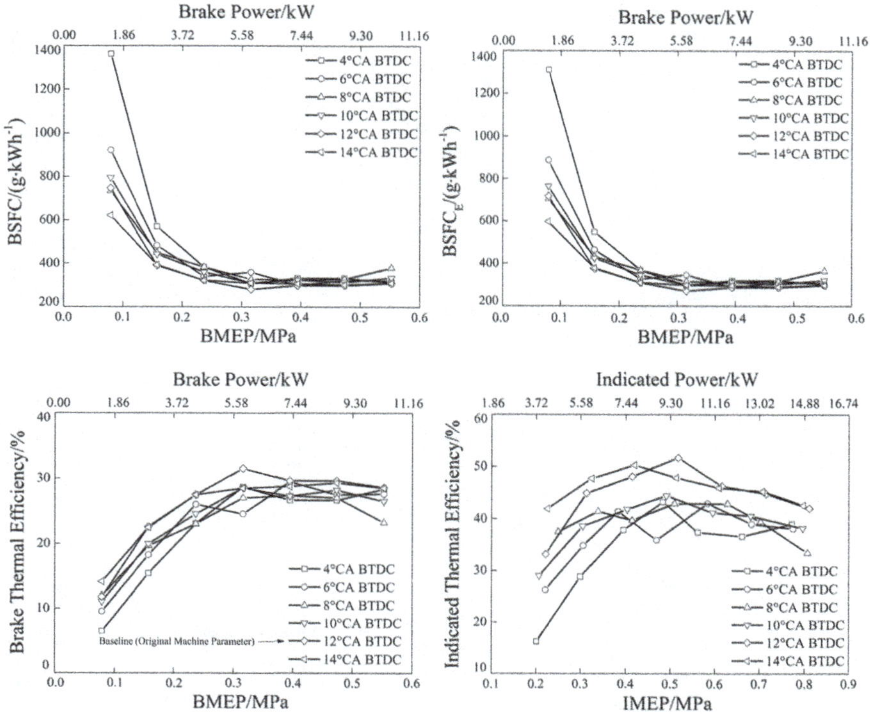

Fig. 7 Effects of fuel injection timing on fuel economy characteristics

ther away from TDC when fuel injection timing was delayed, emissions of unburned HC and CO were higher, and burning efficiency was lower. However, with increasing of load, when brake mean effective pressure (BMEP) of the engine was greater than 0.4 MPa, differences of BSFC and $BSFC_E$ among different kinds of fuel injection timings became small. This is mainly because, as fuel injection timing delayed, ignition delay period was extended, more premixed mixture was formed before ignition, percentage of premixed combustion in the whole combustion process was higher, heat releasing was more concentrated, and work-efficiency was higher. Instead, if fuel injection timing was advanced, start timing of combustion occurred before TDC, the combustion process did more negative work, the engine operated unstably, and these caused similar fuel consumption.

For the consideration of engine emissions and efficiency, 12 °CA BTDC fuel injection timing is the best choice, which means that the original machine fuel injection timing is just the best fuel injection timing for SOB30 PPCI combustion mode.

4.3 *Effects of EGR on the PPCI Combustion Mode*

In order to further reduce NO_x emission of the PPCI engine, researches of hot EGR which means EGR without cooling at low load and Quasi-EGR using CO_2 addition in intake air at moderate load have been developed in this paper.

There are several ways to influence engine ignition and combustion through EGR: firstly, the thermal effects to intake charge. Hot EGR can increase intake temperature after mixing with cold intake air before being sucked into the engine, which is good for the charge inside cylinder to achieve compression ignition; secondly, the diluting effects to the whole charge. There are three aspects of diluting effects: The oxygen concentration is reduced; higher specific heat capacity of CO_2 and water vapor than air lead to lower temperature at the end of compression stroke; the ingredients in EGR gas can take part in chemical reaction partially which may cause spontaneous combustion and ignition timing changes.

Considering high intake temperature may cause reduction in intake charge, research of hot EGR has been developed under low operating load: 0.158 MPa BMEP. Intake temperature was controlled in the range: 65 ± 5 °C. Calculation method of EGR in this paper is CO_2 concentration method:

$$EGR = \frac{CO_{2in}}{CO_{2out}} \times 100\% \tag{3}$$

CO_{2in} CO_2 concentration in intake gas; volume %
CO_{2out} CO_2 concentration in exhaust gas; volume %.

As shown in Fig. 8, when increased hot EGR ratio, unburned HC emission increased slightly, CO emission increased sharply, NO_x emission reduced obviously, and soot emission increased. When EGR ratio increased, cylinder temperature was lowered, complete combustion degree decreased and which caused higher HC emis-

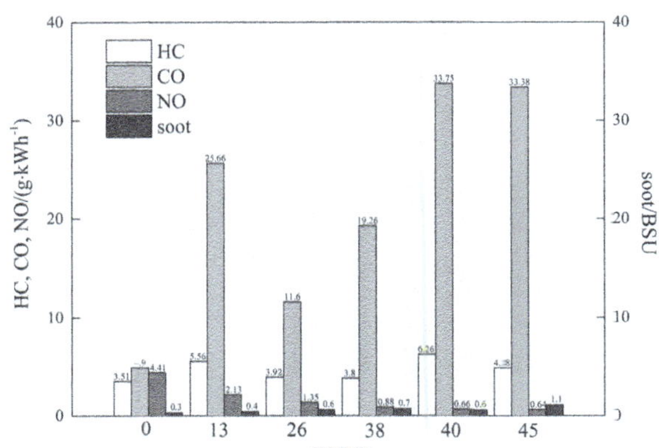

Fig. 8 Effects of hot EGR on emissions

sion. Because of delayed start timing of combustion and combustion phase center, residence time of CO in cylinder became short and CO emission increased.

Great effects of hot EGR on reducing NO_x emission were the result of many ways as mentioned above: First of all, engine exhaust gas contained CO_2, CO, NO_x, UHC, which could lower oxygen concentration in cylinder; secondly, engine exhaust gas contained polyatomic gases H_2O and CO_2, which increased specific heat capacity of mixtures, then cylinder combustion temperature reduced; thirdly, engine exhaust gas contained NO_x, which inhibited the formation of NO_x.

Although EGR could delay start timing of combustion and combustion phase center, but start timing of combustion and combustion phase center was still close to TDC as high temperature of intake gas, which resulted in higher percentage of diffusion combustion, and local excessively rich mixture zones in cylinder still existed. With EGR rate increasing, start timing of combustion and combustion phase delayed, the maximum combustion temperature decreased, soot oxidation rate reduced, so soot emission increased.

Figure 9 gives bar charts of fuel economic parameters of SOB30 fuel vary with different kinds of hot EGR ratios. Before hot EGR reaching up to 40%, fuel consumption and thermal efficiency remained basically unchanged, and there was a little drop of thermal efficiency when EGR ratio exceeded 40%. This is because, introducing EGR into the cylinder reduced the temperature in cylinder, but it also made the ignition delay period longer, so the formation of the mixture became more before fuel burning, the proportion of premixed combustion increased, heat release was more concentrated, work-efficiency was higher. To some extent, these made up for the decrease of cylinder temperature and oxygen concentration.

From above, we can see that hot EGR could reduce NO_x emission as much as 86% while the fuel consumption increased only 10%. Take all emissions and fuel consumption into consideration, 26% hot EGR at the load of 0.158 MPa BMEP is the best: HC emission is 3.9 g/kWh, CO emission is 11.6 g/kWh, NO_x emission is 1.3 g/kWh, soot emission is 0.6BSU, the equivalent brake specific fuel consumption is 419 g/kWh, and the brake thermal efficiency is 20% (37% indicated thermal efficiency). Compared to operation condition without hot EGR, HC emission increased

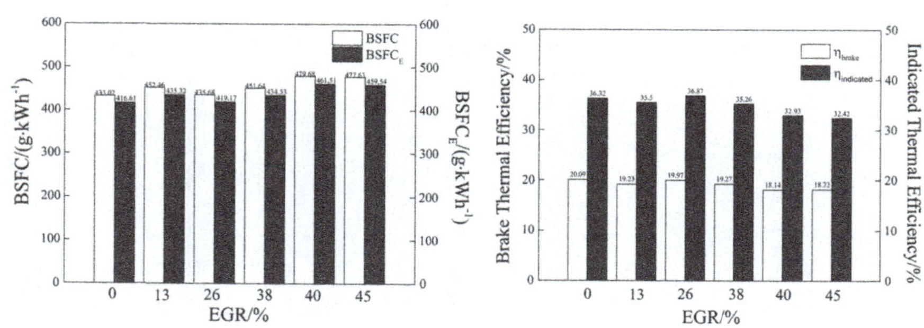

Fig. 9 Effects of hot EGR on fuel economy characteristics

10%, CO and soot emissions doubled, NO_x emission reduced 70%, the fuel consumption remained unchanged.

The principle of EGR to reduce NO_x is mainly about the inert gas such as CO_2 in the exhaust gas increases specific heat capacity of the mixture, and it reduces the oxygen concentration in the mixture at the same time. Thereby, the reaction rate and combustion temperature are decreased, which inhibit the formation of NO_x. Because the components of the engine emissions are very complex, single inert gas to simulate EGR can be used in the basic research of EGR. CO_2 is one of the main components of the engine exhaust gas, which has the nature of the inert gas, not involved in the formation and decomposition reaction of NO_x. In addition, CO_2 is a greenhouse gas, so in this paper we use CO_2 as an EGR simulation gas to achieve the purpose of reduce the impact on the atmosphere, and also reduce NO_x emission.

Research of Quasi-EGR has been developed under moderate operating load: 0.395 MPa BMEP. Intake temperature was controlled in the range 35 ± 1 °C. Calculation method of Quasi-EGR in this paper is CO_2 concentration method as mentioned above.

As shown in Fig. 10, when increased Quasi-EGR ratio, unburned HC emission changed little, CO emission increased sharply, NO_x emission reduced obviously, and soot emission changed little.

Even Quasi-EGR ratio increased, cylinder temperature was still high enough to allow HC oxidation under moderate load. CO emission increased because of delayed start timing of combustion and combustion phase center, residence time of CO in cylinder became short. Quasi-EGR also showed great effects on reducing NO_x emission. Firstly, intake CO_2 decreased oxygen concentration in cylinder; secondly, CO_2 increased specific heat capacity of mixtures, then cylinder combustion temperature reduced. Although Quasi-EGR could cause a lower oxygen concentration relatively, but as the engine was operating under moderate load the cylinder temperature was not as high as high load. So soot emission changed little and remained in low level.

Figure 11 gives bar charts of fuel economic parameters of SOB30 fuel vary with different kinds of Quasi-EGR ratios.

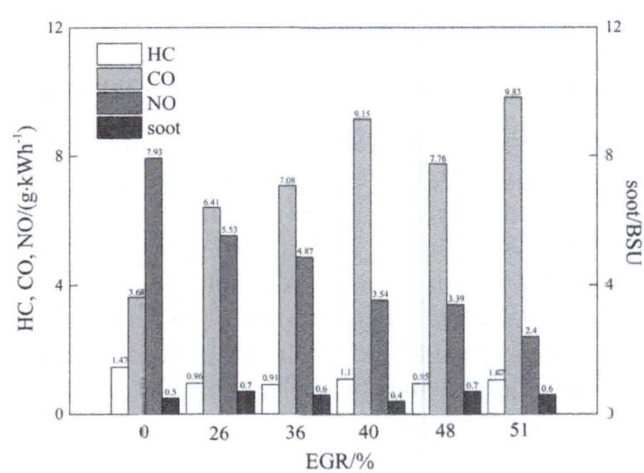

Fig. 10 Effects of Quasi-EGR on emissions

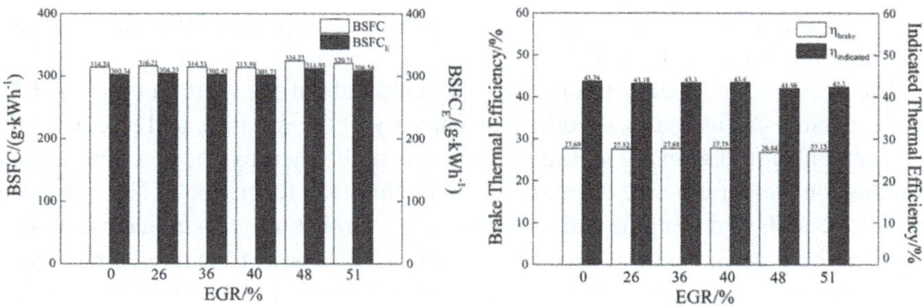

Fig. 11 Effects of Quasi-EGR on fuel economy characteristics

Fuel consumption and thermal efficiency remained basically unchanged under different Quasi-EGR ratios. This is because, introducing Quasi-EGR into the cylinder reduced the oxygen concentration in cylinder, but it also made the ignition delay period longer, so the formation of the mixture became more before fuel burning, the proportion of premixed combustion increased, heat release was more concentrated, and work-efficiency was higher. To some extent, these made up for the decrease of oxygen concentration.

From above, we can see that Quasi-EGR could reduce NO_X emission as much as 70% while the fuel consumption increased only 2%. Take all emissions and fuel consumption into consideration, 40% Quasi-EGR at the load of 0.395 MPa BMEP is the best: HC emission is 1.1 g/kWh, CO emission is 9.1 g/kWh, NO_X emission is 3.5 g/kWh, soot emission is 0.4BSU, the equivalent brake specific fuel consumption is 301 g/kWh, and the brake thermal efficiency is 28% (43% indicated thermal efficiency). Compared to operation condition without Quasi-EGR, HC emission reduced 27%, CO emission doubled, NO_X emission reduced 56%, soot emission reduced 20%, and the fuel consumption remained unchanged (even a little drop: 0.2%).

5 Conclusions

Based on the investigations carried out the following conclusions can be arrived:

(1) Compared to intake gas should be heated to above 50 °C to realize compression ignition of $93^{\#}$ gasoline, 10% SOB addition blend fuel can achieve stably operation of engine under the condition of 35 °C intake gas, which can improve the engine's operating stability and economy characteristics under low loads.

(2) SOB can reduce HC and CO emissions of gasoline PPCI dramatically, and also its soot emission can be controlled nearly zero with proper biodiesel fraction. For the consideration of engine performance and emissions, 30% SOB addition into gasoline may be the best choice. Compared to pure gasoline at 0.316 MPa BMEP load, HC emission reduced 56%, CO emission reduced 82%, NO emis-

sion increased 60%, soot emission increased 0. 3BSU (still very small), and equivalent brake-specific fuel consumption reduced 11%.

(3) Fuel injection timing is one of the main parameters affecting the engine performance. With the optimal fuel injection timing, both the efficiency and emissions of the engine could be improved. Delaying fuel injection timing can reduce NO_x; however, this will result in increased HC, CO, and soot emissions. The more advanced the fuel injection timing, the lower the fuel consumption. For the consideration of engine performance and emissions, 12 °CA BTDC fuel injection timing is the best choice, which means that the original machine fuel injection timing is just the best fuel injection timing for SOB30 fuel.

(4) Hot EGR could reduce NO_x emission as much as 86% while the fuel consumption increased only 10%. Take all emissions and fuel consumption into consideration, 26% hot EGR at the load of 0.158 MPa BMEP is the best. Quasi-EGR could reduce NO_x emission as much as 70% while the fuel consumption increased only 2%. Take all emissions and fuel consumption into consideration, 40% Quasi-EGR at the load of 0.395 MPa BMEP is the best.

References

1. Yang Y, Dec J, Dronniou N et al (2012) Boosted HCCI combustion using low-octane gasoline with fully premixed and partially stratified charges. SAE Paper, 2012-01-1120
2. Manente V, Johansson B, Tunestal P (2009) Partially premixed combustion at high load using gasoline and ethanol, a comparison with diesel. SAE Paper, 2009-01-0944
3. Manente V (2010) Gasoline partially premixed combustion. Lund Institute of Technology, Sweden
4. Okude K et al (1907) Premixed compression ignition (PCI) combustion for simultaneously reduction NO_x and soot in diesel engine. SAE Paper 2004-01-1907
5. Kalghatgi G, Risberg P, Angstrom H (2006) Advantages of fuels with high resistance to auto-ignition in late-injection, low-temperature, compression ignition combustion. SAE Paper, 2006-04-3385
6. Kalghatgi G, Risberg P, Angstrom H (2007) Partially pre-mixed auto-ignition of gasoline to attain low smoke and low NO_x at high load in a compression ignition engine and comparison with a diesel fuel. SAE Paper, 2007-01-0006
7. Adams CA, Loeper P, Krieger R et al (2013) Effects of biodiesel–gasoline blends on gasoline direct-injection compression ignition (GCI) combustion. Fuel 111:784–790
8. Sun W, Du J, Guo L et al (2015) The effects of gasoline/diesel blend fuels on the combustion and ultra-fine particulate emission of compression ignition engine. 2015SAECCE, Shanghai
9. Li Q, Tan L, Huang G et al (2016) Experimental research on PPCI combustion and emissions with DTBP/gasoline blend. Chin Intern Combust Engine Eng 37(3):6–12
10. Sheng M, Guo D, Zhang D (2002) Study on preparation of bio-diesel with soybean oil. China Oils Fats 27(1):70–72
11. Freedman B, Pryde EH, Mounts TL (1984) Variables affecting the yields of fatty esters from transesterified vegetable oils. JAOCS 61(10):1638–1643

The Deblocking Function Optimal Design for a Continuous Variable Valve Lift System

Mingxi Liang, Cheng Liu, Wenxin Cai, Cheng Meng, Jian Wan and Liu Hong

Abstract In order to optimize the deblocking characteristics for a hydraulic motor-driven CVVL system, we abstracted the working process of internal shaft actuator and investigated the parameters which could significantly affect system deblocking characteristics by simulation and test methods. The results show that: (1) Increasing the locking pin clearance can extend the time length of rotation movement under the same unlocking speed, which is able to improve system deblocking performance. (2) Replacing step unlocking signal by slope unlocking signal will restrain rotation and unlocking movement simultaneously, while rotation movement is affected more obviously due to the differences in oil chamber volume. (3) Stick issue has been solved after lower limit of locking pin clearance was adjusted to 0.5°, and the signal slope was well restricted. The optimal design method mentioned in this paper can be popularized in similar systems and brings more conveniences on the deblocking study and critical parameters definition.

Keywords Continuous variable valve lift · CVVL · Deblocking characteristics · Optimal design

In recent years, the over-expanded cycle technology represented by Miller and Atkinson cycle is widely used in small-displacement turbocharged engines. In order to obtain different sets of ideal valve lifts and durations simultaneously, continuous variable valve lift (CVVL) is refocused by more and more OEMs [1].

Typically, the VEL system applied in Nissan VQ series gasoline engine is able to drive eccentric cam and changes the swing of the output cam by electric motor, which enables the valve lifts and durations adjusting continuously between 1.3 and 8.4 mm and 20 and 260 °CA. The Valvetronic system applied in BMW N series gasoline engine, using electric motor as its power source, is able to change the foot swing of the middle push rod by rotating its sector gear, which enables the valve lifts and durations adjusting continuously between 0.2 and 9.7 mm and 0 and 250 °CA. [2]. Besides, the Multi-Air system equipped in Fiat Tigershark engine and

M. Liang (✉) · C. Liu · W. Cai · C. Meng · J. Wan · L. Hong
Dongfeng Motor Corporation Technical Center, 430058 Wuhan, China
e-mail: liangmingx@dfmc.com.cn

China SAE (ed.), *Proceedings of China SAE Congress 2018: Selected Papers*,
Lecture Notes in Electrical Engineering 574,
https://doi.org/10.1007/978-981-13-9718-9_36

Freevalve system equipped in Qoros Qamfree engine provide brand-new solutions for valve actuation. They flexibly control the valve movement by compressed oil and compressed gas instead of the traditional camshaft.

In the development process of a next-generation three-cylinder engine, VCR and CVVL technologies have been used to improve thermal efficiency and achieve particularly low fuel consumption in the engine part load conditions. For the sake of meeting the special requirements of valve lift characteristics, a cam-in-cam shaft and its followers are implemented at the intake side, to provide more possibilities for intake valve movement.

1 Hydraulic-Driven CVVL System

1.1 Structure and Working Principle

The CVVL system consists of a cam-in-cam shaft, upper and lower rocker arms and actuators for both internal and external shaft. The moving lobes in the middle are connected with the internal shaft, and the fixed lobes connected with external shaft are symmetrically arranged on both sides of the moving one. Rollers in upper rocker arm are contacted both with the moving and the fixed cam lobes surface. When camshaft is rotating, the displacement vectors created by two sets of rollers accumulate at the joint of the upper and lower rocker arms, and finally drive the lower rocker arm to swing around HLA pivot cyclically, which realize the opening and closing motion of valves on the intake side (Fig. 1).

Valve movement: Different from traditional valve train, the valve movement of this CVVL system is controlled separately by its moving and fixed cam lobes. During the opening section, moving cam remains outputting maximum lift, but the fixed cam is gradually rotating from base circle to high lift area. The lift change enables lower rocker arms to move downward and open the intake valves. On the contrary, during the closing section, fixed cam remains outputting maximum lift, while the moving

Fig. 1 Cam-in-cam CVVL system

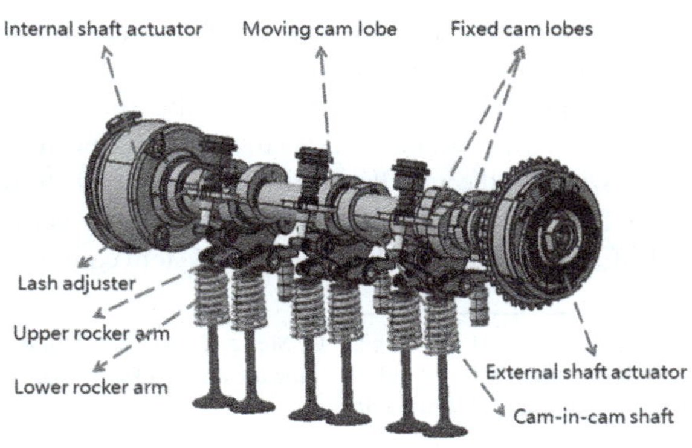

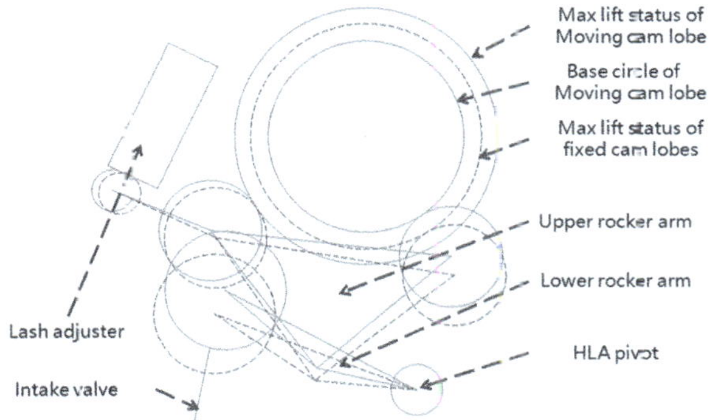

Fig. 2 Skeleton layout of CVVL system

cam is gradually rotating from high lift area to base circle. Spring force pushes the lower rocker arms to move upward and close the intake valves.

Change of valve lifts and durations: Change of valve lifts and durations can be achieved by adjusting the relative angle between moving and fixed cam lobes. As described above, the opening and closing section appear alternately. When the closing section appears before the valve fully opened, the system is not able to reach its theoretical maximum lift. On the other hand, when closing section appears after the valve fully opened, the system is able to reach its theoretical maximum lift and even keeps maximum lift for a period of time (Fig. 2).

2 CVVL Deblocking Issue

2.1 Phenomenon and Reason Analysis

During P&E development period, initial calibration is carried out on the CVVL system. Engineers found that the internal shaft actuator was easily locked at zero position when engine speed was lower than 1200 rpm. As a result, the rotor of an internal shaft actuator could not follow the PWM signal to the target phase, and the system loses its ability to real-time change valve lift and duration as shown in Fig. 3.

Base on former development experience, the issue may be caused by the failure of rotor unlocking. To solve this issue, we pay great attention to investigate CVVL hardware and working conditions.

After inspection, oil and gas can flow smoothly in actuator internal oil gallery. The unlocking pressure is less than 0.6 bar and the locking pin clearance is within the range of 0.3°–0.7°, which completely satisfy the definition on drawing. The core and sleeve do not get stuck, and the functions of OCV are not affected in this prototype. The investigation of sensor signal also shows that oil pressure at OCV P port is much

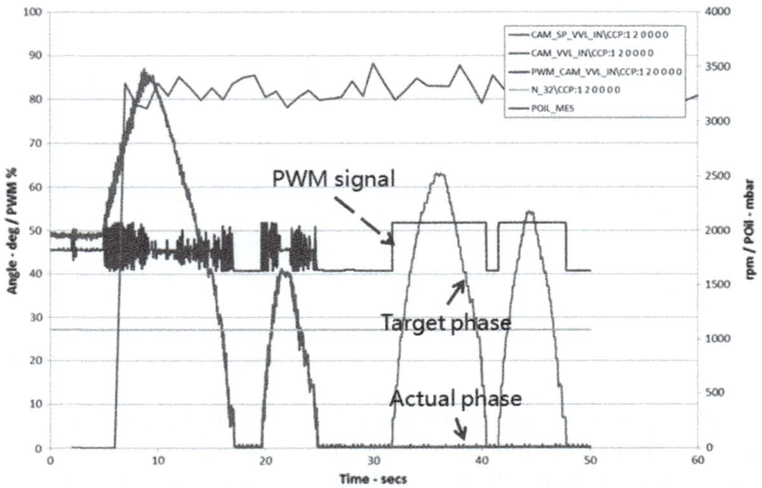

Fig. 3 Locking issue at 1000 rpm

larger than locking pin unlock pressure, while the PWM signal works well during engine operation at the same time. There is no objective condition for this deblocking issue.

From the analysis of the dynamic working process of locking pin, there are two independent movements during the process of unlocking. Unlock movement: the oil pressure acting at locking pin top surface overcomes the spring force, compresses the spring and pushes the locking pin move toward spring seat. Rotation movement: pin side wall moves to the edge of pin hole, due to the rotation of rotor vanes.

The end mark of unlock movement is the movement that the top surface of the locking pin completely sinks under rotor end surface, and the system loses its ability to lock at the initial position. Pin gets rigid contact with pin hole. Locking pin clearance reducing to 0 marks the end of rotation movement. If unlock movement completes before the rotation movement, CVVL mechanism can be unlocked smoothly and realizes basic functions. However, when the rotation movement completes before the unlock movement, two parts contact and at the time if locking pin is not in the locking position completely, the friction will prevent the locking pin moving downward and bring potential stuck risk to the system (Fig. 4).

According to the analysis above, the finish timing of the two movements directly determines whether the rotor can be unlocked successfully or not. Therefore, calculating the finish timing of both the aforementioned movements and optimizing critical parameters that significantly affect deblocking characteristics is of great importance to the trouble shooting.

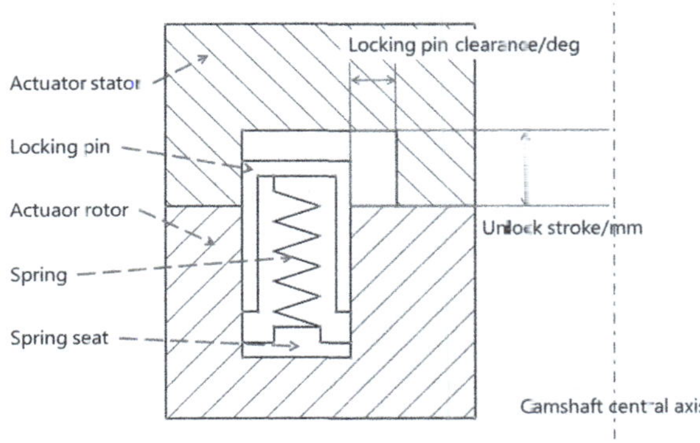

Fig. 4 CVVL system deblocking process

2.2 *Modeling of Unlock Movement*

AMESim is a mature 1D simulation software applied in multiple disciplines such as mechanical, electrical and hydraulic. By simplifying the internal shaft actuator with matched 3 position 4 way solenoid, sub-models in hydraulic, mechanical and signal libraries are used to establish the system model and calculate the characteristics of these two movements [3].

Figure 5 shows the sub-system model for unlock movement parameters calculating. The unlocking oil path of internal shaft actuator is connected to OCV A port. A check valve is arranged between OCV A port and locking pin chamber. Under the pressure from oil pump, locking pin overcomes the spring and the friction force to move downward. In the model, a limit block is set for the locking pin. Calculation stops automatically as soon as the locking pin displacement exceeds the unlocking stroke (Table 1).

In the model, the oil pressure data is directly related to the engine speed and oil temperature obtained by rig test. OCV can receive different driving signals according to demand. The maximum and hold point flow rate of OCV are set based on the actual product specifications. Due to the existence of check valve and hydraulic chamber

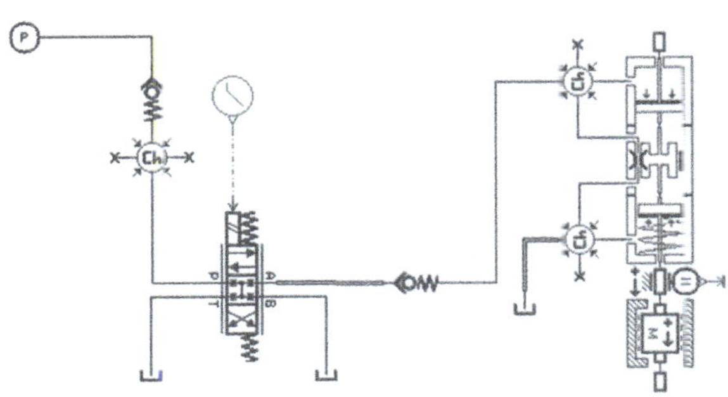

Fig. 5 Sub-model for unlock movement

Table 1 Settings of unlock movement sub-model

Oil pressure	According to engine speed and oil temperature
Pin diameter	φ8 mm
Spring stiffness	0.5 N/mm
Pin mass	3 g
Unlock stroke	3.3 mm

model, the time length for check valve opening and chamber full filled will also be reflected in the output results.

2.3 Modeling of Rotation Movement

The internal shaft actuator is essentially a hydraulic motor system controlled by a closed-loop signal. As shown in Fig. 6, size of rotor vanes, camshaft torque, hydraulic torque and spring torque are taken into consideration in the study of rotation movement characteristic. What we concern most is the beginning of rotation movement. During this time, the locking pin clearance still exists. The pin does not obstruct the relative angular motion between the rotor and the stator. In this model, the normal open signal has been implemented in the clutch interface, so the rotation of the actuator rotor will not be restricted [4] (Table 2).

We must notice that the valve movement of this CVVL system is controlled separately by its moving and fixed cam lobes. The cam torques act on the internal and external shaft is different with the traditional valve train. Valve spring is compressed during the opening section, and at this moment, the lift change only appears at fixed cam lobes. So, the valve spring resistance torque mainly acts at the external shaft. Valve spring is elongated during closing section, and at this moment, the lift change only appears at moving cam lobes. So, the valve spring assistance torque mainly acts at the internal shaft. Therefore, in a whole working cycle, spring torque always promotes internal shaft rotation and hinders external shaft rotation.

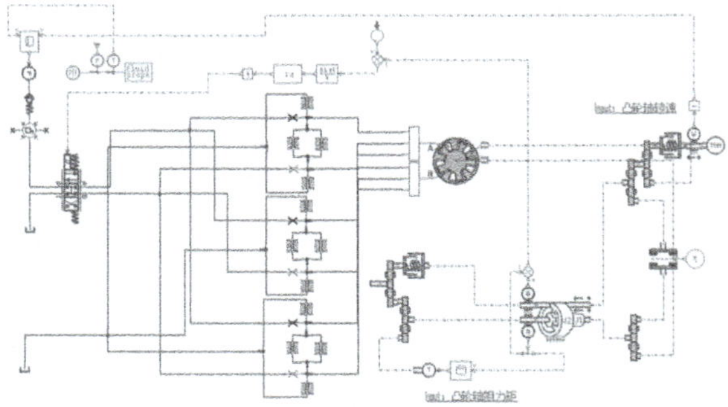

Fig. 6 Sub-model for rotation movement

Table 2 Settings of rotation movement sub-model

Number of vanes	3
Vane length	25 mm
Vane width	14 mm
Camshaft torque	According to engine speed
Spring stiffness	0.018 Nm/°
Inertia of rotor and internal shaft	15.5 kgmm^2
Type of signal	Step/Slope
Maximum flow rate of OCV	4.7 L/min

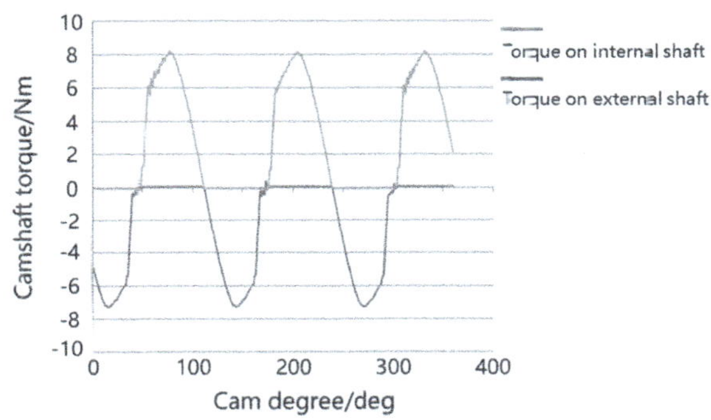

Fig. 7 Camshaft torque at 1000 rpm

Figure 7 shows the calculated camshaft torque on the internal and external shaft at 1000 rpm. For the rotor of internal shaft actuator, positive area of the camshaft torque is able to push the rotor apart from initial position together with hydraulic torque, while the negative area of the camshaft torque has an opposite effect. Compared with the common VVT, there is no negative torque applied on internal shaft. Camshaft torque only accelerates the rotation movement apart from zero position [5].

2.4 Calculation Results and Analysis

In order to find the relationship between stuck phenomenon and the timing of aforementioned two movements, we picked the working condition of 1000 rpm with high failure rate during the test, to investigate the influence of locking pin clearance and unlock signal on CVVL deblocking function under different oil pressures.

The calculation results are shown below. The *X*-axis is a time in ms. *Y*-axis on the left is rotation angular displacement in degree, and the *Y*-axis on the right is pin unlock stroke in mm. Unlock movement and rotation movement are, respectively, marked by dashed lines and solid lines. Computer will record the ending timing of unlock movement as $t1$ and the ending timing of rotation movement as

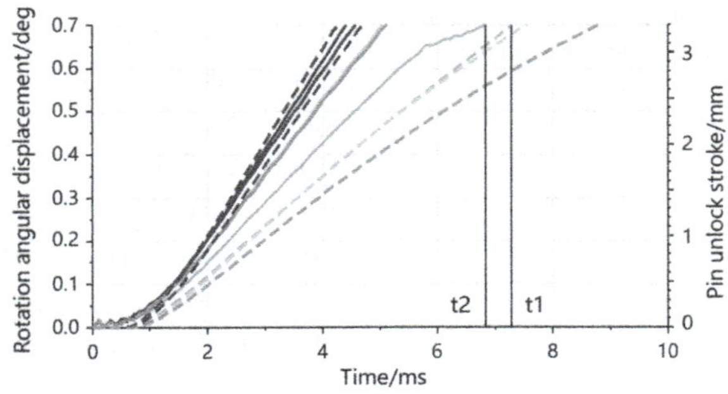

Fig. 8 0.7° locking pin clearance + step signal

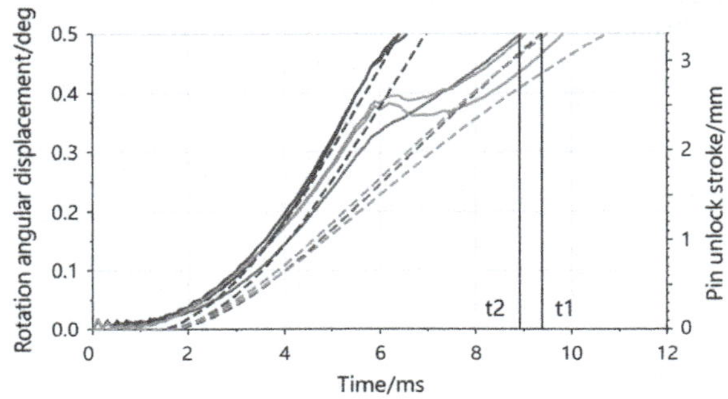

Fig. 9 0.5° locking pin clearance + slope signal

*t*2. If $t1 - t2 < 0$, unlocking process will go smoothly, otherwise the system will be under high risk of blocking (Figs. 8 and 9).

2.5 *Influence of Locking Pin Clearance*

Locking pin clearance is the relative angle between locking pin side wall and pin hole side wall inside the actuator. This lash is majorly determined by parts manufacturing and assembly tolerances. The clearance value has a significant impact on the ending timing of rotation movement.

We can know from the calculation results that when nominal locking pin clearance is 0.5°, the system has blocking risk at 1000 rpm no matter with the oil temperature and pressure. If the nominal clearance is increasing to 0.7°, part of the timing difference curve enters into safety unlocking area at approximately 90 °C oil temperature. And if the nominal clearance is increased to 0.9°, in simulation results, the CVVL system can be successfully unlocked under all working conditions (Fig. 10).

We also learned from the results that, due to the lack of resistance effect from camshaft torque, the rotation movement speed of internal shaft actuator reached 160–200 °CA/s at 1000 rpm, which was much faster than conventional VVT phaser.

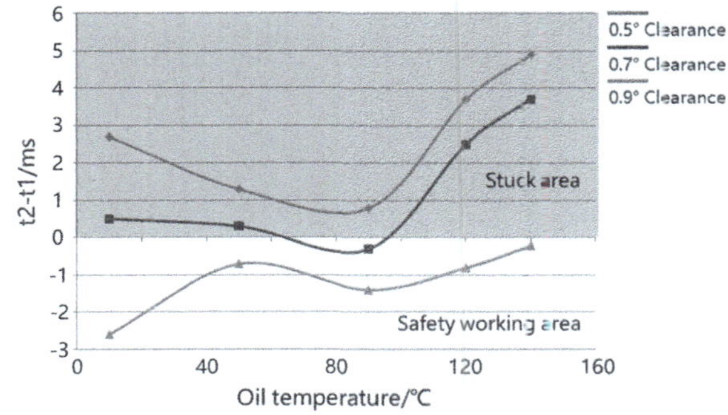

Fig. 10 Impact of locking pin clearance

That is one of the reasons why VVT phaser works well but internal shaft actuator locks at its initial position under a similar level of locking pin clearance.

Increasing the locking pin clearance can prevent the CVVL from blocking. But too large clearance will bring negative influences to the actuator, especially in the engine cold start condition, while the oil pressure does not completely established yet. And the collision between locking pin and the pinhole will cause abnormal noise, even deform the pin and lead to serious reliability problems.

2.6 *Influence of Unlocking Signal*

In the early development stage of this project, in order to guarantee rapid responsive ability, step signal was used to control actuator to realize deblocking function. Step signals can quickly open OCV A port so that the oil is allowed to flow into oil chamber and locking pin hole in a relatively short time. If the oil flowing into the locking pin hole and oil chamber is reduced simultaneously, both unlock movement and rotation movement velocity will be affected. However, the volume of unlock oil chamber is much smaller than advance/retard oil chamber. Rotation movement will be restricted more obviously [6]. In order to verify the inference, we chose different slope signals to calculation variation in flow rate and movement ending timing.

We can know from the calculation results that under 0.5° locking pin clearance, part of the timing difference curve enters into safety unlocking area at approximately 90 °C oil temperature and 1000 rpm when slope 2000 signal is input to OCV interface. Continue reducing the slope of the unlocking signal to less than 1500, the system can be successfully unlocked under all working conditions (Fig. 11).

The smoother the unlocking signal is, the larger time difference the curve deviates from zero line and so is the safety factor of unlocking. As long as the OCV flow area is limited, rotation movement is affected obviously by the differences in oil chamber volume.

Unlock process only accounts for a very low time proportion to total rotation movement. Reducing the flow rate at the unlock process has no obvious influence

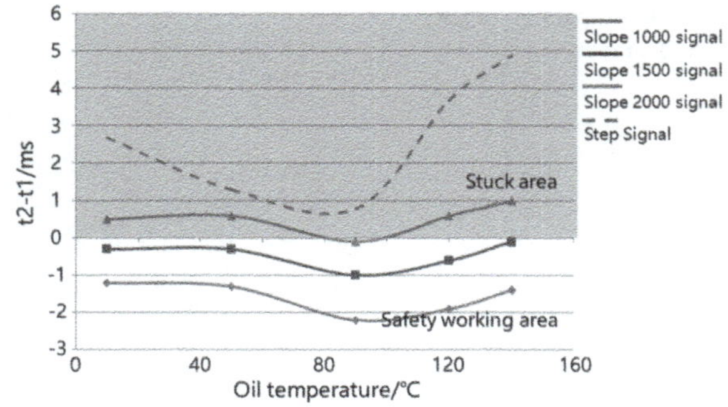

Fig. 11 Impact of unlocking signal

on overall response performance. Calculation results show that, in the condition of 2.3 bar oil pressure and 90 °C oil temperature, the response time length of 40° adjustment range only increased by 4.6 ms. The effect of slope signal on overall response speed can be neglected.

3 Test Verification

3.1 Optimal Design for Deblocking Function

For the sake of solving unlock problem essentially, optimizations both in hardware and software are carried out in this CVVL system. In terms of hardware, the sub-components manufacturing tolerance of the internal shaft actuator were modified. Hence, the lower limit of locking pin clearance raised from 0.3° to 0.5° to avoid the situation that too small clearance affects deblocking performance. In terms of software, step unlocking signal was replaced by slope unlocking signal to achieve better unlock reliability under extreme working conditions.

3.2 Deblocking Function Verification

As shown in Fig. 12, the test bench consists of motor, oil external control device, cylinder head assembly with CVVL and sensors to record phase of the internal and external shaft. During operation, the motor acts as engine crankshaft to drive the intake camshaft upon to target speed. Oil external control device heats and presses the oil into cylinder head oil path directly. Solenoid for internal shaft actuator receives PWM signal from ECU, converses it into spool displacement and controls CVVL system to achieve basic functions. Since the locking pin inside the actuator is covered by stator and cover plates, and it is not easy to measure the movement characteristics

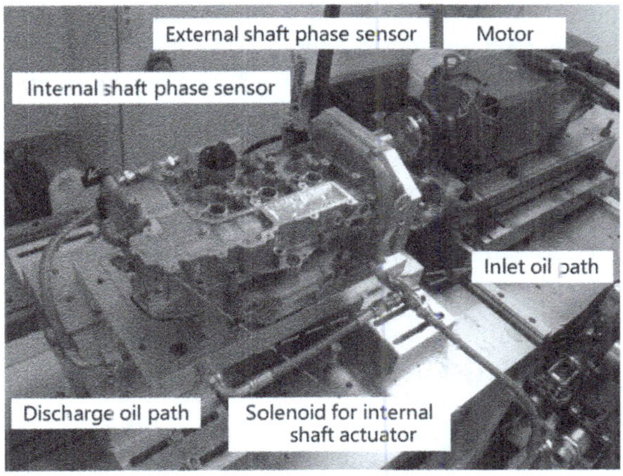

Fig. 12 Function verification test rig

of locking pin directly. Therefore, in this test, we judge whether actuator unlocked or not through checking the relative angular position of internal and external shaft.

We can learn from simulation results that it has no matter with the oil temperature; this deblocking issue appears when the locking pin clearance is lower than 0.5° and step signal is used. Increasing the gap value and optimizing the signal type are able to improve the system deblocking phenomenon. We tried to reproduce this phenomenon by test. Tables 3 and 4 show the comparison between simulation and test results. Results at the front are obtained by calculating, and results at back are obtained by

Table 3 Simulation and test results at 1000 rpm, 0.5° locking pin clearance

	St	Sl1000	Sl1500
10 °C	NG/NG	OK/OK	OK/OK
50 °C	NG/OK	OK/OK	OK/OK
90 °C	NG/OK	OK/OK	OK/OK
120 °C	NG/NG	OK/OK	OK/OK
140 °C	NG/NG	OK/OK	OK/OK

Table 4 Simulation and test results at 1000 rpm, 0.7° locking pin clearance

	St	Sl1000	Sl1500
10 °C	NG/OK	OK/OK	OK/OK
50 °C	NG/OK	OK/OK	OK/OK
90 °C	OK/OK	OK/OK	OK/OK
120 °C	NG/NG	OK/OK	OK/OK
140 °C	NG/NG	OK/OK	OK/OK

measuring. In each condition, we tested five times. Any failure during testing will be recorded by NG as result.

It can be seen that in Table 3, test results of 50 and 90 °C oil temperature are inconsistent with simulation results. And similar phenomena appear in Table 4 as well. The deviation may be caused by the difference between simulation inputs and actual parameters, especially the settings of Q-I characteristics and system damping. Results of major conditions show good consistence. Deblocking issue is solved properly after the application of slope unlocking signals.

4 Conclusions

In a hydraulic-driven continuously variable valve lift system, due to the asymmetrical torque on internal shaft and unreasonable locking pin clearance, the system locks at initial position frequently under specific conditions. For the sake of trouble shooting and design optimizing, we abstracted the working process of internal shaft actuator and investigated the parameters which could significantly affect system deblocking characteristics by simulation and test methods. The results show that:

(1) Increasing the locking pin clearance can extend the time length of rotation movement under the same unlocking speed, which is able to improve system deblocking performance.
(2) Replacing step unlocking signal by slope unlocking signal will restrain rotation and unlocking movement at the same time, while rotation movement is affected obviously due to the differences in oil chamber volume.
(3) Stick issue has been solved after the lower limit of locking pin clearance was adjusted to 0.5°, and the signal slope was well restricted.

Not limited to the CVVL system mentioned above, in practical use, excessive load driven by camshaft and asymmetric in opening and closing cam profile will also imbalance the actuator response speed on both directions, which brings similar deblocking risk to hydraulic motor system. The optimal design methods mentioned in this paper can be popularized in similar systems and will bring more conveniences on the deblocking study and critical parameters definition.

References

1. Wan Y (2013) Reducing part load pumping loss and improving thermal efficiency through high compression ratio over-expanded cycle. SAE Paper, 2013-01-1744
2. Wu D (2010) Simulation and research on fully variable mechanical driven valve lift control system. Nanjing University of Aeronautics and Astronautics, Nanjing
3. Liang Q (2016) Hydraulic system Amesim computer simulation advanced tutorial. Mechanical Industry Press

4. Liu C (2017) Establishment and application of dynamic characteristics prediction model for variable valve timing system. In: Proceedings of SAE-China congress 2017: selected papers, 2017APAC-CIT-027
5. Zhang Y (2016) Camshaft load torque calculation and virtual check control strategy. Intern Combust Engine Power Plant 33-2-2016
6. Hou G (2006) Engineering fluid mechanics. Mechanical Industry Press

Research on the Design Method and Vibration Reduction Performance of Dual-Mass Flywheel

Haipeng Luo, Guangquan Wu, Qi Wu, Wenfeng Zhan, Jiang Yang and Lin Xu

Abstract This study introduces the operating principle of dual-mass flywheel (DMF) and establishes the dynamical equations of transmission system based on the theory of forced vibration and vibration reduction. Coupling with empirical parameters, the rotary inertia distribution of DMF and the torsional stiffness design of DMF are also established. The second-order torsional vibration natural frequency (8.5 Hz) and other main order frequencies of engine under idle speed are proved to be deviated from the excitation frequencies through the modal analysis. The test shows that the resonance appears at around 255 r/min under starting condition, which matches well with the simulation. Under idle speed, the engine angular acceleration fluctuation decreases by 83.5% by using DMF. Under driving condition, the angular acceleration of gearbox input shaft is much lower than that of engine, while under driving WOT of second gear condition, the gearbox angular acceleration is only 141 rad/s^2 compared with 1947 rad/s^2 of the engine maximum angular acceleration, and the two angular accelerations of the transmission system with CTD, both around 430–930 rad/s^2, have no great difference. The result shows that the method of designing DMF is reliable, providing a new design concept of DMF.

Keywords Internal combustion engine · DMF · Vibration reduction · Rotary inertia · Torsional stiffness · Transmission system

1 Introduction

As people have more and more stringent requirements on automobile comfort [1, 2], the torsional vibration and noise of the vehicle transmission system have received extensive attention. And the vibration is mainly caused by engine combustion and reciprocating inertia force [3, 4], which is then transmitted to the whole vehicle through the gearbox and driveshaft. Because the conventional clutch torsional damper

H. Luo (✉) · G. Wu · Q. Wu · W. Zhan · J. Yang · L. Xu
GAC Automotive Research & Development Center, 668 East of Jinshan Road, 511434 Panyu District, Guangzhou, People's Republic of China
e-mail: luohaipeng@gacrnd.com

China SAE (ed.), *Proceedings of China SAE Congress 2018: Selected Papers*, Lecture Notes in Electrical Engineering 574,
https://doi.org/10.1007/978-981-13-9718-9_37

(CTD) integrates the friction plate and damping spring, it is necessary to balance the friction torque and the vibration reduction performance, which makes the compression space of damping spring small, resulting in a large spring stiffness, limiting vibration reduction performance of the engine, especially three-cylinder engine with poor rotational speed stability. And the multi-speed gearbox also requires higher engine speed stability, so the CTD has been difficult to meet the increasing NVH requirements. With the rapid development of the torsional vibration damper, the dual-mass flywheel (DMF) came into being.

DMF can greatly increase the compression space of the damping spring by placing it inside, which effectively reduces spring stiffness and achieves good vibration reduction performance. In addition, by adjusting the spring stiffness and the DMF rotary inertia to change the torsional characteristics of the vehicle transmission system, the vehicle could obtains good NVH performance. Because the DMF plays an important role in reducing the torsional vibration and the impact of the vehicle transmission system, it has become an important component of the engine. Therefore, a lot of studies have been conducted on structure design, performance analysis and vehicle behavior of DMF at home and abroad [5–14]. In this paper, research and analysis are carried out on the design of DMF performance parameter, and also its vibration reduction characteristics are verified and analyzed in combination with the test results.

2 Working Principle and Torsional Vibration Mathematical Model of DMF

2.1 Working Principle of Dual-Mass Flywheel

The DMF is essentially a "low-pass filter" for engine torque that absorbs and isolates high-frequency torque fluctuations during torque transfer, allowing low-frequency torque transmission. This paper takes the dual-mass flywheel-circumferential spring (DMF-CS)-type torsional vibration damper as research object to study, which structure is mainly decomposed into primary flywheel, secondary flywheel and damping spring equipment. As shown in Fig. 1, the primary flywheel is manufactured by stamping and connected to the crankshaft for engine starting and transferring the engine rotational torque. The secondary flywheel connected with the clutch is cast and then machined. The arc spring and the slide rail are circumferentially installed in the punching groove of the primary flywheel, and the lubricating grease is filled between them to reduce wear. Besides, the arc springs are often embedded into two groups, so the DMF has two different torsional stiffnesses to cope with different operating conditions of the engine. Meanwhile, the primary and secondary flywheels rotate relatively through sliding or rolling bearing. Finally, the engine torque is transferred from the primary flywheel to the secondary flywheel through the drive plate compressing arc spring and then transmitted to the vehicle through the gearbox.

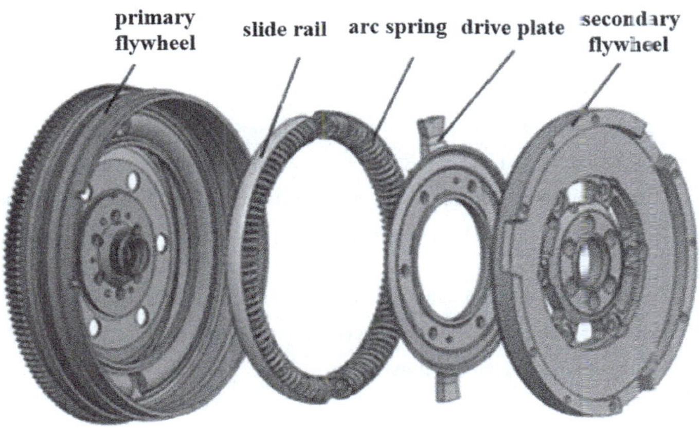

Fig. 1 Dual-mass flywheel structure

2.2 *Torsional Vibration Mathematical Model of Dual-Mass Flywheel*

For the saloon car, the first thing to consider in the design of the torsional vibration damper is to reduce and control the torsional vibration and noise under the idle condition [5, 7, 15]. Therefore, when determining the rotary inertia and stiffness of the torsional damper, it is necessary to ensure that the lowest order elastic torsional vibration frequency of the powertrain under idle condition is lower than and far away from the excitation frequency of engine idle speed [5, 7, 15].

So in the design of dual-mass flywheel, the torsional vibration characteristics of the transmission system are calculated with engine idle condition as the model in this paper, and the system is cut into several rotational masses; thus, the torsional vibration characteristic calculation and analysis model could be simplified as a nine-degree-of-freedom (9 DOF) system subjected to the output torque of engine as shown in Fig. 2.

In Fig. 2, T_1, T_2, T_3 and T_4, respectively, are the explosive torques of the four cylinders to the crankshaft, and the remaining parameters are shown in Table 1.

Fig. 2 Torsional vibration characteristic calculation and analysis model during engine idling

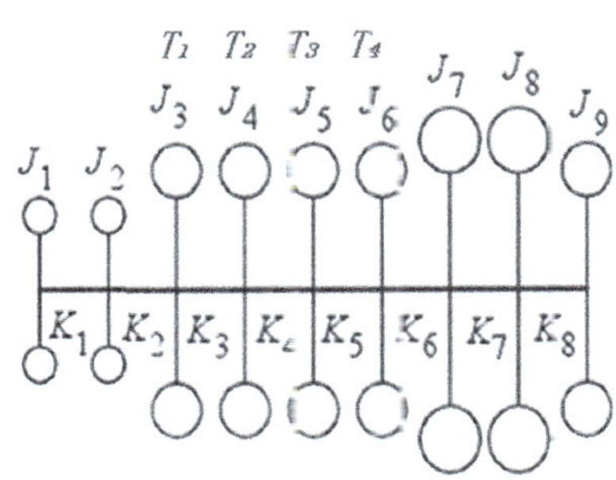

Table 1 Parameter of torsional vibration characteristic calculation and analysis model during engine idling

Parameter	Specification
J_1	Rotary inertia of outer ring of torsional vibration damper(TVD)
J_2	Rotary inertia of TVD hub, etc.
J_3	Rotary inertia of crank and connecting rod mechanism of the first engine cylinder
J_4	Rotary inertia of crank and connecting rod mechanism of the second engine cylinder
J_5	Rotary inertia of crank and connecting rod mechanism of the third engine cylinder
J_6	Rotary inertia of crank and connecting rod mechanism of the fourth engine cylinder
J_7	Rotary inertia of primary flywheel
J_8	Rotary inertia of secondary flywheel and clutch assembly
J_9	Rotary inertia of gearbox input shaft and constant mesh gears, etc
K_1	Torsional stiffness of TVD
K_2	Torsional stiffness of TVD hub, crankshaft front end, etc.
K_3	Torsional stiffness of crank and connecting rod mechanism of the first engine cylinder
K_4	Torsional stiffness of crank and connecting rod mechanism of the second engine cylinder
K_5	Torsional stiffness of crank and connecting rod mechanism of the third engine cylinder
K_6	Torsional stiffness of crank and connecting rod mechanism of the fourth engine cylinder
K_7	Torsional stiffness of DMF during idling
K_8	Torsional stiffness of gearbox input shaft

The transmission system dynamic equation under engine idle condition can be described as Eq. (1), and in order to analyze the nature torsional vibration characteristics of the transmission system with DMF under idle condition, it can be evolved into Eq. (2) through the system matrix method.

$$\begin{cases} J_1\ddot{\theta}_1 + K_1(\theta_1 - \theta_2) = 0 \\ J_2\ddot{\theta}_2 - K_1(\theta_1 - \theta_2) + K_2(\theta_2 - \theta_3) = 0 \\ J_3\ddot{\theta}_3 - K_2(\theta_2 - \theta_3) + K_3(\theta_3 - \theta_4) = T_1 \\ J_4\ddot{\theta}_4 - K_3(\theta_3 - \theta_4) + K_4(\theta_4 - \theta_5) = T_2 \\ J_5\ddot{\theta}_5 - K_4(\theta_4 - \theta_5) + K_5(\theta_5 - \theta_6) = T_3 \\ J_6\ddot{\theta}_6 - K_5(\theta_5 - \theta_6) + K_6(\theta_6 - \theta_7) = T_4 \\ J_7\ddot{\theta}_7 - K_6(\theta_6 - \theta_7) + K_7(\theta_7 - \theta_8) = 0 \\ J_8\ddot{\theta}_8 - K_7(\theta_7 - \theta_8) + K_8(\theta_8 - \theta_9) = 0 \\ J_9\ddot{\theta}_9 - K_8(\theta_8 - \theta_9) = 0 \end{cases} \tag{1}$$

$$[J][\ddot{\theta}] + [K][\theta] = [0] \tag{2}$$

where $[J]$ is the rotary inertia matrix, $[K]$ is the torsional stiffness matrix, $[\theta]$ is the angular rotation matrix, and $[\ddot{\theta}]$ is the angular acceleration matrix.

Besides, each matrix can be written as:

$$[J] = \begin{bmatrix} J_1 & & & \\ & J_2 & & \\ & & \ddots & \\ & & & J_9 \end{bmatrix}$$

$$[K] = \begin{bmatrix} K_1 & -K_1 & & & & & \\ -K_1 & K_1+K_2 & -K_2 & & & & \\ & -K_2 & K_2+K_3 & -K_3 & & & \\ & & \ddots & \ddots & \ddots & & \\ & & & \ddots & \ddots & \ddots & \\ & & & & \ddots & \ddots & \ddots \\ & & & & -K_7 & K_7+K_8 & -K_8 \\ & & & & & -K_8 & K_8 \end{bmatrix}$$

$$[\theta] = [\theta_1\ \theta_2\ \cdots \theta_9]^{\mathrm{T}},\ [\ddot{\theta}] = [\ddot{\theta}_1\ \ddot{\theta}_2\ \cdots\ \ddot{\theta}_9]^{\mathrm{T}}.$$

Therefore, in order to analyze and solve the modal of the transmission system with DMF during idling, it can be summarized as the problem of solving the eigenvalue and eigenvector of free vibration; the necessary and sufficient condition for the nonzero solution of the nature frequency Eq. (2) is that the system matrix determinant is equal to 0, so the equation can be written as:

$$\left|[J]^{-1}[K] - W^2[E]\right| = 0 \tag{3}$$

where W is expressed as the torsional frequency of the multi-mass torsional vibration system, and $[E]$ is expressed as an identity matrix.

The eigenvalue of the characteristic Eq. (3) is the natural frequency value of each order of the torsional vibration system, and the eigenvector is the amplitude of each rotational mass, so that the modal of the engine idle condition can be obtained.

3 The Design Method of Dual-Mass Flywheel

The performance of DMF is primarily optimized by distributing the rotary inertia of primary and secondary flywheel and the design of DMF stiffness. Nevertheless, in the design and matching process of DMF parameters, different manufacturers, scientific research institutions and related universities have different calculation methods. In this paper, the two-stage torsional stiffness DMF of one 4-stroke, 4-cylinder gasoline engine matching MT gearbox is studied, and the design of key performance parameters such as inertia and stiffness is studied and analyzed.

3.1 Distribution Principle of Rotary Inertia of Dual-Mass Flywheel

In order to firstly ensure that the transmission system has good vibration reduction effect during engine idling, the parameter design principle of DMF is analyzed and studied under such condition, and while designing and distributing the rotary inertia of DMF, the system model during idling could be simplified as a two-DOF torsional vibration system. Thus, the natural frequency calculation formula of this condition can be described as Eq. (4), and the relation between ratios of the two rotary inertia and natural frequency is shown in Fig. 3.

$$f = \frac{1}{2\pi} \times \sqrt{\frac{K_{s1}(J_e + J_t)}{J_e J_t}} \tag{4}$$

where K_{s1} is the first-order stiffness of DMF; J_e is the rotary inertia of engine side, which contains TVD, crank and connecting rod mechanism, primary flywheel; J_t is the rotary inertia of gearbox side, which contains secondary flywheel, clutch assembly and gearbox shaft, etc.

Generally, when the rotary inertia of the primary and secondary flywheel is distributed theoretically, the lowest resonance speed is often to be pursued, so that the system could be farthest away from the idle speed region. As can be seen from Fig. 3, when the spring stiffness of DMF is fixed, the natural frequency of the transmission system under idle condition depends on the two rotary inertias J_e and J_t.

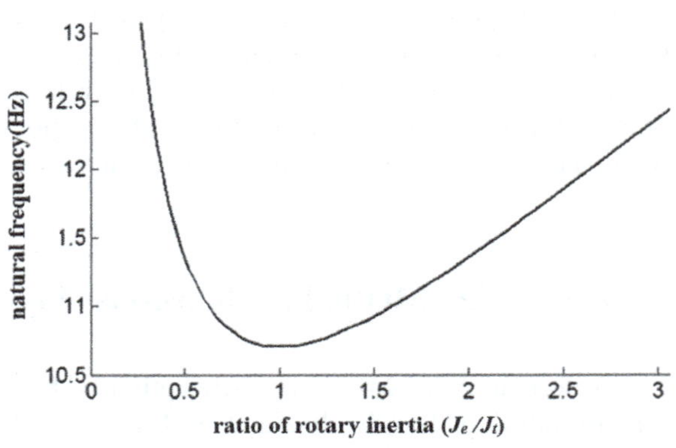

Fig. 3 Relation between ratios of rotary inertia and natural frequency

Furthermore, only when the ratio of J_e and J_t, is equal to 1, the natural frequency f is the smallest, i.e.:

$$J_e = J_t \tag{5}$$

Since J_e contains rotary inertia of TVD J_1 and J_2, rotary inertia of crank and connecting rod mechanism J_3–J_6, and rotary inertia of primary flywheel $J_{s1}(J_7)$, i.e.:

$$J_e = \sum_{i=1}^{6} J_i + J_{s1} \tag{6}$$

And J_t contains rotary inertia of secondary flywheel J_{s2}, rotary inertia of clutch assembly J_{21} and J_{22}, rotary inertia of gearbox shaft J_{23}, etc.

$$J_t = J_8 + J_9 = J_{s2} + J_{21} + J_{22} + J_{23} \tag{7}$$

References [9, 15] proposed that the total rotary inertia of DMF should be equal to the rotary inertia of single mass flywheel (SMF) of the primary engine, but in practical engineering, some engines do not have a SMF powertrain configuration. Besides, because of the limited space for putting DMF, and also to ensure good vibration reduction performance by using DMF, all these reasons result in the total rotary inertia of DMF which is not usually equal to SMF, and most of the former is larger. Because the ratio of SMF rotary inertia to the total engine rotary inertia is usually between 0.8 and 0.9 [16], while the ratio of primary flywheel rotary inertia J_{s1} is slightly lower than the lower limit or in the lower limit region by database comparison. This paper takes 0.78, namely:

$$\psi = J_{s1} / \sum_{i=1}^{7} J_i = 0.78 \tag{8}$$

3.2 *Design Principle of Torsional Stiffness of Dual-Mass Flywheel*

For two-stage torsional stiffness DMF, the first-stage stiffness K_{s1} is usually used in engine ignition, idling and some low-load conditions, while the second-stage stiffness K_{s1} is used in most driving conditions, and the design of DMF torsional stiffness combines the calculation and distribution of the critical torque T_t, the limiting torque T_{max}, the free rotation angle θ_f, the critical rotation angle θ_t and the limiting rotation angle $\theta_{\max}$; meanwhile, some of the above parameters are based on experience and process structure.

In order to ensure that the natural frequency of the powertrain is far away from the excitation frequency under idle condition and obtains good vibration reduction effect, the value of DMF K_{s1} is usually very small in design, while the excitation frequency f_e of four-cylinder engine at different rotation speeds is written as Eq. (9):

$$f_e = \frac{n}{60} \cdot \frac{\upsilon \cdot z}{\tau} \tag{9}$$

where n, v, z and τ, respectively, are engine rotation speed, harmonic number, number of cylinders and number of strokes. For four-stroke, four-cylinder engine, if the value of harmonic number v is an integer multiple of 2, it means that the harmonic number is the main harmonic. Moreover, the lowest main harmonic ($v = 2$) is the most important harmonic component of excitation torque that causes the torsional vibration of the whole vehicle transmission system.

In addition, according to the forced vibration theory, the excitation angular frequency w has the following relation with the natural angular frequency w_n:

$$\eta = \frac{1}{\sqrt{\left(1 - \frac{w^2}{w_n^2}\right)^2 + 4\xi^2 \frac{w^2}{w_n^2}}} \tag{10}$$

where η denotes the magnification coefficient and ς is the ratio of system damping to critical damping coefficient. According to the theory, only when $w/w_n > 2^{0.5}$, the natural frequency of the lowest order elastic torsional vibration could avoid the second main harmonic of the torsional vibration system under idle condition, and then the gearbox can obtain good vibration reduction effect. Through the modal analysis of vehicle idle condition, the first-order modal is rigid body modal, which is not used as a reference, so combined with Eq. (9), the relationship between second-order natural frequency f_2 and engine idle rotation speed n_d could be obtained:

$$f_2 \langle \frac{2n_d}{60\sqrt{2}} \tag{11}$$

After determining f_2, the first-stage stiffness K_{s1} of DMF can be obtained by solving the characteristic Eq. (3), the solution of second-stage stiffness K_{s2} can be obtained by the parameters such as the limiting torque $T_{\max}$, the critical rotation angle θ_t, the limiting rotation angle $\theta_{\max}$ and the critical torque T_t, and the equation of second-stage stiffness K_{s2} can be defined as:

$$K_{s2} = (T\max - T_t)/(\theta\max - \theta_t) \tag{12}$$

Among them, the DMF limiting torque $T_{\max}$ is related to the engine maximum torque T_e. In order to ensure the reliability of the engine torque transmission, $T_{\max}$

is equal to T_e multiplied by a torque safety factor κ, which can be determined by referring to the clutch safety factor [6], for truck, $\kappa = 1.7–2.25$, and for salocn car, $\kappa = 1.2–1.3$. Eventually, based on the current process technology, manufacturing accuracy and improvement of the working life of DMF, the safety factor κ is set to 1.4.

The DMF limiting angle θ_{max} is generally in the range of 30°–70°, and meanwhile, the limiting angle θ_{max} should be as large as possible to meet the requirements of torque transmission, so as to reduce the DMF stiffness and reduce the vibration of the transmission system. But in the actual design process of DMF, it also depends on the available space arrangement and the manufacturer's process technology. Summing up the above analysis, the limiting angle θ_{max} is set to 64°.

The calculation of the critical torque T_t is determined by the acceleration load of transmission system at the vehicle starting condition, which can be obtained by vehicle kinematics Eq. (13):

$$\frac{I_g I_o T_t \xi}{r} = \delta ma + mg\mu + \frac{C_D A}{21.15} \cdot V^2 \tag{13}$$

where I_g, I_o, ξ, r, m, a, δ, μ, C_D, A, V are, respectively, gearbox ratio, vehicle final drive ratio, mechanical efficiency of transmission system, tire rolling radius, full load mass of vehicle, vehicle acceleration, conversion coefficient of vehicle rotation mass, rolling resistance coefficient, air drag coefficient, windward area, relative speed.

Therefore, there is Eq. (14) for the DMF critical angle θ_t, in which the maximum free rotation angle θ_f usually does not exceed 15°, and the angle θ_f is preliminarily set as 6° in this paper.

$$\theta_t = T_t/K_{s1} + \theta_f/2 \tag{14}$$

The rotary inertia distribution and torsional stiffness design of DMF are preliminarily determined according to the above principles, but some parameters also need to be adjusted and improved during engine and vehicle operation.

4 Example Design and Vibration Reduction Performance Analysis of Dual-Mass Flywheel

This paper takes a four-stroke, four-cylinder engine as research object to study, in which idle speed $n_d = 730$ r/min, and the maximum torque $T_e = 195$ N m. Since the basic damping torque F_1 of DMF is usually designed at 2–20 N m, F_1 is preliminarily set as 5 N m according to the experience. The other key parameters of DMF are designed and matched according to the above principles, and f_2 must satisfy the inequality (11); meanwhile, in order to improve and optimize the vibration reduction effect, f_2 is also limited to avoid the excitation of the lowest harmonic ($v = 1$) torsional vibration. As the 2nd order natural frequency f_2 of the transmission system is usually in 6–15 Hz [13] under idle condition, and in order to avoid and stay away from the 1st and 2nd harmonics torsional vibration during idling. So combined with the above factors, f_2 is eventually set as 8.5 Hz, i.e.: $f_2 = 8.5$ Hz, and the resonance speed $n_r = 8.5 \times 60/2 = 255$ r/min. In addition, the first-stage torsional stiffness K_{s1} can be solved by bringing f_2 to the characteristic Eq. (3), which value is 2.5, i.e.: $K_{s1} = 2.5$ N m/(°). The performance parameters of the designed and matched DMF are reported in Table 2, and the real DMF object is shown in Fig. 4.

Figure 5 shows the result of modal analysis of engine idle condition, in which the rotary inertia J_9 and torsional stiffness K_8 of the transmission input shaft are, respectively, 0.00241 kg m^2 and 12,661 N m/rad. It can be seen from Fig. 5 that the second-order natural frequency is 8.5 Hz, which is the preset target value, and the third-, fourth- and fifth-order natural frequencies are greater than 220 Hz, so all the main order natural frequencies are far from the second, fourth and sixth main harmonic excitation frequencies which are, respectively, 24.3, 48.7 and 73 Hz; as a result, the idle condition would not cause resonance.

Because the resonance speed 255 r/min could not be reflected in the common working range of the engine, it should be validated at engine starting condition. Therefore, the engine crankshaft speed and the transmission input shaft speed are both monitored in this paper, and the curve of monitored engine starting speed is

Table 2 Performance parameters of dual-mass flywheel

Parameter	Symbol	Unit	Value
Rotary inertia of primary flywheel	J_{s1}	kg m^2	0.0752
Rotary inertia of secondary flywheel	J_{s2}	kg m^2	0.0454
First-stage torsional stiffness	K_{s1}	N m/(°)	2.5
Second-stage torsional stiffness	K_{s2}	N m/(°)	7.5
Free rotation angle θ_f	θ_f	(°)	6
Critical rotation angle	θ_t	(°)	40
Limiting rotation angle	θ_{max}	(°)	64

Fig. 4 Dual-mass flywheel object

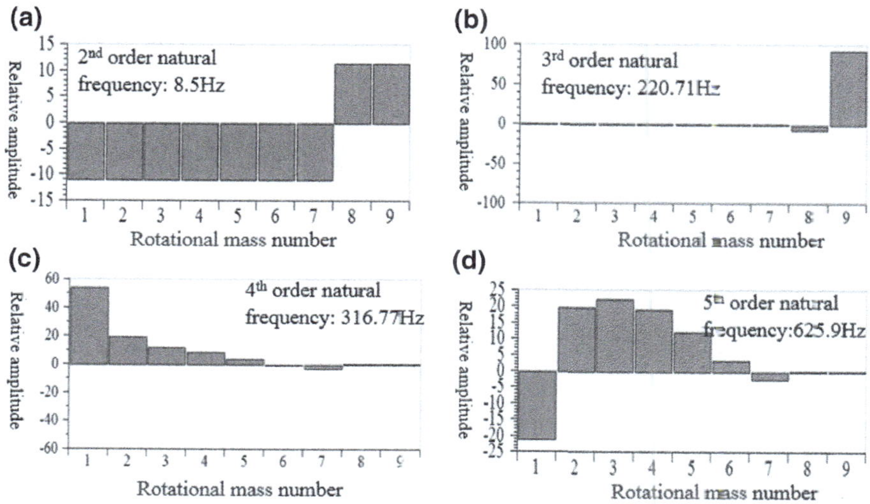

Fig. 5 Different modals under engine idle condition

shown in Fig. 6. It can be seen that the engine produces resonance near the speed 255 r/min, and the phenomenon is proved to be consistent with the theoretical design.

Through the actual measurement of the rotational speed and angular acceleration during idling, the results are, respectively, shown in Figs. 7 and 8. It can be seen that engine rotational speed and angular acceleration fluctuations are attenuated and filtered by DMF, and they have become more stable while reaching the gearbox input shaft. The engine rotational speed and angular acceleration fluctuation decrease by

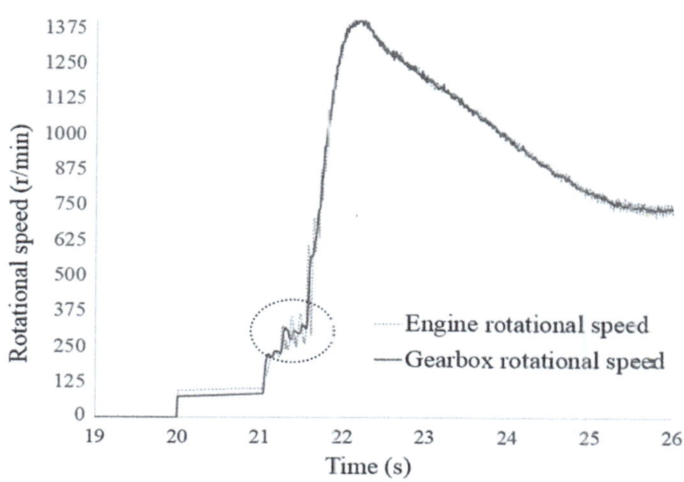

Fig. 6 Monitored rotational speed under engine starting condition

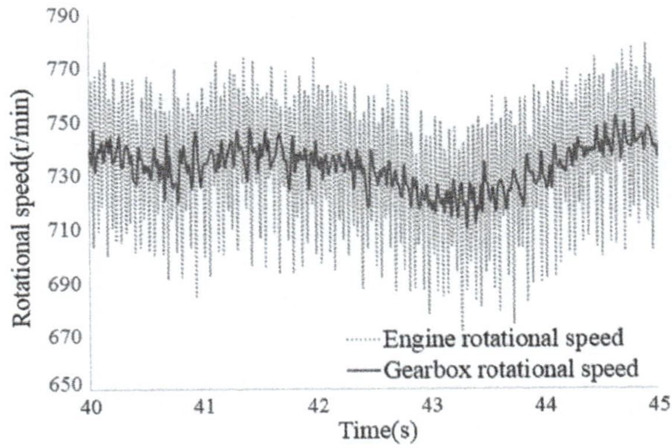

Fig. 7 Monitored rotational speed under engine idle condition

Fig. 8 Monitored angular acceleration under engine idle condition

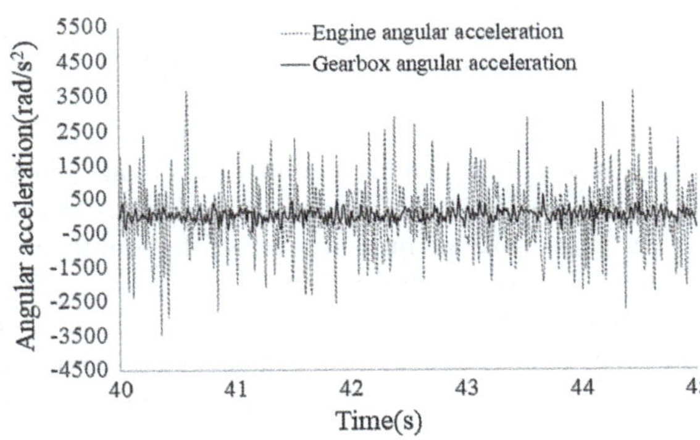

77 and 83.5%, respectively, and it can be proved that the designed DMF has excellent vibration reduction performance under idle condition.

The vibration reduction effect of DMF under the vehicle driving condition is also analyzed. By comparing and analyzing the second-order angular acceleration of engine and gearbox, the vibration reduction effect of transmission system with DMF is very significant, and the angular acceleration of gearbox is far lower than engine. Meanwhile, in order to compare and analyze the vibration reduction effect of DMF and CTD, this paper takes driving WOT of second gear condition of vehicle as an example. Results are shown in Fig. 9, the gearbox angular acceleration with DMF is always below 250 rad/s^2, and also it shows the maximum angular acceleration of engine is 1947 rad/s^2 at 2400 r/min while the gearbox angular acceleration is 141 rad/s^2, proving the good vibration reduction effect of DMF.

However, the angular acceleration of engine and gearbox, belonged to the original SMF transmission system with CTD, is all in the range of 430–930 rad/s^2 and has no great difference, which is located in the middle of the two curves of transmission system with DMF.

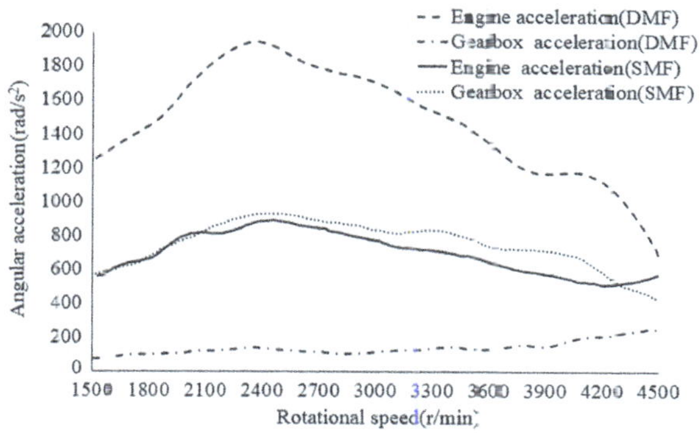

Fig. 9 Comparison of second-order angular acceleration between DMF and CTD under vehicle driving condition

5 Conclusion

(1) Based on the theory of forced vibration and vibration reduction, the dynamical equations of transmission system have been established, and coupling with empirical parameters, the rotary inertia distribution of DMF and the torsional stiffness design of DMF are also established.

(2) Through the modal analysis of engine idle condition, the second-order natural frequency 8.5 Hz and other higher-order natural frequencies are far from the second, fourth and sixth main harmonic excitation frequencies which, respectively, are 24.3, 48.7 and 73 Hz, avoiding the resonance of the transmission system.

(3) The engine takes place resonance near the speed 255 r/min under engine starting condition, which is consistent with the theoretical calculation, proving the method of designed DMF reliable.

(4) DMF has excellent vibration reduction effect under engine idle condition,and the vibration reduction rate of angular acceleration is up to 83.5%.

(5) Under vehicle driving condition, the angular acceleration of engine and gearbox, belonged to the SMF transmission system with CTD, has no great difference, while the vibration reduction effect of transmission system with DMF is very significant, and the angular acceleration of gearbox is far lower than engine. When under driving WOT of second gear condition, the engine maximum angular acceleration is 1947 rad/s^2 compared with 141 rad/s^2 of gearbox angular acceleration at that time, whereas the two angular accelerations of the transmission system with CTD are both around 430–930 rad/s^2.

Acknowledgement Supported by National Key Research and Development Plan: Performance Optimization of Multi-mode Driven Plug-in Saloon Car (2017YFB0103300).

References

1. Perera MSM, Theodossiades S, Rahnejat H (2007) A multi-physics multi-scale approach in engine design analysis. Proc Inst Mech Eng Part K J Multi-body Dyn 221(3):335–348
2. Koronias G, Theodossiades S, Rahnejat H et al (2011) Axle whine phenomenon in light trucks: a combined numerical and experimental investigation. Proc Inst Mech Eng Part D J Automobile Eng 225(7):885–894
3. Tangasawi O (2011) Non-linear vibro-impact phenomeron belying transmission idle rattle. Drive Syst Tech 222(10):1909–1923
4. Rahnejat MDLC, Theodossiades S, King P (2011) Transmission drive rattle with thermo-elastohydrodynamic impacts: numerical and experimental investigations. Int J Powertrains 1(2):137
5. Li W (2009) Research on the method and the performance of isolation of torsional vibration of DMF-CS based on the vehicle power train. Jilin University
6. Lei Chen, Mingran Deng, Zhengfeng Jiang (2012) Optimization method of performance parameters of dual mass flywheel. J Intern Combust Engine 3:277–283
7. Zhenhua Lv, Tao Chen (2006) Design of the elastic characteristics of torsional damper with double mass flywheel and circum ferential spring. Automot Eng 28(1):73–77
8. Zhao X (2009) Structural analysis and performance study of the torsional vibration system of dual mass flywheel. Chongqing University
9. Shi WK, Long Y, Lu YD (2009) Parameter design and optimization of multi-stage nonlinear dual mass flywheel. J Vib Shock 28(5):92–96
10. Wei G, Wang S, Liu Y (2017) The control strategy research of engine start with hybrid clutch. Chin J Automot Eng 7(4):270–279
11. Galvagno E, Velardocchia M, Vigliani A et al (2015) Experimental analysis and model validation of a dual mass flywheel for passenger cars. Sae Technical Papers
12. Gupta K, Choudhary A, Bidre R (2017) NVH performance improvement study using a dual mass flywheel (DMF), inertia ring type tuned torsional vibration damper (TVD) and single mass flywheel (SMF) in a front engine and rear wheel driveline architecture. In: Noise and vibration conference and exhibition
13. Li G, Hu J, Qin D (2008) Study on natural torsional vibration characteristics of dual mass flywheel-radial spring type torsional vibration damper. Chin Mech Eng 19(15):1800–1805
14. Walter A, Lingenfelser C, Kiencke U et al (2008) Cylinder balancing based on reconstructed engine torque for vehicles fitted with a dual mass flywheel (DMF). SAE Int J Passeng Cars—Mech Syst 1(1):810–819
15. Jiang Z, Chen L (2009) Study on the design method of circumferential arc spring dual mass flywheel. Chin Intern Combust Engine Eng 30(6):31–36
16. Yuan Z (2012) Internal combustion engine design. Version 2. China Machine Press, pp 327–232

The Influence of Methanol Mass Ratio and Compression Ratio on Combustion Characteristics of Dual-Fuel Engine

Peng Li, Jianjun Zhu and Wenjie Wu

Abstract A study was conducted in a dual-fuel engine to explore the influence of methanol mass ratio and compression ratio on the combustion characteristics and a set of methanol injection device was installed on a single cylinder diesel engine. The results show that, with the increase of the methanol mass ratio, the in-cylinder pressure gradually decreases, the pressure rise rate and the heat release rate increase firstly and then decrease, the start of combustion gradually postpone and combustion duration gradually short. The decrease of the compression ratio causes the increase of the peak heat release rate, but it has complex effects on the peak in-cylinder pressure, the peak pressure rise rate, start of combustion and combustion duration. When the compression ratio is 15.9, the peak in-cylinder pressure and the peak pressure rise rate are highest, the start of combustion is earliest and the combustion duration is longest.

Keywords Dual-fuel engine · Methanol · Biodiesel · Compression ratio · Combustion

1 Introduction

In order to solve the problem of air pollution caused by combustion diesel fuel, it is urgent to study the new combustion mode and alternative fuel. The dual-fuel combustion mode is widely studied as a new type of combustion mode. The dual-fuel combustion model is used high cetane rating fuel to ignite high octane rating fuel. The premixed gas is formed in the cylinder, and a large scale of ignition occurs at the same time, which accelerates the combustion rate. Thus, the combustion can achieve higher equal volume degree and thermal efficiency. Moreover, the lean mixture can be used at high compression ratio to significantly reduce the particulate matter emissions of the engine [1–3].

P. Li · J. Zhu (✉) · W. Wu
Taiyuan University of Technology, Taiyuan, China
e-mail: 18734868516@163.com

China SAE (ed.), *Proceedings of China SAE Congress 2018: Selected Papers*, Lecture Notes in Electrical Engineering 574,
https://doi.org/10.1007/978-981-13-9718-9_38

As an alternative fuel with high oxygen content, methanol is the clean energy which was first proposed to be applied to the market. China's coal resources are relatively abundant, and high-sulphur and low-quality coal resources can be used as raw materials for produce methanol. Using methanol as a substitute fuel in internal combustion engine is not only conducive to the treatment of high-sulphur and inferior coal, but also can alleviate the energy crisis and air pollution and can guarantee the long-term supply of resources. It is of great strategic significance to China's energy security [4].

To produce biodiesel from waste cooking oil, on the one hand, can alleviate the demand pressure of petrochemical energy, and on the other hand, it can prevent waste cooking oil from returning to 'dining-table'. In the 1980s, the large-scale research and application of biodiesel began in the world. At present, Chinese enterprises have mastered the technology of producing biodiesel from waste food oil, and its conversion efficiency is up to 93.6% [5].

The cetane rating of methanol is very low and difficult to direct compression ignition on diesel engine. However, the cetane rating of biodiesel is very high and can be easily applied in diesel engine. The injection of biodiesel into the cylinder to ignite the methanol injected into the intake port, and this method can control the combustion process flexibly by controlling different fuel concentration. Moreover, the latent heat of vaporization of methanol is very high, which is conducive to reducing the temperature in combustion chamber and reducing the emission of nitrogen oxides and particulate matter of the engine [6].

2 Test Equipment and Methods

2.1 Test Equipment

Table 1 is a comparison of the physical and chemical properties of fuel used for the test. The engine used in the test is modified on a CY25 single cylinder diesel engine.

Table 1 Physical and chemical properties of fuel

Fuel	0# diesel	Biodiesel	Methanol
Oxygen content (%)	0	10.9	50
Cetane rating	40–55	59	3
Octane rating (RON)	–	–	114
Density (g ml^{-1}) (20 °C)	0.84	0.885	0.788
Kinematic viscosity (mm^2 s^{-1}) (60 °C)	2.77	2.41	1.00
Latent heat of vaporization (kJ kg^{-1})	270	320.2	1178
Low heat value (MJ kg^{-1})	42.5	39.2	19.7
Stoichiometric air–fuel ratio	14.3	12.7	6.45
Autogenous ignition temperature (°C)	200–220	–	470

On the basis of not changing other devices, a set of methanol supply device is installed on the intake port to ignite the homogeneous mixture of methanol by biodiesel fuel supplied by the fuel system on original diesel engine. The Kistler 6125C11 cylinder pressure sensor and the 4618A2 type charge amplifier are used to obtain the combustion pressure in the cylinder, and the data are collected and analysed by the DEWE-800-CA-SE combustion analyser of the DEWETRON company. Table 2 is the operating parameters of the test engine.

2.2 Experimental Method

In this experiment, engine is stabilized running on a certain working condition. By adjusting the pulse width of methanol and the value opening of biodiesel, the maximum methanol mass ratio under this condition is found. Then the injection pulse width of methanol is gradually reduced and the valve opening of biodiesel is increased and mass ratio of methanol was reduced. The data of different methanol mass ratio under this working condition are recorded. Change the engine operating conditions, repeat the above operations and record the methanol mass ratio data under different conditions. Then replace the piston, repeat the above operation and record the methanol mass ratio data of different compression ratios. The change of the compression ratio of the piston was achieved by mechanically cutting the upper edge region of the combustion chamber to increase the volume of the combustion chamber.

The mass ratio of methanol(r) is defined as the mass ratio of methanol consumption per cycle to total fuel consumption per cycle.

Table 2 Diesel engine parameters

Diesel engine model	CY25
Diesel engine type	Four stroke, single cylinder
Bore (mm)	115
Stoke (mm)	120
Combustion chamber shape	ω
Injection pressure (MPa)	20
Fuel delivery advance angle (°CA)	22
Displacement (L)	1.25
Compression ratio	17:1
Rated power (kW/r min^{-1})	14.7/2200
Maximum torque (N m/r min^{-1})	70/1600

3 Effect of Methanol Mass Ratio on Combustion Characteristics

3.1 In-Cylinder Pressure

As shown in Fig. 1, with the increase of methanol mass ratio, the in-cylinder pressure decreases gradually, and the position of crankshaft angle corresponding to the peak pressure is gradually leaning back, and the change of speed has no obvious effect on in-cylinder pressure change trend. This is because that the latent heat of vaporization of methanol is relatively high, which will absorb a lot of heat in the combustion chamber, resulting in a reduction of the temperature in the combustion chamber. With the increase of methanol mass ratio, the in-cylinder temperature gradually decreases, which inhibits the atomization and evaporation of biodiesel in the combustion chamber. Moreover, the chemical reaction rate of methanol premixed gas combustion model is mainly related to the combustion chamber temperature. With the reduction of the combustion chamber temperature, the chemical reaction speed decreases gradually and the fuel combustion rate slows down. Therefore, with the increase of methanol mass ratio, the in-cylinder pressure gradually decreases.

3.2 Pressure Rise Rate

As seen from Fig. 2, with the increase of the mass ratio of methanol, the pressure rise rate increases first and then decreases, and the position of crankshaft angle corresponding to the peak pressure rise rate also gradually delays, and the speed change has no obvious effect on the pressure rise rate change trend. For the biodiesel ignition methanol premixed gas combustion mode, the pressure rise rate is closely related to the number of combustible mixture formed in the ignition delay [7]. When methanol mass ratio is relatively small, the combustion in combustion chamber is

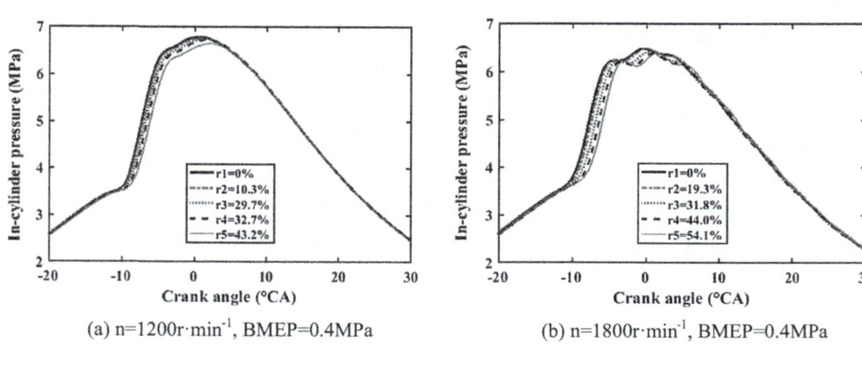

(a) n=1200r·min^{-1}, BMEP=0.4MPa (b) n=1800r·min^{-1}, BMEP=0.4MPa

Fig. 1 Effect of methanol mass ratio on in-cylinder pressure

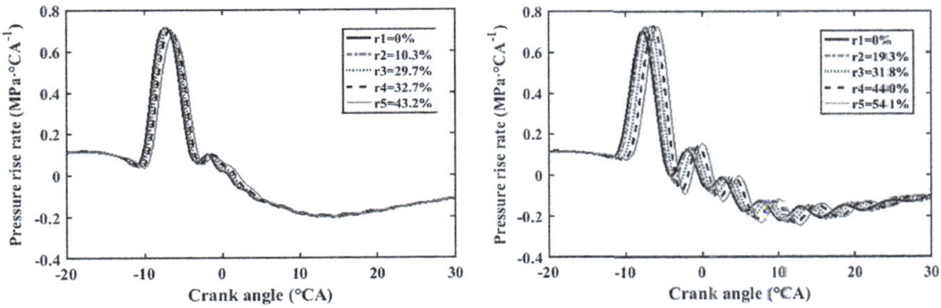

(a) n=1200r·min^{-1},BMEP=0.4MPa (b) n=1800r·min^{-1}, BMEP=0.4MPa

Fig. 2 Effect of methanol mass ratio on pressure rise rate

mainly diffusion combustion. With the increase of the amount of methanol injection, the number of methanol premixed gas in the combustion chamber increases and the combustion reaction speed is accelerated, thus pressure rise rate increases gradually. When the mass ratio of methanol increases to a certain limit, the amount of methanol is more and the amount of heat absorbed by evaporation is more, which will obviously reduce the temperature in combustion chamber and affect the combustion speed of the fuel, so the pressure rise rate begins to decrease gradually.

3.3 Heat Release Rate

As shown in Fig. 3, with the increase in methanol mass ratio, the heat release rate rises first and then decreases, and the position of crankshaft angle corresponding to the peak heat release rate is gradually delayed. The speed change has no obvious influence on the heat release rate change trend. The peak heat release rate is related to the combustion rate of the fuel, the temperature of the combustion chamber, the

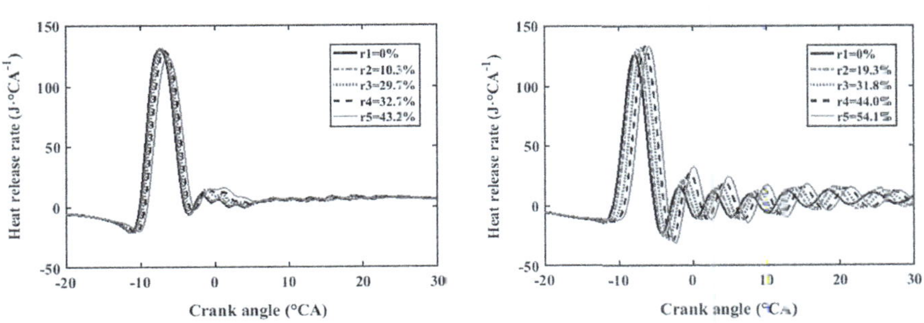

(a) n=1200r·min^{-1}, BMEP=0.4MPa (b) n=1800r·min^{-1}, BMEP=0.4MPa

Fig. 3 Effect of methanol mass ratio on heat release rate time of start of combustion and the heat loss in the combustion chamber. Under the condition of lower methanol mass ratio, the combustion in the combustion chamber is mainly diffusion combustion. With the increase of the methanol mass ratio, the number of methanol premixed gas in the combustion chamber increases, the proportion of premixed combustion increases gradually, the rate of combustion reaction accelerates, and the heat release rate increases gradually. When the methanol mass ratio is increased to a certain limit, the amount of methanol premixed gas in the combustion chamber is too much, and the methanol vaporization will lead to a sharp decrease in the temperature of the combustion chamber, which affects the combustion rate of the fuel, so the heat release rate reduces gradually.

3.4 Start of Combustion and Combustion Duration

The start of combustion determines the distance between the heat release rate curve and the top dead center and has a very important influence on the cyclic thermal efficiency, the pressure rise rate and the maximum combustion pressure. As shown in Fig. 4, with the increase of methanol mass ratio, the start of combustion is gradually postponed, but speed has no obvious effect on start of combustion change trend. This is because that the latent heat of vaporization of methanol is higher, and it will absorb a lot of heat after into the combustion chamber, which leads to the decrease of the temperature and pressure in the combustion chamber, which leads to the biodiesel atomize effect become worse. With increasing methanol mass ratio, the more obvious the temperature drop in the combustion chamber, the biodiesel atomize effect in the cylinder becomes worse. With increasing methanol mass ratio, the injection amount of biodiesel is gradually reduced and the oxygen concentration near the biodiesel is gradually reduced, so the biodiesel in the combustion chamber needs more time to ignite methanol premixed gas, and the start of combustion will be postponed gradually.

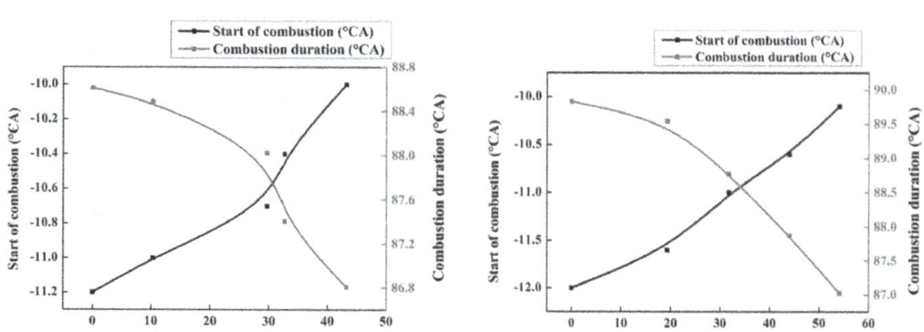

(a) n=1200r·min^{-1}, BMEP=0.4MPa (b) n=1800r·min^{-1}, BMEP=0.4MPa

Fig. 4 Effect of methanol mass ratio on the start of combustion and combustion duration

The length of the combustion duration is partly a reflection of the theoretical isobaric heat release and expansion ratio, which has a huge impact on the cycle thermal efficiency of the engine and the generation of emissions. With the increase of methanol mass ratio, the combustion duration decreases gradually, and the speed change has no obvious effect on the combustion duration change trend. For conventional diesel engines, combustion duration is related to fuel injection advance angle, the speed of combustible mixture formed during the diffusion combustion period and the degree of perfection of combustion. However, for biodiesel fuel ignite methanol premixed gas combustion mode, combustion duration is mainly related to the combustion rate of mixed gas [8], and the combustion rate of the mixture is mainly related to the mixture gas concentration and the combustion chamber temperature. With the increase of the mass ratio of methanol, the concentration of methanol premixed gas in the combustion chamber increases. Although the vaporization of methanol will lead to the decrease in the temperature of the combustion chamber, the effect of the increase of premixture concentration on combustion duration is greater than the effect of combustion chamber temperature on combustion duration. Therefore, with the increase of the mass ratio of methanol, the duration of combustion is gradually shortened.

4 Effect of Compression Ratio on Combustion Characteristics

4.1 *Peak Pressure Release Rate*

As shown in Fig. 5, it can be seen that at high speed and low speed, the maximum in-cylinder pressure is the biggest when the compression ratio is 15.9, and is the smallest when the compression ratio is 15.4. The cause of this phenomenon is: when the com-

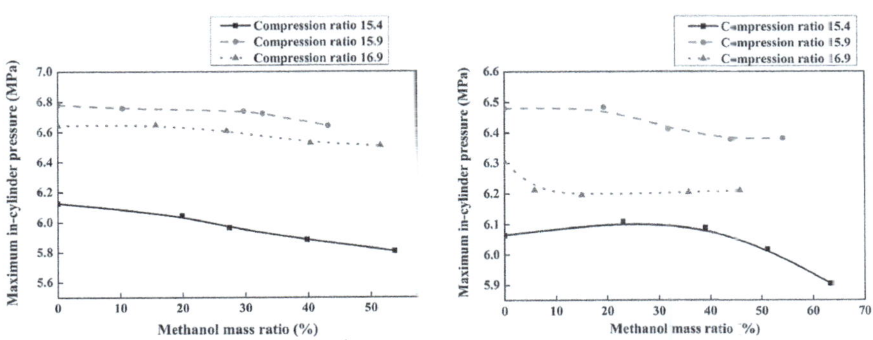

(a) n=1200r·min^{-1}, BMEP=0.4MPa (b) n=1800r·min^{-1}, BMEP=0.4MPa

Fig. 5 Effect of compression ratio on maximum in-cylinder pressure pression ratio is reduced from 16.9 to 15.9, the volume of the combustion chamber is increased and the intensity of eddy current movement is weakened, which reduces loss of the heat dissipation of combustible mixture, the reduction of temperature due to reduction of compression ratio is suppressed and is beneficial to full mixing of combustible mixture in combustion chamber, and beneficial to increase combustion rate of fuel, and the start of combustion is ahead, which causes the maximum explosion pressure to rise. When compression ratio continues to decrease to 15.4, the volume of combustion chamber continues to increase. The pressure and temperature of the combustion chamber continue to decrease, which inhibits atomization and evaporation of biodiesel and methanol in the combustion chamber. It is possible that more methanol does not form a uniform mixture in the combustion chamber. The space distribution of combustible mixture in the combustion chamber becomes extremely uneven, and the flame formed by combustion of biodiesel may straight ignite the unblended methanol fuel. This phenomenon causes the combustion rate to be faster, and the flame has not spread to the edge of combustion chamber and the combustion may have stopped. Thus, may leads to that some fuel is not completely burned at combustion chamber edge. The maximum burst pressure in the cylinder is obviously reduced under the combined action of various factors.

4.2 *Peak Pressure Rise Rate*

As shown in Fig. 6, when the compression ratio is 15.9, the peak pressure rise rate is the highest. When the compression ratio is 16.9, the pressure rise rate is the lowest. The rate of pressure rise is closely related to the start of combustion and the temperature in the combustion chamber. With the reduction of compression ratio, the volume of the combustion chamber increases, and the pressure and temperature of combustion chamber decrease. When the compression ratio decreases from 16.9 to 15.9, the volume of the combustion chamber increases and the pressure in cylinder decreases.

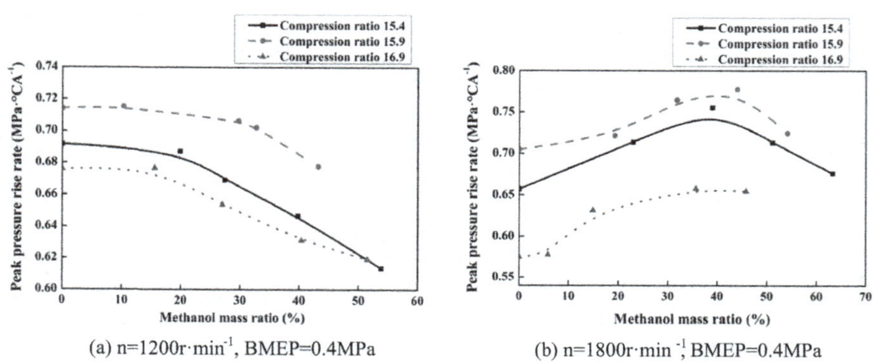

Fig. 6 Effect of compression ratio on peak pressure rise rate

At this time, the eddy current in combustion chamber may promote the evaporation and mixing of methanol, which makes more methanol premixed gas formed before biodiesel fire, and biodiesel can quickly ignite more methanol premixture after the ignition, which leads to the peak pressure rise rate increase. When the compression ratio continues to decrease to 15.4, the volume of the combustion chamber increases, the eddy current in combustion chamber is weakened, the temperature of combustion chamber is reduced and the lower temperature inhibits fuel to atomize and evaporate, resulting the peak pressure rise rate decrease.

4.3 Peak Heat Release Rate

It can be seen from Fig. 7 that with the decrease of compression ratio, the peak heat release rate increases gradually. With the reduction of compression ratio, the volume of the combustion chamber increases, and the pressure and temperature of the combustion chamber decrease. When the compression ratio is reduced from 16.9 to 15.9, the volume of combustion chamber increases. At this time, the eddy current in combustion chamber may promote the evaporation and mixing of methanol, so that there are more methanol premixture in combustion chamber before biodiesel start fire, and biodiesel can quickly ignite more methanol premixture after the ignition, thus the peak heat release rate increases. When the compression ratio continues to decrease to 15.4, the combustion chamber space is further increased and the start of combustion is delayed, resulting in more time for methanol to form premixture in the combustion chamber. The number of methanol premixture formed during the ignition delay increases, resulting in biodiesel can ignite more methanol premixed gas faster after ignition. Therefore, the peak heat release rate has been further improved.

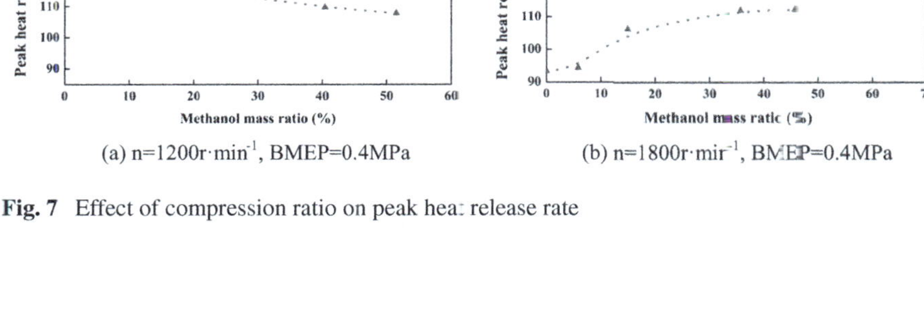

(a) n=1200r·min^{-1}, BMEP=0.4MPa (b) n=1800r·min^{-1}, BMEP=0.4MPa

Fig. 7 Effect of compression ratio on peak heat release rate

4.4 Start of Combustion and Combustion Duration

The changing trend of the start of combustion is shown in Fig. 8. At low speed, when the compression ratio is 15.9, the start of combustion of dual-fuel engine is the earliest, and when the compression ratio is 16.9, the start of combustion is closest to top dead center. At high speed, the start of combustion is earliest when compression ratio is 15.9, when compression ratio is 15.4, the start of combustion is closest to top dead center. The start of combustion of dual-fuel engine is mainly determined by the ignition time of biodiesel which ignites methanol, and the critical ignition temperature of biodiesel is related to combustion chamber pressure and excess air coefficient. With decrease of the pressure in combustion chamber, the higher of critical temperature of ignition of fuel; the excess air coefficient is closer to 1, the lower the critical temperature of ignition of fuel. When the compression ratio is reduced from 16.9 to 15.9, the volume of combustion chamber increases and the air content in combustion chamber increases, resulting in increased excess air coefficient and approaching to 1, lead to the critical temperature of ignition of fuel decreases and the start of combustion is ahead of time. When the compression ratio is reduced to 15.4, although the critical temperature of fuel ignition will be reduced by the increase of the excess air coefficient, the pressure and temperature of the combustion chamber decline more intense at this time. At this moment, the temperature and pressure are the dominant factors, which seriously affect the ignition of biodiesel, which leads to the start of combustion delay.

The trend of combustion duration change is shown in Fig. 9. At low speed and high speed, when the compression ratio is 15.9 that the combustion duration is the longest and when the compression ratio is 16.9 that the combustion duration is the shortest. For the mode of methanol premixed gas ignited by biodiesel, the combustion duration is mainly controlled by the burning rate of combustible mixture, that is to say, it is mainly related to the temperature in combustion chamber and the concentration of combustible mixture. When the compression ratio is reduced from 16.9 to

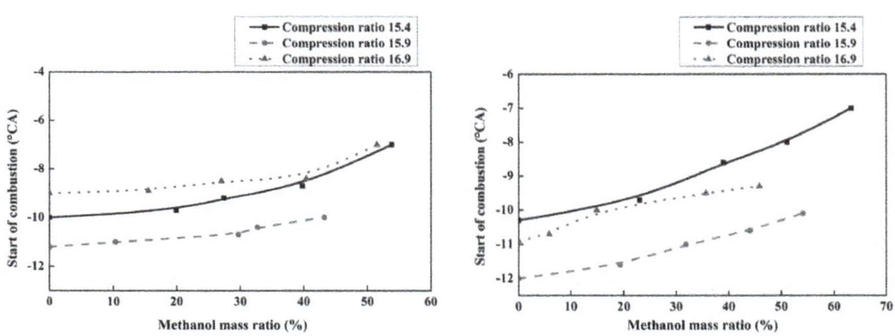

(a) n=1200r·min^{-1}, BMEP=0.4MPa (b) n=1800r·min^{-1}, BMEP=0.4MPa

Fig. 8 Effect of compression ratio on start of combustion

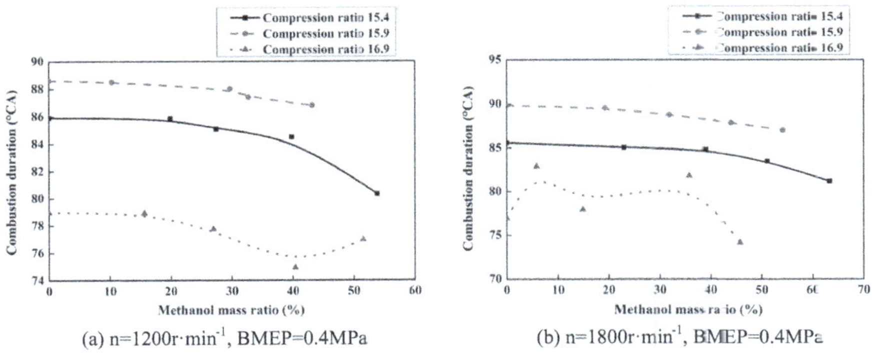

Fig. 9 Effect of compression ratio on combustion duration

15.9, the temperature and pressure in the combustion chamber will be reduced, the volume of the combustion chamber is increased, the start of combustion is advanced and the concentration of mixture is reduced. Although the premixed gas formed by methanol is more perfect at this time, the effect of atomization and evaporation of biodiesel becomes worse because the decrease of the temperature in combustion chamber, which makes diffusion combustion duration prolonged, thus the combustion duration was prolonged. When the compression ratio continues to decrease to 15.4, the volume of combustion chamber continues to increase, the temperature and pressure in combustion chamber continue to decrease, leading the atomization and evaporation of methanol and biodiesel become more difficult, the start of combustion is greatly delayed, the fuel space distribution in the combustion chamber is extremely uneven, and the flame formed by combustion of biodiesel may be straight ignite the unblended methanol fuel. This phenomenon causes the combustion rate to be faster, and the flame has not spread to the edge of combustion chamber and the combustion may have stopped, so the combustion duration is obviously shortened.

5 Conclusion

(1) With the increase of methanol mass ratio, the pressure in the cylinder gradually decreases, and the pressure rise rate and heat release rate increase first and then decrease, and the position of crankshaft angle corresponding to the maximum pressure, the peak pressure rise rate and the peak heat release rate is gradually lagged, the start of combustion is gradually delayed and the combustion duration is gradually shortened.

(2) The decrease of compression ratio will lead to increase of the peak heat release rate, but the influence on the maximum burst pressure, the peak pressure rise rate, the start of combustion and the combustion duration is complicated. When compression ratio decreases from 16.9 to 15.9, the peak pressure and peak

pressure rise rate in cylinder will increase, and the start of combustion will be advanced and the combustion duration will be prolonged. When the compression ratio decreases from 15.9 to 15.4, the peak pressure and peak pressure rise rate in cylinder will decrease, and the start of combustion will be delayed and the combustion duration will be shortened. Therefore, when the compression ratio is 15.9, the combustion state of engine is the most ideal for the mode of methanol premixed gas ignited by biodiesel.

References

1. Geng P, Yao C, Wang Q et al (2015) Effect of DMDF on the PM emission from a turbo-charged diesel engine with DDOC and DPOC. Appl Energy 148:449–455
2. Geng P, Yao C, Wei L et al (2014) Reduction of PM emissions from a heavy-duty diesel engine with diesel/methanol dual fuel. Fuel 123(1):1–11
3. Wang Q, Wei L, Wang P et al (2015) Investigation of operating range in a methanol fumigated diesel engine. Fuel 140:164–170
4. Mingchen Xu, Qi Huiqing (2014) Analysis on methanol fueled vehicle industry and the development countermeasures. On Econ Prob 12:63–67
5. Kong JH, Yu DD, Huang JP et al (2016) Synthesis of biodiesel from waste cooking oil catalyzed by compound basic ionic liquid. China Oils Fats
6. Zhen X, Wang Y (2015) An overview of methanol as an internal combustion engine fuel. Renew Sustain Energy Rev 52:477–493
7. Wang J, Shuai S (2011) Automotive engine fundanmentals. Tsinghua University Press, Beijing
8. An Q (2015) Experimental study on the engine with premixed methanol charge induced ignition by f-t diesel. Taiyuan University of Technology

Mechanical Properties Research on Cylinder Head Based on Simulation of Secondary Dendrite Arm

Zhengfeng Jia, Tao Liu, Qingqiang Zeng, Huibin Qing and Huixian Shen

Abstract Firstly, the qualitative relationship is verified by test data between secondary dendrite arm spacing (SDAS) and tensile strength (σ_b), and the quantitative equation is $\sigma_b = 641.75(\text{SDAS})^{-0.28}$, deduced by the linear regression method. Meanwhile, Furer–Wunderlin formula is equivalently transformed to $\text{SDAS} = A^* \times (T_s)^{1/3}$ by modifying the constant part to A^*. Then, the method is deeply studied to get SDAS rapidly and accurately. The first step is to obtain actual local solidification time (T_s) by calibrating solidification temperature by casting simulation and measurement. The second step is to calculate SDAS by modified theoretical formula ($\text{SDAS} = A^* \times (T_s)^{1/3}$). The third step is to calibrate proper A^* by SDAS test data. Finally, the research method is applied to the same kind of aluminum alloy cylinder head development, the head passed the reliability test one time and reduced test cost-effectively and improved the quality significantly. One of the study results is that SDAS and tensile strength (σ_b) have quantitative relationship: $\sigma_b = 641.75(\text{SDAS})^{-0.28}$. SDAS sample is obtained easier than tensile strength specimen and breakthrough the limitation of the tensile strength specimen from the product. SDAS can be as an indicator to evaluate tensile strength in the early stage of product development. The other is that the method to get SDAS is innovative and feasible by combining with cast simulation and modified theoretical formula ($\text{SDAS} = A^* \times (T_s)^{1/3}$), but it is necessary to calibrate two key parameters of local solidification time T_s and A^*, and the proper A^* is ranged from 6.5 to 7.5 for aluminum alloy cylinder head.

Keywords Secondary dendrite arm spacing · Aluminum alloy cylinder head · Local solidification time · Tensile strength · Casting simulation · Test · Calibration

Z. Jia (✉) · T. Liu · Q. Zeng · H. Qing · H. Shen
Powertrain R&D Institute, Changan Automobile, Chongqing 401120, China
e-mail: jiazf@changan.com.cn

China SAE (ed.), *Proceedings of China SAE Congress 2018: Selected Papers*, Lecture Notes in Electrical Engineering 574,
https://doi.org/10.1007/978-981-13-9718-9_39

1 Introduction

1.1 Background

With the continuous improvement of computer numerical simulation technology, the numerical simulation of flow and temperature field during cast filling and solidification is becoming more and more perfect. The function of commercial casting CAE software is gradually enhanced, so that, casting simulation is very popularly applied in engineering and has achieved significant economic and social benefits in the production [1]. With the automotive requirements of lightweight and low fuel consumption, more and more auto parts use aluminum alloy casting; for example, aluminum alloy cylinder head has been widely used in automotive business due to anti-stretch, anti-fatigue, corrosion resistance, high specific strength, and other excellent performances.

1.2 Significance of This Study

Mechanical properties of the product under certain casting condition can be predicted by samples from the product. However, the structure of the aluminum alloy cylinder head is complicated and the shape is irregular. Samples acquirement is hard in such area, and mechanical properties of the body can not be obtained. The qualitative relationship between secondary dendrite arm spacing (SDAS) and mechanical properties has been studied in the literature [2–5]. In conclusion, the method is feasible to predict mechanical properties indirectly by SDAS test but there are limitations to obtain SDAS sample in the early product development. Other methods to get SDAS rapidly and accurately may be worthy of study, such as simulation, but there is little research on this aspect.

1.3 Contents of This Study

Firstly, the qualitative relationship is verified by test data between secondary dendrite arm spacing and tensile strength, and quantitative equation is deduced by the linear regression method. Furer–Wunderlin formula is equivalently transformed by modifying constant part to A^*. Then, the method is deep studied to get SDAS rapidly and accurately including calibration of two key parameters of local solidification time T_s and A^*. Finally, the research method is applied to the same kind of aluminum alloy cylinder head development.

2 Theoretical Study on SDAS

Secondary dendrite arm spacing (SDAS) refers to the distance between the secondary dendrite arms in the aluminum alloy microstructure.

2.1 Study on the Relationship Between SDAS and Tensile Strength

The conclusion is the smaller SDAS of the aluminum alloy test bar, the better mechanical properties, which has been confirmed by many experimental data and also proved by Flemings and others' research results [6]. In this paper, the tensile strength and SDAS data are obtained by tests from cylinder head body, as shown in Table 1.

According to Table 1 data, the relationship between SDAS and tensile strength is as shown in Fig. 1.

As can be seen from Fig. 1, the strength of aluminum alloy reduced with SDAS increased. The results anastomosis with literature [6]. The quantitative equation of tensile strength and SDAS is deduced by the linear regression method.

$$\sigma_b = 641.75(\mathrm{SDAS})^{-0.28} \tag{1}$$

Formula (1), SDAS can predict tensile strength for aluminum alloy cylinder head. SDAS sample is obtained easier than tensile strength specimen and breakthrough the limitation of the tensile strength specimen from the product. SDAS can be as an indicator to evaluate tensile strength in the early stage of product development.

Table 1 Test data

	*s*1	*s*2	*s*3	*s*4	*s*5
SDAS (μm)	27.78	36.81	40.36	41.77	51.18
Tensile strength (MPa)	246.9	240.3	227.2	235.4	206.0

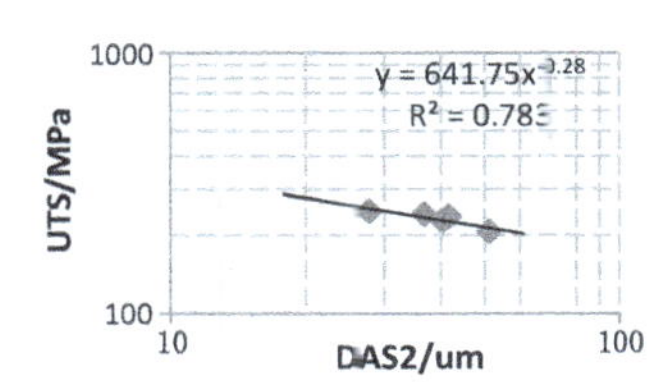

Fig. 1 Relationship between SDAS and strength

2.2 Study on Relationship Between SDAS and Solidification Time T_S

Furer–Wunderlin formula [7] of SDAS and solidification time T_S is:

$$\mathrm{SDAS} = 5.5(A \times \mathrm{T_S})^{1/3} \tag{2}$$

Which SDAS—secondary dendrite arm spacing
T_S—local solidification time
A—rough coefficient

$$A = \frac{\cdot D_L L_n (C_L/C_0)}{m_L(1-k)(C_0 - C_L}$$

In the formula, L_n: Gibbs–Thompson coefficient; D_L: liquid solute diffusion coefficient; C_L: liquid concentration; C_0: the original concentration of the molten alloy; m_L: liquid line slope; k: equilibrium distribution coefficient).

In Eq. (2), A is a constant, 5.5 is a constant, will be converted to A^*, then Eq. (2) is converted into Eq. (3)

$$\mathrm{SDAS} = A^* \times (T_S)^{1/3} \tag{3}$$

In formula (3): SDAS is the secondary dendrite arm spacing, T_S is the local solidification time, A^* is a constant obtained by combining the constant part, which relates to alloy composition, cooling conditions, etc.

The secondary dendrite arm space has a quantitative relationship with tensile strength and solidification time through the study of 2.1 and 2.2.

3 Simulation Study on SDAS

3.1 Model and Boundary Description

In order to calculate the temperature field accurately, some complicated factors in the casting process must be taken into account in the solving conditions of the numerical equation. For example, the initial conditions, boundary conditions, latent heat, thermophysical properties of casting, and molds vary with various factors, as well as the handling of complex geometries and so on.

The finite difference method is suitable and effective in the flow and temperature field simulation during casting process. Cylinder head, mold, cooling pipe, sand core, and all other parts use hexahedral grid. Meanwhile, cylinder head casting grid should be fined. FDM model is shown in Fig. 2.

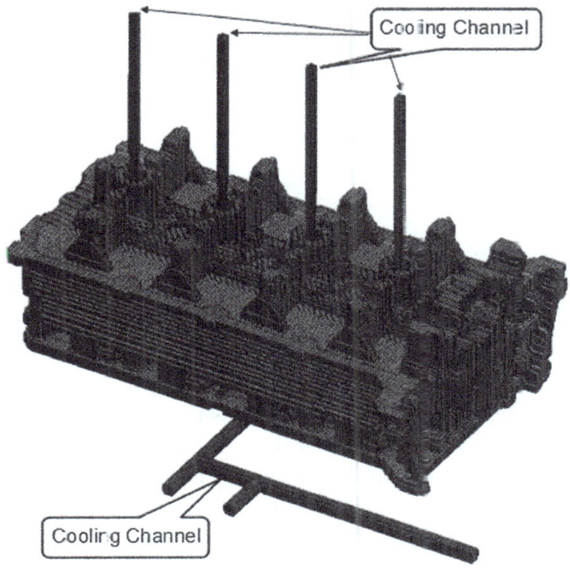

Fig. 2 FDM mode

Table 2 Material and initial temperature

Parts	Cast	Top	Side	Bottom	Riser	Core
Material	AC4B	H13	H13	H13	H13	Coated resin
Initial temp (°C)	700	350	350	400	570	30

Cylinder head using low-pressure casting process for the actual production, the material for cylinder head is AC4B, a kind of aluminum–silicon alloy. The actual process, cooling conditions, and pressure curve should be taken into account. The materials used in the analysis and the initial temperature are shown in Table 2.

3.2 Calibration of Casting Temperature

Before the calibration, the initial temperature, opening and closing time of cooling pipe, casting process parameters, and the entire process must be investigated and confirmed on-site. Ensure that the impact factors of the temperature field are taken into account.

Boundary convective heat transfer coefficient and interface heat transfer coefficient are difficult to obtain through test, manual, or literature. In the temperature field simulation, the boundary heat transfer coefficient is hypothetical, and the casting and mold temperature is to be sought. Since the temperature field is easily measured by experiment, it is now necessary to deduce the boundary heat transfer coefficient, that

Table 3 Temperature results before and after calibration

Points	Measured (°C)	Before calib (°C)	After calib (°C)	Error (°C)	Error (%)
Top1	345	117.4	341.7	−3.3	−0.96
Bot2	487	280.5	492	5	1.03
Bot3	503	297	506.6	3.6	0.72
Bot4	503	300.2	511.1	8.1	1.61
Bot5	489	290.4	507.6	18.6	3.80
Bot6	458	245	469.1	11.1	2.42
Bot7	467	221.2	454.1	−12.9	−2.76

is, to measure a number of points temperature from the casting system. The error between the calculated and measured temperature gradually decreases by several rounds of simulation and calibration and get the proper convection and interface heat transfer coefficient. The target of temperature field calibration is achieved, results are shown in Table 3, the maximum error is 18.6 °C, and the error is 3.8%, within 5%.

3.3 *Simulation of Local Solidification Time* T_S

The precise local solidification time T_S is gotten by the precise temperature field after calibration. The local solidification time results of cylinder head cut section are shown in Fig. 3.

Fig. 3 Local solidification time

3.4 Calibration of Constant A*

The dendrite arm spacing is calculated by dendrite method [5]. SDAS test is carried out at 1–6 in Fig. 4, according to conventional metallographic sample preparation method for grinding, polishing, and corrosion.

As far as possible, to find five measurement areas, in the representative of magnification field of 50 times or 100 times, the number of secondary dendrites is preferably at least 10, at least not less than 5, and the total length between each dendrite is measured, divided by the number of dendrites contained in this length range, SDAS is obtained as shown in Fig. 5.

A^* is calibrated by SDAS test data related to alloy composition, cooling conditions, etc. The calibration A^* coefficient ranges from 6.5 to 7.5, and the results are shown in Table 4.

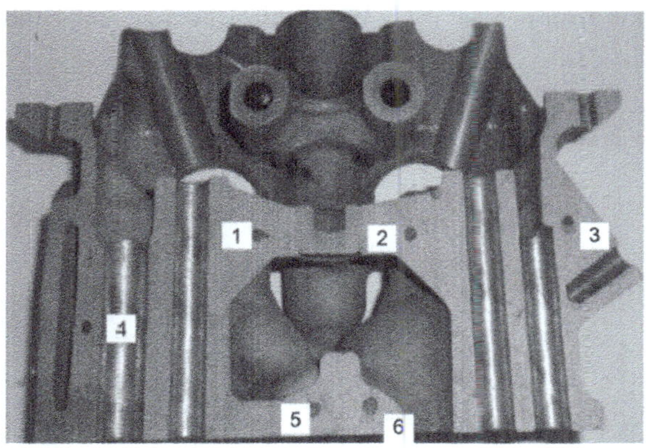

Fig. 4 SDAS samples of cylinder head section

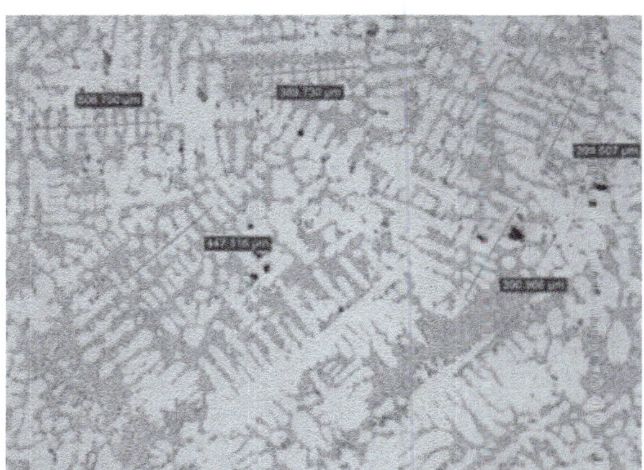

Fig. 5 SDAS test

Table 4 SDAS results

Sample	Test (μm)	Simulation (μm)	Error (μm)	Error (%)
1	47.3	47.8	0.5	1.06
2	47.7	47.8	0.1	0.21
3	28.5	30	1.5	5.26
4	31.9	33	1.1	3.45
5	29.7	31.7	2	6.73
6	31	32.8	1.8	5.81

By formula (3), SDAS $= A^* \times (T_S)^{1/3}$ SDAS simulation result after calibration is shown in Fig. 6.

The SDAS simulation results obtained after calibration of A^* coefficient are in good agreement with the test results. Figure 7 shows the maximum error is 2 μm, the error percent is 6.73%, within 10%, reaching calibration target of A^* coefficient.

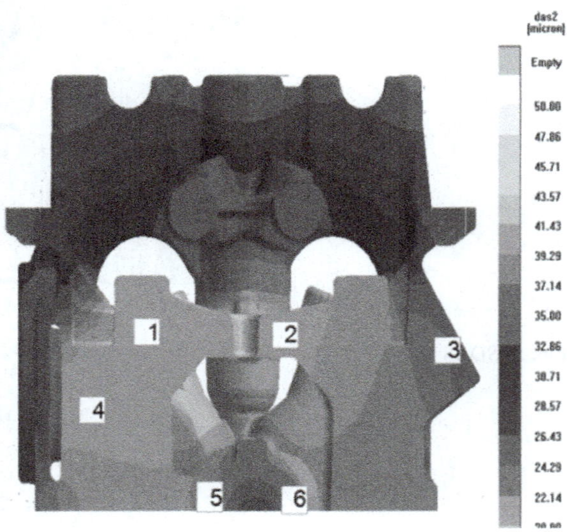

Fig. 6 SDAS

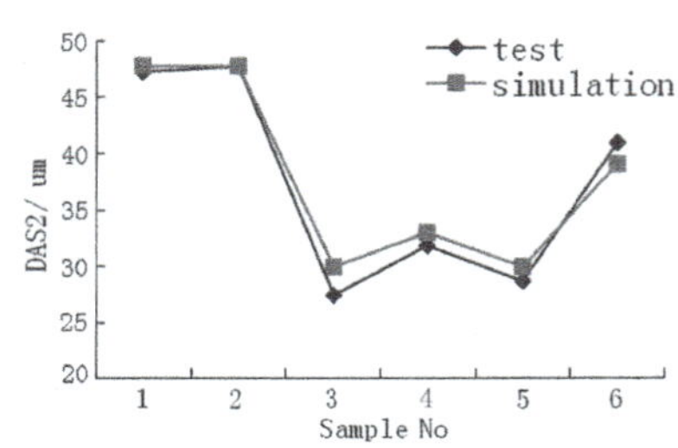

Fig. 7 SDAS results by simulation and test

4 Application to Other Aluminum Alloy Cylinder Head Development

4.1 SDAS Technical Requirements of Head

The research method is applied to the same kind of aluminum alloy cylinder head development, SDAS results are shown in Fig. 8, and SDAS ranges from 20 to 48 μm as shown in Table 5. It is used as one mechanical performance indicator written to technical requirements, SDAS proposed for the key areas of the cylinder head: the combustion chamber ≤30 μm, the middle plate ≤40 μm, the top of the cylinder head, and other areas ≤50 μm.

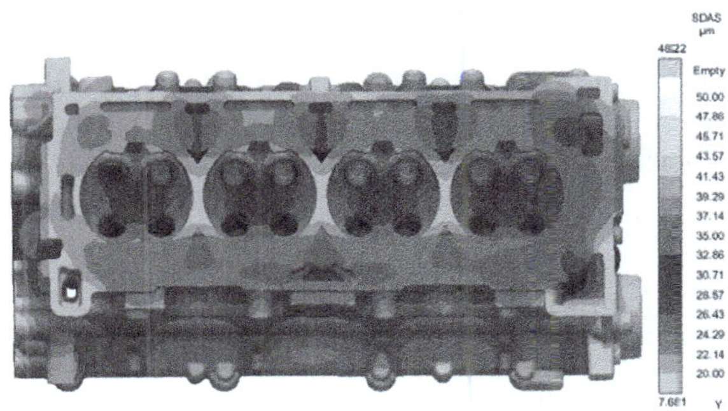

Fig. 8 SDAS simulation results

Table 5 SDAS simulation results and technical requirements

Key areas	Simulation (μm)	Average (μm)	Technical requirements (μm)
Combustion chamber	28.861	29.93	30
	30.104		
	30.078		
	30.674		
Middle plate	37.838	38.03	40
	38.186		
	38.705		
	37.373		
Top (Maximum) and other areas	45.861	47.08	50
	48.065		
	48.116		
	47.972		
	47.982		
	44.49		

Table 6 SDAS and tensile strength test data

	SDAS (μm)	Tensile strength (MPa)
Combustion chamber	29.3	305.3
	31.2	286
Middle plate	47.1	268.4
	47.9	264.1
Top	48.9	258

4.2 *Test Results*

After the cylinder head product samples come out, SDAS test data of key parts of cylinder head are shown in Table 6. The test data basically meet the development requirements and ensure the achievement of the mechanical performance objectives.

The development of the cylinder head passed the reliability of verification once time, ensuring the reliability of the development of the cylinder head, at the same time, effectively reducing the test cost and improving the product quality and development efficiency.

5 Conclusion

(1) The qualitative relationship is verified by test data between secondary dendrite arm spacing (SDAS) and tensile strength (σ_b), and the quantitative equation $\sigma_b = 641.75(\text{SDAS})^{-0.28}$ is deduced by the linear regression method. SDAS sample is obtained easier than tensile strength specimen and break the limitation of the tensile strength specimen from the product. SDAS can be as an indicator to evaluate tensile strength in the early stage of product development.

(2) The method to get SDAS is innovative and feasible by combining with cast simulation and modified Furer–Wunderlin theoretical formula ($\text{SDAS} = A^* \times (T_s)^{1/3}$) by transforming constant part to A^*, but it is necessary to calibrate two key parameters of local solidification time T_s and A^*, and the proper A^* calibrated is ranged from 6.5 to 7.5 for aluminum alloy cylinder head.

(3) The research method is applied to the same kind of aluminum alloy cylinder head development, the head passed the reliability test one time and reduced test cost-effectively and improved the quality significantly. It is proven that the method to obtain SDAS is useful and reliable especially in the early product development.

References

1. Liu B, Xu Q, Xion S, etc. (2003) Research progress of multi-scale simulation of casting process. J Mech Eng 10: 53–63
2. Zhang B, Wang S, Wang Y (2008) Study on secondary dendrite arm spacing of gravity casting aluminum alloy engine cylinder head. Foundry Technol 29(11):1546–1549
3. Li C, Guo T, Li R (2004) Study on secondary dendrite arm spacing. Casting 53(12):1011–1014
4. Zhang G, Jing T, Liu B, et al (2002) Numerical simulation of microstructure of aluminum alloy casting using macro-micro coupled method. Proc ASME-IMECE, CD-No. 2002-32903. 2002, NewOrleans, USA
5. Zhang B, Garro M, Tagliano C (2003) Dendrite arm spacing in aluminum alloy cylinder heads produced by gravity semi2 permanent mold. Published Metallurgical Sci Technol 21(1):3–9
6. Zhang B, Garro M, Giglio A (2004) Effect of dendrite arm spacing on mechanical properties of aluminum alloy cylinder heads and engine blocks. SAE Int 2:101–110
7. Feurer U, Wunderlin R (1977) Metal solidification. DGM Fachber, Stuttgart

Analysis of Energy Absorption Characteristics of Typical High-Strength Steel Based on Drop Test

Wei Zhang, Chunguang Li, Jianwei Yang, Lixian Liu, Qing Chen and Fulin Wei

Abstract The drop test is an important method to analyze the energy absorption characteristics of materials, which causes the sample to collapse axially and absorb energy by initial energy impact. Hat-shaped beam is a typical structure of energy absorber of front longitudinal beam of automobile. It is widely used for its lightweight, high strength, and good energy absorption. Select four typical high-strength steels: HC340LAD + Z, HC420/DP780D + Z, HC550/DP980D + Z and TRIP690, the cap girder structure is obtained through stamping and welding, and the test is taken based on the drop test. According to the test results, evaluation indexes such as energy absorption ratio, load ratio, maximum load, and average load are obtained, and the energy absorption characteristics of the materials are compared and analyzed. The results show that the energy absorption characteristics of different materials are quite different, and the energy absorption ratio of the materials tends to decrease gradually as the material strength increases.

Keywords Drop test · Energy absorption characteristics · High-strength steel · Thin-walled beam

1 Introduction

The collapse test is mainly used to observe the absorbing effect of metal thin wall material under impact environment. The thin-walled beam structure of sheet metal obtained by stamping and welding is widely used in energy-absorbing elements due to its lightweight, high strength, and good energy-absorbing effect and mainly absorbs energy through axial crush deformation mode [1]. The axial crush test passes the drop hammer test bench, and the drop hammer sets certain initial energy to apply axial load on the sample to obtain its deformation mode, dynamic force mode, dynamic

W. Zhang (✉) · C. Li · J. Yang · L. Liu · Q. Chen · F. Wei
Application Technology Department, Shougang Research Institute of Technology, Beijing 10043, China
e-mail: zlg1314@yeah.net

China SAE (ed.), *Proceedings of China SAE Congress 2018: Selected Papers*, Lecture Notes in Electrical Engineering 574,
https://doi.org/10.1007/978-981-13-9718-9_40

force-displacement curve, energy-displacement curve, and energy-time images used to examine the energy-absorbing properties of typical components [2].

In this paper, select four typical high-strength steels: HC340LAD + Z, HC420/DP780D + Z, HC550/DP980D + Z, and TRIP690, the cap girder structure is obtained through stamping and welding, and the test is taken based on the drop test. The dynamic force-displacement curve, energy-displacement curve, and energy-time images are taken through the drop test. According to the test results, evaluation indexes such as energy absorption ratio, load ratio, maximum load, and average load are obtained, and the energy absorption characteristics of the materials are compared and analyzed.

2 The Test Sample

(1) Experiment material

The steel used in this test was HC340LAD + Z, HC420/DP780D + Z, HC550/DP980D + Z, and TRIP690. The thicknesses were all 1.0 mm. The basic mechanical properties of the material are shown in Table 1.

(2) Sample size

The sample adopts a rectangular cross section. The rectangular cross-section energy-absorbing beam has a simple structure and strong energy absorption ability. It is a typical form of the automobile longitudinal beam [3, 4]. In order to ensure axial crushing smoothly, the sample size must meet the following requirements: section aspect ratio $a/b > 0.67$, aspect ratio $t/b > 0.016$, aspect ratio $l_o/b < 15$, the cross-sectional dimensions are shown in Fig. 1. The typical sample dimensions are shown in Table 2.

Therefore, according to the requirements of the above sample size, according to the material thickness of the test, the sample size is selected as shown in Table 3.

According to the sample size, the sheet is punched or bent into a U shape, and the U-shaped flange and the web are spot welded to obtain a thin-walled beam structure. The spot welding pitch is 30 mm, the welding core diameter is 6 mm, and the welding point is 15 mm from the edge [5].

Table 1 Basic mechanical properties

Material	Specified plastic elongation strength $R_{p0.2}$/MPa	Tensile strength R_m/MPa	Elongation after fracture A/%
HC340LAD + Z	345	445	30
HC420/DP780D + Z	492	832	19.5
HC550/DP980D + Z	615	971	14
TRIP690	426	692	28.5

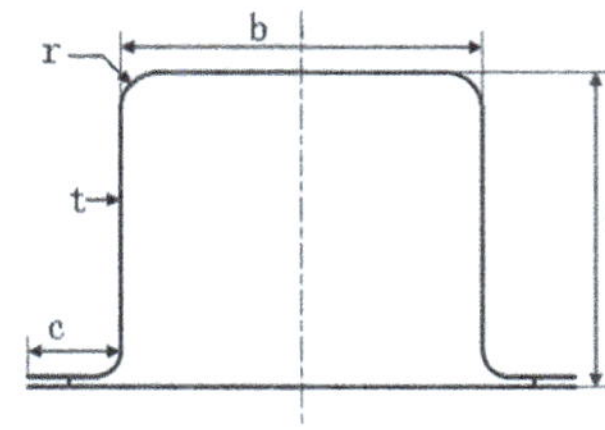

Fig. 1 Sample section

Table 2 Typical sample size

Feature	Value		
	1	2	3
Sheet thickness (*t*/mm)	1.2–2.0	1.5–3.5	2.0–4.0
Rectangular section height (*a*/mm)	60	80	120
Rectangular section width (*b*/mm)	70	80	120
Flange width (*c*/mm)	25	30	30
Corner radius (*r*/mm)	7	8	10
Length (l_0/mm)	300	400	500

Table 3 Sampling test specimen size

Characteristic	Value
Sheet thickness (*t*/mm)	1.0
Rectangular section height (*a*/mm)	50
Rectangular section width (*b*/mm)	30
Flange width (*c*/mm)	25
Corner radius (*r*/mm)	7
Length (l_0/mm)	300

3 Drop Weight Crush Test

3.1 Test Bench

The drop hammer crush test was conducted on a high-speed drop hammer test bench. The test equipment was designed to increase the height of the drop hammer (maximum height of 5 m) and increase the weight of the drop hammer can increase initial energy (maximum weight of 234 kg, including hammer weight of 34 kg). In addition, the test equipment can achieve a maximum impact speed of 15 m/s by configuring the spring-assisted energy storage device. The test bench is shown in Fig. 2.

Throughout the test process, in order to achieve high-precision, high-response, high-frequency acquisition of strain gauge voltage signals, ultra-dynamic strain gauges are used, and high-speed cameras are used for image acquisition. Satisfy the

Fig. 2 Drop hammer crush test bench

requirement of sampling frequency for this experiment, and can accurately record the deformation of the test piece at each moment in the test, recorded high-quality pictures ensure the accuracy of post-processing software's deformation processing of test pieces [6].

When using a high-speed drop hammer test stand, in order to improve the measurement accuracy of the loading force of the tube, at the same time reduce the repetitive work such as strain gauges, independently developed load cells and fixtures. The upper part of the sensor is fixed with a strain gauge, and an electric signal can be output when there is a loading force [7] as shown in Fig. 3.

The upper end of the sensor is fixed with a cap-type pipe fitting with a strain gauge in the middle, which can output the electrical signal when loaded. The quasi-static calibration of the sensor is required before the test, that is, the relationship between the sensor electrical signal and the force data is established. The sensor itself is independent of the strain rate during the elastic phase. When measuring the force, the loading force must not exceed the yield force of the sensor. After determining the calibration coefficients, this coefficient can be used in the dynamic test to obtain the real force data in the dynamic test.

Fig. 3 Drop hammer crush test fixture and sensor

3.2 Testing Process

The size of the test sample is measured, and the external dimensions are used to determine the influencing factors of the energy absorption effect of the sample and to evaluate the energy absorption effect of the sample. Since it is difficult to directly measure the impact force of the test piece during the experiment, it is necessary to design the test fixture and the base design. Therefore, when the strain gauge is pasted on the base, an elastic deformation section may occur. The whole base can be approximated as a rigid body, so the strain gauge can be used to measure the impact force of the specimen. Because the sensor used in the experiment is a voltage signal, and a force signal is needed in the energy calculation process, it is necessary to calibrate the sensor, measure the voltage signal and the force signal, and find out the corresponding relationship before the test.

Two sets of data were collected during the test which are obtained by a high-speed camera and hyperdynamic strainmeter. And the time—displacement data and time—force data is obtained, respectively. During the experiment, high-speed video is used to collect test images, and mark points are added to the lowest weight block before the test. Through the processing of special image processing software in the later stage, the time-displacement data of the specimen during the test is obtained. The time-force data is obtained through the strain gauges pasted on the pedestal, and the voltage signal after the gain is read by the hyperdynamic strain gauge, and then the time-force data is obtained according to the sensor calibration voltage–force relationship parameters before the test [8].

3.3 Data Processing

Test data processing includes three parts: test calibration data processing, experimental image data processing, and test energy data processing.

Test calibration data processing: during the calibration process, time—force and time—voltage data were collected. Since the two sets of data are read on different instruments and are not triggered at the same time, the time axes of the two sets of data are not aligned. Therefore, the importance of obtaining relational parameters lies in the time axis alignment and interpolation of the two sets of data.

Test image data processing and the experimental image processing are realized under the professional image processing software Vic-2D. The main way to achieve this is to use the marker points added before the test for tracking calculations. Calculate the deformation, deformation speed, and deformation acceleration of the test piece during the test. The high-speed camera test image is shown in Fig 4.

Test energy data processing: Calculate through the data obtained from the above two-step test, the triggering method used in the test is manual triggering and triggering at the same time. So, the two sets of data timelines acquired are synchronized, because the frequency of the two is different, the time nodes of the read data are inconsistent.

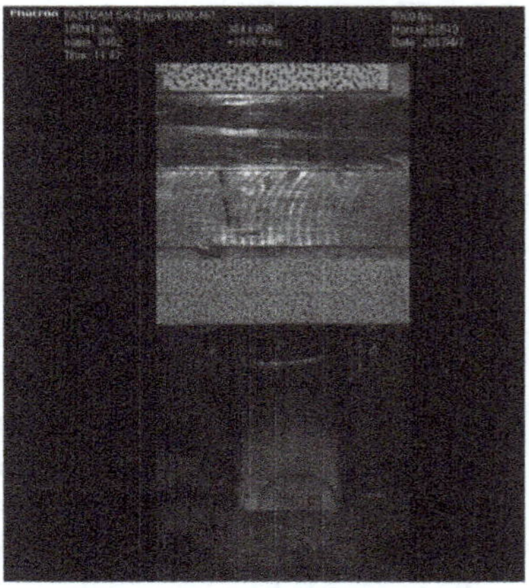

Fig. 4 High-speed camera captures displacement change images

Therefore, interpolation calculations are needed to interpolate the two sets of data to the same time axis. The starting point of the impact is determined from the force image and the displacement image, and the energy absorbed by the test piece during impact is calculated according to the $W = FS$ integral and drawn as needed [9]. Each image is shown in Fig. 5.

The test specimens are subjected to dynamic impact tests at speeds of 5, 8, and 10 m/s, respectively, and three tests are performed for each material at each rate. The deformation mode of the sample is shown in Fig. 6.

Through the post-processing of the data is collected during the test, the dynamic force-displacement curves and energy-displacement curves of the test are shown in Fig. 7.

4 Comparison of Energy Absorption Characteristics

4.1 Evaluation Index

The main parameters for evaluating the energy absorption properties of thin-walled specimens include the following:

(1) Average load F_{mean} and Peak load F_{max}

As shown in Fig. 8, the average load F_{mean} reflects the overall load level; In order to maximize the absorption of impact energy through plastic deformation, the average load of the structure should be as large as possible within a certain range. Peak load F_{max} reflects initial overload or maximum overload condition. Because the impact force is transmitted to the occupant through the body in the form of deceleration, excessive deceleration can cause serious injury to the occupant, so the maximum peak load should be limited [10] (Fig. 9).

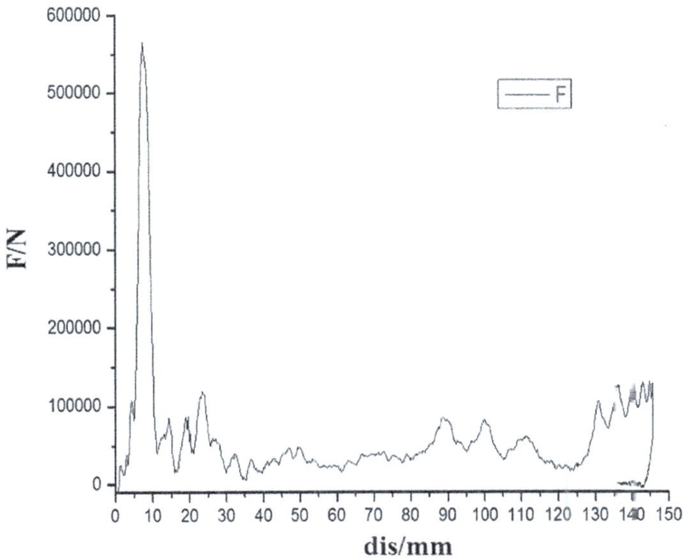

(a) Force-displacement curve

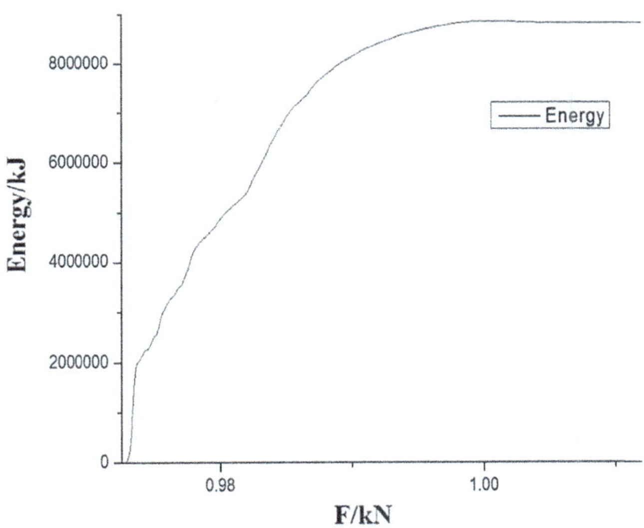

(b) Energy-time curve

Fig. 5 Test curve processing

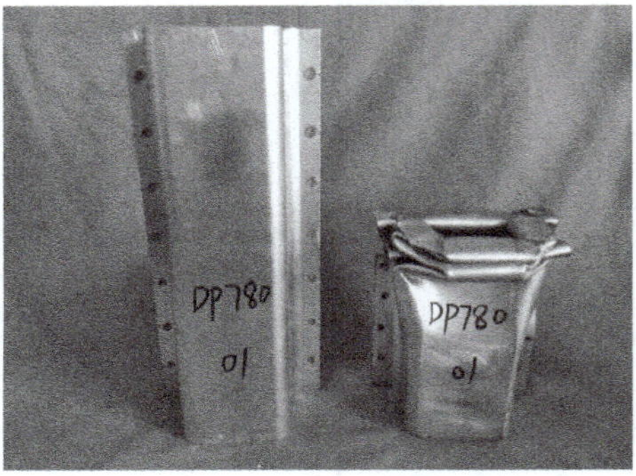

Fig. 6 Comparison of specimens before and after crush test

(2) Total energy absorption E_k and Energy absorption ratio H

The energy absorbed during the entire crush deformation process can be obtained by integrating the load-displacement history curve, i.e.,:

$$E_k = \int_{S_i}^{S_b} F\mathrm{d}S = F_{\mathrm{mean}}(S_b - S_i) = F_{\mathrm{mean}}\Delta l \quad (1)$$

In the formula, S_b and S_i are the final failure point and initial failure point, respectively; F is the forward load; F_{mean} is the average load; Δl is the effective failure length.

The energy absorption ratio H refers to the ratio of the total energy absorbed by the tube to the initial energy, i.e.,:

$$H = \frac{E_k}{E_f} \quad (2)$$

In the formula, E_f is the input initial energy.

(3) Specimen deformation Δh

Specimen deformation Δh refers to the amount of collapse of square tubes and rectangular tubes during crushing. In order to maximize the absorption of impact energy through plastic deformation, the collapse of the structure should be as large as possible within a certain range.

(4) Load ratio

The load ratio is the ratio of the average load to the maximum load, which is $F_{\mathrm{mean}}/F_{\mathrm{max}}$. This ratio reflects the fluctuation of the load. The larger the ratio is, the smaller the fluctuation of the load. When the ratio is 1.0, the theoretical optimal load ratio is achieved.

Fig. 7 Test results

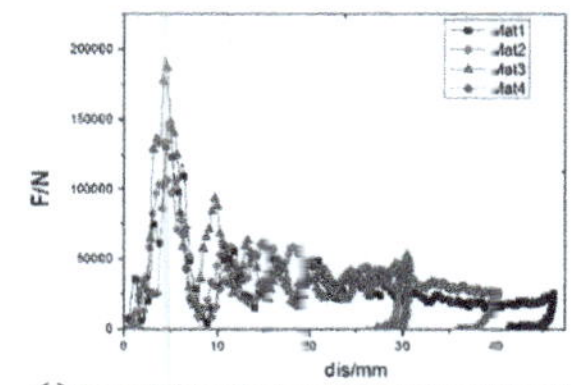

(a) Dynamic force-displacement curve at a velocity of 5m/s

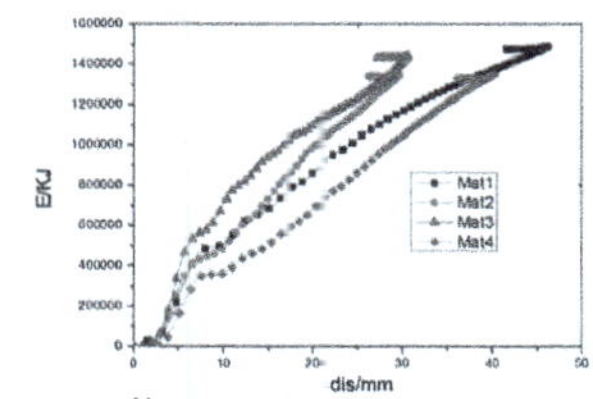

(b) Energy-displacement curve at a rate of 5m/s

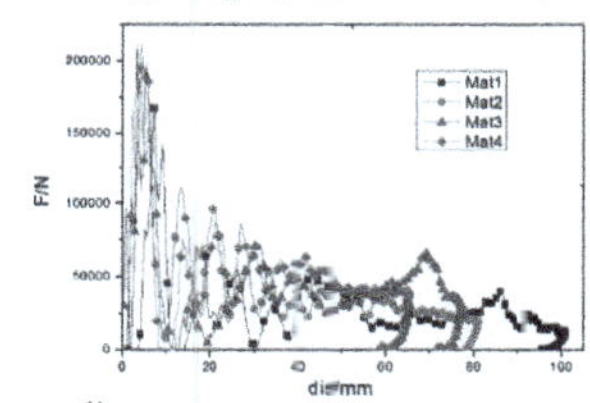

(c) Dynamic force-displacement curve at a rate of 8m/s

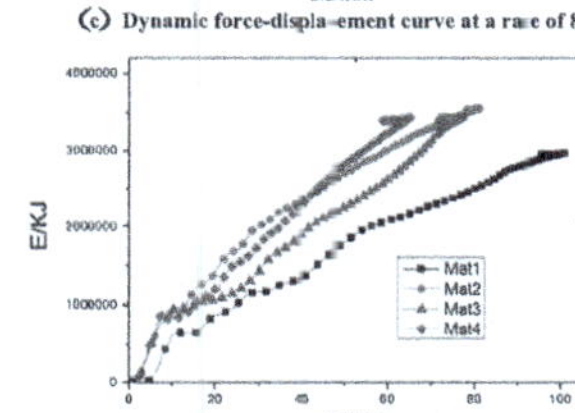

(d) Energy-displacement curve at a rate of 8m/s

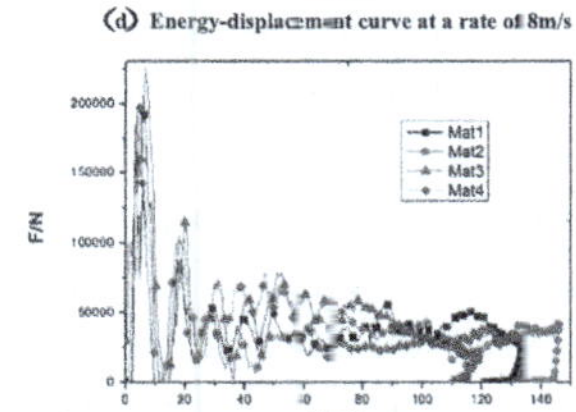

(e) Dynamic force-displacement curve at a velocity of 10m/s

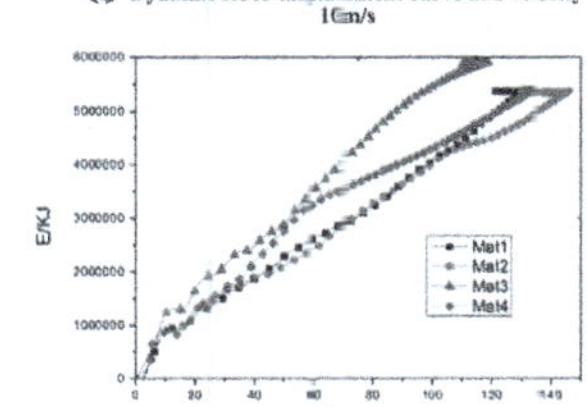

(f) Energy-displacement curve at a rate of 10m/s

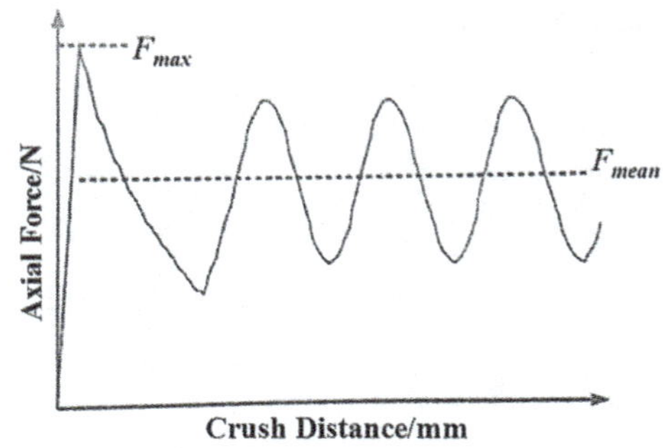

Fig. 8 Crush distance and axial force curve

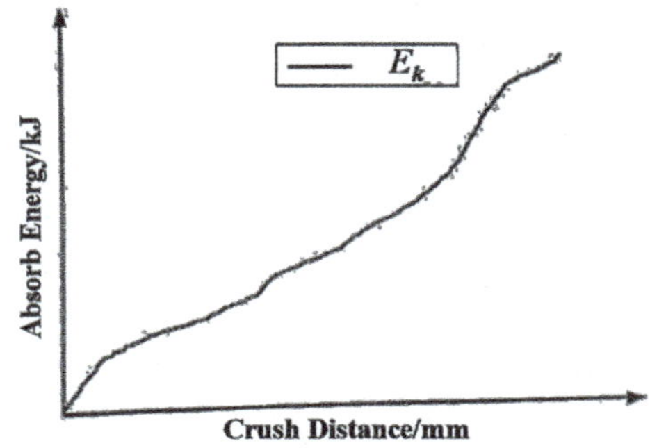

Fig. 9 Crush distance and absorption energy curve

4.2 Result Analysis

According to the test results, combined with the evaluation index, the data parameters of the material energy absorption characteristics are shown in Table 4.

According to the evaluation index of the material structure performance, the analysis of the test results shows that.

The energy absorption of the HC550/DP980D + Z is relatively low. Its peak load and mean load are larger than other material structures, but its load ratio is also greater than other material structures. The peak load of HC420/DP780D + Z is larger, the average load is relatively small, and its load ratio is smaller than other material structures. TRIP690 load, load ratio, energy absorption ratio are slightly greater than HC340LAD + Z; comparison shows that the absorption ratio of materials decreases with the increase of material strength.

Table 4 Material energy absorption characteristic parameters

Specimen	Test speed (m/s)	Absorb energy (J)	Peak load (F_{max}/N)	Deformation (mm)	Average force (F_{mean}/N)	Energy absorption ratio	Load ratio
HC340LAD + Z	4.86	1493.47	149,429.58	46.18	32,340.19	0.944	0.2164
	4.91	1403.23	175,384.06	38.22	36,714.54	0.869	0.2093
	7.93	3374.85	177,412.20	84.76	39,816.54	0.801	0.2244
	7.92	2970.56	167,641.96	100.94	29,428.96	0.707	0.1755
	9.90	5403.17	191,780.66	133.53	40,464.09	0.823	0.2110
HC420/DP780D + Z	5.09	1242.42	150,035.05	31.57	39,354.45	0.716	0.2623
	4.92	1444.39	194,539.57	30.59	47,217.71	0.891	0.2427
	8.13	3448.56	211,347.65	77.19	44,676.25	0.779	0.2114
	8.01	3722.45	284,098.21	62.15	59,894.61	0.866	0.2108
	10.98	5932.65	225,429.85	118.70	49,980.20	0.734	0.2217
HC550/DP980D + Z	5.00	1375.78	211,520.86	18.64	73,807.94	0.821	0.3489
	4.99	1417.81	219,008.02	16.73	84,746.56	0.850	0.3869
	8.01	3267.56	220,077.00	56.41	57,925.19	0.760	0.2632
	9.10	4006.18	219,927.91	74.57	53,723.75	0.722	0.2443
	10.80	5492.63	230,806.33	107.92	50,895.38	0.703	0.2205
TRIP690	4.97	1354.82	111,048.45	40.13	33,760.78	0.819	0.3040
	4.95	1353.44	192,063.20	29.05	46,590.02	0.824	0.2426
	7.81	3435.93	212,360.98	65.25	52,657.93	0.834	0.2480
	7.88	3415.38	166,848.16	82.07	41,615.45	0.821	0.2494
	9.85	5389.26	196,614.53	133.80	40,278.47	0.829	0.2049
	9.94	5568.89	184,745.64	111.63	49,887.04	0.841	0.2700

5 Conclusion

The comparison between the test results of a typical high-strength steel thin-walled beam with a drop hammer is as follows:

(1) The designed test fixture and sensor mounting equipment are simple in structure and can meet the requirements of the test.
(2) HC550/DP980D + Z load is relatively large, which reflects the fluctuation of the load.
(3) The energy absorption characteristics of different materials are significantly different, and the energy absorption ratio gradually decreases as the material strength increases.

References

1. Fang J, Zhou YD, Wang L (2012) A method to determine the features of energy absorption for automotive steels during dynamic deformation. J Plast Eng 19(1):92–96
2. Strnadel B, Ferfecki P, Židlík P (2013) Statistical characteristics of fracture surfaces in high-strength steel drop weight tear test specimens. Eng Fract Mech 112–113(11):1–13
3. Fan TQ (2016) Lightweight research of the threshold beam used high strength steel based on components test. In: The international conference on advanced high strength steel and press hardening, pp 693–698
4. Zhang W, Li CG, Yang JW et al (2017) Analysis on dynamic mechanical properties of high strength steel based on thin-walled beam crushing. Forging Stamping Technol 42(10):174–179
5. Lian CW, Chen XP, Yu NF (2017) Mechanical properties on high elongation steel QP under high strain rate. Forging Stamping Technol 42(7):164–168
6. Ma N, Park T, Kim D et al (2010) Evaluation of charpy impact test performance for advanced high-strength steel sheets based on a damage model. Int J Mater Form 3(1):183–186
7. Jongvivatsakul P, Attachaiyawuth A, Pansuk W (2016) A crack-shear slip model of high-strength steel fiber-reinforced concrete based on a push-off test. Constr Build Mater 126(6):924–935
8. Yu ZS, Dong WP, Ma CW et al (2015) Resistance spot weldability of hot-formed ultra high strength steel based on virtual and physical experiment. Appl Mech Mater 750(4):244–250
9. Sin SR, Yang SM, Yu HS et al (2008) Fatigue analysis for a multi-lap spot welded joint of high strength steel using quasi static tensile-shear test. Int J Automot Technol 9(1):81–86
10. Umemoto M, Hao X, Yasuda T et al (2002) Formation and annealing behavior of nanocrystalline steels produced by ball drop test. Mater Trans 43(10):2536–2542

Research on Comprehensive Performance of the PTFE Alloy Materials

Bo Lin, Gongqi Jia and Zhi Li

Abstract The PTFE was used as the main body, and the PEEK was used as the organic filler, using mechanical mixing—cold forming—prepared by the sintering process with PTFE alloy materials. The best formulation of PTFE/PEEK alloy sealing material was determined by test and analysis. The addition of the PEEK organic fillers improved the thermal and mechanical properties of the alloy system and showed good rigidity of high-temperature mechanics. This paper provides a theoretical basis for the application of PTFE alloy in the field of high-performance composite materials.

Keywords PTFE · Alloy · Thermal property · High-temperature mechanical rigidity

Since the 1990s, the world automobile, especially the Chinese automobile industry, has continued to develop, and automobile production continues to grow [1]. With the development of the automotive industry, people have a higher requirement for the performance of automotive parts and materials, and alloy materials have attracted more and more attention. Among auto parts, PTFE sealing materials are widely used in body parts, doors, seats, sunroofs, engines and fuel tanks. PTFE is an engineering plastic with excellent comprehensive performance and is known as the "Plastic King." It has the advantages of high- and low-temperature resistance, superior chemical stability, thermal stability, non-combustion, minimal water absorption, low friction coefficient and no toxicity [2]. At present, polytetrafluoroethylene is usually modified by filler modification, surface modification and blend modification. These modifications are mostly aimed at improving the wear resistance of polytetrafluoroethylene and improving its mechanical properties. These are few studies on the composite thermal and thermodynamic properties of composites [3]. For the thermal properties of polymers, blending between different polymers is one of the traditional and effective methods to improve their thermal properties [4].

B. Lin (✉) · Z. Li
Brilliance-auto Research Design Center Body, Shenyang 110142, Liaoning, China
e-mail: linbo_23@163.com

G. Jia
Shanghai Wanchao Car Sunroof CO., LTD., Shanghai 201807, China

China SAE (ed.), *Proceedings of China SAE Congress 2018: Selected Papers*, Lecture Notes in Electrical Engineering 574,
https://doi.org/10.1007/978-981-13-9718-9_41

In this study, the mechanical blending cold press forming-sintering process was used to systematically study the blending system of polytetrafluoroethylene (PTFE)/polyetheretherketone (PEEK) by different amounts of organic filler PEEK. Thermal performance and dynamic mechanical properties. The filling of organic materials improves the heat resistance, creep resistance, compression resistance, compression, bending and wear resistance of PTFE [5]. This ideal PTFE alloy material has good performance at high temperature and can be used as an advanced composite material.

1 Test Method

1.1 Experimental Materials

The main raw materials are polytetrafluorothylene—PTFE(Shandong Dongyue Polymer Material Co., Ltd.), polyetheretherketone—PEEK(Changzhou Guangcheng Plastic Co., Ltd).

1.2 Main Instruments and Equipment

Flat-plate vulcanizing machine: XLB 400 × 400 × 2E, Qingdao third rubber machinery factory; Maofu furnace: sm-2.8-12, Shenyang Industrial Electric Furnace Factory; Vicat Softening Temperature Tester: XWB-300C, Shanghai Precision science Instrument co., Ltd. Cantilever beam impact test machine: gt-7100-mi, Taiwan High-speed Rail Technology co., Ltd. Plastic Rockwell hardness tester: XHR-150, Shanghai material testing machine factory; High-speed universal grinder: 30B, Tianjin TEST Instrument co., Ltd. Scanning electron microscopy (SEM) test: After the cross sections of the blended alloy impact sample was gold-plated in a vacuum, the shape of the impact section was observed with a JSM-6360l Scanning Electron Microscope. Dynamic Mechanical Thermal Analyzer: DMTA-IV type, American rheological Scientific Instrument Company.

1.3 Sample Preparation

The dry mixing method is used to weigh PTFE powder and different parts of the organic filler PEEK, and put them into the high-speed universal grinder. After being stirred evenly, they were put into the self-made mold and then cold-pressed by the plate vulcanizing machine (pre-pressure: 10 MPa; dampening time: 2 min), demoulding to obtain pre-formed samples, then placed in the Moffe furnace for sintering (heating method: first heat up to 330 ± 5 °C, heat for 60 min, then heat to 380 ± 5 °C, insulation for 60 min; cooling method: natural cooling to room temperature with the furnace), after grinding and processing, to obtain PTFE alloy material products.

1.4 Performance Test

The notched impact performance test is carried out according to GB/t1043-79. The Vicat softening temperature is tested according to GB1633-79; the dynamic thermal mechanical analysis is carried out according to GB3356-82; the friction and wear test is carried out with reference to the plate-type friction testing machine. Test method: first fix the jig on the test bench, then put the sample into the fixture, fix the electric drill and place it directly above the sample and give a constant pressure of 29.4 n (3000 g). Turn on the power switch, the grinding wheel rotates at 200 r/min, remove the sample after grinding for 30 min and weigh the mass before and after the friction with an electronic balance with an accuracy of 0.0001. The wear amount is used as the index to evaluate the wear performance of materials in wear test. The notch impact section of the alloy was observed by scanning electron microscopy (SEM). The impact morphology and structural characteristics of the composite were analyzed.

2 Results and Discussion

2.1 Analysis of Mechanical Properties of PTFE/PEEK Alloy Materials

Figure 1 is a plot of PEEK content versus impact properties for PTFE/PEEK alloys. It can be seen from Fig. 1 that with the increase of PEEK content, the impact strength

Fig. 1 Effect of PEEK amount on impact PTFE/PEEK alloys

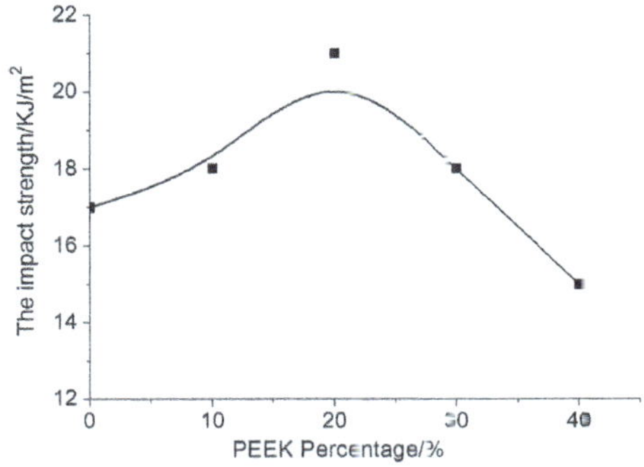

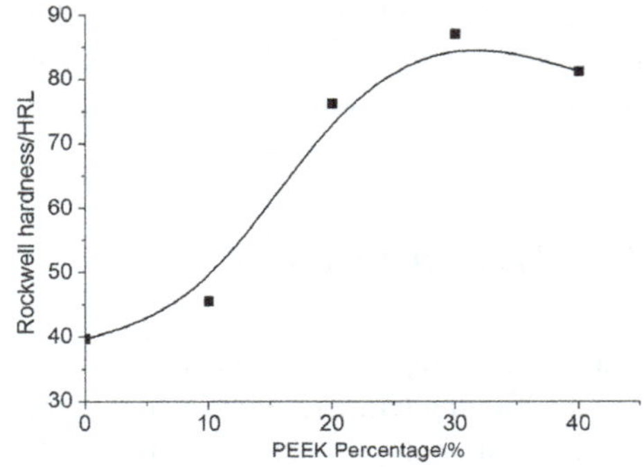

Fig. 2 Effect of PEEK amount on impact PTFE/PEEK alloys

of PTFE/PEEK alloy increases first and then decreases. When the PEEK content is 20 phr, the impact strength of the PTFE/PEEK alloy is increased to 21 kJ/m^2, and the toughness of the alloy material is improved.

When the content of PEEK is small, the impact strength of the alloy can be improved. This is because the interface bonding between PEEK and PTFE is strong, which is beneficial to the transition of the material from brittle to tough, so that PEEK is a rigid particle to PTFE. The matrix produces a toughening effect. When the content exceeds 20 phr, the impact is lowered because the excessive addition of PEEK causes a corresponding change in the composition of the interface layer, which makes it from dense to loose, which is not conducive to the transfer of stress.

Figure 2 is a plot of PEEK content versus PTFE alloy hardness. It can be seen from the figure that as the PEEK content increases, the hardness of the alloy material first increases and then decreases. When the PEEK content is 30 phr, the hardness of the PTFE/PEEK alloy material reaches a maximum of 87.0 HRL, which is 47.3 HRL higher than that of pure PTFE (39.7 HRL). However, due to excessive PEEK content, the hardness of the alloy material is reduced.

The addition of PEEK increases the hardness of PTFE/PEEK alloy. The properly proportioned hard particle PEEK exhibits rigidity and uniform dispersion in the system, which increases the contact area between the two phases and improves the viscosity between the phases. In the formation of a stable interface layer, the system is relatively dense, the hardness is improved; but when the PEEK content is too much, the composition of the PEEK and PTFE two-phase interface layer changes correspondingly, making it from dense to loose, which is not conducive to stress transmission.

2.2 *Thermal Performance Analysis of PTFE/PEEK Alloy Materials*

Figure 3 is a plot of PEEK content versus Vicat softening temperature for PTFE/PEEK alloys. It can be seen from the figure that after the polymer blending,

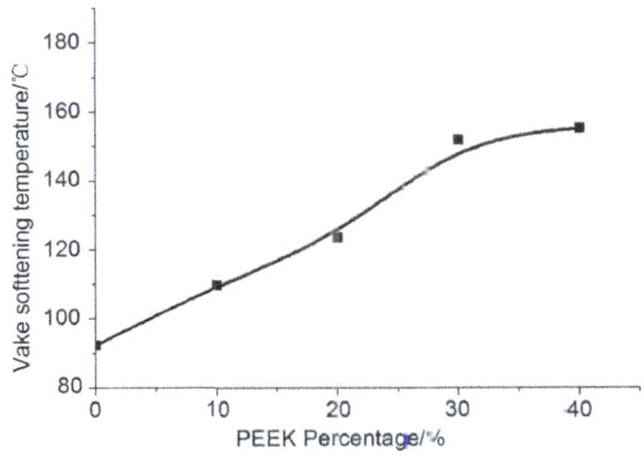

Fig. 3 Effect of PEEK amount on Vica softening temperature PTFE/PEEK alloys

the Vicat softening temperature of the PEEK/PTFE alloy material is significantly improved, and the heat resistance is remarkably improved. As the PEEK content increases, the Vicat softening temperature increases. PEEK reaches 155.2 °C at 40 phr, which is 62.9 °C higher than pure PTFE (92.3 °C).

PEEK is a high-temperature-resistant thermoplastic resin with a melting point of 334 °C, and PEEK acts as a physical crosslink in the resin matrix, which limits the thermal motion of the polymer chain and reduces the elastic deformation and plasticity of the matrix at the same temperature. Deformation increases the Vicat softening temperature of the material. However, within a certain range, the addition of a large amount of PEEK will weaken the original force between the polymer chains. Therefore, as the amount of PEEK increases, the increase of the Vicat softening temperature of the alloy tends to be moderate. In addition, the increase in crystallinity also improves the heat resistance of the alloy material. PEEK is a semi-crystalline polymer, and the increase in crystallinity is advantageous for the improvement of the heat resistance of the alloy.

2.3 Analysis of Friction and Wear Properties of PTFE/PEEK Composites

Figure 4 is a plot of PEEK content versus PTFE/PEEK alloys wear. It can be seen from the figure that after the alloy system is blended, the wear amount of the PTFE/PEEK alloy is greatly reduced, and the wear resistance performance is remarkably improved.

The wear nature of PTFE is that the macromolecular chain slips or breaks under the action of external force, so that the PTFE is pulled out of the crystallization zone and formed into a sheet shape to transfer to the surface of the grinding member, causing adhesive wear. PTFE itself has plasticity. After adding the hard powdered PEEK, it also plays a preferential role in the load.

The surface of PEEK can adsorb large molecular chains. The surface of one particle can adsorb multiple macromolecular chains, and the macromolecular chains adsorbed by the particles are entangled with each other, so that the particles adsorbing

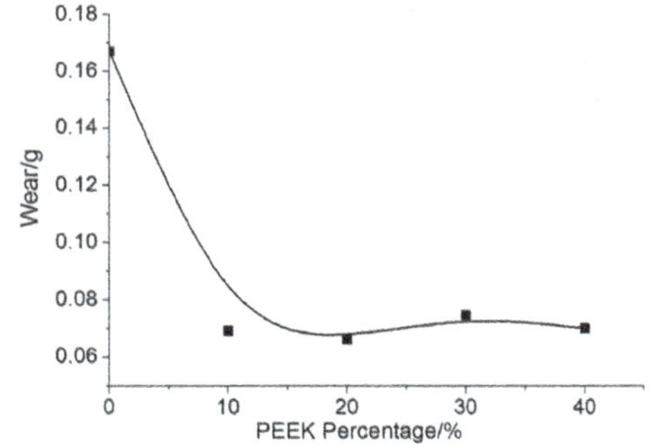

Fig. 4 Effect of PEEK content on wear of PTFE/PEEK alloy

the macromolecular chains are stressed, the PEEK particles can be used is uniform, so that the PTFE macromolecular chain in the blend system is less likely to slip and break, thus improve wear resistance.

2.4 *Dynamic Thermo Mechanical Analysis of PTFE/PEEK Alloy*

(1) Storage modulus E'

Figure 5 is a plot of PEEK content versus PTFE/PEEK alloys temperature storage modulus. It can be seen from the figure that the PEEK content in the PTFE alloy is 20%. It can be seen that the addition of PEEK has a great influence on the storage modulus of the PTFE alloy material with temperature. In the range of normal temperature to 270 °C, the storage modulus of PEEK/PTFE alloy material decreases rapidly with the increase of temperature. Compared with pure PTFE, the storage modulus increases to different degrees and goes to high temperature. When the temperature

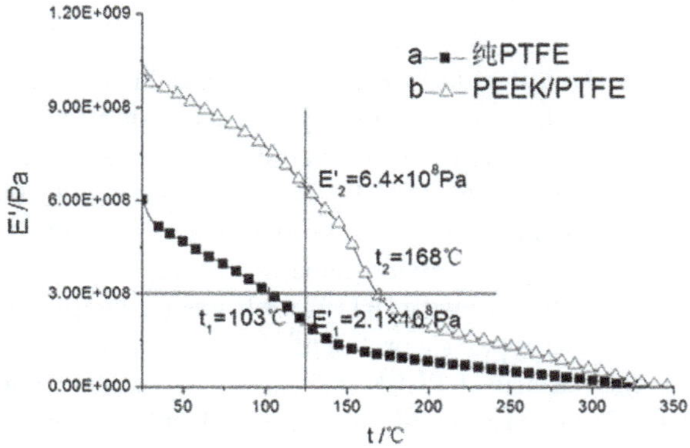

Fig. 5 Pure PTFE and PTFE/PEEK alloys on temperature energy storage modulus curve

is 125 °C, the storage modulus of PTFE/PEEK alloy is more than three times that of pure PTFE, showing good high-temperature mechanical rigidity.

As a semi-crystalline high polymer, PEEK is dispersed in the amorphous state in the form of crystallites. The hard crystallites in the amorphous state play the dual role of filler particles and cross-linking bridges, so that the mold of the alloy system amount is increased. On the other hand, the main chain of PEEK is made up of a benzene ring connected by a ketone group and an ether bond. It does not contain any flexible fatty segments. Even after T_g, the chain is not easily rearranged by conformation. Stress is released, which also makes its modulus jump less. Therefore, the storage modulus of the composite material is increased to varying degrees, and the material and high-temperature rigidity are significantly improved.

(2) Mechanical loss tanδ

Figure 6 is a plot of PEEK content versus temperature versus loss for PTFE/PEEK alloys. It can be seen from the figure that after the blend is added, the loss peak intensity of PTFE/PEEK alloy material is significantly lower than that of pure PTFE in the range of room temperature to 250 °C, which is due to the aromatic heterocyclic ring of PEEK blend. The structure destroys the linear structure of PTFE, hindering the free movement of its molecules, making its glass transition more difficult and requiring higher energy. It can be seen that the addition of the blend improves the disadvantage of the large loss of pure PTFE material in a certain temperature range.

The mechanical loss tanδ represents the ratio of energy loss to energy storage for each time period of temperature change, and the temperature corresponding to the peak value is the glass transition temperature T_g'. In Fig. 6, two loss peaks appear at about 131 °C and about 167 °C, respectively, corresponding to the two glass transition temperatures of PTFE and PEEK. The T_g' peak height indicates that the material segment relaxation requires more energy, and the T_g' peak is narrow, indication that the material segment movement is less dispersed and the segment relaxation process is shorter. It can be seen that the high-temperature mechanical properties of the PEEK/PTFE alloy are significantly improved.

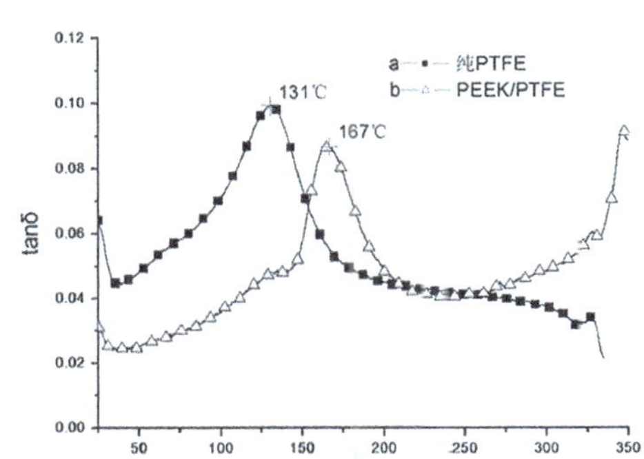

Fig. 6 Pure PTFE and PTFE/PEEK alloys on temperature-loss curve

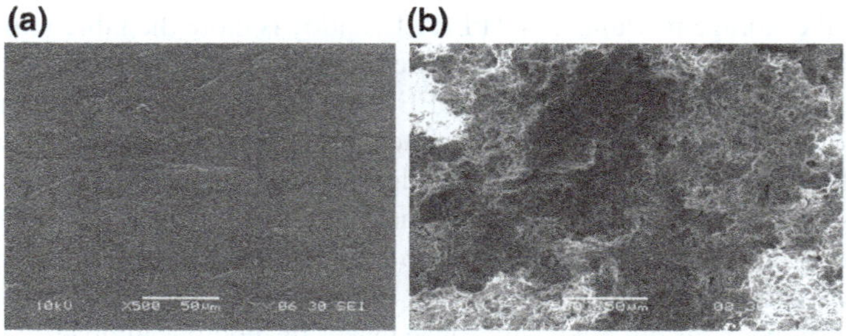

Fig. 7 SEM of pure PTFE (**a**) and PTFE/PEEK alloys (**b**)

2.5 Microstructure Analysis of PTFE/PEEK Alloy

Figure 7 shows the crystal morphology of the notched impact section of the PTFE/PEEK alloy sample. (a) is the impact profile of pure PTFE. The section is smooth and flat, and it is a homogeneous system, showing a typical brittle fracture morphology.

It can be seen from (b) that when the amount of PEEK added is 20%, the impact fracture surface of PTFE/PEEK alloy has large undulations, and some micro-pits are visible. The fracture surface is in a torn state and is a ductile fracture. The combination of PEEK and PTFE is better, and the shear lip has a relatively large shear lip, which indicates that a certain plastic deformation has occurred before the fracture, and the crack propagation speed is relatively slow. As a rigid supporting phase in the system, PEEK increases the contact area between the two phases and forms a stable interfacial layer, which is beneficial to the transition of the material from brittle to tough. This rigid particle of PEEK produces a toughening effect on the PTFE matrix. It can be seen that the addition of PEEK in a certain range is beneficial to the improvement of the mechanical properties of the blending system, and the physical and mechanical properties are superior.

3 Conclusion

(1) After adding the PEEK blend, the Vicat softening temperature of PEEK/PTFE alloy system is increased. When the PEEK content is 40 phr, the Vicat softening temperature is increased from 92.3 to 155.2 °C; on the other hand, PEEK is a semi-crystalline polymer. The improvement of crystallinity is beneficial to the heat resistance of the alloy system; the dynamic thermo mechanical analysis shows that when the temperature is 125 °C, the storage modulus of PEEK/PTFE alloy material is three times that of pure PTFE, showing good mechanical rigidity of high temperature.

(2) When PEEK is blended into PTFE, the hardness, impact strength and wear resistance of the alloy system are significantly improved. The hardness increased form 39.7HRL to 81.0HRL. When the PEEK content was 20 phr, the impact strength of the alloy system increased to 21 kJ/m^2; the wear amount decreased by 58%. PTFE/PEEK alloy material is a ductile fracture, the PEEK dispersion is uniform, the bonding between the two phases is enhanced, and PEEK plays an auxiliary toughening effect.

References

1. Zhong M (2007) Research, development and progress of polypropylene automobile bumper materials. Automot Technol Mater 12(2):53–55
2. Wang C (2007) Production status and modification progress of PTFE. Guizhou Petrochem 32(04):16–18
3. Zhang Y, Li P et al (2006) Principle and process of polymer blending modification. Mechanical Industry Press, Beijing, pp 267–269
4. Wu P, Zhang L (1984) Principle and process of polymer blending modification, vol 22. Light Industry Publishing, Beijing, pp 179–184
5. Pu Z, Liu J et al (2003) Tribological properties of polyphony ester/polyimide filled polytetrafluoroethylene. J Mater Sci Eng 21(6):851–854

Cracking Mechanisms of Hot-Dip Galvannealed Coatings on CP1000 Steel

Huasai Liu, Libin Liu, Haiquan Wang, Yun Han, Chunqian Xie and Yinghua Jiang

Abstract In recent years, ultra-high-strength steels are used as structural components more and more. For automobile industry and for different carmakers, different coating layers are chosen for corrosion resistance, such as galvanized (GI) and galvannealed (GA) layers. In this study, a hot-dip galvannealed complex phase 1000 MPa grade (CP1000) steel was chosen as the model material. Different GA parameters were adopted to get different Fe–Zn coating microstructures on CP1000, and bending and tensile loading were loaded on CP1000. Microstructure of the coating layers was investigated using SEM, and the cracking mechanisms of the coating layers were also observed using SEM. It was found that, with increasing GA time, ζ phase decreased on the surface of the coating layers, and more and more δ phase was formed from ζ phase on the surface. Furthermore, with increasing GA temperature, the formation velocity of δ phase got faster and faster. Bending tests showed that cracks could be found at the interface between the coating layers and CP1000 steel, especially when GA temperature was 550 °C. Cracks in δ phase, which were perpendicular to the interface, could also be found. Tensile tests showed that when the tensile strain was low enough, interface delamination could not be found. At the interface, when the GA temperature was 510 °C, cracks perpendicular to the interface could also be found.

Keywords Cracking mechanisms · Hot-dip galvannealed · CP1000

H. Liu (✉) · L. Liu · H. Wang · Y. Han · C. Xie · Y. Jiang
Shougang Group Co., Ltd. Research Institute of Technology, 69# Yangzhuang Street, Shijingshan District, Beijing 100043, People's Republic of China
e-mail: hsliu@alum.imr.ac.cn; hsliu@shougang.com.cn

Beijing Key Laboratory of Green Recyclable Process for Iron & Steel Production Technology, 69# Yangzhuang Street, Shijingshan District, Beijing 100043, People's Republic of China

L. Liu · H. Wang
University of Science & Technology Beijing, 30# Xueyuan Load, Haidian District, Beijing 100083, People's Republic of China

China SAE (ed.), *Proceedings of China SAE Congress 2018: Selected Papers*, Lecture Notes in Electrical Engineering 574,
https://doi.org/10.1007/978-981-13-9718-9_42

1 Introduction

In recent years, for reducing the entire car weight and improve vehicle safety, advanced high steels (AHSS) are used as structural components more and more in automotive industry [1–3]. Among these AHSS steels, dual-phase (DP) steel is used most, but with increasing the tensile strength, the DP steels show a higher sensitivity to edge fracture and a lower hole expansion ratio, which will lead to poor bendability and stretch-flangeability of the steels. The innovation of complex-phase (CP) steels which contain bainite, martensite, and ferrite is viewed as a key point to solve the problem, for CP steels have the minimum strength difference between constituent phases [4]. Furthermore, corrosion of the materials is also a big problem for automobiles. For automobile industry and for different carmakers, different coating layers are chosen for corrosion resistance, such as galvanized (GI) and galvannealed (GA) layers [5–7].

Compared with GI coatings, steels with GA coatings have better corrosion resistance and weldability. However, the brittle intermetallic Fe–Zn coatings on GA steels are the disadvantages during the sheet forming process, for powdering of GA coatings will influence the quality of products. Alpas et al. [5] performed a draw bead test on Al-killed and Ti-added low carbon steels with GA coatings and found that, with increasing Fe content of the coating layers, the amount of powdering will increase. There are several Fe–Zn intermetallic compounds formed on the coating layer of GA steels according to the Fe–Zn binary phase diagram [8], and different GA parameters will influence the formation the Fe–Zn coatings with different Fe content.

In this study, several galvannealing parameters were adopted to get different Fe–Zn coating microstructures on a hot-dip galvannealed complex phase 1000 MPa grade (CP1000) steel, and bending and tensile loading were loaded on CP1000. Microstructure of the coating layers was investigated using SEM, and the cracking mechanisms of the coating layers were also observed using SEM.

2 Experimental Procedure

The cold-rolled CP1000 steel samples (200 × 110 × 1 mm) were galvannealed in a hot-dip process simulator (HDPS), and the chemical composition of the steel in wt% was 0.156 C, 1.508 Mn, 0.376 Si, and 0.037 Al. The CP1000 steel was annealed at intercritical temperature and cooled down to 455 °C at a speed of 20 °C/s; then, the steel was dipped into the liquid zinc bath for 3 s, and the Al content in the zinc bath was about 0.15% in wt%. Isothermal annealing treatments were carried out at different temperatures and for different time to get different microstructures of the Fe–Zn coating layers; the detailed galvannealing parameters were listed in Table 1. The galvannealed steel was cooled down to room temperature using pressurized N_2 gas uniformly.

Table 1 Galvannealing parameters for CP1000

Galvannealing temperature, (°C)	480	510	550
Galvannealing time, (s)	20, 60, 360	20, 60, 360	20, 60, 360

Three-point bending tests were performed on the galvannealed CP1000 using a tensile tester (Instron5985) under displacement control at a rate of 0.2 mm/min. The dimensions of the specimen were 2 × 6 mm (width × length). The span between two supporting points was 4 mm. Tensile test was also tested using Instron5985 on CP1000.

The microstructure of the coating layers was investigated using a scanning electron microscope (SEM, ZEISS EVO18) equipped with Oxford energy dispersive X-ray spectrometer (EDS) operating at 20 kV, and the cracking mechanisms of the coating layers were also observed using SEM.

3 Results and Discussion

3.1 Microstructure of the Coatings

Figure 1 shows the microstructure of the coating layers of galvannealed CP1000 under different parameters. When the galvanized parameter is 480 °C × 20 s, there is very little intermetallic phase formed on the surface of CP1000 as shown in Fig. 1a. With increasing alloying reaction time, both pure zinc and ζ phase can be found on the surface of the coatings as shown in Fig. 1b. When reaction time is 360 s, the surface of the coating layer is mostly δ phase with a small amount of ζ phase as shown in Fig. 1c. When galvannealing temperature increases to 510 °C, holding for 20 s, pure zinc still occupies most of the surface of the coatings, a few ζ phases can also be found as shown in Fig. 1d. When galvannealing time reaches 60 s, there are full of intermetallic phases, among which δ phase takes up most of the surface area as shown in Fig. 1e. When galvannealing temperature is 550 °C, alloying reaction happens in a short time as shown in Fig. 1g–i, δ phase can be found on the surface of the coatings, and it is a little hard to find any ζ phase among them.

Figure 2 shows the relationship between Fe content of Zn–Fe coatings and galvannealing time. It is seen that Fe content in Zn–Fe coatings is only ~2.6% when galvannealing time is 20 s at both 480 and 510 °C, but there is an explosive increase of the Fe content of Zn–Fe coatings when galvannealing time is from 20 to 60 s at 510 °C. When galvannealing time is 20 s at 550 °C, Fe content of Zn–Fe coatings is ~11.8%, which means at high temperature there exists an explosive alloy reaction for a short time and it is very clear that high-temperature galvannealing is not fit for industrial production.

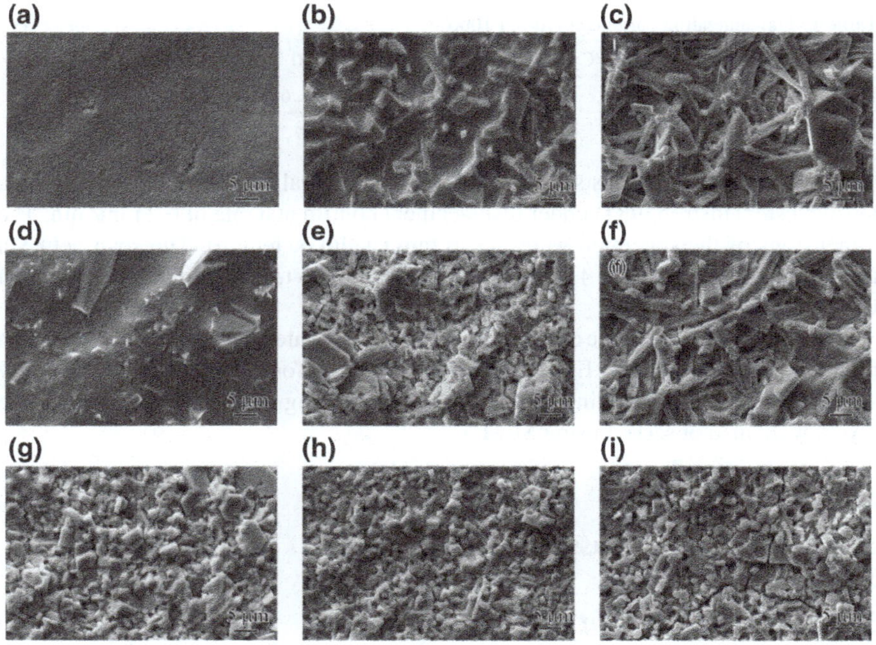

Fig. 1 SEM microstructure of the coating of galvannealed CP1000 under different parameters: **a** 480 °C × 20 s, **b** 480 °C × 60 s, **c** 480 °C × 360 s, **d** 510 °C × 20 s, **e** 510 °C × 60 s, **f** 510 °C × 360 s, **g** 550 °C × 20 s, **h** 550 °C × 60 s, and **i** 550 °C × 360 s

Fig. 2 Relationship between Fe content of Zn–Fe coatings and galvannealing time

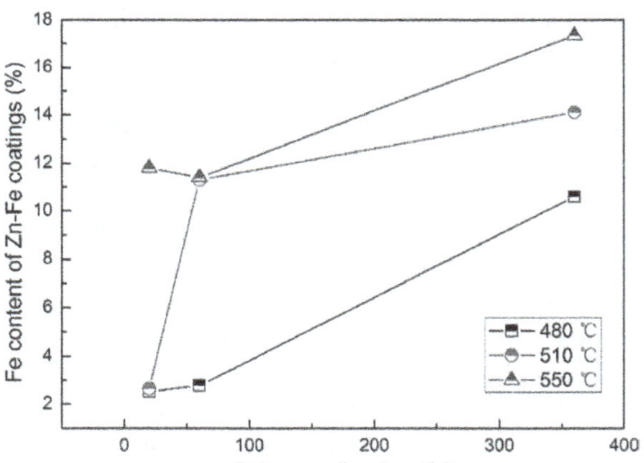

3.2 *Fracture Behaviors of the Zn–Fe Coatings*

Cross-sectional view of the galvannealed CP1000 under different galvannealing parameters is shown in Fig. 3. When CP1000 is galvannealed at 480 °C for 20 s, the microstructure of Zn–Fe coatings consists of pure zinc layer, ζ phase, δ phase, and Γ phase as shown in Fig. 3a. The surface of the coatings is pure, and in some areas, ζ phase of the surface can also be seen, δ phase is under the zinc layer, and Γ phase can be found at the interface. When galvannealing time reaches 360 s, the surface is covered with ζ phase and δ phase, and Γ phase can also be seen at the interface between the coating and the steel as shown in Fig. 3b. When galvannealing temperature is 550 °C, few ζ phases can be found, and the surface is covered with δ phase, it is very hard to distinguish ζ phase on the SEM pictures. Moreover, Γ phase can be found at the interface, Γ_1 phase above Γ phase can also be seen as shown in Fig. 3c, d.

Figure 4 shows the cross-sectional view of the galvannealed CP1000 after bending test. When the CP1000 is galvannealed at 480 °C for 360 s, cross-sectional view shows that delamination could be found at the interface between the coating layers and the CP1000 steel, but good bonding areas can also be found as shown in Fig. 4a. Especially when GA temperature reaches 550 °C, fully delamination of the interface

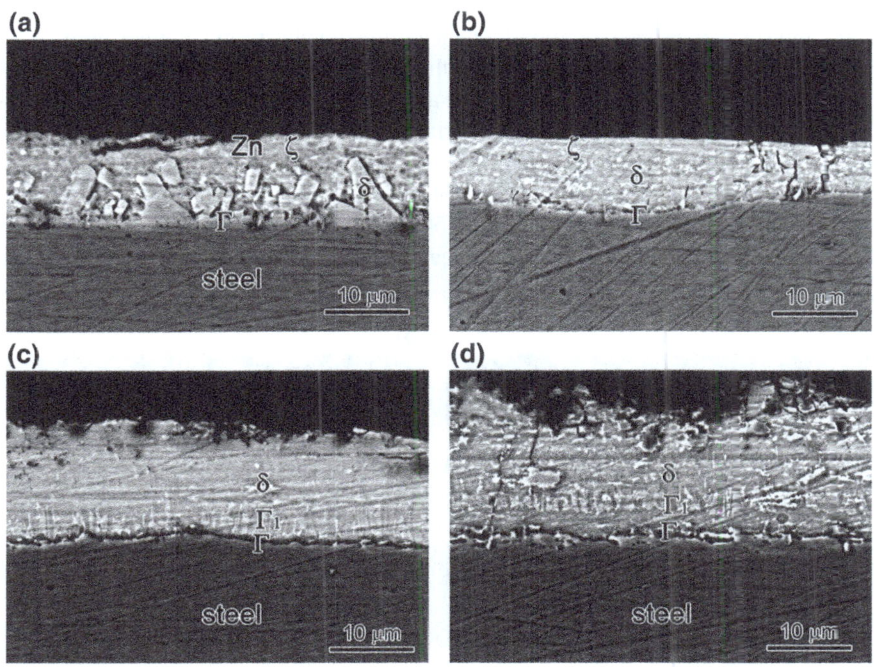

Fig. 3 Cross-sectional view of the galvannealed CP1000 under different galvannealing parameters: **a** 480 °C × 20 s, **b** 480 °C × 360 s, **c** 550 °C × 20 s, and **d** 550 °C × 360 s

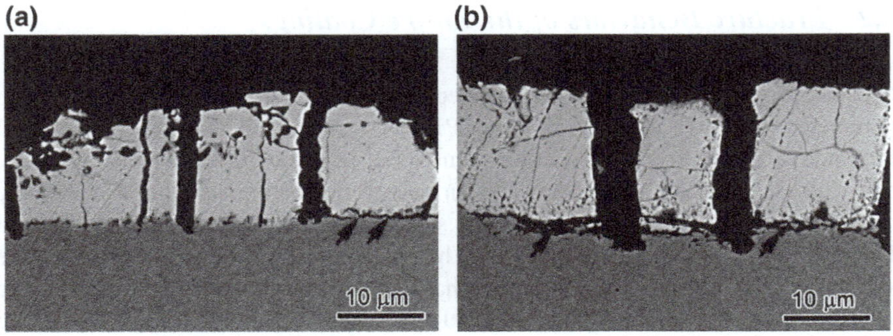

Fig. 4 Cross-sectional view of the galvannealed CP1000 after bending test: **a** 480 °C × 360 s, and **b** 550 °C × 360 s

happens as shown in Fig. 4b. Besides that cracks in δ phase, which are perpendicular to the interface, could also be found in Fig. 4a, b. Crack nucleation and propagation can be seen at the surface of CP1000 substrate under bending stress.

Figure 5 shows the galvannealed CP1000 tested under tensile strength. It is clear that when the tensile strain is 1%, there is no interface delamination as shown in Fig. 5a. But when tensile stain reaches 5%, interface delamination could be found as the red arrows shown in Fig. 5b. Besides interface delamination, cracks perpendicular to the interface could also be found whenever the tensile strain is low or high.

Compared with pure zinc, it is clear that intermetallic phases on the surface of galvannealed sheets are much more hard and brittle, and during sheet forming process, the powdering of GA coatings will influence the corrosion of products. When the surface layer contains a few ζ phases, only a small amount of powdering occurs, which means the GA coatings consisting of δ phase as the main constituent with a few ζ phases on the surface could provide a good fracture resistance [5, 7]. In this study, when the galvannealed CP1000 was annealed at 510 °C for 60 s, the microstructure

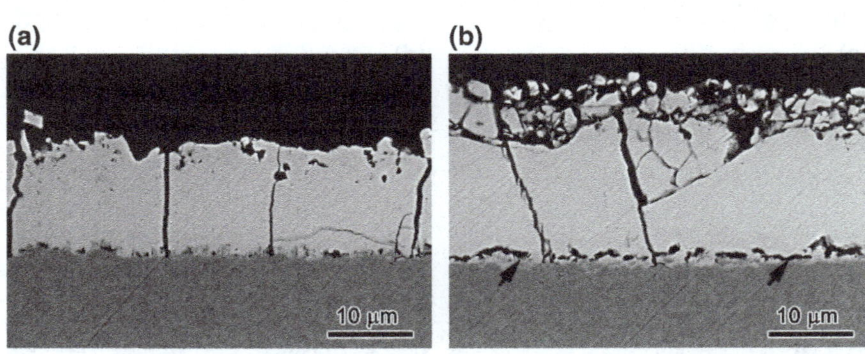

Fig. 5 Cross-sectional view of the CP1000 galvannealed at 510 °C for 60 s after tensile test at the strain of 1 and 5% (Color figure online)

of the coatings has a few ζ phase above δ phase, and this microstructure is consistent with previous research, so this galvannealing parameter should be adopted for industrial production in the future.

4 Conclusions

(1) It was found that, with increasing GA time, ζ phase decreased on the surface of the coating layers, and more and more δ phase was formed from ζ phase on the surface. Furthermore, with increasing GA temperature, the formation velocity of δ phase got faster and faster.
(2) Bending tests showed that cracks could be found at the interface between the coating layers and CP1000 steel, especially when GA temperature was 550 °C. Cracks in δ phase, which were perpendicular to the interface, could also be found.
(3) Tensile tests showed that when the tensile strain was low enough, interface delamination could not be found. At the interface, when the GA temperature was 510 °C, cracks perpendicular to the interface could also be found.

Acknowledgements The authors would like to thank Dr. Guangrui Jiang for his helpful discussion on this paper, and Mr. Bo Sun for three-point bending and tensile test. This work was supported by the National Key R&D Program of China (2017YFB0304401).

References

1. Nakada N, Arakawa Y, Park K-S, Tsuchiyama T, Takaki S (2012) Mater Sci Eng A 553:128
2. Ingarao G, Lorenzo RD, Micari F (2009) Mater Des 30:4421
3. Schemmann L, Zaefferer S, Raabe D, Friedel F, Mattissen D (2015) Acta Mater 95:386
4. Karelova A, Krempaszky C, Werner E, Tsipouridis P, Hebesberger T, Pichler A (2009) Steel Res Int 80:71
5. Alpas AT, Inagaki J (2000) ISIJ Int 40:172
6. Song GM, Vystavel T, van der Pers N, De Hosson JTM, Sloof WG (2012) Acta Mater 60:2973
7. Long JM, Haynes DA, Hodgson PD (2011) Mater Forum 27:832
8. Kato T, Hong MH, Nunome K, Sasaki K, Kuroda K, Saka H (1998) Thin Solid Films 319:132

The Influence of Combined Hygrothermal Environment and Load on the Performance of Adhesively Bonded CFRP–Aluminum Alloy Joints

Jingxin Na, Wenlong Mu, Guofeng Qin and Wei Tan

Abstract To provide reference and guidance for the design of the adhesive structures of the CFRP–aluminum alloy in the actual service environments, the aging laws of the adhesive joints under the hygrothermal environment and the load were studied in this paper, and a prediction method considering different aging conditions was proposed based on the cohesive zone model. The structural adhesive Araldite® 2015 was selected to fabricate the CFRP–aluminum alloy butt joint, and the durability tests were carried out under the condition of hygrothermal cycle, combined hygrothermal cycle and static load, and combined hygrothermal cycle and alternating load. The residual strength and failure modes of the joint after aging were analyzed. The results showed that the strength of joints decreased markedly after 60 h of hygrothermal cycle. The load further aggravates the aging of the joint, the adverse effect of dynamic alternating load on the failure strength of joint is more obvious than that of the static load; at the same time, the failure mode of joint has changed after aging, and compared to the selected CFRP, adhesive performance decreased more obviously; The damage factors under different aging conditions were defined, combined with the failure criterion obtained by the test, the classical double linear traction–displacement cohesion rule is modified, and the residual strength of the adhesive joint after aging is predicted by numerical simulation analysis.

Keywords Carbon FRP · Adhesive joints · Hygrothermal cycle · Load · Failure prediction

---

Funded by the National Natural Science Foundation of China (51775230) and Graduate Innovation Fund of Jilin University (Project No. 101832018C198).

J. Na · W. Mu (✉) · G. Qin · W. Tan
State Key Laboratory of Automotive Simulation and Control,
Jilin University, Changchun 130022, China
e-mail: mwh0032@163.com

Teachers College for Vocational and Technical Education, Guangxi Normal University, Guilin 541004, China
e-mail: qinguofengjlu@163.com

China SAE (ed.), *Proceedings of China SAE Congress 2018: Selected Papers*,
Lecture Notes in Electrical Engineering 574,
https://doi.org/10.1007/978-981-13-9718-9_43

1 Introduction

With the continuous improvement of residents' income and travel demand, the automobile industry has entered a stage of rapid development, which also brings some problems such as environmental pollution and excessive energy consumption. One of the effective ways to solve the above problems is to realize vehicle lightweight by applying new materials [1]. New materials such as aluminum alloy and carbon fiber reinforced polymer (CFRP), which with good properties, offering the possibility of manufacturing lightweight cars. In view of the requirements of complex performance and cost control, it is an inevitable trend of future development to use appropriate materials in different parts of vehicle and realize the mixing of multiple materials [2]. Therefore, the connection of different material (especially the CFRP and metal) is one of the key technologies for new material application. As a new type of structural connection technology, adhesive bonding technology has played an increasingly important role in the automobile industry for joining both metallic and nonmetallic components. It can be used to connect dissimilar materials without damaging the structure and ensure sufficient strength. In addition, it has the advantages of large load-bearing area, uniform stress distribution, good fatigue performance, and a better lightweight effect [3].

In the actual service process, the adhesive structure of vehicle is often subjected to the combined degradation action of environment and load. Because of their polymeric nature, the properties of the adhesive and resin matrix of FRP considerably change under different environment, especially after the hygrothermal exposure. Meanwhile, under the action of continuous loading, the mechanical properties of adhesive structure also may present attenuation, which will threat the safety and integrity of structure [4–6]. Therefore, it is important to investigate the effect of various factors (humidity, temperature and load) on the performance of CFRP–Al joints, and establish corresponding failure prediction method.

Many scholars have studied the durability of adhesively bonded joints under the action of humidity, temperature, load, and their coupling. The moisture absorption process of single-lap joint which fabricated with epoxy adhesive was studied by Mubashar et al. [7], and it was found that the strength of joints decreases significantly with the increase of exposure time. Han et al. tested the residual strength of single-lap joints which immersed (some loaded and some unloaded) in deionised water at 50 °C for six months. The results showed that the degradation of adhesive specimens caused by moisture absorption was more obvious than the sustained load [8]. Cai selected two kind of flexible adhesive to bond single-lap joints and butt joints, and the mechanical properties of joints were tested after ten days of hygrothermal exposure with constant load. It was concluded that the load not only affects the residual strength of joints but also changes the failure mode [9]. Koricho et al. studied the influence of bond length and thickness on the strength and failure mode of adhesive joints in different hygrothermal environments, and the aging of adhesive was further analyzed by Fourier-Transform Infrared (FTIR) spectra, thermogravimetric and DSC test [10]. In [11], the environmental aging test of double-cantilever beam joints was carried out at 60 °C–95% relative humidity (RH) and 60 °C–82% RH conditions. It was found that the failure strength, impact performance, and energy release rate of joints in high-temperature–humidity all change obviously. Nguyen et al. [12] investigated the degradation of adhesively bonded steel/CFRP joints at elevated temperatures with different load level and concluded that residual strength of joints is related to temperature change and exposure time. Agarwal et al. studied the influence of combined mechanical and environmental loads on the behavior of CFRP strengthened steel structures. The residual strength of joints is related to the glass transition temperature (T_g) of the adhesive and humidity [13].

The existing research about the durability of adhesive system mainly focused on single-lap joints which shear stress is dominating in the adhesive layer. However, the environmental effect on the mechanical properties of joint under tensile stress is seldom considered. Meanwhile, compare with the numerous lectures on the adverse effect of temperature, moisture, and load individual, there is still a lack of comprehensive understanding about the influence of combined environment and load. In this paper, the adhesive-bonded CFRP–aluminum alloy joints were fabricated with epoxy adhesive Araldite® 2015. Four scenarios (25 °C/50% RH, hygrothermal cycle, combined hygrothermal cycle and static load/alternating load) were defined to study the durability of joints. The failure strength and failure mode under each condition were analyzed and the variation law of joints properties was studied. On this basis, the damage factor was introduced to modify the bilinear traction–separation cohesive zone model (CZM). The failure loads of the butt joints under different working conditions were simulated, and the results of test and simulation were compared, which proved the effectiveness of the prediction method.

2 Experimental Process

2.1 Experimental Materials

The 6005A aluminum alloy widely used in automobile industry was selected. The material properties are shown in Table 1. CFRP laminate was made using twill prepreg which the substrates and fibers are YPH-23 and T300, respectively. The numbers of layer were eight and all spread in 0° direction. The thickness of CFRP laminate was about 2 mm. A two-component epoxy adhesive—Araldite® 2015 which widely used in vehicle structural bonding was selected. The properties of CFRP and 2015 are shown in Tables 2 and 3.

Table 1 Material properties of 6005A aluminum alloy

Density (kg/m^3)	Poisson's ratio	Young's modulus (Mpa)
2730	0.33	71,000

Table 2 Material properties of CFRP

$E_1 = 60 \pm 7\,\text{GPa}$	$\nu_{12} = 0.2$	$G_{12} = 4 \pm 0.5\,\text{GPa}$
$E_2 = 60 \pm 7\,\text{GPa}$	$\nu_{13} = 0.3$	$G_{13} = 4 \pm 0.5\,\text{GPa}$
$E_3 = 8 \pm 0.5\,\text{GPa}$	$\nu_{23} = 0.3$	$G_{23} = 4 \pm 0.5\,\text{GPa}$

Table 3 Properties of the adhesive Araldite® 2015

Young's modulus (Mpa)	Shear modulus (Mpa)	Poisson's ratio
1850	560	0.33

2.2 *Joint Fabrication*

The adhesively bonded CFRP–aluminum alloy butt joint was designed and fabricated in this paper, and its geometric dimension is shown in Fig. 1. The thickness of the adhesive layer was 0.2 mm. In order to prevent tearing at the boundary of the CFRP laminate, the cross-dimension of CFRP was slightly larger than that of aluminum alloy. To improve the utilization rate of the joint and the uniformity of the adhesive fixture, the butt joint was not only used for tensile test, as shown in Fig. 2a, but also can be tested with the auxiliary device to obtain initial shear failure strength for simulation, as shown in Fig. 2b. The calculation formulas of tensile stress σ and shear stress τ of joints are as follows:

$$\sigma = F/S \tag{1}$$

$$\tau = F/S \tag{2}$$

where F is the failure load and S indicates the adhesive area.

All adhesive joints were manufactured according to the unified process. First, the adhesive surfaces of aluminum alloy and CFRP were polished with sandpaper, and then acetone was used to remove grease and dust from the adhesive surface of aluminum and CFRP. A two-component caulking gun was used to mix the adhesive

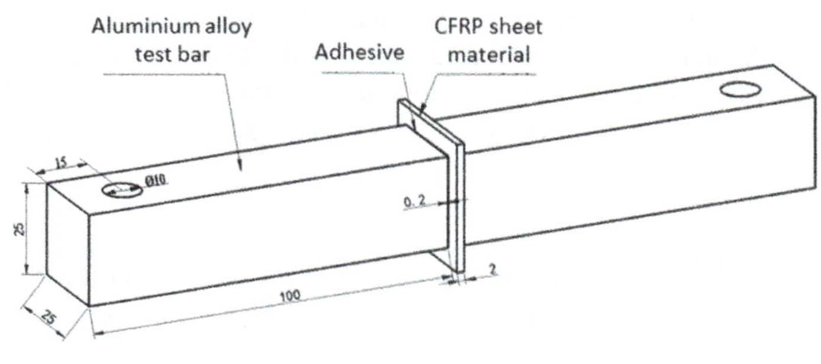

Fig. 1 Geometries and dimensions of joints (dimensions in mm)

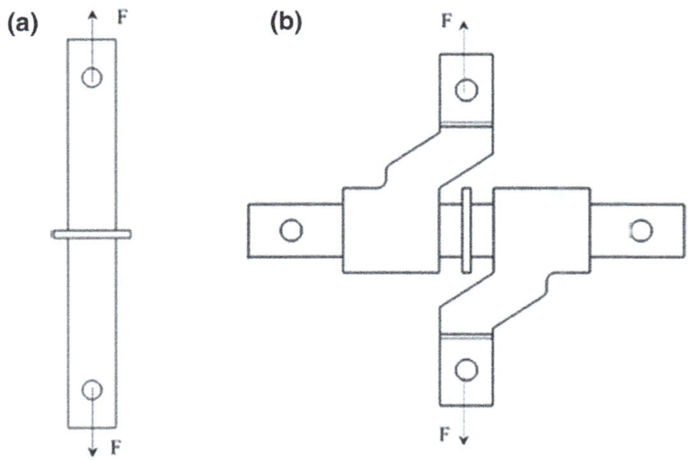

Fig. 2 Diagram of testing principle **a** tensile test; **b** shear test

and then adhesive was gelatinized on the surface of the aluminum alloy. The thickness of the adhesive layer was controlled by glass beads with a diameter of 0.2 mm. The joint fabrication was finished by using the adhesive fixture (as shown in Fig. 3). The base and cover plate were respectively used to support and fix the adhesive joint, and the sheathed screw was used to press the adhesive layer to ensure the thickness. After 24 h, the bonded joint was removed from the fixture and cured in an environmental chamber for 2 h at 80 °C to complete the entire joint manufacturing process.

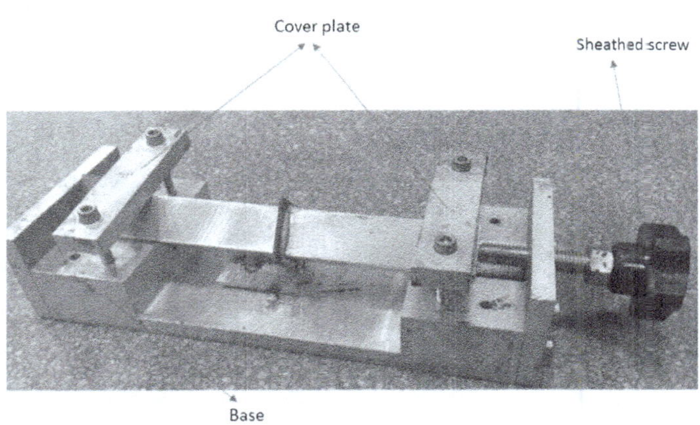

Fig. 3 Model of adhesive fixture

2.3 Test Method

To test the durability of adhesive joints under different aging conditions, the dynamic test device capable of performing hygrothermal cycle and load coupling was designed and processed. The test device was consisted of fatigue loading device, temperature and humidity environment chamber, display, control system, and hydraulic oil pump. The test device is shown in Fig. 4. By adopting this structure, the test loads of bond joints can be guaranteed to be equal, and the non-axial loads can be eliminated. Thus, the accelerated aging test that combined hygrothermal cycle and load can be carried out for multiple joints and the mechanical properties of the bonded structure under different service conditions can be obtained.

In order to study the influence of hygrothermal cycle and load on the performance of the adhesive joint, three aging conditions, including hygrothermal cycle (HC), hygrothermal cycle and static load (HC + SL), hygrothermal cycle and alternating load (HC + AL),were selected to test the durability of the joint. According to the standard DIN6701, the hygrothermal cycle spectrum was formulated, as shown in Fig. 5. One-cycle time was 12 h. The magnitude of static sustain load is 625 N, the maximum load of alternating load is also 625 N, and the load ratio (minimum load/maximum load) is $R = 0.1$ with a frequency 5 Hz.

Fig. 4 Test device

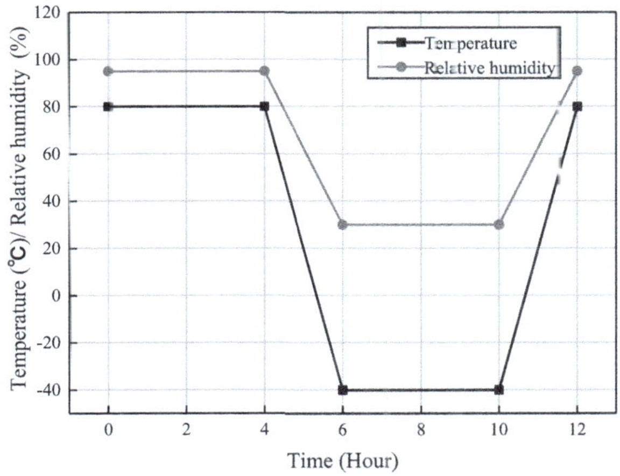

Fig. 5 One experimental period of hygrothermal cycles

The hygrothermal cycle test was carried out on the cured joint according to the hygrothermal cycle spectrum, and the combined hygrothermal cycle and static/alternating load test were performed using the test device. After 60 h (5 cycles) aging test, the joints were taken out and cooled to room temperature Four repeats were conducted for each group. Mechanical properties of the joints were measured by universal electronic testing machine, and cross-universal joints were used to eliminate the influence of nonaxial force. The tensile tests were conducted at a constant speed of 1 mm/min and the variation curves of the test load with displacement were recorded by the testing machine. The failure strength of joints can be obtained by dividing the failure load by the adhesive area.

3 Results and Discussion

3.1 Failure Strength

The tensile failure strength of adhesive joint after 60 h (5 cycles) of different aging conditions is shown in Fig. 6. It can be seen that after aging, the tensile failure strength of joints decreases significantly. The strength of joints decreased by 13.13% compared with that of the unaged joint under the hygrothermal environment, while the respective strength of joints was 16.93 and 19.26% less than that of the unaged joints under the combined condition of hygrothermal cycle + static load and hygrothermal cycle + alternating load. The results indicating that the hygrothermal environment has significant influence on the performance of joint, and the existence of load aggravates the strength attenuation, meanwhile compare with the static load, the alternating load was more detrimental.

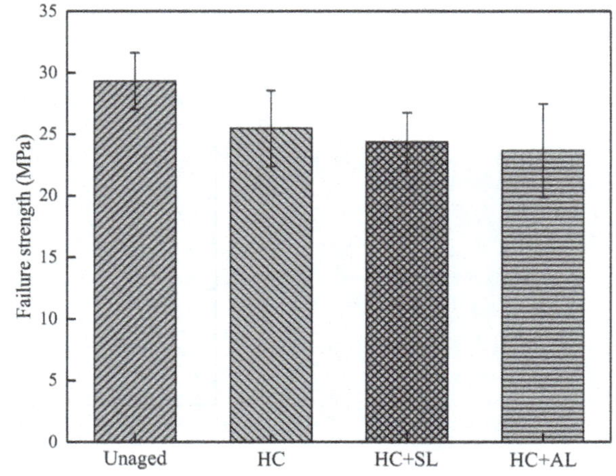

Fig. 6 Failure strength of adhesive joints under different aging conditions

3.2 *Failure Mode*

The Representative fracture surfaces of adhesively bonded joints under different conditions are presented in Fig. 7. It can be seen that the failure modes have changed obviously after aging. The CFRP fiber tear was observed in the unaged joint, indicating that the adhesive strength is close to that of CFRP epoxy resin under the initial condition. However, after hygrothermal cycle, combined hygrothermal cycle and load aging, the failure mode more often presents cohesive failure. This can be

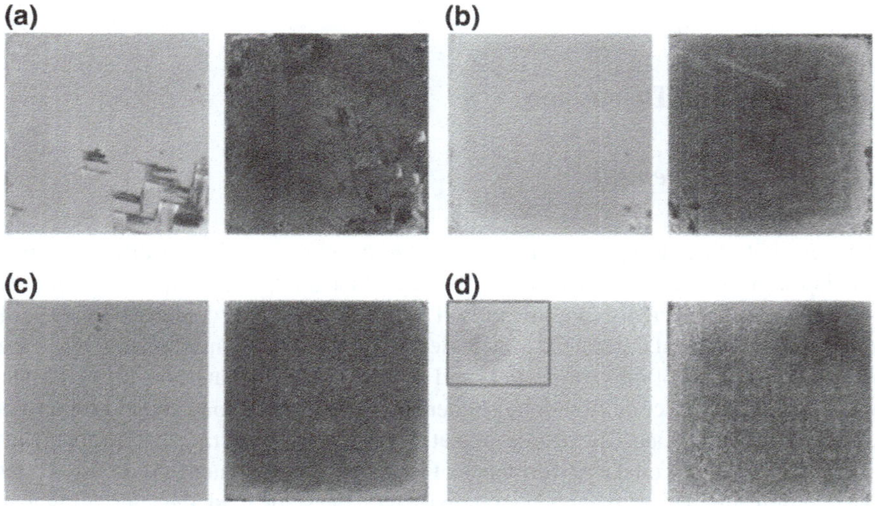

Fig. 7 Representative fracture surfaces of adhesively bonded joints: **a** unaged; **b** HC; **c** HC + SL; **d** HC + AL

explained that the selected adhesive is more sensitive to the hygrothermal environment compare with CFRP, and the reduction of mechanical properties of Araldite® 2015 adhesive is more significance under the action of hygrothermal environment. After the exposure of combined hygrothermal cycle and alternating load, local interface damage occurred on the fracture surface of the joint (marked with red box), and the similar phenomenon was observed in literature [14, 15].

4 Numerical Simulation

4.1 *Definition of Failure Criteria and Damage Factor*

It is necessary to provide valid failure criterion for the finite element simulation of adhesion joint by using cohesive zone model. The quadratic stress failure criterion is widely used to predict the failure of bonding structures, as shown in Eq. (3), the σ and τ respective represents the tensile stress and shear stress, and N, S are the failure strength in the normal and shear directions.

$$\left(\frac{\tau}{S}\right)^2 + \left(\frac{\sigma}{N}\right)^2 = 1 \tag{3}$$

The failure criterion formula obtained from the experimental data is shown in Eq. (4).

$$\left(\frac{\tau}{26.5}\right)^2 + \left(\frac{\sigma}{29.3}\right)^2 = 1 \tag{4}$$

In order to predict the failure load of the aged adhesive joint, the damage factor D is defined, and the calculation formula is shown in Eq. (5).

$$D = \frac{S_d}{S_f} \tag{5}$$

where S_d is the average residual strength of aged joints and S_f is the average quasistatic failure strength of unaged joint.

4.2 *Finite Element Modeling*

The simulation analysis was conducted by using software package ABAQUS® and geometric nonlinearity was considered. The three-dimensional finite element model of the adhesive joint is shown in Fig. 8. Although there is a layer of adhesives on both sides of CFRP in the actual testing process, their stress conditions are the same, so

Fig. 8 Finite element model of butt joint

only one layer of adhesives is established in the finite element modeling, while the other side of the CFRP was directly connected to aluminum alloy with shared mesh nodes. The size of the grid is controlled by distributing seeds on the geometric model (the size is 0.5 mm), and the grid is generated by automatic grid division technology. In the numerical model, CFRP was modeled using continuum shell element (SC8R). The aluminum alloy was modeled from 3D stress element (C3D8R) and the adhesive layer was built using cohesive element (COH3D8). One end of the joint is constrained by a fixed constraint, and the other end subjected to horizontal displacement and the degrees of freedom in other directions are constrained, as shown in Fig. 8.

In this paper, the classical bilinear traction–displacement law is adopted, and the initial failure of the cohesive element is determined by the quadratic stress failure criterion. When the initial failure is reached, the stiffness of the adhesive begins to soften, and the final separation is evaluated by the energy law, as shown in formula (6).

$$\frac{G_{\mathrm{I}}}{G_{\mathrm{I}C}} + \frac{G_{\mathrm{II}}}{G_{\mathrm{II}C}} = 1 \tag{6}$$

where the G_{I} G_{II} represents the energy release rates and $G_{\mathrm{I}C}$ $G_{\mathrm{II}C}$ is the fracture toughness.

The simulation parameters of aluminum alloy and CFRP are respectively shown in Tables 1 and 2, and the parameters of adhesive are shown in Table 4. The failure strength of modes I and II is obtained by experimental tests in this paper. Other parameters refer to existing studies [8].

4.3 Simulation Results

The experimental and simulated load–displacement curves of the unaged butt joint are shown in Fig. 9. It can be seen that the simulation curve match well with the curve

Table 4 Simulation parameters of adhesive

Parameter Type	Number
Young modulus, E (GPa)	1.82
Shear modulus, G (GPa)	0.56
Mode I failure strength, N (MPa)	25.46
Mode II failure strength, S (MPa)	24.98
Mode I fracture toughness, G_{IC} (N/mm)	0.43
Mode II fracture toughness, G_{IIC} (N/mm)	4.7

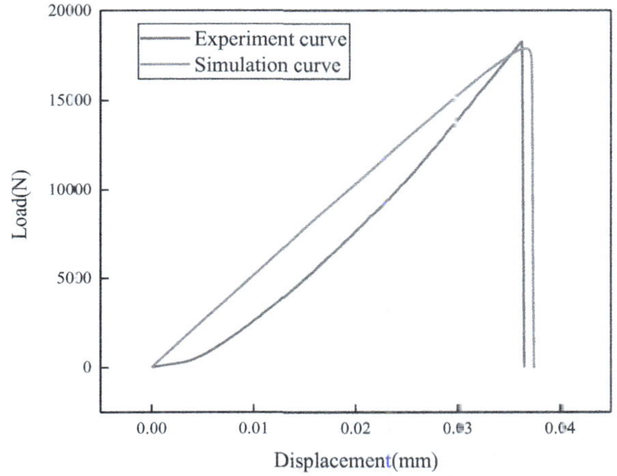

Fig. 9 Test and simulation load-displacement curve of butt joint

obtained by the test, in which the relative error values for the maximum failure load and failure displacement are 2.4 and 3.5%, respectively, indicating that the simulation model is validity.

To simplicity, it is assumed that the parameters of adhesive cohesive zone model are degraded in proportion with the damage factors. The simulation parameters of adhesive were modified according to the damage factors under different aging conditions by multiplying the original parameters with the factor, and the failure strength of the joints after aging was calculated. The experimental and simulated failure strength under different aging conditions is shown in Fig. 10. The maximum relative error of the simulation strength is no more than 3.2%, which indicates that the prediction model established in this paper has certain accuracy and can predict the failure strength of joints after aging under different conditions.

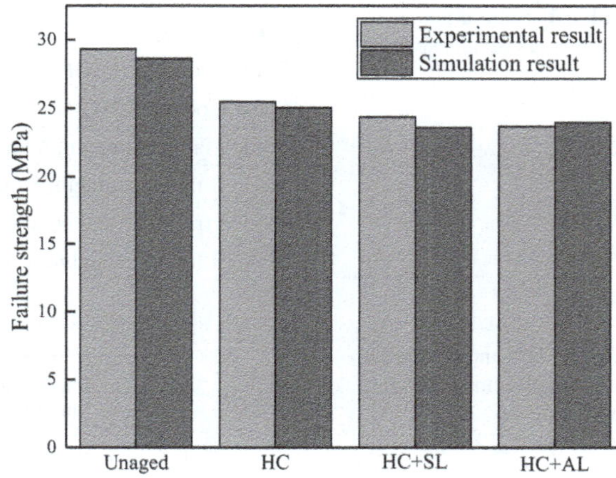

Fig. 10 Results of experiment and simulation

5 Conclusions

In this paper, the aging conditions of hygrothermal cycle, combined hygrothermal cycle and static/alternating load were established respectively according to the actual service environment of adhesive structure in vehicle. The relevant equipment was designed and processed and the accelerated aging tests of CFRP–aluminum alloy joint were carried out. Based on the experimental data, the durability of joint under different aging conditions was investigated and the failure modes of joint were analyzed to explain the fracture mechanism. The damage factors that consider different aging conditions were defined and the prediction model of failure strength of adhesive joints after aging was established. Quadratic stress failure criterion obtained by experiment was included in this model. The following conclusions were drawn:

(1) The hygrothermal environment has a significant influence on the performance of CFRP–aluminum alloy bonding joints, and the load further aggravated the aging of joints. Compared with static load, alternating load has more influence on joint failure strength;
(2) The failure modes of adhesive joint changed after aging. The unaged joints present a degree of CFRP fiber tear, and cohesive failure was observed for the hygrothermal aging joints. However, local interface failure is more likely to occur in the boundary of adhesive area when load exist, especially under the action of alternating load;
(3) The constitutive of the cohesive zone model was modified by introducing the damage factor. Combine with the secondary stress failure criterion obtained from experimental data, the failure process of aged joints can be predicted and the prediction accuracy is satisfactory for engineering application.

References

1. Feng M (2006) Development and applications of new materials in automotive lightweighting technologies. Automot Eng 28(3):213–220
2. Sakundarini N, Taha Z, Abdul-Rashid SH et al (2013) Optimal multi-material selection for lightweight design of automotive body assembly incorporating recyclability. Mater Des 50(17):846–857
3. Silva LFMD, Öchsner, A, Adams RD (2011) Handbook of adhesion technology. NOVA. The Univ Newcastle's Digital Repository. University of Newcastle
4. Nassar S, Sakai K (2015) Effect of cyclic heat, humidity, and joining methods on the static and dynamic performance of lightweight-material single lap joints. J Manufact Sci Eng 137(5)
5. Zhang F, Yang X, Wang HP et al (2013) Durability of adhesively-bonded single lap–shear joints in accelerated hygrothermal exposure for automotive applications. Int J Adhes Adhes 44(44):130–137
6. Shenoy V, Ashcroft IA, Critchlow GW et al (2009) Strength wearout of adhesively bonded joints under constant amplitude fatigue. Int J Fatigue 31(5):820–830
7. Mubashar A, Ashcroft IA, Critchlow GW et al (2009) Moisture absorption–desorption effects in adhesive joints. Int J Adhes Adhes 29(8):751–760
8. Han X, Crocombe AD, Anwar SNR et al (2014) The strength prediction of adhesive single lap joints exposed to long term loading in a hostile environment. Int J Adhes Adhes 55:1–11
9. Cai L (2016) Research on the effect of load on weatherability of two kinds of automobile adhesive. Jilin University
10. Koricho EG, Verna E, Belingardi G et al (2016) Parametric study of hot-melt adhesive under accelerated ageing for automotive applications. Int J Adhes Adhes 68:169–181
11. Ameli A, Datla NV, Azari S et al (2012) Prediction of environmental degradation of closed adhesive joints using data from open-faced specimens. Compos Struct 94(2):779–786
12. Nguyen TC, Bai Y, Al-Mahaidi R et al (2012) Time-dependent behaviour of steel/CFRP double strap joints subjected to combined thermal and mechanical loading. Compos Struct 94(5):1826–1833
13. Agarwal A, Foster SJ, Hamed E (2016) Testing of new adhesive and CFRP laminate forsteel-CFRP joints under sustained loading and temperature cycles. Compos Part B Eng 99:235–247
14. Jen YM, Ko CW (2010) Evaluation of fatigue life of adhesively bonded aluminum single-lap joints using interfacial parameters. Int J Fatigue 32(2):330–340
15. Pang J, Du Y, Wu K et al (2013) Fatigue analysis of adhesive joints under vibration loading. J Adhes 89(12):899–920

Shear Stress Analysis and Optimization of Dual-Zone Adhesive-Bonded Steel/Al Joints

Xiaokai Chen, Dong Fu, Ziyu Guo and Mengqiang Li

Abstract As one of the fundamental joints, adhesive joint can be applied to hybrid joints for the complex structures. To improve the material utilization of adhesive, the stress distribution of dual-zone adhesive-bonded joints was studied and optimized. Considering the elastic deformation of adhesives, the analytical models for shear stress of single- and dual-zone adhesive joint were deduced. The analytical model of single-zone adhesive joint for steel/Al was validated by FEA model and test data. Based on the analytical model, the shear stress distribution of dual-zone adhesive-bonded steel/Al joints was optimized. Results of optimization show that the stress level in dual-zone adhesive is higher than the traditional adhesive joint structure, which means the mechanical potential of adhesive can be realized better. The adhesive usage amount in dual-zone structure can be reduced 11.7% by keeping the overlapping length constant and the strength of joints not decreasing. In addition, the adhesive length close to lower stiffness material is longer than the other side which can remedy the defect of stiffness imbalance of adherends.

Keywords Steel/Al adhesive-bonded joints · Shear stress analytical model · Structure optimization

1 Introduction

In order to meet the lightweight design requirements, the lightweight materials, such as magnalium, engineering plastics, and composite materials, have been widely utilized for automobile structures. The traditional joints like welding and riveting cannot meet the design requirements, so adhesive bonded has become another alternative joint method. Adhesive bonded can significantly improve stiffness of joint and joint efficiency which is defined as the ratio of joint's shear strength to strength of the weaker adherend [1]. And it can also reduce stress concentration caused by mechanical connection or spot welding.

X. Chen · D. Fu (✉) · Z. Guo · M. Li
Beijing Institute of Technology, Beijing 100081, China
e-mail: 1563445454@qq.com

China SAE (ed.), *Proceedings of China SAE Congress 2018: Selected Papers*,
Lecture Notes in Electrical Engineering 574,
https://doi.org/10.1007/978-981-13-9718-9_44

Single-lap joint, one of the most common joint type, is the basic research model of other complex structure of hybrid joint. The shear stress distribution of arbitrary size of single-lap joint can easily get from numerical analytical model of shear stress distribution. It can also help us to figure out the structure's failure mechanism which will produce preliminary design direction for future study.

Up till now, research on the analytical model of single-lap joint is still relatively rare. Volkersen et al. figured out the shear-lag model considering the elastic effect of adherend for the first time [2]. Goland established the G-R model based on one-dimensional beam model which considered the influence of bending moment [3]. Ojalvo et al. utilized an analytical method to optimize the size of adhesive in single and double lap joint whose target was to make shear stress distributed evenly [4]. Peng Yue et al. developed a mixed-modulus bonded single-lap joint analytical model based on linear elasticity theory by neglecting the adherends shear strains [5].

However, for single zone of adhesive, the stress in the middle part is quite low, so designers could use dual zone of adhesive to overcome the defect of low material utilization. In this paper, we deduced the analytical model of single and dual zone of adhesive in single-lap joint. And on the foundation of the analytical model, we optimized the physical dimension of adhesive in dual-zone adhesive of single-lap joint.

2 The Establishment of Shear Stress Analytical Model of Adhesive Joint

2.1 Model Assumption

Focusing on single-lap adhesive joint with dissimilar materials, the formula of shearing stress in overlapping area could be deduced by considering the shear-lag effect in this area. The specific derivation process is as follows.

Figure 1 shows the simplified schematic of single-lap joint. The boundary condition is that one end is fully fixed and a uniformly distributed load $P = \overline{p} \cdot A$ is applied on the opposite end. Considering the influence of adherend's thickness, the shearing stress in the direction of thickness is assumed to be linear [6]. In addition, due to the short width in overlapping area, we presume that the stress in width is constant [7].

2.2 The Shear Stress Analytical Model

After analyzing the material element force diagram in Fig. 1, the displacement along $\times$ direction of 2 adherend are derived in a form:

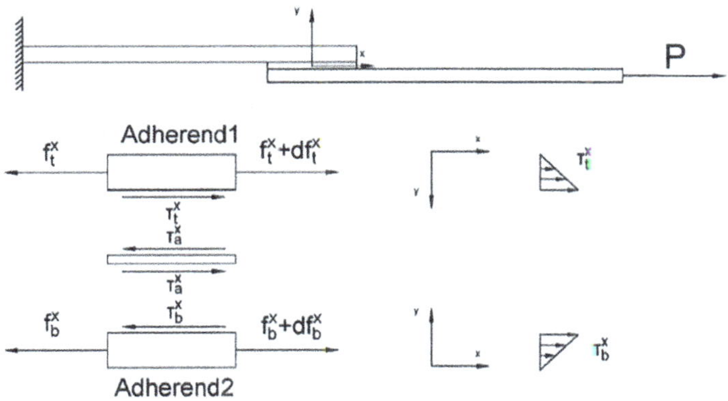

Fig. 1 Simplified schematic and material element force diagram of single-lap joint

$$u_t = u_t^f + u_t^\tau,\ u_b = u_b^f - u_b^\tau \tag{1}$$

where u_t^f and u_b^f are, respectively, the displacement along × direction of 2 adherend caused by normal stress; u_t^τ and u_b^τ are, respectively, the displacement along × direction of 2 adherend caused by shear stress.

In local coordinates $y_i'(i = 1, 2)$:

$$u_t^\tau(y_1') = \int_0^{y_1'} \gamma_t \mathrm{d}y_1', u_b^\tau(y_2') = \int_0^{y_2'} \gamma_t \mathrm{d}y_2' \tag{2}$$

where γ_t, γ_b are, respectively, shear strain of 2 adherend. With Hook's law for shear, we can get:

$$\gamma_t = \frac{\tau_t^x}{G_1},\ \gamma_b = \frac{\tau_b^x}{G_2} \tag{3}$$

According to the assumption of shear stress distribution in adherends, the shear stress in joint surface is equal to the stress τ_a^x in adhesive and the value in free interface is zero, so:

$$\tau_t^x(y_1') = \frac{\tau_a^x y_1'}{t_1},\ \tau_b^x(y_2') = \frac{\tau_a^x y_2'}{t_2} \tag{4}$$

From (1) to (3) and (4) follows:

$$u_t = u_t^f + \frac{\tau_a^x y_1'^2}{2G_1 t_1},\ u_b = u_b^f - \frac{\tau_a^x y_2'^2}{2G_2 t_2} \tag{5}$$

The resultant force of material element in 2 adherends is:

$$f_t^x = E_1 a t_1 \frac{\mathrm{d}u_t^f}{\mathrm{d}x} + \frac{E_1 a t_1}{6G_1}\frac{\mathrm{d}\tau_a^x}{\mathrm{d}x},\ f_b^x = E_2 a t_2 \frac{\mathrm{d}u_b^f}{\mathrm{d}x} - \frac{E_2 a t_2}{6G_2}\frac{\mathrm{d}\tau_a^x}{\mathrm{d}x} \tag{6}$$

With Hook's law for shear, the shear stress in adhesive is:

$$\tau_a^x = G_3\gamma_a = \frac{(u_b|_{y_2'=t_2} - u_t|_{y_1'=t_1})G_3}{t_3} \tag{7}$$

From (5), (6), and (7) follows:

$$\left(\frac{t_1}{3G_1} + \frac{t_2}{3G_2} + \frac{t_3}{G_3}\right)\frac{\mathrm{d}\tau_a^x}{\mathrm{d}x} + \frac{f_t^x}{E_1 t_1 a} - \frac{f_b^x}{E_2 t_2 a} = 0 \tag{8}$$

From the force analysis along × direction of material element, we can deduce:

$$\mathrm{d}f_t^x + \tau_a^x a \cdot \mathrm{d}x = 0, \mathrm{d}f_b^x - \tau_a^x a \cdot \mathrm{d}x = 0, P = f_t^x + f_b^x \tag{9}$$

From Formula (8) and (9), a linear differential equation with constant coefficient of the second order can be established:

$$\frac{\mathrm{d}^2 f_t^x}{\mathrm{d}x^2} - \omega^2 f_t^x = -\frac{3PG_1G_2G_3}{E_2t_2(t_1G_2t_3 + G_1t_2G_3 + 3G_1G_2t_3)} \tag{10}$$

where $\omega = -\sqrt{\frac{3G_1G_2G_3(E_1t_1+E_2t_2)}{E_1E_2t_1t_2(3G_1G_2t_3+t_1G_2G_3+G_1t_2G_3)}}$.

By solving the Eq. (10), the force f_t^x is:

$$f_t^x = A\mathrm{e}^{\omega x} + B\mathrm{e}^{-\omega x} + \frac{PE_1t_1}{E_1t_1 + E_2t_2} \tag{11}$$

According to the boundary condition: $\sigma_t^x|_{x=-\frac{L}{2}} = \frac{f_t^x}{t_1 a} = \frac{P}{t_1 a}$, $\sigma_t^x|_{x=\frac{L}{2}} = \frac{f_t^x}{t_1 a} = 0$, we can get the coefficient A and B in Eq. (11):

$$A = \frac{P\left[E_2t_2\mathrm{e}^{-\frac{1}{2}L\omega} + E_1t_1\mathrm{e}^{\frac{1}{2}L\omega}\right]}{(E_2t_2 + E_1t_1)(\mathrm{e}^{-L\omega} - \mathrm{e}^{L\omega})},\ B = -\frac{P\left[E_1t_1\mathrm{e}^{-\frac{1}{2}L\omega} + E_2t_2\mathrm{e}^{\frac{1}{2}L\omega}\right]}{(E_2t_2 + E_1t_1)(\mathrm{e}^{-L\omega} - \mathrm{e}^{L\omega})}$$

Hence, the expression of shear stress in adhesive is:

$$\tau(x) = -\frac{1}{a}\frac{\mathrm{d}f_t^x}{\mathrm{d}x} = -\frac{\omega}{a}\left(A\mathrm{e}^{\omega x} - B\mathrm{e}^{-\omega x}\right) \tag{12}$$

3 Verification of the Expression of Analytical Model

The validity of Formula (12) can be confirmed by finite element simulation. Considering the common material used in vehicle's body, we choose aluminum alloy AA6111 and mild steel St12 as adherends, and the basic mechanical parameters of these 2 materials are shown in Table 1[8].

We choose Ergo 7200 as the adhesive, and the basic parameters are displayed in Table 2.

Using finite element software ABAQUS to build model, the parameters of structure are as follows. Three different magnitudes of overlapping length are selected which are $L = 10$ mm, 15 mm, 20 mm. The value of overlapping width is $a = 25$ mm. And considering the size in automobile's body, the thickness of aluminum alloy, mild steel, and adhesive is, respectively, $t_1 = 0.8$ mm, $t_2 = 0.8$ mm, $t_3 = 0.1$ mm.

The boundary condition and loading mode of the model are as follows. The left-hand side is fully fixed, and the right-hand side is applied with a uniformly distributed load of 1500 N. By simulation in ABAQUS, we can acquire the stress distribution along the direction of overlapping. The comparison result between simulation and theory is shown in Fig. 2.

From the comparison result, we can conclude that both of the 2 sets of curves are parabolic with concave upside and they as a whole fit well along the direction of lap length. All the curves show the stress in middle area are close to zero, and the spot of maximum stress in adhesive layer appears in the end of aluminum whose stiffness is lower.

Although the simulation results have a certain error with the theoretical results at the edge of the adhesive layer, in the actual production process of the adhesive joint, there will be an adhesive fillet at the edge which will generate stress concentration to compensate for the error in the simulation result. Therefore, the effect of this error in

Table 1 Basic mechanical parameters of adherends

Adherends	Elasticity modulus (E/GPa)	Poisson's ratio	Yield strength (σ_s/MPa)	Tensile strength (τ_b/MPa)
Mild steel St12	210	0.3	186	330
Aluminum alloy AA6111	70	0.33	227	317

Table 2 Basic parameters of adhesive

Adhesive	Elasticity modulus (E/GPa)	Shear modulus (G/GPa)	Poisson's ratio	Crack I		Crack II	
				G_{IC} (N/m)	σ_c (MPa)	G_{IIC} (N/m)	σ_c (MPa)
Ergo 7200	2.875	0.96	0.33	80	8	320	18

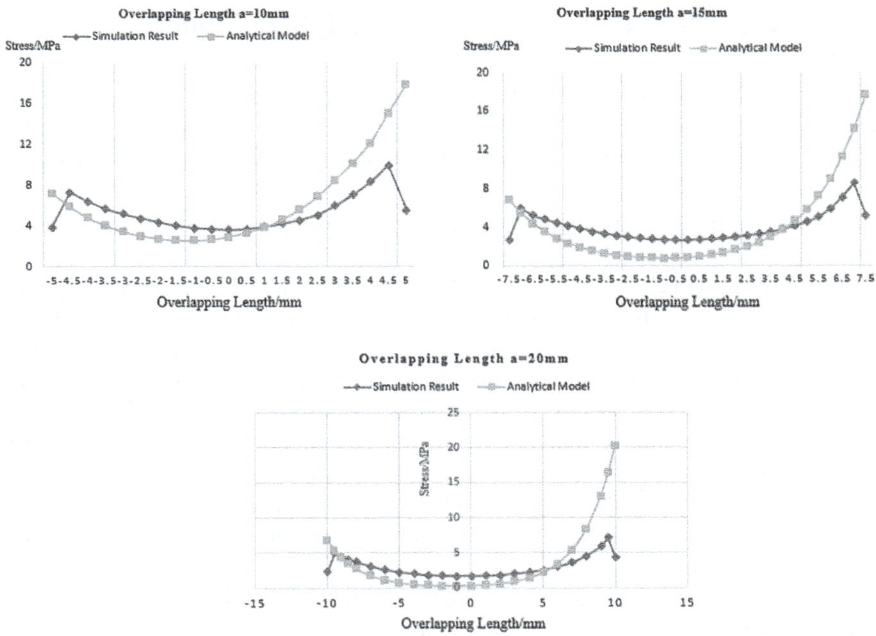

Fig. 2 Comparison result between simulation and theory

the actual test will be further reduced. In addition, considering the drop in the ends, the stress in the ends is equal to the figure in $\pm 90\% \frac{L}{2}$.

4 Strength Test Verification of Analytical Model

For the verification of Eq. (12), the theoretical failure load can be obtained through giving shear strength of adhesive and then compared with the failure load in the test. For the single-lap joint specimen in this article, the failure mode is cohesive failure. Therefore, the failure load can be predicted by judging whether the maximum shear stress reaches the shear strength. The failure criterion is:$\tau(x)\text{max} \geq \tau_r$, where τ_r is the shear strength of the adhesive layer.

In this paper, the experimental data are selected in the literature [9]. And the constraints and loading methods are the same. The adherends combination in the literature can be classified into three types: mild steel + mild steel; mild steel + aluminum alloy; aluminum alloy + aluminum alloy, as shown in Fig. 3.

With Eq. (12), the failure load is:

$$P\text{max} = \frac{\tau_r}{-\frac{\omega}{a}(A_1 e^{\omega x} - B_1 e^{-\omega x})} \tag{13}$$

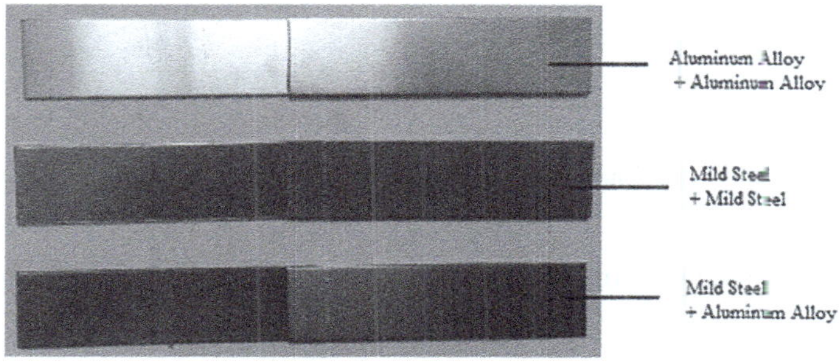

Fig. 3 Specimens with different adherend combination [9]

Table 3 Comparison of shear-lag model and test values for failure load of adhesive specimens

Type of specimen	Theoretical results/*N*	Test results/*N*	Relative error (%)
Mild Steel + mild steel	4575.3	4486.4	1.98
Mild steel + aluminum alloy	3233.1	3899.3	17.1
Aluminum alloy + aluminum alloy	3654.4	3948.7	7.5

where τ_r is the failure load; $A_1 = \frac{E_2t_2e^{-\frac{1}{2}L\omega}+E_1t_1e^{\frac{1}{2}L\omega}}{(E_2t_2+E_1t_1)(e^{-L\omega}-e^{L\omega})}$; $B_1 = -\frac{E_1t_1e^{-\frac{1}{2}L\omega}+E_2t_2e^{\frac{1}{2}L\omega}}{(E_2t_2+E_1t_1)(e^{-L\omega}-e^{L\omega})}$.

Substitute the material parameters to Formula (13), and compare the theoretical results with test results which are shown in Table 3.

From the comparison results, it can be seen that the precision of the model is within an acceptable range, although the relative error of the heterogeneous materials joint is slightly higher than that of the homogenous materials. Due to the difference in surface treatment between the two materials and the difference in bonding ability of the adhesive and the two materials, the relative error of the joint of the dissimilar materials will be slightly greater than that of the same material. Hence, the subsequent design of the adhesive layer can be optimized by using the analytical model.

5 The Establishment of Shear Stress Analytical Model of Dual-Zone Adhesive

According to the verification of finite element analysis and test results, we can use the conclusion above to establish the shear stress analytical model of dual-zone adhesive. Now, the structure of adhesive is changed to dual zone, and the parameters like overlapping length *L*, overlapping width *a*, and adhesive layer's thickness t_3 keep the same which are shown in Fig. 4.

Although the adhesive layer's structure is changed, it can be easily seen from above that the force analysis of the microelements does not change. Therefore, the

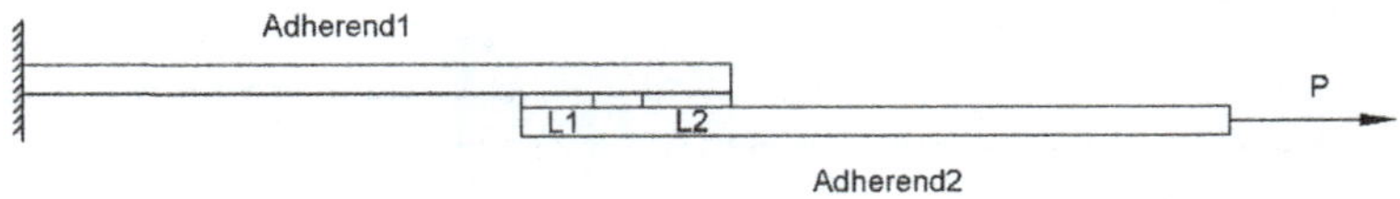

Fig. 4 Sketch of dual-zone adhesive layer structure

second-order constant-coefficient differential equation of the adhesive layer remains:

$$\frac{d^2 f_t^x}{dx^2} - \omega^2 f_t^x = \frac{-3PG_1G_2G_3}{E_2t_2(t_1G_2G_3 + 3G_1G_2t_3 + G_1t_2G_3)} \tag{14}$$

Therefore, the average normal stress of the upper adherend is:

$$\overline{\sigma}_t^x = \frac{C}{t_1a}e^{\omega x} + \frac{D}{t_1a}e^{-\omega x} + \frac{PE_1}{a(E_1t_1 + E_2t_2)} \tag{15}$$

The force analysis is shown in Fig. 5. The boundary conditions of the adhesive layer on the left side are:

$$\overline{\sigma}_t^x\big|_{x=-\frac{l_1}{2}} = \frac{P}{t_1a},\ \overline{\sigma}_t^x\big|_{x=\frac{l_1}{2}} = \frac{1}{t_1a}(P - \tau_1) \tag{16}$$

where:

$$\tau_1 = \int_{-\frac{l_1}{2}}^{\frac{l_1}{2}} \tau_a^x \cdot a \cdot dx = -\int_{-\frac{l_1}{2}}^{\frac{l_1}{2}} \omega(Ce^{\omega x} - De^{-\omega x})dx$$
$$= -(Ce^{\frac{\omega l_1}{2}} + De^{-\frac{\omega l_1}{2}} - Ce^{-\frac{\omega l_1}{2}} - De^{\frac{\omega l_1}{2}}) \tag{17}$$

From (15), (16), and (17) follows:

$$C_1 = \frac{PE_2t_2 \cdot e^{\frac{\omega l_1}{2}}}{E_1t_1 + E_2t_2},\ D_1 = 0$$

By analyzing the force of the adhesive layer on the right side in Fig. 5, the boundary conditions are:

Fig. 5 Force analysis sketch of dual-zone adhesive layer

$$\overline{\sigma}_t^x\Big|_{x=-\frac{l_2}{2}} = \frac{1}{t_1 a}(P - \tau_1),\ \overline{\sigma}_t^x\Big|_{x=\frac{l_2}{2}} = 0 \tag{18}$$

From (15), (17), and (18) follows:

$$C_2 = -\frac{P \cdot (E_1 t_1 e^{\frac{\omega l_2}{2}} + E_2 t_2 e^{\omega l_1} e^{-\frac{\omega l_2}{2}})}{(E_1 t_1 + E_2 t_2)(e^{\omega l_2} - e^{-\omega l_2})},$$

$$D_2 = \frac{P \cdot (E_1 t_1 e^{-\frac{\omega l_2}{2}} + E_2 t_2 e^{\omega l_1} e^{\frac{\omega l_2}{2}})}{(E_1 t_1 + E_2 t_2)(e^{\omega l_2} - e^{-\omega l_2})}$$

The shear stress distribution model of a bi-region adhesive joint can be obtained as:

$$\tau_1(x) = -\frac{\omega}{a} \cdot \frac{P E_2 t_2 \cdot e^{\frac{\omega l_1}{2}}}{E_1 t_1 + E_2 t_2} e^{\omega x} \tag{19}$$

$$\tau_2(x) = \frac{\omega}{a}\left(\frac{P \cdot (E_1 t_1 e^{\frac{\omega l_2}{2}} + E_2 t_2 e^{\omega l_1} e^{-\frac{\omega l_2}{2}})}{(E_1 t_1 + E_2 t_2)(e^{\omega l_2} - e^{-\omega l_2})} e^{\omega x} + \frac{P \cdot (E_1 t_1 e^{-\frac{\omega l_2}{2}} + E_2 t_2 e^{\omega l_1} e^{\frac{\omega l_2}{2}})}{(E_1 t_1 + E_2 t_2)(e^{\omega l_2} - e^{-\omega l_2})} e^{-\omega x}\right) \tag{20}$$

where $x \in \left[-\frac{l}{2}, \frac{l}{2}\right]$.

6 Optimum Structural Design of Dual-Zone Adhesive Layer

Nowadays, under the trend of automotive lightweight, the designer's ideal goal is to minimize the weight as much as possible. For the single-lap joint in this work, the purpose of reducing the usage of adhesive is achieved by optimizing the length and thickness of the two adhesive layers under the condition of ensuring the connection strength.

Firstly, the failure load of single-zone adhesive needs to be calculated. From the above Eq. (12) and simulation results, the maximum shear strength is:

$$\begin{aligned}\tau(x)\max &= \tau\left(0.9 \cdot \frac{L}{2}\right) \\ &= -\frac{P\omega}{a} \cdot \frac{E_1(e^{0.95L\omega} + e^{-0.95L\omega}) + E_2(e^{0.05L\omega} + e^{-0.05L\omega})}{(E_1 + E_2)(e^{-L\omega} - e^{L\omega})}\end{aligned} \tag{21}$$

In order to make sure the adhesive layer not destroyed, the maximum shear stress should meet:$\tau(x)\max \le \tau_r$. And for simplifying the calculation process, we substitute the parameters of adhesive and adherends into Formula (21). The material parameters are presented in Tables 1 and 2. Other physical dimension is: overlapping length $L = 30\,\text{mm}$, overlapping width $a = 25\,\text{mm}$, the thickness of aluminum alloy, mild steel, and adhesive layer $t_1 = t_2 = 0.8\,\text{mm}$, $t_3 = 0.1\,\text{mm}$. So the external load is:

$$P \leq -\frac{a}{\tau_r \omega} \cdot \frac{(E_1 + E_2)(\mathrm{e}^{-L\omega} - \mathrm{e}^{L\omega})}{E_1(\mathrm{e}^{0.95L\omega} + \mathrm{e}^{-0.95L\omega}) + E_2(\mathrm{e}^{0.05L\omega} + \mathrm{e}^{-0.05L\omega})} = \frac{121.2}{\tau_r} \tag{22}$$

Then on the condition of keeping materials, overlapping length, width, and thickness of adherend the same, we change single-zone adhesive to dual-zone adhesive. And the design variables are 2 pieces of adhesive's length and thickness of adhesive layer.

From the simulation results, we can easily know the position of maximum shear stress appears in the end close to aluminum alloy. And the maximum shear stress $\tau(x)_{\max}$ also needs to meet: $\tau(x)_{\max} \leq \tau_r$. So the external load is:

$$P \leq \frac{25}{\tau_r \omega} \cdot \frac{28\mathrm{e}^{l_2\omega} - 28\mathrm{e}^{-l_2\omega}}{21(\mathrm{e}^{0.95l_2\omega} + \mathrm{e}^{-0.95l_2\omega}) + 7\mathrm{e}^{\omega l_1}(\mathrm{e}^{-0.05l_2\omega} + \mathrm{e}^{0.05l_2\omega})} \tag{23}$$

Comparing Eq. (22) with (23), if the structure's strength does not decline when the number of adhesive layer changes from 1 to 2, it needs to meet:

$$\frac{25}{\tau_r \omega} \cdot \frac{28\mathrm{e}^{l_2\omega} - 28\mathrm{e}^{-l_2\omega}}{21(\mathrm{e}^{0.975l_2\omega} + \mathrm{e}^{-0.975l_2\omega}) + 7\mathrm{e}^{\omega l_1}(\mathrm{e}^{-0.025l_2\omega} + \mathrm{e}^{0.025l_2\omega})} \geq \frac{121.2}{\tau_r} \tag{24}$$

Hence, this problem can be transferred to a single objective optimization problem:

$$\begin{cases} \text{Min}: \ V = (l_1 + l_2) \cdot t_3 \\ \text{s.t. } P' \geq P_0 \\ l_1 + l_2 < L = 30 \\ 0 \leq t_3 \leq 0.2 \\ l_1, \ l_2 \geq 4\,\text{mm} \\ \text{Design variables: } l_1, \ l_2, \ t_3 \end{cases}$$

where l_1, l_2, t_3, V, P are, respectively, the left piece of adhesive's length, the right piece of adhesive's length, the thickness of adhesive layer, the volume of adhesive, and the failure load. And equation $P' \geq P_0$ refers to Formula (24).

After the progress of optimization of Multi-Island GA, the results show in Table 4.

From the optimization results, we can conclude that the adhesive length close to aluminum which has lower elasticity modulus is $l_2 = 21.9\,\text{mm}$, while the length close to mild steel is $l_1 = 4.6\,\text{mm}$. It is obvious that the adhesive length close to the lower elasticity modulus adherend is higher than the other side, which can remedy the defect of stiffness imbalance of adherends. Although the usage of adhesive decline

Table 4 Optimization results

Parameters	l_1/mm	l_2/mm	t_3/mm	V	P
Pre-optimization	15	15	0.1	75	$\frac{121.2}{\tau_r}$
Post-optimization	4.6	21.9	0.12	66.25	$\frac{121.43}{\tau_r}$

11.7% which takes up a small part for one specimen's cost, in actual mass production, the decreased cost amount of adhesive is reasonable.

7 Conclusion

In this paper, the shear stress distribution discipline is analyzed, and based on this, the optimization focused on steel/Al adhesive joint is carried out. The main achievements are:

1. On the basis of mechanics of materials, the shear stress distribution in single-zone adhesive is acquired by analyzing microelement of adhesive joints. And the validity of this analytical model is confirmed by FEA model and test data.
2. By selecting the same microelement of adhesive joints, the analytical model of dual-zone adhesive along the direction of lap length is obtained based on the results of single-zone adhesive.
3. After optimization, under condition of keeping the overlapping length constant and the strength of joints not decreasing, the usage amount of adhesive can reduce 11.7%, and the production cost can be also reduced.
4. Due to the stiffness imbalance of adherends, the adhesive length near to lower stiffness adherend is longer in dual-zone adhesive joint.

Acknowledgements This project is supported by National Natural Science Foundation of China (51675044) and National Key Research and Development Plan (2017YFB0103704).

References

1. Zhao B, Li ZH (2009) Elastic stress analysis of single-lap adhesive-bonded joints. Chin J Appl Mech, 608–613
2. Sun DX, You M (2008) Research on the mechanical behavior of coaxially bonded single lap joint. China Elastometrics, 18–21
3. Goland M, Reissner E (1944) The stresses in cemented joints. J Appl Mech ASME 11(1):A17–A27
4. Ojalvo IU (1985) Optimization of bonded joints. AIAA J 23(10):1578–1582
5. Yue P, Zhao B (2008) Stress analysis and adhesive shear stress optimization for mixed-modulus bonded single-lap joints. Beijing Adhesion Society, Beijing, pp 901–915
6. Adams RD, Peppiatt NA (1973) Effect of poisson' s ratio strains in adherends on stresses of an idealized lap joints. J Strain Anal 8(2):134–139

7. Wang J, Zhang C (2009) Three parameter elastic foundation model for analysis of adhesively bonded joints. Int J Adhes Adhes 29:495–502
8. Li L (2011) Research on static strength analysis and strength prediction of adhesively bonded single lap joint for automotive. Jilin University, Jilin, p 75
9. Qin ZX (2015) Research on the performance of clinch-bonded hybrid joints between steel and aluminum alloy sheets and its application on auto body. South China University of Technology, Guangdong, p 28

The Study on Lightweight and Cost Control in Vehicle Stamping SE of Material Technology

Wencai Xie, Xueshuang Wang, Hong Zheng, Zhuang Fu and Mingqi Xu

Abstract Automotive simultaneous engineering (SE) plays an important role in the whole cycle of automobile product development, shortening the product development greatly and reducing the cost. As a basic discipline, the major of material has provided technical support in the work of stamping simultaneous engineering and promoted the development of stamping SE. This paper introduces the characteristics of stamping SE and the role which is played by material engineers in various stages of stamping SE. Also, the paper introduces the emphasis on knowledge accumulation, the establishment of material database, material lightweight proposal, and material cost reduction proposal, etc.

Keywords Simultaneous engineering · Materials · Lightweight · Cost reduction

Preface

With the increasing of market competition, in order to reduce the cost and save the development cycle, the development of automotive simultaneous engineering (SE) has attracted more and more attention from the auto factory. The SE in the automobile design cycle involves multiple majors, and each major has the characteristics of synchronization, binding, coordination, and consistency. The most important major of SE is stamping SE. The development of stamping SE effectively reduces the later project or design change, greatly shortens the time of product development cycle, and saves the cost. However, the development of stamping SE cannot be separated from the technical support of professional material engineers. For example, in the product planning stage, the material selection scheme based on vehicle model cost and quality positioning was proposed for BOM evaluation and optimization. In the product definition stage, it assists to achieve the vehicle lightweight target and reduces the cost proposal of vehicle materials, etc. In the product design and validation stage, it reduces the material variety and specification according to the BOM inspection results, and improves the material utilization rate. In the production stage, the material

W. Xie · X. Wang (✉) · H. Zheng · Z. Fu · M. Xu
FAW JieFang Automotive CO., LTD, ChangChun, JiLin, China
e-mail: wangxueshuang@rdc.faw.com.cn

China SAE (ed.), *Proceedings of China SAE Congress 2018: Selected Papers*, Lecture Notes in Electrical Engineering 574,
https://doi.org/10.1007/978-981-13-9718-9_45

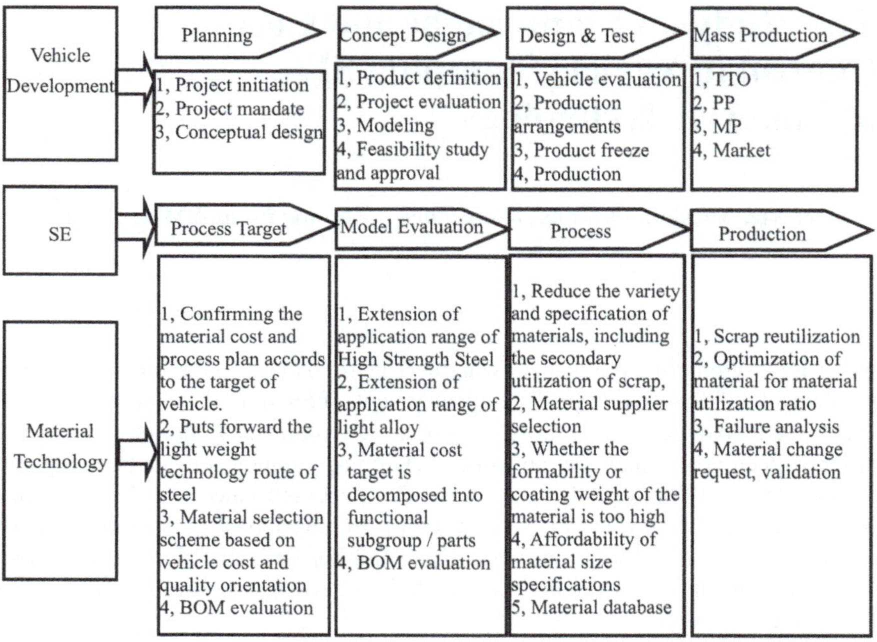

Fig. 1 Main role of material technology in the process of SE

arrangement optimization, subsequent material change application, verification, and so on. Therefore, in all the above stages, material engineers can play an important role and become an indispensable part. Figure 1 shows the technical support of material technology in the SE process. This paper will introduce how to effectively carry out the above work for the material major in the process of automotive SE, discuss its working methods, and finally summarize how to accumulate data files and establish corresponding technical specifications and standards.

1 Main Technical Support and Existing Problems of Material Specialty in Stamping SE

In the process of stamping SE development, material engineers are required to provide basic data and technical support relating to metal sheet materials, mainly including summarizing and analyzing the material of the benchmark vehicle, proposing material selection scheme based on the cost and quality positioning of the model according to the benchmark vehicle, reviewing and optimizing material BOM, reducing material variety and specification, considering the purchasing ability of material coil width and thickness specification, and selecting the material supplier. Also, it can improve material utilization rate, scrap reutilization, etc. Material performance database supports the stamping CAE calculation, mainly including the basic properties of the material (mechanical properties, tensile stress–strain curve, the value of n and r, etc.), process performance (expanding rate, forming limit and rebound characteristics, welding, painting, etc.), service performance (dynamic mechanical properties, fatigue characteristics, delayed fracture, etc.), failure analysis of parts (road test, etc.).

In the processing of stamping SE development, the material engineer can propose the material cost reduction proposal according to the requirements of the vehicle project: According to the material cost target of the vehicle, the material cost reduction target can be decomposed into functional subgroups/parts to assist the realization of the vehicle product cost reduction target.

In the processing of stamping SE development, the material engineer can propose the material lightweight proposal according to the requirements of the whole vehicle project: the technical route of sheet metal parts under the premise of performance guarantee, cost control and feasible forming can be proposed; According to the target positioning of the development vehicle model, the application range of high strength steel and light alloy can be expanded to promote the realization of the lightweight target of the design product.

In the development of stamping SE, it is still a difficult problem for material engineers that how to play a better role. First, the materials engineers and stamping SE engineers cannot work well in previous BOM evaluation, so the material properties cannot give the correct plan. For example, in the process of lightweight, the high-strength steel replaced the mild steel. Both of the material engineers and the stamping SE engineers need to confirm formability and springback characteristic to put forward the accurate lightweight scheme. Second, the data accumulation of the basic performance of steel, especially the forming limit of the trimming edge under different historical strain conditions, cannot be well connected with the data required for stamping CAE analysis.

2 Technical Reserve of Material Specialty in Future Stamping SE

In order to achieve the above work, the material major needs to make relevant technical reserves, establish the knowledge accumulation of relevant database required in the process of development, so as to provide better technical support for stamping simultaneous engineering.

2.1 Establishment of Material Basic Performance Database

In order to accurately calculate the stamping CAE, it is necessary to provide whole material basic data to comprehensively evaluate the material performance. The establishment of material basic performance database is urgent to be solved. The database content (Table 1, FC340/590DP) should mainly include material (materials of different steel factory) brand, thickness, tensile strength, yield strength, percentage elongation rate, *n*, *r*, *K*, FLC, stress–strain curve, etc.

Building the material performance database, you need to consider the different steel factories, types of different materials and different strength levels, different materials thickness, coating category (such as electroplating, hot dip, pure zinc, or zinc alloy) and under the conditions of different thickness of coating, materials tensile strength, yield strength, elongation, the value of *n* and *r*, and the coefficient of friction of different coating state range, thus we can build the material performance range of optional window, in the hoping of better reflects the material reality; it is hoped that the applicable scope of steel can be classified accurately: which is suitable for one-time drawing or multiple times drawing; which is suitable for flanging; which is quality sensitive for hedged cutting surface; At the same time, it is necessary to consider the forming characteristics and yield criteria of the developed steel newly, and it is necessary to establish the material parameter documents.

In the process of establishing the performance database of sheet metal forming limit (FLC), the following situations should be considered:

The flanging limit after trimming. Especially, the high-strength steel and thick steel ($\geq$3 mm) are trimmed. We should know how to take into accounting the metallographic structure, shear hardening, tearing band, burrs and dimensional effect of materials, and how to consider the forming limit deviation which is caused by the change of strain path;

The influence of strain gradient in local area for forming limit and the influence of local curvature for forming limit;

The forming limit of thick steel, especially the forming limit of edge parts. At present, there are still some problems in the prediction of springback deformation of thick steel in the calculation of stamping CAE. The prediction of springback is inaccurate and the performance of the material fluctuates greatly. The accurate strain range of hardening curve (which can be obtained accurately) is limit;

Flip limit. Hole turned limit with strain history; hole turned limit which is based on diameter–thickness ratio; surface elongation turned limit; how to establish the correct test method for evaluation;

We should know how to establish a standard test method to measure the limit of flanging by compression with contour-thickness ratio that is considered.

2.2 *The Material Cost Reduction Proposals and Knowledge Accumulation*

The cost of white body generally accounts for 10–15% of the total cost of the vehicle, and the cost of steel accounts for about 75% of the total cost of white body [1, 2]. Therefore, the material cost reduction plays an important role in SE.

Table 1 Main contents of material performance database

Material	FC340/590DP	Thickness	1.5 mm	Factory	Bao steel	
Technical specifications	TS (MPa)	YS (MPa)	Percentage elongation (%)	*K*	*n*	*r*
	≥590	340–440	≥21	–	≥0.14	–
90% confidence interval statistical of upper limit	TS (MPa)	YS (MPa)	Percentage elongation (%)	*K*	*n*	*r*
	640	400	26.2	982.7	0.147	0.90
90% confidence interval statistical of lower limit	TS (MPa)	YS (MPa)	Percentage elongation (%)	*K*	*n*	*r*
	610	360	23.1	890.5	0.135	0.86
Hole expansion rate	–	Cold bending, 180°		$d = 0.5a$		
Stress–strain curve (including original data, at least five pieces of each)		FLC (including original data, at least three pieces of each)				

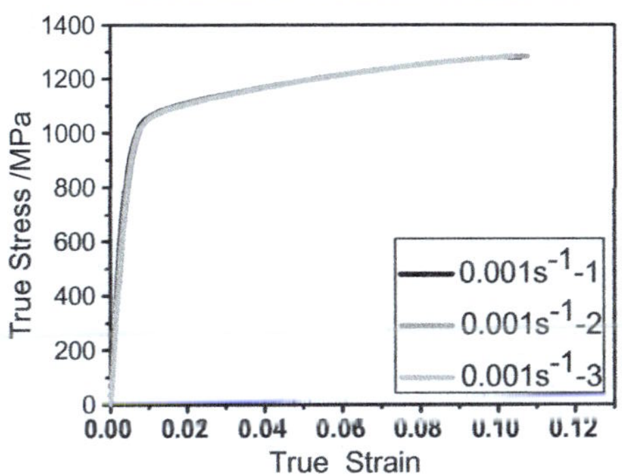

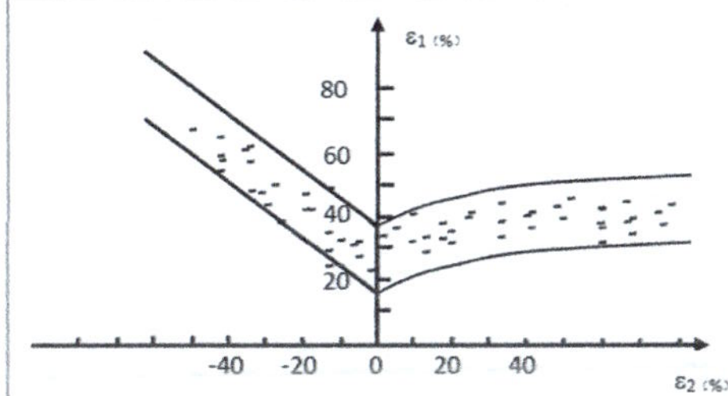

First, we should build the principle of material selection. The material engineers can determine the preliminary material selection plan according to the target cost positioning. The general principle is to make full use of the performance and cost advantages of different types of steel, select suitable materials for each part of the body, and take the performance and cost of the body as the basic starting point to achieve the goal of light weight of the body. The material selection of the new vehicle can be considered from the following aspects:

(1) In order to satisfy the requirements of the part's performance and formability, we should select steel with low cost as far as possible. The selection sequence is: hot-rolled steel (1.8–3.0 mm preferred), cold-rolled steel (low-alloy high-strength steel, dual-phase steel, QP steel), coated steel (hot-dip galvanizing preferred). In addition, priority should be given to hot-rolling instead of cold-rolling.
(2) Considering the availability of materials, we should select the types of materials available from domestic steel to ensure the stability of materials supply in low cost; also should select the common material, thickness specification, should avoid the unused material brand, that the thickness specification is neither too thin nor too thick. We should build the database of common specifications and thickness of cold-rolled steel, hot-rolled steel, hot-galvanized steel, and electric galvanizing steel else.
(3) We should establish the merging principle of brand number and thickness. In the same vehicle model, the combination of material number and thickness specification are best controlled less than 70 types. Small parts should choose large parts of scrap reutilization or set cutting technology.

Second, we should improve the material utilization rate and reduce cost. Generally the weight of the white body is from 230 to 450 kg, the material utilization rate is 55–65%, so, a vehicle needs steel 354–820 kg, every 1% increasing material utilization ratio, will reduce the materials consumption 7–15 kg, the cost savings of about ¥60 to ¥120. So we can know that, improving the body material utilization is an important way to decrease the cost of car body [2]. First of all, according to the safety star level requirements, modeling characteristics, and material selection of the development model, the appropriate material utilization target is set. For the three-box car with 5-star safety requirements, the ideal material utilization target of the body is generally between 63 and 65%. If it is a two-compartment car with 5-star safety requirements, the ideal material utilization target of the body is generally between 61 and 63% [2]. Improving material utilization can be implemented through the following ways:

(1) Optimization the selection of blanking process. When the blanking parts adopt the technology of uncoiling blanking and uncoiling swing shearing, the choice of coil width should be made according to the technology. Figure 2 is a schematic diagram of uncoiling pendulum shear of a part. Table 2 shows the optimization principle of width of different types of steel.

Figure 3 is a schematic diagram of different layout processes, which are limited by the length of the part. The material brand of the part is St280, when the part of

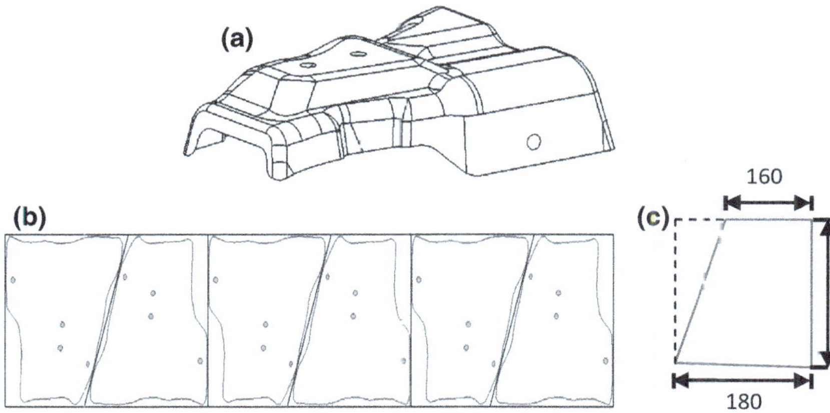

Fig. 2 Schematic diagram of coil pendulum shear. **a** Part. **b** Layout diagram. **c** Shear sheet

Table 2 Optimization principles of coil width of different types of steel

Steel	First	Second	Third
Galvanized	1200–1499	1000–1199	800–999, 1500–1699
Cold (≤440 MPa)	1000–1499	600–999, 1500–1699	–
AHSS (>440 MPa)	900–1300		

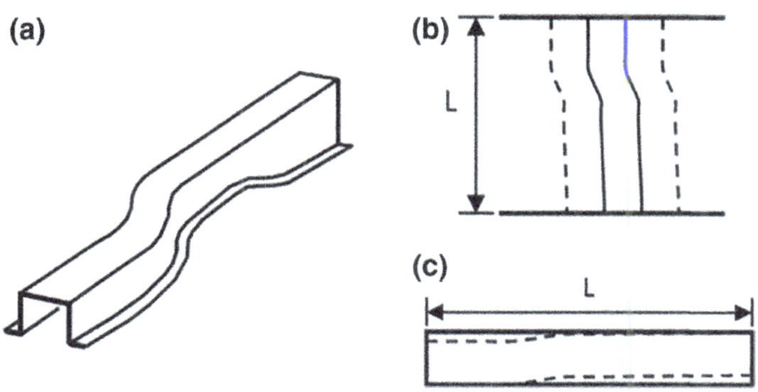

Fig. 3 Blank layout. **a** Part. **b** Length direction. **c** Width direction

the billet length is $L = 1300$ mm, can adopt the way of Fig. 3b to carry on the layout, satisfies the requirement of the length of the common roll width, but when the parts of billet length is $L = 1800$ mm, you need to adopt the way of Fig. 3c to carry on the layout, in order to improve the material utilization, achieve the goal of optimal layout.

Table 3 shows the utilization rate of materials when several different discharging processes are used. So we can know from Table 3, the material utilization rate of quadrangular blanking is the lowest and the material utilization rate of arc blanking is increased by 6.86% (Fig. 4).

(2) Optimized the product structure design. It is necessary to optimize material utilization rate with complex part, the part's appearance, and partial appearance. As shown below, the design of a car and the partial protrusion ears of the part affect the material utilization rate. In the SE process, the metal designer put

Table 3 Material utilization rate of different discharging modes

Mode	Quadrangle	Trapezium	Camber
Material utilization rate (%)	56.63	57.67	63.49

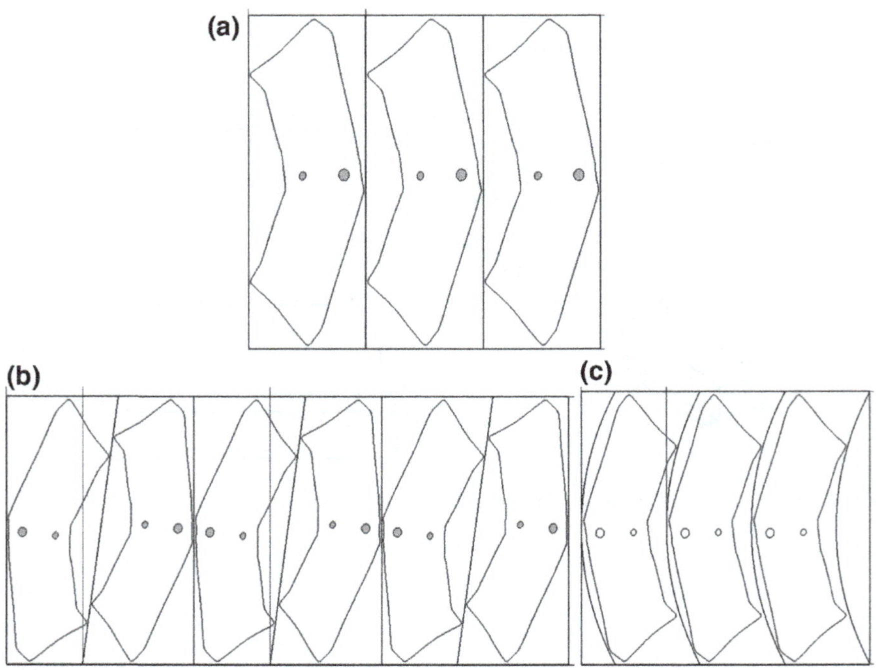

Fig. 4 Schematic diagram of different blanking modes. **a** Quadrangle. **b** Trapezium. **c** Camber

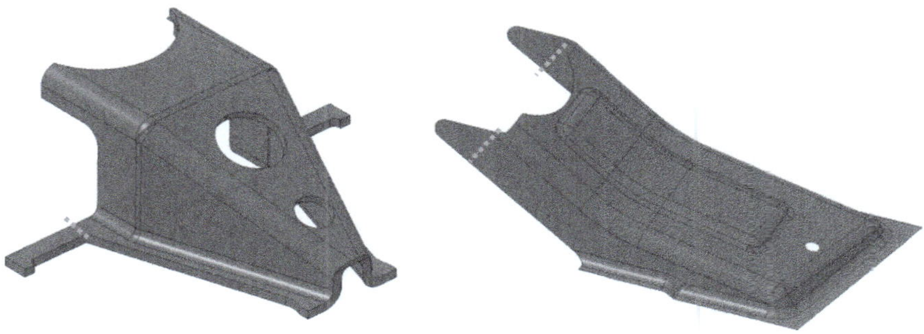

Fig. 5 Modification of part boundary to improve material utilization

forward the ECR report, suggesting the designer to change the structure, which can save the material utilization rate and reduce the cost (Fig. 5).

(3) Optimization product process. From the perspective of die, according to the specific shape characteristics of part, selective use of forming process, open drawing process, left and right parts are combined drawing process and symmetric process of two parts of one mold to improve material utilization rate (Table 4).
(4) Scrap reutilization technology. Use the scrap generated in the production of large products to make other parts [2] (Fig. 6).

2.3 Material Lightweight Proposal and Knowledge Accumulation

Material engineers can propose lightweight materials during the product benchmarking and finish the feasibility report together with the cooperation of the stamping engineers. The main contents of knowledge accumulation include first, establish the lightweight database of high-strength steel, mainly including the mechanical performance, stamping performance, and elastic energy of high-strength steel. At the same time, we should know the level of lightweight of foreign and internal vehicles, the level of accumulation according to different vehicles of high strength steel application examples, in automotive covering parts, components, security classification, technology roadmap to develop high strength steel application, put forward feasible premise to ensure performance, cost control, forming the lightweight high strength steel parts technical route, gradual thinning lightweight proposal in all the application of autonomous vehicles; second, the joint design and stamping professionals will optimize the structure of stamping parts (opening process holes, reducing weight holes, etc.) to achieve the goal of lightweight.

In the lightweight process, it is recommended to introduce the results of stamping hardening and blanking of stamping CAE analysis into structural CAE and safety

Table 4 Material utilization rate of the parts is improved according to merge the left and right parts

Part	Forming process	Die	Material utilization	Improvement
	Drawing separated	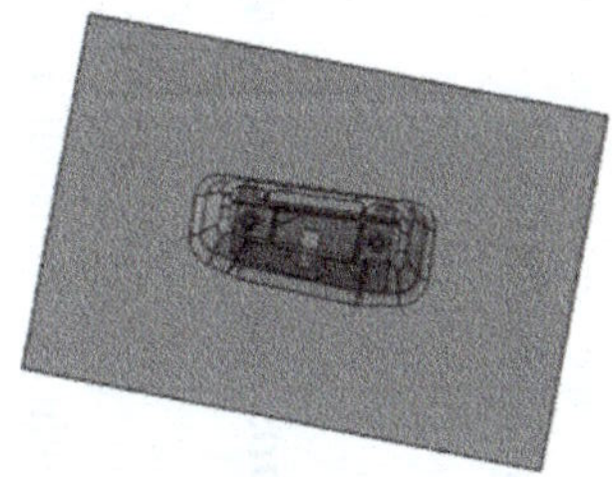	Part weight (single): 0.265 kg Blank weight (single): 1.243 kg Material utilization rate: 21.11%	Material utilization rate↑12.39%
	Drawing combined	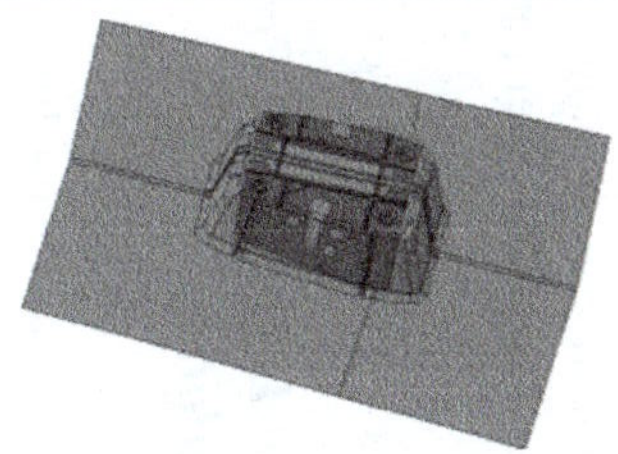	Part weight (single): 0.265 kg Blank weight (single): 0.791 kg Material utilization rate: 33.50%	

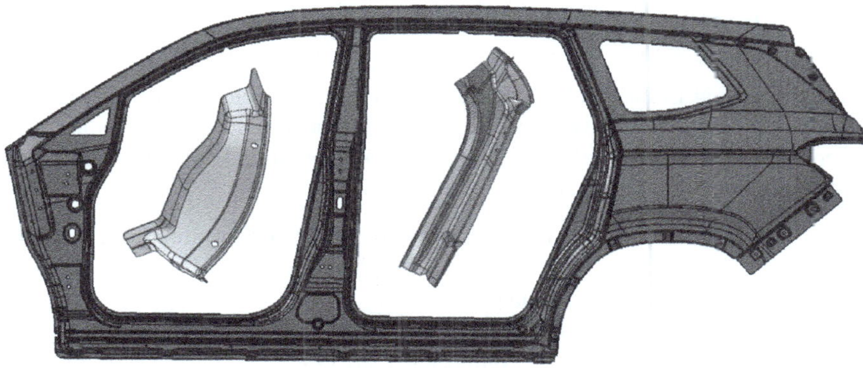

Fig. 6 Improve material utilization rate through scrap reutilization process

CAE analysis, so as to improve the calculation accuracy and reduce design redundancy.

The lightweight proposal and the cost reduction proposal of materials should be carried out before or in parallel with the structural CAE, so as to avoid excessive design changes and CAE calculation, adding work and wasting time and cost.

According to the above analysis, the material engineers need to establish the following technical specifications and standards in the work of stamping SE (Table 5).

Table 5 Technical specifications

No.	File name	Content
1	Material performance database	Basic mechanical properties of materials, the influence of strain history and trim condition on FLC, stress–strain curve
2	Common material cost bases on stamping parts	Material cost of different steel factory
3	General material specifications and thickness optimization principles for stamping parts	Material trademark, thickness, coating, roll width, etc.
4	General methods of improving material utilization rate	Product structure requirements, layout, blanking, etc.
5	Lightweight technology route	Cold forming of high-strength steel, hot forming of high-strength steel, lightweight alloy (aluminum alloy, magnesium alloy, etc.)
6	General methods of material cost reduction	Material optimization, structural optimization, etc.
7	Data accumulation of new steel	Basic forming characteristics, yield criterion, material model, and failure criterion of new steel

3 Conclusion

The development of automobile SE saves the product development cycle, reduces the research cost, and greatly promotes the development of automobile industry. As a member of the SE team, the material engineers play an indispensable part in the implementation of stamping SE:

(1) Establish the material basic performance database and forming limit database, provide the required basic data for stamping CAE analysis, and ensure the accuracy of stamping CAE analysis;
(2) Put forward material lightweight proposal and cost reduction proposal in the initial stage of development, promote the realization of vehicle lightweight target and cost target, establish the knowledge accumulation of material lightweight and cost proposal, and improve product design quality;
(3) Establish professional technical specifications and standards for materials, make good technical reserves, and cooperate with EVI of the steel factory to join promote the development of stamping SE.

References

1. Zhao W, Zhu K, Zhao Y (2015) Application of simultaneous engineering in the development of autonomous vehicle. Automot Pract Technol 11:144–146
2. Wang D, Chen H (2017) Application research of simultaneous engineering in the early development of car body. Shanghai Automot 2:47–52

Application of Shared Concepts in 3D Virtual Assembly

Shuaitao Zhang, Junjun Wang, Ning Tian, Hongguang Zhou, Tao Chen, Fei Liu and Shuang Liu

Abstract At present, an important direction of intelligent manufacturing is the product's Digital Twin, "Digital Twin" means virtual three-dimensional models exactly like real objects, so as to help enterprises optimize, simulate and test in the virtual environment before actually putting into production. If the enterprises have a number of product variants and part type and adopt traditional 3D model assembly method, it will occupy a lot of human resources, and limited by a lot of "constraint command", it is difficult to quickly modify the 3D model of product variant. This is also one of the important reasons that many companies are difficult to achieve the Digital Twin for their full range of products. In this paper, take powertrain as an example use the shared concepts to introduce the 3D virtual assembly method of a platform. This method can build the accurate and real-time Digital Twin with the help of PDM software. The innovation is: (1) the "without Piles Located on" mode of space positioning method for CAD model is implemented; (2) the platform method for the skeleton model is set up; and (3) the platform management standards for 3D model assembly are set up. It is especially suitable for large-scale virtual assembly models such as automobile and powertrain.

Keywords Engine · Platform · Variant · Constraint command · Virtual assembly · Skeleton

Technical Paper

Preface

As we all know, bike sharing is booming in China, there are some excellent bike-sharing company such as Mobike and OFO, their bike-sharing system has outstanding advantage: the shared bikes can be parked anywhere with GPS systems, and the shared bikes do not need to be parked into the piles. So shared bicycle perfectly solves the

S. Zhang (✉) · J. Wang · N. Tian · H. Zhou · T. Chen · F. Liu · S. Liu
DFM Corporation Technical Center, Wuhan 430058, Hubei Province, China
e-mail: zhangst@dfmc.com.cn

China SAE (ed.), *Proceedings of China SAE Congress 2018: Selected Papers*, Lecture Notes in Electrical Engineering 574,
https://doi.org/10.1007/978-981-13-9718-9_46

problem of the "last kilometre travel", thus showing vigorous vitality and replacing the traditional system of piled public bicycles.

In the field of 3D virtual assembly, there was a similar situation: in the traditional 3D virtual assembly, the part's 3D models are constrained by each other. If the designer wants to complete the assembly of the total product's 3D model, it needs to adopt a large number of "constraint commands" between the part's 3D models. We can imagine the following shortcoming: First, if the total product has 700 parts to the assembly, it will probably bring on thousands of serial "constraint commands". Once a front "constraint command" has been wrong, the next several hundred "constraint commands" may have been wrong, so it greatly reduces the reliability of the total product's 3D models; second, if the product is the basic product, the basic product will derive a lot of product variants (such as upgrading product models and different configuration product models), and each new product variant's 3D model will have to repeat a large number of components and parts assembly(even if most parts are common) and then repeat a large number of "constraint orders", leading to longer research and development cycles of the product variant assembly.

The "constraint command" between part's models is just like the "pile" of a public bicycle, and it greatly limits the convenience of multiple product variant assemblies to share the same parts. Because of the complicated operation and poor stability of the "constraint command", the traditional 3D virtual assembly method will be disastrous for the design efficiency in the full range of products, and it can hardly realize the multi-person collaborative virtual assembly design.

In order to solve the above pain points, a convenient 3D virtual assembly method is presented through the shared concept.

1 Introduction of the 3D Virtual Assembly Method Which Based on Platform

1.1 Shortcomings of Traditional Bottom-Up 3D Virtual Assembly Method

The shortcomings of the traditional bottom-up 3D virtual assembly methods are introduced by the following case: as shown in Fig. 1, A powertrain platform includes 50 powertrain product variants (used for carrying 50 vehicle variants, respectively), and each powertrain product variant's 3D model includes more than 700 part's 3D models, and each part's 3D model needs to use three "constraint command" to complete the 3D virtual assembly. In the extreme case, the powertrain platform contains $50 * 700 = 35{,}000$ parts in total, and the 3D models of these parts need to be installed $50 * 700 * 3 = 105{,}000$ times (that is, 105,000 "constraint commands" needed), the workload is huge, and the "constraint commands" will influence each other. If only a few employees are responsible for the work, it will not only be very tired, but also need a lot of work time to seriously drag the overall development progress of the

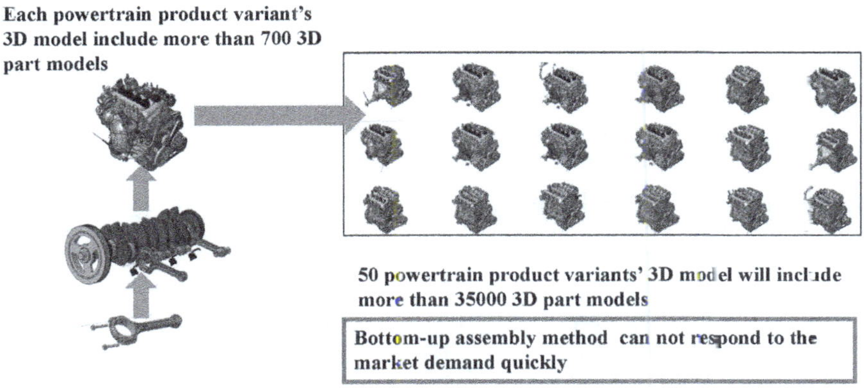

Fig. 1 Bottom-up assembly method with mutual constraints between part models

project. Moreover, it is difficult to modify the powertrain product variant's 3D model quickly and cannot respond to the market demand quickly, which results in a long research and development cycle for the subsequent modification of the project.

According to the above analysis, the traditional 3D virtual assembly method is not suitable for the large-scale 3D virtual assembly models of the product series, especially the large development projects, such as automobiles and engines.

1.2 The Principle and Innovation of 3D Virtual Assembly Method Which Based on Platform

This thesis mainly introduces a convenient 3D virtual assembly method which based on the platform. The method is described as an example of the platform design of powertrain, in this powertrain platform, a top-down 3D virtual assembly method is adopted. First, a skeleton reference model is established for the powertrain platform's unique coordinate. Then all the 3D models of product variants and parts' spatial location will accord to this skeleton reference model as shown in Fig. 2.

The method has the following innovation points:

(1) The "without Piles Located on" mode of space positioning method for 3D model is implemented: cancelling the traditional "parking pile" (corresponding to cancelling the "constraint command" between part's models). In the specific 3D virtual assembly process, the part's model is suspended in a correct space position, rather than be constrained on the other models; thus, it can avoid the huge workload of establishing "constraint command".
(2) The platform method for the skeleton model is set up: break the "regional limit" (corresponding to sharing the position of the parts in the whole powertrain platform, not only in a single powertrain product variant). In the whole platform, the spatial location of the part models and component models is carried out by

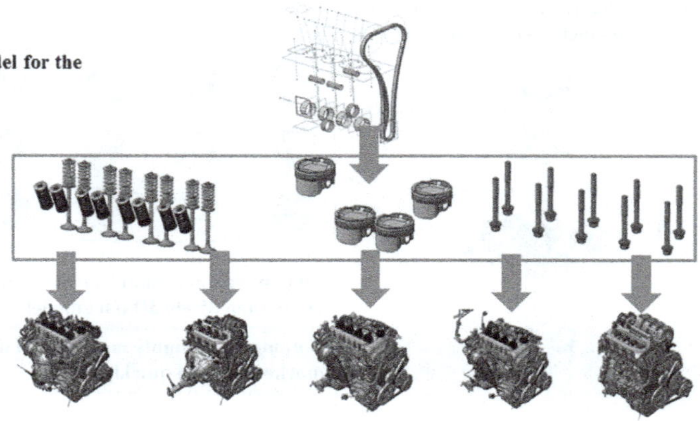

Fig. 2 Top-down 3D virtual assembly method which based on the platform

the unified coordinate system for different product variations under the same product platform, in general, most parts' space position is the same, because the position information of the 3D model is "without Piles Located on", the part's 3D model can be "parked" to the correct position of any product variant in a moment, and the part's 3D model does not need to be constrained repeated in each product variant.

(3) The platform management standard for 3D model assembly is set up: by using PDM system to manage the above product platforms, product variants, components and parts models, we set up a series of 3D virtual assembly standards and rules so that the method can be used efficiently in the PDM system, and the integrated management of product BOM and product assembly model is now available.

Then, taking the application of 3D virtual assembly of a powertrain platform project in PDM and CATIA as an example, the above three innovations are introduced, respectively.

2 The Implementation of "Without Piles Located on" Mode of Space Positioning Method for Cad Model

2.1 The Difference Between "with Piles Located on" Mode and "Without Piles Located on" Mode

As we all know, the biggest advantage of sharing bicycles compared to the previous public bicycles is "without Piles Located on", and the shared bikes can be parked anywhere with GPS systems.

There are similar scenes in the 3D virtual assembly field. "Constraint commands" between part's models just like "piles" of previous public bicycle system. If they can avoid "constraint orders" between part's models, it is equivalent to achieve the "no pile" mode.

The interpretations are as follows:

The "with Piles Located on" mode refers to the traditional bottom-up 3D virtual assembly mode; the part's models are constrained by each other. For example, the part A's model and part B's model are assembled by each other by using "constraint commands" such as "axis alignment" and "surface fitting", and then the part C's model is assembled to the part A or part B with the similar "constraint command". By using the same method, the other part's models are also assembled in turn; finally, we can finish the whole powertrain's virtual assembly.

The "without Piles Located on" mode refers to the top-down 3D virtual assembly mode; the space position of all the part's models is designated by skeleton model, instead of using "constraint commands" between part's models.

As stated in the preface, the "with Piles Located on" method is unsuitable for large-scale 3D virtual assembly models which include many product variants.

In fact, the top-down design concept of "without Piles Located on" method is not a new concept. However, there are few cases in which the top-down concept is applied to the series of complex products such as automobiles and powertrains. The thesis which named "Application of Top-Down Strategy in Product Series Design" [1] introduced a case of a series of simple products (drawer), but the complexity of the case is not comparable to the automobile industry.

This thesis introduces the method of "without Piles Located on" model space positioning. It adopts the concept of top-down design and combines the concept of platform design, which greatly extends the scope of application. The scope of application includes the whole life cycle stage, such as conceptual design stage, detailed design stage, and design modification stage.

2.2 *Introduction of "Without Piles Located on" Space Positioning Method*

Method introduction:

First, the skeleton model is established. It is used to provide a unified coordinate position.

Then put all the part's models in the right position. When the position is placed, it is necessary to refer to the skeleton model as shown in Fig. 3.

Finally, as the position of all part's models has been placed correctly, the position of the parent levels such as subassembly, assembly, and product variant also be kept in sync. It avoids the need to establish a large number of "constraint commands" between the part's models in the traditional assembly mode and avoids a lot of "constraint command" errors in the subsequent design changes.

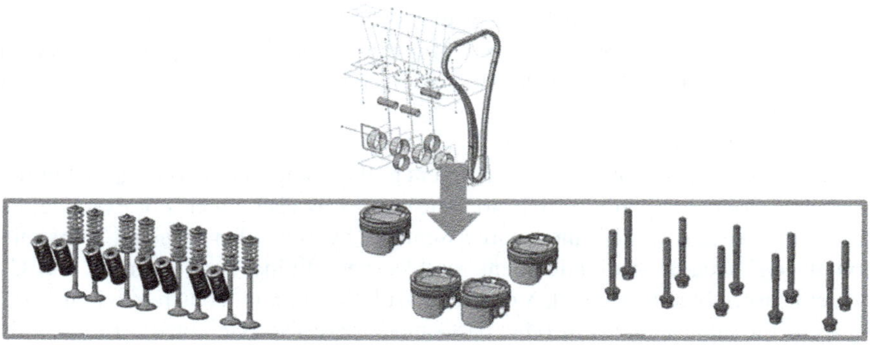

Fig. 3 "Without piles located on" space positioning method

"Without Piles Located on" space positioning method cancelled the traditional "parking pile" (corresponding to cancelling the "constraint command" between part's models), which greatly improves the stability of the whole powertrain products, and is especially suitable for the series of large-scale product development projects.

3 The Establishment of the Platform Method for the Skeleton Model

3.1 Shortcomings of Non-platform Management

Before the emergence of a shared bicycle with GPS systems, many cities have their own system of public bicycles which using parking pile, but the national common bicycle system cannot be achieved because of regional restrictions. For example, if you have a magnetic stripe card for Beijing's public bicycle system, you can only use Beijing's public bicycle; it cannot be used in Shanghai's public bicycle system.

Such situation is not convenient for people working across regions, such as business trips and travel.

There are similar pain points in the 3D virtual assembly field. For example, during the development of large product projects, such as automobile and powertrain, the same product platform will derive a lot of product variants, and the differences between product variants are very small. It is the same for the variety, quantity, and relative space position of most of the parts (or components) in each product variant. For example, a powertrain product platform derived too many product variants; the differences between these product variants are some parts just like wire harness, and small bracket. However, in the traditional 3D virtual assembly, many parts need to be repeatedly installed in different product variants because of the use of bottom-up assembly methods in which the part's models are constrained by each other.

As mentioned above, if a product platform derived to 50 product variants, a lot of virtual assembly works need to be repeated 49 times. If each product variant's virtual assembly work needs ten working days, it will waste 490 working days for the whole product platform, which is a great waste to the project progress and personnel configuration.

The above problems are particularly prominent when using PDM software to manage product variants. In particular, when it needs to achieve the Digital Twin for full range of products with the PDM software, it means a huge amount of work and risk (a lot of companies in China are unable to achieve digital twins in a full range of products; one of the reasons is here).

Based on the above pain points, our company has established the platform method for the skeleton model.

3.2 Introduction to the Platform Method for the Skeleton Model

Method introduction:

On the basis of the "without Piles Located on" space positioning method of the last section, the skeleton model serves all product variants of the same product platform (instead of serving the single product variant), that is, the skeleton model's coordinates are used by all the product variants, the components and the parts which belong to the same product platform.

As shown in Fig. 2 of the previous section, for the powertrain platform, the skeleton model's coordinates are used by all the powertrain product variants, the components and the parts. As shown in Fig. 4, in the skeleton model, the 3D features of the platform, such as points, lines, and surfaces, are parameterized, and the parameters can be modified directly, which can drive 3D features and be convenient to use.

As shown in Fig. 5, our company has realized the Digital Twin work of many platforms and variants in the field of powertrain 3D model management. With the help of PDM software, the goal of real-time updating of the whole digital model for multiple platforms and variants is realized. It has made a qualitative leap in the virtual assembly field of large-scale serial products.

The advantages of the platform method for the skeleton model: in most cases, the 3D model of the same part(or component) only needs to be positioned once, which can satisfy the spatial position demand of all the product variants, and no longer needs to be repeatedly installed into all the 3D digital models of the product variants.

It can be seen that because the part's designers have positioned the models of parts in the correct position, if the BOM structure of the product variant has been built in the PDM, the whole powertrain model of the product variant can be automatically generated with the aid of PDM, and the designer does not need to worry about the construction of those powertrain's product variants which belong to the same platform.

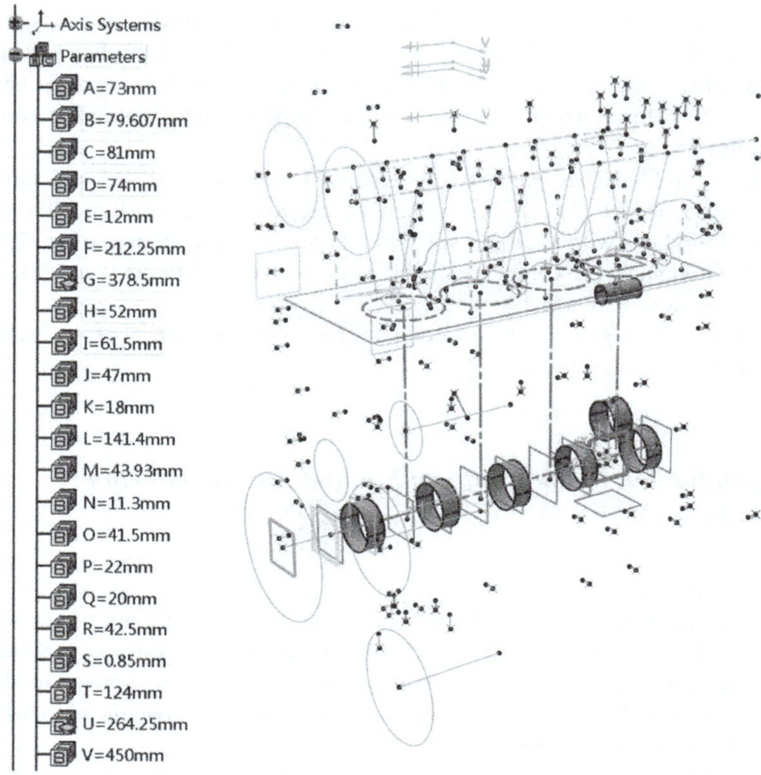

Fig. 4 Parameterized management of platform's skeleton model

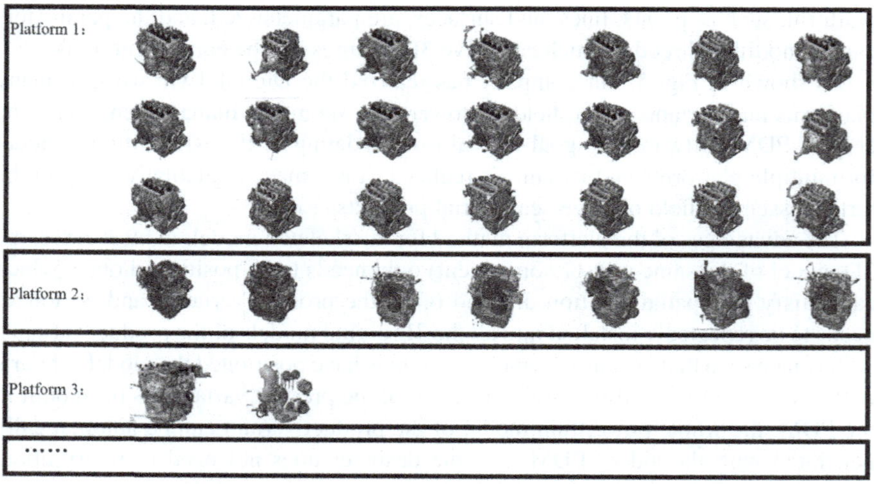

Fig. 5 Achievements in platform management for Digital Twins

In summary, this advantage can save a large number of times and improve the stability of the product variant's model, and it is good for rapid changing of the product variant's model. It is especially suitable for large-scale product development projects such as automobiles and powertrains.

4 The Establishment of the Platform Management Standards for 3D Model Assembly

After using PDM software to manage all digital models, the uniqueness and real-time performance of the digital model can be realized.

When PDM to combine with the 3D virtual assembly method which based on the platform, the most difficult and important thing is setting standards. The reasons are as follows:

(1) In PDM software, for every product variant, its BOM and product variant's model correspond to each other. However, the automobile or powertrain has a very complex BOM, and there are many scenes that the product variant's BOM is not consistent with the product variant's model. For example, the procurement and logistics links need various part numbers, but these part numbers only exist in the product variant's BOM; meanwhile, these part numbers are not needed in the product variant's model. Take the piston as an example: we need to buying two kinds of pistons for the powertrain assembly, one is bigger, and the other one is smaller, the bigger piston needs a part number, and the smaller piston also needs a part number. But when we assembly a three-cylinder engine, we need to choose three pistons from these two kinds of pistons, maybe we choose three bigger pistons, maybe we choose three smaller pistons, maybe we choose one bigger piston and two smaller pistons, maybe we choose another group, so it is complex.
(2) The relative position of several parts will change to meet different product variants. Such as some sensors, we know that the water temperature sensor may be fixed on the cylinder block or on the cylinder head.
(3) The blank parts are used in the procurement and logistics links, so it will appear in the BOM of product variant, but the blank parts cannot appear in the model of product variant. Such as cylinder head blank, we know that cylinder head blank is processed into the cylinder head, and then the cylinder head is assembled on the engine.
(4) PDM software and CATIA software are finished software, the internal structure solidified, if not redevelop the software, blind standards are likely to be impossible to achieve, and any standard must be tested in the software.
(5) The division of labour must be considered and enable the part designer and product variant designer to work each other to ensure that the final powertrain (or the vehicle) model assembly can be completed and continuously updated.

From the above several typical reasons, we can see the importance of theory combined with actual operation.

By defining relevant standards, the author solved the above problems. By using PDM, the unified management of the powertrain's product variant models was realized (as shown in Fig. 6). It solved a series of DMU layout checking problems caused by the scattered data storage. It was very convenient to transfer the 3D model data between different departments.

The long-term and efficient implementation of the standards must be built on a continuous, stable personnel architecture system. Figure 7 shows an example of the

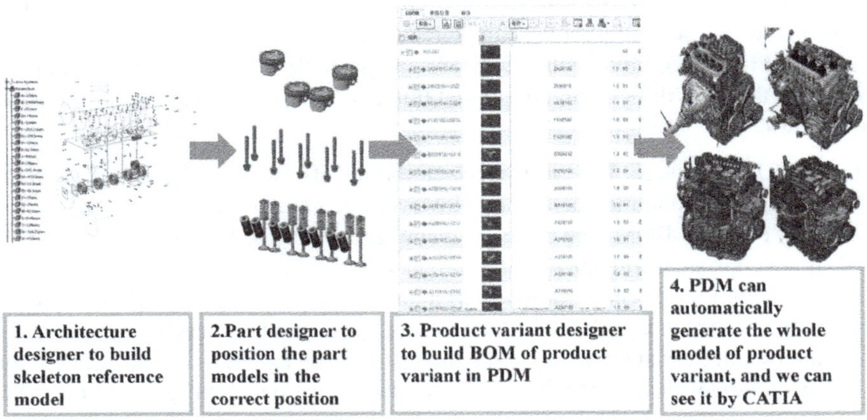

Fig. 6 Diagram of management for the whole powertrain product variants 3D data in PDM and CATIA software

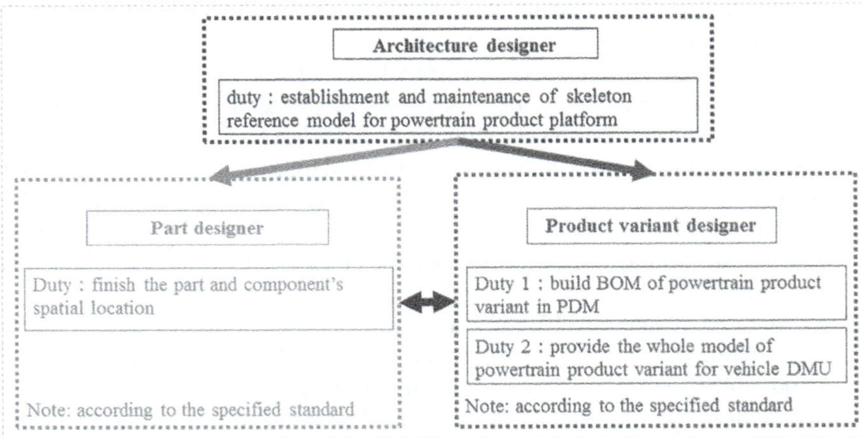

Fig. 7 Diagram of the personnel architecture system for the top-down 3D virtual assembly method which based on the platform

personnel architecture set-up in the implementation of the project. In the process of building the system, the most important thing is to carry out standardized training and practical exercises to the relevant designers, so that they can establish a platform of 3D virtual assembly concept and subvert the traditional concept of assembly in the mind.

In the PDM software, the "model-EBOM" integrated design concept has been adopted, and the barriers between the CATIA model structure tree and the EBOM structure tree have been cleared up, and the precise Digital Twins are realized, which is very beneficial to the different department collaborative development in the cross-field.

5 Conclusion

By using shared concepts to introduce the 3D virtual assembly method of a platform, this method can build the accurate and real-time Digital Twin with the help of PDM software.

This method has been applied to many projects, and the method is mature. This method is especially suitable for large series of virtual assembly models such as automobiles and powertrains.

It is hoped that it can help other colleagues and can learn from this method for the digital model management of the large complex assembly, such as the hybrid powertrain or the vehicle.

Reference

1. Haiping Liu, Jianping Rao (2011) Application of top-down strategy in product series design. Electron Mech Eng 27(5):57–60

Vehicle Doors Leakage Issue Control and Solve

Zhiqiang Xu, Li Tao, Jinchao Ruan and Tianxin Yang

Abstract Leakage issue in production line significantly influences assembly shop FTC and leads to rework cost and risks. For the leakage defect cars, research on rivets' damage force, rivet sequence, and also use complete car CMM measurement methods can find out the main reason that results in the defect, next use short-term solution to temporarily control the batch issue. At last, optimize the unqualified supplier parts and change rivet sequence to totally solve this issue. Engineering verification shows the whole methods and process are stable and effective.

Keywords Doors leakage · Cross-check experiment · Short-term solution · Long-term solution

1 Introduction

The overall perception of a vehicle's quality is significantly influenced by its function performance [1, 2]. The car with very good function performance can not only bring excellent customer satisfaction, but also enhance the enterprise and products' competitiveness [3]. With the fast development of vehicle industry, automotive systems are becoming more and more complex, integrating a large number of functions and features. Nearly, all the automotive enterprises, especially for luxury brands, from preliminary development to mass production, seriously focus on function performance and invest a lot of manpower and financial costs in this aspect.

In this article, we will research the vehicle's function performance. To control and solve leakage issue in assembly line is one of the main contents of function performance. When a car is offline from assembly shop, immediately will be test in shower room. Daily production quality reports show in PT1 stage (before mass production), about 20% cars' doors will have leakage defect, which is a batch priority two issues

Z. Xu (✉) · L. Tao · J. Ruan · T. Yang
Beijing Benz Automotive Co., Ltd.,
Beijing Economic-Technological Development Area (BDA),
No. 8 Boxing Road, Beijing 100176, People's Republic of China
e-mail: xuzhq@bbac.com.cn

China SAE (ed.), *Proceedings of China SAE Congress 2018: Selected Papers*,
Lecture Notes in Electrical Engineering 574,
https://doi.org/10.1007/978-981-13-9718-9_47

(NOK, and not acceptable). If not absolutely controlled and solved, this defect will seriously affect doors' electrical system, decrease FTC (First Time Capability) in the mass production stage. Finally, the leakage cars must be blocked and reworked, which will lead to huge money and time loss.

Therefore, for this defect, we take doors leakage issue as an example, make cross-check experiments, check rivets damage force, and use complete car CMM measurement to find out the root cause. After that, temporarily control this batch issue through the short-term solution. Finally, suppliers optimize the related parts of the dimension, and assembly shop adjusts rivets sequence, which can totally solve this issue. Engineering verification shows the whole solving process is stable and effective.

2 Analysis Process

2.1 Find and Confirm Leakage Area

In PT stage(production test), before mass production, when the complete cars are offline from the assembly line, all of them will have 40 min shower test, in order to check whether there is leakage issue or not. The whole shower process is including 15 min normal shower, 10 min high-pressure normal shower, 10 min high pressure and large flow shower, and 5 min drying process.

After the Q-sensors find the leakage car, we need to find and confirm the leakage point. First, we disassemble the door interior panel and then put the car into the shower room again. About 17 min later, we find the leakage point, and water leaks from door module rivet's edge area. Although the rivets have very good fitting, still has leakage issue, the leakage area is shown in Fig. 1.

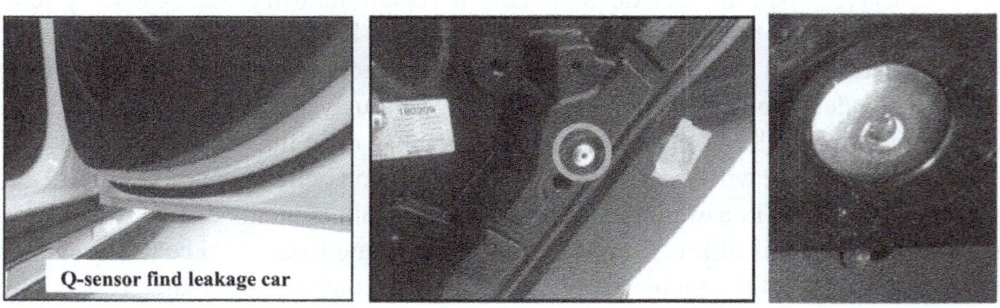

Fig. 1 Water leaks from rivet area

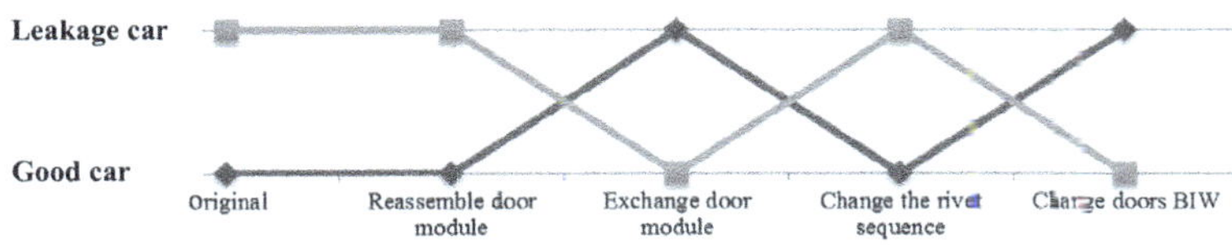

Fig. 2 Cross-check results

2.2 Cross-check Experiment

After the doors' leakage area is found, we use to cross-check experiment to find out the main contributor for this issue. Cross-check method is very commonly used when analysis the main contributors to the defect car. It is not very accurate, but quite fast and convenient to qualitatively analyze the issues [4, 5].

Prepare two cars, one good car, and one defect car, do the cross-check experiments according to the following steps. The cross-check results are shown in Fig. 2.

- Reassemble the door module panel, leakage car still leaks, good car still no leakage, so assembly has no influence on this issue;
- Exchange the two cars' door module panel, the leakage defect goes with the door module panel, so the door module panel has influence on this issue;
- Change the rivet sequence, the leakage defect also goes with rivet sequence, and so this is also rivets sequence related;
- Exchange the two cars' doors body in white, the issue goes with the doors body in white.

From the cross-check results, we can see the door module supplier part, doors body in white, and rivets' sequence are the main contributors to this leakage issue.

2.3 Rivet Dimension and Damage Force Research

After the cross-check, we will continue to research on the rivets' dimension and damage force. The most important dimension is rivet diameter and length.

Randomly select ten rivets from assembly line and from different batches, use the caliper to check the dimension, the dimensions are shown in Table 1, all the rivets dimension are within tolerance.

Next, randomly select three rivets to check the damaged force. In 2D drawings, the specification of damage force is above 4000 N. From the rivets' test machine, the results show all the rivets are above 10,000 N, so the rivets' damage force is qualified. The results are shown in Fig. 3 and Table 2.

Table 1 Rivets dimension

No.	Length ($16^{-0}_{+1.0}$)	Diameter ($6.4^{-0.15}_{+0.08}$)	Qualified?	
1	16.78	6.34	Y	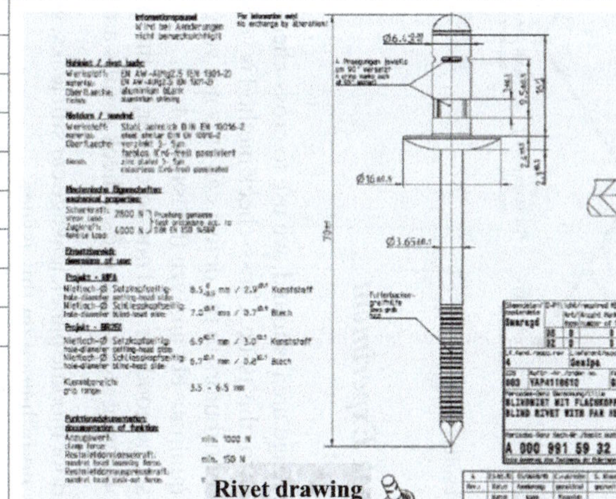
2	16.97	6.37	Y	
3	16.91	6.34	Y	
4	16.93	6.35	Y	
5	16.86	6.35	Y	
6	19.93	6.36	Y	
7	16.91	6.36	Y	
8	16.92	6.33	Y	
9	16.87	6.33	Y	
10	16.88	6.34	Y	

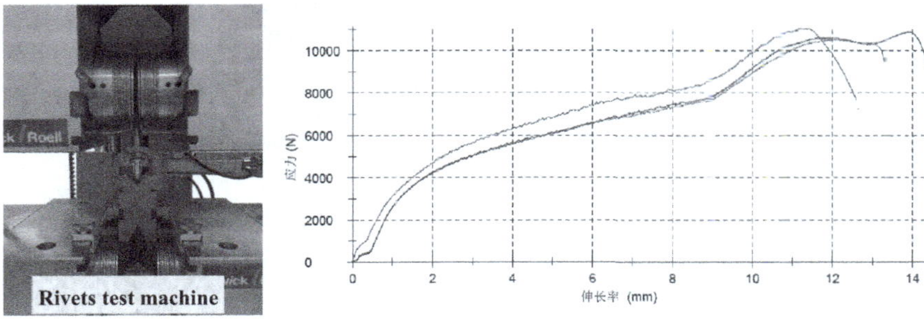

Fig. 3 Rivets' damage force results

Table 2 Rivets' damage force results

Label	Sample no.	Damage force (KN)
▬	1	11.04
▬	2	10.51
▬	3	10.88

2.4 CMM Measurement

Complete car CMM measurement is only used for analysis measurement. It is totally different from body-in-white series audit measurement. And use this method, we can get the precise deviation and contribution of each factor to this issue, all the measurements are based on CAD models and Gauss square least method [6].

Prepare another leakage car and make sure it is in the original status, which means no reassemble, no cross-check and so on. The measurement reports are shown in Fig. 4.

- In door alignment, doors body-in-white holes used to fix and locate door module is 1.5 mm deviation (negative contribution);
- In door module part alignment, location hole also has −1.3 mm deviation (negative contribution), out of tolerance ±1.0 mm;
- Check the rivet sequence in working instructions, there are no clear requirements for the rivets' sequence except for the two location ones.

Finally, we can get the deviation and contribution of all the related factors, which are shown in Table 3.

From the cross-check results and CMM measurement results we can summarize the root cause for the leakage issue.

- Door module part location hole has 1.3 mm deviation—Supplier part issue.
- Doors body in white has 1.4 mm deviation—Press shop issue.

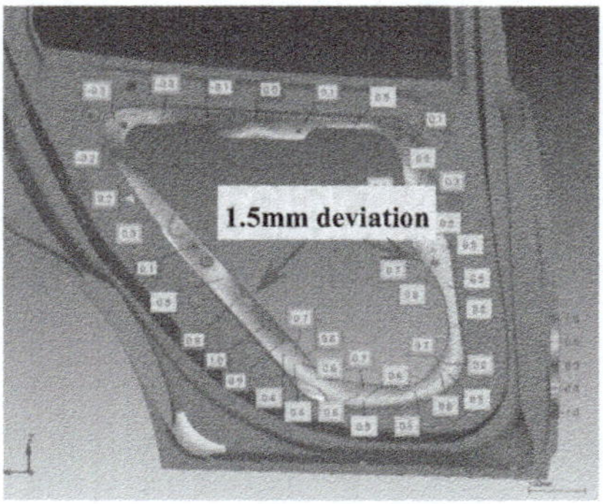

a) **Left door BIW deviation**

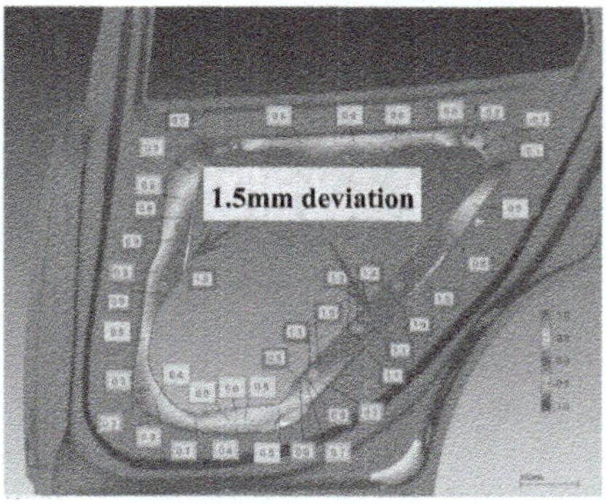

b) **Right door BIW deviation**

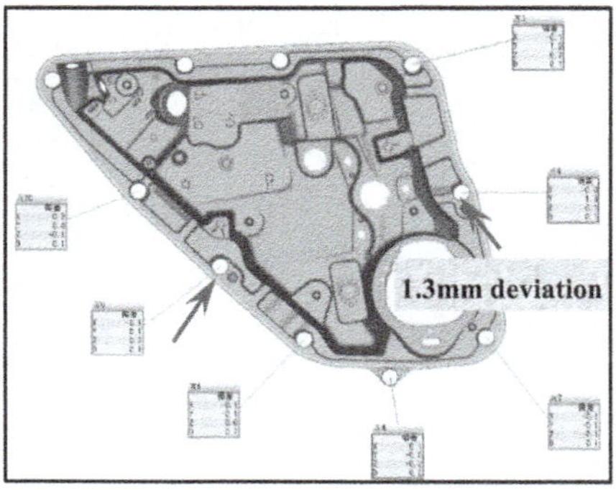

c) **Left door module deviation**

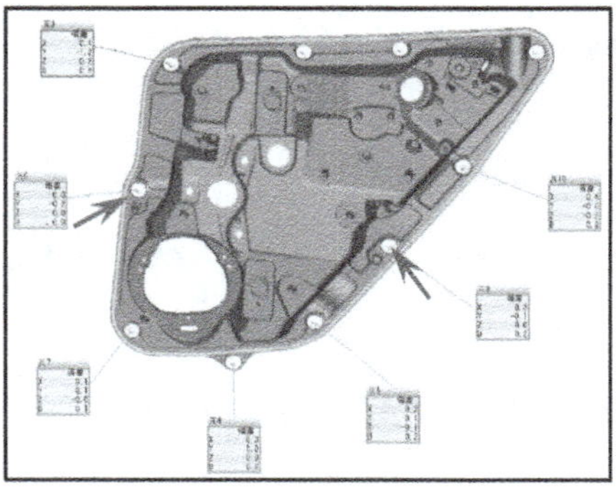

d) **Right door module deviation**

Fig. 4 CMM measurement reports

Table 3 Deviation and contribution of all the factors

	Door module part/mm	Doors body in white/mm	Assembly
Deviation/mm	1.3	1.4	Wrong sequence
Contribution	Negative	Negative	Negative
Tolerance	±1.0	±1.5	–

- Wrong rivets sequence—Assembly issue.

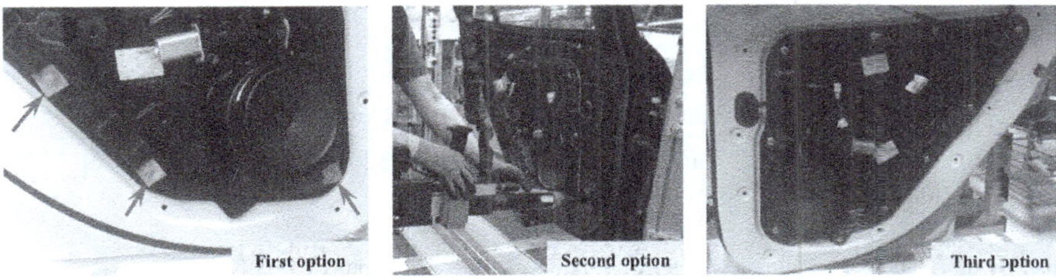

Fig. 5 Three options of short-term solutions

Table 4 Comparison and evaluation of three short-term solutions

	Advantages	Disadvantages	Effect	Approve?
First tryout	Low cost	Corruption risks, not easy to carry out	Good	N
Second tryout	Easy to implement	Middle cost	Good	Y
Third tryout	Definitely no leakage car	High cost, take long time	Good	N

3 Short-Term Solution

According to the root cause analysis results, we need to work with R&D and production to propose a short-term solution, in order to temporarily control the leakage issue. For the short-term solution, in most cases, we use some materials or optimize some process or sorting the parts and so on to temporality stop and solve the defects. For this issue, we add glue on the leakage rivets area, which is shown in Fig. 5.

- First option: add waterproof tape on three lower rivets, which is implemented on finish car;
- Second option: add sealing glue on three lower rivets, which is implemented on door assembly line;
- Third option: rework all the leakage cars, which mean to remove rivets, change door module, and rivet again.

After fully discuss and evaluate with R&D, production and quality department (Table 4), we choose the second one as the short-term solution, which is released by R&D, valid for three months.

4 Long-Term Solution

For the long-term solution, we should ensure it is stable and effective. Not only don't affect other parts' fitting status, but also has small money and time costs. When summarized the CMM reports, cross-check results and verified short-term solutions

Table 5 Evaluation of long-term solutions

Options	Evaluation			Decision
	Time	Costs/RMB	Risks	
(a) Press shop optimize doors press part dimension	3 months(tooling, inventory production and release)	320,000	High	N
(b) Optimize door module dimension and rivets sequence	3 weeks	No cost	No risks	Y
(c) Add water proof tape on rivets, need engineering change	The remaining life cycle of the car, 3 years	450,000	High	N
(d) Make short-term solution as long-term solution	The remaining life cycle of the car, 3 years	1,200,000	No risks	N

in PT2 stage, we proposed the available options of a long-term solution, the risks and costs for the four options are shown in Table 5.

Option (a): Optimize the doors body-in-white deviation;
Option (b): Optimize door module dimension and rivets' sequence;
Option (c): Add water proof tape on rivets until EOP (end of production);
Option (d): Make short-term solution as long-term solution until EOP (end of production).

After discussed with R&D, production and quality departments, we choose option (b) as the long-term solution, which will implement for the whole lifecycle of the car.

5 Engineering Verification

Since February 14, 2018, the new model of the car was in PT2 (production test) stage, totally there are 78 cars. When we use the short-term solution for this defect, only four leakage cars were found, leakage rate was greatly reduced, the short-term solution is stable and effective, and four leakage cars are reworked.

Next, from April 9, 2018, the new model of the car was in PT3 stage. Totally, the assembly will produce 176 cars, for which, the door module supplier has optimized the parts dimension, and assembly changed the rivets sequence, which is shown in Fig. 6. After the shower test, no leakage car was found, and the long-term solution is effective and stable.

Figure 7 shows the defect rate greatly dropped from PT1 to PT3, and we can see the leakage issue is absolutely solved. Then in mass production, we are sure that the customers will get qualified cars.

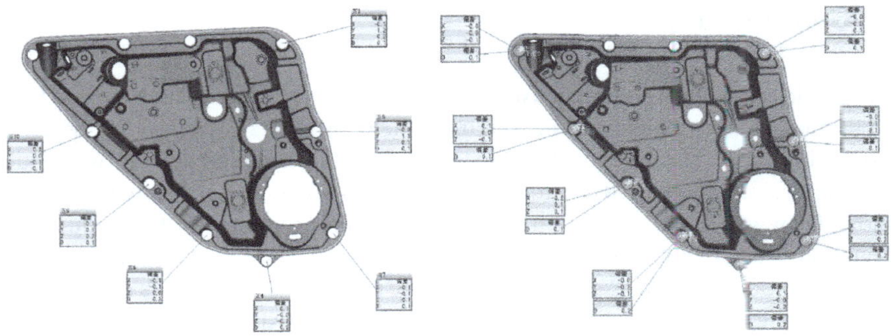

a) Supplier optimized door module part

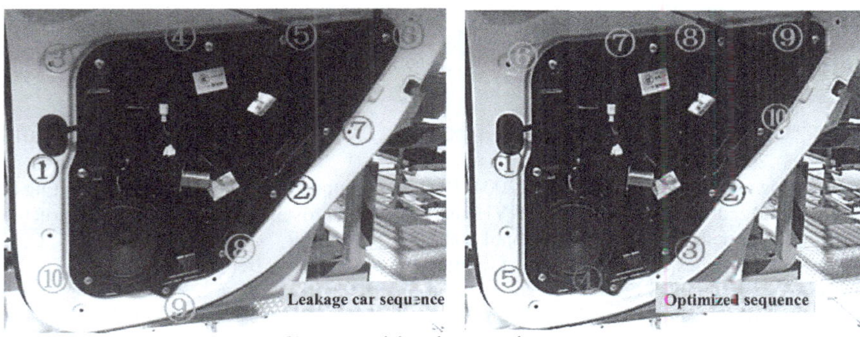

b) Assembly change rivets sequence

Fig. 6 Long-term solution

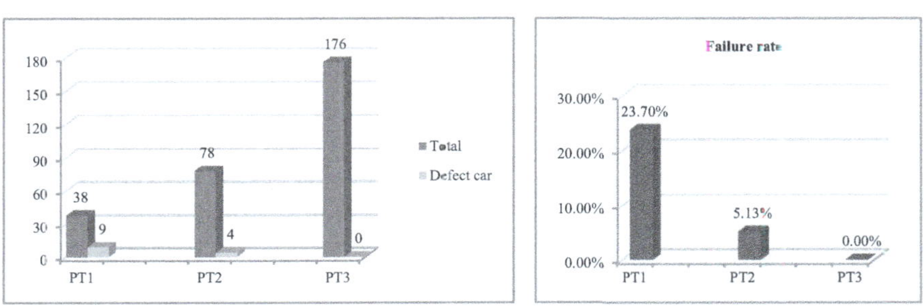

Fig. 7 Defect rate reduced

6 Conclusion

The article presented a comprehensive process based on rivets' research, cross-check experiments and complete car measurement methods to deal with leakage issues of production cars. Therefore, for the dimension deviation-coupled defects, we should find out the vehicle body deviation, assembly deviation, and parts dimension deviation. Then aim at the main contributors, take actions to temporarily eliminate the deviation. At last, based on the enough tryouts and verification, optimize the related parts and production process to absolutely solve the defects. The results of this article show the whole methods and process is stable and effective and can be widely used in vehicle mass production.

References

1. 杜克强, 张海华. "车门密封胶条漏水因素以及解决方法", 装备制造技术, 2013, 11(208):201–204
2. Govindswamy K, Hartwig M, Alt N (2004) Designing sound to build character. Automot Eng Int
3. Azadi S, Azadi M, Zahedi F (2009) NVH analysis and improvement of a vehicle body structure using DOE method. J Mech Sci Technol 2009(23):2980–2989
4. Liu WQ (2005) Experiments design. Tsinghua University Press, Beijing
5. Cui W, Li X, Zhou S (2007) Investigation on process parameters of electro spinning system through orthogonal experimental design. J Appl Polym Sci 2006(103):3105–3112
6. Li M, Fei L (2012) Geometrical measurement technology and applications. Standards Press of China, Beijing

Microstructures and Mechanical Properties of Butt Joint of A356/6005A Dissimilar Aluminum Alloy by Pulse MIG Welding

Guangshan Hu, Yun Cheng, Mingzhu Zhang and Yi Zhang

Abstract The microstructures and mechanical properties of the butt joint between cast A356 aluminum alloy and wrought 6005A aluminum alloy welded using ER5356 welding wire by the method of DC pulse metal-inert arc gas (MIG) welding were investigated by optical microscopy (OM), scanning electron microscopy (SEM), equipped with energy dispersive spectrometry (EDS) and tensile property tests. The fracture occurred at the heat effected zone and weld zone mainly for the A356/6005A welded joint. Its ultimate tensile strength, elongation and welded joint strength coefficient were 219 MPa, 6.21% and 0.7, respectively. And the welding strength coefficient was about 0.7. The fusion zone was characterized by dendritic structure and composed of α-Al, Mg_2Si and Mg_5Al_8 phases. The fusion zones in the A356 and 6005A alloys showed different microstructure characteristic. For the fusion zone in the A356 side, its width was over 500 μm and consisted of island-shaped (α-Al + Al–Si) eutectic and coarse Si particles. The fusion zone in the 6005A side which exhibited transition microstructures and consisted of (α-Al + Al–Mg–Fe–Si + Al–Fe–Si) mixed phases was about 80 μm in width.

Keywords A356/6005A joint · Metal-inert arc gas welding · Microstructures · Mechanical properties

1 Introduction

As the automotive industry fast developing, the possession quantity of cars in China increases year by year. In 2017, the auto manufacture and sale volumes respective are 29.015 million and 28.879 million, respective growing 3.2 and 3.0% compared with last year, and ranked first in the world for the consecutive nine years [1]. The current development tendency of automotive is lightening, safety, comfort, low-cost

G. Hu (✉) · Y. Cheng · M. Zhang · Y. Zhang
Zotye Automotive Engineering Research Institute, Hangzhou 310018, China
e-mail: beixue10@163.com

G. Hu
Zhejiang University, Hangzhou 310027, China

China SAE (ed.), *Proceedings of China SAE Congress 2018: Selected Papers*, Lecture Notes in Electrical Engineering 574,
https://doi.org/10.1007/978-981-13-9718-9_48

and environmental care. The lightweight materials such as aluminum (Al) alloys, magnesium alloys and composite materials have been widely used for weight and carbon emission reduction [2, 3].

Since the low density, high specific strength, excellent formability and almost completely recycled, Al alloys are characterized as "Green metals" and have been applied in the automobile industry widely [4]. The Al alloys have been applied to produce engine cover, door inner and outer plate, steering knuckle, auxiliary frame and so on [5, 6]. To demonstrate their technical prowess, some international automotive OEMs have developed the mass-produced all Al car bodies, such as Audi's fourth-generation A8, Cadillac CT6, Range Rover and Jaguar XF. Zotye Automotive Engineering Research Institute has independently developed an all Al alloy vice-frame which partial structures are made of wrought 6005A Al alloy and cast A356 Al alloy. The 6005A and A356 alloys are welded by pulse metal-inert arc gas (MIG). MIG welding is a potentially beneficial fastening process for automobile manufacturing as it reduces weight as well as lower cost. The pulse MIG welding is usually used in automobile Al alloys components welded. As expected, intensive researches have pay attention to the pulse MIG welding in wrought and cast Al alloys. Nie et al. have investigated the microstructure evolution and mechanical properties of MIG welded 6061/A356 Al alloys dissimilar joints. For the 6061/A356 butt joint, grain boundary liquation and segregation occurred in the partially melted zone (PMZ) on 6061 Al alloy side, and brittle Fe-rich phases were observed in partially melted zone on A356 Al alloy side. The ultimate tensile strength of the joint reached 235 MPa, which is about 83% of that of T6 treated 6016 alloy. In addition, the joint failed mainly in PMZ and heat-affected zone (HAZ) on A356 Al alloy side through mixed fracture mode with quasi-cleavage and dimples on the fracture surface [7, 8]. The researchers in CRRC Qingdao Sifang CO., LTD. have studied the welding performance of 6005A wrought Al alloy and found that the welding parameters exerted remarkably influence on the grain sizes and distribution characteristic of second phases in the welded zone [9]. However, phase composition of the 6005A joints was unchanged during the welding process. The research on MIG welding between 6005A wrought Al alloy and A356 cast Al alloy is limited. In this paper, the butt joints of 6005A/A356A alloys have been prepared using the pulse MIG welding, and the microstructure characteristic, alloying element distribution and mechanical properties of the welding joints filled by ER5356 welding wire were investigated. This research aims to improve the welding quality of Al alloy auto parts, to promote the application of Al profiles and castings in automobile chassis, longitudinal beam and beam structure, etc., and to reduce the lightweight coefficient.

2 Materials and Experimental Procedures

2.1 Experimental Materials

Wrought 6005A and cast A356 Al alloys in T6 state were used as the base metal in this test, and the specimen size was 100 × 40 × 5 mm. The yield strength (YS) and ultimate tensile strength (UTS) of 6005A-T6 alloy is 255 and 292 MPa, respectively. The YS and UTS of A356-T6 alloy are 300 and 315 MPa, respectively. Modification treatment and T6 heat treatment (the artificial aging) can effectively spheroidize the primary Si phase and eutectic Si in Al–Si cast alloy and then improve the mechanical properties. Figure 1a, b is the microstructures of A356 alloy and 6005A alloy, respectively. A large number of spheroidized Si particles disperse in the matrix of A356 alloy. Mg_2Si phase could be completely dissolved into the matrix in solid solution treatment and precipitated very fine particles in the T6 treatment, so it was hard to be observed [10]. The particles in the matrix of 6005A alloy were mainly Al(Fe,Mn)Si phase which widely existed in Al–Mg–Si wrought alloys. ER5356 welding wire with a diameter of 1.2 mm was used as the filler in the welding experiment. The chemical composition of the base metals and the welding wire are shown in Table 1.

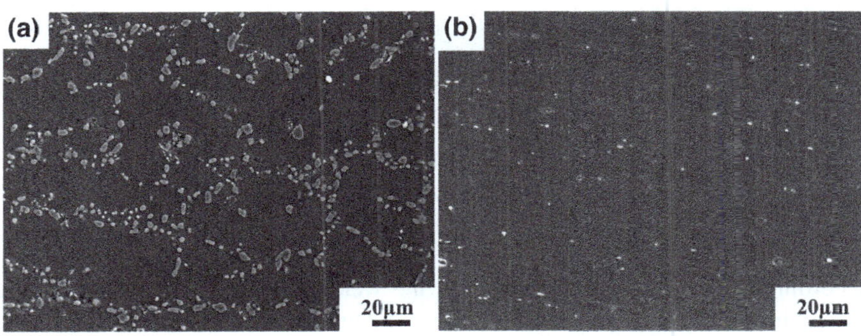

Fig. 1 Microstructures of T6 treated Al alloys: **a** A356, **b** 6005A

Table 1 Chemical composition of base metals and welding wire (mass fraction, %)

Materials	Si	Mg	Mn	Fe	Cu	Cr	Zn	Al
6005A	0.822	0.647	0.278	0.133	0.136	0.063	0.023	Bal.
A356	6.86	0.294	0.002	0.166	0.004	0.005	0.005	Bal.
ER5356	0.21	4.71	0.07	0.160	0.008	0.075	0.007	Bal.

Table 2 Welding parameters of the experimental materials

Welding current (A)	Welding voltage (V)	Feeding rate (m/min)	Welding speech (m/min)	Gas flow (L/min)
170–180	19–21	7	0.7	20

2.2 *Experimental Procedures*

Welding specimens were prepared with a 45° V groove and 1 mm root face. Before welding, the sample was polished with wire brush and sandpaper to wipe off the oxidation layer of the surface, then cleaned with acetone and dried to remove the residual moisture. The TPS 400i digital MIG welding machine (Fronius, Austria) and ABB robot have been used in MIG welding to prepare the joint. The base metals (A356 alloy and 6005A alloy) were welded by direct current pulse MIG welding to prepared the butt joints, and the protective gas was 99.99% argon (Ar) gas with a gas flow of 20 L/min. The MIG welding parameters were summarized in Table 2. The microstructure analysis was performance by OLYMPUS GX51 optical microscope (OM), HITACHI SUI510 scanning electron microscope (SEM) with X-ray energy dispersive spectrometer (EDS). The tensile direction was perpendicular to the weld seam. The tensile experiment was conducted via the MTS E45.305 electronic universal testing machine with a tensile speed of 2 mm/min at room temperature.

3 Results and Discussion

3.1 *Microstructures*

According to the heat input variation in the welding process, the joint could be divided into HAZ, fusion zone (FZ) and weld zone (WZ) [11]. Figure 2 shows the metallographic analysis of different zones of A356/6005A alloy welded joint. It can be seen from Fig. 2b that WZ exhibited typical characteristics of as-cast dendrite structure, which composes of α-Al, Mg_2Si and Mg_5Al_8 phases mainly. In the welding

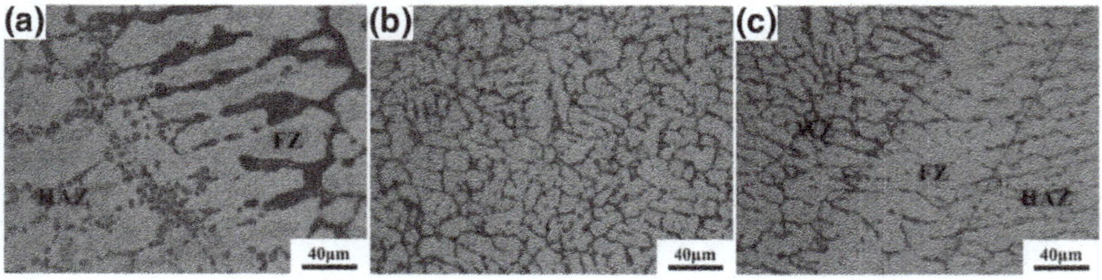

Fig. 2 Metallographic analysis of different zones of the welded joint: **a** A356 side, **b** weld zone, **c** 6005A side

process, the welding wire is melted into the welding pool as the temperature of WZ reached 700–800 °C. Subsequently, the welding pool starts to rapid cooling and then crystallizes, leading to the heterogeneous diffusion of solute atoms. Furtherly, the concentration difference of solute atoms is exacerbated, because of the different solidification sequence of different phases, a typical dendritic structure is formed during the solidification. Since the alloying element Mg in the ER5356 welding wire could refine the WZ grains, the microstructures of melting zone were relatively fine and uniform [12]. FZ is the mixed microstructures of welding wire and base metal. The columnar crystals which form in the way of epitaxial solidification along the direction of heat dissipation range near the welded side [13, 14]. As shown in Fig. 2a, the compound content in FZ in the side of A356 alloy was significantly higher than that of the side of WZ and 6005A alloy. The FZ width in the side of 6005A profile is about 80 μm, and it is less than that of A356 alloy. It is attributed to the eutectic temperature of Al–Mg–Si wrought alloys (595 °C) is higher than that of Al–Si cast alloys (577 °C). Figure 2c demonstrates that the grains in HAZ of 6005A alloy have been coarsened obviously.

SEM and EDS are carried out to further analyze the microstructure characteristics of different zones in welded joints. Based on Fig. 3a, b, the FZ width in the side of A356 alloy is more than 500 μm, and it is mainly composed of a large number of insular compounds and a small number of coarser granular compounds. The insular compounds are composed of lamellar structure, bony structure and fine granular compounds, exhibiting a characteristic of eutectic structure. EDS analysis suggests that the contents of Al, Si and Mg in coarse granular compounds (arrow 1 in Fig. 3b) are 4.836, 91.753 and 1.012 at.%, respectively. Therefore, the granular compound is Si phase. The formation of coarse Si particle is mainly attributed to: (1) The unmelted Si particles in A356 base metal enter into the FZ by the effects of electromagnetic agitation and heat convection; (2) the Si atoms in melted A356 base material and welding wire solidify and precipitate in the welding process. The FZ in the 6005A section side is characterized by the transitional structure and consisted of dendrite crystals and granular compounds. Figure 4a, b is the enlargement of the fusion zone in Fig. 3a, c, respectively, and some Mg atoms are dissolved into the matrix (as shown in Fig. 4b middle point 5). As shown in Fig. 4a, the FZ matrix in the side of A356 alloy is α-Al, and the island compounds are mainly composed of Al and Si elements, forming Al–Si eutectic structure. It can be seen from Fig. 4b that Al

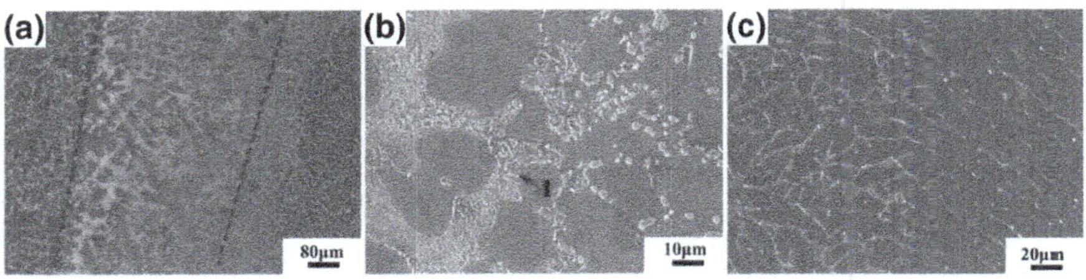

Fig. 3 SEM analysis of the different zones of the welded joint: **a**, **b** A356 side, **c** 6005A side

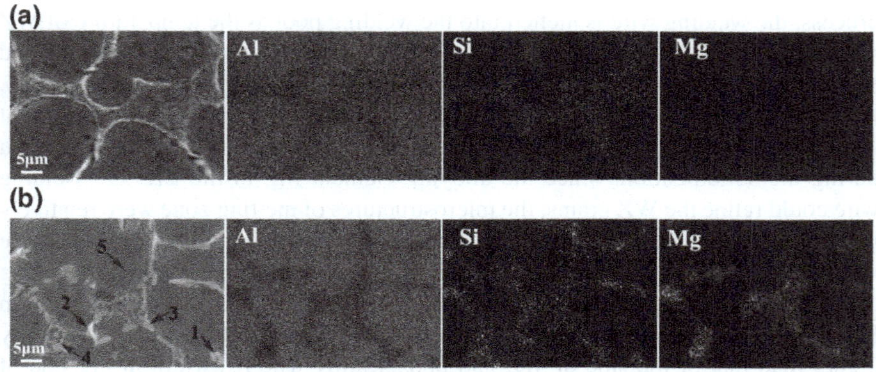

Fig. 4 SEM and EDS analysis of the fusion zone of the welded joint: **a** A356 side, **b** 6005A side

Table 3 EDS analysis results of the points in Fig. 4

Points	Chemical composition (atomic percent, %)				
	Al	Si	Mg	Mn	Fe
1	78.827	15.241	4.77	0.216	0.946
2	81.877	14.943	2.714	0.342	0.124
3	74.721	11.763	12.647	0.056	0.785
4	60.396	18.619	20.697	0.077	0.212
5	97.18	0.219	2.161	0.166	0.273

elements are mainly distributed on the FZ matrix in the side of 6005A alloy, resulting the formation of α-Al, but the Mg and Si elements mainly existed in the dendrites. Compared to the uniform distribution of Si in dendrites, the concentration of Mg element in gray dendrites is low, but in bright white dendrites is high, even higher than the content of Si. EDS analysis results of marked areas in Fig. 4b have been summarized in Table 3. Based on Al–Mg–Si ternary phase diagram and previous studies, it is speculated that the gray dendrites in FZ in the side of 6005A alloy are Al–Mg–Fe–Si phase, while the bright white dendrites are mainly Al–Fe–Si phase, which could be attributed to the diffusion of Fe atoms in the base material. Therefore, the FZ in the side of 6005A alloy is the (α-Al + Al–Mg–Fe–Si + Al–Fe–Si) mixed structures.

3.2 Mechanical Properties

The tensile testing was conducted on six groups of A35656/6005A alloy butt joints at room temperature. The tensile engineering stress–strain curves are shown in Fig. 5, and the experimental results are summarized in Table 4. The HAZ and WZ are the

Fig. 5 Engineering stress–strain curves of the welded joints

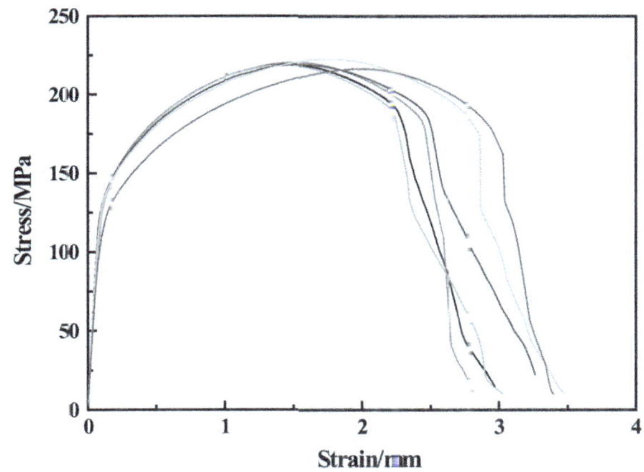

Table 4 Mechanical properties of the welded joints

No.	UTS (MPa)	Average (MPa)	Elongation (%)	Average (%)	Fracture location
1	219	219	6.49	6.21	WZ
2	220		5.92		WZ
3	220		5.60		HAZ
4	219		6.03		WZ
5	216		6.27		WZ
6	222		6.95		HAZ

mainly fracture failure position of the joints, suggesting these two zones are the weakest areas for the butt joints. The average UTS and elongation of the joints are 219 MPa and 6.21%, respectively. Compared with the base material, the UTS of the joints is 73 and 70% of A356 and 6005A alloys, respectively. As a result, the welding strength coefficient is about 0.7. According to the ISO standard 15624-2, the UTS ($Rm(w)$) of welded joint of 4 mm 6005A alloy in T6 state must be satisfied:

$$Rm(w) = Rm(BM) \times T \tag{1}$$

where $Rm(BM)$ is the required minimum UTS of base metal, and T is the utilization factor. 6005A alloy in T6 state is treated using the artificial aging and T corresponded to a value of 0.7. According to the design requirements, the strength requirement of 6005A alloy in T6 state is ≥310 MPa, then $Rm(w)$ is 217 MPa, which is lower than the strength of the joint of A356/6005A alloy. Therefore, the welded joint of A356/6005A alloy has better mechanical properties and meets the design requirements.

Figure 6 shows the SEM morphology of the tensile fracture surface of the A35656/6005A joint. The pores and dimples with different sizes are evenly distributed on the tensile fracture surface, as indicated in Fig. 6a. The existence of pores

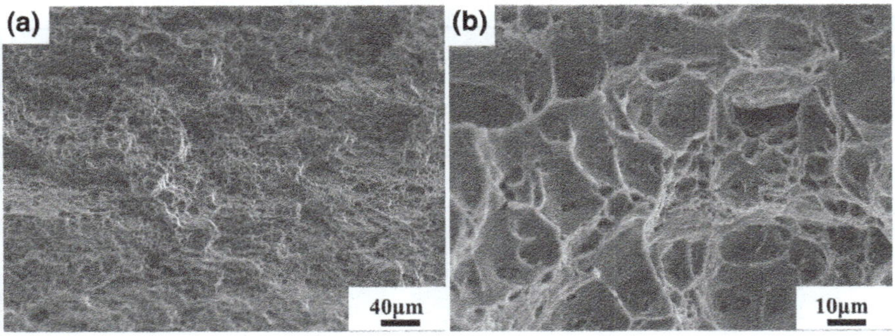

Fig. 6 SEM morphology of the tensile fracture of the joint

reduces the effective bearing area of the cross section of the joints, resulting in the reduction of the strength and plasticity. In the loading process, the pores can promote the initiation and expansion of cracks and make WZ became into the weakest area of the joint. It is observed from Fig. 6b that small holes distribute in the dimple. In the welding process, the temperature of WZ exceeded the solid solution temperature of the low-melting point phase, such as Mg_2Si phase, so these phases dissolved into the matrix, leaving small holes. Therefore, the strength and plasticity of the joint are lower than that of the base metals.

4 Conclusions

(1) The MIG welding butt joint of A356/6005A alloy had excellent mechanical properties, and the UTS and elongation rate of the joint were 219 MPa and 6.21%, respectively. Its welding strength coefficient was about 0.7.

(2) The weld zone of A356/6005A alloy joint showed a typical as-cast dendritic structure, which mainly consisted of α-Al, Mg_2Si and Mg_5Al_8 phases. The width of the fusion zone in the side of 6005A alloy was about 80 μm, while the width in the side of A356 alloy was over 500 μm. The structures in the fusion zone in the side of A356 alloy were mainly composed of insular (α-Al + Al–Si) eutectic phase and coarser Si particles. The fusion zone in the 6005A alloy side presented the characteristics of transitional structures, which were composed of the mixed structure of (α-Al + Al–Mg–Fe–Si + Al–Fe–Si)phase.

(3) The heat-affected zone and weld zone were the weakest areas of A356/6005A alloy welding joint, and there were many pores and dimples with different sizes on the fracture surface.

References

1. Cui H (2017) Application of aluminum alloys connection technology in automobile manufacture. Mod Compon 6:20–23
2. An C, Han Z, Yu D, Shizhan Z, Yujia Y (2018) A study on material selection for Multi-material autobody components based on PSI method. Automot Eng 40(2):239–244
3. Gong Y, Wang K, Zhang Z (2017) New energy vehicles lightweight approach and its evaluation. Automobile Appl Technol 1:5–6
4. Gao Y (2018) Technical schemes and implementation examples of automobile lightweight. Chin J Automot Eng 8(1):1–9
5. Huang Y, Huang Y, Zhou C, Cui L (2017) Finite element analysis on hydro-mechanical deep-draw forming for the outer panel of aluminum-alloy car door. Automot Eng 39(4):486–490
6. Shi D, Gao P, Wei Y, Wang X, Liu P (2016) Contact corrosion analysis of forged aluminum steering knuckle and steel bushing. Automobile Technol Mater 12:36–38
7. Nie F, Dong H, Chen S, Li P, Wang L, Zhao Z, Li X, Zhang H (2016) Microstructure and mechanical properties of pulse MIG Welded 6061/A356 aluminum alloy dissimilar butt joints. J Mater Sci Technol 3:1–10
8. Nie F, Dong H, Li P, Zhao Z, Wang L, Zhang H (2016) Lap joining of dissimilar aluminum alloys between wrought 6061 and cast with pulse MIG welding. J Mech Eng 52(24):65–71
9. Dan C, Shi X (2017) Comparative study for welding process of 6005 aluminum. Weld Technol 46(7):50–53
10. Ran G, Zhou E, Wang Y, Zho S (2007) Microstructure and tensile properties of cast A356 aluminum alloy. Heat Treat Met 32(3):13–18
11. Yang Z, Yang S, Jiang Y, Wang Y (2017) Microstructure and mechanical properties of laser welding joints of 7075 aluminum alloy with filler wire. Mater Rev 31(12):60–63
12. Jie Y (2009) The study of mechanical properties and microstructure evolution of Al alloy weld in double-pulsed MIG welding. Hunan University, Changsha
13. Xue J, Li Y, Chen H, Zhu Z (2018) Effects of heat input on wettability, interface microstructure and properties of Al/steel butt joint in laser-metal inert-gas hybrid welding-brazing. J Mater Process Technol 255:47–54
14. Peng X, Cao X, Duan Y, Chen J, Guofu X, Yin Z (2014) Microstructures and properties of MIG welded joint of 7020 aluminum alloy. Chin J Nonferrous Metals 24(4):912–918

Experimental Study on Bolt Tightening of Torsional Vibration Damper

Fan Dong, Mingrui Wang, Xingtao Sheng and Qing Sun

Abstract In engines, torsional vibration is inevitably caused by the fluctuation of the engine torque. Therefore, the establishment of torsional vibration damper is the necessary measure to diminish the torsional vibration. In this paper, an experimental study was carried out in a gasoline engine to analyze the bolt tightening of torsional vibration damper. The influence of the friction coefficient of the bolt of the torsional vibration damper, the material properties of the torsional vibration damper, and the axial force calibration were studied. The results show that the changes in the friction coefficient of the bolt and the material of the torsional vibration damper have great impacts on the bolt torque. It is effective to improve the bolt tightening state by checking the axial force and establishing suitable tightening parameters.

Keywords Torsional vibration damper · Friction coefficient · Bolt tightening

1 Introduction

Torsional vibration damper is an important part of the engine, which is installed in the front of the engine crankshaft. It is connected to the crankshaft rigidly, and the rubber ring between the hub and shell is used as the damping medium. When the engine is at work, the hub rotates along with the crankshaft, driving the shell and the inertia ring rotating synchronously through the rubber ring [1, 2]. When the torsional vibration occurs, the damper vibrates along with the crankshaft simultaneously, and meanwhile, the inertial shell and the body with a certain rotational inertia still rotate at a constant speed. Thus, relative movement occurs between the shell and the inertia body, and vibration energy is absorbed through the damping effect of the rubber ring. This damper can reduce torsional vibration and noise in the engine, improve the emission level, reduce the fuel consumption, prevent the crankshaft from fracture, and protect the cylinder body as well [3–6].

F. Dong (✉) · M. Wang · X. Sheng · Q. Sun
Dongfeng Motor Corporation Technical Center, Wuhan, China
e-mail: dongfan@dfmc.com.cn

China SAE (ed.), *Proceedings of China SAE Congress 2018: Selected Papers*, Lecture Notes in Electrical Engineering 574,
https://doi.org/10.1007/978-981-13-9718-9_49

Torsional vibration damper is often connected with the engine crankshaft by bolts, and hence the tightening torque of the bolt is also one of the significant technological parameters. To replace the bolts of the damper not only affects the production takt but also increases the labor intensity of the operators. How to better control the bolt tightening torque has been one of the important indicators of quality assurance in the process of production [7–11]. As basic fastening pieces, the bolt tightening has direct influences on the safety and reliability of the torsional vibration damper which works in the vibratory environments. Various studies have been performed on the energy dissipation and damping of bolted joints, failure, and fatigue of structural joints, and loosening mechanisms [12]. According to the load applied, four different types of loads are applied to the bolt connections: axial tensile load, transverse or shear load, torsion load, and prying load. These loads can cause vibration which may lead to the bolt loosening and increase maintenance or failure. After experiencing enough vibration, the friction force of the bolt decreases, and then the nut starts to fall off. Finally, the clamping force completely disappears. Bolt loosening mechanism has been studied extensively in a lot of previous studies [13–17].

In the 1950s, Goodier and Sweeney [18] took the lead in studying the loosening mechanism of the bolt tightening under the axial load. The study showed that the increase of inload resulted in the bolt threads move in radial direction and the nut threads move out radial direction. The radial slippage occurs between the contact threads affected by this effect. Based on the static equilibrium conditions, this theory predicted the loosening torsion during loading and the tightening torsion during unloading which has a loosening torsion in each cycle. This work is further studied in [19]. Basava and Hess examined the variation of the clamping force in a single-bolt assembly model caused by the axial vibration. Specifically, the effects of vibration level and the initial preload on clamping force were also studied. They pointed out that when the bolt loosening or tightening occurs, the changes of clamping force are usually instantaneous and stable value over time. The maximum loss of clamping force caused by the axial vibration is about 52.9%, while the increase of clamping force is as much as 83.4%.

Ibrahim and Pettit [20] provided an overview of the problems pertaining to structural dynamics of bolted joints. The energy dissipation of bolted joints, the linear and non-linear identification of the dynamic characteristics of the joints, the uncertainty and relaxation of parameters, and the active control of the pre-tightening force of the joints are emphatically studied in this paper. This paper also discussed the design of fully and partially restrained joints, the sensitivity analysis of the joint parameter variations, and fatigue prediction of metal and composite joints. In Reference [21], the author studied the influences of the fatigue behavior and the failure modes of the aluminum-alloy-bolted joints from three factors theoretically. An experimental study was carried out on 100 aluminum symmetric double-butt joints to analyze the properties and the mechanisms of the fatigue. They proposed a rough idea of the scatter and pointed out the width of the scatter bands not only depends on the intrinsic scatter of each series but also affects by the parameters of different series. At last, they figured out that the fatigue behavior of the bolted aluminum joints is similar to that of the aluminum-welded structures, but the detailed analysis standards were not given.

The objective of the present study is to investigate the bolt tightening of the torsional vibration damper. The paper is organized according to the following sequence: The numerical calculation of the bolt tightening is described in Sect. 2. Section 3 presents the influence of the bolt friction coefficient, the torsional vibration damper material, and the tightening parameters along with the discussion of the experiment results. Conclusions are provided in Sect. 4.

2 Numerical Calculation

The tightened threaded connection must be assured so that the torsional vibration damper is compressed and the bolt is tensile. In order to ensure the tightening quality of the torsional vibration damper, the tightening axial force of bolts should be checked first. The numerical calculation is based on the AVL algorithm. The equation used to calculate the equivalent radius of the bolt frictional force is introduced as follows:

$$r = \frac{2}{3} \times \frac{r_1^3 - r_2^3}{r_1^2 - r_2^2} \tag{1}$$

where r_1 and r_2 are the outer and inner radii of the friction surface, respectively, and r is the equivalent radius of the frictional force of the bolt. Then, the preload required by the bolt can be calculated based on the nominal diameter of the bolt, and the equation employed is

$$Q = \frac{k(M_1+M_2)}{f \cdot r} \tag{2}$$

where M_1 is friction torque that the belt acted to torsional vibration damper, M_2 the friction torque of the chain wheel, k the safety factor, and f the friction coefficient. One can calculate the required preload for the tightening bolt is 92 kN.

3 Results and Discussion

3.1 Bolt Friction Coefficient

In the assembly process of the torsional vibration damper, the preload will be different even applying same technique in same bolt, which leads to occurrence of thread elongation and even fracture. Insufficient preload will affect production efficiency.

The tightening torque is associated not only with the diameter, material, surface roughness, lubrication, heat treatment, nut-supporting surface and internal chamfer size of the bolt itself but also, most importantly, with the friction coefficient of the threaded surface [22, 23]. To analyze the influence of the friction coefficient, the composition of the frictional force on the two friction surfaces has to be identified. Under the effect of the external load, the contact stress between two contacted surfaces is very large. If the surfaces are clean and pure, then atoms (molecules and ions) on one surface will attract atoms on the other one, producing a strong adhesive force. The phenomenon of adhesion on the metal surfaces is called cold welding. It is particularly remarkable on the surfaces with high-degree finish under high vacuum. Frictional force is usually divided into two types, i.e., the static frictional force and the dynamic frictional force. In general, the static frictional force is greater than the dynamic one in the same condition. The ratio of the frictional force to the normal load is called the friction coefficient.

To elevate the friction coefficient of the bearing will increase the frictional resistance of the bearing and reduce the net torque. If the frictional resistance of the bearing becomes high enough to keep the net torque negative during the entire loading cycle, then the bolt will not rotate loosely [24]. In the tightening process of the torsional vibration damper, the friction coefficient consists of two parts: (1) the friction coefficient of the bolt; (2) the surface roughness of the torsional vibration damper [25–28]. Figure 1 shows the relationship between the tightening torque of the damper bolt and the axial force, in which μ is the friction coefficient of the bolt. According to the figure, one can see that reducing the friction coefficient can have a positive impact on the axial force and the tightening torque. The extension strength when the non-proportional elongation of 0.2% is also shown in this figure, which is

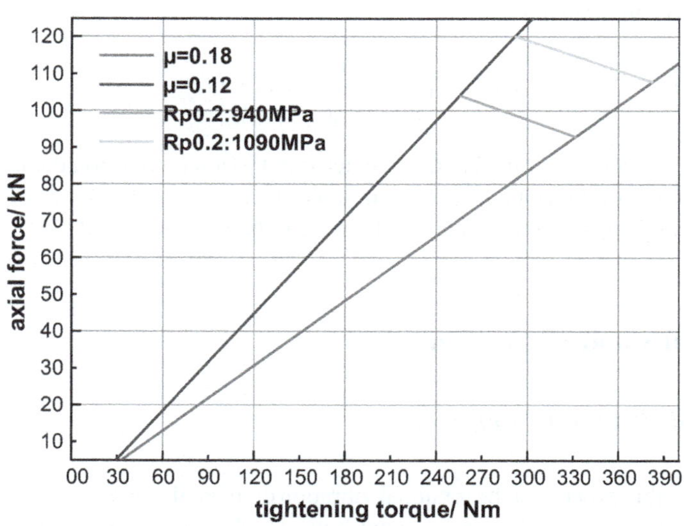

Fig. 1 Relationship between the tightening torque and the axial force

Table 1 Measurement of friction coefficient of tightening bolt

Batch	Friction coefficient (0.12–0.18)					
1	0.131	0.145	0.133	0.131	0.147	0.145
2	0.141	0.135	0.133	0.141	0.137	0.135
3	0.123	0.127	0.123	0.130	0.126	0.123

tested at 940 and 1090 MPa. The friction coefficients of the bolts in the following axial experiment are also tested in the present study. Shown in Table 1, there are three batches of friction coefficient for the bolts. One can see that all of them are within the range of qualification although they are close to the lower limit.

3.2 *Analysis of Damper Material*

The material property of the torsional vibration damper is another factor that influences the friction coefficient. The effects of the damper material can be divided into two sections: (1) the structure and hardness of the damper; (2) the roughness of the pressure bearing surface. The metallographic structure of the damper is normally the ferrite or the ferrite + pearlite matrices. The strength and hardness of the pearlite matrix are significantly higher than those of the ferrite matrix while its plasticity and toughness are lower. If the torsional vibration damper is not hard enough, it will collapse when the bolts are tightened, which is the main reason for the false moment. Hence, the pearlite matrix in the damper is suggested to be less than 25%, while the carbide matrix is less than 3% (Fig. 2).

The hardness is tested in the present study, which is measured by Brinell hardness. The influence mechanism of the surface roughness is by changing the friction coefficient of the surface. A smaller roughness may lead to a larger tightening torque and axial force. The analysis results are shown in Table 2.

3.3 *Analysis of the Tightening Parameters*

In the process of bolt tightening, there are four methods at the present time: torque method, torque-angle method, yield point method, and extension method. The torque-angle method is to tighten the threaded parts to the initial torque first and then turn the parts to the required angle. In this paper, the bolt tightening of the torsional vibration damper is controlled by the torque-angle method, which can obtain a larger preload and its numerical value can be fixed near its mean value. The current parameter is 41 Nm + 105°. In order to check the accuracy of the tightening technology, the axial force calibration of bolts under different tightening parameters is carried out.

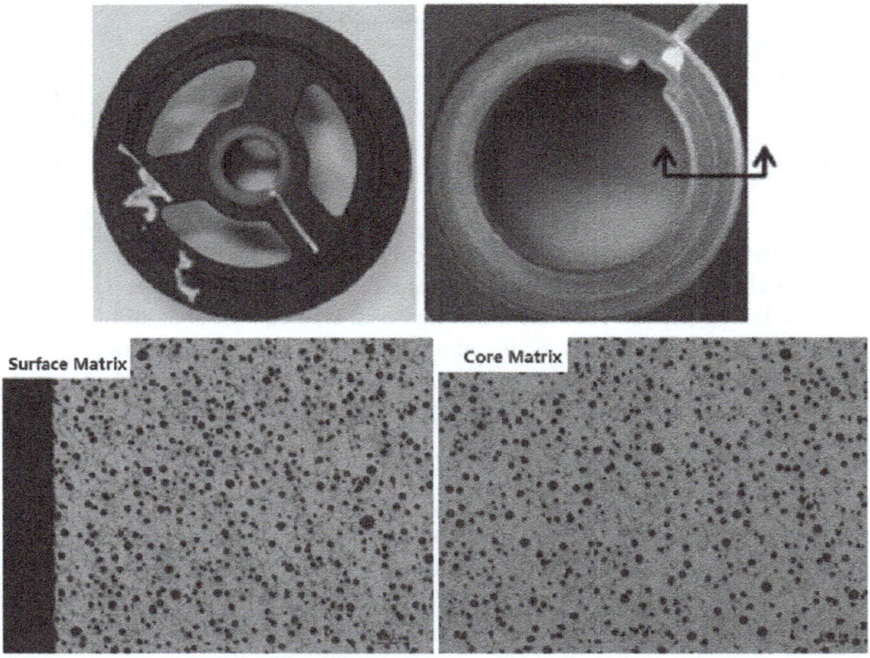

Fig. 2 Matrix structure of the torsional vibration damper

Table 2 Results of analysis on material of torsional vibration damper

	Results	Datum
Metallographic structure	Nodularity: level 1 Graphite size: level 7 Matrix structure: ferrite	Nodularity: Levels 1–4 Matrix structure: ferrite matrix or ferrite + pearlite matrices, pearlite ≤25%, carbide ≤3%
Hardness/HB	167/166/172/170/172 Average value: 169	160–210
Roughness of pressure-bearing surface Ra/μm	1.74/1.76/1.87/2.06 Average value: 1.86	1.6

In this calibration process, four different tightening parameters, i.e., 41 Nm + 100°, 41 Nm + 105°, 41 Nm + 110° and 41 Nm + 115°, are selected. When tightening the bolt, technicians keep the torque unchanged first and then rotate the bolt to the desired angles. Every bolt is measured three times, and the average value is calculated. Five bolts are measured for each set of tightening parameter, and the coating method is used to conduct the measurement. Figure 3 shows each measurement of each bolt for different sets of tightening parameters. The tightening torque increases apparently

Fig. 3 Tightening torque at different tightening parameters

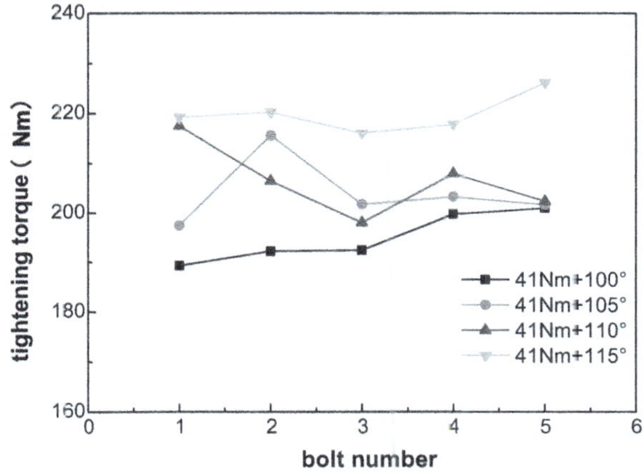

as the tightening angle increases. Under the condition of 41 Nm + 115°, the average tightening torque is about 219.86 Nm.

Figure 4 shows the axial force of each bolt for each set of parameters. Under the condition of 41 Nm + 115°, the average axial force of these five bolts is 91.84 kN, which is closest to 92 kN, the required axial force. Meanwhile, the average tightening torque is also the largest. However, under other conditions, the axial force does not meet the numerical results.

Fig. 4 Axial force at different tightening parameters

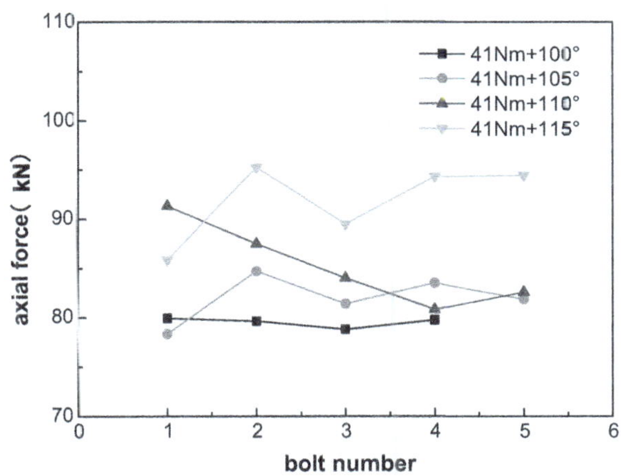

4 Conclusions

A series of experiments on the bolt tightening of torsional vibration damper were preformed in this study. Through the numerical calculation according to the AVL algorithm, it was concluded that under the current boundary conditions, the axial force must be 92 kN to meet the demand. The friction coefficient of the bolt also has a great impact on the tightening of the torsional vibration damper. The smaller friction coefficient leads to a lower tightening torque, while the larger friction coefficient has the opposite effect. A smaller friction coefficient is favorable for the tightening of the torsional vibration damper. The material of the torsional vibration damper can influence the axial force by changing the friction coefficient of the friction pair. Lower hardness of the torsional vibration damper may cause a collapse when tightening, then producing a false moment in bolt. The roughness of the pressure bearing surface of the torsional vibration damper has the same effect as the friction coefficient of the bolt. At last, a suitable tightening parameter can improve the tightening torque and the axial force effectively.

References

1. Hwang BC, Kim C, Bae WB (2009) A study of structural analysis and torsional characteristic of the sleeve spring-type torsional vibration damper. Korean Soc Precis Eng 26(2):94–100
2. Jolly MR, Bender JW, Carlson JD (2000) Properties and applications of commercial magneto-rheological fluids. J Intell Mater Syst Struct 10(1):5–13
3. Carlson D, Catanzarite DM, Clair KASt (1996) Commercial magneto-rheological fluid devices. Int J Mod Phys 10(23–24):2857–2863
4. Yang M, Shima S (1988) Simulation of pyramid type three roll bending process. Int J Mech Sci 30(12):877–886
5. Lee CK, Jeon HJ (1989) A study on the design of the torsional vibration viscous damper for the crankshaft and developing of its performance simulation computer program. J Korean Soc Mar Eng 13(1):77–96
6. Kim JS, Jei YG, Jeong JH, Jeon HJ (1994) A theoretical study on the dynamic characteristic of damping flexible coupling. J Korean Soc Mar Eng 18(1):11–22
7. Shin YS, Iverson JC, Kim KS (1991) Experimental studies on damping characteristics of bolted joints for plates and shells. J Pressure Vessel Technol 113:402–408
8. Hua M, Baines K, Cole IM (1999) Continuous four-roll plate bending: a production process for the manufacture of single seamed tubes of large and medium diameters. Int J Mach Tools Manuf 39:905–935
9. Esteban J, Rogers CA (2000) Energy dissipation through joints: theory and experiments. Comput Struct 75:347–359
10. Hanss M, Oexl S, Gaul L (2002) Simulation and analysis of structural joint models with uncertainties. In: Proceedings of the international conference on structural dynamics modeling-test, 165–174
11. Hanss M, Oexl S, Gaul L (2002) Identification of a bolted-joint model with fuzzy parameters loaded normal to the contact interface. Mech Res Commun 29:177–187
12. Izumi S, Take T, Kimura M, Sakai S (2007) Self-loosening analysis of bolt-nut tightening system subjected to axial loading by three-dimensional finite element method. Nihon Kikai Gakkai Ronbunshu A Hen/Trans Jpn Soc Mech Eng Part A 73(732):869–876

13. Sakai T (2011) Mechanism for a bolt and nut self-loosening under repeated bolt axial tensile load. J Solid Mech Mater Eng 5(11):627–639
14. Atzori B, Lazzarin P, Quaresimin M (1997) A re-analysis on fatigue data of aluminum alloy bolted joints. Int J Fatigue 19:579–588
15. Birch RS, Alves M (2000) Dynamic failure of structural joint systems. Thin-Walled Struct 36:137–154
16. Junker GH (1969) New criteria for self-loosening faster under vibration. Eur Res Eng 78:314–335
17. Jiang Y, Zhang M, Park TW, Lee CH (2004) An experimental study of self-loosening of bolted joints. Int J Mech Mater Des 126(5):925
18. Goodier JN, Sweeney RJ (1945) Loosening by vibration of threaded fastenings. Mech Eng 67:798–802
19. Basava S, Hess DP (1998) Bolted joint clamping force variation due to axial vibration. J Sound Vib 210(2):255–265
20. Ibrahim RA, Pettit CL (2005) Uncertainties and dynamic problems of bolted joints and other fasteners. J Sound Vibr 279(3–5):857–936
21. Lazzarin P, Milani V, Quaresimin M (1997) Scatter bands summarizing the fatigue strength of aluminum alloy bolted joints. Int J Fatigue 19:401–407
22. Hosokawa S, Sato S, Miyata C, Tsumura T (1987) Contraction deformation of bolt shank in threaded connection. J Jpn Soc Precis Eng 53(11):1726–1732
23. Hess DP, Sudhirkashyap SV (1996) Dynamic analysis of threaded fasteners subjected to axial vibration. J Sound Vib 193(5):1079–1090
24. Bickford JH (1998) Hand book of bolts and bolted joints. The Chemical Rubber Company Press, Florida
25. Bickford JH (2007) Introduction to the design and analysis of bolted joints, 4th edn. The Chemical Rubber Company Press, Florida
26. Nassar SA, Barber GC, Zuo D (2004) Bearing friction torque in bolted joints. A S L E Trans 48(1):67–75
27. Karamiş MB, Selçuk B (1993) Analysis of the friction behavior of bolted joints. Wear 166(1):73–83
28. Juvinall RC, Marshek KM (2011) Fundamentals of machine component design, 4th edn. Wiley, New York

The Heat Treatment Deformation Law and the Improvement of Accuracy of the Transmission Gear

Peng Sun, Di Pang, Xuechun Qi and Ningming Luo

Abstract Transmission is one of the core assembly units of automobile, and its performance has a decisive effect on the overall performance of automobile And gear accuracy has the biggest influence on the performance of transmission, and to improve gear accuracy is the key point to improve the performance of transmission. But the heat treatment deformation of gear has become the tiger on the road (means the obstacle on the road) to improve gear accuracy, especially for the shaved gear which would not have finish machining after heat treatment. The heat treatment deformation quantity can directly influence gear accuracy, and as the factor of heat treatment is complicated and sophisticated, the deformation can be diminished as far as possible, but cannot be eliminated, usually we can try to diminish the deformation and analyze and summarize the deformation law to make a reverse machining, we do both at the same time to improve gear accuracy. This thesis through the analysis and researches on deformation law of gear heat treatment, tested and verified by gear shaved in FAW shaft and gear manufacturing center, finally made a summarization on the deformation law of gear heat treatment, and with this law, we can direct the machining of white parts, make a reverse modification to counteract the heat treatment deformation finally to achieve the goal of improving finished gear accuracy. Here are the researches done:

1. The influence of heat treatment deformation on gear common normal line;
2. The influence of heat treatment deformation on gear involute;
3. The influence of heat treatment deformation on gear helical line;
4. The influence of heat treatment deformation on gear tooth spacing;

Through this research, the gear deformation law of heat treatment is summarized, and under the direction of this law, we can make a reverse modification machining to improve gear accuracy, and then to improve the performance of transmission.

P. Sun (✉) · D. Pang
FAW Jiefang Automobile Workshop Transmission Department, Changchun 130011, China
e-mail: sp851104813@126.com

X. Qi · N. Luo
FAW Jiefang Shaft-Gear Plant of Transmission Department, Changchun 130011, China

China SAE (ed.), *Proceedings of China SAE Congress 2018: Selected Papers*,
Lecture Notes in Electrical Engineering 574,
https://doi.org/10.1007/978-981-13-9718-9_50

Keywords Transmission · Gear · Deformation · Law · Shaving modification · Gear accuracy

1 Introduction

Transmission is one of the core assembly units of automobile, and its performance has a decisive effect on the overall performance of automobile. And gear accuracy has the biggest influence on the performance of transmission; to improve gear accuracy is the key point to improve the performance of transmission. But the heat treatment deformation of gear has become the tiger on the road (means the obstacle on the road) to improve gear accuracy, especially for the shaved gear which would not have finish machining after heat treatment. The heat treatment deformation quantity can directly influence gear accuracy.

Although finish machining can correct the deformed parts, but it rise the product cost a lot. The factor of heat treatment deformation of gear is complicated and sophisticated, and the main reason can be divided into two parts: the first one is the volumetric deformation caused by the transformation of the volume of the parts during heat treatment phase transition; the second one is the shape distortion caused by the inhomogeneity plastic deformation under the comprehensive effects of complicated stress during heat treatment.

Many reasons for the two kinds of deformation, for example, difference of rise and fall of each point of gear affected by the homogeneity of material; the release of inner residual stress of parts during heat treatment, the heat stress during heat treatment, dead weight of parts, the bracing and clamping of parts in furnace when heating and cooling, etc. No deformation of parts can be attributed to one reason; it is a result of comprehensive impact, and every part must have several different kinds of deformation factors. So the strategy is to decrease the deformation and at the same time, make experiments and analyze the deformation law, and make a reverse modification to counteract the heat treatment deformation finally to achieve the goal of improving product accuracy.

This thesis through the analysis and researches on deformation law of gear heat treatment, tested and verified by gear shaved in FAW shaft and gear manufacturing center, finally made a summarization on the deformation law of gear heat treatment, and with this law, we can direct the machining of white parts and make a reverse modification to counteract the heat treatment deformation finally to achieve the goal of improving finished gear accuracy.

2 The Influence of Heat Treatment on Common Normal Line of Gear

2.1 Analysis and Prediction of the Common Normal Line Transformation

The heat treatment deformation of the common normal line of gear is mainly caused by the volumetric deformation caused by the transformation of the volume of the parts during heat treatment phase transition. The major component of gear material is pearlite of white parts; after heat treatment, the surface changed into martensite, the inner part changed into a mixture of pearlite, martensite, etc, as martensite has larger volume than pearlite, shown as Chart 1, so the volume of the gear become larger. Since the inner part of gear material is a mixture of different components, the deformation quantity related to the component proportion, the specific deformation cannot be decided because of uncertainty of the component proportion, so a test is necessary.

2.2 Test and Verify

Contrastive detection of common normal of five kinds of gears produced by FAW axle tooth center before and after heating, among this five kinds of products: No.1–No.4, have different shape and size, but the same material; No.5–No.4, same size, different material, test result shown in Chart 2, we can see:

组　织	wc(%)	室温下的比体积 / (cm³。g⁻¹)
奥氏体	0~2	0.1212+0.0033(C%)
马氏体	0~2	0．1271+0．0025(C%)
铁素体	0~0．02	0．1271
渗碳体	6．7±0．2	0．130±0．001
ε -碳化物	8.5±0.7	0．140±0．002
石墨	100	0.451
铁素体+渗碳体	0~2	0.1271+0.0005(C%
低碳马氏体+ε -碳化物	0~2	0．1277+0．0015(C%-0．25)
铁素体+ε -碳化物	0~2	0．1271+0．0015(C%)

Chart 1 Volume proportion of each component of steel

序号	零件材料	零件模数	外径	厚度	零件状态	1	2	3	4	5	6	7	8	9	10	均值	变形量
1	FAS3420	2.96	129	47	热前	31.8	31.8	31.8	31.8	31.8	31.8	31.8	31.8	31.8	31.8	31.8	0.039
					热后	31.8	31.8	31.8	31.8	31.8	31.8	31.8	31.8	31.8	31.8	31.8	
2	FAS3420	2.95	145	53	热前	40.5	40.5	40.5	40.5	40.5	40.5	40.5	40.5	40.5	40.5	40.5	0.048
					热后	40.5	40.5	40.5	40.5	40.5	40.5	40.5	40.5	40.5	40.5	40.5	
3	FAS3420	2.5	164	35	热前	49.6	49.6	49.6	49.6	49.7	49.6	49.6	49.6	49.6	49.7	49.6	0.043
					热后	49.7	49.7	49.7	49.7	49.7	49.7	49.7	49.7	49.7	49.7	49.7	
4	FAS3420	2.56	181	35	热前	50.2	50.2	50.2	50.2	50.2	50.2	50.2	50.2	50.2	50.2	50.2	0.04
					热后	50.2	50.2	50.2	50.2	50.2	50.2	50.2	50.2	50.2	50.2	50.2	
5	20CrMnTiH	2.56	181	35	热前	50.2	50.2	50.2	50.2	50.2	50.2	50.2	50.2	50.2	50.2	50.2	0.046
					热后	50.2	50.2	50.2	50.2	50.2	50.2	50.2	50.2	50.2	50.2	50.2	

Chart 2 Statistics of the common normal line transformation of different gear products (white and black)

(1) for the FAS3420H and 20CrMnTiH gear, the common normal line become larger;
(2) For parts with the same material and different sizes, the heat treatment deformation is different. The deformation is positively correlated with the size, and the deformation is about 0.5% of the volume.
(3) For parts with same shape and size, different material, the deformation differs, materials have better hardenability characteristic have relatively larger deformation, but the difference is quite less.

2.3 The Countermeasure Against the Transformation of the Common Normal Line

In order to have a perfect common normal line value for black parts, the processed size needs to be smaller than the finished parts, and the reduction quantity is about 0.5% of the volume.

3 The Influence of Heat Treatment on Gear Involute

3.1 Analysis and Prediction of the Deformation of Tooth Profile

The involute is the deformation of tooth profile, and it is the result of a comprehensive effect of the heat stress and structural stress during quenching process. The heat stress and structural stress are a couple of opposite stress influenced by the heat treatment condition and the material composition. They are different in size; the final deformation is decided by the greater one. The deformation law of heat stress is where the cooling speed is fast it has a reduction in size, where the cooling speed is slow and it has an increase in size. So, under the influence of heat stress, the tooth top become thinner, that is, the $fH\alpha$ of the tooth has a negative increase, the $C\alpha$ increases. The deformation under the influence of the structural stress is in contrary to the heat treatment; meanwhile, the residual stress of cutting also has a certain effect on the deformation of tooth. Especially, for the shaved parts, because the principle of shaving, the middle part of the teeth has larger squeezing pressure to have larger residual tress, the residual stress released during the heat treatment, making the profile crown increase. The final deformation quantity is the difference between the three stresses. But which one has a larger effect, should be decided under certain experiments.

3.2 Test and Verify

Five kinds of products different in module of FAW shaft and gear manufacturing center were chosen, profile of white and black parts is tested, and the statistics of the results are shown in Chart 3 and Picture 1, we can see:

(1) the heat stress and the residual stress of cutting has a lager effect than the structural stress on the deformation of the involute of gear.
(2) The deformation law is the tooth top thinner, that is, $fH\alpha$ decreases, about 7 μm, the crown $C\alpha$ increases about 4 μm.
(3) For the gears with the same parameters, better hardenability characteristic material has a relatively larger deformation.

3.3 The Countermeasure Against the Transformation of Involute

In order to improve the accuracy of gear products, a reverse modification to the involute machining-gear shaving is necessary, that is, to say, the $fH\alpha$ of white parts

序号	零件材料	零件模数	外径	厚度	零件状态	精度项	1	2	3	4	5	6	7	8	9	10	均值	变形量
1	FAS3420	2.96	129	47	热前	fHα	6	8	7	5	8	6	8	7	6	4	6.5	
						Cα	8	3	5	5	5	4	7	4	7	7	5.5	
					热后	fHα	1	2	2	3	2	1	2	1	1	0	1.5	**−5**
						Cα	10	5	8	5	8	5	9	6	10	9	7.5	**2**
2	FAS3420	2.95	145	53	热前	fHα	6	5	8	7	6	5	7	8	8	3	6.3	
						Cα	4	5	6	5	5	4	6	7	2	3	4.7	
					热后	fHα	0	−2	−1	0	0	−3	−1	2	3	−1	−0.3	**−6.6**
						Cα	10	9	8	10	6	7	9	7	9	10	8.5	**3.8**
3	FAS3420	2.5	164	35	热前	fHα	10	10	12	11	9	12	13	9	10	11	10.7	
						Cα	5	3	4	7	6	3	8	5	6	8	5.5	
					热后	fHα	1	3	3	2	1	4	4	6	3	3	3	**−7.7**
						Cα	10	9	8	12	13	9	13	10	10	13	10.7	**5.2**
4	FAS3420	2.56	181	35	热前	fHα	9	10	10	11	8	12	11	12	11	10	10.4	
						Cα	5	3	4	7	6	3	8	5	6	5	5.2	
					热后	fHα	2	4	3	2	4	1	1	2	7	8	3.4	**−7**
						Cα	10	9	8	12	13	9	13	10	10	13	10.7	**5.5**
5	20CrMnTiH	2.56	181	35	热前	fHα	16	15	14	15	13	14	14	17	16	15	14.9	
						Cα	3	5	7	7	5	4	7	6	8	8	6	
					热后	fHα	2	3	4	1	2	5	3	2	3	3	2.8	**−12.1**
						Cα	13	15	11	12	14	9	16	14	10	13	12.7	**6.7**

Chart 3 Statistics of the deformation of profile of different gear products for white and black parts

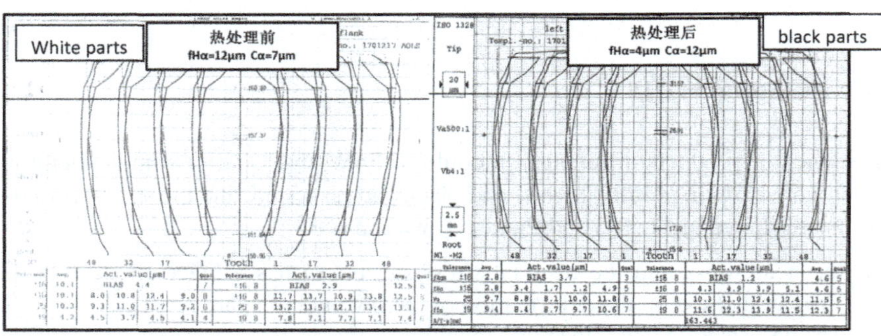

Picture 1 Deformation of profile of different gear products for white and black parts

should be 7 μm larger than the finished parts and the crown $C\alpha$ should be 4 μm less than the finished parts.

4 The Influence of Heat Treatment on Gear Helical Line

The factor that influences the transformation of helical line is quite complicated, such as heat stress, structural stress, gear shape, heat treatment clamping method, and self-weight.

These factors act on the gear heat treatment process at the same time. Under different heat treatment conditions, the strength of each factor varies. Under the combined action of different factors, the helix of gear will change differently.

4.1 *Transformation Law of Gear Helical Line in Different Ratio of Diameter to Thickness*

The gear shape directly affects the transformation of gear helical line, *I* here introduces the concept of ratio of diameter to thickness and a ratio of gear diameter to gear thickness ($K = D/L$), and when K increases, the transformation increases. Because when K increases, the gear tends to be metal sheet, and the character of metal sheet is it has a weak resistance to the centripetal force; in the process of heat treatment, it is much easer to transform under the effect of stress.

4.1.1 Test and Verify

Six kinds of gear with different K values were chosen to make a comparative detection of white and black parts, and the result is shown in Chart 4, we can see, when K increases, the deformation of $fH\beta$ increases, but the $C\beta$ (crown) has nothing to do with K value, it has a steady deformation quantity of about 3 μm. Picture 2 is the detection report of Max$K$ and Min$K$ gear

4.1.2 Countermeasure Against the Effect of *K* Value on Helical Line

K value is decided by the structure of parts, so we should pay attention to the K value of the product when designing it, cut the K value as much as possible, but when it has to be big, we can enhance the symmetry to the full, choose better material, and choose better heat treatment method to reduce the deformation of helical line (we suggest for gears K value >5 and accuracy request higher than 7°, to grind the gear teeth).

序号	零件材料	零件模数	外径	厚度	K值	零件状态	精度项	1	2	3	4	5	6	7	8	9	10	均值	变形量
1	FAS3420	2.95	145	53.446	**2.7111**	热前	fHα	4	1	1	2	1	2	4	2	6	4	2.7	
							Cα	15	18	14	18	16	18	15	14	13	14	16	
						热后	fHα	6	2	6	1	6	2	10	2	6	4	4.5	**1.8**
							Cα	15	12	14	14	13	14	12	14	12	10	13	**-2.5**
2	FAS3420	2.961	129	47.3	**2.7297**	热前	fHα	2	1	2	4	2	4	1	2	6	4	2.8	
							Cα	16	17	15	14	16	17	16	17	15	14	16	
						热后	fHα	7	3	6	1	5	10	6	2	6	4	5	**2.2**
							Cα	11	12	14	12	13	13	12	10	12	11	12	**-3.7**
3	FAS3420	2.555	124	35.17	**3.5388**	热前	fHα	1	2	1	2	1	4	2	2	6	4	2.5	
							Cα	15	16	14	15	16	17	15	14	13	14	15	
						热后	fHα	6	10	4	2	3	10	4	2	5	4	5	**2.5**
							Cα	11	12	10	11	11	12	11	9	9	10	11	**-4.3**
4	FAS3420	2.555	129	32.45	**3.9753**	热前	fHα	2	1	5	2	1	2	4	2	6	4	2.9	
							Cα	16	17	15	15	16	17	15	14	13	14	15	
						热后	fHα	4	5	8	2	3	6	4	11	5	6	5.4	**2.5**
							Cα	12	11	12	12	13	14	12	11	12	10	12	**-3.3**
5	FAS3420	2.498	164	34	**4.8118**	热前	fHα	5	4	2	6	7	2	4	2	6	4	4.2	
							Cα	15	16	14	15	16	17	15	14	13	14	15	
						热后	fHα	17	5	26	16	14	20	10	19	6	4	14	**9.5**
							Cα	10	12	14	12	13	15	12	13	12	10	12	**-2.6**
6	FAS3420	2.555	181	34.95	**5.1645**	热前	fHα	6	1	3	2	1	5	4	4	6	4	3.6	
							Cα	16	12	10	12	13	15	14	13	10	10	13	
						热后	fHα	40	29	6	15	19	10	9	39	6	19	19	**15.6**
							Cα	12	9	10	10	10	10	14	13	9	8	11	**-2**

Chart 4 Statistics of the deformation of helical line (*fHα*)

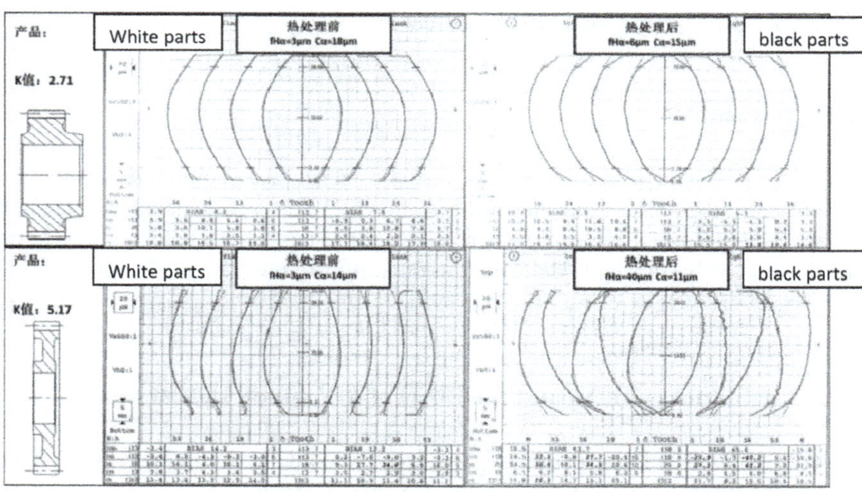

Picture 2 The detection report of Max*K* and Min*K* gear

4.2 The Deformation Law of Symmetric Gear Helical Line with Relatively Small K Value

4.2.1 Analysis and Prediction

When K value is relatively small and symmetric, the main factor that affects deformation of helical line is the interaction of heat stress and structural stress. For the crown of the helical line, under the influence of heat stress, the deformation trend is the crown increase and the two flanks shrink; under the influence of structural stress, the deformation trend is the crown reduces, and the two flanks bow and twisted. Which stress can influence more needs to be verified though experiment.

4.2.2 Test and Verify

A kind of gear with symmetrical shape and smaller K value (shown in Picture 3) was selected to compare the helix of 10 gears before and after heat treatment. The test results are shown in Chart 5 and Picture 4. The drum shape of helix is reduced by 4 microns, and the structural stress plays a leading role in the helix deformation.

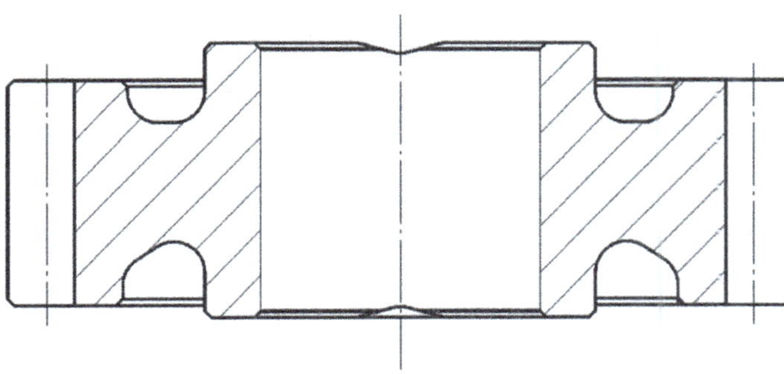

Picture 3 Gear (idle gear)

序号	零件材料	零件模数	外径	厚度	K值	零件状态	精度项	1	2	3	4	5	6	7	8	9	10	均值	变形量
1	FAS3420	2.555	124	35.17	3.5388	热前	Cα	15	16	14	15	16	17	15	14	13	14	15	−4.3
						热后	Cα	11	12	10	11	11	12	11	9	9	10	11	

Chart 5 Statistics of the gear (idle gear) helical line crown deformation of white and black parts

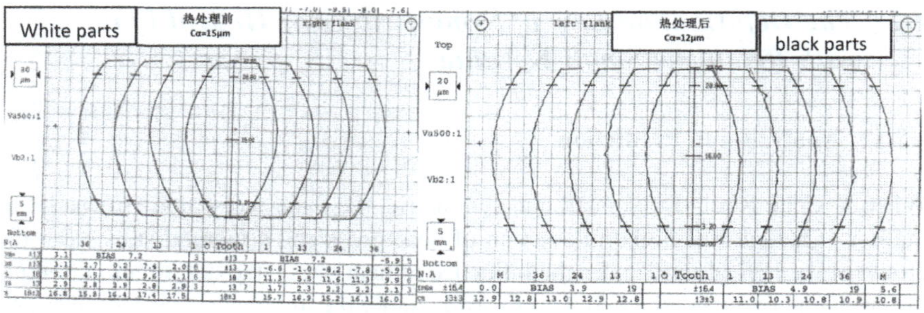

Picture 4 Report pictures

4.2.3 Countermeasure Against the Deformation of Helical Line in This Condition

Compared to the crown request of the finished pars, the crown of white parts should increase 3 μm, to have an optimal crown for the finished parts.

4.3 The Deformation Law of Helical Line of Asymmetric Gear with Relatively Large K *Value*

4.3.1 Analysis and Prediction of Deformation

When gear's *K* value is large, and asymmetric, the shape of the gear plays a leading role in the deformation of helical line. The volume is increase caused by the phase transition, the volume of the side with more entity expands larger than the other side, and this causes the gear bend to the center of the other side, so the helical line has a cone angle. At the same time, because the cooling speed of the side with more entities is slower than the other side when quenching, under the effect of heat stress, the inner material of the gear pressed to the side with more entities also can cause the helical line's cone angle. These two factors have the same influence on the cone angle deformation of the helical line, also the interaction of heat stress and structural stress can change the crown of helical line.

4.3.2 Test and Verify

A gear produced in shaft and gear manufacturing center with large *K* value but asymmetric was chosen, first gear (Picture 5), ten pieces, a contrast detection of the helical line of white and black parts was made, result shown in Chart 6 and Picture 6, the helical line has a cone angle, the quantity is about 22 μm on average, the crown decreased 3 μm, the structural stress plays a leading role on the deformation of the

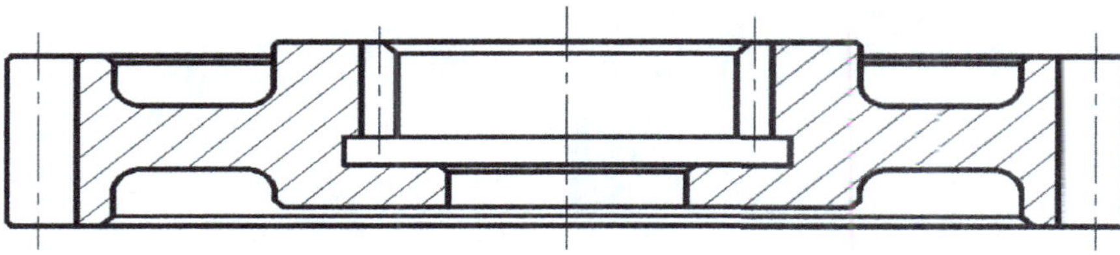

Picture 5 First gear picture

零件材料	零件模数	外径	厚度	K值	零件状态	精度项	1	2	3	4	5	6	7	8	9	10	均值	变形量
FAS3420	4.46	218	35	6.22	热前	fHα	6	1	3	2	1	5	4	4	6	4	3.6	
						Cα	16	12	10	12	13	15	14	13	10	10	13	
					热后	fHα	25	26	27	28	29	24	23	27	19	22	25	21.4
						Cα	12	9	10	10	10	10	9	13	9	8	10	−2.5

Chart 6 Statistics of the deformation quantity of white and black parts

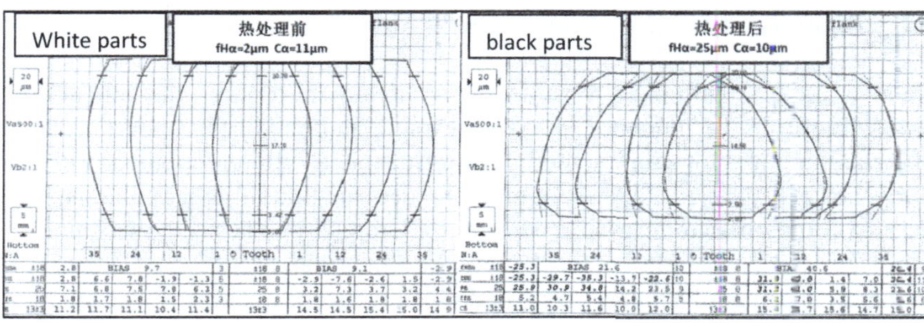

Picture 6 Contrast report of the first gear helical line deformation

helical line, the forming of the stress may have two reasons, one is the homogeneity of the material organization is poor, the other one is when clamping in heat treatment process the part get partial acting forces.

4.3.3 Countermeasure Against the Deformation of Helical Line

Inverse the helical line cone angle of white parts to make a countermeasure modification when machining, in the heat treatment process it can counteract the cone angle of helical line, to get optimal helical line, and at the same time, make the crown 2 μm larger than the request of finished parts, to get optimal crown.

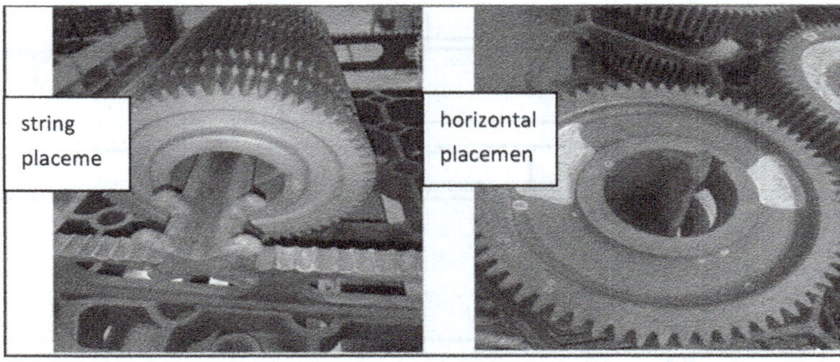

Picture 7 String and leveling

4.4 The Deformation Law of Helical Line in Different Ways of Heat Treatment Location Mode

4.4.1 Analysis and Expectations of the Deformation

As above, for gear with lager *K* value, in heat treatment process, the helical line is easy to deform in the influence of external force, and one of the important factors is the clamping and location method in heat treatment, different methods affect the homogeneity of the parts, and then affect the deformation of helical line. Now, we have two common clamping ways, shown in Picture 7, one is string placement and the other is horizontal placement. String placement has less support point, while horizontal placement has more. We can dope out that for horizontal placement, the deformation law can be better than string placement. But at the same time, for the horizontal placement, the cooling speed on the two sides can be different, one faster one slower, under the effect of heat stress, the material of the side which cool faster were pushed to the other side, and this cause the helical line have a cone angle. Under the effect of gravity, the part creeps downward, and also causes the tilt deformation of helical line

4.4.2 Test and Verify

Twenty gears (see Picture 7) with larger *K* and symmetrical shape (eliminating shape influence) were selected, 10 gears were heat treated in series and 10 gears were heat treated in paperback. The spirals before and after heat treatment were compared and tested. The results were shown in Pictures 8 and 9.

(1) comparing with string placement, horizontal placement has larger average deformation value
(2) comparing with horizontal placement, the deformation of string placement is in a mess and irregular

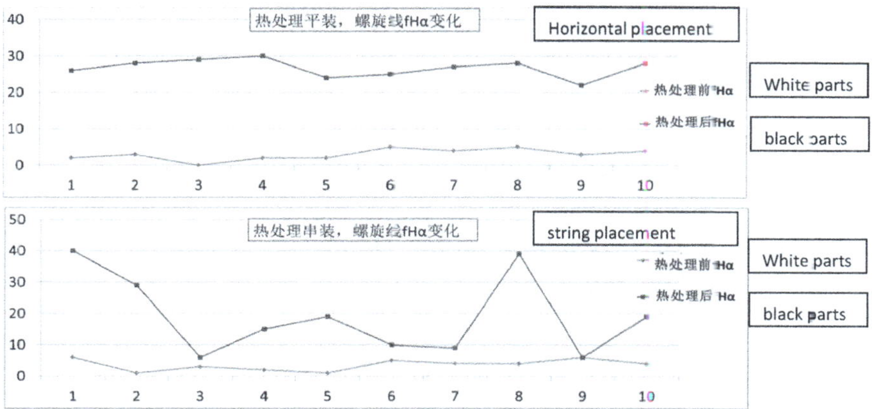

Picture 8 Statistics of deformation of the fcurth gear white and black

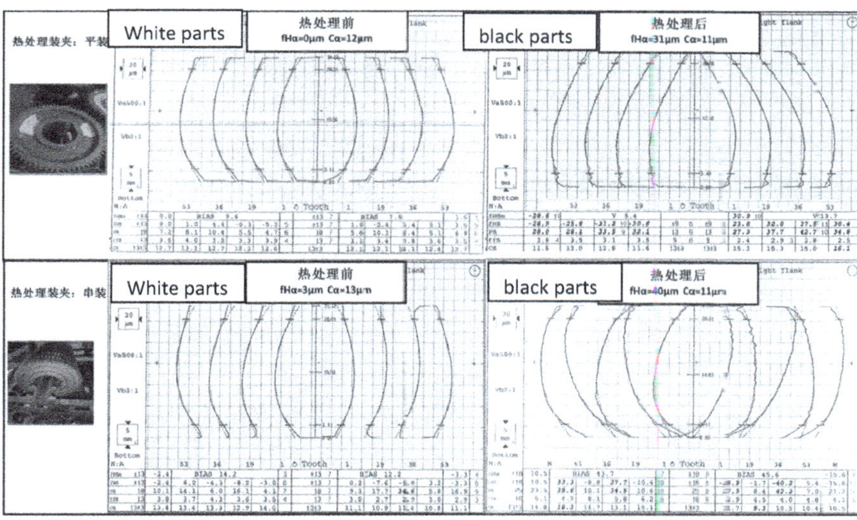

Picture 9 Report of deformation of the fourth gear white and black

(3) comparing with string placement, the deformation of horizontal placement is steady and regular.

4.4.3 Countermeasure Against the Helical Line Deformation

Because of the better regularity of heat treatment flat-mounted deformation, the gear with large K value adopts flat-mounted heat treatment method, and before heat treatment, the helix is processed into reverse cone shape to offset the deformation of heat treatment cone, and the ideal helix inclination deviation is obtained.

5 The Effects of Heat Treatment on Tooth Spacing (Fp)

5.1 Analysis and Prediction on the Deformation Law of Tooth Spacing

As gear is rotational parts, symmetry on circumference, the deformation caused by the structural stress and heat stress is the same on each tooth, so the tooth spacing will not change. Tooth spacing is the change of tooth distance on circumference, and the change in size caused by phase transition is the same for each gear, that is, the deformation of tooth spacing is caused by the inhomogeneity of material, but the effect can be little by average. The other factor affects tooth spacing is the clamping and locationing method, and for string placement, the tooth spacing get larger because the gear turn to be ellipse under gravitation.

5.2 Test and Verify of the Deformation Law of Tooth Spacing

Twenty parts of shaft and gear manufacturing center were chosen, 10 horizontal placed, 10 string placed, contrast detection were made, reports are shown in Pictures 10 and 11, and we can see:

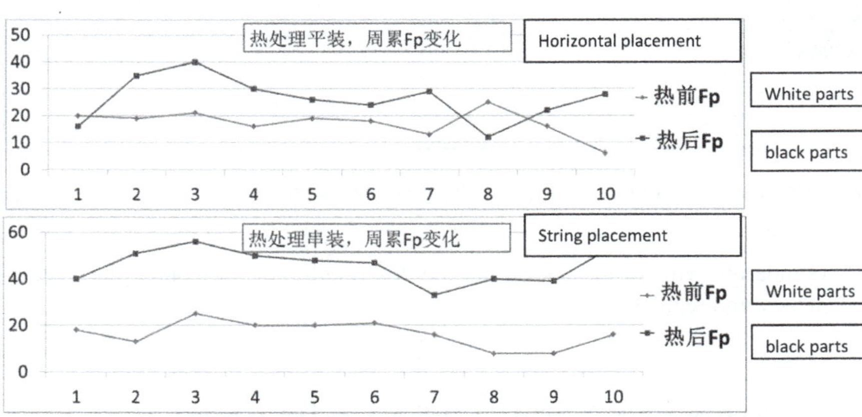

Picture 10 Statistics of the contrast of tooth spacing deformation of string placement and horizontal placement

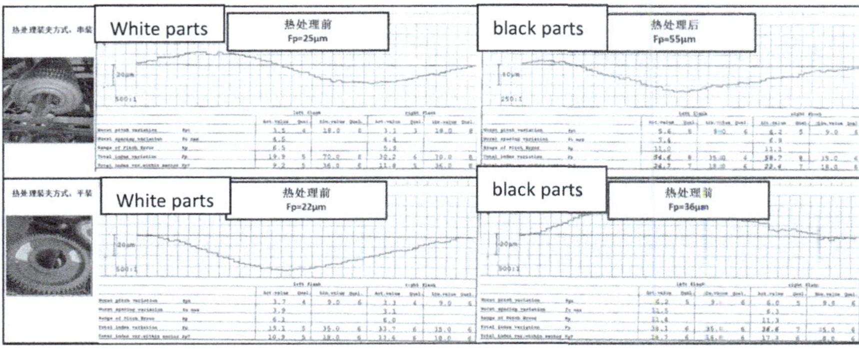

Picture 11 Statistics of tooth spacing test report

(1) The location method has a great influence on the tooth spacing deformation in heat treatment
(2) Horizontal placement is better than string placement in heat treatment

5.2.1 Countermeasure Against Tooth Spacing Deformation

For heavier gears, we suggest horizontal placement method, to diminish the tooth spacing deformation and improve the tooth spacing accuracy.

6 Conclusion

Through the research and summarize, the gear deformation law is found, applying the deformation law in the gear manufacturing of shaft and gear manufacturing center, and the gear accuracy improved a lot. The accuracy of 12TA transmission gear promoted one grade, and the percent of pass of gears improved 15%.

References

1. 许洪基. 齿轮手册[M]. 机械工业出版社, 2013. Xu hongji Gear manual. China Machine Press
2. 潘健生. 热处理手册[M]. 机械工业出版社, 2013, 8. Pan jiansheng Heat treatment manual. China Machine Press

Analysis of the Influence of the Cylinder Body on the Position of the Base Face System

Wang Hong and Sun Lin

Abstract Diesel engine cylinder's main bearing cover is tightened by torque Angle method, the bolt is subjected to high preload, and the crankcase wall is thin, which leads to deformation and displacement of the cylinder's bottom face. As the processing of the bottom face system of the cylinder block was completed before the main bearing cover was assembled, the position of the bottom face system changed after the cover was installed, which finally showed that the position of the bottom face system was qualified in the process inspection, but it was out of tolerance in the finished product inspection. In this paper, the stress and deformation of 36D cylinder are calculated by finite element method. The actual displacement direction and displacement amount of the base face system were verified by analyzing the data collected by the process test. In this way, the machining process is compensated and the position of the bottom face of the cylinder block product is out of tolerance.

Keywords Cylinder block · Pre-tightening force · Deformation · Finite element analysis · Degree of base alignment

1 Introduction

The 36D cylinder of FAWDE is gantry structure with thin side wall of the crankcase skirt, so the force is easy to deform, resulting in displacement of the bottom surface. The process of producing 36D cylinder blocks is first to process the bottom face system and then to assemble the main bearing cover. Therefore, it is very important to study the influence of the assembly of the main bearing cover on the position of the bottom face system to ensure the quality of the product. In this paper, force analysis is carried out on the cylinder block through finite element method to calculate the displacement direction and displacement amount of the bottom surface, and then the data collected is analyzed through the process test to find the displacement amount of each screw hole on the bottom surface. Therefore, the NC processing

W. Hong (✉) · S. Lin
FAWDE, Wuxi 214000, Jiangsu Province, China
e-mail: 565000440@qq.com

China SAE (ed.), *Proceedings of China SAE Congress 2018: Selected Papers*,
Lecture Notes in Electrical Engineering 574,
https://doi.org/10.1007/978-981-13-9718-9_51

program is compensated to eliminate the displacement of the main bearing cover after assembly, so as to ensure that the position of finished screw holes meets the product requirements.

2 Total Structure and Force Analysis of Cylinder

Figure 1 shows the structure of the cylinder block after assembly of the main shaft cover.

According to the picture, after the main bearing cover is installed, the force on the cylinder can be divided into two forces, one is the tension of the bolt, the other is the pressure of the spindle cover, both of which are equal to the pre-tightening force of the bolt.

Each cylinder body is installed with a total of 7 main bearing covers, each bearing cover is tightened with two M16 × 2–5 h strong bolts, and the center distance between the two adjacent bolts is 127 mm. Bolt tightening torque Angle method: tighten torque of 90 N m + 210°. Through calculations and measurements of bolt tightening torque eventually get after the Angle rotated 210°, eventually tightening torque T material 300 N m.

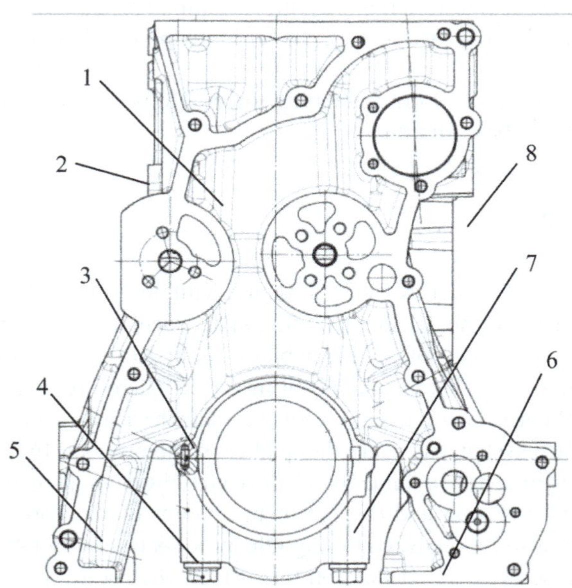

Fig. 1 36D Cylinder block structure. 1—cylinder block; 2—side of oil pump; 3—positioning pin of bearing cap; 4—strong bolt; 5—crankcase skirt hem; 6—cylinder body underside; 7—main bearing cover; 8—camshaft side

According to the formula:

$$T = F'\left[\tan(\varphi + \rho')\frac{d_2}{2} + \frac{u_n}{3}\left(\frac{D_0^3 - d_0^3}{D_0^2 - d_0^2}\right)\right]$$

F'—Pre-tightening force;
φ—Thread hade;
ρ'—Thread deputy equivalent friction, $\rho' = \arctan(1.15\mu)$, where μ is the friction factor of thread pair;
d_2—Thread pitch diameter (mm);
u_n—Friction factor of support surface;
D_0—Outer diameter of thread support surface (mm);
d_0—The diameter of the connector or washer (mm);

The formula can be simplified as follows:

$$T = k_1 F' d$$

where d—nominal diameter; Tightening torque factor is related to thread size, thread parameter, thread pair and friction factor of support surface, etc. Its value can be found by Table 1.

M16 rough thread is used in the 36D cylinder, and the thread is lubricated by machine oil, so it can be estimated by the following test: $T = 0.19F'd$.

So the prestressing force $F' = \frac{T}{0.19 \times d}$, Substitute data into:

$$F' = \frac{300}{00.19 \times 0.016} \approx 98.68\,\text{kN}$$

Table 1 Value reference table

Surface coating	No coating	Cadmium plated	Tu molybdenum base grease	Coated light crude	Oil and phosphate
Average	0.199	0.186	0.137	0.123	0.19
Maximum	0.267	0.250	0.16	0.15	0.23
Minimum	0.158	0.106	0.10	0.099	0.15

3 Finite Element Analysis Simulation

3.1 Establishment of Three-Dimensional Model of Cylinder Block

Figure 2 shows the solid model of 36D.

Because of the complex cylinder structure, the finite element mesh is very large and the calculation and analysis time is long. Therefore, the cylinder block is simplified to remove the sub-surface, unrelated hole series, chamfer, support surface, convex platform, inner cavity, etc., only the simplified model of the cylinder block required in this paper is retained, as shown in Fig. 3.

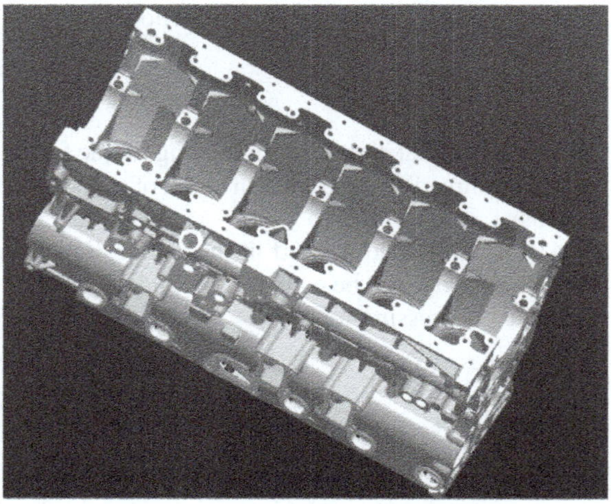

Fig. 2 Solid model

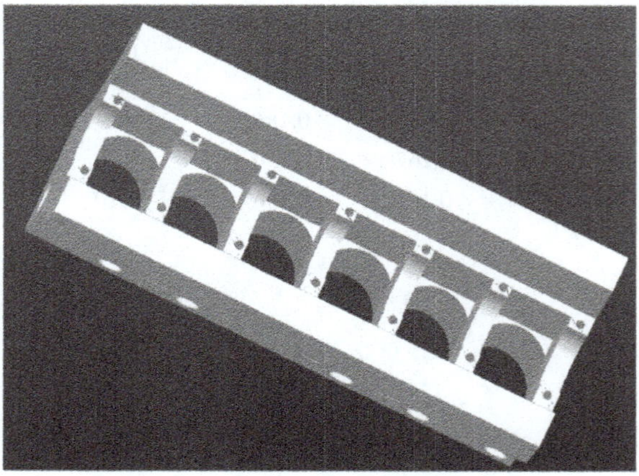

Fig. 3 Simplified model

3.2 Loading Mode and Constraint of Cylinder Block Force

The pre-tightening force of bolt exists in the axial tensile stress of bolt connection. The bearing cover bolt analyzed in this paper is not subjected to the working load, so the pre-tightening force is equal to the tensile stress of bolt. Analysis is carried out with the bearing cover as the reference origin: the cylinder body is subjected to the tensile stress of the bolt to it, and the bearing cover is equivalent to the constraint of the cylinder body, which limits the displacement of the cylinder body. Therefore, the axial load is applied in the cylinder's screw hole, and the constraint is applied on the joint surface of the cylinder's bearing cover to fix its movement in the X and Y directions. For the convenience of calculation, the upper surface of the cylinder is restrained simultaneously. The cylinder block is treated according to nonlinear contact analysis.

3.3 Calculation Results and Analysis

It can be seen from the stress and displacement diagram that the joint surface of the bearing cover and the cylinder block is the place where the stress is most concentrated after the bearing cover is installed, and the maximum displacement of the cylinder block has occurred on the bottom surface, reaching 0.026 mm. The change direction is the rise of the bottom to the left and right sides, that is, the deformation away from the bearing cover, which has a great impact on the Y direction of the bottom face system, while the Z direction of the front and back ends is unchanged, as shown in Fig. 4.

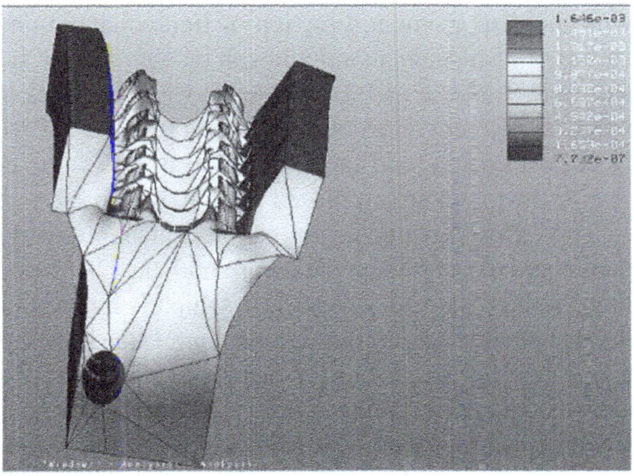

Fig. 4 Finite element analysis results

4 Process Test Verification

According to the finite element analysis, the deformation of the bottom surface is large, and the deformation of the bottom face system seriously affects its position. In order to compensate the position coordinates of the hole system and guarantee the product quality during processing, the displacement of each hole should be determined.

4.1 *Determine the Test Program*

In this paper, the machining center of the bottom face system is adopted in the machining center of Dalian Machine Factory, and the positioning surface and pin hole of the side of the CAM are used as the positioning datum of the cylinder body. In order to ensure the accuracy of the measurement, the three-coordinate detector is used to detect the position coordinates of all the bottom face systems before and after the main bearing cover installation with the CAM side as the reference (to ensure the consistency between the process reference and the test reference). Through the calculation and analysis, the specific value of the change of the position degree of the base face system can be obtained.

4.2 *Collection and Analysis of Test Data*

By collecting the data of 10 groups of 5 workpieces, it was found that the *Z* direction of the base face system did not change, while the *Y* direction changed greatly (consistent with the displacement direction of finite element analysis). Therefore, the specific values of displacement amount can be obtained by subtracting the *Y* coordinates, as shown in Tables 2 and 3. In order to visually check the *y*-direction displacement of holes in different areas of the base surface, the OP70 process is used to process the base face system, and the number of all holes, the corresponding position and the size of the offset are all placed on Fig. 5.

The following conclusions can be drawn intuitively from Fig. 5: since the process datum and measurement datum are both the CAM side, the position of the CAM side face system is almost unchanged, which is ignored in this paper. The oil pump side displacement is larger, through the pump side hole separately, get Fig. 6 line chart (*Y* coordinate for the size of the displacement, *X* coordinate for the hole number), you can see the bottom of the cylinder block face is a position from the back end face to face before change is bigger and bigger (36 d after the cylinder block face organic oil

Table 2 Oil pump side face system

Hole	Work piece number/difference (Y/mm)					Value (mm)
	GS974	GS955	GS1002	GS1006	GS1019	
#2110	−0.1324	−0.1352	−0.1265	−0.1319	−0.1258	−0.130
#2111	−0.0862	−0.0796	−0.0717	−0.0778	−0.0686	−0.077
#2120	−0.1254	−0.1302	−0.1254	−0.1262	−0.1215	−0.126
#2121	−0.1173	−0.126	−0.1204	−0.1201	−0.1144	−0.120
#2122	−0.1172	−0.1203	−0.1121	−0.1132	−0.1153	−0.116
#2123	−0.1056	−0.1133	−0.1079	−0.1031	−0.1073	−0.107
#2124	−0.102	−0.1119	−0.1067	−0.1033	−0.1035	−0.105
#2125	−0.0988	−0.1074	−0.0999	−0.0986	−0.0993	−0.101
#2126	−0.0956	−0.1039	−0.0984	−0.0995	−0.0979	−0.099
#2127	−0.0941	−0.1023	−0.0975	−0.101	−0.0927	−0.098
#2128	−0.0992	−0.1017	−0.0949	−0.0978	−0.0929	−0.097
#2129	−0.0926	−0.0936	−0.0853	−0.0923	−0.0875	−0.090
#2130	−0.0899	−0.0874	−0.0831	−0.0884	−0.0818	−0.086
#2131	−0.0883	−0.0829	−0.0747	−0.0801	−0.074	−0.080
#2150	−0.1307	−0.1316	−0.1295	−0.1277	−0.1189	−0.128
#2151	−0.1199	−0.1228	−0.1241	−0.1208	−0.1167	−0.121
#2152	−0.1138	−0.1207	−0.1079	−0.112	−0.1106	−0.113
#2153	−0.108	−0.1155	−0.1087	−0.112	−0.1069	−0.110
#2154	−0.108	−0.1133	−0.1029	−0.105	−0.1053	−0.107
#2155	−0.1035	−0.1029	−0.1002	−0.0975	−0.0956	−0.100
#2156	−0.0956	−0.1005	−0.0964	−0.0968	−0.092	−0.096
#2157	−0.0972	−0.0998	−0.093	−0.1035	−0.0903	−0.097
#2158	−0.0938	−0.0938	−0.0901	−0.0923	−0.0883	−0.092
#2159	−0.088	−0.0923	−0.0843	−0.0893	−0.0789	−0.087
#2160	−0.0901	−0.0866	−0.0801	−0.0823	−0.0772	−0.083
#2161	−0.0848	−0.0791	−0.0703	−0.0762	−0.0644	−0.075

pump installation lumen, skirt edge thickness is greater than the front end, at the rear end is better than former transverse rigidity), displacement of 0.075–0.130 mm, the average displacement of about 0.1 mm. Since the cylinder can be simply viewed as a symmetrical body, the unilateral displacement of the base surface is about 0.05 mm.

Table 3 Camshaft side face system

Hole	Work piece number/difference (*Y*/mm)					Value (mm)
	GS974	GS955	GS1002	GS1006	GS1019	
#2162	0.0293	0.0153	0.024	0.0193	0.0227	0.022
#2163	0.0271	0.0089	0.0263	0.0198	0.0191	0.020
#2164	0.0223	0.009	0.0192	0.0128	0.0144	0.016
#2165	0.0139	0.0047	0.0142	0.0062	0.0203	0.012
#2166	0.0156	0.0008	0.0088	0.0047	0.0079	0.008
#2167	0.009	−0.0026	0.0064	0.0006	0.0042	0.004
#2168	0.0019	−0.0073	−0.0018	−0.0017	−0.0029	−0.002
#2169	0	−0.0122	−0.0102	−0.0111	−0.0021	−0.007
#2170	−0.0036	−0.0151	−0.0143	−0.0122	−0.0106	−0.011
#2171	−0.0068	−0.0185	−0.0123	−0.0133	−0.0121	−0.013
#2172	−0.0096	−0.0182	−0.0194	−0.0151	−0.0149	−0.015
#2173	−0.0107	−0.0249	−0.0223	−0.0186	−0.0152	−0.018
#2132	0.0295	0.0114	0.026	0.0192	0.0221	0.022
#2133	0.0204	0.0108	0.0232	0.0176	0.0216	0.019
#2134	0.0236	0.0105	0.0197	0.0143	0.0193	0.017
#2135	0.0195	0.006	0.0169	0.0124	0.0162	0.014
#2136	0.0133	0.0027	0.011	0.0079	0.0123	0.009
#2137	0.0104	−0.0009	0.0059	0.0033	0.0071	0.005
#2138	0.0055	−0.0048	0.0006	−0.0006	0.0019	0.001
#2139	0.0025	−0.008	−0.0041	−0.0041	−0.0013	−0.003
#2140	−0.0028	−0.0131	−0.0086	−0.0101	−0.0065	−0.008
#2141	−0.0058	−0.0165	−0.0131	−0.0132	−0.0095	−0.012
#2142	−0.0077	−0.018	−0.0154	−0.0164	−0.0136	−0.014
#2143	−0.0115	−0.0225	−0.0201	−0.0173	−0.0146	−0.017
#2144	−0.0142	−0.0288	−0.0286	−0.0235	−0.018	−0.023

4.3 Experimental Verification After Compensating NC Program

The NC program of OP70 process is manually compensated (screw hole #2120–#2131, #2150–#2161 position coordinates in NC program are corrected by 0.1 mm). Five workpieces are selected to be processed with the coordinates after compensation, and the position degree of the bottom face system is detected after the finished product is offline (the required position degree should be found 0.4). As shown in Table 4, the adjusted position of the base face system meets the product requirements.

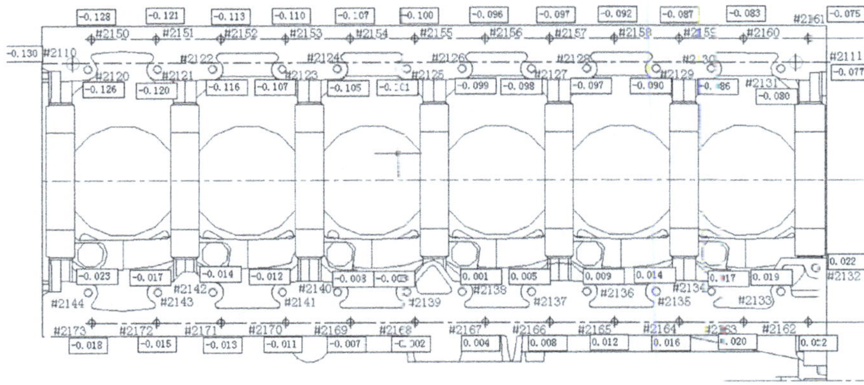

Fig. 5 Detailed displacement of the hole system

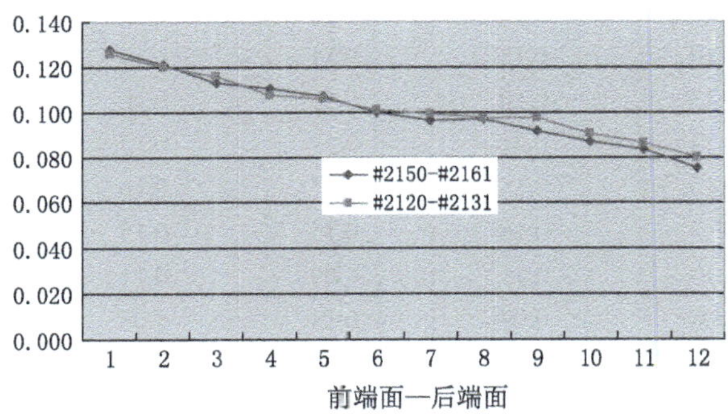

Fig. 6 Displacement broken line

Table 4 Location of oil pump side face system

Hole	Work piece number/position (product requirements ϕ 0.4)				
	GS1114	GS1117	GS1118	GS1119	GS1129
#2120	0.14	0.14	0.13	0.15	0.11
#2121	0.15	0.09	0.09	0.09	0.06
#2122	0.10	0.09	0.11	0.10	0.08
#2123	0.04	0.08	0.04	0.08	0.03
#2124	0.03	0.06	0.05	0.05	0.06
#2125	0.07	0.06	0.08	0.07	0.05
#2126	0.09	0.08	0.11	0.13	0.06
#2127	0.08	0.04	0.04	0.10	0.03
#2128	0.11	0.05	0.10	0.08	0.06
#2129	0.10	0.02	0.08	0.04	0.05
#2130	0.06	0.07	0.08	0.06	0.04
#2131	0.13	0.11	0.10	0.10	0.11
#2150	0.14	0.14	0.15	0.12	0.12
#2151	0.10	0.11	0.10	0.10	0.10
#2152	0.11	0.13	0.12	0.11	0.14
#2153	0.10	0.18	0.11	0.14	0.09
#2154	0.13	0.10	0.13	0.12	0.10
#2155	0.14	0.21	0.17	0.15	0.15
#2156	0.13	0.15	0.15	0.16	0.12
#2157	0.13	0.16	0.15	0.15	0.12
#2158	0.10	0.11	0.09	0.11	0.11
#2159	0.08	0.09	0.14	0.08	0.06
#2160	0.09	0.06	0.05	0.13	0.10
#2161	0.13	0.10	0.06	0.11	0.12

5 Conclusion

The following conclusions can be drawn from the analysis:

(1) Through finite element calculation, it is concluded that the deformation direction of the cylinder is consistent with the actual displacement direction, and the displacement amount is 48% error due to the fact that the calculation is convenient for the simplification of the cylinder block model.
(2) After the cylinder body is installed with the main bearing cover, it will cause displacement of the bottom face system and move about away from the main shaft cover 0.05 mm. Taking the side of the pump as the reference calculation, the CAM side face system moves 0.10 mm away from the side of the pump.
(3) The NC program of OP70 process was manually compensated (the machining coordinates of screw holes #2120–#2131, #2150–#2161 in NC program were corrected by 0.1 mm) and verified. The problem of the location of the bottom face system of the cylinder block is solved.
(4) Through the research on the size and direction of the displacement of the bottom face system behind the main bearing cover of the cylinder block, it provides a reference for the future research and analysis of the deformation in the machining process of the cylinder block.

Case Design and Implementation of Hypothesis Test Analyses for Attribute Measurement Systems

Shuxian Zhao, Yuanyuan Qi, Yanling Liu and Bin Fu

Abstract This paper is written by the author on the basis of a thorough understanding of the Measurement Systems Analysis (MSA) reference manual and summarizing its successful experience in the production practice of Weichai Power Co., Ltd. The typical attribute measurement system (go/no go gage) commonly used in the production site was chosen, and the consistency is determined by the *Kappa* value–Cross-tabulation method. This paper elaborated the detailed operation flow of the attribute MSA systematically and detailed introduce the specific design and implementation of each process, including the sample choosing, the determination of the random sequence of data collection, calculation, analysis, and the conclusion.

Keywords Measurement system analysis (MSA) · Attribute measurement systems · Hypothesis test analyses · *Kappa* value–Cross-tabulation method

1 Introduction to Attribute Measurement Systems

Attribute measurement systems are the class of measurement systems where the measurement value is one of a finite number of categories. For example, go/no go gage, which is the most commonly used in the batch production process of machinery manufacturing industry, is a typical representative. Unlike a variable gage, this attribute gage cannot indicate how good or how bad a part is, but only that the part is accepted or rejected. This inspection method will make two types of errors to products near the specification limit boundary (gray area as shown in Fig. 1): discard truth or accept false.

As shown in Fig. 1, if process capability index is not high enough, the parts from the process may be randomly distributed in the areas of Figures I, II, and III. The parts in area I are defined "very bad" parts, and the parts in area III are defined "very good" parts, and the rest ones in gray area II are defined the critical parts. With "qualified parts" for the null hypothesis H_0, in statistical trials, either to reject or

S. Zhao (✉) · Y. Qi · Y. Liu · B. Fu
Weichai Power Co., LTD, Weifang 261061, Shandong, China
e-mail: zhaosx@weichai.com

China SAE (ed.), *Proceedings of China SAE Congress 2018: Selected Papers*, Lecture Notes in Electrical Engineering 574,
https://doi.org/10.1007/978-981-13-9718-9_52

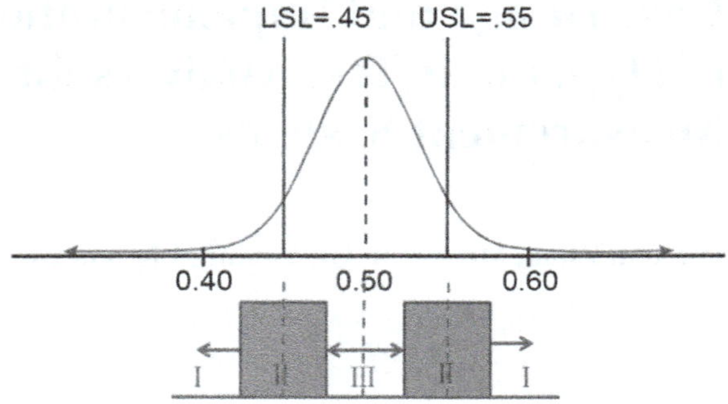

Fig. 1 Counting inspection and process capability

accept the null hypothesis, to achieve the correct results is impossible with certain error, especially for the components in area II. The following four possible results occur in the hypothesis trials, as shown in Table 1:

(1) Type I error: If the null hypothesis that is true is rejected, the error committed is called type I error. Under the condition that hypothesis trial is true, the trial statistics just in a small probability rejection area, the null hypothesis rejected, type I error committed, the probability of which is equal to the significance level of alpha, so we called them alpha errors in statistics. We can control the probability of error by controlling the significance level.
(2) Type II error: If the null hypothesis that is false is accepted, the error committed is called type II error. The probability of type II error committed is equal to the significance level of β, so we called them β errors in statistics.

In view of the above reasons, in order to reduce the two types of errors as far as possible in the statistical trial process, the trial capability and judgment consistency of the attribute measurement system must be determined in advance. In general, carrying out the hypothetical trial analysis–Cross-tabulation method is used to analyze and determine the attribute measurement system [1].

This paper will discuss the analysis of attribute measurement system by taking aperture (Φ10H7) test as an example.

Table 1 Four possible outcomes of hypothesis trial analyses

Hypothesis classes	Accept H_0	Reject H_0
Null hypothesis H_0 is right	Right judgment	Reject truth error (Type I error or "α" errors)
Null hypothesis H_0 is wrong	Accept false errors (Type II error or "β" errors)	Right judgment

2 Design and Analysis of MSA Case of Attribute Measurement System

2.1 Measurement System Overview

Measurement task: Diameter measurement of a hole in a diesel engine block, with requirements for Φ10H7;
Measuring equipment: Plain limit gauge (Φ10H7), which is qualified with JJG 343-2012 "Verification Regulation of Plain Limit Gauges;"
Test samples: According to the requirements of the MSA manual, 30–50 samples should be taken to evaluate attribute measurement systems with the Cross-tabulation method. Figures of a representative sample of 30 holes cover the area I, II, III. In this case, namely sample points, usually collected in the production process from MSA in advance, extend the lower limit and upper limit for product specification. After selection, 30 samples were marked as 01–30, respectively. Note that the number should be hidden to meet the requirements for "blind measurements" in the process of data collection [2]. Blind measurements are measurements obtained in the actual measurement environment by an operator who does not know that an assessment of the measurement system is being conducted. Properly administered, tests based on blind measurements are usually not contaminated by the well-known Hawthorne effect.
Appraiser: Three trained and qualified appraisers, numbered *A*, *B* and *C*, respectively;
Measurement method: To measure the aperture by plain limit gauge. If GO-side go through the hole and NO-GO-side not go, parts are qualified, mark them as "1;" if GO-side not go through the hole or NO-GO-side go through, mark them as "0."
Environmental conditions: The temperature range: (16–30)°C.

2.2 Data Collection

(1) Determine the order of measurement. In order to ensure that the process variation distributed randomly in the process of data collection, the random order of data collection should be designed in advance. Three appraisers measure 30 sample points for 270 times, with every appraiser measuring each sample for 3 times. Determine the order randomly, referring to "Fig. 2 determining the measurement order in Excel:" The first column in the figure is listed as the original column, the number of the evaluator and sample input, a total of 270 rows. For example, *A*01 is the sample No. 01 measured by appraiser *A*, and three identical *A*01 means that appraiser *A* measures sample 01 three times. In the first row of the second column, input "=RAND()," enter and pull down to 270 rows, generate the random number column with the function of Excel, and then use the sorting function to generate the new order (see the green area on the right of the figure)

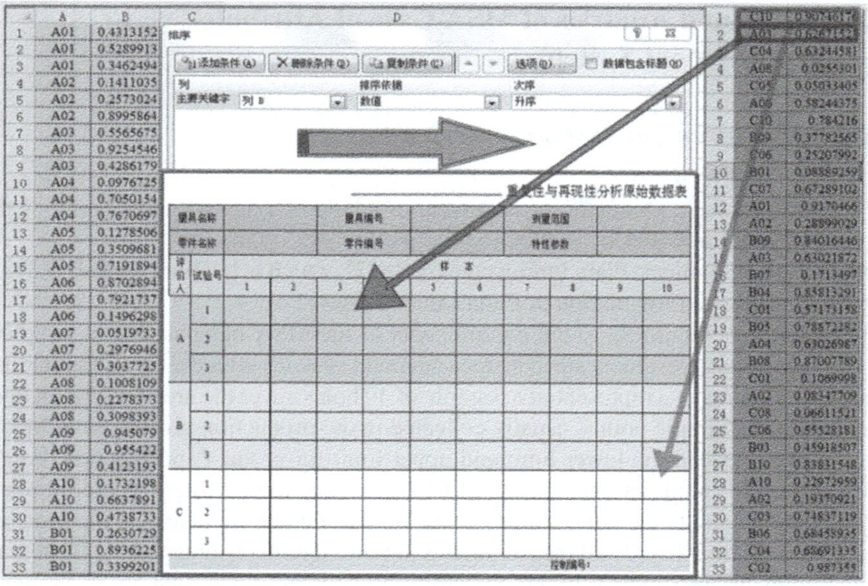

Fig. 2 Determine the measurement order by excel

column, which is called the execution column. Field data collection shall be conducted in the first column (i.e., execution column) of the green area. If according to the execution column generated in this table, the first time is to trial the 10th sample by appraiser *C*, the second time is to trial the 3rd sample by appraiser *A*, and the third time is to trial the 4th sample by appraiser *C*. In the same way, 270 data were collected.

(2) Then, three appraisers carry out the trial in the way of blind measurement in the order randomly above, and summarize data obtained in Table 2.

(3) Determine the characteristic figure of each sample according to the data obtained from the attribute measurement systems choosing the sample. If it meets the specification requirements, it is denoted as "1" in the column of "Ref;" otherwise, it is denoted as "0." The data obtained are filled in Table 2.

2.3 Statistics and Calculation of Data

(1) The "1" and "0" data in the "hypothetical test analysis original data summary Table 2" were statistically calculated according to the evaluator $A * B$, $B * C$, $A * C$ and A * Ref, B * Ref, C * Ref, and fill in the "hypothetical test analysis original data summary Table 3." For example, column $A * B$: $0 * 0$ represents the rejection of both of them. There are 5 in the statistics. $1 * 0$ means A to

Table 2 Data collection sheet of hypothesis trial analyses

序号	*A*-1	*A*-2	*A*-3	*B*-1	*B*-2	*B*-3	*C*-1	*C*-2	*C*-3	Ref
1	1	1	1	1	1	1	1	1	1	**1**
2	1	1	1	1	1	1	1	1	1	**1**
3	1	1	1	1	1	1	1	1	1	**1**
4	1	1	1	1	1	1	1	1	1	**1**
5	0	0	0	0	1	0	0	0	0	**0**
6	1	1	1	1	1	1	1	1	1	**1**
7	1	1	1	1	1	1	1	1	1	**1**
8	1	1	1	1	1	1	1	1	1	**1**
9	1	1	1	1	1	1	1	1	1	**1**
10	0	0	0	0	0	0	0	0	0	**0**
11	1	1	1	1	1	1	1	1	1	**1**
12	1	1	1	1	1	1	1	1	1	**1**
13	1	1	1	1	1	1	1	1	1	**1**
14	1	1	1	1	1	1	1	1	1	**1**
15	1	1	1	1	1	1	1	1	1	**1**
16	1	1	1	1	1	1	1	1	1	**1**
17	1	1	1	1	1	1	1	1	1	**1**
18	1	1	1	1	1	1	1	1	1	**1**
19	1	1	1	1	1	1	1	1	1	**1**

(continued)

Table 2 (continued)

序号	*A*-1	*A*-2	*A*-3	*B*-1	*B*-2	*B*-3	*C*-1	*C*-2	*C*-3	Ref
20	1	1	1	1	1	1	1	1	1	**1**
21	1	1	1	1	1	1	1	1	1	**1**
22	1	1	1	1	1	1	1	1	1	**1**
23	1	1	1	1	1	1	1	1	1	**1**
24	1	1	1	1	1	1	1	1	1	**1**
25	1	1	1	1	1	1	1	1	1	**1**
26	1	1	1	1	1	1	1	1	1	**1**
27	1	1	1	1	1	1	1	1	1	**1**
28	1	1	1	1	1	1	1	1	1	**1**
29	1	1	1	1	1	1	1	1	1	**1**
30	1	1	1	1	1	1	1	1	1	**1**

Table 3 Summary table of hypothetical test analysis data

*	$A * B$	$B * C$	$A * C$	A * Ref	B * Ref	C * Ref
0 * 0	5	5	6	6	5	6
1 * 0	0	1	0	0	1	0
0 * 1	1	0	0	0	0	0
1 * 1	84	84	84	84	84	84

accept and B to reject, there are 0 statistics; *0 * 1* means A rejects and B accepts. There is 1 in the statistics. 1 * 1 means $A * B$, both are judged and accepted, and there are 84 statistics. Fill in the table corresponding to the statistical results. The remaining columns are counted in the same way. Ref is the decision result of the reference value.

(2) According to the statistical results in Table 3, the Cross-tabulation of A and B, B and C, and A and C are, respectively, established, and A Cross-tabulation of A and Ref, B and Ref, C and Ref. Taking the Cross-tabulation of A and B (Table 4) as an example, the expectation in the table is calculated, respectively, after the observation values are filled in, which corresponds to the expectation calculation in the Cross-tabulation.

The specific calculation is as follows:
Each of the 30 samples was judged three times, that is, each was judged 90 times. Among the 90 times evaluated by A evaluator, 6 times rejected, and 84 times received. Among the 90 times evaluated by B evaluator, 5 rejected and 85 received, then:

$$\text{The proportion of reviewer } A\text{'s rejection: } P_{A0} = \frac{6}{90} \tag{1}$$

$$\text{The proportion of reviewer } B\text{'s rejection: } P_{B0} = \frac{5}{90} \tag{2}$$

$$\text{The proportion received by evaluator } A\text{: } P_{A1} = \frac{84}{90} \tag{3}$$

Table 4 $A * B$ cross-tabulation

$A * B$			B		Total	
			0	1		
A	0	Count	5	1	6	P_o and P_e:
		Expected count	0.33	5.67	6.0	
	1	Count	0	84	84	P_o=0.99
		Expected count	4.67	79.33	84.0	P_e=0.89
Total		Count	5	85	90	
		Expected count	5.00	85.00	90.00	

$$\text{The proportion received by evaluator } A\text{: } P_{B1} = \frac{85}{90} \tag{4}$$

Since the observation processes of two appraisers are independent, the expected probability is calculated as follows:

The probability that the appraiser *A* and *B* jointly reject:

$$p(A0 \cap B0) = P_{A0} \cdot P_{B0} = \frac{(6 \times 5)}{90} \approx 0.33 \tag{5}$$

The probability that the appraiser *A* and *B* jointly receive:

$$p(A1 \cap B1) = P_{A1} \cdot P_{B1} = \frac{(84 \times 85)}{90} \approx 79.33 \tag{6}$$

The probability that the appraiser *A* rejects and *B* receives:

$$p(A0 \cap B1) = P_{A0} \cdot P_{B1} = \frac{(6 \times 85)}{90} \approx 5.67 \tag{7}$$

The probability that the appraiser *A* receives and *B* rejects:

$$p(A1 \cap B0) = P_{A1} \cdot P_{B0} = \frac{(84 \times 5)}{90} \approx 4.67 \tag{8}$$

(3) Calculate P_o and P_e:

P_o = The sum of diagonal elements for observational statistics; means the rate that the evaluator observes consistently.
P_e = The sum of diagonal elements for the value of expectation; means the rate that people expect or accidentally agree

Here, the formula:

$$P_o = \frac{A0B0 + A1B1}{\text{total}} = \frac{5 + 84}{90} \approx 0.99 \tag{9}$$

$$P_e = \frac{p(A0 \cap B0) + p(A1 \cap B1)}{\text{total}} = \frac{0.33 + 79.33}{90.00} \approx 0.89 \tag{10}$$

Then, fill in Table 4 with the above data. Use the same method to construct and complete the below tables, including Table 5 *B* * *C* Cross-tabulation, Table 6 *A* * *C* Cross-tabulation, Table 7 *A* * Ref Cross-tabulation, Table 8 *B* * Ref Cross-tabulation and Table 9 *C* * Ref Cross-tabulation.

(4) Calculate the value of *Kappa*.

Kappa is defined to measure the ratio of the agreement between the evaluations of appraisers after eliminating accidental agreement. A value of 1 indicates perfect

Table 5 $B * C$ cross-tabulation

$B * C$			C		Total	P_o and P_e:
			0	1		$P_o = 0.99$
B	0	Count	5	0	5	$P_e = 0.89$
		Expected count	0.33	4.67	5	
	1	Count	1	84	85	
		Expected count	5.67	79.33	85	
Total		Count	6	84	90	
		Expected count	6.00	84.00	90.00	

Table 6 $A * C$ cross-tabulation

$A * C$			C		Total	P_o and P_e:
			0	1		$P_o = 1.00$
A	0	Count	6	0	6	$P_e = 0.88$
		Expected count	0.40	5.60	6	
	1	Count	0	84	84	
		Expected count	5.60	78.40	84	
Total		Count	6	84	90	
		Expected count	6.00	84.00	90.00	

Table 7 A * Ref cross-tabulation

A * Ref			Ref		Total	P_o and P_e:
			0	1		$P_o = 1.00$
A	0	Count	6	0	6	$P_e = 0.88$
		Expected count	0.40	5.60	6.0	
	1	Count	0	84	84	
		Expected count	5.60	78.40	84.0	
Total		Count	6	84	90	
		Expected count	6.00	84.00	90.00	

Table 8 B * Ref cross-tabulation

B * Ref			Ref		Total	P_o and P_e:
			0	1		$P_o = 0.99$
B	0	Count	5	0	5	$P_e = 0.89$
		Expected count	0.33	4.67	5.0	
	1	Count	1	84	85	
		Expected count	5.67	79.33	85.0	
Total		Count	6	84	90	
		Expected count	6.00	84.00	90.00	

Table 9 C * Ref cross-tabulation

C * Ref			Ref		Total	P_o and P_e: $P_o = 1.00$ $P_e = 0.88$
			0	1		
C	0	Count	6	0	6	
		Expected count	0.40	5.60	6.0	
	1	Count	0.00	84.00	84	
		Expected count	5.60	78.40	84.0	
Total		Count	6	84	90	
		Expected count	5.00	84.00	90.00	

agreement. A value of 0 indicates the agreement is no better than chance. This parameter was firstly proposed by Cohen in 1960, and it is a reliability index that measures the level of the agreement between measurement results. *Kappa* is only available for tables in which both variables use the same category values and both variables have the same number of categories.

The value of *Kappa* is calculated by the below formula.

$$Kappa = (P_o - P_e)/(1 - P_e) \tag{11}$$

We can calculate the *Kappa* between the evaluations of appraisers according to Formula (11).

$$A * B\text{: } Kappa = \frac{0.99 - 0.89}{1 - 0.89} \approx 0.90 \tag{12}$$

$$B * C\text{: } Kappa = \frac{0.99 - 0.89}{1 - 0.89} \approx 0.90 \tag{13}$$

$$A * C\text{: } Kappa = \frac{1.00 - 0.88}{1 - 0.88} \approx 1.00 \tag{14}$$

By the same method, we make out the *Kappa* between each appraiser and the reference decision.

$$A * \text{Ref: } Kappa = \frac{1.00 - 0.88}{1 - 0.88} \approx 1.00 \tag{15}$$

$$B * \text{Ref: } Kappa = \frac{0.99 - 0.89}{1 - 0.89} \approx 0.90 \tag{16}$$

$$C * \text{Ref: } Kappa = \frac{1.00 - 0.88}{1 - 0.88} \approx 1.00 \tag{17}$$

Table 10 *Kappa* summary

Cross-category	$A * B$	$B * C$	$A * C$	$A *$ Ref	$B *$ Ref	$C *$ Ref
Kappa	0.90	0.90	1.00	1.00	0.90	1.00
Conclude	Good	Good	Good	Good	Good	Good

2.4 Data Analysis and Determination

The criterion of *Kappa*:

If the value of *Kappa* is greater than 0.75, it indicates good to excellent agreement.
If the value of *Kappa* is located between 0.4 and 0.75, it indicates general agreement. Comprehensive evaluation of measurement system should be made according to the importance of process and maintenance cost of measuring equipment. It may be acceptable in the conditions of customer recognition.
If the value of *Kappa* is less than 0.4, it indicates poor agreement. We should analyze the whole measurement system to improve, such as repairing measure equipment or training appraisers.

Then, fill the above data into Table 10.

In the example, all the values of *Kappa* are greater than 0.75, it indicates that all the appraisers show good agreement between each other and the reference decision in the measurement system.

If the final calculation result is less than 0.75, it means that there are some problems in the process to determine the agreement of system and we need to analyze the process from the following reason to formulate corrective measures.

(1) the measuring equipment may be worn out of tolerance and should be calibrated again;
(2) the skills of appraisers are inconsistent, so it is necessary to train and strengthen their detection skills;
(3) the evaluation standards and methods of measurement are not detailed enough and prone to ambiguity, so we need to refine the document [3].

3 Conclusion

Based on the author's understanding and full mastery of the reference manual of "Measurement Systems Analysis (MSA)," this paper creatively applies it to the productive practice of the research and development process of the company's engine. According to the inherent characteristics of the attribute measurement system, the author used the hypothesis test analyses–Cross-tabulation method, the specific operation process of the analysis of the typical attribute measurement system is designed.

Specifically include: the selection of sample pieces, the determination of the random sequence of data collection, data collection, calculation, analysis, and the conclusion, etc.

With the development of technology and the increasing automation of the automobile manufacturing industry, the classification of the measurement system is becoming a more and more detailed division. Many special online measurement systems have been used more and more in the production line. All of these have put forward more and more requirements for the analysis of the measurement system. Only by understanding and mastering more statistical tools and relevant theoretical knowledge, and paying close attention to the production practice in the enterprise, can we truly improve the ability to discover and solve problems, and make the measurement system analysis the statistical analysis tool really serve the process, not only in the form.

References

1. AIAG (2010) Measurement systems analysis (MSA) reference manual, 4th edn, 6:131
2. Zhao S, Yin S, Ju L et al (2006) Application of measurement system analysis in automobile manufacturing industry (to the end). Chin Metrol 4:65
3. Zhao S (2015) Application research of measurement system analysis (MSA) in engine development (master's thesis). Tianjin University, 50

Research on the Materials and Production Process of Low Odor and Environmental Automotive Sealing Strips

Shuchen Wei and Yi Zhu

Abstract To improve the air quality in an automotive cab, after analyzing each impact from materials and production process of automotive sealing strips, we confirm the harmful ingredients in each additive and can provide some solutions to produce low odor and environmental automotive sealing strips.

Keywords Automotive sealing strips · Odor · Additives · Production process · Attachments

As consumers are paying more attention to the air quality inside the vehicle while buying cars, the odor of the air inside the vehicle becomes one of the most important indicators to evaluate the air quality in the automotive cab. The materials of most of the interior trims are high molecular polymers, which contain various unsaturated bonds and unstable functional groups. The small molecule VOCs (volatile organic compounds) can gradually escape from the interior trims' surfaces into the air of an automotive cab and make bad smelling. A short-term breathing in the harmful substances can make passengers feel dizzy, nausea and cause headaches. And a long-term breathing in the harmful substances leads to harmful substances deposition in the human body which can damage visceral organs, nervous system and even cause cancer. So improving the air quality in the automotive cab to protect the passengers' health has become an important responsibility of automotive enterprises.

The materials that affect the air quality inside the automotive cab are plastics, rubber, fabrics, leathers, adhesives, and paint. The main rubber contributor producing the VOCs and odor into the air of an automotive cab is The sealing strips, which mainly include outer door sealing strips of four doors, inner door sealing strips of four doors, window glass guides of four doors, and backdoor sealing strip (or trunk sealing strip). Most of these strips are made of rubber, which takes EPDM (Ethylene Propylene Diene Monomer) as the main material. Some models of window glass guides, especially some Japanese cars', are made of thermoplastic elastomers (TPV-EPDM/PP) as well. By drawing some non-conjugated diene monomers (3–10%) into

S. Wei (✉) · Y. Zhu
FAW, Changchun, China
e-mail: 516605469@qq.com

China SAE (ed.), *Proceedings of China SAE Congress 2018: Selected Papers*, Lecture Notes in Electrical Engineering 574,
https://doi.org/10.1007/978-981-13-9718-9_53

the copolymer of ethylene and propylene, we can make this common rubber type—EPDM. EPDM is a good material for sealing strips that has excellent heat resistance and weather resistance. The raw rubber and the tens of additives in the formulation of EPDM products can both affect the odor performance of sealing strips. It is not an easy job to improve the odor performance of sealing strips as there are lots of impacts that may change the production process and the material performances. We will analyze the impacts one by one in the following words.

1 Raw Rubber and Additives

1.1 The Raw Rubber

In the formulation of automotive sealing strip materials, the content of EPDM is generally 30–50%.

1.1.1 The Composition of Raw Rubber

Automotive sealing strips are usually composite structured that made up mainly by dense EPDM, sponge EPDM, metal keel, and so on. The sponge EPDM material is mainly found in outer door sealing strips and trunk sealing strips. The Mooney viscosity of raw rubber to make sponge EPDM material is usually 60–70. The raw rubber contains 45–60% ethylene and 9–10% ENB (ethylene norbornene). And its molecular structure is CLCB (controllable long-chain branching).

The dense EPDM material can be mainly found in inner door sealing strips and window glass guides. The Mooney viscosity of raw rubber to make dense EPDM material is usually 80. The raw rubber contains about 55% ethylene and 5.5% ENB. And its molecular structure is also CLCB. This high-Mooney-viscosity rubber material is suitable to make dense EPDM with Shore A hardness of 50–80.

1.1.2 How the Raw Rubber Influences the Odor Performance of Sealing Strips

Mooney viscosity reflects the molecular weight distribution, plasticity, and fluidity of rubber. The rubber material with high Mooney viscosity has the characteristics of high molecular weight and wide distribution range. Although it increases the difficulty of the mixing and extrusion process, it is beneficial to improve the emission characteristics and odor performance of the extruded product. And the product has higher strength and better comprehensive performance after the vulcanization process.

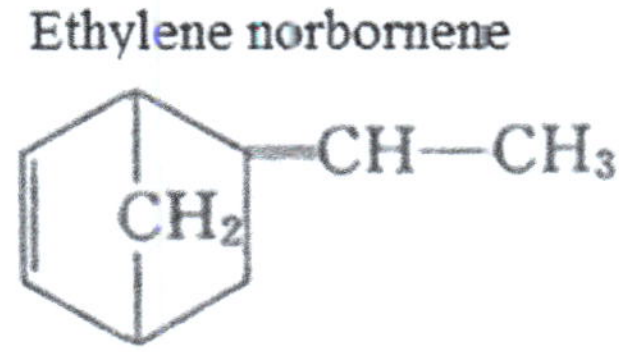

Fig. 1 Monomer structural formula of ENB

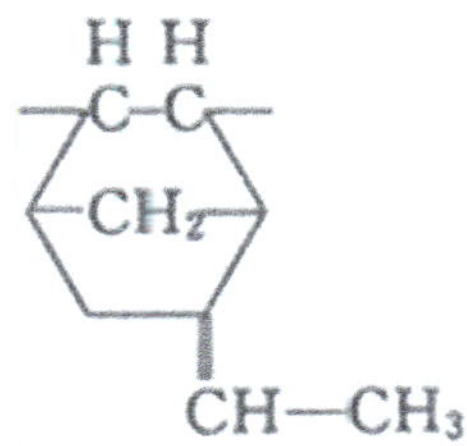

Fig. 2 Structural formula of ENB-EPDM

The main factor affecting the odor performance of sealing strips in the raw rubber is the content of the third monomer, which is ENB. As far as possible, the dense EPDM should use raw rubber with lower ENB content. The sponge rubber is limited by the foaming process, and the raw rubber with higher ENB content is currently selected. ENB is a polymerization catalyst for EPDM. Although it will penetrate into the water during the vulcanization process, there is still a certain amount of residue on the product, which is unavoidable. Therefore, the use of low ENB raw rubber can improve the odor performance of the sealing strip product. The monomer structural formula of ENB is shown in Fig. 1, and the structural formula of ENB-EPDM is shown in Fig. 2.

1.2 The Reinforcing Agent

The reinforcing agent generally used in automotive EPDM sealing strip products is carbon black. In rubber products, carbon black consumption generally accounts for 40–50% of rubber consumption, ranking second in the world's rubber industry raw material consumption. In order to improve the odor performance of the sealing strips, carbon black derived from natural gas should be used.

Ordinary carbon black contains a certain amount of PAHs (polycyclic aromatic hydrocarbons). The cracking and volatilization of PAHs will produce a large amount of pollutants such as benzene series, which has an unpleasant smell and is harmful to the human body, which seriously affects the air quality inside the vehicle. In the 2005/69/EC Directive, the EU stipulates the content of PAHs in products. Although there are no relevant regulations in China, the odor performance of EPDM is also closely related to the high content of PAHs. Therefore, in order to improve the environmental performance of the sealing strip products, environmentally friendly carbon blacks with a low PAHs content of less than 5% PAHs, or even carbon blacks without PAHs may be used.

1.3 The Filler

EPDM is a kind of rubber with very good filling properties. It can form products with good strength and elasticity when the rubber content is relatively low. Filler is a kind of inorganic mineral filler added to reduce cost, improve rubber's processing performance and products' strength, also known as a compatibilizer or inert filler. Fillers for automotive rubber sealing strips are generally selected from calcium carbonate, kaolin, clay, or wollastonite. There are many automotive sealing strip factories using calcium carbonate as filler. Light calcium carbonate or heavy calcium carbonate can be used, which both have little effect on the environmental performance of the product. From the viewpoint of low odor, it is mainly desired that the filler material has the characteristics of higher purity, appropriate fineness, uniform particle size distribution, lower moisture content, no heavy metal, non-toxic, and tasteless. New materials such as active mineral powder or nanomaterials also work very well.

1.4 The Soften Agent

In the formulation of automotive rubber sealing strip, the softener is added to improve the plasticity, fluidity, and adhesion of the rubber compound and to promote the dispersion of the compounding agent, which is beneficial to the rubber mixing and subsequent processing. The softener commonly used in automotive rubber sealing strips is paraffin oil, also known as white oil. Different grades of paraffin oil have different flash points, and the content of impurities and PAHs are different. The paraffin oil with high flash point and less impurities has better environmental performance. In order to achieve the purpose of reducing the odor of automotive sealing strips, medical grade paraffin oil can be used, which with the PAHs content is 0 or only 1–2 benzene rings of aromatic hydrocarbons (no PAHs).

1.5 The Vulcanizing Agent

The vulcanizing agent commonly used in automotive rubber sealing strips is sulfur. Due to the unpleasant smell generated during the vulcanization process, which will remain in the product, a vulcanization system with lower sulfur content and higher accelerator amount can be used.

The peroxide vulcanization process also produces the smell of vulcanizing agent, and the processes in the vulcanization process such as salt bath are likely to cause secondary pollution to the product. Therefore, the odor of the peroxide-cured sealing strip product does not show a significant advantage over the sulfur-cured one.

1.6 The Active Agent

In order to promote the cross-linking of rubber molecules and adjust the activity of the vulcanization system, producers add active agents into EPDM products, such as stearic acid, zinc oxide, and calcium oxide. Stearic acid reacts with zinc oxide to form zinc stearate. Calcium oxide is the indispensable carrier for EPDM that accelerates the vulcanization speed. However, since zinc oxide and calcium oxide products are liable to contain heavy metals such as lead or their compounds, and hydrolysis of stearic acid may cause paint staining, it is preferable to minimize the amount of both and to make the reaction as complete as possible during the process. Selecting high zinc content active environmentally friendly zinc oxide, increasing the activity and amount of calcium oxide to react a part of the nitrosamine produced by the accelerator can reduce the odor and the VOCs of the product.

1.7 The Accelerator

The vulcanization system of the rubber sealing strip is mostly a sulfur vulcanization system and various accelerators need to be added in, and the amount is relatively large. Some toxic and harmful substances contained in the accelerator, such as nitrosamines, which can cause cancer, seriously affect the environmental performance of the sealing strip and produce an abnormal odor. Therefore, the choices of vulcanization accelerators should be low toxicity and odor products. And comprehensively consider the residue of the accelerator after the vulcanization reaction and the formation of harmful small molecule volatile substances. On the basis of ensuring that the sealing strip can be vulcanized properly, the accelerator should be reacted completely with the volatile substances.

Previously used types of accelerators such as guanidines, thioureas, and aldehydes have certain toxicity and are basically no longer used. The DTDM is 4,4′-dithiodimorpholine, which usually produces toxic, harmful substances, and unpleasant smells during vulcanization, should be restricted. Most kinds of the thiurams and thiocarbamates also contain nitrosamines, which have been restricted in Europe. Only a few varieties, such as TBzTD, ZBEC, and ZDTP, are environmentally friendly accelerators. There are also some environmentally friendly accelerators in the series of thiazoles, sulfenamides, and alkyl thiophosphates, such as MBT, MBTS, ZMBT, and ZBOP. An environmentally friendly accelerator vulcanization system can be used instead of the currently widely used non-environmental accelerator vulcanization system to eliminate the formation of volatile nitrosamines and improve the odor and environmental performance of the sealing strip product. An environmentally friendly accelerator vulcanization system can be used instead of the currently widely used non-environmental accelerator vulcanization system to eliminate the formation of volatile nitrosamines and improve the scent and environmental performance of the seal. At present, the formulation of the environmentally friendly accelerator vulcanization system has matured, and it is feasible for the producer to prepare the comprehensive accelerator by himself or systematically supplied by the accelerator suppliers. At present, there are several kinds of granular accelerator masterbatch products to replace the powders, which can make further environmental improvement for the reaction sites and the sealing strip products.

1.8 The Foaming Agent

Thermal decomposition type organic chemical foaming agent is generally used in the foaming process of sponge EPDM rubber sealing strips. OBSH (diphenylsulfonylmethyl) and AC (azodicarbonamide) are commonly used foaming agents, either alone or in combination. The OBSH has a low decomposition temperature and a high decomposition rate. As the decomposition products of OBSH are non-polluting water and nitrogen, so it is widely used. The AC produces ammonia, which can affect the odor performance of the product, during the foaming process. Reducing or eliminating the use of AC can reduce the odor of products. Figure 3 shows the comparison of VOCs' emissions before and after the formulation improvement (methods 1.1 to 1.8).

2 Production Process Analysis of Sealing Strips

2.1 Extrusion and Vulcanization

Automotive sealing strips undergo different high-temperature environments from extrusion to vulcanization, such as preforming, shaping, coating drying, microwaves, and high-temperature vulcanization processes. High temperature will cause the cracking of PAHs and other toxic and harmful substances in the polymer chain, resulting in the continuous deposition of small molecular organic compounds, which contain harmful functional groups, from the surface of the sealing strip product. Therefore, appropriately reducing the process temperature by about 10 °C can reduce the cracking of some high-boiling polymer chains, thereby reducing the formation of toxic and harmful substances and reducing the product's odor and VOC emission.

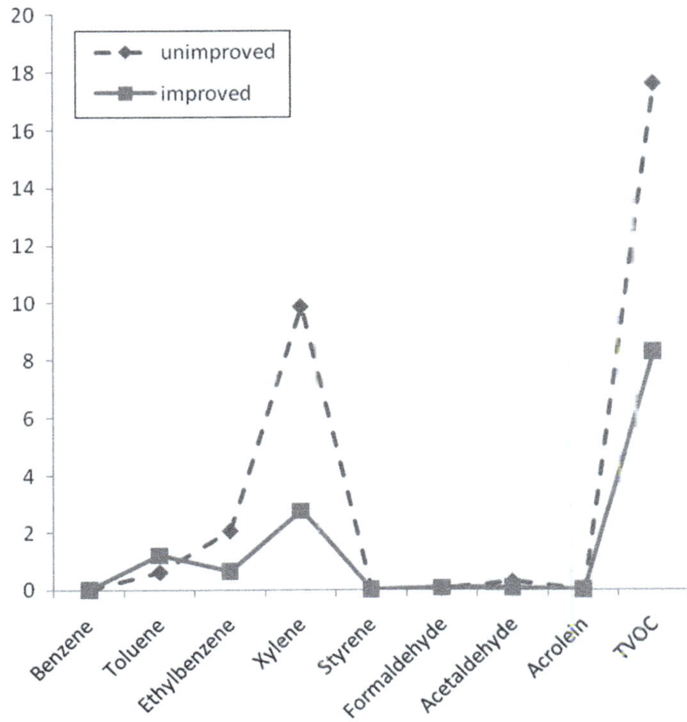

Fig. 3 Comparison of benzene series' and aldehydes' emissions before and after the formulation improvement (unit: mg/m^3)

2.2 *The Flocking Process*

The window glass guide is mostly flocked. The flocking glue used in the flocking process is mostly solvent-based glue, which is not environmentally friendly. Products produced by the surface coating process can exhibit better wear resistance than the flocked ones. And the surface coated product, is more environmentally friendly, emits less VOCs and harmful gases. Therefore, the use of PU coating or silicone coating instead of the flocking process is also the trend of the development of wear-resistant layer materials for window glass guides.

2.3 *The Post-treatment Process for Sealing Strip Products*

By adding a hot air exchange convection post-treatment process to the extruded sealing strip product, the odor and VOC emission of the product can be effectively reduced.

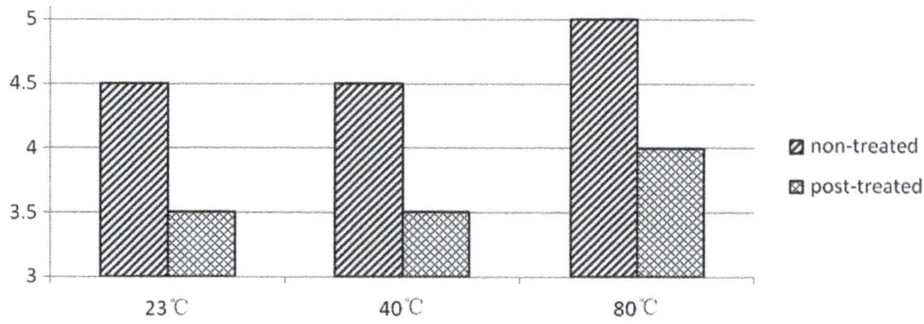

Fig. 4 Comparison of odor performances before and after hot air exchange post-treatment process of door sealing strips (unit: level)

Heating the air around the product can increase the saturated vapor pressure of organic volatiles from the sealing strip, remove low-boiling organic volatiles, and reduce the content of high-boiling organic volatiles. At the same time, the air exchange, continuously reduce the concentration of organic volatiles in the air around the product, and accelerate the emission of organic volatiles on the surface of the product. There several kinds of devices for sealing strip products that can achieve the function, such as steam heating convection deodorization device, electric heating convection deodorization device, and microwave convection deodorization device.

The post-treatment process needs to strictly control the heating temperature. If the temperature is too high, it will affect the mechanical properties of the sealing strip product and affect the stability of the sealing strip cornering process. If the temperature is too low, the deodorizing effect will not be obvious. It is recommended to use the process condition of '60–80 °C/4–6 h' to ensure product performance and production cycle (Fig. 4).

3 Other Factors Affecting Odor Performance

3.1 The Odor Concealer and Deodorant

At present, there are products such as odor concealers and deodorants for reducing the smell of nonmetallic materials. The main principle of the odor concealer is to cover the pungent odor with a specific scent. Deodorants are mainly divided into two types: physical adsorption type and chemical reaction type.

3.1.1 The Odor Concealer

The odor of the product itself can be covered by adding odor concealer to the raw material or by painting a specific scented concealer to the surface of the shaped product. Thereby we can achieve the purpose of reducing the pungent odor perception of nonmetallic products. Due to the unique rubber smell of sealing strip products, the odor generated after mixing with the concealer is more disgusting, so the effect of the odor concealer on the odor performance of sealing strips is not obvious.

3.1.2 Physical Adsorption Type Deodorant

By adding some porous macromolecular compounds into raw materials to adsorb small molecules of volatile organic compounds produced in the production process, the physical adsorption type deodorant can reduce the VOC emission of products at a certain degree, especially hard plastic products. However, since the rubber is a viscous fluid, this type of deodorant cannot be thoroughly mixed during the mixing process, and the flowing high-temperature rubber will block the pores, so the effect of the physical adsorption type deodorant on rubber products is not obvious.

3.1.3 Chemical Reaction Type Deodorant

By adding and thoroughly mixing the deodorant into the production process to chemically react with the small molecule volatile organic compounds produced in the production process in order to reduce the VOC emission of the product. The product of the chemical reaction is harmless. However, this type of deodorant cannot be thoroughly mixed during the rubber seal mixing process, so the deodorant has no obvious effect on the rubber product.

3.2 The Attachments

3.2.1 The Adhesive Tape

In the automotive sealing strip products, window glass guides and outer door sealing strips often use double-sided tapes, and even some models of door sealing strips are fixed to the door all by double-sided tapes. Solvent-based adhesives are generally applied to the tapes, and although the bonding performance is good, the odor is very irritating. Reducing the amount of double-sided tape from the designing stage, such as install the strip by buckles, can effectively reduce the odor of the product.

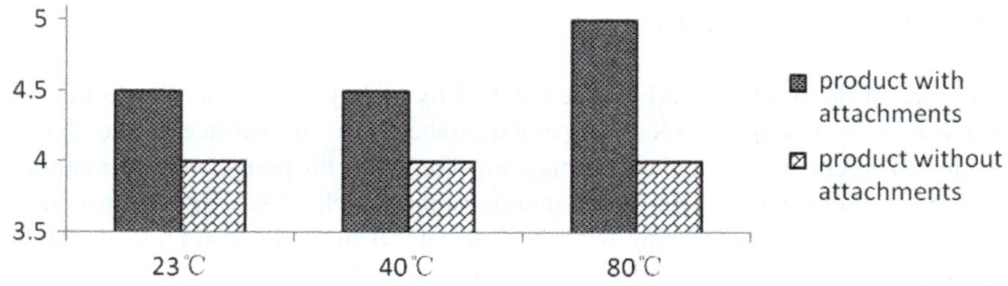

Fig. 5 Comparison of odor performances with and without attachments of outer door sealing strips (unit: level)

3.2.2 The Sponge Pad

In order to reduce the stress on the window glass guide by the window glass and prolong the life of the product, the sponge pad is generally used. The sponge pad material is generally EPDM, EVA, or PUR, of which EPDM material is the most. The sponge pad is installed by a double-sided pressure-sensitive adhesive film, which is generally provided directly by the tape supplier, and is mostly a non-environmentally friendly material. Therefore, improving the materials of the sponge pad can reduce the odor of the window glass guide product. Figure 5 shows the comparison of odor performances with and without attachments of sealing strips.

3.3 The Coating

An inner door sealing strip product is surface spray coated after the vulcanization process to improve wear resistance and reduce surface gloss. The coating will increase the xylene content of the product. By adjusting the rubber material formulation, the surface of the sealing strip can also have a matte effect, thereby reducing the amount of coating and improving the environmental performance of the product.

3.4 Storage and Transportation

After the products are formed, they should be placed in the warehouse for ventilation at room temperature. It should be noted that there is a gap between the products and they should not be stacked. Use clean boxes with venting holes to hold the products during transportation and storage. The product should be laid in layers in the box, separated by corrugated cardboards, and supported by spacers. During the storage and transportation process, the surface air exchange of the product can be increased by the above means, and the low-boiling point volatile organic compounds with strong activity on the surface of the product are further dispersed.

4 Conclusions and Cost Analysis

The methods to improve the odor and VOC performances of EPDM sealing strip products should be chosen from the following aspects:

(1) The selection of raw rubber types;
(2) The adjustment of additive types and proportions;
(3) The improvement of process conditions;
(4) Choosing an environmentally friendly vulcanization system;
(5) Choosing a proper post-treatment process;
(6) A clean storage and transportation environment.

If the EPDM sealing strips are improved by the methods described in this paper, the material and process costs will increase by 5–15%. If the labor, storage, and transportation costs are counted, the final products' price of each vehicle will increase by 10–20%.

4 Conclusions from Case Study

[illegible]

Study of the Influence of Structural Adhesive on Durability of Vehicle Body

Chao Qi, Xianhong Mao, Hua Wang and Binjiao Deng

Abstract Based on the stiffness results of the tensile shear joint of structural adhesive, a reasonable modeling method of structural adhesive was obtained. A finite element (FE) model of the thin peeling specimen was established on this method. Then, the FE analysis and the fatigue test results of the peeling specimen were compared. Finally, spot-welding fatigue analysis of SUV vehicle body was conducted, and the structural adhesive was used to optimize the spot welding.

Keywords Structural adhesive · Fatigue · Spot-welding

1 Introduction

Structural adhesive possesses the characters of high strength, resistance to peeling, durability to impact, and easy to process. It is widely used in the connection of metal, ceramic, plastic, rubber, wood, etc. in recent years, as structural adhesive has been more and more used in the structural design of the body in white (BIW) by foreign auto companies, domestic auto companies begin to use it for structure connection and strengthening. Applying structural adhesive at the critical locations of the vehicle body could remarkably improve the torsional rigidity with little mass increased. According to data of the industry, the application of structural adhesive could increase the body modal by 1.5–2.5 Hz and the body torsional stiffness by 10–20% [1].

Electric resistance welding is commonly used for connection on the vehicle body. However, there is no scientific arrangement design for welding, and it is usually arranged by experience. Thus, the failure of the welding spot occurs frequently. The structural adhesive connection is of great engineering significance for improving the fatigue behavior of the spot welding. Besides, the investigation of the durability of the structural adhesive is imperative.

C. Qi (✉) · X. Mao · H. Wang · B. Deng
Changan Auto Global R&D Centre, Changan Automobile Co. Ltd.,
Chongqing 401120, China
e-mail: visionaryqc@126.com

China SAE (ed.), *Proceedings of China SAE Congress 2018: Selected Papers*,
Lecture Notes in Electrical Engineering 574,
https://doi.org/10.1007/978-981-13-9718-9_54

2 The Calibration of FE Model of the Structural Adhesive

Structural adhesive is flexible gelatinoids. The connecting section area is usually controlled by the diameter of the adhesive. During the connecting process, the adhesive would be spread uniformly to the connecting section. To simulate the behavior of the adhesive by FE method, it is required to calibrate the FE model with joint rigidity test.

2.1 The Rigidity Calibration of the Structural Adhesive Modeling Method

An FE model of a tensile shear joint specimen was built. The sheet metal thickness was 1.5 mm and was modeled by shell element. The structural adhesive was modeled by RBE3 + CHXA + RBE3 element. The modeling method is shown in Fig. 1a. The structural adhesive's elastic modulus was 1830 MPa, and its Poisson's ratio was 0.34. Since the structural adhesive is uniformly filled in the connecting section, CHEXA

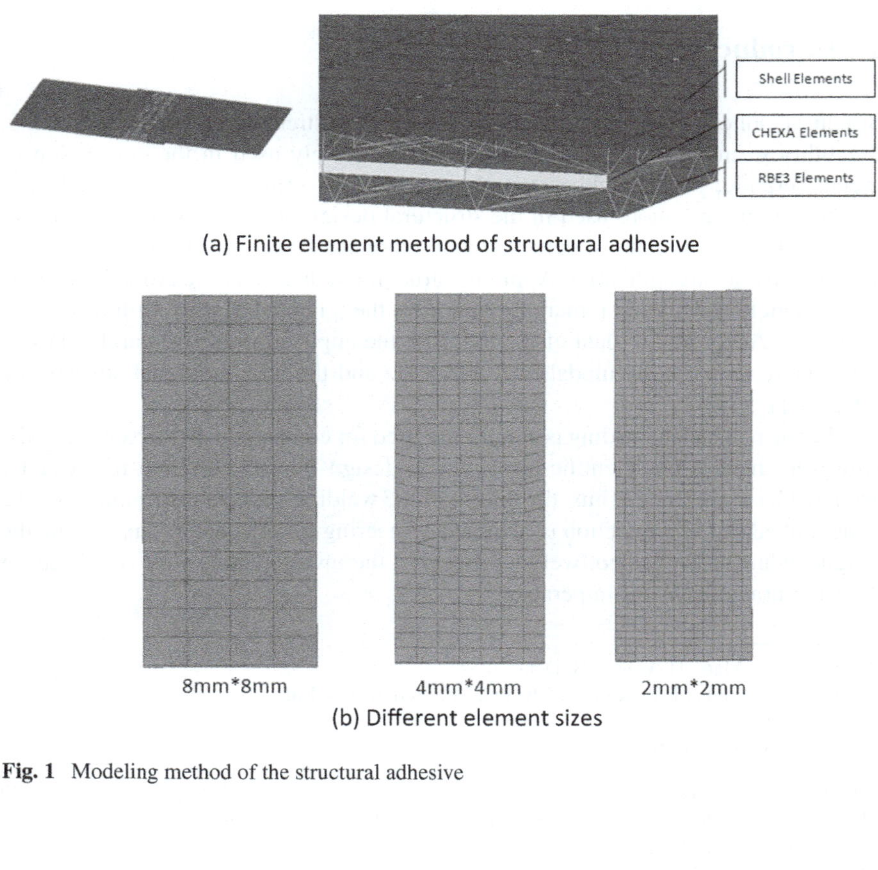

(a) Finite element method of structural adhesive

(b) Different element sizes

Fig. 1 Modeling method of the structural adhesive

Table 1 Stiffness results of various schemes

	1 layer			3 layer		
	2 * 2	4 * 4	8 * 8	2 * 2	4 * 4	8 * 8
0.1 mm	61,824.6	62,417.53	63,150.93	61,821.78	62,410.64	63,150.73
	0.73%	1.70%	2.89%	0.73%	1.69%	2.89%
0.2 mm	61,184.57	61,652.7	62,248.63	61,175.14	61,640.27	62,248.17
	0.31%	0.45%	1.42%	0.33%	0.43%	1.42%
0.3 mm	60,705.77	61099.67	61 604.5	60,689.48	61,082.9[illegible]	61,603.67
	1.09%	0.45%	0.37%	1.12%	0.48%	*0.37%*
0.4 mm	60,304.5	60,647.3	61,089.48	60,281.56	60,626.82	61,088.25
	1.75%	1.19%	0.47%	1.78%	1.22%	0.47%
Test	61,376					

element with thickness of 0.1, 0.2, 0.3, and 0.4 mm was used, respectively, in the analyses. Besides, for each different thickness, 1 or 3 layers of CHEXA element were used, respectively. Additionally, in the plane direction of the adhesive, mesh size of 2 mm * 2 mm, 4 mm * 4 mm, and 8 mm * 8 mm was used, respectively (see Fig. 1b).

Comparing the joint rigidity analysis results of the structural adhesive with different thickness, layers, and mesh size gave the conclusions that:

(1) The rigidity of the structural adhesive with the thickness of 0.1 mm was 62,417 N/mm, and it is 60,647 N/mm for the thickness of 0.4 mm. The thinner the element, the high the rigidity.
(2) The rigidity of the structural adhesive with mesh size of 8 mm * 8 mm was the highest.
(3) As shown in Table 1, the modeling method with thickness of 0.2 mm and mesh size of 2 mm * 2 mm had the closest results to the test, and the error was 0.31%. When using the thickness of 0.3 mm and mesh size of 8 mm * 8 mm, the error was 0.37%.
(4) Since the mesh size of common NVH and fatigue analysis models is 5–8 mm, consider the analysis accuracy and efficiency, the modeling method with structural adhesive thickness of 0.3 mm and mesh size of 8 mm * 8 mm was recommended. The rigidity calibration results are shown in Fig. 2.

2.2 *The Fatigue Calibration of the Structural Adhesive Modeling Method*

Comparing with the tensile shear joint, the peeling joint is more critical for the durability performance of the structural adhesive. Thus, the peeling joint was used to

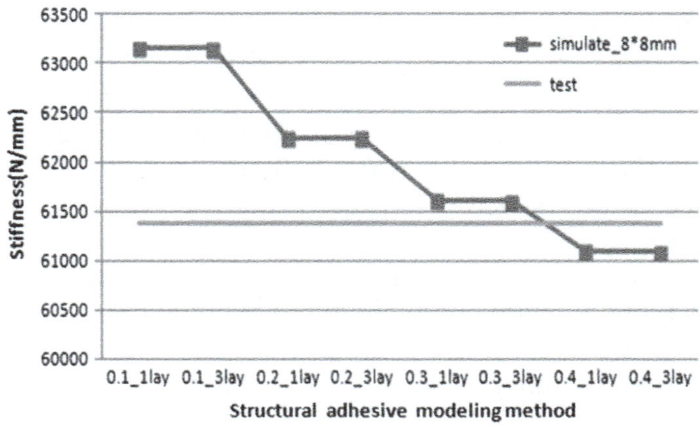

Fig. 2 Rigidity calibration results of the structural adhesive tensile shear joint

conduct the fatigue calibration test. The test load cycle ratio was 0, and the sinusoidal wave was used. Load amplitude of 0–280, 0–330, 0–370, and 0–440 N was used, respectively, for the test. The thickness of the sheet specimen was 0.7 and 1.2 mm. For each group of load cycles, 5 joint specimens were tested, and the cycle times of each group were recorded. The test results are shown in Fig. 3.

An FE model of the peeling joint specimen was built. According to rigidity calibration result, the thickness of the structural adhesive elements was 0.3 mm. A stress analysis was conducted with a load of 440 N, and the analysis results were imported into the fatigue analysis software. Combining the SN curve of the structural adhe-

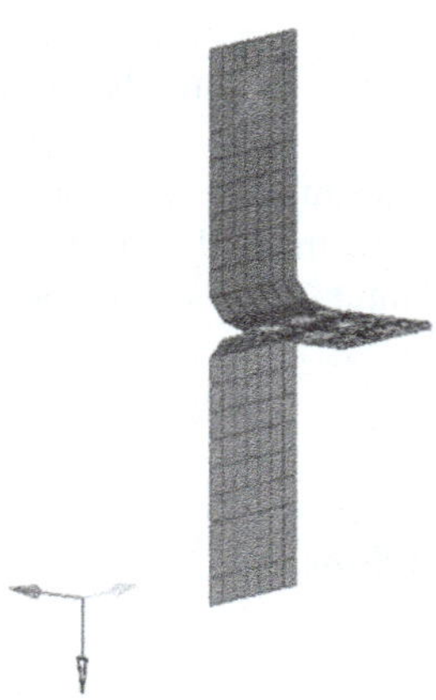

number of sample	force/N	cycles	crack or not
1	330	1277630	crack
2	330	1433620	crack
3	330	1148730	crack
4	330	1558560	crack
5	330	1047430	crack
6	370	254300	crack
7	370	305340	crack
8	370	277560	crack
9	370	341460	crack
10	370	283660	crack
11	400	80420	crack
12	400	63780	crack
13	400	73240	crack
14	400	93710	crack
15	400	99523	crack
16	440	26890	crack
17	440	152213	crack
18	440	46230	crack
19	440	28570	crack
20	440	35440	crack

Fig. 3 Results of the fatigue calibration test

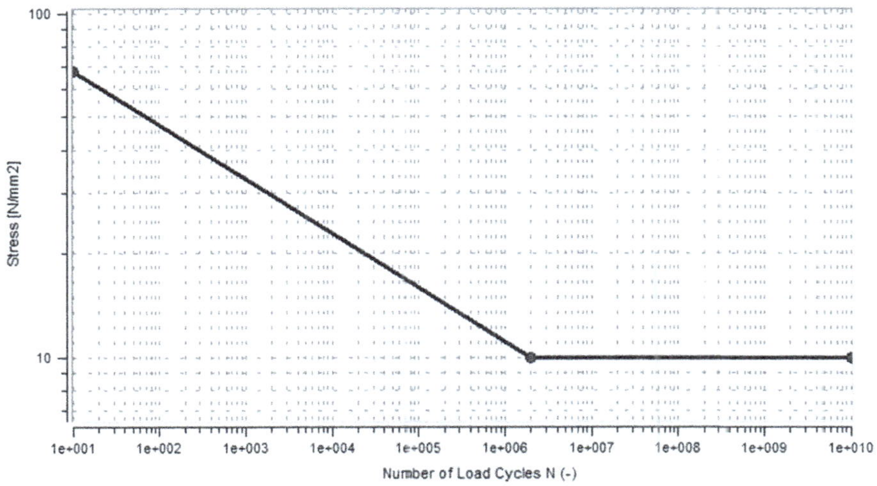

Fig. 4 Fatigue life curve of the structural adhesive

sive, the fatigue life curve of the adhesive was obtained (see Fig. 4). Considering the material plasticity of the structural adhesive, the stress was amended by Neuber method. Considering the influence of the average stress, the curve was amended by Haigh curve (see Fig. 5).

Applying the load cycle of 0–440 N showed that the structural adhesive had the least fatigue life in the peeling joint. The analysis result is shown in Fig. 6, and the minimum fatigue life was 8212. The average fatigue life of the test was 30,468 The proportion of the analysis result to the test result was 3.7. This analysis result could

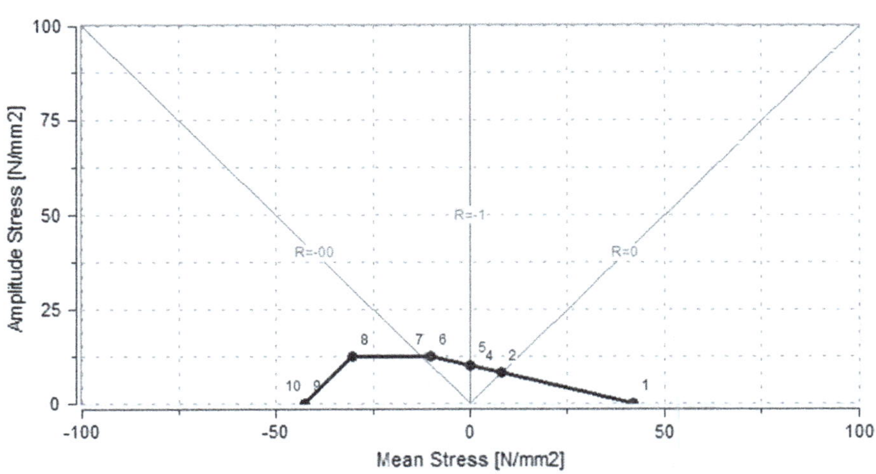

Fig. 5 Haigh curve of the structural adhesive

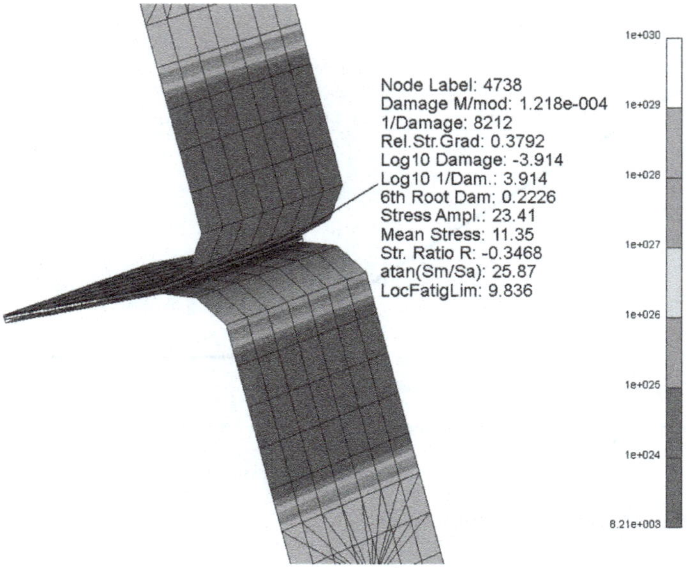

Fig. 6 Fatigue life analysis results of the structural adhesive

be used in the earlier stage of the engineering developing process to avoid the crack problem caused by the unreasonable arrangement of the structural adhesive.

3 The Utilization of the Structural Adhesive in BIW

Structural adhesive is mostly used at the critical joints and the low rigidity parts like the wheelhouse panel. The usage of structural adhesive on BIW could increase the bending and torsional rigidity by 10–20% and the modal by 2–3 Hz. Structural adhesive could also be used to connect the parts that are hard to apply spot welding and ensure the durability performance. Applying the structural adhesive at the spot of crack risk could improve the durability performance of the connection zone.

3.1 *Body Spot Welding Fatigue Analysis Model*

The connection of the spot welding was modeled by 1D CBar element. The CBar elements divided the connected sheet metal into two rounds of mesh areas. The inner round simulated the welding core, and the sheet metal and the 1D elements' elastic modulus were 8.4E6. The outer round simulated the heat affected zone, and the sheet metal's elastic modulus was 2.1E6. The fatigue analysis model is shown in Fig. 7.

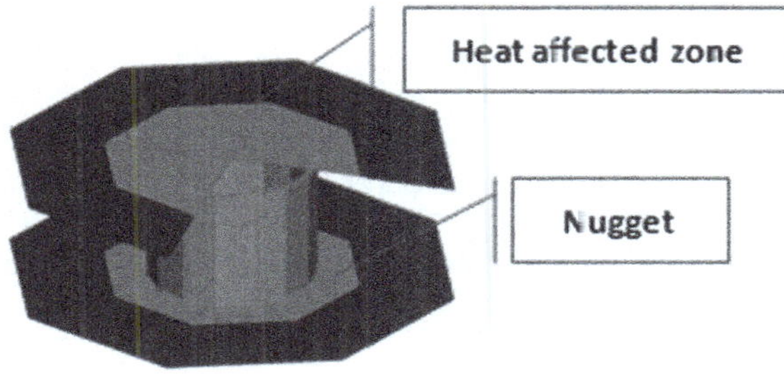

Fig. 7 Fatigue analysis model of the welding spot

In practical, the failure of the spot welding usually occurs between the heat affected zone and the welding core as a fatigue tear. Thus, the fatigue analysis read the element stress of the heat affected zone and combining with the spot welding SN curve to acquire the spot welding fatigue life. In engineering, this analysis method is called the fatigue analysis based on the element structural stress [2].

3.2 The Body Welding Spot Fatigue Analysis of a SUV

Based on the six-dimensional force acquired by proving ground, the random load at the joints of the body and the chassis was obtained via the multi-body dynamics load resolving. The body was excited by unit load at the joints, and the element stress around the welding spots was acquired by the inertia release method. The stress results and the random load were imported into the fatigue analysis software, combining with the welding spot SN curve, the fatigue damage of the body welding spots were obtained [3].

The body welding spot analysis was conducted for an SUV. The result showed that, the fatigue damage of two welding spots at the connection area of the rear cross member and the floor panel were 3.41 and 1.34, respectively. These two welding spots had the risk of crack in proving ground. The analysis result is shown in Fig. 8.

3.3 The Welding Spots Optimization with Structural Adhesive

For the welding spots with crack risk, the structural adhesive was applied around the connection section. Practically, the structural adhesive was applied before the welding, and the heat released during the welding would burn it out. Hence, when building the structural adhesive model, the welding section should be avoided to match the real condition. The structural adhesive modeling arrangement is shown in Fig. 9.

The analysis result of the model with structural adhesive showed that the welding spot fatigue damage was 0.04, which was 85 times of the original ones. The analysis

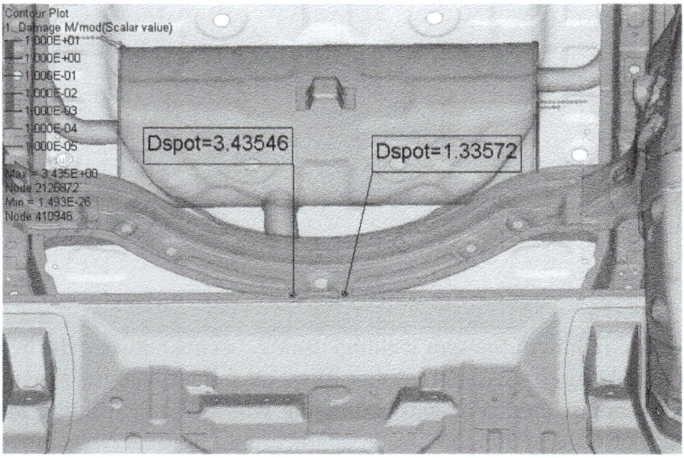

Fig. 8 Welding spot fatigue analysis result of a SUV

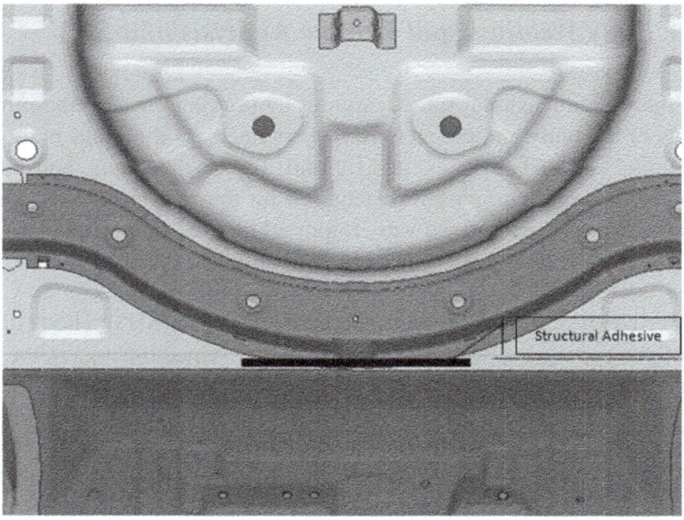

Fig. 9 Structural adhesive modeling arrangement

result of the optimized structural was shown in Fig. 10. The fatigue damage of the structural adhesive was 1.3E−5, which fulfilled the durability requirement. Thus, the usage of structural adhesive could not only improve the durability performance of the weak area of the vehicle body, but also increase the welding spot fatigue life.

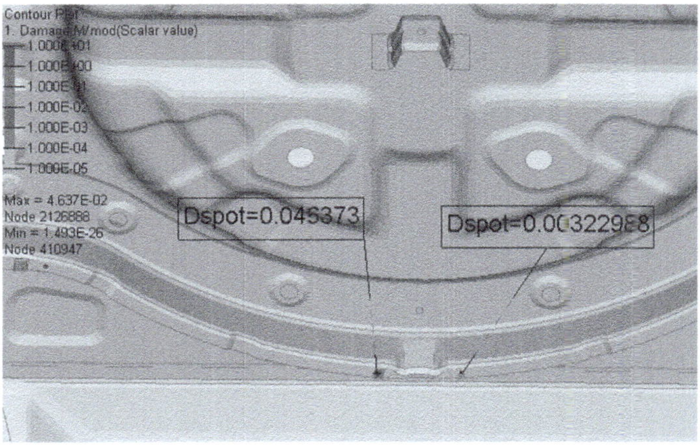

Fig. 10 Analysis result of the optimized structural

4 Conclusion

This paper built structural adhesive models with different thickness, mesh size, and layers. The analysis results of the models with the three variables were compared with the proving ground results. The comparison showed that the model with mesh size of 8 * 8 mm and thickness of 0.3 mm could fulfill the accuracy requirement as well as the vehicle analysis efficiency requirement. The calibrated modeling method was used to conduct a peeling joint fatigue analysis. The dispersion of the FE analysis and the test was 3.7, which indicated the analysis was able to avoid risk in the earlier stages of the engineering process. A body welding spot fatigue analysis was conducted for an SUV, and the structural adhesive was used to optimize the vehicle body. The optimized structural improved the welding spot fatigue life by 35 times.

References

1. Li C, Xu Y, Ji Y et al, Application of structural adhesive in body NVH development. In: The fourth annual conference of China CAE engineering analysis technology
2. Brenner CE, Unger B, Gaier C, Sternwender G, Fatigue assessment of welding seams and spot joints based on FEA. SAE Paper 2000-01-0780
3. Long H, Hu Y et al (2016) Fatigue life prediction for the spot welds of body-in-white based on random loadings. Automot Eng 19(S1):56–58

A Research on Vibration Performance of Light-Duty Truck Doors Based on Road Load Spectrum

Xin Yan, Di Jiang, Yangyang Bai, Weitian Yi, Zhenyu Guo and Zhaochuan Jiang

Abstract Based on the vibration problem of some users' light truck door frame, road load spectrum in similar pavement condition is obtained and extracted as the target signals. Including the doors, the multi-body dynamics model of cab suspension system is built so as to do virtual load iteration to get the real displacement driver of the cabin. By fixing the elastic element system features, more accurate system dynamic response is acquired. By means of structure optimization on the key points of the door, the door frame dynamic response in all kinds of harsh conditions can meet user's requirements. At the same time, the scientific and reasonable method of product forward design is provided for the door system.

Keywords Vibration of door system · Road load spectrum · Virtual load iteration · Structure optimization

1 Introduction

In recent years, with the development of the logistics industry, the demand of light truck used for the carriage of long and short distance is increasing. In addition to the performance of safety, reliability, and energy conservation, the users also have some requests for ride comfort and harshness performance. Therefore, a lot of light truck makers begin to pursue the elaborating design based on user perception of products. The door system belongs to the end of the multi-level vibration isolation system for vehicle, and the vibration characteristics directly affect the driver's subjective feeling. Unreasonable structure of design will lead to serious vibration and noise problem which may cause the users' complaints and claims. Some new arrivals light trucks in Qingdao, Taizhou, and Shenyang districts users came up the problem of door frame vibration in the condition of 80–110 km/h speed on highway road and

X. Yan (✉) · Y. Bai · W. Yi · Z. Guo
FAW Jiefang Auto Co., Ltd. Commercial Vehicle Development Institute, Changchun City, China
e-mail: yanxin@rdc.faw.com.cn

D. Jiang · Z. Jiang
General R&D Institute of FAW, Changchun City, China

China SAE (ed.), *Proceedings of China SAE Congress 2018: Selected Papers*,
Lecture Notes in Electrical Engineering 574,
https://doi.org/10.1007/978-981-13-9718-9_55

highway exit buffer road. The doors flogged so that the visible light can been seen in some position. And there were some related phenomenon such as rain and air leakage. In order to solve this problem, the project team put forward the research of vibration characteristic on door system.

There are many reference literatures on the vibration characteristics of door system. Literature [1] is based on the theoretical and experimental modal analysis. And it takes modal frequency of the door as the optimization goal and the plate thickness as the design variables so as to optimize the dynamic performance of the door. Literature [2] aims at the coach door in the abnormal vibration problems and analyzes the basic theory of vibration and does some experiment research. Based on structure topology optimization method, literature [3] makes optimization design for spot weld joint layout of the door to solve harshness problem.

The present research are more based on modal test and analysis technology. And this research mainly focuses on the dynamics analysis of the complex system of light truck cab, suspension, and the door. It aims at the measured road load spectrum, and takes the method of virtual load iteration to obtain the dynamics response of the door in actual working condition. Hereby, the vibration problem on the door frame structure can be solved to meet users' requirement.

2 Road Load Spectrum Test and Signal Analysis

There are two main purposes for road load spectrum test, one is to get the door's system dynamic response signal, on the other hand, the load signals of cab and suspension system can be obtained as the input for subsequent dynamics analysis and virtual load iteration.

The sensors placement of door system is shown in Fig. 1. One relative displacement sensor lay out near the rear-view mirror fixed point of the door frame which is

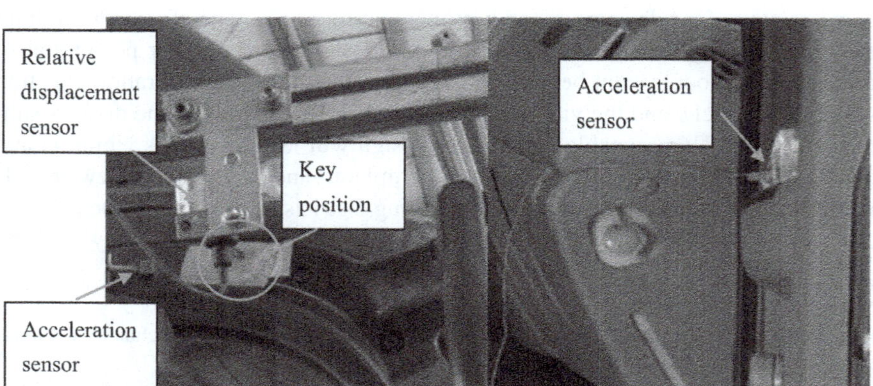

Fig. 1 Sensors placement of door system

called the key position to extract relative vibration displacement between door and body. Two acceleration sensors are, respectively, fixed up on the door frame and the door hinge fixed end, and the measured signal is used to calculate frequency response function of the transmission path from the body to the door.

The sensors layout of cab and suspension system is shown in Fig. 2. The acceleration sensors are fixed on the upper and lower brackets of the cab suspension, and some of the measured data are adapted as the target signal of cab suspension system virtual iteration, the others are used to modify the vibration isolation characteristic parameters of cab suspension system. The relative displacement sensor is mounted along the vertical direction between the upper and lower brackets, and the measured data will be used as the kinematics monitoring signals of simulation model to ensure the transfer function accuracy of cab and suspension system.

Based on the users' feedback survey, it is found that 80% of users perceived the door vibration problem on highway road with the speed of 80–110 km/h, and 20% of users on highway exit buffer road. So, we choose high-speed circular runway road and washboard road as compared for the assessment of test road in order to repeat the users' problem maximally.

Circular runway road test is running, respectively, in 80, 90, 100, 110 km/h speed, and washboard road speed is 55 km/h. At least five cycle signals data should be measured under the steady speed in each working condition. And then we choose one single cycle signal data with the lowest value of crest factor as the preprocessing data.

Fig. 2 Sensors placement of cab suspension system

$$\text{Crest factor} = \text{Max}(|\text{Maximum Value}|, |\text{Minimum Value}|)/\text{Standard Deviation} \quad (1)$$

Signal data check is done to ensure the rationality of the signal, such as signal overrun confirmation and direction calibration. And we preprocess the signals such as Butterworth filter, offset or drift correction, peak burr removing, etc., and finally obtain the signal that can be used for post-processing. Table 1 is shown for the post-processing data signal statistics of door frame key position.

As shown in Table 1, it is found that the displacement extreme value of the door frame key position in runway 100 and 110 km/h speed conditions is larger than the pre-compression theory value of sealing strip (6 mm). And this will produce the leak problem between the body and the door. It will be more serious if coupled with assembly deviation (0–2 mm) and the wind load tensor caused by negative pressure (0–1 mm). The vibration acceleration extreme value and standard deviation of washboard road is larger, but it is always the in-phase vibration between the door and the door, so the displacement extreme value and standard deviation is not so large. It can also illustrate that the excitation frequency of washboard road (20 Hz) is away from *Y*-direction vibration frequency of the door. With the increase of speed, excitation frequency in runway conditions is more and more close to *Y* vibration frequency of the door.

The power spectral density (PSD) analysis of the input and output signal in runway working condition is shown in Fig. 3, and there are two excitation source, source 1 is caused by the unbalance of tire dynamic incentives, which belongs to the narrowband signal with side-lobe energy decaying quickly. At this time the door end is withstand forced vibration response significantly. Source 2 is the runway road surface characteristics mixing a certain order modal frequency of frame system (lateral vibration modal frequency of body is always higher, generally greater than 40 Hz). With the wider frequency band, the fundamental modal frequency of the door is coupled with the excitation frequency so as to cause the resonance phenomenon. Figure 4 shows the frequency response function (FRF) from the body to the door along *Y*-direction.

Table 1 Statistics of door frame key position

Working condition	Displacement extreme value (mm)	Displacement standard deviation (mm)	Acceleration extreme value (*g*)	Acceleration standard deviation (*g*)
Runway 80 km/h	3.1	0.42	1.15	0.3
Runway 90 km/h	4.1	0.55	1.2	0.34
Runway 100 km/h	6.36	0.79	1.25	0.36
Runway 110 km/h	7.18	0.91	1.6	0.46
Washboard road	3.98	0.48	3.66	1.48

Fig. 3 PSD analysis

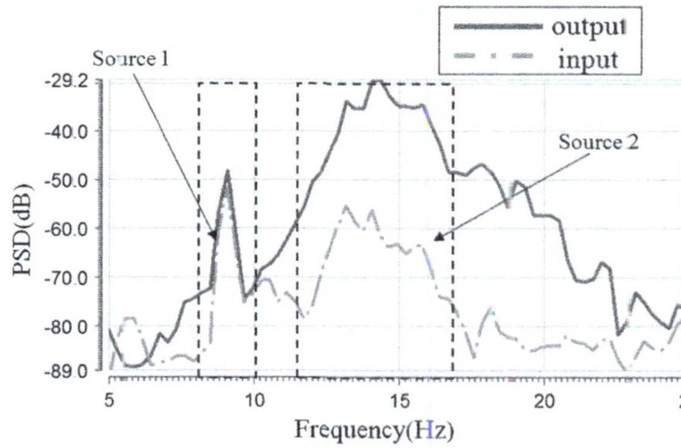

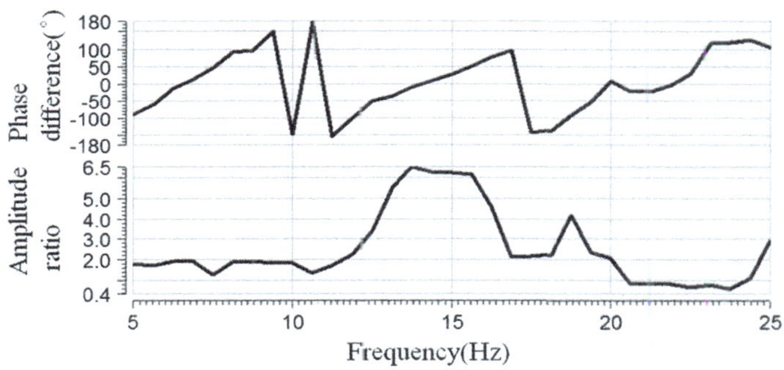

Fig. 4 FRF analysis

It is visible that there is the highest amplitude ratio around 13.5 Hz which can be recognized as the first-order modal frequency of the door along the lateral direction.

Conventional modal analysis method for vibration problem research generally adapts frequency-avoiding design. The goal of structure modal frequency is to avoid the main excitation frequency. The range of common road excitation frequency is about 6–22 Hz. Four-cylinder and four-stroke engine's idle speed excitation frequency range is about 23–27 Hz. So, the range of main excitation frequency is 6–27 Hz. And this door system is mounted with the large rear-view mirror more than 5 kg so as to come up with local low-frequency mode (13.5 Hz). If the modal frequency of the structure is too low (less than 6 Hz), its stiffness value is too low in the same way, and the subjective evaluation requirement cannot be meet; at the same time, it is difficult to make the modal frequency value of door structure higher than the 27 Hz. So we adopt the way to reduce vibration response amplitude of the target. And this research attempts to simulate real state of the kinematics and dynamics of the system and make improvement for the structure to reduce vibration response at the key position to achieve the established target requirements. So, it is necessary to build a multi-body dynamics model of the system firstly.

3 Building of Multi-body Dynamics System Model

For the sake of simulating the kinematics and dynamics characteristics of the door system, it is necessary to build a complex multi-body dynamic system model with the cab and suspension assembly including the door itself. It is needed to identify the system hard point coordinates, the parts joint form, mass and moment of inertia parameters, stiffness damping characteristics of elastic element, load driver input, and sensors setting, etc.

The structures such as the door, the whole trim-body, stabilizer bar, and mount bracket adopt flexible modeling which should ensure that mass and moment of inertia value of the model is consistent with the measured one. The door hinge and lock's constraint mode should be the same as the actual situation. The first-order modal vibration mode of the door system simulation is shown in Fig. 5. The modal frequency of lateral vibration for the door is 13.8 Hz which is relatively consistent with the one from the measured road load spectrum (13.5 Hz). It proves the high accuracy of the flexible model. At the same time, it also can be found that the point of largest modal displacement is where the rear-view mirror fixed on the door frame, and this position is same to the users' feedback about the vibration problem.

The bushing forces are established in cab suspension mount point with six-direction stiffness and damping values. Normally, we pay more attention to the low-range value of excitation frequency (less than 40 Hz), so in this research, the measured static nonlinear stiffness value is adapted, and the test methods are shown in Fig. 6. The stiffness value is compound of working stiffness area and rigidity limited area. For the tilting condition properties of the light truck cab, a pre-torque is setting in the front suspension bushing. The door sealing strip and limit block are input with measured stiffness values as shown in Fig. 7. The door sealing strip adapts

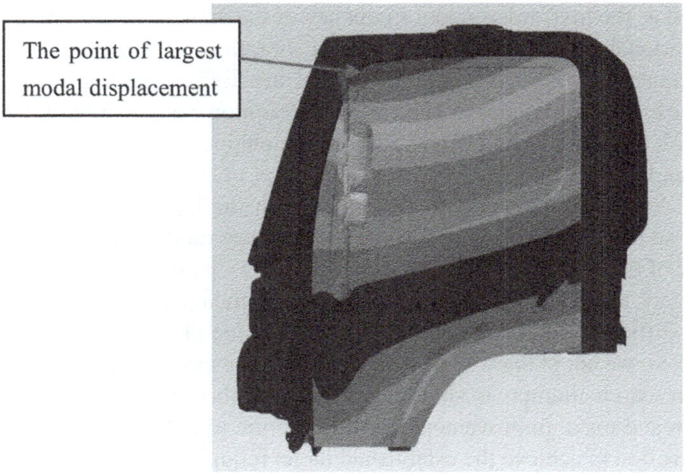

Fig. 5 First-order modal vibration mode of the door system

Fig. 6 Test of rubber bushing

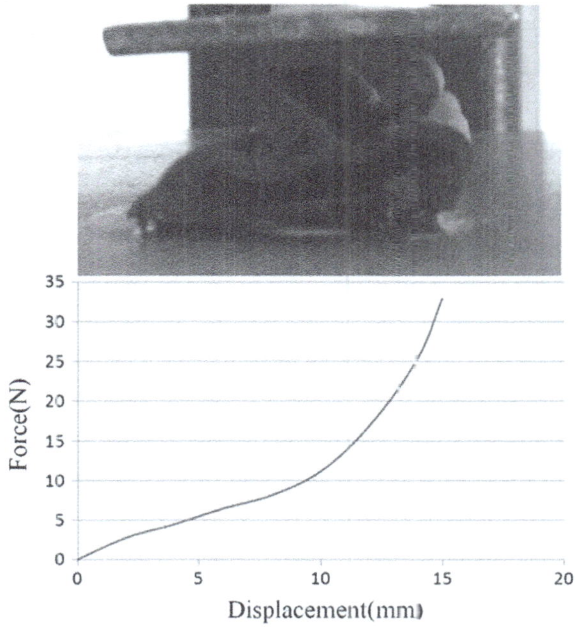

Fig. 7 Test of sealing strip

bushing unit for simulation which sets a unit in every 200 mm. At the same time, it is necessary to set pre-compression load for the door sealing strip. We connect suspension mount bracket and its fixture to the six degrees of freedom vibration table, and define three linear drives (*X*/*Y*/*Z*) and three rotation drives (*RX*/*RY*/*RZ*) at the vibration table center.

Multi-body system dynamic model is shown in Fig. 8, because damping value of suspension rubber bushing is estimated, it is needed to be corrected. We drive the vibration table in *XYZ* direction with the sine sweep signal, the amplitude is 2 mm and the frequency range is from 0 to 40 Hz. And then we extract the acceleration response of upper and lower brackets of the cab suspension to calculate the transmission amplitude ratio as the vibration isolation characteristics. Then, we ensure that the simulation curve is similar to the test one by modifying the damping parameters. The simulation and test consistency comparing curves are shown in Fig. 9. The revised

Fig. 8 Multi-body system dynamic model

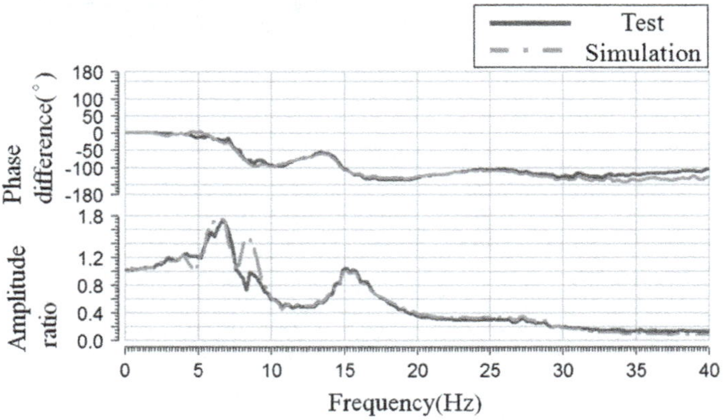

Fig. 9 Simulation and test consistency curves

model ensures that the transfer function precision of cab suspension system is enough which can be used in virtual load iteration subsequently.

4 Virtual Load Iteration

This research is to emulate the vibration response at key position of the door under the actual road load. But we cannot acquire the external drive input of the door by testing. So, we need to use the method of virtual simulation. The main methods in the world are:

(1) Virtual proving ground method: The path is to build the multi-body dynamics system model in vehicle level and the digital road surface model. There are three difficulties for that. First of all, it is difficult to simulate the dynamic characteristics of complex tire. Second, the digital road surface model costs too much and its universality is not so good; Third, truck model has multi-level vibration isolation chains (tire–chassis–suspension–cab–door), and performance correction for each level of vibration isolation chains is very complicated.

(2) Virtual load iteration method: Also known as half-analytical method, the core content is to reversely calculate the input drive by output response. By measuring the object's dynamic response signals such as acceleration and displacement or strain signals as the target, by building multi-body system dynamic model as the transfer function, we can get the drive of external motivation. If the linear degree of system is high, the drive can be obtained by only one-time solving in theory. But, generally, the system is nonlinear so that there is error between the response and the target, so it is needed to do iterations to ensure the consistency for the target and response. This process is also referred to as virtual load iteration, and this research is using this way to solve the problem.

The process of virtual load iteration is:

(1) Generate white pink noise (WPN) u_{noise} that is input to the multi-body dynamics model for getting response y_{noise} and calculate the transfer function H,

$$H = y_{\text{noise}}/u_{\text{noise}} \tag{2}$$

(2) Make the road load testing data signals y_{target} as the target and obtain the first drive u_1 by the inverse transfer function H^{-1},

$$u_1 = H^{-1} y_{\text{target}} \tag{3}$$

Drive the system model by u_1 to get the first response y_1, obtain the RMS values of comparison error of y_1 and y_{target}, adjust error gain a and feedback rms_1 to H^{-1} so as to get u_2,

$$u_2 = u_1 + aH^{-1}\text{rms}_1 \tag{4}$$

3) Repeat iterations to get the final drive u_{n+1},

$$u_{n+1} = u_n + aH^{-1}\text{rms}_n \tag{5}$$

make sure that rms_n is tend to zero, the target and response signal achieve high consistency in time domain, frequency domain, and rain-flow domain in an acceptable level, and then terminate the iterations, so the final drive u_{n+1} is the goal.

It is mentioned above that the vibration table has six drive input channels. Based on theory of iteration algorithm, the channel number of target should be not less than the drive input. In this research, we define a total of 12 acceleration sensors(four position, three direction) on the cab near upper bracket of suspension as iteration channels, four relative displacement sensors between the cab and suspension as monitoring channels, one relative displacement sensor at key position of the door as validation channel. We use the test data on the circular runway with 110 km/h speed as the target signal source, and start the virtual load iteration.

The results of RMS error values for each simulation and test channel are shown in Table 2, the RMS error value for each iteration channel is less than 10%, which can meet the requirement of the iteration and ensure the accuracy of the cab's dynamic response; four monitoring channels can also meet the requirement (less than 15%); But the RMS error of validation channel (relative displacement of key position) is 25% which is far away from requirement(less than 5%). It means that the elastic

Table 2 RMS error of iteration channels, monitoring channels, and validation channel, %

Iteration channels	RMS error	Iteration channels	RMS error	Monitoring/validation channel	RMS error
Left front accelera-tion—*X*	5.8	Left rear accelera-tion—*X*	9.3	Left front relative displacement	10.5
Left front accelera-tion—*Y*	2.9	Left rear accelera-tion—*Y*	2.9	Right front relative displacement	7.9
Left front accelera-tion—*Z*	5.1	Left rear accelera-tion—*Z*	7.7	Left rear relative displacement	12.4
Right front accelera-tion—*X*	6.2	Right rear accelera-tion—*X*	8.4	Right rear relative displacement	8.6
Right front accelera-tion—*Y*	3.5	Right rear accelera-tion—*Y*	3.6	Relative displacement of key position	25
Right front accelera-tion—*Z*	4.7	Right rear accelera-tion—*Z*	7.2		

Table 3 Key statistics results of simulation and test signals in validation channel

Validation channel	RMS (mm)	Average (mm)	Extreme (mm)
Test	0.91	1.25	7.18
Simulation (original)	0.68	1.11	5.7
Simulation (revised)	0.88	1.22	7.15

element parameters of the transmission path from the body to the door need correction. It is found that the structural modal damping ratio at the first-order modal frequency of the door is the main influence factor. The flexible processing for the door structure adopts value of software by default-damping ratio of 8%, and actually there may be difference. By the experimental analysis, this value is adjusted to 3.2% to meet the requirement in the validation channel. The key statistics result is shown in Table 3, and the revised simulation result has high consistency with the measured one in terms of root mean square value, average value, and extreme value which ensure the model's accuracy of each link. So, the model and drive can be used as the basis of structural improvement and optimization for the door.

5 Improvement Analysis of Door System Structure

According to the first-order modal vibration mode of the door and experimental analysis for structure improvement, it is found that the stiffness of door hinge fixed end is insufficient which is the main reason of the door frame vibration problem. So, we improve the door system structure in two ways. One is to increase the stiffness of the door hinges. We optimize the shape for the door hinge and augment its thickness from 4.5 to 5.5 mm, although the hinge weight increases about 60%. Also, we increase the area enclosed by hinge bolt fixed points. The second way is to add a reinforcement plate on door hinge activity end to establish the relationship between the window frame and door inner panels so as to significantly increase the lateral stiffness of the door. The final improved structure is shown in Fig. 10.

After that, the new dynamic solution is updated to obtain the response of the key position on the door. The pre-compression theory value of sealing strip is 6 mm, and if the assembly process can lead to the maximum static tensor 2 mm, the negative pressure of the wind load can cause maximum displacement 1 mm, the maximum permitted target displacement caused by vibration should be less than 3 mm. The simulation extreme value of the improved structure is 2.71 mm which can fulfill the requirement of the target. The analysis of frequency response function is shown in Fig. 11, the door's focused frequency has been changed from 13.5 to 15 Hz, and the peak value of transmission amplitude rate is significantly reduced. The test results of three trial-manufacture prototype trucks with improved structure are shown

Fig. 10 Improvement of the door structure

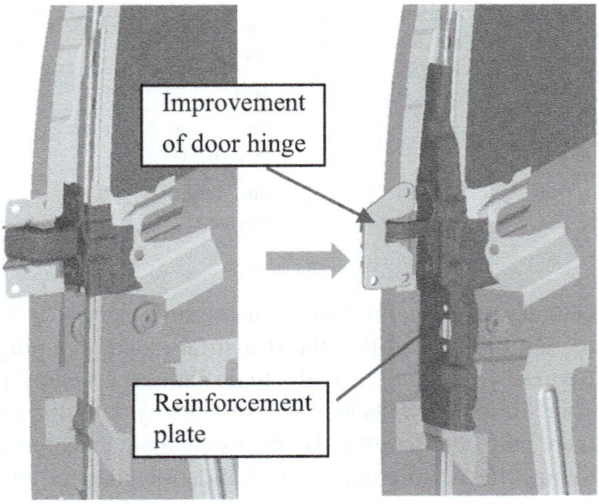

Fig. 11 Comparison of frequency response function curves

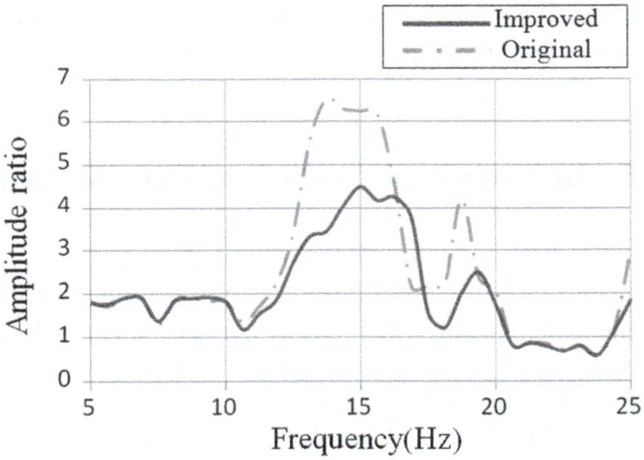

in Table 4. The test values of the key position has the high consistency with the simulation result, and they can satisfy the requirement of the target value (3 mm) at the same time, so the vibration problem of the door gets effectively solved.

Table 4 Comparison of simulation and test extreme value of key position on the improved door

Simulation (mm)	Test-A (mm)	Test-B (mm)	Test-C (mm)
2.71	2.49	2.65	2.77

6 Conclusion

(1) The road load spectrum of the light truck door system is tested and analyzed from excitation, response, and transfer path to get the reason of abnormal vibration on the key position of the door.
(2) By virtual load iteration method, the system external drive is obtained which can reflect the real dynamics and kinematics response characteristics of the door system. The vibration response is significantly reduced by structure improvement so that the problem of the user is effectively solved.
(3) This research provides a scientific and reasonable method to design the NVH performance for light truck door system in forward direction which can shorten the development cycle immensely.

References

1. Chen Y (2011) Research of dynamic characteristic of car door based on mode method. Automob Technol 5:25–29
2. Zhang J (1995) Experimental research on abnormal vibration of door of medium passenger car. Automob Res Dev 2:40–44
3. Cheng Z (2016) Analysis abnormal sound of the passenger door and optimize points. Mach Des Manuf (6):229–235

Conclusion

References

Analysis and Optimization of Vehicle Interior Noise Caused by Tire Excitation

Jianghua Deng, Jianying Sun and Aofei Li

Abstract Tire is one of the main excitation sources of automobile vibration and noise, and it is the main factor of road noise and other abnormal noises in vehicle. In this paper, the road noise caused by the tire force transfer characteristic, the resonant band problem of the interior noise caused by the tire cavity mode, and the problem of the beat frequency caused by the tire uniformity are analyzed and studied. The problems of vehicle interior noise are improved by means of reasonable selection of tire, acoustic treatment of tire cavity, suspension dynamic absorber, and other related structural parts' optimization. The results show that the reasonable control of the tire structural transmission characteristics and modal peak response can effectively reduce the vehicle noise level and improve the road noise. The optimization of tire uniformity and the modal avoidance of the tire and other accessory structures can avoid the problem of abnormal noise caused by beat frequency.

Keywords Tire · Structure transfer function · Cavity modal · Tire uniformity · Beat frequency

Preface

Tire is one of the main noise sources of automobiles, and the vehicle interior noise caused by tires can be divided into two kinds: the airborne noise and the structural borne noise. The airborne noise is related to the tire pattern form and body sound-package performance, and the structural borne noise is related to the modal distribution characteristics of the tire, suspension system, and body system.

The performance of vehicle interior noise caused by tire excitation is various, mainly distributed in 70–300 Hz frequency band. This noise is related to road surface excitation, vibration characteristic of suspension system, vibration characteristic of body structure and vibration transfer characteristic, and modal distribution of tire itself [1]. In the industry and academia, there are many researches on tire excitation induced road noise, which are analyzed from tire itself and transmission path, respec-

J. Deng (✉) · J. Sun · A. Li
China Automotive Technology and Research Center Co., Ltd., Tianjin, China
e-mail: dengjianghua@catarc.ac.cn

China SAE (ed.), *Proceedings of China SAE Congress 2018: Selected Papers*,
Lecture Notes in Electrical Engineering 574,
https://doi.org/10.1007/978-981-13-9718-9_56

tively [2, 3]. In addition, the noise caused by tire excitation will also be reflected in the internal resonance under the WOT condition, the abnormal sound under the constant speed condition, and so on, which seriously affects the driving experience and comfort in the car.

In this paper, several forms of vehicle interior noise caused by tire are studied, and the influence of tire vibration transmission characteristics, modal distribution, and tire uniformity on the interior noise is considered, and the problem is improved effectively by optimizing the tire itself or related accessories.

1 Noise Response Theory Based on Modal Superposition

The noise generated by the road excitation can be expressed by the road surface excitation force and the transfer function of the wheel center to the vehicle [4], as follows:

$$P(\omega) = T(\omega)^{\mathrm{T}} F(\omega) \tag{1}$$

In the formula, $P(\omega)$ is the interior noise, $F(\omega)$ is the road load force, ω is different frequency point, $T(\omega)$ is the force-sound transfer function of the wheel center to the vehicle, which can be expressed as:

$$T(\omega) = T_{\text{tire}}(\omega) T_{\text{susp}}(\omega) T_{\text{body}}(\omega) \tag{2}$$

In the formula, $T_{\text{tire}}(\omega)$ is the vibration transfer function of tire tread to the wheel center, $\frac{\text{m}}{\text{s}^2}/\text{N}$; $T_{\text{susp}}(\omega)$ is the vibration transfer function between the wheel center to the suspension and the body mounting point, $\frac{\text{m}}{\text{s}^2}/\text{N}$; $T_{\text{body}}(\omega)$ is the vibration transfer function of the suspension body installation point to the vehicle, Pa/N.

It can be seen from the above that the vibration response of the wheel center may be expressed in the following:

$$a(\omega) = T_{\text{tire}}(\omega) F(\omega) \tag{3}$$

In the formula, a is the wheel center vibration acceleration.

The vibration displacement response of the wheel center can be obtained by linear superposition of the tire modes:

$$u = \mu R(\omega) \tag{4}$$

In the formula, u is the displacement of the wheel center, and the μ is a matrix composed of the tire modal vectors, and $R(\omega)$ is a vector composed of modal participation factors. The acceleration vector of the wheel center can be expressed as

$$a = -\omega^2 \cdot \mu \cdot R(\omega) = M(\omega) R(\omega) \tag{5}$$

In the form, $M(\omega)$ represents the modal transfer vector. From the above, Formula (1) can be expressed as

$$P(\omega) = T_{\text{susp}}(\omega)T_{\text{body}}(\omega)^{\mathrm{T}}(M(\omega)R(\omega)) \tag{6}$$

By Formula (6), it is found that the vehicle interior noise caused by tires and the linear superposition of different modal vibrations of the tire have corresponding relationship. That is, the modal distribution of tire can directly affect the performance of the noise in the vehicle.

2 Analysis of Vehicle Interior Noise Caused by tire

2.1 *Influence of Tire Structure Transfer Characteristics*

2.1.1 Problem Description

In the course of tire selection, it is found that when tire A is used, the noise level of the vehicle is higher than the target value when driving at a constant 60 km/h condition on the rough road, and the road noise is too large. Through the analysis of vehicle noise, the problem frequency is mainly expressed in the 80–200 Hz, corresponding to the 80, 100, 200 Hz central band noise energy, as shown in Fig. 1. This band noise energy is the main reason of the interior noise on constant speed conditions.

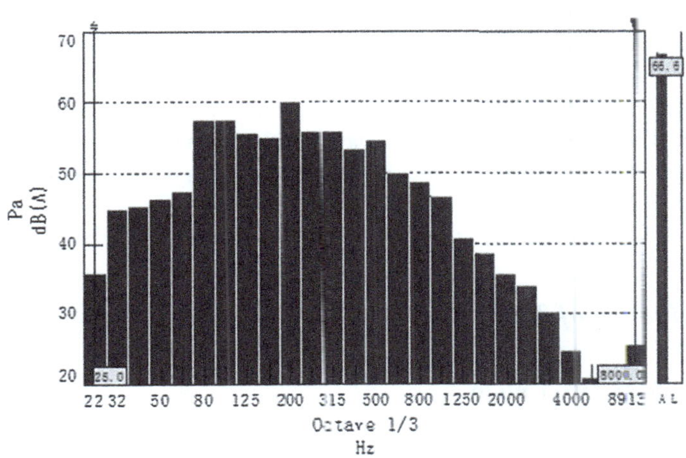

Fig. 1 Uniform 60 km/h noise spectrum

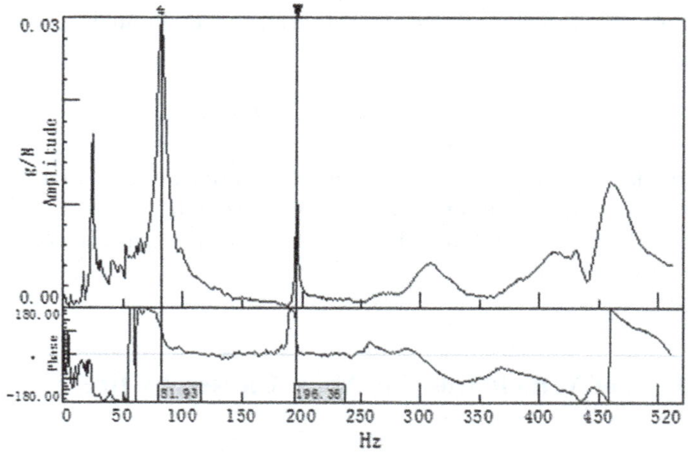

Fig. 2 Transfer function of tire force

2.1.2 Problem Analysis

Through the analysis of the noise spectrum in the vehicle, this problem is mainly expressed as medium- and low-frequency noise, which should be related to suspension and tire structure characteristics. Through the test of force transfer function of wheel center to the body mounting point and wheel center to tire skin, it is found that the three main noise peak problems are directly related to the dynamic transmission characteristics of tire, and the force transfer function is shown in Fig. 2.

The analysis shows that there are two distinct peaks in the dynamic transfer characteristics of the tire itself in 82 and 198 Hz. Through the tire modal test, the two peaks, respectively, for the first order of the tire modal flexibility and first-order empty state modes, the modal response peak directly affects the size of the interior noise level. Therefore, it is necessary to pay attention to the peak response level of the dynamic transmission characteristics of tire in the corresponding mode frequency in the tire selection.

2.2 Influence of Tire Cavity Model

2.2.1 Problem Description

In the process of acceleration, there is a 220-Hz resonant band in the interior noise, as shown in Fig. 3. Moreover, it is coupled with the fourth-order excitation of the engine around 3400 rpm, resulting in obvious peak noise (Fig. 4), which seriously affects the performance of the whole vehicle Noise Vibration Harshness (NVH).

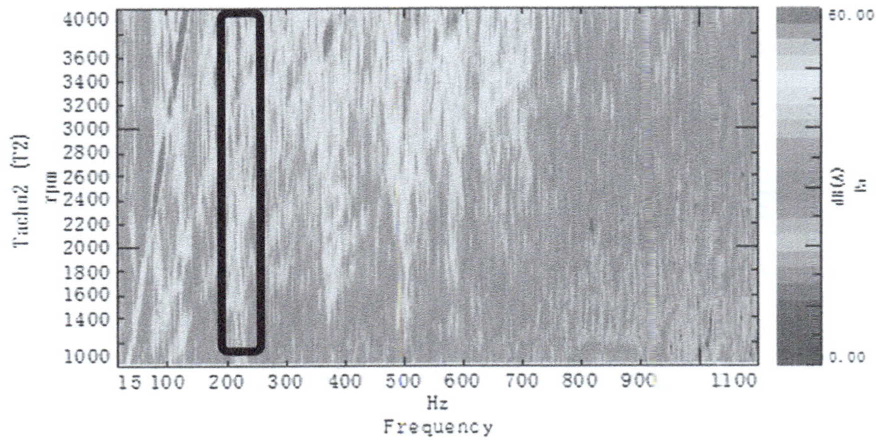

Fig. 3 Interior noise in acceleration

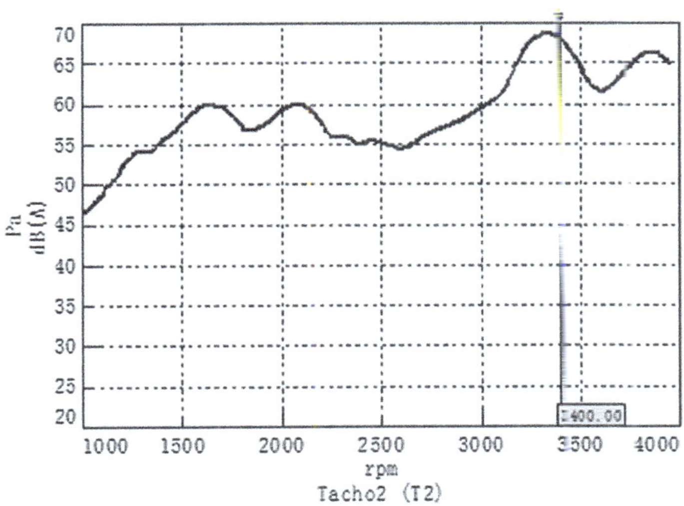

Fig. 4 Interior fourth-order noise in acceleration

2.2.2 Problem Analysis

Through the identification of the noise transfer path, there is a peak value of 220 Hz in the transfer function from wheel center to interior noise, as shown in Fig. 5. Further investigation confirmed that the peak is caused by cavity mode of the tire.

It can be seen from the above analysis that the response peak of the tire cavity mode can be optimized by reducing the response peak of the tire cavity mode itself or decreasing the transfer path response peak of the tire to the vehicle body.

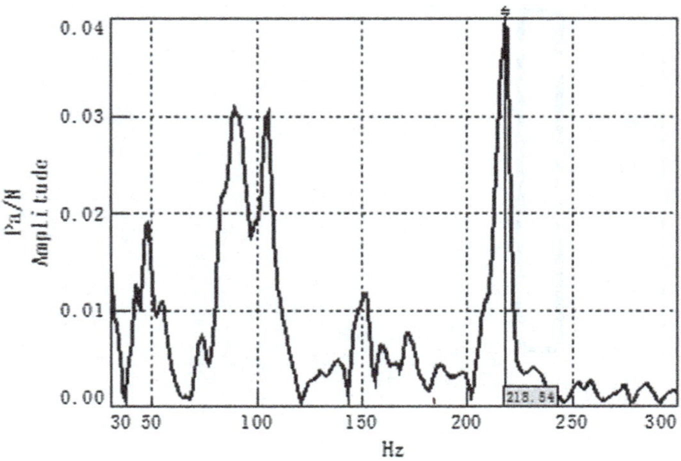

Fig. 5 Sound transfer function from wheel center to driver's right ear

2.3 *Influence of Tire Uniformity*

The uniformity of tire is the unroundness of tire, which usually includes the nonuniformity of geometrical size, nonuniformity of mass distribution, and nonuniformity of rigid distribution. Tire inhomogeneity can cause unbalanced excitation when the tire rotates at high speed, which is mainly manifested as different order excitation, and the corresponding order is related to uniform distribution of the tire, such as first-order dynamic imbalance excitation and eighth-order nonuniformity excitation.

2.3.1 Problem Description

When a car travels at a constant speed of 110 km/h in five gears (corresponding to the engine speed of 3200 rpm), there is obvious periodic "poo-poo" in the subjective feeling inside the car. According to the constant speed test in five gears of 3200 rpm, there are two adjacent noise peaks (108 and 115 Hz) in the car, as shown in Fig. 6.

2.3.2 Problem Analysis

It can be seen from the above analysis that the subjective "poo-poo" sound should be the beat frequency phenomenon generated by two adjacent frequencies, with the beat frequency of 7 Hz. The production of the beat is coupled with two noise sources with adjacent frequency, and the second order of engine is 108 Hz at 3200 rpm. Therefore, there should be a noise source independent of the engine, with frequency of 115 Hz. Through the test of the chassis dynamometer, the vehicle was running at

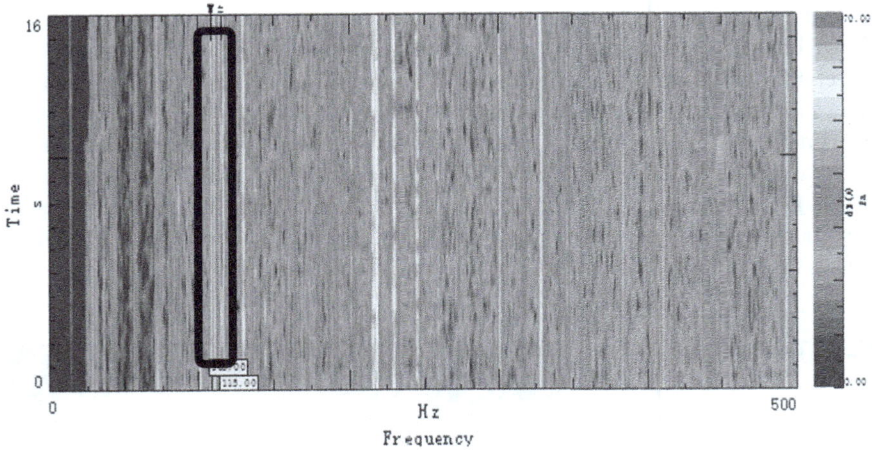

Fig. 6 Interior noise color map in the problematic condition

a constant speed of 110 km/h when the engine was turned off, and the spectrum of interior noise was measured as shown in Fig. 7.

As can be seen from Fig. 7, in the engine key off condition, there is a noise peak of order in the car corresponding to the rotation frequency of the tire, and the peak value of order 8 is obvious, corresponding to around 115 Hz. It is known that the tire specification of the current prototype is 195/55 R16. Under the condition of corresponding speed, the eighth-order noise is about 115 Hz, which is consistent with the frequency of problems in the vehicle.

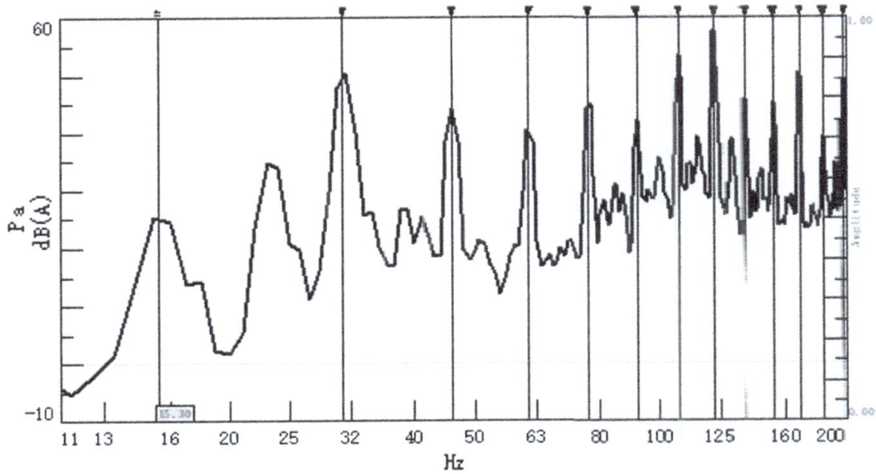

Fig. 7 Interior noise spectrum at constant speed

In the process of tire forming, the tire forming drum is made of eight drum tiles, as shown in Fig. 8. This forming process causes the problem of tire uniformity and forms the eighth-order rotation excitation under this working condition. Further analysis shows that the tire has an elastic mode of 116 Hz, as shown in Fig. 9. That is, under the condition of this problem, the tire mode is stimulated by the eighth-order rotary excitation, which then leads to the 115 Hz resonant belt in the car, and the peak noise of 108 Hz is induced by the second-order excitation of the engine after the vibration absorber is applied on the driving half shaft. The combined action of the two sources causes the beat frequency, which leads to the periodic "poo-poo" in the car.

Fig. 8 Tire forming drum

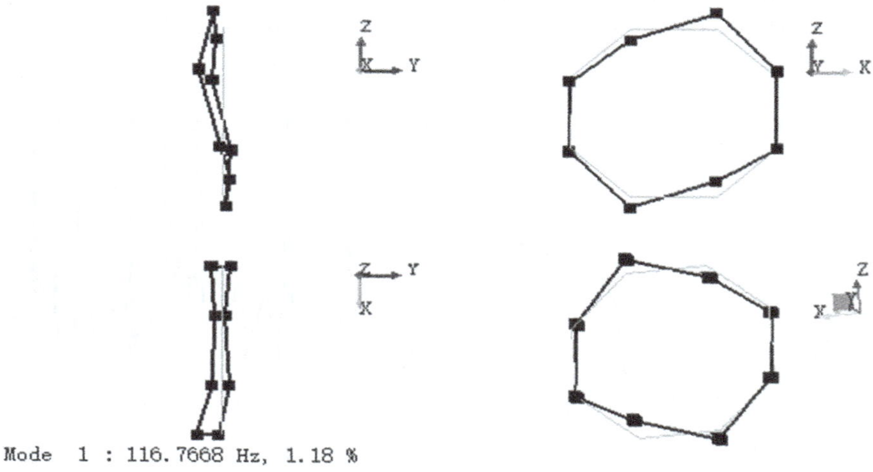

Fig. 9 Tire mode

3 Tire-Induced Interior Noise Optimization

3.1 *Structural Transfer Characteristic Optimization*

In view of the interior noise problem caused by the tire's force transfer characteristics, the tire's force transfer function is focused on the tire selection. After selecting tire B, the comparison between the force transfer function and tire A is shown in Fig. 10.

As can be seen from Fig. 10, the peak transmission frequency of B tire is higher than 80 Hz, and the amplitude is significantly reduced. At the same time, the high-frequency peak value above 300 Hz is also significantly lower than that of tire A. After the adoption of tire B, the interior noise in the rough road with constant speed of 60 km/h is significantly improved than that of tire A, especially at the central frequency band of 80 and 100 Hz, as shown in Fig. 11.

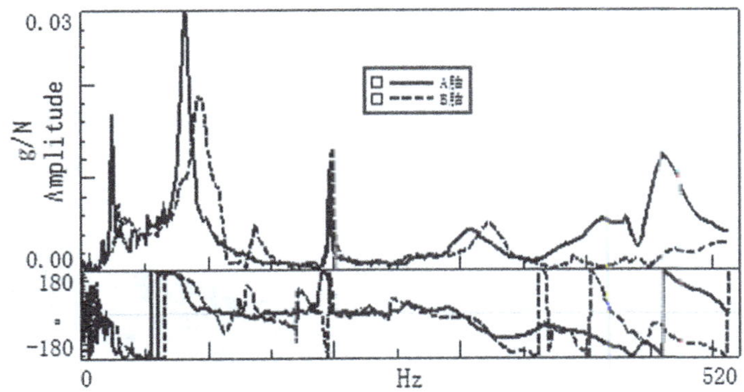

Fig. 10 Comparison of tire force transfer functions between A and B

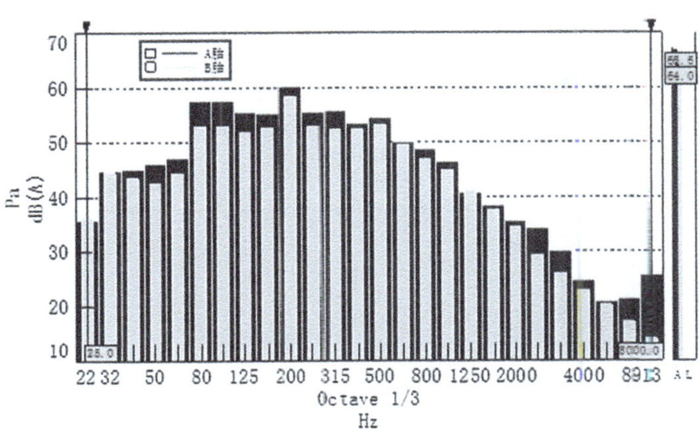

Fig. 11 Comparison of vehicle interior noise with different tires

3.2 Noise Optimization Induced by Tire Modal

There are two ways to deal with the interior noise problem caused by tire cavity mode. One is to optimize the tire cavity mode (mainly for peak response), and the other is to optimize the path.

3.2.1 Cavity Mode Optimization

As the tire specification and model are determined, the cavity structure and size cannot be changed, that is, the modal frequency is difficult to be modified, so the damping is optimized to reduce the amplitude of modal frequency response. The acoustic material is wrapped around the rim to improve the cavity reverberation, i.e., the acoustic cavity damping, as shown in Fig. 12. After adopting this scheme, the interior resonance noise peak value can be significantly improved, as shown in Fig. 13.

Fig. 12 Acoustic material is wrapped around the rim

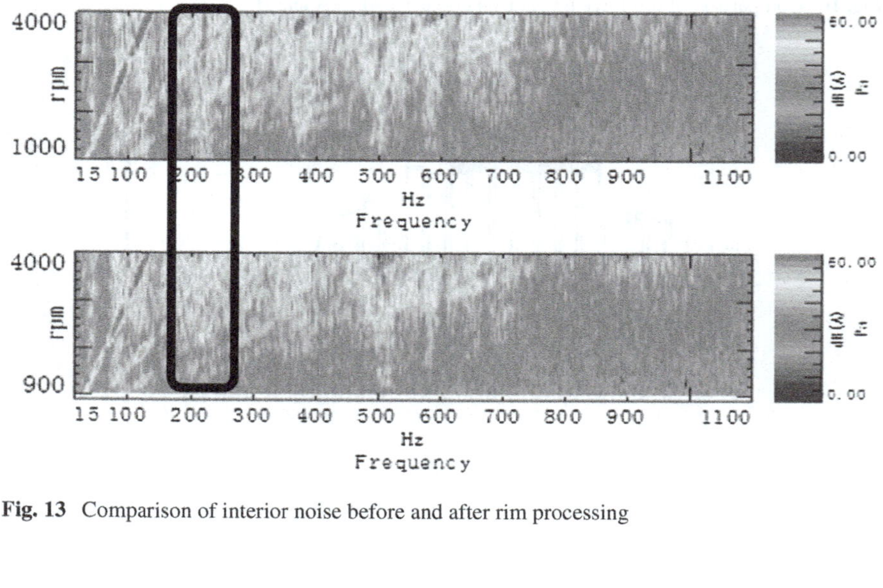

Fig. 13 Comparison of interior noise before and after rim processing

3.2.2 Path Optimization

The tire cavity mode is excited and transmitted to the vehicle through the suspension system in the form of structural borne sound. Therefore, the transmission path of the suspension system can be optimized to reduce the transmission of structural borne sound. A dynamic vibration absorber [5] is installed at the lower end of the suspension vibration damping spring (the frequency of the absorber is 220 Hz), as shown in Fig. 14. After the scheme is applied, the interior noise decreases sharply at the corresponding center frequency, as shown in Fig. 15.

Fig. 14 Optimization of transmission path by dynamic vibration absorber

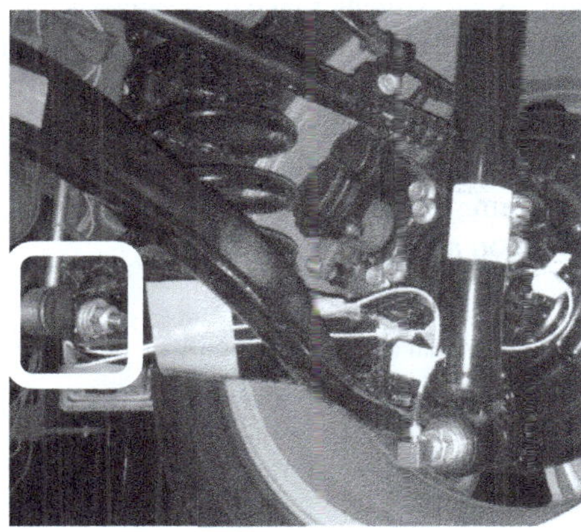

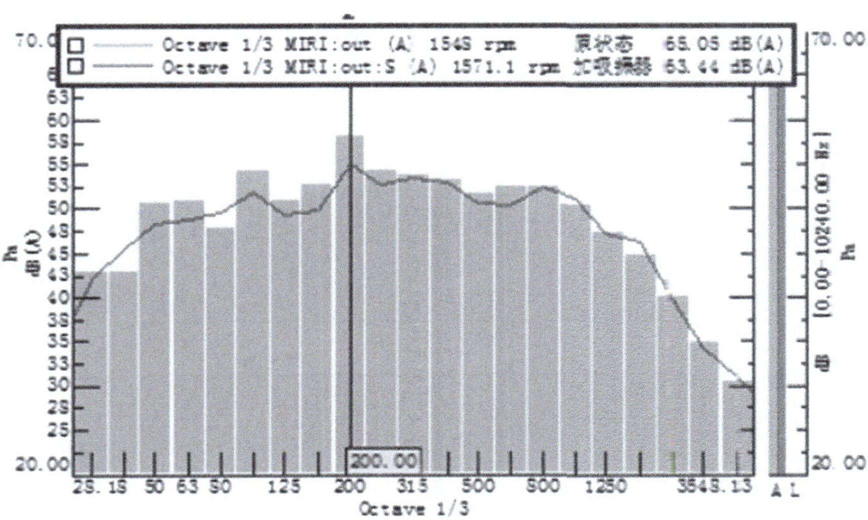

Fig. 15 Comparison of interior noise after the implementation of the dynamic vibration absorber

3.3 Optimization of Beat Frequency Problem

The uniformity of tires is mainly related to the manufacturing process of tires, starting with semi-finished tire components, forming process, curing process, etc. [6, 7]. In addition, the noise problem caused by tire uniformity is also related to tire structure mode and other adjacent accessory mode. Through finite element analysis, the current tire model is optimized. See Table 1 for the variation of tire mode under different schemes.

It can be seen from Table 1 that scheme 4 can significantly improve the frequency avoidance of tire mode by optimizing the beam angle. Through the tire molding process adjustment, make tire materials more uniform, thereby reducing tire dynamic balance, promote tire uniformity. At the same time, scheme 4 is adopted for tire mode optimization, and tire damping is changed by adjusting the tread formula. Finally, the interior noise can be significantly improved, with the 115-Hz resonance peak value basically eliminated and subjective "poo-poo" eliminated, as shown in Fig. 16. The remaining 108-Hz peak value can be processed by optimizing the semi-axle dynamic vibration absorber.

Table 1 Modal optimization of tire

Order	Test values (Hz)	Scheme 1	Scheme 2	Scheme 3	Scheme 4	Scheme 5
		Original (Hz)	Reduce carcass angle (Hz)	Increase carcass angle (Hz)	Beam angle optimization (Hz)	Carcass and beam angle optimization (Hz)
1	94	87.9	84.6	86.9	81.1	77.4
2	116.7	111.7	108.3	110.3	97.7	104.3

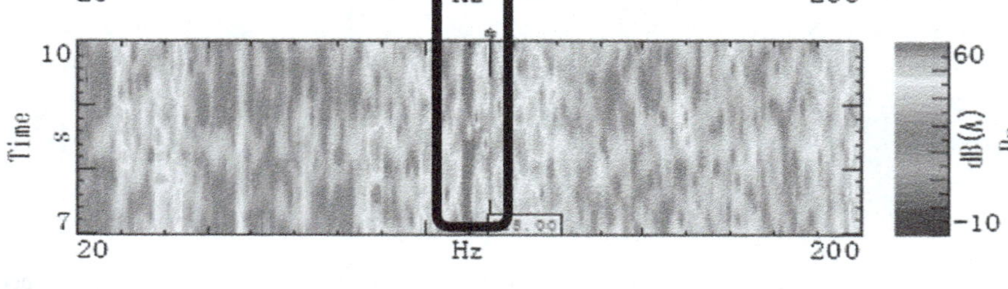

Fig. 16 Comparison of vehicle interior noise before and after optimization

4 Conclusion

In this paper, the different forms of vehicle interior noise caused by tires are analyzed, including road-noise, acceleration noise, and beat frequency. The corresponding tire NV problem was optimized to improve the interior noise of the vehicle. Through this analysis, the following conclusions can be drawn:

(1) It is particularly important to carry out reasonable matching and selection of tires, which should focus on the force transfer characteristics and modal response amplitude of tires. Only by effectively controlling and optimizing its frequency distribution and amplitude response, the risk of NVH problem in the later stage can be reduced. The optimization of the force transfer function and modal response can be achieved by adjusting the ratio of the tire compound, the angle of the tire body, and the angle of the beam.
(2) After the tire selection is determined, the cavity mode frequency has been fixed, and the optimization of the cavity mode can be taken from the reduction of amplitude response. Adding sound absorbing material in the cavity and adding dynamic vibration absorber on the rim or suspension path should be a good choice.
(3) The uniformity of tires not only affects the operation stability and driving safety of vehicles but also makes a great contribution to the performance of the whole vehicle NVH. The risk of NVH problems can be effectively reduced through the optimization of production and processing technology.
(4) The modal avoidance of tires (including structural mode and cavity mode) and other adjacent components are also a concern. Effective frequency avoidance can not only reduce road noise caused by structural mode coupling but also avoid possible beat frequency caused by different excitation sources.

References

1. Ichikawa.K, Research on mechanism of change in suspension transfer force in relation to low frequency road noise. SAE paper: 2015-01-0667
2. Yu X, Min F, Wen W (2013) Structural transfer path analysis of tire/road noise. Automot Eng 35(11):1030–1034
3. Chen L, Yang T, Yang L (2002) Optimization for structure parameters of low noise tread patterns. Tire Ind 22(12):720–728
4. Tsuji H, Maruyama S, Onishi K, Reciprocal measurements of the vehicle transfer function for road noise. SAE paper: 2015-01-2241
5. Shen W, Yi B, Lu W (2010) The application of dynamic vibration absorber to the reduction of in vehicle road noise. Automot Eng 32(8):690–692
6. Zhao S, Deng S, Ding H (2014) Influential factors on tire noise and design method of low-noise tire. Tire Ind 34(2):76–80
7. Ran X, Huang Z, Bin L (2017) Effect of eighth harmonic of radial force on noise of tire at high speed. Tire Ind 37(1):8–12

Diagnosis of Road-Induced Drumming Noise of Passenger Car Based on Multiple Coherence Method

Shichao Gong, Bo Peng, Daijun Chen, Yongqiang Tang and Pengfei Wang

Abstract Due to the fact that the vehicle suspension system is complicated, how to identify the structure-borne path of road noise efficiently and accurately is the key point of road noise diagnosis. In this research, a transfer path diagnosis approach for road noise diagnosis is proposed. In terms of methodology, the proposed method is based on the multiple coherence method, and it also needs to take into account the mechanical transfer paths of road noise. Therefore, the numerical implementation of multiple coherence method is presented first and then the classification of mechanical transfer paths of road noise. As for the practical application, first of all, the presented approach is used to investigate the transfer paths of a 79 Hz road-induced drumming problem of a vehicle. And then the mechanism for above-mentioned problem is predicted combining the modal results of relevant components. Furthermore, the predicted result is validated by adding a tuned 78 Hz damper to the lower control arm on both sides of the rear suspension. The experimental result indicates that the peak value at 79 Hz corresponding to road-induced drumming noise reduces by 10 dB(A), which proves the correctness of the inference for drumming noise and further demonstrates the validity of the proposed diagnosis approach.

Keywords Drumming noise · Multiple coherence · Modal analysis · Tuned damper

1 Introduction

The NVH (noise, vibration, and harshness) performance of a vehicle is an indirect reflection of the vehicle design level and manufacturing process, which will affect the market competitiveness and user experience of the vehicle. The vehicle's NVH performance is mainly composed of powertrain noise, wind noise, electromechanical system noise, and road noise. As the powertrain is changed from conventional

S. Gong (✉) · B. Peng · D. Chen · Y. Tang · P. Wang
State Key Laboratory of Vehicle NVH and Safety Technology, Chongqing, China
e-mail: scgong@163.com

Changan Auto Global R&D Center, Changan Automobile Co., Ltd., Chongqing, China

China SAE (ed.), *Proceedings of China SAE Congress 2018: Selected Papers*,
Lecture Notes in Electrical Engineering 574,
https://doi.org/10.1007/978-981-13-9718-9_57

internal combustion engine to electric motor, the contribution of powertrain noise in vehicle interior noise is gradually decreasing, and road noise (tire/road noise) will be more prominent. The development of road noise performance is becoming more and more important for improving the ride comfort of passenger cars. The road noise is divided into low-frequency road noise, middle-frequency road noise, and high-frequency road noise according to the frequency range. The low-frequency road noise is the noise generated by the interaction between tire and road surface, which is transmitted to the body through suspension, subframe and other structures, resulting in the interaction between the body sheet metal vibration and the acoustic cavity mode of the compartment. The phase difference between the left and right wheels of the front axle as well as that of the rear axle varies from 0° to 180° in real time, and also, there is a time delay between the front and rear axle determined by the wheelbase and speed. The mechanism of low-frequency road noise determines that road noise is a multi-reference partial correlation vibration and noise problem. How to determine the structural path of road noise problem efficiently and accurately is the difficulty of road noise diagnosis.

In literature [1, 2], inverse matrix method combined with modal recognition method is applied to carry out the research on road noise, and good results are obtained. Yang and Chu conduct research on the multiple coherence decomposition of road noise based on the multiple coherence theory, and the decomposition results are relatively accurate [3]. Kim et al. [4] identify the critical paths of knocking noise through transfer path analysis method and adjust bushing stiffness and shock absorber damping of the corresponding path, which eventually solves both knocking noise and improve the ride comfort.

In this paper, first of all, the numerical implementation of multiple coherence method and the path classification approach of road noise problem are introduced, and then the primary path and secondary path identification of a 79 Hz drumming noise is carried out with the proposed diagnosis approach. Combining the principle of low-frequency road noise development and both the analysis results of the suspension structural mode and the acoustic cavity mode of the passenger cabin, the mechanism of the road-induced drumming problem is deduced. The solution of adding tuned damper is adopted to solve the road-induced drumming problem according to the actual situation of project development. Experimental results show that the tuned damper solution can reduce the peak value of 79 Hz by 10 dB(A), which shows that the proposed method is correct.

2 Path Diagnosis of Road Noise Problem Based on Multiple Coherence Method

Random excitation of road profile is the source of road noise, and the unacceptable road noise problem of passengers is more due to the amplification effect of the path on the vibration caused by road excitation during the transmission process. How to accurately and efficiently investigate the critical path is the key to the diagnosis of road noise problem. Actually, the connection between the suspension and the body is extremely complex, which takes a lot of time for all paths investigation. Therefore, this paper employs the multi-coherence analysis method to conduct the hierarchical troubleshooting, which can improve the efficiency greatly.

2.1 Numerical Implementation of Multiple Coherence Method

As for path diagnosis of road noise problem, the multiple coherence analysis method can be used to analyze the proportion of one path or several paths in the interior road noise [5]. Multiple coherence coefficients γ_{XY}^2 between multiple input signals $X_1(f), X_2(f), \ldots, X_M(f)$ and output signal $Y(f)$ can be defined as:

$$\gamma_{XY}^2(f) = \frac{P_{XY}^H(f) P_{XX}^+(f) P_{XY}(f)}{P_{YY}(f)} \tag{1}$$

where $(\cdot)^H$ represents conjugate transpose, $(\cdot)^\dagger$ represents generalized inverse. $P_{YY}(f)$ is the self-power spectral density of output signal $Y(f)$, $P_{\mathbf{XX}}(f)$ is the cross-power spectral density matrix of the input signals, and $P_{\mathbf{XY}}(f)$ is the cross-power spectral density vector between multiple input signals and output signals. The definition given in Eq. (1) is implemented by Welch algorithm [6] during numerical computation, where the calculation method of each parameter is expressed as:

$$\mathbf{P}_{XY}(f_n) = \frac{1}{K}\sum_{i=1}^{K} \mathbf{X}_i^*(f_n) * Y_i(f_n) \tag{2}$$

$$P_{YY}(f_n) = \frac{1}{K}\sum_{i=1}^{K} Y_i^*(f_n) * Y_i(f_n) \tag{3}$$

$$\mathbf{P}_{\mathbf{XX}}(f_n) = \frac{1}{K}\sum_{i=1}^{K} \mathbf{X}_i^*(f_n) * \mathbf{X}_i(f_n) \tag{4}$$

$$\mathbf{X}_i(f_n) = [X_{i1}(f_n), X_{i2}(f_n), \ldots, X_{iM}(f_n)]^{\mathrm{T}} \tag{5}$$

where $(\cdot)^{\mathrm{T}}$ and $(\cdot)^*$,respectively, represent transposition and conjugate; $Y_i(f_n)$ is the ith discrete periodic spectrum of the output signal, and there are K in total. $X_{im}(f_n)$ is the ith periodic spectrum of the m input signal.

Until now, the numerical solution of Eq. (1) can be realized by using Eqs. (2–5).

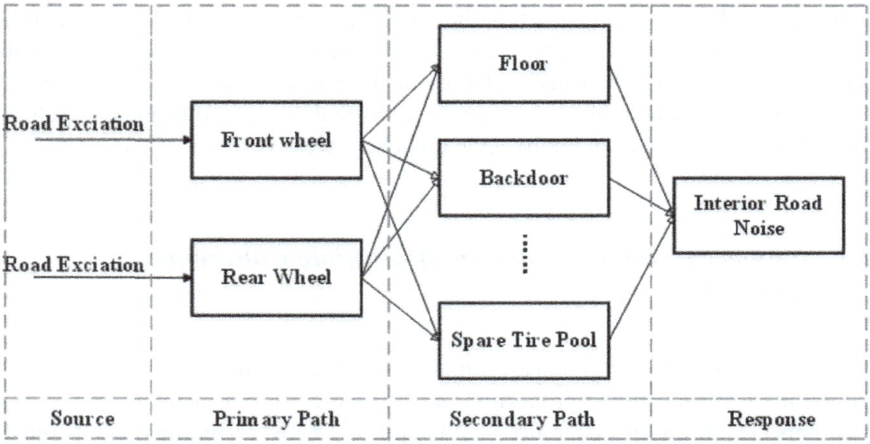

Fig. 1 Transmission of road noise

2.2 *Hierarchical Investigation for Road Noise Problem*

As shown in Fig. 1, when a car is driving on the road, four wheels vibrate under the excitation of road surface profile, and the vibration is transmitted to the car through suspension and car body to generate noise. The wheels are defined as the primary path of road noise transmission, while the suspension or body is the secondary path of road noise transmission. In order to simplify the test, the proportion of the noise transferred by the primary path in the road noise is analyzed through multi-coherence method, and the principal wheel in the front wheel and the rear wheel is identified. Then, the measurement points are arranged on the secondary path corresponding to the principal wheel identified, and then the critical paths belong to the secondary path are analyzed so as to simplify the test and analysis process.

3 Path Diagnosis and Mechanism Analysis of Drumming Problem

3.1 *Description of Road-Induced Drumming Problem*

A passenger car has obvious drumming phenomenon in the front row of the passenger compartment when driving at the speed of 40–60 km/h on coarse road surface and common cement road surface. The measured sound signal is subjectively evaluated with high pass, low pass, bandstop filtering and so on through sound playback system,

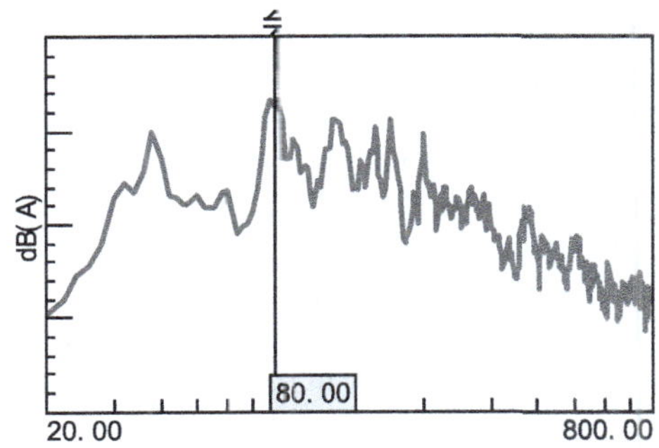

Fig. 2 Spectrum diagram of interior front row noise (FFT)

and the problem frequency band is determined to be below 100 Hz. The result of narrow-band spectrum analysis shows that the main frequency of the problem is 78–80 Hz (Fig. 2).

3.2 *Hierarchical Investigation of Drumming Noise Critical Paths*

In order to quickly identify whether the principal path of the drumming problem is the front wheel or the rear wheel, this paper first applied the multi-coherence method to analyze the coherence coefficient of the noise at the front row of the vehicle and the vibration of the front/rear knuckle, and the results are shown in Fig. 3. It can be seen from the results of the coherence coefficient between noise at front row of the

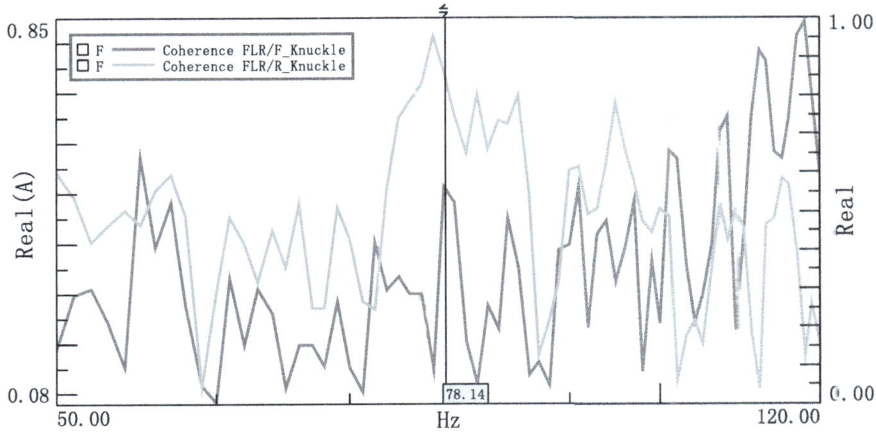

Fig. 3 Correlation between vehicle front noise and front/rear knuckle

vehicle and the front/rear knuckle that the coherence coefficient of the interior noise in the frequency range of 78–80 Hz and the vibration of the rear knuckle is about 0.8, and that of the front knuckle is about 0.5. According to the analysis results of the coherence coefficient of the primary path, it can be determined that the main path of the 78–80 Hz drumming problem is the rear wheel and the rear suspension assembly.

After determining the primary path, the range of road-induced drumming transmission path is reduced to the position of the rear suspension parts and their attachment points to the underbody. As can be seen from the structural diagram of multi-link rear suspension (Fig. 4), the transfer path in rear suspension includes the rear longitudinal trailing arm mounting bush, the rear shock absorber mounting seat, the rear coil spring mounting point, and the front and rear installation points of rear subframe.

The vibration of the rear suspension knuckle, as well as that of the selected path above, was tested and recorded under the condition of 60 kph constant speed on coarse road for multi-coherence analysis. The vibration signal of the rear knuckle was selected as the source reference signal, and the coherence coefficients between the vibration of the above-mentioned 5 * 3 paths and vibration of the rear knuckle were analyzed (Fig. 5).

At 79 Hz frequency point, the coherence coefficient between the x- and y-direction of rear shock absorber top mount attachment point, the *x*- and *y*-direction of rear coil spring mounting seat local point, and the *x*-direction of rear subframe front attachment point vibration with vibration of rear knuckle rank top six and are greater than 0.84. The results of the coherence coefficient show that the above six paths are the critical path of 79 Hz road noise problem. According to the "source-path-response" model in the NVH domain, the vehicle interior response is determined by the excitation force at attachment point and the acoustic transfer function from the attachment point to the interior. In order to understand whether the problem is caused by the suspension system or the body system, it is necessary to study the suspension and the body characteristics.

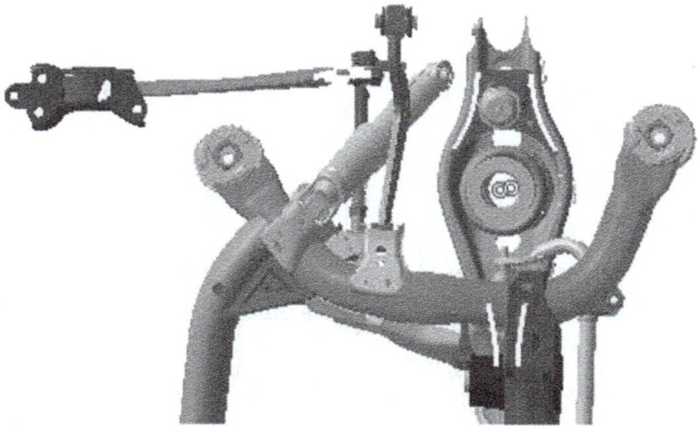

Fig. 4 Structural diagram of multi-link rear suspension

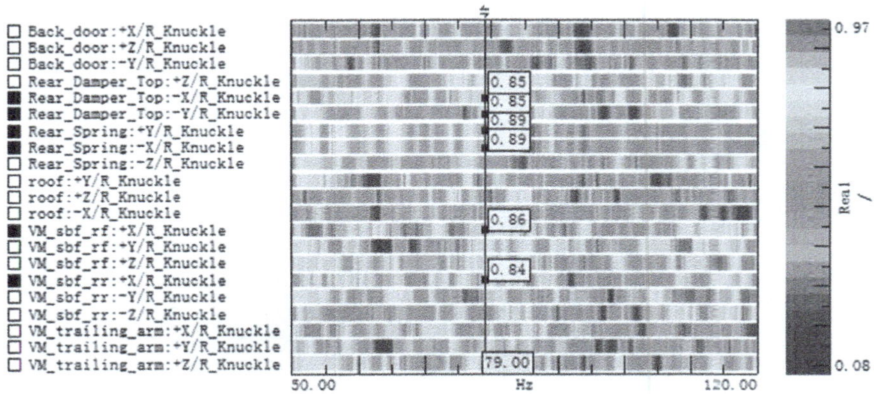

Fig. 5 Multiple coherence analysis results of the rear suspension path

3.3 Mechanism Analysis of Drumming Problem

The road noise problem in frequency band of 20–100 Hz is mainly solved by reasonable modal alignment and sufficient vibration isolation. The key point of modal alignment is separation of the main mode of suspension assembly, with integral and local mode of vehicle body as well as longitudinal acoustic mode of passenger compartment. The larger the frequency separation margin between suspension assembly mode and cabin acoustic cavity mode is, the smaller the interior sound response is. Vibration isolation means reducing the stiffness of suspension bushing and subframe bushing for the purpose of increasing the ratio of mounting point dynamic stiffness, which then increases vibration isolation capacity.

According to the results of multiple coherence analysis of secondary path investigation in Sect. 3.2, it is inferred that the rear suspension assembly may have a mode with a frequency of about 79 Hz. According to the above-mentioned 20–100 Hz low-frequency road noise avoidance method, the suspension mode and the acoustic cavity mode of the passenger compartment need to be identified first.

The finite element model of the rear suspension assembly was established according to the 3D model of the rear suspension and the performance parameters of the elastic components, and modal analysis was carried out. The simulation result showed that the rear suspension assembly has a longitudinal mode of 79 Hz (Fig. 6). In order to verify the accuracy of the simulation result of suspension mode, experimental test of suspension assembly mode based on hammer impact method was carried out. During the test, the vehicle was unloaded and placed on a bridge-type hoist system stationary and horizontally. Generally, the left and right side suspension assembly of an independent rear suspension are symmetrical structure with *XZ* plane as the symmetric plane. The right side suspension assembly was selected for the modal test. A total of 16 tri-axial vibration sensors were mounted on the right suspension assembly. The center of the right wheel was selected as the driving point, and the fre-

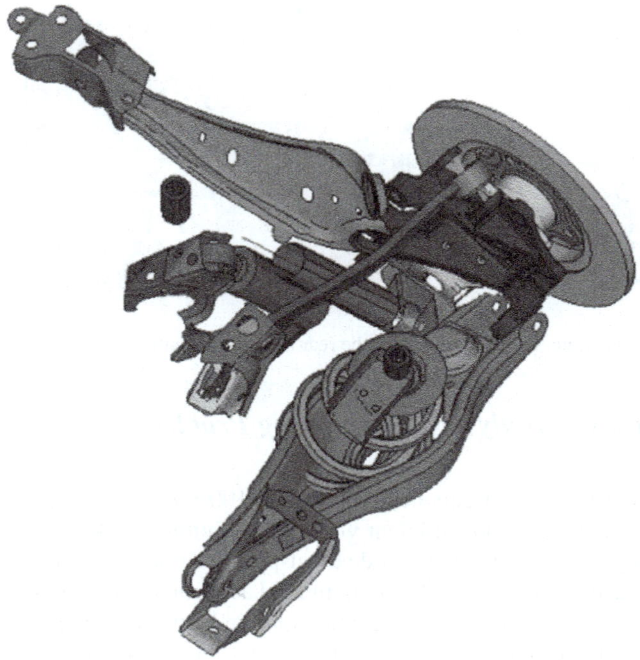

Fig. 6 Longitudinal mode of rear suspension assembly (CAE, 79 Hz)

quency response functions were recorded, respectively, when impacting the driving point toward the longitudinal, vertical, and transverse direction. The experimental modal result (Fig. 7) shows that the longitudinal mode frequency of the rear suspension assembly is 78 Hz, which is in good agreement with the simulation result. In addition, the test result of constraint mode of the rear suspension components shows that the constraint mode at 78–80 Hz frequency band does not exist in the rear frame and the suspension link.

The volume sound source was used as the excitation source, and the microphone in the cabin was used as the acquisition sensor to test the acoustic mode of the cabin. The result of acoustic cavity mode test shows that the second-order longitudinal acoustic mode of the passenger compartment is 76.7 Hz (Fig. 8).

Based on the finite element analysis result of the rear suspension mode and the acoustic cavity mode test results of the passenger cabin, it can be inferred that the root cause of the low-frequency drumming problem of 78–80 Hz is the coupling between the longitudinal motion mode of the rear suspension assembly and the acoustic cavity mode of the passenger cabin. In this paper, the mechanism of road-induced drumming noise is that the rear suspension vibrates fore-aft at frequency corresponding to the

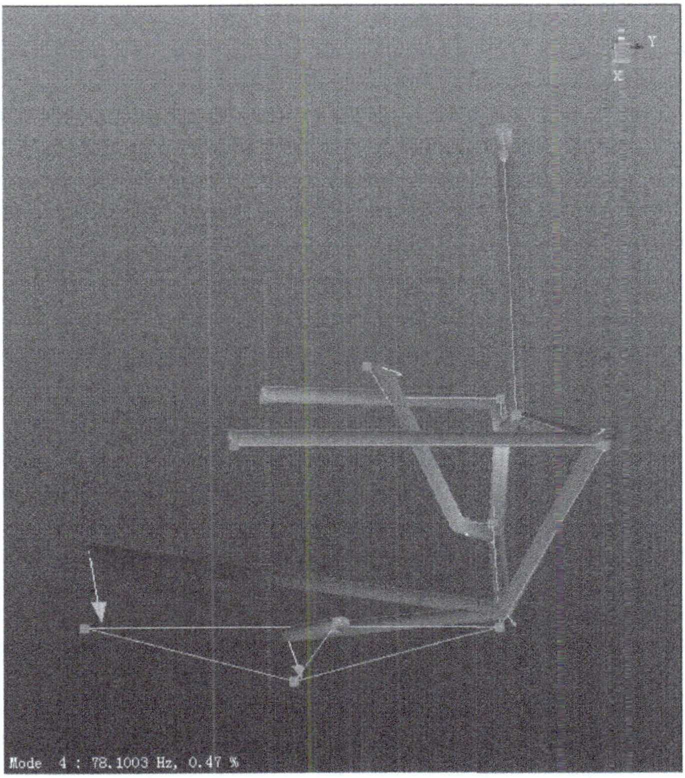

Fig. 7 Longitudinal mode of rear suspension assembly (test, 78 Hz)

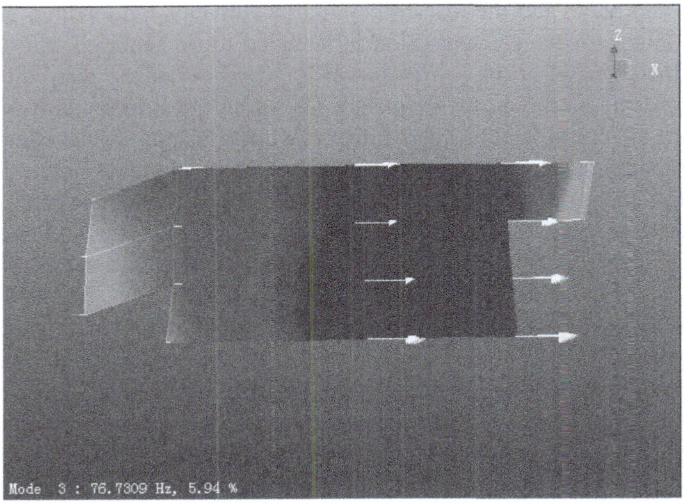

Fig. 8 Second-order longitudinal acoustic mode of passenger compartment (76.7 Hz)

longitudinal mode under the excitation of road surface roughness of coarse road. The forced vibration energy of rear suspension transmits to the car underbody through the rear coil spring, rear shock absorber and the rear subframe attachment points, and then to car back door through the hinge, back door lock, sealing strip, buffer block structure, which force the back door to vibrate globally and locally in longitudinal direction. The second-order longitudinal acoustic cavity mode of the passenger cabin is excited by the longitudinal vibration of the back door, which generates obvious drumming phenomenon.

3.4 Optimization Method of Drumming Problem

The acoustic sensitivity of a trimmed body is characterized by the vibro-acoustic transfer function from the attachment point to the interior response point. The acoustic cavity mode of passenger compartment is an important factor to determine the acoustic sensitivity of the vehicle. The peak value of suspension vibration spectrum often exists at the suspension mode frequency under random road excitation, so it is necessary to adjust the suspension mode frequency to avoid the frequencies at which the peak values of the trimmed body NTF appear.

The acoustic cavity mode of the passenger compartment is determined by the shape and size of the passenger cabin space. The acoustic cavity mode nodes and modal frequencies are difficult to change once the trimmed body design is completed. In the middle and late stage of a project, the size of the whole vehicle has been fixed, so the plan to change the shape and size of the passenger compartment is not feasible.

As shown in Fig. 7, in the 79 Hz modal shape of rear suspension assembly, the combination of the rear lower control arm and the rear coil spring has the maximum vibration amplitude, and the rear coil spring has obvious longitudinal bending. If sufficient modal frequency separation is intended to be achieved, it is necessary to adjust the stiffness of the lower control arm bushing or the rear coil spring to increase or decrease the modal frequency. Changing the stiffness of lower control arm bushing affects the handling performance, while adjusting spring stiffness affects the deviation frequency of the sprung mass, thus affecting the primary ride comfort performance. In the middle and late stage of the project, a lot of tuning and verification work needs to be carried out when the spring stiffness and lower control arm bushing stiffness are changed. However, the project development cycle and cost don't allow this.

For low-frequency NVH problem, in addition to strategies such as reasonable mode alignment and vibration isolation, there are also measures such as placing attachment points at modal nodes and adding tuned absorbers.

The suspension system not only supports the weight of body structure, but also realizes the complex kinematics function and the vibration and noise reduction function. Taking into account the development cycle, cost and the influence on the performance of suspension, the solution of adding tuned absorber is adopted in this project (as shown in Fig. 9). According to the principle of tuned damper, a kind of

Fig. 9 Suspension with tuned mass damper

disk-like tuned damper with an axial (normal direction of disk) frequency of 79 Hz was designed, and its mass was 450 g. There is a through hole on the disk type iron core. While assembled to the rear suspension, a bolt passes through the center hole of the iron core and locks the damper like a pressing metal gasket.

After a tuned damper was added to the rear lower control arm of both the left and right side, the optimization effect of interior noise in front row on coarse road is shown in Fig. 10. The peak value of interior noise spectrum at 79 Hz decreased by 10 dB(A). The subjective evaluation showed that the drumming problem in front row improved significantly.

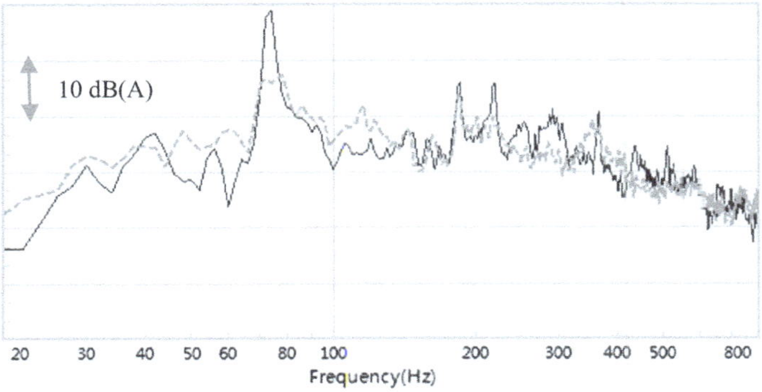

Fig. 10 Road noise effect with tuned damper on rear lower control arm

4 Conclusion

In the practice of road noise development, how to identify the transfer path of low-frequency road noise quickly and accurately is the foundation of solving the problem. This paper proposes and applies the method of road noise path diagnosis based on the multiple coherence theory, determines the critical path of interior drumming problem at 79 Hz quickly and accurately, and infers the mechanism of 79 Hz road-induced drumming noise based on the modal results of related parts. Finally, the solution of adding one-tuned damper to the left and right rear lower control arm is adopted when taking into consideration of factors such as the project development cycle and development cost. After implementing the solution of adding tuned damper to the rear lower control arms, the on-road test result showed that the amplitude of interior drumming noise reduced by 10 dB(A). The result shows the correctness of the inference-concerning mechanism of road-induced drumming problem and also proves the proposed road noise path diagnosis method is correct and effective.

References

1. Shivle S, Arora G (2006) Methodology of road noise analysis and improvement strategy for passenger cars. SAE Technical Paper 2006-01-1094
2. Yu X, Min F, Wen W el al (2013) Structural transfer path analysis of tire/road noise. Automot Eng 35(11):1030–1034
3. Yang Y, Chu Z (2015) Multiple coherence decomposition method for road noise of a car. J Vib Shock 34(19):31–36
4. Kim B, Bae K, Chang S, Ryu J et al (2012) Optimal rear suspension design for the improvement of ride comfort and suspension noise. SAE Technical Paper 2012-01-0975
5. Zafeiropoulos N, Ballatore M, Moorhouse A, Mackay A (2015) Active control of structure-borne road noise based on the separation of front and rear structural road noise related dynamics. SAE Int J Passeng Cars Mech Syst 8(3). https://doi.org/10.4271/2015-01-2222
6. Peter DW (1967 June) The use of fast fourier transform for the estimation of power spectra: a method based on time averaging over short, modified periodograms. IEEE Trans Audio Electroacoust AU-15(2):70–73

A Fault Identification Method of Rear Axle Bearing Under Lateral Dynamic Load of Vehicle

Xin Wan, Jun Zhang, Zhongming Xu, Mi Shen and Zhao Yang

Abstract As an important component of automobile, rolling bearing has a great impact on the safety of vehicle and occupants. At present, most of the literatures have studied the damaged bearing on the bench, and there is little research on the bearing failure resulted from the change of the vehicle posture. In this paper, taking the abnormal noise of the rear axle hub bearing induced by the axial load of rear axle in steering condition as an example, a method of fault identification and data analysis for the rolling bearing in the whole vehicle state is presented. First of all, the subjective evaluation of vehicles is carried out and the fault location is analyzed with the transfer path of the abnormal noise. Then, an objective test is designed to collect the vibration acceleration data of the rear axle bearings and the data is analyzed by wavelet packet. The appropriate wavelet base function and the number of decomposition level are selected to decompose the signal into different frequency bands. The frequency band containing the most information of bearing fault is determined by calculating the energy distribution of frequency bands. Envelope analysis for the fault frequency band is used to extract the characteristic frequency of the fault bearing. The correctness of the analysis results is verified by disassembling the hub bearing, and it also shows that this method can be effectively used to judge the fault location and identify the fault type of rolling bearing under the condition of the whole vehicle.

Keywords Rolling bearing · Lateral dynamic load · Fault identification · Wavelet packet

X. Wan · Z. Xu · Z. Yang
Automotive Collaborative Innovation Center of Chongqing University, Chongqing, China

X. Wan · J. Zhang (✉) · M. Shen · Z. Yang
Chongqing Changan Auto Oushang R&D Center, Chongqing, China
e-mail: zj_zmkm@126.com

China SAE (ed.), *Proceedings of China SAE Congress 2018: Selected Papers*,
Lecture Notes in Electrical Engineering 574,
https://doi.org/10.1007/978-981-13-9718-9_58

1 Introduction

As a type of rolling bearing, the hub bearing is an important component of the rear axle of car, mainly used to reduce the friction loss [1]. In addition to being a connecting part, the rolling bearing also has to bear certain loads, which makes it work under alternating stress for a long time. At the same time, because of the problems in the production process and installation, the bearing may experience imperceptible damage due to such problems, like mechanical stress and wear and so on during operation. Under the axial force in steering condition, the working condition of bearing is even worse, causing more serious damages. At this time, the bearing can not only produce annoying abnormal noise, but also directly affect the service life of the vehicle and the safety of the passengers. Therefore, in order to determine the location and cause of the bearing failure in the whole vehicle state, it is particularly important to identify the fault of rolling bearing in an effective way.

Research on fault identification methods of rolling bearings at home and abroad mainly focuses on the following aspects. The eigenvalue method [2] preliminarily judges the bearing fault by calculating the mean value, root-mean-square value, peak value, kurtosis, pulse factor, and other characteristic values of the bearing vibration acceleration signal in time domain. The envelope analysis [3] utilizes the Hilbert transform to separate the amplitude and phase of signal to extract bearing fault information. The resonance demodulation method [4] modulates the low-frequency signal containing bearing fault information to the sensor's higher natural frequency, and then the fault characteristic frequencies are extracted through filtering and demodulation method. Wavelet analysis [5], based on wavelet transform, can decompose the signal to multi-level and multi-band segments, which is beneficial to screen out the fault frequency band of bearing. The empirical mode decomposition method [6] can be well applied to nonlinear, non-stationary bearing fault signals and adaptively separates the signals into independent components. The local mean decomposition method [7] can adaptively decompose the multi-component vibration signal into a series of single-component vibration signals according to the characteristics of the signal, realizing the demodulation of the bearing fault characteristic signal.

This paper, taking the abnormal noise of the rear axle hub bearing induced by the axial load of rear axle in steering condition as an example, finds out the abnormal component as the rear axle hub bearings through transfer path analysis. An objective test is designed to collect the vibration acceleration data of the rear axle bearings. Then, the wavelet packet analysis and frequency band energy analysis are used to extract the abnormal frequency band. The characteristic fault frequencies are successfully extracted by the envelope analysis of the reconstructed signals of the abnormal frequency bands. The analysis results are verified by disassembling the abnormal bearing, and the improvement of the hub bearing press-mounting process is put forward. While the improved scheme is applied to press the hub bearing, no abnormal noise appeared.

2 Analysis of the Abnormal Noise of Vehicle

2.1 Description of the Problem

When a vehicle was evaluated for NVH (Noise ,Vibration and Harshness) performance, it was found that certain abnormal noise, whose frequency changes significantly with the vehicle speed, occurred in the rear axle while driving at 60 km/h or more. This noise appeared on the rear axle of the car in the same direction as the steering, while no audible abnormality appeared in straight line driving. This car is prepositioning engine and rear wheel drive, with integral rear axle and spiral spring shock damper.

2.2 Location Analysis of the Abnormal Noise

Since the abnormal noise changes significantly with the speed of the vehicle, the sound source component is a rotating member. The rear axle, as shown in Fig. 1, is composed of the final drive gear, the final drive bearing (tapered roller bearing), the differential gear, the half shaft spline, and the half shaft bearing (tapered roller bearing) and hub bearing (self-aligning ball bearings). Due to the abnormal noise appearing in either the left or the right, the components related to the final drive and differential are excluded. It is most common to produce an abnormal noise in rolling bearing of the rest components, so it is important to consider whether the rolling bearing is damaged. In summary, it is presumed that the damaged parts may be half shaft bearings or hub bearings, and the abnormal noise analysis process is shown in Fig. 2.

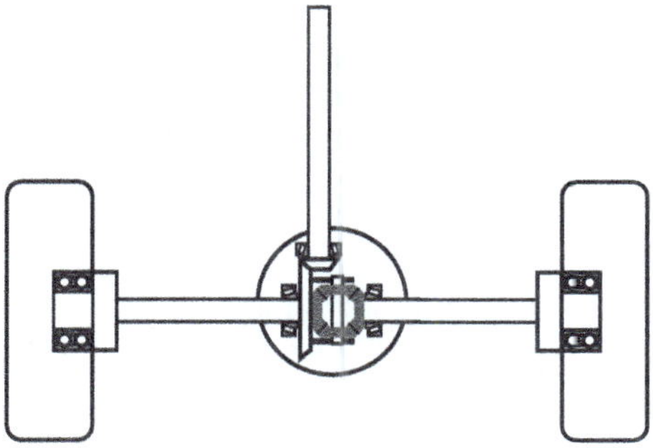

Fig. 1 Sketch of rear axle

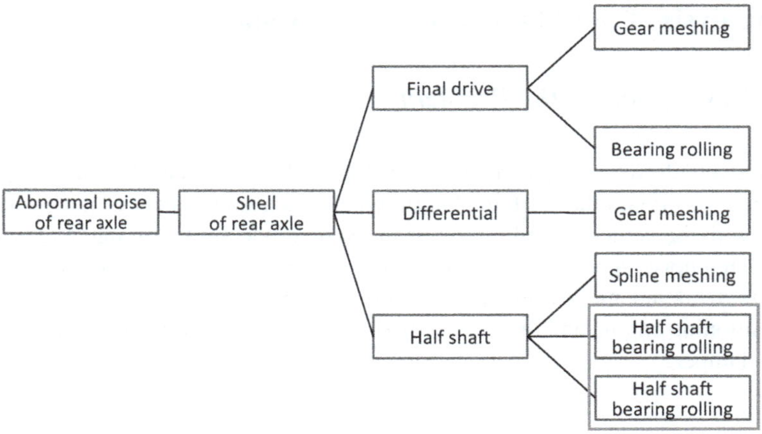

Fig. 2 Analysis process of abnormal noise

3 Experiment and Measurement System

Based on the analysis results above, a test was designed for the abnormal vehicle. Triaxial accelerometers were arranged on the left, middle, and right side of the rear axle, as shown in Fig. 3. Two vehicles were prepared for the test, one with abnormal noise and another without. Test conditions include 60–70 km/h straight and steering running on horizontal straight road.

The structure of the test system is shown in Fig. 4. The accelerometers collect the vibration signals of the rear axle bearings; vehicle speed and engine speed signals are collected through the OBD interface, the LMS data acquisition front-end, and test. Lab, the data analysis software, completes the acquisition and analysis of the data.

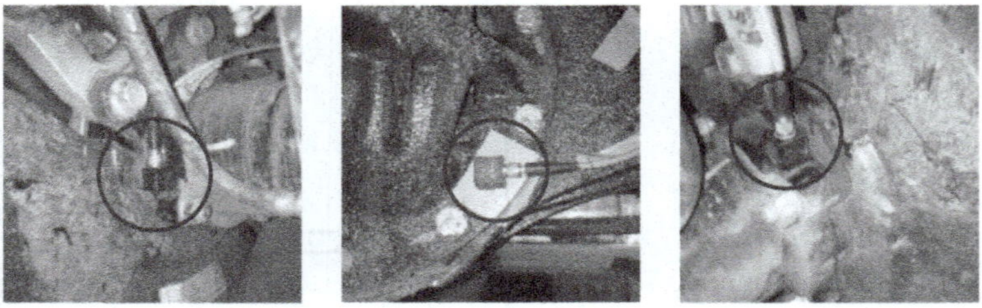

Fig. 3 Location of triaxle accelerometer

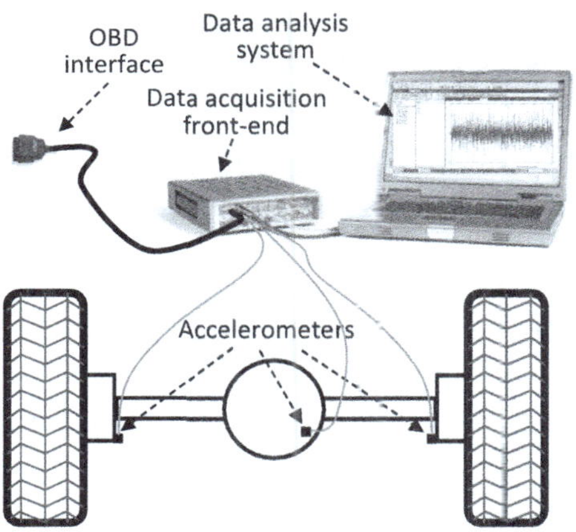

Fig. 4 Structure of measurement system

4 Analysis of Test Data

4.1 Fault Characteristic Frequencies of Rolling Bearing

When the rolling bearing is loaded, the stiffness will change as the variety of the number of loaded rolling elements at different positions, which will be responsible for the vibration of the axle center. The bearing will also vibrate accordingly when the bearing's inner ring, outer ring, rolling element, or cage is damaged. The vibration frequencies resulting from bearing damage are related to the geometric parameters, called the fault characteristic frequencies of rolling bearing. Figure 5 is the structural sketch of a rolling bearing. d is the diameter of the rolling elements, D is the pitch diameter of the bearing, D_1 is the diameter of the inner ring raceway, D_2 is the diameter of the outer ring raceway, and α is the contact angle.

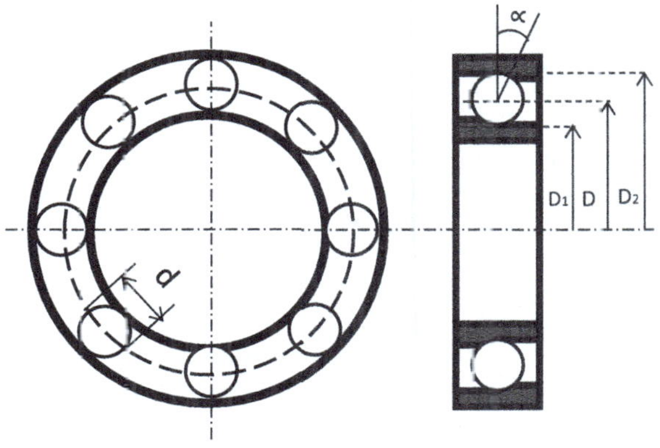

Fig. 5 Structure of a rolling bearing

When the hub bearing and the half shaft bearing are working, the outer ring is fixed, the inner ring is rotated, and the fault characteristic frequencies are calculated as follows [8, 9].

(1) Rotational frequency:

$$f_r = \frac{n}{60} \tag{1}$$

(2) Defect on inner race (ball pass frequency of inner race, BPI):

$$f_{BPI} = \frac{Z}{2}\left(1 + \frac{d}{D}\cos\alpha\right)f_r \tag{2}$$

(3) Defect on outer race (ball pass frequency of inner race, BPO):

$$f_{BPO} = \frac{Z}{2}\left(1 - \frac{d}{D}\cos\alpha\right)f_r \tag{3}$$

(4) Defect on rolling element (ball spin frequency, BS):

$$f_{BS} = \frac{D}{d}\left[1 - \left(\frac{d}{D}\right)^2\cos^2\alpha\right]f_r \tag{4}$$

(5) Defect on cage (fundamental train frequency, FT):

$$f_{FT} = \frac{1}{2}\left(1 - \frac{d}{D}\cos\alpha\right)f_r \tag{5}$$

where n represents the rotational speed of inner race, and Z is the number of rolling elements.

Table 1 shows the geometric parameters of the rear axle hub and half shaft bearing. According to Table 1, the corresponding fault characteristic frequencies are calculated, as shown in Table 2.

Table 1 Geometric parameters of the rear axle bearings

	Hub bearing	Half shaft bearing
Diameter of inner ring raceway D_1 (mm)	24.2	17.8
Diameter of outer ring raceway D_2 (mm)	35.4	13.5
Diameter of rolling element d (mm)	5.6	4.3
Pitch diameter D (mm)	29.8	31.4
Number of rolling element n	14	20
Contact angle cosα	1	0.81

Table 2 Fault characteristic frequencies of the rear axle bearings

	Vehicle speed (km/h)	Rotational frequency f_r (/Hz)	Characteristic frequency (Hz)			
			f_{BPI}	f_{BPO}	f_{BS}	f_{FT}
Hub bearing	68.5	9.3	77.3	52.6	23.5	3.8
Half shaft bearing			103.2	82.5	33.2	4.1

4.2 Wavelet Packet Analysis

Wavelet packet is a further extension of wavelet. When a signal is decomposed by wavelet analysis method, it will be decomposed into a low-frequency part and a high-frequency part, each of which occupies half bandwidth of original signal [10, 11]. Afterward, the low-frequency part is decomposed again into two bands of the same bandwidth. Therefore, the wavelet decomposition is only decomposed on the low-frequency part of the signal. While the wavelet packet analysis is applied to decompose the signal, the low-frequency and high-frequency parts will be decomposed at the same time, which realizes the refinement of the high-frequency portion of the signal.

The definition of subspace W_j^n is the closure space of function $w_n(x)$, while W_j^{2n} is the closure space of function $w_{2n}(x)$ and makes $w_n(x)$ satisfy the following two scale equations.

$$w_{2n}(x) = \sqrt{2} \sum_{k=0,\ldots,2N-1} h(k) w_n(2x - k) \tag{6}$$

$$w_{2n+1}(x) = \sqrt{2} \sum_{k=0,\ldots,2N-1} g(k) w_n(2x - k) \tag{7}$$

where $g(k)$ and $h(k)$ are filter coefficients, and $g(k) = (-1)^k h(1-k)$, that is the two coefficients are also orthogonal. The sequence $\{w_n(x)\}$ constructed by the above equations is called an orthogonal wavelet packet determined by the basis function $w_n(x) = \phi(x)$. When $n = 0$, $w_0(x) = \phi(x)$ is the scaling function and $w_1(x) = \psi(x)$ is the wavelet basis function.

Let $\{w_n(x)\}$ be the wavelet packet family of $h(k)$, and $g_j^n(x) \in W_j^n$, then $g_j^n(x)$ can be expressed as the following equation.

$$g_j^n(x) = \sum_l d_l^{j,n} w_n(2^j x - l) \tag{8}$$

where l represents different wavelet packet coefficient, j is the level number of wavelet packet coefficient, and $d_l^{j,n}$ is wavelet packet coefficient.

Wavelet packet decomposition algorithm:

$$d_l^{j,2n} = \sum_k a_{k-2l} d_k^{j+1,n} \tag{9}$$

$$d_l^{j,2n+1} = \sum_k b_{k-2l} d_k^{j+1,n} \tag{10}$$

where a_{k-2l} and b_{k-2l} are reconstruction filter coefficients.

The reconstructed signal is composed of $\{d_l^{j,2n}\}$ and $\{d_l^{j,2n+1}\}$, that is

$$d_l^{j+1,n} = \sum_k (h_{l-2k} d_k^{j,2n} + g_{l-2k} d_k^{j,2n+1}) \tag{11}$$

When bearing is damaged, the energy distribution in frequency domain is different among different fault positions and fault types. Wavelet packet decomposition can adaptively decompose the signal into a series of frequency bands of the same bandwidth. Therefore, by selecting appropriate wavelet base function and the number of decomposition level, and then extracting the wavelet packet decomposition coefficients of the specified band, the signal of the specified band can be reconstructed, thereby realizing signal filtering. Define the energy of each band as E_i.

$$E_i = \int |x_i(t)|^2 \mathrm{d}t \tag{12}$$

4.3 Hilbert Envelope Analysis

Envelope analysis is based on the Hilbert transformation. It is an important tool in the analysis of rolling bearing signals. It converts the one-dimensional time-domain function into a two-dimensional analytic function, realizing the demodulation of amplitude and phase [12, 13].

There is a signal $x(t)$, and its Hilbert transformation is $\hat{x}(t)$, then

$$\hat{x}(t) = \frac{1}{\pi} \int_{-\infty}^{\infty} \frac{x(\tau)}{t-\tau} \mathrm{d}\tau = \frac{1}{\pi} \int_{-\infty}^{\infty} \frac{x(t-\tau)}{\tau} \mathrm{d}\tau = x(t) * \frac{1}{\pi t} \tag{13}$$

Define the analytic function $z(t)$, then

$$z(t) = x(t) + j\hat{x}(t) = A(t)\mathrm{e}^{j\varphi(t)} \tag{14}$$

where $A(t) = \sqrt{x(t)^2 + \hat{x}(t)^2}$, called the envelope of $x(t)$.

4.4 The Fault Identification Method Based on Wavelet Packet, Envelope, and Energy Distribution

The core of applying the fault identification method based on wavelet packet, envelope, and energy distribution is to decompose the signal into different frequency bands by appropriate wavelet base function and number of decomposition level and find out the band containing the bearing fault information to reconstruct signal. Finally, the envelope analysis of the reconstructed signal is used to extract the characteristic frequencies. The processes are shown in Fig. 6.

Table 3 is the subjective evaluation of the abnormal noise of vehicle No. 1 and No. 2 in straight and steering conditions. Vehicle No. 1 is a normal vehicle, while vehicle No. 2 is the one with abnormal noise, and Fig. 7 shows the corresponding vibration acceleration signals of the rear axle bearings. It can be seen from Fig. 7 that there is impact phenomenon in the signals of vehicle No. 2 in straight and steering conditions, and it is more obvious in steering condition.

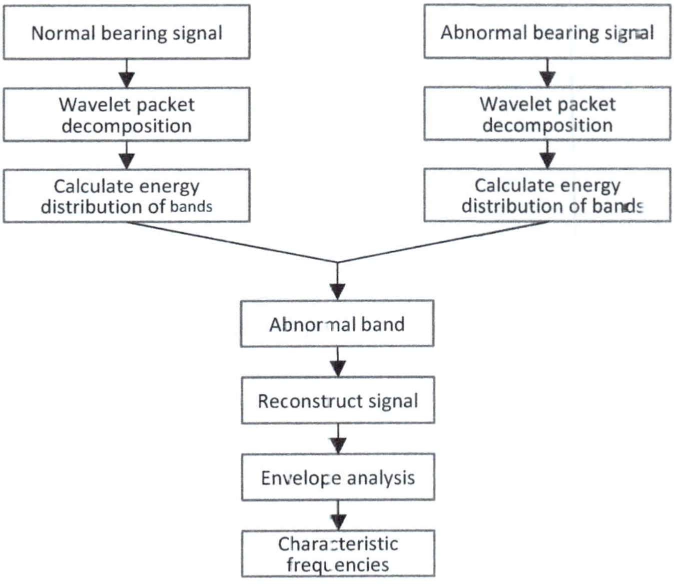

Fig. 6 Processes of the bearing fault identification

Table 3 Subjective evaluation of the abnormal noise of the rear axle

	Straight	Steering
Vehicle No. 1 (normal)	No abnormal noise	No abnormal noise
Vehicle No. 2 (abnormal)	No abnormal noise	Obvious abnormal noise

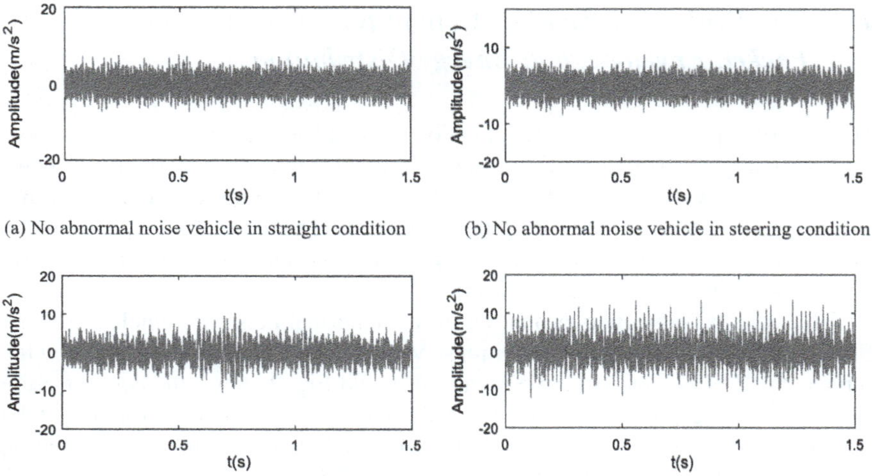

(a) No abnormal noise vehicle in straight condition (b) No abnormal noise vehicle in steering condition

(c) Abnormal noise vehicle in straight condition (d) Abnormal noise vehicle in steering condition

Fig. 7 Signals of the rear axle bearings

With "db9" selected as the wavelet base function and 4 as the decomposition level, each of the four sets of data above is decomposed into sixteen frequency bands by wavelet packet method. The energy of each band of the four sets of data is calculated separately to obtain the energy distribution, as shown in Fig. 8. It can be seen from Fig. 8 that the signal energy of the no abnormal noise vehicle is mainly concentrated in frequency band 1, while part of the signal energy of the abnormal noise appears in band 2. It is worth noting that almost half of the energy of the abnormal noise vehicle in steering condition is concentrated in band 2, indicating that band 2 is the one containing considerable information about the bearing fault.

It can be seen from the spectrums of band 2 of the abnormal noise vehicle, as shown in Fig. 9, that there are a large number of harmonics and frequency modulated phenomena in steering condition, while a small in straight condition. By means of extracting the wavelet packet coefficients of band 2 of the abnormal noise vehicle in straight and steering conditions, the time-domain signals are reconstructed, as shown in Fig. 10. Compared with the original time-domain signal (Fig. 7), the impact

Fig. 8 Energy distribution of the four sets of data

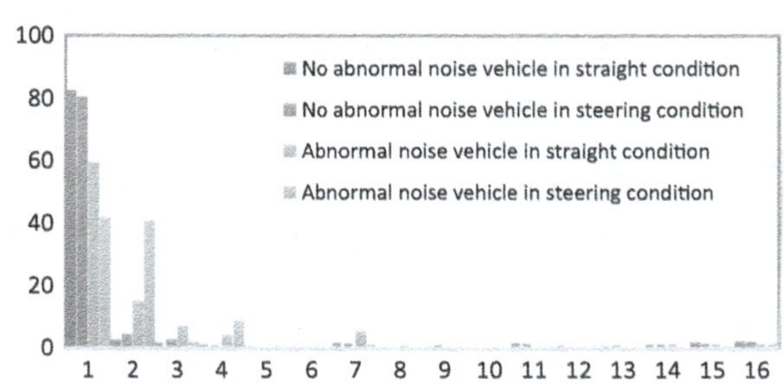

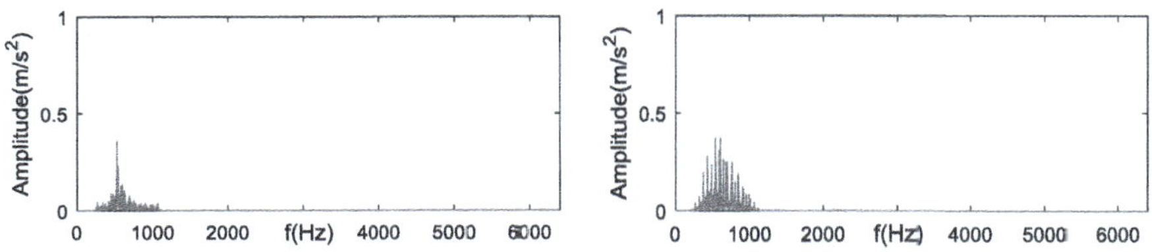

(a) Straight (b) Steering

Fig. 9 Spectrums of band 2 of the abnormal noise vehicle

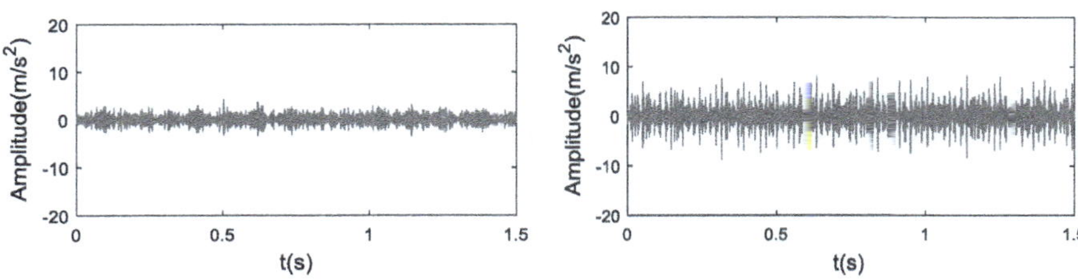

(a) Straight (b) Steering

Fig. 10 Reconstructed signals by wavelet packet coefficients of band 2 of the abnormal noise vehicle

and periodic components of the reconstructed signals are more obvious, and the information irrelevant to the bearing fault is greatly reduced.

The impact and periodic components of the reconstructed signals contain the information about the characteristic frequencies of the bearings, but they are modulated to high frequencies. Therefore, envelope analysis and FFT are performed on the reconstructed signals to obtain the envelope spectrums of the reconstructed signals, as shown in Fig. 11. There are spectral lines of 9.3, 53.3, and 78 Hz in Fig. 11a, 53.3 Hz and its double, triple, and quadruple frequencies, and 75.3 Hz in Fig. 11b. Among them, 9.3 Hz is the rotational frequency of the inner ring, 53.3 Hz is very close to the outer ring characteristic frequency of the hub bearing ($f_{BPO} = 52.6$ Hz), and 78, 75.3 Hz is very close to the inner ring characteristic frequency of the hub

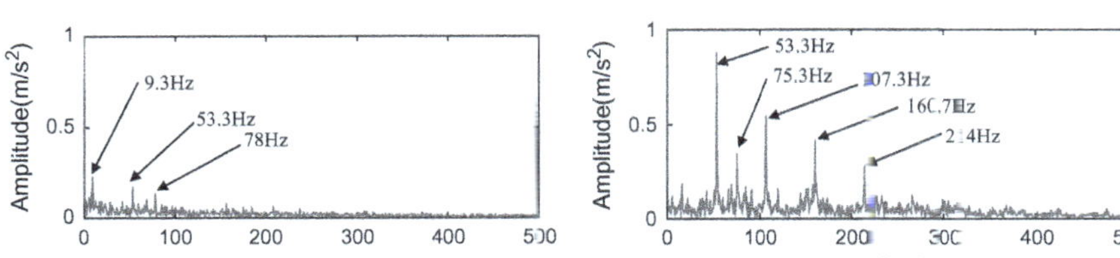

(a) Straight (b) Steering

Fig. 11 Envelope spectrums of the reconstructed signals of band 2 of the abnormal noise vehicle bearing ($f_{BPI} = 77.3$ Hz). Taking into account of the bearing wear, dimension errors and calculation errors and so on, it can be considered that 53.3 Hz is the outer ring characteristic frequency of the hub bearing, and 78, 75.3 Hz is the inner ring characteristic frequency of the hub bearing. Therefore, it can be confirmed that the damaged component of the rear axle is the hub bearing.

In the envelope spectrums of the reconstructed signals of band 2 of the abnormal noise vehicle, as shown in Fig. 11a, the weak energy of f_{BPI} and f_{BPO} of the hub bearing suggests that the damage has already existed, although there is no obvious noise in straight condition. In Fig. 11b, there are obvious spectrum lines of f_{BPI}, f_{BPO}, and harmonics of the hub bearing, indicating the severe damage under the axial force in steering condition.

Combining the energy distribution of the signals in frequency domain (Fig. 8) and the envelope spectrums of reconstructed signals (Fig. 11), it can be inferred that the hub bearing has been slightly damaged in the early stage due to the installation or certain quality problem. In the driving condition, the initial damage is rapidly amplified under the axial force induced by the change of the vehicle posture, such as steering, producing the audible abnormal noise.

5 Disassembly of the Hub Bearing and the Improvement of Process

The hub bearing was removed from the bearing seat and disassembled. It was found that there were obvious scratches on the inner and outer ring raceways (as shown in Fig. 12), which verified the above analysis results. Inspecting the process of the hub bearing assembly found that the bearing was pressed into the seat through the inner ring end face. Since the seat and the outer ring are interference fit, the process of pressing the inner ring end face will transfer the axial force from the inner ring to the outer ring through the rolling elements, which causes the damage in press-mounting process. The initial minor damage gradually increases under the axial load during the steering and finally produces a significant abnormal noise.

Therefore, it is necessary to improve the press-mounting process of the hub bearing to ensure that the end faces of the inner and outer rings are evenly stressed to avoid

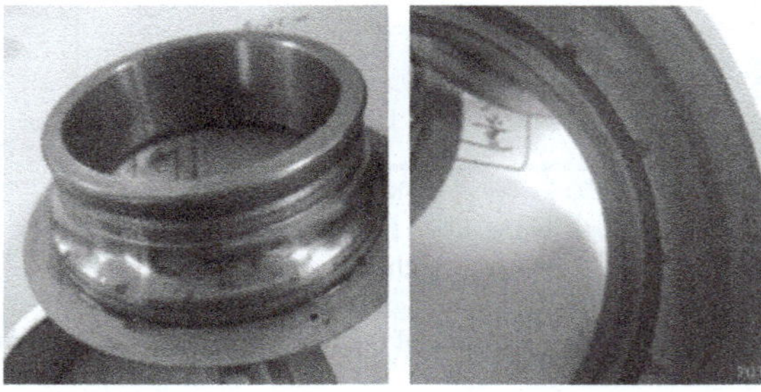

Fig. 12 Disassembly of the hub bearing

press-mounting damage. The press-mounting fixture is improved, and Fig. 13 is the comparison between before and after the improvement. While the improved scheme is applied, no abnormal noise occurred. With the fault identification method above, the vibration signal of the process-improved hub bearing under the same condition is processed (as shown in Fig. 14). In Fig. 14, there are no spectrum lines of the characteristic frequencies appear, indicating that the scheme works in avoiding the press-mounting damage and meets the requirements for use.

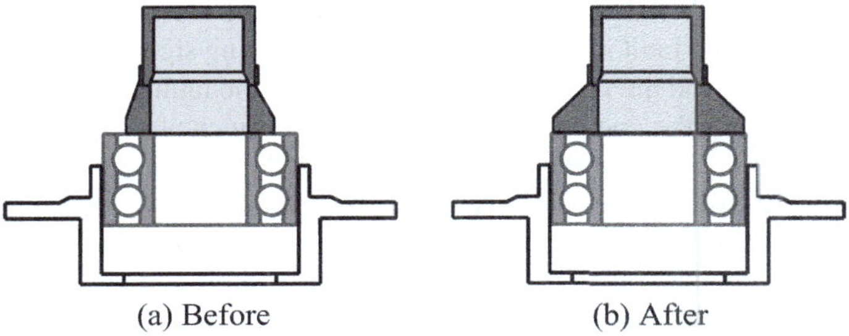

Fig. 13 Sketch of the press-mounting fixture

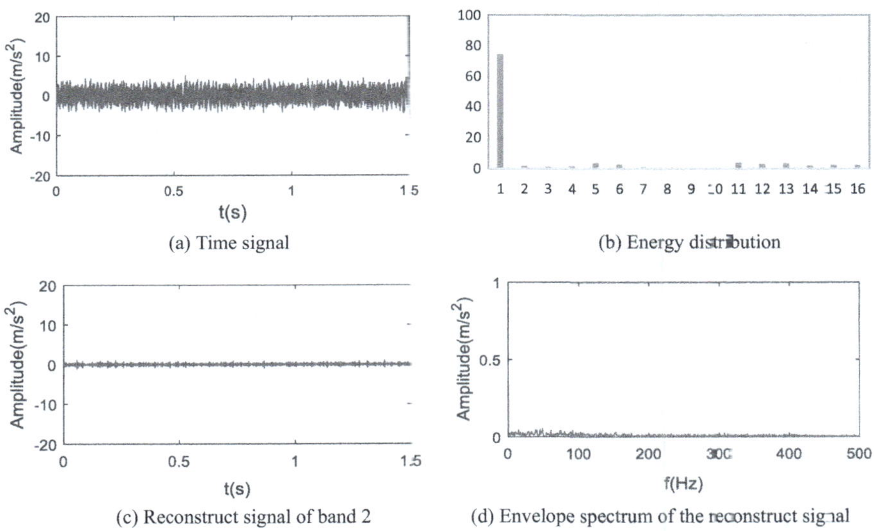

Fig. 14 Bearing signal of rear axle with improved process

6 Conclusion

Rolling bearings are important components of automobiles. When damage occurs, they are generally in the state of the whole vehicle, and it is inconvenient to disassemble and inspect. At this moment, rapid and efficient location and fault identification of damaged bearings can undoubtedly reduce significant time and economic costs. In this paper, the abnormality of the rear axle hub bearing induced by the change of the vehicle posture is taken as an example, and a fault identification method for the rolling bearing in the whole vehicle state is proposed. The wavelet packet analysis is applied to the bearing fault identification. The fault bearing signal is decomposed by the wavelet packet, and the frequency band containing the fault information is determined by analyzing the energy distribution of the frequency bands to reconstruct the bearing fault signal, and the fault characteristic frequencies were successfully extracted by the envelope analysis. The analysis results were verified by disassembling the fault bearing. Meanwhile, it also shows that this method can effectively identify and predict the fault of rolling bearings in the whole vehicle state. The cause of the damage is analyzed by the type of bearing damage, and the defect whose root cause results from the press-mounting process is found. Finally, the bearing press-mounting process was improved by the design of the press-mounting fixture. The improved scheme ensures that the inner and outer ring end faces are evenly stressed when the bearing is pressed, avoiding the press-mounting damage. After adopting the improved scheme, the problem of abnormal noise of the rear axle hub bearing was solved.

References

1. Yang L (2012) Early failure analysis & experimental research of automobile wheel hub bearing. South China University of Technology
2. Hu ZY, Hu JX, Xie LY et al (2016) Review on signal processing for rolling bearing vibrations. Chin J Constr Mach 14(6)
3. Zhao YH, Chen H, Hou YD et al (2012) Method for rolling bearing diagnosis based on HHT. Bearing 12:48–51
4. Liu JZ, Ding XW, Wang GC (2007) Adaptive resonance demodulation method and its application to fault diagnosis of freight car rolling bearings. J Vib Shock 26(1):38–41
5. Zhou XY, Ye YZ (2006) Application of wavelet analysis to fault diagnosis. Control Eng China 13(1):70–73
6. Ali JB, Fnaiech N, Saidi L et al (2015) Application of empirical mode decomposition and artificial neural network for automatic bearing fault diagnosis based on vibration signals. Appl Acoust 89(3):16–27
7. He T, Lin YZ, Gao PG et al (2011) Application of local mean decomposition in gear fault diagnosis. J Vib Shock 30(6):196–201
8. Xu X, Zhao M, Lin J et al (2016) Envelope harmonic-to-noise ratio for periodic impulses detection and its application to bearing diagnosis. Measurement 91:385–397
9. Chen XM, Yu DJ, Luo JS (2012) Envelope demodulation method based on resonance-based sparse signal decomposition and its application in roller bearing fault diagnosis. J Vib Eng 25(6):628–636

10. Wang Y, Xu G, Lin L et al (2015) Detection of weak transient signals based on wavelet packet transform and manifold learning for rolling element bearing fault diagnosis. Mech Syst Signal Process 54–55:259–276
11. Nikolaou NG, Antoniadis IA (2009) Rolling element bearing fault diagnosis using wavelet packets. Coal Mine Mach 35(3):197–205
12. Dong JN, Shen YJ, Yang SP (2005) Study and prospect on signal process technique of bearing fault diagnosis. Bearing 01:43–47
13. Yang Y, Yu DJ, Cheng JS (2005) Roller bearing fault diagnosis based on Hilbert marginal spectrum. J Vib Shock (1):72–74 + 13

Curb Vehicle Mass Target Formulation Based on Regression Analysis

Xiaoya Jiang, Jing Li, Fan Zeng, Qiong Yang and Huijuan Yu

Abstract A method of curb vehicle weight (CVW) target setting based on linear regression analysis algorithm was presented in this research. The curb mass target model based on unitary linear recursive analysis and multivariate linear recursive analysis methods is built by analyzing the effects of vehicle performance factors. The vehicle mass target is generally defined by engineers according to the benchmark vehicles, which is not comprehensive and able to meet the development requirements. The method raised by this paper based on the large data processing and regression analysis algorithm takes a good advantage in establishing vehicle mass target.

Keywords Curb vehicle mass target · Vehicle mass model · Regression analysis

1 Introduction

Curb vehicle weight (CVW) is one of the key performance indicators in vehicle development process, which has an important impact on vehicle performance. Firstly, CVW is one of the main factors that affect fuel consumption. The vehicle fuel economy could be increased by about 5.5%, while 10% vehicle weight was reduced [1, 2]. Secondly, increasing vehicle weight would result in decreasing power performance under the same conditions [3]. And a heavier vehicle weight will lead to a longer acceleration time. Meanwhile, the vehicle inertia will also be increased, which means that the braking distance would be significantly extended when driving at high speed.

Engineers usually estimate CVW according to the density data of several similar benchmark vehicle models in working out the weight target. However, the weight accuracy and market competitiveness evaluated from the traditional method cannot be judged effectively since less data are used to support goal formulation, and the density and volume are merely considered in calculation. The long story of automobile

X. Jiang (✉) · J. Li · F. Zeng · Q. Yang · H. Yu
SAIC Motor Technical Center, Shanghai 201804, China
e-mail: jiangxiaoya@saicmotor.com

China SAE (ed.), *Proceedings of China SAE Congress 2018: Selected Papers*,
Lecture Notes in Electrical Engineering 574,
https://doi.org/10.1007/978-981-13-9718-9_59

industry so far has helped manufacturer accumulate a large amount of vehicle weight data. But these data have not been effectively utilized in the process of vehicle performance development.

In this paper, the regression analysis algorithm was used in the CVW target formulation process. The CVW mathematical model involved with vehicle type characteristics was established by regression analysis algorithm based on the large number of vehicle weight data. The reasonability and accuracy of the mathematical model were evaluated through residual distribution analysis. The weight target calculated from regression analysis model is more comprehensive and reliable since the vehicle characteristics have been taken into account synthetically.

2 Regression Analysis

The CVW target formulation process proposed in this paper is based on regression analysis algorithm. Regression analysis is one of the major branches in statistics, which is an important means in data analysis. Researchers can predict the range and expectation of the dependent variable as the independent variable takes different values by establishing a regression model between both variables. According to the characteristics of the established model, regression analysis can be divided into linear regression analysis and nonlinear regression analysis. The functional relation between the independent variable and the dependent variable is linear in the linear regression analysis model, while nonlinear in the nonlinear regression analysis model. However, some nonlinear regression analysis can be converted into linear regression analysis by means of parameter transformation. In addition, according to the number of independent variables, regression analysis can be divided into unary regression analysis and multiple regression analysis. The functional model between single independent variable and dependent variable is studied in the unitary regression analysis. Multiple regression analysis is widely applied since more than one dependent variable need to be considered in practice [4].

Volume, as one of the factors that affect CVW, has the most significant impact on weight. Therefore, volume was used as the independent variable to establish the relationship with weight based on the unary regression analysis. Then, multiple factors such as volume and wheelbase are regarded as independent variables to acquire a more accurate CVW prediction model based on multiple regression analysis. In order to employ linear regression algorithm, the following assumptions need to be satisfied:

(1) The independent variable X_i is deterministic and fixed in repeated sampling;
(2) Random error has zero mean, homoscedasticity, and autocorrelation;
(3) Random error is not related to the independent variable;
(4) Random error has normal distribution of zero mean, autocorrelation, and zero covariance.

3 CVW Mathematical Model

3.1 Selection and Processing of Independent Variable

Independent variables and dependent variables need to be determined in regression analysis. In this paper, the engine type is a categorical variable with two values: turbocharged engine and non-turbocharged engine. Dummy variables are usually assigned to 0, −1, and 1, which represent different categories in handling categorical [5]. Therefore, the engine type could be assigned as:

$$X = \begin{cases} 1 & \text{Turbocharged engine} \\ 0 & \text{Non-turbocharged engine} \end{cases} \tag{1}$$

According to the above assignment method, some descriptive categorical variables such as engine type and driving type can be considered in the model to improve the accuracy.

Volume density method is used commonly in CVW target formulation:

$$W = \rho V \tag{2}$$

where W is the curb vehicle weight, V is the vehicle volume, and ρ is the average vehicle density. The vehicle volume has a direct effect on the CVW. In addition, the type and displacement of the engine and the driving type of the vehicle are the important factors in considering the CVW target formulation. In order to clarify the correlation among variables, conducting correlation analysis is necessary.

Correlation coefficient is a statistical index to reflect the correlation degree of variables which can be given by Eq. (3):

$$\rho_{XY} = \frac{\mathrm{Cov}(X, Y)}{\sqrt{D(X)}\sqrt{D(Y)}} \tag{3}$$

where $\mathrm{Cov}(X, Y)$ represents the covariance of variables X and Y, and $D(X)$ and $D(Y)$ are variance of variables X and Y, respectively, which can be written as Eqs. (4) and (5):

$$\mathrm{Cov}(X, Y) = E[(X - E[X])(Y - E[Y])] \tag{4}$$

$$\mathrm{Cov}(X, Y) = E(XY) - E(X)E(Y) \tag{5}$$

The closer the correlation coefficient is to 1, the greater the correlation between variables X and Y. If the correlation coefficient is larger than 1, X is positively correlated with Y. Therefore, the independent variables of the model can be determined by calculating the correlation coefficient.

3.2 Unary Regression Analysis Mathematical Model

One of the independent variables, which have been determined in analysis process, is selected to establish a unary regression analysis mathematical model for preliminary estimation. The scatter plot of independent and dependent variables is acquired based on the abstracted sample data in the unary regression analysis. According to the trend of the scatter plot, the functional relationship between the two variables can be determined roughly. Here, an example is used to illustrate the analysis procedure. Once the linear relationship between the independent variable and the CVW is clear, the functional formula could be assumed as Eq. (6):

$$\hat{y} = a\hat{x} + b \tag{6}$$

where $\hat{y}$ and $\hat{x}$ are dependent and independent variables, respectively, and a and b are model parameters which can be calculated from analysis results. Some data points have little effect on the result of regression model, while some have great effect. Therefore, the bad points need to be eliminated. The definition criteria of bad points are according to the leverage ratio of residuals calculated from the initial model. Leverage ratio can be used to characterize the impact of the data points on the overall model. The point with high leverage ratio is bad point that needs to be deleted. The leverage ratio can be calculated through Eq. (7). The point is considered to be a high leverage point when the leverage ratio is greater than $6/n$.

$$h_i = \frac{1}{n} + \frac{(\delta_i - \overline{\delta})^2}{\sum (\delta_i - \overline{\delta})^2} \tag{7}$$

where n is sample total number, $\delta_i = y - \hat{y}$ is the corresponding residual of dependent variable i, and $\overline{\delta} = \frac{1}{n}\sum (y_i - \hat{y})$ is the mean value of residual of all dependent variables.

Once the a and b have been calculated according to the principle of least square method, the credibility of the model could be analyzed and evaluated. The four basic hypotheses of the regression analysis described in Part 1 are employed mainly for evaluation model. And the model can be evaluated and further revised by the plot of residuals and the coefficient of determination.

The coefficient of determination R^2 is the ratio of the explained sum of squares (ESS) to the total sum of squares (TSS). The R^2 can be calculated by Eq. (8). The R^2 is used to illustrate the goodness of fit of the model to a certain extent. The closer the R^2 is to 1, the closer the ESS to the TSS is, and the higher the fitting degree of the regression model is. It is generally accepted that the coefficient R^2 is greater than 0.8, and the model can be accepted.

$$R^2 = \frac{\text{ESS}}{\text{TSS}} = 1 - \frac{\text{RSS}}{\text{TSS}} \tag{8}$$

where ESS is explained sum of squares, TSS is total sum of squares, and RSS is residual sum of squares. ESS, TSS, and RSS can be written as:

$$\begin{cases} \text{ESS} = \sum_{i=1}^{n} (\hat{y}_i - \overline{y})^2 \\ \text{RSS} = \sum_{i=1}^{n} (y_i - \hat{y}_i)^2 \\ \text{TSS} = \sum_{i=1}^{n} (y_i - \overline{y})^2 \end{cases} \tag{9}$$

$$\text{TSS} = \text{ESS} + \text{RSS} \tag{10}$$

The residuals and residual distributions maps can be utilized to verify the basic assumptions of linear regression based on the studentized residual error. If the regression model accords with the basic hypothesis, the studentized residual error will be approximately normal distribution, and about 95% will fall in the interval [−2, 2]. The studentized residual error can be calculated by Eq. (11):

$$r_i = \frac{\delta_i}{\hat{\sigma}\sqrt{1 - h_i}} \tag{11}$$

where h_i can be written as:

$$\begin{cases} \delta_i = y_i - \hat{y}_i \\ \hat{\sigma}^2 = \sum_{i=1}^{n} \delta_i^2/(n - k - 1) \end{cases} \tag{12}$$

3.3 *Multiple Regression Analysis Mathematical Model*

Engineers need to consider the influence of other factors on CVW such as fuel consumption, wheelbase, and so on in practical projects. Therefore, the unary regression model is not enough to acquire an accurate and reasonable result. Multiple linear regression model is used to analyze the influence of multiple independent variables. Assuming that there are k independent variables, multiple linear regression analysis model is given by Eq. (13):

$$y = \beta_0 + \beta_1 x_1 + \beta_2 x_2 + \cdots + \beta_k x_k + e \tag{13}$$

where β_0 is a constant term, and $\beta_1, \beta_2, \ldots, \beta_k$ are partial regression coefficients representing the average change of the dependent variable under the condition that the independent variable changes one unit, while the other independent variables are unchanged. e is the random error of affecting the dependent variable y after removing k independent variables.

The multiple regression model obtained from the sample estimation can be written as Eq. (14):

$$\hat{y} = b_0 + b_1x_1 + b_2x_2 + \cdots + b_kx_k \tag{14}$$

where$\hat{y}$ represents the estimated value of the dependent variable y by the sample, b_0 is a constant term, and $b_1, b_2, \ldots, b_k$ are the partial regression coefficients calculated from the sample.

Assuming that the sample size is n, the vector and matrix are introduced as Eq. (15):

$$b = \begin{pmatrix} b_0 \\ b_1 \\ \cdots \\ b_k \end{pmatrix}, \quad Y = \begin{pmatrix} y_1 \\ y_2 \\ \cdots \\ y_n \end{pmatrix}, \quad X = \begin{bmatrix} 1 & x_{11} & x_{21} & \cdots & x_{k1} \\ 1 & x_{12} & x_{22} & \cdots & x_{k2} \\ 1 & x_{13} & x_{23} & \cdots & x_{k3} \\ \cdots & \cdots & \cdots & \cdots & \cdots \\ 1 & x_{1n} & x_{2n} & \cdots & x_{kn} \end{bmatrix} \tag{15}$$

b can be calculated through the principle of least square method:

$$b = (X^{\mathrm{T}}X)^{-1}X^{\mathrm{T}}Y \tag{16}$$

The established multiple regression model can be evaluated from three aspects: the coefficient of determination, the F test, and the normal distribution test of residual error. The coefficient of determination has been introduced, and the F value of the variance test is given by Eq. (17). Since more independent variables are contained in multiple regression model, it is more difficult to get the studentized residual errors. Therefore, the relative residuals δ_i/y_i were directly used for analysis.

$$F = \frac{\mathrm{ESS}/k}{\mathrm{RSS}/(n-k-1)} \tag{17}$$

4 Calculation and Verification

Here, an instance was used to introduce concretely the CVW target formulation process according to the multiple regression analysis model. Selecting one vehicle randomly with five doors and five seats, SUV, forward-drive, engine displacement 2.0 L, non-turbocharging, and its parameter configuration are shown in Table 1.

Table 1 Vehicle configuration

Configuration description	Parameter
Class	Compact SUV
Length (mm)	4732
Width (mm)	1854
Height (mm)	1680
Transmission case	5MT
Wheelbase (mm)	2710
Front track (mm)	1554
Rear track (mm)	1570
Fuel	Gasoline

4.1 The Selection and Processing of Independent Variables

Many factors can affect the CVW such as volume, engine displacement, and wheelbase. Eight independent variables which have obvious influence on the CVW are selected in this paper, as shown in Table 2.

Since the independent variable contains continuous variables and qualitative variables, here, the reliability of the final model is evaluated only from data analysis, while the actual meaning of the model is not considered. In order to facilitate the calculation, the CVW is logarithmic converted; that is, the dependent variable is ln(CVW).

The capacity of sample is 3263. According to Eq. (1), the correlation coefficient between each variable and the CVW is calculated. The absolute value is obtained which is shown in Table 3.

As can be seen from Table 3, the correlation coefficient of vehicle volume (X_1) is the largest, 0.8552, which verifies that the volume has the greatest impact on the CVW. The correlation coefficient of rear track (X_4) and wheelbase (X_7) is smaller, 0.7885 and 0.7852, respectively. The correlation coefficients of hybrids (X_3) and forward-drive (X_5) are smallest, 0.1231 and 0.2814, respectively. Therefore, X_8 and

Table 2 Meaning and value of each variable

	Meaning	Value
X_1	Volume (m^3)	Length × width × height
X_2	Engine displacement (L)	–
X_3	Turbocharging	Yes: 1, no: 0
X_4	Rear track (m)	–
X_5	Forward-drive	Yes: 1, no: 0
X_6	Rear-drive	Yes: 1, no: 0
X_7	Wheelbase (m)	–
X_8	Hybrids	Yes: 1, no: 0

Table 3 Correlation coefficient between each variable and CVW

Independent variable	Correlation coefficient
X_1	0.8552
X_2	0.6943
X_3	0.4109
X_4	0.7885
X_5	0.2814
X_6	0.5986
X_7	0.7852
X_8	0.1231

Table 4 Independent variables in multiple regression analysis

Independent variable	Meaning	Comments
X_1	Volume/m^3	Length × width × height
X_2	Engine displacement/L	–
X_3	Turbocharging	Yes: 1, no: 0
X_4	Rear track/m	–
X_5	Rear-drive	Yes: 1, no: 0
X_6	Wheelbase/m	–

X_5 are removed from independent variables in the model calculation. The revised independent variables are shown in Table 4.

4.2 *Unary Regression Analysis Model of CVW and Volume*

In this paper, 1839 samples are extracted to establish a preliminary model. The residual leverage ratio of each data is calculated, as shown in Fig. 1. The residual leverage ratio $h > 6/1839 = 0.003263$ is regarded as a high leverage point and deleted. After the high leverage point is removed, the sample size is $n = 1776$.

According to the least square principle, the model parameters a and b are calculated, respectively. The results are $a = 154.95$ and $b = -610.54$. The regression model and data scatter plot are shown in Fig. 2. The regression model between CVW and volume can be written as Eq. (18).

$$\mathrm{CVW} = 154.95V - 610.54 \tag{18}$$

The coefficient of determination R^2 of the model is calculated to be $0.8243 > 0.8$. According to Eq. (18), the studentized residual error map could be drawn, as

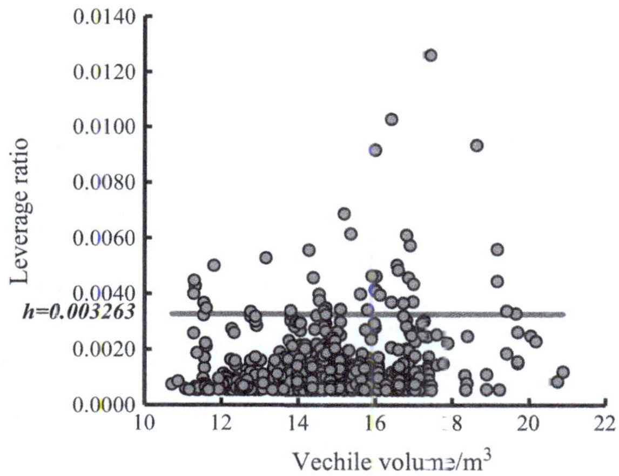

Fig. 1 Scatter plot of leverage ratio and vehicle volume

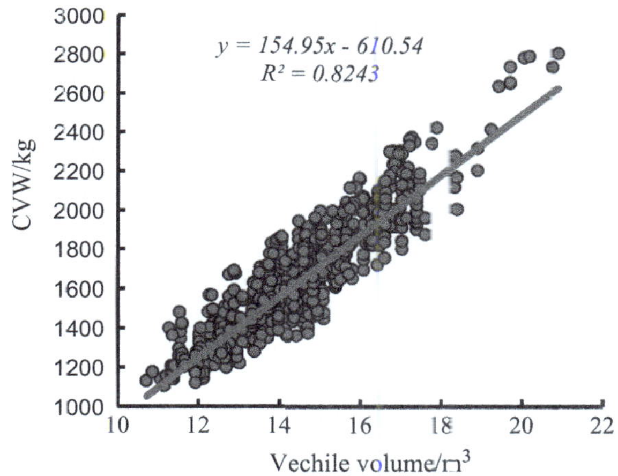

Fig. 2 Regression analysis of CVW and vehicle volume

shown in Fig. 3. The studentized residual error distribution map is shown in Fig. 4. As can be seen from Fig. 3, the mean value of residuals is 0, and the distribution of residuals is uniform, which indicates that the random error is not related to the independent variable (volume), and there is no heteroscedasticity in the regression model. In addition, there is no obvious outlier in the residual distribution, which indicates that the high leverage points are removed effectively and completely. It can be seen from Fig. 4 that residuals are approximately normal distribution. According to the data statistics, we can see that there are 1679 sample points in the range of [−2, 2], accounting for 94.54% of the total number of samples, which indicates that the regression model accords with the basic hypothesis of linear regression. Therefore, the regression model in Eq. (18) can be accepted.

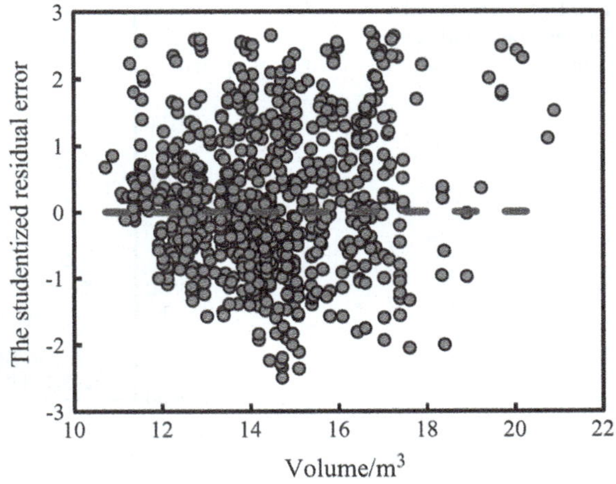

Fig. 3 Studentized residual error diagram

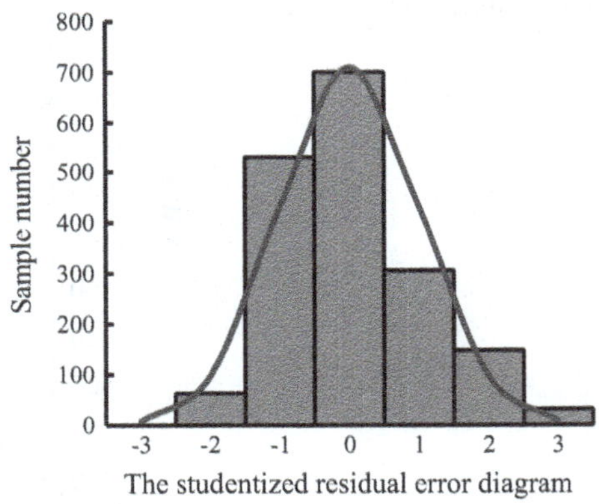

Fig. 4 Distribution of studentized residual error

According to the above one element model, the initial weight target of the case vehicle can be calculated to be CVW = 1673.24 kg.

4.3 Multiple Regression Analysis Model

The samples are selected and bad points are removed, the total capacity is $n = 3263$, and the number of independent variables is $k = 6$. Equation (19) shows the calculated partial regression coefficient b:

$$b = \begin{pmatrix} 5.2173 \\ 0.0422 \\ 0.0540 \\ 0.0547 \\ 0.7416 \\ 0.0831 \\ 0.0820 \end{pmatrix} \quad (19)$$

Therefore, the regression model estimated is as follows:

$$y = 5.217 + 0.042\,x_1 + 0.054\,x_2 + 0.055\,x_3 + 0.741\,x_4 + 0.083\,x_5 + 0.082\,x_6 \quad (20)$$

The estimated value of the CVW is:

$$\mathrm{CVW} = \mathrm{e}^y \quad (21)$$

The coefficient of determination and F value of the regression model (Eq. 21) are calculated, respectively. The coefficient of determination of the model is 0.9097, while $F = 4095$, which shows that the regression model has a high credibility. Figure 5 is the residual error distribution map of the multiple regression model. It can be seen that the residual distribution of the model is approximately consistent with the standard normal distribution model, and the relative residuals are mainly concentrated in the range of $[-0.1, 0.1]$, which indicates that the model can be accepted.

According to the model, the weight target CVW = 1531.21 kg can be calculated by substituting the vehicle parameters.

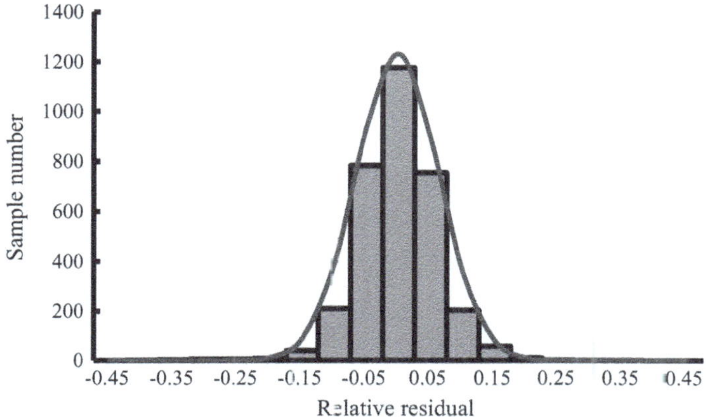

Fig. 5 Distribution of relative residual

5 Conclusion

The regression analysis method is employed to establish the CVW target formulation model including the unary regression model of CVW and volume as well as the multiple regression model of CVW and six independent variables. We can select the unary regression model or multiple regression model according to the demands in the actual project development process. Compared with the previous method of CVW target formulation method, the method described in this paper is supported by a large number of data, and the weight target based on vehicle characteristics is more comprehensive and accurate, which is of great significance to the development of the vehicle performance.

References

1. Wang H, Chen J (2009) Lightweight structure and lightweight material for automobile body. Peking University Press
2. Pagerit S, Sharer P, Rousseau A (2006) Fuel economy sensitivity to vehicle mass for advanced vehicle powertrains. SAE Technical Paper
3. Jihui Z (2016) Automobile performance analysis and new energy vehicle technology. China Water & Power Press
4. Gao H (2012) Optimal model construction in some kinds of commonly used nonlinear regression analysis and their intelligent realization by SAS. Academy of Military Medical Sciences
5. Zhang J, Li H (2005) The assignment of qualitative variables in regression analysis. Evid Based Med (03):169–171

The Research of Friction Coefficient of Seatbelt's Through Ring and Webbing Based on Orthogonal Test Matrix

Guangyi Huang, Chengjing Zhou and Hongji Chen

Abstract With the in-depth understanding of vehicle's safety knowledge and the increase of seatbelt's wearing rates, people pay more and more attention to seatbelt's wearing comfort and rapid retraction performance. Also, the Chinese market's complaint about seatbelt is increasing. From the IQS result, the main part of seatbelt's IQS complaint in Chinese market is about seatbelt's retraction performance, and it shows an upward trend. The friction between seatbelt's through ring and webbing is an important factor that influences the performance of seatbelt's retraction, but there is no enough research result or selection standard about the friction coefficient of the through ring and webbing. In this paper, we analyzed the layout of seatbelt and the main factors affecting webbing's retraction, set orthogonal test matrix for the combination of different through rings and webbings, obtained the friction coefficients, and analyzed its influence to the webbing's pull-out and retract force. This work will provide a lot of convenience for the development of follow-up vehicles.

Keywords Seatbelt · Retraction force · Through ring · Webbing · Retraction performance

1 Introduction

With people's in-depth understanding of vehicle's knowledge, customers have more and more attractive requirements for vehicle's comfort and convenience. At the same time, because of the requirements of national laws and people's awareness of accidents, the wearing rate of seat belts is increasing day by day [1]. Therefore, the attention to seatbelt in Chinese market is also increasing. So seatbelt's other properties except such as convenience, comfort, retract performance have become a very important factor that affecting vehicle's charms.

In recent years, the Chinese market customers are very concerned about whether the seatbelt can be retracted to the initial location after it is released. From the

G. Huang (✉) · C. Zhou · H. Chen
Dongfeng Nissan Passenger Vehicle Company, Guangzhou, China
e-mail: huangguangyi@dfl.com.cn

China SAE (ed.), *Proceedings of China SAE Congress 2018: Selected Papers*, Lecture Notes in Electrical Engineering 574,
https://doi.org/10.1007/978-981-13-9718-9_60

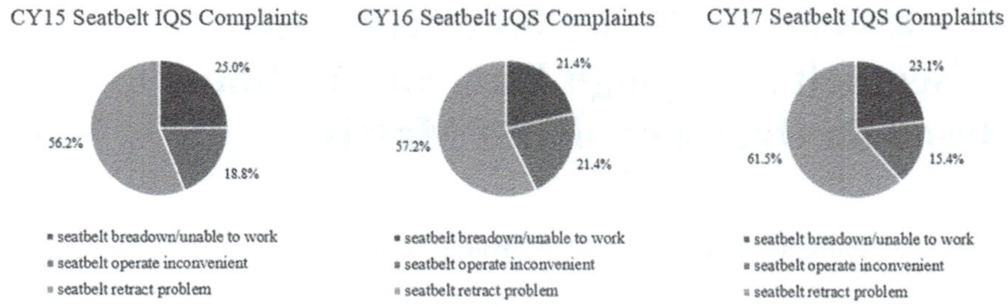

Fig. 1 Seatbelt IQS complaints in recent years

results of IQS [2], in 2015, the ratio of complaints about seatbelt retracted slowly or do not retract is 56.2% in all of the seatbelt's IQS complaints. In 2016, this ratio reached 57.2%. In 2017, it reached 61.5%, as shown in Fig. 1. It can be seen that complaint about seatbelt's poor retract performance is a major complaint item among customers in the Chinese market, and the uptrend is obvious. The design and study of the seatbelt's retract performance have become an important and urgent topic for all OEMs.

At present, in the digital stage of the seatbelt development, there is a clear requirement for the maximum retract force of the seatbelt, and this force should be measured in real vehicle condition. But at this stage, the seatbelt supplier cannot measure the retract force exactly according to the real vehicle condition, and the force can only be measured on the test bed. Under the influence of the friction between the through ring and webbing, the test bed cannot fully simulate the real state of the vehicle and other factors, and it often appears that the retract force measured in the test bed meets the design requirements, but the retract force is not enough after the seatbelt is installed to the vehicle. And in seatbelt's design and development stage, the friction force and friction coefficient between the through ring and webbing are not fully considered, and also there is no good criteria for the through ring and webbing's selection.

In this paper, the seatbelt's layout and the hindering factors of the seatbelt's retract are analyzed firstly. And then, the orthogonal test matrix is set to measure the retract force/pull-out force, so the friction force and coefficient between the through ring and webbing can be obtained by calculating; the relationship between friction coefficient and retract/pull put force is also analyzed in this paper. This results can provide a selection basics for the through ring and webbing in the follow-up vehicles' seatbelt development.

2 Seatbelt Assembly

2.1 *Seatbelt Assembly Introduce*

Seatbelt is an important part of the vehicle's restraint system, which mainly realizes the function of restraining the occupants on the seats and reducing the occupant's injury when the vehicle collides. There are many types of seatbelt, a three point-type 4 seatbelts is usually used for passenger vehicle. All of the seatbelt mentioned below belong to this type.

The seatbelt assembly mainly includes three parts: the retractor assembly, the buckle assembly, and the shoulder adjuster assembly, as shown in Fig. 2. The components of the retractor assembly include retractor, through ring, webbing, tongue, and ANCH plate. The webbing is rolled around the retractor's spindle, and the webbing passes through the through ring and tongue after being drawn out from the spindle and is connected with the ANCH plate. After the tongue is released from the buckle, under the effect of coil spring's retract force, the webbing overcomes the friction with through ring and other resistance in the path, and the tongue returns to its original stop position.

The object of this paper is the friction coefficient and friction force between the through ring and webbing.

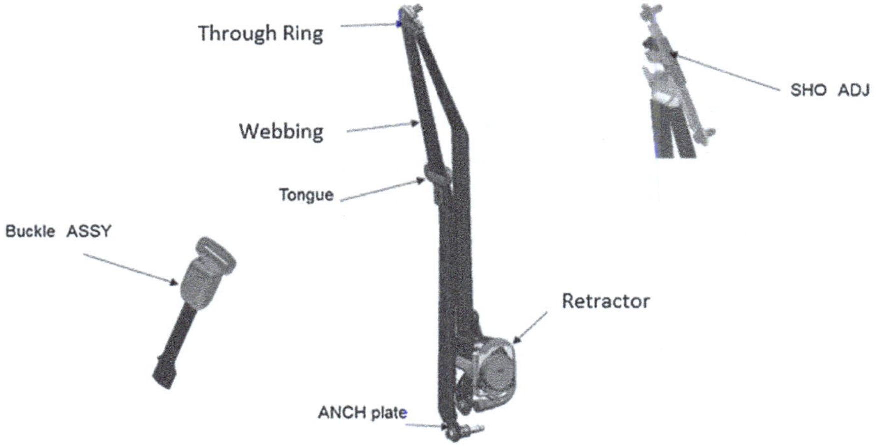

Fig. 2 Seatbelt assembly components

2.2 Classification of the Through Ring and Webbing

There are various forms of the through ring. There are often differences between through rings used by different OEMs and suppliers. From the relationship with B-pillar trim, the through ring can be divided into two forms: built-in (inside B-pillar trim) and exposed (outside B-pillar trim). From the material of the through ring, it can be divided into pure PA, pure POM, POM + insert, metal with chromium plating, metal, and other forms, as shown in Fig. 3. The friction force of through rings with different materials and structures is quite different.

The classification of the webbing is much more complicated; different manufacturers have different standers and forms. Generally speaking, according to the different weaving methods of webbing, it can be divided into four forms: no stripes, five stripes, seven stripes, and American standard seven stripes, as shown in Fig. 4. And the same striped webbing also has different knitting methods; the main knitting methods are regular twill weave, change twill weave, and cross-twill weave, as shown in Fig. 5. Different knitting methods have different contact areas with through ring, and the friction force is different.

Because of the diversity of through rings and webbings, OEMs often face difficult choices when developing seatbelt. Besides the appearance, the friction between the through rings and webbings is the key factor that should be considered. At present, the friction force or coefficient between the through ring and webbing has no well choice standard; generally, it is based on experience to judge. For example, it is considered that the friction force of metal through ring is small, and the smooth through ring's friction force is smaller than the diagonal through ring's force.

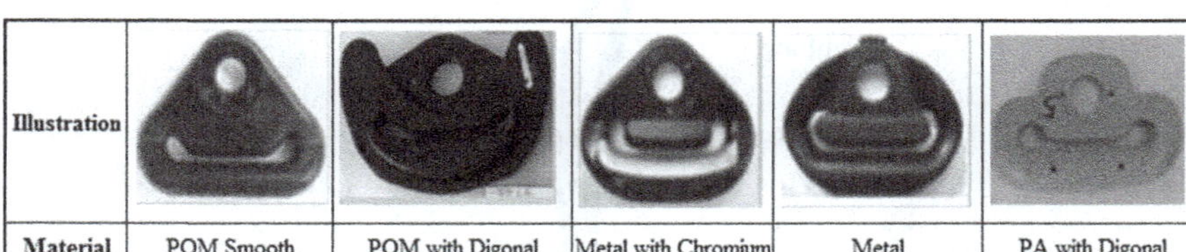

Fig. 3 Through rings with different materials and structures

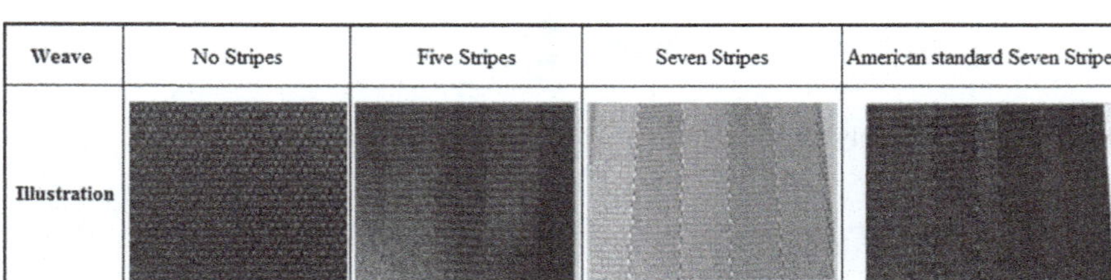

Fig. 4 Webbing with different stripes

Knitting method	Regular twill weave	Change twill weave	Cross twill weave
Features	The longitude and latitude lines are arranged regularly, with uniform density and large contact area.	Between regular twill weave and cross twill weave	The latitude and longitude lines are irregular, the ribbon is soft, and the contact area is the smallest.
Internal organization	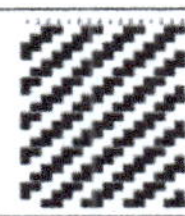		
Practicality amplification	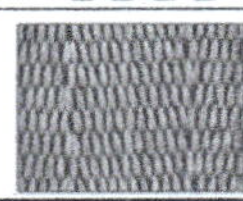		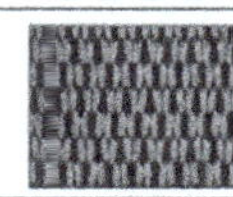

Fig. 5 Webbing with different knitting methods

The purpose of this paper is to analyze the friction coefficient of the combination of different kinds of through rings and webbings by setting the orthogonal test matrix, so that the optimum combination of the through ring and webbing can be selected in the development process of the seatbelt.

3 Seatbelt Retract Obstacles Analysis

This chapter analyzes the layout and environment parts of seatbelt and confirms the main friction effects on the path of the seatbelt's retraction; the purpose is prepared for set up the test bed to measure the friction force and calculate the friction coefficient between the through ring and webbing.

3.1 Main Environmental Parts of the Seatbelt

The front and rear seatbelts are all made of three point-type 4 with similar structure. The matching mode of the through ring and webbing is the same; the difference is the location of the installation. Because of the space limitation, this paper takes the front seatbelt as an example to illustrate.

The retractor assembly is fixed on the B-pillar sheet metal and threshold by the bolts that are used for fastening the upper/lower ANCH plate and the retractor bracket, as shown in Fig. 6. It can be seen from Fig. 6, the main environmental parts of the seatbelt assembly include B-pillar sheet metal, B-pillar trim, threshold, and seat.

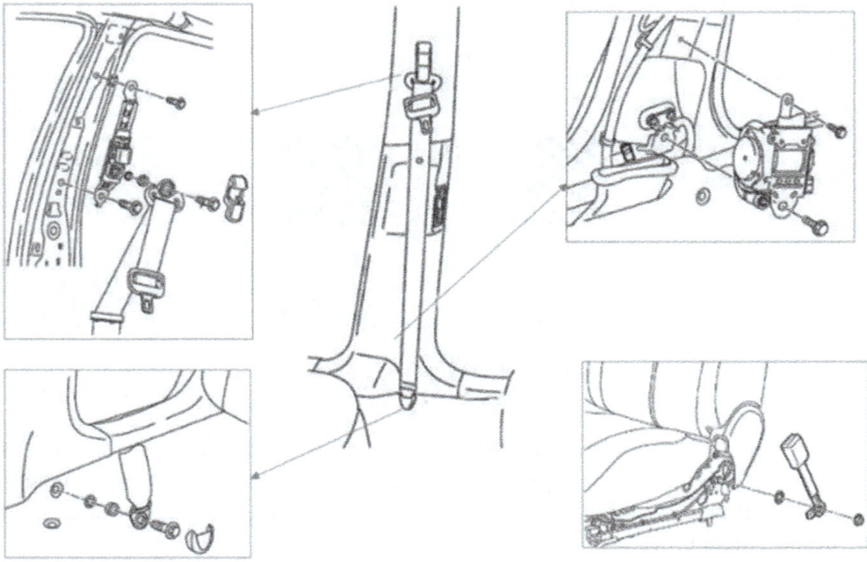

Fig. 6 Installation of seatbelt

3.2 *Analysis of the Retraction Path and Friction Effect of the Seatbelt*

This section analyzes the retraction path of the seatbelt to determine the main friction effects of the webbing's retraction and to prepare for setting up the test bed. Take the front seatbelt as an example at this paper; the webbing drawn out from the retractor, through the gap between the B-pillar sheet metal and the trim, around the through ring, passes through the gap between the B-pillar trim and the seat and finally fixed on the lower ANCH plate which is fixed on the threshold, as shown in Fig. 7.

There is a clearance requirement between webbing and B-pillar sheet metal and B-pillar trim to avoid the friction when the seatbelt is developed and designed. The threshold is below the webbing and will not rub with the webbing. The factors affecting the retraction of the seatbelt are as shown in Table 1.

3.3 *The Force Analysis of Seatbelt Retraction*

When the tongue released from the buckle, the coil spring provides the power for webbing's retraction, so that it can overcome the friction between the through ring and webbing and retract to the initial position. The role of the through ring can be equivalent to fixed pulley.

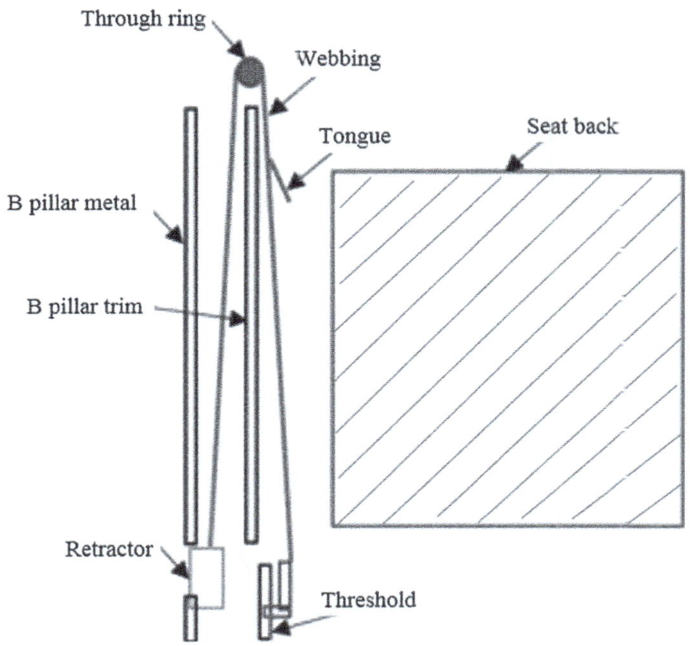

Fig. 7 Path of the webbing

Table 1 Affecting factors on webbing retraction

No.	Affecting factors	Design requirements	Friction with webbing
1	Clearance between webbing and B-pillar metal	W/	×
2	Clearance between webbing and B-pillar trim	W/	×
3	Clearance between seat and B-pillar trim	W/	×
4	Webbing's friction coefficient	W/O	√
5	Through ring's friction coefficient	W/O	√
6	Tongue weight	W/	×

It is assumed that the clearance between the webbing and B-pillar trim or other environmental parts meets the design requirements that means there is no friction between the webbing and the environmental parts; at the same time, ignoring the weight of the webbing, the webbing's retraction process can be simplified as the physical model shown in Fig. 8. In the figure, F_j is the coil spring force, F_f is the internal friction force of coil spring, F_n is the positive pressure of the webbing to through ring, F_m is the sliding friction force between the through ring and webbing,

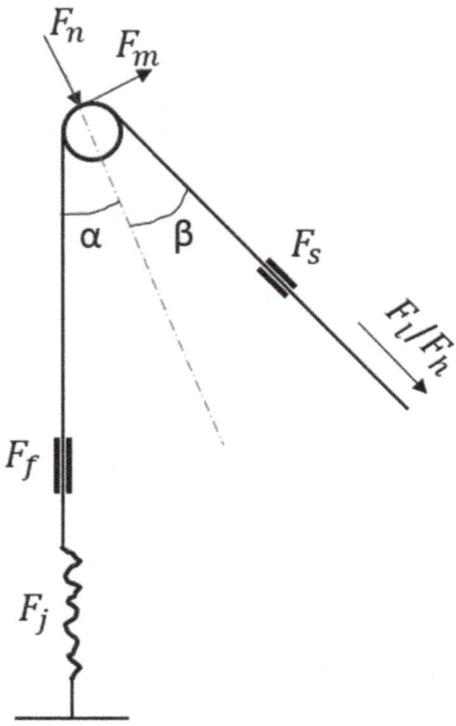

Fig. 8 Force diagram of the seatbelt retraction

F_s is the force produces by the tongue weight, and F_l and F_h is the pulled out and retracted force of the webbing.

From the above analysis, it can be seen that the risk of friction between webbing and B-pillar metal/trim and other environmental parts are very small, because the clearance has design requirements. As the webbing passes through the through ring and contact directly with it, and when the webbing retracts, it has a clear relative motion with the through ring; therefore, the obstruction of webbing retraction is mainly due to the friction between the webbing and the through ring. When constructing the test bed, the influence of other factors is ignored, mainly considering the friction between the through ring and webbing.

4 The Friction Coefficient Analysis

4.1 Orthogonal Test Matrix Setting of Through Ring and Webbing

In this section, the common used through ring and webbing are selected to set the orthogonal test, and the friction force under different combination conditions is mea-

Table 2 Orthogonal test matrix of the through ring and webbing

	No stripes	Five stripes	Seven stripes	Seven stripes II
POM smooth	1	2	3	4
POM with diagonal	5	6	7	8
POM with steel insert	9	10	11	12
POM with chrome-plating insert	13	14	15	16

Table 3 Precondition of the test

No.	Test one: calculation of friction coefficient	Test two: pull-out/retract force measurement
①	Setting up a test bed according to the actual vehicle layout	Setting up a test bed according to the actual vehicle layout
②	Replacing the coil spring with the weight block, the mass is 1 kg	Record the pull-out/retract force
③	Vertical installation of weight block, the distance to through ring is 600 mm	–
④	The angle of the ribbon is 45°	–
⑤	The weight block moves at a constant speed, and the speed is 30 mm/s	–
⑥	Record the pulling force F	–

sured on the test bed, so that the friction coefficient of the through ring and webbing is obtained by calculation.

The experiment is divided into two parts—Test one: the friction coefficient test of through ring and webbing's combination. The purpose is to measure the pull-out force F; then, the friction coefficient can be calculated. Test two: test of pull-out/retract force with different through rings and webbings combinations. The purpose is to analyze the relationship between the friction coefficient and pull-our/retract force of webbing. Sixteen groups of tests were carried out with different through rings and webbing combinations, as shown in Table 2. The specific requirements and conditions of the test are shown in Table 3.

4.2 *Analysis of the Test Result*

Pulling force F of the through rings with different materials combined with webbings with different stripes is measured by means of test 1, as shown in Table 4. The through ring and webbing are equivalent to the fixed pulley model, ignoring the self-weight of the webbing; the friction force between the through ring and webbing equals pulling force F minus gravity of the weight block, so the friction coefficient μ can be calculated, as shown in Table 5. The combination of the POM with chrome-plating

Table 4 Pulling force F in combination of through ring and webbing (Unit: N)

	No stripes	Five stripes	Seven stripes	Seven stripes II
POM smooth	14.22	15.48	15.72	15.41
POM with diagonal	14.35	15.94	16.51	16.06
POM with steel insert	14.11	17.35	16.51	16.06
POM with chrome-plating insert	13.89	15.57	16.16	15.27

Table 5 Friction coefficient μ in combination of through ring and webbing

	No stripes	Five stripes	Seven stripes	Seven stripes II
POM Smooth	0.199	0.243	0.251	0.241
POM with diagonal	0.204	0.258	0.276	0.262
POM with steel insert	0.195	0.274	0.273	0.260
POM with chrome-plating insert	0.187	0.246	0.265	0.236

insert through ring and webbing of no stripes has minimum friction coefficient, but the combination of the POM with diagonal through ring and seven stripes webbing has the maximum friction coefficient. The friction coefficient of different through rings and webbings can be obtained by this method, which can be used as the basis for selecting through rings and webbings.

Analyse the friction coefficient between different through rings and webbings. From Fig. 9, it can be seen that the POM with chrome-plating insert through ring and no stripes webbing has the minimum friction coefficient, but the friction coefficient is maximum when it combine with seven stripes webbing. According to traditional experience, the friction coefficient of POM with chrome-plating insert through ring should be the smallest no matter what kind of webbing. But from the experimental results, it is not exactly. Besides the through ring, the friction coefficient also related

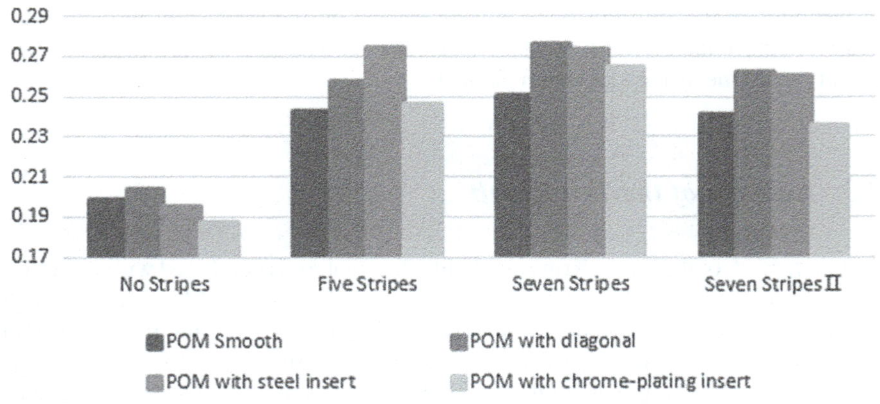

Fig. 9 Comparison of friction coefficient between different through rings and webbings

to the type of webbing. Optimal combination of through ring and webbing should be selected according to friction coefficient.

Analysis of pull-out force of different through ring and webbing combinations. The pull-out force measured by the method of test two is shown in Table 6. It can be seen from Fig. 10, the POM with chrome-plating insert through ring and no stripe webbing's combination has the minimum pull-out force; the POM with diagonal through ring and seven stripes webbing's combination has the maximum pull-out force, corresponding to the minimum/maximum value of friction coefficient. Comparing with Figs. 9 and 10, it can be seen the variation trend of friction coefficient and pull-out force is consistent.

Analysis of retract force of different through ring and webbing combinations. The retract force measured by the method of test two is shown in Table 7. It can

Table 6 Pull-out force of different through ring and webbing (Unit: N)

	No stripes	Five stripes	Seven stripes	Seven stripes II
POM smooth	3.67	3.988	4.06	3.988
POM with diagonal	3.705	4.116	4.359	4.146
POM with steel insert	3.641	4.282	4.192	4.046
POM with chrome-plating insert	3.584	4.017	4.17	3.941

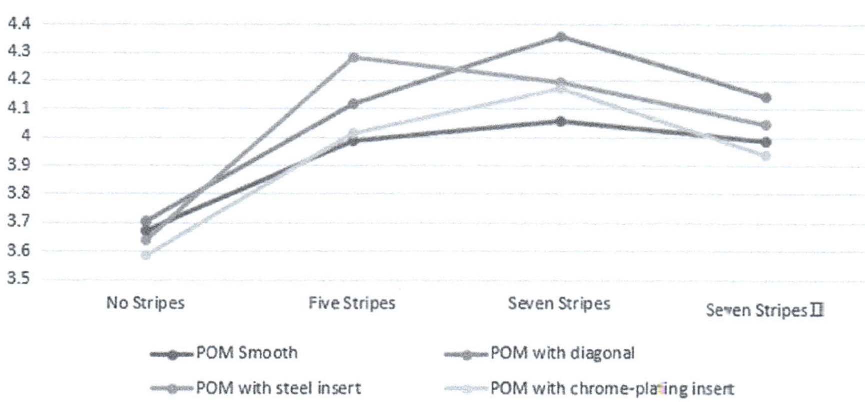

Fig. 10 Diagram of different through ring and webbing's pull-out force

Table 7 Retract force of different through ring and webbing (Unit: N)

	No stripes	Five stripes	Seven stripes	Seven stripes II
POM smooth	1.754	1.600	1.551	1.607
POM with diagonal	1.770	1.596	1.404	1.571
POM with steel insert	1.810	1.503	1.432	1.501
POM with chrome-plating insert	1.863	1.579	1.500	1.700

be seen from Fig. 11, the POM with chrome-plating insert through ring and no stripe webbing's combination has the maximum retract force; the POM with diagonal through ring and seven stripes webbing's combination has the minimum retract force, corresponding to the minimum/maximum value of friction coefficient. Comparing with Figs. 9 and 11, it can be seen the variation trend of friction coefficient and retract force is opposite.

Analyse the relationship of the friction coefficient and the pull-out/retract force. Theoretically, the friction coefficient is proportional to the pull-out force and inversely proportional to the retract force. A diagram of the relationship between friction coefficient and pull-out/retract force is shown in Fig. 12. It can be seen from the diagram: ① When the friction coefficient is maximum, the pull-out force is maximum and the retract force is minimum; when the friction coefficient is minimum, the pull-out force is minimum and the retract force is maximum; ② The pull-out force increases with the increase of friction coefficient and the retract force decreases with

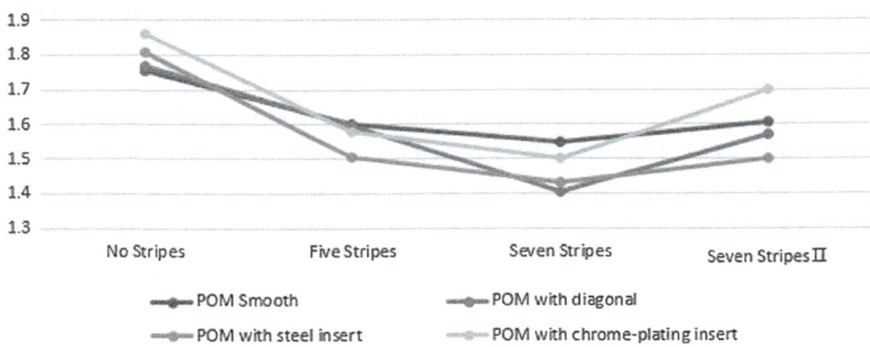

Fig. 11 Diagram of different through ring and webbing's retract forces

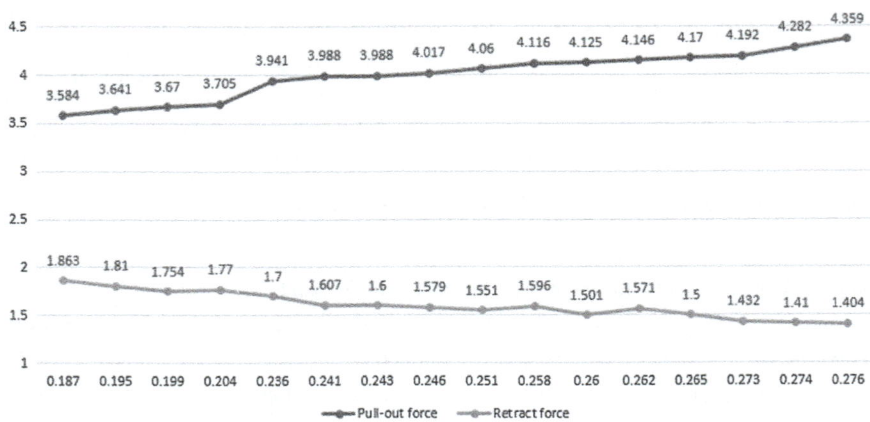

Fig. 12 Relationship between the friction coefficient and the pull-out/retract force

the increase of friction coefficient; ③ The relationship between friction coefficient and pull-out/retract force is consistent with the theoretical situation.

Based on the orthogonal test, the friction coefficient and the pull-out/retract force of the through ring and the webbing are obtained. In order to make the seatbelt easy to pull-out and retract quickly when it is released, the through ring and the webbing combination which can make the pulling force smaller and the retract force greater should be selected. From the above results, the smaller the friction coefficient between through ring and webbing, the smaller the pull-out force and the greater the retract force of ribbon, which is beneficial to the ease of pull-out and the rapidity retract of webbing.

5 Conclusions

This paper introduced seatbelt's retraction issue which is the main item of seatbelt's IQS complaint in the first place and then explained the friction between the through ring and the webbing which is an important factor that affects webbing's retraction, but at present, the selection of through ring and webbing is basically according to experience. Finally, according to the layout of seatbelt, the through ring and webbing which in common used were selected to set the orthogonal test matrix, and the test bed simulated to the real vehicle was established; the friction coefficient between the through ring and webbing was obtained by measurement and calculation. On the basis of those facts, the following conclusions can be reached,

(1) The friction between the through ring and webbing is an important factor that affects seatbelt's retraction performance. Full consideration should be paid to this friction in seatbelt's design and development stage.
(2) Through orthogonal test, webbing's pull-out and retract force, and the friction coefficient of commonly used through rings and webbings are obtained.
(3) According to the research results of this paper, the friction coefficient of different through rings and webbings can be fully compared in the design and development stage of the seatbelt, and the combination of the through rings and webbings with smaller friction coefficient can be selected to reduce the friction force of the through ring and webbing, so that the friction hindrance is reduced when the webbing is pulled out or retracted; thus, the seatbelt can be pulled out with small force and retracted rapidly. To a certain extent, it can solve IQS complaint about seatbelt's retract performance and improve the charm of vehicle.

References

1. Zhang Q (2010) Survey on the seatbelt's fasten situation of Chinese customers. World Car
2. JD Power customer survey. 2015–2017

Application of Color Variance with Angular Variation Effect in the Vehicle Color Exploitation

Huajie Xun, Chaofu Jiang, Xiaoqiang Qin, Zhongqiong Xu and Yu Xiao

Abstract The article is about an innovative method for color designing. Through magnifying the color variance with angular variation effect, we scheme out a new type of vehicle colors, which show different colors depend on light and movement.

Keywords Vehicle color exploitation · Color variance with angular variation effect · Gradient color

1 Introduction

As we know, the effective pigment used on vehicle paint has shown obviously anisotropy. Based on the specialty, most of the vehicle colors show different optical properties depend on angle of light, we name the phenomenon to be color variance with angular variation effect. For the convenience of study, we define a series of characteristic parameters to represent the color variance with angular variation effect. Based on the test range by the universal colorimeter for auto industry, we make the definition as follows. We define the lowest hue of the six test angles as shown in Fig. 1 to be hue of lower limit, the highest hue to be hue of upper limit, the hue of forty-five degree to be hue of front view, the numerical difference between the lowest and highest hue to be hue variation section, the numerical difference between the lowest and highest lightness to be lightness variation section, the numerical difference between the lowest and the highest chroma to be chroma variation section.

In traditional knowledge, the color variance with angular variation effect is harmful to color difference control. Because of this effect, the color difference between vehicle body and exteriors will be more obvious, and the control of color consistency will be more difficult. In the vehicle color exploitation, we usually focus on decreasing the color variance with angular variation effect. On the other hand, we find that it

H. Xun (✉) · C. Jiang · X. Qin · Z. Xu · Y. Xiao
Changan Automobile Co., Ltd., Chongqing, China
e-mail: xunhuajie@126.com

China SAE (ed.), *Proceedings of China SAE Congress 2018: Selected Papers*,
Lecture Notes in Electrical Engineering 574,
https://doi.org/10.1007/978-981-13-9718-9_61

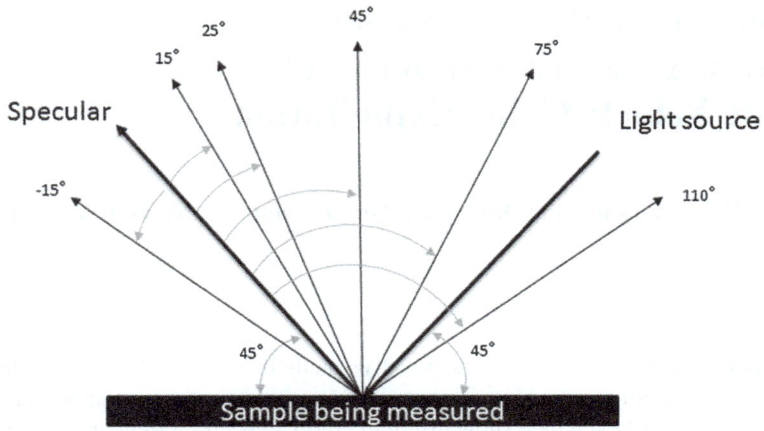

Fig. 1 Schematic table of measuring angles by BYK-mac i 23 mm 7030 colorimeter

will occur an interesting phenomenon when the color variance with angular variation effect is up to a certain degree. We will get different visual sense as the auto-body modeling variation.

2 Conceptual Design of the Gradient Color

The Soul Red of M Company is a classical application of the color variance with angular variation. We have made a digital comparative analysis between the Soul Red and Red of C Company. The results are as follows. The lightness variance section of Soul Red is 89.3% higher than the Red of C Company. The chroma variance section is 108.6% higher than the Red of C Company. The hue variance section is 144.3% higher than the Red of C Company. The difference between these characteristic values led to the visual sensible difference between the Soul Red and Red of C Company. Combined with the visual sensible compare of real vehicles, we find that the layered sense of the Soul Red is mainly due to the higher lightness variance section and chroma variance section. As the curvature change of the auto-body modeling surface, the lightness and chroma of the color will change obviously. So we get the feel of changing between light and shadow on the vehicle (Figs. 2, 3 and 4).

On the enlightenment of the Soul Red, we get a new way for vehicle color designing. Through extend the hue variance section of the colors by using a new type of advanced pearlescent pigment, we achieve the target that the paint coat will show different colors in the different view. We name this type of vehicle colors to be gradient color. The most obvious characteristic of the gradient color is that the hue of the color will change obviously as the change of the test angles. Reflected in the vehicle body, we will get color changing as the curvature change of the auto-body modeling surface (Fig. 5).

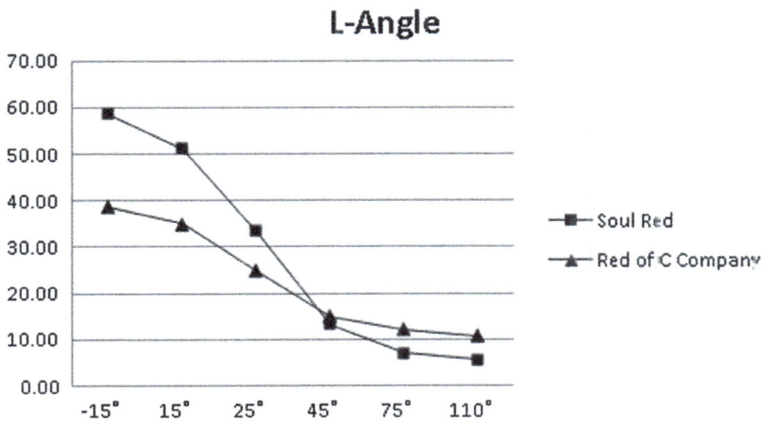

Fig. 2 Lightness variance section comparison between Soul Red and Red of C Company

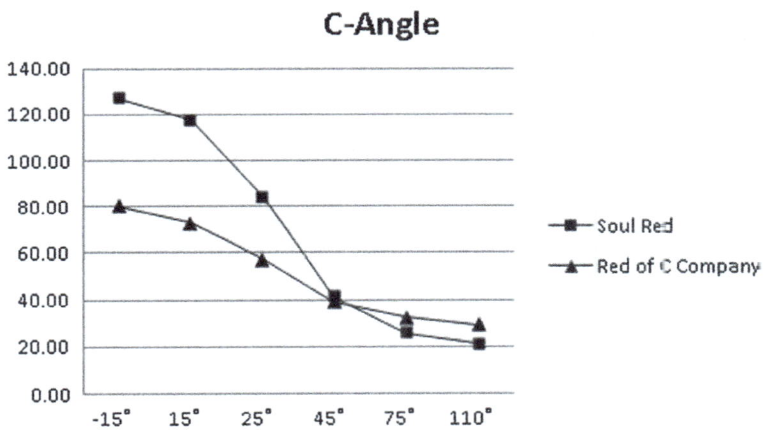

Fig. 3 Chroma variance section comparison between Soul Red and Red of C Company

Compared with the international normal color specification system and visual sense of color samples, we define the colors that the hue variation section is above ninety to be gradient color. As the most important index of gradient color, the hue variation section signifies the ability of color changing. The bigger the index is, the greater color difference between different angles will be. Such as a gradient color we designed, the hue variation section is 96.15. Another index hue of front view signifies the color we perceived in the front side of the sample. It decides the basic color of the sample and serves as the basis of color naming. As the color mentioned above, the hue of front view is 256.89, belong to blue section, so we name the color to be "XX blue." In addition, the hue of upper limit and hue of lower limit signify extension base on the basic color. The hue of lower limit of the sample is 195.82.

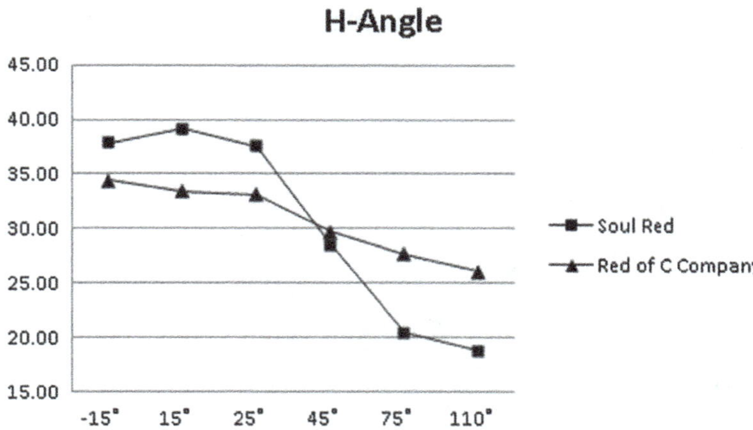

Fig. 4 Hue variance section comparison between Soul Red and Red of C Company

Fig. 5 Schematic table of the international normal color specification

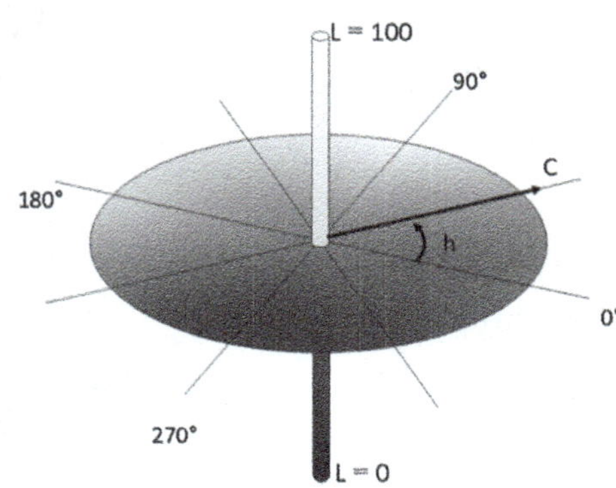

The hue of upper limit of the sample is 291.97. The color variance section we felt is from bluish green to dark blue. It just likes the sea and sky merged into one. So we name the innovative color to be sea sky blue (Fig. 6).

After the specimen has been designed, we try three different kinds of primer such as white primer, gray primer, and black primer as the basic layer. We find that the sample with black primer has the best visual effect in the three samples. The sample shows more sense of depth of the gradient color itself, because of the black primer reflecting less light and decreasing the interference from the basic layer. So we get better color-changing effect with black primer. Finally, we choose black primer as the matching primer of the gradient color.

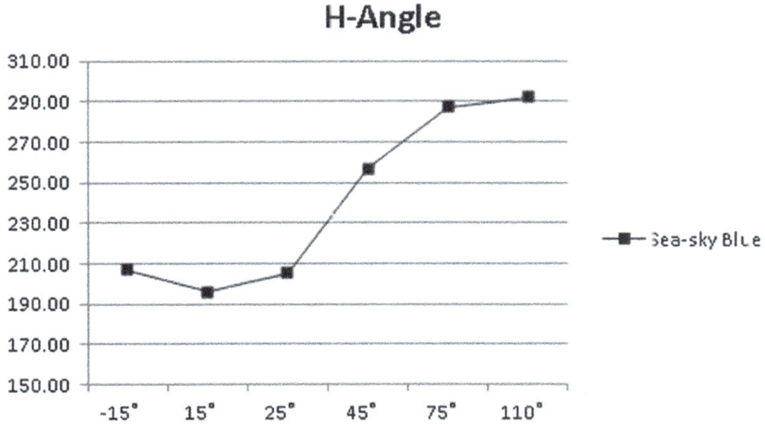

Fig. 6 Hue curve of sea sky blue

3 Coating Film Thickness Control Range Design of the Gradient Color

Because of the color variance of the gradient color, we need to consider different angles for coating film thickness designing. For design simplification, we choose three important angles of the gradient color for analysis. They are the angles of the hue of front view, the hue of lower limit, and the hue of upper limit. Firstly, we test the black and white opacity. The result is 11 Âμm. Secondly, we make a series of gradient thickness samples and test the color characteristic values of them. The results are shown in Table 1.

Table 1 Color characteristic values of the gradient thickness samples

Thickness	Angle	*L*	*C*	*H*	*E*
11	−15	54.91	65.55	206.88	223.86
	15	52.22	64.16	195.56	212.34
	25	34.69	42.21	206.12	213.24
	45	15.96	24.74	259.82	261.48
	75	10.66	27.83	285.79	287.34
	110	8.79	28.06	289.68	291.17
13	−15	54.67	65.75	206.74	223.73
	15	52.06	64.67	195.57	212.46
	25	35.02	43.19	205.64	213.02
	45	16.26	25.00	258.38	260.10
	75	10.57	28.43	286.58	288.18

(continued)

Table 1 (continued)

Thickness	Angle	L	C	H	E
	110	8.58	28.85	291.14	292.69
15	−15	54.22	65.07	206.88	223.55
	15	51.69	64.19	195.82	212.46
	25	35.29	43.61	205.38	212.90
	45	16.64	25.09	256.89	258.65
	75	10.58	28.76	287.03	288.66
	110	8.48	29.39	291.97	293.57
18	−15	54.20	65.44	206.76	223.54
	15	51.10	63.57	196.24	212.52
	25	34.44	43.92	205.86	213.29
	45	16.26	25.17	254.36	256.12
	75	10.44	28.37	286.14	287.73
	110	8.41	29.00	292.14	293.70
22	−15	52.62	63.58	207.15	222.99
	15	50.41	63.00	196.20	212.14
	25	35.27	44.06	205.06	212.68
	45	17.32	25.23	252.82	254.67
	75	10.81	28.79	286.28	287.93
	110	8.47	29.82	292.36	294.00
24	−15	45.75	56.04	209.49	221.63
	15	45.02	56.63	198.52	211.29
	25	34.74	44.37	205.02	212.62
	45	18.69	26.52	243.57	245.72
	75	11.55	28.88	283.66	285.36
	110	8.51	30.93	292.98	294.73

In order to simplify analysis further, we define a new characteristic index $E = \sqrt{L^2 + C^2 + H^2}$, then use the index to assess the relationship between color character and film thickness. We use the quadratic regression model of the Minitab analysis software to get the relationships of the three angles. They are expressed as $y = ax^2 + bx + c$. Then we make the derived function of these functions as $y' = 2ax + b$. Because we need to investigate the severity rate of this value, but not the direction of changing. We make the $Y' = |2ax + b|$ as the analysis index. We define this index of the hue of front view to be Y'_{HFV}, the hue of lower limit to be Y'_{HLL}, the hue of upper limit to be Y'_{HUL}. We select the maximum value among the Y'_{HFV}, Y'_{HLL} and Y'_{HUL} of the same film thickness and define it to be $Y'_{\max} = \mathrm{MAX}(Y'_{\mathrm{HFV}}, Y'_{\mathrm{HLL}}, Y'_{\mathrm{HUL}})$. Then we sort the $Y'_{\max}$s from smallest to bigger, which is beginning with the black and white opacity value. Based on the production control level of our coating process, we decide five micrometers as the coating film thickness control range. Combined with the thicknesses sorted by the Y'_{max}s, we get the coating film thickness control range finally.

For example, based on the regression equations of the color sea sky blue, we get

$$Y'_{HFV} = |-0.15226x + 1.663|,$$
$$Y'_{HLL} = |-0.03244x + 0.5046|,$$
$$Y'_{HUL} = |-0.0433x + 0.9808|.$$

Then we make a table of Y'_{max} versus thickness and sort them as follows. Based on the sequence of Y'_{max}, the film thickness sequence is 13, 12, 14, 11, 15, and so on. So we set 11–15 Âµm as the coating film thickness control range (Figs. 7, 8, 9 and Table 2).

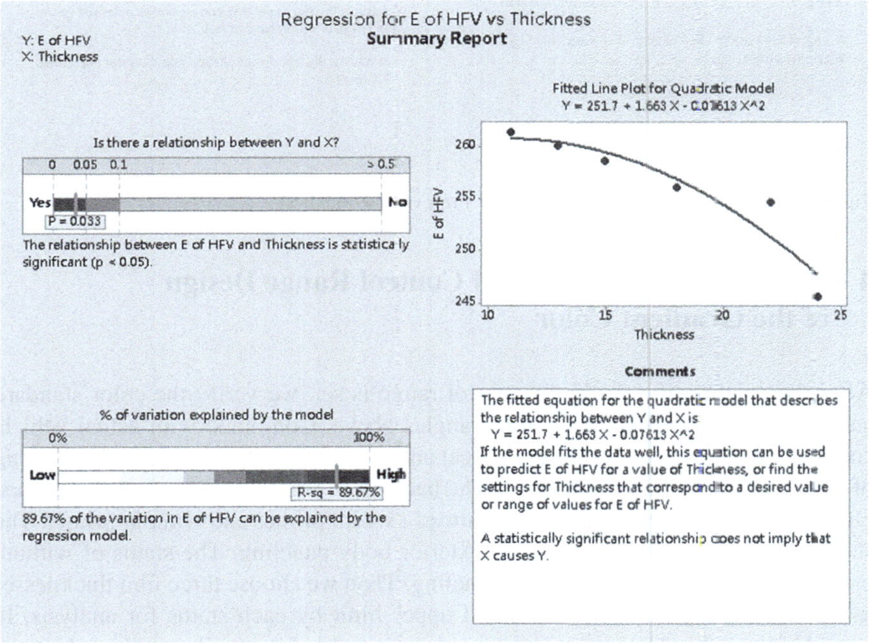

Fig. 7 Regression for *E* of the hue of front view versus thickness

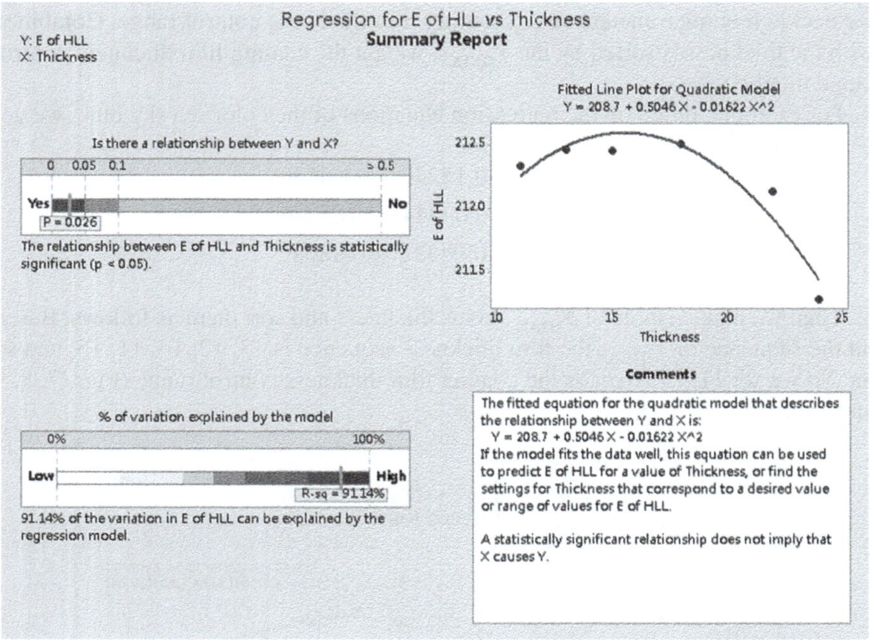

Fig. 8 Regression for *E* of the hue of lower limit versus thickness

4 Color Standard Value and Control Range Design of the Gradient Color

After the coating film thickness control range is set, we verify the color standard value and tolerance also in the three angles above. Combined with actual vehicle coating, we choose two backings as steel and plastic to make samples. The backing of steel represents the vehicle body. The backing of plastic represents the exteriors. On the backing of steel, we set two statuses as with primer and without primer. The status of with primer represents the exterior body paneling. The status of without primer represents the interior body paneling. Then we choose three film thicknesses as the midcourt line, lower limit, and upper limit by each status for analysis. In consideration of vehicle manufacture color control habit, we choose *L*, *a*, *b* as the color standard system. The design of experiments is shown in Table 3.

We make samples as shown in Table 3, and test the *L*, *a*, *b* values of them in the three angles. Then we get mean and standard deviation of these values with the Minitab analysis software. Next, we make the probability distribution plot based on the sigma level of the process. Finally, we get the mean and standard deviation of the color for process control (Table 4).

For example, the test results of the color sea sky blue are shown in Table 5.

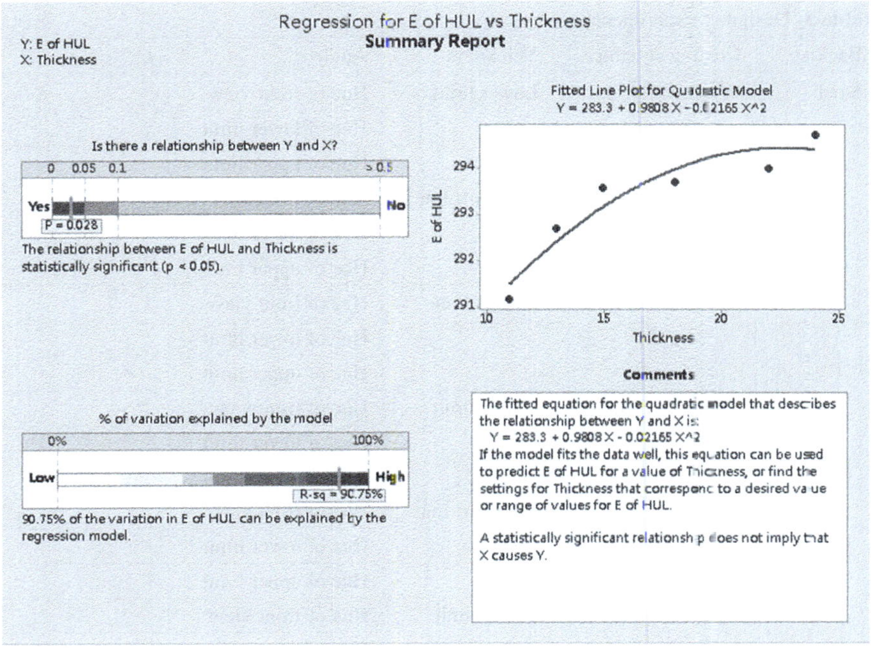

Fig. 9 Regression for *E* of the hue of upper limit versus thickness

Table 2 Sort of $Y'_{\max}$ versus thickness

Thickness	$Y'_{\max}$
13	0.4179
12	0.4612
14	0.4686
11	0.5045
15	0.6209
16	0.7732
17	0.9254
18	1.0777
19	1.2300
20	1.3822
21	1.5345

Table 3 Design of experiments

Backing	Coating structure	Thickness	Angle	*L*	*A*	*B*
Steel	With primer	Lower limit	Hue of front view			
			Hue of lower limit			
			Hue of upper limit			
		Midcourt line	Hue of front view			
			Hue of lower limit			
			Hue of upper limit			
		Upper limit	Hue of front view			
			Hue of lower limit			
			Hue of upper limit			
	Without primer	Lower limit	Hue of front view			
			Hue of lower limit			
			Hue of upper limit			
		Midcourt line	Hue of front view			
			Hue of lower limit			
			Hue of upper limit			
		Upper limit	Hue of front view			
			Hue of lower limit			
			Hue of upper limit			
Plastic	–	Lower limit	Hue of front view			
			Hue of lower limit			
			Hue of upper limit			
		Midcourt line	Hue of front view			
			Hue of lower limit			
			Hue of upper limit			
		Upper limit	Hue of front view			
			Hue of lower limit			
			Hue of upper limit			

Table 4 Contrast of the sigma level, probability of imperfection, and process quality control level

Sigma level	Probability of imperfection (ppm)	Process quality control level
6σ	3.4	Perfect
5σ	230	Excellent
4σ	6210	Good
3σ	66,800	Common
2σ	308,000	Bad
σ	690,000	Out of control

Table 5 Test results of the color sea sky blue

Backing	Coating structure	Thickness	Angle	*L*	*A*	*b*
Steel	With primer	11	45	15.96	−5.19	−24.35
			15	52.22	−61.81	−17.21
			110	8.79	9.33	−26.42
		13	45	16.26	−5.42	−24.49
			15	52.06	−61.92	−17.36
			110	8.58	10.40	−25.91
		15	45	16.64	−5.59	−24.43
			15	51.69	−61.76	−17.58
			110	8.48	10.99	−27.25
	Without primer	11	45	16.94	−5.47	−24.03
			15	52.97	−61.83	−17.94
			110	10.03	11.09	−28.35
		13	45	17.02	−5.81	−24.28
			15	52.64	−61.96	−18.02
			110	9.95	11.36	−28.67
		15	45	17.13	−6.35	−24.38
			15	52.37	−62.08	−18.13
			110	9.88	11.52	−28.88
Plastic	–	11	45	18.02	−6.21	−25.17
			15	53.81	−62.41	−18.16
			110	9.83	11.06	−28.61
		13	45	18.49	−6.42	−25.43
			15	53.47	−62.53	−18.37
			110	10.12	11.47	−28.93
		15	45	18.65	−6.63	−25.64
			15	53.14	−62.74	−18.52
			110	10.34	11.63	−29.13

Through the analysis by the Minitab analysis software, we get the mean and standard deviation of the color sea sky blue for process control base on four sigma level as shown in Table 6 (Figs. 10, 11, 12, 13, 14, 15, 16, 17 and 18).

Table 6 Mean and standard deviation values of the color sea sky blue for process control

Angle	Type	L	A	B
45	Mean	17.2	−5.9	−24.7
45	Standard deviation	14.6–19.8	−7.3 to −4.5	−26.3 to −23.1
15	Mean	52.7	−62.1	−17.9
15	Standard deviation	50.8–54.6	−63.1 to −61.1	−19.2 to −16.7
110	Mean	9.6	11.0	−28.1
110	Standard deviation	7.6 to 11.5	9.4 to 12.6	−30.9 to −25.4

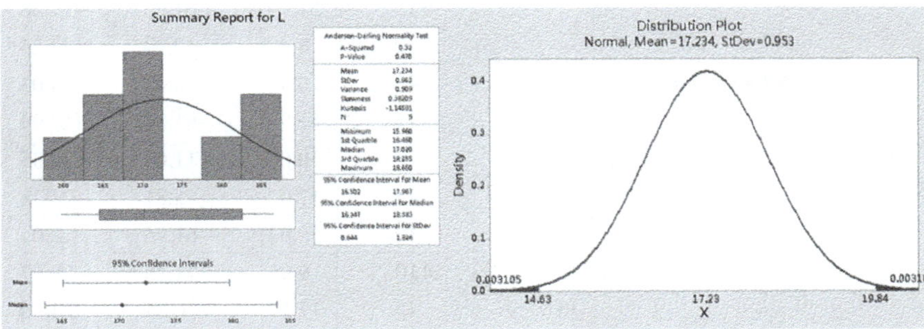

Fig. 10 Mean and standard deviation analysis of *L* in the hue of Front view base on four sigma

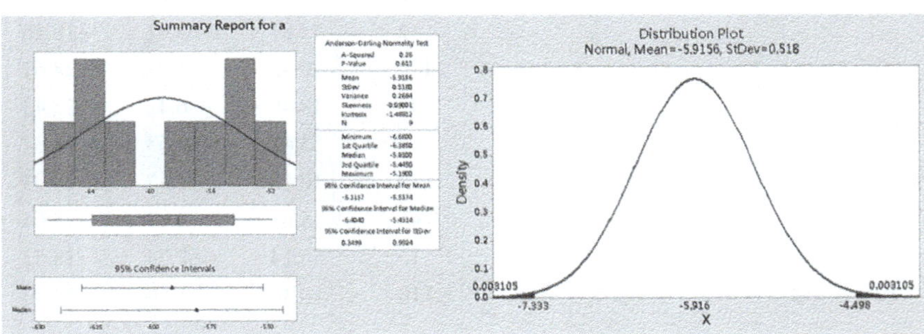

Fig. 11 Mean and standard deviation analysis of *a* in the hue of front view base on four sigma

5 Recoatability of the Gradient Color

The color-changing effect of the gradient color is mainly base on the color-changing pearlescent pigment. In order to ensure the color-changing effect, we must make the colored paint as translucent. So we must repaint the primer before the recoat of the gradient color.

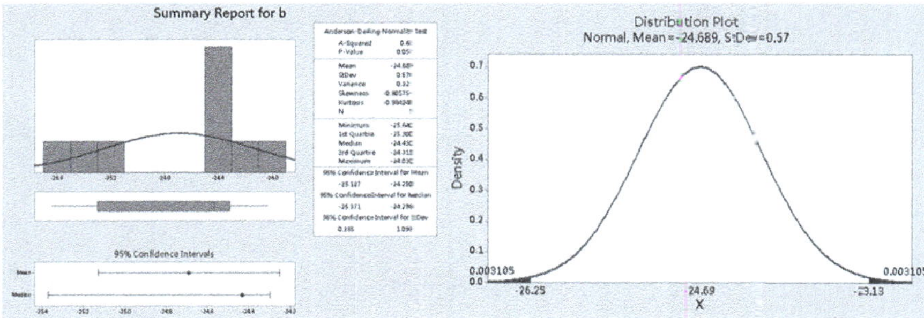

Fig. 12 Mean and standard deviation analysis of *b* in the hue of front view base on four sigma

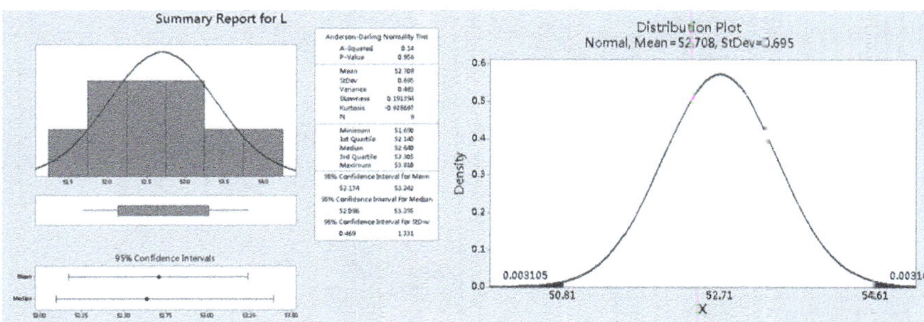

Fig. 13 Mean and standard deviation analysis of *L* in the hue of lower limit base on four sigma

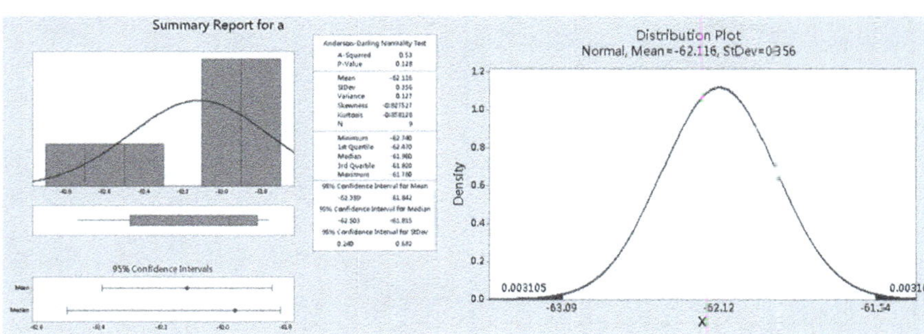

Fig. 14 Mean and standard deviation analysis of *a* in the hue of lower limit base on four sigma

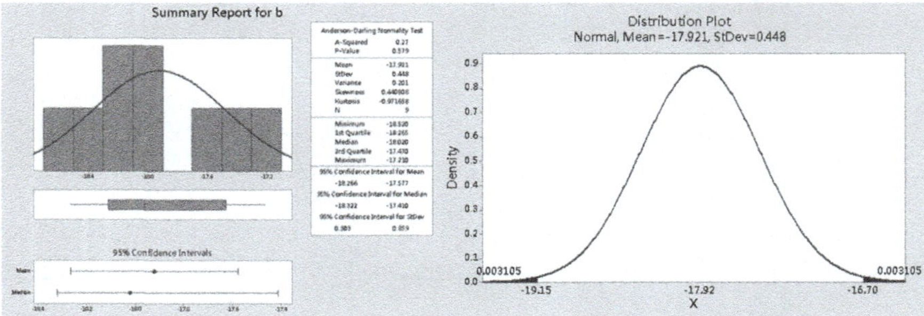

Fig. 15 Mean and standard deviation analysis of *b* in the hue of lower limit base on four sigma

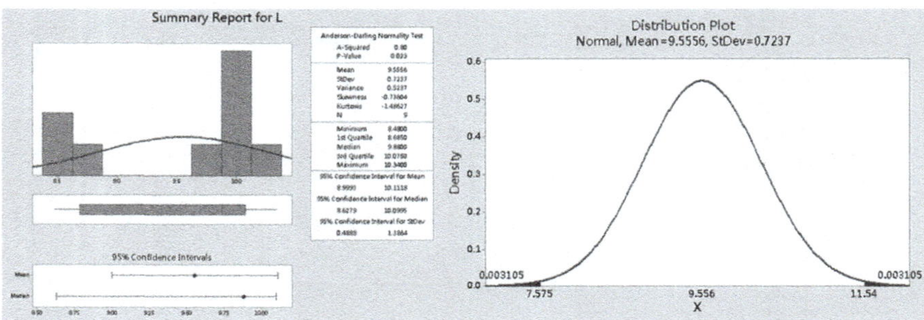

Fig. 16 Mean and standard deviation analysis of *L* in the hue of upper limit base on four sigma

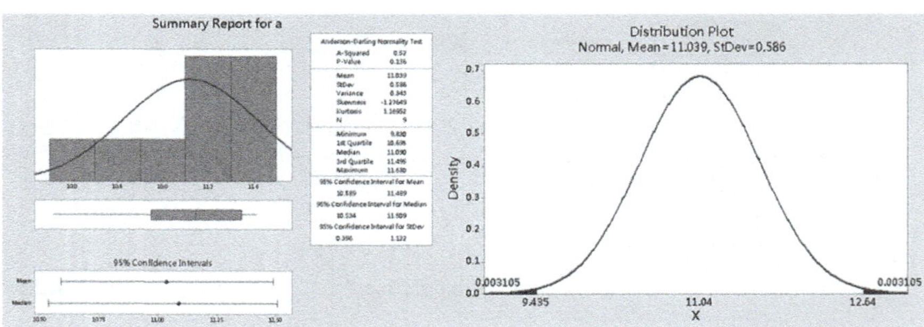

Fig. 17 Mean and standard deviation analysis of *a* in the hue of upper limit base on four sigma

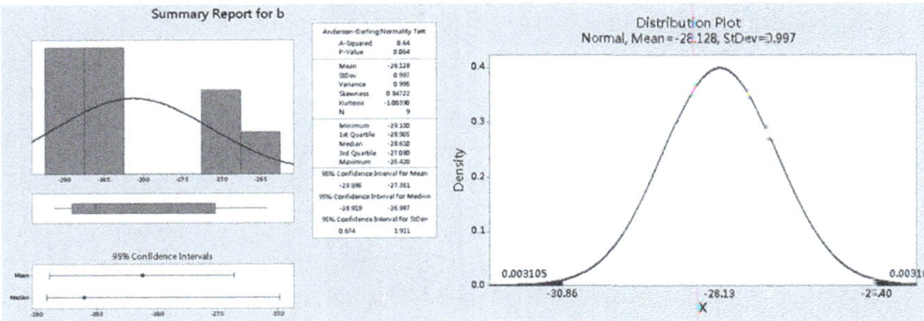

Fig. 18 Mean and standard deviation analysis of *b* in the hue of upper limit base on four sigma

6 Conclusion

The gradient color is a new direction for vehicle color design. It will offer a special colorful world for the customers of the vehicle. In the future, the paint of the vehicle will become the carrier of individuation and art. It will bring us more beauty of the world.

The Process and Application of 3D Collision Checking for New Paint Shop Projects

Chao Wang, Ge Zhang and Xingyu Jiang

Abstract This paper introduces HLS and the standard process for 3D collision checking used by FAW-Volkswagen during preliminary planning stage of new paint shops. Taking a new paint shop project as an example, it introduces the actual result of 3D collision checking.

Keywords Navisworks · Microstation · DGN · NWD · HLS · 3D collision checking

1 Introduction

At present, due to the large number of equipment installed in new paint shops, the space in some area of the plant is too critical, and there will be a lot of collisions between different equipment, utility pipes, HVAC Ducts, etc., in the design phase. If these collisions cannot be solved before the construction phase, there will be a lot of modifications by installation on site. Therefore, in the design phase of a new paint shop, it is necessary to collect 3D drawings of the equipment of each module and to check the collisions between them. Then, the supplier for each module should modify the 3D drawings of their own equipment according to the modifications' plan, until all of the collisions are solved, to make sure all of the equipment will be installed smoothly on site.

C. Wang · G. Zhang · X. Jiang (✉)
FAW-Volkswagen, Changchun 130011, China
e-mail: xingyu.jiang@faw-vw.com

C. Wang
e-mail: chao.wang.pl@faw-vw.com

China SAE (ed.), *Proceedings of China SAE Congress 2018: Selected Papers*, Lecture Notes in Electrical Engineering 574,
https://doi.org/10.1007/978-981-13-9718-9_62

2 HLS

Before the suppliers design the 3D drawings of equipment of each module for a new paint shop, the company should present the standard for 3D drawings and 3D collision checking. The standard used by FAW-Volkswagen is called Hallen-Layout-System (HLS), which contains the subjects as follows:

- Define the origin for the new paint shop, design the grids file for the new paint shop, and then all of the suppliers should design their 3D drawings of equipment based on the grids' file.

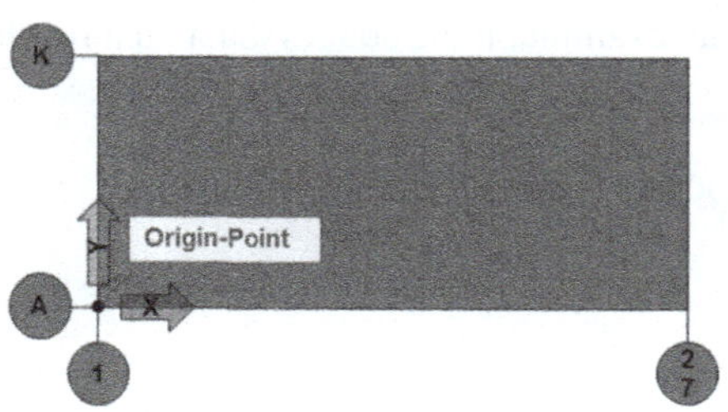

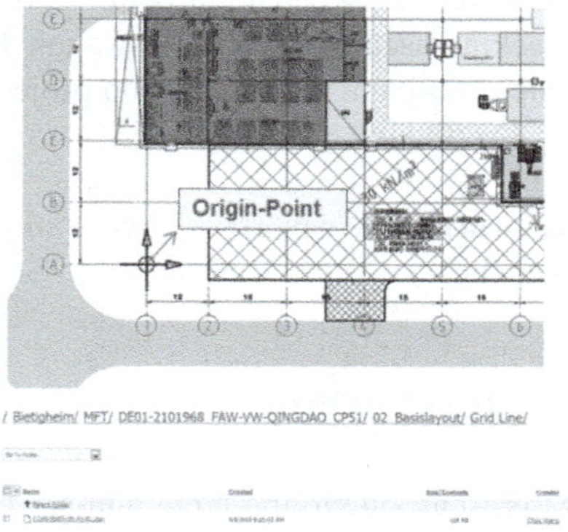

Example of grids' file:

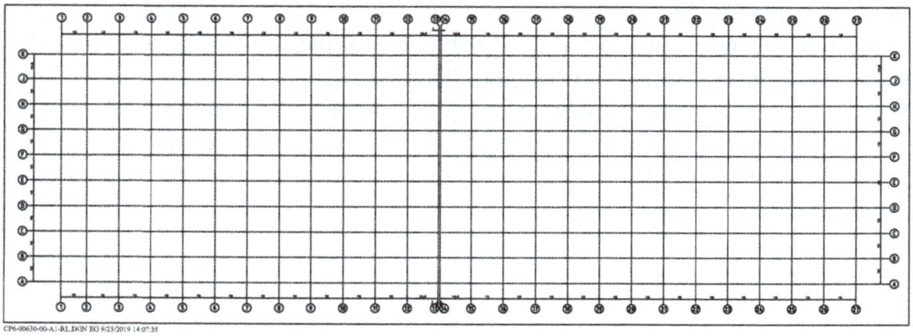

- File naming rules for each supplier to design 3D drawings:

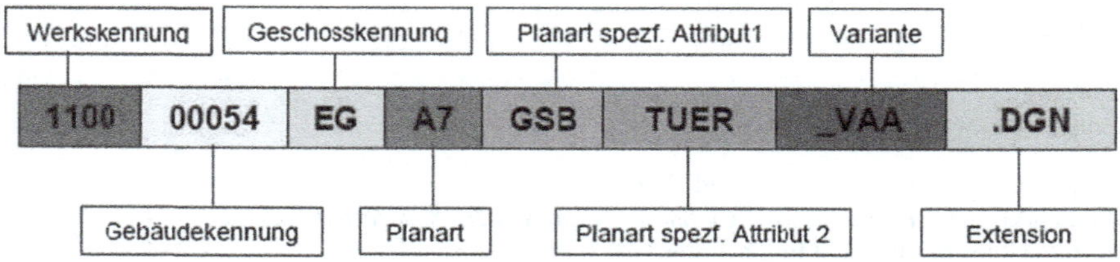

File name naming fules for a new paint shop			
Nr.	Deutsch	English	
1	Werkskennung	Factory code	C400
2	Gebäudekennung	Shop code	00230
3	Geschosskennung	Floor code	See floor code（Geschosskennung）
4	Planart	Equipment type	See equipment type（Planart）
5	Planart spezf. Attribut 1	Equipment type special attribute 1	See Equipment type special attribute 1 & 2
6	Planart spezf. Attribut 2	Equipment type special attribute 2	（Planart spezf. Attribut 1 und Planart spezf.
7	Variante	Variante	See Variante
8	Extension	Extension	.DGN

- Floor code rules:

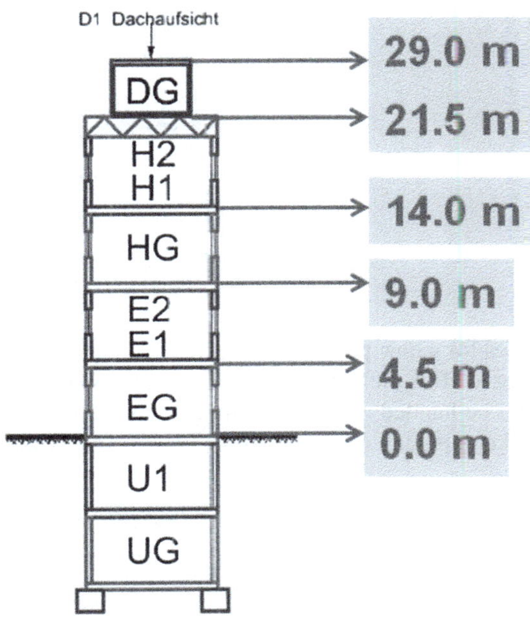

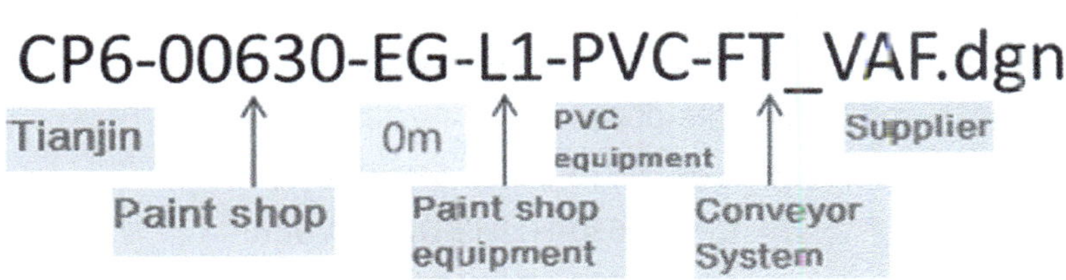

The upper example shows the name of the drawing of PVC conveyor system for a paint shop with the supplier's name VAF.

- Create the HLS file name list for 3D drawings of a new paint shop
  Create all of the file names for all 3D drawings of all modules according to file naming rules, and then, the HLS file name list for 3D drawings for a new paint shop will be created. And then for each file name, we will know the equipment

it contains; for each equipment, we can also find the relevant file name of the 3D drawings in the list.

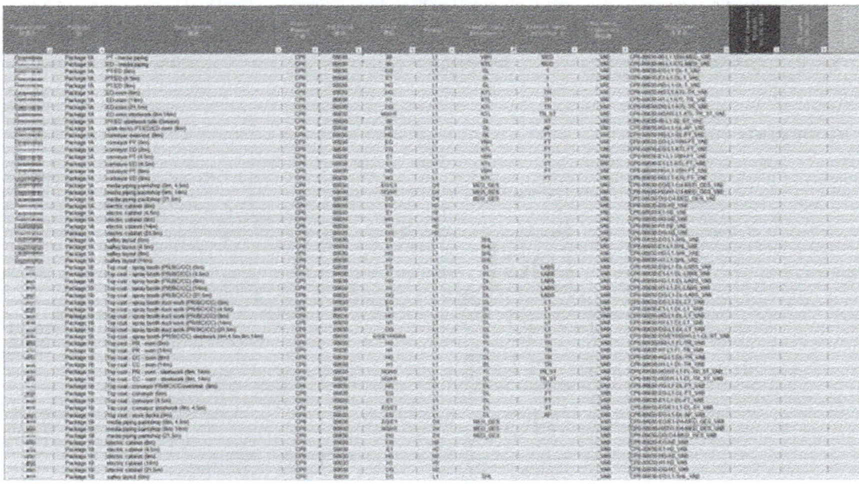

- Create HLS files for a new paint shop
  Then according to HLS file name list, the company will create every single HLS file, which will be handed over to all suppliers as the basis for designing of 3D drawings of all equipment.

CP6-00630-E1-H2-PVC_VAA.dgn
CP6-00630-E1-L1-PVC-APP_VAA.dgn
CP6-00630-EG-D4-MED_GES_VAA.dgn
CP6-00630-EG-H2-DL_VAA.dgn
CP6-00630-EG-H2-PVC_VAA.dgn
CP6-00630-EG-L1-DL-APP_VAA.dgn
CP6-00630-EG-L1-PVC-APP_VAA.dgn
CP6-00630-EG-L1-SHL_PVC_VAA.dgn
CP6-00630-H1-H2-DL_VAA.dgn
CP6-00630-H1-L1-DL-APP_VAA.dgn
CP6-00630-HG-H2-DL_VAA.dgn
CP6-00630-HG-L1-DL-APP_VAA.dgn
CP6-00630-HG-L1-SHL_DL_VAA.dgn

- Establish a server for data exchange, and create folders which will be used for file exchange.

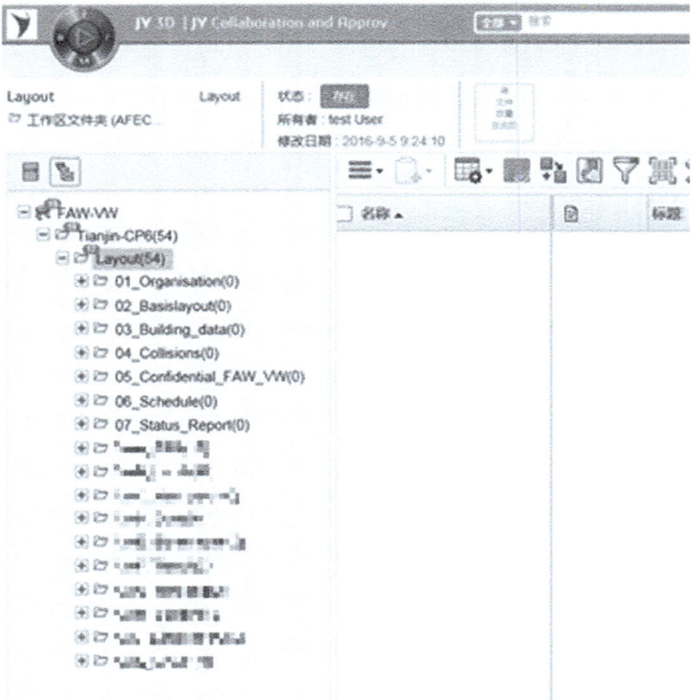

File folders will be classified with two sorts:

(a) Public file folder: The company and all suppliers can upload/download files. For example: Folder "01_Organisation" contains the organization of the project; Folder "02_Basislayout" contains 2D Layout, safety Layout, maintenance Layout, and so on; Folder "03_Building_data" contains drawings of building requirements; Folder "04_Collisions" contains 3D drawing files and 3D collision lists/reports, etc.

(b) Private file folder: Only certain supplier and the company can upload/download files.
For example: Folder "VAA_XXX" contains 3D drawings files and relevant files of robots from the supplier VAA

For each folder, different access rights for each supplier should be set.

- Setting of server access rights
  The access rights to each folder for each supplier should be set by the company, and access rights' table will be as follows:

Right to access DURR's server

Folder / Supplier	01_Organisation	02_Basislayout	03_Building_data	04_Collisions	[illegible]	[illegible]	[illegible]	[illegible]	[illegible]	[illegible]	[illegible]	[illegible]	[illegible]	[illegible]	[illegible]
[illegible]	r	r	r	r	r/w/d										
[illegible]	r	r	r	r		r/w/d									
[illegible]	r	r	r	r			r/w/d								
[illegible]	r	r	r	r				r/w/d							
[illegible]	r	r	r	r					r/w/d						
[illegible]	r	r	r	r						r/w/d					
[illegible]	r	r	r	r							r/w/d				
[illegible]	r/w/d	r/w/d	r/w/d	r/w/d	r/w/d	r/w/d	r/w/d	r/w/d	r/w/d	r/w/d	r/w/d	r/w/d	r/w/d	r/w/d	r/w/d
[illegible]	r	r	r	r									r/w/d		
[illegible]														r/w/d	
[illegible]	r/w/d	r/w/d	r/w/d	r/w/d	r/w/d	r/w/d	r/w/d	r/w/d	r/w/d	r/w/d	r/w/d	r/w/d	r/w/d	r/w/d	r/w/d

r Read
w Write
d Delete

- Process for 3D collision checking
  Each supplier designs 3D drawings of equipment according to HLS requirement from the company and then upload them to the data exchange server according to agreed format (DGN file from Microstation); then, the company will download all of the 3D drawings and check the data format, and if there is no problem, the company will transform/compress the data into NWD (Navisworks) format; then the company will perform 3D collision checking, and afterward, the company will create the 3D collision list/report; later, the company will upload the 3D file (NWD format) and the 3D collision list/report to the server; then each supplier will download the 3D file(NWD format) and the 3D collision list/report, and check the relevant collisions, then will modify relevant 3D drawings, and then upload the updated 3D drawings to the data exchange server again. Until then, it forms a complete closed loop for 3D collision checking. Generally, the company and all suppliers will agree to complete one-time 3D collision checking and 3D drawings updating within 1–2 weeks.

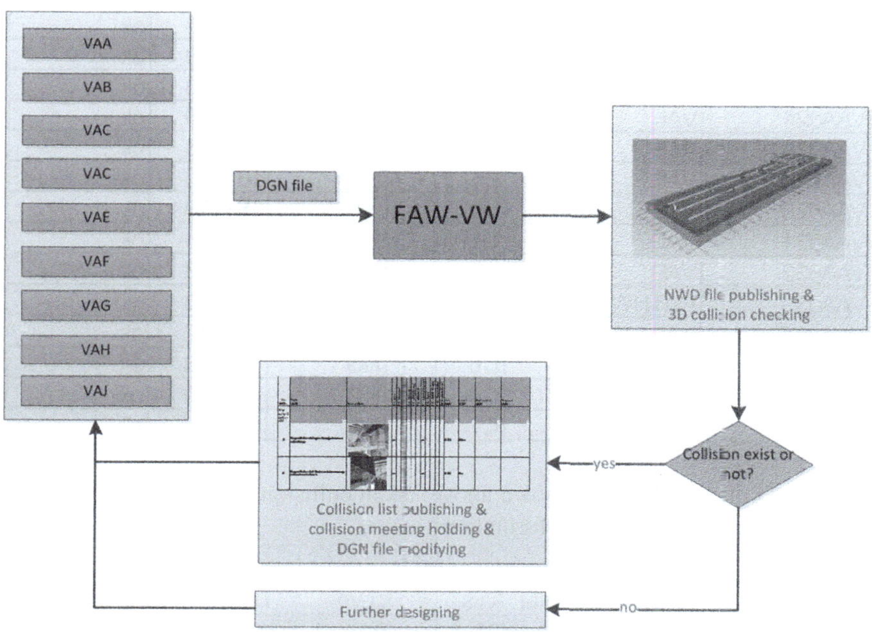

- 3D collision list
  3D collision list consists of actual collisions' information; actual collisions are generated from within building structure, utility pipes, HVAC ducts, equipment, etc.

Collision Poir	Relevant suppli	Axis	Level	Date	Status
VAA001	Y	C7	HG	161128	ok
VAA002	VAB	ALL	HG	161128	ok
VAA004	PATHWAY	4E	EG	161128	ok
VAA005	VAB	C8	H1	161128	ok
VAA006	VAB	G8	H1	161128	ok
VAA007	VAB	C6-7	E1	161128	ok
VAA008	VABVAG	ALL	E1	161128	ok
VAA009	VAF	H12	E1	161128	ok
VAA010	VAF	G16	E1	161128	ok
VAA011	VAH	H17	EG	161128	ok
VAA012	VAF	H16	EG	161128	ok
VAA014	VAF	G17	EG	161128	ok
VAA015	VAF	H17	E1	161128	ok
VAA016	VAF	G17	EG	161128	ok
VAA018	VAF	J16	EG	161128	ok
VAA020	VAF	J17	EG	161128	ok
VAA021	VAE	J17	EG	161128	ok
VAA022	VAE	J17	EG	161128	ok
VAA023	VAF	J17	EG	161128	ok
VAA024	VAF	J16	EG	161128	ok
VAA026	VAB	E8	EG	161128	ok
VAA028	VAB	E5	H1	161128	ok
VAA029	VAB	H5	H1	161128	ok
VAA031	VAD	G17	EG	161128	ok
VAA032	VAB	F6	H1	161128	ok
VAA033	PATHWAY	J17	EG	161206	ok
VAA034	VAE	J18	EG	161212	ok
VAA034	Y	C6	HG	161228	ok
VAA035	VAE	C6	HG	161228	ok
VAA036	Y	C7	E1	161228	ok

Sample of collision list is as shown above, taken collision VAA007 as an example, and it includes the following information:

Collision Point	Relevant supplier	Axis	Level	Date	Status
VAA007	VAB	C6-7	E1	161128	ok
Collision Number	Relevant Supplier is VAB	Collision Position	Level of Collision	Found Time	Status, OK means closed

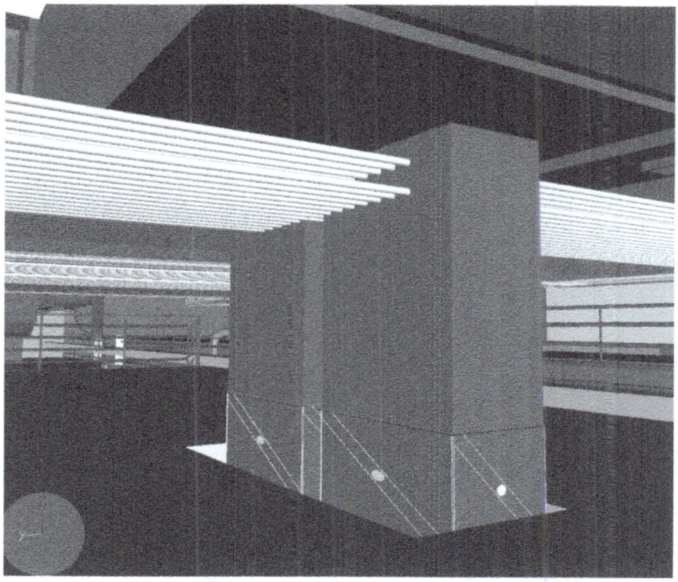

This collision is between the pipes from the supplier VAA and the air duct from the supplier VAB, its position is around axis C6-7, its level is 4.5 m, and this collision was found in November 28, 2016, until now it is closed. Every time when by collision checking, the company should publish this collision list together with the 3D file (NWD format).

- HLS layer naming rules

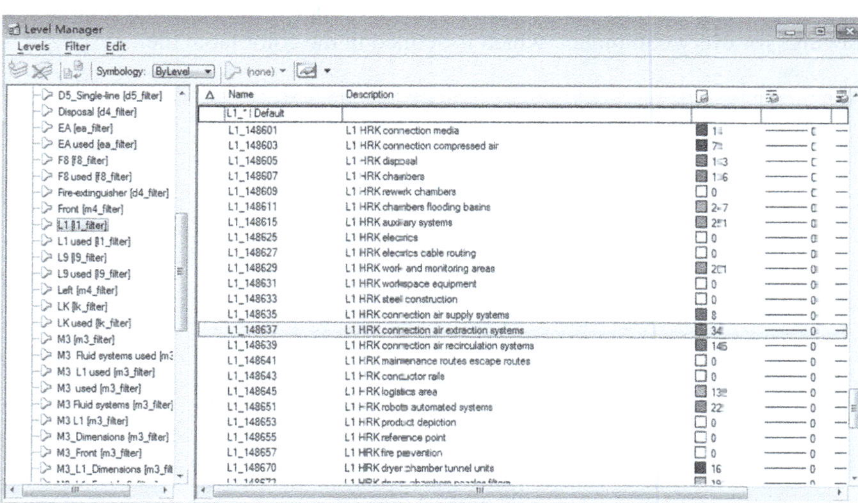

The company should specify the name of each layer for the equipment in each 3D file, and the content, color, line shape, line width, etc., of each layer for the equipment should be regulated.

- HLS color table

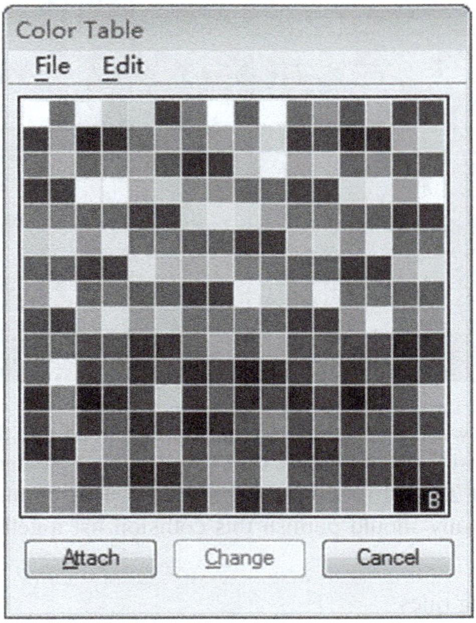

The company should specify the colors which all suppliers can use to design the 3D drawings for every equipment, which is defined in the color table file.

- Unit settings

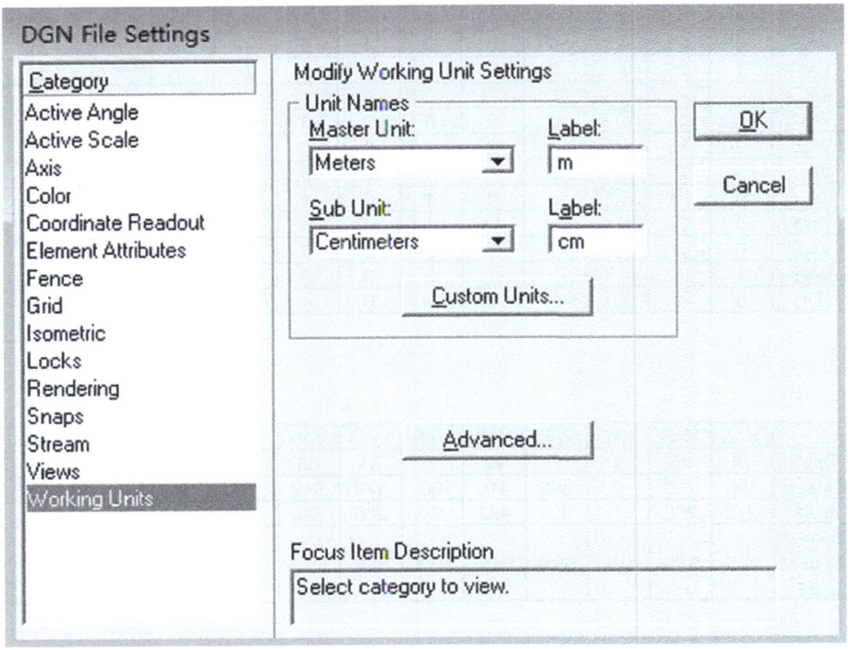

The company should set the unit for all 3D drawings from each supplier

- Other requirements
    (a) All of the dynamic spatial information for equipment/car body, etc., should be included in 3D drawings (should be represented as blocks).
    (b) The size for each 3D file from each supplier should be less than 50 MB.

3 The Application of 3D Collision Checking Used in a New Paint Shop Project

After collision checking for a new paint shop project, nearly 3000 collisions were found. Compared with another paint shop project, 90% collisions' solving time was shortened from 10 months to 5 months, and 90% collisions of the new paint shop project were closed before equipment installation, then the project progress was guaranteed, and the planning fees and fees for modification of the equipment on site were reduced. The statistics information of 3D collision checking for the new paint shop project is shown as follows:

	VAA	VAB	VAC	VAD	VAE	VAF	VAG	VAH	VAJ	VAK	VDU	SUM
VAA	0	13		0	0	0	1	0	0	0		14
VAB	1	25		2	2	1	0	5	0	0		36
VAC								1				1
VAD	0	2		0	2	0	0	3	0	0		7
VAE	0	15		1	2	1	0	22	1	0		42
VAF	0	2		0	1	1	0	3	0	0		7
VAG	0	21		5	0	1	0	4	0	0		31
VAH	0	22		10	11	1	0	17	7	0		68
VAJ	0	0		3	0	0	0	0	1	0		4
VAK	1	0		0	1	2	0	2	0	0		6
VDU	0	1		0	0	0	0	4	0	0	0	5
											SUM	221

	VAA	VAB	VAC	VAD	VAE	VAF	VAG	VAH	VAJ	VAK	VDU	SUM
Open	14	36	1	7	42	7	31	68	4	6	5	221
Closed	198	720	0	598	410	96	174	512	10	9	0	2727
SUM	212	756	1	605	452	103	205	580	14	15	5	2948
Solved Percent age	93%	95%	0%	99%	91%	93%	85%	88%	71%	60%	0%	93%

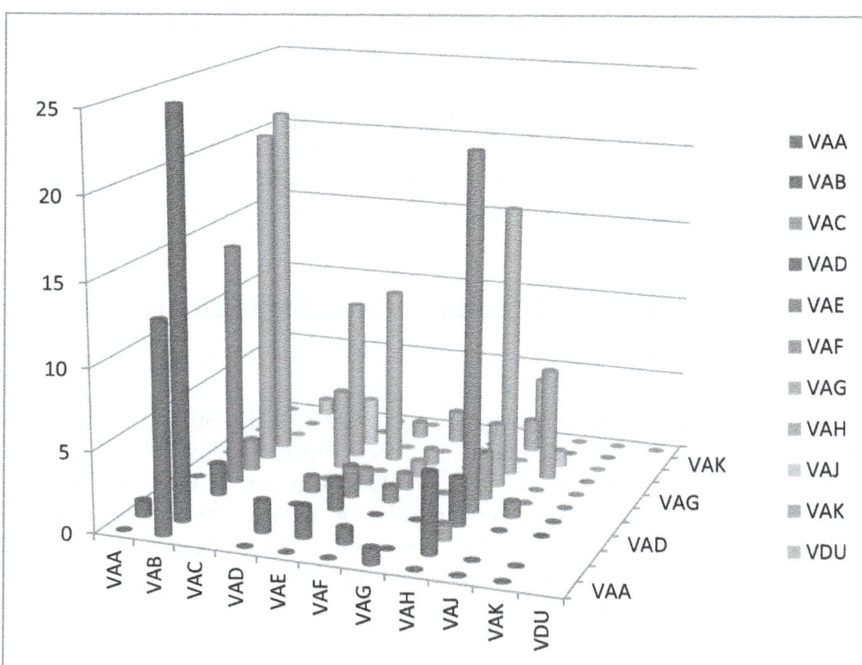

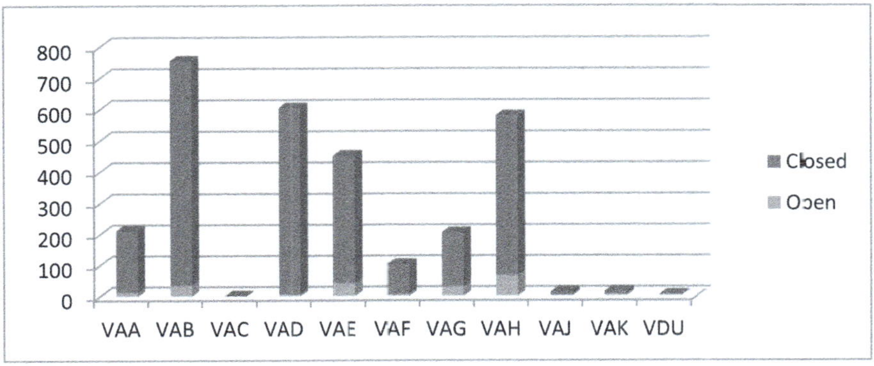

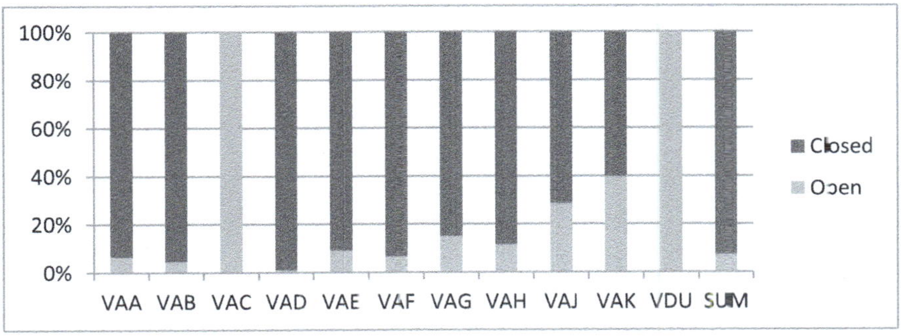

4 Summary

Taken the new paint shop project, it is necessary to carry out 3D collision checking in the preliminary planning stage of new paint shops; most of the collisions can be avoided before equipment installation by 3D collision checking, then a lot of potential equipment modifications on site and the potential delay of the project regarding equipment modifications on site can also be avoided; through 3D collision checking for the new paint shop project, it shows that HLS can meet the requirements for 3D collision checking for new paint shops. And now FAW-Volkswagen has written standard documents of HLS for new paint shops, which will be the basis for planning new paint shops in the preliminary planning stage.

Monocular Vision-Based Real-Time Vehicle Detection at Container Terminals

Zijian Liu, Tianlei Zhang, Bei He, Yu Liu, Li Sun and Wenyang Tang

Abstract We present a practical approach to vehicle detection at container terminals based on a single camera and prevailing convolutional neural network models in computer vision domain. Aiming at container terminal scenarios, we introduce a specialized data labelling strategy for network training, as well as an optimized setting of crucial hyperparameters, leading to a significant improvement on results. Our solution achieves 83% precision with 90% recall for semitrailer trunks within 30 m ahead of the vehicle-mounted monocular camera, at a speed of 32 frames per second (FPS) on a Nvidia Titan X for 416 × 416 image input, also providing more alternatives of fairly easy speed/accuracy trade-off. Compared to traditional lidar-based vehicle detection method for autonomous driving, our solution is more robust for particular container terminal scenarios while still maintaining a real-time performance by completely eliminating the region proposal stage.

Keywords Vehicle detection · Convolutional neural network · Autonomous driving · Container terminal

1 Introduction

Perception system is a complicated component of autonomous driving technology, but computer vision-based solution might handle this difficulty properly. In computer vision domain, convolutional neural network (CNN) has become an irresistible approach to tackle several difficult challenges such as image classification, object detection, and semantic segmentation since AlexNet [1]. On canonical object detection datasets like PASCAL VOC [2] and COCO [3], CNN-based detectors supersede

Z. Liu (✉) · Y. Liu
Beihang University, Beijing, China
e-mail: lzj1995@126.com

T. Zhang · B. He
Beijing Trunk Technology Co., Ltd., Beijing, China

L. Sun · W. Tang
Tianjin Port (Group) Co., Ltd., Tianjin, China

China SAE (ed.), *Proceedings of China SAE Congress 2018: Selected Papers*, Lecture Notes in Electrical Engineering 574,
https://doi.org/10.1007/978-981-13-9718-9_63

the classic DPM [4] method and dominate the benchmark results. These detectors can be divided into two categories according to different pipelines. Models using region proposal technique combined with a convolutional feature extractor achieved state-of-the-art accuracy, including the R-CNN series [5–7] and other variants based on Faster R-CNN [7]. While other detectors with slightly worse accuracy are less computationally expensive by removing proposal generation, running in real time like SSD [8] and YOLO [9].

All aforementioned approaches are flexible and generalize well to various scenarios including autonomous driving. As shown in KITTI [10] benchmark, 2D object detection models for autonomous driving have similar architecture compared with solutions for general detection problems, except for a few works introducing additional stereoimages or point clouds. However, all these methods are designed for datasets which heavy vehicles rarely appear and they are required to detect all kinds of cars as well as pedestrians simultaneously. Compared to usual driving circumstances, traffic condition at container terminals is certainly different. Heavy semitrailer trunk used for carrying containers is the most principal type of vehicles at container terminals along with the low probability of pedestrian occurrences. We noticed that these characteristics could be utilized for optimizing vehicle detection at container terminals in both data labelling process and network training phase.

We propose a specialized data labelling strategy to locate ground-truth bounding boxes of semitrailer trunks in container terminals scenarios, which result in a better feature extraction. We choose proposal-free SSD [8] model to realize real-time detection, with our particular setting of aspect ratios for default boxes. In addition, we compared our method to conventional lidar-based detectors and demonstrate the insight into why vision-based detectors are indispensable in container terminal scenarios.

2 Prior Work

2.1 Faster R-CNN

Faster R-CNN [7] is the backbone of current leading works in object detection benchmarks. It proposed the revolutionary RPN (region proposal network), which takes over the proposal generation process from selective search [11]. RPN shares convolutional features with the detection network, saving plenty of time in region proposal to accelerate inference. Another remarkable innovation from Faster R-CNN is the concept of translation-invariant anchors, which reduce the model size and enhance the performance for objects of multiple scales and aspect ratios. Though Faster R-CNN is much faster than its predecessors, it still struggles to meet a real-time requirement.

2.2 YOLO

YOLO [9] frame object detection as a regression problem and completely remove region proposal to achieve a real-time detection rate. YOLO adopt a single neural network to predict bounding boxes and class probabilities simultaneously instead of combining multiple modules together. The major limitation of YOLO is smaller objects are more likely to be neglected. Moreover, it is unstable when objects presented in unusual aspect ratios.

2.3 SSD

Assimilating the idea of MultiBox [12], SSD [8] proposed default boxes as prediction candidates, which are similar to anchors from Faster R-CNN while without the proposal stage. SSD runs in real time and obtains competitive performance with state-of-the-art models. It also locates small objects more accurately compared to YOLO thanks to generating default boxes in multiscale feature maps rather than a single-scale feature map.

3 Vehicle Detection at Container Terminals

As mentioned previously, solutions for common detection problems generalize well to autonomous driving. After a comparison between multiple prevalent models, we choose SSD as the basic framework to detect vehicles at container terminals for its real-time robust performance to different object aspect ratios, considering that the occurrences of semitrailer trunks always vary drastically in aspect ratios within images.

For the monotonousness of objects presenting in our case, knowing whether there is an object lying ahead is more important than figuring out what exactly the object is. In other words, detectors are expected to get extremely accurate localization and convincible objectness score rather than specific information with respect to every class. Thus, based on SSD model, we conduct our own data labelling strategy which is specialized for container terminal scenario. We also modify the aspect ratios setting of default boxes as well as the weight hyperparameter of loss function.

3.1 Labelling Strategy

We tried several labelling strategies. One obvious way is to label all semitrailer trunks as one class regardless of whether they are loaded with containers or unloaded.

Fig. 1 Forms left to right, one-class labelling, two-classes labelling, subpart labelling. Different color denotes different class

Another straightforward way is to separate loaded trunks from unloaded trunks. Both strategies are reasonable, but inconsistent in feature extraction. We expect to get somewhat consistent feature maps if two objects belong to the same class; otherwise, it will lead to a considerable classification loss. The one-class labelling strategy apparently gives distinct difference in feature maps between loaded and unloaded trunks in the same class. As for two-classes labelling, it is also hard to distinguish loaded trunks from unloaded ones with some specific angles, especially when trunks are truncated. Besides, the feature map of a semitrailer trunk with a single trailer should also be different from the one with multiple trailers (Fig. 1).

Our labelling strategy, named subpart labelling, locates fine-grained ground-truth bounding boxes with respect to subparts of semitrailer trunks, meaning that we divide one semitrailer trunk into a tractor part and trailer parts. The benefit of this strategy is that feature maps of the congener will be more consistent, which may accelerate the training convergence, and it also easily handles the situation when vehicle is truncated. The inspiration comes from the fact that semitrailer trunk is a combination of a tractor unit and semitrailers. In detail, we label the tractor unit as one class, with labelling the loaded trailer and the unloaded trailer as two other classes. Since we divide semitrailer trunks into several parts with subpart labelling, the detector only needs to give a single detection result of either part of the trunk, which obviously decreases the difficulty.

We believe our labelling strategy also adapts to object that is a combination of several separable parts and larger than common ones. This labelling strategy outperforms two other mentioned methods according to our experiments.

3.2 Picking Hyperparameters

We look into some crucial hyperparameters in model training phase to make SSD model more suitable for our case.

The original loss function of SSD is computed as:

$$L(x, c, l, g) = 1/N(L_{\text{conf}}(x, c) + \alpha L_{\text{loc}}(x, l, g)) \tag{1}$$

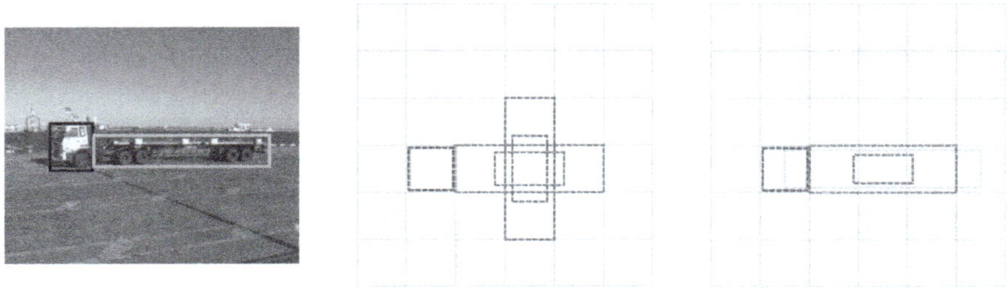

Fig. 2 There might be no default box being able to match a particular ground-truth bounding box if using original aspect ratios setting of SSD. We modified these hyperparameters to fit the shape of trunks

SSD model set the weight term α to one, which makes predicted localization and predicted classification to give an equal contribution to loss function.

Since accurate localization gives more valuable information than classification in our scenario, we set α bigger than one to emphasize localization loss which is a smooth L_1 loss [6] computed as:

$$\text{smooth}_{L_1}(x) = \begin{cases} 0.5x^2 & \text{if } |x| < 1 \\ |x| - 0.5 & \text{otherwise} \end{cases} \quad (2)$$

We choose $\alpha = 1.2$ by cross-validation in the end.

Another set of hyperparameters we tried to play with are aspect ratios of default boxes. In [8], it gives five different aspect ratios as, $\{1, 2, 3, \frac{1}{2}, \frac{1}{3}\}$. But in container terminal scenarios, we found that most ground-truth bounding boxes of semitrailer trunks have aspect ratios below one even if they present in different angles. Thus, using the original setting of aspect ratios might neglect some detection target among bounding box candidates (Fig. 2). According to this, we set aspect ratios of default boxes as, $\{\frac{3}{2}, 1, \frac{1}{2}, \frac{1}{3}, \frac{1}{5}\}$, which definitely fit the shape of semitrailer trunks better.

3.3 Acceleration and Trade-off

Howard et al. [13] introduced the width multiplier and the resolution multiplier to efficiently make the trade-off between speed and accuracy. We adopt the similar idea to accelerate the detection pipeline. Width multiplier reduces channels of each layer by using less convolutional kernels, meaning that we have less parameter to train, leading to a faster convergence. While resolution multiplier shrinks the size of input image and the internal representation of every layer is subsequently reduced by the same multiplier. Both multipliers reduce the computation cost in a quadratic factor. In other words, if we set width multiplier equal to α $(0 < \alpha < 1)$ and resolution multiplier equal to β $(0 < \beta < 1)$, the computation expense will be reduced by roughly $(\alpha\beta)^2$.

We apply the resolution multiplier to our model and the results are in 4.1 part. We get much less latency time of inference phase, without a significant decay of detection accuracy.

4 Experiments

4.1 Results

We conduct our methods on datasets of container terminal scenarios, and the results indicate that both subpart labelling strategy and modified aspect ratios of default boxes could boost the model performance.

We compared results from three mentioned labelling strategies. The precision–recall curves are in Fig. 3. With a recall of 90%, our subpart labelling strategy achieves a precision of 83% compared with 79% of two-classes labelling and 76% of one-class labelling (with our customized aspect ratios of default boxes).

As for our modified aspect ratios setting of default boxes, we also found that it will create a better outcome than the original one in container terminal scenarios. We compared the results with speed/accuracy trade-off in Table 1.

We applied other prevailing models like YOLO and Faster R-CNN to our datasets with subpart labelling as well. As for YOLO, subpart labelling seems not working, giving us a worse result than general labelling methods. Looking into the architecture of YOLO, it can be explained by its deterioration of performance when it comes to small objects. It is obvious that subpart labelling strategy separates a bigger ground-truth box into small ones, which can lead to a problem for models that are less

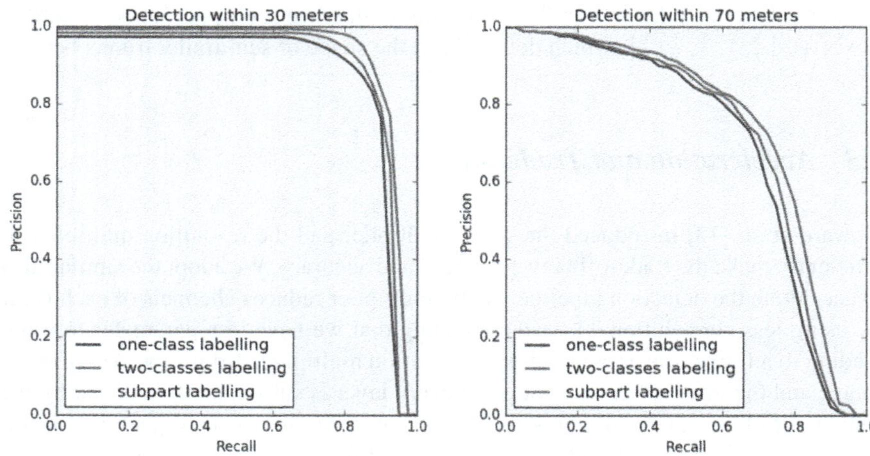

Fig. 3 The precision–recall curves of three different labelling strategies within 30 m distance and within 70 m distance

Table 1 Precision results when using different aspect ratios of default boxes with speed/accuracy trade-off at 90% recall within 30 m distance

	Labelling strategy	SSD's original aspect ratios of default boxes (%)	Our customized aspect ratios of default boxes (%)
416 × 416 input size (32 FPS)	One-class labelling	73.2	76.2
	Two-classes labelling	75.6	79.[illegible]
	Subpart labelling	79.7	**83.7**
300 × 300 input size (45 FPS)	One-class labelling	68.3	71.8
	Two-classes labelling	70.5	73.9
	Subpart labelling	72.9	**76.4**

Bold values represent best result

sensitive to small objects. We believe that is the reason why our method is not a good option for YOLO.

Not surprisingly, Fast R-CNN yields an even more robust result compared to SDD with subpart labelling strategy. In contrast to YOLO, its robustness gives the credit to the superb performance of detecting small objects. But of course, accompanied by a computationally intensive RPN, this model is struggled to meet the real-time requirement, which makes it less competitive compared to SSD in autonomous driving scenario. We list detection results of all models mentioned above in Table 2 given 300 × 300 input size, both SSD and Faster R-CNN use a VGG [14] base model while YOLO combines with its original customized base architecture.

According to Table 2, we should point out that the subpart labelling strategy might slightly increase detection time. That makes sense because this strategy generally generates more bounding boxes with respect to each subpart. But apparently, this tiny deterioration is affordable compared to considerable precision improvement.

Table 2 Results of SSD, YOLO, and Faster R-CNN

	mAP (%)	Processing time per frame (ms)
SSD with general labelling	73.9	**21.6**
SSD with subpart labelling	76.4	22.4
YOLO with general labelling	69.2	24.9
YOLO with subpart labelling	63.2	27.1
Faster R-CNN with general labelling	76.8	187.4
Faster R-CNN with subpart labelling	**79.3**	191.2

Bold values represent best result

4.2 Data Augmentation

Our model doesn't apply a feature resampling procedure to make the system more reliable for small objects like Faster R-CNN. Instead, we choose to use a customized data augmentation strategy, which could enlarge the entire datasets by random cropping and present more solid performance for small objects.

Cropping produces more training examples for our model, making the model more sensitive to objects with different sizes and shapes. We firstly use the entire image for training, followed by patches of original image. These patches overlap ground-truth bounding boxes with ratios of {0.1, 0.3, 0.5, 0.7, 0.9}. At last, we randomly sample patches that also used for training. Following these three steps, we get abundant training examples with ground-truth bounding boxes of various sizes, especially those small bounding boxes, which could enhance the overall performance by a large margin.

The insight into this data augmentation strategy is that cropped patches amplify details of the original image, which could be regarded as a zoom-in operation and it is crucial for the detection of small objects. On top of that, we also need a zoom-out operation to present an overall sketch, which means bigger objects are narrowed to small ones, boosting the chances for a robust detection of small objects that can be seen as sketches of targets. To implement this idea, we resize the original image to a smaller size and randomly place it in a canvas filled with mean values.

Both approaches contribute a lot to the improvement of detection results, small objects in particular, which is also beneficial for our subpart labelling strategy. In our experiments, we get a steady 2–3% enhancement of mAP by applying data augmentation.

4.3 Lidar or Camera?

Lidars and cameras both play a vital part in perception system of autonomous driving technology. Lidars are irreplaceable when accurate distance information is required and they function well in a caliginous environment. But lidars could be incompetent in container terminal scenario for vehicle detection when trunks are unloaded. If a trunk is unloaded with containers, the point clouds of the trailer part could be extremely sparse, which will lead to a low recall of detection results. In this case, the trailer part is hard to be detected (Fig. 4). Considering that trunks in container terminals usually have an uncommon long length, it might bring about a dangerous situation.

We conducted an vehicle detection experiment on point clouds datasets of container terminals in bird's eye view from two Velodyne's VLP-16 3D LiDARs, based on the state-of-the-art MV3D [15] model. We proved that our vision-based approach using monocular camera is more robust compared to lidar-based detection method which also contains a CNN kernel. Furthermore, before the training phase, the

Fig. 4 These are point cloud in bird's eye view and image data collected at the same time. The unloaded semitrailer trunk in image is obvious to be detected. However, the point cloud of the trailer part is extremely sparse, which brings difficulty to the lidar-based detector for a robust detection

labelling process for point clouds is much tougher than labelling images since the message given by point clouds is much more abstract, which may also bring adverse effect to detection results.

As for images captured by cameras, dense pixels reveal more abundant and more transparent information. Cameras function like human's eyes, which makes the labelling process more straightforward. Also, because of the considerable bulky volume of vehicles, vision-based detectors are able to give reliable detection results within a long distance.

5 Conclusion

We introduce a practical approach to detect heavy vehicles at container terminals. Our method uses a monocular camera based on an optimized SSD convolutional neural network model, which is specialized for container terminal scenarios to get more accurate detection results. We adopt a novel labelling strategy called subpart labelling with considered choice of hyperparameters such as weight term of loss function and aspect ratios of default boxes. Both tricks enhance the performance of our model. We believe that techniques proposed in this paper should generalize well to specific scenarios similar to container terminals, where detection objectives are homogeneous and universally have a bulky volume.

We later compare our method with lidar-based detectors and explain the insight into the robustness of vision-based method at container terminals. But still, lidars are irreplaceable with respect to accurate distance estimation and general obstacle detection. For future work ,we hope to combine cameras and lidars, meaning using both image data and point cloud data simultaneously to construct a stronger model, which will be a new trend in the perception system design of autonomous driving.

References

1. Krizhevsky A, Sutskever I, Hinton GE (2012) ImageNet classification with deep convolutional neural networks[C]. In: International conference on neural information processing systems. Curran Associates Inc., pp 1097–1105
2. Everingham M, Gool LV, Williams CKI et al (2010) The pascal visual object classes (VOC) challenge. Int J Comput Vis 88(2):303–338
3. Lin T, Maire M, Belongie SJ et al (2014) Microsoft COCO: common objects in context. In: European conference on computer vision, pp 740–755
4. Felzenszwalb P, Mcallester D, Ramanan D (2008) A discriminatively trained, multiscale, deformable part model. In: Computer vision and pattern recognition, 2008. CVPR 2008. IEEE conference on. IEEE, pp 1–8
5. Girshick RB, Donahue J, Darrell T et al (2014) Rich feature hierarchies for accurate object detection and semantic segmentation. In: Computer vision and pattern recognition, pp 580–587
6. Girshick R (2015) Fast R-CNN[C]. In: IEEE international conference on computer vision. IEEE Computer Society, pp 1440–1448
7. Ren S, Girshick R, Girshick R et al (2017) Faster R-CNN: towards real-time object detection with region proposal networks. IEEE Trans Pattern Anal Mach Intell 39(6):1137–1149
8. Liu W, Anguelov D, Erhan D et al (2016) SSD: single shot MultiBox detector[C]. In: European conference on computer vision. Springer, Cham, pp 21–37
9. Redmon J, Divvala S, Girshick R et al (2016) You only look once: unified, real-time object detection[C]. In: IEEE conference on computer vision and pattern recognition. IEEE Computer Society, pp 779–788
10. Geiger A (2012) Are we ready for autonomous driving? The KITTI vision benchmark suite[C]. In: IEEE conference on computer vision and pattern recognition. IEEE Computer Society, pp 3354–3361
11. Uijlings JRR, Sande KEAVD, Gevers T et al (2013) Selective search for object recognition. Int J Comput Vis 104(2):154–171
12. Erhan D, Szegedy C, Toshev A et al (2014) Scalable object detection using deep neural networks[C]. In: IEEE conference on computer vision and pattern recognition. IEEE Computer Society, pp 2155–2162
13. Howard AG, Zhu M, Chen B et al (2017) MobileNets: efficient convolutional neural networks for mobile vision applications
14. Simonyan K, Zisserman A (2014) Very deep convolutional networks for large-scale image recognition. Comput Sci
15. Chen X, Ma H, Wan J et al (2017) Multi-view 3D object detection network for autonomous driving. IEEE CVPR

Chassis Control in Zero Radius Steer Mode of Four-Wheel-Independently Actuated Unmanned Ground Vehicle in Remote Control Mode

Taipeng Wang, Yuzhuang Zhao, Jun Ni and Sizhong Chen

Abstract This paper describes a heavy-class four-wheel-independently actuated unmanned ground vehicle, which is able to achieve zero radius steer (ZRS) based on the independent steered technique. The major novelties of the unmanned ground vehicle are introduced, especially the mechanisms to achieve independent steer. To enhance the performance of the vehicle to negotiate the complex terrains, a chassis controller for ZRS is proposed. The chassis controller includes a yaw moment controller and a tyre slip ratio controller. The yaw moment is determined by the ground command station, which is remotely controlled by the human operator. The tyre traction forces are distributed according to the tyre vertical load, and a sliding mode slip ratio controller is applied. Finally, the experiments based on the unmanned ground vehicle developed by the authors are conducted to verify the performance of the proposed controllers.

Keywords Unmanned ground vehicle · Steer-by-wire · Chassis control · Dynamics control

1 Introduction

The unmanned ground vehicle (UGV) and heavy-class wheeled mobile robot (WMR) are fast emerging based on the achievements of the environment perception, sensor, intelligence, and electric control techniques [1–3]. The UGV and WMR are majorly applied in military or other special-used applications to take the place of human beings, such as rescue, reconnaissance, search, combat, irrigation, and delivery. Many countries, including China, USA, Russia, and Germany, have started the researches of the military UGV to greatly change the way of the war on the land in the future. Several military UGVs developed in different countries are shown in Fig. 1.

The developments of steer-by-wire and drive-by-wire techniques enable the UGV and WMR to achieve advanced structure configuration and performance. With steer-

T. Wang · Y. Zhao · J. Ni (✉) · S. Chen
Beijing Institute of Technology, Beijing, China
e-mail: nijun_bit@163.com

China SAE (ed.), *Proceedings of China SAE Congress 2018: Selected Papers*, Lecture Notes in Electrical Engineering 574,
https://doi.org/10.1007/978-981-13-9718-9_64

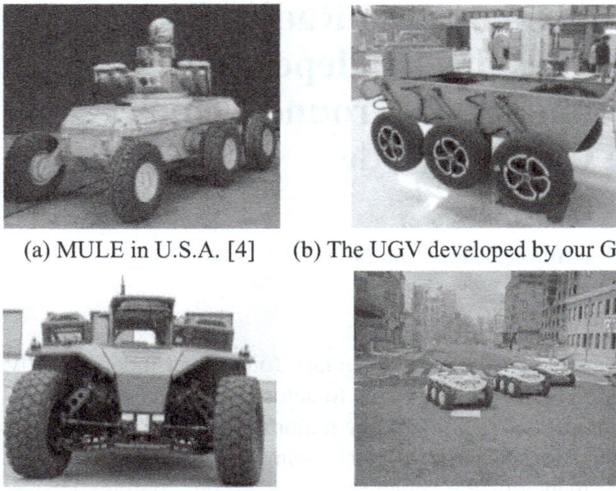

(a) MULE in U.S.A. [4] (b) The UGV developed by our Group [5]

(c) Gecko in Germany [4] (d) The UGV Proposed by Russia [6]

Fig. 1 Several military UGVs in different countries

by-wire technique, each wheel of the UGV has the potential to be independently steered, which could greatly enhance the mobility of the UGV. This technique is called independent steered technique. In the area of the passenger car, the researches of the steer-by-wire techniques have attracted lots of concentrations [7, 8]. However, owing to the relatively weaker requirements of mobility to handle different terrains, the independent steered technique is not widely applied in the area of passenger car. Owing to the higher requirements of mobility for military UGV, the independent steered technique should be focused on in the area of military UGV to improve the ability to handle different complex terrains.

The most significant advantage of the independent steered technique is that the zero radius steer (ZRS) can be achieved. The ZRS could greatly enhance the mobility of the military UGV to negotiate various complex terrains and some narrow spaces. However, the research of the dynamics and control in ZRS of UGV is seldom reported. To this end, this paper focuses on the dynamics and control in ZRS of four-wheel-independently actuated UGV. In this paper, a chassis controller, including the yaw moment distribution module and tyre slip control module, is proposed for the UGV in ZRS. Finally, based on the UGV developed by the authors, the controllers are verified through the experimental tests.

2 Comparison of Mechanisms to Achieve Pivot Steer

In above section, it has been stated that the ability to achieve pivot steer is important to improve the mobility of a military or special-used UGV. Commonly, there are several following mechanisms to achieve pivot steer. The advantages and disadvantages will be discussed in this section.

2.1 *Six-Wheeled Skid Steered UGV*

The six-wheeled UGV is a common configuration of military used UGV, which uses six wheels to drive the vehicle. The famous six-wheeled UGV includes the MULE shown in Fig. 1 and Crusher developed in USA [4]. A unique technique of the six-wheeled UGV is that it uses skid steered technique to steer the vehicle. In other words, there is no particular steer mechanism for these vehicles, such as Ackermann mechanisms. It uses the traction force difference and the speed difference between left and right side wheels to steer the vehicle. Figure 1b shows a six-wheeled skid steered UGV developed by the authors. The readers can find more details in [5].

The skid steered technique enables the vehicle to achieve pivot steer. However, there are some drawbacks of skid steered technique. First, since the skid steered technique uses the speed and force differences between left and right wheels to achieve pivot steer, the large tyre slip and skid ratio are unavoidable. Therefore, the tyres' life is greatly degenerated. Second, owing to the unavoidable tyre slip, the pivot steer trajectory is very difficult to control. The position of the center of gravity (C.G.) fluctuates in the pivot steer maneuver. These drawbacks are discussed in our previous work [5].

2.2 *Omnidirectional Steer Mechanisms for Light-Class Robots*

In the area of light-class robots, the omnidirectional steer technique is widely focused on. Figure 2 shows typical mechanisms of omnidirectional steer technique, including Mecanumm, roller wheels, and transformable suspensions [9].

The roller wheel uses several rollers on the circumference of a single wheel, which are located perpendicular to the direction of rotation of the wheel. The wheel generates thrust force along the wheel's rotation direction and slides passively along the lateral direction by using the rollers. The working principle of the Mecanumm wheel is very similar to that of the roller wheels. The difference is that the rollers in the Mecanumm wheel are aligned 45° to the wheel plane, which aims at generating angular forces. The transformable wheel suspension is a direct approach to achieve

(a) Roller Wheel (b) Mecanumm Wheel

(c) Transformable Suspension

Fig. 2 Typical kinds of omnidirectional steer mechanisms [9]

omnidirectional steer. It adjusts the direction of the control arm of the suspension to change the steer angle of the wheels to adjust the moving direction.

However, the roller wheel, Mecanumm wheel, and transformable suspension cannot be applied for heavy-class UGVs, since the loading capacity of these three mechanisms is very small. Therefore, they can only be applied for the light-class robots and cannot be applied for the heavy-class UGVs. Therefore, the appropriate steering mechanisms for the heavy-class UGVs should be particularly designed and investigated.

3 Development of an Independent Steered Four-Wheel-Independently Actuated UGV

The authors designed an independent steered four-wheel-independently actuated UGV, which is named as unmanned ground carrier (UGC) [10, 11]. This project is supported by the central military commission of China. The major specifications of the UGC are given in Table 1.

The UGC owns following novelties: First, the main idea of the UGV is the 'Carrier' idea. In other words, the UGC is not an individual UGV. It is a mother-with-child type UGV, which is a vehicle platform to hold other rotorcrafts and small carrier-based systems. Based on the collaborative control techniques between the UGV and rotorcrafts, the ability and performance to conduct difficult missions are significantly enhanced. Figure 3 gives the overview picture of the UGC. The mother vehicle is given in Fig. 3a. Following carrier-based systems are carried on the UGC: (1) an armed rotorcraft, which is shown in Fig. 3b. It is able to finish the missions of search,

Table 1 Specifications of the UGC platform

Specifications	Values
Mass	1200 kg
Wheelbase	2.2 m
Track width	1.6 m
Height	1.4 m
Tyre radius	0.44 m
Maximum power	250 kW
Battery capacity	15 kWh
Highest speed	120 km/h

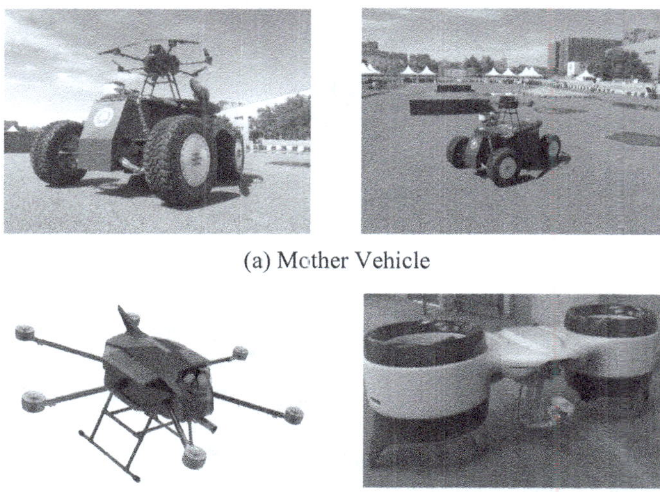

(a) Mother Vehicle

(b) Equipped Armed Aircraft (c) Equipped Twin Duct Aircraft

Fig. 3 Unmanned ground carrier developed by the Authors

reconnaissance, recuse, and light combat. The load capacity of the armed rotorcraft is 35 kg, and its maximum endurance time is 60 min, (2) a twin ducted aircraft with robotic arm, which is shown in Fig. 3c. It is able to finish the missions of grabbing or manipulating in the narrow and closed environments, (3) a small tracked robot, which is located inside the platform. The door at the front of the platform is able to be remotely controlled open or closed to send out the small tracked robot (Fig. 4).

Second, the mother vehicle platform adopts electrified technique. It is powered by a LiFePo4 battery pack that peaks 250 kW. Each wheel of the platform is steered/driven/braked independently, which leads to four-wheel-independently actuated technique. Four wheels are driven directly by the in-wheel-driven motors, respectively. The max power of the motor is approximately 60 kW. The wheel is independently steered by a steer servomotor. The steer mechanism is given in Fig. 5.

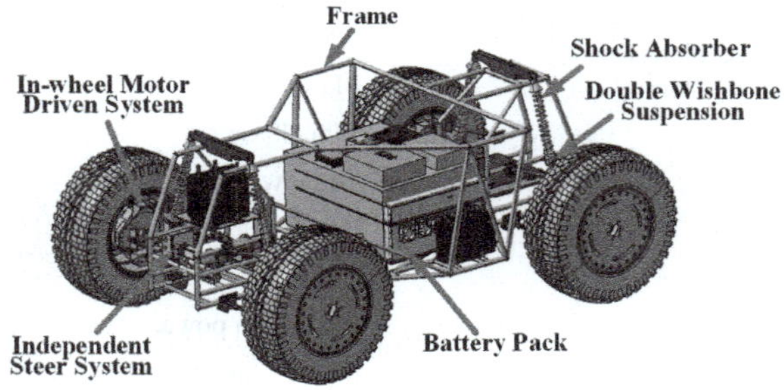

Fig. 4 Configuration of the platform

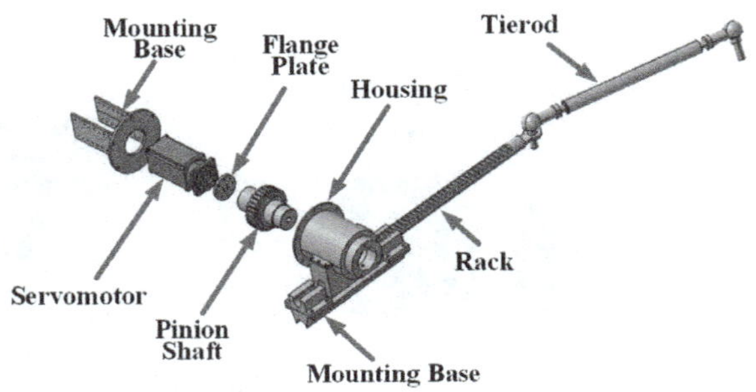

Fig. 5 Mechanisms of the steer system

Compared to the mechanisms shown in Sect. 2, this independent steer mechanism has following advantages: (1) Compared to skid steered technique, this mechanism allows the steering wheel to be steered during the pivot steer such that the vehicle turning center is located in the middle of the platform, which can greatly reduce the tyre slip and skid ratio. (2) With high power and high torque steer servomotor, the steer mechanism has high load capacity to negotiate the heavy UGC. The high power steer servomotor also enables the high-speed response of the steering maneuver.

With independent steered technique, the UGC is able to achieve ZRS as Fig. 6 shows. In ZRS, the steering angles of each left and right side wheels are opposite.

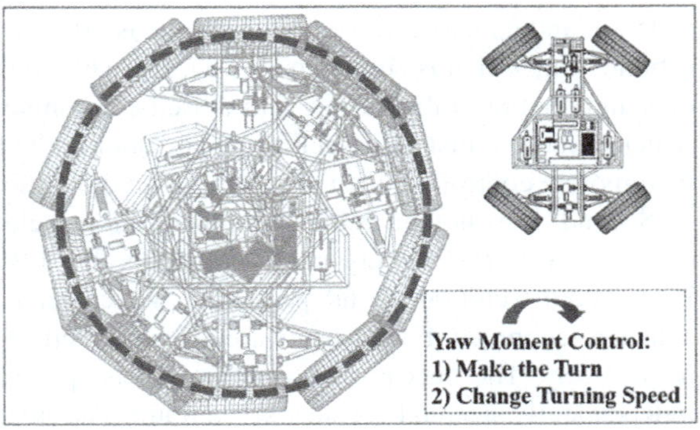

Fig. 6 Illustration of platform behavior in ZRS

The behavior of the platform in ZRS can be seen in Fig. 6. The zero radius turning center is located at the center of the platform to avoid the tyre slip. In addition, since each wheel is driven by an in-wheel-driven motor, the yaw moment can be generated based on the independent traction forces provided by the in-wheel motors. Therefore, the yaw moment can be applied to yaw the platform and change the turning speed as the human operator wants.

4 Yaw Moment Control and Type Slip Ratio Control in ZRS

The UGC is majorly manipulated by the ground command station in remote control mode, which will be shown in the final section. On the command station locate several control levers to control the platform, rotorcrafts, and the robots. The human operator can manipulate the control levers to select the ZRS knob and control the platform to perform ZRS. The UGC is supposed to work in the field environments, where the road conditions are very complex and difficult to handle. Therefore, the chassis control in ZRS includes two modules: yaw moment control and tyre slip ratio control modules.

After the ZRS knob is selected, the steering wheels are placed as Fig. 7 shown. The input value of the yaw moment is in proportion to the lever position:

$$L = K M_{\text{des}} \tag{1}$$

where K is the parameter of control gain. L is the control lever position on the command station. M_{des} is the yaw moment.

The total yaw moment needs to distribute to four wheels. Based on the tyre friction principle, the tyre adhesive capability is in approximate linear proportion to the tyre vertical load. Therefore, after the value of input yaw moment is determined, the tyre traction forces are distributed based on the vertical load acting on the tyres as Fig. 7

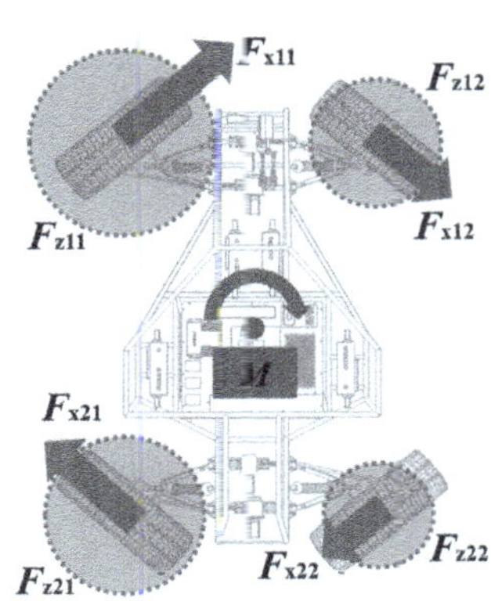

Fig. 7 Illustration of chassis control in ZRS

shows. The relationship between M_{des} and F_{xij} is described as:

$$F_{xij}(k) = \frac{M_{\text{des}}(k-1)}{\sqrt{\frac{B^2}{4} + \frac{L^2}{4}}} \frac{F_{zij}(k-1)}{\sum_{i,j=1}^{2} F_{zij}(k-1)} \tag{2}$$

where k is the sampling step. B is the track width. L is the wheelbase. F_{zij} is the vertical load acting on four tyres, respectively.

When the UGC works on the muddy or field roads in complex environments, the wheel slip increases considerably to degenerate the overall performance. Therefore, the wheel slip control is important to handle the tough terrain. The dynamic model of each steering wheel is:

$$T_{ij}(t) = \frac{I_w \dot{\omega}_{ij}(t) + F_{xij}(t) R_t + F_{Rij}(t)}{i_p} \tag{3}$$

where T_{ij} is the input value of torque of the in-wheel-driven motors. ω_{ij} is the rotation speed of wheels. F_{Rij} is the resistance drag force of wheels. I_w is the wheel inertia. R_t is the radius of tyre. i_p is the planetary gear transmission ratio, which is assembled in the in-wheel motor-driven system.

The state space of the tyre slip ratio dynamics model is shown in Eq. (4). The details of the manipulation to obtain this model are neglected in this paper due to the space limitation.

$$\dot{S}(t) = AS(t) + Bu(t) + C(t) \tag{4}$$

where:

$$S(t) = \left[s_{11}(t) \; s_{12}(t) \; s_{21}(t) \; s_{22}(t) \right]^T \tag{5}$$

$$u(t) = \left[T_{11}(t) \; T_{12}(t) \; T(t)_{21} \; T_{22}(t) \right]^T \tag{6}$$

$$A = \begin{bmatrix} -\dot{\omega}_{11}(t)/\omega_{11}(t) & & & \\ & -\dot{\omega}_{12}(t)/\omega_{12}(t) & & \\ & & -\dot{\omega}_{21}(t)/\omega_{21}(t) & \\ & & & -\dot{\omega}_{22}(t)/\omega_{22}(t) \end{bmatrix} \tag{7}$$

$$B = \begin{bmatrix} -i_p/\omega_{11}(t) m_w R^2 & & & \\ & -i_p/\omega_{12}(t) m_w R^2 & & \\ & & -i_p/\omega_{21}(t) m_w R^2 & \\ & & & -i_p/\omega_{22}(t) m_w R^2 \end{bmatrix} \tag{8}$$

$$C(t) = \begin{bmatrix} -\frac{I_w\dot{\omega}_{11}(t)+F_{R11}(t)(1+R)}{\omega_{11}(t)m_wR^2} + \frac{\dot{\omega}_{11}(t)}{\omega_{11}(t)} & & & \\ & -\frac{I_w\dot{\omega}_{12}(t)-F_{R12}(t)(1+R)}{\omega_{12}(t)m_wR^2} + \frac{\dot{\omega}_{12}(t)}{\omega_{12}(t)} & & \\ & & -\frac{I_w\dot{\omega}_{21}(t)+F_{R21}(t)(1+R)}{\omega_{21}(t)m_wR^2} + \frac{\dot{\omega}_{21}(t)}{\omega_{21}(t)} & \\ & & & -\frac{I_w\dot{\omega}_{22}(t)+F_{R22}(t)(1+R)}{\omega_{22}(t)m_wR^2} + \frac{\dot{\omega}_{22}(t)}{\omega_{22}(t)} \end{bmatrix} \quad (9)$$

where $u(t)$ is considered as the control input parameters. $C(t)$ is considered as the disturbance that causes by the tyre behavior or road condition.

The target of the wheel slip controller is to find the control input $u(t)$ to handle $S(t)$ to reach the desirable value, which is predefined in the controller. Generally, the desired value of the tyre slip ratio is selected as 0.1–0.2 to obtain desirable tyre behavior. This value range is widely used in the field of vehicle dynamics control.

The sliding mode technique is used in this paper to design the controller. The sliding surface is defined as $\lambda = S(t) - S_\text{d}$. S_d is the desired value of the tyre slip ratio. Differentiate the sliding surface:

$$\dot{\lambda} = GS(t) + Hu(t) + L(t) - S_d \quad (10)$$

A continuous law $u_\text{eq}(t)$ can be obtained:

$$u_\text{eq}(t) = \begin{bmatrix} T_{11\text{eq}}(t) \\ T_{12\text{eq}}(t) \\ T_{21\text{eq}}(t) \\ T_{22\text{eq}}(t) \end{bmatrix} = \begin{bmatrix} \frac{F_{R11}(t)\omega_{11}(t)+F_{R11}(t)R\omega_{11}(t)+I_w\dot{\omega}_{11}(t)\omega_{11}(t)+m_w\dot{\omega}_{11}(t)u_{x11}(t)}{i_p\omega_{11}(t)} \\ \frac{F_{R12}(t)\omega_{12}(t)+F_{R12}(t)R\omega_{12}(t)+I_w\dot{\omega}_{12}(t)\omega_{12}(t)+m_w\dot{\omega}_{12}(t)u_{x12}(t)}{i_p\omega_{12}(t)} \\ \frac{F_{R21}(t)\omega_{21}(t)+F_{R21}(t)R\omega_{21}(t)+I_w\dot{\omega}_{21}(t)\omega_{21}(t)+m_w\dot{\omega}_{21}(t)u_{x21}(t)}{i_p\omega_{21}(t)} \\ \frac{F_{R22}(t)\omega_{22}(t)+F_{R22}(t)R\omega_{22}(t)+I_w\dot{\omega}_{22}(t)\omega_{22}(t)+m_w\dot{\omega}_{22}(t)u_{x22}(t)}{i_p\omega_{22}(t)} \end{bmatrix} \quad (11)$$

The sliding relationship is designed to decrease the chattering:

$$\frac{1}{2}\frac{d}{\text{d}t}S^2 \leq -\varepsilon|S| \quad (12)$$

where ε is a positive scalar.

If the condition given in Eq. (12) is used, the trajectories of the dynamics system are controlled to achieve the surface within a known time period. The time period is smaller than the value of $|S(t = 0|/\eta$. Generally, following law with a discontinuous part should be used to meet the above condition:

$$u(t) = u_{\text{eq}}(t) - K_s \operatorname{sgn}(S(t)) \tag{13}$$

The K_s can be calculated to obtain the relationship in Eq. (12). To reduce the chattering through the $S = 0$, take the place of sgn($S(t)$) by the approximation term sat($S(t)/\Phi$):

$$\begin{aligned} u(t) &= u_{\text{eq}}(t) - K_s \operatorname{sat}\left(\frac{S(t)}{\phi}\right) \\ &= \begin{bmatrix} \frac{F_{R11}(t)\omega_{11}(t)(1+R)+I_w\dot{\omega}_{11}(t)\omega_{11}(t)+m_w\dot{\omega}_{11}(t)u_{x11}(t)}{i_p\omega_{11}(t)} \\ \frac{F_{R12}(t)\omega_{12}(t)(1+R)+I_w\dot{\omega}_{12}(t)\omega_{12}(t)+m_w\dot{\omega}_{12}(t)u_{x12}(t)}{i_p\omega_{12}(t)} \\ \frac{F_{R21}(t)\omega_{3}(t)(1+R)+I_w\dot{\omega}_{21}(t)\omega_{21}(t)+m_w\dot{\omega}_{21}(t)u_{x21}(t)}{i_p\omega_{21}(t)} \\ \frac{F_{R22}(t)\omega_{22}(t)(1+R)+I_w\dot{\omega}_{22}(t)\omega_{22}(t)+m_w\dot{\omega}_{22}(t)u_{x22}(t)}{i_p\omega_{22}(t)} \end{bmatrix} - K_s \operatorname{sat}\begin{bmatrix} \frac{s_{11}(t)}{\phi} \\ \frac{s_{12}(t)}{\phi} \\ \frac{s_{21}(t)}{\phi} \\ \frac{s_{22}(t)}{\phi} \end{bmatrix} \end{aligned} \tag{14}$$

where sat is the saturation term. Φ is the thickness of the layer of the boundary.

The $\omega_{ij}(t)$ is feedback by the information of the in-wheel motor. The $F_{Rij}(t)$ is obtained by multiplying the resistance drag coefficient and the vertical load on the tyre. The $u_{xij}(t)$ can be obtained by the calculation with the vehicle speed, lateral speed, and the yaw velocity of the vehicle.

5 Experimental Verification

In this section, the proposed chassis controller is verified by the experiments. The UGC is controlled remotely by the human operator as Fig. 8 shows. The human operator controls the rotorcrafts, mother platform, and the small robot, based on the command station. The telecommunicating frequency is 915 MHz. The digital cameras are equipped on the rotorcrafts, mother platform, and the small robot, which help to determine the next moves.

5.1 Individual Experiment for the Tyre Slip Ratio Controller

To verify the proposed wheel slip controller, an individual experiment is conducted. The platform is placed on a muddy road in the filed environment. Both the cases of uncontrolled and under control are done to compare the tyre slip ratio values to demonstrate the performance of the wheel slip controller. In each experiment, the maximum motor torque is applied to accelerate the platform, and the tyre slip ratio

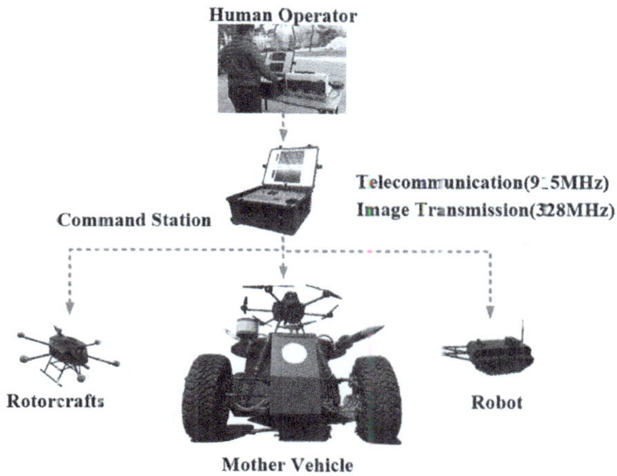

Fig. 8 Telecommunication between rotorcrafts, platform, and robot

is calculated based on the vehicle speed and the wheel rotation speed. The vehicle speed is feedback by the differential GPS/INS.

Figure 9a shows the value of slip ratio without control. The tyre slip ratio value is as high as 1. The average of the slip value is approximately 0.7. The slip ratio value reaches 1 indicates the wheel rotation speed is much larger than the value of vehicle speed, which means the extreme skid and slip occurs. The adhesive capability of tyre is largely reduced when high tyre slip occurs. Consequently, the driving and acceleration performance of the vehicle is largely decreased. Figure 9b shows the value of motor torque under tyre slip ratio controller. The torque of each motor is allocated based on the tyre vertical load and modified to obtain desired slip ratio value. Figure 9c shows the value of tyre slip when tyre slip ratio control is applied to the vehicle. The slip ratio is reduced largely due to the application of the tyre slip ratio controller. It can be seen that the average of the slip value is approximately 0.5. The slip ratio controller designed in this paper successfully reduces the value of tyre slip ratio and enhances the driving performance in the off road. Before the experiments in ZRS, the individual experiment is conducted to clearly show the performance of the proposed tyre slip ratio controller (Fig. 10).

5.2 Experiments in ZRS

To verify the tyre traction force distributing controller and the tyre slip controller in ZRS, both the tests on sand road and the tests on pavement road are done. In the experiments, the ZRS knob is selected. The maximum yaw moment (maximum torque of motor) is used. In sand road case, the traction forces are allocated based on the vertical load distribution and the tyre slip control is switched on. In the pavement road case, the traction forces are equally allocated and the slip control is switched off.

Fig. 9 Tyre slip ratio comparison

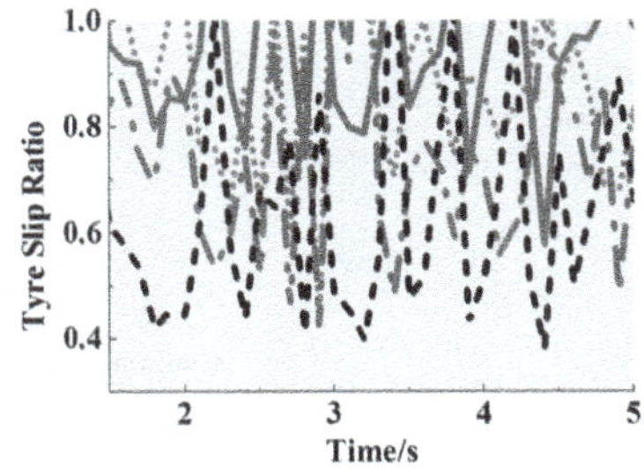

(a) Tyre Slip Ratio Response without Slip Ratio Control

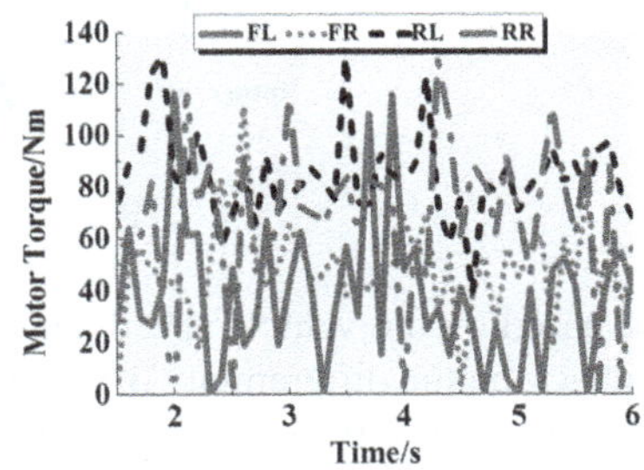

(b) Motor Torque with Slip Ratio Control

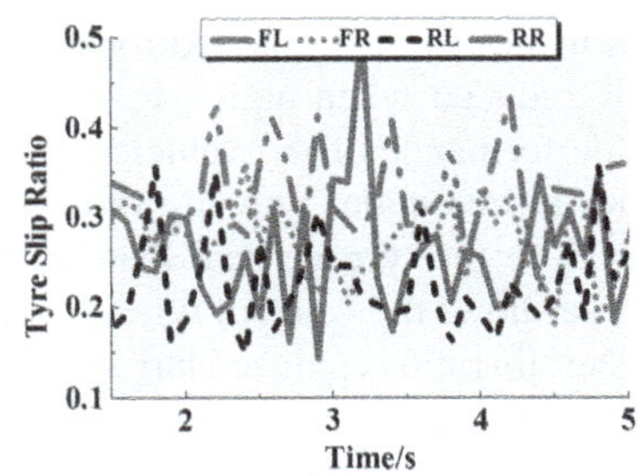

(c) Tyre Slip Ratio Response with Slip Ratio Control

Fig. 10 Experiment site

(a) ZRS on Sand Road Condition (With Control)

(b) ZRS on Pavement Road Condition (No Control)

Fig. 11 Experiments in ZRS

According to Fig. 11, the UGC with control smoothly finishes the ZRS maneuver. The turn center of the vehicle is located in the center of the vehicle. The turn trace in the experiment is quietly smooth. In addition, the moving trace of each steering wheel matches to each other well, which shows no tyre skid or tyre slip. The C.G. position of the vehicle is stable in the ZRS maneuver. According to Fig. 11b, the performance in ZRS without control is worse. The vehicle turn center is not in the middle of the vehicle. The moving trace of each steering wheel is not same to each other, which shows great tyre skid and slip. It causes the great movement of the C.G. of the vehicle. It can be known that the proposed tyre traction forces distributing approach and the tyre slip controller enhance the performance in ZRS successfully.

6 Conclusion

This paper introduced the UGC platform, which is designed by the authors. A unique advantage of the UGC is that it is able to perform ZRS based on the independent steered technique. To enhance the performance of the UGC to negotiate the complex terrains, a chassis controller for ZRS is proposed. Finally, the experiments based on the UGC validate the performance and efficiency of the proposed controller.

References

1. Albert YS, Leung YW, Chu X (2016) Autonomous-vehicle public transportation system: scheduling and admission control. IEEE Trans Intell Transp Syst 17(5):1210–1226
2. Wang R, Jing H, Hu C et al (2016) Robust H∞ path following control for autonomous ground vehicles with delay and data dropout. IEEE Trans Intell Transp Syst 17(7):2042–2049
3. Kayacan E, Ramon H, Saeys W (2016) Robust trajectory tracking error model-based predictive control for unmanned ground vehicles. IEEE/ASME Trans Mechatron 21(2):806–814
4. https://en.wikipedia.org/wiki/Unmanned_ground_vehicle
5. Ni J, Hu J (2016) Dynamic modeling and experimental validation of skid steered vehicle in the pivotal steer condition. Proc Inst Mech Eng, Part D: J Automobile Eng 231(2):1–16
6. http://v.youku.com/v_show/id_XNjcyOTAwOTk2.html
7. Ni J, Hu J, Xiang C (2017) Envelope control for four-wheel-independently-actuated autonomous ground vehicle through AFS/DYC integrated control. IEEE Trans Veh Tech. https://doi.org/10.1109/tvt.2017.2723418
8. Zhang Z, Zhang X, Pan H et al (2017) A novel steering systems for a space-saving 4WS4WD electric vehicle: design, modeling and road tests. IEEE Trans Intell Trans Syst 18(1):114–127
9. Genya I, Iagnemma K, Overholt J et al (2015) Design, development, and mobility evaluation of an omnidirectional mobile robot for rough terrain. J Field Robot 32(6):880–896
10. Ni J, Hu J, Xiang C Design and advanced robust chassis dynamics control for X-by-wire unmanned ground vehicle, Morgan & Claypool publisher
11. Ni J, Hu J, Xiang C (2017) Control configured vehicle design and implementation on X-by-wire electric vehicle, IEEE Trans Veh Tech. https://doi.org/10.1109/tvt.2018.2805886

Construction of F-PSG Analysis Model Based on Automobile Modeling Design Language

Jingfeng Shao, Jianliang Shi, Leqing Huang, Feijiao Huo and Shaojie Gao

Abstract With the continuous development of economy and society, the rapid progress in science and technology, and the upgrading living standards of modern people, automobiles are no longer tools just used for transportation, but something more and more attached to people's needs for diversified and individualized designs. However, different designers produce different styles. Design is something very subjective, and the language itself is vague. Therefore, for a long time design style has been described in an abstract way. The absence of scientific description on design makes the design language of automobiles an obstacle, which will reduce the effect and efficiency of analysis and communication between designers, or between designers and other related staffs. The existing analysis method of automobiles' design language is mainly understood and adopted by users with professional background knowledge, so it is difficult to use the method to deliver design concept to collaborators and consumers for their lack of specialized design knowledge. Given that, based on the technique of semantic cognition, a new analytical method, the F-PSG analytical method is proposed in this paper, which hopefully will be a solution to the problem.

Keywords Automotive design language communication · Semantic cognitive method · F-PSG analysis model

1 Foreword

With the advancement of technology and the improvement of people's living standards, cars that were originally expensive have already entered the homes of ordinary people. Although the advancement of technology has brought about a reduction in the cost of automobiles, the performance and appearance of automobiles have been iter-

J. Shao · J. Shi · L. Huang (✉) · F. Huo · S. Gao
SAIC Motor Corporation Limited Technology Center Co, Shanghai, China
e-mail: HuangLeQing@saicmotor.com

J. Shao
Tongji University, Shanghai, China

China SAE (ed.), *Proceedings of China SAE Congress 2018: Selected Papers*, Lecture Notes in Electrical Engineering 574,
https://doi.org/10.1007/978-981-13-9718-9_65

ated and enriched with diversified needs from consumers. This has led to the design of automobile products increasingly requiring different professionals to participate in cooperation, but aesthetic personality and ambiguity of the different design language make the original complex and professional design language an obstacle to effective analysis, and causing difficulties in the communication of design development.

2 Overview on Automotive Styling Design Language

2.1 Aesthetic Rules and Preference

Aesthetics is a shallow sense of feeling or emotion, which is a direct response to beauty and ugliness (Bai Hanzhi and Chen Chuyun 2009). Through research, it is found that aesthetic law of common style exists, such as the linear elements commonly used in design, conveying feelings of the tall and solemn. Horizontal lines bring in a sense of stability and balance, while inclined lines produce a strong and dynamic feeling. Curves present different degrees of acceleration according to its curvature. Although aesthetic commonality exists, individual aesthetic preferences always exist at the same time. Studies have shown that individual aesthetic preferences and the ability to associate form with beauty are determined by both innate and acquired factors. In addition to the innate aesthetic sensitivity, professional designers have also experienced acquired professional training and work, which endowed designers with more skills in professional judgment and communication, in terms of design aesthetics. In addition to professional training, relevant variables that contribute to aesthetic preferences also include:

(1) Intrinsic external motivation. The aesthetic experience is related to the intrinsic motivation; thus if the product is regarded as a tool for completing a certain task, judged from the exterior design, its practical aesthetics will be much more appreciated [1].
(2) Individual differences in age, gender, personality, mood, interest, income level, etc., will affect people's interest of the product and the way of its visual information processed [1].
(3) Application scenarios. Different application scenarios are the most common factors affecting individual aesthetic judgment, so in the user's empathy test, the tested objects must be arranged in the same scene.
(4) Different cultural and customary factors formed by different geographical features.
(5) Visual personality/language personality. Visual consumers and linguistic consumers have different aesthetic judgments on products. The visual type is sensitive to the whole and prefers organic forms. The linguistic type is sensitive on specific aspects and is good at paying attention to independent details. Visual type is more likely to be attracted to different aspects of the form than the language type.

(6) The time length of design observation (Creusen and Schoormans 1998).
(7) Subjective attributes of color/shape. For example, red will produce a feeling of enthusiasm and sharp objects will bring in insecurity.
(8) Aesthetic fatigue. When exposed to a certain form for a long time, the aesthetic preference shifts to a newer form [1], resulting in a popular aesthetic.
(9) Brand preference. The preference for a certain brand will cause consumers' preference for design styling to a certain extent. Thus during consumer research, the logo of the test product must be removed.

2.2 Analysis and Communication Process of Automobile Design

Although some designers can comprehend and produce judgments based on experience or perception, there is still a need for a concise expression of the design style. If the design is still in the communication field of designers or professionals, it is difficult to obtain recognition and support from inside and especially outside of company's related parties. From the design work itself, car design is competitive from the first initial sketch, but the subsequent design work is based on teamwork. This involves not only communication between professional designers, but also the collaboration of internal engineers and communication to external parties. The design object expands from the car itself to its surroundings and service experience. The problem that the design needs to solve is much more complicated, and thus, professional teams need to participate in a diversified collaborative design process. From the perspective of the design communication process, car design carries semantic communication functions of design, manufacture, production, sales, market, and consumer. Due to the complexity of car design itself and aesthetic differences of the individual, there are obstacles in terms of communication and consensus building.

When the design is not expressed in a common language, non-design professionals are likely to have negative emotions, as Mozota discovered in his research:

(1) Pride: "I know what a good design is."
(2) Despise: "What is the function of the design tool, considering the position of the design in the company?"
(3) Fear: Differences in beauty or design perspectives.
(4) Disregard: Designers did not receive an education on the importance of design assessment [1].

In design reviews, we find it difficult to communicate from the designer to its consumer without transforming design into a common language. Therefore, automotive design requires a much more general analytical method to break barriers of communication between the professionals and audiences.

2.3 Automotive Styling Language Concept

The automotive design language has comprehensive visual information of high technology and complexity, which is an aesthetic expression that combines art, design, and technology perfectly. It is based on the overall thinking of user experience, mainly used in automotive exterior and interior design.

(1) Exteriors include proportions, poses, surface features, material crafts, colors, exterior trims, etc.
(2) Interiors include driving space, interior human–machine interaction (HMI), surface features, material craftsmanship, color, interior trim, etc.

The design language needs to convey three kinds of semantic information:

(1) Aesthetic. The aesthetic is evaluated whether the design of car meets the consumer's aesthetic taste and identity expression, or whether it can be quickly accepted by the target group, since the design has a direct impact on the consumer's purchase choice.
(2) Functional. The design rule of "form follows function" is also applicable to automobile design. The relationship between vehicle structure and design is mutually influential. For example, sedans and SUVs are two models based on different usage scenarios, and hence, the structure of the vehicles is also significantly different. So that the design language should be able to accurately express the core characteristics of different models, to meet the functional needs of users for marketing.
(3) Brand recognition. Automobile styling design is an important means for enterprises to participate in market competition. How to stand out from the crowd and be better identified, recognized, and generated by its consumers is the goal that major auto manufacturers have been pursuing [5].

3 Research on Analysis Methods of Automobile Styling Language

At present, the analysis methods of automotive design languages mainly include the following:

3.1 Semantics Recognition Method

At present, semantic recognition is the most common design communication method, through which forms the psychological intention from the actual styling images, and constitutes the semantic feature through vocabulary. Semantics refers to meaning,

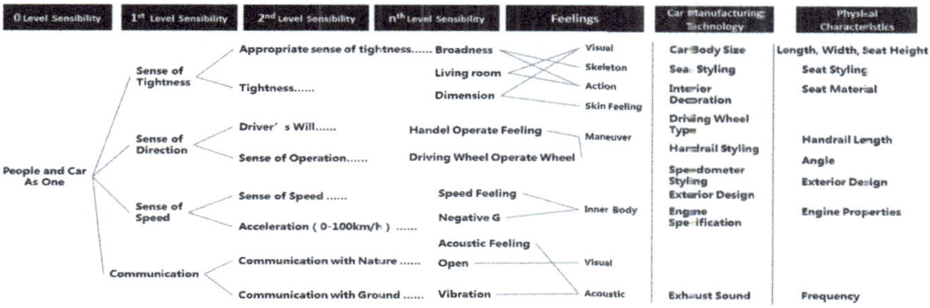

Fig. 1 Nagamachi, Sansei. Tale of Kansei Engineering [J]. 1995. Classification based on the perceptual level of new products, in order to establish product sensory structure to obtain design details

intention, and conceptual idea of form (natural, human, abstract) under certain conditions. The product itself constitutes a completed symbolic system and becomes a symbolic carrier for conveying information and meaning, expressing functional descriptive semantics, operational indicative semantics, and symbolic semantics [2]. Starting from the Japanese Kansei Engineering (sensibility engineering), Mazda, Mitsubishi, Toyota, Honda, Fiat, Mercedes-Benz, and other automotive manufacturers have applied semantic cognition to the development of automotive design, helping designers understand the target group's expectations for design. Through the factor analysis, a link between intentional semantics and styling features is established [3], as shown in Fig. 1. In the process of design and implementation, intuitive and logical design analysis, and communication are realized through this hierarchical connection.

3.2 *Image Board Method*

Image board method is a re-use of speech recognition tools, whose interface consists of two parts: "word tag" and "graphics". It assists the designer to map the design theme in the multidimensional semantic space, in a way that combines the design semantics with the visualization, which is used to help designers locate clear design positioning [6]. Taking the Roewe RX8 as an example, the three-pair adjectives of "luxury-coarse," "hard-round," "steady-dynamic" are applied to define the semantic space of its design goal, as shown in Fig. 2.

3.3 *Line-Shape Analysis Method*

The line-shape analysis method is to grasp the core characteristics of the whole vehicle styling, through analyzing key lines that forms shapes, potency, and space

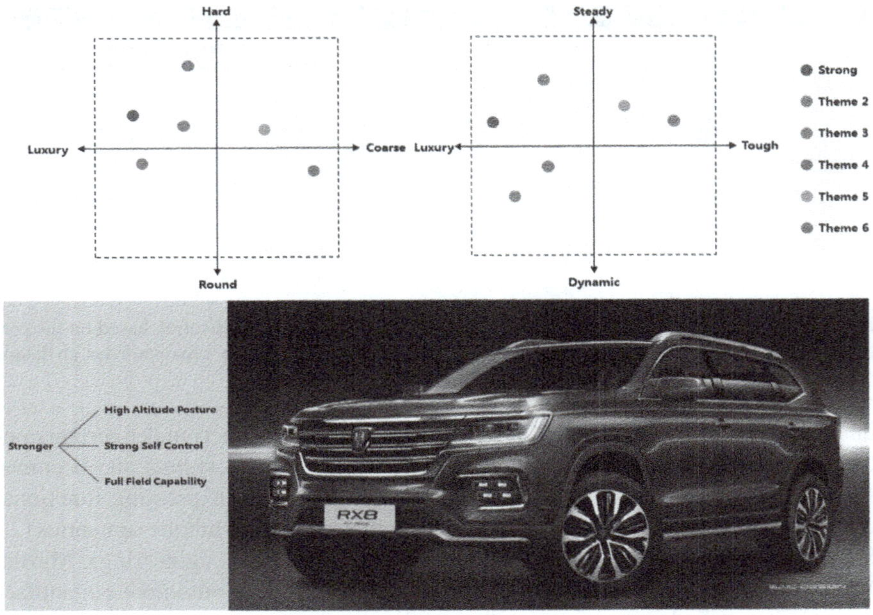

Fig. 2 Automotive analysis case based on intentional image board tools

of the product, extracting the key lines including waistline, contour line, fractal line, section line, and other lines related to its branding features. The key lines build the core structure of the overall shape, forming the basic dynamic potency of the whole vehicle product styling.

It helps to find the styling regular pattern by analyzing and comparing rhythm of each key line and the relative relationship between them, especially grasping the important proportional correlation formed by the ratio of the line spacing [5].

3.4 Diagram Analysis Method

Diagram analysis method is an analytical method that uses sketches and diagrams to help designers think. As a common design analysis and communication method for designers, this model combines both mind mapping and fast expression, expressing the derivation process and thinking process in design. Compared with the textual thinking or expression, it benefits from utilizing the concrete and intuitive image to describe the fleeting and ambiguous conceptual relationship in the design process, forming visual logic framework that is easy to understand.

3.5 Digital Modeling Analysis Method

The digital modeling analysis method is based on the principle of establishing a three-dimensional model of a digital surface (such as the creation of a NURBS surface). The surface can be regarded as the scanning trajectory of the line, and such the edge line of the surface and the key section line in the middle determine the size, spatial location, potency, curvatures of the surface [5]. The function of the digital model is involved in the design process of the car, which is mainly reflected in the surface stitching technology and surface curvature quality control, ensuring the enhancement of the quality of the car body, containing a certain meaning of life [4]. With the continuous improvement of the level of intelligence, on one hand, digital modeling tools are more involved in the form discussion of the early design stage. On the other hand, the progress and development of VR/AR technology, digital modeling technology will be much more used in the design review and releasing stage. The digital virtual reality center established by SAIC Motor Design Center uses VR technology to present the real-time rendered car model in front of the viewer. In addition to reviewing the proportional appearance, the door of the virtual model can be opened for viewers to experience its interior design (Fig. 3).

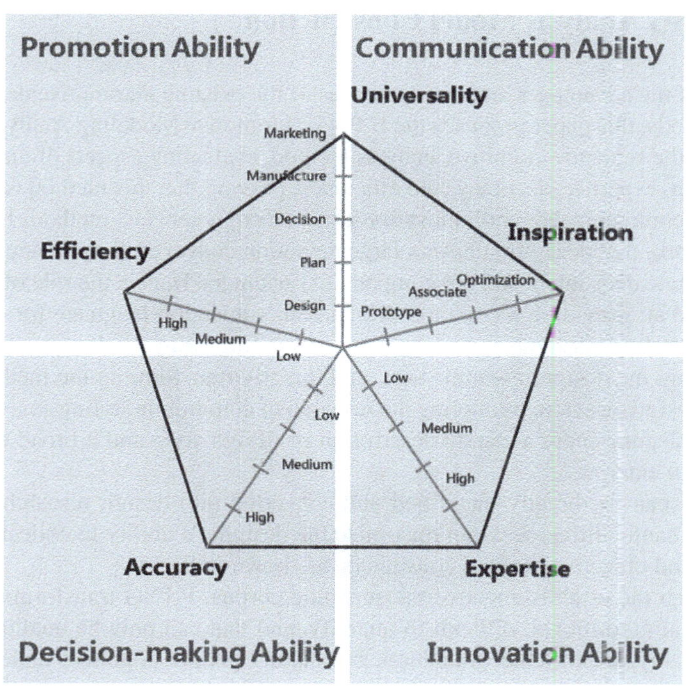

Fig. 3 Attribute map of automotive styling analysis method

Through in-depth interviews with personnel of design, manufacturing, production, sales, and marketing and consumers, the application of several modeling design analysis methods is concluded as follows. According to the functions of several analytical methods, the semantics recognition method is the most versatile (applicable to most wide range audience) and efficient analysis method, which is consistent with the theoretical research results, as shown in Table 1. By establishing the relationship between semantics and design elements, an extensive and efficient communication can be carried out to implement design solutions. It should be noted that the analysis method achieves different functional purposes according to relevant usage scenarios, implementation conditions, and scope of application; hence, there is no difference between good and bad among these methods. However, besides semantic recognition method, we also see that the other four analytical methods are mainly applicable to those with professional knowledge, such as designers, while communication and collaboration with partners, sellers, and other people who do not have professional design background are often required from design to final product, with which becomes difficult to convey the design concept and image accurately to these audiences.

4 F-PSG Analysis Model Construction

In view of the advantages and disadvantages of the existing automotive design analysis methods, this paper proposes the F-PSG Automotive Modeling Analysis Model based on the semantic cognitive analysis method, evaluating aspects of universality, inspiration, expertise, accuracy, and efficiency, proving that this method is currently the most communicable and innovation-driven design analysis method. In the past design work, the ideas often have a large deviation conveyed to the designers from the corporate decision makers or company's consumers. Hence, the role of F-PSG is positioned at: providing a common design communication language for enterprise decision makers, designers, and its consumers; following the brand connotation, undertaking the design research stage, and directly transforming the method into a tool for design research; promoting the research of automobile styling analysis methods; establishing more accurate description of design style and a broad consensus throughout enterprise.

F-PSG can be directly quantified and converted into design research tools. F-PSG's semantic differences can maximize the designer's ability to collect the most realistic and objective ideas of consumers, as shown in Fig. 4.

Through the establishment of the semantic corpus, F-PSG transforms aesthetic feeling that is irrational, difficult to quantify, and that can only be qualitative, into hierarchically quantifiable evaluation, correlating people cognitive understanding of certain automotive exterior styling with specific design aspect, facilitating the adjustment of automotive design development.

Table 1 Comparison of analysis methods

Analysis method	Realize criteria	Applicable range	Advantages	Disadvantages	Analysis function	Capacity level
Semantics recognition method	Intent semantic library; establish a connection between intentional semantics and styling features	Designers, planners, engineers, marketers, decision makers, operators, consumers	A generally applicable analytical language; wide range of applicable subjects; concise semantics cause less ambiguity, high accuracy	Need numerous perceptual linguistic data to build an intent semantic library	Communication ability; promotion ability	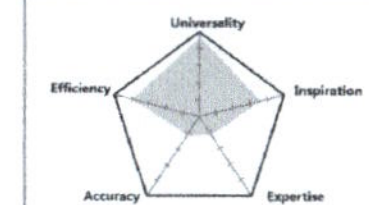
Image board method	Word tag; graphics	Designers, product planners	Graphic images improve association efficiency	Easy to cause positioning distraction due to aesthetic preference; single semantic spatial dimension; Only suitable for the development phase of the design process	Innovation ability	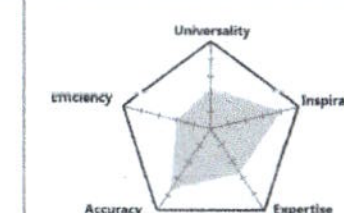

(continued)

Table 1 (continued)

Analysis method	Realize criteria	Applicable range	Advantages	Disadvantages	Analysis function	Capacity level
Line-shape analysis method	Key lines abstraction; construction of core architecture	Designers, decision makers, operators, consumers	Help non-professional understanding overall feeling and attitude of a car design	Lack in detail feature and surface feature analysis	Communication ability	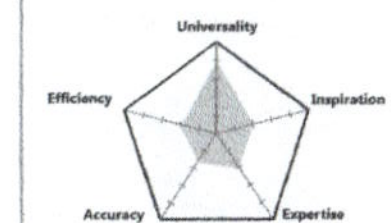
Diagram analysis method	Quick sketch, process analysis	Designers	More intuitive use of the thinking process of car design; the sketches and diagrams are more focused and specific	Only applicable for design professionals, lack in universality	Innovation ability	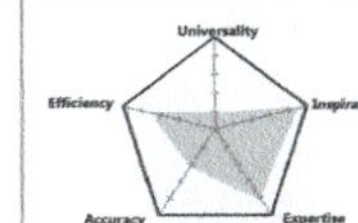
Digital modeling analysis method	Surface analysis tool, virtual presentation technology	Designers, decision maker	Convert design features into quantifiable digital modeling features; design expression is most complete among other methods	Lack of "meaning" expression in design review stage	Decision-making ability	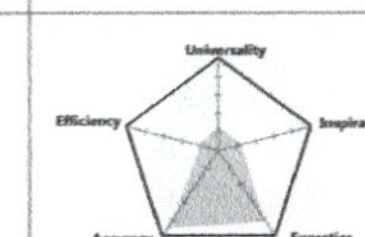

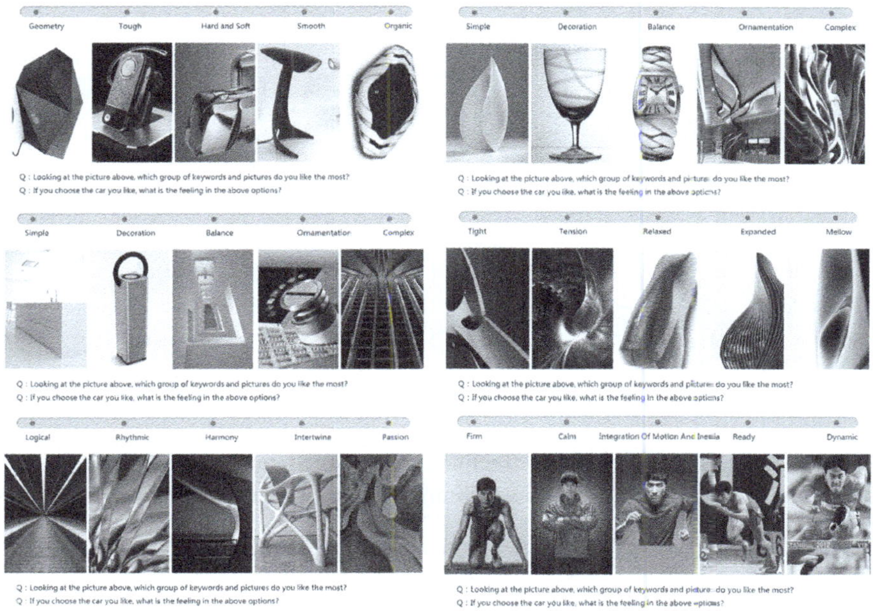

Fig. 4 Actual survey questionnaires guided by F-PSG modeling style analysis method

4.1 Establishment of Semantics Corpus for Automobile Styling and Design

When car styling design is presented, audience will have different intentions and associations. Through interviews and desktop research, 120 descriptions of vehicles emotional vocabulary are collected. The construction of the semantic corpus is based on two aspects: On the one hand, from respondents' description on multiple vehicle style pictures, the emotional descriptive vocabulary (corpus) is collected and refined by the interviewees through guidance; on the other hand, emotional vocabulary about car design style is collected through Web searching. Respondents' screening range includes multiple enterprise levels and external related personnel: 40 car owners, 10 senior car designers, 2 marketers, 2 engineers, 10 car sales staff, 2 enterprise decision makers. We set up same interview scene for the respondents, using black and white photographs and other measures to reduce the aesthetic preferences, trying to clarify the design image.

The vocabulary is classified and screened by cluster analysis and semantic difference method, and finally 15 pairs of perceptual vocabulary are collected: Restrained-Untrammeled, Avant-Garde-Conservative, Young-Mature, Modern-Traditional, Strong-Flexible, Logical-Sensual, Pure-Complex, Succinct-

Table 2 Questionnaire on semantic difference

Semantic vocabulary	Score							Semantic vocabulary
Calm	−3	−2	−1	0	1	2	3	Exaggerated
Restraining	−3	−2	−1	0	1	2	3	Untrammeled
Logical	−3	−2	−1	0	1	2	3	Sensual
Pure	−3	−2	−1	0	1	2	3	Complex
……	−3	−2	−1	0	1	2	3	……
Geometry	−3	−2	−1	0	1	2	3	Organic
Technology	−3	−2	−1	0	1	2	3	Sexy
Static	−3	−2	−1	0	1	2	3	Dynamic
Aggressive	−3	−2	−1	0	1	2	3	Defense

Decoration, Geometry-Organic, Technology-Sexy, Static-Dynamic, Aggressive-Defense, Personal-Public, Luxurious-Simple. We use semantic difference questionnaire as shown in Table 2.

4.2 *Screening of Semantic Vocabulary*

The questionnaire divides the feelings into seven levels (−3 to 0 to 3), two ends of which indicate the opposite feelings; therefore, the representative semantic vocabulary is screened. Using cluster and factor analysis, 15 pairs of perceptual vocabulary are clustered and selected. The basic purpose of factor analysis is to use few factors to describe the relationship between many indicators, that is, to attribute several closely related indicators to same class. Each type of clustered indicators becomes a factor that reflects most of the information from the original material. The research found that five factors that affected the semantic cognition of styling design, corresponded to the five major aspects of the automotive styling, as shown in Fig. 5.

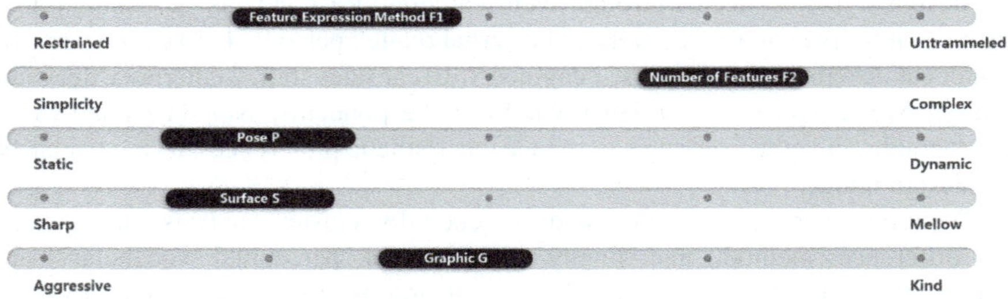

Fig. 5 Construction F-PSG model

4.3 Establish the Correlation Between Intentional Semantics and Automotive Styling

The F-PSG analysis model determines the type and scope of the automotive style by disassembling the perceptive car design language and splitting it into several dimensions that can be quantified. The so-called F-PSG is feature, proportion, surface, and graphic. The feature n is divided into two dimensions; namely the expression F1 represents a single feature, and the F2 represents the number of features. Therefore, the styling style can be subdivided into five dimensions, as shown in Fig. 5. F-PSG expresses the design style by the semantic differences of the car design and correlative combinations of the design levels.

(1) Feature Expression: Restrained (Rational) and External (Sensible)

The designer's method of presentation will determine the responsive relationship between the overall design of the car and its detail design features. Introverted (rational) design presentation: Volkswagen design presents a strict sense of order, where a clear relationship of its detail design features, and comparison of physical form and positions are developed. Lexus did not follow this rigorous comparison, but adopted a radical and exaggerated presentation.

(2) Number of Features: Pure and Complex

Simplicity does not depend on the number of design features. The simplicity to the extreme is "minimalism": The audience does not feel the trace of design, but are much satisfied with aesthetic and their functional requirements. Minimalism is to remove redundant design features as much as possible and to extract a more uniform and bright style of design. On the contrary, using more design features to express the design will bring in a feeling complexity. For example, respondents generally believe that the DS5 has a strong sense of ornamentation.

(3) Posture Feature: Static and Dynamic

Car design should have appropriate body proportion and corresponding posture characteristics. Generally speaking, from the side view of the car, the proportion of the body that meets the public's aesthetic habits is the distribution of golden ratio of the wheelbase of the car and the total length of the car. In addition to this, the lengths of the front and rear overhangs, the difference in the inclination angle of the windshield and rear window, the inclination angle of the waistline, and curvature change can all change the posture of the vehicle body. Generally, the tilt will produce a sense of dynamics and instability, and the horizontal will produce a feeling of static and stability. In addition, a lower the center of gravity of the form, which is closer to the center of the support surface, shall bring in a stronger the sense of stability.

(4) Surface Feature: Sharp and Mellow

In the expression of the surface features of the vehicle body, the respondents believe that the rounded and mellow surface will bring in a noble and elegant subjective feeling, while the curved surface features such as BMW's classic anti-concave and Lexus skirt line design will bring in a sense of sharpness and power. Using digital modeling tools to analyze surface features, you can use the curvature of the control line to evaluate whether the surface is smooth or discontinuous. Designers generally use curvature acceleration to express surface features. The correlation of posture ratio and surface characteristics of the car are equivalent to the relationship of bones and flesh. The shaping of the surface is crucial to the shaping of the car, presenting the creature attributes of automobiles.

(5) Graphic Feature: Aggressive and Kind

Graphic features include not only the front of the car, but also the graphical features of other faces. The front face of the car has obvious emotional recognition features as the visual center of the vehicle design. In particular, the combination of detail features such as grille, headlights, and lower air intake makes it easier for people to associate with facial expressions and hence deepen their impression of car's emotion. The study found that people's feelings about the expression of the vehicle can produce two extremes (the different front faces of the car can bring in exact opposite feeling): Respondents feel that the face of Lexus has an offensive, fierce look, but for golf and Fiat their front face feelings can be summarized as affinity, kind, and cute.

As mentioned above, the car's design styling is subdivided by these dimensions. Linking subjective feelings specifically to a certain aspect of automotive design facilitates iterations of design review and design semantics. However, the feeling that the design needs to convey is holistic, so it is necessary to integrate these five dimensions into an integrity to form a unified design style. In the process of integration, dimensions with no particular features need to be diluted, and here, only the most special dimension within the whole style is described and summarized. Therefore, the analysis level of the entire F-PSG modeling style analysis model is shown in Fig. 6.

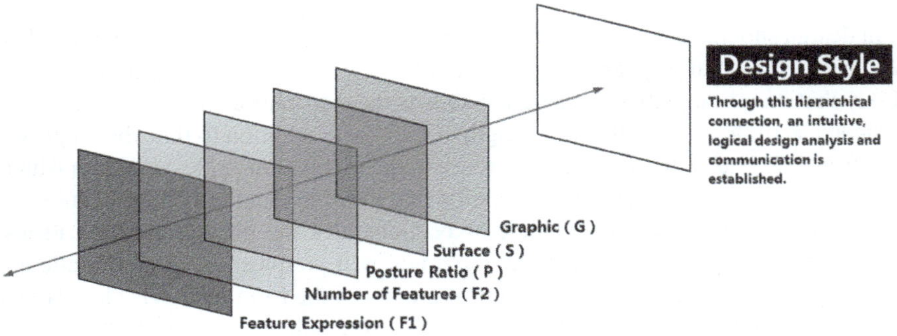

Fig. 6 Analysis levels of F-PSG model

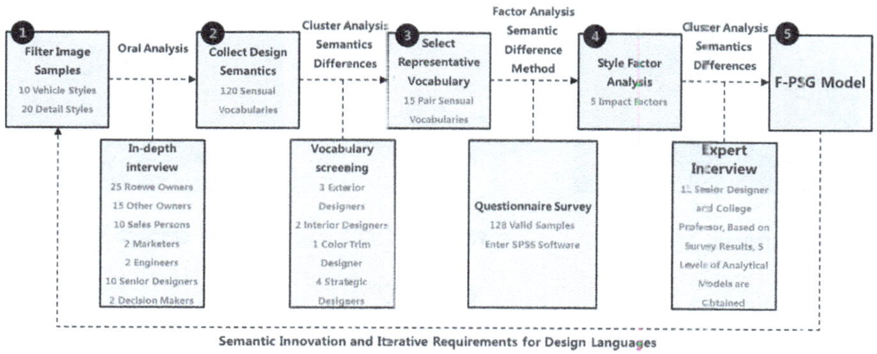

Fig. 7 Basic maintenance mechanism of F-PSG

4.4 Maintenance of F-PSG Model

The accuracy of F-PSG is relative. The semantic perception of automobile design is based on people's emotional needs and functional needs, which is both influenced by congenital factors and acquired environment. The basic semantic library needs to be iteratively updated according to social aesthetics and technology development, exploring the semantic feelings brought by the latest design forms to relevant personnel. The periodic semantic library change is the basic maintenance mechanism of F-PSG, as shown in Fig. 7.

5 Conclusion

A complete style experience includes many levels of intentional semantics. The intentional perception of car design is combined from all levels and is subjected to the whole, as shown in Fig. 8. The presentational connection between the intentional semantics and the design style forms the F-PSG design style analysis method In the process of design and implementation, it effectively promotes the design communication, so that the design style can be achieved with a broader consensus within the enterprise, which will be ultimately presented to its consumers.

The establishment of Roewe brand family design language "Rhythm Design" fully utilizes the advantages of the F-PSG analysis model and enhances the communication quality of design research, development, execution, and control of various development scenarios, improving the communication efficiency of design and development. Moreover, relevant personnel inside and outside the enterprise has formed a broader consensus in the process of design communication. The Roewe brand design language system is a set of family design language system based on the scientific and rigorous research systems, careful research on users and the market, and core of the Roewe brand appeal. After repeated market research and repeated argumen-

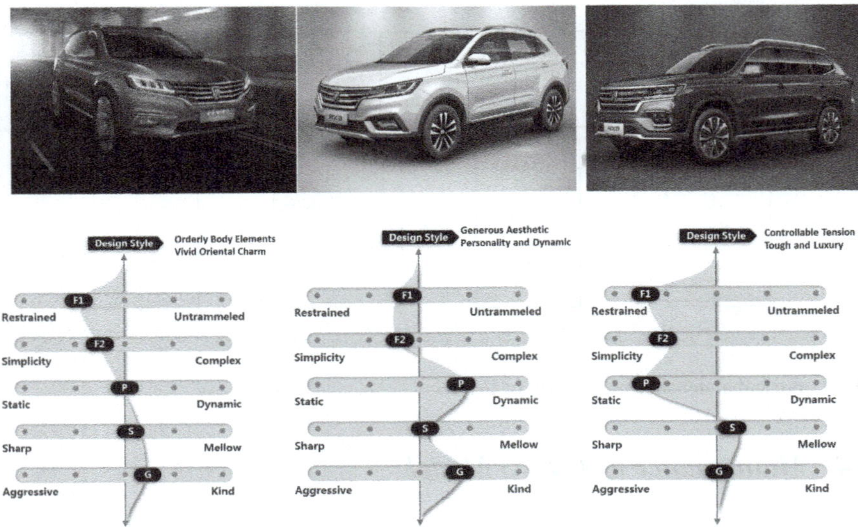

Fig. 8 Application of F-PSG on Roewe design language

tation, Roewe "Rhythm Design" fully conveys the product concept of "confidence and elegance" to its users and public society. In the current rapid development of China's economy, it conveys positive energy to users and public society. Since 2015 when the Roewe Rhythm Design concept was released, the design language has been applied to the brand concept car Vision-R and the well sold super internet SUV Roewe RX5,etc.,receiving success in the market. "Rhythm Design" has comprehensively enhanced the brand value of the Roewe and has become the most valuable company wealth during its brand reformation and improvement.

References

1. Borjade Mozota B (2003) Design management: using design to build brand value and corporate innovation [M]. Allworth Press
2. Cai Y (2009) A preliminary study of product semantics in automobile modeling features. Art and Design: Theor 12:212–214
3. Nagamachi S (1995) The story of Kansei engineering [J]
4. Ming S, Jin J (2013) Research on automobile modeling method from sketch to initial digital model. Packaging Engineering 6:59–63
5. Wang B, Luo J, Zhu R (2010) Line-type analysis method for automobile modeling design. Automotive Engineering 32(6):470–476
6. Zou F, Zhu Y (2010) Establishing image to perfection—on image shaping and processing of modern automobile modeling. Art & Design 12:88–89

Lifecycle Assessment and Prediction of Proton Exchange Membrane Fuel Cell Vehicles for 2020

Yisong Chen, Zhensen Ding, Jiahui Liu and Jinqiu Ma

Abstract Currently, fuel cell vehicles have attracted the attention of China and even the world, which perhaps will become the commanding point of technical competition in the automotive industry. By modeling with GaBi software, the prediction and quantitative analysis are made on lifecycle energy-saving and emission reduction for fuel cell vehicles in China in 2020 based on the 2020 target of China and the US Department of Energy 2020 plan. Finally, sensitivity analysis was conducted for selected key factors. The result shows that the total resource depletion, energy consumption, and environmental impact for fuel cell vehicles of China in its complete life cycle were 0.354 kg Sb-Eq., 7.38 × 105 MJ and 4.48 × 104 kg CO_2 eq. In 2020, the impact of the power generation on the environment will decrease as the proportion of thermal power generation in the power structure declines, and it will provide a possibility for large-scale electrolysis of water to produce hydrogen.

Keywords Fuel cell vehicles · Lifecycle assessment · DOE 2020 target · Energy depletion and environmental impact

1 Introduction

In recent years, the rapid development of automobile industry has exerted an increasing influence on resources, energy, and environment in China. According to statistics, 31% of environmental impacts such as climate change and haze in China are caused by automobiles. In addition, the degree of dependence on foreign crude oil has been above 65% in China and may reach as high as 70% in 2020. Faced with the dual pressure of resources and the environment, the state has intensively issued a number of policies to support the development of new energy vehicles and encourage the green consumption among consumers [1, 2].

Some foreign scholars have carried out some research on the resource and energy depletion of fuel cell vehicles. For example, Sara Evangelisti of London University

Y. Chen · Z. Ding (✉) · J. Liu · J. Ma
School of Automobile, Chang'an University, Xi'an 710064, China
e-mail: 571130429@qq.com

China SAE (ed.), *Proceedings of China SAE Congress 2018: Selected Papers*,
Lecture Notes in Electrical Engineering 574,
https://doi.org/10.1007/978-981-13-9718-9_56

et al. [3] proposed a comprehensive evaluation method for fuel cell vehicles focusing on the production process, and the comparative analysis was made between it and the production, use, and abandonment process of BEV and ICEV. At the same time, the sensitivity analysis of key components of the fuel cell stack and fuel cell system was carried out. The results show that reducing the environmental impact of the fuel cell vehicles in manufacturing phase is still a challenge. Christian Bauer of London University et al. [4] put forward a comprehensive lifecycle assessment method based on the new vehicle simulation framework and made a comparative analysis of the lifecycle environmental impacts between fuel cell vehicles and pure electric vehicles, traditional fuel vehicles and hybrid vehicles, diesel vehicles and natural gas vehicles. The results show that hydrogen source is of great significance for fuel cell vehicles to save energy and reduce emissions. Compared with foreign countries, domestic research is mainly concentrated in the field of hydrogen source infrastructure evaluation. For example, Ju et al. [5], from the Automotive Industry Development Research Institute of China Automotive Technology and Research Center, conducted a full lifecycle comprehensive evaluation of the Beijing fuel cell bus demonstration project by using the full lifecycle GREET energy analysis method. The results show that the hydrogen production scheme of natural gas centralized is optimal.

In summary, the lifecycle assessment in the field of new energy vehicles started relatively late in China. Although there has been lifecycle assessment for hydrogen source infrastructure in recent years, lifecycle assessment of proton exchange membrane fuel cell vehicle (hereinafter referred to as fuel cell vehicle) has not been found. Based on the 2020 target of China's fuel cell technology and the US Department of Energy 2020 requirements for fuel cell cars, this paper compares the technical characteristics of fuel cell cars in 2020 with the plug-in hybrid Prius, which constructs a lifecycle evaluation model from the production of raw materials to the maintenance stage. The evaluation results can provide a quantitative reference basis for the formulation of industrial policy and enterprise technical strategy planning of fuel cell vehicles.

2 Evaluation Method

2.1 System Boundaries and Functional Units

Firstly, the research objectives, system boundaries, and functional units need to be clarified. The purpose of this paper is to explore the impact of resources, energy consumption, and environmental emissions at all stages of proton exchange membrane fuel cells' life cycle in 2020 and to make reference for decision-making by enterprises, research institutions, or government departments. Therefore, the study covers the entire life cycle of fuel cell vehicles, including raw material acquisition, component manufacturing, assembly, use, maintenance, discarding, and recycling.

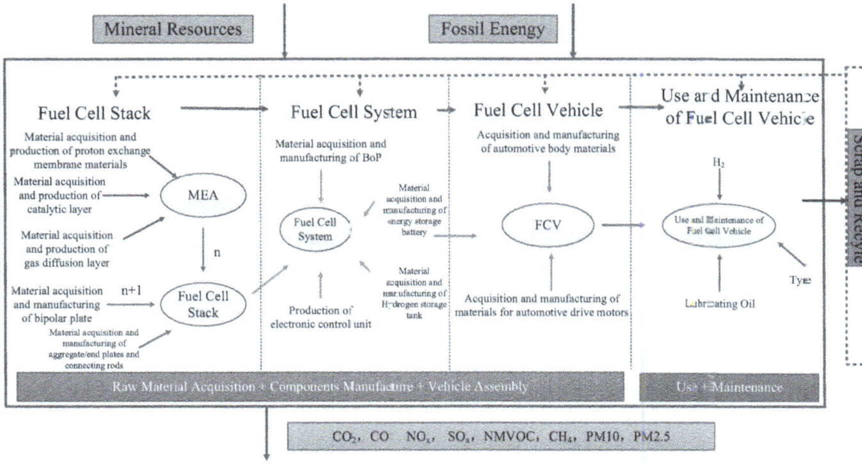

Fig. 1 System boundary of proton exchange membrane fuel cell vehicle lifecycle evaluation

The lifecycle evaluation system boundary of the proton exchange membrane fuel cell vehicle constructed in this paper is shown in Fig. 1. This paper uses the proton exchange membrane light fuel cell vehicle with the technology of 2020 to drive 150,000 km on the Chinese road as a functional unit [3].

2.2 *Evaluation Objects and Data Sources*

As this paper is intended to predict and evaluate fuel cell vehicles under the 2020 technical background, there is no actual mass production model at present. Therefore, on the basis of the study of the target and trend of fuel cell vehicles in 2020 in this paper, the other data that analogies with the Prius PHV will be secondary treated, in addition to the power system. The hydrogen consumption per 100 km of the selected model is based on the cyclical conditions in Beijing's semi-urban cities [3]. The actual scene data required for each stage of fuel cell vehicles in this paper mainly comes from domestic and foreign references [3–5], a small part of which comes from enterprise research, while background data mainly refers to the upstream process data mainly from the data of GaBi7 (GaBi ts) database in 2017.

2.3 *Impact Assessment Indicators and Calculation Methods*

Impact assessment refers to the process of transforming inventory analysis results into potential environmental impacts based on selected impact indicators and evaluation models. It is designed to help people understand the impact of product systems on certain important environmental (including resources and energy) characteristics [6].

This paper, aiming at the characteristics of new energy vehicles, especially fuel cell vehicles, selects ADP(e), ADP (f), GWP, AP, EP, POCP, and ODP, which are used to measure resources and energy depletion and environmental impact. Unit is kg, Sb-Eq, MJ, kg, CO_2-Eq, kg, SO_2-Eq, kg, Phosphate-Eq, kg, Ethene-Eq, kg, and R11-Eq, respectively. The characterization factors which derived from the CML model are developed by Leiden University in the Netherlands [6].

3 Model Constructions

This paper uses the GaBi ts software system platform developed by the University of Stuttgart in Germany to build a model. The real-world data and background data necessary for model construction mainly come from the references and GaBi database. In this paper, the proton exchange membrane light fuel cell vehicle is improved on the basis of the ANL [7] classification method. Thus, the whole fuel cell vehicle is divided into six parts: automobile main body, high-pressure hydrogen storage tank, motor and control system, converters, fuel cell stacks, and energy storage batteries.

3.1 Raw Material Acquisition Stage

The raw material acquisition process refers to the conversion of mineral resources into raw materials for vehicles by processing, which can be used to manufacture automotive parts. As mentioned above, this paper divides the entire fuel cell vehicle into six parts. The material composition data of each part comes from references [3–5], and the hydrogen tank adopts the most advanced non-metallic liner fiber full winding (IV) Type 70 MPa hydrogen bottle [8].

After obtaining the material composition list (BOM table) of the fuel cell vehicle in the above six parts, the raw material acquisition stage can be modeled in the GaBi ts software, which is shown in Fig. 2.

3.2 Component Manufacture Stage

Component manufacturing refers to the process of converting the obtained raw materials into parts for automotive assembly according to function. The fact is that his paper predicts the lifecycle impact of fuel cell vehicles in 2020 in the absence of mass production models. Other parts which analogy in the more advanced plug-in hybrid cars Toyota Prius NWH35 manufacturing process are obtained after the secondary processing according to the structural differences between fuel cell vehicles

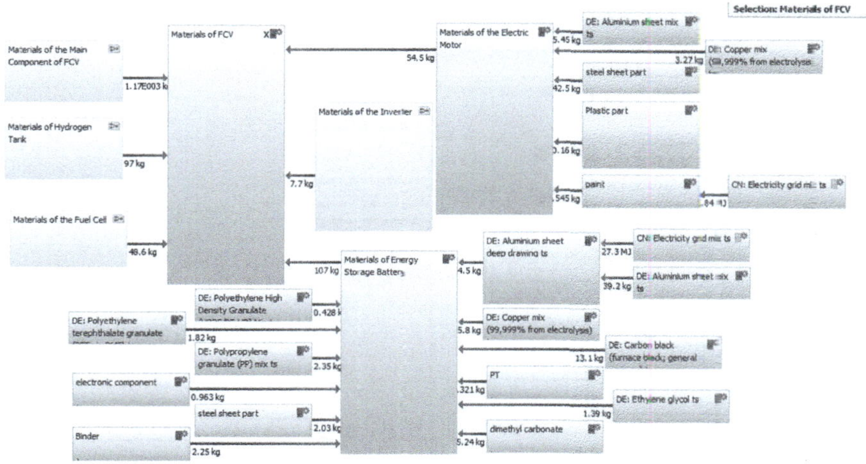

Fig. 2 Raw material acquisition stage mode 2.2 component manufacturing stage

and plug-in hybrid, except for the entire power system. Since the data of a small structural part manufacturing phase is difficult to obtain, it is assumed that the minimum impact of the fuel cell vehicle life cycle in 2020 is studied in this paper. The results obtained in this paper are estimated based on the development of various technologies and the updating of data in 2020.

After acquiring the energy consumption values of each part in the component manufacturing stage, it is modeled in the GaBi ts software, which is shown in Fig. 3.

3.3 *Assembly Stage of the Vehicle*

According to the data of vehicle assembly stage, this paper refers to the research of the automotive assembly industry at the University of California, Berkeley. The main energy consumption of the automobile assembly plant comes from painting, air-conditioning system and lighting, heating, material handling, welding, workshop compressed air, and other processes or equipment [9, 10]. Except the coal used in the heating process in the assembly plant, all other processes used electric energy. The total consumption of electric energy is 6.86 MJ/kg, and thermal energy is 2.03 MJ/kg. The GaBi model is shown in Fig. 4.

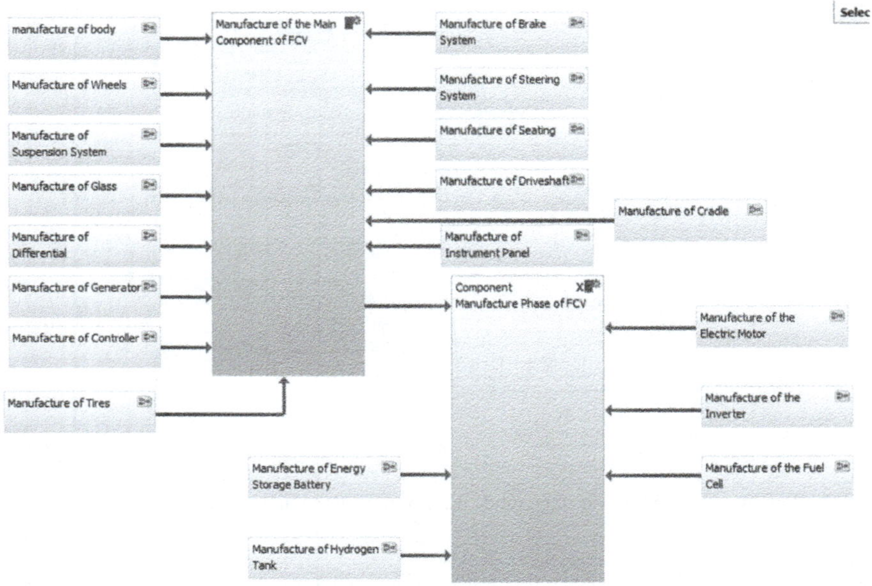

Fig. 3 Component manufacturing stage model

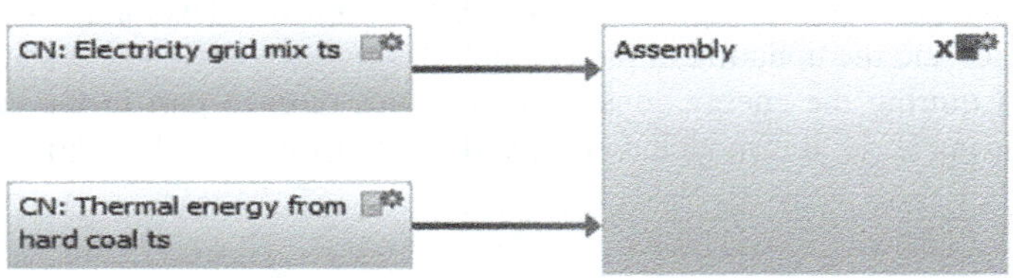

Fig. 4 Model of vehicle assembly stage

3.4 Use Stage

Fuel cell vehicles can be put into use after the assembly is completed. This paper assumes that the vehicle is driven in Beijing semi-urban cycle condition, and the average hydrogen consumption of fuel cell vehicle per 100 km in 2020 is 0.85 kg [3].

The hydrogen production process in this paper comes from the GaBi ts database, in which hydrogen is produced a diverse hybrid of technologies, including steam reforming, water electrolysis, and steam cracking ethylene.

The GaBi model for the usage phase is shown in Fig. 5.

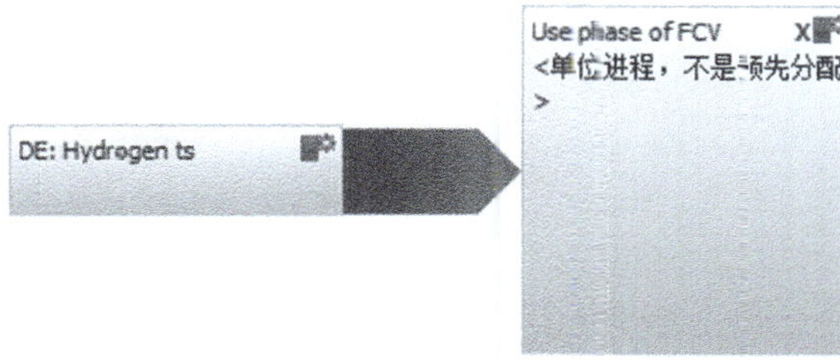

Fig. 5 Model of vehicle usage stage

3.5 Maintenance Phase

During the whole service life of the automobile, the parts need to be replaced. This paper only considers the environmental impact of the manufacturing process of the replaced parts, excluding light parts such as air filter and brake pads due to their less impact on the environment, which makes results close to the minimum impact of the 2020 fuel cell vehicle life cycle. According to the references [13], a car needs to change its tires every 62,500 km, so a fuel cell car needs to change its tires twice in the whole service life. It is assumed that the fuel cell degenerates no more than 15% after using 150,000 km [11] which do not need replacing. The replacement of other parts, including a small amount of lubricating fluid, has little impact on the environment, which is not considered in this paper.

3.6 Discard and Recycle Stage

The discard and recycle phase refers to the recycling of cars that have reached the end of their useful life. The main purpose of the recycling process is to disassemble and reuse the components that still have value in the car to eliminate the energy and materials consumed in the process of manufacturing new parts. Another purpose is to recycle and treat environmentally harmful parts of the car. The references [12–15] introduced the recovery process of fuel cell vehicles, but there is no professional solution for fuel cell vehicle recycling, so this paper will simulate other new energy vehicles for modeling.

The main difference between fuel cell vehicle and other new energy vehicle recycling models is that the recovery of platinum on the fuel cell stack and carbon fiber coating on the hydrogen storage tank. However, there is no relevant data on the carbon fiber recycling process, so this modeling does not take this process into account. The GaBi model of the recovery phase is shown in Fig. 6.

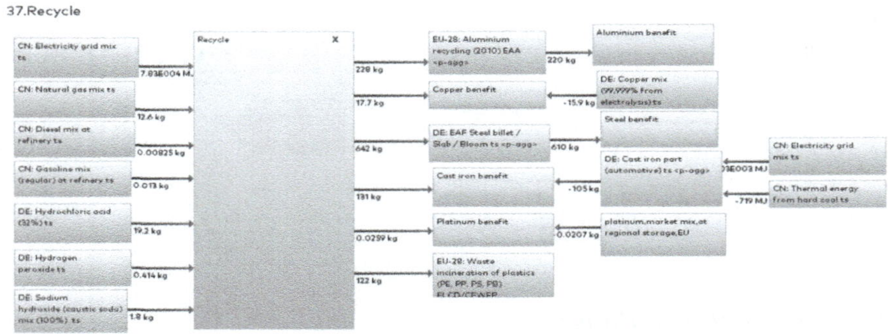

Fig. 6 Recovery phase model

4 Impact Evaluation and Interpretation of Results

4.1 Lifecycle Energy Consumption Emissions and Characteristic Results

In terms of energy consumption, this paper mainly considers the consumption of non-renewable energy (raw coal, crude oil, natural gas) and major emissions. The energy consumption and emission are shown in Table 1.

As can be seen from Table 5, the fuel cell vehicle will consume most raw coal in 2020, followed by natural gas. The main reason is that a large amount of electric energy is consumed in the recovery of various vehicle parts, while the proportion of thermal power generation in China is too large and the hydrogen production process consumes a large amount of natural gas. Carbon dioxide emissions are much higher than other emissions so that GWP were higher in each stage. This is mainly because the hydrogen production processed by mixing technology and consumed a lot of electric energy, while China is mainly based on coal-fired electricity. Therefore, a large amount of carbon dioxide is generated. Secondly, the raw material acquisition stage also consumed vast amount of fossil energy and produced a lot of carbon dioxide because of platinum and other precious metals' mining.

In this paper, the lifecycle assessment of fuel cell vehicles in 2020 is predicted by using GaBi ts software, and internationally widely used CML2001 evaluation method is cited as well. Seven influencing factors were analyzed, which was mentioned before. The characterization results are as shown in Table 2.

Table 1 Energy consumption and emissions table for each stage of the life cycle

	Raw material production stage	Component manufacturing stage	Vehicle assembly	Use stage	Maintenance stage	Scrap recovery stage	Total
Crude oil (kg)	415.807	7.7537	13.300	39.923	864.831	83.879	1425.493
Raw coal (kg)	1737.992	533.2841	998.404	542.385	192.252	6619.593	10,623.911
Natural gas (kg)	508.503	20.943	24.990	8362.717	52.537	239.421	9209.111
CO_2 (kg)	11,334.163	1563.797	2640.422	3264.667	2312.668	17,571.174	38,686.890
CO (kg)	25.826	1.7292	3.248	4.617	0.282	25.071	60.771
NO_X (kg)	12.737	3.3266	5.958	7.149	0.643	39.848	69.662
SO_X (kg)	124.940	5.1117	7.271	7.234	0.967	−10.740	134.784
NMVOC (kg)	5307.341	361.3259	0.721	3.091	0.238	−2110.725	3561.992
CH_4 (kg)	15.424	3.6008	7.226	53.571	1.882	51.867	133.571
PM10 (kg)	1.847	1.0707	2.194	1.090	0.056	14.439	20.697
PM2.5 (kg)	0.861	0.8597	1.760	0.976	0.029	11.571	16.058

Table 2 Characteristic results of environmental impacts of fuel cell vehicles in 2020

	ADP(e) (kg, Sb-Eq)	ADP(f) (MJ)	GWP (kg, CO_2-Eq)	AP(kg, SO_2-Eq)	EP (kg, Phosphate-Eq)	POCP (kg, Ethene-Eq)	ODP(kg, R11-Eq)
I	0.406	1.62E+05	2.31E+04	308	3.76	19.7	2.70E−06
II	3.06E−04	2.42E+04	2.51E+03	10.7	0.745	1.44	4.61E−10
III	2.19E−04	2.80E+04	2.80E+03	11.8	0.846	1.64	5.16E−10
IV	1.73E−03	3.85E+05	4.58E+03	12.3	1.02	2.99	3.32E−10
V	2.38E−05	4.43E+04	2.61E+03	1.49	0.0943	0.196	9.55E−12
VI	−0.0548	9.48E+04	9.27E+03	3.76	2.91	1.92	7.63E−07
VII	0.3535	738,300	44,870	348.05	9.3753	27.886	3.46E−06

Note I—raw material acquisition phase; II—parts manufacturing phase; III—assembly phase; IV—use phase; V—maintenance phase; VI—recycling phase; VII—full life cycle

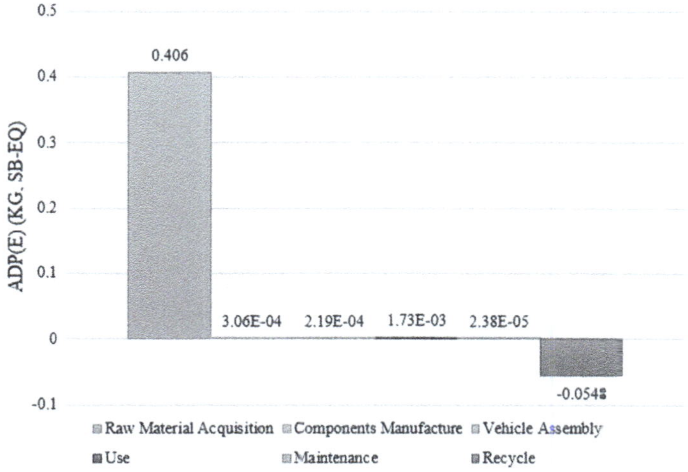

Fig. 7 Consumption of mineral resources

4.2 *Mineral Resource Consumption*

The GaBi software is used to the lifecycle model of fuel cell vehicle in 2020 with the calculation method of balance table. According to the calculation results, the mineral resource consumption of each stage can be obtained as shown in Fig. 7.

It can be seen from Fig. 7 that the raw material acquisition stage consumes the most mineral resources. This is mainly because the acquisition of various automobile raw materials at this stage requires a large number of mineral resources as the material premise. For example, precious metal mineral resources such as platinum are needed in the production process of fuel cells, and the acquisition of lithium battery raw materials consumes a large number of mineral resources such as lithium ore and rocks.

4.3 *Fossil Energy Consumption*

In GaBi software, the fossil energy consumption of the model is calculated in the balance table. After the calculation is completed, the results can be obtained from the software, as shown in Fig. 8.

Figure 8 shows that fossil energy consumption of fuel cell vehicles occurs mainly in the use of stage because the car used in the whole stage (150,000 km) consumes a large amount of hydrogen gas. This article assumes that the current widely used hybrid hydrogen technology is selected and consumes large amounts of electricity and natural gas, while thermal power is the primary of our country's power structure. So, first of all, it ought to seek efficient and energy-saving hydrogen production

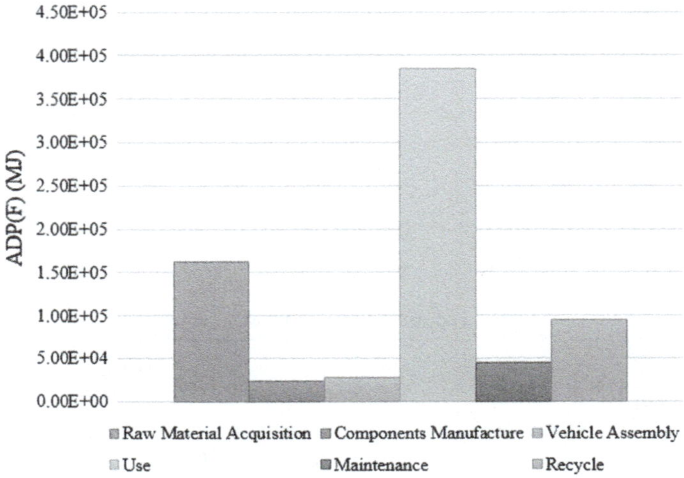

Fig. 8 Fossil energy consumption

technology in order to reduce consumption of fossil energy. Secondly, it is also to improve the fuel efficiency of the battery's internal chemical reaction of hydrogen and oxygen and cooperate with the use of lightweight techniques. For example, seeking better materials reduces the consumption of hydrogen and the weight of the larger parts, such as fuel cells and hydrogen tanks. Besides, it is still important to improve China's power structure, reduce the proportion of coal power, and use more clean energy to generate electricity.

4.4 Impacts of Environmental Emissions

In order to better explain the relative importance of various environmental impact types, the CML2001 method was used to normalize and quantify the above five environmental impact types. The normalized baseline value is derived from GaBi database, and the weight is derived from references [16]. The normalized and quantitative results of the five environmental impact types are shown in Table 3 and Fig. 9.

It can be seen from Table 3 that the five environmental impact types of fuel cell vehicles in the five stages from large to small are GWP (40.73%), AP (38.39%), POCP (20.00%), EP (0.77%), and ODP (0.001%).

As can be seen from Fig. 9, the environmental impact of fuel cell vehicles in 2020 is mainly concentrated in the stage of raw material acquisition, followed by the stage of recycling and use. The potential value of acidification is the highest in the acquisition stage of raw materials, mainly because the acquisition process of raw materials such as platinum, steel, and aluminum consumes a large amount of fossil energy and electric energy, resulting in a large number of acidified gases SO_2 and

Table 3 Normalized and quantified results of five environmental impact types

	Raw material production stage	Component manufacturing stage	Vehicle assembly	Use stage	Maintenance stage	Scrap recovery stage	Total
GWP	1.52E−10	1.65E−11	1.84E−11	3.01E−11	1.72E−12	6.09E−11	2.79E−10
AP	2.33E−10	8.09E−12	8.92E−12	9.30E−12	1.13E−12	2.84E−12	2.63E−10
EP	2.11E−12	4.19E−13	4.76E−13	5.74E−13	5.30E−14	1.64E−12	5.27E−12
POCP	9.69E−11	7.08E−12	8.07E−12	1.47E−11	9.64E−13	9.44E−12	1.37E−10
ODP	3.27E−15	5.58E−19	6.24E−19	4.02E−19	1.16E−16	9.23E−16	4.31E−15
Total	4.84E−10	3.21E−11	3.59E−11	5.47E−11	3.87E−12	7.48E−11	6.85E−10

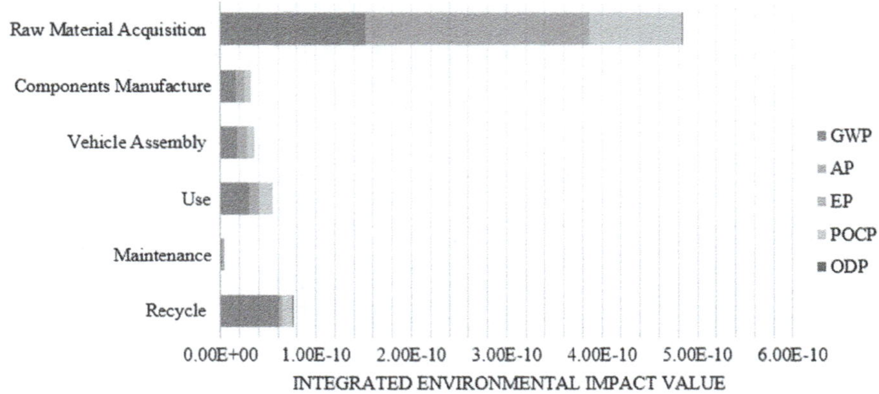

Fig. 9 Comprehensive value of environmental impact at each stage of the life cycle

NO_X. In the recovery stage, GWP has the highest potential value, mainly because a large amount of electric energy is consumed in the recovery process. In China, coal burning is the main form of thermal power generation, resulting in a large amount of carbon dioxide. The maximum GWP in the use stage is mainly due to the fact that the hydrogen production technology is not mature, resulting in the generation of large quantities of carbon dioxide.

5 Sensitivity Analysis

Sensitivity analysis is an indeterminate analysis technique that studies the degree of influence of a certain change on a certain key indicator or a set of key indicators from the perspective of quantitative analysis. In this paper, the single-factor sensitivity analysis method is used to improve the power structure.

This paper selects the power structure as the key factor and improves the power structure according to the 10% reduction of thermal power ratio successively in order to observe its impact on the environment, as shown in Table 4 [17]. Considering that the geographical distribution of water resources in China is quite different, it is assumed that the proportion of hydropower will not change. Nuclear energy is an ideal clean energy, but considering the frequent accidents and long construction period, it is assumed that the proportion of nuclear power generation will not change. Therefore, this paper only improves the proportion of thermal power, wind power, and photovoltaic power generation, and the GaBi model of the improved power structure will not be shown due to the space limitation. The comparison results of environmental impact under different power structures calculated by GaBi model are shown in Table 5, and the comparison results of comprehensive environmental impact of power structure improvement are shown in Fig. 10.

Table 4 Current power structure and improved power structure

	Current (%)	Improvement 1 (%)	Improvement 2 (%)	Improvement 3 (%)
Coal electricity	71.60	61.60	51.60	41.60
Hydropower	19.71	19.71	19.71	19.71
Nuclear power	3.56	3.56	3.56	3.56
Wind power	4.02	9.02	14.02	19.02
Solar power	1.11	6.11	11.11	16.11
Total	100	100	100	100

Table 5 Comparison results of environmental impacts under different power structures

	Current (Coal electricity 71.60%)	Improvement 1(Coal electricity 61.60%)	Improvement 2 (Coal electricity 61.60%)	Improvement 3 (Coal electricity 41.60%)
GWP	1.24E+05	1.08E+05	9.09E+04	7.43E+04
AP	525	455	384	313
EP	36.3	31.5	26.6	21.8
POCP	50	43.2	36.4	29.7
ODP	4.06E−08	4.36E−08	4.65E−08	4.94E−08
Integrated environmental impact value	1.48E−09	1.28E−09	1.08E−09	8.83E−10

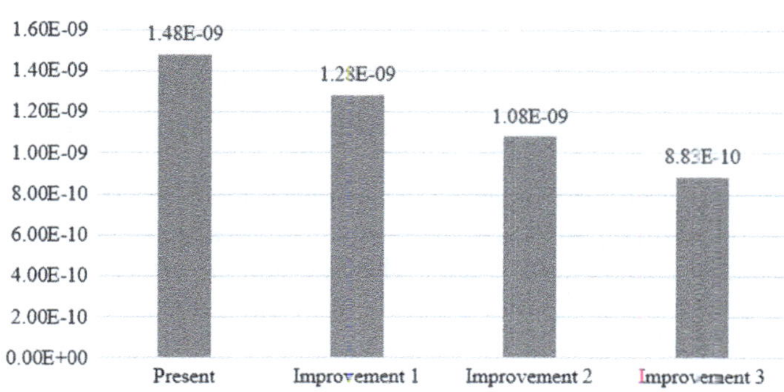

Fig. 10 Comprehensive environmental impact after power structure improvement

As can be seen from Fig. 10, through sensitivity analysis of the power structure, it is found that with the improvement of the power structure, the ratio of thermal power decreases by 10% successively. And the comprehensive value of the whole lifecycle environmental impact of the proton exchange membrane fuel cell vehicle also decreases gradually. Whether the power structure is clean or not has a great impact on fuel cell vehicles. China should pay attention to the adjustment of the power structure to achieve the goal of energy conservation and emission reduction.

6 Conclusions

In this paper, the following four important conclusions are drawn from the whole lifecycle quantitative analysis on the resource, energy, and environment impact of the proton exchange membrane fuel cell vehicle in 2020 and the sensitivity analysis of the power structure:

(1) In 2020, proton exchange membrane fuel cell vehicles will consume the most resources in the raw material acquisition stage and consume the most fossil energy in the use stage. In addition to the consumption of iron, aluminum, and other mineral resources in this stage, the consumption of platinum and other precious metal mineral resources is also an important influencing factor. The proportion of natural gas consumption will be the largest in fossil energy consumption, which is closely related to the main use of natural gas for industrial hydrogen production and the "uncleanliness" of China's power grid. Therefore, it would be an effective method to accelerate the development of alternative material for fuel cell vehicle critical materials, particularly platinum. At the same time, increasing the proportion of clean energy in Chinese electric power structure will greatly reduce the consumption of mineral resources and fossil energy.

(2) The proportion of GWP emission of fuel cell vehicles in 2020 is higher in all stages. The highest emission is in the raw material acquisition stage, recycling stage, and use stage, mainly due to the emission of carbon dioxide. The production process of platinum has produced a large amount of CO_2, so it is still an urgent problem to reduce the content of platinum in the catalyst (platinum load). In addition, compressing hydrogen consumes a large amount of electric energy, so that the increase of carbon emissions caused by the high proportion of coal electricity in China is highlighted again.

(3) Sensitivity analysis shows that with the decrease of the proportion of thermal power generation in the power structure and the increase of the proportion of clean energy, the influence of the power generation process on the environment is also gradually reduced and the possibility of large-scale electrolysis of water to make hydrogen is provided.

References

1. Yan G, Hongchao L, Bin GUO (2014) Review on key issues of product life cycle assessment. Comput Integr Manuf Syst 20(05):1141–1148
2. Yan G, Hongchao L, Qian Z et al (2014) Method and application on product life cycle assessment. Comput Integr Manuf Syst 20(03):661–665
3. Evangelisti S, Tagliaferri C, Dan JLB et al (2017) Life cycle assessment of a polymer electrolyte membrane fuel cell system for passenger vehicles. J Clean Prod 142:4339–4355
4. Bauer C, Hofer J, Althaus HJ et al (2015) The environmental performance of current and future passenger vehicles: life cycle assessment based on a novel scenario analysis framework. Appl Energy 157(3):871–883
5. Ju W, Kewei Y, Dan YU (2013) Fuel cell bus demonstration in Beijing and Shanghai. Automobile Technol 10:19–22 (in Chinese)
6. Universiteit Leiden. CML-IA Characterisation Factors[EB/OL]. [2016-09-05]. http://www.leidenuniv.nl/cml/ssp/databases/cmlia/cmlia.zip, [2017-06-19]
7. Burnham A, Wang M, Wu Y (2006) Development and application of GREET 2.7-the transportation vehicle-cycle model. Argonne National Laboratory, Chicago
8. Lee YD, Ahn KY, Morosuk T et al (2015) Environmental impact assessment of a solid-oxide fuel-cell-based combined-heat-and-power-generation system. Energy 79:455–466
9. Ming H, Baolian Y (2016) Key technology of fuel cell. Sci Technol Rev 34(6):52–61 (in Chinese)
10. Kai HU (2015) Research on product carbon label evaluation model based on LCA[D]. Zhejiang University of Technology, Hangzhou (in Chinese)
11. Ahmadi P, Kjeang E (2016) Realistic simulation of fuel economy and life cycle metrics for hydrogen fuel cell vehicles. Int J Energy Res 41(5)
12. Yang G, Jia W, Yifang Z et al (2014) Environmental impact study on recycling and utilization of vehicle power battery. Automobile Parts 20:41–43 (in Chinese)
13. Shuhua J, Han H, Yisong C (2016) Research on carbon emissions in manufacturing process of lithium ion battery for vehicles[C]. In: The 7th energy economics and management academic annual meeting, 2016.1014-16, Nanjing, Jiangsu(in Chinese)
14. Juan L (2015) Life cycle assessment and analysis of BEV and FV power system[D]. Hunan University, (in Chinese)
15. Duclos L, Lupsea M, Mandil G, Svecov L, Thivel P-X, Laforest V (2017) Environmental assessment of proton exchange membrane fuel cell platinum catalyst recycling. J Clean Prod 142:2618–2628
16. Standardization Administration of PRC (2008) GB/T 24044-2008[S]. Standards Press of China, Beijing
17. Ministry of industry development and environment, cec. List of basic data of power statistics in 2016 [DB/OL] (2017). http://www.cec.org.cn/d/file/guihuayutongji/tongjxinxi/niandushuju/2018-03-21/59a0eae449c914c2353153683e77d5b9.xls, [2018-7-20]

Analysis on Contact Strength of Needle Roller Bearing of Transmission and Effect of Surface Modification

Yong Chen, Kai Li, Libin Zang, Yang Zheng, Sen Jia, Huidong Zhou, Miao Yu and Boren Xue

Abstract Needle roller bearing is one of the most widely used parts in transmission; however, it often fails due to wear. In order to investigate the mechanism of fatigue damage, the three-dimensional model based on the sixth shift needle roller bearing of a six-speed manual transmission was established with Romax Designer. Then, the distribution of the contact stresses on bearing raceways was simulated under different loads. It is found that the stress concentration caused by the deflection of the needle rollers will increase the contact stress and lead to the indentations on the raceway surface. The composite surface modification of the roller raceway was employed, and the loading cycle test was carried out on the transmission test rig. The results show that the composite surface modification technique can improve the fatigue intensity of the contact surface and effectively prevent the indentation. Relevant experimental results provide technical guidance and theoretical basis for the optimization of needle roller bearings' raceway strength.

Keywords Needle roller bearing · Contact stress · Deflection · Indentation · Modification treatment

1 Introduction

The needle roller bearing is the key basic component of the automobile transmission. It will be subjected to complicated alternating external load during operation, and there is an error in the process of design and manufacture, which makes the needle roller bearing easily damaged [1]. The more serious the "asymmetrical loading effect" [2] produced when the needle runs, the more obvious the indentation on the raceway surface is, the more serious the impact on the bearing's lifetime and reliability is [3]. Large bias load distance not only causes the light load side of roller out of contact

Y. Chen · K. Li (✉) · L. Zang · Y. Zheng · S. Jia · H. Zhou · M. Yu
Hebei University of Technology, Tianjin, China
e-mail: firstlikai@163.com

B. Xue
Hangzhou Welltest Metal Science & Technology Co. Ltd, Hangzhou, China

China SAE (ed.), *Proceedings of China SAE Congress 2018: Selected Papers*, Lecture Notes in Electrical Engineering 574,
https://doi.org/10.1007/978-981-13-9718-9_67

with the raceway, which leads to contact length reduction, but also makes contact stress of the heavy load side too large, which causes the roller to fail faster than the bearing [4, 5]. Yang et al. [6] used finite element method to analyze the bearing under bias load. It was concluded that the stress concentration became more and more obvious with the increase of bias load distance. Wang et al. [7] analyzed the distribution of contact stress and surface stress field in the process of increasing the skew angle of needle roller. Chen et al. [8] used fast Fourier transform (FFT) method and conjugate gradient method to solve the contact stress of skewed roller contact pairs. In recent years, because of the continuous improvement of surface hardening technique and the application of new surface strengthening technique [9], the bearing capacity and lifetime of bearings have been greatly improved. Jin et al. [10] used ion implantation technique to inject large amounts of Mo^+ and N^+ into the bearing surface, which effectively improves the bearing's life. Wang [11] adopted nano-modification technique to strengthen the bearing surface, which greatly improved the bearing performance. Bouzakis et al. [12] presented that the technological parameters of coatings directly affect the fatigue life of rolling bearings through comparative tests. Lei et al. [13] carried out laser hardening treatment on bearing's raceways and raised the hardened layer of the raceways to 0.8–0.9 mm.

In order to settle the indentation of the needle roller bearing of a six-speed manual transmission, the contact stress distribution of the needle roller bearing was analyzed based on the Hertzian theory, and combined with dynamic simulation analysis, it is revealed that stress concentration on raceway surface is the main cause of indentation. The composite surface modification technique, including thermal spraying and shot peening, and cyclic loading test were carried out for raceway surface. It is proved that increasing the fatigue ultimate strength of the raceway can prevent the indentation.

2 Contact Analysis of Needle Roller Bearing

2.1 *The Contact Stress of Two Cylinders with Parallel Axes*

When the two cylinders are in contact with parallel axes, there is a straight contact line before the external force is applied [14]. When the force P shown in Fig. 1a is applied to the two cylinders, the contact area shows elastic deformation and becomes a rectangular surface with a width of 2b, and the distributing disciplinarian of the unit pressure on the contact surface is shown in Fig. 1b. The pressure on the X-axis is the largest, which is presented by P_0. The distributing disciplinarian of the pressure on other points of the contact surface which is presented by P_y is semi-elliptical shaped, so its value can be expressed as $P_y = P_0\sqrt{1 - \frac{y^2}{b^2}}$. In fact, the volume of the semi-elliptic cylinder is equal to the total force P, so it can be calculated as $P = \frac{\pi b}{2} L P_0$, where L represents the length of the contact line along the axial direction.

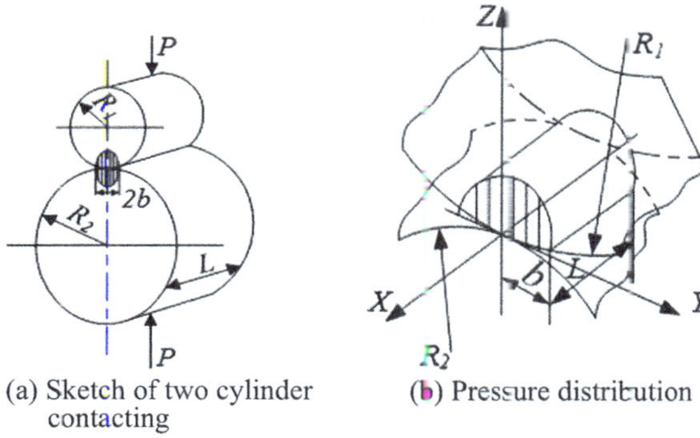

Fig. 1 Contact state and pressure distribution of two cylinders

The maximum unit pressure can be calculated as

$$P_0 = \sigma_H = \frac{2P}{\pi bL} \tag{1}$$

It is known from Hertzian formula that the half width of Hertzian contact region b can be given as [15]

$$b = \sqrt{\frac{4P}{\pi L} \cdot \frac{\frac{1-\mu_1^2}{E_1} + \frac{1-\mu_2^2}{E_2}}{\frac{1}{R_1} + \frac{1}{R_2}}} \tag{2}$$

where R_1 and R_2 represent the radius of the bases of two contacted cylinders, respectively. E_1, E_2 and μ_1, μ_2 represent Young's modulus of elasticity and Poisson ratio of two cylinders, respectively.

Replacing b with Eq. (2), and Eq. (1) will become,

$$\sigma_H = \sqrt{\frac{P}{\pi L} \cdot \frac{\frac{1}{R_1} + \frac{1}{R_2}}{\frac{1-\mu_1^2}{E_1} + \frac{1-\mu_2^2}{E_2}}} \tag{3}$$

If the materials of the two cylinders are all steel, E_1 and E_2 can be represented by E, the Poisson ratio is $\mu_1 = \mu_2 = 0.3$, the curvature sum R is written as $\frac{1}{R} = \frac{1}{R_1} + \frac{1}{R_2}$, so the maximum contact stress can be expressed as

$$\sigma_H = 0.418\sqrt{\frac{PE}{LR}} \tag{4}$$

If two cylinders are in internal contact, the axes of the two cylinders are on the same side of the contact surface. The curvature sum R is calculated as $\frac{1}{R} = \frac{1}{R_1} - \frac{1}{R_2}$.

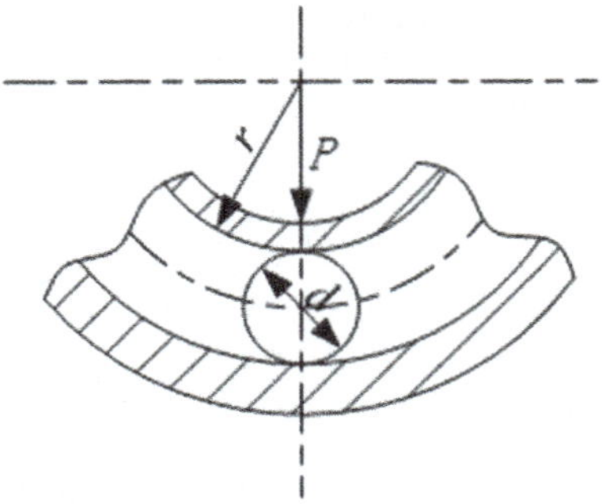

Fig. 2 Contact state of roller bearing

2.2 Contact Stress Between Roller and Raceway of Bearing

As shown in Fig. 2, the contact between roller and inner ring of the bearing is equivalent to external contact of two cylinders, and the contact between roller and outer ring of the bearing is equivalent to internal contact of two cylinders.

The curvature sum for external contact R can be expressed as $\frac{1}{R} = \frac{1}{R_1} + \frac{1}{R_2} = \frac{2}{d} + \frac{1}{r}$, where d and r represent diameter of the roller and contact radius of inner ring, respectively. According to Eq. (4), the maximum contact stress between roller and inner ring can be obtained by

$$\sigma_H = 0.418\sqrt{\frac{PE}{L}\left(\frac{2}{d} + \frac{1}{r}\right)} \tag{5}$$

The curvature sum for internal contact R can be expressed as $\frac{1}{R} = \frac{1}{R_1} - \frac{1}{R_2} = \frac{2}{d} - \frac{1}{R_b}$, where R_b represents contact radius of outer ring. According to Eq. (4), the maximum contact stress between roller and outer ring can be obtained by

$$\sigma_H = 0.418\sqrt{\frac{PE}{L}\left(\frac{2}{d} - \frac{1}{R_b}\right)} \tag{6}$$

where L is the active length of the roller in Eqs. (5) and (6).

2.3 The Influence of Needle Roller Deflection on the Contact Stress

Figure 3 shows the deflection of needle roller. In the radial plane, there is an angle β between the axis of the needle roller and the raceway line of ring. The elastic deformation of the needle roller and the raceway surface is generated under the acting force. The maximum deformation generated by compression is represented

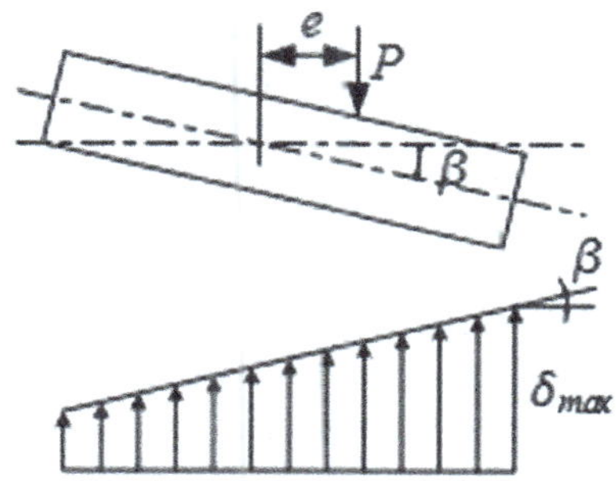

Fig. 3 Sketch of needle roller deflection

as $\delta_{\max}$. Based on the moment balance principle, bending moment is $W = Pe$, where e represents the distance between acting force P and mid-point of the needle roller as shown in Fig. 3. With the increase of the skew angle, the difference between the maximum contact stress at both ends of the needle is also enlarged. The contact stress at one end is 2 times that of the other end when β is 0.04° obtained from Ref. [7], which results in the needle roller suffering from severe local wear.

3 Simulation Process and Result Analysis

The working condition of the needle roller bearing is complicated. The load and the stress produced on each component will change over time. The loaded rolling elements are controlled by the raceway and synchronized with the movement of the ring; however, the unloaded rolling elements fall behind the ring because of the damping [16].

According to the characteristics of rolling bearing's movement, the assumptions in the simulation process are made as follows:

(1) Since the influence of the chamfering on the internal stress distribution of the bearing can be neglected, the physical model does not contain chamfering.
(2) The influence of radial clearance, axial clearance and oil film is not considered.

3.1 *The Establishment of the Transmission Model*

The physical model of the manual transmission is established based on Romax Designer. In order to minimize the work in the process of modeling and assembling, the model should be established as simplified as possible under the prerequisite of the correct position of each component. The needle roller bearing with the sixth gear shows the most serious indentation, so the model contains only the sixth gear components and essential shafts, gears and bearings. The model is mainly composed of the components on the sixth gear transmission route, including the input shaft, the output shaft, the sixth gear pairs, the needle roller bearings and other parts on shafts. Figure 4 clearly shows assembly relationship among the needle roller bearing and the gear and the sleeve. The radial force generated by the gear is not located in the axial symmetry plane of the needle roller, resulting in the deflection of the needle.

For the accuracy of the simulation results, the parameters of structure and material of the simulation model must be consistent with actuality [17]. The structural parameters of the needle roller bearing are given in Table 1. The material composition and parameters of transmission components are given in Table 2.

Fig. 4 Assembly sketch of components

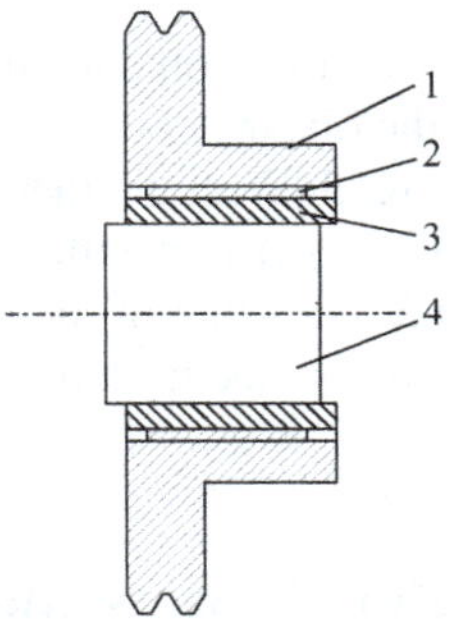

1—driving gear 2—needle roller bearing

3—shaft sleeve 4—shaft

Table 1 Structure parameters of needle roller bearing for 6th gear

Parameters	Specification
Cage width/mm	23.80
Roller diameter/mm	2.5
Roller length/mm	20.30
Number of roller paths	1
Number of rollers per path	30
Inner diameter/mm	37
Outer diameter/mm	42

Table 2 Material parameters of components

	Gear	Shaft sleeve	Roller	Cage
Material	20CrMoH	GCr15	GCr15	Nylon
Elastic modulus/Pa	2.1×10^{11}	2.19×10^{11}	2.19×10^{11}	2.62×10^{9}
Poisson's ratio	0.278	0.3	0.3	0.34
Density/(kg/m^3)	7.84×10^{3}	7.83×10^{3}	7.83×10^{3}	1.12×10^{3}

Table 3 Load cases of simulation

Item	Specification
Torque/(N m)	170
	190
	220
Speed/(r/min)	4000
Working temperature/°C	80

3.2 *Loads Condition*

Contact load is the primary cause of fatigue failure of transmission components [18]. The operating parameters of simulation are given in Table 3. The maximum input torque of the transmission is 220 N · m. Considering the limit and normal working conditions of transmission, three kinds of torque which are 170 N · m, 190 N · m and 220 N · m, respectively, are chosen, the speed of the input shaft is 4000 r/min, and the working temperature is 80 °C.

3.3 *Result Discussion and Analysis*

Figure 5 shows the circumferential distribution of the contact stress on the inner and outer raceways of the needle roller bearing under different loads, in which the red line represents the inner raceway and the blue line represents the outer raceway.

It is seen from Fig. 5 that the contact stress between roller and inner ring is bigger than that between roller and outer ring. The reason is that the curvature radii of the inner and outer rings are different, and the contact area between roller and inner ring is smaller. With the increase of load, the location of the bearing area does not significantly change, while the number of loaded roller changed from 6 to 7 when

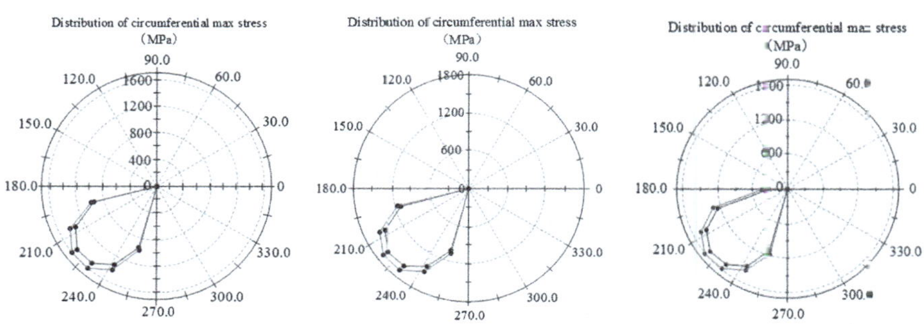

(a) T=170N·m (b) T=190N·m (c) T=220 N · m

Fig. 5 Circumferential distribution of contact stress on inner and outer raceway the torque changed from 170 N · m and 190 N · m to 220 N · m, indicating that the increase of load enlarges the range of the bearing area.

Figures 6 and 7 show the contact stress distribution contours of raceways under different loads. Compared to Fig. 5 which only depicted the maximum contact stress distribution of each roller contacted with the raceway, the contours can also demonstrate the stress distribution of the same roller.

By comparing Figs. 6 and 7, it can be drawn that there is no significant change in contact region and maximum contact stress location on raceway under different loads, which is corresponding to the results shown in Fig. 5. However, the contact

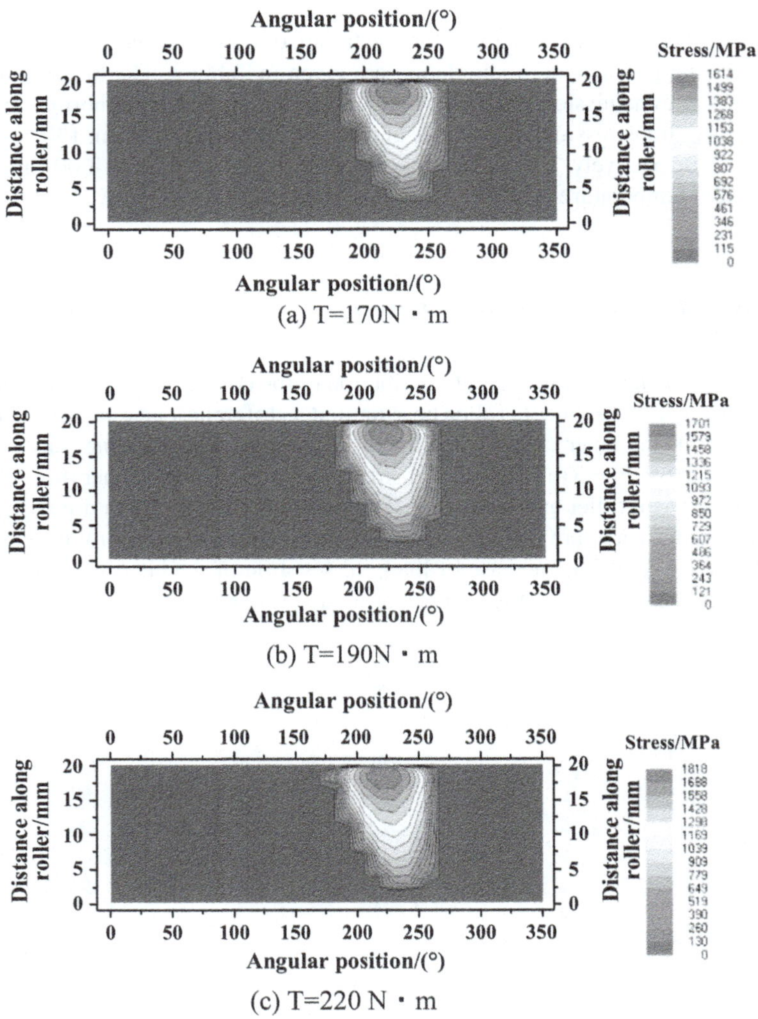

Fig. 6 Contours of stress of inner race under different loads

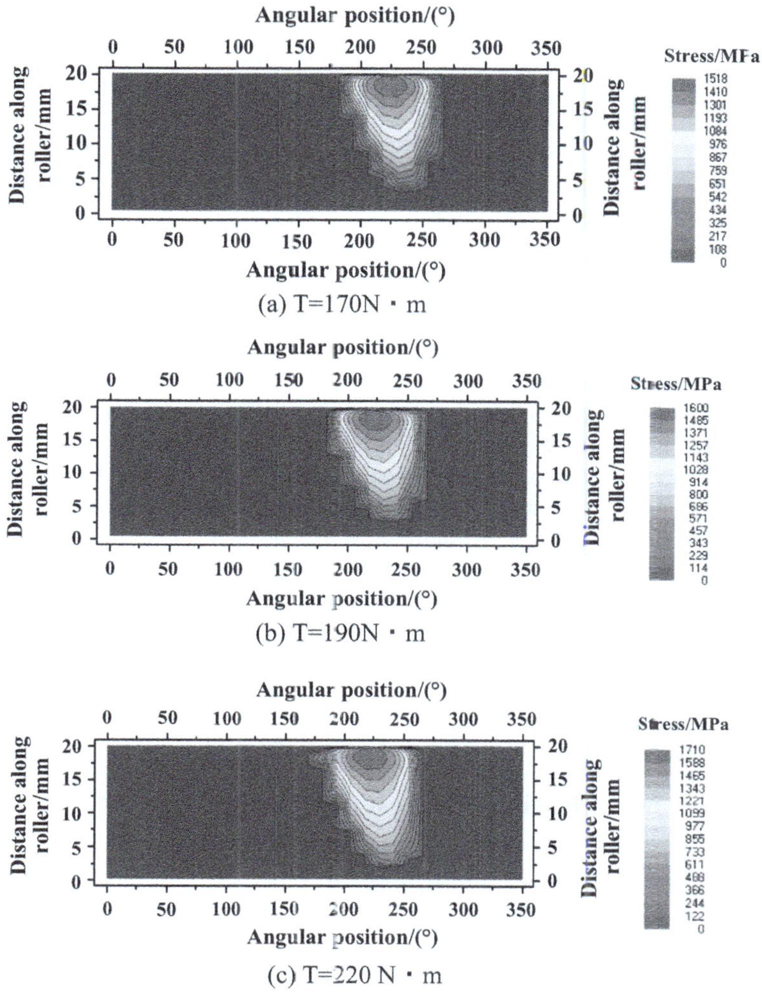

Fig. 7 Contours of stress of outer race under different loads

stresses on raceways increase with load, the maximum contact stresses of the needle rollers contacted with the inner ring and the outer ring are 1614 and 1518 MPa for the 170 N m load case, respectively. For the 190 N m case, they are 1701 and 1600 MPa and for the 220 N m case, they are 1818 and 1710 MPa, respectively. In addition, the area generated by the maximum contact stress is elliptical, and the stress decreases around the stress center. The maximum contact stress at the stress concentration is about 1.7–2 times of that on the central region, and the length of the stress concentration is about 11–19% of the contact length. More importantly, the maximum contact stress is not located in the axial symmetry plane of the roller bearing. This is because the needle roller bearing is deflected during operating, which

Table 4 Comparison of the simulated and theoretical values of contact stress

Torque/(N m)		Simulated values/MPa	Theoretical values/MPa	Relative error/%
170	Inner race	1614	1510.2	6.87
	Outer race	1518	1416.5	7.17
190	Inner race	1701	1599.0	6.38
	Outer race	1600	1500.0	6.67
220	Inner race	1818	1723.0	5.51
	Outer race	1710	1616.3	5.80

leads to the stress concentration. In addition, when the contact stress exceeds the yield limit of the material, the indentation which has a great influence on the life of the roller bearing is generated on raceway surface.

Table 4 shows the comparison between the simulated values and the theoretical values based on the Hertzian theory. The values listed in Table 4 clearly demonstrate that the simulated values of maximum stress are bigger than the theoretical values whether in external contact or in internal contact under different loads, and the relative error of both is within a reasonable range. Hence, Table 4 validates the accuracy of the three-dimensional model of the transmission. Additionally, Hertzian contact theory is based on many assumptions, and the theoretical values can serve as a reference for accurately revealing the real mechanical condition of bearings.

4 Surface Modification and Test of Bearing

4.1 Composite Surface Modification

Based on the composite surface modification technique, the composite material composed of molybdenum disulfide and hard solid particles was impacted on the raceway surfaces under high temperature and pressure, which changed the structure of the matrix surface and improved the surface fatigue intensity by applying a kind of hard coating on the substrate surface.

4.2 Fatigue Test of Bearing

The transmission is a manual transmission with six gears, of which 4, 5, 6 are high speed gears and are used for longer time compared with 1, 2, 3 low speed gears, so the fatigue test for bearings of 4, 5, 6 gears was carried out. Two sets of shaft sleeves and gears of 4, 5, 6 gears were prepared, one was not treated with composite surface

Table 5 Operating cases of test

Shift	Torque/(N m)	Speed/(r/min)	Number of cycles	Test time/h	Remark
4	220	4000	42	42	Single cycle for 60 min
5	220	4000	25	50	Single cycle for 120 min
6	220	4000	37.5	75	Single cycle for 120 min
Total test time				167	

modification technique, the other, including the outer surfaces of the sleeves and the inner surfaces of the gears, was treated with surface modification. The cyclic load was applied to the transmission on the transmission test rig.

4.2.1 Testing Conditions

(1) The lubricating oil number of the transmission was SAE 75W/90, the volume of lubricating oil was 2.3L, and the oil temperature was controlled at (80 ± 5) °C.
(2) The running-in of transmission was carried out before the experiment, and the specifications were as follows:

   (a) The speed of the input shaft was 3000 r/min;
   (b) The input torque was 55 N m and the running-in time of each forward gear was 1 h;
   c) The input torque was 110 N m and the running-in time of each forward gear was 1 h.

4.2.2 Testing Process

In order to save time and improve the efficiency of test, cyclic loads were applied to 4, 5 and 6 gears, respectively. The values for each load case are shown in Table 5. Most importantly, the three gears were experimented circularly, and it needed to stop for 10 min after a cycle was completed.

4.3 The Experimental Results and Discussion

Figure 8 shows the morphology of the roller bearing inner raceway surface after test. The raceway surface is not treated with composite surface modification technique, and a micrograph of the indentation is shown in the left corner [Fig. 8a]. It is seen that the indentation is very obvious and is much deeper at some contact region. The

Fig. 8 Comparison of surface morphology of raceway

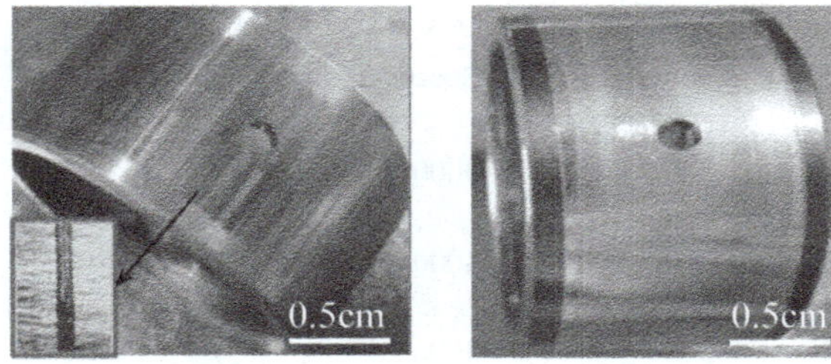

(a) Untreated (b) Treated

raceway surface is treated with composite surface modification technique [Fig. 8b]. It is obvious that there is no indentation generated and the surface maintains smooth.

It is inferred that the reason for the formation of indentation is the needle rollers' reciprocating movement under large contact stress. The indentation will cause vibration, noise and temperature rise, deteriorate the working environment of the roller and shorten its service life. The composite modification technique can form an enhancement layer with a certain thickness on the surface of bearing raceway, which can enhance the hardness and reduce the wear. Additionally, it improves the compressive residual stress of the surface, which can reduce the indentation caused by external force at the contact area and improve the service life of needle roller bearing.

5 Conclusion

(1) A three-dimensional model of a six-speed manual transmission is established, which consists mainly of sixth gear transmission components, and the simulation analysis for the contact stress of the inner and outer rings of the needle bearing is successfully carried out under the load cases that the input speed is 4000 r/min and the input torques are 170, 190 and 220 N m, respectively. The results show that the needle rollers deflected, and the maximum contact stress of the raceway, which is about 1.7–2 times of that on the central contact region, increases with the torque.

(2) The contact stress of the raceway surface is calculated based on the Hertzian contact theory, and the simulated values are compared with the theoretical values. The relative errors are in a reasonable range, which proves that the excessive contact stress is a major cause of the indentation.

(3) The composite surface modification of the needle bearing raceway surfaces is carried out. The results of the fatigue experiment show that the composite surface modification technique can increase the compressive residual stress and hardness of the contact surface and improve the working condition of the bearing. Moreover, the treated surface can prevent formation of the indentation. This paper provides method and theoretical references for the intensity optimization of bearing raceway surface.

Acknowledgements The authors would like to thank the National Key R&D Program of China (2017YFB0102400) and Innovation Project of Graduate Students of Hebei Province (CXZZBS2018036).

References

1. Chen LP (2005) Dynamics analysis of mechanical system and ADAMS application course[M]. Tsinghua University Press, Beijing
2. Wei YG, Ge ZL, Jiang QY (2002) Finite element analysis of cylindrical rollers under nonuniform load used in shaft boxes bearings of high-speed vehicles and unsymmetric modification of the rollers. Lubr Eng 2:9–11
3. Zhang RT, Zhao YX, Cai H (2015) Skewed loading effect on the probabilistic lives of journal bearing of railway wagon. J Mech Eng 51(24):177–183
4. Mao YX, Shen XJ, Chen XY et al (2009) Study on contact stress distribution of roller bearings and roller profile design under misaligned loads. China Mech Eng 20(16):1918–1922
5. Chen XY, Zhou Y, Shen XJ et al (2015) Contact stress analysis and crowing design of friction pair with rollers under misalignment and skewing conditions. J South China Univ Technol (Nat Sci Ed) 6:77–83
6. Yang CH, Chen JB (2017) Analysis on eccentric load stress of tapered roller bearing in railway wagon. Bearing 5:9–11
7. Wang AL, Hong YF, Wang JG (2011) Analysis on contact of needle roller bearings. Bearing (11):1–4
8. Sun DC, Chen XY, Wang ZJ et al (2014) Numerical solution of the contact stress for load offsetting roller pairs based on fast fourier transform and conjugate method. Mach Des Manuf 11:43–46
9. Chen Y, Zang LB, Ju DY et al (2017) Research status and development trend on strengthening technology of high strength automobile gear surface. China Surf Eng 30(1):1–15
10. Jin H, Sakai M, Wu CL et al (1997) A study of the inhibitive effect of Mo on the temper softening of bearing surfaces caused by ion implantation. Wear 209(1–2):193–198
11. Wang Y (2004) Nano-and submicron-structured sulfide self-lubricating coatings produced by thermal spraying. Tribol Lett 17(2):165–168
12. Bouzakis KD, Vidakis N, Lontos A et al (2000) Implementation of low temperature-deposited coating fatigue parameters in commercial roller bearings catalogues. Surf Coat Technol 133:489–496
13. Lei S, Liu QK, Xue ZT et al (2009) Laser surface hardened layers for race surface of bearing rings. Bearing 10:30–33
14. Zhang XS, Wu Z (2000) Calculation of contact stress for mechanical components. Machinery 1:24–26
15. Xu K, Su JX, Zhou YD et al (2014) Theoretical formula and finite element analysis between gear pairs with line contact and point contact. J Mech Transm 8:77–81
16. Zhang LL, Gao X, Tan NL et al (2007) Simulation and analysis on rolling bearings based on ANSYS/LS-DYNA. J Mach Des 24(9):62–65
17. Shi MQ (2004) Study on multi-contacts with ADAMS. Comput Eng Appl 40(29):220–222
18. Zhang CB (2009) The studay of rolling contact fatigue failure and tests on super hard coating parts. Yanshan University

Development of Longitudinal Slope Estimation Algorithm for Premium SUV

Jian Zhang, Wang Yu, Liu Ye and Wang Kang

Abstract In this paper, a longitudinal slope estimation algorithm for a premium SUV is developed. First, the longitudinal dynamic vehicle model is established, and two longitudinal slope estimation methods are proposed. Furthermore, the longitudinal acceleration is corrected by suspension displacements from semi-active suspension system. Then, two Luenberger observers for longitudinal slope estimation are proposed based on variable of vehicle information. Then, the algorithm prototype is proposed by MATLAB/SIMULINK, and the code is generated and embedded into the vehicle ECU. Finally, the experiment system is established and the performance of the proposed algorithm is verified by road test.

Keywords Premium SUV · Longitudinal slope estimation · Luenberger observer · Road test

1 Introduction

Road slope information is widely used in vehicle control system. For electric park brake system (EPB), the slope signal is necessary to calculate required braking force which is applied to the vehicle to achieve reliable stop [1]. For transmission control unit (TCU), the slope signal is required to adjust shifting logic to improve the performance when the vehicle is driving on the roads with slopes [2]

There are many researches on longitudinal slope estimation algorithm. Lei [3] established longitudinal dynamic equation and extended kalman filtering method (EKF) is used for developing the algorithm. Ran [4] proposed a real-time slope estimation method by vehicle longitudinal dynamics and longitudinal acceleration signal. Yang [5] directly measured slope angle by angle sensor which has similar structure to simple pendulum and corrected the result by additional information to improve estimation accuracy.

J. Zhang (✉) · W. Yu · L. Ye · W. Kang
China FAW Group Co. Ltd, Intelligent Connected Vehicle Development Institute, Changchun, China
e-mail: 369419196@qq.com

China SAE (ed.), *Proceedings of China SAE Congress 2018: Selected Papers*, Lecture Notes in Electrical Engineering 574,
https://doi.org/10.1007/978-981-13-9718-9_68

However, the condition of the road on which SUV drives is more complicated, and the requirement of slope estimating is higher. The estimation error of longitudinal acceleration signal-based method is obvious when slope is large because accelerometer is mounted on sprung part of vehicle, and pitch angle should be considered as correction.

For premium SUVs, accurate estimation of longitudinal slope is possible, as a variety of electronic control systems are equipped, and more vehicle state information is available. Two model-based high-accuracy slope estimation methods for premium SUV are proposed in this paper. One method is based on vehicle longitudinal dynamics and uses engine torque, gear position, master cylinder pressure, longitudinal speed signals which can be obtained by EMS, TCU, and ESP. The other method is based on the measured value of accelerometer and is corrected by pitch angle which is calculated by suspension height signals. Two Luenberger observers using these models are designed, and two longitudinal slope estimation algorithms are proposed. The algorithm prototypes are established by MATLAB/SIMULINK, and the code is generated automatically and embedded into the vehicle ECU. The experiment system is established, and the performance of the proposed algorithm is verified by road tests.

2 Longitudinal Slope Estimation Algorithm

2.1 *Vehicle Longitudinal Dynamics*

The vehicle longitudinal dynamics model is:

$$m\dot{v} = F_{\mathrm{t}} - F_{\mathrm{f}} - F_{\mathrm{w}} - F_{\mathrm{i}} \tag{1}$$

where m is the mass, v is the longitudinal velocity, F_{t} is the tractive force, F_{f} is rolling resistance, F_{w} is the air drag force, and F_{i} is the gradient resistance.

The vehicle states, such as longitudinal acceleration, engine torque, can be measured online. According to the configuration of vehicles, two models for longitudinal slope estimation are established as below.

2.2 *Estimation Models*

2.2.1 Model Based on Tractive Force and Driving Resistances

The tractive force F_{t} can be expressed by:

$$F_{\mathrm{t}} = \frac{T_{\mathrm{t}} i_{\mathrm{g}} i_{0} \eta_{\mathrm{T}}}{r} - \frac{2 p_{\mathrm{mc}} (k_{\mathrm{bf}} + k_{\mathrm{br}})}{r} \tag{2}$$

where T_t is the engine torque, i_g is the gear ratio of transmission, i_0 is the gear ratio of final drive, η_T is the efficiency of powertrain system, r is the radius of driving wheel, p_{mc} is the pressure of master cylinder, and k_{bf} and k_{br} are the brake effective factors of front and rear wheel.

The air drag force F_w can be expressed by:

$$F_w = \frac{C_d A \rho v^2}{2} = \frac{C_d A v^2}{21.15} \tag{3}$$

where C_d is the drag coefficient, A is the frontal area.

The rolling resistance F_f can be expressed by:

$$F_f = mgf \cos\alpha \approx mgf \tag{4}$$

where g is the gravitational acceleration, f is the coefficient of rolling resistance, α is the angle of slope.

The gradient resistance F_i can be expressed by:

$$F_i = mg \sin\alpha \approx mgi \tag{5}$$

where i is the longitudinal slope.

2.2.2 Model Based on the Signal of Accelerometer

The longitudinal acceleration signal obtained by accelerometer which is mounted on sprung part of vehicle is not the real longitudinal acceleration of vehicle but contains the slope information when vehicle is driving on the slope road, which can be expressed by:

$$ma_x = F_t - F_f - F_w \tag{6}$$

where a_x is the longitudinal acceleration signal obtained by accelerometer. So (1) can be rewritten as:

$$m\dot{v} = -mgi + ma_x \tag{7}$$

It needs to be emphasized that a_x in (7) does not count the pitch angle of vehicle, and correction with pitch angle should be considered to obtain high-accuracy. Premium SUVs are usually equipped with electronic control air suspension (ECAS), the pitch angle can be calculated by four suspension height signals as below:

$$k_{body} = \frac{(d_{srl} + d_{srr}) - (d_{sfl} + d_{sfr})}{2L} \tag{8}$$

where $d_{\text{sfl}}, d_{\text{sfr}}, d_{\text{srl}}, d_{\text{srr}}$ are the suspension height of four corners, L is the wheelbase. Thus, the corrected a_x can be expressed by:

$$a_x' = a_x - k_{\text{body}} g \tag{9}$$

2.3 Observer Design

The models are expressed in state-space form as:

$$\begin{cases} \dot{\mathbf{x}} = Ax + Bu \\ \mathbf{y} = \mathbf{Cx} \end{cases} \tag{10}$$

where $\mathbf{x} = \begin{bmatrix} v \ i \end{bmatrix}^T, \mathbf{y} = v,$

$$\mathbf{A} = \begin{bmatrix} 0 & -g \\ 0 & 0 \end{bmatrix}, \quad \mathbf{C} = \begin{bmatrix} 1 \ 0 \end{bmatrix}.$$

For model based on tractive force and driving resistances:

$$\mathbf{B} = \begin{bmatrix} 1/m \\ 0 \end{bmatrix}, \quad \mathbf{u} = F_t - F_f - F_w$$

For model based on the measured value of accelerometer:

$$\mathbf{B} = \begin{bmatrix} 1 \\ 0 \end{bmatrix}, \quad \mathbf{u} = a_x'$$

The rank of observable matrix is:

$$\text{rank}\begin{bmatrix} \mathbf{C} \\ \mathbf{CA} \end{bmatrix} = \text{rank}\begin{bmatrix} 1 & 0 \\ 0 & -g \end{bmatrix} = 2 \tag{11}$$

Necessary conditions for the system to be observable are obtained.

Luenberger observer is used to design slope estimation, which can be expressed by:

$$\dot{\hat{\mathbf{x}}} = A\hat{\mathbf{x}} + \mathbf{Bu} + \mathbf{L}e \tag{12}$$

where

$$e = \mathbf{y} - \hat{\mathbf{y}} \tag{13}$$

and (12) is rewritten as:

$$\begin{aligned}\dot{\hat{\mathbf{x}}} &= A\hat{\mathbf{x}} + \mathbf{Bu} + \mathbf{L}(\mathbf{y} - \hat{\mathbf{y}}) \\ &= (\mathbf{A} - \mathbf{LC})\hat{\mathbf{x}} + \mathbf{Bu} + \mathbf{Ly}\end{aligned} \tag{14}$$

The design objective of Luenberger observer is:

$$\lim_{t \to \infty} e = 0 \tag{15}$$

According to the theory of Luenberger observer, the convergence of the observer can be ensured if all eigenvalues of **A-LC** have negative real part [6]. Let λ_1, λ_2 be the eigenvalues, and the feedback matrix can be expressed by:

$$\mathbf{L} = \begin{pmatrix} -\lambda_1 - \lambda_2 \\ \frac{-\lambda_1 \lambda_2}{g} \end{pmatrix} \tag{16}$$

2.4 Estimation Algorithm

The algorithm prototypes are established by MATLAB/SIMULINK, which are shown in Fig. 1. The code is generated and embedded into the vehicle ECU.

Two different algorithm prototypes are programmed. In the following part of this paper, the algorithm based on tractive force and driving resistances is named as algorithm I, and the algorithm based on the measured value of accelerometer is named as algorithm II.

3 Establish of Test System

3.1 Implementation of Slope Estimation Algorithm

Algorithm is embedded into the vehicle ECU which uses MPC5634M microcontroller (80 MHz, 1.5 MB Flash, 94 KB SRAM). Figure 2 shows system architecture and hardware interface. Required signals are obtained by chassis CAN bus.

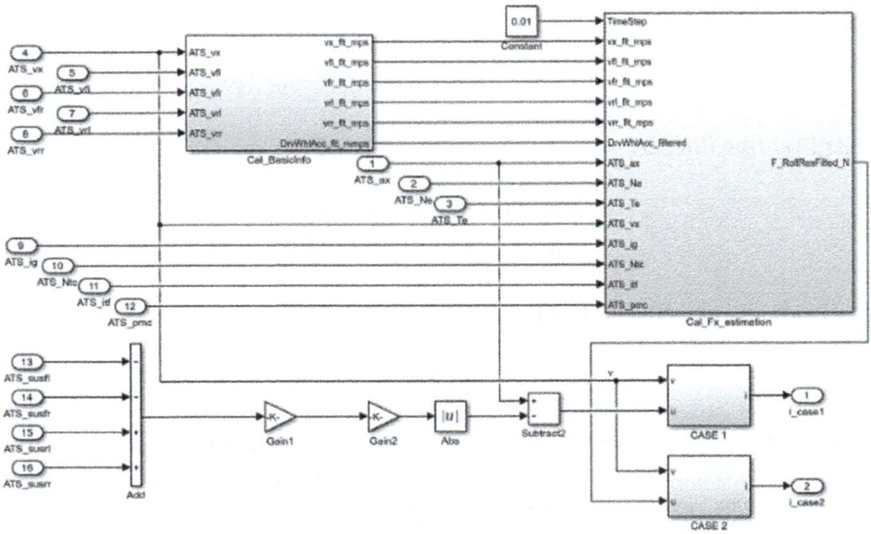

Fig. 1 Algorithm prototypes

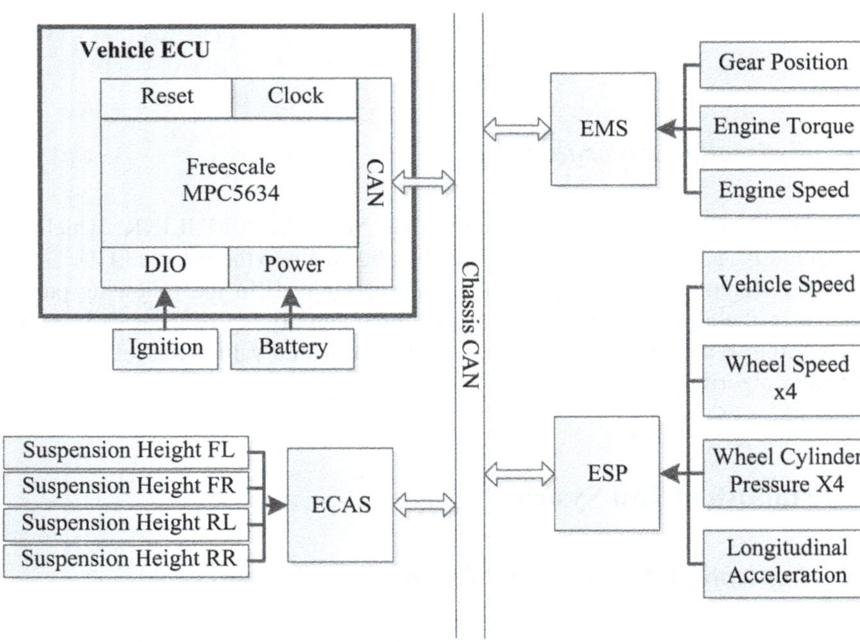

Fig. 2 System architecture and hardware interface

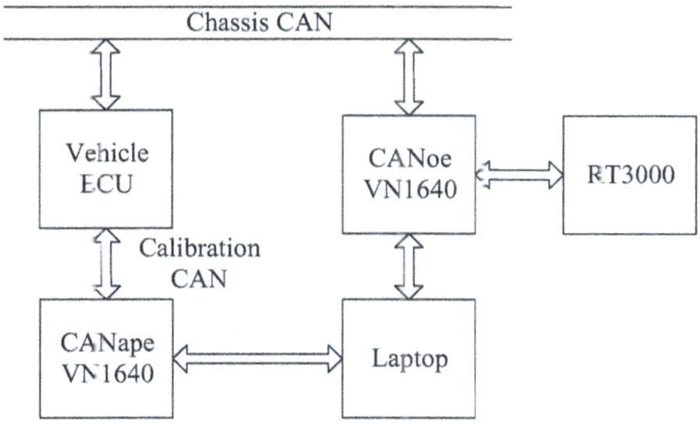

Fig. 3 Architecture of test system

3.2 *Architecture of Test System*

Figure 3 shows the architecture of test system. RT3000 is representing a coupled Global Positioning System/Inertial Navigation System which is manufacture by OxTS, and high-accuracy longitudinal speed, vertical speed, longitudinal acceleration, and vertical acceleration signals are obtained by it to compare with vehicle bus signals and calculate the reference value of slope. Vector CANoe is used as data acquisition system by which chassis CAN messages and RT3000 signals are stored to laptop. Vector CANape is used to flash program for vehicle ECU and online calibration.

4 Verification by Road Test

4.1 *Testing Filed*

The road tests were carried out at Nong'an testing ground of FAW, China, and four standard slopes are selected to verify the estimation algorithm, which are shown in Fig. 4. The nominal slopes of these roads are 8, 12, 20, and 29%, respectively. The Bosch GIM 60 Digital Level is used to check the slope of each road by measuring at several different sampling points. Figure 5 shows that the error between nominal value and measured value is less than 0.5%.

Fig. 4 Testing slopes

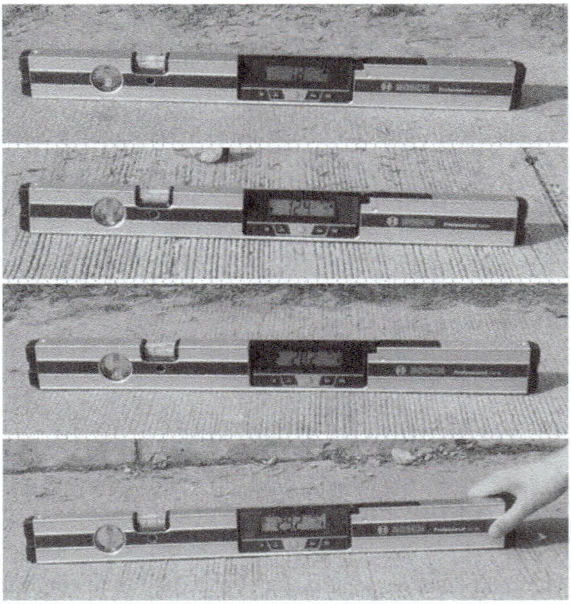

Fig. 5 Check slope

4.2 *Test Results*

4.2.1 Comparison of Observation Results with Different Parameters

First, the tested vehicle is stopped on −12% slope road, that is, $i = -12\%$, $v_x = 0$. In this test, algorithm I is no longer effective because brake force of actuator is larger than the brake force generated on ground, so algorithm II is used to compare the results with different parameters. Three groups of parameters are selected. These parameters are $\lambda_1 = -30$, $\lambda_2 = -50$; $\lambda_1 = -5$, $\lambda_2 = -10$, and $\lambda_1 = -1$, $\lambda_2 = -3$.

Figure 6 shows the results of estimated vehicle speed. The design objective defined by (15) can be evaluated by comparison between the estimated value and measured value of vehicle speed. Figure 7 shows the details of these values. Figure 8 shows the results of estimated slope, and Fig. 9 shows the details. It is seen that quicker convergence speed can be obtained by larger parameters but estimated slope has more fluctuation. Finally, $\lambda_1 = -5$, $\lambda_2 = -10$ are selected.

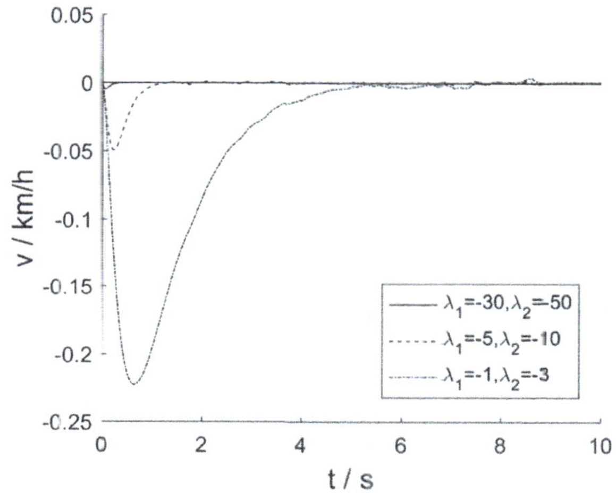

Fig. 6 Estimated vehicle speed

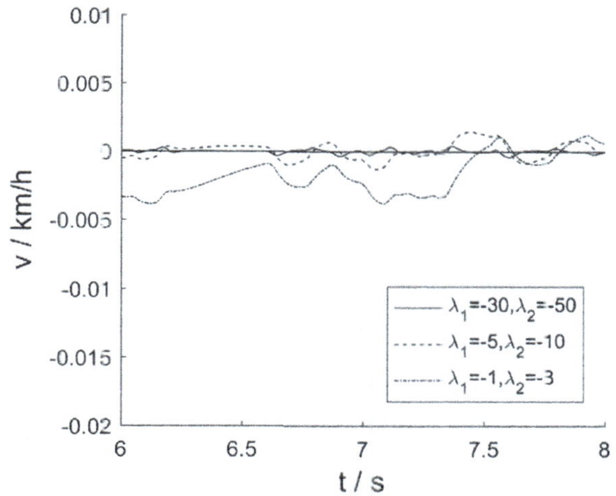

Fig. 7 Estimated vehicle speed (in detail)

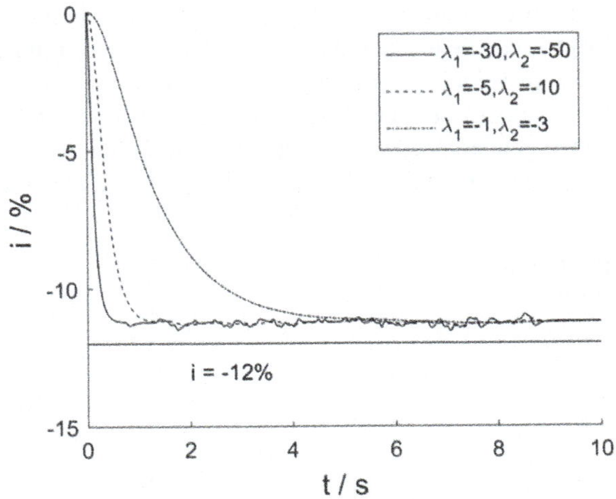

Fig. 8 Estimated slope

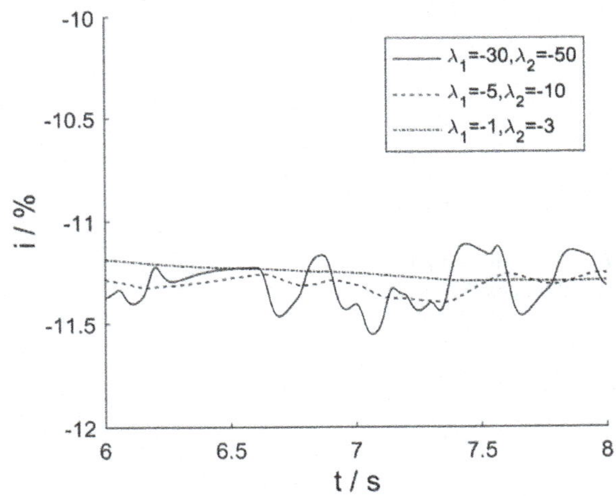

Fig. 9 Estimated slope (in detail)

4.2.2 Effect of Correction with Pitch Angle

Correction with pitch angle is not considered in the previous test, the output of accelerometer is directly used, and the results have a steady error which can be seen in Fig. 8. The effect of correction with pitch angle is shown in Fig. 10. It is seen that the estimated value with correction is very close to the true value.

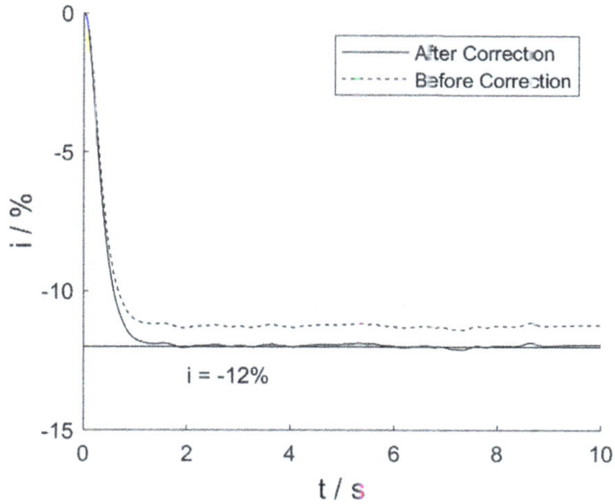

Fig. 10 Effect of correction with pitch angle

4.2.3 Comparison of Observation Results with Different Algorithms

The observation results of two algorithms with same parameters ($\lambda_1 = -5, \lambda_2 = -10$) is compared. The tested vehicle was drove up on 12% slope and then drove down on −29% slope.

Figure 11 shows the result of estimated vehicle speed by different methods, and Fig. 12 shows the details. It is seen that quick convergence speed and high accuracy can be obtained by both methods. Figure 13 shows the result of estimated slope by different methods. The estimated values of both methods are basically same and are close to the reference value. Figure 14 shows the inputs of both methods.

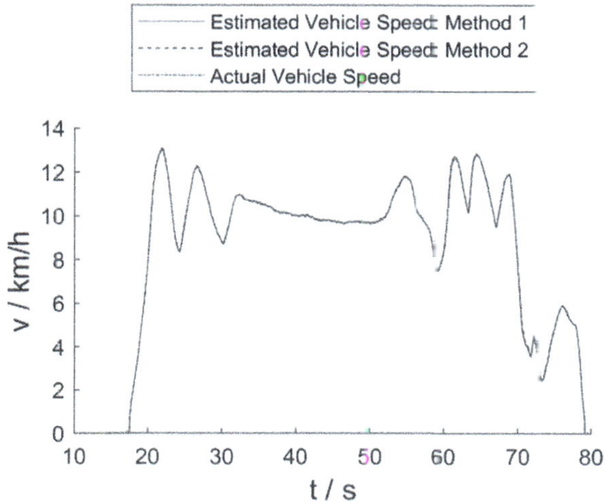

Fig. 11 Estimated vehicle speed by different methods

Fig. 12 Estimated vehicle speed by different methods (in detail)

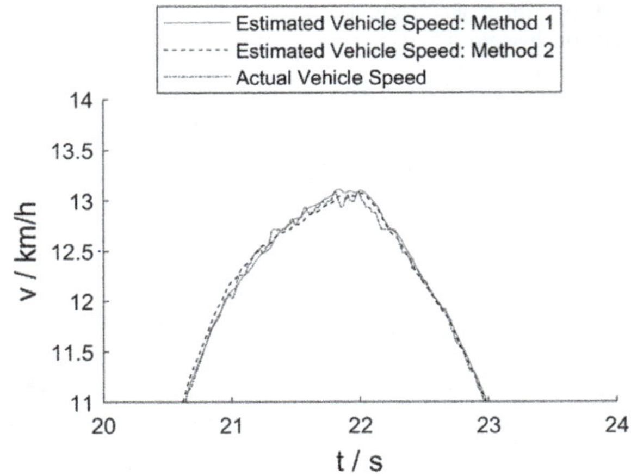

Fig. 13 Estimated slope by different methods

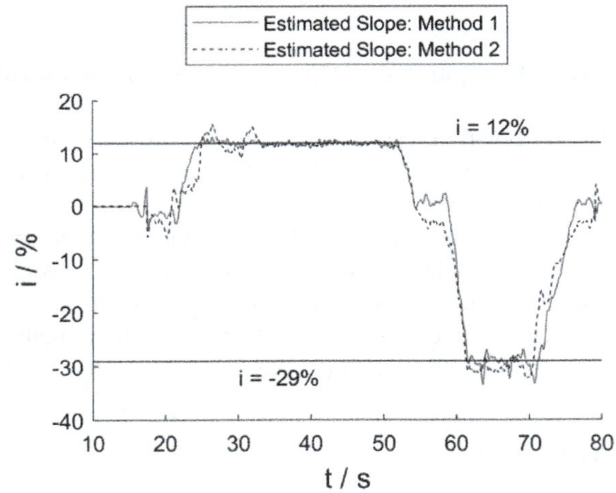

Fig. 14 Input of both methods

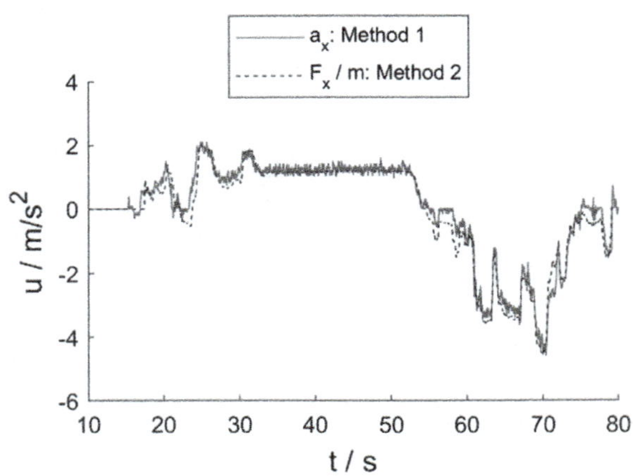

4.2.4 Comparative Analysis of the Two Methods

From the results, high accuracy can be obtained by both methods, but there is still some difference. The longitudinal tire forces cannot be estimated when vehicle is stopped on the slope road, which makes algorithm I no longer available, but algorithm II is still valid. The fluctuation of estimated result of algorithm I is larger than algorithm II because of the estimation error of longitudinal vehicle force. From this viewpoint, algorithm II is better. However, algorithm I is not affected by the orientation of body, while the estimation results of algorithm II should be corrected by suspension heights, but the correction cannot be realized if ECAS is not equipped.

In conclusion, the two methods have their own advantages, respectively. It is possible to obtain more accurate results through mutual correction, and the signal redundancy can be realized in the case of individual sensor failures, which will be studied in the future researches.

5 Conclusion

Two longitudinal slope estimation algorithms based on the Luenberger observer are designed. The algorithms are embedded into the ECU on a premium SUV. The parameters of observer are determined, and the performance of the proposed algorithms is verified by road test. Research in this paper can provide the basis for more accurate and reliable longitudinal slope estimation.

References

1. Wang H, Yuan Q, Zhong H et al (2014) Research on the hill-start auto control for the vehicles with EPB. Trans Beijing Inst Technol 34(4):344–348
2. Li G (2016) Estimation of vehicle mass and road slope and their influence on electric vehicle's shift schedule. Jilin University
3. Lei Y, Fu Y, Liu Ket al (2014) Vehicle mass and road grade estimation based on extended Kalman filter. Trans Chin Soc Agric Mach 45(11):9–13+8
4. Ran X, Li L, Zhao X et al (2019) Real-time recognition algorithm of longitudinal road slope based on vehicle dynamics and acceleration sensor information. J Mech Eng 52(18):98–104
5. Yang Z, Cao C, Su Y (2002) The method of identifying dynamically road grade by means of multi-sensor information fusion. J Chongqing Jiaotong Univ (Soc Sci Ed) 21(2):97–102
6. Yun H, Zhao Y, Zhong Z (2010) On-line SOC estimation of li-ion battery using luenberger closed-loop states observer. Chin Mech Eng 21(20):2505–2509

Scheme Design of Engine Anti-theft System Based on RFID Technology

Chao Rao and Jiawu Ling

Abstract This article for the auto theft crime occurs frequently, and domestic automobile anti-theft system's safety performance is low; this paper proposes an engine anti-theft system scheme based on RFID technology, and it is mainly composed of a highly integrated ignition key, IMMO controller and engine management controller. Key components of engine anti-theft system hardware design and software design in detail, at the same time to multiple test validation of guard against theft system, theory and practice fully proves that the system safe and reliable, cost-effective, have broad application prospects, and application in bulk in some SUV vehicle models.

Keywords Automotive electronics · Engine anti-theft system · IMMO · RFID

1 Introduction

With the popularity of automobiles and the rapid development of automotive electronic technology, car theft cases also occur frequently, and car theft has become a major public hazard in the world [1]. In order to effectively protect the owner's property safety, all vehicle manufacturers are constantly improving anti-theft technology. Especially with the progress of microelectronics technology, vehicle anti-theft technology has developed towards the direction of electronization, automation, intelligence and networking. The traditional simple mechanical anti-theft system has unable to effectively prevent car theft and to meet people's needs. The current advanced anti-theft measures are to use the immobilizer (IMMO) on the car to control the engine starting by matching the password of the chip in the key with the security controller and the password in the engine controller to achieve the purpose of anti-theft. For cars equipped with engine anti-theft locking system, even if the car driver opens the door, he cannot start the engine and drive the car. The core technology of the engine anti-theft system is radio frequency identification (RFID), which is a non-contact automatic identification technology [2]. The target object is

C. Rao (✉) · J. Ling
Dongfeng Motor Corporation Technology Center, Wuhan, China
e-mail: super_2612@163.com

China SAE (ed.), *Proceedings of China SAE Congress 2018: Selected Papers*,
Lecture Notes in Electrical Engineering 574,
https://doi.org/10.1007/978-981-13-9718-9_69

automatically identified by RF signal and acquired relevant data. RFID technology uses radio frequency transmission and can read chip data through external materials to achieve non-contact operation. Communication data is encrypted by encryption algorithm to realize data security storage, management and communication. In the early 1990s, the practical technology for data acquisition was a breakthrough in the field of radio frequency identification; in 1994, the Ford Motor Co was first equipped with an RFID-based automobile anti-theft system on its European ESCORT vehicles; in 1995, the car burglar system based on the control ignition system became the standard of all domestic cars sold in Europe. The anti-theft system was installed as standard configuration, the rate of vehicle theft was decreased by 90% and gradually extended to all parts of the world, at present, China has been gradually popularized in low and medium cars.

2 Class of Automobile Anti-theft System

Nowadays, the automobile anti-theft system has developed from the early mechanical type to the electronic password, remote alarm and other high-tech products. According to the structure and function division, it can be divided into four main categories: mechanical anti-theft system, electronic burglar-proof system, chip-based anti-theft system and network-based anti-theft system. They have their own advantages and disadvantages, but the development direction of vehicle anti-theft is to develop with chip and network with higher intelligence [3].

2.1 Mechanical Anti-theft System

This is the earliest automobile anti-theft system, currently rarely used alone, usually used with electronic, chip-based anti-theft system, in order to obtain the best anti-theft effect. The main principle of the mechanical anti-theft system is to lock the clutch, brake, throttle, steering wheel or transmission gear and other important parts of the car to achieve the purpose of anti-theft. The advantage is that the principle is simple and the price is relatively cheap. The disadvantages are as follows: the assembly and disassembly are very troublesome, the security is not good and the alarm information cannot be transmitted when the vehicle was stolen. These limitations lead to the gradual replacement of the electronic anti-theft system.

2.2 Electronic Anti-theft System

With the rapid development of automotive electronic technology, electronic anti-theft system based on electronic control unit has been widely used in automobiles.

The electronic anti-theft system can monitor the movement of the door, the trunk, the engine hatch cover and the ignition key through the communication between the electronic controllers and also has the sound and light alarm functions. Once the vehicle is completely locked, once the system detects the abnormal movements of the monitoring objects, such as the door or the trunk, the alarm signals will be sent through the burglar horn and the external lights, and the engine will be prevented from starting. The electronic anti-theft system has the advantages of complete functions and simple operation. The disadvantage is that it is easy for the stolen vehicle to use the receiver or scanner to steal the wireless signal sent by the remote controller and then decode the car's anti-theft system through decoding.

2.3 Chip-Type Anti-theft System

The principle of chip-type anti-theft system is to control the start and operation of the engine by locking the circuit and the oil circuit of the engine, so as to achieve the purpose of burglary prevention. The anti-theft system embeds a chip with unique ID and password in the ignition key. When the key hits the ON or START gear, the system first authenticates the ID and password of the key, and only the authentication can start the car. This ensures that even if the thief stole the key and entered the car, it cannot be started and opened. The key authentication process is realized by low-frequency wireless communication, and the low-frequency signal attenuation is very fast. It can effectively prevent the authentication information from being leaked out of the vehicle and be scanned. The key is embedded in digitalized passwords with very low bit rate, which ensures the high security of the anti-theft system.

2.4 Network Anti-theft System

Its principle is to realize vehicle location, interception and other functions through network, so as to achieve the purpose of theft prevention afterwards. Most of the networks used are satellite positioning and tracking systems (GPS). This kind of anti-theft system is usually used in combination with electronic burglar security system and chip-type anti-theft system. The advantage is that it can pass the alarm information and the location of the alarm vehicle to the alarm centre silently through the GPS system (or other network systems). It has high security and powerful function, which breaks through the limits of distance and covers a wide range. The disadvantages are that the reaction speed is relatively slow, it cannot be used for real-time anti-theft, the network signal has blind area (a few areas that cannot receive the signal), the cost is very high, and it is not popular in the market at present.

Through the above analysis, the comprehensive consideration of cost and safety, the electronic burglar system and the chip-based anti-theft system are more suitable for the current development level. It is the most widely used anti-theft system at present. It has high cost performance and high security performance. With the continuous improvement of the level of electronic technology and network technology, the future car anti-theft system will develop towards safety, simplicity, intelligence and diversification of functions.

3 Scheme Design of Engine Anti-theft System

3.1 Composition Scheme of Anti-theft System

The basic structure of the engine anti-theft system is shown in Fig. 1.

From the above picture, it is known that the engine anti-theft system mainly consists of three parts: ignition key, anti-theft controller (including anti-theft coil, base station controller, body controller) and engine controller.

Ignition key: inside is embedded with transceiver chip TRP (Transponder), and a low-frequency antenna communicates with the anti-theft controller. The transceiver chip TRP does not have its own power supply. The energy required for its work is provided by the base station antenna coil in the anti-theft controller. This design can reduce the volume of the transceiver chip TRP and make the ignition key more convenient to carry and use.

Anti-theft controller: it usually includes anti-theft coil, base station controller and body controller. The anti-theft coil is usually installed on the ignition lock. It is a low-frequency wireless communication circuit with the low-frequency antenna driving circuit inside the base station controller, which is responsible for communication

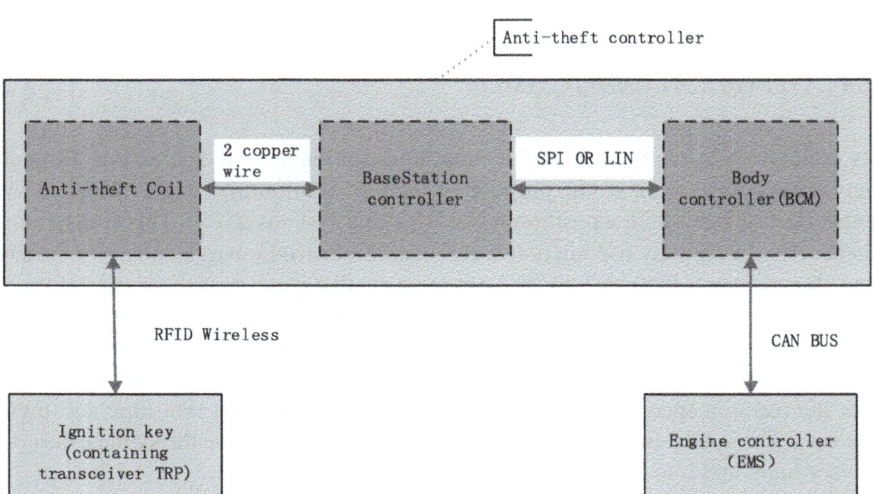

Fig. 1 The basic block diagram of the anti-theft system of the engine

with the ignition key. The body controller mainly carries out the logic operation and control processing of the anti-theft system.

Engine controller: it is mainly responsible for controlling various actions of the engine system and generally communicates with the burglar controller through the CAN bus.

3.2 *Key Information and Storage Definition of Anti-theft System Authentication*

- IDE (TRP Identifier): TRP identification code, data length is 32bit, and each TRP contains a fixed and unique code, written before the factory. Its role is to participate in the algorithm authentication of the key.
- ISK (Immobilizer Secret Key): anti-theft key, the data length is 64bit; it is derived from ESK code, and each car configuration of all the keys (including TRP) contains the same ISK code. Its role is to participate in the algorithm authentication of the key.
- PSW (Password): random password, the length of the data is 24bit, which is produced at random in the key learning process, and the PSW in each TRP is different in theory. Its role is to participate in the algorithm authentication of the key.
- ESK (EMS Communication Cryptology Secret Key): engine controller communication encryption key, the length of the data is 128bit; it is written through the whole vehicle EOL equipment downline matching process; the ESK code in each car is different in theory, and it is transferred through the vehicle VIN code. Its role is to participate in anti-theft controller algorithm authentication.
- PIN (Personal Identification Number): safety code, it is written through the whole vehicle EOL equipment for the downline matching process, and each car's PIN code is not the same in theory; it is transformed through the whole vehicle VIN code. Its function is to carry out the security access authorization code for the after service process of the engine anti-theft system.

  - The transceiver chip TRP stores the fixed unique identification code IDE (TRP Identifier), the anti-theft key ISK (Immobilizer Secret Key), the random password PSW (Password).
  - The anti-theft controller stores or configures the fixed unique identification code IDE, the anti-theft key ISK, the random password PSW, the engine controller's communication encryption key ESK (EMS Communication Cryptology Secret Key) and the security code PIN (Personal Identification).
  - The engine controller is stored or configured with the engine controller communication encryption key ESK and the safety code PIN.

3.3 Hardware Design Scheme of Anti-theft System

The core component of the anti-theft system of the engine is the anti-theft controller. The design of the base station controller and the ignition key (including TRP) plays a decisive role in the safety and reliability of the anti-theft system. At present, there are two main chip vendors in the domestic and foreign popular engine burglar anti-theft system: NXP and TI, as shown in Table 1.

Figure 2 shows the hardware design of the engine anti-theft system used in an SUV model, using the NXP scheme: PCF7941E (integrated remote control and TRP chip) + PCF7991 (the base controller anti-theft coil drive chip) [4, 5]. The hardware design of the anti-theft system mainly includes the ignition key (integrated remote control and TRP function), anti-theft controller (including anti-theft coil, base station controller, body controller) and engine controller. The anti-theft controller communicates with the ignition key through the RFID low-frequency wireless signal and communicates with the engine controller through the CAN bus. The serial port bus LIN is used to communicate between the base station controller and the body controller inside the anti-theft controller.

3.3.1 Design of Ignition Key

The design of the ignition key is NXP's PCF7941E chip, which integrates the UHF emitter (telecontrol radio frequency module) and the LF transceiver to achieve radio frequency emission and transceiver TRP low-frequency communication authentica-

Table 1 Hardware design of traditional anti-theft system

Number	Vendor	Domain of design scheme	Encryption algorithm	Key data length (bit)
1	NXP	Base station, TRP	HITAG2	64
2	TI	Base station, TRP	DST80	80

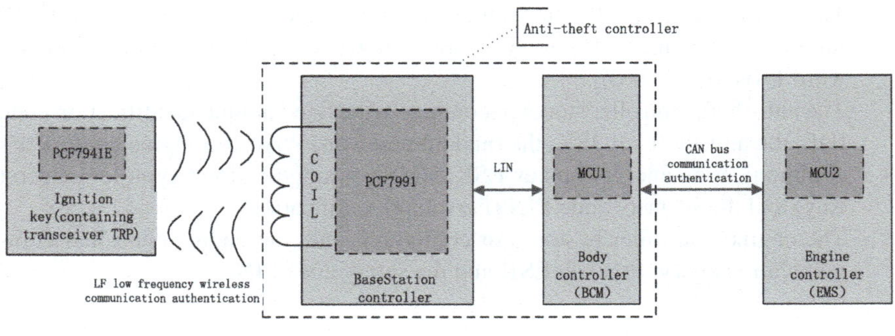

Fig. 2 Hardware design scheme for anti-theft system of engine

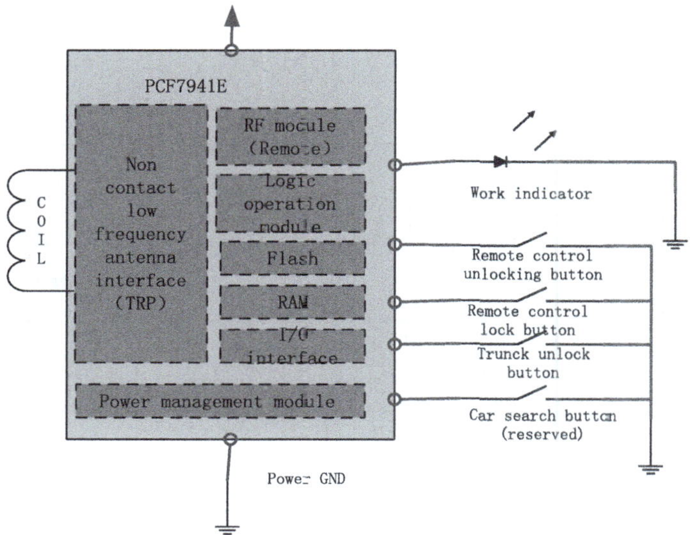

Fig. 3 Schematic diagram of PCF7941E interface design

tion. It uses HITAG2 encryption identification algorithm, uses random numbers, keys and passwords to carry out operation and has the characteristics of high sensitivity and short identification time. PCF7941E also provides the 32bit TRP identifier (IDE) that has been programmed and cured at the factory. Figure 3 is the schematic diagram of the interface design of PCF7941E.

3.3.2 Design of Base Station Controller

The design of the anti-theft base station controller uses the PCF7991 chip of NXP, which realizes the modulation and demodulation of the low-frequency signal by integrating it with the anti-theft coil in the anti-theft base station and authenticates with the auto body controller through the serial port LIN bus. The PCF7991 chip has high integration and provides read and write access to TRP. It can use ASK write operation and AM/PM read operation to realize two-way communication with TRP [6]. Figure 4 is a schematic of the interface design of PCF7991.

3.4 Software Design Scheme for Anti-theft System

The software of the anti-theft system mainly includes two parts: the authentication of the anti-theft system including the ignition key (including the TRP) and the wireless communication authentication of the anti-theft controller, the CAN bus authenti-

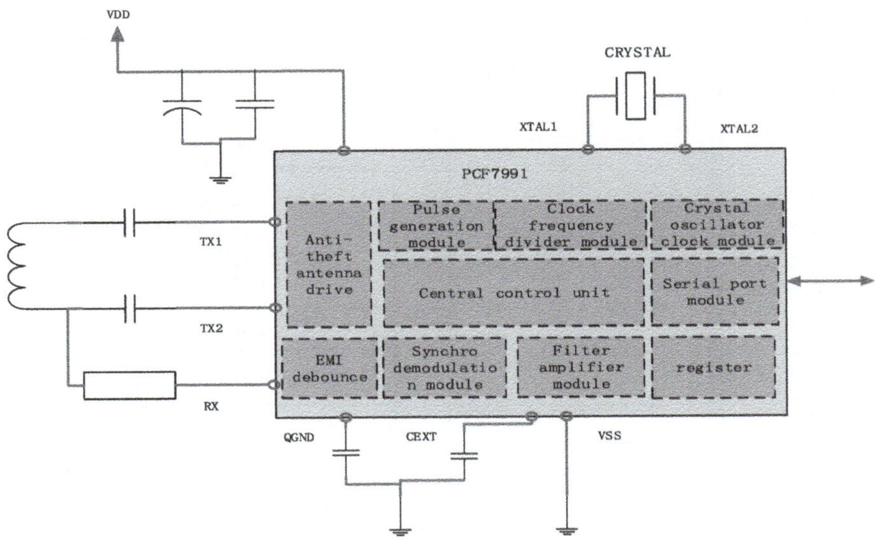

Fig. 4 Schematic diagram of PCF7991 interface design

cation of the anti-theft controller and the engine controller, and the authentication information configuration service of the anti-theft system.

3.4.1 Authentication of Anti-theft System

The process of authentication of the anti-theft system, as shown in Fig. 5, mainly includes level two authentication: first level is the wireless communication authentication of the ignition key (including TRP) and the anti-theft controller authenticates the legitimacy of the TRP. If the first level of authentication fails, the engine will be locked, the vehicle cannot be activated, and only the first level authentication is successfully passed, Only to enter the second level certification; the second level is the CAN bus authentication of the anti-theft controller and the engine controller authenticates the legitimacy of the anti-theft controller. If the second level authentication fails, it will also cause the engine to be locked, unable to start the vehicle, only the second-level certification is also successfully passed, the engine is released, allowed to be allowed. Ignition starts the vehicle [7].

Wireless Communication Authentication Between Ignition Key (Including TRP) and Anti-theft Controller

When the ignition lock is switched to the position of ON, the TRP and the anti-theft controller are the of communication through the RFID low-frequency wireless signal

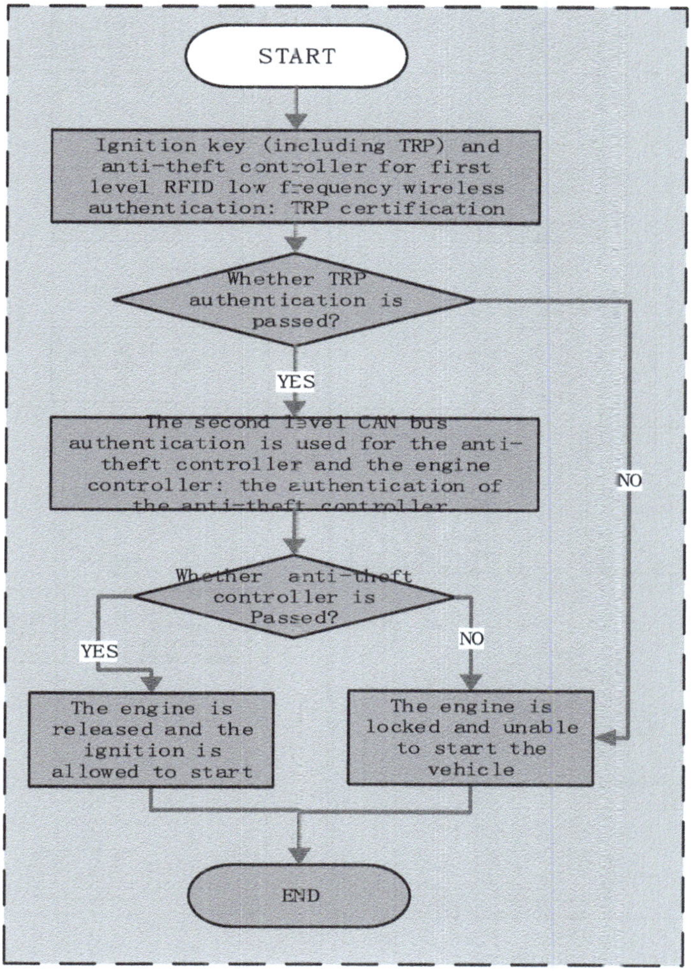

Fig. 5 Authentication process of anti-theft system

of the 125 kHz [8], as in the first level TRP authentication process of Fig. 6. First, the anti-theft controller sends the command to start the authentication, TRP receives this command to the anti-theft controller to feed back the IDE, and the anti-theft controller checks the received IDE with the IDE stored in the configuration service process. If the anti-theft controller uses the configured IDE, ISK and the random number to encrypt it, get the sign. The name Signature then sends the random number and the signature Signature to TRP. The TRP end performs the same encryption algorithm to check the received signature Signature. If consistent, the feedback random password PSW checks the anti-theft controller, and if the agreement is consistent, the

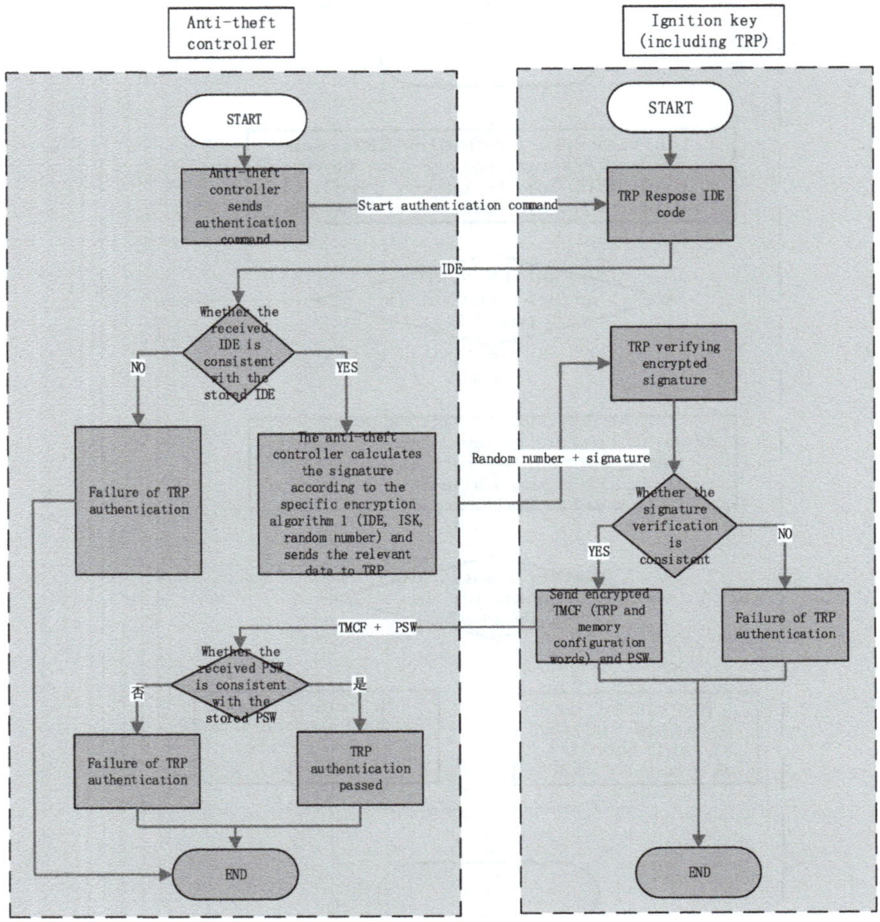

Fig. 6 TRP authentication process

authentication of the TRP is passed. If there is any inconsistency between the above checks and comparisons, the TRP authentication fails.

CAN Bus Authentication Between Anti-theft Controller and Engine Controller

When the ignition lock is switched to the ON, the second-level anti-theft controller, as shown in Fig. 7, is communicated between the anti-theft controller and the engine controller through the CAN bus. First of all, the engine controller randomly generates challenge code to send to the anti-theft controller. If the first level TRP authentication result is passed, the anti-theft controller uses the configured ESK and the received

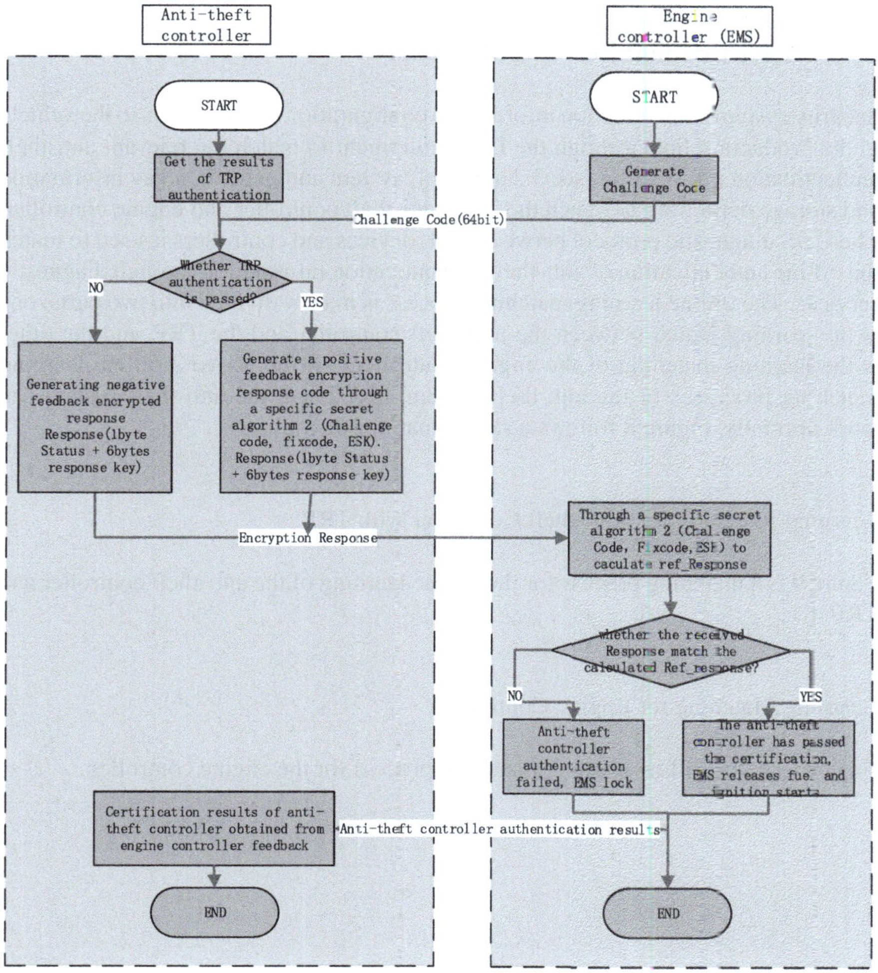

Fig. 7 Anti-theft controller authentication process

challenge code; the anti-theft controller is encrypted with the fixed code FixCode of the engine controller agreement and produces the encrypted response code Response. After that, the arrangement is sent to the engine controller. The transmitter controller ends the same encryption algorithm to check the encrypted response code Response, and if the consistent anti-theft controller authentication passes, the EMS releases the fuel and allows the ignition to start the vehicle; if inconsistent, the anti-theft controller authentication fails, the EMS lock is locked, and the vehicle cannot be activated. At the same time, the engine controller feedback the results of authentication to the anti-theft controller [9].

3.4.2 Anti-theft System Authentication Information Configuration Service (Offline Learning Matching)

Security system authentication information configuration service refers to the vehicle in the production line, through the EOL equipment to match the relevant anti-theft authentication information (see 3.2 anti-theft system authentication key information and storage definition) to match the TRP, anti-theft controller and engine controller. The UDS diagnostic protocol between EOL devices and controllers is used to match the offline authentication of anti-theft authentication information through diagnostic services. The offline learning matching process is mainly divided into two parts: one is the learning match between the anti-theft controller and the TRP, and the other is the learning matching of the engine controller. All the above 2 offline learning matching processes of the anti-theft system are passed, the anti-theft system can work normally. Figure 8 follows a sketch map for the offline.

Learning Matching for Anti-theft Controller with TRP

Figure 9 is a matching process for the offline learning of the anti-theft controller and TRP.

Learning Matching for Engine Controller

Figure 10 is the offline learning matching process for the engine controller.

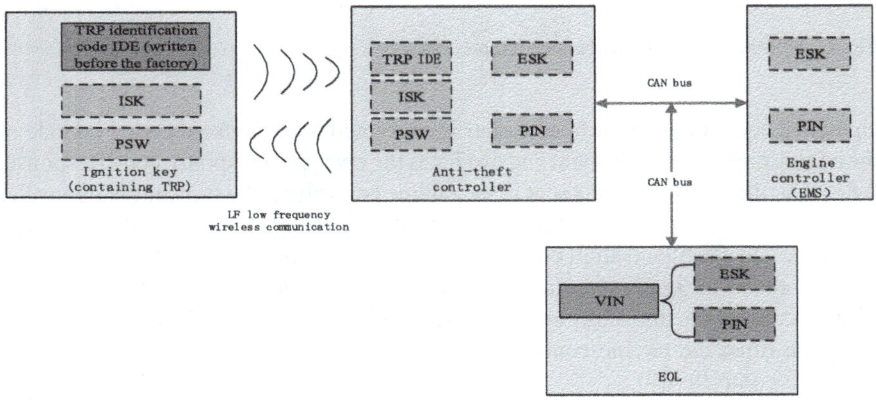

Fig. 8 A sketch map for the learning of the downline of the anti-theft system

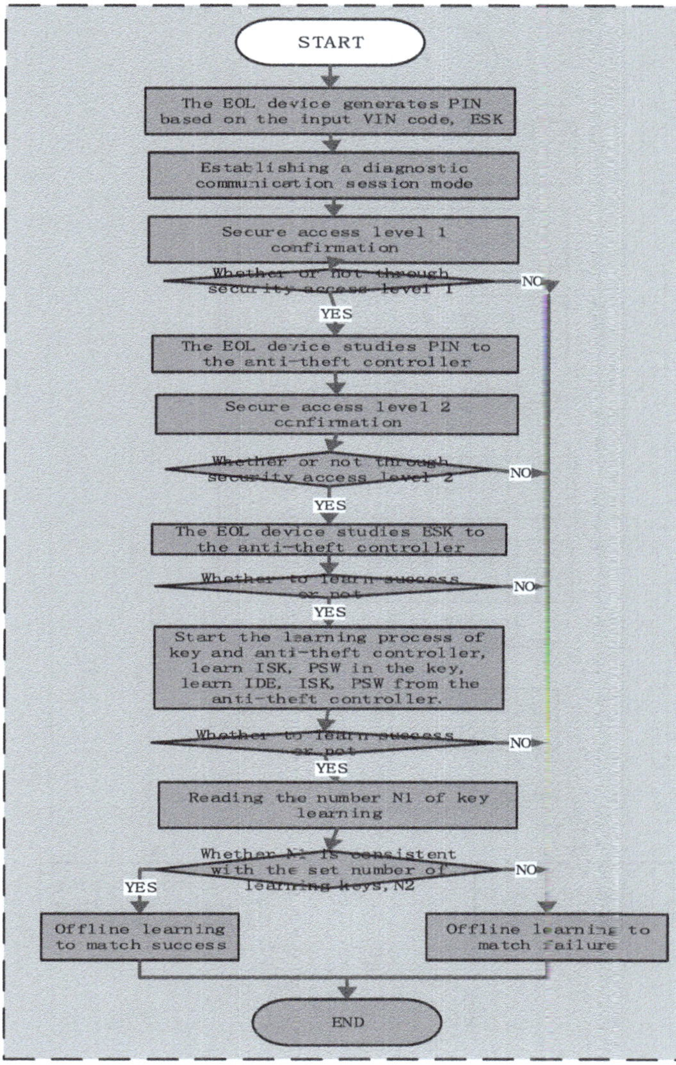

Fig. 9 Matching process of offline learning between anti-theft controller and TRP

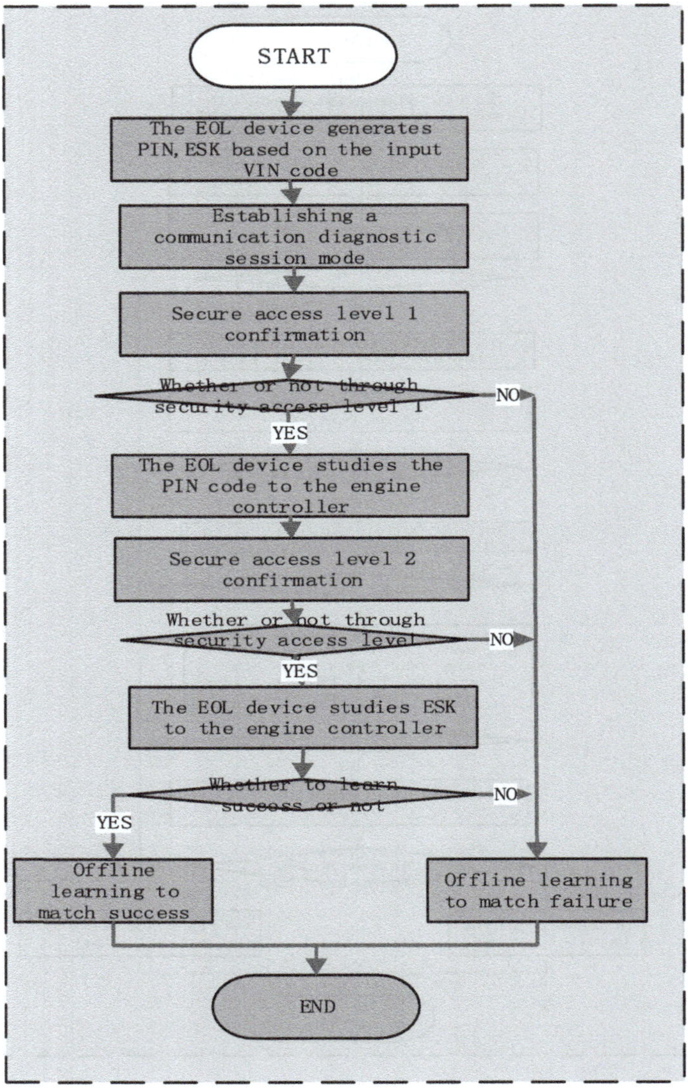

Fig. 10 Offline learning matching process of engine controller

3.5 Innovation of Anti-theft System Scheme

The scheme of engine anti-theft system described in this paper has the following innovations compared with the existing products.

- The hardware design of anti-theft system has high integration and low cost.
- Adopting advanced RFID communication technology and two-level authentication security scheme, it has high safety and reliability.
- Anti-theft system authentication information configuration service process is perfect and standardized.

4 Verification and Conclusion

4.1 Verification of Anti-theft System

The verification of this system mainly includes functional test, bench performance test and vehicle reliability road test.

- **Functional testing**: mainly test verification of the anti-theft system components including TRP, anti-theft controller and engine controller functional logic processing is in line with the design specification requirements.
- **Test bench performance**: electrical performance verification, EMC performance validation, environmental adaptability test and mechanical performance verification. The specific experimental items are shown in Table 2 below, with a total of 48 items, all tested and verified.
- **Vehicle reliability road test**: including static and dynamic testing of all vehicle sections.

This anti-theft system has passed the functional test successfully, the bench performance test and the vehicle reliability road test have been supplied in bulk on an SUV vehicle, and the work is stable and reliable.

4.2 Conclusion

Through the investigation and study of the car burglar-proof system, this paper puts forward a design scheme for the anti-theft system of the engine. At the same time, the main hardware design and software design of the engine anti-theft system are explained in detail. Through the development and application of the anti-theft system, it effectively reduces the cost of the vehicle, improves the security level of the automobile anti-theft system, masters the independent intellectual property rights and has a wide application prospect in the automotive electronics field.

Table 2 Experimental verification project of anti-theft system

Number	Classification	Test Item	Number	Classification	Test Item
1	Electrical performance verification	Abnormal voltage test	25	Verification of EMC performance [10]	Radiation noise measurement
2		Automatic heating measurement	26		Abnormal power supply voltage
3		Insulation impedance test	27		Change of power supply voltage in general voltage control range
4		Dust Test	28		Anti-jamming performance of impulse output inductive load
5		Static current	29		Restarting the anti-jamming performance of pulse
6		Grounding bias	30		Anti-jamming performance of voltage-controlled pulse
7		Power supply voltage performance	31		Measurement of radiofrequency conduction noise at the output end
8		Abnormal power supply voltage performance	32	Verification of environmental adaptability	IMMO durability activation
9		Earthing and connection network positive performance	33		After low temperature storage
10		Long-time overload performance	34		After high temperature storage
11	Verification of EMC performance	Performance of power supply voltage slowly rising and decreasing	35		The limit temperature operating under the general power supply voltage

(continued)

Table 2 (continued)

Number	Classification	Test Item	Number	Classification	Test Item
12		Re initialization test	36		Temperature gradient
13		Micro-interruption performance (transient) of power supply system	37		Determination of function limitation under temperature
14		Performance of pulse 1 or 1BIS and 2A	38		Temperature resistance cycle test
15		Performance of pulse 3A and 3B	39		Endurance test
16		Performance of pulse 4	40		High temperature durability
17		Performance of pulse 5B	41		Thermal shock durability test
18		Performance of power supply voltage fluctuation in vehicular power supply network	42		Humid and corrosion
19		Transient anti-interference performance of signal lines	43		combustion performance
20		Anti-jamming performance of high current injection (BCI)	44	Mechanical performance verification	Free fall
21		Radiation harassment anti-interference test	45		External stress test of equipment
22		Electrostatic discharge performance of equipment without power supply	46		Resonance frequency detection and resonance point 1 h vibration
23		Electrostatic discharge performance under the condition of equipment power supply	47		random vibration
24		Conduction radiation test	48		Mechanical impact test

References

1. Zhang D, Liu Z (2009) Reliability design of a vehicle anti-theft key authentication system. Autom Technol Appl 28(11):61–64,85
2. Teng JL, Long Y (2010) Design of intelligent vehicle security and anti-theft system based on RFID. J Hunan Univ Technol 24(6):80–83
3. Zhang D (2009) Research and development of vehicle anti-theft controller. Harbin Institute of Technology, Harbin
4. PCF7941 Security Transponder and RISC Controller Product Specification. Philips, 2006
5. PCF7991 AT Advanced BaseStation IC. Philips, 1995
6. Shi G, Hu L (2013) Integrated automotive electronic security system design. In: China automotive engineering society annual meeting, Beijing, 2013:113–119
7. Bao YY (2013) Application and development of vehicle anti-theft system. Dev Innov Mech Electr Prod 26(4):132–134
8. Shuhua Y (2008) Design and implementation of vehicle security system. Dalian University of Technology, Dalian
9. Wan D (2018) Analysis of Bao Jun automotive engine anti-theft programming principle. Intern Combust Eng Accessories 12:167–168
10. Zeng XW (2017) Solutions to EMC problems caused by engine anti-theft system. Automot Eng (04):29–30 + 50

Thermal Problem Solution of Vehicle DRL LED Driver

Jiefei Xiong and Yuantao Hou

Abstract The thermal problem of vehicle lamp DRL LED driver is investigated. The root cause of flickering is analyzed, and theoretical way to solve thermal problem is provided. Finally, low-cost solution is applied on this issue and is validated by measurement. This solution can also be applied on other DC-DC headlamp driver projects to solve thermal problem.

Keywords Automotive lamp · Thermal problem · LED driver · Thermal resistance · Heat dissipation

1 Introduction

As automotive industry developed, vehicle lamp plays a more and more important role in vehicle safety area. Especially driving at night, the performance of lamp has a big influence on safety of other vehicles and passerby. In last 10 years, with fast LED lighting technology development and LED cost reduction, more and more vehicle manufacturers choose LED as vehicle lamp light source due to LED's obvious advantage such as long life, small volume, low power dissipation, high reliability, fast response, and flexible styling design comparing to halogen bulb [1, 2]. In LED lamp, an important part is LED driver, which is used to supply suitable voltage and current for LEDs, making LEDs operate as designed. However, there's a critical problem affecting the LED drivers, called thermal problem. It doesn't only affect performance of lamp, but reduce lamp life a lot. In worst case, lighting function is degraded which results to safety accident. In this paper, thermal problem is studied and solved based on a thermal issue—DRL flickering, during lamp development. Summary and suggestions are provided for analyzing thermal problems.

J. Xiong (✉) · Y. Hou
Dongfeng Nissan Technical Centre, Guangzhou 510800, China
e-mail: xiongjiefei@dfl.com.cn

China SAE (ed.), *Proceedings of China SAE Congress 2018: Selected Papers*, Lecture Notes in Electrical Engineering 574,
https://doi.org/10.1007/978-981-13-9718-9_70

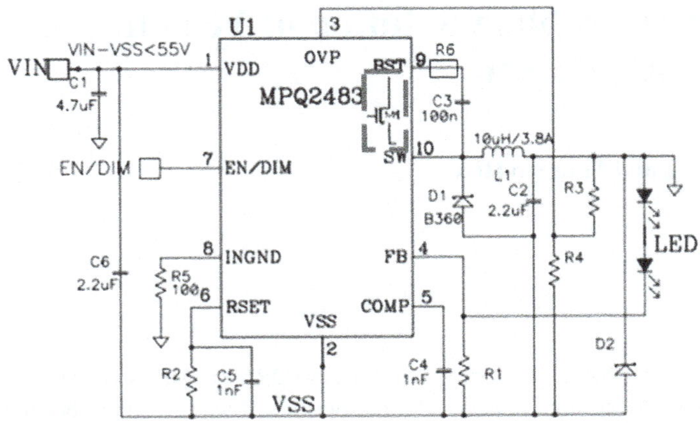

Fig. 1 MPQ2483 application circuit

2 Background and Root Cause Analysis

2.1 DRL Ambient Temperature and Flickering

During development of vehicle headlamp, the DRL operates abnormally—DRL flickers about 0.5 Hz after vehicle operating for a period. For this vehicle, DRL is located in headlamp, which is near the engine, so the internal temperature of headlamp can reach 105 °C during driving (by measurement), while the junction temperature (T_j) of driver IC reaches 150 °C. This DRL driver is designed based on the IC-MPQ2483, which equipped with over-temperature protection (OTP). This IC shuts down the driver when it detects IC's temperature reaches 150 °C to protect IC from temperature rising even failure. Then, the T_j of IC decreases because the driver is shutting down. After T_j decreases below the threshold, the IC goes out of OTP mode and the DRL driver resumes. So repeatedly. As Fig. 1 shows the application circuit of the driver [3].

2.2 Root Cause of IC Thermal Problem

There are two reasons that contribute to the temperature rise on the driver IC.

One reason is heat conduction. The heat sources such as power inductor and rectifying diode near the IC transfer the heat to IC by the PCB copper, which helps to increase IC temperature.

The other reason is the high power dissipation of IC that resulting to self-heating. There's an internal switching MOSFET, which dissipates a lot of power during ON phase and SWITCHING phase of each period, packaged in the IC. This is the main

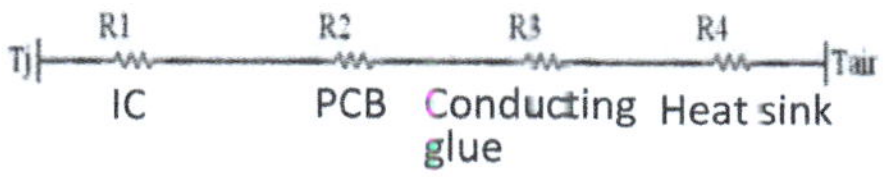

Fig. 2 DRL driver system thermal resistance

heat source of the IC. In addition, power dissipation also exists because of the quiescent current when IC is operating.

The junction temperature T_j of IC can be expressed by,

$$T_j = T_{amb} + R_{thja} * P_{total}$$

where T_{amb} is based on ambient temperature and can't be improved. So thermal resistance R_{thja} and power dissipation of IC P_{total} will be studied in this paper.

2.3 *Equivalent Resistance Model of IC*

See Fig. 2.

Generally, the root of thermal design is to transfer the heat of IC to outside of lamp and reduce the intermediate steps if possible. IC–PCB–heat conducting glue–heat sink–air is the main path to dissipate the power. The heat from IC is conducted to PCB by solder lead and conducted to heat sink by heat conducting glue, and finally, heat is transferred to the air by conduction and convection.

Thermal resistance of IC R_{thja} is,

$$R_{thja} = R1 + R2 + R3 + R4$$

3 Thermal Problem Solution

According to the analysis above, the thermal problem can be solved from two ways: reducing R_{thja} of IC and reducing the power dissipation P.

So these two ways will be tried on this problem.

3.1 *Active Cooling*

As we know, there are 3 ways for thermal transmission: conduction, convection, and radiation. While cooling ways are divided into active cooling and passive cooling. Active cooling contains several kinds of ways such as fan cooling, heat pipe cooling, semi-conductor refrigeration, and liquid cooling, but all of the methods need extra power supply or device. The obvious advantage of active cooling is the high efficiency but on the other hand, additional space and cost for fan is needed, and failure risk should also be considered. Concerning on this project, it is impossible to modify the mechanical objects in late development phase, so active cooling cannot be adopted.

3.2 *Passive Cooling*

Comparing to active cooling, passive cooling is much simpler. It means the heat of IC is transferred to air by heat sink under natural convection, so the surface of heat sink is very significant. The best benefit is simple, low cost, high reliability, and without extra power supply [4].

For DRL driver, following passive cooling methods are adopted to reduce R_{thja} of IC.

1. Increase cooling copper for IC, thicken the copper from 1 to 2 oz;
2. Increase the vias on PCB around IC;
3. Make all vias not filled;
4. Remove solder mask of the copper around IC.

Heat sink is also considered to add but not adopted due to the extra cost. According to the LED headlamp driver heat sink optimization, the higher size of heat sink is, not the better cooling performance [5]. The performance of heat sink is linked with base length, fin height, fin thickness and fin numbers. Following sizes are gotten considering the internal lamp space: base thickness is 1.5 mm, fin height is 12.17 mm, fin thickness is 1.517 mm and fin number is 10.

Measurement (T_s is solder temperature) after modifying PCB layout is as follows:

Condition: 13.5 V, 25 °C[a]	Temperature rise before change (°C)	Temperature rise after change (°C)	Improvement (°C)
Ts of IC	125	97.5	27.5

[a]25 °C is in favor of comparing temperature rising

3.3 *Decrease LED Current*

This way to reduce heat of IC is by reducing power dissipation of internal MOSFET. The power dissipation of internal MOSFET during ON period can be expressed by

$$\mathrm{P} = I_{\mathrm{D}}^{2} \cdot R_{DS(\mathrm{ON})}$$

where I_D is the drain current when MOSFET is ON, which is positive correlated with LED current.

After confirming light distribution meets regulation, the LED current can be reduced 25% from original 800 to 500 mA, and estimated power dissipation is expected to decrease 43.75%.

Measurement after reducing LED current is as follows:

Condition: 13.5 V, 25 °C	Temperature rise before change (°C)	Temperature rise after change (°C)	Improvement (°C)
Ts of IC	125	53	72

3.4 Decrease MOSFET Switching Loss

As Fig. 1 shows previously, "bootstrap circuit" is applied for internal MOSFET to boost GATE voltage (BST pin of MPQ2483) so that MOSFET is surely conducted during each period. When MOSFET is switching, the switching speed is determined by the charging time of $C3$, while the time constant can be gotten by $t = R6 * C3$. According to this equation, the higher $R6$ is, the longer time of GATE voltage rise, and the MOSFET switching time is longer.

This theory can be explained by the MOSFET waveform in Fig. 3. The switching loss of MOSFET during switching phase is,

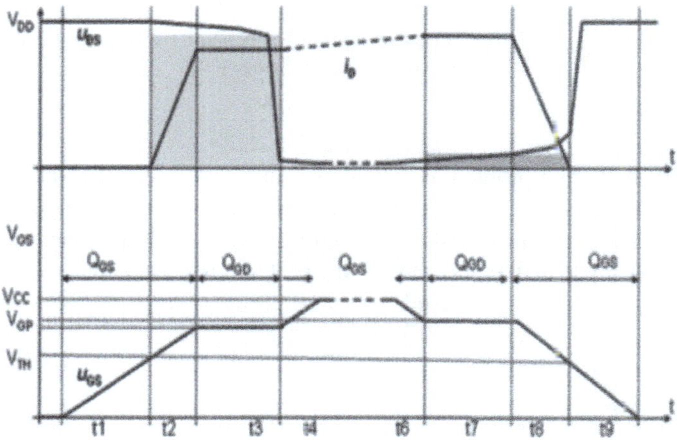

Fig. 3 Waveform of switching MOSFET

$$P = \int_{t2}^{t3} I_D \cdot V_{DS}$$

We can conclude from this equation, when $t2 + t3$ is longer, the switching time is longer, and power loss is more.

Measurement after decrease $R6$ from 100 to 10 is as follows:

Condition: 13.5 V, 25 °C	Temperature rise before change (°C)	Temperature rise after change (°C)	Improvement (°C)
Ts of IC	125	89.6	35.4

It is noted that more noise is generated when accelerating MOSFET switching, so modifying input filter and confirming noise level meet EMC standard is necessary.

3.5 *Results*

Finally, the DRL function light distribution meets regulation with 20% margin and noise test meets vehicle manufacturer EMC standard, after solution 3.2, 3.3 and 3.4 are adopted simultaneously. DRL lights continuously at 105 °C without flickering, and there's much margin to trigger OTP.

The measurement at 13.5 V, 105 °C[1] is as follows:

Condition: 13.5 V, 25 °C	Temperature rise before change (°C)	Temperature rise after change (°C)	Improvement(°C)
Ts of IC	>150	128	>22

The DRL flickering problem is solved even at worst case and without additional cost. Three solutions are combined and all applied in this project, but it is not mandatory to apply simultaneously. All of the solutions can be selected one or more alternatively according to actual project until the problem is solved, meanwhile the other performance of lamp must be cared.

[1] It is test condition in standard.

4 Conclusion

DRL LED driver thermal problem is analyzed and solved in this paper, and following items are explained.

(1) Active cooling. It is high efficient, but additional cost, space and risk is needed. This method is suitable at the beginning of the lamp design, and usually no enough space to install the fan after design freeze.
(2) Passive cooling. Heat exchange happens between PCB copper and air, so it is not as efficient as active cooling. This cooling effect is limited but without increasing cost. This common solution is used in this project.
(3) Reducing power dissipation of IC. The trade-off—between reducing output power and meeting light distribution regulation, between reducing noise level and reducing power dissipation of IC, realizes the optimization of performance and cost.

The thermal problem of vehicle lamp DRL LED driver is investigated based on the DRL flickering issue. The root cause of flickering is analyzed and theoretical way to solve thermal problem is provided. Both of reducing thermal resistance and power dissipation are applied, and low-cost solutions are validated by measurement. This thermal solution is effective and can also be applied on other DC-DC headlamp driver projects, providing ideas to solve thermal problems.

References

1. Yan K-S, Niu P-J, Fu X-S (2008) Light Source of automobile headlamp and its trend. OME Inf 11:36–40
2. Fu J, Li W, Zhou H, Yu H (2010) Applications of LED technology in auto lighting. Chin Illum Eng J 21(3):64–69
3. MPQ2483_r1.05. Monolithic Power Systems, Inc
4. Jing Wang (2014) Implementation and heat dissipation mechanism research of LED headlamp passing beam system[D]. Jiangsu University, Zhenjiang, pp 54–101
5. Dongchun Liu, Yanmei Chen, Jinzhou Lin (2017) Optimization of heatsink for some automotive LED lights[J]. Tianjin Sci Technol 44(11):77–85

Research and Verification of Cooperative Regenerative Braking Function Based on Electrical Brake Booster System

Yanjing Wang, Mingwei Xie and Dihua Yi

Abstract The regenerative braking technology is an effective means to increase the mileage of electric vehicles and reduce the energy consumption per hundred kilometers and is a key issue in the research of electric vehicles. With the development trend of intelligent and electrified vehicles, electrical brake booster system is undoubtedly the best choice for automotive boost brake. This article proposes a control scheme of cooperative regenerative braking function based on a domestic home appliance electrical brake booster system. The program refers to the functional framework and communication interface of the VDA360 standard. The main controller is a vehicle control unit. The motor system and the electrical brake booster system act as actuators. At the same time, based on the wheel hub test and real vehicle road test, the control scheme was verified by multiple conditions. The results show that the cooperative regenerative braking function greatly improves the vehicle energy recovery rate, and the development of this function lays the groundwork and foundation for the subsequent development of intelligent driving functions and active braking functions.

Keywords Electric boost brake system · Cooperative regenerative braking function · Vehicle energy recovery rate

Preface

As the problem of energy shortage becomes more and more prominent, all countries in the world regard new energy vehicles as the focus of future strategies, and pure electric vehicles are the key projects of national research. In recent two years, with the optimization of battery capacity, energy density, and other performance, the cruising

---

First author: Yanjing Wang (1984), female, postgraduate, engaged in electric vehicle control strategy development.

Fund Project: Beijing Science and Technology Commission Project—Research on Chassis and Vehicle Industrialization of Distributed Electric Vehicle.

Y. Wang (✉) · M. Xie · D. Yi
Beijing Electric Vehicle Co., Ltd, Beijing 100176, China
e-mail: wangyanjing@bjev.com.cn

China SAE (ed.), *Proceedings of China SAE Congress 2018: Selected Papers*,
Lecture Notes in Electrical Engineering 574,
https://doi.org/10.1007/978-981-13-9718-9_71

range of pure electric vehicles has been improved. The cruising range of domestic pure electric vehicles can reach 400–500 km, which greatly improves the "mileage anxiety" problem. However, as the national policy for new energy vehicles' subsidy is increasingly demanding the energy consumption per 100 km, domestic host factories are increasingly researching brake energy recovery.

With the transformation of technology, intelligence and electrification are the main trends in the future development of automobiles. The so-called intelligence is to reduce the dependence on the driver, more through the control system to choose the optimal solution. The so-called electrification requires the use of electric energy instead of traditional fossil energy to achieve energy saving and emission reduction. The braking system has always been a key research object of the automobile. The current trend of intelligence and electrification also puts higher requirements and challenges on the braking system. At the same time, with the development of society, the driver demand for braking systems is also increasing, not only requiring it to provide reliable braking safety but also requiring a more comfortable driving experience and a smarter driving mode. In this context, the electrical brake booster system is undoubtedly the best choice for vehicle boost braking [1].

In order to improve the energy recovery rate, a series braking energy recovery (also known as cooperative regenerative braking) control scheme based on the electrical brake booster system is proposed in this paper and is verified based on the wheel hub test and the real vehicle road test.

1 Boost Braking System

1.1 Vacuum Boost Braking System in Traditional Vehicle

The schematic diagram of vacuum boost braking system in the conventional vehicle is shown in Fig. 1 [2].

The vacuum booster is generally located between the brake master cylinder and the brake pedal. For easy installation, it is usually combined with the brake master cylinder in an assembly.

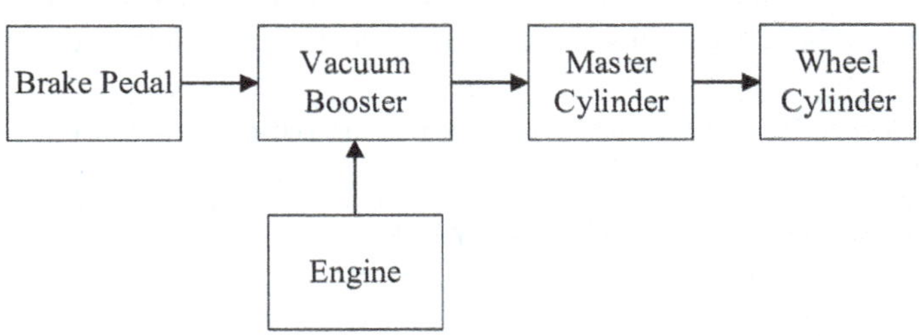

Fig. 1 Schematic diagram of vacuum boost braking system in the traditional vehicle

The vacuum booster is mainly composed of a vacuum servo chamber and a control valve. The vacuum servo chamber is composed of a front shell and a rear shell. A servo chamber diaphragm is clamped between the two chambers, and the servo chamber is divided into a normal pressure chamber and a variable pressure chamber. The vacuum in the normal pressure chamber is usually (60–80) kPa. The vacuum booster utilizes the pressure difference generated by the intake manifold of the automobile engine in the two chambers as a power source for increasing the braking force, and the pressure difference between the two chambers determines the vacuum booster size. When the maximum boosting effect is reached, that is, when the pressure of the variable chamber is the same as the outside air pressure, the assisting effect will become a constant value. The pedal force and the servo force generated by the vacuum booster are directly applied to the piston push rod of the brake master cylinder, thereby increasing the output pressure of the brake master cylinder

1.2 Vacuum Boost Braking System in Electric Vehicle

The schematic diagram of vacuum boost braking system in the electric vehicle is shown in Fig. 2.

In the traditional vehicles, the vacuum of the vacuum booster is provided by the intake manifold of the engine. The power source of the electric vehicle is converted from a conventional engine to an electric motor, and the brake system does not have the original vacuum boost power source. Therefore, in order to ensure the safety of the vehicle and the comfort of driving, it is necessary to add a vacuum power source capable of providing sufficient pressure, that is, a vacuum pump. When the electric vehicle is started, or when driving, the vacuum sensor will detect and monitor the vacuum inside the vacuum tank. The system directly controls the operation and stop of the vacuum pump motor according to the driving state of the vehicle and the

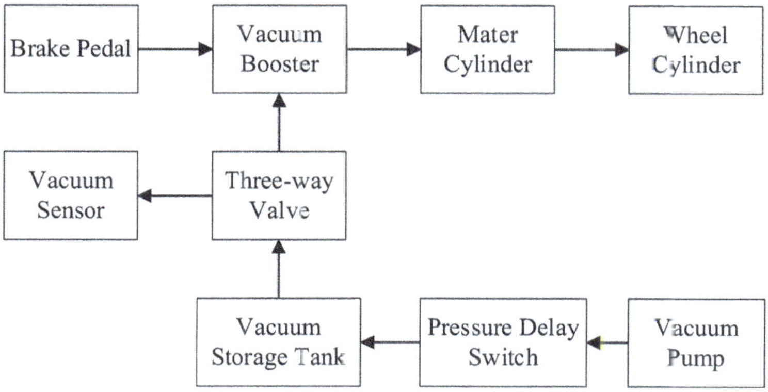

Fig. 2 Schematic diagram of vacuum boost braking system in the electric vehicle

vacuum range within the vacuum tank. If the vacuum is below the lower threshold set by the system, the vacuum pump starts to work. If the vacuum reaches the upper threshold set by the system, the system will turn off the vacuum pump.

1.3 Electrical Boost Braking System in Electric Vehicle

The schematic diagram of electrical boost braking system in the electric vehicle is shown in Fig. 3.

Different from the traditional vacuum booster, the electrical boost braking system does not need any vacuum source. It only relies on electric motor for servo boosting, which can achieve the same pedal feeling as the traditional vacuum boost braking system, so that the driver can better adapt. At the same time, in order to adapt to the different needs of different drivers, the electrical boost braking system can change the boosting force by adjusting the control parameters and thus obtain different pedal feelings [1].

In addition to the basic boost function described above, the electrical boost braking system also achieves braking energy recovery function. The braking system needs to work in conjunction with the regenerative braking system to ensure both braking energy recovery efficiency and a good brake pedal feeling. The traditional vacuum boosters fail to decouple the pedal from the master cylinder, and the brake pedal feeling felt worse. The electric boost braking system can achieve partial decoupling or complete decoupling between the brake pedal and the master cylinder, ensuring that the vehicle has a good pedal feeling while achieving energy recovery.

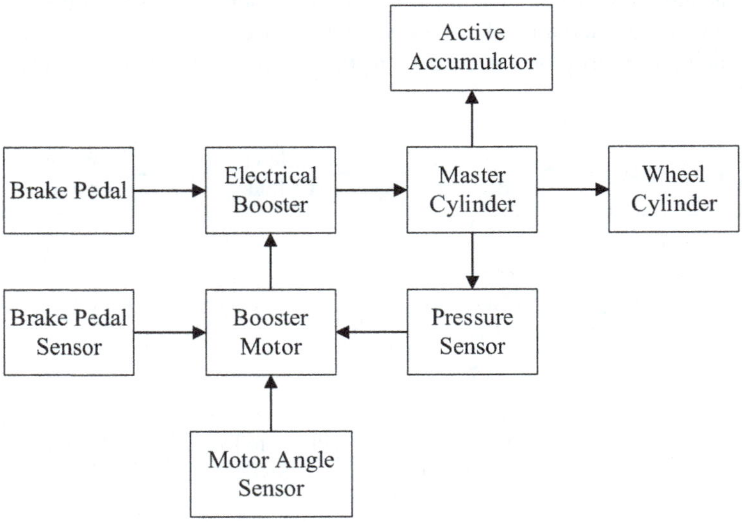

Fig. 3 Schematic diagram of electrical boost braking system in the electric vehicle

In addition, the electric boost braking system has the ability to quickly and comfortably build pressure, which can meet the needs of emergency braking, adaptive cruise control, and other auxiliary driving functions. At the same time, the electric boost braking system can still provide a certain braking force by manpower in the event of failure.

2 Braking Energy Recovery

Braking energy recovery is that the electric motor is in a power-generating state, and the generated electric energy is returned to the power battery during the braking process of the vehicle. At the same time, the feedback torque generated by the motor is applied to the driving shaft to brake the entire vehicle. From the perspective of vehicle energy, it has achieved the purpose of extending the cruising range and improving energy utilization efficiency [3]. However, the pure electric vehicle is faced with the problem that the electric braking force cannot complete the braking task independently. Therefore, the vehicle must be equipped with a mechanical brake system to supplement the mechanical braking force when needed and to complete the braking task together with the electric brake force to ensure braking efficiency and braking stability.

For the electromechanical composite brake system, there are two main contradictions: First, the brake force distribution problem of the front and rear axles, which directly affects the braking stability, is the primary problem to ensure the safety of the vehicle; another problem is the distribution between the electrical brake force and the mechanical brake force applied on the drive axle, and the ratio of the two kinds of brake force affects the braking energy recovery. Therefore, in order to ensure the safe driving of the vehicle while maximizing the recovery of braking energy, it is necessary to properly coordinate the braking force ratio. At present, the composite brake system is mainly divided into two types of parallel and series [4].

2.1 *Parallel Braking Energy Recovery*

The parallel braking energy recovery system completely retains the traditional mechanical brake system but simply superimposes the regenerative braking force on the drive wheels. The mechanical brake system and the regenerative braking system are completely independent. When the electric braking force is big, the deceleration of the whole vehicle is large, and when the electric braking force is small, the deceleration of the whole vehicle is small, that is to say, the deceleration of the whole vehicle directly depends on the electric braking force. The control principle of the parallel brake system is shown in Fig. 4. The disadvantage of parallel braking energy recovery is that the electric motor power cannot be fully utilized, and the braking energy recovery efficiency is low. However, due to the simple structure and low cost,

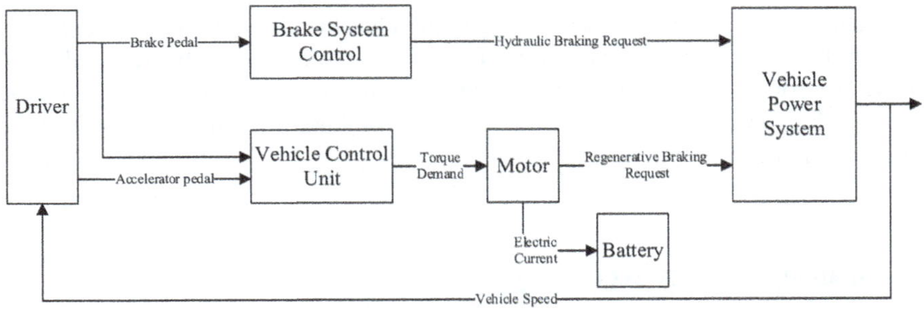

Fig. 4 Schematic diagram of parallel braking energy recovery

the braking energy recovery can be realized with control based on the hardware configuration of the original car, so the current application is extensive.

2.2 *Series Braking Energy Recovery*

Under the braking condition, the series brake distribution controller determines the driver's demand braking force according to the depth of the brake pedal and then determines the hydraulic pressure request and the regeneration torque request according to the electric brake ability sent by the vehicle controller. When the electric braking force generated by the motor is greater than or equal to the driver's required braking force, the driver's required braking force is all provided by the electric braking force; when the electric braking force generated by the motor is less than the driver's required braking force, the mechanical braking system is controlled to complete hydraulic compensation. The control principle of the series brake system is shown in Fig. 5.

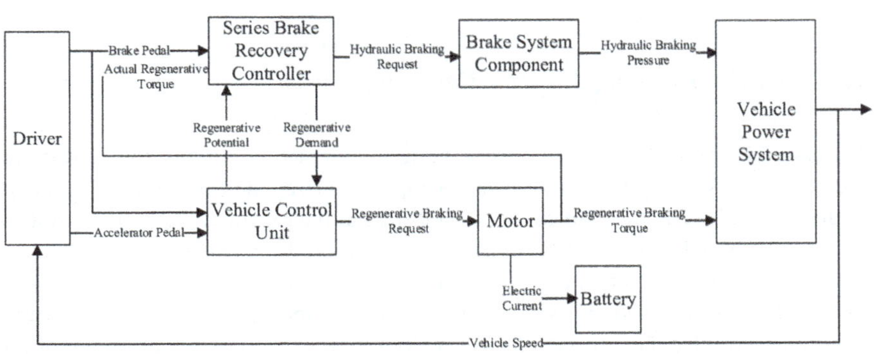

Fig. 5 Schematic diagram of series braking energy recovery

Therefore, the series braking system by controlling the distribution of the required braking force has the following advantages: the electric braking force generated by the motor can be fully utilized, and the energy utilization rate can be improved, and a consistent deceleration can be maintained during the braking process, without that the size of the electric braking force affects the deceleration of the vehicle. However, a highly integrated controller and new hydraulic brake system are required to achieve high precision control.

3 Cooperative Regenerative Braking Function Based on the Electrical Brake Booster

3.1 Function Architecture According to VDA Standard

The VDA 360 standard is a non-binding recommendation of the German Association of the Automotive Industry, which includes the definition and description for the communication interface between an electrical brake booster (eBooster) and an ESC control unit. The functional scope based on the combination of the eBooster and ESC includes basic boost function, cooperative regenerative braking function, external brake request function, and related HMI display function.

Based on the functional architecture and communication interface of the VDA360 standard [5], this paper proposes a new functional architecture which is shown in Fig. 6, that is "vehicle control unit (VCU) master cooperative regenerative braking function, motor control unit (MCU) as the actuator of regenerative deceleration, electrical brake booster (eBooster) as the actuator of hydraulic pressure and the controller for the pedal force compensation". The specific function descriptions are described as follows:

(1) Driver brake request function (DBR function)

In a conventional brake system, the driver brake request is clearly determined via the master cylinder pressure. But in a cooperative regenerative brake system, the master cylinder pressure is no longer equivalent to the driver brake request due to the coordinated control of the brake fluid.

When the driver depresses the brake pedal, the movement of the brake pedal is recorded by the integrated sensors in the electrical brake booster and output the signal sOutputRodDriver. Therefore, the vehicle control unit (VCU) calculates a virtual master cylinder pressure (pMCVirtual) using the brake pedal stroke to pressure characteristic (PV characteristic) and output the total brake request (DriverBrakeRequest).

(2) Brake torque coordinator function (BTC function)

In order to coordinate the brake request (DriverBrakeRequest), the relevant information which the vehicle control unit (VCU) should consider includes the electric motor potential of the regenerative deceleration (RecuBrakeTorqueCap), the

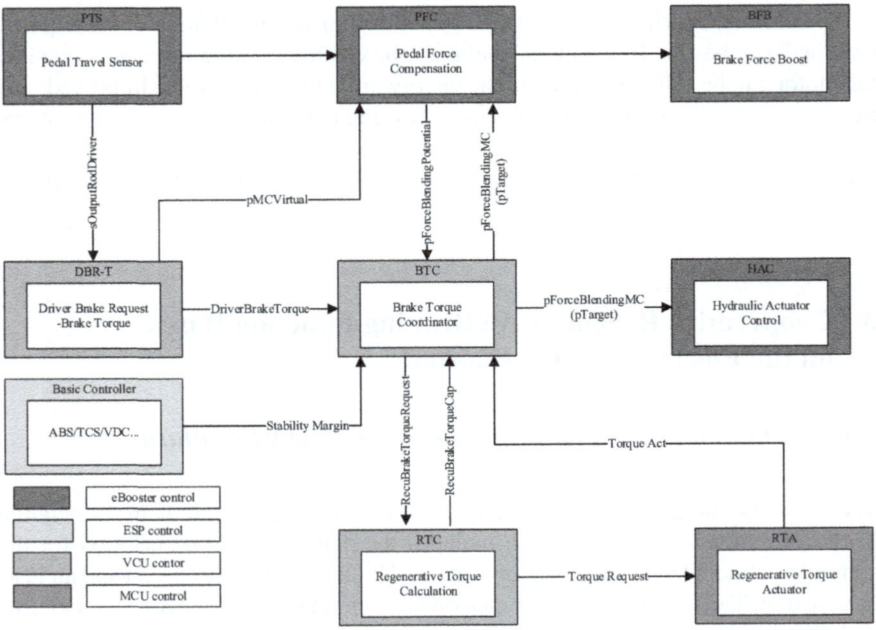

Fig. 6 Functional architecture of the system

eBooster potential for pedal force compensation (pForceBlendingPotential) and the stability conditions of the vehicle. The BTC function determines the target brake torque for the regenerative deceleration to the electric motor. For the residual brake request, BTC function distributes the pressure build-up task to HAC function module. The output signals of this function are the brake torque (RecuBrakeTorqueRequest) and the target pressure (pForceBlendingMC).

(3) Pedal force compensation function (PFC function)

Due to the coordinated cooperation between the hydraulic brakes and the regenerative brake via electric motor, the actual master cylinder pressure in the brake system will no longer matches the target pressure compared with purely hydraulic braking events or the target pressure requested by the driver. To achieve a consistent pedal feeling, the correlation between the brake action and the pedal force must be taken into account. Therefore, the electrical brake booster (eBooster) needs to balance out the difference between the actual pressure and the virtual pressure. This function is known as pedal force compensation.

(4) Hydraulic actuator control function (HAC function)

After the vehicle control unit(VCU) completes the brake request distribution, the master cylinder target pressure signal (pForceBlendingMC) is sent to the HAC function module of the electrical brake booster (eBooster) for hydraulic control, that is, carry out the closed loop control of pressure according to the target pressure and the actual pressure.

(5) Regenerative torque control function (RTC function)

The vehicle control unit (VCU) finally calculates the request torque of the motor taking into account the coast regenerative torque and the brake regenerative torque calculated by the BTC function.

The potential for the regenerative torque depends on many factors, including the vehicle speed, battery maximum charge power, motor temperature limit, the potential for pedal force compensation, and so on, and this maximum regenerative torque is sent to the BTC function.

3.2 Function Assignment and Signal List

Based on Fig. 6, the electrical brake booster (eBooster) realizes three sub-functions of basic boost function, pedal force compensation function, and target hydraulic pressure control function; the vehicle control unit (VCU) realizes three sub-functions of brake request calculation function, regenerative potential calculation function, and brake request coordination function; the motor control unit (MCU) realizes two sub-functions of request torque execution function and working state monitoring function. The functional block diagram and the involved communication interfaces are shown in Fig. 7. The signal list is shown in Table 1.

3.3 Braking Force Distribution Principle

In the distribution of braking force, the following principles are mainly considered:

(1) In order to prevent dangerous side slip from the rear axle lock, the actual front and rear brake force distribution lines (β line) of the vehicle brake system should always be below the ideal brake force distribution line (I line);
(2) For as much energy recovery as possible, the design and control will follow the criteria for the total braking force to be distributed on the front wheels as much as possible under the conditions of the brake regulations (ECE regulations). That is, the total brake force distribution will follow the maximum front axle braking force curve (minimum rear braking force) as defined by the ECE regulations.
(3) f The f line group is the brake force relationship curves between the front and rear axles when the front wheels are not locked and the rear wheels are locked on various φ values (φ is road adhesion coefficient). Therefore, the f line when the φ value is equal to φ_0 (synchronous adhesion coefficient) is selected as the boundary.

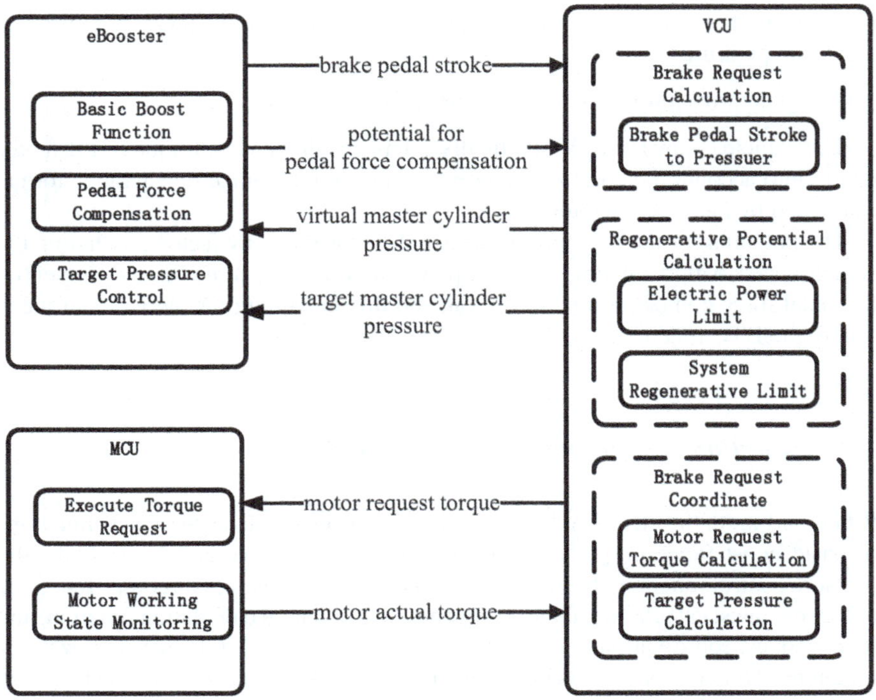

Fig. 7 Function assignment of ECU

Table 1 Signal list of ECU

No	Signal name	Signal description	Send ECU	Receive ECU
1	sOutputRodDriver	brake pedal stroke	eBooster	VCU
2	pForceBlendingPotential	potential for pedal force compensation	eBooster	VCU
3	pMcVirtual	virtual master cylinder pressure	VCU	eBooster
4	pForceBlendingMC	target master cylinder pressure	VCU	eBooster
5	TorqueRequest	motor request torque	VCU	MCU
6	TorqueAct	motor actual torque	MCU	VCU

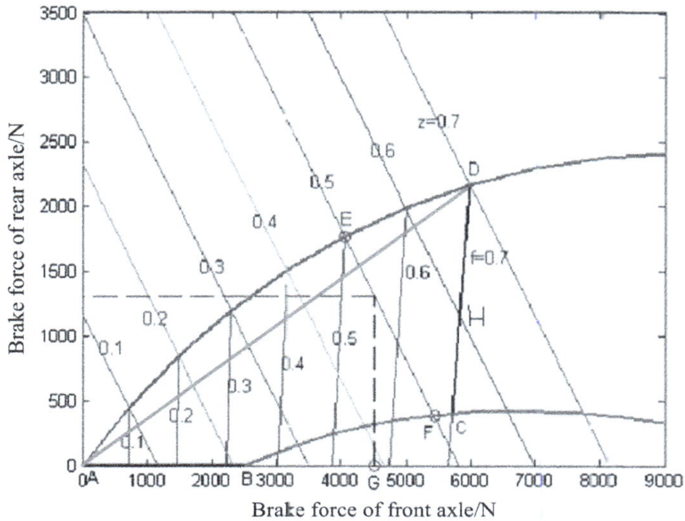

Fig. 8 Schematic diagram of braking force distribution line and related intersection points

According to the above principle, the brake force distribution constraint is set in the area of A-B-C-D-A, as shown in Fig. 8, where point *B* is the intersection of the ECE regulation line and the horizontal axis, point *C* is the intersection of the *f* Line and the ECE regulation line when the φ value is φ_0, and point *D* is the intersection of the ideal brake force distribution curve (*I* line) and the fixed brake force distribution line (β line). In the process of braking force distribution, two intersections are involved, that is, point *F* is the intersection of the total demand braking force and the ECE regulation line, and point *H* is the intersection of the *f* line and the total demand braking force when the φ value is φ_0.

The overall braking force distribution process is shown in Fig. 9 where the variables involved in Fig. 9 are defined as follows:

F_t Request brake force;
F_{t_B} Total brake force of *B* point;
F_{t_C} Total brake force of *C* point;
F_{t_D} Total brake force of *D* point;
F_{t_F} Electrical brake force of F point;
F_{t_H} Electrical brake force of *H* point;
$F_{\mathrm{recup\_max}}$ Maximum electrical brake force;
$F_{\mathrm{recup\_req}}$ Request electrical brake force.

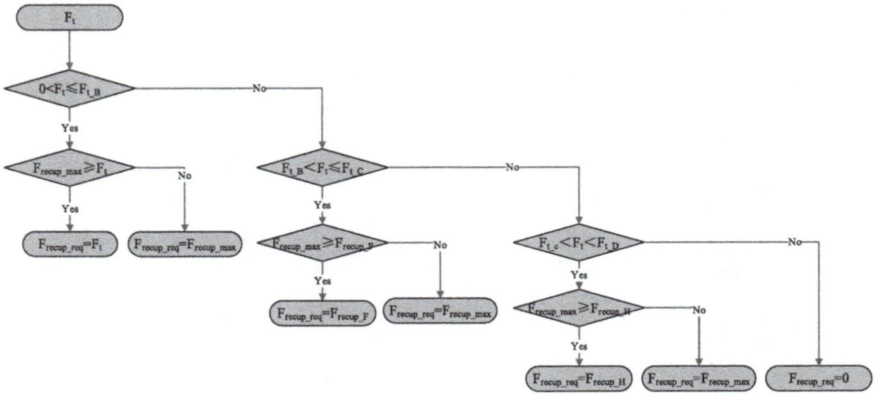

Fig. 9 Schematic diagram of motor demand braking force calculation process

4 Real Vehicle Test

Based on a *C*-class car which has been listed in the company, the electrical brake booster system and vehicle stability control system provided by domestic suppliers are modified, and wheel hub test and vehicle road test are carried out. The basic parameters of the test vehicle are as follows:

(1) Vehicle parameters: vehicle preparation quality is 1613 kg and transmission ratio is 7.817.
(2) Battery parameters: the available energy at room temperature is 42.8 kW.h, and the nominal voltage is 336 V.
(3) Motor parameters: peak power 90 kW, peak torque 260 N.m, peak speed 10,000 rpm.

4.1 Hub Test

Based on the wheel hub test, the recovery rate of cooperative regenerative braking function under NEDC cycle was tested. The formula of energy recovery rate was: battery recovered energy/battery discharged energy, and the energy recovery rate was increased by more than 5% compared with the original vehicle (without electrical brake booster system, using parallel brake recovery system). The data for a single NEDC cycle is shown in Fig. 10, the ECE segment data in the NEDC cycle is shown in Fig. 11, and the EUDC segment data in the NEDC cycle is shown in Fig. 12.

Through Figs. 10, 11, and 12, it can be seen that when the vehicle speed is greater than 8 km/h, the deceleration is completely provided by the motor recovery torque; when the vehicle speed is less than 8 km/h, the motor torque is gradually withdrawn, and the hydraulic pressure gradually intervenes to ensure the deceleration process

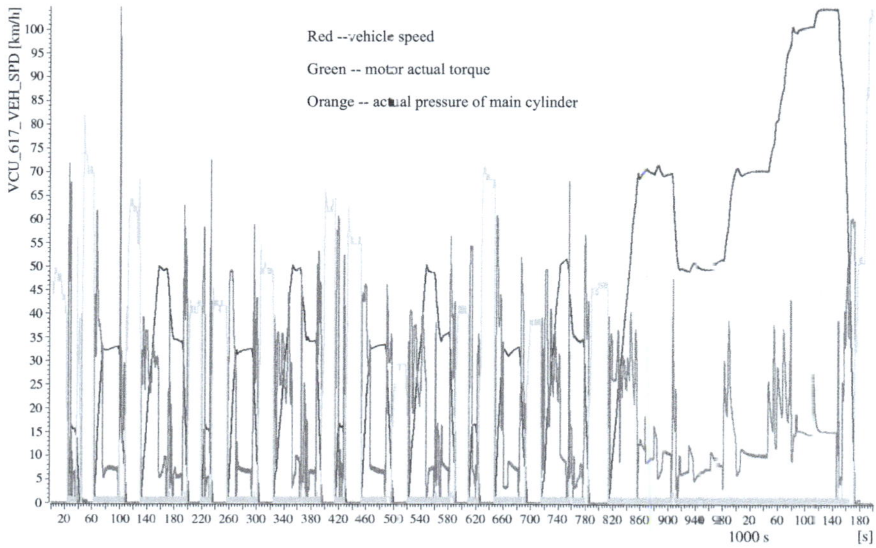

Fig. 10 Data of NEDC cycle (Color figure online)

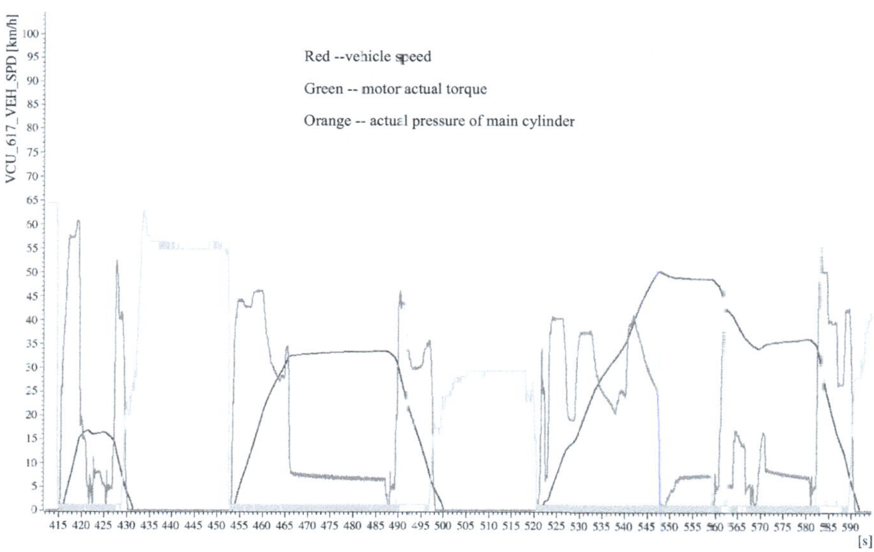

Fig. 11 Data of ECE cycle (Color figure online)

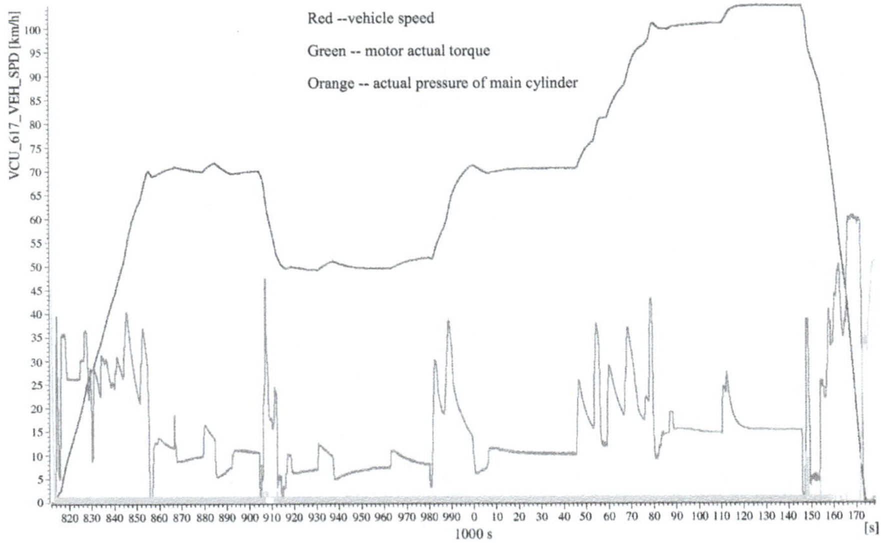

Fig. 12 Data of EUDC cycle (Color figure online)

smooth. After the vehicle is stopped, the motor recovery torque is zero and the vehicle is completely braked by hydraulic pressure. It can be concluded that the cooperative regenerative braking function based on the electrical brake booster system completely utilizes the motor recovery torque of the vehicle to achieve braking and recover all the energy.

4.2 *Road Test*

The results of the real vehicle road test are shown in Figs. 13 and 14. As can be seen from Fig. 13, under the small brake pedal stroke, the electric braking ability meets the driver's braking demand, so all the deceleration can be achieved by the electric braking; from Fig. 14, under the large brake pedal stroke, the electric braking ability cannot meet the driver's braking requirements, so the electric braking and hydraulic braking together to achieve deceleration.

5 Conclusion and Prospect

In this paper, a cooperative regenerative braking function scheme is proposed which is based on the electrical brake booster system and controlled by vehicle control unit (VCU) and has been verified based on the wheel hub test and the real vehicle road

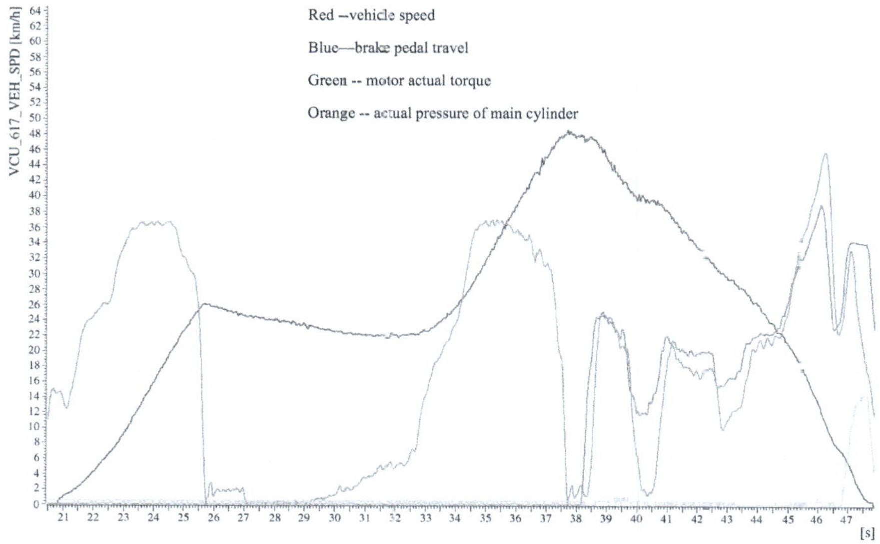

Fig. 13 Data under the small brake pedal stroke (Color figure online)

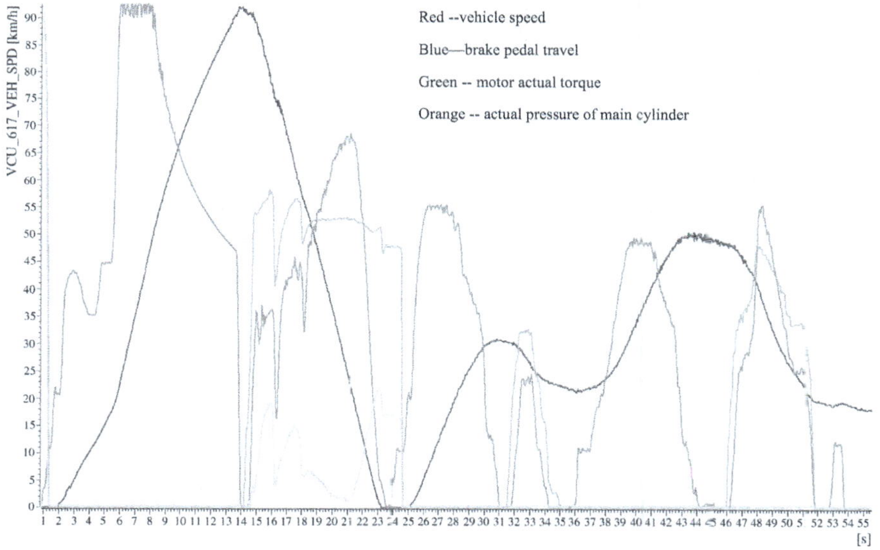

Fig. 14 Data under the big brake pedal stroke (Color figure online)

test. The test results show that when the electric braking ability meets the driving brake demand, the driver's required braking force is all provided by the electric braking force; when the electric braking ability cannot meet the driver's braking demand, the hydraulic brake is replenished in real time to ensure the smoothness of the deceleration process; and the energy recovery rate of the vehicle is greatly improved, which in turn improves the economy of the vehicle. At the same time, in the process of braking force distribution, combined with the electric braking ability and the driver braking demand, the electric braking demand and the hydraulic demand are adjusted in real time, and the electrical brake booster system adjusts the boosting force in real time to ensure the consistency of the pedal feeling. The real vehicle verification effect is good.

The focus of this paper is to realize the development of series braking recovery function based on the characteristics of electric boost braking system. For the first time, the series braking energy recovery function is developed and applied, and the control of series braking energy recovery function is controlled by the vehicle control unit (VCU) for the first time. In the subsequent research and development, it is still necessary to deeply optimize and improve the braking force distribution principle and electro-hydraulic coordination process of the series brake recovery function. At the same time, based on the research of the series braking recovery function and based on the electric boost braking system hardware, it lays the foundation for the development of distributed drive control technology, intelligent driving control technology, brake redundancy, or backup technology.

References

1. Qian Lei (2017) Research on electric power braking system. Jilin University, Master's Dissertation, p 11
2. Sun Dongrui (2016) Research on matching technology of vacuum assisted braking system for electric vehicle. Jilin University, Master's Dissertation, p 14
3. Wen Jing (2017) Research on the longitudinal dynamics braking control of the Pure electric vehicles based on the intelligent driving. Hefei University of Technology, Master's Dissertation, p 20
4. Fang Lei (2014) Research on the regenerative braking control strategy of electric vehicle. Beijing Information S&T University, Master's Dissertation, p 17
5. Recommendation for the implementation of a communication interface between an electrical brake booster and an ESC control unit, p 42

Dual-Redundancy Steering by Wire Control System with High Safety

Junnan Mi, Tong Wang, Zhikai Cai, Xi Chen and Xiaomin Lian

Abstract Steering by wire control system cannot be applied to mass production cars because of the low safety. A structure of dual-redundancy steering by wire control system with high safety and redundancy management method are proposed in this paper. By applying the redundancy theory to the design of steering by wire control system and combining the redundancy management method, when any single random electrical fault occurs, no system failure can be achieved. The whole system has a working mode of "fault-safe-forewarning," thus the safety of steering by wire control system is enhanced, increasing the possibility of applying the system to mass production cars.

Keywords SBW · Safety · Redundancy · Control · Fault-tolerance

1 Introduction

Steering by wire control system (SBW) becomes a hot topic in research for its ability of enhancing vehicle controllability and stability, passive safety, riding comfort and improving driving environment. It is a critical technology for autonomous driving as well. SBW safety is of great importance because lives of passengers are at risk once SBW fails. According to ISO 26262 Road Vehicles—Functional Safety [1], vehicle steering system is at ASIL-D safety level, which is the highest level for safety requirement. Conventional SBW cannot meet the requirement and consequently cannot be applied to mass production cars.

Methods for improving SBW safety have been studied by researchers at home and abroad. Yao [2] in Visteon Corporation, Zheng [3] in Purdue University, Chaaban [4] in France, Onoda [5] in TOYOTA and Zong [6] in Jilin University have conducted research on applying redundancy methods to system design and proposed fault-tolerant control methods. Although these can improve the safety of SBW at individual

J. Mi · T. Wang · Z. Cai · X. Chen · X. Lian (✉)
Tsinghua University, Beijing, China
e-mail: lianxm@tsinghua.edu.cn

China SAE (ed.), *Proceedings of China SAE Congress 2018: Selected Papers*, Lecture Notes in Electrical Engineering 574,
https://doi.org/10.1007/978-981-13-9718-9_72

component level, at the whole system level, the system would still fail once a fault occurs in other single-redundancy parts of the system.

A dual-redundancy steering by wire control system (DRSBW) is proposed in this paper. The system-level dual-redundancy design method and management method are proposed by applying the redundancy theory, which can make the system not-fail when any single random electrical fault occurs. The method can dramatically improve the SBW safety to satisfy the random hardware failure rate requirements of ASIL-D, making it possible to apply SBW to mass production cars.

2 Dual-Redundancy SBW

Dual-redundancy steering by wire control system (DRSBW) is of high safety SBW designed by applying the redundancy theory to the system. The system is designed with dual-redundancy method and redundancy management method is applied to the system management, which works in "fault-safe-forewarning" mode.

2.1 Determination of System Redundancy Number

The redundancy number of DRSBW is 2, which is determined by analysis of comparing conventional SBW with the fly-by-wire system of airplane.

The redundancy number of conventional SBW is 1, in which a single electrical fault may cause the system to fail. Consequently, the system redundancy number should at least be increased to 2. Fly-by-wire system has been applied to most airplanes for many years, and its redundancy number is 3 or 4 verified by the huge amount of aviation data, of which the safety is extremely high.

The fly-by-wire system must keep working in dangerous condition during flight, as once the system failure occurs, the survival probability of passengers is extremely low. In contrast, when a SBW failure occurs, cautious driving, low-speed driving, maintenance nearby or waiting for rescue can be taken to avoid severe safety accidents. As a result, the redundancy number of SBW can be lower than that of fly-by-wire system, built into a dual-redundancy system.

2.2 System Border and Components

The structure of DRSBW is shown in Fig. 1.

The full-line blocks in Fig. 1 are the components of DRSBW. The dual-redundancy steering wheel by wire subsystem (DRSWBW) is connected to steering wheel to provide feel of steering to driver. The dual-redundancy steering gear by wire subsystem (DRSGBW) drives the steering gear to steer the front wheels. The dual-redundancy

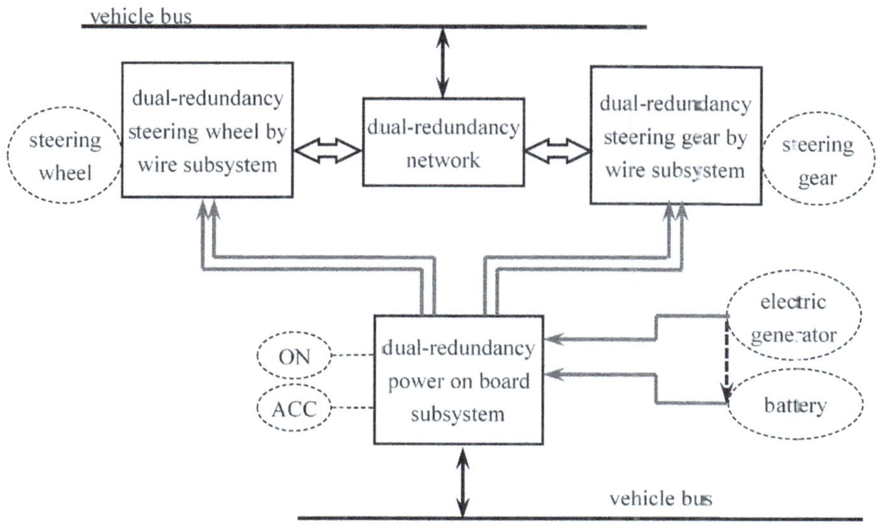

Fig. 1 Structure of DRSBW

network (DRN) is applied to dual-redundancy communication between the two subsystems, which are powered by dual-redundancy power on board subsystem (DRP). The DRP is controlled by both ON and ACC switches, powered by electric generator and battery simultaneously and communicates with vehicle bus.

2.3 Analysis of System Safety

The system safety is analyzed to verify whether it meets the safety requirements. The system fault tree is shown in Fig. 2 [9].

As is shown in Fig. 2, Fussell-Vesely theory is used to find out the minimum cut sets of the fault tree. The failure rate of top event (system failure) is calculated by failure rates of bottom events (components failure). According to calculation results, the probability of system random hardware failure is 2.3×10^{-9}/h, lower than 1×10^{-8}/h that is required in ISO26262. The system safety requirement is satisfied.

2.4 System Redundancy Management

The random single electrical fault no-failure function can be realized through system redundancy management after the system structure shown in Fig. 1 is proposed.

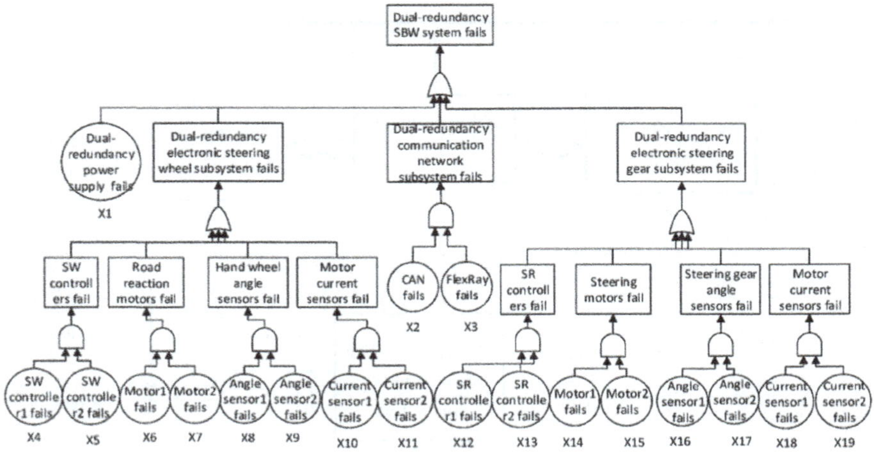

Fig. 2 System fault tree

The system redundancy management consists of four parts, which are redundant data synchronization, redundant data transmission and arbitration, fault diagnosis and forewarning, and system order-reduce and reconstruction. They will be discussed in subsystem sections.

3 Dual-Redundancy Steering Gear by Wire Subsystem

The dual-redundancy steering gear by wire subsystem (DRSGBW) [10, 11] powered by DRP in Fig. 1 can steer the front wheels and communicate with steering wheel subsystem.

3.1 Dual-Redundancy Steering Driver and Sensors

The structure of steering driver and sensors of DRSGBW are shown in Fig. 3.

The structure is shown in Fig. 3. M_{r1} and M_{r2} are dual-redundancy motors, coordinately driving gear and worm mechanism through a pair of gears. Dual-redundancy angle sensors S_{r1} and S_{r2} are put on the gear to measure the angle. The gear is linked to the steering gear to drive the rack and pinion mechanism.

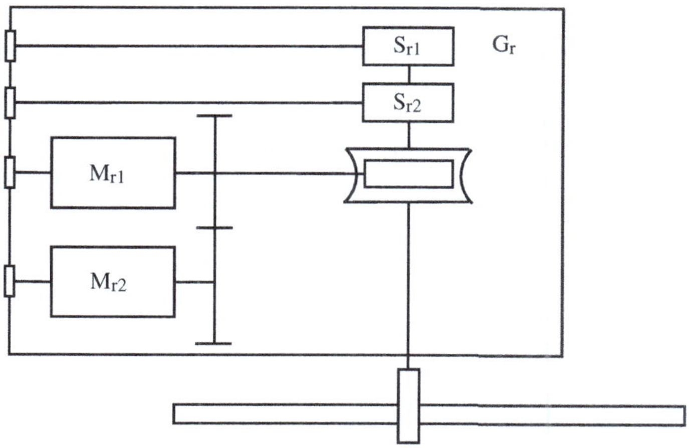

Fig. 3 Dual-redundancy steering driver and sensors

3.2 Dual-Redundancy Controller of DRSGBW

The structure of dual-redundancy controller of steering gear subsystem is shown in Fig. 4.

The full-line blocks in Fig. 4 are DRSGBW controller, in which C_{r1}, C_{r2} and C_{r3} are three control units connected to DRN. C_{r1} and C_{r2} are main control units,

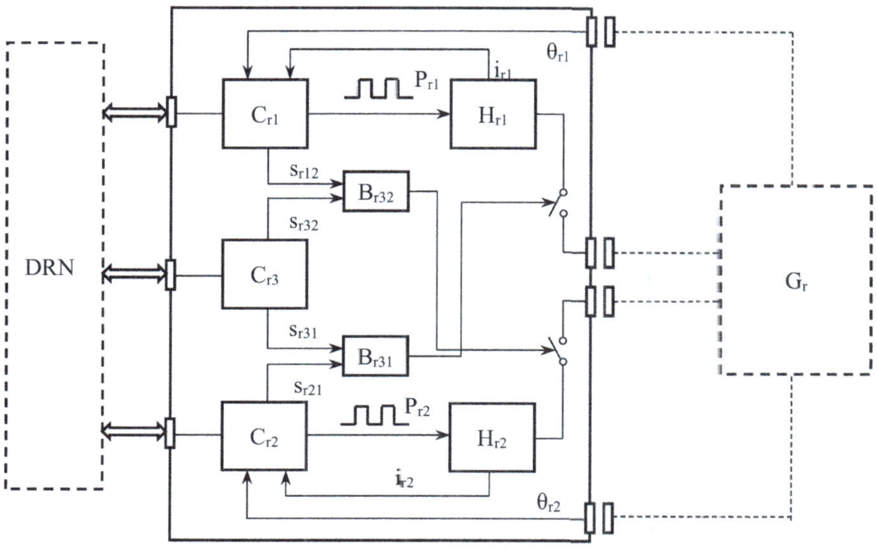

Fig. 4 Steering gear subsystem dual-redundancy controller

controlling the motors to steer front wheels; C_{r3} is a dual-redundancy management control unit. Mutual monitoring relations are established among the three control units to ensure only when C_{r1} and C_{r3} output cutting signals s_{r12}, s_{r32} simultaneously the logical circuit B_{r32} can cut the power of the second drive circuit. The condition is the same with C_{r2} and C_{r3}. The mutual monitoring in dual-redundancy controller is established according to the separation of the three powers to avoid mistaken power cutting conditions.

3.3 Dual-Redundancy Steering Control Algorithm

The control blocks for dual-redundancy steering are shown in Fig. 5.

As shown in Fig. 5, the two main control units receive steering wheel angle signals $\theta_{hi}(i = 1, 2)$ and vehicle speeds v_{ri} separately. The target angles θ_{2di} are calculated by steering ratio blocks K_{ri} according to Eq. (1).

$$\theta_{2di} = f(\theta_{hi}, v_{ri}) \tag{1}$$

In Eq. (1), target angles θ_{2di} are functions of steering wheel angles θ_{hi} and vehicle speeds v_{ri}.

Upper controllers C_{rui} receive target angles θ_{2di} and angles feedback θ_2, which is obtained by fusing the angle sensor signals θ_{2i}. The original target electric currents i_{rui} can be obtained through the PID algorithm in Eqs. (2)–(4).

$$\Delta\theta_{2i}(n) = \theta_{2di}(n) - \theta_2(n) \tag{2}$$

$$\begin{aligned}\Delta i_{rui}(n) = K_{prui}(\Delta\theta_{2i}(n) - \Delta\theta_{2i}(n-1)) + K_{irui}\Delta\theta_{2i}(n) \\ + K_{drui}(\Delta\theta_{2i}(n) - 2\Delta\theta_{2i}(n-1) + \Delta\theta_{2i}(n-2))\end{aligned} \tag{3}$$

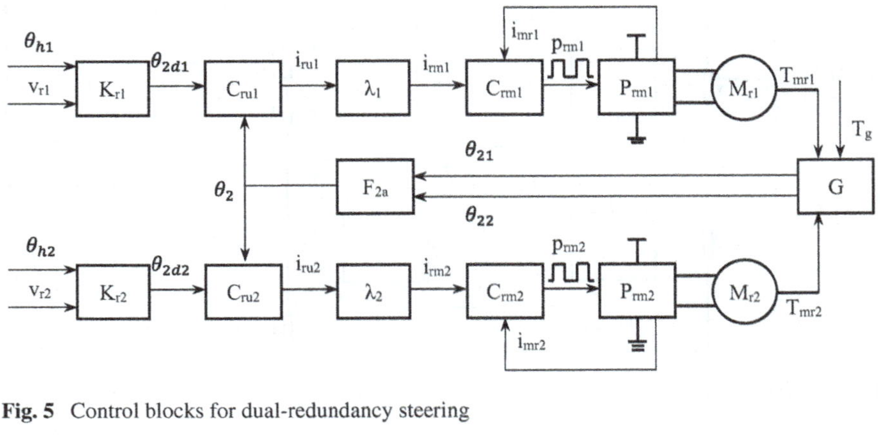

Fig. 5 Control blocks for dual-redundancy steering

$$i_{rui}(n) = i_{rui}(n-1) + \Delta i_{rui}(n) \quad (4)$$

In Eq. (2), the differences between target angles θ_{2di} and angles feedback θ_2 in step n are calculated. The increments of target electric current Δi_{rui} in step n are obtained according to the angle differences $\Delta\theta_{2i}$ of step $n-2$, $n-1$ and n in Eq. (3). The target electric currents i_{rui} of step n are calculated by adding the increments to the target electric current of step $n-1$ in Eq. (4).

The actual target electric currents i_{rmi} are obtained through the torque distribution coefficients λ_i, which is shown in Eq. (5).

$$i_{rmi}(n) = \lambda_i i_{rui}(n) \quad (5)$$

The values of λ_i are in Eq. (6).

$$\lambda_i = \begin{cases} 0 \text{ faults in both drive circuits or local driver circuit} \\ 1 \text{ no fault} \\ 2 \text{ fault in another drive circuit} \end{cases} \quad (6)$$

The function of the coefficients is to double the output torque of the motor in no-fault drive circuit when a fault occurs in one drive circuit, making the system switching process more stable.

The lower controllers calculate the duty cycle through PID algorithm according to the actual target electric currents and currents feedback. Then the controllers generate PWM pulses p_{rmi} to drive circuits P_{rmi} to control the motors to drive load G.

3.4 *Redundancy Management Algorithm for DRSGBW*

The redundancy management algorithm for DRSGBW consists of two parts, which are fault diagnosis, isolation and switch.

In system fault diagnosis, the three control units C_{r1}, C_{r2} and C_{r3} communicate with each other through DRN, estimating whether fault occurs according to angle and electric current message separately. The diagnosis algorithm is shown in Eq. (7).

$$E_{ji}(n) = \begin{cases} E_{Aji}(n) \cup E_{Cji}(n) \cup C_i(n) & (i,j) \in \{(1,2),(1,3),(2,1),(2,3)\} \\ C_i(n) & (i,j) \in \{(3,1),(3,2)\} \end{cases} \quad (7)$$

In Eq. (7), $E_{ji}(n)$ is the state value of control unit i estimated by control unit j. $E_{Aji}(n)$, $E_{Cji}(n)$ and $C_i(n)$ are the angle, electric current and control unit fault state values of unit i estimated by j. The state values are 0 when no fault occurs and 1 when a fault occurs.

The system isolation and switch is performed after the fault diagnosis. The process is as follows.

A system forewarning index $W_{ji}(n)$ is defined as Eq. (8).

$$\begin{aligned} & W_{ji}(n) = E_{ji}(n) \cap E_{ji}(n-1) \cap E_{ji}(n-2) \\ & (i, j) \in \{(1, 2), (1, 3), (2, 1), (2, 3), (3, 1), (3, 2)\} \end{aligned} \tag{8}$$

$W_{ji}(n)$ is the forewarning index that gives the forewarning when controller j identifies three continuous faults in drive circuit i.

A order-reduce fault state value $E_{Rji}(n)$ is defined in Eq. (9).

$$\begin{aligned} & E_{Rji}(n) = E_{Cji}(n) \cup C_i(n) \\ & (i, j) \in \{(1, 2), (1, 3), (2, 1), (2, 3)\} \end{aligned} \tag{9}$$

A fault isolation index $S_{ji}(n)$ satisfies Eq. (10).

$$\begin{aligned} & S_{ji}(n) = E_{Rji}(n) \cap E_{Rji}(n-1) \cap E_{Rji}(n-2) \\ & (i, j) \in \{(1, 2), (1, 3), (2, 1), (2, 3), (3, 1), (3, 2)\} \end{aligned} \tag{10}$$

$S_{ji}(n)$ is the isolation index that gives the isolating index when controller j identifies three continuous order-reduce faults in drive circuit i.

Drive circuit isolation is performed by two control units isolating the third one simultaneously, thus the fault isolation state value σ_i satisfies Eq. (11).

$$\sigma_i = \overline{S_{ji}(n)} \cup S_{3j}(n)(i, j) \in \{(1, 2), (2, 1)\} \tag{11}$$

The value σ_i is 0 or 1, representing the drive circuit i being isolated or not.

The system reduces order when a driver circuit is isolated with normal control unit outputting fault forewarning message to DRN.

4 Dual-Redundancy Steering Wheel by Wire Subsystem

The dual-redundancy steering wheel by wire subsystem (DRSWBW) in Fig. 1 provides steering feel and communicates with vehicle bus and DRSGBW, which is also powered by DRP.

4.1 Dual-Redundancy Steering Feel Driver and Sensors

The structures of steering feel driver and sensors in DRSWBW are shown in Fig. 6.

As the structures shown in Fig. 6, the steering wheel W is connected to mechanical angle limit L by a pair of Cardan joints, which limits the steering wheel angle between $\pm 540°$; M_{h1} and M_{h2} are dual-redundancy steering feel motors, generating steering

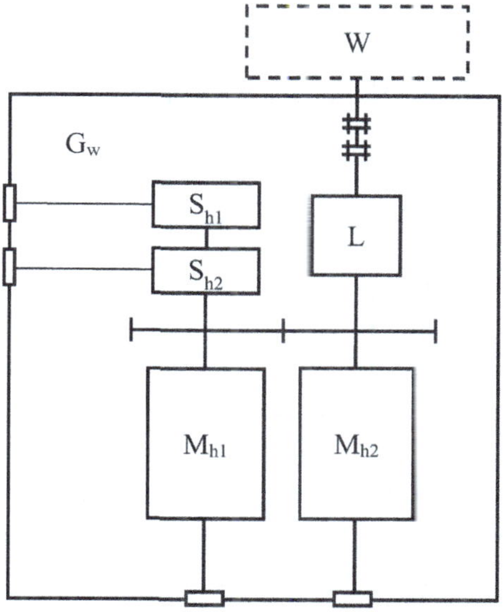

Fig. 6 Dual-redundancy steering feel driver and sensors

feel torques coordinately on steering column through a pair of gears. The steering wheel angle is measured by dual-redundancy angle sensors S_{h1} and S_{h2} on the output axle of M_{h1}.

4.2 Dual-Redundancy Controller of DRSWBW

The structure of dual-redundancy controller of DRSWBW is the same as that of DRSGBW shown in Fig. 4.

4.3 Dual-Redundancy Steering Feel Control Algorithm

The dual-redundancy steering feel control algorithm is shown in Fig. 7.

As is shown in Fig. 7, the steering wheel angle message θ_h, which is the fusion of angle sensor signals $\theta_{hi}(i = 1, 2)$, vehicle speeds v_{hi} and target electric currents of steering motors i_{2di} are received by the two main control units, which calculate the target electric currents of steering feel motors i_{1di} through the steering feel blocks K_{ri}. The steering feel blocks K_{ri} are as Eq. (12).

$$i_{1di} = f(\theta_{hi}, v_{hi}, i_{2di}) \tag{12}$$

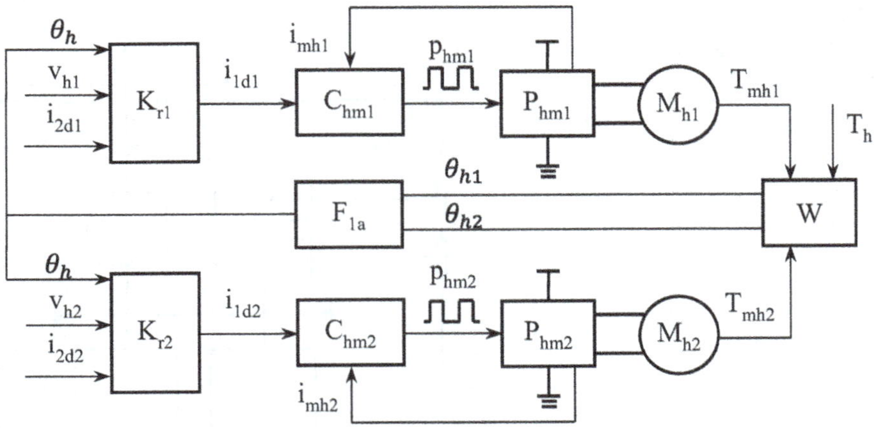

Fig. 7 Control blocks for dual-redundancy steering feel control

In Eq. (12), target electric currents i_{1di} are functions of steering wheel angles θ_{hi}, vehicle speeds v_{hi} and target electric currents of steering motors i_{2di}.

Lower controllers C_{hmi} control the steering feel motors M_{hi} according to target electric currents i_{1di} to generate steering feel torques on the steering wheel W.

4.4 Redundancy Management Algorithm for DRSWBW

The redundancy management algorithm for DRSWBW is the same as DRSGBW introduced in Sect. 2.4.

5 Dual-Redundancy Network

Dual-redundancy network (DRN) [12, 13] is a heterogeneous bus network used for the dual-redundancy communication of subsystems in DRSBW.

5.1 Framework of Dual-Redundancy Network

The framework of DRN is shown in Fig. 8.

In Fig. 8, W_A and W_B are steering feel motor controllers and W_E is the isolation controller in DRSWBW. G_A and G_B are steering motor controllers and G_E is the isolation controller in DRSGBW. Ga is the gateway controller. CAN bus and FlexRay

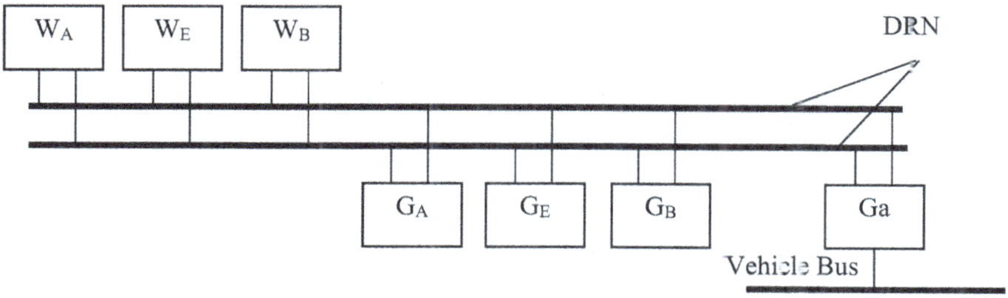

Fig. 8 Framework of DRN

bus combine to form the network, which is a heterogeneous communication network to avoid failures for the same causes and enhance the safety of the network.

During the data transmission, each of the two main controllers sends the same message to CAN and FlexRay once, respectively. For instance, to transmit message M, the messages sent by the two controllers are $M_{F1}(n)$ and $M_{C1}(n)$, $M_{F2}(n)$ and $M_{C2}(n)$, which are to be arbitrated by the message receiving controller.

5.2 Data Arbitration of DRN

The method of data arbitration is as follows.

Compare the 8-byte data of $M_{F1}(n)$, $M_{C1}(n)$, $M_{F2}(n)$ and $M_{C2}(n)$ in Eqs. (13)–(16)

$$M_1 = \sum_{n=0}^{7} |M_{F1}(n) - M_{C1}(n)| \quad (13)$$

$$M_2 = \sum_{n=0}^{7} |M_{F1}(n) - M_{F2}(n)| \quad (14)$$

$$M_3 = \sum_{n=0}^{7} |M_{F2}(n) - M_{C1}(n)| \quad (15)$$

$$M_4 = \sum_{n=0}^{7} |M_{C2}(n) - M_{C1}(n)| \quad (16)$$

According to the result of the comparison, the final message M_f is calculated by Eq. (17).

$$M_f = \begin{cases} M_{F1}(n), M_1 = 0 \\ M_{F2}(n), M_1 \neq 0 \wedge M_2 = 0 \\ M_{C1}(n), M_1 \neq 0 \wedge M_2 \neq 0 \wedge M_3 = 0 \\ M_{C2}(n), M_1 \neq 0 \wedge M_2 \neq 0 \wedge M_3 \neq 0 \wedge M_4 = 0 \\ \text{Safe}, \quad \text{else} \end{cases} \tag{17}$$

Data with high safety can be obtained by the message receiving controller from DRN through the data arbitration method above.

6 Dual-Redundancy Power on Board Subsystem

Dual-redundancy power on board subsystem (DRP) [12, 14] provides dual-redundancy power supplies to other subsystems and is powered by electric generator and battery simultaneously.

6.1 Structure of Power and Subsystem Switch

The structure of DRP is shown in Fig. 9.

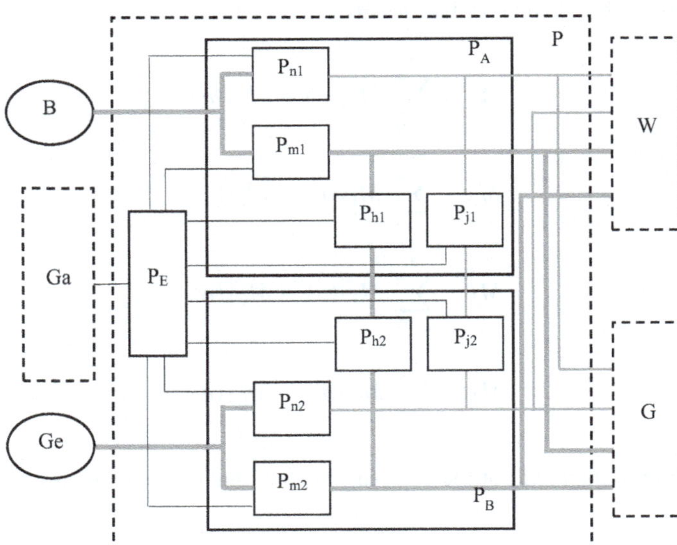

Fig. 9 Structure of DRP

In Fig. 9, B is battery, Ge is electric generator, W is DRSWBW and G represents DRSGBW. P_A and P_B are the first and second power supply sections of DRP, which are exactly the same and offer power simultaneously. The two sections are monitored and controlled by fault detection section P_E. DRP consists of dual-redundancy power electricity channels P_{m1}, P_{m2}, digital electricity channels P_{n1}, P_{n2}, power electricity isolation channels P_{h1}, P_{h2} and digital electricity isolation channels P_{j1}, P_{j2}.

6.2 *Fault Diagnosis and Order-Reduce Forewarning of DRP*

When a fault occurs in a section between P_A and P_B, the section with fault is isolated automatically while another power section offers power normally. The working states of dual-redundancy channels P_{m1}, P_{m2}, P_{n1}, P_{n2}, P_{h1}, P_{h2}, P_{j1}, P_{j2} are monitored by fault detection section P_E, which sends fault forewarning message through gateway controller Ga if a fault occurs.

7 Experiment Verification

A DRSBW hardware-in-loop test bed is designed and manufactured according to the system structure proposed in this paper, as shown in Fig. 10.

The DRSBW hardware-in-loop test bed in Fig. 10 is made up of DRSWBW, DRSGBW, DRN, DRP and fault display.

The function of random single electrical fault causing no-failure is verified by the random failure experiment results on the testbed.

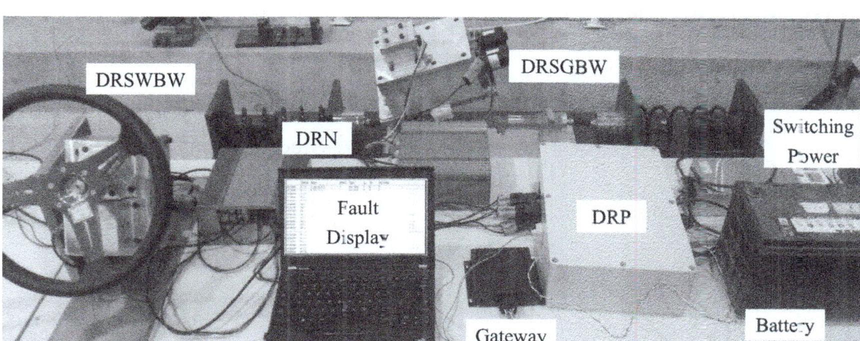

Fig. 10 DRSBW hardware-in-loop test bed

7.1 Angle Sensor Fault Experiment

One of the angle sensors in DRSGBW fails during the steering process and the signals of the steering gear angle sensors are shown in Fig. 11.

As is shown in Fig. 11, the angle sensor 2 fails at 20 s causing the disappearance of its angle signal. The fault is not a system order-reduce fault, thus the system should work normally.

The curves of the steering gear angle tracking are as Fig. 12.

As is shown in Fig. 12, the steering gear tracks the steering wheel close and there is no distinct angle tracking error after the failure, demonstrating that the system works normally.

The system forewarning signal is shown in Fig. 13.

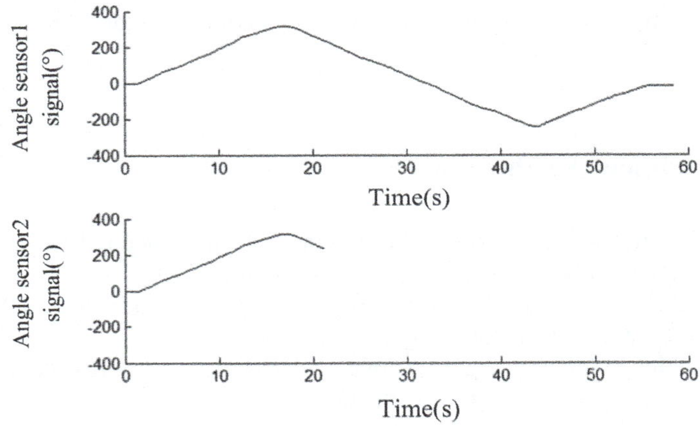

Fig. 11 Steering gear angle sensor signals

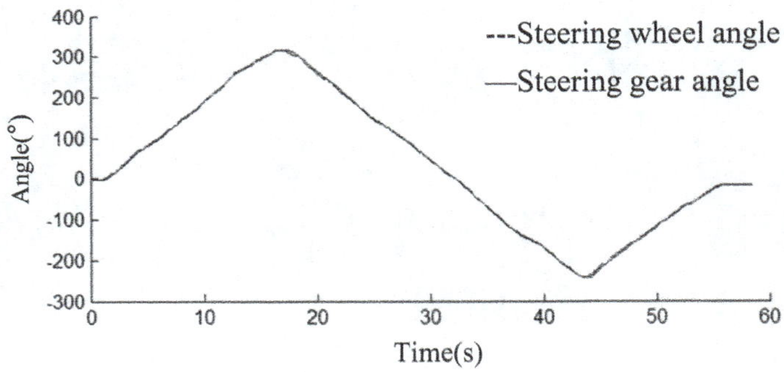

Fig. 12 Steering gear angle tracking curves with angle sensor failure

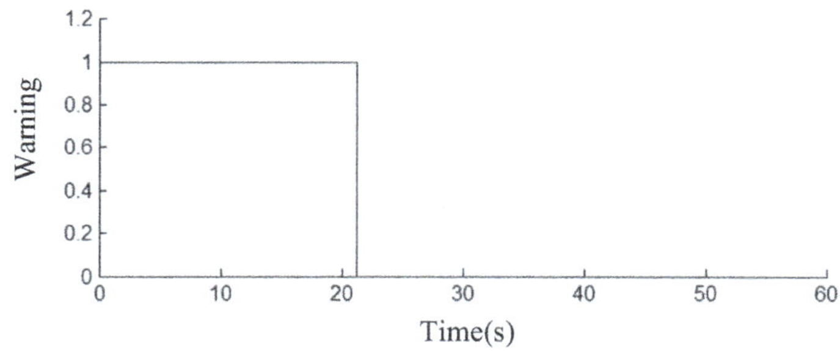

Fig. 13 System forewarning signal with angle sensor failure

In Fig. 13, the warning signal changes from 1 to 0 after the angle sensor failure occurs, showing that the fault forewarning signal is transmitted to vehicle bus. The "fault-safe-forewarning" working mode in angle sensor failing condition is demonstrated by the results.

7.2 *Electricity Sensor Fault Experiment*

One of the electricity sensors in DRSGBW fails during the steering process to study the steering gear angle tracking effect. The results of the electricity sensor signals are shown in Fig. 14.

As shown in Fig. 14, the signal of sensor 2 changes to 0, causing a system order-reduce fault. The drive circuit consisting of sensor 2 is cut and the angle tracking curves are shown in Fig. 15.

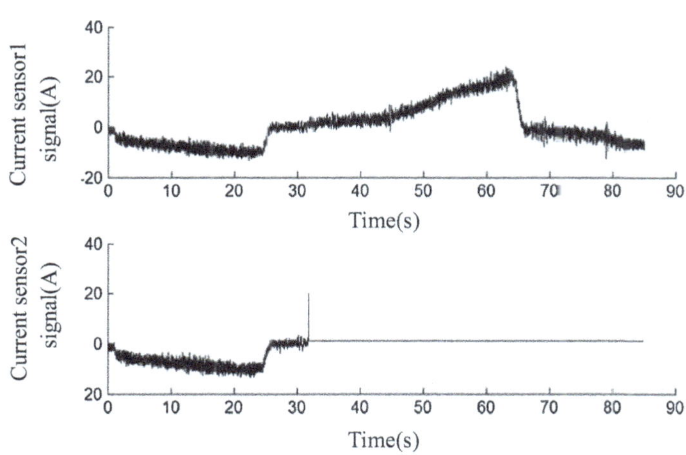

Fig. 14 Results of the electricity sensor signals

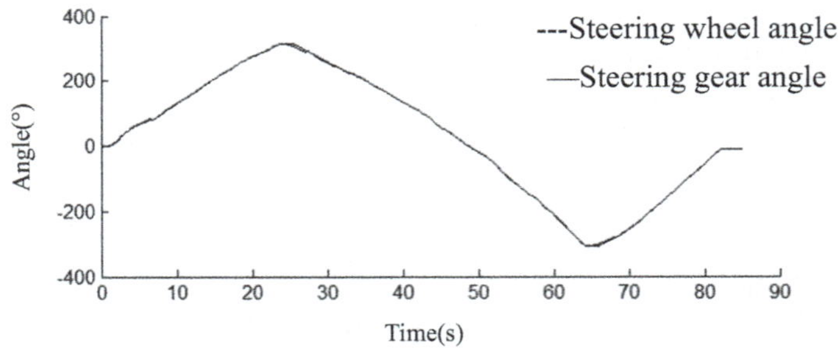

Fig. 15 Steering gear angle tracking curves with electricity sensor failure

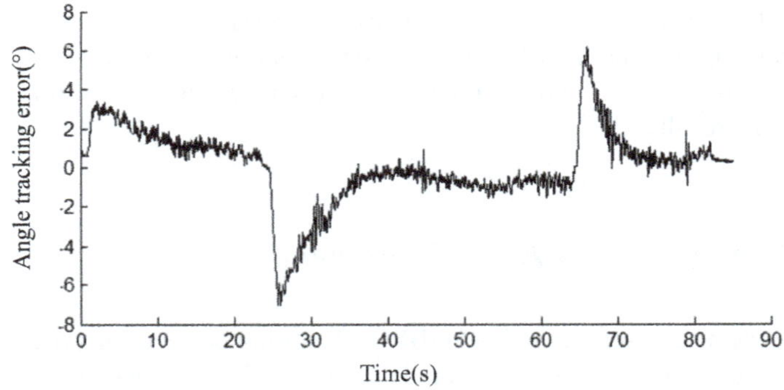

Fig. 16 Steering gear angle tracking error with electricity sensor failure

Figure 15 illustrates that the steering gear tracks the steering wheel angle close. The angle tracking error is shown in Fig. 16.

As shown in Fig. 16, the peak value of angle tracking error occurs when the moment the steering wheel steers inversely. No distinct angle tracking error can be observed when the electricity sensor fails, illustrating that the system switching process is smooth.

The system fault forewarning signal is shown in Fig. 17.

According to Fig. 17, the warning signal changes from 1 to 0 after the electricity sensor failure occurs, showing that the fault forewarning signal is transmitted to vehicle bus. The "fault-safe-forewarning" working mode is demonstrated by the results.

7.3 Network Fault Experiment

The dual-redundancy messages transmitted to node2 are C1, F1 and C2, F2, which are from the two controllers in node1 in DRN. Two of the four messages will disappear

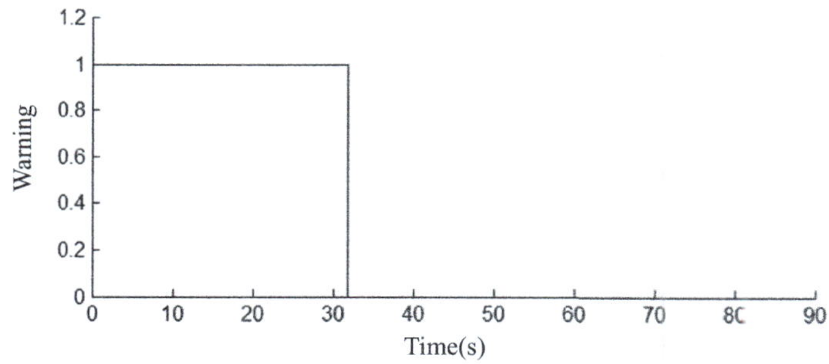

Fig. 17 System forewarning signal with electricity sensor failure

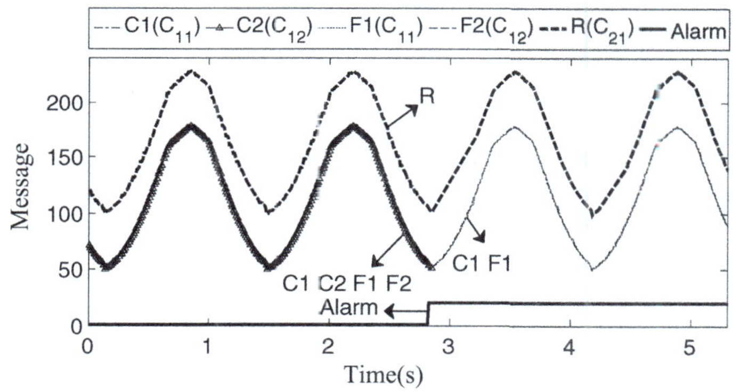

Fig. 18 DRN fault experiment results

when one of the two controllers fails, which means only C1, F1 or C2, F2 can be received by node2. The experiment results are in Fig. 18.

As shown in Fig. 18, the messages in DRN decrease from C1, F1, C2, F2 to C1, F1. *R* curve is the arbitrated data by node2, which is moved upward in order to illustrate the result more clearly. It can be concluded from the curves that no abnormal data occurs in R curve after the fault, demonstrating that the network transmits data normally. A warning signal occurs at the fault point, verifying the "fault-safe-forewarning" working mode of DRN.

7.4 Power Fault Experiment

Faults are injected to power electricity channel and digital electricity channel separately to verify the high safety of DRP. The experiment results are shown in Fig. 19.

In the left plot of Fig. 19, a fault is injected to power electricity channel, causing a small volt decrease to the channel volt U_{p1}, while U_{p1} is still within normal range.

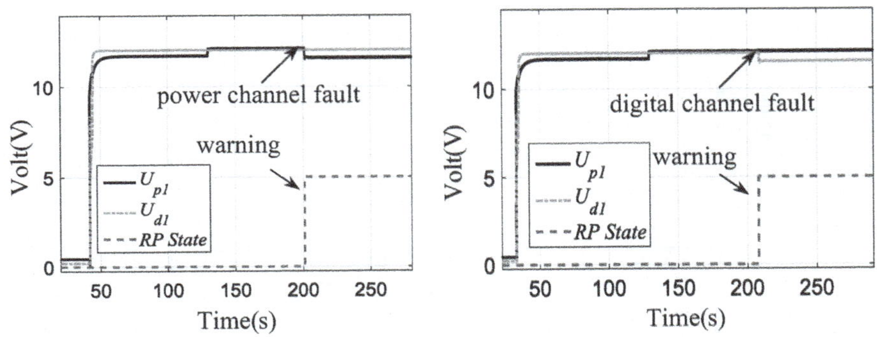

Fig. 19 DRP fault experiment results

The volt of digital channel U_{dl} is not influenced by the fault. A warning occurs when the fault happens. The right plot of Fig. 19 shows a fault injected to digital electricity channel, causing a small volt decrease to the channel volt U_{dl}, while U_{dl} is still within normal range. The volt of power channel U_{pl} is not influenced. A warning is given when the fault occurs. The "fault-safe-forewarning" working mode of DRP is demonstrated by the results.

8 Conclusion

The design method and redundancy management method of DRSBW are proposed in this paper and experiment verifications are performed. The conclusions are as follows:

(1) Random single electrical fault causing no-failure function is achieved in SBW system by applying the dual-redundancy working mode, which is in "fault-safe-forewarning" mode. The system safety is improved significantly.
(2) DRSWBW can provide normal steering feel with random single electrical fault and give fault forewarning of steering wheel subsystem, enhancing the subsystem safety.
(3) DRSGBW can drive front wheels normally with random single electrical fault and give fault forewarning of steering gear subsystem, improving the subsystem safety.
(4) DRN can transmit message normally with random single fault and give fault forewarning of network. The safety of the network in SBW system increased significantly.
(5) DRP can provide power to the system normally with one fault in power circuits and power fault forewarning can be transmitted to vehicle bus, improving the safety of the power supply in SBW.

References

1. ISO 26262 Road vehicles—functional safety—part 5: product development: hardware level (2011)
2. Yao Y, Daugherty B (2007) Control method of dual motor-based steer-by-wire system. SAE Tech Pap
3. Zheng B, Altermare C, Anwar S (2005) Fault tolerant steer-by-wire road wheel control system. In: American control conference, 2005. Proceedings of the 2005. IEEE, pp 1619–1624
4. Chaaban K, Leserf P (2009) Simulation of a steer-by-wire system using FlexRay-based ECU network. In: Advances in computational tools for engineering applications, 2009. ACTEA'09. international conference on. IEEE, pp 21–26
5. Onoda Y, Onuma Y, Goto T et al (2008) Design concept and advantages of steer-by-wire system. SAE Tech Pap
6. Changfu Zong et al. Redundant fault tolerant control method with application to SBW system with dual-motors[P]. China Patent: CN102320325A, 2012-01-18
7. Yao Y (1997) Reliability and redundancy technology. Beihang University Press, Beijing
8. Gao Jinyuan (2005) Fly-by-wire system and active control technology. Beihang University Press, Beijing, pp P8–P9
9. Wang T, Chen X, Cai Z, Mi J, Lian X (2017) Safety reliability analysis of a dual-redundancy steer-by-wire system. In: 19th ASIA Pacific automotive engineering conference. Shanghai
10. Cai Z (2017) Dual-redundancy steering gear system. Tsinghua University Beijing
11. Cai Z et al (2018) Full dual redundancy steer-by-wire steering gear and its redundancy management. Automobile Technol 3:41–46
12. Chen Xi (2017) Vehicle dual redundancy electrical system. Tsinghua University, Beijing
13. Chen Xi et al (2018) A study on the redundant and heterogeneous bus network with high safety and reliability. Automobile Technol 1:38–44
14. Chen X, Wang T, Lian X-M (2017) Vehicle redundant power with high safety and reliability. In: International conference on electrical engineering and automation control

Super Knock Control Logic Design and Verification for Turbocharged Gasoline Direct Injection Engine

Rui Xu, Long Chen, Han Yun Tuo, Wen Tao Xiao, Bai Qi Li and Feng Yuan

Abstract During development of turbocharged gasoline direct injection engine, super knock is a common and dangerous phenomenon. The development of super knock suppression strategy involves in three points. First, super knock control logic design; second, super knock with proper frequency induction on test bench. Third, system which super knock logic embedded verification should be done, including whether the super knock is detected, whether proper steps are taken by the system to suppress super knock. These issues are discussed in the paper. It has been proved by bench test result that the super knock control logic described in the paper can give appropriate protection to turbocharged gasoline direct injection (GDI) engine.

Keywords Super knock · Control logic · Test bench verification

1 Introduction

With the application of direct injection technology and turbocharging technology, and the strengthening level of gasoline engine is constantly improving [1]. Super knock is an abnormal combustion phenomenon caused by pre-ignition. At the same time, the enormous maximum combustion pressure, extremely high combustion temperature and great pressure fluctuation in the cylinder are accompanied. It generally occurs in low speed and heavy load. Super knock is likely to have a destructive effect on the engine [2, 3].

This strategy should be developed in a qualified EMS system to ensure the safety of the engine. Three aspects should be taken into consideration during developing the strategy: First, the strategy design. Second, inducing super knock with appropriate occurrence frequency to meet the strategy verification experiment. Third, the strategy should detect the super knock event, and specified actions could be applied to protect the engine. This paper elaborates the super knock strategy design and verification process, which is based on a turbocharged gasoline direct injection engine with small

R. Xu (✉) · L. Chen · H. Y. Tuo · W. T. Xiao · B. Q. Li · F. Yuan
Dong Feng Motor Corporation Technical Center, Wuhan, China
e-mail: 281079399@qq.com

China SAE (ed.), *Proceedings of China SAE Congress 2018: Selected Papers*,
Lecture Notes in Electrical Engineering 574,
https://doi.org/10.1007/978-981-13-9718-9_73

displacement from Dong Feng Motor Company. It proves that is an effective super knock induction method, and the strategy could be used to protect the engine.

2 The Super Knock Control Strategy Design

2.1 The Super Knock Detection Strategy Design

The non-resonant piezoelectric knock sensor is still used in the super detection, transforming the original signal from knock sensor the vibration signal of engine cylinder into differential voltage signal.

The super knock detection process is shown in Fig. 1. The original analog signal is amplified by gain and converted to digital signal. The gain selection depends on the accuracy of analog to digital conversion, and the signal distortion should be avoided. Low pass filter is used for high-frequency noise filtering. For subsequent bandpass filtering process, it is necessary to reduce the sampling rate of the current signal. Finally, the super knock intensity is obtained after rectification and integration. Then, the characteristic value of background noise and the threshold for super knock detection are obtained by analyzing the data collected under different working conditions. If the super knock intensity exceeds the threshold. This situation can be considered as super knock event.

Due to the mechanism of super knock, it should be noted that the occurrence time and suppression measures of super knock are totally different from those of normal knock, which is shown in Fig. 2. Therefore, it is necessary to set the independent data acquisition window and the independent threshold for super knock detection, and the super knock detection window should be in front of the normal knock detection window. The super knock windows are defined by the start position and window length, both of which are functions of rotational speed and load.

The center frequency of the super knock should be calculated before the subsequent bandpass filtering process, its calculation method is explained by the following Formula 1, the Draper equation [4], $\rho_{m,n}$ is the vibration mode constant, C is the Local sonic speed, and B is the Cylinder diameter. Then the Fourier transform is used to get the center frequency, finally the center frequency is confirmed by verification experiment.

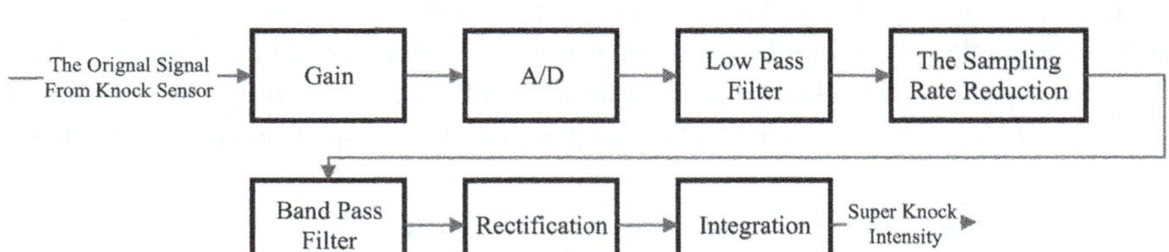

Fig. 1 The super knock signal calculation process

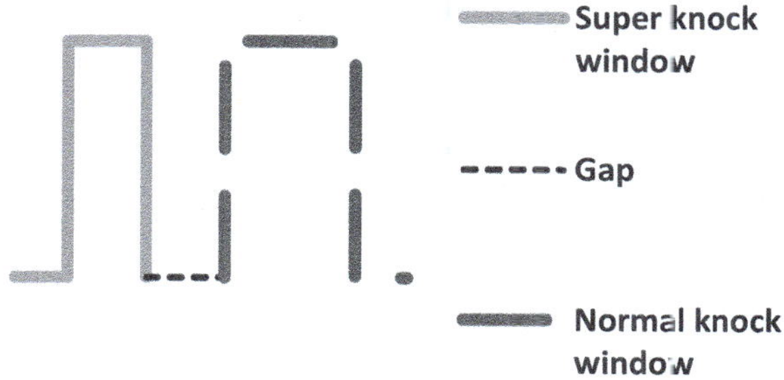

Fig. 2 The diagram for data acquisition window

$$f = \frac{\rho_{m,n} C}{\pi B} \tag{1}$$

2.2 *The Super Knock Suppression Strategy Design*

The super knock suppression logic is shown in Fig. 3, the system continuously monitors the signal from the knock sensor, and calculates the super knock intensity signal and the threshold. Then the system will compare the super knock intensity to the threshold, if the super knock intensity is larger than the threshold, the system identifies the event as a super knock event. Possible measures which will be taken to suppress super knock are fuel enrichment, load reduction, and fuel cut.

Worse situation of super knock will lead to more suppression measures which will be taken. Each suppression measure has a duration, the suppression measure will be canceled if there is no new super knock event happened during its duration. But if a new super knock event happened during its duration, the electronic management system (EMS) will not stop the current suppression measures, and more suppression measures will be taken. The fuel enrichment can obviously reduce the combustion temperature in the cylinder, this is why the fuel enrichment could suppress the super knock. Load reduction greatly reduces the probability of super knock. It should be cautious about the application of fuel cut, which has a great influence on drivability and safety.

3 The Super Knock Induction Method

Inducing super knock with appropriate occurrence frequency is the basis for super knock control strategy verification experiment. Super knock happens with small possibility in some engines under normal working condition. The test engine is working at 1500 revolutions per minute and 80–90% full load, this ensures the engine works under heavy load and without super knock before that super knock induction

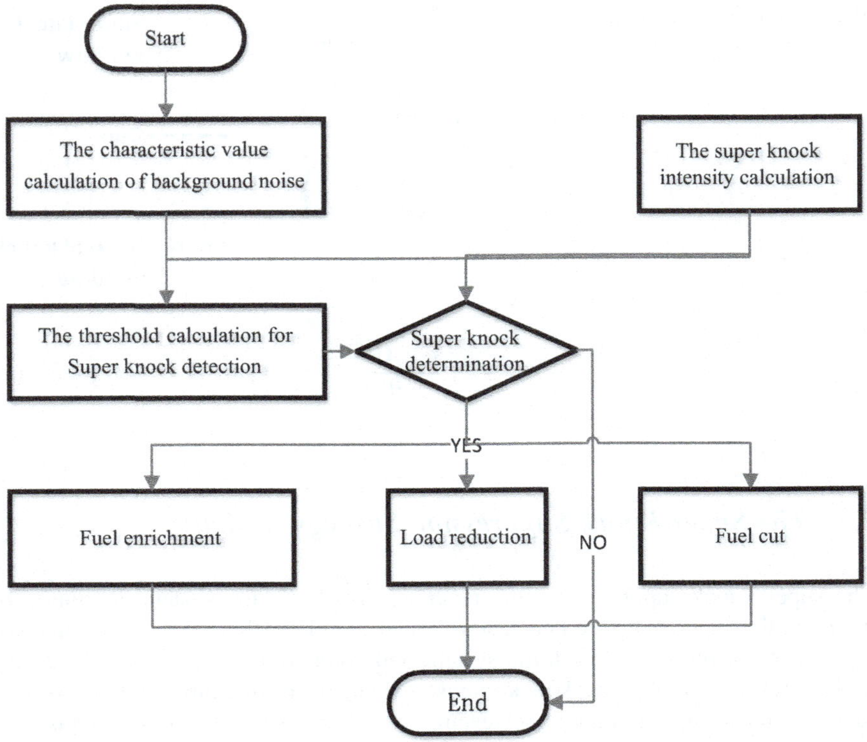

Fig. 3 The Super Knock Suppression Strategy

measures are taken, the intake air temperature is controlled in 50–60 °C, coolant water temperature is controlled in 90–108 °C. In order to get simple experiment process, two ways for super knock induction are chosen by analyzing the super knock mechanism and previous data.

3.1 The Normal Knock Induction Method

Normal knock can significantly increase the combustion temperature in the cylinder, which is helpful for super knock production. It can be predicted that when the super knock happens by this way, which will be close to the ignition angle, this is useful in super knock induction control.

A simple data acquisition plan is applied, whether super knock happens can be determined by signal from knock sensor, in which cylinder the super knock happens can be determined by signal from crankshaft speed sensor and ignition pulse signal

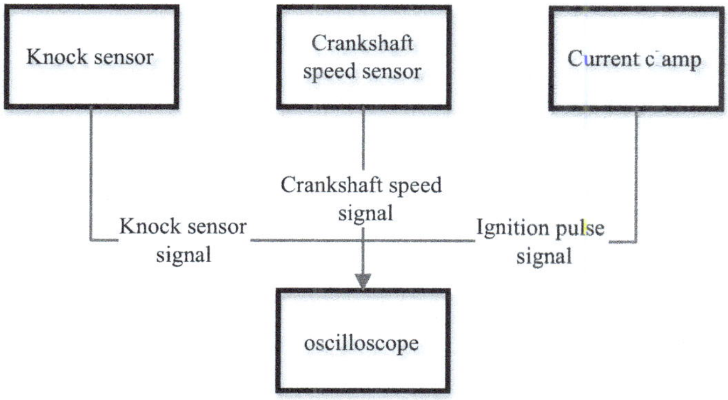

Fig. 4 The simple data acquisition plan

from current clamp, all signals mentioned above are sent into oscilloscope, which records the data continuously. The data acquisition plan is shown in Fig. 4.

The super knock data collected by the plan is shown in Fig. 5, super knock and the ignition event occurs almost at the same time, and super knock has obviously larger intensity than normal knock.

The advantage of this method is which has better control over when the super knock happens and the frequency of the super knock, but continuous strong normal knock could also damage the engine.

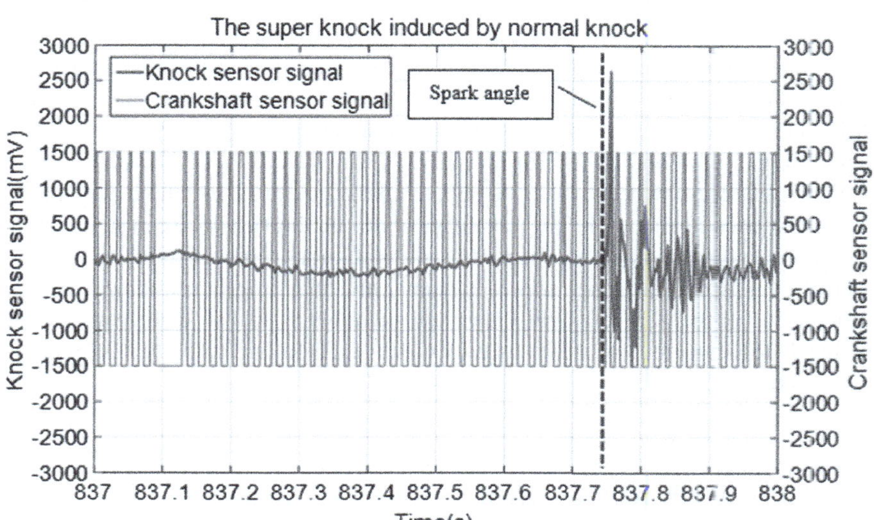

Fig. 5 The super knock induced by normal knock

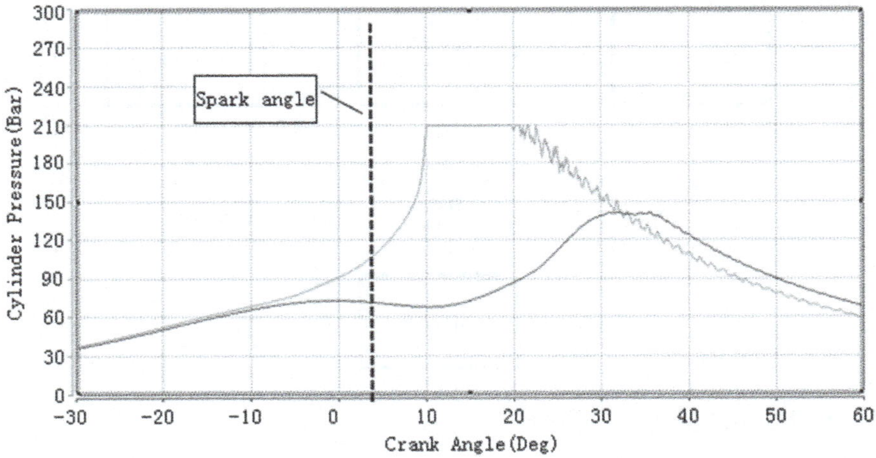

Fig. 6 The super knock induced by inferior fuel (Color figure online)

3.2 Inferior Fuel Induction Method

Inferior fuel can significantly increase the probability of super knock [2, 5]. The method is to mix the engine oil into gasoline in a small proportion, then gradually increase the engine oil until the super knock happens. In this experiment, a combustion analyzer is used to collect the cylinder pressure data.

The super knock induced in the experiment is shown in Fig. 6. Super knock and the ignition event occurs almost at the same time, and the super knock is very close to the compression top dead center. The pressure signal (red) from the cylinder which the super knock happens prematurely deviates from the compression line compared with other cylinder pressure signal (black). The super knock intensity is even greater than the upper limit of the combustion analyzer, which leads to a peak shaving event.

Because it does not need to use the high temperature caused by continuous normal knock to induce super knock. The advantage of this method is which can significantly reduce the engine damage risk caused by normal knock. But this method has poor control over when the super knock happens, and inappropriate mixing ratio can improve the probability of super knock considerably, it is also dangerous for test engine. Due to the previous experience accumulated in this area, finally this induction plan has been adopted.

4 The Super Knock Control Strategy Verification

The super knock detection strategy verification result is shown in Fig. 7. The super knock intensity calculated by the system based on the voltage signal of the super

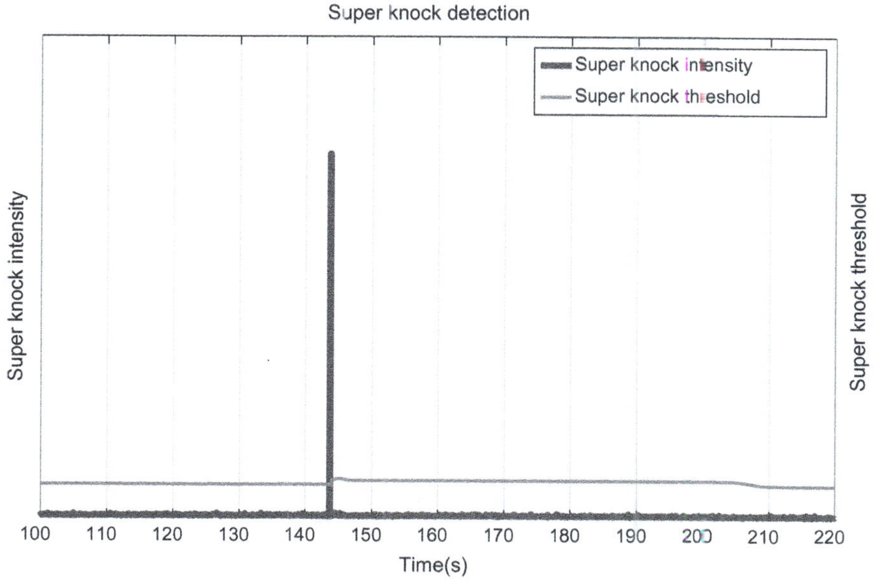

Fig. 7 Super knock detection function verification

knock sensor far exceeds the threshold when the super knock happened, then the system identified the event as a super knock event.

The super knock intensity always stayed under the super knock threshold, it has been shown that there is no omission or false of super knock event report in the system. Super knock is much more dangerous to the engine than normal knock, so the super knock suppression measures are generally more aggressive. Too loose diagnostic criteria will increase the risk of engine damage due to omission of super knock event report. On the contrary, too conservative calibration will greatly affect the engine performance. Therefore, the calibration of super knock detection should be as accurate as possible, and the relevant parameters of super knock suppression strategy should be carefully calibrated and evaluated. The test bench experiment should be carried out based on theoretical calculation and simulation test to ensure that the super control strategy can always works properly.

Fuel enrichment request is sent by the system when a super knock event was detected, which is shown in Fig. 8. When a super knock was detected, the super knock flag is set to true, and the fuel enrichment request is activated simultaneously. When there is no new super knock event is detected for a period of time, the fuel enrichment request fades out.

When a super knock was detected, the super knock flag is set to true, and the load reduction request is activated simultaneously, which is shown in Fig. 9. When there is no new super knock event is detected for a period of time, the load request gradually recovers.

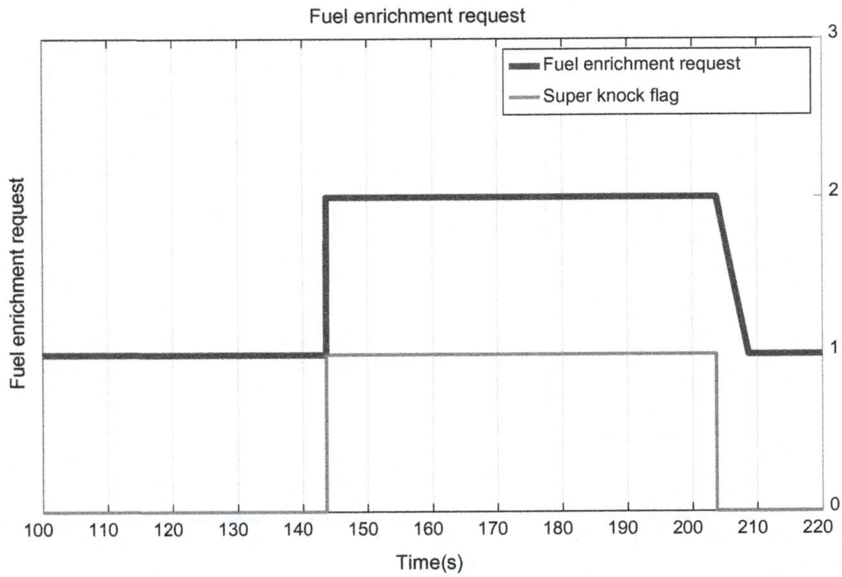

Fig. 8 Fuel enrichment function verification

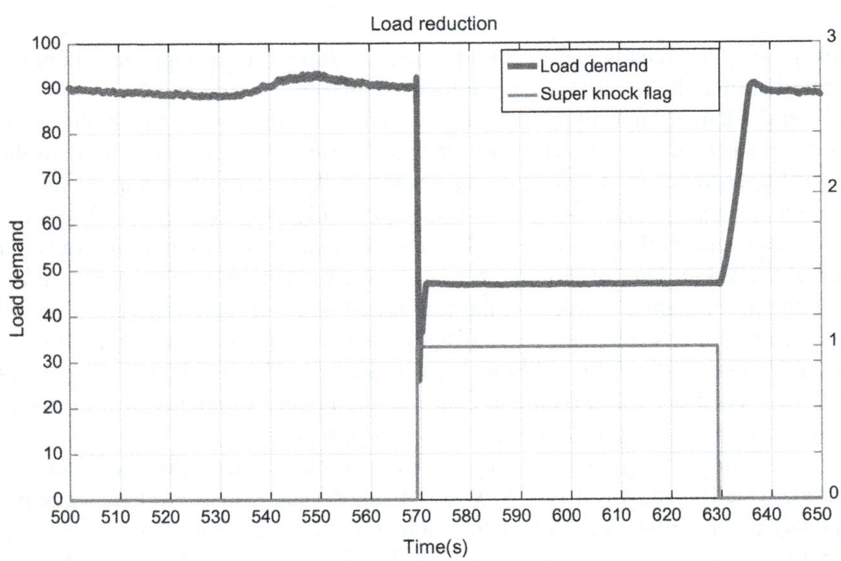

Fig. 9 Load reduction function verification

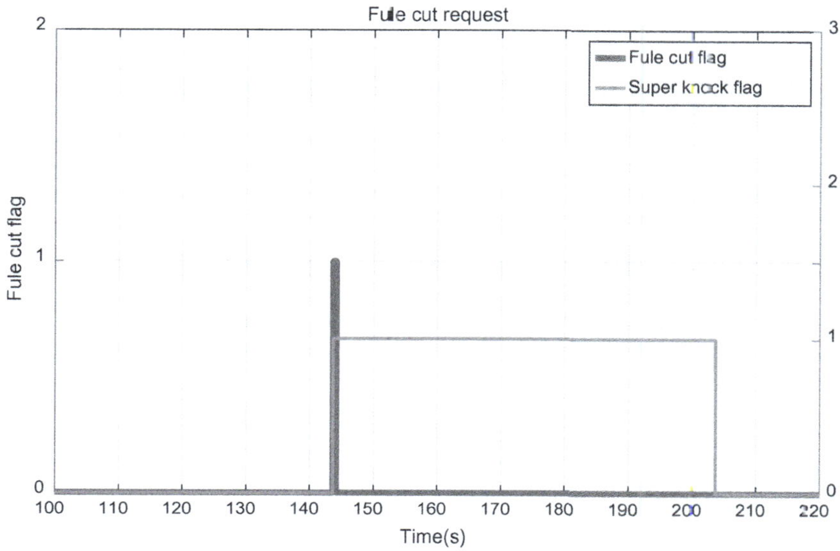

Fig. 10 Fuel cut function verification

Due to the fuel cut function has great influence on drivability and safety, fuel cut with short duration was applied in the experiment. The result is shown in Fig. 10, when a super knock was detected, the super knock flag and the fuel cut flag are set to true simultaneously, after the duration, the fuel cut request is canceled quickly.

With the help of the super knock control strategy mentioned above and careful calibration, the test bench experiment result shows that the EMS can detect super knock event correctly, the following suppression measures applied can ensure the engine works normally and safely.

5 Summary

Now super knock is a great threat to the turbocharged gasoline direct injection (GDI) engine. In the paper, the super knock strategy design and verification processes are discussed, which is based on a turbocharged gasoline direct injection engine with small displacement from Dong Feng Motor Company. This paper elaborates the super knock strategy, then the super knock induction methods are discussed for engine which works with low probability of super knock. In addition, a simple data acquisition plan is applied in the super knock induction experiment. Finally, the test bench verification experiment result is shown in the paper, it proves that the super knock control strategy can detect the super knock correctly, and also ensure the engine works normally and safely.

References

1. Zhou L, Liu Z, Gao Z et al (2014) Internal combustion engine (The third edition). Machinery Industry Press, Beijing
2. Dahnz C, Han K, Spicher U, Magar M et al (2010) Investigations on pre-ignition in highly supercharged SI engines. SAE Int J Eng 3(1):214–224
3. Amann M, Mehta D, Alger T (2011) Engine operating condition and gasoline fuel composition effects on low-speed pre-ignition in high-performance spark ignited gasoline engines. SAE Int J Eng 4(1):274–285
4. Hao Zhou (2013) Investigation on knock analysis of direct injection engine. School of Mechanical Engineering, Tianjin University
5. Wang Z, Qi Y, Liu H, Long Y, Wang J-X (2015) Experimental study on pre-ignition and super-knock in gasoline engine combustion with carbon particle at elevated temperatures and pressures. SAE Technical Paper, 2015-01-0752

Improvement of Sound Quality in Car Based on the Third-Party Sound Effect

Zilong Liang, Zebing Tang, Ying Cai, Jin Yang and Haixia Hu

Abstract Sound quality of cars is an important part of the automobile information and entertainment system for user's experience. The enclosed space of cars combined with the multi-channel stereo audio source can create the sound effect like a moving concert hall. With the development experience of automobile audio system, this paper analyzed the influence factors of the sound quality in low-cost passenger vehicles, and it introduced the classification and technical characteristics of some sound effects, and how to integrate the third-party sound effects to improve the sound quality in car. With the mass production verification, it obtained development method, test, and evaluation criteria, which could provide reference for the sound quality improvement of the audio entertainment system.

Keywords Sound quality · Sound effect · Audio entertainment system · Speaker

1 Preface

With the development of automobile intellectualization and networking, automobile life has become very normal, and the information and entertainment system in car gradually becomes an important entrance of human–machine interaction besides computer, TV, mobile phone, and other intelligent terminals [1]. The sound quality in car is one of the most important indicators for the user experience and the comfort evaluation of the in-car entertainment systems. Most users or even music enthusiasts have some high requirements for the sound quality in car, and they hope to achieve the grand sound effects like being in a concert hall or a cinema with the cars' limited interior space. Based on the user experience of vehicle information entertainment system, this paper analyzed the current situation and deficiencies of the acoustics in low-cost passenger cars, analyzed the main factors which affect sound quality, and emphatically introduced the solution of improving sound quality in car through the third-party sound effect. Through the real case of mass production, this paper

Z. Liang (✉) · Z. Tang · Y. Cai · J. Yang · H. Hu
Dongfeng Motor Technical Center, Wuhan 430058, China
e-mail: liangzl@dfmc.com.cn

China SAE (ed.), *Proceedings of China SAE Congress 2018: Selected Papers*, Lecture Notes in Electrical Engineering 574,
https://doi.org/10.1007/978-981-13-9718-9_74

introduced the principle of the third-party sound effect, the process of development and integration, the objective measurement method, and the subjective evaluation method of sound quality in car.

2 The Deficiency and the Main Influencing Factors of Sound Quality in Car

2.1 The Deficiency of Sound Quality in Car

For the low-cost cars on the market, considering the cost and the generalization, the speaker configuration, specifications selection, and layout scheme are basically similar, and the speakers mainly use low-cost ones. The commercial evaluators think that the sound effect in those vehicles is relatively poorer than other high-end vehicles, and there is a large opportunity for improvement, which mainly manifest as the bass weakness, missing treble details, scattered sound field and muddy overall sound effects and so on.

2.2 The Main Influencing Factors of Sound Quality in Car

The sound quality of audio system in car is related to many factors. Speakers' selection, speakers' layout, and signal processing in the audio system are the main influencing factors for it, besides the space, airtightness, interior materials in car [2].

2.2.1 Selection of Speakers

As the output of the whole car's acoustics, the performance of speakers will directly determine the sound quality. The most medium- and low-end vehicles are equipped with low-cost ordinary speakers, whose performance parameters (such as frequency response and distortion) need to be improved and whose ability of the sound playback and restore is limited. As shown in Fig. 1, it is the frequency response curve of a vehicle's tweeters. Above 13 kHz, the frequency response of the tweeter decays rapidly. If we want to improve sound quality and effect, the high-performance brand loudspeakers should be used, and the number of speakers should be increased. However, the cost is also very high, which need to be configured according to the product positioning.

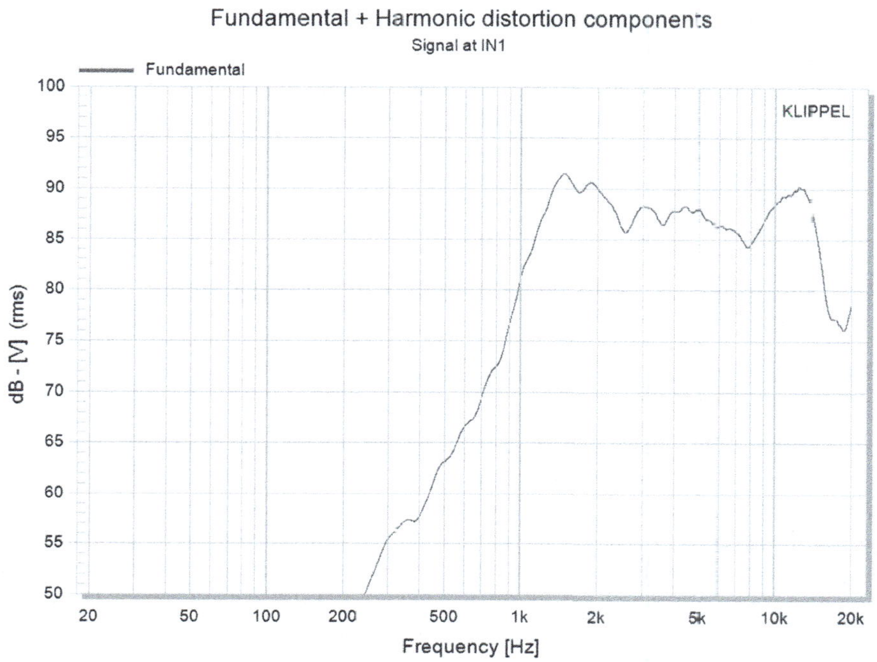

Fig. 1 Frequency response curve of tweeters in a certain mass production vehicle

2.2.2 Layout of Speakers

A high-quality speaker is an important condition for the performance of the vehicle acoustics, but the layout of speakers also plays a decisive role for the sound effects in car. The woofers have a large size, which require more space to layout. They are generally located in the lower part of the door panel, which could form an echo chamber inside the door panel. The tweeters generally have directivity, whose located height should be similar to that of the driver's ear. The central speaker is generally located at the center of the dashboard or the top of the back seats. Subwoofer is generally placed in the trunk because of large amplitude and volume. Due to factors such as the shape and layout restrictions of some vehicles, sometimes the layout of speakers is also difficult to be in the best position (e.g., tweeters are not on a pillar, but placed on the dashboard, resulting in its poor directionality).

2.2.3 Sound Processing of Head Unit

Head unit could be regarded as the sound source, where sound is generated and processed. It also directly affects the overall effect of sound quality in car. The most head units just have general sound effects with frequency equilibrium, such as rock, jazz, pop, custom, and so on. As for the improvement of sound stage and image, a professional sound effect algorithm is needed.

In the case that the whole vehicle hardware configuration and the structure layout has been fixed, the processing algorithm software of third-party sound effect based on DSP is a good solution for sound quality improvement, which has low cost, obvious effect, and convenient implementation.

3 Classification and Characteristics of Sound Effects

3.1 Classification of Sound Effects

According to the processing method of digital audio signal, the sound effects roughly can be divided into four categories: delayed class, modulated class, filtering class, and non-linear processing class. The sound effects of delayed class mainly refer to that the original signals would add the signals delayed with some coefficient in each channel, including fixed delay (such as echoes) and variable delay (such as vibrato and chorus), which can improve thickness and elasticity of sound. The sound effects of modulated class mainly refer to that original signals are modulated to generate mono or multiple-tone chords by enveloping filters, which are used for warning sounds in vehicle. The sound effects of filtering class mainly refer to that digital audio signals are processed by various functional filters to realize frequency equalization, loudness control, tone adjustment, and so on, which is widely used in compensated and modified sound effects. The sound effects of non-linear processing class mainly refer to realize the control of sound stage balance and gain distortion through amplitude expansion in each channel, volume gain control, amplitude limiter, etc. [3].

According to the function, sound effects can be divided into two categories: reductive sound effects and modified sound effects. The former aims to reasonably repair and compensate audio signals' distortion generated in transmission, conversion, amplification, playback, and other processes, which would restore the original characteristics of the audio signal as far as possible. This kind of sound effects is mainly represented by Harman Clari-Fi, Dolby, BBE. The latter aims to sound stage expansion, sound balance enhancement, and sound image positioning, which make the sound scene rich. Those sound effects are generated by delay, filtering, gain control, and other methods on the basis of the original sound signals, which is mainly represented by Arkamys, Yamaha, DTS Cs-auto. This paper will mainly introduce the main characteristics of modified sound effects and its mass production verification.

3.2 *Characteristics of the Modified Sound Effects*

3.2.1 EQ Compensation

The audio signals are processed by DSP chipset and sent to amplifiers and speakers, and then sound is generated and enters into ears through the air. Due to the signal processing, conversion, amplification, transmission, and other reasons, the frequency response of the audio signal is different from the original one, resulting in a decline in sound quality. EQ compensation, through processing of combinative filters, can restore the frequency response characteristics. In the meanwhile, it can achieve a variety of music style effects, such as rock, jazz, popular, and so on, according to the audio style and user preferences.

3.2.2 Sound Stage Enhancement

According to the auditory characteristics of human's ears, using vertical HRTF (Head Related Transfer Function) processing technology, it can achieve the vertical elevation of sound stage for the speakers in a low position to the height of the ears as close as possible.

3.2.3 Sound Image Positioning

In acoustics, the human's ears judge the distance mainly through the loudness of the sound, the reverberation, and the delay time and judge the azimuth through the difference of the left and right loudness. The spatial sense of sound image can be extended by adjusting the loudness and delay of each channel, and the sound image has the certain directivity and position. The best sound image is located in front of the human's ears, as shown in the following Fig. 2, the left and right sound source is equilateral between the human ears, namely $A = B = C$, at least $B = C$, in which the wider of A, the wider of the width of the sound stage. It is a mono sound effect when A is 0.

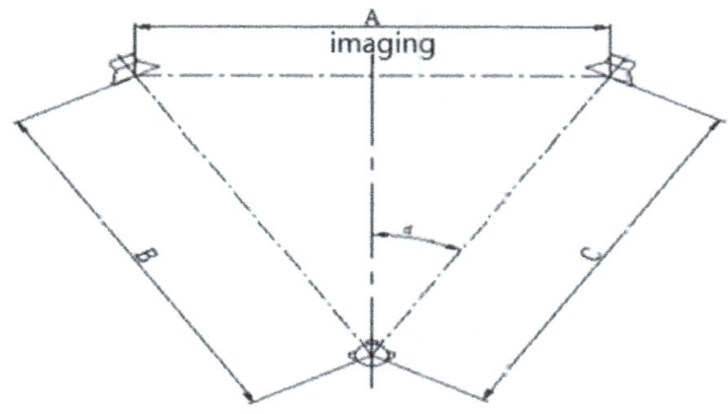

Fig. 2 Schematic diagram of sound image composition

3.2.4 Virtual Bass

From the perspective of psychological acoustics, when the ears receive the low-frequency sound signal that lacks the basic harmonics, the human's auditory system can use the suitable harmonics to supplement the basic frequency signal. That is, by enhancing the second and third harmonics, the psychoacoustic algorithm can simulate the perception of very low frequencies, and listeners can feel lower frequencies that a physical woofer does not have. Software algorithm can be designed to enhance the low-frequency performance of speaker configurations that do not include a physical subwoofer. When a speaker is not capable of reproducing a frequency, it is possible to recreate the feeling of this frequency by playing its harmonics.

3.2.5 Clarity Control

The human ears have different sensitivity to different frequency sounds. For low-frequency and high-frequency sound, the auditory sensitivity is significantly reduced, while more sensitive to the voice of the middle frequency band. With noise control and equalization, the medium frequency can be weakened appropriately, so that listeners can clearly feel the voice, different musical instruments, and other delicate sound details in audio source files.

4 Verification of Mass Production Scheme for Sound Effect a Based on a Third-Party

4.1 Mass Production Scheme

In order to improve the sound quality of in-car sound system, enhance product competitiveness and form unique USPs for vehicle market sales, currently we select sound effect *A* as a verification scheme. Compared with other sound effect schemes, the cost is lower and the effect is better.

The company *A* is one of global leading supplier on software solutions of sound processing and voice enhancement, and it focused on the sound quality improvement of car audio systems. It improves the sound quality of in-car sound system through the mainstream DSP software platform, professional voice calibration tools, and senior Gold Ear acoustics experts. It provides a comprehensive solution for the sound quality of the in-car sound system which can enhance sound stage and image and release more sound details.

The verified vehicle uses NXP TEF6635 as DSP chipset and is equated with the sound effects of V1 (positioning & equalization) and V2 (phantom subwoofer), as shown in the following Table 1.

Table 1 Sound scheme of the third sound effects

Sound scheme	Optimize content	Note
V1	**Equalization**: optimize frequency response and correct the in-car acoustic defects **Sound stage enhancement**: raise and focus on sound stage, construct a virtual center stage in front of the passenger, provide precise positioning of sound in front of passengers	Basic sound stage optimization
V2	**Phantom subwoofer**: Process based on psychoacoustic perception enhancing very low frequencies to create a rendering similar to a physical subwoofer	Phantom subwoofer

There are mainly three kinds of sound effect modes: the all passengers mode, the main driver mode, and the off mode. In the main driver mode, the sound stage is optimized for the driver, which does not consider other passengers, and is suitable for driving alone. The sound stage is located in the center of the instrument board and is very open with spatial sense.

The all passengers mode is designed for all seats including front and back with the overall optimization. The sound stage is located in the front of each passenger. The off mode means that the third-party sound effect processing is removed.

4.2 *Scheme Implementation*

The integrated development of the third-party sound effects follows the development of head unit and even involves the whole life cycle of vehicle development. It mainly includes four phases: the definitions and requirements, engineering development, sound tuning and evaluation, and integration test (Table 2).

In the engineering development phase, the layout of all speakers in car will be carried out according to the recommended standard, which can make the performance of speakers at those best. The audio-path design of DSP needs to be in accordance with the design requirements of the head unit, whose parameters set cannot affect the volume curve and source balance.

In the sound tuning phase, the selected vehicles need to be in a project phase close enough to the final vehicles to be sold to the final customers. The hardware and software of head unit should be stabilized. All speakers must be the final version, the brackets, configuration of speakers, interior type, seating type, door panels, and watertight parts also must be the final condition. All speakers must be functional and aged, because the chemical coating usually is used on the plastic/rubber material

Table 2 Main development process of the third sound effect integration

Requirements and definitions	Engineering development	Sound tuning and evaluation	Integration test
• Product positioning • Definition of function and performance • Selection of DSP software and hardware platform • Design of UI/UE • Development plan	• Speakers layout • DSP audio-path design • Communication integration design • Bench function verification	• Speakers aging • Sound tuning • Objective test • Subjective evaluation	• EOL parameters integration • Bench performance test • Mass production import

parts of the new speakers to ensure speakers' lifetime, which led to tightness of speakers. By aging, the speakers restore softness and toughness.

"Golden Ear" acoustic experts tune sound in final vehicles and calibrate the DSP parameters, which can achieve the desired sound. According to the objective test and subjective evaluation results, the performance of sound effects can be fine-tuned to achieve perfection.

4.3 *Objective Test*

The objective test of sound quality mainly includes the maximum sound pressure level test and the frequency response curve test. The frequency response curve is an important performance parameter of the audio system and the important basis of sound effect design and optimization.

4.3.1 Frequency Response Curve Test Case

As shown in Fig. 3, a 6-speaker vehicle with two 0.75 in. tweeters on the dashboard, two 6.5 in. woofers, and two 6.5 in. full-range speakers in the front and rear door guards. The pink noise at-23dBspl is used as the test source in the frequency response curve test case, and it is measured with a hand-held audio analyzer at the main driver's ears position (Fig. 4).

4.3.2 Test Results Analysis

According to the following test curve in the Fig. 5, in the whole frequency range between 20 Hz–20 kHz, both the low-frequency band ① and the high-frequency band ⑤ decay fast, and the middle frequency band ③ has a sharp peak. After the sound effect processing, the band ① at 20–70 Hz is appropriately enhanced to make

Fig. 3 Speakers layout

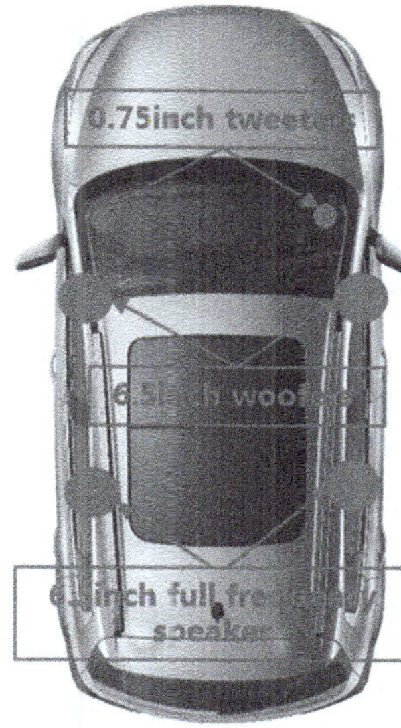

Fig. 4 Objective performance test

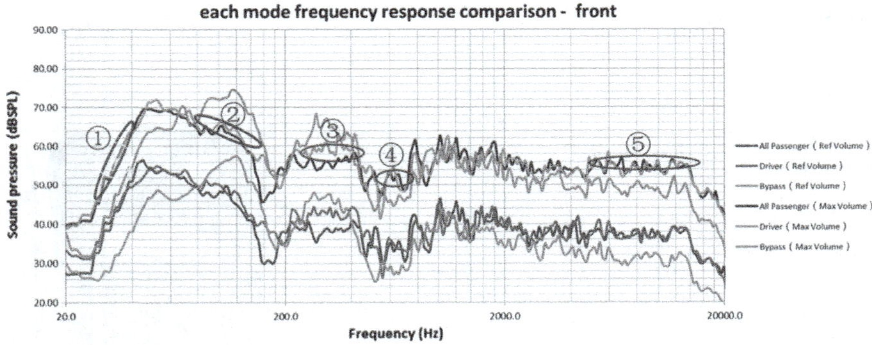

Fig. 5 Test results of frequency response curve

the bass strong and powerful. The band ② at 80–150 Hz spikes properly weakened to balance the entire bass section and make the bass full and natural. The peak of band ③ at medium frequency 200–400 Hz is weakened to make the midrange sound clear and eliminate "vague" feeling. The trough of band ④ at 500–800 Hz is filled to make up the reflection loss in the car. The decline of the band ⑤ at high-frequency 5–20 kHz is appropriately enhanced to make up the loss of tweeters layout.

When the sound effect is on, the frequency response curve is flatter overall than off. The high and bass frequencies are properly compensated, and the midrange frequencies are appropriately weakened, so that the overall sound is clearer, and the bass is more powerful and flexible.

4.4 Subjective Evaluation

The objective frequency response test can verify the overall performance of sound effects, but the overall sound effects also require the subjective evaluation from the professional acoustic evaluators. The following introduces subjective evaluation method of sound quality in car.

4.4.1 Subjective Evaluation Method

We select the driver's position and use the fixed tracks during subjective evaluation. The appraisal team includes the professional "golden ear" and 28 engineers who have the sound appraisal experience. We will record the evaluation results and calculate scores.

Subjective evaluation index below is divided into 4 dimensionalities and 17 indicators, which is quantified by values. The bandwidth and balance mainly evaluate the sound details and balance of each frequency band, which is the main determi-

1.Bandwidth&Balance				3.Imaging			
No.	Item	Max.	Score	No.	Item	Max.	Score
1-1	Treble	10		3-1	Center	5	
1-2	Mid-range	10		3-2	Left	5	
1-3	Mid-bass	10		3-3	Right	5	
1-4	Bass	10		3-4	Stereoscopic	5	
1-5	Balance	10					
2.Staging				4.Preception&Ambience			
No.	Item	Max.	Score	No.	Item	Max.	Score
2-1	Width	5		4-1	Dynamic	3	
2-2	Height	5		4-2	Fresh	3	
2-3	Depth	5		4-3	Airy	2	
2-4	Distance	5		4-4	Body	2	

Fig. 6 Subjective evaluation index

nant of sound effects. This evaluation is relatively simple and accounts for 50%. The staging and imaging are the main evaluation of sound spatial sense, stage sense, and 3D-positioning, which can be evaluated accurately after listening training. It accounts for 20% each evaluation; the perception and ambience mainly evaluate the delicate characteristics of sound, which can be evaluated after professional training. It accounts for 10% (Fig. 6).

4.4.2 Analysis of Subjective Evaluation Results

The subjective evaluation results of vehicle A and other related competing vehicles are seen in Fig. 7. It can be seen from the diagram that the subjective evaluation of sound quality after tuning is significantly better than that before tuning.

Compared to other competing vehicles, the vehicle *M* also configured the third-party sound effects and its tweeters are placed in *A* pillar, whose treble details is better. However, the bass of vehicle *A* is more powerful than vehicle *M*, and the

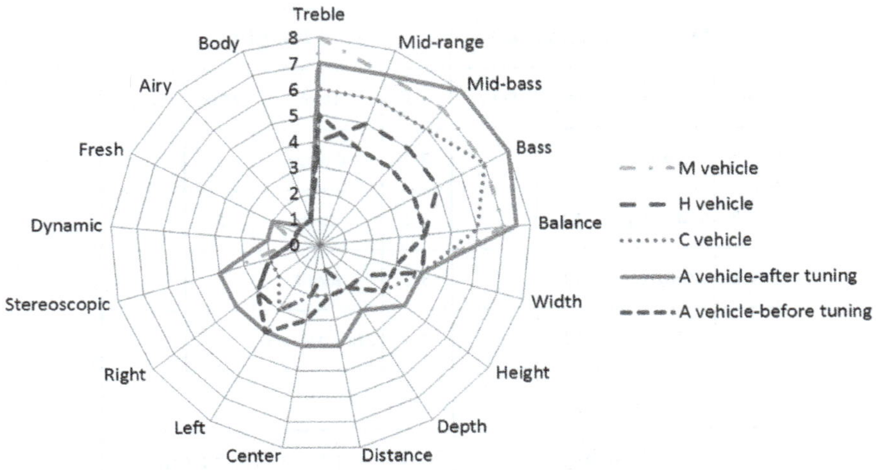

Fig. 7 Subjective evaluation results

overall sound stage and image are comparable between them. Both vehicle *H* and *C* have no special sound effect processing in equalization, sound stage, and sound image, whose sound quality is similar to vehicle *A* without sound effect.

5 Summary

Through objective test and subjective evaluation, the improvement scheme of sound quality in car based on the third-party sound effect has obvious effect. For the middle- and low-end ordinary cars, the cost of hardware does not increase, but excellent sound effect experience is provided, which greatly improves the competitiveness of the vehicles. This paper analyzed the main influencing factors of sound quality in the low-cost cars, and it introduced the method of improving sound quality with the third-party effect, and it introduced the integrating and testing methods of sound effects with the example of mass production vehicles, which is instructive for the other development of the third-party sound effect.

References

1. DBScar (2014) Telematics: battle of fourth screen. Electronic Industry Press. pp 4–15
2. Zhan D (2015) Design and evaluation method of vehicle sound effects. Automot Pract Technol (7)
3. Anghelescu P, Angelescu S, Ionita S (2014) Real-time audio effects with DSP algorithms and direct sound. In: International conference on electronics, computers and artificial intelligence (ECAI), pp 35–40

A Practical Path Planning Strategy for Automatic Parking in Constrained Parking Space

Hui Lu, Kangxi Xu, Huanran Wang, Cheng Qi and Dazhi Wang

Abstract A practical automatic parking strategy of path planning in constrained parking space is stated in this paper. The path planning algorithm is designed by reversing a retrieving path that consists of two steps. First, a sequence of forward and backward moving is implemented from the final parking position, until the vehicle reaches a configuration of states from which it can pull out of the parking slot with one maneuver. Second, a continuous path is generated from this configuration to the vehicle original states. The constraints of parking space and vehicle dynamics are considered in both steps.

Keywords Automatic parking · Motion planning · Path planning · Vehicle motion control

1 Introduction

The automatic parking system can detect the environment and assist the driver to parking the vehicle into appropriate parking slot. An automatic parking system is shown to be able to improve driving comfort and safety by reducing the amount of stress people feel when manually steering for parallel parking and garage parking maneuvers. This system also can be included in the autonomous point-to-point parking system or autonomous valet parking system. The low computational demand, few sensors requirement, predictable, and trustworthy for drivers are important constraints and system requirements for production implementation.

Circular locus method is widely used by the path planning algorithms for automatic parking based on the fundamental of mathematics, i.e., it has been proved that the shortest path between two configurations for the simplified car is made up of line segments connected with tangential circular arcs of minimum radius [1]. Pohl from Volvo [2] proposed a semi-automated parallel parking system; the path planning

H. Lu (✉) · K. Xu · H. Wang · C. Qi · D. Wang
Software Engineering Department, SAIC Motor Corporation Limited
Passenger Vehicle Co., Shanghai, China
e-mail: luhui01@saicmotor.com

China SAE (ed.), *Proceedings of China SAE Congress 2018: Selected Papers*, Lecture Notes in Electrical Engineering 574,
https://doi.org/10.1007/978-981-13-9718-9_75

adopted a two arcs strategy for one maneuver parallel parking. Moshchuk from GM [3] proposed path planning strategies for both one and two turn parking maneuvers, which also consisted of two arcs. Vorobieva from Renault [4] proposed a path planning strategy for automatic parallel parking in tiny spots. First, a path constituting of circle arcs is generated. Then this path is transformed into a continuous-curvature path using clothoid curves.

This paper proposes a practical automatic parking strategy of path planning in constrained parking space and especially focuses on the practical implementation of multiple maneuvers parking strategy considering the multiple constraints of the parking space. The path planning algorithm is designed by reversing a retrieving path that consists of two steps. First, a sequence of forward and backward moving is implemented from the final parking position, until the vehicle reaches a configuration of states from which it can pull out of the parking slot with one maneuver. Second, a continuous path is generated from this configuration to the vehicle original states. Furthermore, the constraints of parking space and vehicle dynamics are considered in each step.

2 Path Planning

The object of path planning algorithm is to generate a traversable trajectory from the original position into the parking slot with acceptable final states.

Figure 1 demonstrates the basic one maneuver strategy adopted in this paper. The main steering segments are two arcs connected by a straight line. The clothoid curve is adopted to smooth the arc and straight line. The vehicle at starting point 0 is assumed to be parallel with final parking orientation, because it is easy to correct the vehicle orientation with a CAC (Clothoid–Arc–Clothoid) or CC (Clothoid–Clothoid) curve [3]. The first segment between point 0 and point 1 is a straight line parallel with the

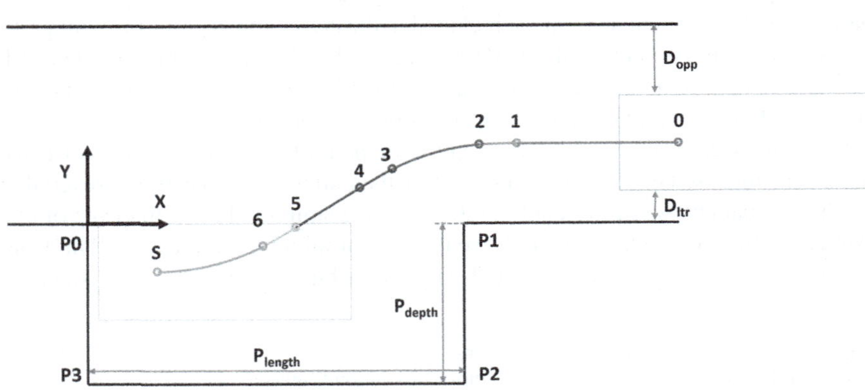

Fig. 1 Basic one maneuver sequence

Table 1 Vehicle configuration

Parameters	Notations	Values
Vehicle length	L_{VL}	4.7 m
Wheelbase	L_{WB}	2.7 m
Front overhang	L_{FO}	0.9
Rear overhang	L_{RO}	1.1
Vehicle width	W_{VW}	1.8 m
Track	W_T	1.6

final parking orientation. The segment between point 1 and point 4 is a CAC curve with a constrained minimum turning radius. The segment between point 4 and point 5 is a straight line to increase the path planning flexibility. The segment between point 5 and point S is a CAC/CA curve to move the vehicle to final parking states.

It is assumed that the stand steering when shift gear is acceptable in the following discussion, because this will significantly improve the performance in tiny parking slot, and it will not make the driver feel uncomfortable. On the other hand, for the parking slot that is big enough to parking in one maneuver without any stand steering, the path planning will be easier to solve with the proposed strategy. It is also assumed that the stand steering is acceptable at the final position, and since this paper focuses on the limited parking space, the segment between point 5 and point S is limited to CA in the following discussion.

Several denotations and assumptions are given here firstly. The configuration and denotations of the used vehicle are shown in Table 1. By establishing a coordinate system fixed on the parking slot corner $P0\{(x_{P0}, y_{P0}) = (0, 0)\}$, the other parking space points and the vehicle configuration will be expressed in this coordinate if there is no further explanation. A parallel parking to the right side is considered in this paper, because it is simple to transfer to the left side parking.

The parking maneuver assumes the four wheels turning around the same instantaneous center without slipping under an Ackerman steering angle δ_f. Hence, the turning radius of the rear axle center is $R_o = L_{WB}/\tan\delta_f$, and the minimum turning radius determined by the vehicle geometry can be expressed as $R_{o\min} = L_{WB}/\tan\delta_{f\max}$. Also the turning radius of each corner of vehicle outline can be easily deduced.

2.1 The Strategy of One Maneuver and Multiple Maneuvers

Considering the constraints of the vehicle and parking space, major objects of the path planning algorithm are to move the vehicle into parking slot with minimum maneuvers of change direction and minimum travel distance. The parking path in this paper is deduced by reversing a retrieving path that consists of two steps, as shown in Fig. 2.

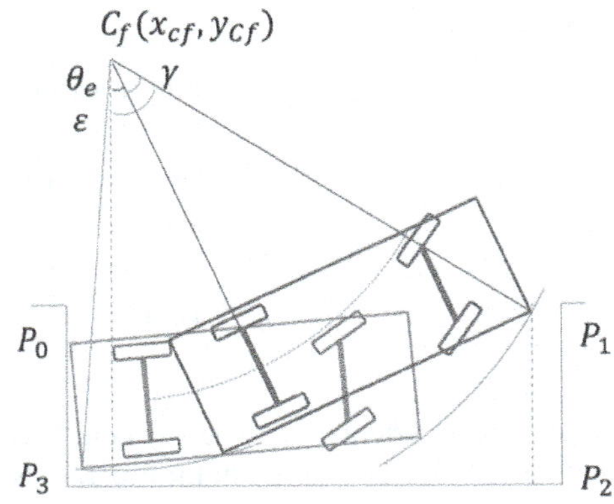

Fig. 2 Moving forward in parking slot

First, a sequence of forward and backward moving is implemented from the final parking position, until the vehicle reaches a configuration of states from which it can pull out of the parking slot with one maneuver. For the automatic parking system, the number of maneuvers influences the driver experience more significantly. Hence, this paper adopted an iteration strategy to find a sequence with acceptable number of maneuvers.

Second, a continuous path is generated from the found configuration to the vehicle original states following the basic one maneuver sequence as shown in Fig. 1.

Given a parking slot P, a vehicle original state V_0, and a vehicle final parking state V_s, the possibility to pull out with one maneuver can be checked first. If it is possible, the continuous path between V_0 and V_s is generated. If it is not possible to pull out with one maneuver, an iteration is implemented from the final parking state. The vehicle is supposed to move forward and backward with allowed minimum turning radius until reach a configuration from which the vehicle can pull out with one maneuver. The first move of the iteration can be forward or backward. Both of them are solved to find a smaller number of maneuvers. Meanwhile, a maximum of iteration numbers is used to give up the parking space that need too many maneuver numbers (Fig. 3).

2.2 *Maneuvers in Parking Slot*

(i) *Moving forward in parking slot and pulling out check*

Given a configuration in the parking slot, $V_s(x_s, y_s, \theta_s)$, the vehicle is supposed to move forward with a constant turning radius until reach the front boundary or to move out of the parking slot, as shown in Fig. 2. Firstly, the one maneuver pulling out is checked by the collision free of front right corner. Meanwhile, the constraints of the constant turning radius need to be considered if the parking slot depth is limited. The obstacle of the inner side of the parking slot is assumed to be a wall in the following

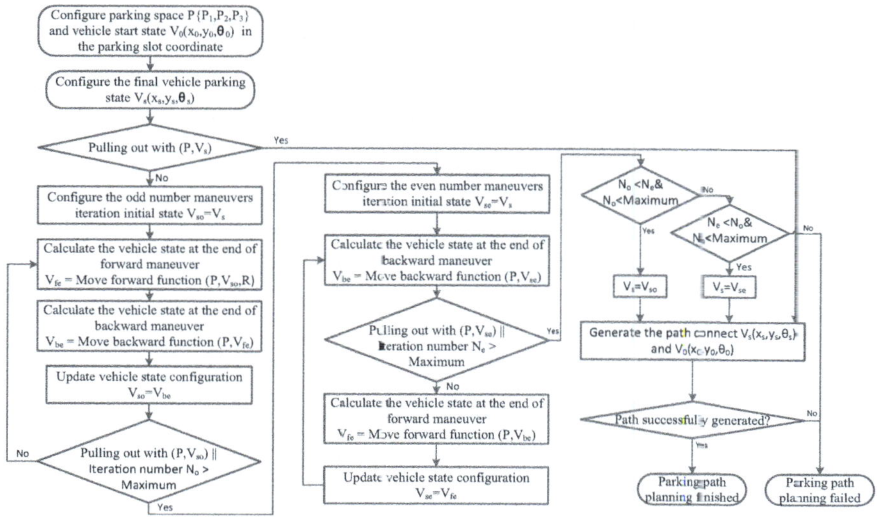

Fig. 3 Strategy of one maneuver and multiple maneuvers

discussion, and the rear right corner collision free is guaranteed. For the curbstone, the same way can be adopted by check the turning radius of the rear right wheel outline.

Denoting R_f as the constrained turning radius of the rear axle center, it can be solved by check the collision of vehicle rear right corner with the inner side obstacle. The turning center location (x_{cf}, y_{cf}) of this forward maneuver can be calculated as below.

$$x_{cf} = x_s - R_f * \sin(\theta_s) \tag{1}$$

$$y_{cf} = y_s + R_f * \cos(\theta_s) \tag{2}$$

Denoting the location of rear right vehicle corner at the $V_s(x_s, y_s, \theta_s)$ as (x_{srr}, y_{srr}), if $x_{srr} < x_{cf}$, then the parking slot depth constraint needs to be checked. The required minimum parking slot depth P_{depth_min} can be deduced by solving P_{depth_f} by letting $R_f = R_{omin}$ following below equations.

$$y_{cf} + P_{depth\_f} = R_{RR} + B_{P23} \tag{3}$$

$$R_{RR} = \left(\left(R_f + \frac{W_{VW}}{2}\right)^2 + L_{RO}^2\right)^{1/2} \tag{4}$$

If the parking slot depth $P_{depth} > P_{depth\_min}$, then the minimum turning radius R_{omin} can be adopted, $R_f = R_{omin}$. If $P_{depth} < P_{depth\_min}$, then the constrained minimum turning radius R_f can be calculated by solving R_f by letting $P_{\text{depth\_}f} = P_{\text{depth}}$ according to the above Eqs. (2)–(4).

Considering the constrained R_f, the pull out can be checked by the collision of vehicle front right corner with P_1. If the distance from the turning center C_f to the parking slot front corner P_1 is bigger than the turning radius of the vehicle front right corner, $R_{\text{FR}} = \left(\left(R_f + \frac{W_{\text{VW}}}{2}\right)^2 + (L_{\text{WB}} + L_{\text{FO}})^2\right)^{1/2}$, plus a safe margin B_{fP1}, then this configuration (x_s, y_s, θ_s) is supposed to be able to move out of parking slot directly, and this configuration will be used to generate the path to connect to the vehicle start position.

$$\left(x_{cf} - x_{P1}\right)^2 + \left(y_{cf} - y_{P1}\right)^2 > R_{\text{FR}} + B_{fP1} \tag{5}$$

If the distance from C_f to P_1 is smaller than $R_{\text{FR}} + B_{fP1}$, then the end configuration of this maneuvers, (x_e, y_e, θ_e), can be deduced with the following equations.

$$\sin\varepsilon = x_{P1} - B_{P12} - x_{cf}/R_{\text{FR}} \tag{6}$$

$$\sin\gamma = L_{\text{WB}} + L_{\text{FO}}/R_{\text{FR}} \tag{7}$$

$$\theta_e = \varepsilon - \gamma \tag{8}$$

The angle ε and γ are shown in Fig. 2. B_{P12} is the safety margin between the front corners of vehicle to line P_1P_2 of parking slot boundary. Then the final location of the rear axle center at the end of this forward maneuvers can be calculated as below.

$$x_e = x_{cf} + R_f \sin\theta_e \tag{9}$$

$$y_e = y_{cf} - R_f \cos\theta_e \tag{10}$$

(ii) ***Moving backward in parking slot***

Given a configuration of vehicle state in the parking slot, $V_s(x_s, y_s, \theta_s)$, a backward maneuver in parking slot assumes the vehicle move backward with a constant turning radius until reach the rear boundary or inner side boundary. For a given turning radius, R_b, the location of the turning center can be calculated as below.

$$x_{cb} = x_s + R_b * \sin(\theta_s) \tag{11}$$

$$y_{cb} = y_s - R_b * \cos(\theta_s) \tag{12}$$

To reach the rear boundary, let the rear left vehicle corner located at the line P_0P_3 with a shift of safety margin B_{P03}, as shown in Fig. 4a. The turning radius of the

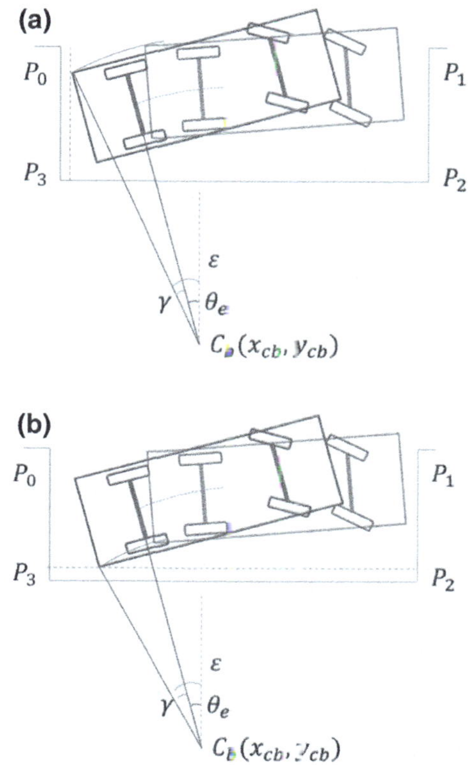

Fig. 4 Moving backward in parking slot

rear left vehicle corner can be deduced as $R_{\text{RL}} = \left(\left(R_b + \frac{W_{\text{VW}}}{2}\right)^2 + L_{\text{RO}}^2\right)^{1/2}$. Then the angles ε_1 γ_1 and $\theta_{\text{e}1}$, as shown in Fig. 4a, can be calculated as below.

$$\sin \varepsilon_1 = x_{cb} - B_{P03}/R_{\text{RL}} \tag{13}$$

$$\sin \gamma = L_{\text{RO}}/R_{\text{RL}} \tag{14}$$

$$\theta_{\text{e}1} = \varepsilon_1 - \gamma \tag{15}$$

To reach the inner side boundary, let the rear right vehicle corner located at the line P_2P_3 with a shift of safety margin, B_{P23}, as shown in Fig. 4b. The turning radius of the rear right vehicle corner can be deduced as $R_{\text{RR}} = \left(\left(R_b - \frac{W_{\text{VW}}}{2}\right)^2 + L_{\text{RO}}^2\right)^{1/2}$. Then the angles, ε_2, γ_2, and $\theta_{\text{e}2}$, as shown in figure, can be calculated as below,

$$\cos \varepsilon_2 = (y_{P3} - y_{cb} + B_{P23})/R_{\text{RR}} \tag{16}$$

$$\sin \gamma_2 = L_{\mathrm{RO}} / R_{\mathrm{RR}} \tag{17}$$

$$\theta_{e2} = \varepsilon_2 - \gamma_2 \tag{18}$$

Then, the final vehicle head angle can be determined as $\theta_e = \min\{\theta_{e1}, \theta_{e2}\}$. The final location of the rear axle center at the end of this backward maneuver can be calculated as below.

$$x_e = x_{cb} - R_b \sin \theta_e \tag{19}$$

$$y_e = y_{cb} + R_b \cos \theta_e \tag{20}$$

2.3 Maneuvers to the Original Point

From the previous section, a configuration $V_s(x_s, y_s, \theta_s)$ is deduced, from which the vehicle can pulling out with one maneuver. This section aims to generate a continuous path that connects $V_s(x_s, y_s, \theta_s)$ and vehicle original position $V_0(x_0, y_0, \theta_0)$ following the basic sequence described in Sect. 2.1.

The constrained turning radius of arc S-6 has been deduced as R_f in the section of moving forward in parking slot. The constrained turning radius of arc 2–3 can be deduced by the constraint of the opposite obstacle. It is compensated to avoid the collision of vehicle front left corner.

As shown in Fig. 5, the clear distance from the obstacle to vehicle outline is denoted as D_{opp}. The required minimum clear distance $D_{\mathrm{opp\_min}}$ can be deduced by solving D_{opp1} by letting $R_1 = R_{o\min}$ following below equations.

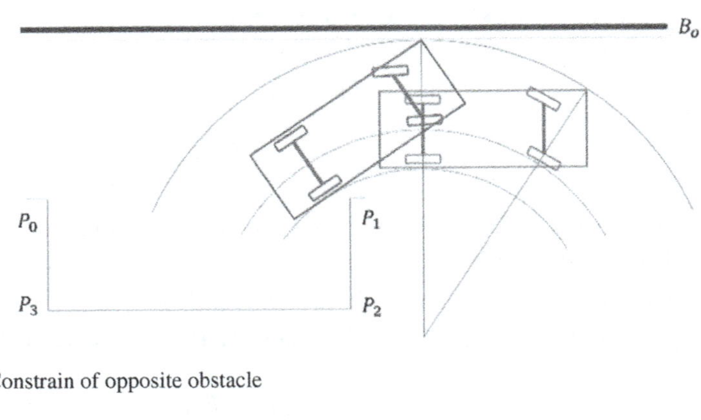

Fig. 5 Constrain of opposite obstacle

$$R_1 + \frac{W_{VW}}{2} + D_{opp1} = R_{FL} + B_o \tag{21}$$

$$R_{FL} = \left(\left(R_1 + \frac{W_{VW}}{2}\right)^2 + (L_{WB} + L_{FO})^2\right)^{1/2} \tag{22}$$

If $D_{opp} > D_{opp\_min}$, then the minimum turning radius R_{omin} can be adopted, $R_1 = R_{omin}$. If $D_{opp} < D_{opp\_min}$, then the constrained minimum turning radius R_1 can be calculated by solving R_1 by letting $D_{opp1} = D_{opp}$ according to the above Eqs. (21) and (22).

To connect the arc S-6 and arc 2–3 with an optional straight line, the vehicle head angle θ_t at the tangency point or tangent line of the two arcs is used as a critical parameter to determine the final path, as shown in Fig. 6.

θ_{tmin} can be deduced by the collision free of the point P_1. By setting a safety margin B_{p1} around P_1, the distance between turning center $C_{1min\theta}$ of arc 2–3 and point P_1 is determined to be the turning radius of vehicle rear right tyre outline.

$$\left(x_{c1min\theta} - x_{p1}\right)^2 + \left(y_{c1min\theta} - y_{p1}\right)^2 = \left(R_1 - \frac{1}{2}W_{VW} - B_{p1}\right)^2 \tag{23}$$

Since $y_{c1min\theta} = x_0 - R_1$, the turning center $C_{1min\theta}$ can be determined. The turning center C_2 can be determined by the given configuration $V_s(x_s, y_s, \vartheta_s)$, as described in Eqs. (1) and (2). Then the tangency line can be obtained.

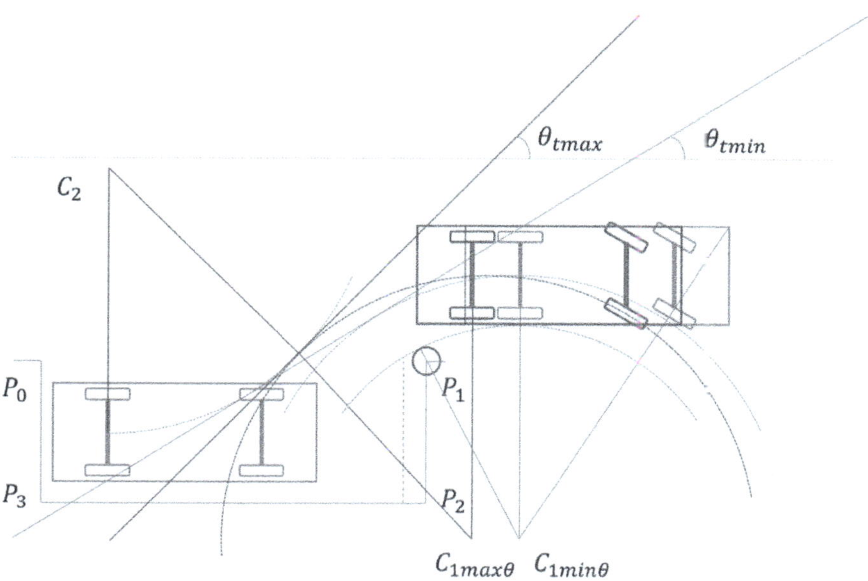

Fig. 6 Basic parking sequence of cycle and line without clothoid

$\theta_{t\max}$ can be deduced by letting the two arcs tangency at one point directly without a straight line segment. This can be thought as moving the turning center of arc 2–3 along the line $y = x_0 - R_1$ until it tangency with arc S-6.

Finally, $\theta_t = (\theta_{t\min} + \theta_{t\max})/2$ is adopted to generate the path connect V_s and V_0 for the consideration of safety and comfort.

The clothoid curves are adopted in this paper to smooth the connection between lines and arcs in order to avoid steering at stop. The clothoid is a curve with curvature $\kappa = 1/R$ varies linearly with the arc length L, $\kappa(L) = \sigma L + \kappa(0)$. σ is the sharpness of the clothoid. A parameter A is usually used, $A^2 = RL$ and $A = 1/\sqrt{\sigma}$. The detail discussion about clothoid can be found in paper [5].

In the parking maneuvers, the maximum steering speed $v_{\delta\max}$ is given by the vehicle hardware system. Then the minimum travel distance to steering from 0 to $\delta_{\max}$ is determined by the vehicle velocity, $L_{\min} = v * \delta_{\max}/v_{\delta\max}$. Then the parameter $A_{\min}^2 = R_{\min}L_{\min}$ is adopted in this paper. For a certain A, given the turning radius, the following values as shown in Fig. 7 can be calculated online ΔX_c, ΔY_c, and $\Delta\theta_c$.

Figure 8 shows the smoothed parking curve with clothoid. To calculate the $\theta_{t\min}$, firstly the turning center $C_{1\min\theta}$ is shifted by inserting clothoid 1–2. With the obtained constrained R_1, the character values of clothoid 1–2 can be deduced, ΔX_{c1}, ΔY_{c1}, and $\Delta\theta_{c1}$.

$$y_{c1\min\theta} = y_0 - \Delta Y_{c1} - R_1 * \cos(\Delta\theta_{c1}) \tag{24}$$

Then $x_{c1\min\theta}$ can be calculated by Eq. (23).

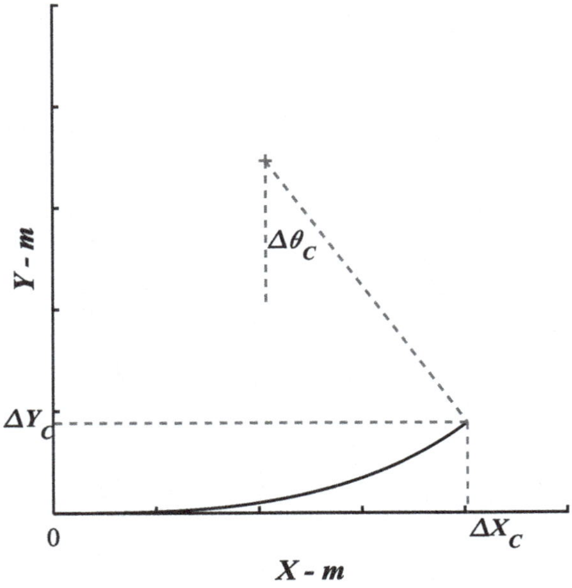

Fig. 7 Clothoid curve

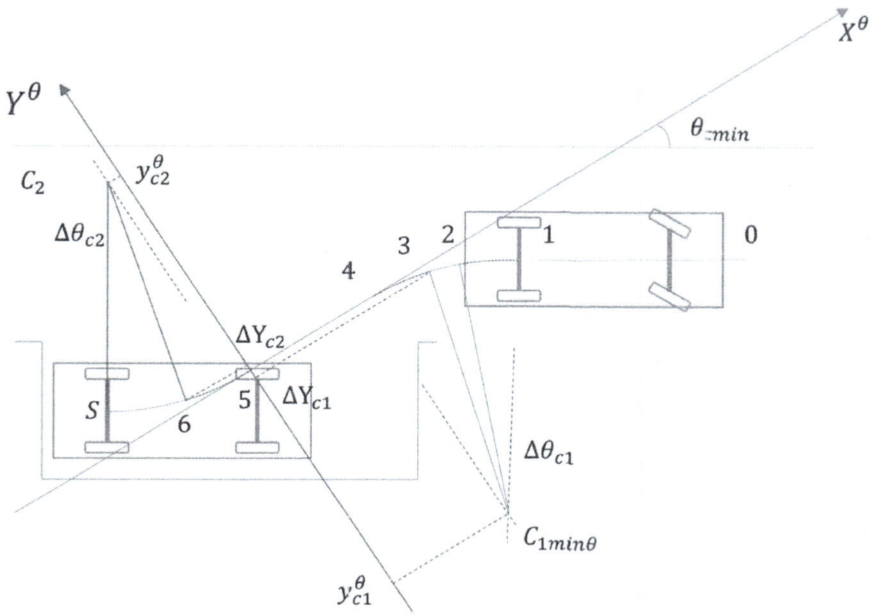

Fig. 8 Basic sequence of smooth parking path with clothoid curves

To calculate $\theta_{t\min}$, a temp coordinate is established on the tangency point 5, the X^θ axis is along the straight line, and the Y^θ axis is perpendicular to the straight line. Then the projection of turning center C_2 and $C_{1\min\theta}$ on this Y^θ axis can be expressed as below, respectively.

$$y^\theta_{c1\min\theta} = -\Delta Y_{c1} - R_1 * \cos(\Delta\theta_{c1}) \tag{25}$$

$$y^\theta_{c2} = \Delta Y_{c2} + R_2 * \cos(\Delta\theta_{c2}) \tag{26}$$

Since the distance between C_2 and $C_{1\min\theta}$ can be obtained by the coordinate value in the coordinate fixed to point $P0$, the angle between line $C_2C_{1\min\theta}$ and X^θ axis and the angle between line $C_2C_{1\min\theta}$ and X axis of the coordinate fixed to $P0$ can be described as below, respectively.

$$\sin\theta^\theta_{c1c2} = \left(y^\theta_{c2} - y^\theta_{c1\min\theta}\right) \Big/ \sqrt{(x_{c2} - x_{c1\min})^2 + (y_{c2} - y_{c1\min})^2} \tag{27}$$

$$\sin\theta_{c1c2} = (y_{c2} - y_{c1\min\theta}) \Big/ \sqrt{(x_{c2} - x_{c1\min})^2 + (y_{c2} - y_{c1\min})^2} \tag{28}$$

Then $\theta_{t\min}$ can be deduced as $\theta_{t\min} = \theta^\theta_{c1c2} - \theta_{c1c2}$. Similar approach can be adopted to calculate $\theta_{t\max}$ with the inserted clothoid segment.

3 Analysis

In this section, we analyze the proposed path planning algorithm with a prototype vehicle configured as Table 1. We consider two approaches to select the final parking position. Approach A, the final parking position is always set to the middle of the parking slot. Approach B, the final parking position is set to close to the rear boundary with a distance of B_{P03} for odd number of maneuvers and close to the front vehicle with a distance of B_{P12} for even number of maneuvers. The difference between A and B only lies on the set of Vs at the beginning of the iteration.

The parameters of the safe margins are set as $B_{p1} = 0.3\,\text{m}$, $B_o = 0.2\,\text{m}$, $B_{P23} = 0.1\,\text{m}$, $B_{P03} = 0.2\,\text{m}$, and $B_{P12} = 0.3\,\text{m}$. The parking slot depth P_{depth} is set to 2.5 m, the opposite clear distance D_{opp} is set to 1.5 m, and the lateral distance D_{ltr} is set to 1 m. Figure 9 shows the total number of maneuvers varying with the different parking slot length. The result of approach A is shown in orange line. It can be found that, for the smallest parking length 5.7 m, the number of maneuvers is still acceptable. Because there are buffers of 0.2 m in front and rear boundary, the travelled space in parking slot is only vehicle length + 0.6 m.

Meanwhile, the blue line is the result of Approach B shifted by 1 along the y axis, because usually it needs a last straight line adjustment to move the vehicle from the position close to boundary to the acceptable position. It can be found that setting the initial states of iteration directly to the final acceptable position can decrease the number of maneuvers in some cases compared with adding one straight line adjustment, which is usually used in the other applications.

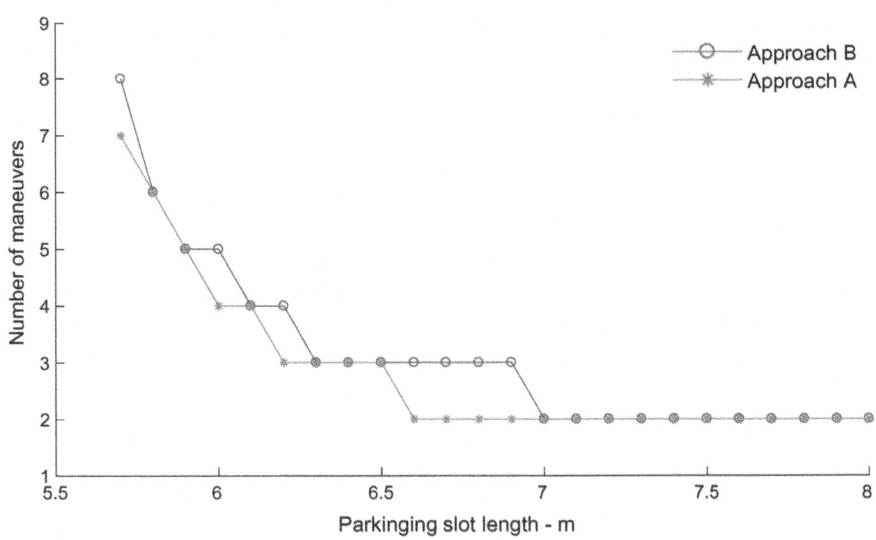

Fig. 9 Total number of maneuvers compared by different parking slot length

The parking slot depth P_{depth} is set to 3 m, the parking slot length P_{length} is set to 7 m, and the lateral distance D_{ltr} is set to 0.6 m. Figure 10 shows the path planning results varying with the different opposite clear distances, D_{opp}, which is set to 1.5, 1, 0.8 m, respectively. It can be found that the turning radius of arc 2–3 is adjusted according to the opposite clear space, and the move in parking slot is not influenced.

The opposite clear distance D_{opp} is set to 1.5 m, the parking slot length P_{length} is set to 6.5 m, and the lateral distance D_{ltr} is set to 0.6 m. Figure 11 shows the path planning results varying with the different parking slot depth, P_{depth}, which is set to 2.5 and 2.3 m, respectively. It can be found that the path in parking slot is adjusted according to the parking slot depth.

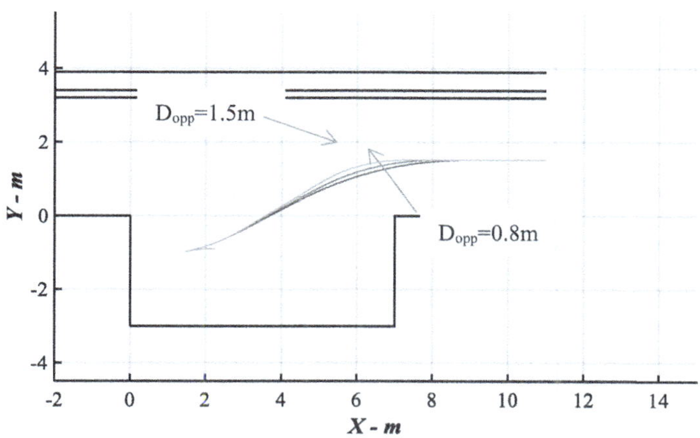

Fig. 10 Parking paths compared by different opposite clear distance

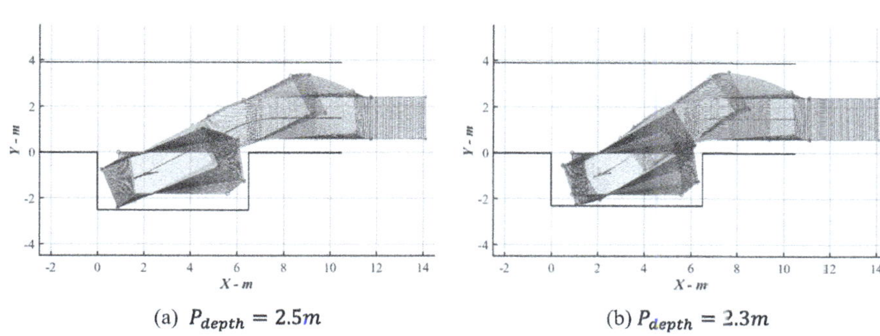

Fig. 11 Parking paths compared by different parking slot depth

4 Conclusions

This paper proposed a strategy of vehicle path planning for automatic parking. An iteration method considering the multiple constraints of tiny parking slot is adopted. Odd number and even number of maneuvers are both solved, and the minimum numbers of maneuvers are implemented. The analysis indicates that the proposed planning algorithm can adapt to various constrained parking space successfully.

The vehicle field test is under preparation. In the future, the online vehicle path regeneration will be included in order to adjust according to the vehicle control error and the online updated environment information.

References

1. Reeds J (1990) Optimal paths for a car that goes both forwards and backwards. Pacific J Math 145(2):367–393
2. Pohl J, Sethsson M, Degerman P et al (2006) A semi-automated parallel parking system for passenger cars. Proc Inst Mech Eng Part D J Automobile Eng 220(220):53–65
3. Moshchuk N, Chen SK (2008) Autonomous parking strategy. In: ASME 2008 international mechanical engineering congress and exposition, pp 195–202
4. Vorobieva H, Glaser S, Minoiu-Enache N et al (2015) Automatic parallel parking in tiny spots: path planning and control. IEEE Trans Intell Transp Syst 16(1):396–410
5. Fraichard T, Scheuer A (2004) From Reeds and Shepp's to continuous-curvature paths. IEEE Trans Rob 20(6):1025–1035

The manufacturer's authorised representative in the EU is Springer Nature Customer Service Centre GmbH, Europaplatz 3, 69115 Heidelberg, Germany. If you have any concerns regarding our products, please contact ProductSafety@springernature.com

Printed and bound by CPI Group (UK) Ltd, Croydon, CR0 4YY

30/09/2026

02225305-0001